Instructor's Solutions Manual

DANIEL S. MILLER Niagara County Community College

Blitzer

PRECALCULUS THIRD EDITION

Volume I

Upper Saddle River, NJ 07458

Acquisitions Editor: Adam Jaworski
Editor-in-Chief: Sally Yagan
Project Manager: Dawn Murrin
Supplements Editor: Christine Whitlock
Executive Managing Editor: Kathleen Schiaparelli
Managing Editor: Nicole M. Jackson
Assistant Managing Editor: Karen Bosch
Production Editor: Traci Douglas
Supplement Cover Designer: Christopher Kossa
Manufacturing Buyer: Ilene Kahn
Manufacturing Manager: Alexis Heydt-Long

Pearson Prentice Hall
Pearson Education, Inc.
Upper Saddle River, NJ 07458

Printed in the United States of America

10 9 8 7 6 5 4 3 2 1

ISBN 0-13-188037-3

Pearson Education Ltd., *London*
Pearson Education Australia Pty. Ltd., *Sydney*
Pearson Education Singapore, Pte. Ltd.
Pearson Education North Asia Ltd., *Hong Kong*
Pearson Education Canada, Inc., *Toronto*
Pearson Educación de Mexico, S.A. de C.V.
Pearson Education—Japan, *Tokyo*
Pearson Education Malaysia, Pte. Ltd.

Table of Contents

Chapter P

Section P.1

Check Point Exercises

1. $8+6(x-3)^2 = 8+6(13-3)^2$
$= 8+6(10)^2$
$= 8+6(100)$
$= 8+600$
$= 608$

2. The year 2000 is 40 years after 1960.
$P = 0.72x^2 + 9.4x + 783$
$= 0.72(40)^2 + 9.4(40) + 783$
$= 0.72(1600) + 376 + 783$
$= 1152 + 376 + 783$
$= 2311$
The equation's value of 2311 models the data in the bar graph quite well.

3. The elements common to $\{3, 4, 5, 6, 7\}$ and $\{3, 7, 8, 9\}$ are 3 and 7.
$\{3,4,5,6,7\} \cap \{3,7,8,9\} = \{3,7\}$

4. The union is the set containing all the elements of either set.
$\{3,4,5,6,7\} \cup \{3,7,8,9\} = \{3,4,5,6,7,8,9\}$

5. $\left\{-9,\ -1.3,\ 0,\ 0.\bar{3},\ \frac{\pi}{2},\ \sqrt{9},\ \sqrt{10}\right\}$

a. Natural numbers: $\sqrt{9}$ because $\sqrt{9} = 3$

b. Whole numbers: $0,\ \sqrt{9}$

c. Integers: $-9,\ 0,\ \sqrt{9}$

d. Rational numbers: $-9,\ -1.3,\ 0,\ 0.\bar{3},\ \sqrt{9}$

e. Irrational numbers: $\frac{\pi}{2},\ \sqrt{10}$

f. Real numbers:
$\left\{-9,\ -1.3,\ 0,\ 0.\bar{3},\ \frac{\pi}{2},\ \sqrt{9},\ \sqrt{10}\right\}$

6. **a.** $\left|1-\sqrt{2}\right|$
Because $\sqrt{2} \approx 1.4$, the number inside the absolute value bars is negative. The absolute value of x when $x < 0$ is $-x$. Thus,
$\left|1-\sqrt{2}\right| = -\left(1-\sqrt{2}\right) = \sqrt{2}-1$

b. $|\pi - 3|$
Because $\pi \approx 3.14$, the number inside the absolute value bars is positive. The absolute value of a positive number is the number itself. Thus,
$|\pi - 3| = \pi - 3.$

c. $\frac{|x|}{x}$
Because $x > 0$, $|x| = x.$
Thus, $\frac{|x|}{x} = \frac{x}{x} = 1$

7. $|-4-(5)| = |-9| = 9$
The distance between -4 and 5 is 9.

8. $7(4x^2+3x)+2(5x^2+x)$
$= 7(4x^2+3x)+2(5x^2+x)$
$= 28x^2+21x+10x^2+2x$
$= 38x^2+23x$

9. $6+4[7-(x-2)]$
$= 6+4[7-x+2)]$
$= 6+4[9-x]$
$= 6+36-4x$
$= 42-4x$

Exercise Set P.1

1. $7+5(10) = 7+50 = 57$

2. $8+6(5) = 8+30 = 38$

3. $6(3)-8 = 18-8 = 10$

4. $8(3)-4 = 24-4 = 20$

5. $8^2+3(8) = 64+24 = 88$

6. $6^2+5(6)=36+30=66$

7. $7^2-6(7)+3=49-42+3=7+3=10$

8. $8^2-7(8)+4=64-56+4=8+4=12$

9. $4+5(9-7)^3=4+5(2)^3$
$=4+5(8)=4+40=44$

10. $6+5(8-6)^3=6+5(2)^3$
$=6+5(8)$
$=6+40=46$

11. $8^2-3(8-2)=64-3(6)$
$=64-18=46$

12. $8^2-4(8-3)=64-4(5)=64-20=44$

13. $\frac{5(x+2)}{2x-14}=\frac{5(10+2)}{2(10)-14}$
$=\frac{5(12)}{6}$
$=5\cdot 2$
$=10$

14. $\frac{7(x-3)}{2x-16}=\frac{7(9-3)}{2(9)-16}=\frac{7(6)}{2}=7\cdot 3=21$

15. $\frac{2x+3y}{x+1}; x=-2, y=4$
$=\frac{2(-2)+3(4)}{-2+1}=\frac{-4+12}{-1}=\frac{8}{-1}=-8$

16. $\frac{2x+y}{xy-2x}$; $x=-2$ and $y=4$
$\frac{2(-2)+4}{(-2)(4)-2(-2)}=\frac{-4+4}{-8+4}=\frac{0}{4}=0$

17. $C=\frac{5}{9}(50-32)=\frac{5}{9}(18)=10$
10°C is equivalent to 50°F.

18. $C=\frac{5}{9}(F-32)=\frac{5}{9}(86-32)=\frac{5}{9}(54)=30$ 30°C is equivalent to 86°F.

19. $h=4+60t-16t^2=4+60(2)-16(2)^2$
$=4+120-16(4)=4+120-64$
$=124-64=60$
Two seconds after it is kicked, the ball's height is 60 feet.

20. $h=4+60t-16t^2$
$=4+60(3)-16(3)^2$
$=4+180-16(9)$
$=4+180-144$
$=184-144=40$
Three seconds after it is kicked, the ball's height is 40 feet.

21. $\{1,2,3,4\}\cap\{2,4,5\}=\{2,4\}$

22. $\{1,3,7\}\cap\{2,3,8\}=\{3\}$

23. $\{s,e,t\}\cap\{t,e,s\}=\{s,e,t\}$

24. $\{r,e,a,l\}\cap\{l,e,a,r\}=\{r,e,a,l\}$

25. $\{1,3,5,7\}\cap\{2,4,6,8,10\}=\{\ \}$
The empty set is also denoted by $\varnothing$.

26. $\{1,3,5,7\}\cap\{-5,-3,-1\}=\{\ \}$ or $\varnothing$

27. $\{a,b,c,d\}\cap\varnothing=\varnothing$

28. $\{w,y,z\}\cap\varnothing=\varnothing$

29. $\{1,2,3,4\}\cup\{2,4,5\}=\{1,2,3,4,5\}$

30. $\{1,3,7,8\}\cup\{2,3,8\}=\{1,2,3,7,8\}$

31. $\{1,3,5,7\}\cup\{2,4,6,8,10\}$
$=\{1,2,3,4,5,6,7,8,10\}$

32. $\{0,1,3,5\}\cup\{2,4,6\}=\{0,1,2,3,4,5,6\}$

33. $\{a,e,i,o,u\}\cup\varnothing=\{a,e,i,o,u\}$

34. $\{e,m,p,t,y\}\cup\varnothing = \{e,m,p,t,y\}$

35. **a.** $\sqrt{100}$

b. $0, \sqrt{100}$

c. $-9, 0, \sqrt{100}$

d. $-9, -\frac{4}{5}, 0, 0.25, 9.2, \sqrt{100}$

e. $\sqrt{3}$

f. $-9, -\frac{4}{5}, 0, 0.25, \sqrt{3}, 9.2, \sqrt{100}$

36. **a.** $\sqrt{49}$

b. $0, \sqrt{49}$

c. $-7, 0, \sqrt{49}$

d. $-7, -0.\overline{6}, 0, \sqrt{49}$

e. $\sqrt{50}$

f. $-7, -0.\overline{6}, 0, \sqrt{49}, \sqrt{50}$

37. **a.** $\sqrt{64}$

b. $0, \sqrt{64}$

c. $-11, 0, \sqrt{64}$

d. $-11, -\frac{5}{6}, 0, 0.75, \sqrt{64}$

e. $\sqrt{5}, \pi$

f. $-11, -\frac{5}{6}, 0, 0.75, \sqrt{5}, \pi, \sqrt{64}$

38. **a.** $\sqrt{4}$

b. $0, \sqrt{4}$

c. $-5, 0, \sqrt{4}$

d. $-5, -0.\overline{3}, 0, \sqrt{4}$

e. $\sqrt{2}$

f. $-5, -0.\overline{3}, 0, \sqrt{2}, \sqrt{4}$

39. 0

40. Answers may vary.

41. Answers may vary.

42. Answers may vary.

43. True; -13 is to the left of -2 on the number line.

44. False; 6 is to the left of 2 on the number line.

45. True; 4 is to the right of -7 on the number line.

46. True; -13 is to the left of -5 on the number line.

47. True; $-\pi = -\pi$

48. True; -3 is to the right of -13 on the number line.

49. True; 0 is to the right of -6 on the number line.

50. True; 0 is to the right of -13 on the number line.

51. $|300| = 300$

52. $|-203| = 203$

53. $|12 - \pi| = 12 - \pi$

54. $|7 - \pi| = 7 - \pi$

55. $\left|\sqrt{2} - 5\right| = 5 - \sqrt{2}$

56. $\left|\sqrt{5} - 13\right| = 13 - \sqrt{5}$

57. $\frac{-3}{|-3|} = \frac{-3}{3} = -1$

58. $\frac{-7}{|-7|} = \frac{-7}{7} = -1$

59. $\left||-3| - |-7|\right| = |3 - 7| = |-4| = 4$

60. $\left||-5| - |-13|\right| = |5 - 13| = |-8| = 8$

61. $|x + y| = |2 + (-5)| = |-3| = 3$

62. $|x - y| = |2 - (-5)| = |7| = 7$

63. $|x| + |y| = |2| + |-5| = 2 + 5 = 7$

64. $|x|-|y|=|2|-|-5|=2-5=-3$

65. $\frac{y}{|y|}=\frac{-5}{|-5|}=\frac{-5}{5}=-1$

66. $\frac{|x|}{x}+\frac{|y|}{y}=\frac{|2|}{2}+\frac{|-5|}{-5}=\frac{2}{2}+\frac{5}{-5}=1+(-1)=0$

67. The distance is $|2-17|=|-15|=15$.

68. The distance is $|4-15|=|-11|=11$.

69. The distance is $|-2-5|=|-7|=7$.

70. The distance is $|-6-8|=|-14|=14$.

71. The distance is $|-19-(-4)|=|-19+4|=|-15|=15$.

72. The distance is $|-26-(-3)|=|-26+3|=|-23|=23$.

73. The distance is $|-3.6-(-1.4)|=|-3.6+1.4|=|-2.2|=2.2$.

74. The distance is $|-5.4-(-1.2)|=|-5.4+1.2|=|-4.2|=4.2$.

75. $6 + (-4) = (-4) + 6$; commutative property of addition

76. $11\cdot(7+4)=11\cdot7+11\cdot4$; distributive property of multiplication over addition

77. $6 + (2 + 7) = (6 + 2) + 7$; associative property of addition

78. $6\cdot(2\cdot3)=6\cdot(3\cdot2)$; commutative property of multiplication

79. $(2 + 3) + (4 + 5) = (4 + 5) + (2 + 3)$; commutative property of addition

80. $7\cdot(11\cdot8)=(11\cdot8)\cdot7$; commutative property of multiplication

81. $2(-8 + 6) = -16 + 12$; distributive property of multiplication over addition

82. $-8(3+11)=-24+(-88)$; distributive property of multiplication over addition

83. $\frac{1}{x+3}(x+3)=1; x\neq-3$, inverse property of multiplication

84. $(x+4)+[-(x+4)]=0$; inverse property of addition

85.
$$\begin{aligned}5(3x+4)-4&=5\cdot3x+5\cdot4-4\\&=15x+20-4\\&=15x+16\end{aligned}$$

86.
$$\begin{aligned}2(5x+4)-3&=2\cdot5x+2\cdot4-3\\&=10x+8-3\\&=10x+5\end{aligned}$$

87.
$$\begin{aligned}5(3x-2)+12x&=5\cdot3x-5\cdot2+12x\\&=15x-10+12x\\&=27x-10\end{aligned}$$

88.
$$\begin{aligned}2(5x-1)+14x&=2\cdot5x-2\cdot1+14x\\&=10x-2+14x\\&=24x-2\end{aligned}$$

89.
$$\begin{aligned}&7(3y-5)+2(4y+3)\\&=7\cdot3y-7\cdot5+2\cdot4y+2\cdot3\\&=21y-35+8y+6\\&=29y-29\end{aligned}$$

90.
$$\begin{aligned}&4(2y-6)+3(5y+10)\\&=4\cdot2y-4\cdot6+3\cdot5y+3\cdot10\\&=8y-24+15y+30\\&=23y+6\end{aligned}$$

91.
$$\begin{aligned}&5(3y-2)-(7y+2)=15y-10-7y-2\\&=8y-12\end{aligned}$$

92.
$$\begin{aligned}&4(5y-3)-(6y+3)\\&=20y-12-6y-3\\&=14y-15\end{aligned}$$

93.
$$\begin{aligned}7-4[3-(4y-5)]&=7-4[3-4y+5]\\&=7-4[8-4y]\\&=7-32+16y\\&=16y-25\end{aligned}$$

94. $6-5[8-(2y-4)]$
$6-5[8-2y+4]$
$=6-5[12-2y]$
$=6-60+10y$
$=10y-54$

95. $8x^2+4-\left[6\left(x^2-2\right)+5\right]$
$=18x^2+4-\left[6x^2-12+5\right]$
$=18x^2+4-\left[6x^2-7\right]$
$=18x^2+4-6x^2+7$
$=18x^2-6x^2+4+7$
$=(18-6)x^2+11=12x^2+11$

96. $4x^2+5-\left[7\left(x^2-2\right)+4\right]$
$=14x^2+5-\left[7x^2-14+4\right]$
$=14x^2+5-\left[7x^2-10\right]$
$=14x^2+5-7x^2+10$
$=14x^2-7x^2+5+10$
$=(14-7)x^2+15$
$=7x^2+15$

97. $-(-14x)=14x$

98. $-(-17y)=17y$

99. $-(2x-3y-6)=-2x+3y+6$

100. $-(5x-13y-1)=-5x+13y+1$

101. $\frac{1}{3}(3x)+[(4y)+(-4y)]=x+0$
$=x$

102. $\frac{1}{2}(2y)+[(-7x)+7x]=y+0=y$

103. $|-6|\ \square\ |-3|$
$6\ \square\ 3$
$6>3$
Since $6>3$, $|-6|>|-3|$.

104. $|-20|\ \square\ |-50|$
$20\ \square\ 50$
$20<50$
Since $20<50$, $|-20|<|-50|$.

105. $\left|\frac{3}{5}\right|\ \square\ |-0.6|$
$|0.6|\ \square\ |\ 0.6|$
$0.6\ \square\ 0.6$
$0.6\ =\ 0.6$
Since $0.6=0.6$, $\left|\frac{3}{5}\right|=|-0.6|$.

106. $\left|\frac{5}{2}\right|\ \square\ |-2.5|$
$|2.5|\ \square\ |-2.5|$
$2.5\ \square\ 2.5$
$2.5\ =\ 2.5$
Since $2.5=2.5$, $\left|\frac{5}{2}\right|=|-2.5|$.

107. $\frac{30}{40}-\frac{3}{4}\ \square\ \frac{14}{15}\cdot\frac{15}{14}$
$\frac{30}{40}-\frac{30}{40}\ \square\ \frac{\cancel{14}}{\cancel{15}}\cdot\frac{\cancel{15}}{\cancel{14}}$
$0\ \square\ 1$
$0<1$
Since $0<1$, $\frac{30}{40}-\frac{3}{4}<\frac{14}{15}\cdot\frac{15}{14}$.

108. $\frac{17}{18}\cdot\frac{18}{17}\ \square\ \frac{50}{60}-\frac{5}{6}$
$\frac{\cancel{17}}{\cancel{18}}\cdot\frac{\cancel{18}}{\cancel{17}}\ \square\ \frac{50}{60}-\frac{50}{60}$
$1\ \square\ 0$
$1>0$
Since $1>0$, $\frac{17}{18}\cdot\frac{18}{17}>\frac{50}{60}-\frac{5}{6}$.

109. $\frac{8}{13} \div \frac{8}{13} \square |-1|$

$\frac{8}{13} \cdot \frac{13}{8} \square 1$

$1 \square 1$

$1 = 1$

Since $1 = 1$, $\frac{8}{13} \div \frac{8}{13} = |-1|$.

110. $|-2| \square \frac{4}{17} \div \frac{4}{17}$

$2 \square \frac{4}{17} \cdot \frac{17}{4}$

$2 \square 1$

$2 > 1$

Since $2 > 1$, $|-2| > \frac{4}{17} \div \frac{4}{17}$.

111. $x-(x+4) = x-x-4 = -4$

112. $x-(8-x) = x-8+x = 2x-8$

113. $6(-5x) = -30x$

114. $10(-4x) = -40x$

115. $5x-2x = 3x$

116. $6x-(-2x) = 6x+2x = 8x$

117. $8x-(3x+6) = 8x-3x-6 = 5x-6$

118. $8-3(x+6) = 8-3x-18 = -3x-10$

119. $N = 17x^2 - 65.4x + 302.2$
$= 17(4)^2 - 65.4(4) + 302.2$
$= 17(16) - 261.6 + 302.2$
$= 272 - 261.6 + 302.2$
≈ 313
The formula models the graph's data very well.

120. $N = 17x^2 - 65.4x + 302.2$
$= 17(3)^2 - 65.4(3) + 302.2$
$= 17(9) - 196.2 + 302.2$
$= 153 - 196.2 + 302.2$
≈ 259
The formula models the graph's data very well.

121. $N = 17x^2 - 65.4x + 302.2$
$= 17(6)^2 - 65.4(6) + 302.2$
$= 17(36) - 392.4 + 302.2$
$= 612 - 392.4 + 302.2$
≈ 522
The formula predicts that there will be 522 U.S. billionaires in 2006.

122. $N = 17x^2 - 65.4x + 302.2$
$= 17(7)^2 - 65.4(7) + 302.2$
$= 17(49) - 457.8 + 302.2$
$= 833 - 457.8 + 302.2$
≈ 677
The formula predicts that there will be 677 U.S. billionaires in 2007.

123. Model 1: $N = -2.04x + 10.24$
$= -2.04(0) + 10.24$
$= 10.24$
Model 2: $N = 0.04x^2 - 3.6x + 11$
$= 0.04(0)^2 - 3.6(0) + 11$
$= 11$
Model 3: $N = 0.76x^3 - 4x^2 + 1.8x + 10.5$
$= 0.76(0)^3 - 4(0)^2 + 1.8(0) + 10.5$
$= 10.5$
Model 3 was the best model for 1999.

124. Model 1: $N = -2.04x + 10.24$
$= -2.04(1) + 10.24$
$= 8.2$
Model 2: $N = 0.04x^2 - 3.6x + 11$
$= 0.04(1)^2 - 3.6(1) + 11$
$= 7.44$
Model 3: $N = 0.76x^3 - 4x^2 + 1.8x + 10.5$
$= 0.76(1)^3 - 4(1)^2 + 1.8(1) + 10.5$
$= 9.06$
Model 3 was the best model for 2000.

125. Model 1: $N = -2.04x + 10.24$
$= -2.04(4) + 10.24$
$= 2.08$
Model 2: $N = 0.04x^2 - 3.6x + 11$
$= 0.04(4)^2 - 3.6(4) + 11$
$= -2.76$

Model 3: $N = 0.76x^3 - 4x^2 + 1.8x + 10.5$
$= 0.76(4)^3 - 4(4)^2 + 1.8(4) + 10.5$
$= 2.34$
Model 3 was the best model for 2003.

126. Model 1: $N = -2.04x + 10.24$
$= -2.04(4) + 10.24$
$= 2.08$
Model 2: $N = 0.04x^2 - 3.6x + 11$
$= 0.04(4)^2 - 3.6(4) + 11$
$= -2.76$
Model 3: $N = 0.76x^3 - 4x^2 + 1.8x + 10.5$
$= 0.76(4)^3 - 4(4)^2 + 1.8(4) + 10.5$
$= 2.34$
Model breakdown occurs for model 2 in 2003 as shown by the negative result.

127. a. $0.05x + 0.12(10,000 - x)$
$= 0.05x + 1200 - 0.12x$
$= 1200 - 0.07x$

b. $1200 - 0.07x = 1200 - 0.07(6000)$
$= \$780$

128. a. $0.06t + 0.5(50 - t)$
$= 0.06t + 25 - 0.5t$
$= 25 - 0.44t$

b. $0.06(20) + 0.5(50 - 20)$
$= 1.2 + 0.5(30)$
$= 1.2 + 15$
$= 16.2$ miles

140. a. False; For example, 1.7 is a rational number and it is not an integer.

b. False; All whole numbers, $\{0, 1, 2, 3, \cdots\}$, are also integers.

c. True; −7.5 is a rational number and it is not positive.

d. False; $-\pi$ is an irrational number that is also negative.

(c) is true.

141. a. False; x has a coefficient of 1.

b. False; $5 + 3(x - 4) = 5 + 3x - 12 = 3x - 7$

c. False; $-x - x = -2x$.

d. True; $x - 0.02(x + 200)$
$= x - 0.02x - 4$
$= x - 0.02x - 0.02(200)$
$= 0.98x - 4$

(d) is true.

142. $\sqrt{2} \approx 1.4$
$1.4 < 1.5$
$\sqrt{2} < 1.5$

143. $-\pi > -3.5$

144. $-\frac{3.14}{2} = -1.57$
$-\frac{\pi}{2} \approx -1.571$
$-1.57 > -1.571$
$-\frac{3.14}{2} > -\frac{\pi}{2}$

145. a. $x = 100$
$\frac{0.5x + 5000}{100}$
$= \frac{5050}{100}$
$= \$50.50$

$x = 1000$
$\frac{0.5(1000) + 5000}{1000}$
$= \frac{5500}{1000}$
$= \$5.50$

$x = 10,000$
$\frac{0.5(10,000) + 5000}{10,000}$
$= \frac{10000}{10000}$
$= \$1$

b. No, they would need make 10,000 clocks for the cost to be \$1 so they can add \$.50.

Section P.2

Check Point Exercises

1. **a.** $\left(2x^3y^6\right)^4 = (2)^4\left(x^3\right)^4\left(y^6\right)^4 = 16x^{12}y^{24}$

b. $\left(-6x^2y^5\right)\left(3xy^3\right)$
$= (-6)\cdot 3\cdot x^2\cdot x\cdot y^5\cdot y^3$
$= -18x^3y^8$

c. $\dfrac{100x^{12}y^2}{20x^{16}y^{-4}} = \left(\dfrac{100}{20}\right)\left(\dfrac{x^{12}}{x^{16}}\right)\left(\dfrac{y^2}{y^{-4}}\right)$
$= 5x^{12-16}y^{2-(-4)}$
$= 5x^{-4}y^6$
$= \dfrac{5y^6}{x^4}$

d. $\left(\dfrac{5x}{y^4}\right)^{-2} = \dfrac{(5)^{-2}(x)^{-2}}{\left(y^4\right)^{-2}}$
$= \dfrac{(5)^{-2}(x)^{-2}}{\left(y^4\right)^{-2}}$
$= \dfrac{5^{-2}x^{-2}}{y^{-8}}$
$= \dfrac{y^8}{5^2x^2}$
$= \dfrac{y^8}{25x^2}$

2. **a.** $-2.6\times10^9 = -2{,}600{,}000{,}000$

b. $3.017\times10^{-6} = 0.000003017$

3. **a.** $5{,}210{,}000{,}000 = 5.21\times10^9$

b. $-0.00000006893 = -6.893\times10^{-8}$

4. $410\times10^7 = \left(4.1\times10^2\right)\times10^7$
$= 4.1\times\left(10^2\times10^7\right)$
$= 4.1\times10^9$

5. **a.** $\left(7.1\times10^5\right)\left(5\times10^{-7}\right)$
$= 7.1\cdot5\times10^5\cdot10^{-7}$
$= 35.5\times10^{-2}$
$= \left(3.55\times10^1\right)\times10^{-2}$
$= 3.55\times\left(10^1\times10^{-2}\right)$
$= 3.55\times10^{-1}$

b. $\dfrac{1.2\times10^6}{3\times10^{-3}} = \dfrac{1.2}{3}\cdot\dfrac{10^6}{10^{-3}}$
$= 0.4\times10^{6-(-3)}$
$= 0.4\times10^9$
$= 4\times10^8$

6. $\dfrac{2.02\times10^{12}}{2.88\times10^8} = \dfrac{2.02}{2.88}\cdot\dfrac{10^{12}}{10^8}$
$\approx 0.7014\times10^4$
≈ 7014
The per capita tax was $7014 in 2002.

7. $S = (1.76\times10^5)[(1.44\times10^{-2}) - r^2]$
$= (1.76\times10^5)[(1.44\times10^{-2}) - 0^2]$
$= 2534.4$
The speed of the blood at the central axis of the artery is 2534.4 centimeters per second.

Exercise Set P.2

1. $5^2\cdot2 = (5\cdot5)\cdot2 = 25\cdot2 = 50$

2. $6^2\cdot2 = (6\cdot6)\cdot2 = 36\cdot2 = 72$

3. $(-2)^6 = (-2)(-2)(-2)(-2)(-2)(-2) = 64$

4. $(-2)^4 = (-2)(-2)(-2)(-2) = 16$

5. $-2^6 = -2\cdot2\cdot2\cdot2\cdot2\cdot2 = -64$

6. $-2^4 = -2\cdot2\cdot2\cdot2 = -16$

7. $(-3)^0 = 1$

8. $(-9)^0 = 1$

9. $-3^0 = -1$

10. $-9^0 = -1$

11. $4^{-3} = \frac{1}{4^3} = \frac{1}{4 \cdot 4 \cdot 4} = \frac{1}{64}$

12. $2^{-6} = \frac{1}{2^6} = \frac{1}{2 \cdot 2 \cdot 2 \cdot 2 \cdot 2 \cdot 2} = \frac{1}{64}$

13. $2^2 \cdot 2^3 = 2^{2+3} = 2^5 = 2 \cdot 2 \cdot 2 \cdot 2 \cdot 2 = 32$

14. $3^3 \cdot 3^2 = 3^{3+2} = 3^5 = 3 \cdot 3 \cdot 3 \cdot 3 \cdot 3 = 243$

15. $(2^2)^3 = 2^{2 \cdot 3} = 2^6 = 2 \cdot 2 \cdot 2 \cdot 2 \cdot 2 \cdot 2 = 64$

16. $(3^3)^2 = 3^{3 \cdot 2} = 3^6 = 3 \cdot 3 \cdot 3 \cdot 3 \cdot 3 \cdot 3 = 729$

17. $\frac{2^8}{2^4} = 2^{8-4} = 2^4 = 2 \cdot 2 \cdot 2 \cdot 2 = 16$

18. $\frac{3^8}{3^4} = 3^{8-4} = 3^4 = 3 \cdot 3 \cdot 3 \cdot 3 = 81$

19. $3^{-3} \cdot 3 = 3^{-3+1} = 3^{-2} = \frac{1}{3^2} = \frac{1}{3 \cdot 3} = \frac{1}{9}$

20. $2^{-3} \cdot 2 = 2^{-3+1} = 2^{-2} = \frac{1}{2^2} = \frac{1}{2 \cdot 2} = \frac{1}{4}$

21. $\frac{2^3}{2^7} = 2^{3-7} = 2^{-4} = \frac{1}{2^4} = \frac{1}{2 \cdot 2 \cdot 2 \cdot 2} = \frac{1}{16}$

22. $\frac{3^4}{3^7} = 3^{4-7} = 3^{-3} = \frac{1}{3^3} = \frac{1}{3 \cdot 3 \cdot 3} = \frac{1}{27}$

23. $x^{-2}y = \frac{1}{x^2} \cdot y = \frac{y}{x^2}$

24. $xy^{-3} = x \cdot \frac{1}{y^3} = \frac{x}{y^3}$

25. $x^0 y^5 = 1 \cdot y^5 = y^5$

26. $x^7 \cdot y^0 = x^7 \cdot 1 = x^7$

27. $x^3 \cdot x^7 = x^{3+7} = x^{10}$

28. $x^{11} \cdot x^5 = x^{11+5} = x^{16}$

29. $x^{-5} \cdot x^{10} = x^{-5+10} = x^5$

30. $x^{-6} \cdot x^{12} = x^{-6+12} = x^6$

31. $(x^3)^7 = x^{3 \cdot 7} = x^{21}$

32. $(x^{11})^5 = x^{11 \cdot 5} = x^{55}$

33. $(x^{-5})^3 = x^{-5 \cdot 3} = x^{-15} = \frac{1}{x^{15}}$

34. $(x^{-6})^4 = x^{-6 \cdot 4} = x^{-24} = \frac{1}{x^{24}}$

35. $\frac{x^{14}}{x^7} = x^{14-7} = x^7$

36. $\frac{x^{30}}{x^{10}} = x^{30-10} = x^{20}$

37. $\frac{x^{14}}{x^{-7}} = x^{14-(-7)} = x^{14+7} = x^{21}$

38. $\frac{x^{30}}{x^{-10}} = x^{30-(-10)} = x^{30+10} = x^{40}$

39. $(8x^3)^2 = 8^2(x^3)^2 = 8^2 x^{3 \cdot 2} = 64x^6$

40. $(6x^4)^2 = (6)^2(x^4)^2 = 6^2 x^{4 \cdot 2} = 36x^8$

41. $\left(-\frac{4}{x}\right)^3 = \frac{(-4)^3}{x^3} = -\frac{64}{x^3}$

42. $\left(-\frac{6}{y}\right)^3 = \frac{(-6)^3}{y^3} = -\frac{216}{y^3}$

43.
$$\begin{aligned}(-3x^2y^5)^2 &= (-3)^2(x^2)^2 \cdot (y^5)^2 \\ &= 9x^{2 \cdot 2}y^{5 \cdot 2} \\ &= 9x^4y^{10}\end{aligned}$$

44.
$$\begin{aligned}(-3x^4y^6)^3 &= (-3)^3(x^4)^3(y^6)^3 \\ &= -27x^{4 \cdot 3}y^{6 \cdot 3} \\ &= -27x^{12}y^{18}\end{aligned}$$

45. $(3x^4)(2x^7) = 3 \cdot 2x^4 \cdot x^7 = 6x^{4+7} = 6x^{11}$

46. $(11x^5)(9x^{12}) = 11 \cdot 9x^5x^{12} = 99x^{5+12} = 99x^{17}$

47.
$$\begin{aligned}(-9x^3y)(-2x^6y^4) &= (-9)(-2)x^3x^6yy^4 \\ &= 18x^{3+6}y^{1+4} \\ &= 18x^9y^5\end{aligned}$$

48.
$$\begin{aligned}(-5x^4y)(-6x^7y^{11}) &= (-5)(-6)x^4x^7yy^{11} \\ &= 30x^{4+7}y^{1+11} \\ &= 30x^{11}y^{12}\end{aligned}$$

49. $\frac{8x^{20}}{2x^4}=\left(\frac{8}{2}\right)\left(\frac{x^{20}}{x^4}\right)=4x^{20-4}=4x^{16}$

50. $\frac{20x^{24}}{10x^6}=\left(\frac{20}{10}\right)\left(\frac{x^{24}}{x^6}\right)=2x^{24-6}=2x^{18}$

51. $\frac{25a^{13}\cdot b^4}{-5a^2\cdot b^3}=\left(\frac{25}{-5}\right)\left(\frac{a^{13}}{a^2}\right)\left(\frac{b^4}{b^3}\right)$
$=-5a^{13-2}b^{4-3}$
$=-5a^{11}b$

52. $\frac{35a^{14}b^6}{-7a^7b^3}=\left(\frac{35}{-7}\right)\left(\frac{a^{14}}{a^7}\right)\left(\frac{b^6}{b^3}\right)$
$=-5a^{14-7}b^{6-3}$
$=-5a^7b^3$

53. $\frac{14b^7}{7b^{14}}=\left(\frac{14}{7}\right)\left(\frac{b^7}{b^{14}}\right)=2\cdot b^{7-14}=2b^{-7}=\frac{2}{b^7}$

54. $\frac{20b^{10}}{10b^{20}}=\left(\frac{20}{10}\right)\left(\frac{b^{10}}{b^{20}}\right)$
$=2b^{10-20}$
$=2b^{-10}$
$=\frac{2}{b^{10}}$

55. $(4x^3)^{-2}=(4^{-2})(x^3)^{-2}$
$=4^{-2}x^{-6}$
$=\frac{1}{4^2x^6}$
$=\frac{1}{16x^6}$

56. $(10x^2)^{-3}=10^{-3}x^{2\cdot(-3)}$
$=10^{-3}x^{-6}$
$=\frac{1}{10^3x^6}$
$=\frac{1}{1000x^6}$

57. $\frac{24x^3\cdot y^5}{32x^7y^{-9}}=\frac{3}{4}x^{3-7}y^{5-(-9)}$
$=\frac{3}{4}x^{-4}y^{14}$
$=\frac{3y^{14}}{4x^4}$

58. $\frac{10x^4y^9}{30x^{12}y^{-3}}=\frac{1}{3}x^{4-12}y^{9-(-3)}$
$=\frac{1}{3}x^{-8}y^{12}$
$=\frac{y^{12}}{3x^8}$

59. $\left(\frac{5x^3}{y}\right)^{-2}=\frac{5^{-2}x^{-6}}{y^{-2}}=\frac{y^2}{25x^6}$

60. $\left(\frac{3x^4}{y}\right)^{-3}=\left(\frac{y}{3x^4}\right)^3$
$=\frac{y^3}{3^3x^{4\cdot 3}}$
$=\frac{y^3}{27x^{12}}$

61. $\left(\frac{-15a^4b^2}{5a^{10}b^{-3}}\right)^3$
$=\left(\frac{-3b^{2-(-3)}}{a^{10-4}}\right)^3$
$=\left(\frac{-3b^5}{a^6}\right)^3$
$=\frac{-27b^{15}}{a^{18}}$

62. $\left(\frac{-30a^{14}b^8}{10a^{17}b^{-2}}\right)^3$
$=\left(\frac{-3b^{8-(-2)}}{a^{17-14}}\right)^3$
$=\left(\frac{-3b^{10}}{a^3}\right)^3$
$=\frac{-27b^{30}}{a^9}$

63. $\left(\dfrac{3a^{-5}b^2}{12a^3b^{-4}}\right)^0 = 1$

64. $\left(\dfrac{4a^{-5}b^3}{12a^3b^{-5}}\right)^0 = 1$

65. $3.8 \times 10^2 = 380$

66. $9.2 \times 10^2 = 920$

67. $6 \times 10^{-4} = 0.0006$

68. $7 \times 10^{-5} = 0.00007$

69. $-7.16 \times 10^6 = -7{,}160{,}000$

70. $-8.17 \times 10^6 = -8{,}170{,}000$

71. $7.9 \times 10^{-1} = 0.79$

72. $6.8 \times 10^{-1} = 0.68$

73. $-4.15 \times 10^{-3} = -0.00415$

74. $-3.14 \times 10^{-3} = -0.00314$

75. $-6.00001 \times 10^{10} = -60{,}000{,}100{,}000$

76. $-7.00001 \times 10^{10} = -70{,}000{,}100{,}000$

77. $32{,}000 = 3.2 \times 10^4$

78. $64{,}000 = 6.4 \times 10^4$

79. $638{,}000{,}000{,}000{,}000{,}000$
$= 6.38 \times 10^{17}$

80. $579{,}000{,}000{,}000{,}000{,}000 = 5.79 \times 10^{17}$

81. $-5716 = -5.716 \times 10^3$

82. $-3829 = -3.829 \times 10^3$

83. $0.0027 = 2.7 \times 10^{-3}$

84. $0.0083 = 8.3 \times 10^{-3}$

85. $-0.00000000504 = -5.04 \times 10^{-9}$

86. $-0.00000000405 = -4.05 \times 10^{-9}$

87. $(3 \times 10^4)(2.1 \times 10^3)$
$= (3 \times 2.1)(10^4 \times 10^3)$
$= 6.3 \times 10^{4+3} = 6.3 \times 10^7$

88. $(2 \times 10^4)(4.1 \times 10^3) = 8.2 \times 10^7$

89. $(1.6 \times 10^{15})(4 \times 10^{-11})$
$= (1.6 \times 4)(10^{15} \times 10^{-11})$
$= 6.4 \times 10^{15+(-11)} = 6.4 \times 10^4$

90. $4(2y-6)+3(5y+10)$
$= 4 \cdot 2y - 4 \cdot 6 + 3 \cdot 5y + 3 \cdot 10$
$= 8y - 24 + 15y + 30$
$= 23y + 6$

91. $(6.1 \times 10^{-8})(2 \times 10^{-4})$
$= (6.1 \times 2)(10^{-8} \times 10^{-4})$
$= 12.2 \times 10^{-8+(-4)}$
$= 12.2 \times 10^{-12} = 1.22 \times 10^{-11}$

92. $(5.1 \times 10^{-8})(3 \times 10^{-4}) = 15.3 \times 10^{-12}$
$= 1.53 \times 10^{-11}$

93. $(4.3 \times 10^8)(6.2 \times 10^4)$
$= (4.3 \times 6.2)(10^8 \times 10^4)$
$= 26.66 \times 10^{8+4}$
$= 26.66 \times 10^{12}$
$= 2.666 \times 10^{13} \approx 2.67 \times 10^{13}$

94. $(8.2 \times 10^8)(4.6 \times 10^4)$
$= 37.72 \times 10^{8+4} = 37.72 \times 10^{12}$
$= 3.772 \times 10^{13} \approx 3.77 \times 10^{13}$

95. $\dfrac{8.4 \times 10^8}{4 \times 10^5} = \dfrac{8.4}{4} \times \dfrac{10^8}{10^5}$
$= 2.1 \times 10^{8-5} = 2.1 \times 10^3$

96. $\dfrac{6.9 \times 10^8}{3 \times 10^5} = 2.3 \times 10^{8-5} = 2.3 \times 10^3$

97. $\dfrac{3.6 \times 10^4}{9 \times 10^{-2}} = \dfrac{3.6}{9} \times \dfrac{10^4}{10^{-2}}$
$= 0.4 \times 10^{4-(-2)}$
$= 0.4 \times 10^6 = 4 \times 10^5$

98. $\frac{1.2\times10^4}{2\times10^{-2}} = 0.6\times10^{4-(-2)} = 0.6\times10^6$
$= (6\times10^{-1})\times10^6 = 6\times10^5$

99. $\frac{4.8\times10^{-2}}{2.4\times10^6} = \frac{4.8}{2.4}\times\frac{10^{-2}}{10^6}$
$= 2\times10^{-2-6} = 2\times10^{-8}$

100. $\frac{7.5\times10^{-2}}{2.5\times10^6} = 3\times10^{-2-6} = 3\times10^{-8}$

101. $\frac{2.4\times10^{-2}}{4.8\times10^{-6}} = \frac{2.4}{4.8}\times\frac{10^{-2}}{10^{-6}}$
$= 0.5\times10^{-2-(-6)}$
$= 0.5\times10^4 = 5\times10^3$

102. $\frac{1.5\times10^{-2}}{5\times10^{-6}} = 0.5\times10^{-2-(-6)}$
$= 0.5\times10^4 = 5\times10^3$

103. $\frac{480,000,000,000}{0.00012} = \frac{4.8\times10^{11}}{1.2\times10^{-4}}$
$= \frac{4.8}{1.2}\times\frac{10^{11}}{10^{-4}}$
$= 4\times10^{11-(-4)}$
$= 4\times10^{15}$

104. $\frac{282,000,000,000}{0.00141} = \frac{2.82\times10^{11}}{1.41\times10^{-3}}$
$= 2\times10^{11-(-3)}$
$= 2\times10^{14}$

105. $\frac{0.00072\times0.003}{0.00024}$
$= \frac{(7.2\times10^{-4})(3\times10^{-3})}{2.4\times10^{-4}}$
$= \frac{7.2\times3}{2.4}\times\frac{10^{-4}\cdot10^{-3}}{10^{-4}} = 9\times10^{-3}$

106. $\frac{66000\times0.001}{0.003\times0.002} = \frac{(6.6\times10^4)(1\times10^{-3})}{(3\times10^{-3})(2\times10^{-3})}$
$= \frac{6.6\times10^1}{6\times10^{-6}} = 1.1\times10^{1-(-6)}$
$= 1.1\times10^7$

107. $\frac{(x^{-2}y)^{-3}}{(x^2y^{-1})^3} = \frac{x^6y^{-3}}{x^6y^{-3}}$
$= x^{6-6}y^{-3-(-3)} = x^0y^0 = 1$

108. $\frac{(xy^{-2})^{-2}}{(x^{-2}y)^{-3}} = \frac{x^{-2}y^4}{x^6y^{-3}}$
$= x^{-2-6}y^{4-(-3)} = x^{-8}y^7 = \frac{y^7}{x^8}$

109. $(2x^{-3}yz^{-6})(2x)^{-5} = 2x^{-3}yz^{-6}\cdot2^{-5}x^{-5}$
$= 2^{-4}x^{-8}yz^{-6} = \frac{y}{2^4x^8z^6} = \frac{y}{16x^8z^6}$

110. $(3x^{-4}yz^{-7})(3x)^{-3} = 3x^{-4}yz^{-7}\cdot3^{-3}x^{-3}$
$= 3^{-2}x^{-7}yz^{-7} = \frac{y}{3^2x^7z^7} = \frac{y}{9x^7z^7}$

111. $\left(\frac{x^3y^4z^5}{x^{-3}y^{-4}z^{-5}}\right)^{-2} = (x^6y^8z^{10})^{-2}$
$= x^{-12}y^{-16}z^{-20} = \frac{1}{x^{12}y^{16}z^{20}}$

112. $\left(\frac{x^4y^5z^6}{x^{-4}y^{-5}z^{-6}}\right)^{-4} = (x^8y^{10}z^{12})^{-4}$
$= x^{-32}y^{-40}z^{-48} = \frac{1}{x^{32}y^{40}z^{48}}$

113. $\frac{(2^{-1}x^{-2}y^{-1})^{-2}(2x^{-4}y^3)^{-2}(16x^{-3}y^3)^0}{(2x^{-3}y^{-5})^2}$
$= \frac{(2^2x^2y^2)(2^{-2}x^8y^{-6})(1)}{(2^2x^{-6}y^{-10})}$
$= \frac{x^{18}y^6}{4}$

114. $\frac{(2^{-1}x^{-3}y^{-1})^{-2}(2x^{-6}y^4)^{-2}(9x^3y^{-3})^0}{(2x^{-4}y^{-6})^2}$
$= \frac{(2^2x^6y^2)(2^{-2}x^{12}y^{-8})(1)}{(2^2x^{-8}y^{-12})}$
$= \frac{x^{26}y^6}{4}$

115. $62.6\text{ million} = 62.6\times10^6 = 6.26\times10^7$
So, 6.26×10^7 people will be 65 and over in 2025.

116. $82.0\text{ million} = 82{,}000{,}000 = 8.2\times10^7$
8.2×10^7 people will be 65 or over in 2050.

117. $131.2-34.9=96.3$
$96.3\text{ million} = 96.3\times10^6 = 9.63\times10^7$ There will be 9.63×10^7 more people 65 and over in the year 2100 than in 2000.

118. $82.0-34.9=47.1$ million more people
$47.1\text{ million} = 47{,}100{,}000 = 4.71\times10^7$
There will be 4.71×10^7 more people 65 and over in the year 2050 than in the year 2000.

119. $20\text{ billion} = 2\times10^{10}$

$$\frac{2\times10^{10}}{2.92\times10^8} = \frac{2}{2.92}\times\frac{10^{10}}{10^8} \approx 0.6849\times10^{10-8} = 0.6849\times10^2 = 6.849\times10^1 \approx 68$$

The average American consumes about 68 hotdogs each year.

120.

$$\frac{6\times10^8}{2.92\times10^8} \approx 2.0548\times10^{8-8} = 2.0548\times10^0 = 2.0548 \approx 2$$

Approximately 2 Big Macs per person would be consumed by each American in a year.

121. $8\text{ billion} = 8\times10^9$

$$\frac{8\times10^9}{3.2\times10^7} = \frac{8}{3.2}\times\frac{10^9}{10^7} = 2.5\times10^{9-7} = 2.5\times10^2 = 250$$

$2.5\times10^2 = 250$ chickens are raised for food each second in the U.S.

122. $127\times3.2\times10^7 = 406.4\times10^7 = 4.064\times10^9$

123.

$$\frac{6.8\times10^{12}}{2.9\times10^8} = \frac{6.8}{2.9}\cdot\frac{10^{12}}{10^8} \approx 2.3448\times10^4 \approx 23{,}448$$

If the national debt was divided evenly among every individual in the U.S., each citizen would have to pay $23,448.

124.

$$\frac{365\text{ days}}{1\text{ year}}\cdot\frac{24\text{ hours}}{1\text{ day}} = 8760\text{ hours/year} = 8.76\times10^3\text{ hours/year}$$

$$\frac{8.76\times10^3\text{ hours}}{1\text{ year}}\cdot\frac{60\text{ minutes}}{1\text{ hour}} = 525.6\times10^3\text{ minutes/year} = 5.256\times10^5\text{ minutes/year}$$

$$\frac{5.256\times10^5\text{ minutes}}{1\text{ year}}\cdot\frac{60\text{ seconds}}{1\text{ minute}} = 315.36\times10^5\text{ seconds/year} = 3.1536\times10^7\text{ seconds/year}$$

There are 3.1536×10^7 seconds in a year.

133. a. False, $4^{-2} = \frac{1}{16} > 4^{-3} = \frac{1}{64}$

b. True, $5^{-2} = \frac{1}{25} > 2^{-5} = \frac{1}{32}$

c. False, $16 = (-2)^4 \neq 2^{-4} = \frac{1}{16}$

d. False. $5^2\cdot5^{-2} = 5^{2-2} = 5^0 = 1$
$2^5\cdot2^{-5} = 2^{5-5} = 2^0 = 1$
1 is not greater than 1

134. The doctor has gathered:

$$2^{-1}+2^{-2} = \frac{1}{2}+\frac{1}{2^2} = \frac{2}{4}+\frac{1}{4} = \frac{3}{4}$$

So, $1-\frac{3}{4}=\frac{1}{4}$ is remaining.

135. $b^A = MN, b^C = M, b^D = N$
$b^A = b^C b^D$
$A = C + D$

136.

$$\frac{70\text{ bts}}{\text{min}}\cdot\frac{60\text{ min}}{\text{hr}}\cdot\frac{24\text{ hrs}}{\text{day}}\cdot\frac{365\text{ days}}{\text{yr}}\cdot 80\text{ yrs}$$

$= 70\cdot60\cdot24\cdot365\cdot80$ beats
$= 2943360000$ beats
$= 2.94336\times10^9$ beats
$\approx 2.94\times10^9$ beats

The heartbeats approximately 2.94×10^9 times over a lifetime of 80 years.

Section P.3

Check Point Exercises

1\. a. $\sqrt{81}=9$

b. $-\sqrt{9}=-3$

c. $\sqrt{\frac{1}{25}}=\frac{1}{5}$

d. $\sqrt{36+64}=\sqrt{100}=10$

e. $\sqrt{36}+\sqrt{64}=6+8=14$

2\. a. $\sqrt{75}=\sqrt{25\cdot 3}=\sqrt{25}\sqrt{3}=5\sqrt{3}$

b. $$\begin{aligned}\sqrt{5x}\cdot\sqrt{10x}&=\sqrt{5x\cdot 10x}\\&=\sqrt{50x^2}\\&=\sqrt{25\cdot 2x^2}\\&=\sqrt{25x^2}\cdot\sqrt{2}\\&=5x\sqrt{2}\end{aligned}$$

3\. a. $\sqrt{\frac{25}{16}}=\frac{\sqrt{25}}{\sqrt{16}}=\frac{5}{4}$

b. $$\begin{aligned}\frac{\sqrt{150x^3}}{\sqrt{2x}}&=\sqrt{\frac{150x^3}{2x}}\\&=\sqrt{75x^2}\\&=\sqrt{25x^2}\cdot\sqrt{3}\\&=5x\sqrt{3}\end{aligned}$$

4\. a. $$\begin{aligned}8\sqrt{13}+9\sqrt{13}&=(8+9)\sqrt{3}\\&=17\sqrt{13}\end{aligned}$$

b. $$\begin{aligned}&\sqrt{17x}-20\sqrt{17x}\\&=1\sqrt{17x}-20\sqrt{17x}\\&=(1-20)\sqrt{17x}\\&=-19\sqrt{17x}\end{aligned}$$

5\. a. $$\begin{aligned}&5\sqrt{27}+\sqrt{12}\\&=5\sqrt{9\cdot 3}+\sqrt{4\cdot 3}\\&=5\cdot 3\sqrt{3}+2\sqrt{3}\\&=15\sqrt{3}+2\sqrt{3}\\&=(15+2)\sqrt{3}\\&=17\sqrt{3}\end{aligned}$$

b. $$\begin{aligned}&6\sqrt{18x}-4\sqrt{8x}\\&=6\sqrt{9\cdot 2x}-4\sqrt{4\cdot 2x}\\&=6\cdot 3\sqrt{2x}-4\cdot 2\sqrt{2x}\\&=18\sqrt{2x}-8\sqrt{2x}\\&=(18-8)\sqrt{2x}\\&=10\sqrt{2x}\end{aligned}$$

6\. a. If we multiply numerator and denominator by $\sqrt{3}$, the denominator becomes $\sqrt{3}\cdot\sqrt{3}=\sqrt{9}=3$. Therefore, multiply by 1, choosing $\frac{\sqrt{3}}{\sqrt{3}}$ for 1.

$$\frac{5}{\sqrt{3}}=\frac{5}{\sqrt{3}}\cdot\frac{\sqrt{3}}{\sqrt{3}}=\frac{5\sqrt{3}}{\sqrt{9}}=\frac{5\sqrt{3}}{3}$$

b. The *smallest* number that will produce a perfect square in the denominator of $\frac{6}{\sqrt{12}}$ is $\sqrt{3}$ because $\sqrt{12}\cdot\sqrt{3}=\sqrt{36}=6$. So multiply by 1, choosing $\frac{\sqrt{3}}{\sqrt{3}}$ for 1.

$$\frac{6}{\sqrt{12}}=\frac{6}{\sqrt{12}}\cdot\frac{\sqrt{3}}{\sqrt{3}}=\frac{6\sqrt{3}}{\sqrt{36}}=\frac{6\sqrt{3}}{6}=\sqrt{3}$$

7\. Multiply by $\frac{4-\sqrt{5}}{4-\sqrt{5}}$.

$$\begin{aligned}\frac{8}{4+\sqrt{5}}&=\frac{8}{4+\sqrt{5}}\cdot\frac{4-\sqrt{5}}{4-\sqrt{5}}\\&=\frac{8(4-\sqrt{5})}{4^2-(\sqrt{5})^2}\\&=\frac{8(4-\sqrt{5})}{16-5}\\&=\frac{8(4-\sqrt{5})}{11}\text{ or }\frac{32-8\sqrt{5}}{11}\end{aligned}$$

8. **a.** $\sqrt[3]{40} = \sqrt[3]{8\cdot 5} = \sqrt[3]{8}\cdot\sqrt[3]{5} = 2\sqrt[3]{5}$

b. $\sqrt[5]{8}\cdot\sqrt[5]{8} = \sqrt[5]{64} = \sqrt[5]{32}\cdot\sqrt[5]{2} = 2\sqrt[5]{2}$

c. $\sqrt[3]{\frac{125}{27}} = \frac{\sqrt[3]{125}}{\sqrt[3]{27}} = \frac{5}{3}$

9. $3\sqrt[3]{81} - 4\sqrt[3]{3}$
$= 3\sqrt[3]{27\cdot 3} - 4\sqrt[3]{3}$
$= 3\cdot 3\sqrt[3]{3} - 4\sqrt[3]{3}$
$= 9\sqrt[3]{3} - 4\sqrt[3]{3}$
$= (9-4)\sqrt[3]{3}$
$= 5\sqrt[3]{3}$

10. **a.** $25^{\frac{1}{2}} = \sqrt{25} = 5$

b. $8^{\frac{1}{3}} = \sqrt[3]{8} = 2$

c. $-81^{\frac{1}{4}} = -\sqrt[4]{81} = -3$

d. $(-8)^{\frac{1}{3}} = \sqrt[3]{-8} = -2$

e. $27^{-\frac{1}{3}} = \frac{1}{27^{\frac{1}{3}}} = \frac{1}{\sqrt[3]{27}} = \frac{1}{3}$

11. **a.** $27^{\frac{4}{3}} = \left(\sqrt[3]{27}\right)^4 = (3)^4 = 81$

b. $4^{\frac{3}{2}} = \left(\sqrt[2]{4}\right)^3 = (2)^3 = 8$

c. $32^{-\frac{2}{5}} = \frac{1}{32^{\frac{2}{5}}} = \frac{1}{\left(\sqrt[5]{32}\right)^2} = \frac{1}{2^2} = \frac{1}{4}$

12. **a.** $\left(2x^{4/3}\right)\left(5x^{8/3}\right)$
$= 2\cdot 5x^{4/3}\cdot x^{8/3}$
$= 10x^{(4/3)+(8/3)}$
$= 10x^{12/3}$
$= 10x^4$

b. $\frac{20x^4}{5x^{3/2}} = \left(\frac{20}{5}\right)\left(\frac{x^4}{x^{3/2}}\right)$
$= 4x^{4-(3/2)}$
$= 4x^{(8/2)-(3/2)}$
$= 4x^{5/2}$

13. $\sqrt[6]{x^3} = x^{3/6} = x^{1/2} = \sqrt{x}$

Exercise Set P.3

1. $\sqrt{36} = \sqrt{6^2} = 6$

2. $\sqrt{25} = \sqrt{5^2} = 5$

3. $-\sqrt{36} = -\sqrt{6^2} = -6$

4. $-\sqrt{25} = -\sqrt{5^2} = -5$

5. $\sqrt{-36}$, The square root of a negative number is not real.

6. $\sqrt{-25}$, The square root of a negative number is not real.

7. $\sqrt{25-16} = \sqrt{9} = 3$

8. $\sqrt{144+25} = \sqrt{169} = 13$

9. $\sqrt{25} - \sqrt{16} = 5 - 4 = 1$

10. $\sqrt{144} + \sqrt{25} = 12 + 5 = 17$

11. $\sqrt{(-13)^2} = \sqrt{169} = 13$

12. $\sqrt{(-17)^2} = \sqrt{289} = 17$

13. $\sqrt{50} = \sqrt{25\cdot 2} = \sqrt{25}\sqrt{2} = 5\sqrt{2}$

14. $\sqrt{27} = \sqrt{9\cdot 3} = \sqrt{9}\sqrt{3} = 3\sqrt{3}$

15. $\sqrt{45x^2} = \sqrt{9x^2\cdot 5}$
$= \sqrt{9x^2}\sqrt{5}$
$= \sqrt{9}\sqrt{x^2}\sqrt{5}$
$= 3|x|\sqrt{5}$

16. $\sqrt{125x^2} = \sqrt{25x^2 \cdot 5}$
$= \sqrt{25x^2}\sqrt{5}$
$= \sqrt{25}\sqrt{x^2}\sqrt{5}$
$= 5|x|\sqrt{5}$

17. $\sqrt{2x} \cdot \sqrt{6x} = \sqrt{2x \cdot 6x}$
$= \sqrt{12x^2}$
$= \sqrt{4x^2} \cdot \sqrt{3}$
$= 2x\sqrt{3}$

18. $\sqrt{10x} \cdot \sqrt{8x} = \sqrt{10x \cdot 8x}$
$= \sqrt{80x^2}$
$= \sqrt{16x^2} \cdot \sqrt{5}$
$= 4x\sqrt{5}$

19. $\sqrt{x^3} = \sqrt{x^2} \cdot \sqrt{x} = x\sqrt{x}$

20. $\sqrt{y^3} = \sqrt{y^2} \cdot \sqrt{y} = y\sqrt{y}$

21. $\sqrt{2x^2} \cdot \sqrt{6x} = \sqrt{2x^2 \cdot 6x}$
$= \sqrt{12x^3}$
$= \sqrt{4x^2} \cdot \sqrt{3x}$
$= 2x\sqrt{3x}$

22. $\sqrt{6x} \cdot \sqrt{3x^2} = \sqrt{6x \cdot 3x^2}$
$= \sqrt{18x^3}$
$= \sqrt{9x^2} \cdot \sqrt{2x}$
$= 3x\sqrt{2x}$

23. $\sqrt{\frac{1}{81}} = \frac{\sqrt{1}}{\sqrt{81}} = \frac{1}{9}$

24. $\sqrt{\frac{1}{49}} = \frac{\sqrt{1}}{\sqrt{49}} = \frac{1}{7}$

25. $\sqrt{\frac{49}{16}} = \frac{\sqrt{49}}{\sqrt{16}} = \frac{7}{4}$

26. $\sqrt{\frac{121}{9}} = \frac{\sqrt{121}}{\sqrt{9}} = \frac{11}{3}$

27. $\frac{\sqrt{48x^3}}{\sqrt{3x}} = \sqrt{\frac{48x^3}{3x}} = \sqrt{16x^2} = 4x$

28. $\frac{\sqrt{72x^3}}{\sqrt{8x}} = \sqrt{\frac{72x^3}{8x}} = \sqrt{9x^2} = 3x$

29. $\frac{\sqrt{150x^4}}{\sqrt{3x}} = \sqrt{\frac{150x^4}{3x}}$
$= \sqrt{50x^3}$
$= \sqrt{25x^2} \cdot \sqrt{2x}$
$= 5x\sqrt{2x}$

30. $\frac{\sqrt{24x^4}}{\sqrt{3x}} = \sqrt{\frac{24x^4}{3x}}$
$= \sqrt{8x^3}$
$= \sqrt{4x^2} \cdot \sqrt{2x}$
$= 2x\sqrt{2x}$

31. $\frac{\sqrt{200x^3}}{\sqrt{10x^{-1}}}$
$= \sqrt{\frac{200x^3}{10x^{-1}}}$
$= \sqrt{20x^{3-(-1)}}$
$= \sqrt{20x^4}$
$= \sqrt{4 \cdot 5x^4}$
$= 2x^2\sqrt{5}$

32. $\frac{\sqrt{500x^3}}{\sqrt{10x^{-1}}} = \sqrt{\frac{500x^3}{10x^{-1}}} = \sqrt{50x^{3-(-1)}}$
$= \sqrt{50x^4} = \sqrt{25 \cdot 2x^4} = 5x^2\sqrt{2}$

33. $7\sqrt{3} + 6\sqrt{3} = (7+6)\sqrt{3} = 13\sqrt{3}$

34. $8\sqrt{5} + 11\sqrt{5} = (8+11)\sqrt{5} = 19\sqrt{5}$

35. $6\sqrt{17x} - 8\sqrt{17x} = (6-8)\sqrt{17x} = -2\sqrt{17x}$

36. $4\sqrt{13x} - 6\sqrt{13x} = (4-6)\sqrt{13x} = -2\sqrt{13x}$

37. $\sqrt{8}+3\sqrt{2}=\sqrt{4\cdot 2}+3\sqrt{2}$
$=2\sqrt{2}+3\sqrt{2}$
$=(2+3)\sqrt{2}$
$=5\sqrt{2}$

38. $\sqrt{20}+6\sqrt{5}=\sqrt{4\cdot 5}+6\sqrt{5}$
$=2\sqrt{5}+6\sqrt{5}$
$=(2+6)\sqrt{5}$
$=8\sqrt{5}$

39. $\sqrt{50x}-\sqrt{8x}=\sqrt{25\cdot 2x}-\sqrt{4\cdot 2x}$
$=5\sqrt{2x}-2\sqrt{2x}$
$=(5-2)\sqrt{2x}$
$=3\sqrt{2x}$

40. $\sqrt{63x}-\sqrt{28x}=\sqrt{9\cdot 7x}-\sqrt{4\cdot 7x}$
$=3\sqrt{7x}-2\sqrt{7x}$
$=(3-2)\sqrt{7x}$
$=\sqrt{7x}$

41. $3\sqrt{18}+5\sqrt{50}=3\sqrt{9\cdot 2}+5\sqrt{25\cdot 2}$
$=3\cdot 3\sqrt{2}+5\cdot 5\sqrt{2}$
$=9\sqrt{2}+25\sqrt{2}$
$=(9+25)\sqrt{2}$
$=34\sqrt{2}$

42. $4\sqrt{12}-2\sqrt{75}=4\sqrt{4\cdot 3}-2\sqrt{25\cdot 3}$
$=4\cdot 2\sqrt{3}-2\cdot 5\sqrt{3}$
$=8\sqrt{3}-10\sqrt{3}$
$=(8-10)\sqrt{3}$
$=-2\sqrt{3}$

43. $3\sqrt{8}-\sqrt{32}+3\sqrt{72}-\sqrt{75}$
$=3\sqrt{4\cdot 2}-\sqrt{16\cdot 2}+3\sqrt{36\cdot 2}-\sqrt{25\cdot 3}$
$=3\cdot 2\sqrt{2}-4\sqrt{2}+3\cdot 6\sqrt{2}-5\sqrt{3}$
$=6\sqrt{2}-4\sqrt{2}+18\sqrt{2}-5\sqrt{3}$
$=20\sqrt{2}-5\sqrt{3}$

44. $3\sqrt{54}-2\sqrt{24}-\sqrt{96}+4\sqrt{63}$
$=3\sqrt{9\cdot 6}-2\sqrt{4\cdot 6}-\sqrt{16\cdot 6}+4\sqrt{9\cdot 7}$
$=3\cdot 3\sqrt{6}-2\cdot 2\sqrt{6}-4\sqrt{6}+4\cdot 3\sqrt{7}$
$=9\sqrt{6}-4\sqrt{6}-4\sqrt{6}+12\sqrt{7}$
$=\sqrt{6}+12\sqrt{7}$

45. $\frac{1}{\sqrt{7}}=\frac{1}{\sqrt{7}}\cdot\frac{\sqrt{7}}{\sqrt{7}}=\frac{\sqrt{7}}{7}$

46. $\frac{2}{\sqrt{10}}=\frac{2}{\sqrt{10}}\cdot\frac{\sqrt{10}}{\sqrt{10}}=\frac{2\sqrt{10}}{10}=\frac{\sqrt{10}}{5}$

47. $\frac{\sqrt{2}}{\sqrt{5}}=\frac{\sqrt{2}}{\sqrt{5}}\cdot\frac{\sqrt{5}}{\sqrt{5}}=\frac{\sqrt{10}}{5}$

48. $\frac{\sqrt{7}}{\sqrt{3}}=\frac{\sqrt{7}}{\sqrt{3}}\cdot\frac{\sqrt{3}}{\sqrt{3}}=\frac{\sqrt{21}}{3}$

49. $\frac{13}{3+\sqrt{11}}=\frac{13}{3+\sqrt{11}}\cdot\frac{3-\sqrt{11}}{3-\sqrt{11}}$
$=\frac{13(3-\sqrt{11})}{3^2-(\sqrt{11})^2}$
$=\frac{13(3-\sqrt{11})}{9-11}$
$=\frac{13(3-\sqrt{11})}{-2}$

50. $\frac{3}{3+\sqrt{7}}=\frac{3}{3+\sqrt{7}}\cdot\frac{3-\sqrt{7}}{3-\sqrt{7}}$
$=\frac{3(3-\sqrt{7})}{3^2-(\sqrt{7})^2}$
$=\frac{3(3-\sqrt{7})}{9-7}$
$=\frac{3(3-\sqrt{7})}{2}$

51. $\frac{7}{\sqrt{5}-2}=\frac{7}{\sqrt{5}-2}\cdot\frac{\sqrt{5}+2}{\sqrt{5}+2}$
$=\frac{7(\sqrt{5}+2)}{(\sqrt{5})^2-2^2}$
$=\frac{7(\sqrt{5}+2)}{5-4}$
$=7(\sqrt{5}+2)$

52. $\frac{5}{\sqrt{3}-1}=\frac{5}{\sqrt{3}-1}\cdot\frac{\sqrt{3}+1}{\sqrt{3}+1}$
$=\frac{5(\sqrt{3}+1)}{(\sqrt{3})^2-1^2}$
$=\frac{5(\sqrt{3}+1)}{3-1}$
$=\frac{5(\sqrt{3}+1)}{2}$

53. $\frac{6}{\sqrt{5}+\sqrt{3}}=\frac{6}{\sqrt{5}+\sqrt{3}}\cdot\frac{\sqrt{5}-\sqrt{3}}{\sqrt{5}-\sqrt{3}}$
$=\frac{6(\sqrt{5}-\sqrt{3})}{(\sqrt{5})^2-(\sqrt{3})^2}$
$=\frac{6(\sqrt{5}-\sqrt{3})}{5-3}$
$=\frac{6(\sqrt{5}-\sqrt{3})}{2}$
$=3(\sqrt{5}-\sqrt{3})$

54. $\frac{11}{\sqrt{7}-\sqrt{3}}=\frac{11}{\sqrt{7}-\sqrt{3}}\cdot\frac{\sqrt{7}+\sqrt{3}}{\sqrt{7}+\sqrt{3}}$
$=\frac{11(\sqrt{7}+\sqrt{3})}{(\sqrt{7})^2-(\sqrt{3})^2}$
$=\frac{11(\sqrt{7}+\sqrt{3})}{7-3}$
$=\frac{11(\sqrt{7}+\sqrt{3})}{4}$

55. $\sqrt[3]{125}=\sqrt[3]{5^3}=5$

56. $\sqrt[3]{8}=\sqrt[3]{2^3}=2$

57. $\sqrt[3]{-8}=\sqrt[3]{(-2)^3}=-2$

58. $\sqrt[3]{-125}=\sqrt[3]{(-5)^3}=-5$

59. $\sqrt[4]{-16}$ is not a real number.

60. $\sqrt[4]{-81}$ is not a real number.

61. $\sqrt[4]{(-3)^4}=|-3|=3$

62. $\sqrt[4]{(-2)^4}=|-2|=2$

63. $\sqrt[5]{(-3)^5}=-3$

64. $\sqrt[5]{(-2)^5}=-2$

65. $\sqrt[5]{-\frac{1}{32}}=\sqrt[5]{-\frac{1}{2^5}}=-\frac{1}{2}$

66. $\sqrt[6]{\frac{1}{64}}=\frac{\sqrt[6]{1}}{\sqrt[6]{2^6}}=\frac{1}{2}$

67. $\sqrt[3]{32}=\sqrt[3]{8\cdot 4}=\sqrt[3]{8}\sqrt[3]{4}=2\cdot\sqrt[3]{4}$

68. $\sqrt[3]{150}$ cannot be simplified further.

69. $\sqrt[3]{x^4}=\sqrt[3]{x^3\cdot x}=x\cdot\sqrt[3]{x}$

70. $\sqrt[3]{x^5}=\sqrt[3]{x^3x^2}=x\sqrt[3]{x^2}$

71. $\sqrt[3]{9}\cdot\sqrt[3]{6}=\sqrt[3]{54}=\sqrt[3]{27\cdot 2}=\sqrt[3]{27}\sqrt[3]{2}=3\sqrt[3]{2}$

72. $\sqrt[3]{12}\cdot\sqrt[3]{4}=\sqrt[3]{48}=\sqrt[3]{8\cdot 6}=2\sqrt[3]{6}$

73. $\frac{\sqrt[5]{64x^6}}{\sqrt[5]{2x}}=\sqrt[5]{\frac{64x^6}{2x}}=\sqrt[5]{32x^5}=2x$

74. $\frac{\sqrt[4]{162x^5}}{\sqrt[4]{2x}}=\sqrt[4]{\frac{162x^5}{2x}}=\sqrt[4]{81x^4}=3x$

75. $4\sqrt[5]{2}+3\sqrt[5]{2}=7\sqrt[5]{2}$

76. $6\sqrt[5]{3}+2\sqrt[5]{3}=8\sqrt[5]{3}$

77. $5\sqrt[3]{16}+\sqrt[3]{54}=5\sqrt[3]{8\cdot 2}+\sqrt[3]{27\cdot 2}$
$=5\cdot 2\sqrt[3]{2}+3\sqrt[3]{2}$
$=10\sqrt[3]{2}+3\sqrt[3]{2}$
$=13\sqrt[3]{2}$

78. $3\sqrt[3]{24}+\sqrt[3]{81}=\sqrt[3]{8\cdot 3}+\sqrt[3]{27\cdot 3}$
$=3\cdot 2\sqrt[3]{3}+3\sqrt[3]{3}$
$=6\sqrt[3]{3}+3\sqrt[3]{3}$
$=9\sqrt[3]{3}$

79. $\sqrt[3]{54xy^3} - y\sqrt[3]{128x}$
$= \sqrt[3]{27 \cdot 2xy^3} - y\sqrt[3]{64 \cdot 2x}$
$= 3y\sqrt[3]{2x} - 4y\sqrt[3]{2x}$
$= -y\sqrt[3]{2x}$

80. $\sqrt[3]{24xy^3} - y\sqrt[3]{81x}$
$= \sqrt[3]{8 \cdot 3xy^3} - y\sqrt[3]{27 \cdot 3x}$
$= 2y\sqrt[3]{3x} - 3y\sqrt[3]{3x}$
$= -y\sqrt[3]{3x}$

81. $\sqrt{2} + \sqrt[3]{8} = \sqrt{2} + 2$

82. $\sqrt{3} + \sqrt[3]{15}$ will not simplify

83. $36^{1/2} = \sqrt{36} = 6$

84. $121^{1/2} = \sqrt{121} = 11$

85. $8^{1/3} = \sqrt[3]{8} = 2$

86. $27^{1/3} = \sqrt[3]{27} = 3$

87. $125^{2/3} = \left(\sqrt[3]{125}\right)^2 = 5^2 = 25$

88. $8^{2/3} = \left(\sqrt[3]{8}\right)^2 = 4$

89. $32^{-4/5} = \dfrac{1}{32^{4/5}} = \dfrac{1}{2^4} = \dfrac{1}{16}$

90. $16^{-5/2} = \dfrac{1}{16^{5/2}} = \dfrac{1}{(\sqrt{16})^5} = \dfrac{1}{4^5} = \dfrac{1}{1024}$

91. $(7x^{1/3})(2x^{1/4}) = 7 \cdot 2x^{1/3} \cdot x^{1/4}$
$= 14 \cdot x^{1/3+1/4}$
$= 14x^{7/12}$

92. $(3x^{2/3})(4x^{3/4}) = 3 \cdot 4x^{2/3} \cdot x^{3/4}$
$= 12 \cdot x^{2/3+3/4}$
$= 12x^{17/12}$

93. $\dfrac{20x^{1/2}}{5x^{1/4}} = \left(\dfrac{20}{5}\right)\left(\dfrac{x^{1/2}}{x^{1/4}}\right)$
$= 4 \cdot x^{1/2-1/4}$
$= 4x^{1/4}$

94. $\dfrac{72x^{3/4}}{9x^{1/3}} = \left(\dfrac{72}{9}\right)\left(\dfrac{x^{3/4}}{x^{1/3}}\right) = 8 \cdot x^{3/4-1/3} = 8x^{5/12}$

95. $\left(x^{2/3}\right)^3 = x^{2/3 \cdot 3} = x^2$

96. $(x^{4/5})^5 = x^{4/5 \cdot 5} = x^4$

97. $(25x^4y^6)^{1/2} = 25^{1/2}x^{4 \cdot 1/2}y^{6 \cdot 1/2} = 5x^2|y|^3$

98. $(125x^9y^6)^{1/3} = 125^{1/3}x^{9/3}y^{6/3} = 5x^3y^2$

99. $\dfrac{\left(3y^{\frac{1}{4}}\right)^3}{y^{\frac{1}{12}}} = \dfrac{27y^{\frac{3}{4}}}{y^{\frac{1}{12}}} = 27y^{\frac{3}{4}-\frac{1}{12}}$
$= 27y^{\frac{8}{12}} = 27y^{\frac{2}{3}}$

100. $\dfrac{\left(2y^{1/5}\right)^4}{y^{3/10}} = \dfrac{2^4\left(y^{1/5}\right)^4}{y^{3/10}}$
$= \dfrac{16y^{4/5}}{y^{3/10}} = 16y^{4/5-3/10} = 16y^{1/2}$

101. $\sqrt[4]{5^2} = 5^{2/4} = 5^{1/2} = \sqrt{5}$

102. $\sqrt[4]{7^2} = 7^{2/4} = 7^{1/2} = \sqrt{7}$

103. $\sqrt[3]{x^6} = x^{6/3} = x^2$

104. $\sqrt[4]{x^{12}} = x^{12/4} = |x|^3$

105. $\sqrt[6]{x^4} = \sqrt[6/2]{x^{4/2}} = \sqrt[3]{x^2}$

106. $\sqrt[9]{x^6} = \sqrt[9/3]{x^{6/3}} = \sqrt[3]{x^2}$

107. $\sqrt[9]{x^6y^3} = x^{\frac{6}{9}}y^{\frac{3}{9}} = x^{\frac{2}{3}}y^{\frac{1}{3}} = \sqrt[3]{x^2y}$

108. $\sqrt[12]{x^4y^8} = |x|^{\frac{4}{12}}|y|^{\frac{8}{12}} = |x|^{\frac{1}{3}}|y|^{\frac{2}{3}} = \sqrt[3]{|x|y^2}$

109. $\sqrt[3]{\sqrt[4]{16} + \sqrt{625}} = \sqrt[3]{2+25} = \sqrt[3]{27} = 3$

110. $\sqrt[3]{\sqrt{\sqrt{169}+\sqrt{9}}+\sqrt{\sqrt[3]{1000}+\sqrt[3]{216}}}$
$= \sqrt[3]{\sqrt{13+3}+\sqrt{10+6}}$
$= \sqrt[3]{\sqrt{16}+\sqrt{16}}$
$= \sqrt[3]{4+4} = \sqrt[3]{8}$
$= 2$

111. $\left(49x^{-2}y^{4}\right)^{-1/2}\left(xy^{1/2}\right)$
$= (49)^{-1/2}\left(x^{-2}\right)^{-1/2}\left(y^{4}\right)^{-1/2}\left(xy^{1/2}\right)$
$= \frac{1}{49^{1/2}}x^{(-2)(-1/2)}y^{(4)(-1/2)}\left(xy^{1/2}\right)$
$= \frac{1}{7}x^{1}y^{-2}\cdot xy^{1/2} = \frac{1}{7}x^{1+1}y^{-2+(1/2)}$
$= \frac{1}{7}x^{2}y^{-3/2} = \frac{x^2}{7y^{3/2}}$

112. $\left(8x^{-6}y^{3}\right)^{1/3}\left(x^{5/6}y^{-1/3}\right)^{6}$
$= 8^{1/3}x^{(-6)(1/3)}y^{(3)(1/3)}x^{(5/6)(6)}y^{(-1/3)(6)}$
$= 2x^{-2}y^{1}x^{5}y^{-2} = 2x^{-2+5}y^{1+(-2)}$
$= 2x^{3}y^{-1} = \frac{2x^3}{y}$

113. $\left(\frac{x^{-5/4}y^{1/3}}{x^{-3/4}}\right)^{-6} = \left(x^{(-5/4)-(-3/4)}y^{1/3}\right)^{-6}$
$= \left(x^{-2/4}y^{1/3}\right)^{-6} = x^{(-2/4)(-6)}y^{(1/3)(-6)}$
$= x^{3}y^{-2} = \frac{x^3}{y^2}$

114. $\left(\frac{x^{1/2}y^{-7/4}}{y^{-5/4}}\right)^{-4} = \left(x^{1/2}y^{(-7/4)-(-5/4)}\right)^{-4}$
$= \left(x^{1/2}y^{-2/4}\right)^{-4} = x^{(1/2)(-4)}y^{(-2/4)(-4)}$
$= x^{-2}y^{2} = \frac{y^2}{x^2}$

115. $d(x) = \sqrt{\frac{3x}{2}}$
$d(72) = \sqrt{\frac{3(72)}{2}}$
$= \sqrt{3(36)}$
$= \sqrt{3}\cdot\sqrt{36}$
$= 6\sqrt{3} \approx 10.4$ miles
A passenger on the pool deck can see roughly 10.4 miles.

116. $r(120) = \sqrt{\frac{3(120)}{2}} = \sqrt{180}$
$= \sqrt{36\cdot 5} = \sqrt{36}\cdot\sqrt{5} = 6\sqrt{5}$ miles

The captain can see $6\sqrt{5} \approx 13.4$ miles.

117. $v = \sqrt{20L}; L = 245$
$v = \sqrt{20\cdot 245} = \sqrt{4900} = 70$

The motorist was traveling 70 miles per hour, so he was speeding.

118. $v = \sqrt{20L}; L = 45$
$v = \sqrt{20\cdot 45} = \sqrt{900} = 30$

The motorist was traveling 30 miles per hour, so she was not speeding.

119. $\frac{7\sqrt{2\cdot 2\cdot 3}}{6} = \frac{7\cdot 2\sqrt{3}}{6} = \frac{14\sqrt{3}}{6} = \frac{7}{3}\sqrt{3}$

120. a. $R_f\frac{\sqrt{c^2-v^2}}{\sqrt{c^2}} = R_f\sqrt{\frac{c^2}{c^2}-\frac{v^2}{c^2}}$
$= R_f\sqrt{1-\left(\frac{v}{c}\right)^2}$

b. $R_f\sqrt{1-\left(\frac{v}{c}\right)^2} = R_f\sqrt{1-\left(\frac{0.9\cancel{c}}{\cancel{c}}\right)^2}$
$= R_f\sqrt{1-(0.9)^2}$
$= R_f\sqrt{1-0.81}$
$= R_f\sqrt{0.19}$
$= 0.44R_f$

When moving at 90% of the speed of light, the aging rate is 0.44 relative to a friend on earth.

$0.44R_f = 44$
$R_f = \frac{44}{0.44} \approx 100$

If you are gone for 44 weeks, approximately 100 weeks have passed for your friend on Earth.

121. a. $C = 35.74 + 0.6215t - 35.74\sqrt[25]{v^4} + 0.4275t\sqrt[25]{v^4}$
$C = 35.74 + 0.6215t - 35.74v^{\frac{4}{25}} + 0.4275tv^{\frac{4}{25}}$

b. $C = 35.74 + 0.6215(25) - 35.74(30)^{\frac{4}{25}} + 0.4275(25)(30)^{\frac{4}{25}}$
$\approx 8° \text{ F}$

122. a. $C - 35.74 + 0.6215t - 35.74\sqrt[25]{v^4} + 0.4275t\sqrt[25]{v^4}$
$C = 35.74 + 0.6215t - 35.74v^{\frac{4}{25}} + 0.4275tv^{\frac{4}{25}}$

b. $C = 35.74 + 0.6215(35) - 35.74(15)^{\frac{4}{25}} + 0.4275(35)(15)^{\frac{4}{25}}$
$\approx 25° \text{ F}$

123. $P = 2l + 2w$
$= 2\left(2\sqrt{20}\right) + 2\left(\sqrt{125}\right)$
$= 4\sqrt{20} + 2\sqrt{125}$
$= 4\sqrt{4 \cdot 5} + 2\sqrt{25 \cdot 5}$
$= 4 \cdot 2\sqrt{5} + 2 \cdot 5\sqrt{5}$
$= 8\sqrt{5} + 10\sqrt{5}$
$= (8 + 10)\sqrt{5} = 18\sqrt{5}$
The perimeter is $18\sqrt{5}$ feet.
$A = lw = 2\sqrt{20} \cdot \sqrt{125}$
$= 2\sqrt{20 \cdot 125} = 2\sqrt{2500}$
$= 2 \cdot 50 = 100$
The area is 100 square feet.

124. $P = 2l + 2w = 2\left(4\sqrt{20}\right) + 2\left(\sqrt{80}\right)$
$= 8\sqrt{20} + 2\sqrt{80} = 8\sqrt{4 \cdot 5} + 2\sqrt{16 \cdot 5}$
$= 8 \cdot 2\sqrt{5} + 2 \cdot 4\sqrt{5} = 16\sqrt{5} + 8\sqrt{5}$
$= 24\sqrt{5}$
The perimeter is $24\sqrt{5}$ feet.
$A = lw = 4\sqrt{20} \cdot \sqrt{80} = 4\sqrt{20 \cdot 80}$
$= 4\sqrt{1600} = 4 \cdot 40 = 160$
The area is 160 square feet.

133. a. false; $(-8)^{\frac{1}{3}} = \sqrt[3]{-8} = -2$ is real.

b. false; $\sqrt{x^2 + y^2} \neq x + y$

c. false; $(8)^{-\frac{1}{3}} = \frac{1}{(8)^{\frac{1}{3}}} = \frac{1}{\sqrt[3]{8}} = \frac{1}{2}$

d. true; $2^{\frac{1}{2}} \cdot 2^{\frac{1}{2}} = 2^{\frac{1}{2}+\frac{1}{2}} = 2^1 = 2$

134. $\left(5 + \sqrt{\boxed{3}}\right)\left(5 - \sqrt{\boxed{3}}\right) = 22$
$25 - \boxed{3} = 22$
$\boxed{3} = 3$

135. $\sqrt{\boxed{25}x^{\boxed{14}}} = 5x^7$

136. $\sqrt{13 + \sqrt{2} + \frac{7}{3 + \sqrt{2}}}$
$= \sqrt{13 + \sqrt{2} + \frac{7}{3 + \sqrt{2}} \cdot \frac{3 - \sqrt{2}}{3 - \sqrt{2}}}$
$= \sqrt{13 + \sqrt{2} + \frac{21 - 7\sqrt{2}}{9 - 2}}$
$= \sqrt{13 + \sqrt{2} + \frac{21 - 7\sqrt{2}}{7}}$
$= \sqrt{13 + \sqrt{2} + 3 - \sqrt{2}}$
$= \sqrt{16}$
$= 4$

137. a. $3^{\frac{1}{2}} \boxed{>} 3^{\frac{1}{3}}$

Calculator Check: $1.7321 > 1.4422$

b. $\sqrt{7}+\sqrt{18} \boxed{>} \sqrt{7+18}$

Calculator Check: $6.8884 > 5$

138. a. $2^{\frac{5}{2}} \cdot 2^{\frac{3}{4}} \div 2^{\frac{1}{4}} = \dfrac{2^{\frac{5}{2}} \cdot 2^{\frac{3}{4}}}{2^{\frac{1}{4}}} = 2^{\frac{5}{2}+\frac{3}{4}-\frac{1}{4}} = 2^3 = 8$

Her son is 8 years old.

b. Son's portion:

$$\frac{8^{-\frac{4}{3}}+2^{-2}}{16^{-\frac{3}{4}}+2^{-1}} = \frac{\dfrac{1}{\left(\sqrt[3]{8}\right)^4}+\dfrac{1}{2^2}}{\dfrac{1}{\left(\sqrt[4]{16}\right)^3}+\dfrac{1}{2}}$$

$$= \frac{\dfrac{1}{2^4}+\dfrac{1}{4}}{\dfrac{1}{2^3}+\dfrac{1}{2}}$$

$$= \frac{\dfrac{1}{16}+\dfrac{1}{4}}{\dfrac{1}{8}+\dfrac{1}{2}}$$

$$= \frac{\dfrac{5}{16}}{\dfrac{5}{8}}$$

$$= \frac{8}{16}$$

$$= \frac{1}{2}$$

Mom's portion:

$$\frac{1}{2}\left(1-\frac{1}{2}\right) = \frac{1}{2}\left(\frac{1}{2}\right) = \frac{1}{4}$$

Section P.4

Check Point Exercises

1. a. $(-17x^3+4x^2-11x-5)+(16x^3-3x^2+3x-15)$

$= (-17x^3+16x^3)+(4x^2-3x^2)+(-11x+3x)+(-5-15)$

$= -x^3+x^2-8x-20$

b. $(13x^2-9x^2-7x+1)-(-7x^3+2x^2-5x+9)$

$= (13x^3-9x^2-7x+1)+(7x^3-2x^2+5x-9)$

$= (13x^3+7x^3)+(-9x^2-2x^2)+(-7x+5x)+(1-9)$

$= 20x^3-11x^2-2x-8$

2. $(5x-2)(3x^2-5x+4)$

$= 5x(3x^2-5x+4)-2(3x^2-5x+4)$

$= 5x\cdot 3x^2-5x\cdot 5x+5x\cdot 4-2\cdot 3x^2+2\cdot 5x-2\cdot 4$

$= 15x^3-25x^2+20x-6x^2+10x-8$

$= 15x^3-31x^2+30x-8$

3. $(7x-5)(4x-3) = 7x\cdot 4x+7x(-3)+(-5)4x+(-5)(-3)$

$= 28x^2-21x-20x+15$

$= 28x^2-41x+15$

4. **a.** $(7x-6y)(3x-y)=(7x)(3x)+(7x)(-y)+(-6y)(3x)+(-6y)(-y)$
$=21x^2-7xy-18xy+6y^2$
$=21x^2-25xy+6y^2$

b. $(2x+4y)^2=(2x)^2+2(2x)(4y)+(4y)^2$
$=4x^2+16xy+16y^2$

5. **a.** $(3x+2+5y)(3x+2-5y)=(3x+2)^2-(5y)^2$
$=9x^2+12x+4-25y^2$
$=9x^2+12x-25y^2+4$

b. $(2x+y+3)^2=(2x+y)^2+2(2x+y)(3)+3^2$
$=4x^2+4xy+y^2+12x+6y+9$
$=4x^2+4xy+12x+y^2+6y+9$

Exercise Set P.4

1. Yes; $2x+3x^2-5=3x^2+2x-5$

2. No; The term $3x^{-1}$ does not have a whole number exponent.

3. No; The form of a polynomial involves addition and subtraction, not division.

4. Yes; $x^2-x^3+x^4-5=x^4-x^3+x^2-5$

5. $3x^2$ has degree 2
$-5x$ has degree 1
4 has degree 0
$3x^2-5x+4$ has degree 2.

6. $-4x^3$ has degree 3
$7x^2$ has degree 2
-11 has degree 0
$-4x^3+7x^2-11$ has degree 3.

7. x^2 has degree 2
$-4x^3$ has degree 3
$9x$ has degree 1
$-12x^4$ has degree 4
63 has degree 0
$x^2-4x^3+9x-12x^4+63$ has degree 4.

8. x^2 has degree 2
$-8x^3$ has degree 3
$15x^4$ has degree 4
91 has degree 0
$x^2-8x^3+15x^4+91$ has degree 4.

9. $(-6x^3+5x^2-8x+9)+(17x^3+2x^2-4x-13)=(-6x^3+17x^3)+(5x^2+2x^2)+(-8x-4x)+(9-13)$
$=11x^3+7x^2-12x-4$

The degree is 3.

10. $(-7x^3+6x^2-11x+13)+(19x^3-11x^2+7x-17) = (-7x^3+19x^3)+(6x^2-11x^2)+(-11x+7x)+(13-17)$
$= 12x^3-5x^2-4x-4$

The degree is 3.

11. $(17x^3-5x^2+4x-3)-(5x^3-9x^2-8x+11) = (17x^3-5x^2+4x-3)+(-5x^3+9x^2+8x-11)$
$= (17x^3-5x^3)+(-5x^2+9x^2)+(4x+8x)+(-3-11)$
$= 12x^3+4x^2+12x-14$

The degree is 3.

12. $(18x^4-2x^3-7x+8)-(9x^4-6x^3-5x+7) = (18x^4-2x^3-7x+8)+(-9x^4+6x^3+5x-7)$
$= (18x^4-9x^4)+(-2x^3+6x^3)+(-7x+5x)+(8-7)$
$= 9x^4+4x^3-2x+1$

The degree is 4.

13. $(5x^2-7x-8)+(2x^2-3x+7)-(x^2-4x-3) = (5x^2-7x-8)+(2x^2-3x+7)+(-x^2+4x+3)$
$= (5x^2+2x^2-x^2)+(-7x-3x+4x)+(-8+7+3)$
$= 6x^2-6x+2$

The degree is 2.

14. $(8x^2+7x-5)-(3x^2-4x)-(-6x^3-5x^2+3) = (8x^2+7x-5)+(-3x^2+4x)+(6x^3+5x^2-3)$
$= 6x^3+(8x^2-3x^2+5x^2)+(7x+4x)+(-5-3)$
$= 6x^3+10x^2+11x-8$

The degree is 3.

15. $(x+1)(x^2-x+1) = x(x^2)-x\cdot x+x\cdot 1+1(x^2)-1\cdot x+1\cdot 1$
$= x^3-x^2+x+x^2-x+1$
$= x^3+1$

16. $(x+5)(x^2-5x+25) = x(x^2)-x(5x)+x(25)+5(x^2)-5(5x)+5(25)$
$= x^3-5x^2+25x+5x^2-25x+125)$
$= x^3+125$

17. $(2x-3)(x^2-3x+5) = (2x)(x^2)+(2x)(-3x)+(2x)(5)+(-3)(x^2)+(-3)(-3x)+(-3)(5)$
$= 2x^3-6x^2+10x-3x^2+9x-15$
$= 2x^3-9x^2+19x-15$

18. $(2x-1)(x^2-4x+3) = (2x)(x^2)+(2x)(-4x)+(2x)(3)+(-1)(x^2)+(-1)(-4x)+(-1)(3)$
$= 2x^3-8x^2+6x-x^2+4x-3$
$= 2x^3-9x^2+10x-3$

19. $(x+7)(x+3) = x^2+3x+7x+21 = x^2+10x+21$

20. $(x+8)(x+5) = x^2+5x+8x+40 = x^2+13x+40$

21. $(x-5)(x+3) = x^2+3x-5x-15 = x^2-2x-15$

22. $(x-1)(x+2) = x^2+2x-x-2 = x^2+x-2$

23. $(3x+5)(2x+1) = (3x)(2x) + 3x(1) + 5(2x) + 5 = 6x^2 + 3x + 10x + 5 = 6x^2 + 13x + 5$

24. $(7x+4)(3x+1) = (7x)(3x) + 7x(1) + 4(3x) + 4(1) = 21x^2 + 7x + 12x + 4 = 21x^2 + 19x + 4$

25. $(2x-3)(5x+3) = (2x)(5x) + (2x)(3) + (-3)(5x) + (-3)(3) = 10x^2 + 6x - 15x - 9 = 10x^2 - 9x - 9$

26. $(2x-5)(7x+2) = (2x)(7x) + (2x)(2) + (-5)(7x) + (-5)(2) = 14x^2 + 4x - 35x - 10 = 14x^2 - 31x - 10$

27. $(5x^2-4)(3x^2-7) = (5x^2)(3x^2) + (5x^2)(-7) + (-4)(3x^2) + (-4)(-7) = 15x^4 - 35x^2 - 12x^2 + 28 = 15x^4 - 47x^2 + 28$

28. $(7x^2-2)(3x^2-5) = (7x^2)(3x^2) + (7x^2)(-5) + (-2)(3x^2) + (-2)(-5) = 21x^4 - 35x^2 - 6x^2 + 10 = 21x^4 - 41x^2 + 10$

29. $(8x^3+3)(x^2-5) = (8x^3)(x^2) + (8x^3)(-5) + (3)(x^2) + (3)(-5) = 8x^5 - 40x^3 + 3x^2 - 15$

30. $(7x^3+5)(x^2-2) = (7x^3)(x^2) + (7x^3)(-2) + (5)(x^2) + (5)(-2) = 7x^5 - 14x^3 + 5x^2 - 10$

31. $(x+3)(x-3) = x^2 - 3^2 = x^2 - 9$

32. $(x+5)(x-5) = x^2 - 5^2 = x^2 - 25$

33. $(3x+2)(3x-2) = (3x)^2 - 2^2 = 9x^2 - 4$

34. $(2x+5)(2x-5) = (2x)^2 - 5^2 = 4x^2 - 25$

35. $(5-7x)(5+7x) = 5^2 - (7x)^2 = 25 - 49x^2$

36. $(4-3x)(4+3x) = 4^2 - (3x)^2 = 16 - 9x^2$

37. $(4x^2+5x)(4x^2-5x) = (4x^2)^2 - (5x)^2 = 16x^4 - 25x^2$

38. $(3x^2+4x)(3x^2-4x) = (3x^2)^2 - (4x)^2 = 9x^4 - 16x^2$

39. $(1-y^5)(1+y^5) = (1)^2 - (y^5)^2 = 1 - y^{10}$

40. $(2-y^5)(2+y^5) = (2)^2 - (y^5)^2 = 4 - y^{10}$

41. $(x+2)^2 = x^2 + 2\cdot x\cdot 2 + 2^2 = x^2 + 4x + 4$

42. $(x+5)^2 = x^2 + 2\cdot x\cdot 5 + 5^2 = x^2 + 10x + 25$

43. $(2x+3)^2 = (2x)^2 + 2(2x)(3) + 3^2 = 4x^2 + 12x + 9$

44. $(3x+2)^2 = (3x)^2 + 2(3x)(2) + 2^2 = 9x^2 + 12x + 4$

45. $(x-3)^2 = x^2 - 2\cdot x\cdot 3 + 3^2 = x^2 - 6x + 9$

46. $(x-4)^2 = x^2 - 2\cdot x\cdot 4 + 4^2 = x^2 - 8x + 16$

47. $(4x^2-1)^2=(4x^2)^2-2(4x^2)(1)+1^2=16x^4-8x^2+1$

48. $(5x^2-3)^2=(5x^2)^2-2(5x^2)(3)+3^2=25x^4-30x^2+9$

49. $(7-2x)^2=7^2-2(7)(2x)+(2x)^2=49-28x+4x^2=4x^2-28x+49$

50. $(9-5x)^2=9^2-2(9)(5x)+(5x)^2=81-90x+25x^2$ or $25x^2-90x+81$

51. $(x+1)^3=x^3+3\cdot x^2\cdot 1+3x\cdot 1^2+1^3=x^3+3x^2+3x+1$

52. $(x+2)^3=x^3+3\cdot x^2\cdot 2+3\cdot x\cdot 2^2+2^3=x^3+6x^2+12x+8$

53. $(2x+3)^3=(2x)^3+3\cdot(2x)^2\cdot 3+3(2x)\cdot 3^2+3^3=8x^3+36x^2+54x+27$

54. $(3x+4)^3=(3x)^3+3(3x)^2\cdot 4+3(3x)\cdot 4^2+4^3=27x^3+108x^2+144x+64$

55. $(x-3)^3=x^3-3\cdot x^3\cdot 3+3\cdot x\cdot 3^2-3^3=x^3-9x^2+27x-27$

56. $(x-1)^3=x^3-3x^2\cdot 1+3x\cdot 1^2-1^3=x^3-3x^2+3x-1$

57. $(3x-4)^3=(3x)^3-3(3x)^2\cdot 4+3(3x)\cdot 4^2-4^3=27x^3-108x^2+144x-64$

58. $(2x-3)^3=(2x)^3-3(2x)^2\cdot 3+3(2x)\cdot 3^2-3^3=8x^3-36x^2+54x-27$

59. $(x+5y)(7x+3y)=x(7x)+x(3y)+(5y)(7x)+(5y)(3y)=7x^2+3xy+35xy+15y^2=7x^2+38xy+15y^2$

60. $(x+9y)(6x+7y)=x(6x)+x(7y)+(9y)(6x)+(9y)(7y)=6x^2+7xy+54xy+63y^2=6x^2+61xy+63y^2$

61. $(x-3y)(2x+7y)=x(2x)+x(7y)+(-3y)(2x)+(-3y)(7y)=2x^2+7xy-6xy-21y^2=2x^2+xy-21y^2$

62. $(3x-y)(2x+5y)=(3x)(2x)+(3x)(5y)+(-y)(2x)+(-y)(5y)=6x^2+15xy-2xy-5y^2=6x^2+13xy-5y^2$

63. $(3xy-1)(5xy+2)=(3xy)(5xy)+(3xy)(2)+(-1)(5xy)+(-1)(2)=15x^2y^2+6xy-5xy-2$
$=15x^2y^2+xy-2$

64. $(7x^2y+1)(2x^2y-3)=(7x^2y)(2x^2y)+(7x^2y)(-3)+(1)2x^2y+(1)(-3)=14x^4y^2-21x^2y+2x^2y-3$
$=14x^4y^2-19x^2y-3$

65. $(7x+5y)^2=(7x)^2+2(7x)(5y)+(5y)^2=49x^2+70xy+25y^2$

66. $(9x+7y)^2=(9x)^2+2(9x)(7y)+(7y)^2=81x^2+126xy+49y^2$

67. $(x^2y^2-3)^2=(x^2y^2)^2-2(x^2y^2)(3)+3^2=x^4y^4-6x^2y^2+9$

68. $(x^2y^2-5)^2=(x^2y^2)^2-2(x^2y^2)(5)+5^2=x^4y^4-10x^2y^2+25$

69. $(x-y)(x^2+xy+y^2) = x(x^2)+x(xy)+x(y^2)+(-y)(x^2)+(-y)(xy)+(-y)(y^2)$
$= x^3+x^2y+xy^2-x^2y-xy^2-y^3$
$= x^3-y^3$

70. $(x+y)(x^2-xy+y^2) = x\left(x^2\right)+x(-xy)+x(y^2)+y(x^2)+y(-xy)+y(y^2)$
$= x^3-x^2y+xy^2+x^2y-xy^2+y^3$
$= x^3+y^3$

71. $(3x+5y)(3x-5y) = (3x)^2-(5y)^2 = 9x^2-25y^2$

72. $(7x+3y)(7x-3y) = (7x)^2-(3y)^2 = 49x^2-9y^2$

73. $(x+y+3)(x+y-3) = (x+y)^2-3^2 = x^2+2xy+y^2-9$

74. $(x+y+5)(x+y-5) = (x+y)^2-5^2 = x^2+2xy+y^2-25$

75. $(3x+7-5y)(3x+7+5y) = (3x+7)^2-(5y)^2 = 9x^2+42x+49-25y^2$

76. $(5x+7y-2)(5x+7y+2) = (5x+7y)^2-2^2 = 25x^2+70xy+49y^2-4$

77. $[5y-(2x+3)][5y+(2x+3)] = (5y)^2-(2x+3)^2 = 25y^2-(4x^2+12x+9) = 25y^2-4x^2-12x-9$

78. $[8y+(7-3x)][8y-(7-3x)] = (8y)^2-(7-3x)^2 = 64y^2-(49-42x+9x^2) = 64y^2-49+42x-9x^2$

79. $(x+y+1)^2 = (x+y)^2+2(x+y)+1 = x^2+2xy+y^2+2x+2y+1$

80. $(x+y+2)^2 = (x+y)^2+2(x+y)(2)+2^2 = x^2+2xy+y^2+4x+4y+4$

81. $(2x+y+1)^2 = (2x+y)^2+2(2x+y)+1 = 4x^2+4xy+y^2+4x+2y+1$

82. $(5x+1+6y)^2 = (5x+1)^2+2(5x+1)(6y)+(6y)^2 = 25x^2+10x+60xy+1+12y+36y^2$

83. $(3x+4y)^2-(3x-4y)^2 = \left[(3x)^2+2(3x)(4y)+(4y)^2\right]-\left[(3x)^2-2(3x)(4y)+(4y)^2\right]$
$= \left(9x^2+24xy+16y^2\right)-\left(9x^2-24xy+16y^2\right)$
$= 9x^2+24xy+16y^2-9x^2+24xy-16y^2$
$= 48xy$

84. $(5x+2y)^2-(5x-2y)^2 = \left[(5x)^2+2(5x)(2y)+(2y)^2\right]-\left[(5x)^2-2(5x)(2y)+(2y)^2\right]$
$= \left(25x^2+20xy+4y^2\right)-\left(25x^2-20xy+4y^2\right)$
$= 25x^2+20xy+4y^2-25x^2+20xy-4y^2$
$= 40xy$

85. $(5x-7)(3x-2)-(4x-5)(6x-1)$
$=\left[15x^2-10x-21x+14\right]-\left[24x^2-4x-30x+5\right]$
$=\left(15x^2-31x+14\right)-\left(24x^2-34x+5\right)$
$=15x^2-31x+14-24x^2+34x-5$
$=-9x^2+3x+9$

86. $(3x+5)(2x-9)-(7x-2)(x-1)$
$=\left(6x^2-27x+10x-45\right)-\left(7x^2-7x-2x+2\right)$
$=\left(6x^2-17x-45\right)-\left(7x^2-9x+2\right)$
$=6x^2-17x-45-7x^2+9x-2$
$=-x^2-8x-47$

87. $(2x+5)(2x-5)\left(4x^2+25\right)$
$=\left[(2x)^2-5^2\right]\left(4x^2+25\right)$
$=\left(4x^2-25\right)\left(4x^2+25\right)$
$=\left(4x^2\right)^2-(25)^2$
$=16x^4-625$

88. $(3x+4)(3x-4)\left(9x^2+16\right)$
$=\left[(3x)^2-4^2\right]\left(9x^2+16\right)$
$=\left(9x^2-16\right)\left(9x^2+16\right)$
$=\left(9x^2\right)^2-(16)^2$
$=81x^4-256$

89. $\dfrac{(2x-7)^5}{(2x-7)^3}=(2x-7)^{5-3}$
$=(2x-7)^2$
$=(2x)^2-2(2x)(7)+(7)^2$
$=4x^2-28x+49$

90. $\frac{(5x-3)^6}{(5x-3)^4} = (5x-3)^{6-4}$

$= (5x-3)^2$

$= (5x)^2 - 2(5x)(3) + (3)^2$

$= 25x^2 - 30x + 9$

91. Model 4 is not in standard form.
$N = 0.09x^2 + 0.01x^3 + 1.1x + 5.64$
Model 4 in standard form:
$N = 0.01x^3 + 0.09x^2 + 1.1x + 5.64$

92. Model 2 is not a polynomial model

93. Model 1:
$N = 1.8x + 5.1$
$N = 1.8(2) + 5.1$
$N = 8.7$
Model 2:
$N = 5.6(1.2)^x$
$N = 5.6(1.2)^2$
$N = 8.064$
Model 3:
$N = 0.17x^2 + 0.95x + 5.68$
$N = 0.17(2)^2 + 0.95(2) + 5.68$
$N = 8.26$

Model 4:
$N = 0.09x^2 + 0.01x^3 + 1.1x + 5.64$
$N = 0.09(2)^2 + 0.01(2)^3 + 1.1(2) + 5.64$
$N = 8.28$

Model 1 best describes the data in 2000.

94. Model 1:
$N = 1.8x + 5.1$
$N = 1.8(0) + 5.1$
$N = 5.1$
Model 2:
$N = 5.6(1.2)^x$
$N = 5.6(1.2)^0$
$N = 5.6$
Model 3:

$N = 0.17x^2 + 0.95x + 5.68$

$N = 0.17(0)^2 + 0.95(0) + 5.68$

$N = 5.68$

Model 4:

$N = 0.09x^2 + 0.01x^3 + 1.1x + 5.64$

$N = 0.09(0)^2 + 0.01(0)^3 + 1.1(0) + 5.64$

$N = 5.64$

Model 3 best describes the data in 1998.

95. Model 3 is the model of degree 2.

$N = 0.17x^2 + 0.95x + 5.68$

$N = 0.17(5)^2 + 0.95(5) + 5.68$

$N = 14.68$

Model 3 best describes the data in 2003 very well.

96. Model 4 is the polynomial model that is not in standard form.

$N = 0.09x^2 + 0.01x^3 + 1.1x + 5.64$

$N = 0.09(4)^2 + 0.01(4)^3 + 1.1(4) + 5.64$

$N = 12.12$

Model 4 best describes the data in 2002 very well.

97.
$$\begin{aligned} x(8-2x)(10-2x) &= x\left(80 - 36x + 4x^2\right) \\ &= 80x - 36x^2 + 4x^3 \\ &= 4x^3 - 36x^2 + 80x \end{aligned}$$

98.
$$\begin{aligned} x(8-2x)(5-2x) &= x\left(40 - 26x + 4x^2\right) \\ &= 40x - 26x^2 + 4x^3 \\ &= 4x^3 - 26x^2 + 40x \end{aligned}$$

99. $(x+9)(x+3) - (x+5)(x+1)$

$= x^2 + 12x + 27 - \left(x^2 + 6x + 5\right)$

$= x^2 + 12x + 27 - x^2 - 6x - 5$

$= 6x + 22$

100. $(x+4)(x+3) - (x+2)(x+1)$

$= x^2 + 7x + 12 - \left(x^2 + 3x + 2\right)$

$= x^2 + 7x + 12 - x^2 - 3x - 2$

$= 4x + 10$

108. $(x+3)(x-1) + ((x+3) - x)(x - (x-1))$

$= (x+3)(x-1) + 3(x - x + 1)$

$= x^2 - x + 3x - 3 + 3$

$= x^2 + 2x$

109. $(2x-1)x(x+3)-x(x-2)x$
$=(2x^2+5x-3)(x+2)-x^2(x-2)$
$=2x^3+5x^2-3x-x^3+2x^2$
$=x^3+7x^2-3x$

110. $(x+5)(2x+1)(x+2)-3\cdot x(x+5)$
$=(2x^2+11x+5)(x+2)-3x^2-15x$
$=2x^3+15x^2+27x+10-3x^2-15x$
$=2x^3+12x^2+12x+10$

111. $(y^n+2)(y^n-2)-(y^n-3)^2$
$=y^{2n}-4-(y^{2n}-6y^n+9)$
$=y^{2n}-4-y^{2n}+6y^n-9$
$=6y^n-13$

Section P.5

Check Point Exercises

1. **a.** $10x^3-4x^2$
$=2x^2(5x)-2x^2(2)$
$=2x^2(5x-2)$

b. $2x(x-7)+3(x-7)$
$=(x-7)(2x+3)$

2. $x^3+5x^2-2x-10$
$=(x^3+5x^2)-(2x+10)$
$=x^2(x+5)-2(x+5)$
$=(x+5)(x^2-2)$

3. **a.** Find two numbers whose product is 40 and whose sum is 13. The required integers are 8 and 5. Thus,
$x^2+13x+40=(x+5)(x+8)$ or $(x+8)(x+5)$

b. Find two numbers whose product is –14 and whose sum is –5. The required integers are –7 and 2. Thus,
$x^2-5x-14=(x-7)(x+2)$ or $(x+2)(x-7)$.

4. Find two First terms whose product is $6x^2$.
$6x^2+19x-7=(6x\quad)(x\quad)$
$6x^2+19x-7=(3x\quad)(2x\quad)$

Find two Last terms whose product is –7.
The possible factors are 1(–7) and –1(7).

Try various combinations of these factors to find the factorization in which the sum of the Outside and Inside products is 19*x*.

Possible Factors of $6x^2+19x-7$	Sum of Outside and Inside Products (Should Equal $19x$)
$(6x+1)(x-7)$	$-42x+x=-41x$
$(6x-7)(x+1)$	$6x-7x=-x$
$(6x-1)(x+7)$	$42x-x=41x$
$(6x+7)(x-1)$	$-6x+7x=x$
$(3x+1)(2x-7)$	$-21x+2x=-19x$
$(3x-7)(2x+1)$	$3x-14x=-11x$
$(3x-1)(2x+7)$	$21x-2x=19x$
$(3x+7)(2x-1)$	$-3x+14x=11x$

Thus, $6x^2+19x-7=(3x-1)(2x+7)$ or $(2x+7)(3x-1)$.

5. Find two First terms whose product is $3x^2$.
$3x^2-13xy+4y^2=(3x\quad)(x\quad)$

Find two Last terms whose product is $4y^2$.
The possible factors are $(2y)(2y)$, $(-2y)(-2y)$, $(4y)(y)$, and $(-4y)(-y)$.

Try various combinations of these factors to find the factorization in which the sum of the Outside and Inside products is $-13xy$.

$3x^2-13xy+y^2=(3x-y)(x-4y)$ or $(x-4y)(3x-y)$.

6. Express each term as the square of some monomial. Then use the formula for factoring A^2-B^2.

a. $x^2-81=x^2-9^2=(x+9)(x-9)$

b. $36x^2-25=(6x)^2-5^2=(6x+5)(6x-5)$

7. Express $81x^4-16$ as the difference of two squares and use the formula for factoring A^2-B^2.
$81x^4-16=(9x^2)^2-4^2=(9x^2+4)(9x^2-4)$

The factor $9x^2-4$ is the difference of two squares and can be factored. Express $9x^2-4$ as the difference of two squares and again use the formula for factoring A^2-B^2.
$(9x^2+4)(9x^2-4)=(9x^2+4)\left[(3x)^2-2^2\right]=(9x^2+4)(3x+2)(3x-2)$

Thus, factored completely,
$81x^4-16=(9x^2+4)(3x+2)(3x-2)$.

8. **a.** $x^2+14x+49=x^2+2\cdot x\cdot 7+7^2=(x+7)^2$

b. Since $16x^2=(4x)^2$ and $49=7^2$, check to see if the middle term can be expressed as twice the product of $4x$ and 7. Since $2\cdot 4x\cdot 7=56x$, $16x^2-56x+49$ is a perfect square trinomial. Thus,

$$\begin{aligned}16x^2-56x+49&=(4x)^2-2\cdot 4x\cdot 7+7^2\\&=(4x-7)^2\end{aligned}$$

9. **a.** $= x^3 + 1^3$
$= (x+1)(x^2 - x \cdot 1 + 1^2)$
$= (x+1)(x^2 - x + 1)$

b. $-8 = (5x)^3 - 2^3$
$= (5x-2)\left[(5x)^2 + (5x)(2) + 2^2\right]$
$= (5x-2)(25x^2 + 10x + 4)$

10. Factor out the greatest common factor.
$3x^3 - 30x^2 + 75x = 3x\left(x^2 - 10x + 25\right)$
Factor the perfect square trinomial.
$3x\left(x^2 - 10x + 25\right) = 3x\left(x-5\right)^2$

11. Reorder to write as a difference of squares.
$x^2 - 36a^2 + 20x + 100$
$= x^2 + 20x + 100 - 36a^2$
$= \left(x^2 + 20x + 100\right) - 36a^2$
$= \left(x+10\right)^2 - 36a^2$
$= \left(x + 10 + 6a\right)\left(x + 10 - 6a\right)$

12. $x\left(x-1\right)^{-\frac{1}{2}} + \left(x-1\right)^{\frac{1}{2}}$
$= \left(x-1\right)^{-\frac{1}{2}}\left[x + \left(x-1\right)^{\frac{1}{2}-\left(-\frac{1}{2}\right)}\right]$
$= \left(x-1\right)^{-\frac{1}{2}}\left[x + \left(x-1\right)\right]$
$= \left(x-1\right)^{-\frac{1}{2}}\left(2x-1\right)$
$= \dfrac{\left(2x-1\right)}{\left(x-1\right)^{\frac{1}{2}}}$

Exercise Set P.5

1. $18x + 27 = 9 \cdot 2x + 9 \cdot 3 = 9(2x+3)$

2. $16x - 24 = 8(2x) + 8(-3) = 8(2x-3)$

3. $3x^2 + 6x = 3x \cdot x + 3x \cdot 2 = 3x(x+2)$

4. $4x^2 - 8x = 4x(x) + 4x(-2) = 4x(x-2)$

5. $9x^4 - 18x^3 + 27x^2$
$= 9x^2(x^2) + 9x^2(-2x) + 9x^2(3)$
$= 9x^2(x^2 - 2x + 3)$

6. $6x^4 - 18x^3 + 12x^2$
$= 6x^2(x^2) + 6x^2(-3x) + 6x^2(2)$
$= 6x^2\left(x^2 - 3x + 2\right)$

7. $x(x+5) + 3(x+5) = (x+5)(x+3)$

8. $x(2x+1) + 4(2x+1) = (2x+1)(x+4)$

9. $x^2(x-3) + 12(x-3) = (x-3)(x^2+12)$

10. $x^2\left(2x+5\right) + 17\left(2x+5\right) = \left(2x+5\right)\left(x^2+17\right)$

11. $x^3 - 2x^2 + 5x - 10 = x^2(x-2) + 5(x-2)$
$= (x^2+5)(x-2)$

12. $x^3 - 3x^2 + 4x - 12 = x^2\left(x-3\right) + 4\left(x-3\right)$
$= \left(x-3\right)\left(x^2+4\right)$

13. $x^3 - x^2 + 2x - 2 = x^2(x-1) + 2(x-1)$
$= (x-1)(x^2+2)$

14. $x^3 + 6x^2 - 2x - 12 = x^2\left(x+6\right) - 2\left(x+6\right)$
$= \left(x+6\right)\left(x^2-2\right)$

15. $3x^3 - 2x^2 - 6x + 4 = x^2(3x-2) - 2(3x-2)$
$= (3x-2)(x^2-2)$

16. $x^3 - x^2 - 5x + 5 = x^2\left(x-1\right) - 5\left(x-1\right)$
$= \left(x-1\right)\left(x^2-5\right)$

17. $x^2 + 5x + 6 = (x+2)(x+3)$

18. $x^2 + 8x + 15 = (x+3)(x+5)$

19. $x^2 - 2x - 15 = (x-5)(x+3)$

20. $x^2 - 4x - 5 = (x-5)(x+1)$

21. $x^2 - 8x + 15 = (x-5)(x-3)$

22. $x^2 - 14x + 45 = (x-5)(x-9)$

23. $3x^2 - x - 2 = (3x+2)(x-1)$

24. $2x^2 + 5x - 3 = (2x-1)(x+3)$

25. $3x^2 - 25x - 28 = (3x-28)(x+1)$

26. $3x^2 - 2x - 5 = (3x-5)(x+1)$

27. $6x^2 - 11x + 4 = (2x-1)(3x-4)$

28. $6x^2 - 17x + 12 = (2x-3)(3x-4)$

29. $4x^2 + 16x + 15 = (2x+3)(2x+5)$

30. $8x^2 + 33x + 4 = (8x+1)(x+4)$

31. $9x^2 - 9x + 2 = (3x-1)(3x-2)$

32. $9x^2 + 5x - 4 = (9x-4)(x+1)$

33. $20x^2 + 27x - 8 = (5x+8)(4x-1)$

34. $15x^2 - 19x + 6 = (3x-2)(5x-3)$

35. $2x^2 + 3xy + y^2 = (2x+y)(x+y)$

36. $3x^2 + 4xy + y^2 = (3x+y)(x+y)$

37. $6x^2 - 5xy - 6y^2 = (3x+2y)(2x-3y)$

38. $6x^2 - 7xy - 5y^2 = (3x-5y)(2x+y)$

39. $x^2 - 100 = x^2 - 10^2 = (x+10)(x-10)$

40. $x^2 - 144 = x^2 - 12^2 = (x+12)(x-12)$

41. $36x^2 - 49 = (6x)^2 - 7^2 = (6x+7)(6x-7)$

42. $64x^2 - 81 = (8x)^2 - 9^2 = (8x+9)(8x-9)$

43. $$\begin{aligned} 9x^2 - 25y^2 &= (3x)^2 - (5y)^2 \\ &= (3x+5y)(3x-5y) \end{aligned}$$

44. $$\begin{aligned} 36x^2 - 49y^2 &= (6x)^2 - (7y)^2 \\ &= (6x+7y)(6x-7y) \end{aligned}$$

45. $$\begin{aligned} x^4 - 16 &= (x^2)^2 - 4^2 \\ &= (x^2+4)(x^2-4) \\ &= (x^2+4)(x+2)(x-2) \end{aligned}$$

46. $$\begin{aligned} x^4 - 1 &= (x^2)^2 - 1^2 = (x^2+1)(x^2-1) \\ &= (x^2+1)(x+1)(x-1) \end{aligned}$$

47. $$\begin{aligned} 16x^4 - 81 &= (4x^2)^2 - 9^2 \\ &= (4x^2+9)(4x^2-9) \\ &= (4x^2+9)[(2x)^2 - 3^2] \\ &= (4x^2+9)(2x+3)(2x-3) \end{aligned}$$

48. $$\begin{aligned} 81x^4 - 1 &= (9x^2)^2 - 1^2 \\ &= (9x^2+1)(9x^2-1) \\ &= (9x^2+1)[(3x)^2 - 1^2] \\ &= (9x^2+1)(3x+1)(3x-1) \end{aligned}$$

49. $x^2 + 2x + 1 = x^2 + 2 \cdot x \cdot 1 + 1^2 = (x+1)^2$

50. $x^2 + 4x + 4 = x^2 + 2 \cdot x \cdot 2 + 2^2 = (x+2)^2$

51. $$\begin{aligned} x^2 - 14x + 49 &= x^2 - 2 \cdot x \cdot 7 + 7^2 \\ &= (x-7)^2 \end{aligned}$$

52. $x^2 - 10x + 25 = x^2 - 2 \cdot x \cdot 5 + 5^2 = (x-5)^2$

53. $$\begin{aligned} 4x^2 + 4x + 1 &= (2x)^2 + 2 \cdot 2x \cdot 1 + 1^2 \\ &= (2x+1)^2 \end{aligned}$$

54. $25x^2 + 10x + 1 = (5x)^2 + 2 \cdot 5x \cdot 1 + 1^2 = (5x+1)^2$

55. $$\begin{aligned} 9x^2 - 6x + 1 &= (3x)^2 - 2 \cdot 3x \cdot 1 + 1^2 \\ &= (3x-1)^2 \end{aligned}$$

56. $64x^2 - 16x + 1 = (8x)^2 - 2 \cdot 8x \cdot 1 + 1^2 = (8x-1)^2$

57. $$\begin{aligned} x^3 + 27 &= x^3 + 3^3 \\ &= (x+3)(x^2 - x \cdot 3 + 3^2) \\ &= (x+3)(x^2 - 3x + 9) \end{aligned}$$

58. $x^3+64=x^3+4^3$
$=(x+4)(x^2-x\cdot 4+4^2)$
$=(x+4)(x^2-4x+16)$

59. $x^3-64=x^3-4^3$
$=(x-4)(x^2+x\cdot 4+4^2)$
$=(x-4)(x^2+4x+16)$

60. $x^3-27=x^3-3^3$
$=(x-3)(x^2+x\cdot 3+3^2)$
$=(x-3)(x^2+3x+9)$

61. $8x^3-1=(2x)^3-1^3$
$=(2x-1)[(2x)^2+(2x)(1)+1^2]$
$=(2x-1)(4x^2+2x+1)$

62. $27x^3-1=(3x)^3-1^3$
$=(3x-1)[(3x)^2+(3x)(1)+1^2]$
$=(3x-1)(9x^2+3x+1)$

63. $64x^3+27=(4x)^3+3^3$
$=(4x+3)[(4x)^2-(4x)(3)+3^2]$
$=(4x+3)(16x^2-12x+9)$

64. $8x^3+125=(2x)^3+5^3$
$=(2x+5)[(2x)^2-(2x)(5)+5^2]$
$=(2x+5)(4x^2-10x+25)$

65. $3x^3-3x=3x(x^2-1)=3x(x+1)(x-1)$

66. $5x^3-45x=5x(x^2-9)=5x(x+3)(x-3)$

67. $4x^2-4x-24=4(x^2-x-6)$
$=4(x+2)(x-3)$

68. $6x^2-18x-60=6(x^2-3x-10)$
$=6(x+2)(x-5)$

69. $2x^4-162=2(x^4-81)$
$=2[(x^2)^2-9^2]$
$=2(x^2+9)(x^2-9)$
$=2(x^2+9)(x^2-3^2)$
$=2(x^2+9)(x+3)(x-3)$

70. $7x^4-7=7(x^4-1)$
$=7[(x^2)^2-1^2]$
$=7(x^2+1)(x^2-1)$
$=7(x^2+1)(x+1)(x-1)$

71. $x^3+2x^2-9x-18=(x^3+2x^2)-(9x+18)$
$=x^2(x+2)-9(x+2)$
$=(x^2-9)(x+2)$
$=(x^2-3^2)(x+2)$
$=(x-3)(x+3)(x+2)$

72. $x^3+3x^2-25x-75=(x^3+3x^2)-(25x+75)$
$=x^2(x+3)-25(x+3)$
$=(x^2-25)(x+3)$
$=(x^2-5^2)(x+3)$
$=(x-5)(x+5)(x+3)$

73. $2x^2-2x-112=2(x^2-x-56)=2(x-8)(x+7)$

74. $6x^2-6x-12=6(x^2-x-2)$
$=6(x-2)(x+1)$

75. $x^3-4x=x(x^2-4)$
$=x(x^2-2^2)$
$=x(x-2)(x+2)$

76. $9x^3-9x=9x(x^2-1)=9x(x-1)(x+1)$

77. x^2+64 is prime.

78. x^2+36 is prime.

79. $x^3+2x^2-4x-8=(x^3+2x^2)+(-4x-8)$
$=x^2(x+2)-4(x+2)=(x^2-4)(x+2)=(x^2-2^2)(x+2)=(x-2)(x+2)(x+2)=(x-2)(x+2)^2$

80. x^3+2x^2-x-2
$=(x^3+2x^2)+(-x-2)=x^2(x+2)-1(x+2)=(x^2-1)(x+2)=(x^2-1^2)(x+2)=(x-1)(x+1)(x+2)$

81. $y^5 - 81y$
$= y(y^4 - 81) = y[(y^2)^2 - 9^2] = y(y^2 + 9)(y^2 - 9) = y(y^2 + 9)(y^2 - 3^2) = y(y^2 + 9)(y + 3)(y - 3)$

82.
$y^5 - 16y$
$= y(y^4 - 16) = y[(y^2)^2 - 4^2] = y(y^2 + 4)(y^2 - 4) = y(y^2 + 4)(y^2 - 2^2) = y(y^2 + 4)(y + 2)(y - 2)$

83. $20y^4 - 45y^2 = 5y^2(4y^2 - 9) = 5y^2[(2y)^2 - 3^2] = 5y^2(2y + 3)(2y - 3)$

84. $48y^4 - 3y^2 = 3y^2(16y^2 - 1) = 3y^2[(4y)^2 - 1^2] = 3y^2(4y + 1)(4y - 1)$

85. $x^2 - 12x + 36 - 49y^2 = (x^2 - 12x + 36) - 49y^2 = (x - 6)^2 - 49y^2 = (x - 6 + 7y)(x - 6 - 7y)$

86. $x^2 - 10x + 25 - 36y^2 = (x^2 - 10x + 25) - 36y^2 = (x - 5)^2 - 36y^2 = (x - 5 + 6y)(x - 5 - 6y)$

87. $9b^2x - 16y - 16x + 9b^2y$
$= (9b^2x + 9b^2y) + (-16x - 16y) = 9b^2(x + y) - 16(x + y) = (x + y)(9b^2 - 16) = (x + y)(3b + 4)(3b - 4)$

88. $16a^2x - 25y - 25x + 16a^2y$
$= (16a^2x + 16a^2y) + (-25y - 25x) = 16a^2(x + y) - 25(x + y) = (x + y)(16a^2 - 25) = (x + y)(4a + 5)(4a - 5)$

89. $x^2y - 16y + 32 - 2x^2$
$= (x^2y - 16y) + (-2x^2 + 32) = y(x^2 - 16) - 2(x^2 - 16) = (x^2 - 16)(y - 2) = (x + 4)(x - 4)(y - 2)$

90. $12x^2y - 27y - 4x^2 + 9$
$= (12x^2y - 27y) + (-4x^2 + 9) = 3y(4x^2 - 9) - 1(4x^2 - 9) = (4x^2 - 9)(3y - 1) = (2x + 3)(2x - 3)(3y - 1)$

91. $2x^3 - 8a^2x + 24x^2 + 72x$
$= 2x(x^2 - 4a^2 + 12x + 36) = 2x\left[(x^2 + 12x + 36) - 4a^2\right] = 2x\left[(x + 6)^2 - 4a^2\right] = 2x(x + 6 - 2a)(x + 6 + 2a)$

92. $2x^3 - 98a^2x + 28x^2 + 98x$
$= 2x(x^2 - 49a^2 + 14x + 49) = 2x\left[(x^2 + 14x + 49) - 49a^2\right] = 2x\left[(x + 7)^2 - 49a^2\right] = 2x(x + 7 - 7a)(x + 7 + 7a)$

93. $x^{\frac{3}{2}} - x^{\frac{1}{2}} = x^{\frac{1}{2}}\left(x^{\frac{3}{2}-\frac{1}{2}}\right) - 1 = x^{\frac{1}{2}}(x - 1)$

94. $x^{\frac{3}{4}} - x^{\frac{1}{4}} = x^{\frac{1}{4}}\left(x^{\frac{3}{4}-\frac{1}{4}} - 1\right) = x^{\frac{1}{4}}\left(x^{\frac{1}{2}} - 1\right)$

95. $4x^{-\frac{2}{3}} + 8x^{\frac{1}{3}} = 4x^{-\frac{2}{3}}\left(1 + 2x^{\frac{1}{3}-\left(-\frac{2}{3}\right)}\right) = 4x^{-\frac{2}{3}}(1 + 2x) = \dfrac{4(1 + 2x)}{x^{\frac{2}{3}}}$

96. $12x^{-\frac{3}{4}}+6x^{\frac{1}{4}}=6x^{-\frac{3}{4}}\left(2+x^{\frac{1}{4}-\left(-\frac{3}{4}\right)}\right)=6x^{-\frac{3}{4}}(2+x)=\dfrac{6(x+2)}{x^{\frac{3}{4}}}$

97. $(x+3)^{\frac{1}{2}}-(x+3)^{\frac{3}{2}}=(x+3)^{\frac{1}{2}}\left[1-(x+3)^{\frac{3}{2}-\frac{1}{2}}\right]=(x+3)^{\frac{1}{2}}\left[1-(x+3)\right]=(x+3)^{\frac{1}{2}}(-x-2)=-(x+3)^{\frac{1}{2}}(x+2)$

98. $\left(x^2+4\right)^{\frac{3}{2}}+\left(x^2+4\right)^{\frac{7}{2}}=\left(x^2+4\right)^{\frac{3}{2}}\left[1+\left(x^2+4\right)^{\frac{7}{2}-\frac{3}{2}}\right]=\left(x^2+4\right)^{\frac{3}{2}}\left[1+\left(x^2+4\right)^2\right]=\left(x^2+4\right)^{\frac{3}{2}}\left(x^4+8x^2+17\right)$

99. $(x+5)^{-\frac{1}{2}}-(x+5)^{-\frac{3}{2}}=(x+5)^{-\frac{3}{2}}\left[(x+5)^{-\frac{1}{2}-\left(-\frac{3}{2}\right)}-1\right]=(x+5)^{-\frac{3}{2}}\left[(x+5)-1\right]=(x+5)^{-\frac{3}{2}}(x+4)=\dfrac{x+4}{(x+5)^{\frac{3}{2}}}$

100. $\left(x^2+3\right)^{-\frac{2}{3}}+\left(x^2+3\right)^{-\frac{5}{3}}=\left(x^2+3\right)^{-\frac{5}{3}}\left[\left(x^2+3\right)^{-\frac{2}{3}-\left(-\frac{5}{3}\right)}+1\right]=\left(x^2+3\right)^{-\frac{5}{3}}\left[\left(x^2+3\right)+1\right]=\dfrac{x^2+4}{(x^2+3)^{5/3}}$

101. $(4x-1)^{\frac{1}{2}}-\frac{1}{3}(4x-1)^{\frac{3}{2}}$

$=(4x-1)^{\frac{1}{2}}\left[1-\frac{1}{3}(4x-1)^{\frac{3}{2}-\frac{1}{2}}\right]=(4x-1)^{\frac{1}{2}}\left[1-\frac{1}{3}(4x-1)\right]=(4x-1)^{\frac{1}{2}}\left[1-\frac{4}{3}x+\frac{1}{3}\right]$

$=(4x-1)^{\frac{1}{2}}\left(\frac{4}{3}-\frac{4}{3}x\right)=(4x-1)^{\frac{1}{2}}\frac{4}{3}(1-x)=\dfrac{-4(4x-1)(x-1)}{3}$

102. $-8(4x+3)^{-2}+10(5x+1)(4x+3)^{-1}=2(4x+3)^{-2}\left[-4+5(5x+1)(4x+3)\right]=\dfrac{2(100x^2+95x+11)}{(4x+3)^2}$

103. $10x^2(x+1)-7x(x+1)-6(x+1)=(x+1)\left(10x^2-7x-6\right)=(x+1)(5x-6)(2x+1)$

104. $12x^2(x-1)-4x(x-1)-5(x-1)=(x-1)\left(12x^2-4x-5\right)=(x-1)(6x-5)(2x+1)$

105. $6x^4+35x^2-6=\left(x^2+6\right)\left(6x^2-1\right)$

106. $7x^4+34x^2-5=\left(7x^2-1\right)\left(x^2+5\right)$

107. $y^7+y=y\left(y^6+1\right)=y\left[\left(y^2\right)^3+1^3\right]=y\left(y^2+1\right)\left(y^4-y^2+1\right)$

108. $(y+1)^3+1=(y+1)^3+1^3=\left[(y+1)+1\right]\left[(y+1)^2-(y+1)+1\right]=(y+2)\left[\left(y^2+2y+1\right)-y-1+1\right]$

$=(y+2)\left(y^2+2y+1-y-1+1\right)=(y+2)\left(y^2+y+1\right)$

109. $x^4-5x^2y^2+4y^4=\left(x^2-4y^2\right)\left(x^2-y^2\right)=(x+2y)(x-2y)(x+y)(x-y)$

110. $x^4-10x^2y^2+9y^4=\left(x^2-9y^2\right)\left(x^2-y^2\right)=(x+3y)(x-3y)(x+y)(x-y)$

111. $(x-y)^4-4(x-y)^2$
$=(x-y)^2\left((x-y)^2-4\right)=(x-y)^2\left((x-y)+2\right)\left((x-y)-2\right)=(x-y)^2(x-y+2)(x-y-2)$

112. $(x+y)^4-100(x+y)^2=(x+y)^2\left((x+y)^2-100\right)=(x+y)^2(x+y-10)(x+y+10)$

113. $2x^2-7xy^2+3y^4=(2x-y^2)(x-3y^2)$

114. $3x^2+5xy^2+2y^4=(3x+2y^2)(x+y^2)$

115. a. $(x-0.4x)-0.4(x-0.4x)=(x-0.4x)(1-0.4)=(0.6x)(0.6)=0.36x$

b. No, the computer is selling at 36% of its original price.

116. a. $(x-0.3x)-0.3(x-0.3x)=(x-0.3x)(1-0.3)=(0.7x)(0.7)=0.49x$

b. No, the computer is selling at 49% of its original price.

117. a. $(3x)^2-4\cdot 2^2=9x^2-16$

b. $9x^2-16=(3x+4)(3x-4)$

118. a. $(7x)^2-4\cdot 3^2=49x^2-36$

b. $49x^2-36=(7x+6)(7x-6)$

119. a. $x(x+y)-y(x+y)$

b. $x(x+y)-y(x+y)=(x+y)(x-y)$

120. a. $x^2+xy+xy+y^2=x^2+2xy+y^2$

b. $x^2+2xy+y^2=(x+y)^2$

121.
$$\begin{aligned}V_{\text{shaded}}&=V_{\text{outside}}-V_{\text{inside}}\\&=a\cdot a\cdot 4a-b\cdot b\cdot 4a\\&=4a^3-4ab^2\\&=4a(a^2-b^2)\\&=4a(a+b)(a-b)\end{aligned}$$

122. $V_{shaded} = V_{outside} - V_{inside}$
$$= a \cdot a \cdot 3a - b \cdot b \cdot 3a$$
$$= 3a^3 - 3ab^2$$
$$= 3a\left(a^2 - b^2\right)$$
$$= 3a(a+b)(a-b)$$

130. **a.** false, $x^3 + 1$ is factorable.

b. false, $x(x - 4) + 3$ does not factor the whole polynomial.

c. false, $x^3 - 64 = (x-4)(x+4x+16)$

d. true

131. $x^{2n} + 6x^n + 8 = \left(x^n + 4\right)\left(x^n + 2\right)$

132. $-x^2 - 4x + 5 = -1\left(x^2 + 4x - 5\right) = -1(x+5)(x-1) = -(x+5)(x-1)$

133. $x^4 - y^4 - 2x^3y + 2xy^3 = \left(x^4 - y^4\right) + \left(-2x^3y + 2xy^3\right) = \left(x^2 - y^2\right)\left(x^2 + y^2\right) - 2xy\left(x^2 - y^2\right)$
$$= \left(x^2 - y^2\right)\left(x^2 + y^2 - 2xy\right) = (x-y)(x+y)\left(x^2 - 2xy + y^2\right) = (x-y)(x+y)(x-y)^2 = (x-y)^3(x+y)$$

134. $(x-5)^{-\frac{1}{2}}(x+5)^{-\frac{1}{2}} - (x+5)^{\frac{1}{2}}(x-5)^{-\frac{3}{2}} = (x-5)^{-\frac{3}{2}}(x+5)^{-\frac{1}{2}}\left[(x-5)^{-\frac{1}{2}-\left(-\frac{3}{2}\right)} - (x+5)^{\frac{1}{2}-\left(-\frac{1}{2}\right)}\right]$
$$= (x-5)^{-\frac{3}{2}}(x+5)^{-\frac{1}{2}}\left[(x-5) - (x+5)\right] = (x-5)^{-\frac{3}{2}}(x+5)^{-\frac{1}{2}}(-10) = \frac{-10}{(x-5)^{3/2}(x+5)^{1/2}}$$

135. $x^2 + bx + 15$, $b = 16$, -16, 8 or -8

136. $b = 0, 3, 4$, or $-c(c+4)$, where $c > 0$ is an integer.

Mid-Chapter P Check Point

1. $(3x+5)(4x-7) = (3x)(4x) + (3x)(-7) + (5)(4x) + (5)(-7)$
$$= 12x^2 - 21x + 20x - 35$$
$$= 12x^2 - x - 35$$

2. $(3x+5) - (4x-7) = 3x + 5 - 4x + 7$
$$= 3x - 4x + 5 + 7$$
$$= -x + 12$$

3. $\sqrt{6} + 9\sqrt{6} = 10\sqrt{6}$

4. $3\sqrt{12} - \sqrt{27} = 3 \cdot 2\sqrt{3} - 3\sqrt{3} = 6\sqrt{3} - 3\sqrt{3} = 3\sqrt{3}$

5. $7x + 3[9 - (2x - 6)] = 7x + 3[9 - 2x + 6] = 7x + 3[15 - 2x] = 7x + 45 - 6x = x + 45$

6. $(8x-3)^2 = (8x)^2 - 2(8x)(3) + (3)^2 = 64x^2 - 48x + 9$

7. $\left(x^{\frac{1}{3}}y^{-\frac{1}{2}}\right)^6 = x^{\frac{1}{3}\cdot 6}y^{-\frac{1}{2}\cdot 6} = x^2y^{-3} = \frac{x^2}{y^3}$

8. $\left(\frac{2}{7}\right)^0 - 32^{-\frac{2}{5}} = 1 - \frac{1}{\left(\sqrt[5]{32}\right)^2} = 1 - \frac{1}{(2)^2} = 1 - \frac{1}{4} = \frac{3}{4}$

9. $(2x-5)-(x^2-3x+1) = 2x-5-x^2+3x-1 = -x^2+5x-6$

10.
$$\begin{aligned}(2x-5)(x^2-3x+1) &= 2x(x^2-3x+1)-5(x^2-3x+1)\\ &= 2x(x^2-3x+1)-5(x^2-3x+1)\\ &= 2x^3-6x^2+2x-5x^2+15x-5\\ &= 2x^3-6x^2-5x^2+2x+15x-5\\ &= 2x^3-11x^2+17x-5\end{aligned}$$

11. $x^3+x^3-x^3\cdot x^3 = 2x^3-x^6 = -x^6+2x^3$

12.
$$\begin{aligned}(9a-10b)(2a+b) &= (9a)(2a)+(9a)(b)+(-10b)(2a)+(-10b)(b)\\ &= (9a)(2a)+(9a)(b)+(-10b)(2a)+(-10b)(b)\\ &= 18a^2+9ab-20ab-10b^2\\ &= 18a^2-11ab-10b^2\end{aligned}$$

13. $\{a,c,d,e\}\cup\{c,d,f,h\} = \{a,c,d,e,f,h\}$

14. $\{a,c,d,e\}\cap\{c,d,f,h\} = \{c,d\}$

15.
$$\begin{aligned}\left(3x^2y^3-xy+4y^2\right)-\left(-2x^2y^3-3xy+5y^2\right) &= 3x^2y^3-xy+4y^2+2x^2y^3+3xy-5y^2\\ &= 3x^2y^3-xy+4y^2+2x^2y^3+3xy-5y^2\\ &= 3x^2y^3+2x^2y^3-xy+3xy+4y^2-5y^2\\ &= 5x^2y^3+2xy-y^2\end{aligned}$$

16. $\frac{24x^2y^{13}}{-2x^5y^{-2}} = -12x^{2-5}y^{13-(-2)} = -12x^{-3}y^{15} = -\frac{12y^{15}}{x^3}$

17. $\left(\frac{1}{3}x^{-5}y^4\right)\left(18x^{-2}y^{-1}\right) = 6x^{-5-2}y^{4-1} = \frac{6y^3}{x^7}$

18. $\sqrt[12]{x^4} = x^{\frac{4}{12}} = \left|x^{\frac{1}{3}}\right| = \left|\sqrt[3]{x}\right|$

19. $[4y-(3x+2)][4y+(3x+2)] = (4y)^2-(3x+2)^2 = 16y^2-(9x^2+12x+4) = 16y^2-9x^2-12x-4$

20. $\begin{aligned}(x-2y-1)^2 &= x(x-2y-1)-2y(x-2y-1)-(x-2y-1)\\ &= x^2-2xy-x-2xy+4y^2+2y-x+2y+1\\ &= x^2-4xy+4y^2-2x+4y+1\end{aligned}$

21. $\dfrac{24\times10^3}{2\times10^6}=\dfrac{24}{2}\cdot\dfrac{10^3}{10^6}=12\times10^{-3}=\left(1.2\times10^1\right)\times10^{-3}=1.2\times\left(10^1\times10^{-3}\right)=1.2\times10^{-2}$

22. $\dfrac{\sqrt[3]{32}}{\sqrt[3]{2}}=\sqrt[3]{\dfrac{32}{2}}=\sqrt[3]{16}=\sqrt[3]{2^4}=2\sqrt[3]{2}$

23. $(x^3+2)(x^3-2)=x^6-4$

24. $(x^2+2)^2=(x^2)^2+2(x^2)(2)+(2)^2=x^4+4x^2+4$

25. $\sqrt{50}\cdot\sqrt{6}=5\sqrt{2}\cdot\sqrt{6}=5\sqrt{2\cdot6}=5\sqrt{12}=5\cdot2\sqrt{3}=10\sqrt{3}$

26. $\dfrac{11}{7-\sqrt{3}}=\dfrac{11}{7-\sqrt{3}}\cdot\dfrac{7+\sqrt{3}}{7+\sqrt{3}}=\dfrac{77+11\sqrt{3}}{49-3}=\dfrac{77+11\sqrt{3}}{46}$

27. $\dfrac{11}{\sqrt{3}}=\dfrac{11}{\sqrt{3}}\cdot\dfrac{\sqrt{3}}{\sqrt{3}}=\dfrac{11\sqrt{3}}{3}$

28. $7x^2-22x+3=(7x-1)(x-3)$

29. x^2-2x+4 is prime.

30. $7x^2-22x+3=(7x-1)(x-3)$

31. $3x^2-4xy-7y^2=(3x-7y)(x+y)$

32. $64y-y^3=y\left(64-y^2\right)=y(8+y)(8-y)$

33. $50x^3+20x^2+2x=2x\left(25x^2+10x+1\right)=2x(5x+1)^2$

34. $x^2-6x+9-49y^2=(x-3)^2-49y^2=\left[(x-3)+7y\right]\left[(x-3)-7y\right]=(x-3+7y)(x-3-7y)$

35. $x^{-\frac{3}{2}}-2x^{-\frac{1}{2}}+x^{\frac{1}{2}}=x^{-\frac{3}{2}}\left(1-2x+x^2\right)=\dfrac{(1-x)^2}{x^{\frac{3}{2}}}$

36. $\left(x^2+1\right)^{\frac{1}{2}}-10\left(x^2+1\right)^{-\frac{1}{2}}=\left(x^2+1\right)^{-\frac{1}{2}}\left[\left(x^2+1\right)-10\right]=\left(x^2+1\right)^{-\frac{1}{2}}\left(x^2-9\right)=\dfrac{(x+3)(x-3)}{\left(x^2+1\right)^{\frac{1}{2}}}$

37. $\left\{-11, -\frac{3}{7}, 0, 0.45, \sqrt{25}\right\}$

38. Since $2-\sqrt{13}<0$ then $\left|2-\sqrt{13}\right|=\sqrt{13}-2$

39. Since $x<0$ then $|x|=-x$. Thus $x^2|x|=-x^2x=-x^3$

40. $120\cdot 2.9\times 10^8=348\times 10^8=3.48\times 10^2\times 10^8=3.48\times 10^{10}$
The total annual spending on ice cream is $\$3.48\times 10^{10}$.

41. $\dfrac{3\times 10^{10}}{7.5\times 10^9}=\dfrac{3}{7.5}\cdot\dfrac{10^{10}}{10^9}=0.4\times 10=4$
A human brain has 4 times as many neurons as a gorilla brain.

42. **a.** Model 1:
$D=236(1.5)^x$
$D=236(1.5)^2$
$D=531$
Model 2:
$D=127x+239$
$D=127(2)+239$
$D=493$
Model 3:
$D=-54x^2+234x+220$
$D=-54(2)^2+234(2)+220$
$D=472$

Model 3 best describes the data in 2004.

b. Model 2 is the polynomial of degree 1:
$D=127x+239$
$D=127(6)+239$
$D=1001$

Model 2 predicts Americans will spend 1001 million dollars on online dating in 2008.

Section P.6

Check Point Exercises

1. a. The denominator would equal zero if $x = -5$, so -5 must be excluded from the domain.

   b. $x^2 - 36 = (x+6)(x-6)$
   The denominator would equal zero if $x = -6$ or $x = 6$, so -6 and 6 must both be excluded from the domain.

2. a. $$\frac{x^3+3x^2}{x+3} = \frac{x^2(x+3)}{x+3} = \frac{x^2(x+3)}{x+3} = x^2,\ x \neq -3$$
   Because the denominator is $x + 3$, $x \neq -3$

   b. $$\frac{x^2-1}{x^2+2x+1} = \frac{(x-1)(x+1)}{(x+1)(x+1)} = \frac{x-1}{x+1},\ x \neq -1$$
   Because the denominator is $(x+1)(x+1), x \neq -1$

3. $$\frac{x+3}{x^2-4} \cdot \frac{x^2-x-6}{x^2+6x+9}$$
$$= \frac{x+3}{(x+2)(x-2)} \cdot \frac{(x-3)(x+2)}{(x+3)(x+3)}$$
$$= \frac{x+3}{(x+2)(x-2)} \cdot \frac{(x-3)(x+2)}{(x+3)(x+3)}$$
$$= \frac{x-3}{(x-2)(x+3)},\ x \neq -2,\ x \neq 2,\ x \neq -3$$
   Because the denominator has factors of $x+2$, $x-2$, and $x+3$, $x \neq -2$, $x \neq 2$, and $x \neq -3$.

4. $$\frac{x^2-2x+1}{x^3+x} \div \frac{x^2+x-2}{3x^2+3}$$
$$= \frac{x^2-2x+1}{x^3+x} \cdot \frac{3x^2+3}{x^2+x-2}$$
$$= \frac{(x-1)(x-1)}{x(x^2+1)} \cdot \frac{3(x^2+1)}{(x+2)(x-1)}$$
$$= \frac{3(x-1)}{x(x+2)},\ x \neq 0,\ x \neq -2,\ x \neq 1$$

5. $\frac{x}{x+1} - \frac{3x+2}{x+1} = \frac{x-3x-2}{x+1}$
$= \frac{-2x-2}{x+1}$
$= \frac{-2(x+1)}{x+1}$
$= -2, x \neq -1$

6. $\frac{3}{x+1} + \frac{5}{x-1}$
$= \frac{3x(x-1)+5(x+1)}{(x+1)(x-1)}$
$= \frac{3x-3+5x+5}{(x+1)(x-1)}$
$= \frac{8x+2}{(x+1)(x-1)}$
$= \frac{2(4x+1)}{(x+1)(x-1)}$
$= \frac{2(4x+1)}{(x+1)(x-1)}$, $x \neq -1$ and $x \neq 1$.

7. Factor each denominator completely.
$x^2 - 6x + 9 = (x-3)^2$
$x^2 - 9 = (x+3)(x-3)$

List the factors of the first denominator.
$x-3,\ x-3$

Add any unlisted factors from the second denominator.
$x-3,\ x-3,\ x+3$

The least common denominator is the product of all factors in the final list.
$(x-3)(x-3)(x+3)$ or $(x-3)^2(x+3)$ is the least common denominator.

8. Find the least common denominator.
$x^2 - 10x + 25 = (x-5)^2$
$2x - 10 = 2(x-5)$

The least common denominator is $2(x-5)(x-5)$.
Write all rational expressions in terms of the least common denominator.

$\frac{x}{x^2-10x+25} - \frac{x-4}{2x-10}$
$= \frac{x}{(x-5)(x-5)} - \frac{x-4}{2(x-5)}$
$= \frac{2x}{2(x-5)(x-5)} - \frac{(x-4)(x-5)}{2(x-5)(x-5)}$

Add numerators, putting this sum over the least common denominator.

$$= \frac{2x-(x-4)(x-5)}{2(x-5)(x-5)}$$

$$= \frac{2x-(x^2-5x-4x+20)}{2(x-5)(x-5)}$$

$$= \frac{2x-x^2+5x+4x-20}{2(x-5)(x-5)}$$

$$- \frac{2x-x^2+5x+4x-20}{2(x-5)(x-5)}$$

$$= \frac{-x^2+11x-20}{2(x-5)(x-5)}$$

$$= \frac{-x^2+11x-20}{2(x-5)^2},\ x \neq 5$$

9. $$\frac{\frac{1}{x}-\frac{3}{2}}{\frac{1}{x}+\frac{3}{4}} = \frac{\frac{2}{2x}-\frac{3x}{2x}}{\frac{4}{4x}+\frac{3x}{4x}},\ x \neq 0$$

$$= \frac{\frac{2-3x}{2x}}{\frac{4+3x}{4x}},\ x \neq \frac{-4}{3}$$

$$= \frac{2-3x}{2x} \div \frac{4+3x}{4x}$$

$$= \frac{2-3x}{2x} \cdot \frac{4x}{4+3x}$$

$$= \frac{2-3x}{4+3x} \cdot \frac{4}{2}$$

$$= \frac{2-3x}{4+3x} \cdot \frac{2}{1}$$

$$= \frac{2(2-3x)}{4+3x},\ x \neq 0 \textit{ and } x \neq \frac{-4}{3}$$

10. Multiply each of the three terms, $\frac{1}{x+7}$, $\frac{1}{x}$, and 7 by the least common denominator of $x(x+7)$.

$$\frac{\frac{1}{x+7}-\frac{1}{x}}{7} = \frac{x(x+7)\left(\frac{1}{x+7}\right)-x(x+7)\left(\frac{1}{x}\right)}{7x(x+7)}$$

$$= \frac{x-(x+7)}{7x(x+7)}$$

$$= \frac{-7}{7x(x+7)}$$

$$= -\frac{1}{x(x+7)},\ x \neq 0,\ x \neq -7$$

11. $$\frac{\sqrt{x}+\frac{1}{\sqrt{x}}}{x} = \frac{\left(\sqrt{x}+\frac{1}{\sqrt{x}}\right)\sqrt{x}}{x\sqrt{x}} = \frac{x+1}{x^{3/2}}$$

12. $$\frac{\sqrt{x+3}-\sqrt{x}}{3}$$

$$= \frac{\sqrt{x+3}-\sqrt{x}}{3} \cdot \frac{\sqrt{x+3}+\sqrt{x}}{\sqrt{x+3}\cdot\sqrt{x}}$$

$$= \frac{\left(\sqrt{x+3}\right)^2-(\sqrt{x})^2}{3\left(\sqrt{x+3}+\sqrt{x}\right)}$$

$$= \frac{x+3-x}{3\left(\sqrt{x+3}+\sqrt{x}\right)}$$

$$= \frac{1}{\sqrt{x+3}+\sqrt{x}}$$

Exercise Set P.6

1. $\frac{7}{x-3}, x \neq 3$

2. $\frac{13}{x+9}, x \neq -9$

3. $\frac{x+5}{x^2-25} = \frac{x+5}{(x+5)(x-5)}, x \neq 5, -5$

4. $\frac{x+7}{x^2-49} = \frac{x+7}{(x+7)(x-7)}, x \neq 7, -7$

5. $\frac{x-1}{x^2+11x+10} = \frac{x-1}{(x+1)(x+10)}, x \neq -1, -10$

6. $\frac{x-3}{x^2+4x-45} = \frac{x-3}{(x+9)(x-5)}, x \neq -9, 5$

7. $\frac{3x-9}{x^2-6x+9} = \frac{3(x-3)}{(x-3)(x-3)}$

$= \frac{3}{x-3}, x \neq 3$

8. $\frac{4x-8}{x^2-4x+4} = \frac{4(x-2)}{(x-2)(x-2)} = \frac{4}{x-2},\ x \neq 2$

9. $\frac{x^2-12x+36}{4x-24} = \frac{(x-6)(x-6)}{4(x-6)} = \frac{x-6}{4}.$

$x \neq 6$

10. $\frac{x^2-8x+16}{3x-12} = \frac{(x-4)(x-4)}{3(x-4)} = \frac{x-4}{3},\ x \neq 4$

11. $\frac{y^2+7y-18}{y^2-3y+2} = \frac{(y+9)(y-2)}{(y-2)(y-1)} = \frac{y+9}{y-1},$

$y \neq 1, 2$

12. $\frac{y^2-4y-5}{y^2+5y+4} = \frac{(y-5)(y+1)}{(y+4)(y+1)} = \frac{y-5}{y+4}, y \neq -4, -1$

13. $\frac{x^2+12x+36}{x^2-36} = \frac{(x+6)^2}{(x+6)(x-6)} = \frac{x+6}{x-6},$

$x \neq 6, -6$

14. $\frac{x^2-14x+49}{x^2-49} = \frac{(x-7)^2}{(x-7)(x+7)}$

$= \frac{x-7}{x+7},$

$x \neq 7, -7$

15. $\frac{x-2}{3x+9} \cdot \frac{2x+6}{2x-4} = \frac{x-2}{3(x+3)} \cdot \frac{2(x+3)}{2(x-2)}$

$= \frac{2}{6} = \frac{1}{3}, x \neq 2, -3$

16. $\frac{6x+9}{3x-15} \cdot \frac{x-5}{4x+6} = \frac{3(2x+3)}{3(x-5)} \cdot \frac{x-5}{2(2x+3)}$

$= \frac{3}{6} = \frac{1}{2}, \quad x \neq 5, -\frac{3}{2}$

17. $\frac{x^2-9}{x^2} \cdot \frac{x^2-3x}{x^2+x-12}$

$= \frac{(x-3)(x+3)}{x^2} \cdot \frac{x(x-3)}{(x+4)(x-3)}$

$= \frac{(x-3)(x+3)}{x(x+4)}, x \neq 0, -4, 3$

18. $\frac{x^2-4}{x^2-4x+4} \cdot \frac{2x-4}{x+2} = \frac{(x+2)(x-2)}{(x-2)^2} \cdot \frac{2(x-2)}{x+2}$

$= 2,$

$x \neq 2, -2$

19. $\frac{x^2-5x+6}{x^2-2x-3} \cdot \frac{x^2-1}{x^2-4}$

$= \frac{(x-3)(x-2)}{(x-3)(x+1)} \cdot \frac{(x+1)(x-1)}{(x-2)(x+2)}$

$= \frac{x-1}{x+2},\ x \neq -2, -1, 2, 3$

20. $\frac{x^2+5x+6}{x^2+x-6} \cdot \frac{x^2-9}{x^2-x-6}$

$= \frac{(x+3)(x+2)}{(x+3)(x-2)} \cdot \frac{(x-3)(x+3)}{(x-3)(x+2)} = \frac{x+3}{x-2},$

$x \neq -3, -2, 2, 3$

21. $\frac{x^3-8}{x^2-4} \cdot \frac{x+2}{3x} = \frac{(x-2)(x^2+2x+4)}{(x-2)(x+2)} \cdot \frac{x+2}{3x}$

$= \frac{x^2+2x+4}{3x}, x \neq -2, 0, 2$

22. $\dfrac{x^2+6x+9}{x^3+27}\cdot\dfrac{1}{x+3}$
$=\dfrac{(x+3)(x+3)}{(x+3)(x^2-3x+9)}\cdot\dfrac{1}{x+3}=\dfrac{1}{x^2-3x+9}$,
$x\neq -3$

23. $\dfrac{x+1}{3}\div\dfrac{3x+3}{7}=\dfrac{x+1}{3}\div\dfrac{3(x+1)}{7}$
$=\dfrac{x+1}{3}\cdot\dfrac{7}{3(x+1)}$
$=\dfrac{7}{9}, x\neq -1$

24. $\dfrac{x+5}{7}\div\dfrac{4x+20}{9}=\dfrac{x+5}{7}\div\dfrac{4(x+5)}{9}$
$=\dfrac{x+5}{7}\cdot\dfrac{9}{4(x+5)}$
$=\dfrac{9}{28}$,
$x\neq -5$

25. $\dfrac{x^2-4}{x}\div\dfrac{x+2}{x-2}=\dfrac{(x-2)(x+2)}{x}\cdot\dfrac{x-2}{x+2}$
$=\dfrac{(x-2)^2}{x}; x\neq 0, -2, 2$

26. $\dfrac{x^2-4}{x-2}\div\dfrac{x+2}{4x-8}=\dfrac{(x-2)(x+2)}{x-2}\div\dfrac{x+2}{4(x-2)}$
$=\dfrac{(x-2)(x+2)}{x-2}\cdot\dfrac{4(x-2)}{x+2}$
$=4(x-2)$,
$x\neq 2, -2$

27. $\dfrac{4x^2+10}{x-3}\div\dfrac{6x^2+15}{x^2-9}$
$=\dfrac{2(2x^2+5)}{x-3}\div\dfrac{3(2x^2+5)}{(x-3)(x+3)}$
$=\dfrac{2(2x^2+5)}{x-3}\cdot\dfrac{(x-3)(x+3)}{3(2x^2+5)}$
$=\dfrac{2(x+3)}{3}, x\neq 3, -3$

28. $\dfrac{x^2+x}{x^2-4}\div\dfrac{x^2-1}{x^2+5x+6}$
$=\dfrac{x(x+1)}{(x-2)(x+2)}\div\dfrac{(x-1)(x+1)}{(x+2)(x+3)}$
$=\dfrac{x(x+1)}{(x-2)(x+2)}\cdot\dfrac{(x+2)(x+3)}{(x-1)(x+1)}$
$=\dfrac{x(x+3)}{(x-2)(x-1)}$,
$x\neq 2, 1, -1, -2, -3$

29. $\dfrac{x^2-25}{2x-2}\div\dfrac{x^2+10x+25}{x^2+4x-5}$
$=\dfrac{(x-5)(x+5)}{2(x-1)}\div\dfrac{(x+5)^2}{(x+5)(x-1)}$
$=\dfrac{(x-5)(x+5)}{2(x-1)}\cdot\dfrac{(x+5)(x-1)}{(x+5)^2}$
$=\dfrac{x-5}{2}, x\neq 1, -5$

30. $\dfrac{x^2-4}{x^2+3x-10}\div\dfrac{x^2+5x+6}{x^2+8x+15}$
$=\dfrac{(x+2)(x-2)}{(x+5)(x-2)}\div\dfrac{(x+2)(x+3)}{(x+3)(x+5)}$
$=\dfrac{(x+2)(x-2)}{(x+5)(x-2)}\cdot\dfrac{(x+3)(x+5)}{(x+2)(x+3)}$
$=1$
$x\neq 2, -2, -3, -5$

31. $\dfrac{x^2+x-12}{x^2+x-30}\cdot\dfrac{x^2+5x+6}{x^2-2x-3}\div\dfrac{x+3}{x^2+7x+6}$
$=\dfrac{(x+4)(x-3)}{(x+6)(x-5)}\cdot\dfrac{(x+2)(x+3)}{(x+1)(x-3)}\cdot\dfrac{(x+6)(x+1)}{x+3}$
$=\dfrac{(x+4)(x+2)}{x-5}$
$x\neq -6, -3, -1, 3, 5$

32. $\dfrac{x^3-25x}{4x^2}\cdot\dfrac{2x^2-2}{x^2-6x+5}\div\dfrac{x^2+5x}{7x+7}$
$=\dfrac{x(x-5)(x+5)}{4x^2}\cdot\dfrac{2(x-1)(x+1)}{(x-1)(x-5)}\cdot\dfrac{7(x+1)}{x(x+5)}$
$=\dfrac{7(x+1)^2}{2x^2}$
$x\neq 0, 1, -1, 5, -5$

33. $\frac{4x+1}{6x+5}+\frac{8x+9}{6x+5}=\frac{4x+1+8x+9}{6x+5}$
$=\frac{12x+10}{6x+5}$
$=\frac{2(6x+5)}{6x+5}=2, x\neq -\frac{5}{6}$

34. $\frac{3x+2}{3x+4}+\frac{3x+6}{3x+4}=\frac{3x+2+3x+6}{3x+4}$
$=\frac{6x+8}{3x+4}$
$=\frac{2(3x+4)}{3x+4}$
$=2$

$x\neq -\frac{4}{3}$

35. $\frac{x^2-2x}{x^2+3x}+\frac{x^2+x}{x^2+3x}=\frac{x^2-2x+x^2+x}{x^2+3x}$
$=\frac{2x^2-x}{x^2+3x}$
$=\frac{x(2x-1)}{x(x+3)}$
$=\frac{2x-1}{x+3}, x\neq 0, -3$

36. $\frac{x^2-4x}{x^2-x-6}+\frac{4x-4}{x^2-x-6}=\frac{x^2-4x+4x-4}{x^2-x-6}$
$=\frac{x^2-4}{(x-3)(x+2)}$
$=\frac{(x-2)(x+2)}{(x-3)(x+2)}$
$=\frac{x-2}{x-3},$

$x\neq -2, 3$

37. $\frac{4x-10}{x-2}-\frac{x-4}{x-2}=\frac{4x-10-(x-4)}{x-2}$
$=\frac{4x-10-x+4}{x-2}$
$=\frac{3x-6}{x-2}$
$=\frac{3(x-2)}{x-2}$
$=3, x\neq 2$

38. $\frac{2x+3}{3x-6}-\frac{3-x}{3x-6}=\frac{2x+3-(3-x)}{3x-6}$
$=\frac{2x+3-3+x}{3x-6}$
$=\frac{3x}{3(x-2)}$
$=\frac{x}{x-2},$

$x\neq 2$

39. $\frac{x^2+3x}{x^2+x-12}-\frac{x^2-12}{x^2+x-12}$
$=\frac{x^2+3x-(x^2-12)}{x^2+x-12}$
$=\frac{x^2+3x-x^2+12}{x^2+x-12}$
$=\frac{3x+12}{x^2+x-12}$
$=\frac{3(x+4)}{(x+4)(x-3)}$
$=\frac{3}{x-3}, x\neq 3, -4$

40. $\frac{x^2-4x}{x^2-x-6}-\frac{x-6}{x^2-x-6}$
$=\frac{x^2-4x-(x-6)}{x^2-x-6}$
$=\frac{x^2-4x-x+6}{x^2-x-6}$
$=\frac{x^2-5x+6}{x^2-x-6}$
$=\frac{(x-2)(x-3)}{(x-3)(x+2)}$
$=\frac{x-2}{x+2}, x\neq -2, 3$

41. $\frac{3}{x+4}+\frac{6}{x+5}=\frac{3(x+5)+6(x+4)}{(x+4)(x+5)}$
$=\frac{3x+15+6x+24}{(x+4)(x+5)}$
$=\frac{9x+39}{(x+4)(x+5)}, x\neq -4, -5$

42. $\frac{8}{x-2}+\frac{2}{x-3}=\frac{8(x-3)+2(x-2)}{(x-2)(x-3)}$
$=\frac{8x-24+2x-4}{(x-2)(x-3)}$
$=\frac{10x-28}{(x-2)(x-3)},$
$x \neq 2, 3$

43. $\frac{3}{x+1}-\frac{3}{x}=\frac{3x-3(x+1)}{x(x+1)}$
$=\frac{3x-3x-3}{x(x+1)}=-\frac{3}{x(x+1)}, x \neq -1, 0$

44. $\frac{4}{x}-\frac{3}{x+3}=\frac{4(x+3)-3x}{x(x+3)}$
$=\frac{4x+12-3x}{x(x+3)}$
$=\frac{x+12}{x(x+3)}$
$x \neq -3, 0$

45. $\frac{2x}{x+2}+\frac{x+2}{x-2}=\frac{2x(x-2)+(x+2)(x+2)}{(x+2)(x-2)}$
$=\frac{2x^2-4x+x^2+4x+4}{(x+2)(x-2)}$
$=\frac{3x^2+4}{(x+2)(x-2)}, x \neq -2, 2$

46. $\frac{3x}{x-3}-\frac{x+4}{x+2}=\frac{3x(x+2)-(x+4)(x-3)}{(x-3)(x+2)}$
$=\frac{3x^2+6x-(x^2+x-12)}{(x-3)(x+2)}$
$=\frac{2x^2+5x+12}{(x-3)(x+2)},$
$x \neq 3, -2$

47. $\frac{x+5}{x-5}+\frac{x-5}{x+5}$
$=\frac{(x+5)(x+5)+(x-5)(x-5)}{(x-5)(x+5)}$
$=\frac{x^2+10x+25+x^2-10x+25}{(x-5)(x+5)}$
$=\frac{2x^2+50}{(x-5)(x+5)}, x \neq -5, 5$

48. $\frac{x+3}{x-3}+\frac{x-3}{x+3}=\frac{(x+3)(x+3)+(x-3)(x-3)}{(x-3)(x+3)}$
$=\frac{x^2+6x+9+x^2-6x+9}{(x-3)(x+3)}$
$=\frac{2x^2+18}{(x-3)(x+3)},$
$x \neq -3, 3$

49. $\frac{4}{x^2+6x+9}+\frac{4}{x+3}=\frac{4}{(x+3)^2}+\frac{4}{x+3}$
$=\frac{4+4(x+3)}{(x+3)^2}=\frac{4+4x+12}{(x+3)^2}=\frac{4x+16}{(x+3)^2},$
$x \neq -3$

50. $\frac{3}{5x+2}+\frac{5x}{25x^2-4}=\frac{3}{5x+2}+\frac{5x}{(5x-2)(5x+2)}$
$=\frac{3(5x-2)+5x}{(5x-2)(5x+2)}$
$=\frac{15x-6+5x}{(5x-2)(5x+2)}$
$=\frac{20x-6}{(5x-2)(5x+2)},$
$x \neq -\frac{2}{5}, \frac{2}{5}$

51. $\frac{3x}{x^2+3x-10}-\frac{2x}{x^2+x-6}$
$=\frac{3x}{(x+5)(x-2)}-\frac{2x}{(x+3)(x-2)}$
$=\frac{3x(x+3)-2x(x+5)}{(x+5)(x-2)(x+3)}$
$=\frac{3x^2+9x-2x^2-10x}{(x+5)(x-2)(x+3)}$
$=\frac{x^2-x}{(x+5)(x-2)(x+3)}$, $x\neq -5, 2, -3$

52. $\frac{x}{x^2-2x-24}-\frac{x}{x^2-7x+6}$
$=\frac{x}{(x-6)(x+4)}-\frac{x}{(x-6)(x-1)}$
$=\frac{x(x-1)-x(x+4)}{(x-6)(x+4)(x-1)}$
$=\frac{x^2-x-x^2-4x}{(x-6)(x+4)(x-1)}$
$=-\frac{5x}{(x-6)(x-1)(x+4)}$,
$x\neq 6, 1, -4$

53. $\frac{4x^2+x-6}{x^2+3x+2}-\frac{3x}{x+1}+\frac{5}{x+2}$
$=\frac{4x^2+x-6}{(x+1)(x+2)}+\frac{-3x}{x+1}+\frac{5}{x+2}$
$=\frac{4x^2+x-5}{(x+1)(x+2)}+\frac{-3x(x+2)}{(x+1)(x+2)}+\frac{5(x+1)}{(x+1)(x+2)}$
$=\frac{4x^2+x-6-3x^2-6x+5x+5}{(x+1)(x+2)}$
$=\frac{x^2-1}{(x+1)(x+2)}$
$=\frac{(x-1)(x+1)}{(x+1)(x+2)}$
$=\frac{x-1}{x+2}; x\neq -2, -1$

54. $\frac{6x^2+17x-40}{x^2+x-20}+\frac{3}{x-4}-\frac{5x}{x+5}$
$=\frac{6x^2+17x-40}{(x+5)(x-4)}+\frac{3}{x-4}-\frac{5x}{x+5}$
$=\frac{6x^2+17x-40+3(x+5)-5x(x-4)}{(x+5)(x-4)}$
$=\frac{6x^2+17x-40+3x+15-5x^2+20x}{(x+5)(x-4)}$
$=\frac{x^2+40x-25}{(x+5)(x-4)}; x\neq -5, 4$

55. $\frac{\frac{x}{3}-1}{x-3}=\frac{3\left[\frac{x}{3}-1\right]}{3[x-3]}=\frac{x-3}{3(x-3)}=\frac{1}{3}$, $x\neq 3$

56. $\frac{\frac{x}{4}-1}{x-4}=\frac{4\left[\frac{x}{4}-1\right]}{4(x-4)}=\frac{x-4}{4(x-4)}=\frac{1}{4}, x\neq 4$

57. $\frac{1+\frac{1}{x}}{3-\frac{1}{x}}=\frac{x\left[1+\frac{1}{x}\right]}{x\left[3-\frac{1}{x}\right]}=\frac{x+1}{3x-1}, x\neq 0, \frac{1}{3}$

58. $\frac{8+\frac{1}{x}}{4-\frac{1}{x}}=\frac{x\left[8+\frac{1}{x}\right]}{x\left[4-\frac{1}{x}\right]}=\frac{8x+1}{4x-1}$, $x\neq 0, \frac{1}{4}$

59. $\frac{\frac{1}{x}+\frac{1}{y}}{x+y}=\frac{xy\left[\frac{1}{x}+\frac{1}{y}\right]}{xy[x+y]}=\frac{y+x}{xy(x+y)}=\frac{1}{xy}$,
$x\neq 0, y\neq 0, x\neq -y$

60. $\frac{1-\frac{1}{x}}{xy}=\frac{x\left[1-\frac{1}{x}\right]}{x(xy)}=\frac{x-1}{x^2y}, x\neq 0, y\neq 0$

61. $\frac{x-\frac{x}{x+3}}{x+2}=\frac{(x+3)\left[x-\frac{x}{x+3}\right]}{(x+3)(x+2)}=\frac{x(x+3)-x}{(x+3)(x+2)}$
$=\frac{x^2+3x-x}{(x+3)(x+2)}=\frac{x^2+2x}{(x+3)(x+2)}$
$=\frac{x(x+2)}{(x+3)(x+2)}=\frac{x}{x+3}, x\neq -2, -3$

62. $\dfrac{x-3}{x-\dfrac{3}{x-2}} = \dfrac{(x-2)[x-3]}{(x-2)\left[x-\dfrac{3}{x-2}\right]} = \dfrac{(x-2)(x-3)}{x(x-2)-3}$

$= \dfrac{(x-2)(x-3)}{x^2-2x-3}$

$= \dfrac{(x-2)(x-3)}{(x-3)(x+1)} = \dfrac{x-2}{x+1}, x \neq 2, 3, -1$

63. $\dfrac{\dfrac{3}{x-2}-\dfrac{4}{x+2}}{\dfrac{7}{x^2-4}} = \dfrac{\dfrac{3}{x-2}-\dfrac{4}{x+2}}{\dfrac{7}{(x-2)(x+2)}}$

$= \dfrac{\left[\dfrac{3}{x-2}-\dfrac{4}{x+2}\right](x-2)(x+2)}{\left[\dfrac{7}{(x-2)(x+2)}\right](x-2)(x+2)}$

$= \dfrac{3(x+2)-4(x-2)}{7}$

$= \dfrac{3x+6-4x+8}{7} = \dfrac{-x+14}{7}$

$= -\dfrac{x-14}{7} \quad x \neq -2, 2$

64. $\dfrac{\dfrac{x}{x-2}+1}{\dfrac{3}{x^2-4}+1} = \dfrac{\dfrac{x}{x-2}+1}{\dfrac{3}{(x-2)(x+2)}+1}$

$= \dfrac{\left[\dfrac{x}{x-2}+1\right](x-2)(x+2)}{\left[\dfrac{3}{(x-2)(x+2)}+1\right](x-2)(x+2}$

$= \dfrac{x(x+2)+(x-2)(x+2)}{3+(x-2)(x+2)}$

$= \dfrac{x^2+2x+x^2-4}{3+x^2-4} = \dfrac{2x^2+2x-4}{x^2-1}$

$= \dfrac{2(x^2+x-2)}{(x-1)(x+1)}$

$= \dfrac{2(x+2)(x-1)}{(x-1)(x+1)} = \dfrac{2(x+2)}{x+1},$

$x \neq 1, -1, 2, -2$

65. $\dfrac{\dfrac{1}{x+1}}{\dfrac{1}{x^2-2x-3}+\dfrac{1}{x-3}} = \dfrac{\dfrac{1}{x+1}}{\dfrac{1}{(x+1)(x-3)}+\dfrac{1}{x-3}}$

$= \dfrac{\dfrac{(x+1)(x-3)}{x+1}}{\dfrac{(x+1)(x-3)}{(x+1)(x-3)}+\dfrac{(x+1)(x-3)}{x-3}}$

$= \dfrac{x-3}{1+x+1}$

$= \dfrac{x-3}{x+2} \quad x \neq -2, -1, 3$

66. $\dfrac{\dfrac{6}{x^2+2x-15}-\dfrac{1}{x-3}}{\dfrac{1}{x+5}+1} = \dfrac{\dfrac{6}{(x+5)(x-3)}-\dfrac{1}{x-3}}{\dfrac{1}{x+5}+1}$

$= \dfrac{\dfrac{6(x+5)(x-3)}{(x+5)(x-3)}-\dfrac{(x+5)(x-3)}{x-3}}{\dfrac{(x+5)(x-3)}{x+5}+(x+5)(x-3)}$

$= \dfrac{6-(x+5)}{(x-3)+(x+5)(x-3)}$

$= \dfrac{6-x-5}{x-3+x^2+2x-15}$

$= \dfrac{1-x}{x^2+3x-18}$

$= \dfrac{1-x}{(x+6)(x-3)} \quad x \neq -6, -5, 3$

67. $$\frac{\frac{1}{(x+h)^2}-\frac{1}{x^2}}{h}=\frac{\frac{x^2(x+h)^2}{(x+h)^2}-\frac{x^2(x+h)^2}{x^2}}{hx^2(x+h)^2}$$
$$=\frac{x^2-(x+h)^2}{hx^2(x+h)^2}$$
$$=\frac{x^2-(x^2+2hx+h^2)}{hx^2(x+h)^2}$$
$$=\frac{x^2-x^2-2hx-h^2}{hx^2(x+h)^2}$$
$$=\frac{-2hx-h^2}{hx^2(x+h)^2}$$
$$=\frac{-h(2x+h)}{hx^2(x+h)^2}$$
$$=-\frac{(2x+h)}{x^2(x+h)^2}$$

68. $$\frac{\frac{x+h}{x+h+1}-\frac{x}{x+1}}{h}=\frac{\frac{(x+h)(x+h+1)(x+1)}{x+h+1}-\frac{x(x+h+1)(x+1)}{x+1}}{h(x+h+1)(x+1)}$$
$$=\frac{(x+h)(x+1)-x(x+h+1)}{h(x+h+1)(x+1)}$$
$$=\frac{x^2+x+hx+h-x^2-hx-x}{h(x+h+1)(x+1)}$$
$$=\frac{h}{h(x+h+1)(x+1)}$$
$$=\frac{1}{(x+h+1)(x+1)}$$

69. $$\frac{\sqrt{x}-\frac{1}{3\sqrt{x}}}{\sqrt{x}}=\frac{\left(\sqrt{x}-\frac{1}{3\sqrt{x}}\right)(3\sqrt{x})}{\sqrt{x}(3\sqrt{x})}$$
$$=\frac{3x-1}{3x}$$
$$=1-\frac{1}{3x},\ x>0$$

70. $$\frac{\sqrt{x}-\frac{1}{4\sqrt{x}}}{\sqrt{x}}=\frac{\left(\sqrt{x}-\frac{1}{4\sqrt{x}}\right)(4\sqrt{x})}{\sqrt{x}(4\sqrt{x})}$$
$$=\frac{4x-1}{4x}$$
$$=1-\frac{1}{4x},\ x>0$$

71. $\dfrac{\frac{x^2}{\sqrt{x^2+2}}-\sqrt{x^2+2}}{x^2}$

$=\dfrac{\left(\frac{x^2}{\sqrt{x^2+2}}-\sqrt{x^2+2}\right)\sqrt{x^2+2}}{x^2\sqrt{x^2+2}}$

$=\dfrac{x^2-(x^2+2)}{x^2\sqrt{x^2+2}}$

$=-\dfrac{2}{x^2\sqrt{x^2+2}}$

72. $\dfrac{\sqrt{5-x^2}+\frac{x^2}{\sqrt{5-x^2}}}{5-x^2}$

$=\dfrac{\left(\sqrt{5-x^2}+\frac{x^2}{\sqrt{5-x^2}}\right)\sqrt{5-x^2}}{\left(5-x^2\right)\sqrt{5-x^2}}$

$=\dfrac{5-x^2+x^2}{\left(5-x^2\right)\sqrt{5-x^2}}$

$=\dfrac{5}{\left(\sqrt{5-x^2}\right)^2\sqrt{5-x^2}}$

$=\dfrac{5}{\sqrt{\left(5-x^2\right)^3}}$

73. $\dfrac{\frac{1}{\sqrt{x+h}}-\frac{1}{\sqrt{x}}}{h}=\dfrac{\left(\frac{1}{\sqrt{x+h}}-\frac{1}{\sqrt{x}}\right)\sqrt{x+h}\sqrt{x}}{h\sqrt{x+h}\sqrt{x}}$

$=\dfrac{\sqrt{x}-\sqrt{x+h}}{h\sqrt{x(x+h)}},\ h\neq 0$

74. $\dfrac{\frac{1}{\sqrt{x+3}}-\frac{1}{\sqrt{x}}}{3}=\dfrac{\left(\frac{1}{\sqrt{x+3}}-\frac{1}{\sqrt{x}}\right)\sqrt{x+3}\sqrt{x}}{3\sqrt{x+3}\sqrt{x}}$

$=\dfrac{\sqrt{x}-\sqrt{x+3}}{3\sqrt{x}\sqrt{x+3}}$

75. $\dfrac{\sqrt{x+5}-\sqrt{x}}{5}=\dfrac{\sqrt{x+5}-\sqrt{x}}{5}\cdot\dfrac{\sqrt{x+5}+\sqrt{x}}{\sqrt{x+5}+\sqrt{x}}$

$=\dfrac{(\sqrt{x+5})^2-(\sqrt{x})^2}{5(\sqrt{x+5}+\sqrt{x})}$

$=\dfrac{x+5-x}{5(\sqrt{x+5}+\sqrt{x})}$

$=\dfrac{1}{\sqrt{x+5}+\sqrt{x}}$

76. $\dfrac{\sqrt{x+7}-\sqrt{x}}{7}=\dfrac{\sqrt{x+7}-\sqrt{x}}{7}\cdot\dfrac{\sqrt{x+7}+\sqrt{x}}{\sqrt{x+7}+\sqrt{x}}$

$=\dfrac{(\sqrt{x+7})^2-(\sqrt{x})^2}{7(\sqrt{x+7}+\sqrt{x})}$

$=\dfrac{x+7-x}{7(\sqrt{x+7}+\sqrt{x})}$

$=\dfrac{1}{\sqrt{x+7}+\sqrt{x}}$

77. $\dfrac{\sqrt{x}+\sqrt{y}}{x^2-y^2}=\dfrac{\sqrt{x}+\sqrt{y}}{x^2-y^2}\cdot\dfrac{\sqrt{x}-\sqrt{y}}{\sqrt{x}-\sqrt{y}}$

$=\dfrac{(\sqrt{x})^2-(\sqrt{y})^2}{(x+y)(x-y)(\sqrt{x}-\sqrt{y})}$

$=\dfrac{x-y}{(x+y)(x-y)(\sqrt{x}-\sqrt{y})}$

$=\dfrac{1}{(x+y)(\sqrt{x}-\sqrt{y})}$

78. $\dfrac{\sqrt{x}-\sqrt{y}}{x^2-y^2}=\dfrac{\sqrt{x}-\sqrt{y}}{x^2-y^2}\cdot\dfrac{\sqrt{x}+\sqrt{y}}{\sqrt{x}+\sqrt{y}}$

$=\dfrac{(\sqrt{x})^2-(\sqrt{y})^2}{(x^2-y^2)(\sqrt{x}+\sqrt{y})}$

$=\dfrac{x-y}{(x+y)(x-y)(\sqrt{x}+\sqrt{y})}$

$=\dfrac{1}{(x+y)(\sqrt{x}+\sqrt{y})},\ x\neq y$

79. $\left(\frac{2x+3}{x+1}\cdot\frac{x^2+4x-5}{2x^2+x-3}\right)-\frac{2}{x+2}=\left(\frac{\cancel{(2x+3)}}{x+1}\cdot\frac{(x+5)\cancel{(x-1)}}{\cancel{(2x+3)}\cancel{(x-1)}}\right)-\frac{2}{x+2}=\frac{x+5}{x+1}-\frac{2}{x+2}$

$=\frac{(x+5)(x+2)}{(x+1)(x+2)}-\frac{2(x+1)}{(x+1)(x+2)}=\frac{(x+5)(x+2)-2(x+1)}{(x+1)(x+2)}=\frac{x^2+2x+5x+10-2x-2}{(x+1)(x+2)}=\frac{x^2+5x+8}{(x+1)(x+2)}$

80. $\frac{1}{x^2-2x-8}\cdot\left(\frac{1}{x-4}-\frac{1}{x+2}\right)=\frac{1}{(x-4)(x+2)}\div\left(\frac{(x+2)}{(x-4)(x+2)}-\frac{(x-4)}{(x-4)(x+2)}\right)$

$=\frac{1}{(x-4)(x+2)}\div\left(\frac{x+2-x+4}{(x-4)(x+2)}\right)=\frac{1}{(x-4)(x+2)}\div\left(\frac{6}{(x-4)(x+2)}\right)=\frac{1}{(x-4)(x+2)}\cdot\frac{(x-4)(x+2)}{6}=\frac{1}{6}$

81. $\left(2-\frac{6}{x+1}\right)\left(1+\frac{3}{x-2}\right)=\left(\frac{2(x+1)}{(x+1)}-\frac{6}{(x+1)}\right)\left(\frac{(x-2)}{(x-2)}+\frac{3}{(x-2)}\right)$

$=\left(\frac{2x+2-6}{x+1}\right)\left(\frac{x-2+3}{x-2}\right)=\left(\frac{2x-4}{x+1}\right)\left(\frac{x+1}{x-2}\right)=\frac{2\cancel{(x-2)}\cancel{(x+1)}}{\cancel{(x+1)}\cancel{(x-2)}}=2$

82. $\left(4-\frac{3}{x+2}\right)\left(1+\frac{5}{x-1}\right)=\left(\frac{4(x+2)}{x+2}-\frac{3}{x+2}\right)\left(\frac{(x-1)}{x-1}+\frac{5}{x-1}\right)$

$=\left(\frac{4x+8-3}{x+2}\right)\left(\frac{x-1+5}{x-1}\right)=\frac{4x+5}{x+2}\cdot\frac{x+4}{x-1}=\frac{(4x+5)(x+4)}{(x+2)(x-1)}$

83. $\frac{y^{-1}-(y+5)^{-1}}{5}=\frac{\frac{1}{y}-\frac{1}{y+5}}{5}$

LCD $=y(y+5)$

$\frac{\frac{1}{y}-\frac{1}{y+5}}{5}=\frac{y(y+5)\left(\frac{1}{y}-\frac{1}{y+5}\right)}{y(y+5)(5)}=\frac{y+5-y}{5y(y+5)}=\frac{5}{5y(y+5)}=\frac{1}{y(y+5)}$

84. $\frac{y^{-1}-(y+2)^{-1}}{2}=\frac{\frac{1}{y}-\frac{1}{y+2}}{2}$

LCD $=y(y+2)$

$\frac{\frac{1}{y}-\frac{1}{y+2}}{2}=\frac{y(y+2)\left(\frac{1}{y}-\frac{1}{y+2}\right)}{y(y+2)(2)}=\frac{y+2-y}{2y(y+2)}=\frac{2}{2y(y+2)}=\frac{1}{y(y+2)}$

85. $\left(\frac{1}{a^3-b^3}\cdot\frac{ac+ad-bc-bd}{1}\right)-\frac{c-d}{a^2+ab+b^2}=\left(\frac{1}{(a-b)(a^2+ab+b^2)}\cdot\frac{a(c+d)-b(c+d)}{1}\right)-\frac{c-d}{a^2+ab+b^2}$

$=\left(\frac{1}{(a-b)(a^2+ab+b^2)}\cdot\frac{(c+d)(a-b)}{1}\right)-\frac{c-d}{a^2+ab+b^2}=\frac{c+d}{a^2+ab+b^2}-\frac{c-d}{a^2+bd+b^2}$

$=\frac{c+d-c+d}{a^2+ab+b^2}=\frac{2d}{a^2+ab+b^2}$

86. $\frac{ab}{a^2+ab+b^2}+\left(\frac{ac-ad-bc+bd}{ac-ad+bc-bd}\div\frac{a^3-b^3}{a^3+b^3}\right)=\frac{ab}{a^2+ab+b^2}+\left(\frac{a(c-d)-b(c-d)}{a(c-d)+b(c-d)}\cdot\frac{a^3+b^3}{a^3-b^3}\right)$

$=\frac{ab}{a^2+ab+b^2}+\left(\frac{(c-d)(a-b)}{(c-d)(a+b)}\cdot\frac{(a+b)(a^2-ab+b^2)}{(a-b)(a^2+ab+b^2)}\right)=\frac{ab}{a^2+ab+b^2}+\frac{a^2-ab+b^2}{a^2+ab+b^2}$

$=\frac{ab+a^2-ab+b^2}{a^2+ab+b^2}=\frac{a^2+b^2}{a^2+ab\div b^2}$

87. a. $\frac{130x}{100-x}$ is equal to

1. $\frac{130\cdot 40}{100-40}=\frac{130\cdot 40}{60}=86.67$,
   when $x=40$
2. $\frac{130\cdot 80}{100-80}=\frac{130\cdot 80}{20}=520$,
   when $x=80$
3. $\frac{130\cdot 90}{100-90}=\frac{130\cdot 90}{10}=1170$,
   when $x=90$

It costs $86,670,000 to inoculate 40% of the population against this strain of flu, and $520,000,000 to inoculate 80% of the population, and $1,170,000,000 to inoculate 90% of the population.

b. For $x=100$, the function is not defined.

c. As x approaches 100, the value of the function increases rapidly. So it costs an astronomical amount of money to inoculate almost all of the people, and it is impossible to inoculate 100% of the population.

88. $\frac{DA}{A+12}$

$\frac{7D}{7+12}-\frac{3D}{3+12}=\frac{7D}{19}-\frac{3D}{15}=D\left(\frac{7}{19}-\frac{3}{15}\right)$

$=\frac{16}{95}\cdot D$

A 7-year-old child needs to take $\frac{16}{95}$ times the adult dose more than a 3-year-old child.

89. a. The crime rate is the number of crimes per person. Thus the rational expression is the number of crimes divided by the total population

$\frac{-0.3t+14}{3.6t+260}$

b. $\frac{-0.3t+14}{3.6t+260}=\frac{-0.3(8)+14}{3.6(8)+260}\approx 0.04$

The crime rate in 2002 was 0.04. That is 4000 per 100,000 people.

c. The rational expression models the crime rate in 2002 fairly well.

90. $\dfrac{2d}{\dfrac{d}{r_1}+\dfrac{d}{r_2}}$

LCD $= r_1r_2$

$$\frac{2d}{\dfrac{d}{r_1}+\dfrac{d}{r_2}} = \frac{r_1r_2(2d)}{r_1r_2\left(\dfrac{d}{r_1}+\dfrac{d}{r_2}\right)}$$

$$= \frac{2r_1r_2d}{r_2d+r_1d}$$

$$= \frac{2r_1r_2d}{d(r_2+r_1)} = \frac{2r_1r_2}{r_2+r_1}$$

If $r_1 = 40$ and $r_2 = 30$, the value of this expression will be

$$\frac{2\cdot 40\cdot 30}{30+40} = \frac{2400}{70}$$

$$= 34\frac{2}{7}.$$

Your average speed will be $34\frac{2}{7}$ miles per hour.

91. $P = 2L + 2W$

$$= 2\left(\frac{x}{x+3}\right)+2\left(\frac{x}{x-4}\right)$$

$$= \frac{2x}{x+3}+\frac{2x}{x+4}$$

$$= \frac{2x(x+4)}{(x+3)(x+4)}+\frac{2x(x+}{(x+3)(}$$

$$= \frac{2x^2+8x+2x^2+6x}{(x+3)(x+4)}$$

$$= \frac{4x^2+14x}{(x+3)(x+4)}$$

92. $P = 2L + 2W$

$$= 2\left(\frac{x}{x+5}\right)+2\left(\frac{x}{x+6}\right)$$

$$= \frac{2x}{x+5}+\frac{2x}{x+6}$$

$$= \frac{2x(x+6)}{(x+5)(x+6)}+\frac{2x(x-}{(x+5)(}$$

$$= \frac{2x^2+12x+2x^2+10x}{(x+5)(x+6)}$$

$$= \frac{4x^2+22x}{(x+5)(x+6)}$$

93. $R = \dfrac{1}{\frac{1}{R_1}+\frac{1}{R_2}+\frac{1}{R_3}}$

$$= \frac{R_1R_2R_3}{\left(\frac{1}{R_1}+\frac{1}{R_2}+\frac{1}{R_3}\right)R_1R_2R_3}$$

$$= \frac{R_1R_2R_3}{R_2R_3+R_1R_3+R_1R_2}$$

$$R(4,\ 8,\ 12) = \frac{4\cdot 8\cdot 12}{8\cdot 12+4\cdot 12+4\cdot 8}$$

$$= \frac{384}{96+48+32}$$

$$= \frac{384}{176}$$

$$= \frac{24}{11}$$

The parallel resistance is $\frac{24}{11}$ ohms.

107. a. false; $\frac{a}{b}+\frac{a}{c}=\frac{ab+ac}{bc}$

b. false; $6+\frac{1}{x}=\frac{6x+1}{x}$

c. false; $\frac{1}{x+3}+\frac{x+3}{2}=\frac{x^2+6x+11}{2(x+3)}$

d. false; $\frac{x^2-25}{x-5}=\frac{(x+5)(x-5)}{x-5}=x+5$

e. true

108. $\frac{1}{x^n-1}-\frac{1}{x^n+1}-\frac{1}{x^{2n}-1}$

$=\frac{x^n+1}{x^{2n}-1}-\frac{x^n-1}{x^{2n}-1}-\frac{1}{x^{2n}-1}$

$=\frac{x^n+1-x^n+1-1}{x^{2n}-1}$

$=\frac{1}{x^{2n}-1}$

109. $\left(1-\frac{1}{x}\right)\left(1-\frac{1}{x+1}\right)\left(1-\frac{1}{x+2}\right)\left(1-\frac{1}{x+3}\right)=\left(\frac{x}{x}-\frac{1}{x}\right)\left(\frac{x+1}{x+1}-\frac{1}{x+1}\right)\left(\frac{x+2}{x+2}-\frac{1}{x+2}\right)\left(\frac{x+3}{x+3}-\frac{1}{x+3}\right)$

$=\left(\frac{x-1}{x}\right)\left(\frac{(x+1)-1}{x+1}\right)\left(\frac{(x+2)-1}{x+2}\right)\left(\frac{(x+3)-1}{x+3}\right)$

$=\frac{x-1}{\cancel{x}}\cdot\frac{\cancel{x}}{\cancel{x+1}}\cdot\frac{\cancel{x+1}}{\cancel{x+2}}\cdot\frac{\cancel{x+2}}{x+3}=\frac{x-1}{x+3}$

110. $(x-y)^{-1}+(x-y)^{-2}=\frac{1}{(x-y)}+\frac{1}{(x-y)^2}=\frac{(x-y)}{(x-y)(x-y)}+\frac{1}{(x-y)^2}=\frac{x-y+1}{(x-y)^2}$

111. It cubes x.

$$\frac{\frac{1}{x}+\frac{1}{x^2}+\frac{1}{x^3}}{\frac{1}{x^4}+\frac{1}{x^5}+\frac{1}{x^6}}=\frac{\frac{x^6}{x}+\frac{x^6}{x^2}+\frac{x^6}{x^3}}{\frac{x^6}{x^4}+\frac{x^6}{x^5}+\frac{x^6}{x^6}}=\frac{x^5+x^4+x^3}{x^2+x+1}=\frac{x^3\left(x^2+x+1\right)}{x^2+x+1}=x^3$$

Section P.7

Check Point Exercises

1.

$$\begin{aligned}
4(2x+1)-29 &= 3(2x-5)\\
8x+4-29 &= 6x-15\\
8x-25 &= 6x-15\\
8x-25-6x &= 6x-15-6x\\
2x-25 &= -15\\
2x-25+25 &= -15+25\\
2x &= 10\\
\frac{2x}{2} &= \frac{10}{2}\\
x &= 5
\end{aligned}$$

Check:

$$\begin{aligned}
4(2x+1)-29 &= 3(2x-5)\\
4[2(5)+1]-29 &= 3[2(5)-5]\\
4[10+1]-29 &= 3[10-5]\\
4[11]-29 &= 3[5]\\
44-29 &= 15\\
15 &= 15 \text{ true}
\end{aligned}$$

The solution set is $\{5\}$.

2.

$$\begin{aligned}
\frac{x-3}{4} &= \frac{5}{14}-\frac{x+5}{7}\\
28\cdot\frac{x-3}{4} &= 28\left(\frac{5}{14}-\frac{x+5}{7}\right)\\
7(x-3) &= 2(5)-4(x+5)\\
7x-21 &= 10-4x-20\\
7x-21 &= -4x-10\\
7x+4x &= -10+21\\
11x &= 11\\
\frac{11x}{11} &= \frac{11}{11}\\
x &= 1
\end{aligned}$$

Check:

$$\begin{aligned}
\frac{x-3}{4} &= \frac{5}{14}-\frac{x+5}{7}\\
\frac{1-3}{4} &= \frac{5}{14}-\frac{1+5}{7}\\
\frac{-2}{4} &= \frac{5}{14}-\frac{6}{7}\\
-\frac{1}{2} &= -\frac{1}{2}
\end{aligned}$$

The solution set is $\{1\}$.

3.

$$\begin{aligned}
\frac{6}{x+3}-\frac{5}{x-2} &= \frac{-20}{x^2+x-6}\\
\frac{6}{x+3}-\frac{5}{x-2} &= \frac{-20}{(x+3)(x-2)}\\
\frac{6(x+3)(x-2)}{x+3}-\frac{5(x+3)(x-2)}{x-2} &= \frac{-20(x+3)(x-2)}{(x+3)(x-2)}\\
6(x-2)-5(x+3) &= -20\\
6x-12-5x-15 &= -20\\
x-27 &= -20\\
x &= 7
\end{aligned}$$

The solution set is $\{7\}$.

4.
$$\frac{1}{x+2}=\frac{4}{x^2-4}-\frac{1}{x-2}$$
$$\frac{1}{x+2}=\frac{4}{(x+2)(x-2)}-\frac{1}{x-2}$$
$$\frac{1(x+2)(x-2)}{x+2}=\frac{4(x+2)(x-2)}{(x+2)(x-2)}-\frac{1(x+2)(x-2)}{x-2}$$
$$x-2=4-(x+2)$$
$$x-2=4-x-2$$
$$x-2=2-x$$
$$2x=4$$
$$x=2$$
2 must be rejected. The solution set is $\{\ \}$.

5.
$$\frac{1}{p}+\frac{1}{q}=\frac{1}{f}$$
$$\frac{1pqf}{p}+\frac{1pqf}{q}=\frac{1pqf}{f}$$
$$qf+pf=pq$$
$$qf-pq=-pf$$
$$q(f-p)=-pf$$
$$\frac{q(f-p)}{f-p}=\frac{-pf}{f-p}$$
$$q=\frac{pf}{p-f}$$

6. $4|1-2x|-20=0$
$$4|1-2x|=20$$
$$|1-2x|=5$$
$1-2x=5$ or $1-2x=-5$
$-2x=4$ $\quad -2x=-6$
$x=-2$ $\quad x=3$
The solution set is $\{-2, 3\}$.

7. **a.** $3x^2-9x=0$
$$3x(x-3)=0$$
$3x=0$ or $x-3=0$
$x=0$ $\quad x=3$
The solution set is $\{0,3\}$.

b. $2x^2+x=1$
$$2x^2+x-1=0$$
$$(2x-1)(x+1)=0$$
$2x-1=0$ or $x+1=0$
$2x=1$ $\quad x=-1$
$$x=\frac{1}{2}$$
The solution set is $\left\{\frac{1}{2},-1\right\}$.

8. **a.** $3x^2-21=0$
$$3x^2=21$$
$$\frac{3x^2}{3}=\frac{21}{3}$$
$$x^2=7$$
$$x=\pm\sqrt{7}$$
The solution set is $\left\{-\sqrt{7},\sqrt{7}\right\}$.

b. $(x+5)^2=11$
$$x+5=\pm\sqrt{11}$$
$$x=-5\pm\sqrt{11}$$
The solution set is $\left\{-5+\sqrt{11},-5-\sqrt{11}\right\}$.

9. $x^2+4x-1=0$

$x^2+4x \quad =1$

$x^2+4x+4=1+4$

$(x+2)^2=5$

$x+2=\pm\sqrt{5}$

$x=-2\pm\sqrt{5}$

10. $2x^2+2x-1=0$

$a=2, b=2, c=-1$

$$x=\frac{-b\pm\sqrt{b^2-4ac}}{2a}$$

$$=\frac{-2\pm\sqrt{2^2-4(2)(-1)}}{2(2)}$$

$$=\frac{-2\pm\sqrt{4+8}}{4}$$

$$=\frac{-2\pm\sqrt{12}}{4}$$

$$=\frac{-2\pm2\sqrt{3}}{4}$$

$$=\frac{2(-1\pm\sqrt{3})}{4}$$

$$=\frac{-1\pm\sqrt{3}}{2}$$

The solution set is $\left\{\frac{-1+\sqrt{3}}{2}, \frac{-1-\sqrt{3}}{2}\right\}$.

11. $3x^2-2x+5=0$

$a=3,\ b=-2,\ c=5$

$b^2-4ac=(-2)^2-4\cdot3\cdot5=4-60=-56$

The discriminant is –56. The equation has two complex imaginary solutions.

12. $\sqrt{x+3}+3=x$

$\sqrt{x+3}=x-3$

$\left(\sqrt{x+3}\right)^2=(x-3)^2$

$x+3=x^2-6x+9$

$0=x^2-7x+6$

$0=(x-6)(x-1)$

$x-6=0$ or $x-1=0$

$x=6 \qquad x=1$

1 does not check and must be rejected.

The solution set is $\{6\}$.

Exercise Set P.7

1. $7x-5=72$

$7x=77$

$x=11$

Check:

$7x-5=72$

$7(11)-5=72$

$77-5=72$

$72=72$

The solution set is $\{11\}$.

2. $6x-3=63$

$6x=66$

$x=11$

The solution set is $\{11\}$.

Check:

$6x-3=63$

$6(11)-3=63$

$66-3=63$

$63=63$

3. $11x-(6x-5)=40$

$11x-6x+5=40$

$5x+5=40$

$5x=35$

$x=7$

The solution set is $\{7\}$.

Check:

$11x-(6x-5)=40$

$11(7)-[6(7)-5]=40$

$77-(42-5)=40$

$77-(37)=40$

$40=40$

4. $5x-(2x-10)=35$

$5x-2x+10=35$

$3x+10=35$

$3x=25$

$x=\frac{25}{3}$

The solution set is $\left\{\frac{25}{3}\right\}$.

Check:

$$5x-(2x-10)=35$$
$$5\left(\frac{25}{3}\right)-\left[2\left(\frac{25}{3}\right)-10\right]=35$$
$$\frac{125}{3}-\left[\frac{50}{3}-10\right]=35$$
$$\frac{125}{3}-\frac{20}{3}=35$$
$$\frac{105}{3}=35$$
$$35=35$$

5. $2x-7=6+x$
$x-7=6$
$x=13$
The solution set is $\{13\}$.

Check:
$2(13)-7=6+13$
$26-7=19$
$19=19$

6. $3x+5=2x+13$
$x+5=13$
$x=8$
The solution set is $\{8\}$.

Check:
$3x+5=2x+13$
$3(8)+5=2(8)+13$
$24+5=16+13$
$29=29$

7. $7x+4=x+16$
$6x+4=16$
$6x=12$
$x=2$
The solution set is $\{2\}$.

Check:
$7(2)+4=2+16$
$14+4=18$
$18=18$

8. $13x+14=12x-5$
$x+14=-5$
$x=-19$
The solution set is $\{-19\}$.

Check:
$13x+14=12x-5$
$13(-19)+14=12(-19)-5$
$-247+14=-228-5$
$-233=-233$

9. $3(x-2)+7=2(x+5)$
$3x-6+7=2x+10$
$3x+1=2x+10$
$x+1=10$
$x=9$
The solution set is $\{9\}$.

Check:
$3(9-2)+7=2(9+5)$
$3(7)+7=2(14)$
$21+7=28$
$28=28$

10. $2(x-1)+3=x-3(x+1)$
$2x-2+3=x-3x-3$
$2x+1=-2x-3$
$4x+1=-3$
$4x=-4$
$x=-1$
The solution set is $\{-1\}$.

Check:
$2(x-1)+3=x-3(x+1)$
$2(-1-1)+3=-1-3(-1+1)$
$2(-2)+3=-1-3(0)$
$-4+3=-1+0$
$-1=-1$

11. $\frac{x+3}{6}=\frac{3}{8}+\frac{x-5}{4}$

$$24\left[\frac{x+3}{6}=\frac{3}{8}+\frac{x-5}{4}\right]$$
$$4x+12=9+6x-30$$
$$4x-6x=-21-12$$
$$-2x=-33$$
$$x=\frac{33}{2}$$

The solution set is $\left\{\frac{33}{2}\right\}$.

12. $\frac{x+1}{4}=\frac{1}{6}+\frac{2-x}{3}$

$12\left[\frac{x+1}{4}=\frac{1}{6}+\frac{2-x}{3}\right]$

$3x+3=2+8-4x$

$3x+4x=10-3$

$7x=7$

$x=1$

The solution set is $\{1\}$.

13. $\frac{x}{4}=2+\frac{x-3}{3}$

$12\left[\frac{x}{4}=2+\frac{x-3}{3}\right]$

$3x = 24 + 4x - 12$

$3x - 4x = 12$

$-x = 12$

$x = -12$

The solution set is $\{-12\}$.

14. $5+\frac{x-2}{3}=\frac{x+3}{8}$

$24\left[5+\frac{x-2}{3}=\frac{x+3}{8}\right]$

$120+8x-16=3x+9$

$8x-3x=9-104$

$5x=-95$

$x=-19$

The solution set is $\{-19\}$.

15. $\frac{x+1}{3}=5-\frac{x+2}{7}$

$21\left[\frac{x+1}{3}=5-\frac{x+2}{7}\right]$

$7x + 7 = 105 - 3x - 6$

$7x + 3x = 99 - 7$

$10x = 92$

$x=\frac{92}{10}$

$x=\frac{46}{5}$

The solution set is $\left\{\frac{46}{5}\right\}$.

16. $\frac{3x}{5}-\frac{x-3}{2}=\frac{x+2}{3}$

$30\left[\frac{3x}{5}-\frac{x-3}{2}=\frac{x+2}{3}\right]$

$18x-15x+45=10x+20$

$3x-10x=20-45$

$-7x=-25$

$x=\frac{25}{7}$

The solution set is $\left\{\frac{25}{7}\right\}$.

17. a. $\frac{1}{x-1}+5=\frac{11}{x-1}\ (x\neq 1)$

b. $\frac{1}{x-1}+5=\frac{11}{x-1}$

$1+5(x-1)=11$

$1+5x-5=11$

$5x-4=11$

$5x=15$

$x=3$

The solution set is $\{3\}$.

18. a. $\frac{3}{x+4}-7=\frac{-4}{x+4}\ (x\neq -4)$

b. $\frac{3}{x+4}-7=\frac{-4}{x+4}$

$3-7(x+4)=-4$

$3-7x-28=-4$

$-7x=21$

$x=-3$

The solution set is $\{-3\}$.

19. a. $\frac{8x}{x+1}=4-\frac{8}{x+1}\ (x\neq -1)$

b. $\frac{8x}{x+1}=4-\frac{8}{x+1}$

$8x=4(x+1)-8$

$8x=4x+4-8$

$4x=-4$

$x=-1 \Rightarrow$ no solution

The solution set is the empty set, $\varnothing$.

20. **a.** $\frac{2}{x-2} = \frac{x}{x-2} - 2\ (x \neq 2)$

b. $\frac{2}{x-2} = \frac{x}{x-2} - 2$

$2 = x - 2(x-2)$

$2 = x - 2x + 4$

$x = 2 \Rightarrow$ no solution

The solution set is the empty set, $\varnothing$.

21. **a.** $\frac{3}{2x-2} + \frac{1}{2} = \frac{2}{x-1}\ (x \neq 1)$

b. $\frac{3}{2x-2} + \frac{1}{2} = \frac{2}{x-1}$

$\frac{3}{2(x-1)} + \frac{1}{2} = \frac{2}{x-1}$

$3 + 1(x-1) = 4$

$3 + x - 1 = 4$

$x = 2$

The solution set is $\{2\}$.

22. **a.** $\frac{3}{x+3} = \frac{5}{2x+6} + \frac{1}{x-2}\ (x \neq -3, x \neq 2)$

b. $\frac{3}{x+3} = \frac{5}{2(x+3)} + \frac{1}{x-2}$

$6(x-2) = 5(x-2) + 2(x+3)$

$6x - 12 = 5x - 10 + 2x + 6$

$-x = 8$

$x = -8$

The solution set is $\{-8\}$.

23. **a.** $\frac{2}{x+1} - \frac{1}{x-1} = \frac{2x}{x^2-1}\ (x \neq 1, x \neq -1)$

b. $\frac{2}{x+1} - \frac{1}{x-1} = \frac{2x}{x^2-1}$

$\frac{2}{x+1} - \frac{1}{x-1} = \frac{2x}{(x+1)(x-1)}$

$2(x-1) - 1(x+1) = 2x$

$2x - 2 - x - 1 = 2x$

$-x = 3$

$x = -3$

The solution set is $\{-3\}$.

24. **a.** $\frac{4}{x+5} + \frac{2}{x-5} = \frac{32}{x^2-25}$; $x \neq 5, -5$

b. $\frac{4}{x+5} + \frac{2}{x-5} = \frac{32}{(x+5)(x-5)}$

$(x \neq 5, x \neq -5)$

$4(x-5) + 2(x+5) = 32$

$4x - 20 + 2x + 10 = 32$

$6x = 42$

$x = 7$

The solution set is $\{7\}$.

25. **a.** $\frac{1}{x-4} - \frac{5}{x+2} = \frac{6}{(x-4)(x+2)}$; $(x \neq -2, 4)$

b. $\frac{1}{x-4} - \frac{5}{x+2} = \frac{6}{x^2 - 2x - 8}$

$\frac{1}{x-4} - \frac{5}{x+2} = \frac{6}{(x-4)(x+2)}$

$(x \neq 4, x \neq -2)$

$1(x+2) - 5(x-4) = 6$

$x + 2 - 5x + 20 = 6$

$-4x = -16$

$x = 4$

The solution set is the empty set, $\varnothing$.

26. **a.** $\frac{1}{x-3} - \frac{2}{x+1} = \frac{8}{(x-3)(x+1)}$; $x \neq -1, 3$

b. $\frac{1}{x-3} - \frac{2}{x+1} = \frac{8}{(x-3)(x+1)}$

$1(x+1) - 2(x-3) = 8$

$x + 1 - 2x + 6 = 8$

$-x + 7 = 8$

$-x = 1$

$x = -1$

The solution set is the empty set, $\varnothing$.

27. $I = Prt$

$P = \frac{I}{rt}$;

interest

28. $C = 2\pi r$

$r = \frac{C}{2\pi}$;

circumference of a circle

29.
$$T = D + pm$$
$$T - D = pm$$
$$\frac{T-D}{m} = \frac{pm}{m}$$
$$\frac{T-D}{m} = p$$
total of payment

30.
$$P = C + MC$$
$$P - C = MC$$
$$\frac{P-C}{C} = M$$
markup based on cost

31.
$$A = \frac{1}{2}h(a+b)$$
$$2A = h(a+b)$$
$$\frac{2A}{h} = a+b$$
$$\frac{2A}{h} - b = a$$
area of trapezoid

32.
$$A = \frac{1}{2}h(a+b)$$
$$2A = h(a+b)$$
$$\frac{2A}{h} = a+b$$
$$\frac{2A}{h} - a = b$$
area of trapezoid

33.
$$S = P + Prt$$
$$S - P = Prt$$
$$\frac{S-P}{Pt} = r;$$
interest

34.
$$S = P + Prt$$
$$S - P = Prt$$
$$\frac{S-P}{Pr} = t;$$
interest

35.
$$B = \frac{F}{S-V}$$
$$B(S-V) = F$$
$$S - V = \frac{F}{B}$$
$$S = \frac{F}{B} + V$$

36.
$$S = \frac{C}{1-r}$$
$$S(1-r) = C$$
$$1 - r = \frac{C}{S}$$
$$-r = \frac{C}{S} - 1$$
$$r = -\frac{C}{S} + 1$$
markup based on selling price

37.
$$IR + Ir = E$$
$$I(R+r) = E$$
$$I = \frac{E}{R+r}$$
electric current

38.
$$A = 2lw + 2lh + 2wh$$
$$A - 2lw = h(2l + 2w)$$
$$\frac{A-2lw}{2l+2w} = h$$
surface area

39.
$$\frac{1}{p} + \frac{1}{q} = \frac{1}{f}$$
$$qf + pf = pq$$
$$f(q+p) = pq$$
$$f = \frac{pq}{p+q}$$
thin lens equation

40.
$$\frac{1}{R} = \frac{1}{R_1} + \frac{1}{R_2}$$
$$R_1R_2 = RR_2 + RR_1$$
$$R_1R_2 - RR_1 = RR_2$$
$$R_1(R_2 - R) = RR_2$$
$$R_1 = \frac{RR_2}{R_2 - R}$$
resistance

41.
$$f = \frac{f_1 f_2}{f_1 + f_2}$$
$$f(f_1 + f_2) = f_1 f_2$$
$$ff_1 + ff_2 = f_1 f_2$$
$$ff_1 - f_1 f_2 = -ff_2$$
$$f_1(f - f_2) = -ff_2$$
$$\frac{f_1(f - f_2)}{f - f_2} = \frac{-ff_2}{f - f_2}$$
$$f_1 = \frac{ff_2}{f_2 - f}$$
focal length

42.
$$f = \frac{f_1 f_2}{f_1 + f_2}$$
$$f(f_1 + f_2) = f_1 f_2$$
$$ff_1 + ff_2 = f_1 f_2$$
$$ff_2 - f_1 f_2 = -ff_1$$
$$f_2(f - f_1) = -ff_1$$
$$\frac{f_2(f - f_1)}{f - f_1} = \frac{-ff_1}{f - f_1}$$
$$f_2 = \frac{ff_1}{f_1 - f}$$
focal length

43. $|x - 2| = 7$

$x - 2 = 7 \quad x - 2 = -7$

$x = 9 \quad x = -5$

The solution set is $\{9, -5\}$.

44. $|x + 1| = 5$

$x + 1 = 5 \quad x + 1 = -5$

$x = 4 \quad x = -6$

The solution set is $\{-6, 4\}$.

45. $|2x - 1| = 5$

$2x - 1 = 5 \quad 2x - 1 = -5$

$2x = 6 \quad 2x = -4$

$x = 3 \quad x = -2$

The solution set is $\{3, -2\}$.

46. $|2x - 3| = 11$

$2x - 3 = 11 \quad 2x - 3 = -11$

$2x = 14 \quad 2x = -8$

$x = 7 \quad x = -4$

The solution set is $\{-4, 7\}$.

47. $2|3x - 2| = 14$

$|3x - 2| = 7$

$3x - 2 = 7 \quad 3x - 2 = -7$

$3x = 9 \quad 3x = -5$

$x = 3 \quad x = -5/3$

The solution set is $\{3, -5/3\}$

48. $3|2x - 1| = 21$

$|2x - 1| = 7$

$2x - 1 = 7$ or $2x - 1 = -7$

$2x = 8 \quad 2x = -6$

$x = 4 \quad x = -3$

The solution set is $\{4, -3\}$

49.
$$2\left|4 - \frac{5}{2}x\right| + 6 = 18$$
$$2\left|4 - \frac{5}{2}x\right| = 12$$
$$\left|4 - \frac{5}{2}x\right| = 6$$
$$4 - \frac{5}{2}x = 6 \quad \text{or} \quad 4 - \frac{5}{2}x = -6$$
$$-\frac{5}{2}x = 2 \qquad -\frac{5}{2}x = -10$$
$$x = -\frac{4}{5} \qquad x = 4$$
The solution set is $\left\{-\frac{4}{5}, 4\right\}$.

50.
$$4\left|1 - \frac{3}{4}x\right| + 7 = 10$$
$$4\left|1 - \frac{3}{4}x\right| = 3$$
$$\left|1 - \frac{3}{4}x\right| = \frac{3}{4}$$
$$1 - \frac{3}{4}x = \frac{3}{4} \quad \text{or} \quad 1 - \frac{3}{4}x = -\frac{3}{4}$$
$$-\frac{3}{4}x = -\frac{1}{4} \qquad -\frac{3}{4}x = -\frac{7}{4}$$
$$x = \frac{1}{3} \qquad x = \frac{7}{3}$$
The solution set is $\left\{\frac{1}{3}, \frac{7}{3}\right\}$.

51. $|x + 1| + 5 = 3$

$|x + 1| = -2$

No solution

The solution set is $\{\ \}$.

52. $|x+1|+6=2$
$|x+1|=-4$ The solution set is { }.

53. $|2x-1|+3=3$
$|2x-1|=0$
$2x-1=0$
$2x=1$
$x=1/2$
The solution set is $\left\{\frac{1}{2}\right\}$.

54. $|3x-2|+4=4$
$|3x-2|=0$
$3x-2=0$
$3x=2$
$x=\frac{2}{3}$
The solution set is $\left\{\frac{2}{3}\right\}$.

55. $x^2-3x-10=0$
$(x-5)(x+2)=0$
$x-5=0$ or $x+2=0$
$x=5$ $\quad x=-2$
The solution set is $\{-2,5\}$.

56. $x^2-13x+36=0$
$(x-4)(x-9)=0$
$x-4=0$ or $x-9=0$
$x=4$ $\quad x=9$
The solution set is $\{4,9\}$.

57. $x^2=8x-15$
$x^2-8x+15=0$
$(x-3)(x-5)=0$
$x-3=0$ or $x-5=0$
$x=3$ $\quad x=5$
The solution set is $\{3,5\}$.

58. $x^2=-11x-10$
$x^2+11x+10=0$
$(x+10)(x+1)=0$
$x+10=0$ or $x+1=0$
$x=-10$ $\quad x=-1$
The solution set is $\{-10,-1\}$.

59. $5x^2=20x$
$5x^2-20x=0$
$5x(x-4)=0$
$5x=0$ or $x-4=0$
$x=0$ $\quad x=4$
The solution set is $\{0,4\}$.

60. $3x^2=12x$
$3x^2-12x=0$
$3x(x-4)=0$
$3x=0$ or $x-4=0$
$x=0$ $\quad x=4$
The solution set is $\{0,4\}$.

61. $3x^2=27$
$x^2=9$
$\sqrt{x^2}=\pm\sqrt{9}$
$x=\pm 3$
The solution set is $\{\pm 3\}$.

62. $5x^2=45$
$x^2=9$
$\sqrt{x^2}=\pm\sqrt{9}$
$x=\pm 3$
The solution set is $\{\pm 3\}$.

63. $5x^2+1=51$
$5x^2=50$
$x^2=10$
$\sqrt{x^2}=\pm\sqrt{10}$
$x=\pm\sqrt{10}$
The solution set is $\{\pm\sqrt{10}\}$.

64. $3x^2-1=47$
$3x^2=48$
$x^2=16$
$\sqrt{x^2}=\pm\sqrt{16}$
$x=\pm 4$
The solution set is $\{\pm 4\}$.

65. $3(x-4)^2 = 15$

$(x-4)^2 = 5$

$\sqrt{(x-4)^2} = \pm\sqrt{5}$

$x - 4 = \pm\sqrt{5}$

$x = 4 \pm \sqrt{5}$

The solution set is $\{4 \pm \sqrt{5}\}$.

66. $3(x+4)^2 = 21$

$(x+4)^2 = 7$

$\sqrt{(x+4)^2} = \pm\sqrt{7}$

$x + 4 = \pm\sqrt{7}$

$x = -4 \pm \sqrt{7}$

The solution set is $\{-4 \pm \sqrt{7}\}$.

67. $x^2 + 6x = 7$

$x^2 + 6x + 9 = 7 + 9$

$(x+3)^2 = 16$

$x + 3 = \pm 4$

$x = -3 \pm 4$

The solution set is $\{-7, 1\}$.

68. $x^2 + 6x = -8$

$x^2 + 6x + 9 = -8 + 9$

$(x+3)^2 = 1$

$x + 3 = \pm 1$

$x = -3 \pm 1$

The solution set is $\{-4, -2\}$.

69. $x^2 - 2x = 2$

$x^2 - 2x + 1 = 2 + 1$

$(x-1)^2 = 3$

$x - 1 = \pm\sqrt{3}$

$x = 1 \pm \sqrt{3}$

The solution set is $\{1+\sqrt{3}, 1-\sqrt{3}\}$.

70. $x^2 + 4x = 12$

$x^2 + 4x + 4 = 12 + 4$

$(x+2)^2 = 16$

$x + 2 = \pm 4$

$x = -2 \pm 4$

The solution set is $\{-6, 2\}$.

71. $x^2 - 6x - 11 = 0$

$x^2 - 6x = 11$

$x^2 - 6x + 9 = 11 + 9$

$(x-3)^2 = 20$

$x - 3 = \pm\sqrt{20}$

$x = 3 \pm 2\sqrt{5}$

The solution set is $\{3+2\sqrt{5}, 3-2\sqrt{5}\}$.

72. $x^2 - 2x - 5 = 0$

$x^2 - 2x = 5$

$x^2 - 2x + 1 = 5 + 1$

$(x-1)^2 = 6$

$x - 1 = \pm\sqrt{6}$

$x = 1 \pm \sqrt{6}$

The solution set is $\{1+\sqrt{6}, 1-\sqrt{6}\}$.

73. $x^2 + 4x + 1 = 0$

$x^2 + 4x = -1$

$x^2 + 4x + 4 = -1 + 4$

$(x+2)^2 = 3$

$x + 2 = \pm\sqrt{3}$

$x = -2 \pm \sqrt{3}$

The solution set is $\{-2+\sqrt{3}, -2-\sqrt{3}\}$.

74. $x^2 + 6x - 5 = 0$

$x^2 + 6x = 5$

$x^2 + 6x + 9 = 5 + 9$

$(x+3)^2 = 14$

$x + 3 = \pm\sqrt{14}$

$x = -3 \pm \sqrt{14}$

The solution set is $\{-3+\sqrt{14}, -3-\sqrt{14}\}$.

75. $x^2+8x+15=0$

$$x=\frac{-8\pm\sqrt{8^2-4(1)(15)}}{2(1)}$$
$$x=\frac{-8\pm\sqrt{64-60}}{2}$$
$$x=\frac{-8\pm\sqrt{4}}{2}$$
$$x=\frac{-8\pm 2}{2}$$

The solution set is $\{-5,-3\}$.

76. $x^2+8x+12=0$

$$x=\frac{-8\pm\sqrt{8^2-4(1)(12)}}{2(1)}$$
$$x=\frac{-8\pm\sqrt{64-48}}{2}$$
$$x=\frac{-8\pm\sqrt{16}}{2}$$
$$x=\frac{-8\pm 4}{2}$$

The solution set is $\{-6,-2\}$.

77. $x^2+5x+3=0$

$$x=\frac{-5\pm\sqrt{5^2-4(1)(3)}}{2(1)}$$
$$x=\frac{-5\pm\sqrt{25-12}}{2}$$
$$x=\frac{-5\pm\sqrt{13}}{2}$$

The solution set is $\left\{\frac{-5+\sqrt{13}}{2},\frac{-5-\sqrt{13}}{2}\right\}$.

78. $x^2+5x+2=0$

$$x=\frac{-5\pm\sqrt{5^2-4(1)(2)}}{2(1)}$$
$$x=\frac{-5\pm\sqrt{25-8}}{2}$$
$$x=\frac{-5\pm\sqrt{17}}{2}$$

The solution set is $\left\{\frac{-5+\sqrt{17}}{2},\frac{-5-\sqrt{17}}{2}\right\}$.

79. $3x^2-3x-4=0$

$$x=\frac{3\pm\sqrt{(-3)^2-4(3)(-4)}}{2(3)}$$
$$x=\frac{3\pm\sqrt{9+48}}{6}$$
$$x=\frac{3\pm\sqrt{57}}{6}$$

The solution set is $\left\{\frac{3+\sqrt{57}}{6},\frac{3-\sqrt{57}}{6}\right\}$

80. $5x^2+x-2=0$

$$x=\frac{-1\pm\sqrt{1^2-4(5)(-2)}}{2(5)}$$
$$x=\frac{-1\pm\sqrt{1+40}}{10}$$
$$x=\frac{-1\pm\sqrt{41}}{10}$$

The solution set is $\left\{\frac{-1+\sqrt{41}}{10},\frac{-1-\sqrt{41}}{10}\right\}$.

81. $4x^2=2x+7$

$$4x^2-2x-7=0$$
$$x=\frac{2\pm\sqrt{(-2)^2-4(4)(-7)}}{2(4)}$$
$$x=\frac{2\pm\sqrt{4+112}}{8}$$
$$x=\frac{2\pm\sqrt{116}}{8}$$
$$x=\frac{2\pm 2\sqrt{29}}{8}$$
$$x=\frac{1\pm\sqrt{29}}{4}$$

The solution set is $\left\{\frac{1+\sqrt{29}}{4},\frac{1-\sqrt{29}}{4}\right\}$.

82. $3x^2 = 6x - 1$

$3x^2 - 6x + 1 = 0$

$x = \frac{6 \pm \sqrt{(-6)^2 - 4(3)(1)}}{2(3)}$

$x = \frac{6 \pm \sqrt{36 - 12}}{6}$

$x = \frac{6 \pm \sqrt{24}}{6}$

$x = \frac{6 \pm 2\sqrt{6}}{6}$

$x = \frac{3 \pm \sqrt{6}}{3}$

The solution set is $\left\{\frac{3+\sqrt{6}}{3}, \frac{3-\sqrt{6}}{3}\right\}$.

83. $x^2 - 4x - 5 = 0$

$(-4)^2 - 4(1)(-5)$
$= 16 + 20$
$= 36$; 2 unequal real solutions

84. $4x^2 - 2x + 3 = 0$

$(-2)^2 - 4(4)(3)$
$= 4 - 48$
$= -44$; 2 complex imaginary solutions

85. $2x^2 - 11x + 3 = 0$

$(-11)^2 - 4(2)(3)$
$= 121 - 24$
$= 97$; 2 unequal real solutions

86. $2x^2 + 11x - 6 = 0$

$11^2 - 4(2)(-6)$
$= 121 + 48$
$= 169$; 2 unequal real solutions

87. $x^2 = 2x - 1$

$x^2 - 2x + 1 = 0$

$(-2)^2 - 4(1)(1)$
$= 4 - 4$
$= 0$; 1 real solution

88. $3x^2 = 2x - 1$

$3x^2 - 2x + 1 = 0$

$(-2)^2 - 4(3)(1)$
$= 4 - 12$
$= -8$; 2 complex imaginary solutions

89. $x^2 - 3x - 7 = 0$

$(-3)^2 - 4(1)(-7)$
$= 9 + 28$
$= 37$; 2 unequal real solutions

90. $3x^2 + 4x - 2 = 0$

$4^2 - 4(3)(-2)$
$= 16 + 24$
$= 40$; 2 unequal real solutions

91. $2x^2 - x = 1$

$2x^2 - x - 1 = 0$

$(2x + 1)(x - 1) = 0$

$2x + 1 = 0$ or $x - 1 = 0$

$2x = -1$

$x = -\frac{1}{2}$ or $x = 1$

The solution set is $\left\{-\frac{1}{2}, 1\right\}$.

92. $3x^2 - 4x = 4$

$3x^2 - 4x - 4 = 0$

$(3x + 2)(x - 2) = 0$

$3x + 2$ or $x - 2 = 0$

$3x = -2$

$x = -\frac{2}{3}$ or $x = -3$

The solution set is $\left\{-\frac{2}{3}, 2\right\}$.

93. $5x^2 + 2 = 11x$

$5x^2 - 11x + 2 = 0$

$(5x - 1)(x - 2) = 0$

$5x - 1 = 0$ or $x - 2 = 0$

$5x = 1$

$x = \frac{1}{5}$ or $x = 2$

The solution set is $\left\{\frac{1}{5}, 2\right\}$.

94.
$$5x^2 = 6 - 13x$$
$$5x^2 + 13x - 6 = 0$$
$$(5x-2)(x+3) = 0$$
$$5x - 2 = 0 \quad \text{or} \quad x + 3$$
$$5x = 2$$
$$x = \frac{2}{5} \quad \text{or} \quad x = -3$$
The solution set is $\left\{-3, \frac{2}{5}\right\}$.

95. $3x^2 = 60$
$$x^2 = 20$$
$$x = \pm\sqrt{20}$$
$$x = \pm 2\sqrt{5}$$
The solution set is $\left\{-2\sqrt{5}, 2\sqrt{5}\right\}$.

96. $2x^2 = 250$
$$x^2 = 125$$
$$x = \pm\sqrt{125}$$
$$x = \pm 5\sqrt{5}$$
The solution set is $\left\{-5\sqrt{5}, 5\sqrt{5}\right\}$.

97.
$$x^2 - 2x = 1$$
$$x^2 - 2x + 1 = 1 + 1$$
$$(x-1)^2 = 2$$
$$x - 1 = \pm\sqrt{2}$$
$$x = 1 \pm \sqrt{2}$$
The solution set is $\left\{1+\sqrt{2}, 1-\sqrt{2}\right\}$.

98.
$$2x^2 + 3x = 1$$
$$2x^2 + 3x - 1 = 0$$
$$x = \frac{-3 \pm \sqrt{3^2 - 4(2)(-1)}}{2(2)}$$
$$x = \frac{-3 \pm \sqrt{9+8}}{4}$$
$$x = \frac{-3 \pm \sqrt{17}}{4}$$
The solution set is $\left\{\frac{-3+\sqrt{17}}{4}, \frac{-3-\sqrt{17}}{4}\right\}$.

99.
$$(2x+3)(x+4) = 1$$
$$2x^2 + 8x + 3x + 12 = 1$$
$$2x^2 + 11x + 11 = 0$$
$$x = \frac{-11 \pm \sqrt{11^2 - 4(2)(11)}}{2(2)}$$
$$x = \frac{-11 \pm \sqrt{121-88}}{4}$$
$$x = \frac{-11 \pm \sqrt{33}}{4}$$
The solution set is $\left\{\frac{-11+\sqrt{33}}{4}, \frac{-11-\sqrt{33}}{4}\right\}$.

100.
$$(2x-5)(x+1) = 2$$
$$2x^2 + 2x - 5x - 5 = 2$$
$$2x^2 - 3x - 7 = 0$$
$$x = \frac{3 \pm \sqrt{(-3)^2 - 4(2)(-7)}}{2(2)}$$
$$x = \frac{3 \pm \sqrt{9+56}}{4}$$
$$x = \frac{3 \pm \sqrt{65}}{4}$$
The solution set is $\left\{\frac{3+\sqrt{65}}{4}, \frac{3-\sqrt{65}}{4}\right\}$.

101. $(3x-4)^2 = 16$
$$3x - 4 = \pm\sqrt{16}$$
$$3x - 4 = \pm 4$$
$$3x = 4 \pm 4$$
$$3x = 8 \text{ or } 3x = 0$$
$$x = \frac{8}{3} \text{ or } x = 0$$
The solution set is $\left\{0, \frac{8}{3}\right\}$.

102. $(2x+7)^2 = 25$
$$2x + 7 = \pm 5$$
$$2x = -7 \pm 5$$
$$2x = -12 \quad \text{or} \quad 2x = -2$$
$$x = 6 \quad \text{or} \quad x = -1$$
The solution set is $\{-6, -1\}$.

103. $3x^2 - 12x + 12 = 0$

$x^2 - 4x + 4 = 0$

$(x-2)(x-2) = 0$

$x - 2 = 0$

$x = 2$

The solution set is $\{2\}$.

104. $9 - 6x + x^2 = 0$

$x^2 - 6x + 9 = 0$

$(x-3)(x-3) = 0$

$x - 3 = 0$

$x = 3$

The solution set is $\{3\}$.

105. $4x^2 - 16 = 0$

$4x^2 = 16$

$x^2 = 4$

$x = \pm 2$

The solution set is $\{-2, 2\}$.

106. $3x^2 - 27 = 0$

$3x^2 = 27$

$x^2 = 9$

$x = \pm 3$

The solution set is $\{-3, 3\}$.

107. $x^2 = 4x - 2$

$x^2 - 4x + 2 = 0$

$x = \dfrac{-b \pm \sqrt{b^2 - 4ac}}{2a}$

$x = \dfrac{-(-4) \pm \sqrt{(-4)^2 - 4(1)(2)}}{2(1)}$

$x = \dfrac{4 \pm \sqrt{8}}{2}$

$x = 2 \pm \sqrt{2}$

The solution set is $\{2 \pm \sqrt{2}\}$.

108. $x^2 = 6x - 7$

$x^2 - 6x + 7 = 0$

$x = \dfrac{-b \pm \sqrt{b^2 - 4ac}}{2a}$

$x = \dfrac{-(-6) \pm \sqrt{(-6)^2 - 4(1)(7)}}{2(1)}$

$x = \dfrac{6 \pm \sqrt{8}}{2}$

$x = 3 \pm \sqrt{2}$

The solution set is $\{3 \pm \sqrt{2}\}$.

109. $2x^2 - 7x = 0$

$x(2x - 7) = 0$

$x = 0$ or $2x - 7 = 0$

$2x = 7$

$x = 0$ or $x = \dfrac{7}{2}$

The solution set is $\left\{0, \dfrac{7}{2}\right\}$.

110. $2x^2 + 5x = 3$

$2x^2 + 5x - 3 = 0$

$x = \dfrac{-5 \pm \sqrt{5^2 - 4(2)(-3)}}{2(2)}$

$x = \dfrac{-5 \pm \sqrt{25 + 24}}{4}$

$x = \dfrac{-5 \pm \sqrt{49}}{4}$

$x = \dfrac{-5 \pm 7}{4}$

$x = -3, \dfrac{1}{2}$

The solution set is $\left\{-3, \dfrac{1}{2}\right\}$.

111. $\frac{1}{x}+\frac{1}{x+2}=\frac{1}{3}; x \neq 0,-2$

$$3x+6+3x=x^2+2x$$
$$0=x^2-4x-6$$
$$x=\frac{-(-4)\pm\sqrt{(-4)^2-4(1)(-6)}}{2(1)}$$
$$x=\frac{4\pm\sqrt{16+24}}{2}$$
$$x=\frac{4\pm\sqrt{40}}{2}$$
$$x=\frac{4\pm2\sqrt{10}}{2}$$
$$x=2\pm\sqrt{10}$$

The solution set is $\{2+\sqrt{10},\ 2-\sqrt{10}\}$.

112. $\frac{1}{x}+\frac{1}{x+3}=\frac{1}{4}; x \neq 0,-3$

$$4x+12+4x=x^2+3x$$
$$0=x^2-5x-12$$
$$x=\frac{-(-5)\pm\sqrt{(-5)^2-4(1)(-12)}}{2(1)}$$
$$x=\frac{5\pm\sqrt{25+48}}{2}$$
$$x=\frac{5\pm\sqrt{73}}{2}$$

The solution set is $\left\{\frac{5+\sqrt{73}}{2},\frac{5-\sqrt{73}}{2}\right\}$.

113. $\frac{2x}{x-3}+\frac{6}{x+3}=\frac{-28}{x^2-9}; x \neq 3,-3$

$$2x(x+3)+6(x-3)=-28$$
$$2x^2+6x+6x-18=-28$$
$$2x^2+12x+10=0$$
$$x^2+6x+5=0$$
$$(x+1)(x+5)=0$$

The solution set is $\{-5,\ -1\}$.

114. $\frac{3}{x-3}+\frac{5}{x-4}=\frac{x^2-20}{x^2-7x+12}; x \neq 3,4$

$$3x-12+5x-15=x^2-20$$
$$0=x^2-8x+7$$
$$0=(x-7)(x-1)$$
$$x=7 \qquad x=1$$

The solution set is $\{1, 7\}$.

115. $\sqrt{3x+18}=x$

$$3x+18=x^2$$
$$x^2-3x-18=0$$
$$(x+3)(x-6)=0$$
$$x+3=0 \qquad x-6=0$$
$$x=-3 \qquad x=6$$
$$\sqrt{3(-3)+18}=-3 \qquad \sqrt{3(6)+18}=6$$
$$\sqrt{-9+18}=-3 \qquad \sqrt{18+18}=6$$
$$\sqrt{9}=-3 \text{ False} \qquad \sqrt{36}=6$$

The solution set is $\{6\}$.

116. $\sqrt{20-8x}=x$

$$20-8x=x^2$$
$$x^2+8x-20=0$$
$$(x+10)(x-2)=0$$
$$x+10=0 \qquad x-2=0$$
$$x=-10 \qquad x=2$$
$$\sqrt{20-8(-10)}=-10 \qquad \sqrt{20-8(2)}=2$$
$$\sqrt{20+80}=-10 \qquad \sqrt{20-16}=2$$
$$\sqrt{100}=-10 \text{ False} \qquad \sqrt{4}=2$$

The solution set is $\{2\}$.

117. $\sqrt{x+3}=x-3$

$$x+3=x^2-6x+9$$
$$x^2-7x+6=0$$
$$(x-1)(x-6)=0$$
$$x-1=0 \qquad x-6=0$$
$$x=1 \qquad x=6$$
$$\sqrt{1+3}=1-3 \qquad \sqrt{6+3}=6-3$$
$$\sqrt{4}=-2 \text{ False} \qquad \sqrt{9}=3$$

The solution set is $\{6\}$.

118.
$$\sqrt{x+10} = x-2$$
$$x+10 = (x-2)^2$$
$$x+10 = x^2-4x+4$$
$$x^2-5x-6=0$$
$$(x+1)(x-6)=0$$
$$x+1=0 \qquad x-6=0$$
$$x=-1 \qquad x=6$$
$$\sqrt{-1+10}=-1-2 \qquad \sqrt{6+10}=6-2$$
$$\sqrt{9}=-3 \quad \text{False} \qquad \sqrt{16}=4$$
The solution set is $\{6\}$.

119.
$$\sqrt{2x+13}=x+7$$
$$2x+13=(x+7)^2$$
$$2x+13=x^2+14x+49$$
$$x^2+12x+36=0$$
$$(x+6)^2=0$$
$$x+6=0$$
$$x=-6$$
$$\sqrt{2(-6)+13}=-6+7$$
$$\sqrt{-12+13}=1$$
$$\sqrt{1}=1$$
The solution set is $\{-6\}$.

120.
$$\sqrt{6x+1}=x-1$$
$$6x+1=(x-1)^2$$
$$6x+1=x^2-2x+1$$
$$x^2-8x=0$$
$$x(x-8)=0$$
$$x-8=0 \qquad x=0$$
$$x=8$$
$$\sqrt{6(0)+1}=0-1 \qquad \sqrt{6(8)+1}=8-1$$
$$\sqrt{0+1}=-1 \qquad \sqrt{48+1}=7$$
$$\sqrt{1}=-1 \quad \text{False} \qquad \sqrt{49}=7$$
The solution set is $\{8\}$.

121.
$$x-\sqrt{2x+5}=5$$
$$x-5=\sqrt{2x+5}$$
$$(x-5)^2=2x+5$$
$$x^2-10x+25=2x+5$$
$$x^2-12x+20=0$$
$$(x-2)(x-10)=0$$
$$x-2=0 \qquad x-10=0$$
$$x=2 \qquad x=10$$
$$2-\sqrt{2(2)+5}=5 \qquad 10-\sqrt{2(10)+5}=5$$
$$2-\sqrt{9}=5 \qquad 10-\sqrt{25}=5$$
$$2-3=5 \quad \text{False} \qquad 10-5=5$$
The solution set is $\{10\}$.

122.
$$x-\sqrt{x+11}=1$$
$$x-1=\sqrt{x+11}$$
$$(x-1)^2=x+11$$
$$x^2-2x+1=x+11$$
$$x^2-3x-10=0$$
$$(x+2)(x-5)=0$$
$$x+2=0 \qquad x-5=0$$
$$x=-2 \qquad x=5$$
$$-2-\sqrt{-2+11}=1 \qquad 5-\sqrt{5+11}=1$$
$$-2-\sqrt{9}=1 \qquad 5-\sqrt{16}=1$$
$$-2-3=1 \quad \text{False} \qquad 5-4=1$$
The solution set is $\{5\}$.

123.
$$\sqrt{2x+19}-8=x$$
$$\sqrt{2x+19}=x+8$$
$$\left(\sqrt{2x+19}\right)^2=(x+8)^2$$
$$2x+19=x^2+16x+64$$
$$0=x^2+14x+45$$
$$0=(x+9)(x+5)$$
$$x+9=0 \quad \text{or} \quad x+5=0$$
$$x=-9 \qquad x=-5$$
-9 does not check and must be rejected.
The solution set is $\{-5\}$.

124. $\sqrt{2x+15}-6=x$

$\sqrt{2x+15}=x+6$

$\left(\sqrt{2x+15}\right)^2=(x+6)^2$

$2x+15=x^2+12x+36$

$0=x^2+10x+21$

$0=(x+3)(x+7)$

$x+3=0 \quad \text{or} \quad x+7=0$

$x=-3 \qquad x=-7$

–7 does not check and must be rejected.

The solution set is $\{-3\}$.

125.

$$\begin{aligned} 25-[2+5y-3(y+2)] &= -3(2y-5)-[5(y-1)-3y+3] \\ 25-[2+5y-3y-6] &= -6y+15-[5y-5-3y+3] \\ 25-[2y-4] &= -6y+15-[2y-2] \\ 25-2y+4 &= -6y+15-2y+2 \\ -2y+29 &= -8y+17 \\ 6y &= -12 \\ y &= -2 \end{aligned}$$

The solution set is $\{-2\}$.

126.

$$\begin{aligned} 45-[4-2y-4(y+7)] &= -4(1+3y)-[4-3(y+2)-2(2y-5)] \\ 45-[4-2y-4y-28] &= -4-12y-[4-3y-6-4y+10] \\ 45-[-6y-24] &= -4-12y-[-7y+8] \\ 45+6y+24 &= -4-12y+7y-8 \\ 6y+69 &= -5y-12 \\ 11y &= -81 \\ y &= -\frac{81}{11} \end{aligned}$$

The solution set is $\left\{-\frac{81}{11}\right\}$.

127. $7-7x=(3x+2)(x-1)$

$7-7x=3x^2-x-2$

$0=3x^2+6x-9$

$0=x^2+2x-3$

$0=(x+3)(x-1)$

$x+3=0 \quad \text{or} \quad x-1=0$

$x=-3 \qquad x=1$

The solution set is $\{-3,1\}$.

128. $10x-1=(2x+1)^2$

$$10x-1=4x^2+4x+1$$
$$0=4x^2-6x+2$$
$$0=2x^2-3x+1$$
$$0=(2x-1)(x-1)$$

$2x-1=0$ or $x-1=0$

$x=\frac{1}{2}$ $\quad x=1$

The solution set is $\left\{\frac{1}{2},1\right\}$.

129. $|x^2+2x-36|=12$

$x^2+2x-36=12$ $\quad x^2+2x-36=-12$

$x^2+2x-48=0$ or $x^2+2x-24=0$

$(x+8)(x-6)=0$ $\quad (x+6)(x-4)=0$

Setting each of the factors above equal to zero gives $x=-8$, $x=6$, $x=-6$, and $x=4$.

The solution set is $\{-8,-6,4,6\}$.

130. $|x^2+6x+1|=8$

$x^2+6x+1=8$ or $x^2+6x+1=-8$

$x^2+6x-7=0$ $\quad x^2+6x+9=0$

$(x+7)(x-1)=0$ $\quad (x+3)(x+3)=0$

Setting each of the factors above equal to zero gives $x=-7$, $x=-3$, and $x=1$.

The solution set is $\{-7,-3,1\}$.

131.

$$\frac{1}{x^2-3x+2}=\frac{1}{x+2}+\frac{5}{x^2-4}$$
$$\frac{1}{(x-1)(x-2)}=\frac{1}{x+2}+\frac{5}{(x+2)(x-2)}$$

Multiply both sides of the equation by the least common denominator, $(x-1)(x-2)(x+2)$. This results in the following:

$$x+2=(x-1)(x-2)+5(x-1)$$
$$x+2=x^2-2x-x+2+5x-5$$
$$x+2=x^2+2x-3$$
$$0=x^2+x-5$$

Apply the quadratic formula:
$a=1 \quad b=1 \quad c=-5$.

$$x=\frac{-1\pm\sqrt{1^2-4(1)(-5)}}{2(1)}=\frac{-1\pm\sqrt{1-(-20)}}{2}$$
$$=\frac{-1\pm\sqrt{21}}{2}$$

The solutions are $\frac{-1\pm\sqrt{21}}{2}$, and the solution set is $\left\{\frac{-1\pm\sqrt{21}}{2}\right\}$.

132. $\frac{x-1}{x-2}+\frac{x}{x-3}=\frac{1}{x^2-5x+6}$

$\frac{x-1}{x-2}+\frac{x}{x-3}=\frac{1}{(x-2)(x-3)}$

Multiply both sides of the equation by the least common denominator, $(x-2)(x-3)$. This results in the following:

$$(x-3)(x-1)+x(x-2)=1$$
$$x^2-x-3x+3+x^2-2x=1$$
$$2x^2-6x+3=1$$
$$2x^2-6x+2=0$$

Apply the quadratic formula:
$a=2 \quad b=-6 \quad c=2$.

$$x=\frac{-(-6)\pm\sqrt{(-6)^2-4(2)(2)}}{2(2)}$$
$$=\frac{6\pm\sqrt{36-16}}{4}=\frac{6\pm\sqrt{20}}{4}$$
$$=\frac{6\pm\sqrt{4\cdot 5}}{4}=\frac{6\pm 2\sqrt{5}}{4}$$
$$=\frac{3\pm\sqrt{5}}{2}$$

The solutions are $\frac{3\pm\sqrt{5}}{2}$, and the solution set is $\left\{\frac{3\pm\sqrt{5}}{2}\right\}$.

133. $\sqrt{x+8}-\sqrt{x-4}=2$

$$\sqrt{x+8}=\sqrt{x-4}+2$$
$$x+8=(\sqrt{x-4}+2)^2$$
$$x+8=x-4+4\sqrt{x-4}+4$$
$$x+8=x+4\sqrt{x-4}$$
$$8=4\sqrt{x-4}$$
$$2=\sqrt{x-4}$$
$$4=x-4$$
$$x=8$$
$$\sqrt{8+8}-\sqrt{8-4}=2$$
$$\sqrt{16}-\sqrt{4}=2$$
$$4-2=2$$

The solution set is $\{8\}$.

134. $\sqrt{x+5}-\sqrt{x-3}=2$

$$\sqrt{x+5}=\sqrt{x-3}+2$$
$$x+5=(\sqrt{x-3}+2)^2$$
$$x+5=x-3+4\sqrt{x-3}+4$$
$$x+5=x+1+4\sqrt{x-3}$$
$$5=1+4\sqrt{x-3}$$
$$4=4\sqrt{x-3}$$
$$1=\sqrt{x-3}$$
$$1=x-3$$
$$x=4$$
$$\sqrt{4+5}-\sqrt{4-3}=2$$
$$\sqrt{9}-\sqrt{1}=2$$
$$3-1=2$$

The solution set is $\{4\}$.

135. Values that make the denominator zero must be excluded.

$$2x^2+4x-9=0$$
$$x=\frac{-b\pm\sqrt{b^2-4ac}}{2a}$$
$$x=\frac{-(4)\pm\sqrt{(4)^2-4(2)(-9)}}{2(2)}$$
$$x=\frac{-4\pm\sqrt{88}}{4}$$
$$x=\frac{-4\pm2\sqrt{22}}{4}$$
$$x=\frac{-2\pm\sqrt{22}}{2}$$

136. Values that make the denominator zero must be excluded.

$$2x^2-8x+5=0$$
$$x=\frac{-b\pm\sqrt{b^2-4ac}}{2a}$$
$$x=\frac{-(-8)\pm\sqrt{(-8)^2-4(2)(5)}}{2(2)}$$
$$x=\frac{8\pm\sqrt{24}}{4}$$
$$x=\frac{8\pm2\sqrt{6}}{4}$$
$$x=\frac{4\pm\sqrt{6}}{2}$$

137. $D=\frac{1}{9}N+\frac{26}{9};\ D=\frac{7}{2}$

$$\frac{7}{2}=\frac{1}{9}N+\frac{26}{9}$$
$$18\left(\frac{7}{2}\right)=18\left(\frac{1}{9}N+\frac{26}{9}\right)$$
$$63=2N+52$$
$$11=2N$$
$$\frac{11}{2}=\frac{2N}{2}$$
$$5.5=N$$

If the high-humor group averages a level of depression of 3.5 in response to a negative life event, the intensity of that event would be 5.5. The solution is the point along the horizontal axis where the graph for the high-humor group has a value of 3.5 on the vertical axis. This corresponds to the point $(5.5, 3.5)$ on the high-humor graph.

138. Substitute 10 for D in the low humor formula. The LCD is 9.

$$10=\frac{10}{9}N+\frac{53}{9}$$
$$9(10)=9\left(\frac{10}{9}N\right)+9\left(\frac{53}{9}\right)$$
$$90=10N+53$$
$$90-53=10N+53-53$$
$$37=10N$$
$$\frac{37}{10}=\frac{10N}{10}$$
$$3.7=N$$

The intensity of the event was 3.7. This is shown as the point (3.7, 10) on the low-humor graph.

139.
$$C = \frac{x + 0.1(500)}{x + 500}$$
$$0.28 = \frac{x + 0.1(500)}{x + 500}$$
$$0.28(x + 500) = x + 0.1(500)$$
$$0.28x + 140 = x + 50$$
$$-0.72x = -90$$
$$\frac{-0.72x}{-0.72} = \frac{-90}{-0.72}$$
$$x = 125$$
125 liters of pure peroxide must be added.

140. a. $C = \frac{x + 0.35(200)}{x + 200}$

b.
$$0.74 = \frac{x + 0.35(200)}{x + 200}$$
$$0.74(x + 200) = x + 0.35(200)$$
$$0.74x + 148 = x + 70$$
$$-0.26x = -78$$
$$\frac{-0.26x}{-0.26} = \frac{-78}{-0.26}$$
$$x = 300$$
300 liters of pure acid must be added.

141. $f(x) = 0.013x^2 - 1.19x + 28.24$
$$3 = 0.013x^2 - 1.19x + 28.24$$
$$0 = 0.013x^2 - 1.19x + 25.24$$
Apply the quadratic formula:
$a = 0.013 \quad b = -1.19 \quad c = 25.24$
$$x = \frac{-(-1.19) \pm \sqrt{(-1.19)^2 - 4(0.013)(25.24)}}{2(0.013)}$$
$$= \frac{1.19 \pm \sqrt{1.4161 - 1.31248}}{0.026}$$
$$= \frac{1.19 \pm \sqrt{0.10362}}{0.026}$$
$$\approx \frac{1.19 \pm 0.32190}{0.026}$$
$$\approx 58.15 \text{ or } 33.39$$
The solutions are approximately 33.39 and 58.15. Thus, 33 year olds and 58 year olds are expected to be in 3 fatal crashes per 100 million miles driven. The function models the actual data well.

142. $f(x) = 0.013x^2 - 1.19x + 28.24$
$$10 = 0.013x^2 - 1.19x + 28.24$$
$$0 = 0.013x^2 - 1.19x + 18.24$$
$a = 0.013 \quad b = -1.19 \quad c = 18.24$
$$x = \frac{-(-1.19) \pm \sqrt{(-1.19)^2 - 4(0.013)(18.24)}}{2(0.013)}$$
$$= \frac{1.19 \pm \sqrt{1.4161 - 0.94848}}{0.026}$$
$$= \frac{1.19 \pm \sqrt{0.46762}}{0.026} \approx \frac{1.19 \pm 0.68383}{0.026}$$
Evaluate the expression to obtain two solutions.
$$x = \frac{1.19 + 0.68383}{0.026} \quad \text{or} \quad x = \frac{1.19 - 0.68383}{0.026}$$
$$x = \frac{1.87383}{0.026} \qquad x = \frac{0.50617}{0.026}$$
$$x \approx 72.1 \qquad x \approx 19$$
Drivers of approximately age 19 and age 72 are expected to be involved in 10 fatal crashes per 100 million miles driven. The formula does not model the data very well. The formula overestimates the number of fatal accidents.

143. For the year 2100, we use $x = 98$.
$$H = 0.083(98) + 57.9$$
$$= 66.034$$
$$L = 0.36\sqrt{98} + 57.9$$
$$\approx 61.464$$
In the year 2100, the projected high end temperature is about $66°$ and the projected low end temperature is about $61.5°$.

144. For the year 2080, we use $x = 78$.
$$H = 0.083(78) + 57.9 \approx 64.4$$
$$L = 0.36\sqrt{78} + 57.9 \approx 61.1$$
In the year 2080, the projected high end temperature is about $64.4°$ and the projected low end temperature is about $61.1°$.

145. Using H:
$$0.083x + 57.9 = 57.9 + 1$$
$$0.083x = 1$$
$$x = \frac{1}{0.083}$$
$$x \approx 12$$
The projected global temperature will exceed the 2002 average by 1 degree in 2014 (12 years after 2002).

Using L:

$$0.36\sqrt{x} + 57.9 = 1 + 57.9$$
$$0.36\sqrt{x} = 1$$
$$\sqrt{x} = \frac{1}{0.36}$$
$$\left(\sqrt{x}\right)^2 = \left(\frac{1}{0.36}\right)^2$$
$$x \approx 8$$

The projected global temperature will exceed the 2002 average by 1 degree in 2010 (8 years after 2002).

146. Using H:

$$0.083x + 57.9 = 57.9 + 2$$
$$0.083x = 2$$
$$x = \frac{2}{0.083}$$
$$x \approx 24$$

The projected global temperature will exceed the 2002 average by 2 degrees in 2026 (24 years after 2002).

Using L:

$$0.36\sqrt{x} + 57.9 = 57.9 + 2$$
$$0.36\sqrt{x} = 2$$
$$\sqrt{x} = \frac{2}{0.36}$$
$$\left(\sqrt{x}\right)^2 = \left(\frac{2}{0.36}\right)^2$$
$$x \approx 31$$

The projected global temperature will exceed the 2002 average by 2 degrees in 2033 (31 years after 2002).

161. **a.** false; $(2x-3)^2 = 25$

$$\sqrt{(2x-3)^2} = \pm\sqrt{25}$$
$$2x - 3 = \pm 5$$

b. false; some quadratics have one number in their solution set.

c. true; some quadratics have one number in their solution set.

d. false; $ax^2 + c = 0$ can be solved using $b = 0$.

(c) is true.

162.

$$\frac{7x+4}{b} + 13 = x$$
$$\frac{7(-6)+4}{b} + 13 = -6$$
$$\frac{-38}{b} = -19$$
$$-19b = -38$$
$$b = 2$$

163.

$$[x-(-3)][x-(5)] = 0$$
$$(x+3)(x-5) = 0$$
$$x^2 - 2x - 15 = 0$$

$$V = C - \frac{C-S}{L}N$$
$$VL = CL - (C-S)N$$
$$VL = CL - CN + SN$$

164.

$$CN - CL = NS - LV$$
$$C(N-L) = NS - LV$$
$$\frac{C(N-L)}{N-L} = \frac{NS-LV}{N-L}$$
$$C = \frac{NS-LV}{N-L} \text{ or } \frac{LV-NS}{L-N}$$

165.

$$s = -16t^2 + v_0 t$$
$$0 = -16t^2 + v_0 t - s$$
$$a = -16,\ b = v_0,\ c = -s$$
$$t = \frac{-v_0 \pm \sqrt{(v_0)^2 - 4(-16)(-s)}}{2(-16)}$$
$$t = \frac{-v_0 \pm \sqrt{(v_0)^2 - 64s}}{-32}$$
$$t = \frac{v_0 \pm \sqrt{v_0^2 - 64s}}{32}$$

Section P.8

Check Point Exercises

1. Let x = the number of football injuries
Let $x + 0.6$ = the number of basketball injuries
Let $x + 0.3$ = the number of bicycling injuries

$$x+(x+0.6)+(x+0.3)=3.9$$
$$x+x+0.6+x+0.3=3.9$$
$$3x+0.9=3.9$$
$$3x=3$$
$$x=1$$

$x=1$

$x+0.6=1+0.6=1.6$

$x+0.3=1+0.3=1.3$

In 2004 there were 1 million football injuries, 1.6 million basketball injuries, and 1.3 million bicycling injuries.

2. Let x = the number of years after 2004 that it will take until Americans will purchase 79.9 million gallons of organic milk.

$$40.7+5.6x=79.9$$
$$5.6x=79.9-40.7$$
$$5.6x=39.2$$
$$x=\frac{39.2}{5.6}$$
$$x=7$$

Americans will purchase 79.9 million gallons of organic milk 7 years after 2004, or 2011.

3. Let x = the computer's price before the reduction.

$$x-0.30x=840$$
$$0.70x=840$$
$$x=\frac{840}{0.70}$$
$$x=1200$$

Before the reduction the computer's price was \$1200.

4. Let x = the width of the court.
Let $x + 44$ = the length of the court.

$$2l+2w=P$$
$$2(x+44)+2x=288$$
$$2x+88+2x=288$$
$$4x+88=288$$
$$4x=200$$
$$x=\frac{200}{4}$$
$$x=50$$
$$x+44=94$$

The dimensions of the court are 50 by 94.

5.
$$(16+2x)(12+2x)=320$$
$$192+56x+4x^2=320$$
$$4x^2+56x-128=0$$
$$x^2+14x-32=0$$
$$(x+16)(x-2)=0$$
$$x+16=0 \quad \text{or} \quad x-2=0$$
$$x=-16 \qquad x=2$$

-16 must be rejected.
The path must be 2 feet wide.

6.
$$a^2+b^2=c^2$$
$$a^2+(50)^2=(130)^2$$
$$a^2+2500=16,900$$
$$a^2=14,400$$
$$a=\pm 120$$

-120 must be rejected.
The tower is 120 yards tall.

7.

$$\overbrace{\frac{5,000,000}{x}}^{\text{The original amount of money per person}} - \overbrace{375,000}^{\text{reduction per winner}} = \overbrace{\frac{5,000,000}{x+3}}^{\text{The new amount of money per person}}$$

$$x(x+3)\left(\frac{5,000,000}{x} - 375,000\right) = x(x+3)\frac{5,000,000}{x+3}$$

$$5,000,000(x+3) - 375,000x(x+3) = 5,000,000x$$

$$5,000,000x + 15,000,000 - 375,000x^2 - 1,125,000x = 5,000,000x$$

$$-375,000x^2 - 1,125,000x + 15,000,000 = 0$$

$$x^2 + 3x - 40 = 0$$

$$(x+8)(x-5) = 0$$

$x+8=0$ or $x-5=0$

$x=-8$ $\quad x=5$

-8 must be rejected. There were 5 people in the original group.

Exercise Set P.8

1. Let x = the number of births (in thousands)
Let $x-229$ = the number of deaths (in thousands).

$$x+(x-229)=521$$
$$x+x-229=521$$
$$2x-229=521$$
$$2x-229+229=521+229$$
$$2x=750$$
$$\frac{2x}{2}=\frac{750}{2}$$
$$x=375$$

There are 375 thousand births and $375-229=146$ thousand deaths each day.

2. Let x = energy percentage used by Russia.
$x+6$ = energy percentage used by China.
$x+16.4$ = energy percentage used by the United States.

$$x+(x+6)+(x+16.4)=40.4$$
$$x+x+6+x+16.4=40.4$$
$$3x+22.4=40.4$$
$$3x=18$$
$$\frac{3x}{3}=\frac{18}{3}$$
$$x=6$$
$$x+6=12$$
$$x+16.4=22.4$$

Thus, Russia uses 6%, China uses 12%, and the United States uses 22.4% of global energy.

3. Let L = the life expectancy of an American man.
y = the number of years after 1900.

$$L = 55 + 0.2y$$
$$85 = 55 + 0.2y$$
$$30 = 0.2y$$
$$150 = y$$

The life expectancy will be 85 years in the year $1900 + 150 = 2050$.

4. Let L = the life expectancy of an American man,
Let y = the number of years after 1900

$$L = 55 + 0.2y$$
$$91 = 55 + 0.2y$$
$$36 = 0.2y$$
$$180 = y$$

The life expectancy will be 91 years in the year 1900 + 180 = 2080.

5.
$$1.7x + 39.8 = 44.9 + 8.5$$
$$1.7x + 39.8 = 53.4$$
$$1.7x = 13.6$$
$$\frac{1.7x}{1.7} = \frac{13.6}{1.7}$$
$$x = 8$$

The number of Americans without health insurance will exceed 44.9 million by 8.5 million 8 years after 2000, or 2008.

6.
$$1.7x + 39.8 = 44.9 + 10.2$$
$$1.7x + 39.8 = 55.1$$
$$1.7x = 15.3$$
$$\frac{1.7x}{1.7} = \frac{15.3}{1.7}$$
$$x = 9$$

The number of Americans without health insurance will exceed 44.9 million by 10.2 million 9 years after 2000, or 2009.

7. Let x = the number of years after 2005

$$13{,}300 + 1000x = 26{,}800 - 500x$$
$$1500x = 13{,}500$$
$$\frac{1500x}{1500} = \frac{13{,}500}{1500}$$
$$x = 9$$

The two colleges will have the same enrollment about 9 years after 2005, or 2014.

$$13{,}300 + 1000(9) = 22{,}300$$

and

$$26{,}800 - 500(9) = 22{,}300$$

The college's enrollments will be 22,300 at that time.

8. Let x = the number of years after 2000

$$10,600,000-28,000x=10,200,000-12,000x$$
$$-16,000x=-400,000$$
$$x=25$$

The countries will have the same population 25 years after the year 2000, or the year 2025.

$$10,200,000-12,000x=10,200,000-12,000(25)$$
$$=10,200,000-300,000$$
$$=9,900,000$$

The population in the year 2025 will be 9,900,000.

9. Let x = the cost of the television set.

$$x-0.20x=336$$
$$0.80x=336$$
$$x=420$$

The television set's price is \$420.

10. Let x = the cost of the dictionary

$$x-0.30x=30.80$$
$$0.70x=30.80$$
$$x=44$$

The dictionary's price before the reduction was \$44.

11. Let x = the nightly cost

$$x+0.08x=162$$
$$1.08x=162$$
$$x=150$$

The nightly cost is \$150.

12. Let x = the nightly cost

$$x+0.05x=252$$
$$1.05x=252$$
$$x=240$$

The nightly cost is \$240.

13. Let x = the annual salary for men whose highest educational attainment is a high school degree.

$$x+0.22x=44,000$$
$$1.22x=44,000$$
$$x\approx 36,000$$

The annual salary for men whose highest educational attainment is a high school degree is about \$36,000.

14. Let x = the annual salary with a high school degree

$$34,000=x+0.26x$$
$$34,000=1.26x$$
$$26984.13 \doteq x$$

The annual salary for women with a high school degree is approximately \$27,000.

15. Let c = the dealer's cost

$$584=c+0.25c$$
$$584=1.25c$$
$$467.20=c$$

The dealer's cost is \$467.20.

16. Let c = the dealer's cost

$$15=c+0.25c$$
$$15=1.25c$$
$$12=c$$

The dealer's cost is \$12.

17. Let w = the width of the field
Let $2w$ = the length of the field

$$P=2(\text{length})+2(\text{width})$$
$$300=2(2w)+2(w)$$
$$300=4w+2w$$
$$300=6w$$
$$50=w$$

If $w=50$, then $2w=100$. Thus, the dimensions are 50 yards by 100 yards.

18. Let w = the width of the swimming pool,
Let $3w$ = the length of the swimming pool

$$P=2(\text{length})+2(\text{width})$$
$$320=2(3w)+2(w)$$
$$320=6w+2w$$
$$320=8w$$
$$40=w$$

If $w = 40$, $3w = 3(40) = 120$.
The dimensions are 40 feet by 120 feet.

19. Let w = the width of the field
Let $2w + 6$ = the length of the field

$$228=6w+12$$
$$216=6w$$
$$36=w$$

If $w=36$, then $2w+6=2(36)+6=78$. Thus, the dimensions are 36 feet by 78 feet.

20. Let w = the width of the pool,
Let $2w - 6$ = the length of the pool
$$P = 2(\text{length}) + 2(\text{width})$$
$$126 = 2(2w-6)+2(w)$$
$$126 = 4w-12+2w$$
$$126 = 6w-12$$
$$138 = 6w$$
$$23 = w$$
Find the length.
$2w-6 = 2(23)-6 = 46-6 = 40$
The dimensions are 23 meters by 40 meters.

21. Let x = the width of the frame.
Total length: $16+2x$
Total width: $12+2x$
$$P = 2(\text{length}) + 2(\text{width})$$
$$72 = 2(16+2x)+2(12+2x)$$
$$72 = 32+4x+24+4x$$
$$72 = 8x+56$$
$$16 = 8x$$
$$2 = x$$
The width of the frame is 2 inches.

22. Let w = the width of the path
Let $40 + 2w$ = the width of the pool and path
Let $60 + 2w$ = the length of the pool and path
$$2(40+2w)+2(60+2w) = 248$$
$$80+4w+120+4w = 248$$
$$200+8w = 248$$
$$8w = 48$$
$$w = 6$$
The width of the path is 6 feet.

23. Let w = the width
Let $w + 3$ = the length
Area = lw
$$54 = (w+3)w$$
$$54 = w^2+3w$$
$$0 = w^2+3w-54$$
$$0 = (w+9)(w-6)$$
$w+9=0$ $\quad$ $w-6=0$
$w=-9$ $\quad$ $w=6$
Disregard –9 because we can't have a negative length measurement. The width is 6 feet and the length is $6+3=9$ feet.

24. Let w = the width
Let $w + 3$ = the width
Area = lw
$$180 = (w+3)w$$
$$180 = w^2+3w$$
$$0 = w^2+3w-180$$
$$0 = (w+15)(w-12)$$
$w+15=0$ $\quad$ $w-12=0$
~~$w=-15$~~ $\quad$ $w=12$
The width is 12 yards and the length is 12 yards + 3 yards = 15 yards.

25. Let x = the length of the side of the original square
Let $x + 3$ = the length of the side of the new, larger square
$$(x+3)^2 = 64$$
$$x^2+6x+9 = 64$$
$$x^2+6x-55 = 0$$
$$(x+11)(x-5) = 0$$
Apply the zero product principle.
$x+11=0$ $\quad$ $x-5=0$
$x=-11$ $\quad$ $x=5$
The solution set is $\{-11, 5\}$. Disregard -11 because we can't have a negative length measurement. This means that x, the length of the side of the original square, is 5 inches.

26. Let x = the side of the original square,
Let $x + 2$ = the side of the new, larger square
$$(x+2)^2 = 36$$
$$x^2+4x+4 = 36$$
$$x^2+4x-32 = 0$$
$$(x+8)(x-4) = 0$$
$x+8=0$ $\quad$ $x-4=0$
~~$x=-8$~~ $\quad$ $x=4$
The length of the side of the original square, is 4 inches.

27. Let x = the width of the path
$$(20+2x)(10+2x) = 600$$
$$200+40x+20x+4x^2 = 600$$
$$200+60x+4x^2 = 600$$
$$4x^2+60x+200 = 600$$
$$4x^2+60x-400 = 0$$
$$4(x^2+15x-100) = 0$$
$$4(x+20)(x-5) = 0$$

Apply the zero product principle.

$4(x+20)=0 \qquad x-5=0$

$x+20=0 \qquad x=5$

$x=-20$

The solution set is $\{-20,5\}$. Disregard -20 because we can't have a negative width measurement. The width of the path is 5 meters.

28. Let x = the width of the path

$$(12+2x)(15+2x)=378$$
$$180+24x+30x+4x^2=378$$
$$4x^2+54x+180=378$$
$$4x^2+54x-198=0$$
$$2(2x^2+27x-99)=0$$
$$2(2x+33)(x-3)=0$$

$2(2x+33)=0 \qquad x-3=0$

$2x+33=0 \qquad x=3$

$2x=-33$

$x=-\frac{33}{2}$ (crossed out)

The width of the path is 3 meters.

29. $(20+2x)(30+2x)-(20)(30)=336$

$$600+100x+4x^2-600=336$$
$$4x^2+100x-336=0$$
$$x^2+25x-84=0$$
$$(x-3)(x+28)=0$$

$x-3=0$ or $x+28=0$

$x=3 \qquad x=-28$

-28 must be rejected.
The width of the path is 3 feet

30. $(10+2x)(12+2x)-(10)(12)=168$

$$120+44x+4x^2-120=168$$
$$4x^2+44x-168=0$$
$$x^2+11x-42=0$$
$$(x-3)(x+14)=0$$

$x-3=0$ or $x+14=0$

$x=3 \qquad x=-14$

-14 must be rejected.
The width of the path is 3 feet

31. $a^2+b^2=c^2$

$$a^2+15^2=20^2$$
$$a^2+225=400$$
$$a^2=175$$
$$a=\pm\sqrt{175}$$
$$a\approx\pm13.2$$

-13.2 must be rejected.
The ladder reaches 13.2 feet up the house.

32. $a^2+b^2=c^2$

$$a^2+10^2=30^2$$
$$a^2+100=900$$
$$a^2=800$$
$$a=\pm\sqrt{800}$$
$$a\approx\pm28.3$$

-28.3 must be rejected.
The building is 28.3 feet tall.

33. $a^2+b^2=c^2$

$$5^2+x^2=(x+1)^2$$
$$x^2+25=x^2+2x+1$$
$$25=2x+1$$
$$24=2x$$
$$x=12$$
$$x+1=13$$

The wire is 13 feet long.

34. $a^2+b^2=c^2$

$$15^2+x^2=(x+4)^2$$
$$x^2+225=x^2+8x+16$$
$$225=8x+16$$
$$209=8x$$
$$x=26\frac{1}{8}$$
$$x+4=30\frac{1}{8}$$

The wire is $30\frac{1}{8}$ feet long.

35. Let x be the width.

$$a^2 + b^2 = c^2$$
$$x^2 + (2x)^2 = 64^2$$
$$x^2 + 4x^2 = 4096$$
$$5x^2 = 4096$$
$$x^2 = \frac{4096}{5}$$
$$x = \pm\sqrt{\frac{4096}{5}}$$
$$x \approx 28.62 \text{ feet}$$
$$2x \approx 57.24 \text{ feet}$$

The distance along the length and width is about 28.62 + 57.24, or about 85.9 feet. A person could save 85.9 – 64, or about 21.9 feet.

36. Let x be the width.

$$a^2 + b^2 = c^2$$
$$x^2 + (3x)^2 = 92^2$$
$$x^2 + 9x^2 = 8464$$
$$10x^2 = 8464$$
$$x^2 = 846.4$$
$$x = \pm\sqrt{846.4}$$
$$x \approx 29.09 \text{ yd}$$
$$3x \approx 87.28 \text{ yd}$$

The distance along the length and width is about 29.09 + 87.28, or about 116.4 yards. A person could save 116.4 – 92, or about 24.4 yards.

37.

$$\overbrace{\frac{20,000,000}{x}}^{\text{The original amount of money per person.}} - \overbrace{500,000}^{\text{reduction per winner}} = \overbrace{\frac{20,000,000}{x+2}}^{\text{The new amount of money per person.}}$$
$$x(x+2)\left(\frac{20,000,000}{x} - 500,000\right) = x(x+2)\frac{20,000,000}{x+2}$$
$$20,000,000(x+2) - 500,000x(x+2) = 20,000,000x$$
$$20,000,000x + 40,000,000 - 500,000x^2 - 1,000,000x = 20,000,000x$$
$$40,000,000 - 500,000x^2 - 1,000,000x = 0$$
$$x^2 + 2x - 80 = 0$$
$$(x+10)(x-8) = 0$$

$x + 10 = 0$ or $x - 8 = 0$

$x = -10$ $\quad$ $x = 8$

–10 must be rejected. There were 8 people in the original group.

38.

$$\frac{480,000}{x} - 32,000 = \frac{480,000}{x+4}$$
$$x(x+4)\left(\frac{480,000}{x} - 32,000\right) = x(x+4)\frac{480,000}{x+4}$$
$$480,000(x+4) - 32,000x(x+4) = 480,000x$$
$$480,000x + 1,920,000 - 32,000x^2 - 128,000x = 480,000x$$
$$1,920,000 - 32,000x^2 - 128,000x = 0$$
$$x^2 + 4x - 60 = 0$$
$$(x+10)(x-6) = 0$$

$x + 10 = 0$ or $x - 6 = 0$

$x = -10$ $\quad$ $x = 6$

–10 must be rejected. There were 6 people in the original group.

39. Let x be the car's average velocity.

$$\overbrace{\frac{300}{x}}^{\text{car's time traveled}} = \overbrace{\frac{180}{x-20}}^{\text{bus's time traveled}}$$

$$300(x-20) = 180x$$
$$300x - 6000 = 180x$$
$$120x = 6000$$
$$x = 50$$
$$x - 20 = 30$$

The average velocity of the car is 50 miles per hour. The average velocity of the bus is 30 miles per hour.

40. Let x be the passenger train's average velocity.

$$\overbrace{\frac{240}{x}}^{\text{passenger train's time traveled}} = \overbrace{\frac{160}{x-20}}^{\text{freight train's time traveled}}$$

$$240(x-20) = 160x$$
$$240x - 4800 = 160x$$
$$80x = 4800$$
$$x = 60$$
$$x - 20 = 40$$

The average velocity of the passenger train is 60 miles per hour. The average velocity of the freight train is 40 miles per hour.

41. Let x be the average velocity on the return trip.

$$\frac{5}{x+9} + \frac{5}{x} = \frac{7}{6}$$
$$6x(x+9)\left(\frac{5}{x+9} + \frac{5}{x}\right) = 6x(x+9)\frac{7}{6}$$
$$30x + 30(x+9) = 7x(x+9)$$
$$30x + 30x + 270 = 7x^2 + 63x$$
$$0 = 7x^2 + 3x - 270$$
$$0 = (x-6)(7x+45)$$

$x - 6 = 0$ or $7x + 45 = 0$

$x = 6$ $\quad x = -\frac{45}{7}$

$-\frac{45}{7}$ must be rejected. The average velocity on the return trip is 6 miles per hour.

42. Let x be the average velocity of the first engine.

$$\frac{140}{x} + \frac{200}{x+5} = 9$$
$$\left(\frac{140}{x} + \frac{200}{x+5}\right) = 9$$
$$x(x+5)\left(\frac{140}{x} + \frac{200}{x+5}\right) = 9x(x+5)$$
$$140(x+5) + 200x = 9x(x+5)$$
$$140x + 700 + 200x = 9x^2 + 45x$$
$$0 = 9x^2 - 295x - 700$$
$$0 = (x-35)(9x-20)$$

$x - 35 = 0$ or $9x + 20 = 0$

$x = 35$ $\quad x = -\frac{20}{9}$

$x + 5 = 40$

$-\frac{20}{9}$ must be rejected. The average velocity of the first engine is 35 miles per hour. The average velocity of the second engine is 40 miles per hour.

43. Let x = number of hours

$35x$ = labor cost

$35x + 63 = 448$

$35x = 385$

$x = 11$

It took 11 hours.

44. Let x = number of hours

$63x$ = labor cost

$63x + 532 = 1603$

$63x = 1071$

$x = 17$

17 hours were required to repair the yacht.

45. Let x = inches over 5 feet

$100 + 5x = 135$

$5x = 35$

$x = 7$

A height of 5 feet 7 inches corresponds to 135 pounds.

46. Let g = the gross amount of the paycheck

Yearly Salary $= 2(12)g + 750$

$$33150 = 24g + 750$$
$$32400 = 24g$$
$$1350 = g$$

The gross amount of each paycheck is $1350.

47. Let x be the number of consecutive hits.

$$\frac{35+x}{140+x}=0.30$$
$$35+x=0.30(140+x)$$
$$35+x=42+0.30x$$
$$350+10x=420+3x$$
$$7x=70$$
$$x=10$$

You must get 10 consecutive hits to increase your batting average to 0.30.

48. Let x be the number of consecutive hits.

$$\frac{30+x}{120+x}=0.28$$
$$30+x=0.28(120+x)$$
$$30+x=33.6+0.28x$$
$$3000+100x=3360+28x$$
$$72x=360$$
$$x=5$$

You must get 5 consecutive hits to increase your batting average to 0.28.

55. Let x be the length of one leg.

$$a^2+b^2=c^2$$
$$x^2+(x+1)^2=\left[12-x-(x+1)\right]^2$$
$$x^2+x^2+2x+1=\left[12-x-x-1\right]^2$$
$$2x^2+2x+1=(11-2x)^2$$
$$2x^2+2x+1=121-44x+4x^2$$
$$0=2x^2-46x+120$$
$$0=x^2-23x+60$$
$$0=(x-3)(x-20)$$
$$x-3=0 \quad \text{or} \quad x-20=0$$
$$x=3 \qquad\qquad x=20$$
$$x+1=4$$
$$12-(3+4)=5$$

20 must be rejected, as it is greater than the perimeter.
The lengths of the sides are 3, 4, and 5.

56. Let x = original price
$x-0.4x=0.6x$ = price after first reduction
$0.6x-0.4(0.6x)$ = price after second reduction

$$0.6x-0.24x=72$$
$$0.36x=72$$
$$x=200$$

The original price was \$200.

57. Let x = woman's age
$3x$ = Coburn's age

$$3x+20=2(x+20)$$
$$3x+20=2x+40$$
$$x+20=40$$
$$x=20$$

Coburn is 60 years old the woman is 20 years old.

58. Let x = correct answers
$26-x$ = incorrect answers

$$8x-5(26-x)=0$$
$$8x-130+5x=0$$
$$13x-130=0$$
$$13x=130$$
$$x=10$$

10 problems were solved correctly.

59. Let x = mother's amount
$2x$ = boy's amount
$\frac{x}{2}$ = girl's amount

$$x+2x+\frac{x}{2}=14{,}000$$
$$\frac{7}{2}x=14{,}000$$
$$x=\$4{,}000$$

The mother received \$4000, the boy received \$8000, and the girl received \$2000.

60. Let x = the number of plants originally stolen
After passing the first security guard, the thief has: $x-\left(\frac{1}{2}x+2\right)=x-\frac{1}{2}x-2=\frac{1}{2}x-2$

After passing the second security guard, the thief has: $\frac{1}{2}x-2-\left(\frac{\frac{1}{2}x-2}{2}+2\right)=\frac{1}{4}x-3$

After passing the third security guard, the thief has: $\frac{1}{4}x-3-\left(\frac{\frac{1}{4}x-3}{2}+2\right)=\frac{1}{8}x-\frac{7}{2}$

Thus, $\frac{1}{8}x-\frac{7}{2}=1$

$$x-28=8$$
$$x=36$$

The thief stole 36 plants.

Section P.9

Check Point Exercises

1. **a.** $[-2,\ 5) = \{x|-2 \le x < 5\}$

b. $[1,\ 3.5] = \{x|1 \le x \le 3.5\}$

c. $[-\infty,\ -1) = \{x|x < -1\}$

2. **a.** Graph $[1,3]$:

Graph $(2,6)$:

To find the intersection, take the portion of the number line that the two graphs have in common.

Numbers in both $[1,3]$ and $(2,6)$:

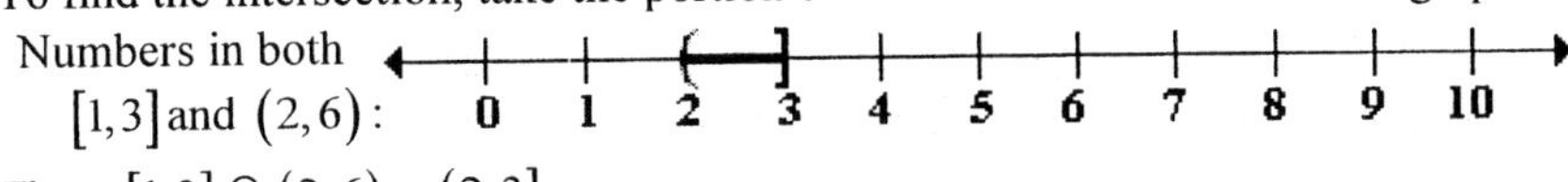

Thus, $[1,3] \cap (2,6) = (2,3]$.

b. Graph $[1,3]$:

Graph $(2,6)$:

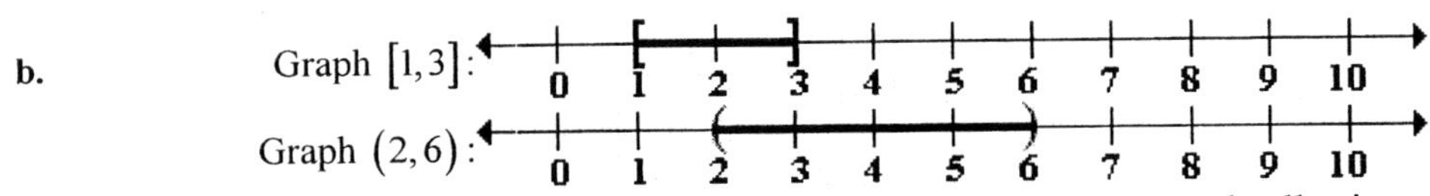

To find the union, take the portion of the number line representing the total collection of numbers in the two graphs.

Numbers in either $[1,3]$ or $(2,6)$ or both:

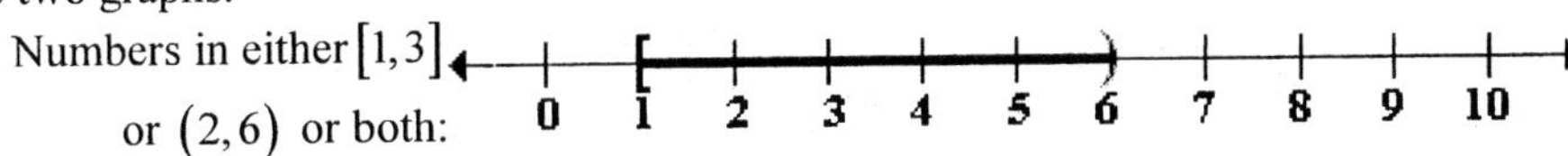

Thus, $[1,3] \cup (2,6) = [1,6)$.

3.
$$2 - 3x \le 5$$
$$-3x \le 3$$
$$x \ge -1$$
The solution set is $\{x|x \ge -1\}$ or $[-1, \infty)$.

4.
$$3x + 1 > 7x - 15$$
$$-4x > -16$$
$$\frac{-4x}{-4} < \frac{-16}{-4}$$
$$x < 4$$
The solution set is $\{x|x < 4\}$ or $(-\infty, 4]$.

5.
$$1 \le 2x + 3 < 11$$
$$-2 \le 2x < 8$$
$$-1 \le x < 4$$
The solution set is $\{x|-1 \le x < 4\}$ or $[-1, 4)$.

6.
$$|x - 2| < 5$$
$$-5 < x - 2 < 5$$
$$-3 < x < 7$$
The solution set is $\{x|-3 < x < 7\}$ or $(-3, 7)$.

7. $-3|5x-2|+20 \ge -19$

$-3|5x-2| \ge -39$

$\frac{-3|5x-2|}{-3} \le \frac{-39}{-3}$

$|5x-2| \le 13$

$-13 \le 5x-2 \le 13$

$-11 \le 5x \le 15$

$\frac{-11}{5} \le \frac{5x}{5} \le \frac{15}{5}$

$-\frac{11}{5} \le x \le 3$

The solution set is

$\left\{x \middle| -\frac{11}{5} \le x \le 3\right\}$ or $\left[-\frac{11}{5}, 3\right]$.

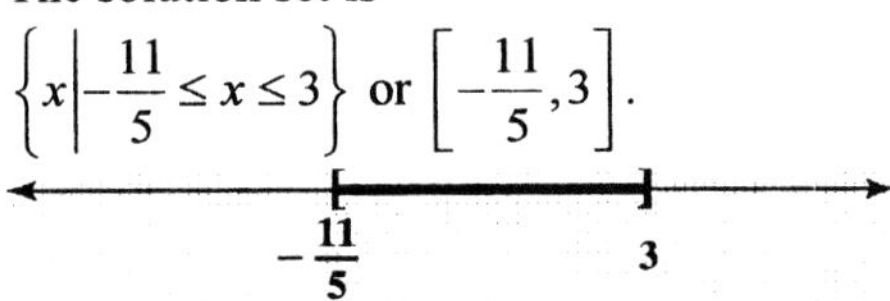

8. $18 < |6-3x|$

$6-3x < -18$ or $6-3x > 18$

$-3x < -24$ $\quad$ $-3x > 12$

$\frac{-3x}{-3} > \frac{-24}{-3}$ $\quad$ $\frac{-3x}{-3} < \frac{12}{-3}$

$x > 8$ $\quad$ $x < -4$

The solution set is $\{x|x < -4 \text{ or } x > 8\}$

or $(-\infty,-4) \cup (8,\infty)$.

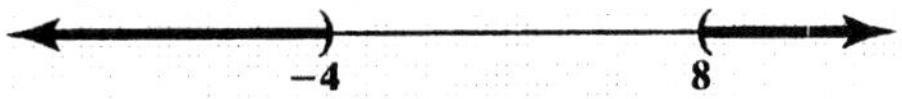

Number line: open at −4 shaded left, open at 8 shaded right.

9. Let x = the number of miles driven in a week.

$260 < 80 + 0.25x$

$180 < 0.25x$

$720 < x$

Driving more than 720 miles in a week makes Basic the better deal.

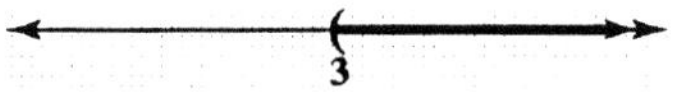

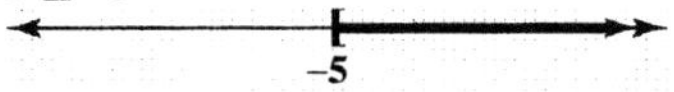

Exercise Set P.9

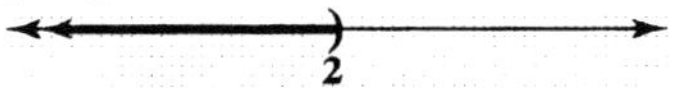

1. $1 < x \le 6$ (number line: 1, 6)

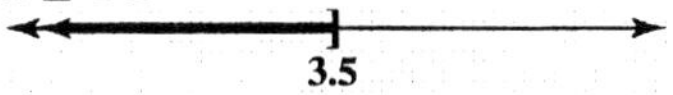

2. $-2 < x \le 4$ (number line: −2, 4)

3. $-5 \le x < 2$ (number line: −5, 2)

4. $-4 \le x < 3$ (number line: −4, 3)

5. $-3 \le x \le 1$ (number line: −3, 1)

6. $-2 \le x \le 5$ (number line: −2, 5)

7. $x > 2$ (number line: 2)

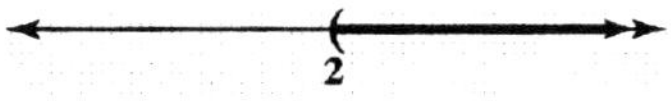

8. $x > 3$ (number line: 3)

9. $x \ge -3$ (number line: −3)

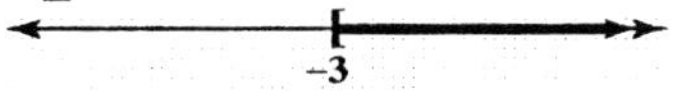

10. $x \ge -5$ (number line: −5)

11. $x < 3$ (number line: 3)

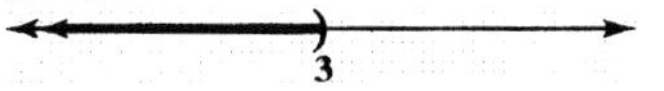

12. $x < 2$ (number line: 2)

13. $x < 5.5$ (number line: 5.5)

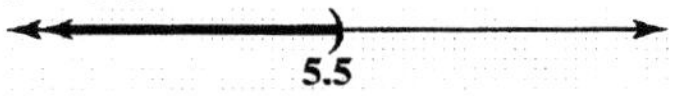

14. $x \le 3.5$ (number line: 3.5)

15. Graph $(-3,0)$:

Graph $[-1,2]$:

To find the intersection, take the portion of the number line that the two graphs have in common.

Numbers in both $(-3,0)$ and $[-1,2]$:

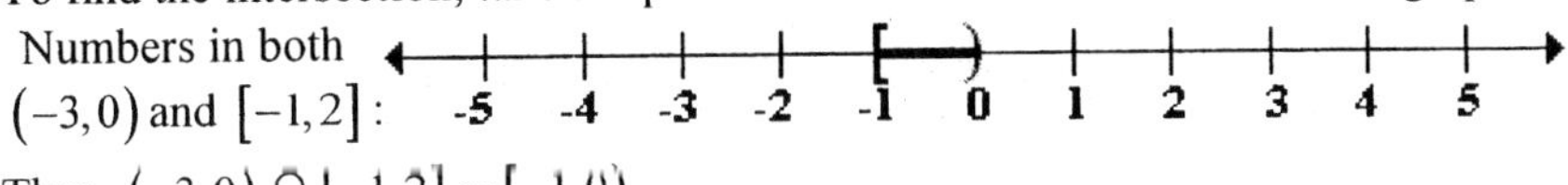

Thus, $(-3,0) \cap [-1,2] = [-1,0)$.

16. Graph $(-4,0)$:

Graph $[-2,1]$:

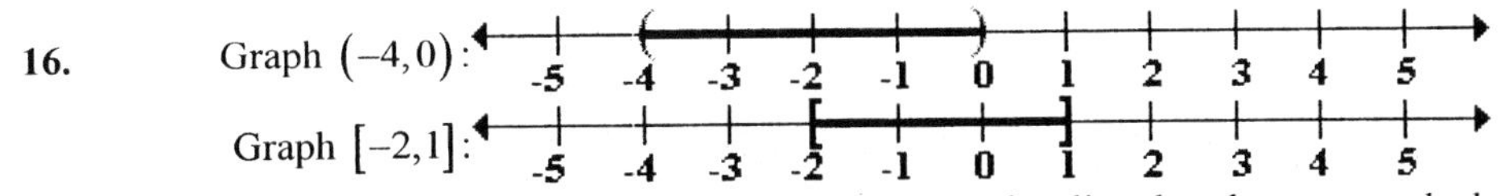

To find the intersection, take the portion of the number line that the two graphs have in common.

Numbers in both $(-4,0)$ and $[-2,1]$:

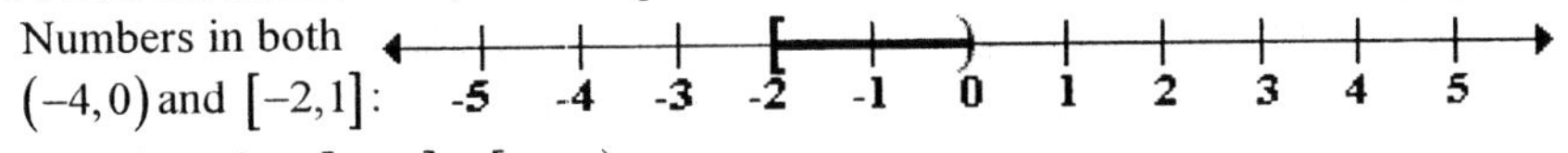

Thus, $(-4,0) \cap [-2,1] = [-2,0)$.

17. Graph $(-3,0)$:

Graph $[-1,2]$:

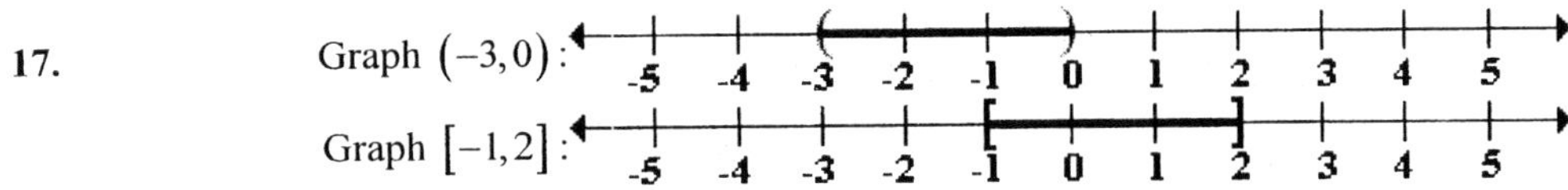

To find the union, take the portion of the number line representing the total collection of numbers in the two graphs.

Numbers in either $(-3,0)$ or $[-1,2]$ or both:

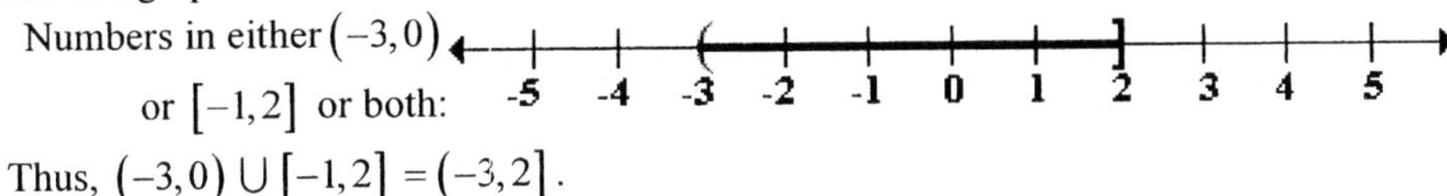

Thus, $(-3,0) \cup [-1,2] = (-3,2]$.

18. Graph $(-4,0)$:

Graph $[-2,1]$:

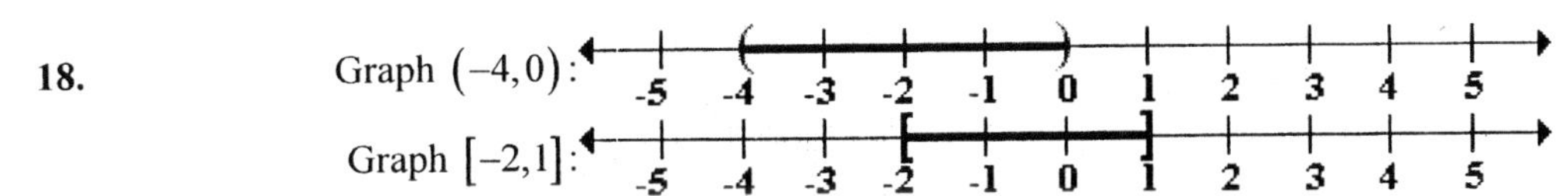

To find the union, take the portion of the number line representing the total collection of numbers in the two graphs.

Numbers in either $(-4,0)$ or $[-2,1]$ or both:

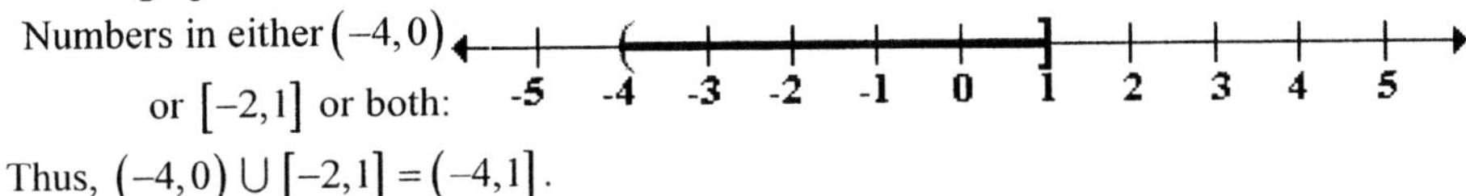

Thus, $(-4,0) \cup [-2,1] = (-4,1]$.

19. Graph $(-\infty,5)$:

Graph $[1,8]$:

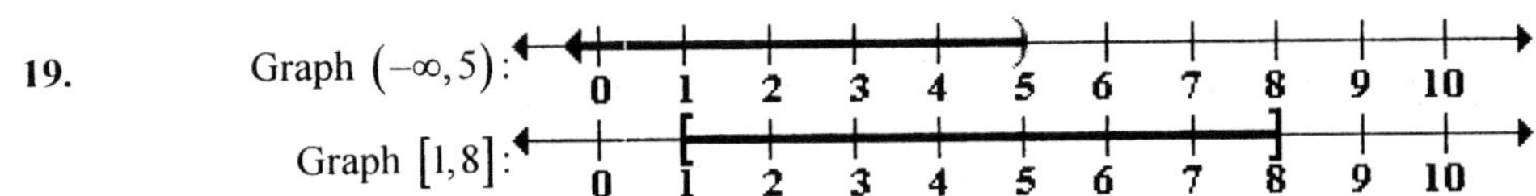

To find the intersection, take the portion of the number line that the two graphs have in common.

Numbers in both $(-\infty,5)$ and $[1,8]$:

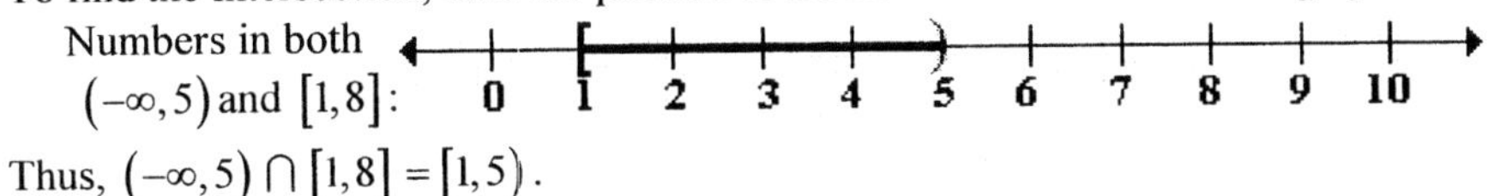

Thus, $(-\infty,5) \cap [1,8] = [1,5)$.

20. Graph $(-\infty,6)$:

Graph $[2,9]$:

To find the intersection, take the portion of the number line that the two graphs have in common.

Numbers in both $(-\infty,6)$ and $[2,9]$:

Thus, $(-\infty,6)\cap[2,9]=[2,6)$.

21. Graph $(-\infty,5)$:

Graph $[1,8]$:

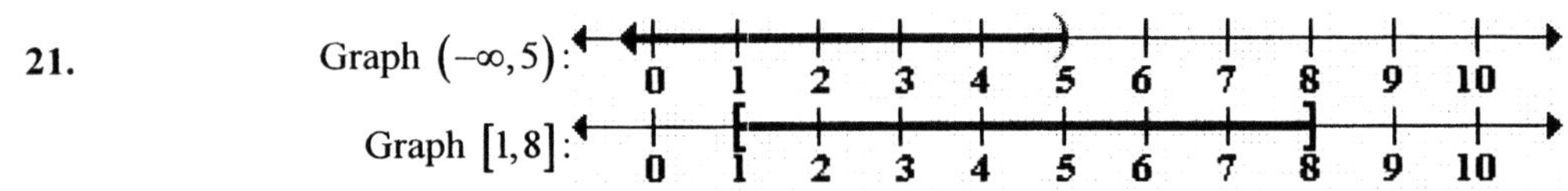

To find the union, take the portion of the number line representing the total collection of numbers in the two graphs.

Numbers in either $(-\infty,5)$ or $[1,8]$ or both:

Thus, $(-\infty,5)\cup[1,8]=(-\infty,8]$.

22. Graph $(-\infty,6)$:

Graph $[2,9]$:

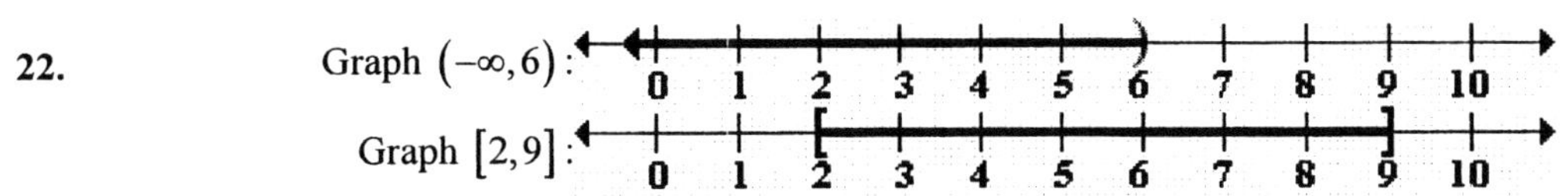

To find the union, take the portion of the number line representing the total collection of numbers in the two graphs.

Numbers in either $(-\infty,6)$ or $[2,9]$ or both:

Thus, $(-\infty,6)\cup[2,9]=(-\infty,9]$.

23. Graph $[3,\infty)$:

Graph $(6,\infty)$:

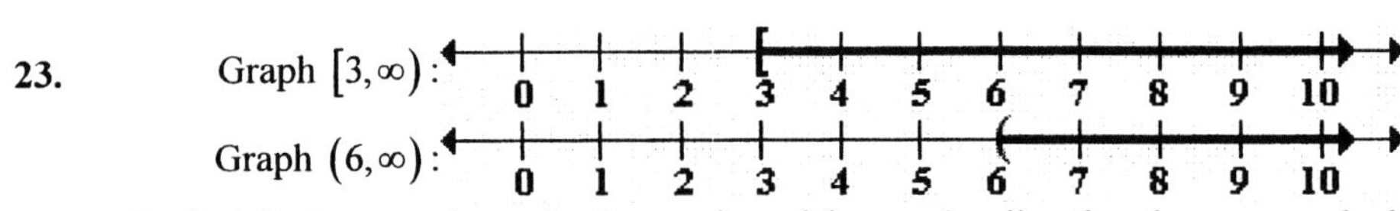

To find the intersection, take the portion of the number line that the two graphs have in common.

Numbers in both $[3,\infty)$ and $(6,\infty)$:

Thus, $[3,\infty)\cap(6,\infty)=(6,\infty)$.

24. Graph $[2,\infty)$:

Graph $(4,\infty)$:

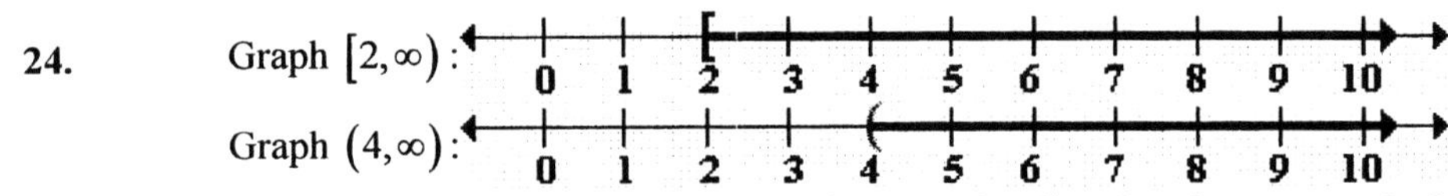

To find the intersection, take the portion of the number line that the two graphs have in common.

Numbers in both $[2,\infty)$ and $(4,\infty)$:

Thus, $[2,\infty)\cap(4,\infty)=(4,\infty)$.

25. Graph $[3, \infty)$: Graph $(6, \infty)$:

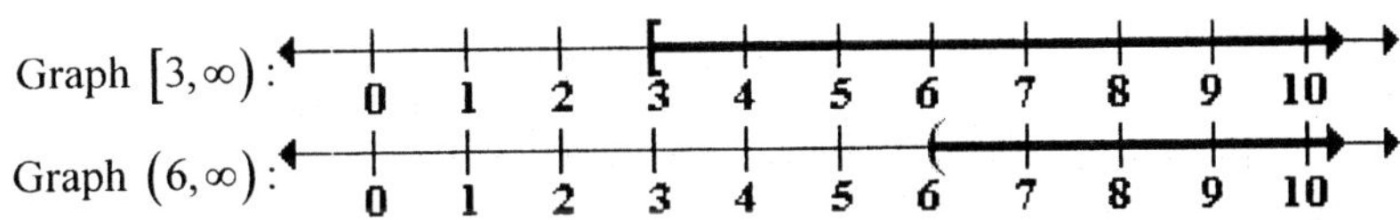

To find the union, take the portion of the number line representing the total collection of numbers in the two graphs.

Numbers in either $[3, \infty)$ or $(6, \infty)$ or both:

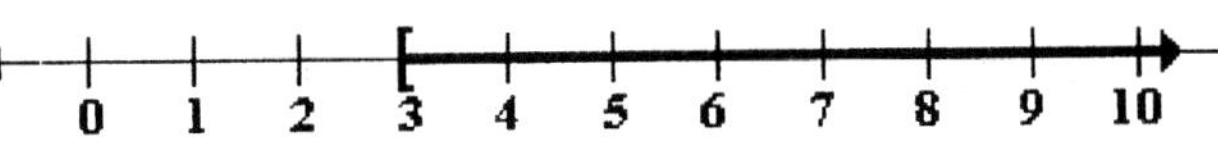

Thus, $[3, \infty) \cup (6, \infty) = [3, \infty)$.

26. Graph $[2, \infty)$: Graph $(4, \infty)$:

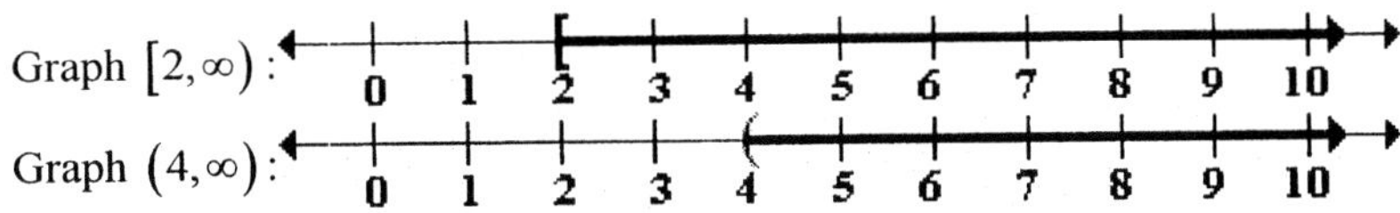

To find the union, take the portion of the number line representing the total collection of numbers in the two graphs.

Numbers in either $[2, \infty)$ or $(4, \infty)$ or both:

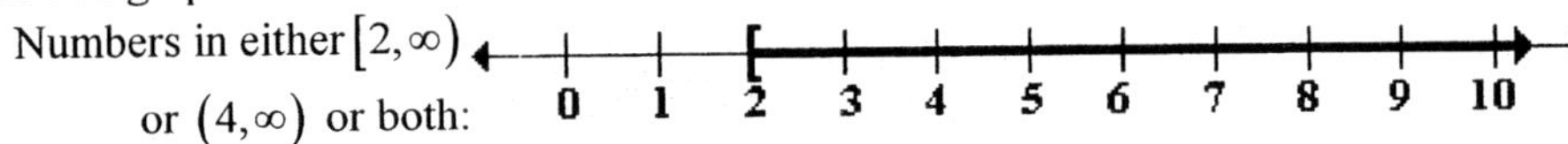

Thus, $[2, \infty) \cup (4, \infty) = [2, \infty)$.

27. $5x + 11 < 26$
$5x < 15$
$x < 3$
The solution set is $\{x \mid x < 3\}$, or $(-\infty, 3)$.

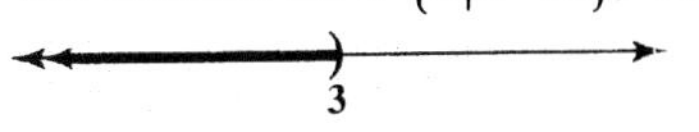

28. $2x + 5 < 17$
$2x < 12$
$x < 6$
The solution set is $\{x \mid x < 6\}$ or $(-\infty, 6)$.

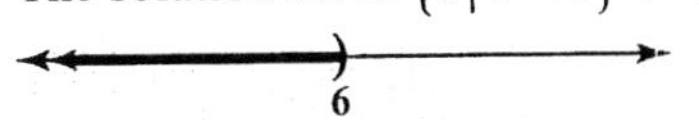

29. $3x - 7 \geq 13$
$3x \geq 20$
$x \geq \frac{20}{3}$
The solution set is $\left\{x \middle| x > \frac{20}{3}\right\}$, or $\left[\frac{20}{3}, \infty\right)$.

$\frac{20}{3}$

30. $8x - 2 \geq 14$
$8x \geq 16$
$x \geq 2$
The solution set is $\{x \mid x > 2\}$ or $[2, \infty)$.

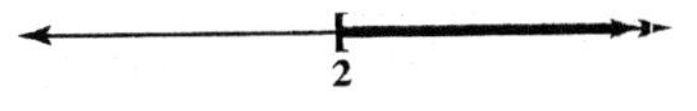

31. $-9x \geq 36$
$x \leq -4$
The solution set is $\{x \mid x \leq -4\}$, or $(-\infty, -4]$.

-4

32. $-5x \leq 30$
$x \geq -6$
The solution set is $\{x \mid x \geq -6\}$ or $[-6, \infty)$.

-6

33. $8x - 11 \leq 3x - 13$
$8x - 3x \leq -13 + 11$
$5x \leq -2$
$x \leq -\frac{2}{5}$
The solution set is $\left\{x \middle| x \leq -\frac{2}{5}\right\}$, or $\left(-\infty, -\frac{2}{5}\right]$.

$-\frac{2}{5}$

34. $18x+45 \le 12x-8$
$18x-12x \le -8-45$
$6x \le -53$
$x \le -\frac{53}{6}$
The solution set is $\left\{x \middle| x \le -\frac{53}{6}\right\}$ or $\left(-\infty, -\frac{53}{6}\right]$.

$-\frac{53}{6}$

35. $4(x+1)+2 \ge 3x+6$
$4x+4+2 \ge 3x+6$
$4x+6 \ge 3x+6$
$4x-3x \ge 6-6$
$x \ge 0$
The solution set is $\{x|x > 0\}$, or $[0, \infty)$.

0

36. $8x+3 > 3(2x+1)+x+5$
$8x+3 > 6x+3+x+5$
$8x+3 > 7x+8$
$8x-7x > 8-3$
$x > 5$
The solution set is $\{x|x > 5\}$ or $(5, \infty)$.

5

37. $2x-11 < -3(x+2)$
$2x-11 < -3x-6$
$5x < 5$
$x < 1$
The solution set is $\{x|x < 1\}$, or $(-\infty, 1)$.

1

38. $-4(x+2) > 3x+20$
$-4x-8 > 3x+20$
$-7x > 28$
$x < -4$
The solution set is $\{x|x < -4\}$ or $(-\infty, -4)$.

−4

39. $1-(x+3) \ge 4-2x$
$1-x-3 \ge 4-2x$
$-x-2 \ge 4-2x$
$x \ge 6$
The solution set is $\{x|x \ge 6\}$, or $[6, \infty)$.

6

40. $5(3-x) \le 3x-1$
$15-5x \le 3x-1$
$-8x \le -16$
$x \ge 2$
The solution set is $\{x|x \ge 2\}$ or $[2, \infty)$.

2

41. $\frac{x}{4}-\frac{3}{2} \le \frac{x}{2}+1$
$\frac{4x}{4}-\frac{4 \cdot 3}{2} \le \frac{4 \cdot x}{2}+4 \cdot 1$
$x-6 \le 2x+4$
$-x \le 10$
$x \ge -10$
The solution set is $\{x|x \ge -10\}$, or $[-10, \infty)$.

−10

42. $\frac{3x}{10}+1 \ge \frac{1}{5}-\frac{x}{10}$
$10\left(\frac{3x}{10}+1\right) \ge 10\left(\frac{1}{5}-\frac{x}{10}\right)$
$3x+10 \ge 2-x$
$4x \ge -8$
$x \ge -2$
The solution set is $\{x|x \ge -2\}$ or $[-2, \infty)$.

−2

43. $1-\frac{x}{2} > 4$
$-\frac{x}{2} > 3$
$x < -6$
The solution set is $\{x|x,-6\}$, or $(-\infty, -6)$.

−6

44. $7-\frac{4}{5}x < \frac{3}{5}$
$-\frac{4}{5}x < -\frac{32}{5}$
$x > 8$
The solution set is $\{x|x > 8\}$ or $(8, \infty)$.

8

45. $\frac{x-4}{6} \ge \frac{x-2}{9} + \frac{5}{18}$

$3(x-4) \ge 2(x-2)+5$

$3x-12 \ge 2x-4+5$

$x \ge 13$

The solution set is $\{x \mid x \ge 13\}$, or $[13, \infty)$.

46.

$\frac{4x-3}{6} + 2 \ge \frac{2x-1}{12}$

$2(4x-3)+24 \ge 2x-1$

$8x-6+24 \ge 2x-1$

$6x+18 \ge -1$

$6x \ge -19$

$x \ge -\frac{19}{6}$

The solution set is $\left\{x \,\middle|\, x \ge \frac{-19}{6}\right\}$ or $\left[-\frac{19}{6}, \infty\right)$.

47. $3[3(x+5)+8x+7]+5[3(x-6)-2(3x-5)] < 2(4x+3)$

$3[3x+15+8x+7]+5[3x-18-6x+10] < 8x+6$

$3[11x+22]+5[-3x-8] < 8x+6$

$33x+66-15x-40 < 8x+6$

$18x+26 < 8x+6$

$10x < -20$

$x < -2$

The solution set is $\{x \mid x < -2\}$ or $[-\infty, -2)$.

−4 −3 −2 −1 −0 1 2 3 4 5 6

48. $5[3(2-3x)-2(5-x)]-6[5(x-2)-2(4x-3)] < 3x+19$

$5[6-9x-10+2x]-6[5x-10-8x+6] < 3x+19$

$5[-7x-4]-6[-3x-4] < 3x+19$

$-35x-20+18x+24 < 3x+19$

$-17x+4 < 3x+19$

$-20x < 15$

$\frac{-20x}{-20} > \frac{15}{-20}$

$x > -\frac{3}{4}$

The solution set is $\left\{x \,\middle|\, x > -\frac{3}{4}\right\}$ or $\left[-\frac{3}{4}, \infty\right)$.

−7 −6 −5 −4 −3 −2 −1 0 1 2 3

49. $6 < x + 3 < 8$
$6 - 3 < x + 3 - 3 < 8 - 3$
$3 < x < 5$
The solution set is $\{x | 3 < x < 5\}$, or (3, 5).

50. $7 < x + 5 < 11$
$7 - 5 < x + 5 - 5 < 11 - 5$
$2 < x < 6$
The solution set is $\{x | 2 < x < 6\}$ or (2, 6).

51. $-3 \le x - 2 < 1$
$-1 \le x < 3$
The solution set is $\{x | -1 \le x < 3\}$, or [–1, 3).

52. $-6 < x - 4 \le 1$
$-2 < x \le 5$
The solution set is $\{x | -2 < x \le 5\}$ or (–2, 5].

53. $-11 < 2x - 1 \le -5$
$-10 < 2x \le -4$
$-5 < x \le -2$
The solution set is $\{x | -5 < x \le -2\}$, or (–5, –2].

54. $3 \le 4x - 3 < 19$
$6 \le 4x < 22$
$\frac{6}{4} \le x < \frac{22}{4}$
$\frac{3}{2} \le x < \frac{11}{2}$
The solution set is $\left\{x \middle| \frac{3}{2} \le x < \frac{11}{2}\right\}$ or $\left[\frac{3}{2}, \frac{11}{2}\right)$.

55. $-3 \le \frac{2}{3}x - 5 < -1$
$2 \le \frac{2}{3}x < 4$
$3 \le x < 6$
The solution set is $\{x | 3 \le x < 6\}$, or [3, 6).

56. $-6 \le \frac{1}{2}x - 4 < -3$
$-2 \le \frac{1}{2}x < 1$
$-4 \le x < 2$
The solution set is $\{x | -4 \ge x < 2\}$ or $[-4, 2)$.

57. $|x| < 3$
$-3 < x < 3$
The solution set is $\{x | -3 < x < 3\}$, or (–3, 3).

58. $|x| < 5$
$-5 < x < 5$
The solution set is $\{x | -5 < x < 5\}$ or (–5, 5).

59. $|x - 1| \le 2$
$-2 \le x - 1 \le 2$
$-1 \le x \le 3$
The solution set is $\{x | -1 \le x \le 3\}$, or [–1, 3].

60. $|x + 3| \le 4$
$-4 \le x + 3 \le 4$
$-7 \le x \le 1$
The solution set is $\{x | -7 \le x \le 1\}$ or [–7, 1].

61. $|2x - 6| < 8$
$-8 < 2x - 6 < 8$
$-2 < 2x < 14$
$-1 < x < 7$
The solution set is $\{x | -1 < x < 7\}$, or (–1, 7).

62. $|3x + 5| < 17$
$-17 < 3x + 5 < 17$
$-22 < 3x < 12$
The solution set is $\left\{x \middle| -\frac{22}{3} < x < 4\right\}$ or $\left(-\frac{22}{3}, 4\right)$.

63. $|2(x - 1) + 4| \le 8$
$-8 \le 2(x - 1) + 4 \le 8$
$-8 \le 2x - 2 + 4 \le 8$
$-8 \le 2x + 2 \le 8$
$-10 \le 2x \le 6$
$-5 \le x \le 3$
The solution set is $\{x | -5 \le x \le 3\}$, or [–5, 3].

64. $|3(x - 1) + 2| \le 20$
$-20 \le 3(x - 1) + 2 \le 20$
$-20 \le 3x - 1 \le 20$
$-19 \le 3x \le 21$
$-\frac{19}{3} \le x \le 7$
The solution set is
$\left\{x \middle| -\frac{19}{3} \le x \le 7\right\}$ or $\left[-\frac{19}{3}, 7\right]$.

65. $\left|\frac{2y+6}{3}\right|<2$

$-2<\frac{2y+6}{3}<2$
$-6<2y+6<6$
$-12<2y<0$
$-6<y<0$

The solution set is $\{x|-6<y<0\}$, or (–6, 0).

66. $\left|\frac{3(x-1)}{4}\right|<6$

$-6<\frac{3(x-1)}{4}<6$
$-24<3x-3<24$
$-21<3x<27$
$-7<x<9$

The solution set is $\{x|-7<x<9\}$ or (–7, 9).

67. $|x|>3$
$x>3$ or $x<-3$

The solution set is $\{x|x>3 \text{ or } x<-3\}$, that is, $(-\infty,-3)$ or $(3,\infty)$.

68. $|x|>5$
$x>5$ or $x<-5$

The solution set is $\{x|x<-5 \text{ or } x>5\}$, that is, all x in $(-\infty,-5)$ or $(5,\infty)$.

69. $|x-1|\geq 2$
$x-1\geq 2$ or $x-1\leq -2$
$x\geq 3$ $\quad$ $x\leq -1$

The solution set is $\{x|x\leq -1 \text{ or } x\geq 3\}$, that is, $(-\infty,-1]$ or $[3,\infty)$.

70. $|x+3|\geq 4$
$x+3\geq 4$ or $x+3\leq -4$
$x\geq 1$ $\quad$ $x\leq -7$

The solution set is $\{x|x\leq -7 \text{ or } x\geq 1\}$, that is, all x in $(-\infty,-7]$ or $[1,\infty)$.

71. $|3x-8|>7$
$3x-8>7$ or $3x-8<-7$
$3x>15$ $\quad$ $3x<1$
$x>5$ $\quad$ $x<\frac{1}{3}$

The solution set is $\left\{x\middle|x<\frac{1}{3} \text{ or } x>5\right\}$, that is, $\left(-\infty,\frac{1}{3}\right)$ or $(5,\infty)$.

72. $|5x-2|>13$
$5x-2>13$ or $5x-2<-13$
$5x>15$ $\quad$ $5x<-11$
$x>3$ $\quad$ $x<-\frac{11}{5}$

The solution set is $\left\{x\middle|x<\frac{-11}{5} \text{ or } x>3\right\}$, that is, all x in $\left(-\infty,\frac{-11}{5}\right)$ or $(3,\infty)$

73. $\left|\frac{2x+2}{4}\right|\geq 2$

$\frac{2x+2}{4}\geq 2$ or $\frac{2x+2}{4}\leq -2$
$2x+2\geq 8$ $\quad$ $2x+2\leq -8$
$2x\geq 6$ $\quad$ $2x\leq -10$
$x\geq 3$ $\quad$ $x\leq -5$

The solution set is $\{x|x\leq -5 \text{ or } x\geq 3\}$, that is, $(-\infty,-5]$ or $[3,\infty)$.

74. $\left|\frac{3x-3}{9}\right|\geq 1$

$\frac{3x-3}{9}\geq 1$ or $\frac{3x-3}{9}\leq -1$
$3x-3\geq 9$ $\quad$ $3x-3\leq -9$
$3x\geq 12$ $\quad$ $3x\leq -6$
$x\geq 4$ $\quad$ $x\leq -2$

The solution set is $\{x|x\leq -2 \text{ or } x\geq 4\}$, or $(-\infty,-2]$ or $[4,\infty)$.

75. $\left|3-\frac{2}{3}x\right|>5$

$3-\frac{2}{3}x>5$ or $3-\frac{2}{3}x<-5$
$-\frac{2}{3}x>2$ $\quad$ $-\frac{2}{3}x<-8$
$x<-3$ $\quad$ $x>12$

The solution set is $\{x|x<-3 \text{ or } x>12\}$, that is, $(-\infty,-3)$ or $(12,\infty)$.

76. $\left|3-\frac{3}{4}x\right|>9$

$3-\frac{3}{4}x>9$ or $3-\frac{3}{4}x<-9$

$-\frac{3}{4}x>6 \qquad -\frac{3}{4}x<-12$

$x<-8 \qquad x>16$

$\{x \mid x<-8 \text{ or } x>16\}$, that is all x in $(-\infty,-8)$ or $(16,\infty)$.

77. $3|x-1|+2\geq 8$

$3|x-1|\geq 6$

$|x-1|\geq 2$

$x-1\geq 2$ or $x-1\leq -2$

$x\geq 3 \qquad x\leq -1$

The solution set is $\{x \mid x\leq 1 \text{ or } x\geq 3\}$, that is, $(-\infty,-1]$ or $[3,\infty)$.

78. $5|2x+1|-3\geq 9$

$5|2x+1|\geq 12$

$|2x+1|\geq \frac{12}{5}$

$2x+1\geq \frac{12}{5} \qquad 2x+1\leq -\frac{12}{5}$

$2x\geq \frac{7}{5}$ or $2x\leq -\frac{17}{5}$

$x\geq \frac{7}{10} \qquad x\leq -\frac{17}{10}$

The solution set is $\left\{x \,\middle|\, x\leq -\frac{17}{10} \text{ or } x\geq \frac{7}{10}\right\}$.

79. $-2|x-4|\geq -4$

$\frac{-2|x-4|}{-2}\leq \frac{-4}{-2}$

$|x-4|\leq 2$

$-2\leq x-4\leq 2$

$2\leq x\leq 6$

The solution set is $\{x \mid 2\leq x\leq 6\}$.

80. $-3|x+7|\geq -27$

$\frac{-3|x+7|}{-3}\leq \frac{-27}{-3}$

$|x+7|\leq 9$

$-9\leq x+7\leq 9$

$-16\leq x\leq 2$

The solution set is $\{x \mid -16\leq x\leq 2\}$.

81. $-4|1-x|<-16$

$\frac{-4|1-x|}{-4}>\frac{-16}{-4}$

$|1-x|>4$

$1-x>4 \qquad 1-x<-4$

$-x>3$ or $-x<-5$

$x<-3 \qquad x>5$

The solution set is $\{x \mid x<-3 \text{ or } x>5\}$.

82. $-2|5-x|<-6$

$-2|5-x|<-6$

$\frac{-2|5-x|}{-2}>\frac{-6}{-2}$

$|5-x|>3$

$5-x>3 \qquad 5-x<-3$

$-x>-2$ or $-x<-8$

$x<2 \qquad x>8$

The solution set is $\{x \mid x<2 \text{ or } x>8\}$.

83. $3\leq |2x-1|$

$2x-1\geq 3 \qquad 2x-1\leq -3$

$2x\geq 4$ or $2x\leq -2$

$x\geq 2 \qquad x\leq -1$

The solution set is $\{x \mid x\leq -1 \text{ or } x\geq 2\}$.

84. $9\leq |4x+7|$

$4x+7\geq 9$ or $4x+7\leq -9$

$4x\geq 2 \qquad 4x\leq -16$

$x\geq \frac{2}{4} \qquad x\leq -4$

$x\geq \frac{1}{2}$

The solution set is $\left\{x \,\middle|\, x\leq -4 \text{ or } x\geq \frac{1}{2}\right\}$.

85. $5 > |4-x|$ is equivalent to $|4-x| < 5$.

$-5 < 4-x < 5$

$-9 < -x < 1$

$\frac{-9}{-1} > \frac{-x}{-1} > \frac{1}{-1}$

$9 > x > -1$

$-1 < x < 9$

The solution set is $\{x | -1 < x < 9\}$.

86. $2 > |11-x|$ is equivalent to $|11-x| < 2$.

$-2 < 11-x < 2$

$-13 < -x < -9$

$\frac{-13}{-1} > \frac{-x}{-1} > \frac{-9}{-1}$

$13 > x > 9$

$9 < x < 13$

The solution set is $\{x | 9 < x < 13\}$.

87. $1 < |2-3x|$ is equivalent to $|2-3x| > 1$.

$2-3x > 1$ or $2-3x < -1$

$-3x > -1$ $\quad -3x < -3$

$\frac{-3x}{-3} < \frac{-1}{-3}$ $\quad \frac{-3x}{-3} > \frac{-3}{-3}$

$x < \frac{1}{3}$ $\quad x > 1$

The solution set is $\left\{x \middle| x < \frac{1}{3} \text{ or } x > 1\right\}$.

88. $4 < |2-x|$ is equivalent to $|2-x| > 4$.

$2-x > 4$ or $2-x < -4$

$-x > 2$ $\quad -x < -6$

$\frac{-x}{-1} < \frac{2}{-1}$ $\quad \frac{-x}{-1} > \frac{-6}{-1}$

$x < -2$ $\quad x > 6$

The solution set is $\{x | x < -2 \text{ or } x > 6\}$.

89. $12 < \left|-2x + \frac{6}{7}\right| + \frac{3}{7}$

$\frac{81}{7} < \left|-2x + \frac{6}{7}\right|$

$-2x + \frac{6}{7} > \frac{81}{7}$ or $-2x + \frac{6}{7} < -\frac{81}{7}$

$-2x > \frac{75}{7}$ $\quad -2x < -\frac{87}{7}$

$x < -\frac{75}{14}$ $\quad x > \frac{87}{14}$

The solution set is $\left\{x \middle| x < -\frac{75}{14} \text{ or } x > \frac{87}{14}\right\}$,

that is, $\left(-\infty, -\frac{75}{14}\right)$ or $\left(\frac{87}{14}, \infty\right)$.

90. $1 < \left|x - \frac{11}{3}\right| + \frac{7}{3}$

$-\frac{4}{3} < \left|x - \frac{11}{3}\right|$

Since $\left|x - \frac{11}{3}\right| > -\frac{4}{3}$ is true for all x,

the solution set is $\{x | x \text{ is any real number}\}$ or $(-\infty, \infty)$.

91. $4 + \left|3 - \frac{x}{3}\right| \ge 9$

$\left|3 - \frac{x}{3}\right| \ge 5$

$3 - \frac{x}{3} \ge 5$ or $3 - \frac{x}{3} \le -5$

$-\frac{x}{3} \ge 2$ $\quad -\frac{x}{3} \le -8$

$x \le -6$ $\quad x \ge 24$

The solution set is $\{x | x \le -6 \text{ or } x \ge 24\}$, that is, $(-\infty, -6]$ or $[24, \infty)$.

92. $\left|2 - \frac{x}{2}\right| - 1 \le 1$

$\left|2 - \frac{x}{2}\right| \le 2$

$-2 \le 2 - \frac{x}{2} \le 2$

$-4 \le -\frac{x}{2} \le 0$

$8 \ge x \ge 0$

The solution set is $\{x | 0 \le x \le 8\}$ or $[0, 8]$.

93. $y \ge 4$
$1-(x+3)+2x \ge 4$
$1-x-3+2x \ge 4$
$x-2 \ge 4$
$x \ge 6$
The solution set is $[6,\infty)$.

94. $y \le 0$
$2x-11+3(x+2) \le 0$
$2x-11+3x+6 \le 0$
$5x-5 \le 0$
$5x \le 5$
$x \le 1$
The solution set is $(-\infty,1]$.

95. $y \le 4$
$7-\left|\frac{x}{2}+2\right| \le 4$
$-\left|\frac{x}{2}+2\right| \le -3$
$\left|\frac{x}{2}+2\right| \ge 3$

$\frac{x}{2}+2 \ge 3$ or $\frac{x}{2}+2 \le -3$
$x+4 \ge 6$ $\quad$ $x+4 \le -6$
$x \ge 2$ $\quad$ $x \le -10$
The solution set is $(-\infty,-10] \cup [2,\infty)$.

96. $y \ge 6$
$8-|5x+3| \ge 6$
$-|5x+3| \ge -2$
$-(-|5x+3|) \le -(-2)$
$|5x+3| \le 2$
$-2 \le 5x+3 \le 2$
$-5 \le 5x \le -1$
$\frac{-5}{5} \le \frac{5x}{5} \le \frac{-1}{5}$
$-1 \le x \le -\frac{1}{5}$
The solution set is $\left[-1,-\frac{1}{5}\right]$.

97. Let x be the number.
$|4-3x| \ge 5$ or $|3x-4| \ge 5$

$3x-4 \ge 5$ $\quad$ $3x-4 \le -5$
$3x \ge 9$ or $3x \le -1$
$x \ge 3$ $\quad$ $x \le -\frac{1}{3}$

The solution set is $\left\{x \mid x \le -\frac{1}{3} \text{ or } x \ge 3\right\}$ or $\left(-\infty,-\frac{1}{3}\right] \cup [3,\infty)$.

98. Let x be the number.
$|5-4x| \le 13$ or $|4x-5| \le 13$

$-13 \le 4x-5 \le 13$
$-8 \le 4x \le 18$
$-2 \le x \le \frac{9}{2}$
The solution set is $\left\{x \mid -2 \le x \le \frac{9}{2}\right\}$ or $\left[-2,-\frac{2}{9}\right]$.

99. $(0,4)$

100. $[0,5]$

101. passion $\le$ intimacy or intimacy $\ge$ passion

102. commitment $\ge$ intimacy or intimacy $\le$ commitment

103. passion $<$ commitment or commitment $>$ passion

104. commitment $>$ passion or passion $<$ commitment

105. 9, after 3 years

106. After approximately $5\frac{1}{2}$ years

107. $3.1x + 25.8 > 63$

$3.1x > 37.2$

$x > 12$

Since x is the number of years after 1994, we calculate 1994+12=2006. 63% of voters will use electronic systems after 2006.

108. $-2.5x + 63.1 < 38.1$

$-2.5x < 25$

$x > 10$

$1994 + 10 = 2004$

In years after 2004, fewer than 38.1% of U.S. voters will use punch cards or lever machines.

109. $28 \le 20 + 0.40(x - 60) \le 40$

$28 \le 20 + 0.40x - 24 \le 40$

$28 \le 0.40x - 4 \le 40$

$32 \le 0.40x \le 44$

$80 \le x \le 110$

Between 80 and 110 ten minutes, inclusive.

110. $15 \le \frac{5}{9}(F - 32) \le 35$

$\frac{9}{5}(15) \le \frac{9}{5}\left(\frac{5}{9}(F - 32)\right) \le \frac{9}{5}(35)$

$9(3) \le F - 32 \le 9(7)$

$27 \le F - 32 \le 63$

$59 \le F \le 95$

The range for Fahrenheit temperatures is $59°\text{F}$ to $95°\text{F}$, inclusive or $[59°\text{F}, 95°\text{F}]$.

111. $\left|\frac{h - 50}{5}\right| \ge 1.645$

$\frac{h - 50}{5} \ge 1.645$ or $\frac{h - 50}{5} \le -1.645$

$h - 50 \ge 8.225$ $\quad$ $h - 50 \le -8.225$

$h \ge 58.225$ $\quad$ $h \le 41.775$

The number of outcomes would be 59 or more, or 41 or less.

112. $50 + 0.20x < 20 + 0.50x$

$30 < 0.3x$

$100 < x$

Basic Rental is a better deal when driving more than 100 miles per day.

113. $15 + 0.08x < 3 + .12x$

$12 < 0.04x$

$300 < x$

Plan A is a better deal when driving more than 300 miles a month.

114. $1800 + 0.03x < 200 + 0.08x$

$1600 < 0.05x$

$32000 < x$

A home assessment of greater than $32,000 would make the first bill a better deal.

115. $2 + 0.08x < 8 + 0.05x$

$0.03x < 6$

$x < 200$

The credit union is a better deal when writing less than 200 checks.

116. $2x > 10,000 + 0.40x$

$1.6x > 10,000$

$\frac{1.6x}{1.6} > \frac{10,000}{1.6}$

$x > 6250$

More than 6250 tapes need to be sold a week to make a profit.

117. $3000 + 3x < 5.5x$

$3000 < 2.5x$

$1200 < x$

More then 1200 packets of stationary need to be sold each week to make a profit.

118. $265 + 65x \le 2800$

$65x \le 2535$

$x \le 39$

39 bags or fewer can be lifted safely.

119. $245 + 95x \le 3000$

$95x \le 2755$

$x \le 29$

29 bags or less can be lifted safely.

120. Let $x =$ the grade on the final exam.

$\frac{86 + 88 + 92 + 84 + x + x}{6} \ge 90$

$86 + 88 + 92 + 84 + x + x \ge 540$

$2x + 350 \ge 540$

$2x \ge 190$

$x \ge 95$

You must receive at least a 95% to earn an A.

121. a. $\frac{86+88+x}{3} \geq 90$

$\frac{174+x}{3} \geq 90$

$174+x \geq 270$

$x \geq 96$

You must get at least a 96.

b. $\frac{86+88+x}{3} < 80$

$\frac{174+x}{3} < 80$

$174+x < 240$

$x < 66$

This will happen if you get a grade less than 66.

122. Let x = the number of hours the mechanic works on the car.

$226 \leq 175+34x \leq 294$

$51 \leq 34x \leq 119$

$1.5 \leq x \leq 3.5$

The man will be working on the job at least 1.5 and at most 3.5 hours.

123. Let x = the number of times the bridge is crossed per three month period

The cost with the 3-month pass is $C_3 = 7.50 + 0.50x$.

The cost with the 6-month pass is $C_6 = 30$.

Because we need to buy two 3-month passes per 6-month pass, we multiply the cost with the 3-month pass by 2.

$2(7.50+0.50x) < 30$

$15+x < 30$

$x < 15$

We also must consider the cost without purchasing a pass. We need this cost to be less than the cost with a 3-month pass.

$3x > 7.50+0.50x$

$2.50x > 7.50$

$x > 3$

The 3-month pass is the best deal when making more than 3 but less than 15 crossings per 3-month period.

132. a. False; $|2x-3| > -7$ is true for any x because the absolute value is 0 or positive.

b. False; $2x > 6, x > 3$

3.1 is a real number that satisfies the inequality.

c. True; $|x-4| > 0$ is not satisfied only when $x = 4$. Since 4 is rational, all irrational numbers satisfy the inequality.

d. False

(c) is true.

133. Because $x > y$, $y - x$ represents a negative number. When both sides are multiplied by $(y-x)$ the inequality must be reversed.

134. a. $|x-4| < 3$

b. $|x-4| \geq 3$

135. Model 1:

$|T-57| < 7$

$-7 < T-57 < 7$

$50 < T < 64$

Model 2:

$|T-50| < 22$

$-22 < T-50 < 22$

$28 < T < 72$

Model 1 describes a city with monthly temperature averages ranging from 50 degrees to 64 degrees Fahrenheit. Model 2 describes a city with monthly temperature averages ranging from 28 degrees to 72 degrees Fahrenheit.

Model 1 describes San Francisco and model 2 describes Albany.

Chapter P Review Exercises

1. $3+6(x-2)^3 = 3+6(4-2)^3$
$= 3+6(2)^3$
$= 3+6(8)$
$= 3+48$
$= 51$

2. $x^2 - 5(x-y) = 6^2 - 5(6-2)$
$= 36-5(4)$
$= 36-20$
$= 16$

3. $S = 0.015x^2 + x + 10$
$S = 0.015(60)^2 + (60) + 10$
$= 0.015(3600) + 60 + 10$
$= 54 + 60 + 10$
$= 124$

4. $A = \{a,b,c\} \quad B = \{a,c,d,e\}$
$\{a,b,c\} \cap \{a,c,d,e\} = \{a,c\}$

5. $A = \{a,b,c\} \quad B = \{a,c,d,e\}$
$\{a,b,c\} \cup \{a,c,d,e\} = \{a,b,c,d,e\}$

6. $A = \{a,b,c\} \quad C = \{a,d,f,g\}$
$\{a,b,c\} \cup \{a,d,f,g\} = \{a,b,c,d,f,g\}$

7. $A = \{a,b,c\} \quad C = \{a,d,f,g\}$
$\{a,d,f,g\} \cap \{a,b,c\} = \{a\}$

8. **a.** $\sqrt{81}$

b. $0, \sqrt{81}$

c. $-17, 0, \sqrt{81}$

d. $-17, -\frac{9}{13}, 0, 0.75, \sqrt{81}$

e. $\sqrt{2}, \pi$

f. $-17, -\frac{9}{13}, 0, 0.75, \sqrt{2}, \pi, \sqrt{81}$

9. $|-103| = 103$

10. $\left|\sqrt{2}-1\right| = \sqrt{2}-1$

11. $\left|3-\sqrt{17}\right| = \sqrt{17}-3$ since $\sqrt{17}$ is greater than 3.

12. $|4-(-17)| = |4+17| = |21| = 21$

13. $3 + 17 = 17 + 3$;
commutative property of addition.

14. $(6\cdot 3)\cdot 9 = 6\cdot(3\cdot 9)$;
associative property of multiplication.

15. $\sqrt{3}(\sqrt{5}+\sqrt{3}) = \sqrt{15}+3$;
distributive property of multiplication over addition.

16. $(6\cdot 9)\cdot 2 = 2\cdot(6\cdot 9)$;
commutative property of multiplication.

17. $\sqrt{3}(\sqrt{5}+\sqrt{3}) = (\sqrt{5}+\sqrt{3})\sqrt{3}$;
commutative property of multiplication.

18. $(3\cdot 7)+(4\cdot 7) = (4\cdot 7)+(3\cdot 7)$;
commutative property of addition.

19. $5(2x-3)+7x = 10x-15+7x = 17x-15$

20. $\frac{1}{5}(5x)+[(3y)+(-3y)]-(-x) = x+[0]+x = 2x$

21. $3(4y-5)-(7y+2) = 12y-15-7y-2 = 5y-17$

22. $8-2[3-(5x-1)] = 8-2[3-5x+1]$
$= 8-2[4-5x]$
$= 8-8+10x$
$= 10x$

23. $E = 10x+166$
$E = 10(20)+166 = 366$

$E = 0.04x^2 + 9.2x + 169$
$E = 0.04(20)^2 + 9.2(20) + 169 = 369$
The actual number was 368 so the better formula was $E = 0.04x^2 + 9.2x + 169$.

24. $(-3)^3(-2)^2 = (-27)\cdot(4) = -108$

25. $2^{-4}+4^{-1}=\frac{1}{2^4}+\frac{1}{4}$
$=\frac{1}{16}+\frac{1}{4}$
$=\frac{1}{16}+\frac{4}{16}$
$=\frac{5}{16}$

26. $5^{-3}\cdot 5=5^{-3}5^1=5^{-3+1}=5^{-2}=\frac{1}{5^2}=\frac{1}{25}$

27. $\frac{3^3}{3^6}=3^{3-6}=3^{-3}=\frac{1}{3^3}=\frac{1}{27}$

28. $(-2x^4y^3)^3=(-2)^3(x^4)^3(y^3)^3$
$=(-2)^3x^{4\cdot 3}y^{3\cdot 3}$
$=-8x^{12}y^9$

29. $(-5x^3y^2)(-2x^{-11}y^{-2})$
$=(-5)(-2)x^3x^{-11}y^2y^{-2}$
$=10\cdot x^{3-11}y^{2-2}$
$=10x^{-8}y^0$
$=\frac{10}{x^8}$

30. $(2x^3)^{-4}=(2)^{-4}(x^3)^{-4}$
$=2^{-4}x^{-12}$
$=\frac{1}{2^4x^{12}}$
$=\frac{1}{16x^{12}}$

31. $\frac{7x^5y^6}{28x^{15}y^{-2}}=\left(\frac{7}{28}\right)(x^{5-15})(y^{6-(-2)})$
$=\frac{1}{4}x^{-10}y^8$
$=\frac{y^8}{4x^{10}}$

32. $3.74\times 10^4=37,400$

33. $7.45\times 10^{-5}=0.0000745$

34. $3,590,000=3.59\times 10^6$

35. $0.00725=7.25\times 10^{-3}$

36. $(3\times 10^3)(1.3\times 10^2)=(3\times 1.3)\times(10^3\times 10^2)$
$=3.9\times 10^5$
$=390,000$

37. $\frac{6.9\times 10^3}{3\times 10^5}=\left(\frac{6.9}{3}\right)\times 10^{3-5}$
$=2.3\times 10^{-2}$
$=0.023$

38. $\frac{10^9}{10^6}=10^{9-6}=10^3$

It would take 10^3 or 1000 years to accumulate \$1 billion.

39. $(2.9\times 10^8)\times 150$
$=(2.9\times 10^8)\times(1.5\times 10^2)$
$=(2.9\times 1.5)\times(10^8\times 10^2)$
$=4.35\times 10^{10}$

The total annual spending on movies is $\$4.35\times 10^{10}$.

40. $\sqrt{300}=\sqrt{100\cdot 3}=\sqrt{100}\cdot\sqrt{3}=10\sqrt{3}$

41. $\sqrt{12x^2}=\sqrt{4x^2\cdot 3}=\sqrt{4x^2}\cdot\sqrt{3}=2|x|\sqrt{3}$

42. $\sqrt{10x}\cdot\sqrt{2x}=\sqrt{20x^2}$
$=\sqrt{4x^2}\cdot\sqrt{5}$
$=2x\sqrt{5}$

43. $\sqrt{r^3}=\sqrt{r^2}\cdot\sqrt{r}=r\sqrt{r}$

44. $\sqrt{\frac{121}{4}}=\frac{\sqrt{121}}{\sqrt{4}}=\frac{11}{2}$

45. $\frac{\sqrt{96x^3}}{\sqrt{2x}}=\sqrt{\frac{96x^3}{2x}}$
$=\sqrt{48x^2}$
$=\sqrt{16x^2}\cdot\sqrt{3}$
$=4x\sqrt{3}$

46. $7\sqrt{5}+13\sqrt{5}=(7+13)\sqrt{5}=20\sqrt{5}$

47. $2\sqrt{50}+3\sqrt{8}=2\sqrt{25\cdot 2}+3\sqrt{4\cdot 2}$
$=2\cdot 5\sqrt{2}+3\cdot 2\sqrt{2}$
$=10\sqrt{2}+6\sqrt{2}$
$=16\sqrt{2}$

48. $4\sqrt{72}-2\sqrt{48}=4\sqrt{36\cdot 2}-2\sqrt{16\cdot 3}$
$=4\cdot 6\sqrt{2}-2\cdot 4\sqrt{3}$
$=24\sqrt{2}-8\sqrt{3}$

49. $\dfrac{30}{\sqrt{5}}=\dfrac{30}{\sqrt{5}}\cdot\dfrac{\sqrt{5}}{\sqrt{5}}=\dfrac{30\sqrt{5}}{5}=6\sqrt{5}$

50. $\dfrac{\sqrt{2}}{\sqrt{3}}=\dfrac{\sqrt{2}}{\sqrt{3}}\cdot\dfrac{\sqrt{3}}{\sqrt{3}}=\dfrac{\sqrt{6}}{3}$

51. $\dfrac{5}{6+\sqrt{3}}=\dfrac{5}{6+\sqrt{3}}\cdot\dfrac{6-\sqrt{3}}{6-\sqrt{3}}$
$=\dfrac{5(6-\sqrt{3})}{36-3}$
$=\dfrac{5(6-\sqrt{3})}{33}$

52.
$\dfrac{14}{\sqrt{7}-\sqrt{5}}=\dfrac{14}{\sqrt{7}-\sqrt{5}}\cdot\dfrac{\sqrt{7}+\sqrt{5}}{\sqrt{7}+\sqrt{5}}$
$=\dfrac{14(\sqrt{7}+\sqrt{5})}{7-5}$
$=\dfrac{14(\sqrt{7}+\sqrt{5})}{2}$
$=7(\sqrt{7}+\sqrt{5})$

53. $\sqrt[3]{125}=5$

54. $\sqrt[5]{-32}=-2$

55. $\sqrt[4]{-125}$ is not a real number.

56. $\sqrt[4]{(-5)^4}=\sqrt[4]{625}=\sqrt[4]{5^4}=5$

57. $\sqrt[3]{81}=\sqrt[3]{27\cdot 3}=\sqrt[3]{27}\cdot\sqrt[3]{3}=3\sqrt[3]{3}$

58. $\sqrt[3]{y^5}=\sqrt[3]{y^3y^2}=y\sqrt[3]{y^2}$

59. $\sqrt[4]{8}\cdot\sqrt[4]{10}=\sqrt[4]{80}=\sqrt[4]{16\cdot 5}=\sqrt[4]{16}\cdot\sqrt[4]{5}=2\sqrt[4]{5}$

60. $4\sqrt[3]{16}+5\sqrt[3]{2}=4\sqrt[3]{8\cdot 2}+5\sqrt[3]{2}$
$=4\cdot 2\sqrt[3]{2}+5\sqrt[3]{2}$
$=8\sqrt[3]{2}+5\sqrt[3]{2}$
$=13\sqrt[3]{2}$

61. $\dfrac{\sqrt[4]{32x^5}}{\sqrt[4]{16x}}=\sqrt[4]{\dfrac{32x^5}{16x}}=\sqrt[4]{2x^4}=x\sqrt[4]{2}$

62. $16^{1/2}=\sqrt{16}=4$

63. $25^{-1/2}=\dfrac{1}{25^{1/2}}=\dfrac{1}{\sqrt{25}}=\dfrac{1}{5}$

64. $125^{1/3}=\sqrt[3]{125}=5$

65. $27^{-1/3}=\dfrac{1}{27^{1/3}}=\dfrac{1}{\sqrt[3]{27}}=\dfrac{1}{3}$

66. $64^{2/3}=(\sqrt[3]{64})^2=4^2=16$

67. $27^{-4/3}=\dfrac{1}{27^{4/3}}=\dfrac{1}{(\sqrt[3]{27})^4}=\dfrac{1}{3^4}=\dfrac{1}{81}$

68. $(5x^{2/3})(4x^{1/4})=5\cdot 4x^{2/3+1/4}=20x^{11/12}$

69. $\dfrac{15x^{3/4}}{5x^{1/2}}=\left(\dfrac{15}{5}\right)x^{3/4-1/2}=3x^{1/4}$

70. $(125\cdot x^6)^{2/3}=(\sqrt[3]{125x^6})^2$
$=(5x^2)^2$
$=25x^4$

71. $\sqrt[6]{y^3}=(y^3)^{1/6}=y^{3\cdot 1/6}=y^{1/2}=\sqrt{y}$

72. $(-6x^3+7x^2-9x+3)+(14x^3+3x^2-11x-7)=(-6x^3+14x^3)+(7x^2+3x^2)+(-9x-11x)+(3-7)$
$=8x^3+10x^2-20x-4$

The degree is 3.

73. $(13x^4-8x^3+2x^2)-(5x^4-3x^3+2x^2-6)=(13x^4-8x^3+2x^2)+(-5x^4+3x^3-2x^2+6)$
$=(13x^4-5x^4)+(-8x^3+3x^3)+(2x^2-2x^2)+6$
$=8x^4-5x^3+6$

The degree is 4.

74. $(3x-2)(4x^2+3x-5)=(3x)(4x^2)+(3x)(3x)+(3x)(-5)+(-2)(4x^2)+(-2)(3x)+(-2)(-5)$
$=12x^3+9x^2-15x-8x^2-6x+10$
$=12x^3+x^2-21x+10$

75. $(3x-5)(2x+1)=(3x)(2x)+(3x)(1)+(-5)(2x)+(-5)(1)$
$=6x^2+3x-10x-5$
$=6x^2-7x-5$

76. $(4x+5)(4x-5)=(4x^2)-5^2=16x^2-25$

77. $(2x+5)^2=(2x)^2+2(2x)\cdot 5+5^2=4x^2+20x+25$

78. $(3x-4)^2=(3x)^2-2(3x)\cdot 4+(-4)^2=9x^2-24x+16$

79. $(2x+1)^3=(2x)^3+3(2x)^2(1)+3(2x)(1)^2+1^3=8x^3+12x^2+6x+1$

80. $(5x-2)^3=(5x)^3-3(5x)^2(2)+3(5x)(2)^2-2^3=125x^3-150x^2+60x-8$

81. $(x+7y)(3x-5y)=x(3x)+(x)(-5y)+(7y)(3x)+(7y)(-5y)$
$=3x^2-5xy+21xy-35y^2$
$=3x^2+16xy-35y^2$

82. $(3x-5y)^2=(3x)^2-2(3x)(5y)+(-5y)^2$
$=9x^2-30xy+25y^2$

83. $(3x^2+2y)^2=(3x^2)^2+2(3x^2)(2y)+(2y)^2$
$=9x^4+12x^2y+4y^2$

84. $(7x+4y)(7x-4y)=(7x)^2-(4y)^2$
$=49x^2-16y^2$

85. $(a-b)(a^2+ab+b^2)$
$=a(a^2)+a(ab)+a(b^2)+(-b)(a^2)$
$+(-b)(ab)+(-b)(b^2)$
$=a^3+a^2b+ab^2-a^2b-ab^2-b^3$
$=a^3-b^3$

86. $[5y-(2x+1)][5y+(2x+1)]$
$= (5y)^2 - (2x+1)^2$
$= 25y^2 - (4x^2+4x+1)$
$= 25y^2 - 4x^2 - 4x - 1$

87. $(x+2y+4)^2$
$= (x+2y+4)(x+2y+4)$
$= x(x+2y+4) + 2y(x+2y+4) + 4(x+2y+4)$
$= x^2 + 2xy + 4x + 2xy + 4y^2 + 8y + 4x + 8y + 16$
$= x^2 + 4xy + 4y^2 + 8x + 16y + 16$

88. $15x^3 + 3x^2 = 3x^2 \cdot 5x + 3x^2 \cdot 1$
$= 3x^2(5x+1)$

89. $x^2 - 11x + 28 = (x-4)(x-7)$

90. $15x^2 - x - 2 = (3x+1)(5x-2)$

91. $64 - x^2 = 8^2 - x^2 = (8-x)(8+x)$

92. $x^2 + 16$ is prime.

93. $3x^4 - 9x^3 - 30x^2 = 3x^2(x^2 - 3x - 10)$
$= 3x^2(x-5)(x+2)$

94. $20x^7 - 36x^3 = 4x^3(5x^4 - 9)$

95. $x^3 - 3x^2 - 9x + 27 = x^2(x-3) - 9(x-3)$
$= (x^2 - 9)(x-3)$
$= (x+3)(x-3)(x-3)$
$= (x+3)(x-3)^2$

96. $16x^2 - 40x + 25 = (4x-5)(4x-5)$
$= (4x-5)^2$

97. $x^4 - 16 = (x^2)^2 - 4^2$
$= (x^2+4)(x^2-4)$
$= (x^2+4)(x+2)(x-2)$

98. $y^3 - 8 = y^3 - 2^3 = (y-2)(y^2+2y+4)$

99. $x^3 + 64 = x^3 + 4^3 = (x+4)(x^2 - 4x + 16)$

100. $3x^4 - 12x^2 = 3x^2(x^2 - 4)$
$= 3x^2(x-2)(x+2)$

101. $27x^3 - 125 = (3x)^3 - 5^3$
$= (3x - 5)[(3x)^2 + (3x)(5) + 5^2]$
$= (3x - 5)(9x^2 + 15x + 25)$

102. $x^5 - x = x(x^4 - 1)$
$= x(x^2 - 1)(x^2 + 1)$
$= x(x - 1)(x + 1)(x^2 + 1)$

103. $x^3 + 5x^2 - 2x - 10 = x^2(x + 5) - 2(x + 5)$
$= (x^2 - 2)(x + 5)$

104. $x^2 + 18x + 81 - y^2 = (x^2 + 18x + 81) - y^2$
$= (x + 9)^2 - y^2$
$= (x + 9 - y)(x + 9 + y)$

105. $16x^{-3/4} + 32x^{1/4} = 16x^{-3/4}\left(1 + 2x^{1/4 - (-3/4)}\right)$
$= 16x^{-3/4}(1 + 2x)$
$= \dfrac{16(1 + 2x)}{x^{3/4}}$

106. $(x^2 - 4)(x^2 + 3)^{\frac{1}{2}} - (x^2 - 4)^2(x^2 + 3)^{\frac{3}{2}}$
$= (x^2 - 4)(x^2 + 3)^{\frac{1}{2}}\left[1 - (x^2 - 4)(x^2 + 3)\right]$
$= (x - 2)(x + 2)(x^2 + 3)^{\frac{1}{2}}\left[1 - (x - 2)(x + 2)(x^2 + 3)\right]$
$= (x - 2)(x + 2)(x^2 + 3)^{\frac{1}{2}}(-x^4 + x^2 + 13)$

107. $12x^{-\frac{1}{2}} + 6x^{-\frac{3}{2}} = 6x^{-\frac{3}{2}}(2x + 1) = \dfrac{6(2x + 1)}{x^{\frac{3}{2}}}$

108. $\dfrac{x^3 + 2x^2}{x + 2} = \dfrac{x^2(x + 2)}{x + 2} = x^2,\ x \neq -2$

109. $\dfrac{x^2 + 3x - 18}{x^2 - 36} = \dfrac{(x + 6)(x - 3)}{(x + 6)(x - 6)} = \dfrac{x - 3}{x - 6},$
$x \neq -6, 6$

110. $\dfrac{x^2 + 2x}{x^2 + 4x + 4} = \dfrac{x(x + 2)}{(x + 2)^2} = \dfrac{x}{x + 2},$
$x \neq -2$

111. $$\frac{x^2+6x+9}{x^2-4}\cdot\frac{x+3}{x-2}=\frac{(x+3)^2}{(x-2)(x+2)}\cdot\frac{x+3}{x-2}$$
$$=\frac{(x+3)^3}{(x-2)^2(x+2)},$$
$x \neq 2, -2$

112. $$\frac{6x+2}{x^2-1}\div\frac{3x^2+x}{x-1}$$
$$=\frac{2(3x+1)}{(x-1)(x+1)}\div\frac{x(3x+1)}{x-1}$$
$$=\frac{2(3x+1)}{(x-1)(x+1)}\cdot\frac{x-1}{x(3x+1)}$$
$$=\frac{2}{x(x+1)},$$
$x \neq 0, 1, -1, -\frac{1}{3}$

113. $$\frac{x^2-5x-24}{x^2-x-12}\div\frac{x^2-10x+16}{x^2+x-6}$$
$$=\frac{(x-8)(x+3)}{(x-4)(x+3)}\div\frac{(x-2)(x-8)}{(x+3)(x-2)}$$
$$=\frac{x-8}{x-4}\cdot\frac{x+3}{x-8}$$
$$=\frac{x+3}{x-4},$$
$x \neq -3, 4, 2, 8$

114. $$\frac{2x-7}{x^2-9}-\frac{x-10}{x^2-9}=\frac{2x-7-(x-10)}{x^2-9}$$
$$=\frac{x+3}{(x+3)(x-3)}$$
$$=\frac{1}{x-3},$$
$x \neq 3, -3$

115. $$\frac{3x}{x+2}+\frac{x}{x-2}=\frac{3x}{x+2}\cdot\frac{x-2}{x-2}+\frac{x}{x-2}\cdot\frac{x+2}{x+2}$$
$$=\frac{3x^2-6x+x^2+2x}{(x+2)(x-2)}$$
$$=\frac{4x^2-4x}{(x+2)(x-2)}$$
$$=\frac{4x(x-1)}{(x+2)(x-2)},$$
$x \neq 2, -2$

116. $$\frac{x}{x^2-9}+\frac{x-1}{x^2-5x+6}$$
$$=\frac{x}{(x-3)(x+3)}+\frac{x-1}{(x-2)(x-3)}$$
$$=\frac{x}{(x-3)(x+3)}\cdot\frac{x-2}{x-2}+\frac{x-1}{(x-2)(x-3)}\cdot\frac{x+3}{x+3}$$
$$=\frac{x(x-2)+(x-1)(x+3)}{(x-3)(x+3)(x-2)}$$
$$=\frac{x^2-2x+x^2+2x-3}{(x-3)(x+3)(x-2)}$$
$$=\frac{2x^2-3}{(x-3)(x+3)(x-2)}$$
$x \neq 3, -3, 2$

117. $\dfrac{4x-1}{2x^2+5x-3}-\dfrac{x+3}{6x^2+x-2}$
$=\dfrac{4x-1}{(2x-1)(x+3)}-\dfrac{x+3}{(2x-1)(3x+2)}$
$=\dfrac{4x-1}{(2x-1)(x+3)}\cdot\dfrac{3x+2}{3x+2}$
$-\dfrac{x+3}{(2x-1)(3x+2)}\cdot\dfrac{x+3}{x+3}$
$=\dfrac{12x^2+8x-3x-2-x^2-6x-9}{(2x-1)(x+3)(3x+2)}$
$=\dfrac{11x^2-x-11}{(2x-1)(x+3)(3x+2)},$
$x\neq\dfrac{1}{2},-3,-\dfrac{2}{3}$

118. $\dfrac{\frac{1}{x}-\frac{1}{2}}{\frac{1}{3}-\frac{x}{6}}=\dfrac{\frac{1}{x}-\frac{1}{2}}{\frac{1}{3}-\frac{x}{6}}\cdot\dfrac{6x}{6x}$
$=\dfrac{6-3x}{2x-x^2}$
$=\dfrac{-3(x-2)}{-x(x-2)}$
$=\dfrac{3}{x},$
$x\neq 0, 2$

119. $\dfrac{3+\frac{12}{x}}{1-\frac{16}{x^2}}=\dfrac{3+\frac{12}{x}}{1-\frac{16}{x^2}}\cdot\dfrac{x^2}{x^2}$
$=\dfrac{3x^2+12x}{x^2-16}$
$=\dfrac{3x(x+4)}{(x+4)(x-4)}$
$=\dfrac{3x}{x-4},$
$x\neq 0, 4, -4$

120. $\dfrac{3-\frac{1}{x+3}}{3+\frac{1}{x+3}}=\dfrac{3-\frac{1}{x+3}}{3+\frac{1}{x+3}}\cdot\dfrac{x+3}{x+3}$
$=\dfrac{3(x+3)-1}{3(x+3)+1}$
$=\dfrac{3x+9-1}{3x+9+1}$
$=\dfrac{3x+8}{3x+10},$
$x\neq -3,-\dfrac{10}{3}$

121 . $\dfrac{\sqrt{25-x^2}+\frac{x^2}{\sqrt{25-x^2}}}{25-x^2}$
$=\dfrac{\left(\sqrt{25-x^2}+\frac{x^2}{\sqrt{25-x^2}}\right)\sqrt{25-2x^2}}{(25-x^2)\sqrt{25-x^2}}$
$=\dfrac{25-x^2+x^2}{(25-x^2)\sqrt{25-x^2}}$
$=\dfrac{25}{\sqrt{(25-x^2)^3}}$
$=\dfrac{25}{\sqrt{(25-x^2)^3}}\cdot\dfrac{\sqrt{25-x^2}}{\sqrt{25-x^2}}$
$=\dfrac{25\sqrt{25-x^2}}{(25-x^2)}$
$=\dfrac{25\sqrt{25-x^2}}{(5-x)^2(5+x)^2}$

122. $1-2(6-x)=3x+2$
$1-12+2x=3x+2$
$-11-x=2$
$-x=13$
$x=-13$
The solution set is $\{-13\}$.
This is a conditional equation.

123. $2(x-4)+3(x+5)=2x-2$
$2x-8+3x+15=2x-2$
$5x+7=2x-2$
$3x=-9$
$x=-3$
The solution set is $\{-3\}$.
This is a conditional equation.

124. $2x - 4(5x + 1) = 3x + 17$
$$2x - 20x - 4 = 3x + 17$$
$$-18x - 4 = 3x + 17$$
$$-21x = 21$$
$$x = -1$$
The solution set is $\{-1\}$.
This is a conditional equation.

125. $x \neq 1, x \neq -1$
$$\frac{1}{x-1} - \frac{1}{x+1} = \frac{2}{x^2-1}$$
$$\frac{1}{x-1} - \frac{1}{x+1} = \frac{2}{(x+1)(x-1)}$$
$$x + 1 - (x-1) = 2$$
$$x + 1 - x + 1 = 2$$
$$2 = 2$$
The solution set is all real numbers except 1 and −1.

126. $x \neq -2, x \neq 4$
$$\frac{4}{x+2} + \frac{2}{x-4} = \frac{30}{(x+2)(x-4)}$$
$$4(x-4) + 2(x+2) = 30$$
$$4x - 16 + 2x + 4 = 30$$
$$6x - 12 = 30$$
$$6x = 42$$
$$x = 7$$
The solution set is $\{7\}$.

127. $-4|2x+1| + 12 = 0$
$$-4|2x+1| = -12$$
$$|2x+1| = 3$$
$$2x + 1 = 3 \quad \text{or} \quad 2x + 1 = -3$$
$$2x = 2 \qquad 2x = -4$$
$$x = 1 \qquad x = -2$$
The solution set is $\{-2, 1\}$.

128. $2x^2 - 11x + 5 = 0$
$$(2x-1)(x-5) = 0$$
$$2x - 1 = 0 \quad x - 5 = 0$$
$$x = \frac{1}{2} \text{ or } x = 5$$
The solution set is $\left\{\frac{1}{2}, 5\right\}$.

129. $(3x+5)(x-3) = 5$
$$3x^2 + 5x - 9x - 15 = 5$$
$$3x^2 - 4x - 20 = 0$$
$$x = \frac{4 \pm \sqrt{(-4)^2 - 4(3)(-20)}}{2(3)}$$
$$x = \frac{4 \pm \sqrt{16+240}}{6}$$
$$x = \frac{4 \pm \sqrt{256}}{6}$$
$$x = \frac{4 \pm 16}{6}$$
$$x = \frac{20}{6}, \frac{-12}{6}$$
$$x = \frac{10}{3}, -2$$
The solution set is $\left\{-2, \frac{10}{3}\right\}$.

130. $3x^2 - 7x + 1 = 0$
$$x = \frac{7 \pm \sqrt{(-7)^2 - 4(3)(1)}}{2(3)}$$
$$x = \frac{7 \pm \sqrt{49-12}}{6}$$
$$x = \frac{7 \pm \sqrt{37}}{6}$$
The solution set is $\left\{\frac{7+\sqrt{37}}{6}, \frac{7-\sqrt{37}}{6}\right\}$.

131. $x^2 - 9 = 0$
$$x^2 = 9$$
$$x = \pm 3$$
The solution set is $\{-3, 3\}$.

132. $(x-3)^2 - 24 = 0$
$$(x-3)^2 = 24$$
$$\sqrt{(x-3)^2} = \pm\sqrt{24}$$
$$x - 3 = \pm 2\sqrt{6}$$
$$x = 3 \pm 2\sqrt{6}$$

133.
$$\frac{2x}{x^2+6x+8}=\frac{x}{x+4}-\frac{2}{x+2}$$
$$\frac{2x}{(x+4)(x+2)}=\frac{x}{x+4}-\frac{2}{x+2}$$
$$\frac{2x(x+4)(x+2)}{(x+4)(x+2)}=(x+4)(x+2)\left(\frac{x}{x+4}-\frac{2}{x+2}\right)$$
$$2x=x(x+2)-2(x+4)$$
$$2x=x^2+2x-2x-8$$
$$0=x^2-2x-8$$
$$0=(x+2)(x-4)$$

$x+2=0$ or $x-4=0$

$x=-2$ $\quad x=4$

–2 must be rejected. The solution set is $\{4\}$.

134. $\sqrt{8-2x}-x=0$
$$\sqrt{8-2x}=x$$
$$\left(\sqrt{8-2x}\right)^2=x^2$$
$$8-2x=x^2$$
$$0=x^2+2x-8$$
$$0=(x+4)(x-2)$$

$x+4=0$ or $x-2=0$

$x=-4$ $\quad x=2$

–4 must be rejected. The solution set is $\{2\}$.

135. $\sqrt{2x-3}+x=3$
$$\sqrt{2x-3}=3-x$$
$$2x-3=9-6x+x^2$$
$$x^2-8x+12=0$$
$$x^2-8x=-12$$
$$x^2-8x+16=-12+16$$
$$(x-4)^2=4$$
$$x-4=\pm 2$$
$$x=4+2$$
$$x=6,2$$

The solution set is $\{2\}$.

136. $vt+gt^2=s$
$$gt^2=s-vt$$
$$\frac{gt^2}{t^2}=\frac{s-vt}{t^2}$$
$$g=\frac{s-vt}{t^2}$$

137.
$$T=\frac{A-P}{\text{Pr}}$$
$$Pr(T)=Pr\frac{A-P}{\text{Pr}}$$
$$PrT=A-P$$
$$PrT+P=A$$
$$P(rT+1)=A$$
$$P=\frac{A}{1+rT}$$

138.
$$x^2=2x-19$$
$$x^2-2x+19=0$$
$$b^2-4ac=(-2)^2-4(1)(19)=-72$$

$-72<0$, thus the equation has no real solutions.

139. $9x^2-30x+25=0$
$$b^2-4ac=(-30)^2-4(9)(25)=0$$

$b^2-4ac=0$, thus the equation has one repeated real solution.

140. Let x = the number of calories in Burger King's Chicken Caesar.
$x+125$ = the number of calories in Taco Bell's Express Taco Salad.
$x+95$ = the number of calories in Wendy's Mandarin Chicken Salad.
$$x+(x+125)+(x+95)=1705$$
$$3x+220=1705$$
$$3x=1485$$
$$x=495$$
$$x+125=495+125=620$$
$$x+95=495+95=590$$

There are 495 calories in the Chicken Caesar, 620 calories in the Express Taco Salad, and 590 calories in the Mandarin Chicken Salad.

141. Let x = the number of years after 1970.
$$P=-0.5x+37.4$$
$$18.4=-0.5x+37.4$$
$$-19=-0.5x$$
$$\frac{-19}{-0.5}=\frac{-0.5x}{-0.5}$$
$$38=x$$

If the trend continues only 18.4% of U.S. adults will smoke cigarettes 38 years after 1970, or 2008.

142. Let x = the original price of the phone
$48 = x - 0.20x$
$48 = 0.80x$
$60 = x$
The original price is \$60.

143. Let x = the amount sold to earn \$800 in one week
$800 = 300 + 0.05x$
$500 = 0.05x$
$10{,}000 = x$
Sales must be \$10,000 in one week to earn \$800.

144. Let w = the width of the playing field,
Let $3w - 6$ = the length of the playing field
$P = 2(\text{length}) + 2(\text{width})$
$340 = 2(3w - 6) + 2w$
$340 = 6w - 12 + 2w$
$340 = 8w - 12$
$352 = 8w$
$44 = w$
The dimensions are 44 yards by 126 yards.

145. b. Check some points to determine that $y_1 = 14{,}100 + 1500x$ and $y_2 = 41{,}700 - 800x$. Since $y_1 = y_2 = 32{,}100$ when $x = 12$, the two colleges will have the same enrollment in the year $2007 + 12 = 2019$. That year the enrollments will be 32,100 students.

146.
$A = lw$
$15 = l(2l - 7)$
$15 = 2l^2 - 7l$
$0 = 2l^2 - 7l - 15$
$0 = (2l + 3)(l - 5)$
$l = 5$
$2l - 7 = 3$
The length is 5 yards, the width is 3 yards.

147. Let x = height of building
$2x$ = shadow height
$x^2 + (2x)^2 = 300^2$
$x^2 + 4x^2 = 90{,}000$
$5x^2 = 90{,}000$
$x^2 = 18{,}000$
$x \approx \pm 134.164$
Discard negative height.
The building is approximately 134 meters high.

148. $(10 + 2x)(16 + 2x) = 280$
$160 + 52x + 4x^2 = 280$
$4x^2 + 52x - 120 = 0$
$x^2 + 13x - 30 = 0$
$(x + 15)(x - 2) = 0$
$x + 15 = 0$ or $x - 2 = 0$
$x = -15$ $\quad x = 2$
−15 must be rejected. The width of the frame is 2 inches.

149.
$$\frac{1500}{x} + 100 = \frac{1500}{x - 4}$$
$$x(x-4)\left(\frac{1500}{x} + 100\right) = x(x-4)\frac{1500}{x-4}$$
$$1500(x - 4) + 100x(x - 4) = 1500x$$
$$1500x - 6000 + 100x^2 - 400x = 1500x$$
$$15x - 60 + x^2 - 4x = 15x$$
$$x^2 - 4x - 60 = 0$$
$$(x + 6)(x - 10) = 0$$
$x + 6 = 0$ or $x - 10 = 0$
$x = -6$ $\quad x = 10$
−6 must be rejected. There were originally 10 people.

150. $\{x \mid -3 \le x < 5\}$

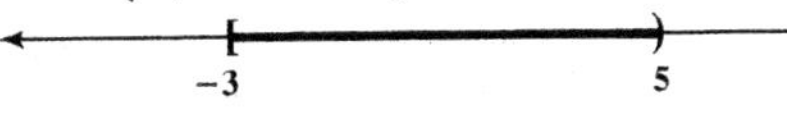

151. $\{x \mid x > -2\}$

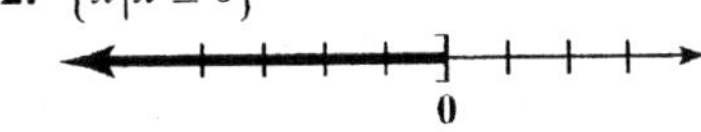

152. $\{x \mid x \le 0\}$

153. Graph $(-2,1]$:

Graph $[-1,3)$:

To find the intersection, take the portion of the number line that the two graphs have in common.

Numbers in both $(-2,1]$ and $[-1,3)$:

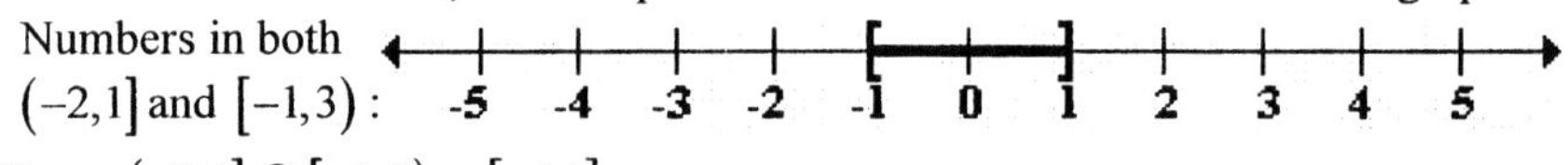

Thus, $(-2,1] \cap [-1,3) = [-1,1]$.

154. Graph $(-2,1]$:

Graph $[-1,3)$:

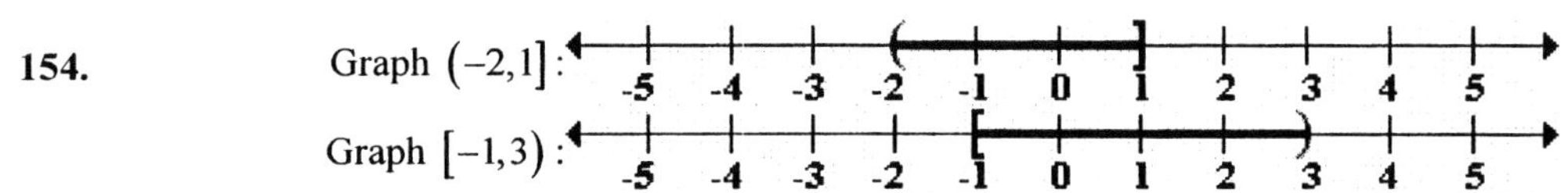

To find the union, take the portion of the number line representing the total collection of numbers in the two graphs.

Numbers in either $(-2,1]$ or $[-1,3)$ or both:

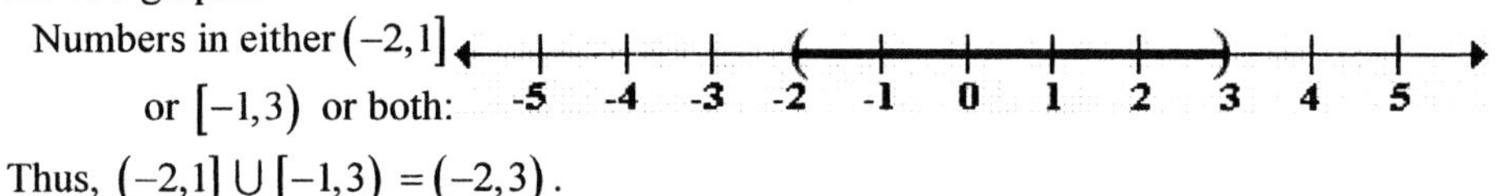

Thus, $(-2,1] \cup [-1,3) = (-2,3)$.

155. Graph $[1,3)$:

Graph $(0,4)$:

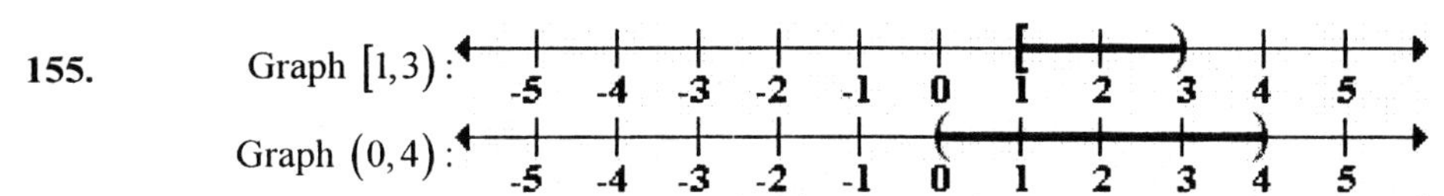

To find the intersection, take the portion of the number line that the two graphs have in common.

Numbers in both $[1,3)$ and $(0,4)$:

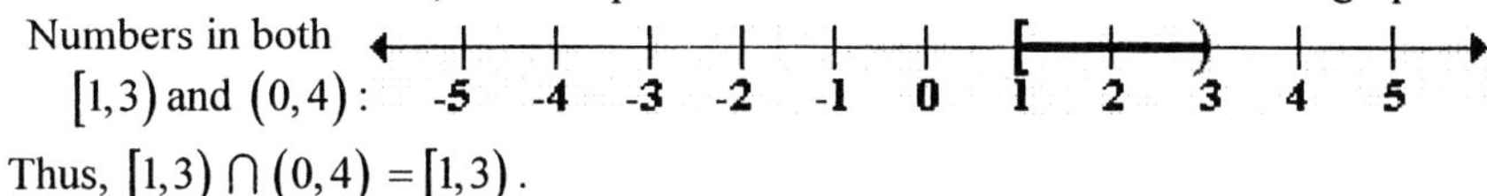

Thus, $[1,3) \cap (0,4) = [1,3)$.

156. Graph $[1,3)$:

Graph $(0,4)$:

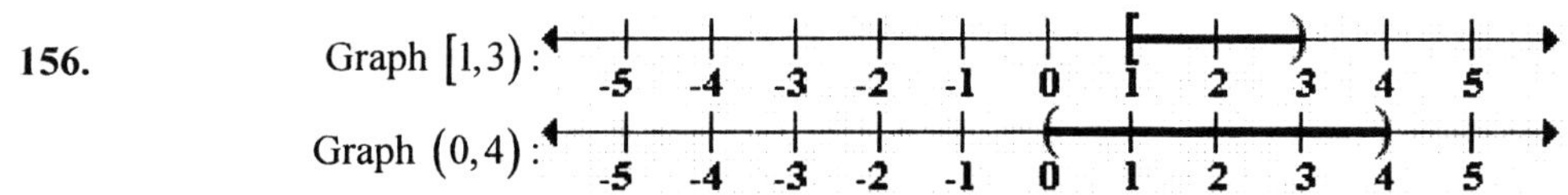

To find the union, take the portion of the number line representing the total collection of numbers in the two graphs.

Numbers in either $[1,3)$ or $(0,4)$ or both:

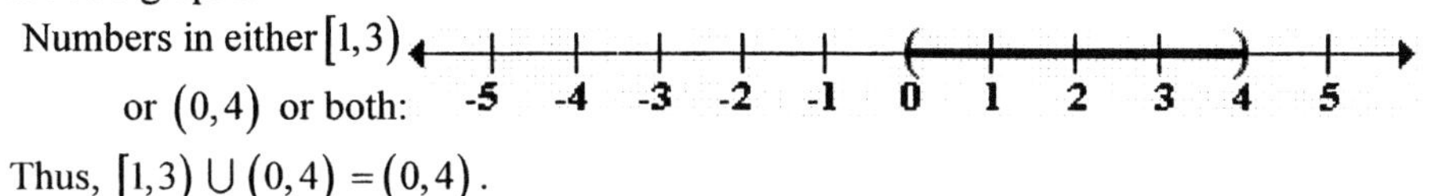

Thus, $[1,3) \cup (0,4) = (0,4)$.

157. $-6x + 3 \le 15$

$-6x \le 12$

$x \ge 2$

-2

The solution set is $[-2, \infty)$.

158. $6x - 9 \ge -4x - 3$

$10x \ge 6$

$x \ge \frac{3}{5}$

$\frac{3}{5}$

The solution set is $\left[\frac{3}{5}, \infty\right)$.

159. $\frac{x}{3}-\frac{3}{4}-1>\frac{x}{2}$

$12\left(\frac{x}{3}-\frac{3}{4}-1\right)>12\left(\frac{x}{2}\right)$

$4x-9-12>6x$

$-21>2x$

$-\frac{21}{2}>x$

$-\frac{21}{2}$

The solution set is $\left(-\infty,-\frac{21}{2}\right)$.

160. $6x+5>-2(x-3)-25$

$6x+5>-2x+6-25$

$8x+5>-19$

$8x>-24$

$x>-3$

-3

The solution set is $(-3,\infty)$.

161. $3(2x-1)-2(x-4)\geq 7+2(3+4x)$

$6x-3-2x+8\geq 7+6+8x$

$4x+5\geq 8x+13$

$-4x\geq 8$

$x\leq -2$

-2

The solution set is $[-\infty,-2)$.

162. $7<2x+3\leq 9$

$4<2x\leq 6$

$2<x\leq 3$

$(2, 3]$

2 3

The solution set is $[2,3)$.

163. $|2x+3|\leq 15$

$-15\leq 2x+3\leq 15$

$-18\leq 2x\leq 12$

$-9\leq x\leq 6$

-9 6

The solution set is $[-9,6]$.

164. $\left|\frac{2x+6}{3}\right|>2$

$\frac{2x+6}{3}>2 \quad \frac{2x+6}{3}<-2$

$2x+6>6 \quad 2x+6<-6$

$2x>0 \quad 2x<-12$

$x>0 \quad x<-6$

-6 0

The solution set is $(-\infty,-6)$ or $(0,\infty)$.

165. $|2x+5|-7\geq -6$

$|2x+5|\geq 1$

$2x+5\geq 1$ or $2x+5\leq -1$

$2x\geq -4 \quad 2x\leq -6$

$x\geq -2$ or $x\leq -3$

-3 -2

The solution set is $(-\infty,-3]$ or $[-2,\infty)$.

166. $-4|x+2|+5\leq -7$

$-4|x+2|\leq -12$

$|x+2|\geq 3$

$x+2\geq 3$ or $x+2\leq -3$

$x\geq 1 \quad x\leq -5$

The solution set is $(-\infty,-5]\cup[1,\infty)$.

-5 1

167. $0.20x+24\leq 40$

$0.20x\leq 16$

$\frac{0.20x}{0.20}\leq\frac{16}{0.20}$

$x\leq 80$

A customer can drive no more than 80 miles.

168. $80\leq\frac{95+79+91+86+x}{5}<90$

$400\leq 95+79+91+86+x<450$

$400\leq 351+x<450$

$49\leq x<99$

A grade of at least 49% but less than 99% will result in a B.

Chapter P Test

1. $5(2x^2-6x)-(4x^2-3x)=10x^2-30x-4x^2+3x$
$=6x^2-27x$

2. $7+2[3(x+1)-2(3x-1)]$
$=7+2[3x+3-6x+2]$
$=7+2[-3x+5]$
$=7-6x+10$
$=-6x+17$

3. $\{1,2,5\}\cap\{5,a\}=\{5\}$

4. $\{1,2,5\}\cup\{5,a\}=\{1,2,5,a\}$

5. $\dfrac{30x^3y^4}{6x^9y^{-4}}=5x^{3-9}y^{4-(-4)}=5x^{-6}y^8=\dfrac{5y^8}{x^6}$

6. $\sqrt{6r}\cdot\sqrt{3r}=\sqrt{18r^2}=\sqrt{9r^2}\cdot\sqrt{2}=3r\sqrt{2}$

7. $4\sqrt{50}-3\sqrt{18}=4\sqrt{25\cdot 2}-3\sqrt{9\cdot 2}$
$=4\cdot 5\sqrt{2}-3\cdot 3\sqrt{2}$
$=20\sqrt{2}-9\sqrt{2}$
$=11\sqrt{2}$

8. $\dfrac{3}{5+\sqrt{2}}=\dfrac{3}{5+\sqrt{2}}\cdot\dfrac{5-\sqrt{2}}{5-\sqrt{2}}$
$=\dfrac{3(5-\sqrt{2})}{25-2}$
$=\dfrac{3(5-\sqrt{2})}{23}$

9. $\sqrt[3]{16x^4}=\sqrt[3]{8x^3\cdot 2x}$
$=\sqrt[3]{8x^3}\cdot\sqrt[3]{2x}$
$=2x\sqrt[3]{2x}$

10. $\dfrac{x^2+2x-3}{x^2-3x+2}=\dfrac{(x+3)(x-1)}{(x-2)(x-1)}=\dfrac{x+3}{x-2},$
$x\neq 2,\ 1$

11. $\dfrac{5\times10^{-6}}{20\times10^{-8}}=\dfrac{5}{20}\cdot\dfrac{10^{-6}}{10^{-8}}=0.25\times10^2=2.5\times10^1$

12. $(2x-5)(x^2-4x+3)$
$=2x^3-8x^2+6x-5x^2+20x-15$
$=2x^3-13x^2+26x-15$

13. $(5x+3y)^2=(5x)^2+2(5x)(3y)+(3y)^2$
$=25x^2+30xy+9y^2$

14. $\dfrac{2x+8}{x-3}\div\dfrac{x^2+5x+4}{x^2-9}$
$=\dfrac{2(x+4)}{x-3}\div\dfrac{(x+1)(x+4)}{(x-3)(x+3)}$
$=\dfrac{2(x+4)}{x-3}\cdot\dfrac{(x-3)(x+3)}{(x+1)(x+4)}$
$=\dfrac{2(x+3)}{x+1},$
$x\neq 3,-1,-4,-3$

15. $\dfrac{x}{x+3}+\dfrac{5}{x-3}$
$=\dfrac{x}{x+3}\cdot\dfrac{x-3}{x-3}+\dfrac{5}{x-3}\cdot\dfrac{x+3}{x+3}$
$=\dfrac{x(x-3)+5(x+3)}{(x+3)(x-3)}$
$=\dfrac{x^2-3x+5x+15}{(x+3)(x-3)}$
$=\dfrac{x^2+2x+15}{(x+3)(x-3)},\ x\neq 3,-3$

16. $\dfrac{2x+3}{x^2-7x+12}-\dfrac{2}{x-3}$
$=\dfrac{2x+3}{(x-3)(x-4)}-\dfrac{2}{x-3}$
$=\dfrac{2x+3}{(x-3)(x-4)}-\dfrac{2}{x-3}\cdot\dfrac{x-4}{x-4}$
$=\dfrac{2x+3-2(x-4)}{(x-3)(x-4)}$
$=\dfrac{2x+3-2(x-4)}{(x-3)(x-4)}$
$=\dfrac{2x+3-2x+8}{(x-3)(x-4)}$
$=\dfrac{11}{(x-3)(x-4)},$
$x\neq 3,\ 4$

17. $\frac{1-\frac{x}{x+2}}{1+\frac{1}{x}} = \frac{\left(1-\frac{x}{x+2}\right)(x+2)x}{\left(1+\frac{1}{x}\right)(x+2)x}$
$= \frac{x(x+2)-x^2}{x(x+2)+(x+2)}$
$= \frac{x^2+2x-x^2}{(x+1)(x+2)}$
$= \frac{2x}{x^2+3x+2}, x \neq 0$

18. $\frac{2x\sqrt{x^2+5}-\frac{2x^3}{\sqrt{x^2+5}}}{x^2+5}$
$= \frac{\left(2x\sqrt{x^2+5}-\frac{2x^3}{\sqrt{x^2+5}}\right)\sqrt{x^2+5}}{(x^2+5)\sqrt{x^2+5}}$
$= \frac{2x(x^2+5)-2x^3}{(x^2+5)\sqrt{x^2+5}}$
$= \frac{2x^3+10x-2x^3}{(x^2+5)\sqrt{x^2+5}}$
$= \frac{10x}{\sqrt{(x^2+5)^3}}$

19. $x^2-9x+18=(x-3)(x-6)$

20. $x^3+2x^2+3x+6=x^2(x+2)+3(x+2)$
$=(x^2+3)(x+2)$

21. $25x^2-9=(5x)^2-3^2=(5x-3)(5x+3)$

22. $36x^2-84x+49=(6x)^2-2(6x)\cdot 7+7^2$
$=(6x-7)^2$

23. $y^3-125=y^3-5^3=(y-5)(y^2+5y+25)$

24. $(x^2+10x+25)-9y^2$
$=(x+5)^2-9y^2$
$=(x+5-3y)(x+5+3y)$

25. $x(x+3)^{-\frac{3}{5}}+(x+3)^{\frac{2}{5}}$
$=(x+3)^{-\frac{3}{5}}\left[x+(x+3)\right]$
$=(x+3)^{-\frac{3}{5}}(2x+3)=\frac{2x+3}{(x+3)^{\frac{3}{5}}}$

26. $-7, -\frac{4}{5}, 0, 0.25, \sqrt{4}, \frac{22}{7}$ are rational numbers.

27. $3(2+5)=3(5+2)$;
commutative property of addition

28. $6(7+4)=6\cdot 7+6\cdot 4$
distributive property of multiplication over addition

29. $0.00076=7.6\times 10^{-4}$

30. $27^{-\frac{5}{3}}=\frac{1}{27^{\frac{5}{3}}}=\frac{1}{\left(\sqrt[3]{27}\right)^5}=\frac{1}{(3)^5}=\frac{1}{243}$

31. $2\left(6.3\times 10^9\right)=12.6\times 10^9=1.26\times 10^{10}$

32. **a.** Model 2 describes data for men and Model 1 describes data for women.

b. $E=0.18t+65$
$E=0.18(50)+65$
$=74$
The model predicts that the life expectancy for men in 2000 was 74 years. This fits the data in the graph fairly well.

c. $E=0.17t+71$
$88=0.17t+71$
$17=0.17t$
$\frac{17}{0.17}=\frac{0.17t}{0.17}$
$100=t$
The life expectancy for women will reach 88 years 100 years after 1950, or 2050.

33. $7(x-2)=4(x+1)-21$
$7x-14=4x+4-21$
$7x-14=4x-17$
$3x=-3$
$x=-1$
The solution set is $\{-1\}$.

34. $\frac{2x-3}{4}=\frac{x-4}{2}-\frac{x+1}{4}$
$2x-3=2(x-4)-(x+1)$
$2x-3=2x-8-x-1$
$2x-3=x-9$
$x=-6$
The solution set is $\{-6\}$.

35. $\frac{2}{x-3}-\frac{4}{x+3}=\frac{8}{(x-3)(x+3)}$

$2(x+3)-4(x-3)=8$

$2x+6-4x+12=8$

$-2x+18=8$

$-2x=-10$

$x=5$

The solution set is $\{5\}$.

36. $2x^2-3x-2=0$

$(2x+1)(x-2)=0$

$2x+1=0$ or $x-2=0$

$x=-\frac{1}{2}$ or $x=2$

The solution set is $\left\{-\frac{1}{2},2\right\}$.

37. $(3x-1)^2=75$

$3x-1=\pm\sqrt{75}$

$3x=1\pm5\sqrt{3}$

$x=\frac{1\pm5\sqrt{3}}{3}$

The solution set is $\left\{\frac{1-5\sqrt{3}}{3},\frac{1+5\sqrt{3}}{3}\right\}$.

38. $x(x-2)=4$

$x^2-2x-4=0$

$x=\frac{-b\pm\sqrt{b^2-4ac}}{2a}$

$x=\frac{2\pm\sqrt{(-2)^2-4(1)(-4)}}{2}$

$x=\frac{2\pm2\sqrt{5}}{2}$

$x=1\pm\sqrt{5}$

The solution set is $\{1-\sqrt{5},1+\sqrt{5}\}$.

39. $\sqrt{x-3}+5=x$

$\sqrt{x-3}=x-5$

$x-3=x^2-10x+25$

$x^2-11x+28=0$

$x=\frac{11\pm\sqrt{11^2-4(1)(28)}}{2(1)}$

$x=\frac{11\pm\sqrt{121-112}}{2}$

$x=\frac{11\pm\sqrt{9}}{2}$

$x=\frac{11\pm3}{2}$

$x=7$ or $x=4$

4 does not check and must be rejected.

The solution set is $\{7\}$.

40. $\sqrt{8-2x}-x=0$

$\sqrt{8-2x}=x$

$\left(\sqrt{8-2x}\right)^2=(x)^2$

$8-2x=x^2$

$0=x^2+2x-8$

$0=(x+4)(x-2)$

$x+4=0$ or $x-2=0$

$x=-4$ $\quad x=2$

–4 does not check and must be rejected.

The solution set is $\{2\}$.

41. $\left|\frac{2}{3}x-6\right|=2$

$\frac{2}{3}x-6=2$ $\quad \frac{2}{3}x-6=-2$

$\frac{2}{3}x=8$ $\quad \frac{2}{3}x=4$

$x=12$ $\quad x=6$

The solution set is $\{6, 12\}$.

42. $-3|4x-7|+15=0$

$-3|4x-7|=-15$

$|4x-7|=5$

$4x-7=5$ or $4x-7=-5$

$4x=12$ $\quad 4x=2$

$x=3$ $\quad x=\frac{1}{2}$

The solution set is $\left\{\frac{1}{2},3\right\}$

43. $\frac{2x}{x^2+6x+8}+\frac{2}{x+2}=\frac{x}{x+4}$

$\frac{2x}{(x+4)(x+2)}+\frac{2}{x+2}=\frac{x}{x+4}$

$\frac{2x(x+4)(x+2)}{(x+4)(x+2)}+\frac{2(x+4)(x+2)}{x+2}=\frac{x(x+4)(x+2)}{x+4}$

$2x+2(x+4)=x(x+2)$

$2x+2x+8=x^2+2x$

$2x+8=x^2$

$0=x^2-2x-8$

$0=(x-4)(x+2)$

$x-4=0$ or $x+2=0$

$x=4$ $\quad x=-2$ (rejected)

The solution set is $\{4\}$.

44. $3(x+4)\ge 5x-12$

$3x+12\ge 5x-12$

$-2x\ge -24$

$x\le 12$

The solution set is $(-\infty,12]$.

12

45. $\frac{x}{6}+\frac{1}{8}\le\frac{x}{2}-\frac{3}{4}$

$4x+3\le 12x-18$

$-8x\le -21$

$x\ge\frac{21}{8}$

The solution set is $\left[\frac{21}{8},\infty\right)$.

$\frac{21}{8}$

46. $-3\le\frac{2x+5}{3}<6$

$-9\le 2x+5<18$

$-14\le 2x<13$

$-7\le x<\frac{13}{2}$

The solution set is $\left[-7,\frac{13}{2}\right)$.

-7 $\quad \frac{13}{2}$

47. $|3x+2|\ge 3$

$3x+2\ge 3$ or $3x+2\le -3$

$3x\ge 1$ $\quad 3x\le -5$

$x\ge\frac{1}{3}$ $\quad x\le -\frac{5}{3}$

The solution set is $\left(-\infty,-\frac{5}{3}\right]\cup\left[\frac{1}{3},\infty\right)$.

$-\frac{5}{3}$ $\quad \frac{1}{3}$

48. $V=\frac{1}{3}lwh$

$3V=lwh$

$\frac{3V}{lw}=\frac{lwh}{lw}$

$\frac{3V}{lw}=h$

$h=\frac{3V}{lw}$

49. $y-y_1=m(x-x_1)$

$y-y_1=mx-mx_1$

$-mx=y_1-mx_1-y$

$\frac{-mx}{-m}=\frac{y_1-mx_1-y}{-m}$

$x=\frac{y-y_1}{m}+x_1$

50.
$$R = \frac{as}{a+s}$$
$$R(a+s) = as$$
$$Ra + Rs = as$$
$$Ra - as = -Rs$$
$$a(R-s) = -Rs$$
$$\frac{a(R-s)}{R-s} = \frac{-Rs}{R-s}$$
$$a = \frac{Rs}{s-R}$$

51.
$$43x + 575 = 1177$$
$$43x = 602$$
$$x = 14$$
The system's income will be \$1177 billion 14 years after 2004, or 2018.

52.
$$B = 0.07x^2 + 47.4x + 500$$
$$1177 = 0.07x^2 + 47.4x + 500$$
$$0 = 0.07x^2 + 47.4x - 677$$
$$0 = 0.07x^2 + 47.4x - 677$$
$$x = \frac{-b \pm \sqrt{b^2 - 4ac}}{2a}$$
$$x = \frac{-(47.4) \pm \sqrt{(47.4)^2 - 4(0.07)(-677)}}{2(0.07)}$$
$x \approx 14$, $x \approx -691$ (rejected)
The system's income will be \$1177 billion 14 years after 2004, or 2018.

53. The formulas model the data quite well.

54. Let $x =$ the number of books in 2002.
Let $x + 62 =$ the number of books in 2003.
Let $x + 190 =$ the number of books in 2004.
$$(x) + (x + 62) + (x + 190) = 2598$$
$$x + x + 62 + x + 190 = 2598$$
$$3x + 252 = 2598$$
$$3x = 2346$$
$$x = 782$$
$$x + 62 = 844$$
$$x + 190 = 972$$
The number of books in 2002, 2003, and 2004 were 782, 844, and 972 respectively.

55.
$$29700 + 150x = 5000 + 1100x$$
$$24700 = 950x$$
$$26 = x$$
In 26 years, the cost will be \$33,600.

56.
$$l = 2w + 4$$
$$A = lw$$
$$48 = (2w + 4)w$$
$$48 = 2w^2 + 4w$$
$$0 = 2w^2 + 4w - 48$$
$$0 = w^2 + 2w - 24$$
$$0 = (w + 6)(w - 4)$$

$w + 6 = 0$ $\quad$ $w - 4 = 0$
$w = -6$ $\quad$ $w = 4$
$$2w + 4 = 2(4) + 4 = 12$$
width is 4 feet, length is 12 feet

57.
$$24^2 + x^2 = 26^2$$
$$576 + x^2 = 676$$
$$x^2 = 100$$
$$x = \pm 10$$
The wire should be attached 10 feet up the pole.

58. Let $x =$ the original selling price
$$20 = x - 0.60x$$
$$20 = 0.40x$$
$$50 = x$$
The original price is \$50.

59.
$$\frac{600,000}{x} - 6000 = \frac{600,000}{x+5}$$
$$x(x+5)\left(\frac{600,000}{x} - 6000\right) = x(x+5)\frac{600,000}{x+5}$$
$$600,000(x+5) - 6000x(x+5) = 600,000x$$
$$600,000x + 3,000,000 - 6000x^2 - 30,000x = 600,000x$$
$$-6000x^2 - 30,000x + 3,000,000 = 0$$
$$x^2 + 5x - 500 = 0$$
$$(x+25)(x-20) = 0$$

$x + 25 = 0$ or $x - 20 = 0$

$x = -25$ $\quad x = 20$

–25 must be rejected. There were originally 20 people.

60. Let x = the number of local calls
The monthly cost using Plan A is $C_A = 25$.
The monthly cost using Plan B is $C_B = 13 + 0.06x$.
For Plan A to be better deal, it must cost less than Plan B.

$C_A < C_B$

$25 < 13 + 0.06x$

$12 < 0.06x$

$200 < x$

$x > 200$

Plan A is a better deal when more than 200 local calls are made per month.

Chapter 1

Section 1.1

Check Point Exercises

1.

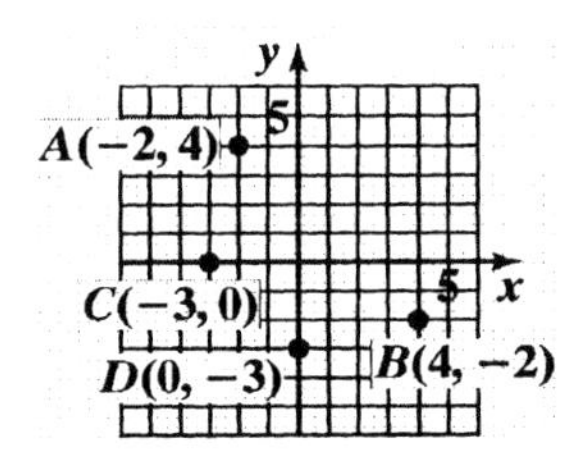

2.

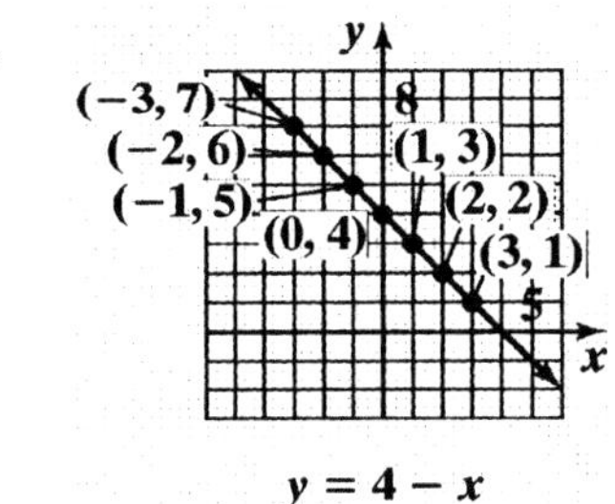

$y = 4 - x$

$x = -3, y = 7$

$x = -2, y = 6$

$x = -1, y = 5$

$x = 0, y = 4$

$x = 1, y = 3$

$x = 2, y = 2$

$x = 3, y = 1$

3.

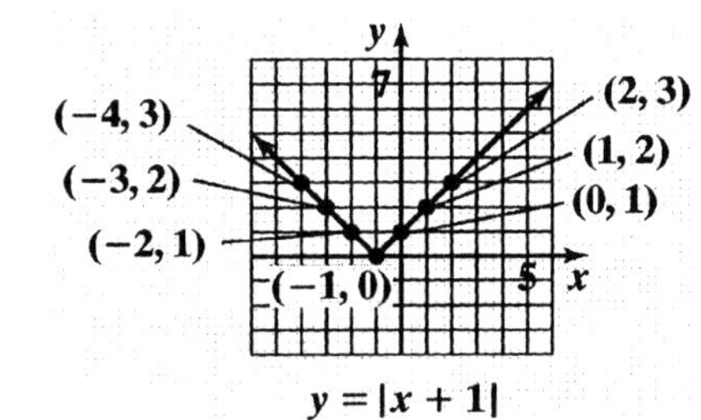

$y = |x + 1|$

$x = -4, y = 3$

$x = -3, y = 2$

$x = -2, y = 1$

$x = -1, y = 0$

$x = 0, y = 1$

$x = 1, y = 2$

$x = 2, y = 3$

4. The meaning of a $[-100, 100, 50]$ by $[-100, 100, 10]$ viewing rectangle is as follows:

$[\underbrace{-100}_{\text{minimum } x\text{-value}}, \underbrace{100}_{\text{maximum } x\text{-value}}, \underbrace{50}_{\text{distance between } x\text{-axis tick marks}}]$

by

$[\underbrace{-100}_{\text{minimum } y\text{-value}}, \underbrace{100}_{\text{maximum } y\text{-value}}, \underbrace{10}_{\text{distance between } y\text{-axis tick marks}}]$

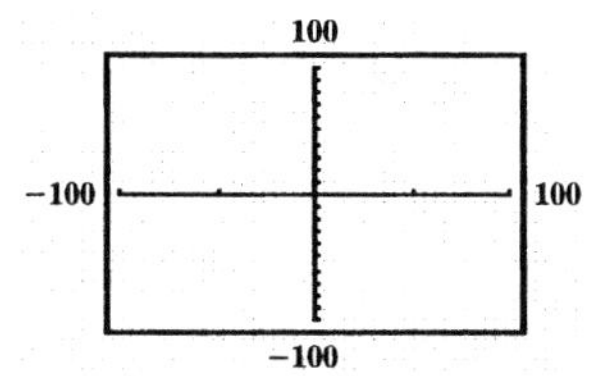

5. **a.** The graph crosses the x-axis at (–3, 0). Thus, the x-intercept is –3. The graph crosses the y-axis at (0, 5). Thus, the y-intercept is 5.

b. The graph does not cross the x-axis. Thus, there is no x-intercept. The graph crosses the y-axis at (0, 4). Thus, the y-intercept is 4.

c. The graph crosses the x- and y-axes at the origin (0, 0). Thus, the x-intercept is 0 and the y-intercept is 0.

6. The number of federal prisoners sentenced for drug offenses in 2003 is about 57% of 159,275. This can be estimated by finding 60% of 160,000.

$$N \approx 60\% \text{ of } 160{,}000$$
$$= 0.60 \times 160{,}000$$
$$= 96{,}000$$

Exercise Set 1.1

1.

y
5
(1, 4)
5 x

2.

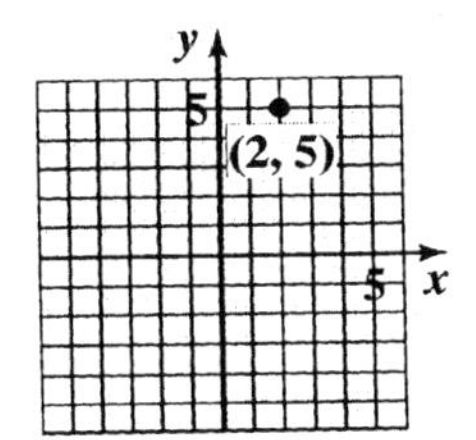

3.

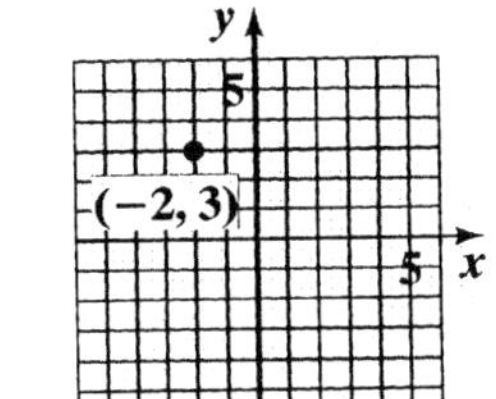

4.

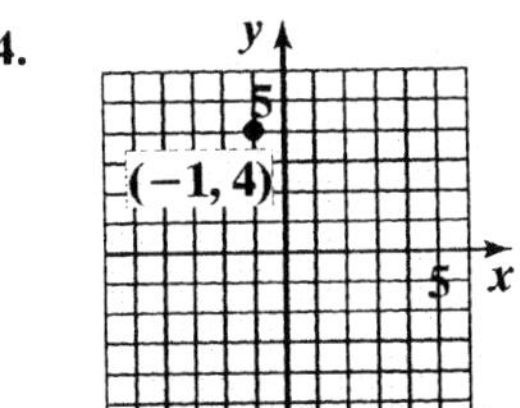

5.

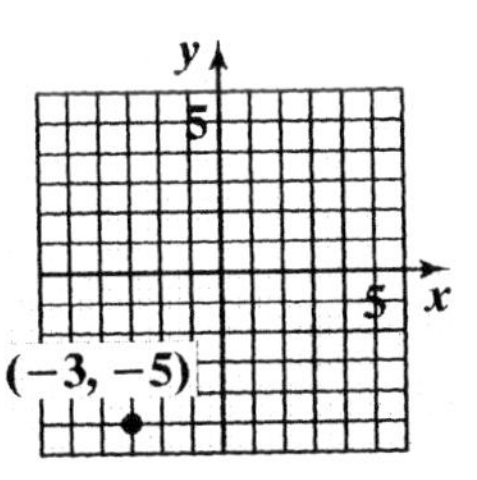

6.

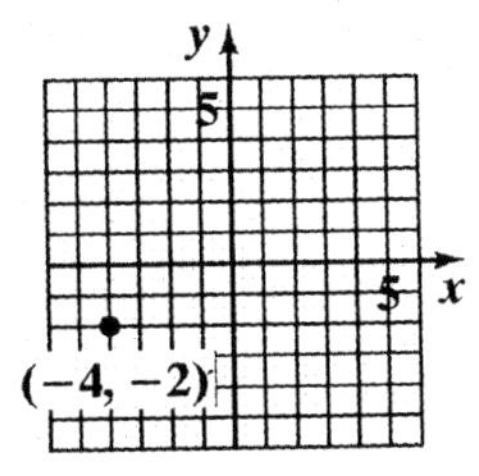

7.

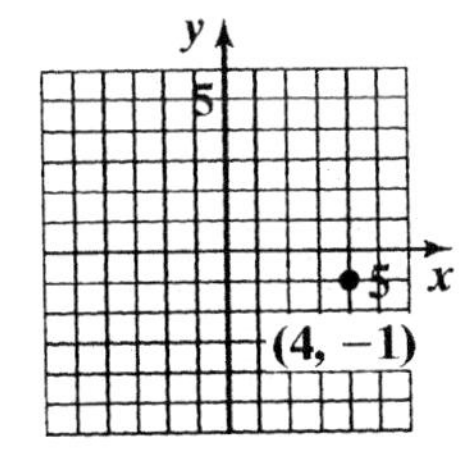

8.

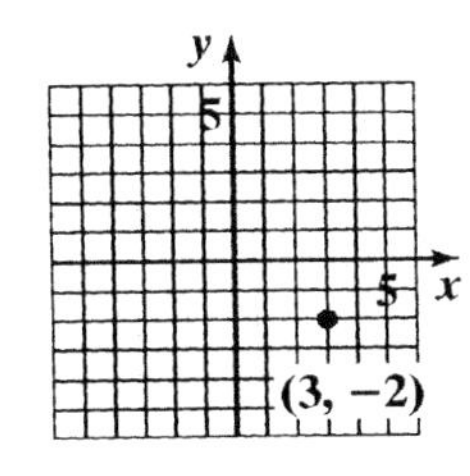

9.

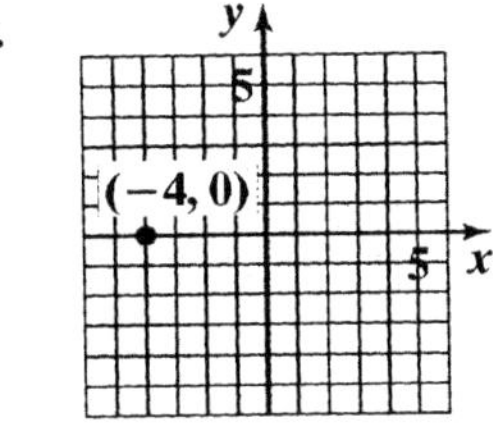

10.

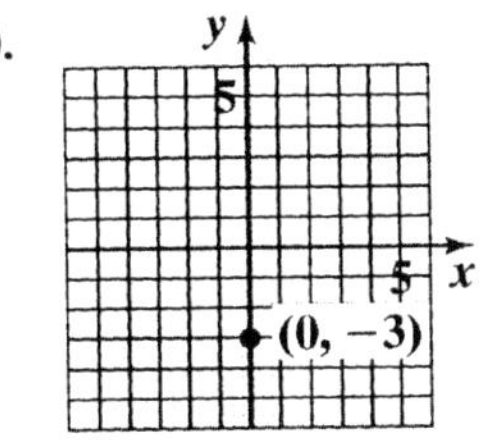

11.

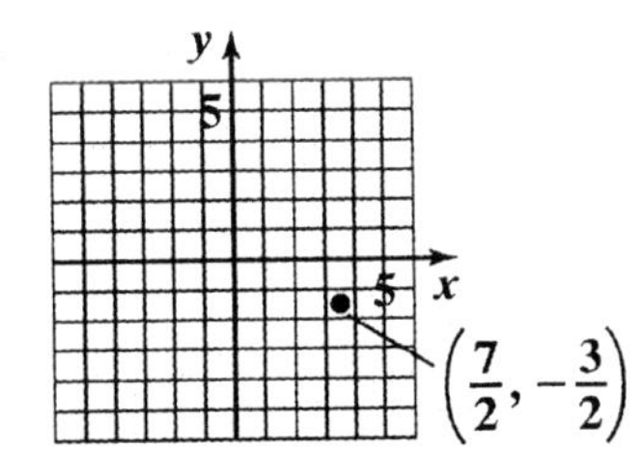

12.

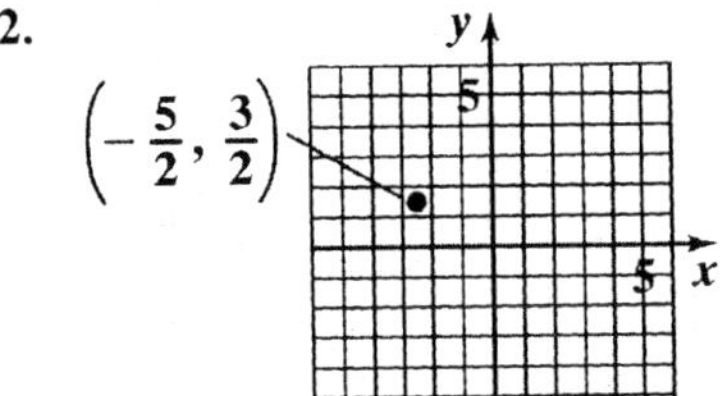

13. 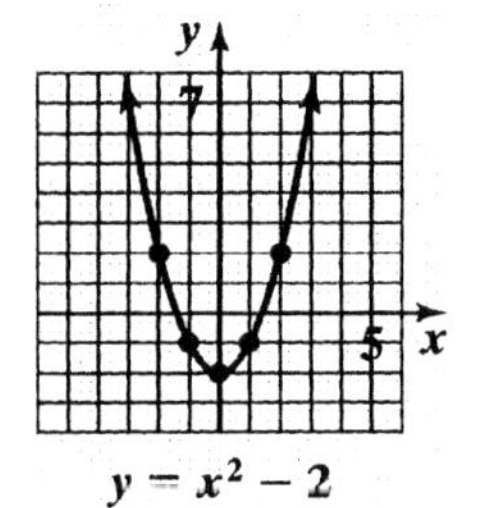

$x = -3, y = 7$

$x = -2, y = 2$

$x = -1, y = -1$

$x = 0, y = -2$

$x = 1, y = -1$

$x = 2, y = 2$

$x = 3, y = 7$

14. 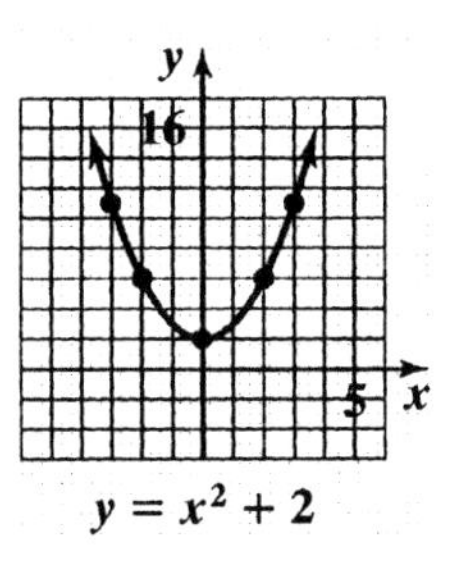

$x = -3, y = 11$
$x = -2, y = 6$
$x = -1, y = 3$
$x = 0, y = 2$
$x = 1, y = 3$
$x = 2, y = 6$
$x = 3, y = 11$

15. 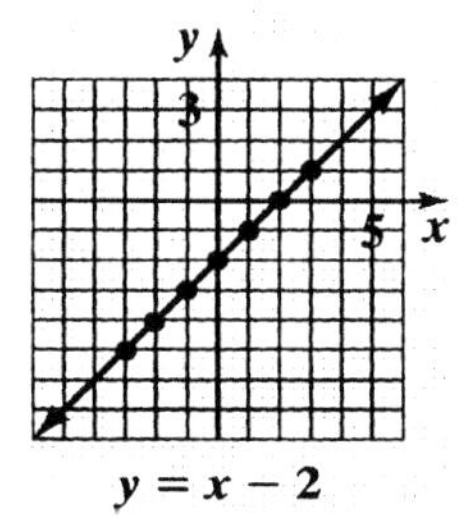

$x = -3, y = -5$

$x = -2, y = -4$

$x = -1, y = -3$

$x = 0, y = -2$

$x = 1, y = -1$

$x = 2, y = 0$

$x = 3, y = 1$

16. 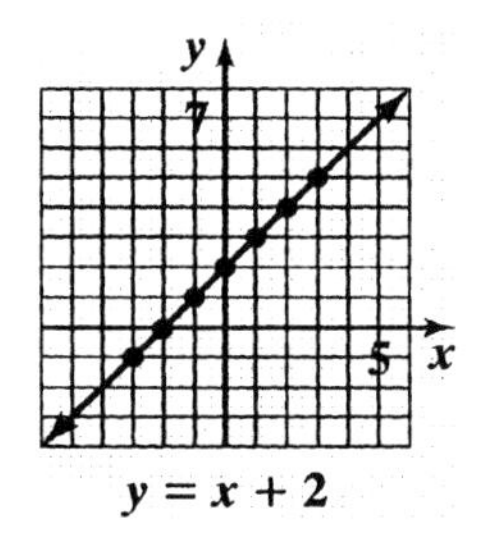

$x = -3, y = -1$
$x = -2, y = 0$
$x = -1, y = 1$
$x = 0, y = 2$
$x = 1, y = 3$
$x = 2, y = 4$
$x = 3, y = 5$

17.

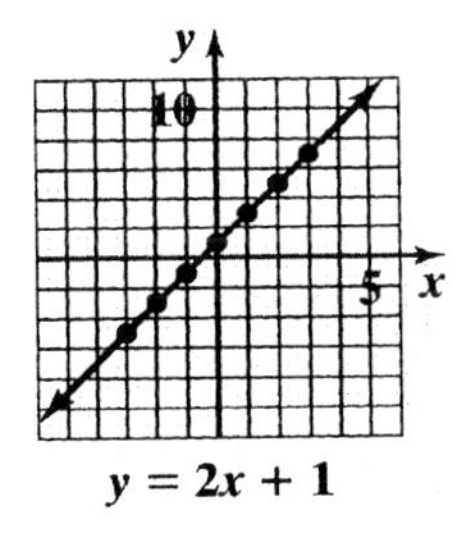

$y = 2x + 1$

$x = -3, y = -5$

$x = -2, y = -3$

$x = -1, y = -1$

$x = 0, y = 1$

$x = 1, y = 3$

$x = 2, y = 5$

$x = 3, y = 7$

18.

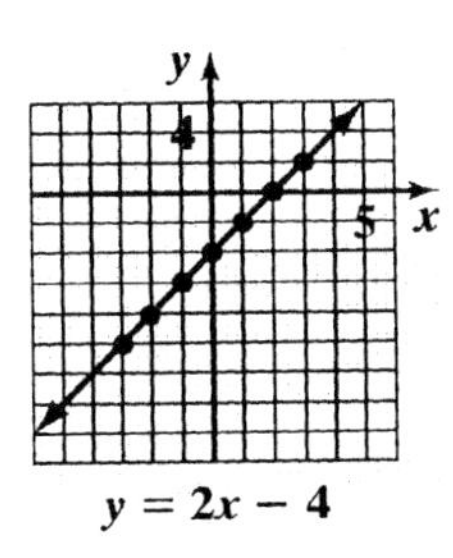

$y = 2x - 4$

$x = -3, y = -10$
$x = -2, y = -8$
$x = -1, y = -6$
$x = 0, y = -4$
$x = 1, y = -2$
$x = 2, y = 0$
$x = 3, y = 2$

19.

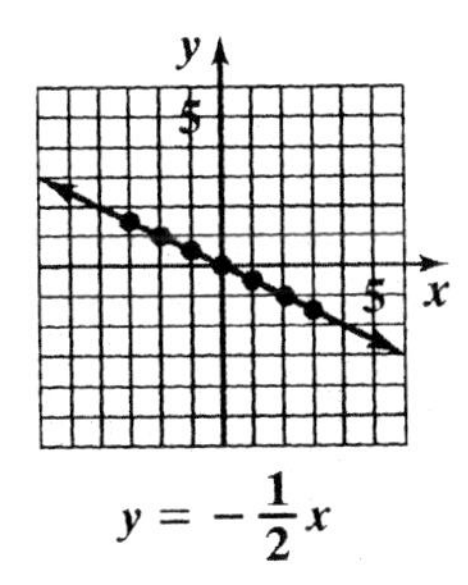

$y = -\frac{1}{2}x$

$x = -3, y = \frac{3}{2}$

$x = -2, y = 1$

$x = -1, y = \frac{1}{2}$

$x = 0, y = 0$

$x = 1, y = -\frac{1}{2}$

$x = 2, y = -1$

$x = 3, y = -\frac{3}{2}$

20.

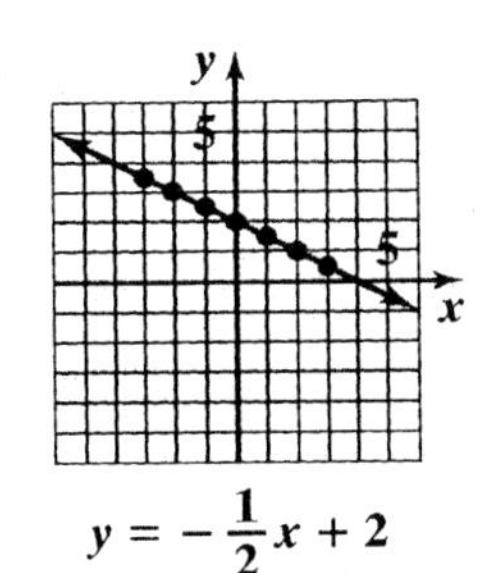

$y = -\frac{1}{2}x + 2$

$x = -3, y = \frac{7}{2}$

$x = -2, y = 3$

$x = -1, y = \frac{5}{2}$

$x = 0, y = 2$

$x = 1, y = \frac{3}{2}$

$x = 2, y = 1$

$x = 3, y = \frac{1}{2}$

21.

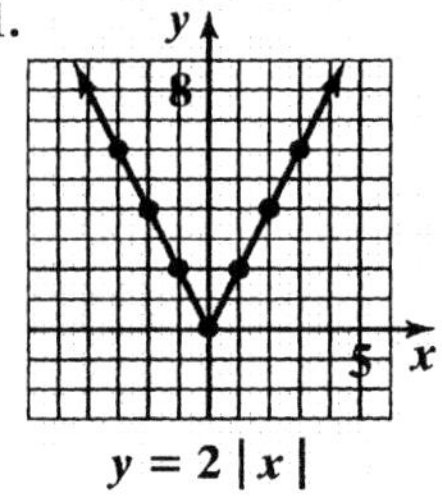

$x = -3, y = 6$

$x = -2, y = 4$

$x = -1, y = 2$

$x = 0, y = 0$

$x = 1, y = 2$

$x = 2, y = 4$

$x = 3, y = 6$

22.

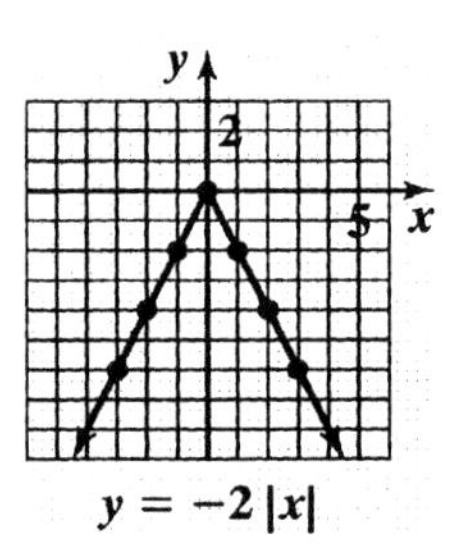

$x = -3, y = -6$

$x = -2, y = -4$

$x = -1, y = -2$

$x = 0, y = 0$

$x = 1, y = -2$

$x = 2, y = -4$

$x = 3, y = -6$

23.

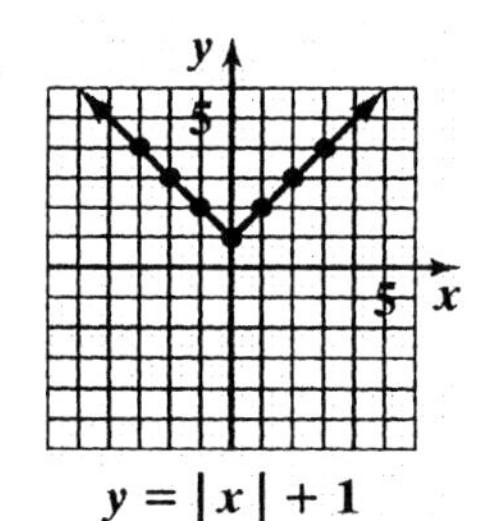

$x = -3, y = 4$

$x = -2, y = 3$

$x = -1, y = 2$

$x = 0, y = 1$

$x = 1, y = 2$

$x = 2, y = 3$

$x = 3, y = 4$

24.

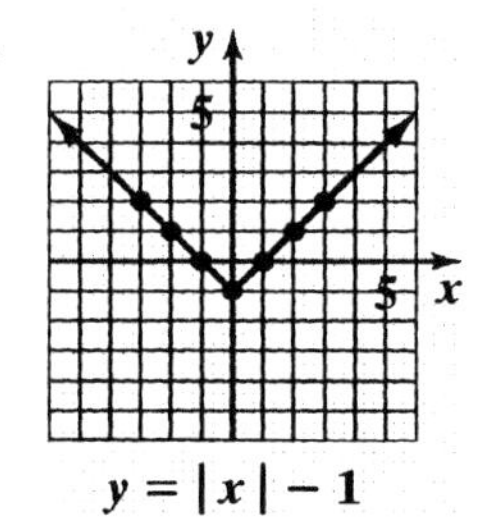

$x = -3, y = 2$

$x = -2, y = 1$

$x = -1, y = 0$

$x = 0, y = -1$

$x = 1, y = 0$

$x = 2, y = 1$

$x = 3, y = 2$

25.

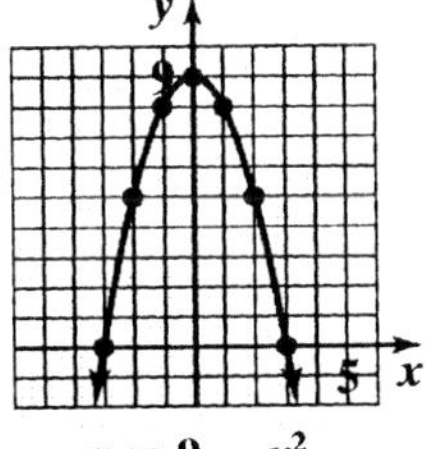

$y = 9 - x^2$

$x = -3, y = 0$

$x = -2, y = 5$

$x = -1, y = 8$

$x = 0, y = 9$

$x = 1, y = 8$

$x = 2, y = 5$

$x = 3, y = 0$

26.

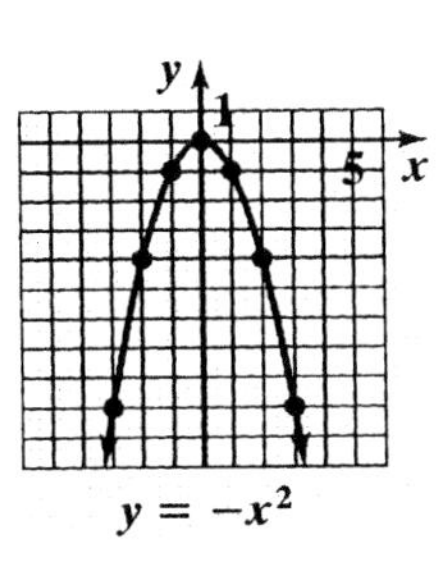

$y = -x^2$

$x = -3, y = -9$

$x = -2, y = -4$

$x = -1, y = -1$

$x = 0, y = 0$

$x = 1, y = -1$

$x = 2, y = -4$

$x = 3, y = -9$

27.

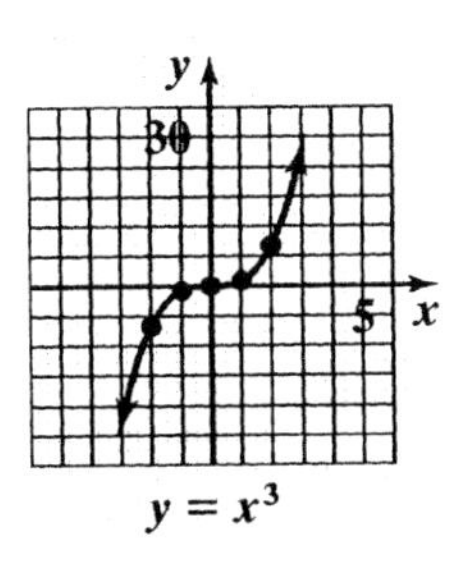

$y = x^3$

$x = -3, y = -27$

$x = -2, y = -8$

$x = -1, y = 1$

$x = 0, y = 0$

$x = 1, y = 1$

$x = 2, y = 8$

$x = 3, y = 27$

28.

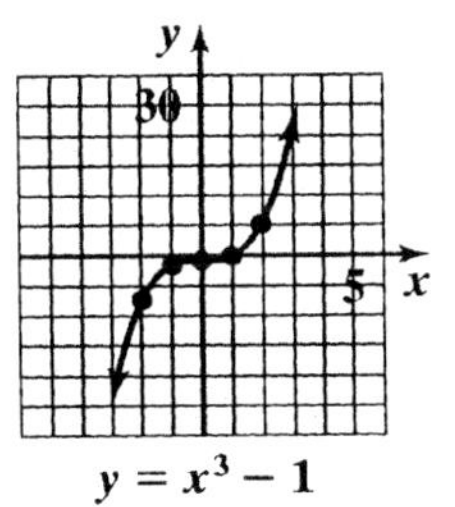

$y = x^3 - 1$

$x = -3, y = -28$

$x = -2, y = -9$

$x = -1, y = -2$

$x = 0, y = -1$

$x = 1, y = 0$

$x = 2, y = 7$

$x = 3, y = 26$

29. (c) x-axis tick marks –5, –4, –3, –2, –1, 0, 1, 2, 3, 4, 5; y-axis tick marks are the same.

30. (d) x-axis tick marks –10, –8, –6, –4, –2, 0, 2, 4, 6, 8, 10; y-axis tick marks –4, –2, 0, 2, 4

31. (b); x-axis tick marks –20, –10, 0, 10, 20, 30, 40, 50, 60, 70, 80; y-axis tick marks –30, –20, –10, 0, 10, 20, 30, 40, 50, 60, 70

32. (a) x-axis tick marks –40, –20, 0, 20, 40; y-axis tick marks –1000, –900, –800, –700, . . . , 700, 800, 900, 1000

33. The equation that corresponds to Y_2 in the table is (c), $y_2 = 2 - x$. We can tell because all of the points $(-3,5)$, $(-2,4)$, $(-1,3)$, $(0,2)$, $(1,1)$, $(2,0)$, and $(3,-1)$ are on the line $y = 2 - x$, but all are not on any of the others.

34. The equation that corresponds to Y_1 in the table is (b), $y_1 = x^2$. We can tell because all of the points $(-3,9)$, $(-2,4)$, $(-1,1)$, $(0,0)$, $(1,1)$, $(2,4)$, and $(3,9)$ are on the graph $y = x^2$, but all are not on any of the others.

35. No. It passes through the point $(0,2)$.

36. Yes. It passes through the point $(0,0)$.

37. $(2,0)$

38. $(0, 2)$

39. The graphs of Y_1 and Y_2 intersect at the points $(-2, 4)$ and $(1, 1)$.

40. The values of Y_1 and Y_2 are the same when $x = -2$ and $x = 1$.

41. **a.** 2; The graph intersects the x-axis at (2, 0).

b. –4; The graph intersects the y-axis at (0,–4).

42. **a.** 1; The graph intersects the x-axis at (1, 0).

b. 2; The graph intersects the y-axis at (0, 2).

43. **a.** 1, –2; The graph intersects the x-axis at (1, 0) and (–2, 0).

b. 2; The graph intersects the y-axis at (0, 2).

44. **a.** 1, –1; The graph intersects the x-axis at (1, 0) and (–1, 0).

b. 1; The graph intersect the y-axis at (0, 1).

45. **a.** –1; The graph intersects the x-axis at (–1, 0).

b. none; The graph does not intersect the y-axis.

46. **a.** none; The graph does not intersect the x-axis.

b. 2; The graph intersects the y-axis at (0, 2).

47.

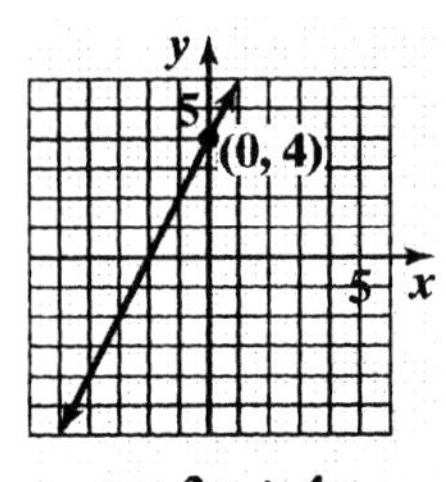

$y = 2x + 4$

48.

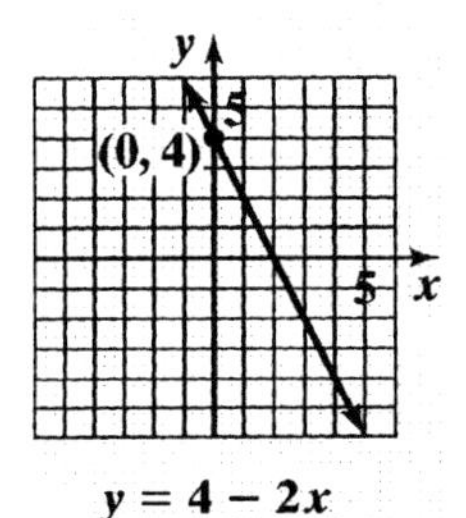

$y = 4 - 2x$

49.

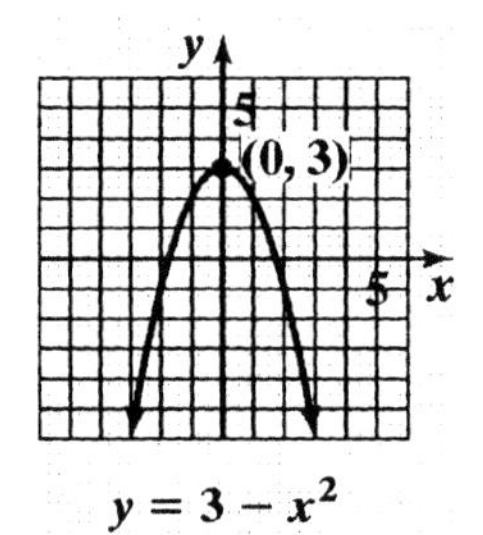

$y = 3 - x^2$

50.

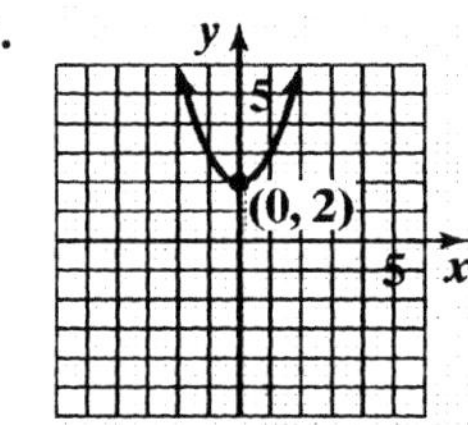

$y = x^2 + 2$

51.

x	(x, y)
–3	(–3, 5)
–2	(–2, 5)
–1	(–1, 5)
0	(0, 5)
1	(1, 5)
2	(2, 5)
3	(3, 5)

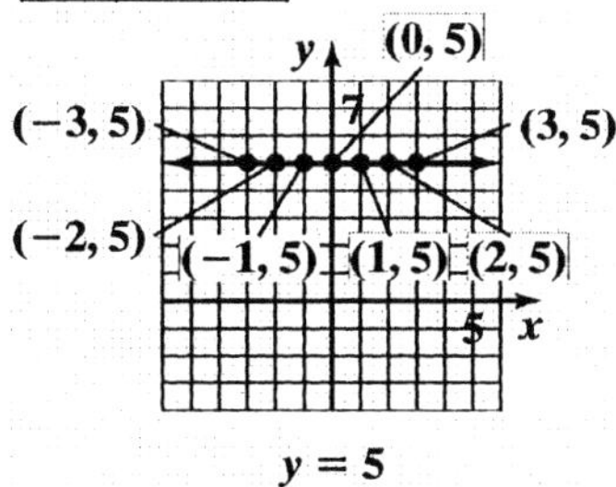

$y = 5$

52.

x	(x, y)
-3	$(-3, -1)$
-2	$(-2, -1)$
-1	$(-1, -1)$
0	$(0, -1)$
1	$(1, -1)$
2	$(2, -1)$
3	$(3, -1)$

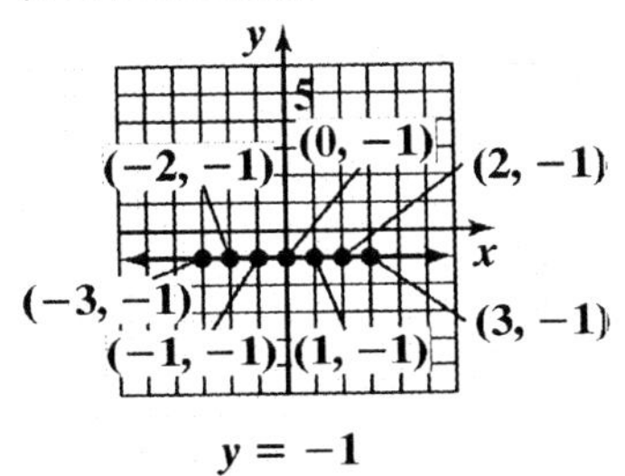

$y = -1$

53.

x	(x, y)
-2	$\left(-2, -\frac{1}{2}\right)$
-1	$(-1, -1)$
$-\frac{1}{2}$	$\left(-\frac{1}{2}, -2\right)$
$-\frac{1}{3}$	$\left(-\frac{1}{3}, -3\right)$
$\frac{1}{3}$	$\left(\frac{1}{3}, 3\right)$
$\frac{1}{2}$	$\left(\frac{1}{2}, 2\right)$
1	$(1, 1)$
2	$\left(2, \frac{1}{2}\right)$

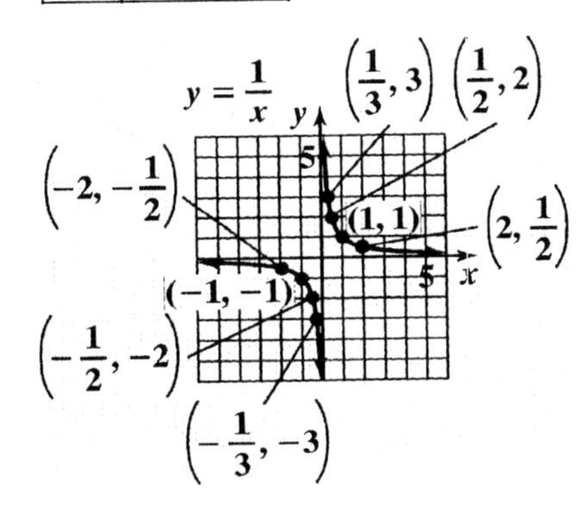

54.

x	(x, y)
-2	$\left(-2, \frac{1}{2}\right)$
-1	$(-1, 1)$
$-\frac{1}{2}$	$\left(-\frac{1}{2}, 2\right)$
$-\frac{1}{3}$	$\left(-\frac{1}{3}, 3\right)$
$\frac{1}{3}$	$\left(\frac{1}{3}, -3\right)$
$\frac{1}{2}$	$\left(\frac{1}{2}, -2\right)$
1	$(1, -1)$
2	$\left(2, -\frac{1}{2}\right)$

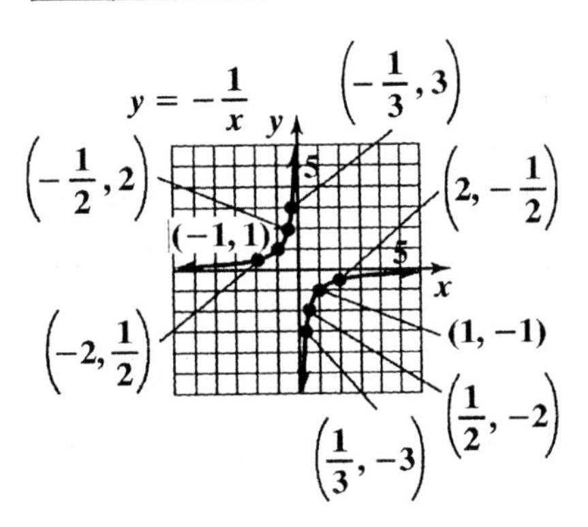

55. There were approximately 65 democracies in 1989.

56. There were $120 - 40 = 80$ more democracies in 2002 than in 1973.

57. The number of democracies increased at the greatest rate between 1989 and 1993.

58. The number of democracies increased at the slowest rate between 1981 and 1985.

59. There were 49 democracies in 1977.

60. There were 110 democracies in 1997.

61. $R = 165 - 0.75A;\ \ A = 40$

$R - 165 - 0.75A = 165 - 0.75(40)$

$= 165 - 30 = 135$

The desirable heart rate during exercise for a 40-year old man is 135 beats per minute. This corresponds to the point (40, 135) on the blue graph.

62. $R = 143 - 0.65A;\ \ A = 40$

$R - 143 - 0.65A = 143 - 0.65(40)$

$= 143 - 26 = 117$

The desirable heart rate during exercise for a 40-year old woman is 117 beats per minute. This corresponds to the point (40, 117) on the red graph.

63. **a.** At birth we have $x = 0$.

$y = 2.9\sqrt{x} + 36$

$= 2.9\sqrt{0} + 36$

$= 2.9(0) + 36$

$= 36$

According to the model, the head circumference at birth is 36 cm.

b. At 9 months we have $x = 9$.

$y = 2.9\sqrt{x} + 36$

$= 2.9\sqrt{9} + 36$

$= 2.9(3) + 36$

$= 44.7$

According to the model, the head circumference at 9 months is 44.7 cm.

c. At 14 months we have $x = 14$.

$y = 2.9\sqrt{x} + 36$

$= 2.9\sqrt{14} + 36$

≈ 46.9

According to the model, the head circumference at 14 months is roughly 46.9 cm.

d. The model describes healthy children.

64. **a.** At birth we have $x = 0$.

$y = 4\sqrt{x} + 35$

$= 4\sqrt{0} + 35$

$= 4(0) + 35$

$= 35$

According to the model, the head circumference at birth is 35 cm.

b. At 9 months we have $x = 9$.

$y = 4\sqrt{x} + 35$

$= 4\sqrt{9} + 35$

$= 4(3) + 35$

$= 47$

According to the model, the head circumference at 9 months is 47 cm.

c. At 14 months we have $x = 14$.

$y = 4\sqrt{x} + 35$

$= 4\sqrt{14} + 35$

≈ 50

According to the model, the head circumference at 14 months is roughly 50 cm.

d. The model describes severe autistic children.

71. $y = 45.48x^2 - 334.35x + 1237.9$

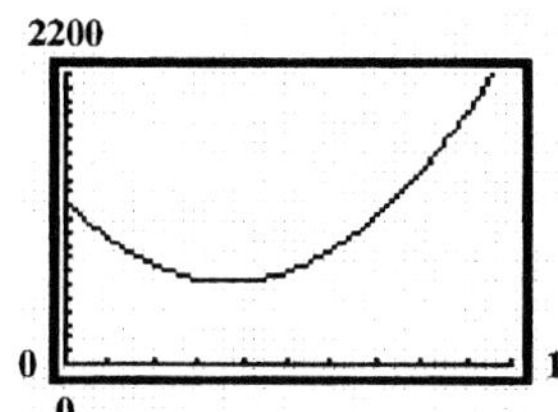

The discharges decreased from 1990 to 1994, but started to increase after 1994. The policy was not a success.

72. **a.** False; (x, y) can be in quadrant III.

b. False; when $x = 2$ and $y = 5$, $3y - 2x = 3(5) - 2(2) = 11$.

c. False; if a point is on the x-axis, $y = 0$.

d. True; all of the above are false. (d) is true.

73. (a)

74. (d)

75. (b)

76. (c)

77. (b)

78. (a)

Section 1.2

Check Point Exercises

1. The domain is the set of all first components: $\{5, 10, 15, 20, 25\}$. The range is the set of all second components: $\{12.8, 16.2, 18.9, 20.7, 21.8\}$.

2. **a.** The relation is not a function since the two ordered pairs (5, 6) and (5, 8) have the same first component but different second components.

b. The relation is a function since no two ordered pairs have the same first component and different second components.

3. **a.** $2x + y = 6$

$y = -2x + 6$

For each value of x, there is one and only one value for y, so the equation defines y as a function of x.

b. $x^2 + y^2 = 1$

$y^2 = 1 - x^2$

$y = \pm\sqrt{1 - x^2}$

Since there are values of x (all values between -1 and 1 exclusive) that give more than one value for y (for example, if $x = 0$, then $y = \pm\sqrt{1 - 0^2} = \pm 1$), the equation does not define y as a function of x.

4. **a.** $f(-5) = (-5)^2 - 2(-5) + 7$

$= 25 - (-10) + 7$

$= 42$

b. $f(x+4) = (x+4)^2 - 2(x+4) + 7$

$= x^2 + 8x + 16 - 2x - 8 + 7$

$= x^2 + 6x + 15$

c. $f(-x) = (-x)^2 - 2(-x) + 7$

$= x^2 - (-2x) + 7$

$= x^2 + 2x + 7$

5.

x	$f(x) = 2x$	(x, y)
-2	-4	$(-2, -4)$
-1	-2	$(-1, -2)$
0	0	$(0, 0)$
1	2	$(1, 2)$
2	4	$(2, 4)$

x	$g(x)=2x-3$	(x,y)
-2	$g(-2)=2(-2)-3=-7$	$(-2,-7)$
-1	$g(-1)=2(-1)-3=-5$	$(-1,-5)$
0	$g(0)=2(0)-3=-3$	$(0,-3)$
1	$g(1)=2(1)-3=-1$	$(1,-1)$
2	$g(2)=2(2)-3=1$	$(2,1)$

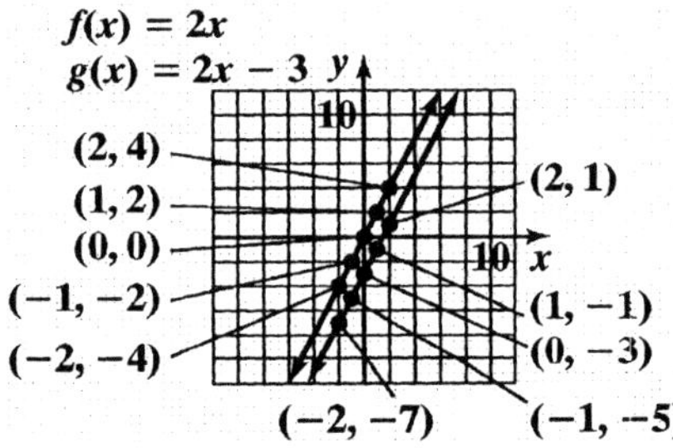

The graph of g is the graph of f shifted down 3 units.

6. The graph (c) fails the vertical line test and is therefore not a function. y is a function of x for the graphs in (a) and (b).

7. **a.** $f(10) \approx 16$ **b.** $x \approx 8$

8. **a.** Domain $= \{x \mid -2 \le x \le 1\}$ or $[-2,1]$.
Range $= \{y \mid 0 \le y \le 3\}$ or $[0,3]$.

b. Domain $= \{x \mid -2 < x \le 1\}$ or $(-2,1]$.
Range $= \{y \mid -1 \le y < 2\}$ or $[-1,2)$.

c. Domain $= \{x \mid -3 \le x < 0\}$ or $[-3,0)$.
Range $= \{y \mid y = -3,-2,-1\}$.

Exercise Set 1.2

1. The relation is a function since no two ordered pairs have the same first component and different second components. The domain is {1, 3, 5} and the range is {2, 4, 5}.

2. The relation is a function because no two ordered pairs have the same first component and different second components The domain is {4, 6, 8} and the range is {5, 7, 8}.

3. The relation is not a function since the two ordered pairs (3, 4) and (3, 5) have the same first component but different second components (the same could be said for the ordered pairs (4, 4) and (4, 5)). The domain is {3, 4} and the range is {4, 5}.

4. The relation is not a function since the two ordered pairs (5, 6) and (5, 7) have the same first component but different second components (the same could be said for the ordered pairs (6, 6) and (6, 7)). The domain is {5, 6} and the range is {6, 7}.

5. The relation is a function because no two ordered pairs have the same first component and different second components The domain is
$\{3, 4, 5, 7\}$ and the range is $\{-2, 1, 9\}$.

6. The relation is a function because no two ordered pairs have the same first component and different second components The domain is
$\{-2, -1, 5, 10\}$ and the range is $\{1, 4, 6\}$.

7. The relation is a function since there are no same first components with different second components. The domain is $\{3, 2, 1, 0\}$ and the range is $\{-3, -2, -1, 0\}$

8. The relation is a function since there are no ordered pairs that have the same first component but different second components. The domain is $\{-7, -5, -3, 0\}$ and the range is $\{-7, -5, -3, 0\}$.

9. The relation is not a function since there are ordered pairs with the same first component and different second components. The domain is $\{1\}$ and the range is $\{4, 5, 6\}$.

10. The relation is a function since there are no two ordered pairs that have the same first component and different second components. The domain is
$\{4, 5, 6\}$ and the range is $\{1\}$.

11. $x + y = 16$
$y = 16 - x$
Since only one value of y can be obtained for each value of x, y is a function of x.

12. $x + y = 25$
$y = 25 - x$
Since only one value of y can be obtained for each value of x, y is a function of x.

13. $x^2 + y = 16$
$y = 16 - x^2$
Since only one value of y can be obtained for each value of x, y is a function of x.

14. $x^2 + y = 25$
$y = 25 - x^2$
Since only one value of y can be obtained for each value of x, y is a function of x.

15. $x^2 + y^2 = 16$
$y^2 = 16 - x^2$
$y = \pm\sqrt{16 - x^2}$
If $x = 0$, $y = \pm 4$.
Since two values, $y = 4$ and $y = -4$, can be obtained for one value of x, y is not a function of x.

16. $x^2 + y^2 = 25$
$y^2 = 25 - x^2$
$y = \pm\sqrt{25 - x^2}$
If $x = 0$, $y = \pm 5$.
Since two values, $y = 5$ and $y = -5$, can be obtained for one value of x, y is not a function of x.

17. $x = y^2$

$y = \pm\sqrt{x}$

If $x = 1$, $y = \pm 1$.

Since two values, $y = 1$ and $y = -1$, can be obtained for $x = 1$, y is not a function of x.

18. $4x = y2$

$y = \pm\sqrt{4x} = \pm 2\sqrt{x}$

If $x = 1$, then $y = \pm 2$.

Since two values, $y = 2$ and $y = -2$, can be obtained for $x = 1$, y is not a function of x.

19. $y = \sqrt{x+4}$

Since only one value of y can be obtained for each value of x, y is a function of x.

20. $y = -\sqrt{x+4}$

Since only one value of y can be obtained for each value of x, y is a function of x.

21. $x + y^3 = 8$

$y^3 = 8 - x$

$y = \sqrt[3]{8-x}$

Since only one value of y can be obtained for each value of x, y is a function of x.

22. $x + y^3 = 27$

$y^3 = 27 - x$

$y = \sqrt[3]{27-x}$

Since only one value of y can be obtained for each value of x, y is a function of x.

23. $xy + 2y = 1$

$y(x+2) = 1$

$y = \dfrac{1}{x+2}$

Since only one value of y can be obtained for each value of x, y is a function of x.

24. $xy - 5y = 1$

$y(x-5) = 1$

$y = \dfrac{1}{x-5}$

Since only one value of y can be obtained for each value of x, y is a function of x.

25. $|x| - y = 2$

$-y = -|x| + 2$

$y = |x| - 2$

Since only one value of y can be obtained for each value of x, y is a function of x.

26. $|x| - y = 5$
$$-y = -|x| + 5$$
$$y = |x| - 5$$
Since only one value of y can be obtained for each value of x, y is a function of x.

27. **a.** $f(6) = 4(6) + 5 = 29$

b. $f(x + 1) = 4(x + 1) + 5 = 4x + 9$

c. $f(-x) = 4(-x) + 5 = -4x + 5$

28. **a.** $f(4) = 3(4) + 7 = 19$

b. $f(x + 1) = 3(x + 1) + 7 = 3x + 10$

c. $f(-x) = 3(-x) + 7 = -3x + 7$

29. **a.**
$$\begin{aligned} g(-1) &= (-1)^2 + 2(-1) + 3 \\ &= 1 - 2 + 3 \\ &= 2 \end{aligned}$$

b.
$$\begin{aligned} g(x+5) &= (x+5)^2 + 2(x+5) + 3 \\ &= x^2 + 10x + 25 + 2x + 10 + 3 \\ &= x^2 + 12x + 38 \end{aligned}$$

c.
$$\begin{aligned} g(-x) &= (-x)^2 + 2(-x) + 3 \\ &= x^2 - 2x + 3 \end{aligned}$$

30. **a.**
$$\begin{aligned} g(-1) &= (-1)^2 - 10(-1) - 3 \\ &= 1 + 10 - 3 \\ &= 8 \end{aligned}$$

b.
$$\begin{aligned} g(x+2) &= (x+2)^2 - 10(8+2) - 3 \\ &= x^2 + 4x + 4 - 10x - 20 - 3 \\ &= x^2 - 6x - 19 \end{aligned}$$

c.
$$\begin{aligned} g(-x) &= (-x)^2 - 10(-x) - 3 \\ &= x^2 + 10x - 3 \end{aligned}$$

31. **a.**
$$\begin{aligned} h(2) &= 2^4 - 2^2 + 1 \\ &= 16 - 4 + 1 \\ &= 13 \end{aligned}$$

b.
$$\begin{aligned} h(-1) &= (-1)^4 - (-1)^2 + 1 \\ &= 1 - 1 + 1 \\ &= 1 \end{aligned}$$

c. $h(-x) = (-x)^4 - (-x)^2 + 1 = x^4 - x^2 + 1$

d. $h(3a) = (3a)^4 - (3a)^2 + 1$
$= 81a^4 - 9a^2 + 1$

32. a. $h(3) = 3^3 - 3 + 1 = 25$

b. $h(-2) = (-2)^3 - (-2) + 1$
$= -8 + 2 + 1$
$= -5$

c. $h(-x) = (-x)^3 - (-x) + 1 = -x^3 + x + 1$

d. $h(3a) = (3a)^3 - (3a) + 1$
$= 27a^3 - 3a + 1$

33. a. $f(-6) = \sqrt{-6+6} + 3 = \sqrt{0} + 3 = 3$

b. $f(10) = \sqrt{10+6} + 3$
$= \sqrt{16} + 3$
$= 4 + 3$
$= 7$

c. $f(x-6) = \sqrt{x-6+6} + 3 = \sqrt{x} + 3$

34. a. $f(16) = \sqrt{25-16} - 6 = \sqrt{9} - 6 = 3 - 6 = -3$

b. $f(-24) = \sqrt{25-(-24)} - 6$
$= \sqrt{49} - 6$
$= 7 - 6 = 1$

c. $f(25-2x) = \sqrt{25-(25-2x)} - 6$
$= \sqrt{2x} - 6$

35. a. $f(2) = \dfrac{4(2)^2 - 1}{2^2} = \dfrac{15}{4}$

b. $f(-2) = \dfrac{4(-2)^2 - 1}{(-2)^2} = \dfrac{15}{4}$

c. $f(-x) = \dfrac{4(-x)^2 - 1}{(-x)^2} = \dfrac{4x^2 - 1}{x^2}$

36. a. $f(2) = \dfrac{4(2)^3 + 1}{2^3} = \dfrac{33}{8}$

b. $f(-2) = \dfrac{4(-2)^3 + 1}{(-2)^3} = \dfrac{-31}{-8} = \dfrac{31}{8}$

c. $f(-x) = \frac{4(-x)^3 + 1}{(-x)^3} = \frac{-4x^3 + 1}{-x^3}$

or $\frac{4x^3 - 1}{x^3}$

37. a. $f(6) = \frac{6}{|6|} = 1$

b. $f(-6) = \frac{-6}{|-6|} = \frac{-6}{6} = -1$

c. $f(r^2) = \frac{r^2}{|r^2|} = \frac{r^2}{r^2} = 1$

38. a. $f(5) = \frac{|5+3|}{5+3} = \frac{|8|}{8} = 1$

b. $f(-5) = \frac{|-5+3|}{-5+3} = \frac{|-2|}{-2} = \frac{2}{-2} = -1$

c. $f(-9-x) = \frac{|-9-x+3|}{-9-x+3}$

$= \frac{|-x-6|}{-x-6} = \begin{cases} 1, \text{ if } x < -6 \\ -1, \text{ if } x > -6 \end{cases}$

39.

x	$f(x) = x$	(x, y)
-2	$f(-2) = -2$	$(-2, -2)$
-1	$f(-1) = -1$	$(-1, -1)$
0	$f(0) = 0$	$(0, 0)$
1	$f(1) = 1$	$(1, 1)$
2	$f(2) = 2$	$(2, 2)$

x	$g(x) = x + 3$	(x, y)
-2	$g(-2) = -2 + 3 = 1$	$(-2, 1)$
-1	$g(-1) = -1 + 3 = 2$	$(-1, 2)$
0	$g(0) = 0 + 3 = 3$	$(0, 3)$
1	$g(1) = 1 + 3 = 4$	$(1, 4)$
2	$g(2) = 2 + 3 = 5$	$(2, 5)$

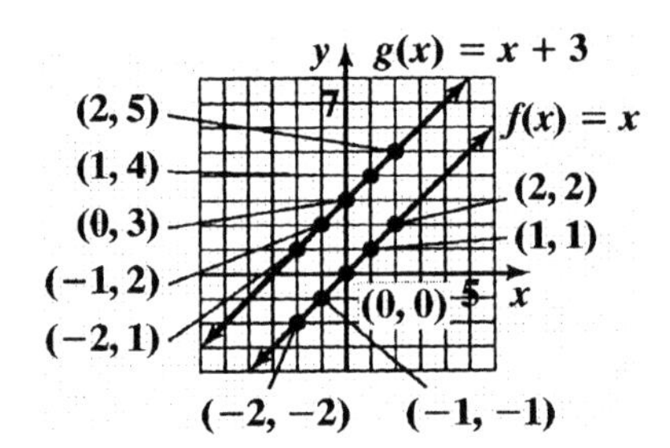

The graph of g is the graph of f shifted up 3 units.

40.

x	$f(x)=x$	(x,y)
–2	$f(-2)=-2$	$(-2,-2)$
–1	$f(-1)=-1$	$(-1,-1)$
0	$f(0)=0$	$(0,0)$
1	$f(1)=1$	$(1,1)$
2	$f(2)=2$	$(2,2)$

x	$g(x)=x-4$	(x,y)
–2	$g(-2)=-2-4=-6$	$(-2,-6)$
–1	$g(-1)=-1-4=-5$	$(-1,-5)$
0	$g(0)=0-4=-4$	$(0,-4)$
1	$g(1)=1-4=-3$	$(1,-3)$
2	$g(2)=2-4=-2$	$(2,-2)$

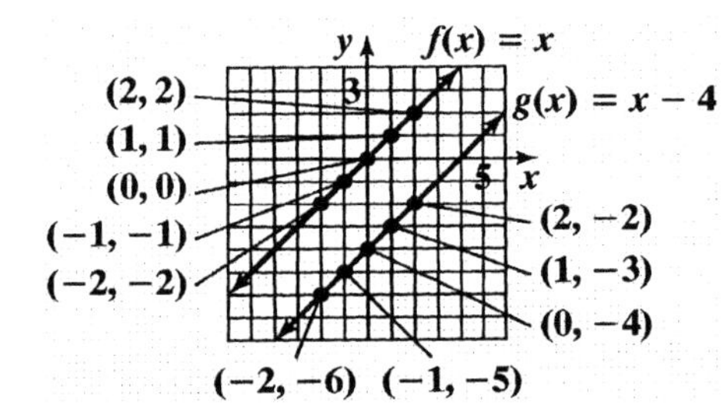

The graph of g is the graph of f shifted down 4 units.

41.

x	$f(x)=-2x$	(x,y)
–2	$f(-2)=-2(-2)=4$	$(-2,4)$
–1	$f(-1)=-2(-1)=2$	$(-1,2)$
0	$f(0)=-2(0)=0$	$(0,0)$
1	$f(1)=-2(1)=-2$	$(1,-2)$
2	$f(2)=-2(2)=-4$	$(2,-4)$

x	$g(x) = -2x - 1$	(x, y)
–2	$g(-2) = -2(-2) - 1 = 3$	$(-2, 3)$
–1	$g(-1) = -2(-1) - 1 = 1$	$(-1, 1)$
0	$g(0) = -2(0) - 1 = -1$	$(0, -1)$
1	$g(1) = -2(1) - 1 = -3$	$(1, -3)$
2	$g(2) = -2(2) - 1 = -5$	$(2, -5)$

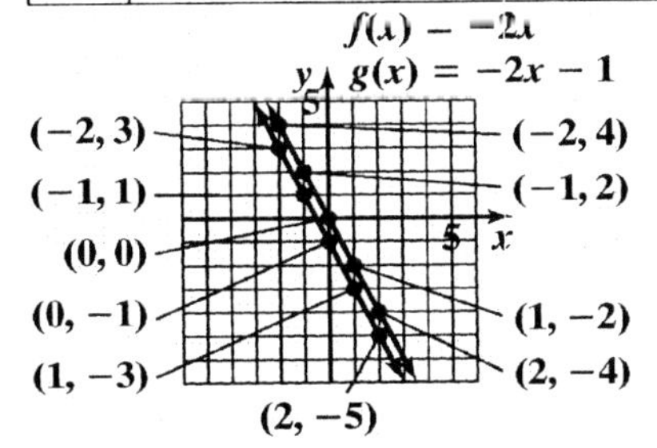

The graph of g is the graph of f shifted down 1 unit.

42.

x	$f(x) = -2x$	(x, y)
–2	$f(-2) = -2(-2) = 4$	$(-2, 4)$
–1	$f(-1) = -2(-1) = 2$	$(-1, 2)$
0	$f(0) = -2(0) = 0$	$(0, 0)$
1	$f(1) = -2(1) = -2$	$(1, -2)$
2	$f(2) = -2(2) = -4$	$(2, -4)$

x	$g(x) = -2x + 3$	(x, y)
–2	$g(-2) = -2(-2) + 3 = 7$	$(-2, 7)$
–1	$g(-1) = -2(-1) + 3 = 5$	$(-1, 5)$
0	$g(0) = -2(0) + 3 = 3$	$(0, 3)$
1	$g(1) = -2(1) + 3 = 1$	$(1, 1)$
2	$g(2) = -2(2) + 3 = -1$	$(2, -1)$

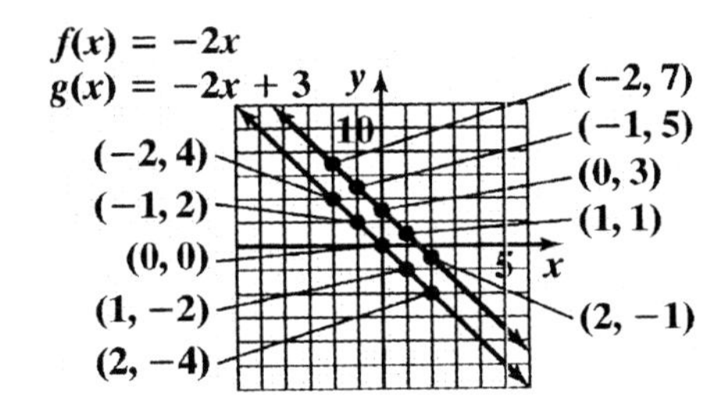

The graph of g is the graph of f shifted up 3 units.

43.

x	$f(x)=x^2$	(x,y)
–2	$f(-2)=(-2)^2=4$	$(-2,4)$
–1	$f(-1)=(-1)^2=1$	$(-1,1)$
0	$f(0)=(0)^2=0$	$(0,0)$
1	$f(1)=(1)^2=1$	$(1,1)$
2	$f(2)=(2)^2=4$	$(2,4)$

x	$g(x)=x^2+1$	(x,y)
–2	$g(-2)=(-2)^2+1=5$	$(-2,5)$
–1	$g(-1)=(-1)^2+1=2$	$(-1,2)$
0	$g(0)=(0)^2+1=1$	$(0,1)$
1	$g(1)=(1)^2+1=2$	$(1,2)$
2	$g(2)=(2)^2+1=5$	$(2,5)$

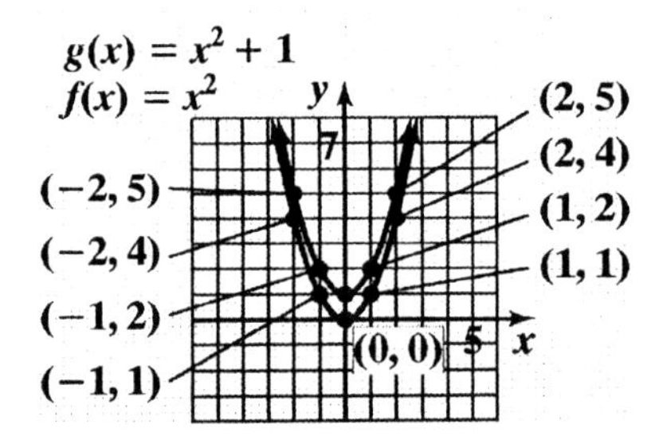

The graph of g is the graph of f shifted up 1 unit.

44.

x	$f(x)=x^2$	(x,y)
–2	$f(-2)=(-2)^2=4$	$(-2,4)$
–1	$f(-1)=(-1)^2=1$	$(-1,1)$
0	$f(0)=(0)^2=0$	$(0,0)$
1	$f(1)=(1)^2=1$	$(1,1)$
2	$f(2)=(2)^2=4$	$(2,4)$

x	$g(x)=x^2-2$	(x,y)
–2	$g(-2)=(-2)^2-2=2$	$(-2,2)$
–1	$g(-1)=(-1)^2-2=-1$	$(-1,-1)$
0	$g(0)=(0)^2-2=-2$	$(0,-2)$
1	$g(1)=(1)^2-2=-1$	$(1,-1)$
2	$g(2)=(2)^2-2=2$	$(2,2)$

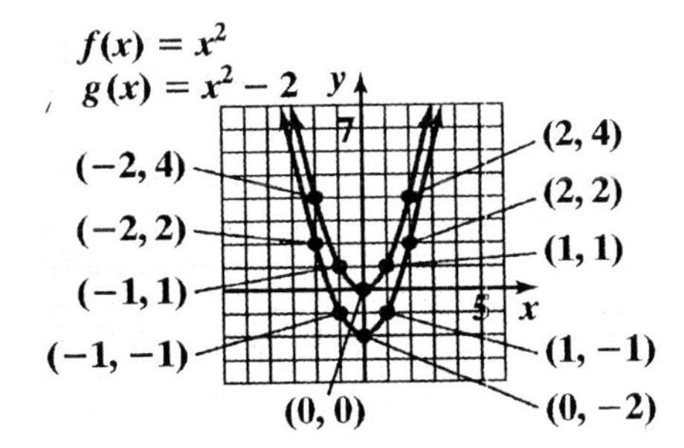

The graph of g is the graph of f shifted down 2 units.

45.

x	$f(x)=\|x\|$	(x,y)
−2	$f(-2)=\|-2\|=2$	$(-2,2)$
−1	$f(-1)=\|-1\|=1$	$(-1,1)$
0	$f(0)=\|0\|=0$	$(0,0)$
1	$f(1)=\|1\|=1$	$(1,1)$
2	$f(2)=\|2\|=2$	$(2,2)$
x	$g(x)=\|x\|-2$	(x,y)
−2	$g(-2)=\|-2\|-2=0$	$(-2,0)$
−1	$g(-1)=\|-1\|-2=-1$	$(-1,-1)$
0	$g(0)=\|0\|-2=-2$	$(0,-2)$
1	$g(1)=\|1\|-2=-1$	$(1,-1)$
2	$g(2)=\|2\|-2=0$	$(2,0)$

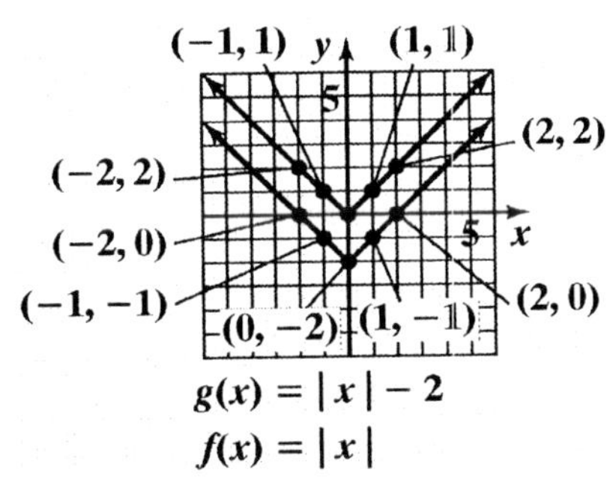

The graph of g is the graph of f shifted down 2 units.

46.

x	$f(x)=\|x\|$	(x,y)
−2	$f(-2)=\|-2\|=2$	$(-2,2)$
−1	$f(-1)=\|-1\|=1$	$(-1,1)$
0	$f(0)=\|0\|=0$	$(0,0)$
1	$f(1)=\|1\|=1$	$(1,1)$
2	$f(2)=\|2\|=2$	$(2,2)$

x	$g(x)=\lvert x\rvert+1$	(x,y)
-2	$g(-2)=\lvert -2\rvert+1=3$	$(-2,3)$
-1	$g(-1)=\lvert -1\rvert+1=2$	$(-1,2)$
0	$g(0)=\lvert 0\rvert+1=1$	$(0,1)$
1	$g(1)=\lvert 1\rvert+1=2$	$(1,2)$
2	$g(2)=\lvert 2\rvert+1=3$	$(2,3)$

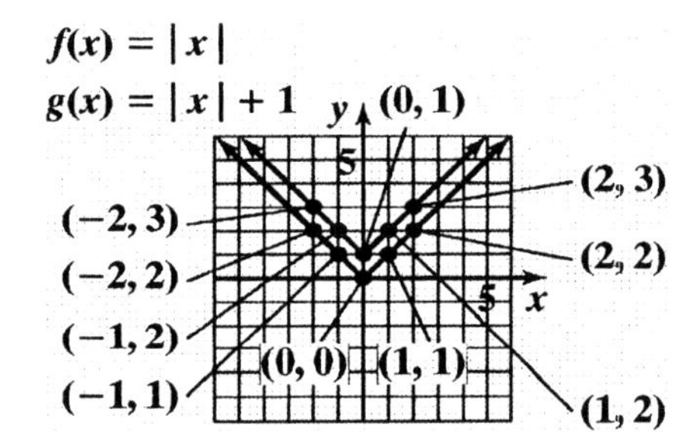

The graph of g is the graph of f shifted up 1 unit.

47.

x	$f(x)=x^3$	(x,y)
-2	$f(-2)=(-2)^3=-8$	$(-2,-8)$
-1	$f(-1)=(-1)^3=-1$	$(-1,-1)$
0	$f(0)=(0)^3=0$	$(0,0)$
1	$f(1)=(1)^3=1$	$(1,1)$
2	$f(2)=(2)^3=8$	$(2,8)$

x	$g(x)=x^3+2$	(x,y)
-2	$g(-2)=(-2)^3+2=-6$	$(-2,-6)$
-1	$g(-1)=(-1)^3+2=1$	$(-1,1)$
0	$g(0)=(0)^3+2=2$	$(0,2)$
1	$g(1)=(1)^3+2=3$	$(1,3)$
2	$g(2)=(2)^3+2=10$	$(2,10)$

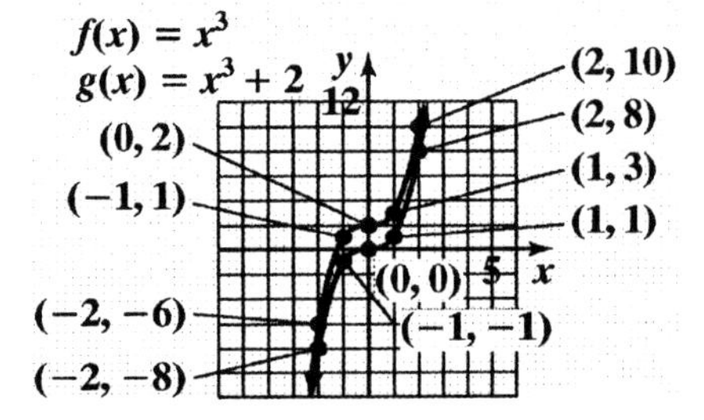

The graph of g is the graph of f shifted up 2 units.

48.

x	$f(x)=x^3$	(x,y)
−2	$f(-2)=(-2)^3=-8$	$(-2,-8)$
−1	$f(-1)=(-1)^3=-1$	$(-1,-1)$
0	$f(0)=(0)^3=0$	$(0,0)$
1	$f(1)=(1)^3=1$	$(1,1)$
2	$f(2)=(2)^3=8$	$(2,8)$

x	$g(x)=x^3-1$	(x,y)
−2	$g(-2)=(-2)^3-1=-9$	$(-2,-9)$
−1	$g(-1)=(-1)^3-1=-2$	$(-1,-2)$
0	$g(0)=(0)^3-1=-1$	$(0,-1)$
1	$g(1)=(1)^3-1=0$	$(1,0)$
2	$g(2)=(2)^3-1=7$	$(2,7)$

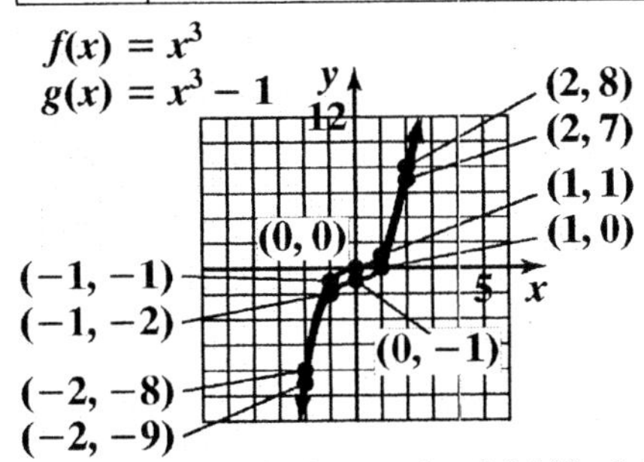

The graph of g is the graph of f shifted down 1 unit.

49.

x	$f(x)=3$	(x,y)
−2	$f(-2)=3$	$(-2,3)$
−1	$f(-1)=3$	$(-1,3)$
0	$f(0)=3$	$(0,3)$
1	$f(1)=3$	$(1,3)$
2	$f(2)=3$	$(2,3)$

x	$g(x)=5$	(x,y)
−2	$g(-2)=5$	$(-2,5)$
−1	$g(-1)=5$	$(-1,5)$
0	$g(0)=5$	$(0,5)$
1	$g(1)=5$	$(1,5)$
2	$g(2)=5$	$(2,5)$

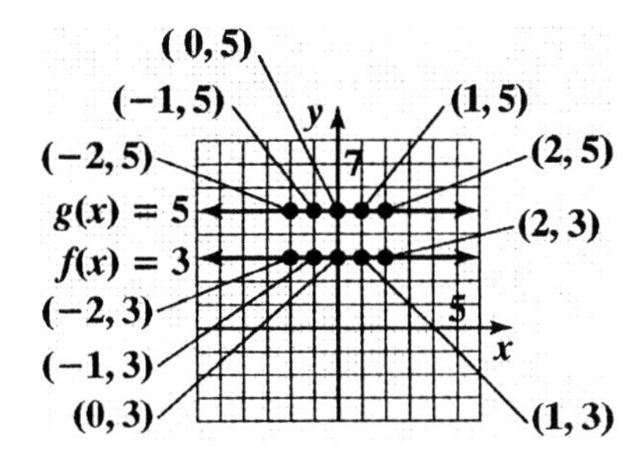

The graph of g is the graph of f shifted up 2 units.

50.

x	$f(x) = -1$	(x, y)
-2	$f(-2) = -1$	$(-2, -1)$
-1	$f(-1) = -1$	$(-1, -1)$
0	$f(0) = -1$	$(0, -1)$
1	$f(1) = -1$	$(1, -1)$
2	$f(2) = -1$	$(2, -1)$

x	$g(x) = 4$	(x, y)
-2	$g(-2) = 4$	$(-2, 4)$
-1	$g(-1) = 4$	$(-1, 4)$
0	$g(0) = 4$	$(0, 4)$
1	$g(1) = 4$	$(1, 4)$
2	$g(2) = 4$	$(2, 4)$

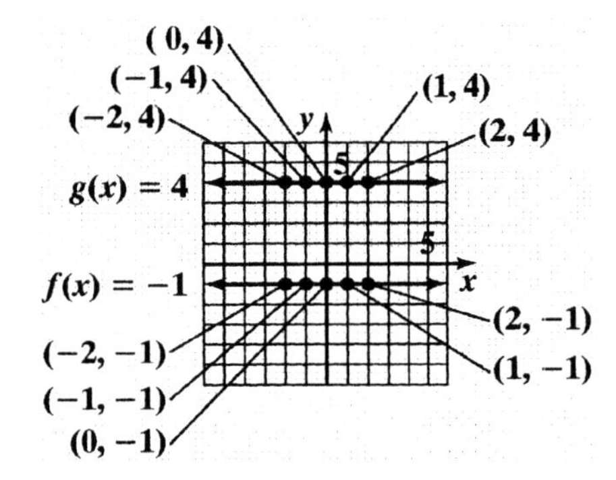

The graph of g is the graph of f shifted up 5 units.

51.

x	$f(x)=\sqrt{x}$	(x,y)
0	$f(0)=\sqrt{0}=0$	$(0,0)$
1	$f(1)=\sqrt{1}=1$	$(1,1)$
4	$f(4)=\sqrt{4}=2$	$(4,2)$
9	$f(9)=\sqrt{9}=3$	$(9,3)$

x	$g(x)=\sqrt{x}-1$	(x,y)
0	$g(0)=\sqrt{0}-1=-1$	$(0,-1)$
1	$g(1)=\sqrt{1}-1=0$	$(1,0)$
4	$g(4)=\sqrt{4}-1=1$	$(4,1)$
9	$g(9)=\sqrt{9}-1=2$	$(9,2)$

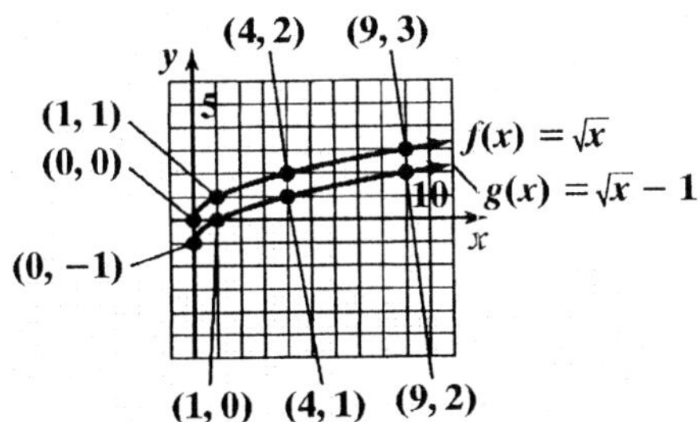

The graph of g is the graph of f shifted down 1 unit.

52.

x	$f(x)=\sqrt{x}$	(x,y)
0	$f(0)=\sqrt{0}=0$	$(0,0)$
1	$f(1)=\sqrt{1}=1$	$(1,1)$
4	$f(4)=\sqrt{4}=2$	$(4,2)$
9	$f(9)=\sqrt{9}=3$	$(9,3)$

x	$g(x)=\sqrt{x}+2$	(x,y)
0	$g(0)=\sqrt{0}+2=2$	$(0,2)$
1	$g(1)=\sqrt{1}+2=3$	$(1,3)$
4	$g(4)=\sqrt{4}+2=4$	$(4,4)$
9	$g(9)=\sqrt{9}+2=5$	$(9,5)$

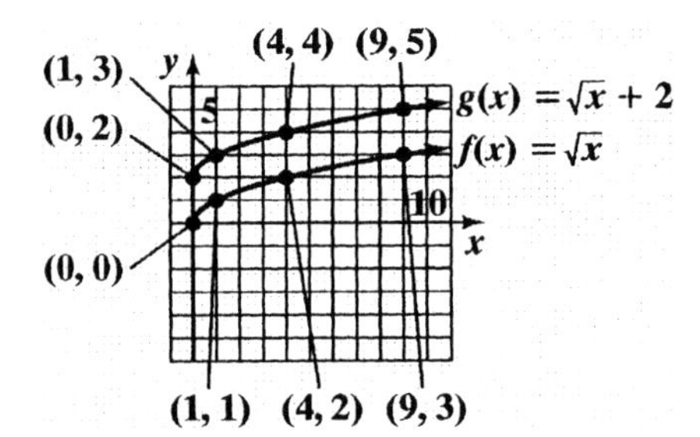

The graph of g is the graph of *f* shifted up 2 units.

53.

x	$f(x)=\sqrt{x}$	(x,y)
0	$f(0)=\sqrt{0}=0$	$(0,0)$
1	$f(1)=\sqrt{1}=1$	$(1,1)$
4	$f(4)=\sqrt{4}=2$	$(4,2)$
9	$f(9)=\sqrt{9}=3$	$(9,3)$

x	$g(x)=\sqrt{x-1}$	(x,y)
1	$g(1)=\sqrt{1-1}=0$	$(1,0)$
2	$g(2)=\sqrt{2-1}=1$	$(2,1)$
5	$g(5)=\sqrt{5-1}=2$	$(5,2)$
10	$g(10)=\sqrt{10-1}=3$	$(10,3)$

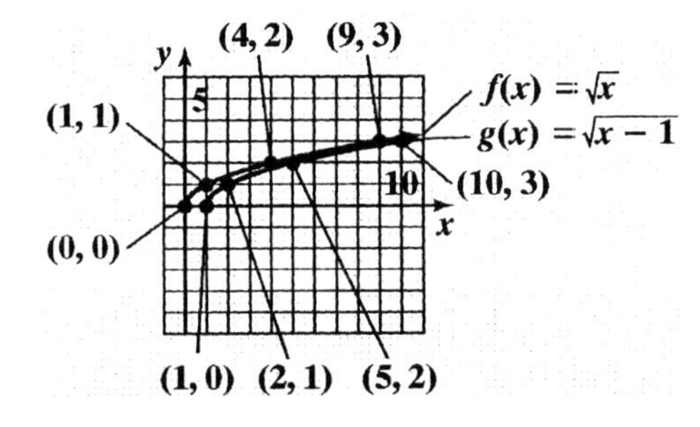

The graph of g is the graph of *f* shifted right 1 unit.

54.

x	$f(x)=\sqrt{x}$	(x,y)
0	$f(0)=\sqrt{0}=0$	$(0,0)$
1	$f(1)=\sqrt{1}=1$	$(1,1)$
4	$f(4)=\sqrt{4}=2$	$(4,2)$
9	$f(9)=\sqrt{9}=3$	$(9,3)$

x	$g(x)=\sqrt{x+2}$	(x,y)
–2	$g(-2)=\sqrt{-2+2}=0$	$(-2,0)$
–1	$g(-1)=\sqrt{-1+2}=1$	$(-1,1)$
2	$g(2)=\sqrt{2+2}=2$	$(2,2)$
7	$g(7)=\sqrt{7+2}=3$	$(7,3)$

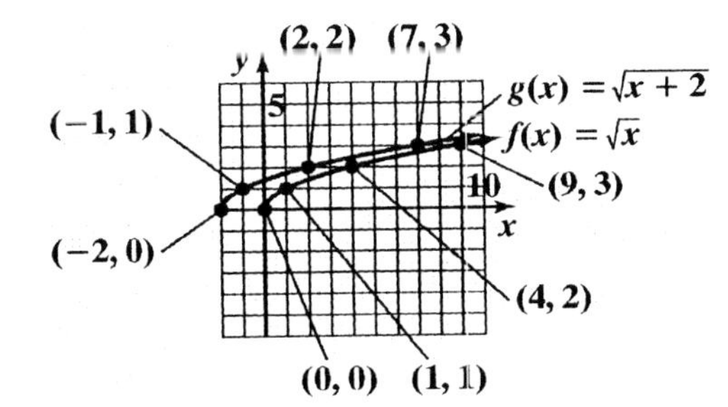

The graph of g is the graph of f shifted left 2 units.

55. function

56. function

57. function

58. not a function

59. not a function

60. not a function

61. function

62. not a function

63. function

64. function

65. $f(-2)=-4$

66. $f(2)=-4$

67. $f(4)=4$

68. $f(-4)=4$

69. $f(-3)=0$

70. $f(-1)=0$

71. $g(-4)=2$

72. $g(2)=-2$

73. $g(-10)=2$

74. $g(10)=-2$

75. When $x=-2$, $g(x)=1$.

76. When $x=1$, $g(x)=-1$.

77. **a.** domain: $(-\infty,\infty)$

b. range: $[-4,\infty)$

c. x-intercepts: –3 and 1

d. y-intercept: –3

e. $f(-2)=-3$ and $f(2)=5$

78. **a.** domain: $(-\infty,\infty)$

b. range: $(-\infty,4]$

c. x-intercepts: –3 and 1

d. y-intercept: 3

e. $f(-2)=3$ and $f(2)=-5$

79. **a.** domain: $(-\infty, \infty)$

b. range: $[1, \infty)$

c. x-intercept: none

d. y-intercept: 1

e. $f(-1) = 2$ and $f(3) = 4$

80. **a.** domain: $(-\infty, \infty)$

b. range: $[0, \infty)$

c. x-intercept: -1

d. y-intercept: 1

e. $f(-4) = 3$ and $f(3) = 4$

81. **a.** domain: $[0, 5)$

b. range: $[-1, 5)$

c. x-intercept: 2

d. y-intercept: -1

e. $f(3) = 1$

82. **a.** domain: $(-6, 0]$

b. range: $[-3, 4)$

c. x-intercept: -3.75

d. y-intercept: -3

e. $f(-5) = 2$

83. **a.** domain: $[0, \infty)$

b. range: $[1, \infty)$

c. x-intercept: none

d. y-intercept: 1

e. $f(4) = 3$

84. **a.** domain: $[-1, \infty)$

b. range: $[0, \infty)$

c. x-intercept: -1

d. y-intercept: 1

e. $f(3) = 2$

85. **a.** domain: $[-2, 6]$

b. range: $[-2, 6]$

c. x-intercept: 4

d. y-intercept: 4

e. $f(-1) = 5$

86. **a.** Domain: $[-3, 2]$

b. Range: $[-5, 5]$

c. x-intercept: $-\frac{1}{2}$

d. y-intercept: 1

e. $f(-2) = -3$

87. **a.** domain: $(-\infty, \infty)$

b. range: $(-\infty, -2]$

c. x-intercept: none

d. y-intercept: -2

e. $f(-4) = -5$ and $f(4) = -2$

88. **a.** domain: $(-\infty, \infty)$

b. range: $[0, \infty)$

c. x-intercept: $\{x \mid x \le 0\}$

d. y-intercept: 0

e. $f(-2) = 0$ and $f(2) = 4$

89. **a.** domain: $(-\infty, \infty)$

b. range: $(0, \infty)$

c. x-intercept: none

d. y-intercept: 1.5

e. $f(4) = 6$

90. **a.** domain: $(-\infty,1)\cup(1,\infty)$

b. range: $(-\infty,0)\cup(0,\infty)$

c. x-intercept: none

d. y-intercept: -1

e. $f(2)=1$

91. **a.** domain: $\{-5,-2,0,1,3\}$

b. range: $\{2\}$

c. x-intercept: none

d. y-intercept: 2

e. $f(-5)+f(3)=2+2=4$

92. **a.** domain: $\{-5,-2,0,1,4\}$

b. range: $\{-2\}$

c. x-intercept: none

d. y-intercept: -2

e. $f(-5)+f(4)=-2+(-2)=-4$

93.
$$g(1)=3(1)-5=3-5=-2$$
$$f(g(1))=f(-2)=(-2)^2-(-2)+4$$
$$=4+2+4=10$$

94.
$$g(-1)=3(-1)-5=-3-5=-8$$
$$f(g(-1))=f(-8)=(-8)^2-(-8)+4$$
$$=64+8+4=76$$

95.
$$\sqrt{3-(-1)}-(-6)^2+6\div(-6)\cdot 4$$
$$=\sqrt{3+1}-36+6\div(-6)\cdot 4$$
$$=\sqrt{4}-36+-1\cdot 4$$
$$=2-36+-4$$
$$=-34+-4$$
$$=-38$$

96.
$$|-4-(-1)|-(-3)^2+-3\div 3\cdot -6$$
$$=|-4+1|-9+-3\div 3\cdot -6$$
$$=|-3|-9+-1\cdot -6$$
$$=3-9+6=-6+6=0$$

97.
$$f(-x)-f(x)$$
$$=(-x)^3+(-x)-5-(x^3+x-5)$$
$$=-x^3-x-5-x^3-x+5=-2x^3-2x$$

98. $f(-x)-f(x)$
$=(-x)^2-3(-x)+7-(x^2-3x+7)$
$=x^2+3x+7-x^2+3x-7$
$=6x$

99. **a.** $\{(\text{U.S., }80\%),(\text{Japan, }64\%),$
$(\text{France, }64\%),(\text{Germany, }61\%),$
$(\text{England, }59\%),(\text{China,}47\%)\}$

b. Yes, the relation is a function. Each element in the domain corresponds to only one element in the range.

c. $\{(80\%,\text{ U.S.}),(64\%,\text{ Japan}),$
$(64\%,\text{ France}),(61\%,\text{ Germany}),$
$(59\%,\text{ England}),(47\%,\text{ China})\}$

d. No, the relation is not a function. 64% in the domain corresponds to both Japan and France in the range.

100. **a.** {(EL,1%), (L,7%), (SL,11%),
(M,52%), (SC,13%), (C,13%),
(EC,3%)}

b. Yes, the relation in part (a) is a function because each ideology corresponds to exactly one percentage.

c. {(1%,EL),(7%,L),(11%,SL), (52%,M),(13%,SC),(13%,C), (3%,EC)}

d. No, the relation in is not a function because 13% in the domain corresponds to two ideologies, SC and C, in the range.

101. $W(16)=0.07(16)+4.1$
$=1.12+4.1=5.22$
In 2000 there were 5.22 million women enrolled in U.S. colleges.
(2000,5.22)

102. $M(16)=0.01(16)+3.9=0.16+3.9=4.06$
In 2000, there were 4.06 million men enrolled in U.S. colleges. This is represented by the point (2000, 4.06) on the graph.

103. $W(20)=0.07(20)+4.1$
$=1.4+4.1=5.5$
$M(20)=0.01(20)+3.9$
$=0.2+3.9=4.1$
$W(20)-M(20)=5.5-4.1=1.4$
In 2004, there will be 1.4 million more women than men enrolled in U.S. colleges.

104. $W(25) = 0.07(25) + 4.1 = 1.75 + 4.1 = 5.85$

$M(25) = 0.01(25) + 3.9 = 0.25 + 3.9 = 4.15$

$W(25) - M(25) = 5.85 - 4.15 = 1.7$

In 2009, there will be 1.7 million more women than men enrolled in U.S. colleges.

105. a. According to the graph, women's earnings were about 73% of men's in 2000.

b. $P(x) = 0.012x^2 - 0.16x + 60$

$P(40) = 0.012(40)^2 - 0.16(40) + 60 = 72.8$

According to the function, women's earnings were about 72.8% of men's in 2000.

c. $\frac{27,355}{37,339} \approx 0.733 = 73.3\%$

The answers in parts (a) and (b) model the actual data quite well.

106. a. According to the graph, women's earnings were about 76% of men's in 2003.

b. $P(x) = 0.012x^2 - 0.16x + 60$

$P(43) = 0.012(43)^2 - 0.16(43) + 60 = 75.3$

According to the function, women's earnings were about 75.3% of men's in 2003.

c. $\frac{30,724}{40,668} \approx 0.755 = 75.5\%$

The answers in parts (a) and (b) model the actual data quite well.

107. $C(x) = 100,000 + 100x$

$C(90) = 100,000 + 100(90) = \$109,000$

It will cost $109,000 to produce 90 bicycles.

108. $V(x) = 22,500 - 3200x$

$V(3) = 22,500 - 3200(3) = \$12,900$

After 3 years, the car will be worth $12,900.

109.

$$T(x) = \frac{40}{x} + \frac{40}{x+30}$$

$$\begin{aligned} T(30) &= \frac{40}{30} + \frac{40}{30+30} \\ &= \frac{80}{60} + \frac{40}{60} \\ &= \frac{120}{60} \\ &= 2 \end{aligned}$$

If you travel 30 mph going and 60 mph returning, your total trip will take 2 hours.

110. $S(x) = 0.10x + 0.60(50 - x)$

$S(30) = 0.10(30) + 0.60(50 - 30) = 15$

When 30 mL of the 10% mixture is mixed with 20 mL of the 60% mixture, there will be 15 mL of sodium-iodine in the vaccine.

121. a. false; the domain of *f* is [–4, 4]

b. false; the range of *f* is [–2, 2)

c. true; $f(-1) - f(4) = 1 - (-1) = 2$

d. false; $f(0) < 1$ x

(c) is true.

122. $f(a+h) = 3(a+h)+7 = 3a+3h+7$

$$f(a) = 3a+7$$

$$\frac{f(a+h)-f(a)}{h}$$

$$= \frac{(3a+3h+7)-(3a+7)}{h}$$

$$= \frac{3a+3h+7-3a-7}{h} = \frac{3h}{h} = 3$$

123. Answers may vary.
An example is {(1,1),(2,1)}

124. It is given that $f(x+y) = f(x)+f(y)$ and $f(1) = 3$.
To find $f(2)$, rewrite 2 as 1 + 1.

$$f(2) = f(1+1) = f(1)+f(1)$$
$$= 3+3 = 6$$

Similarly:

$$f(3) = f(2+1) = f(2)+f(1)$$
$$= 6+3 = 9$$
$$f(4) = f(3+1) = f(3)+f(1)$$
$$= 9+3 = 12$$

While $f(x+y) = f(x)+f(y)$ is true for this function, it is not true for all functions. It is not true for $f(x) = x^2$, for example.

Section 1.3

Check Point Exercises

1. a.

$$f(x) = -2x^2 + x + 5$$
$$f(x+h) = -2(x+h)^2 + (x+h) + 5$$
$$= -2(x^2 + 2xh + h^2) + x + h + 5$$
$$= -2x^2 - 4xh - 2h^2 + x + h + 5$$

b. $\dfrac{f(x+h)-f(x)}{h}$

$=\dfrac{-2x^2-4xh-2h^2+x+h+5-\left(-2x^2+x+5\right)}{h}$

$=\dfrac{-2x^2-4xh-2h^2+x+h+5+2x^2-x-5}{h}$

$=\dfrac{-4xh-2h^2+h}{h}$

$=\dfrac{h\left(-4x-2h+1\right)}{h}$

$=-4x-2h+1$

2. $C(t)=\begin{cases}20 & \text{if } 0\le t\le 60\\ 20+0.40(t-60) & \text{if } t>60\end{cases}$

a. Since $0\le 40\le 60$, $C(40)=20$
With 40 calling minutes, the cost is \$20.
This is represented by $(40,20)$.

b. Since $80>60$, $C(80)=20+0.40(80-60)=28$
With 80 calling minutes, the cost is \$28.
This is represented by $(80,28)$.

3. The function is increasing on the interval $(-\infty,-1)$, decreasing on the interval $(-1, 1)$, and increasing on the interval $(1, \infty)$.

4. **a.** $f(-x)=(-x)^2+6=x^2+6=f(x)$
The function is even.

b. $g(-x)=7(-x)^3-(-x)=-7x^3+x=-f(x)$
The function is odd.

c. $h(-x)=(-x)^5+1=-x^5+1$
The function is neither even nor odd.

Exercise Set 1.3

1. $\frac{f(x+h)-f(x)}{h}$
$=\frac{4(x+h)-4x}{h}$
$=\frac{4x+4h-4x}{h}$
$=\frac{4h}{h}$
$=4$

2. $\frac{f(x+h)-f(x)}{h}$
$=\frac{7(x+h)-7x}{h}$
$=\frac{7x+7h-7x}{h}$
$=\frac{7h}{h}$
$=7$

3. $\frac{f(x+h)-f(x)}{h}$
$=\frac{3(x+h)+7-(3x+7)}{h}$
$=\frac{3x+3h+7-3x-7}{h}$
$=\frac{3h}{h}$
$=3$

4. $\frac{f(x+h)-f(x)}{h}$
$=\frac{6(x+h)+1-(6x+1)}{h}$
$=\frac{6x+6h+1-6x-1}{h}$
$=\frac{6h}{h}$
$=6$

5. $\frac{f(x+h)-f(x)}{h}$
$=\frac{(x+h)^2-x^2}{h}$
$=\frac{x^2+2xh+h^2-x^2}{h}$
$=\frac{2xh+h^2}{h}$
$=\frac{h(2x+h)}{h}$
$=2x+h$

6. $\frac{f(x+h)-f(x)}{h}$
$=\frac{2(x+h)^2-2x^2}{h}$
$=\frac{2(x^2+2xh+h^2)-2x^2}{h}$
$=\frac{2x^2+4xh+2h^2-2x^2}{h}$
$=\frac{4xh+2h^2}{h}$
$=\frac{h(4x+2h)}{h}$
$=4x+2h$

7. $\frac{f(x+h)-f(x)}{h}$
$=\frac{(x+h)^2-4(x+h)+3-(x^2-4x+3)}{h}$
$=\frac{x^2+2xh+h^2-4x-4h+3-x^2+4x-3}{h}$
$=\frac{2xh+h^2-4h}{h}$
$=\frac{h(2x+h-4)}{h}$
$=2x+h-4$

8. $\frac{f(x+h)-f(x)}{h}$
$=\frac{(x+h)^2-5(x+h)+8-(x^2-5x+8)}{h}$
$=\frac{x^2+2xh+h^2-5x-5h+8-x^2+5x-8}{h}$
$=\frac{2xh+h^2-5h}{h}$
$=\frac{h(2x+h-5)}{h}$
$=2x+h-5$

9. $\frac{f(x+h)-f(x)}{h}$
$=\frac{2(x+h)^2+(x+h)-1-(2x^2+x-1)}{h}$
$=\frac{2x^2+4xh+2h^2+x+h-1-2x^2-x+1}{h}$
$=\frac{4xh+2h^2+h}{h}$
$=\frac{h(4x+2h+1)}{h}$
$=4x+2h+1$

10. $\frac{f(x+h)-f(x)}{h}$
$=\frac{3(x+h)^2+(x+h)+5-(3x^2+x+5)}{h}$
$=\frac{3x^2+6xh+3h^2+x+h+5-3x^2-x-5}{h}$
$=\frac{6xh+3h^2+h}{h}$
$=\frac{h(6x+3h+1)}{h}$
$=6x+3h+1$

11. $\frac{f(x+h)-f(x)}{h}$
$=\frac{-(x+h)^2+2(x+h)+4-(-x^2+2x+4)}{h}$
$=\frac{-x^2-2xh-h^2+2x+2h+4+x^2-2x-4}{h}$
$=\frac{-2xh-h^2+2h}{h}$
$=\frac{h(-2x-h+2)}{h}$
$=-2x-h+2$

12. $\frac{f(x+h)-f(x)}{h}$
$=\frac{-(x+h)^2-3(x+h)+1-(-x^2-3x+1)}{h}$
$=\frac{-x^2-2xh-h^2-3x-3h+1+x^2+3x-1}{h}$
$=\frac{-2xh-h^2-3h}{h}$
$=\frac{h(-2x-h-3)}{h}$
$=-2x-h-3$

13. $\frac{f(x+h)-f(x)}{h}$
$=\frac{-2(x+h)^2+5(x+h)+7-(-2x^2+5x+7)}{h}$
$=\frac{-2x^2-4xh-2h^2+5x+5h+7+2x^2-5x-7}{h}$
$=\frac{-4xh-2h^2+5h}{h}$
$=\frac{h(-4x-2h+5)}{h}$
$=-4x-2h+5$

14. $\dfrac{f(x+h)-f(x)}{h}$

$= \dfrac{-3(x+h)^2 + 2(x+h) - 1 - (-3x^2 + 2x - 1)}{h}$

$= \dfrac{-3x^2 - 6xh - 3h^2 + 2x + 2h - 1 + 3x^2 - 2x + 1}{h}$

$= \dfrac{-6xh - 3h^2 + 2h}{h}$

$= \dfrac{h(-6x - 3h + 2)}{h}$

$= -6x - 3h + 2$

15. $\dfrac{f(x+h)-f(x)}{h}$

$= \dfrac{-2(x+h)^2 - (x+h) + 3 - (-2x^2 - x + 3)}{h}$

$= \dfrac{-2x^2 - 4xh - 2h^2 - x - h + 3 + 2x^2 + x - 3}{h}$

$= \dfrac{-4xh - 2h^2 - h}{h}$

$= \dfrac{h(-4x - 2h - 1)}{h}$

$= -4x - 2h - 1$

16. $\dfrac{f(x+h)-f(x)}{h}$

$= \dfrac{-3(x+h)^2 + (x+h) - 1 - (-3x^2 + x - 1)}{h}$

$= \dfrac{-3x^2 - 6xh - 3h^2 + x + h - 1 + 3x^2 - x + 1}{h}$

$= \dfrac{-6xh - 3h^2 + h}{h}$

$= \dfrac{h(-6x - 3h + 1)}{h}$

$= -6x - 3h + 1$

17. $\dfrac{f(x+h)-f(x)}{h} = \dfrac{6-6}{h} = \dfrac{0}{h} = 0$

18. $\dfrac{f(x+h)-f(x)}{h} = \dfrac{7-7}{h} = \dfrac{0}{h} = 0$

19. $\dfrac{f(x+h)-f(x)}{h}$

$= \dfrac{\frac{1}{x+h} - \frac{1}{x}}{h}$

$= \dfrac{\frac{x}{x(x+h)} + \frac{-(x+h)}{x(x+h)}}{h}$

$= \dfrac{\frac{x-x-h}{x(x+h)}}{h}$

$= \dfrac{\frac{-h}{x(x+h)}}{h}$

$= \dfrac{-h}{x(x+h)} \cdot \dfrac{1}{h}$

$= \dfrac{-1}{x(x+h)}$

20. $\dfrac{f(x+h)-f(x)}{h}$

$= \dfrac{\frac{1}{2(x+h)} - \frac{1}{2x}}{h}$

$= \dfrac{\frac{x}{2x(x+h)} - \frac{x+h}{2x(x+h)}}{h}$

$= \dfrac{\frac{-h}{2x(x+h)}}{h}$

$= \dfrac{-h}{2x(x+h)} \cdot \dfrac{1}{h}$

$= \dfrac{-1}{2x(x+h)}$

21. $\dfrac{f(x+h)-f(x)}{h}$

$= \dfrac{\sqrt{x+h}-\sqrt{x}}{h}$

$= \dfrac{\sqrt{x+h}-\sqrt{x}}{h} \cdot \dfrac{\sqrt{x+h}+\sqrt{x}}{\sqrt{x+h}+\sqrt{x}}$

$= \dfrac{x+h-x}{h\left(\sqrt{x+h}+\sqrt{x}\right)}$

$= \dfrac{h}{h\left(\sqrt{x+h}+\sqrt{x}\right)}$

$= \dfrac{1}{\sqrt{x+h}+\sqrt{x}}$

22. $\dfrac{f(x+h)-f(x)}{h}$

$= \dfrac{\sqrt{x+h-1}-\sqrt{x-1}}{h}$

$= \dfrac{\sqrt{x+h-1}-\sqrt{x-1}}{h} \cdot \dfrac{\sqrt{x+h-1}+\sqrt{x-1}}{\sqrt{x+h-1}+\sqrt{x-1}}$

$= \dfrac{x+h-1-(x-1)}{h\left(\sqrt{x+h-1}+\sqrt{x-1}\right)}$

$= \dfrac{x+h-1-x+1}{h\left(\sqrt{x+h-1}+\sqrt{x-1}\right)}$

$= \dfrac{h}{h\left(\sqrt{x+h-1}+\sqrt{x-1}\right)}$

$= \dfrac{1}{\sqrt{x+h-1}+\sqrt{x-1}}$

23. **a.** $f(-2) = 3(-2) + 5 = -1$

b. $f(0) = 4(0) + 7 = 7$

c. $f(3) = 4(3) + 7 = 19$

24. **a.** $f(-3) = 6(-3) - 1 = -19$

b. $f(0) = 7(0) + 3 = 3$

c. $f(4) = 7(4) + 3 = 31$

25. **a.** $g(0) = 0 + 3 = 3$

b. $g(-6) = -(-6 + 3) = -(-3) = 3$

c. $g(-3) = -3 + 3 = 0$

26. **a.** $g(0) = 0 + 5 = 5$

b. $g(-6) = -(-6 + 5) = -(-1) = 1$

c. $g(-5) = -5 + 5 = 0$

27. **a.** $h(5) = \dfrac{5^2-9}{5-3} = \dfrac{25-9}{2} = \dfrac{16}{2} = 8$

b. $h(0) = \dfrac{0^2-9}{0-3} = \dfrac{-9}{-3} = 3$

c. $h(3) = 6$

28. **a.** $h(7) = \dfrac{7^2-25}{7-5} = \dfrac{49-25}{2} = \dfrac{24}{2} = 12$

b. $h(0) = \dfrac{0^2-25}{0-5} = \dfrac{-25}{-5} = 5$

c. $h(5) = 10$

29. **a.** increasing: $(-1, \infty)$

b. decreasing: $(-\infty, -1)$

c. constant: none

30. **a.** increasing: $(-\infty, -1)$

b. decreasing: $(-1, \infty)$

c. constant: none

31. **a.** increasing: $(0, \infty)$

b. decreasing: none

c. constant: none

32. **a.** increasing: $(-1, \infty)$

b. decreasing: none

c. constant: none

33. **a.** increasing: none

b. decreasing: $(-2, 6)$

c. constant: none

34. **a.** increasing: (–3, 2)

b. decreasing: none

c. constant: none

35. **a.** increasing: $(-\infty, -1)$

b. decreasing: none

c. constant: $(-1, \infty)$

36. **a.** increasing: $(0, \infty)$

b. decreasing: none

c. constant: $(-\infty, 0)$

37. **a.** increasing: $(-\infty, 0)$ or $(1.5, 3)$

b. decreasing: $(0, 1.5)$ or $(3, \infty)$

c. constant: none

38. **a.** increasing: $(-5, -4)$ or $(-2, 0)$ or $(2, 4)$

b. decreasing: $(-4, -2)$ or $(0, 2)$ or $(4, 5)$

c. constant: none

39. **a.** increasing: (–2, 4)

b. decreasing: none

c. constant: $(-\infty, -2)$ or $(4, \infty)$

40. **a.** increasing: none

b. decreasing: (–4, 2)

c. constant: $(-\infty, -4)$ or $(2, \infty)$

41. **a.** $x = 0$, relative maximum = 4

b. $x = -3, 3$, relative minimum = 0

42. **a.** $x = 0$, relative maximum = 2

b. $x = -3, 3$, relative minimum = –1

43. **a.** $x = -2$, relative maximum = 21

b. $x = 1$, relative minimum = –6

44. **a.** $x = 1$, relative maximum = 30

b. $x = 4$, relative minimum = 3

45. $f(x) = x^3 + x$
$f(-x) = (-x)^3 + (-x)$
$f(-x) = -x^3 - x = -(x^3 + x)$
$f(-x) = -f(x)$, odd function

46. $f(x) = x^3 - x$
$f(-x) = (-x)^3 - (-x)$
$f(-x) = -x^3 + x = -(x^3 - x)$
$f(-x) = -f(x)$, odd function

47. $g(x) = x^2 + x$
$g(-x) = (-x)^2 + (-x)$
$g(-x) = x^2 - x$, neither

48. $g(x) = x^2 - x$
$g(-x) = (-x)^2 - (-x)$
$g(-x) = x^2 + x$, neither

49. $h(x) = x^2 - x^4$
$h(-x) = (-x)^2 - (-x)^4$
$h(-x) = x^2 - x^4$
$h(-x) = h(x)$, even function

50. $h(x) = 2x^2 + x^4$
$h(-x) = 2(-x)^2 + (-x)^4$
$h(-x) = 2x^2 + x^4$
$h(-x) = h(x)$, even function

51. $f(x) = x^2 - x^4 + 1$
$f(-x) = (-x)^2 - (-x)^4 + 1$
$f(-x) = x^2 - x^4 + 1$
$f(-x) = f(x)$, even function

52. $f(x) = 2x^2 + x^4 + 1$
$f(-x) = 2(-x)^2 + (-x)^4 + 1$
$f(-x) = 2x^2 + x^4 + 1$
$f(-x) = f(x)$, even function

53. $f(x)=\frac{1}{5}x^6-3x^2$

$f(-x)=\frac{1}{5}(-x)^6-3(-x)^2$

$f(-x)=\frac{1}{5}x^6-3x^2$

$f(-x)=f(x)$, even function

54. $f(x)=2x^3-6x^5$

$f(-x)=2(-x)^3-6(-x)^5$

$f(-x)=-2x^3+6x^5$

$f(-x)=-(2x^3-6x^5)$

$f(-x)=-f(x)$, odd function

55. $f(x)=x\sqrt{1-x^2}$

$f(-x)=-x\sqrt{1-(-x)^2}$

$f(-x)=-x\sqrt{1-x^2}$

$=-\left(x\sqrt{1-x^2}\right)$

$f(-x)=-f(x)$, odd function

56. $f(x)=x^2\sqrt{1-x^2}$

$f(-x)=(-x)^2\sqrt{1-(-x)^2}$

$f(-x)=x^2\sqrt{1-x^2}$

$f(-x)=f(x)$, even function

57. The graph is symmetric with respect to the y-axis. The function is even.

58. The graph is symmetric with respect to the origin. The function is odd.

59. The graph is symmetric with respect to the origin. The function is odd.

60. The graph is not symmetric with respect to the y-axis or the origin. The function is neither even nor odd.

61. **a.** Domain: $(-\infty,\infty)$

b. Range: $[-4,\infty)$

c. x-intercepts: 1, 7

d. y-intercept: 4

e. $(4,\infty)$

f. $(0,4)$

g. $(-\infty,0)$

h. $x=4$

i. $y=-4$

j. $f(-3)=4$

k. $f(2)=-2$ and $f(6)=-2$

l. neither ; $f(-x)\neq x$, $f(-x)\neq -x$

62. **a.** Domain: $(-\infty,\infty)$

b. Range: $(-\infty,4]$

c. x-intercepts: –4, 4

d. y-intercept: 1

e. $(-\infty,-2)$ or $(0,3)$

f. $(-2,0)$ or $(3,\infty)$

g. $(-\infty,-4]$ or $[4,\infty)$

h. $x=-2$ and $x=3$

i. $f(-2)=4$ and $f(3)=2$

j. $f(-2)=4$

k. $x=-4$ and $x=4$

l. neither ; $f(-x)\neq x$, $f(-x)\neq -x$

63. a. Domain: $(-\infty,3]$

b. Range: $(-\infty,4]$

c. x-intercepts: –3, 3

d. $f(0)=3$

e. $(-\infty,1)$

f. $(1,3)$

g. $(-\infty,-3]$

h. $f(1)=4$

i. $x=1$

j. positive; $f(-1)=+2$

64. a. Domain: $(-\infty,6]$

b. Range: $(-\infty,1]$

c. zeros of f: –3, 3

d. $f(0)=1$

e. $(-\infty,-2)$

f. $(2,6)$

g. $(-2,2)$

h. $(-3,3)$

i. $x=-5$ and $x=5$

j. negative; $f(4)=-1$

k. neither

l. no; f(2) is not greater than the function values to the immediate left.

65. $f(1.06)=1$

66. $f=(2.99)=2$

67. $f\left(\frac{1}{3}\right)=0$

68. $f(-1.5)=-2$

69. $f(-2.3)=-3$

70. $f(-99.001)=-100$

71. $\sqrt{f(-1.5)+f(-0.9)}-[f(\pi)]^2+f(-3)\div f(1)\cdot f(-\pi)$
$=\sqrt{1+0}-[-4]^2+2\div(-2)\cdot 3$
$=\sqrt{1}-16+(-1)\cdot 3$
$=1-16-3$
$=-18$

72. $\sqrt{f(-2.5)-f(1.9)}-[f(-\pi)]^2+f(-3)\div f(1)\cdot f(\pi)$
$\sqrt{f(-2.5)-f(1.9)}-[f(-\pi)]^2+f(-3)\div f(1)\cdot f(\pi)$
$=\sqrt{2-(-2)}-[3]^2+2\div(-2)\cdot(-4)$
$=\sqrt{4}-9+(-1)(-4)$
$=2-9+4$
$=-3$

73. $30+0.30(t-120)=30+0.3t-36=0.3t-6$

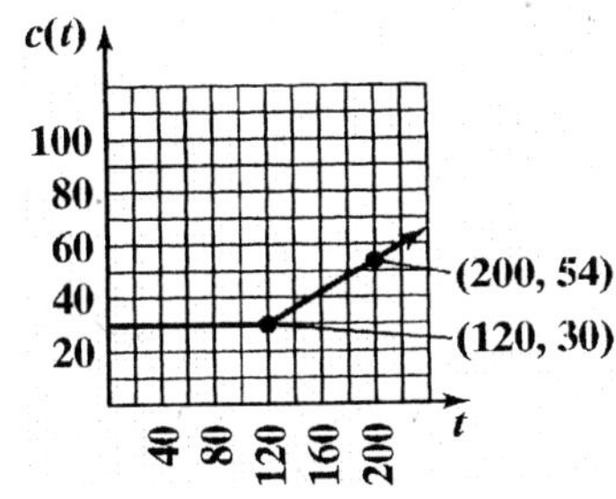

74. $40 + 0.30(t-200) = 40 + 0.3t - 60 = 0.3t - 20$

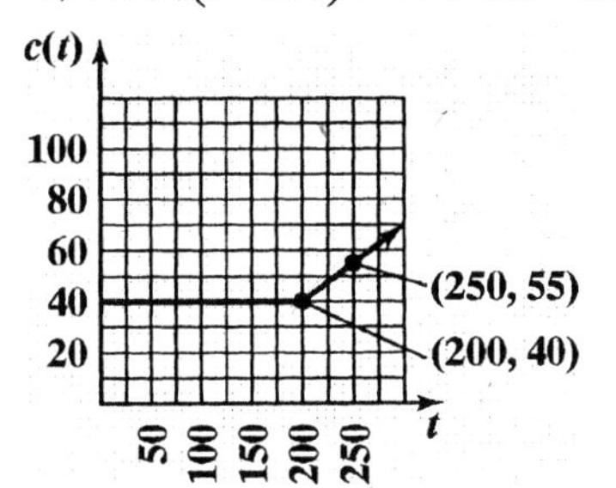

75. $C(t) = \begin{cases} 50 & \text{if } 0 \le t \le 400 \\ 50 + 0.30(t-400) & \text{if } t > 400 \end{cases}$

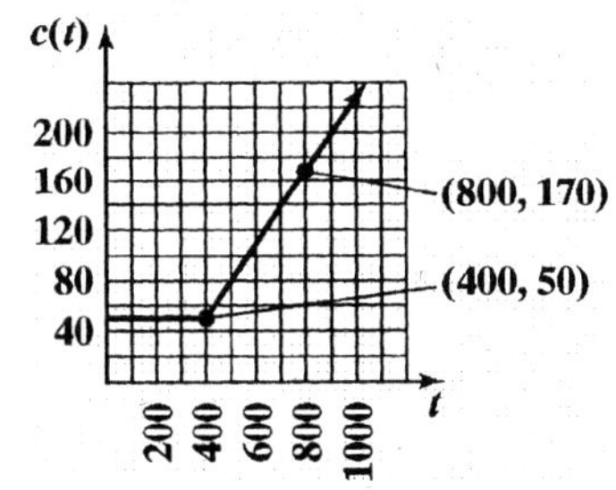

76. $C(t) = \begin{cases} 60 & \text{if } 0 \le t \le 450 \\ 60 + 0.35(t-450) & \text{if } t > 450 \end{cases}$

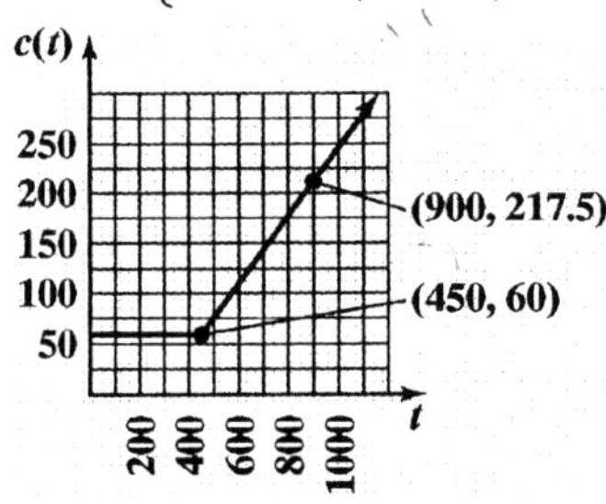

77. $f(60) \approx 3.1$
In 1960, about 3.1% of the population were Jewish-Americans.

78 $f(100) \approx 2.2$
In 2002, about 2.2% of the population were Jewish-Americans.

79. $x \approx 19$ and $x \approx 64$
In 1919 and 1964, about 3% of the population were Jewish-Americans.

80. $x = 14$ and $x \approx 90$, In 1914 and 1990, about 2.5% of the population were Jewish-Americans.

81. In 1940, the maximum of 3.7% of the population were Jewish-American.

82. In 1900, about 1.4% of the population were Jewish-American.

83. Each year corresponds to only 1 percentage.

84. The percentage of Jewish-Americans in the population increased until 1940 and decreased since then.

85. Increasing: (45, 74)
Decreasing: (16, 45)
The number of accidents occurring per 50 million miles driven increases with age starting at age 45, while it decreases with age starting at age 16.

86. $x = 45$ and $f(45) = 190$
The fewest number of accidents per 50 million miles driven occurs at age 45.

87. Answers may vary. An example is 16 and 74 year olds will have 526.4 accidents per 50 million miles.

88. $f(x) = 0.4x^2 - 36x + 1000$
$f(50) = 0.4(50)^2 - 36(50) + 1000 = 200$
50 year old drivers have 200 accidents per 50 million miles.
This is represented on the graph as the point $(50, 200)$.

89. $f(30) = 61.9(30) + 132 = 1989$ cigarettes per adult
This describes the actual data quite well.

90. $f(80) = -2.2(80)^2 + 256(80) - 3503 = 2897$ cigarettes per adult
This describes the actual data quite well.

91. The maximum occurred in 1960. Graph estimates will vary near 4100.
$f(50) = -2.2(50)^2 + 256(50) - 3503 = 3797$ cigarettes per adult
The function does not describe the actual data reasonably well.

92. The minimum occurred in 1910. Graph estimates will vary near 100.
$f(0) = 61.9(0) + 132 = 132$ cigarettes per adult
The function describes the actual data reasonably well.

93.

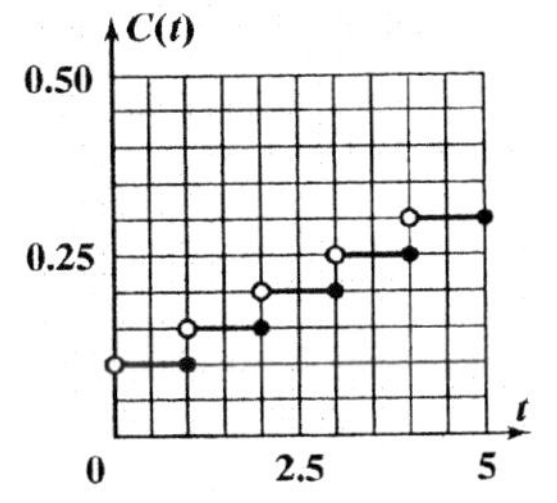

94.

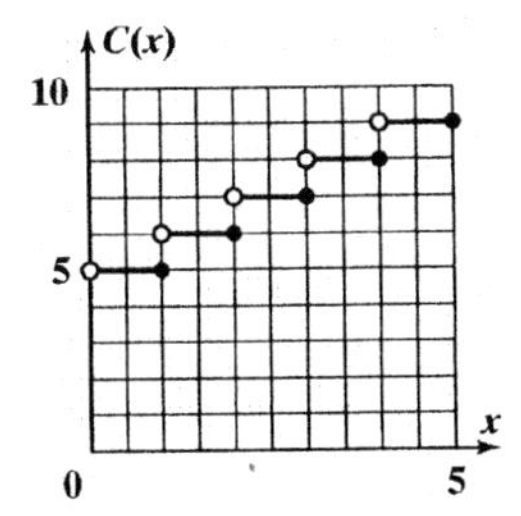

103.

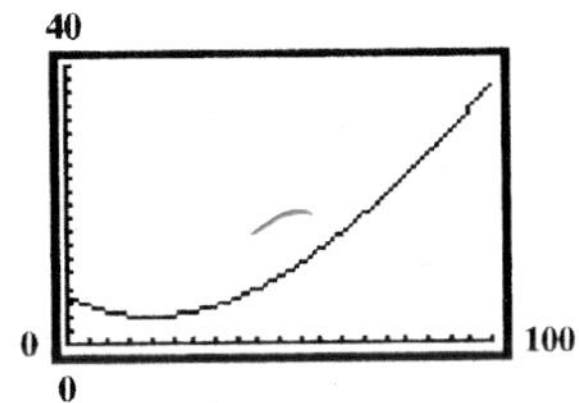

The number of doctor visits decreases during childhood and then increases as you get older. The minimum is (20.29, 3.99), which means that the minimum number of doctor visits, about 4, occurs at around age 20.

104

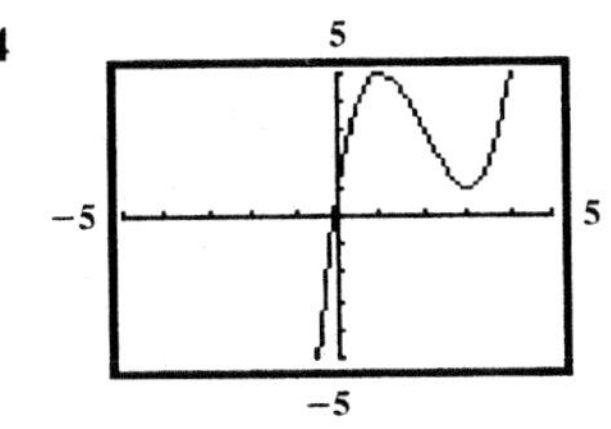

Increasing: $(-\infty, 1)$ or $(3, \infty)$
Decreasing: (1, 3)

105.

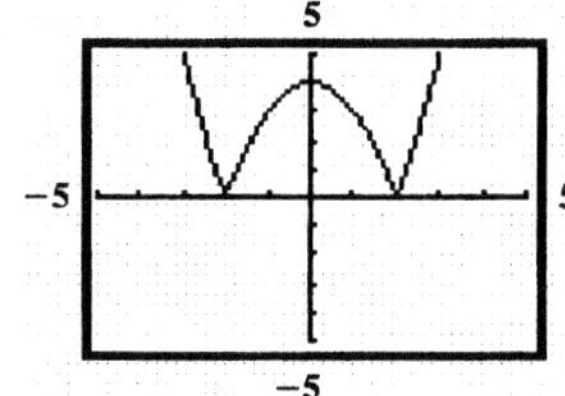

Increasing: (–2, 0) or (2, ∞)
Decreasing: (–∞, –2) or (0, 2)

106.

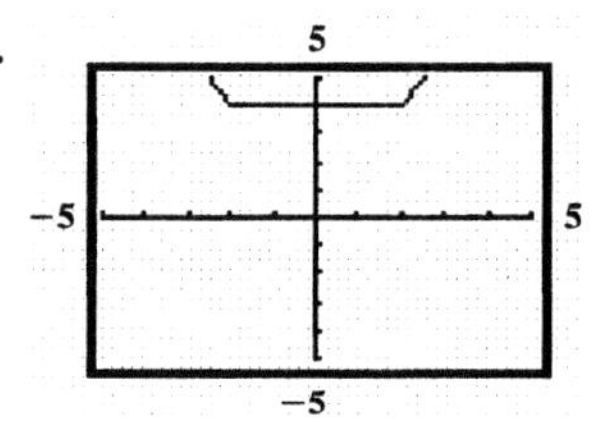

Increasing: $(2, \infty)$

Decreasing: $(-\infty, -2)$

Constant: (–2, 2)

107.

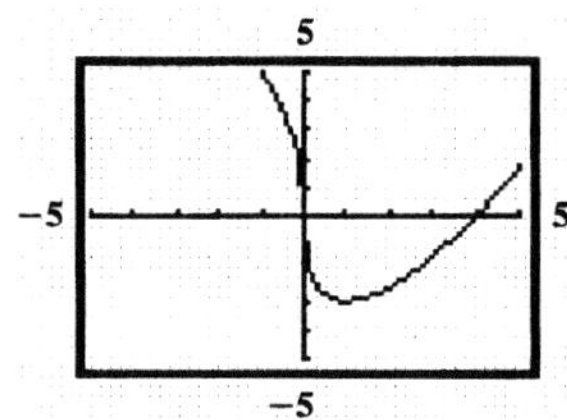

Increasing: (1, ∞)
Decreasing: (–∞, 1)

108.

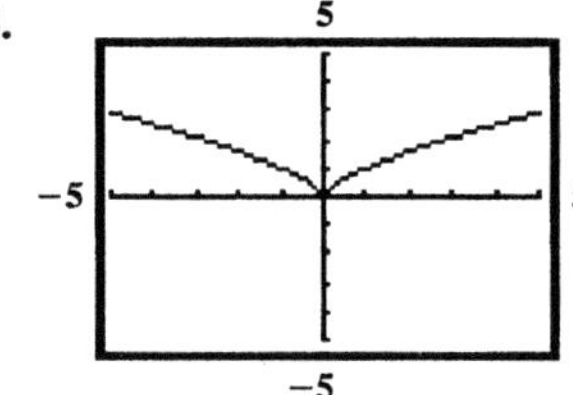

Increasing: $(0, \infty)$

Decreasing: $(-\infty, 0)$

109.

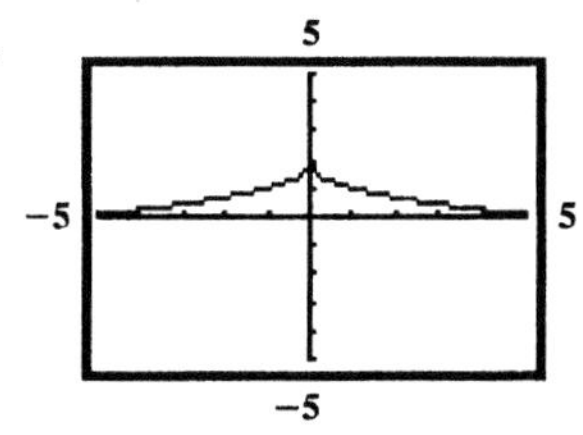

Increasing: (–∞, 0)
Decreasing: (0, ∞)

110. a.

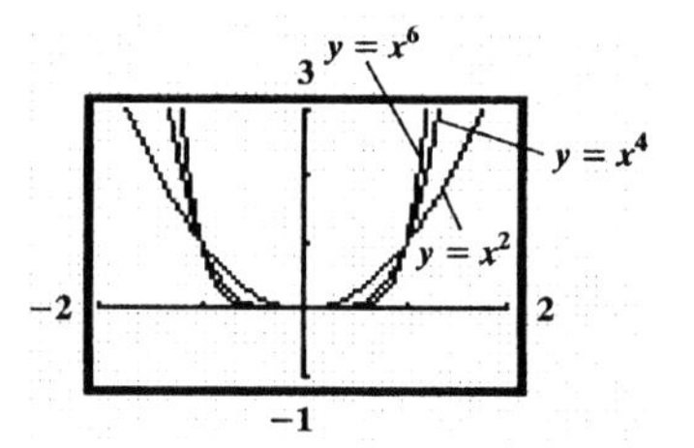

b.

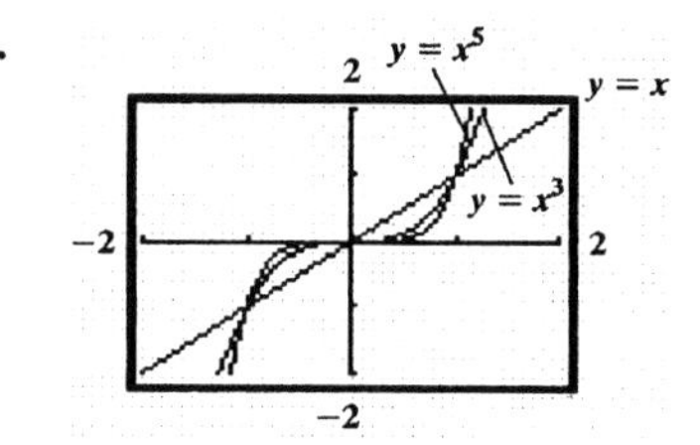

c. Increasing: (0, ∞)
Decreasing: (–∞, 0)

d. $f(x) = x^n$ is increasing from (–∞, ∞) when n is odd.

e.

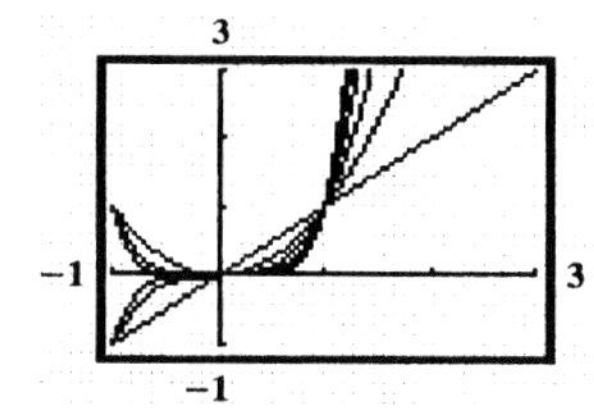

113. a. h is even if both f and g are even or if both f and g are odd.
f and g are both even:

$$h(-x) = \frac{f(-x)}{g(-x)} = \frac{f(x)}{g(x)} = h(x)$$

f and g are both odd:

$$h(-x) = \frac{f(-x)}{g(-x)} = \frac{-f(x)}{-g(x)} = \frac{f(x)}{g(x)} = h(x)$$

b. h is odd if f is odd and g is even or if f is even and g is odd.
f is odd and g is even:

$$h(-x) = \frac{f(-x)}{g(-x)} = \frac{-f(x)}{g(x)} = -\frac{f(x)}{g(x)} = -h(x)$$

f is even and g is odd:

$$h(-x) = \frac{f(-x)}{g(-x)} = \frac{f(x)}{-g(x)} = -\frac{f(x)}{g(x)} = -h(x)$$

114.

Weight at least	Cost
0 oz.	\$0.37
1	0.60
2	0.83
3	1.06
4	1.29

Section 1.4

Check Point Exercises

1. a. $m = \dfrac{-2-4}{-4-(-3)} = \dfrac{-6}{-1} = 6$

   b. $m = \dfrac{5-(-2)}{-1-4} = \dfrac{7}{-5} = -\dfrac{7}{5}$

2. $$y - y_1 = m(x - x_1)$$
$$y - (-5) = 6(x - 2)$$
$$y + 5 = 6x - 12$$
$$y = 6x - 17$$

3. $m = \dfrac{-6-(-1)}{-1-(-2)} = \dfrac{-5}{1} = -5$,

   so the slope is –5. Using the point (–2, –1), we get the point slope equation:
$$y - y_1 = m(x - x_1)$$
$$y - (-1) = -5[x - (-2)]$$
$$y + 1 = -5(x + 2). \text{Solve the equation for } y:$$
$$y + 1 = -5x - 10$$
$$y = -5x - 11.$$

4. The slope m is $\frac{3}{5}$ and the y-intercept is 1, so one point on the line is (1, 0). We can find a second point on the line by using the slope $m = \frac{3}{5} = \frac{\text{Rise}}{\text{Run}}$: starting at the point (0, 1), move 3 units up and 5 units to the right, to obtain the point (5, 4).

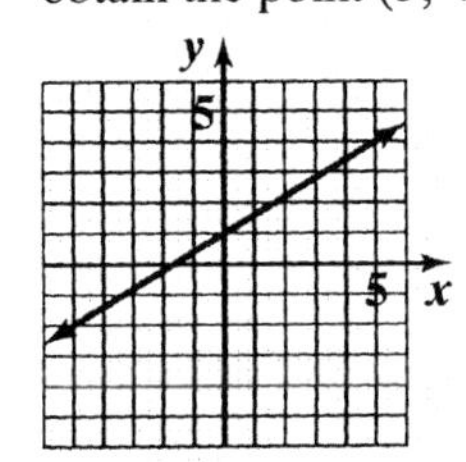

$f(x) = \dfrac{3}{5}x + 1$

5. All ordered pairs that are solutions of $x = -3$ have a value of x that is always –3. Any value can be used for y.

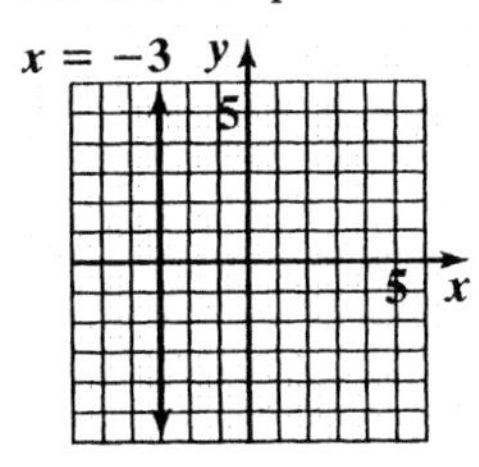

6. $3x+6y-12=0$

$$6y=-3x+12$$
$$y=\frac{-3}{6}x+\frac{12}{6}$$
$$y=-\frac{1}{2}x+2$$

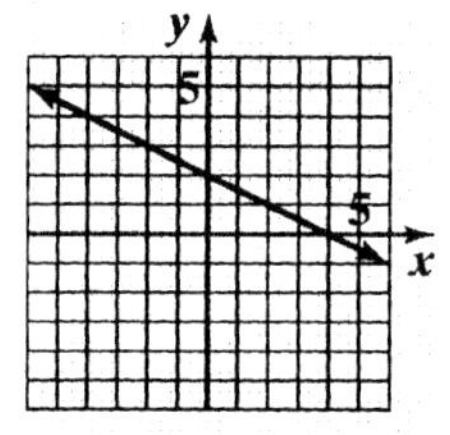

$3x + 6y - 12 = 0$

The slope is $-\frac{1}{2}$ and the y-intercept is 2.

7. Find the x-intercept:

$$3x-2y-6=0$$
$$3x-2(0)-6=0$$
$$3x-6=0$$
$$3x=6$$
$$x=2$$

Find the y-intercept:

$$3x-2y-6=0$$
$$3(0)-2y-6=0$$
$$-2y-6=0$$
$$-2y=6$$
$$y=-3$$

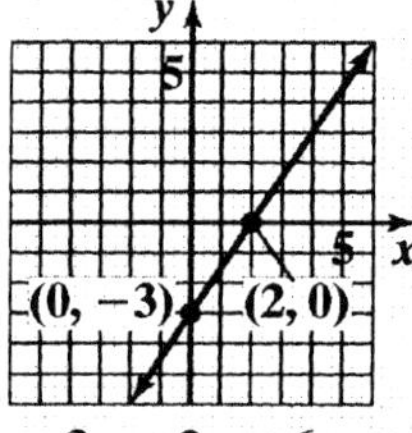

$3x - 2y = 6$

8. $m=\dfrac{\text{Change in } y}{\text{Change in } x}=\dfrac{32.8-30.0}{20-10}=\dfrac{2.8}{10}=0.28$

$$y-y_1=m(x-x_1)$$
$$y-30.0=0.28(x-10)$$
$$y-30.0=0.28x-2.8$$
$$y=0.28x+27.2$$

2020 is 50 years after 1970.

$$y=0.28x+27.2$$
$$y=0.28(50)+27.2$$
$$y=41.2$$

In 2020 the median age is expected to be 41.2.

Exercise Set 1.4

1. $m=\dfrac{10-7}{8-4}=\dfrac{3}{4}$; rises

2. $m=\dfrac{4-1}{3-2}=\dfrac{3}{1}=3$; rises

3. $m=\dfrac{2-1}{2-(-2)}=\dfrac{1}{4}$; rises

4. $m=\dfrac{4-3}{2-(-1)}=\dfrac{1}{3}$; rises

5. $m=\dfrac{2-(-2)}{3-4}=\dfrac{0}{-1}=0$; horizontal

6. $m=\dfrac{-1-(-1)}{3-4}=\dfrac{0}{-1}=0$; horizontal

7. $m=\dfrac{-1-4}{-1-(-2)}=\dfrac{-5}{1}=-5$; falls

8. $m=\dfrac{-2-(-4)}{4-6}=\dfrac{2}{-2}=-1$; falls

9. $m=\dfrac{-2-3}{5-5}=\dfrac{-5}{0}$ undefined; vertical

10. $m=\dfrac{5-(-4)}{3-3}=\dfrac{9}{0}$ undefined; vertical

11. $m=2,\ x_1=3,\ y_1=5$;
point-slope form: $y-5=2(x-3)$;
slope-intercept form: $y-5=2x-6$
$$y=2x-1$$

12. point-slope form: $y-3=4(x-1)$;
$m=4, x_1=1, y_1=3$;
slope-intercept form: $y=4x-1$

13. $m=6,\ x_1=-2,\ y_1=5$;
point-slope form: $y-5=6(x+2)$;
slope-intercept form: $y-5=6x+12$
$$y=6x+17$$

14. point-slope form: $y+1=8(x-4)$;
$m=8, x_1=4, y_1=-1$;
slope-intercept form: $y=8x-33$

15. $m = -3,\ x_1 = -2,\ y_1 = -3;$
point-slope form: $y + 3 = -3(x + 2)$;
slope-intercept form: $y + 3 = -3x - 6$
$y = -3x - 9$

16. point-slope form: $y + 2 = -5(x + 4)$;
$m = -5, x_1 = -4, y_1 = -2;$
slope-intercept form: $y = -5x - 22$

17. $m = -4,\ x_1 = -4,\ y_1 = 0;$
point-slope form: $y - 0 = -4(x + 4)$;
slope-intercept form: $y = -4(x + 4)$
$y = -4x - 16$

18. point-slope form: $y + 3 = -2(x - 0)$
$m = -2, x_1 = 0, y_1 = -3;$
slope-intercept form: $y = -2x - 3$

19. $m = -1,\ x_1 = \frac{-1}{2},\ y_1 = -2;$
point-slope form: $y + 2 = -1\left(x + \frac{1}{2}\right);$
slope-intercept form: $y + 2 = -x - \frac{1}{2}$
$y = -x - \frac{5}{2}$

20. point-slope form: $y + \frac{1}{4} = -1(x + 4);$
$m = -1, x_1 = -4, y_1 = -\frac{1}{4};$
slope-intercept form: $y = -x - \frac{17}{4}$

21. $m = \frac{1}{2},\ x_1 = 0,\ y_1 = 0;$
point-slope form: $y - 0 = \frac{1}{2}(x - 0);$
slope-intercept form: $y = \frac{1}{2}x$

22. point-slope form: $y - 0 = \frac{1}{3}(x - 0);$
$m = \frac{1}{3}, x_1 = 0, y_1 = 0;$
slope-intercept form: $y = \frac{1}{3}x$

23. $m = -\frac{2}{3},\ x_1 = 6,\ y_1 = -2;$
point-slope form: $y + 2 = -\frac{2}{3}(x - 6);$
slope-intercept form: $y + 2 = -\frac{2}{3}x + 4$
$y = -\frac{2}{3}x + 2$

24. point-slope form: $y + 4 = -\frac{3}{5}(x - 10);$
$m = -\frac{3}{5}, x_1 = 10, y_1 = -4;$
slope-intercept form: $y = -\frac{3}{5}x + 2$

25. $m = \frac{10 - 2}{5 - 1} = \frac{8}{4} = 2;$
point-slope form: $y - 2 = 2(x - 1)$ using $(x_1,\ y_1) = (1,\ 2)$, or $y - 10 = 2(x - 5)$ using $(x_1,\ y_1) = (5,\ 10)$;
slope-intercept form: $y - 2 = 2x - 2$ or
$y - 10 = 2x - 10,$
$y = 2x$

26. $m = \frac{15 - 5}{8 - 3} = \frac{10}{5} = 2;$
point-slope form: $y - 5 = 2(x - 3)$ using $(x_1, y_1) = (3, 5)$, or $y - 15 = 2(x - 8)$ using $(x_1, y_1) = (8, 15)$;
slope-intercept form: $y = 2x - 1$

27. $m = \frac{3-0}{0-(-3)} = \frac{3}{3} = 1$;

point-slope form: $y - 0 = 1(x + 3)$ using $(x_1, y_1) = (-3, 0)$, or $y - 3 = 1(x - 0)$ using $(x_1, y_1) = (0, 3)$; slope-intercept form: $y = x + 3$

28. $m = \frac{2-0}{0-(-2)} = \frac{2}{2} = 1$;

point-slope form: $y - 0 = 1(x + 2)$ using $(x_1, y_1) = (-2, 0)$, or $y - 2 = 1(x - 0)$ using $(x_1, y_1) = (0, 2)$;
slope-intercept form: $y = x + 2$

29. $m = \frac{4-(-1)}{2-(-3)} = \frac{5}{5} = 1$;

point-slope form: $y + 1 = 1(x + 3)$ using $(x_1, y_1) = (-3, -1)$, or $y - 4 = 1(x - 2)$ using $(x_1, y_1) = (2, 4)$; slope-intercept form: $y + 1 = x + 3$ or

$$y - 4 = x - 2$$
$$y = x + 2$$

30. $m = \frac{-1-(-4)}{1-(-2)} = \frac{3}{3} = 1$;

point-slope form: $y + 4 = 1(x + 2)$ using $(x_1, y_1) = (-2, -4)$, or $y + 1 = 1(x - 1)$ using $(x_1, y_1) = (1, -1)$
slope-intercept form: $y = x - 2$

31. $m = \frac{6-(-2)}{3-(-3)} = \frac{8}{6} = \frac{4}{3}$;

point-slope form: $y + 2 = \frac{4}{3}(x + 3)$ using $(x_1, y_1) = (-3, -2)$, or $y - 6 = \frac{4}{3}(x - 3)$ using $(x_1, y_1) = (3, 6)$;

slope-intercept form: $y + 2 = \frac{4}{3x} + 4$ or

$$y - 6 = \frac{4}{3}x - 4,$$
$$y = \frac{4}{3}x + 2$$

32. $m = \frac{-2-6}{3-(-3)} = \frac{-8}{6} = -\frac{4}{3}$;

point-slope form: $y - 6 = -\frac{4}{3}(x + 3)$ using $(x_1, y_1) = (-3, 6)$, or $y + 2 = -\frac{4}{3}(x - 3)$ using $(x_1, y_1) = (3, -2)$;

slope-intercept form: $y = -\frac{4}{3}x + 2$

33. $m = \frac{-1-(-1)}{4-(-3)} = \frac{0}{7} = 0$;

point-slope form: $y + 1 = 0(x + 3)$ using $(x_1, y_1) = (-3, -1)$, or $y + 1 = 0(x - 4)$ using $(x_1, y_1) = (4, -1)$;
slope-intercept form: $y + 1 = 0$, so

$$y = -1$$

34. $m = \dfrac{-5-(-5)}{6-(-2)} = \dfrac{0}{8} = 0$;

point-slope form: $y + 5 = 0(x + 2)$ using $(x_1, y_1) = (-2, -5)$, or $y + 5 = 0(x - 6)$ using $(x_1, y_1) = (6, -5)$;

slope-intercept form: $y + 5 = 0$, so

$$y = -5$$

35. $m = \dfrac{0-4}{-2-2} = \dfrac{-4}{-4} = 1$;

point-slope form: $y - 4 = 1(x - 2)$ using $(x_1, y_1) = (2, 4)$, or $y - 0 = 1(x + 2)$ using $(x_1, y_1) = (-2, 0)$;

slope-intercept form: $y - 9 = x - 2$, or

$$y = x + 2$$

36. $m = \dfrac{0-(-3)}{-1-1} = \dfrac{3}{-2} = -\dfrac{3}{2}$

point-slope form: $y + 3 = -\dfrac{3}{2}(x - 1)$ using $(x_1, y_1) = (1, -3)$, or $y - 0 = -\dfrac{3}{2}(x + 1)$ using $(x_1, y_1) = (-1, 0)$;

slope-intercept form: $y + 3 = -\dfrac{3}{2}x + \dfrac{3}{2}$, or

$$y = -\frac{3}{2}x - \frac{3}{2}$$

37. $m = \dfrac{4-0}{0-\left(-\frac{1}{2}\right)} = \dfrac{4}{\frac{1}{2}} = 8$;

point-slope form: $y - 4 = 8(x - 0)$ using $(x_1, y_1) = (0, 4)$, or $y - 0 = 8\left(x + \frac{1}{2}\right)$ using $(x_1, y_1) = \left(-\frac{1}{2}, 0\right)$; or

$y - 0 = 8\left(x + \frac{1}{2}\right)$

slope-intercept form: $y = 8x + 4$

38. $m = \dfrac{-2-0}{0-4} = \dfrac{-2}{-4} = \dfrac{1}{2}$;

point-slope form: $y - 0 = \dfrac{1}{2}(x - 4)$ using $(x_1, y_1) = (4, 0)$,

or $y + 2 = \dfrac{1}{2}(x - 0)$ using $(x_1, y_1) = (0, -2)$;

slope-intercept form: $y = \dfrac{1}{2}x - 2$

39. $m = 2$; $b = 1$

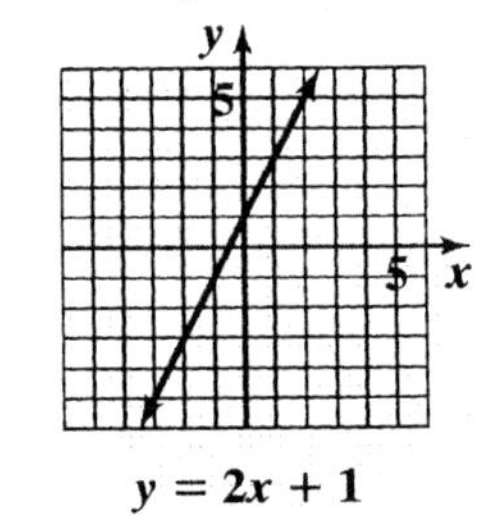

$y = 2x + 1$

40. $m = 3$; $b = 2$

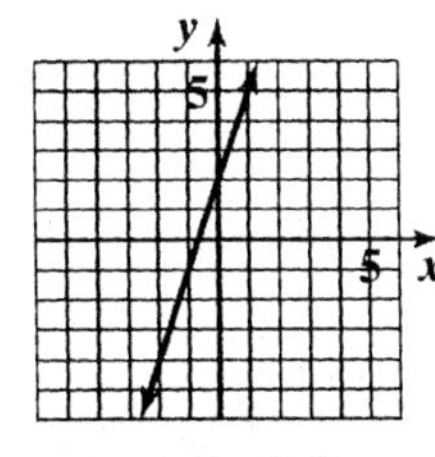

$y = 3x + 2$

41. $m = -2;\ b = 1$

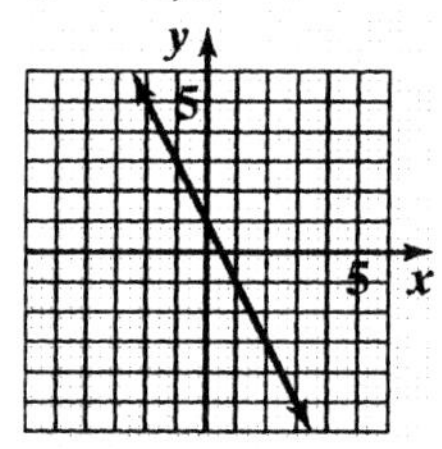

$f(x) = -2x + 1$

42. $m = -3;\ b = 2$

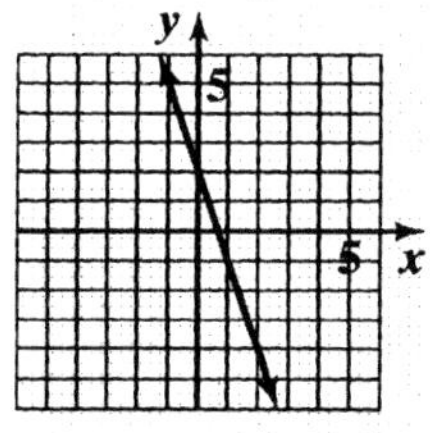

$f(x) = -3x + 2$

43. $m = \frac{3}{4};\ b = -2$

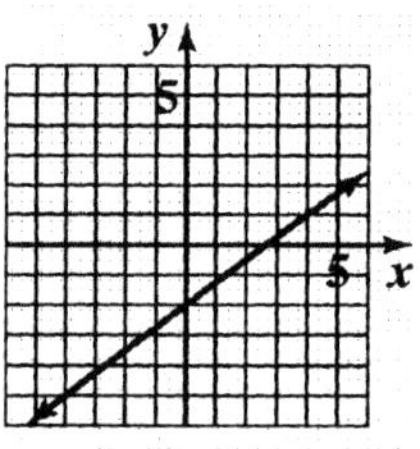

$f(x) = \frac{3}{4}x - 2$

44. $m = \frac{3}{4};\ b = -3$

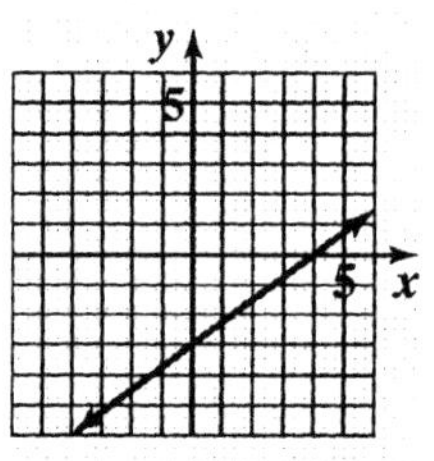

$f(x) = \frac{3}{4}x - 3$

45. $m = -\frac{3}{5};\ b = 7$

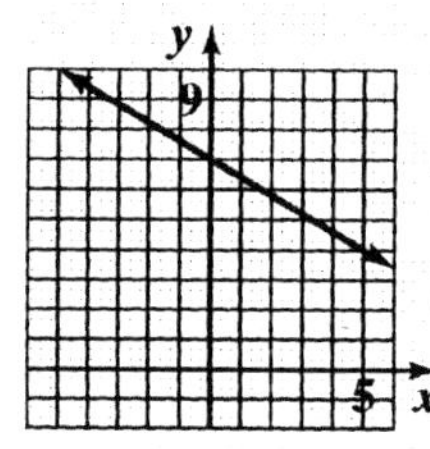

$y = -\frac{3}{5}x + 7$

46. $m = -\frac{2}{5};\ b = 6$

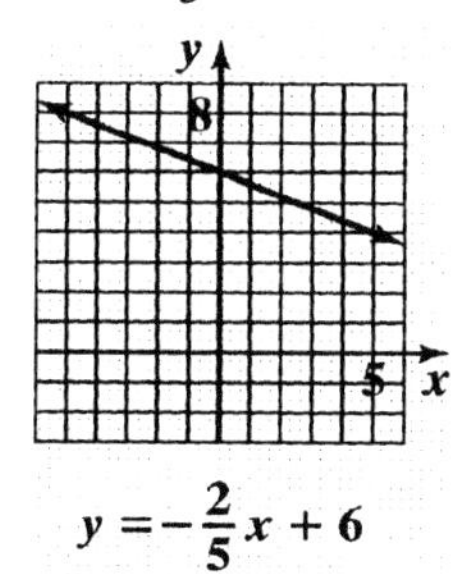

$y = -\frac{2}{5}x + 6$

47. $m = -\frac{1}{2};\ b = 0$

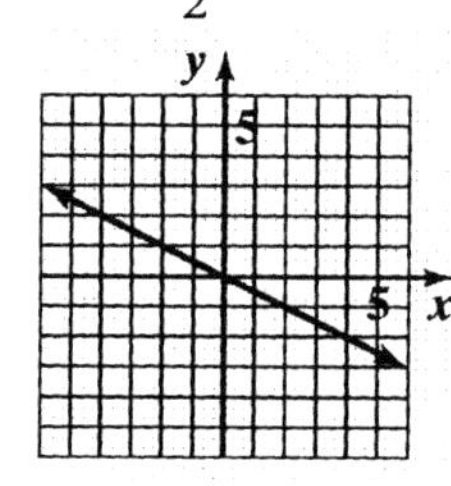

$g(x) = -\frac{1}{2}x$

48. $m = -\frac{1}{3};\ b = 0$

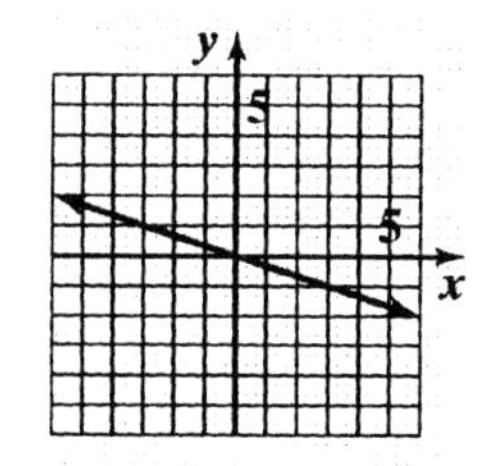

$g(x) = -\frac{1}{3}x$

49.

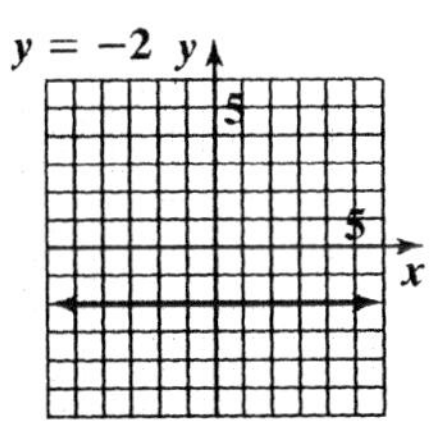

50.

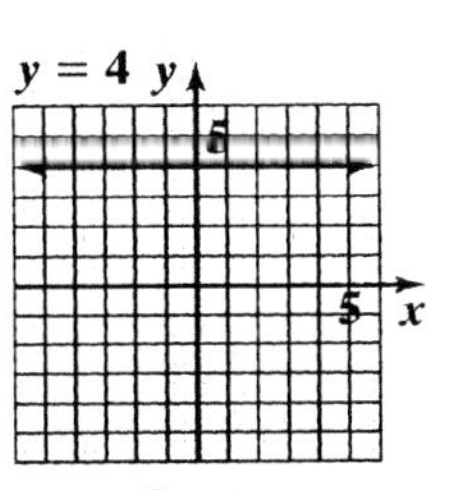

51.

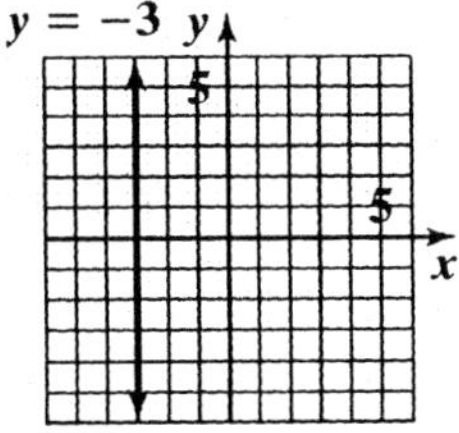

52.

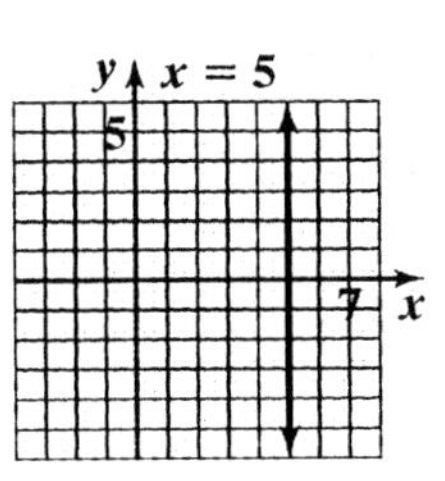

53.

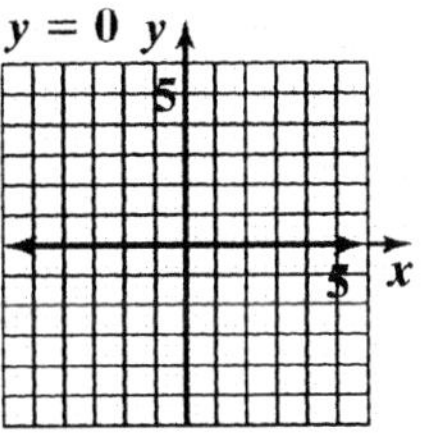

54.

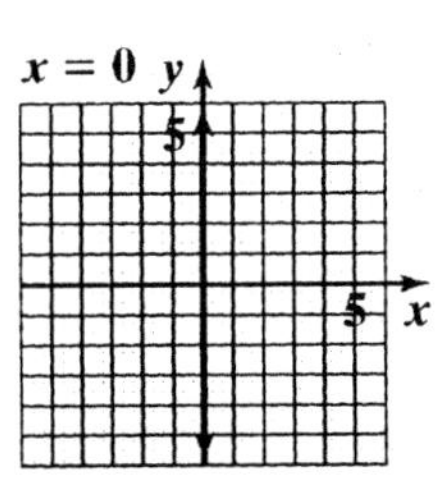

55.

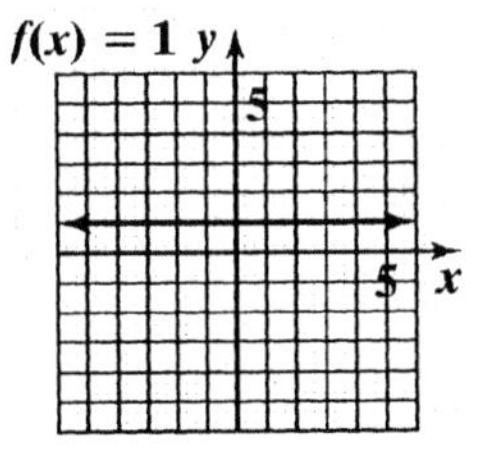

56.

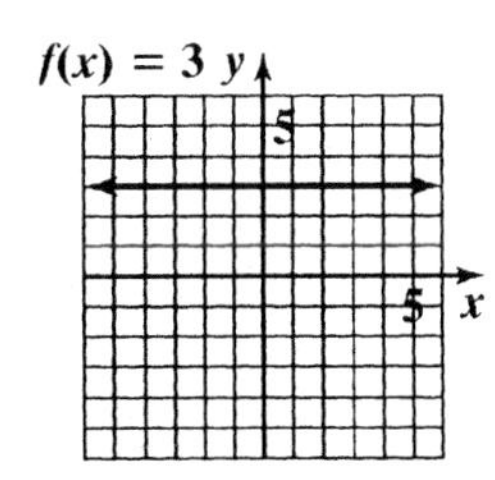

57. $3x - 18 = 0$

$3x = 18$

$x = 6$

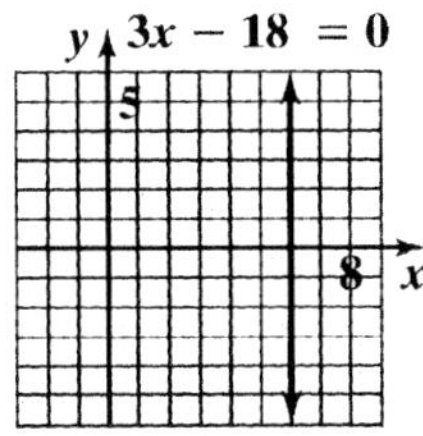

58. $3x + 12 = 0$

$3x = -12$

$x = -4$

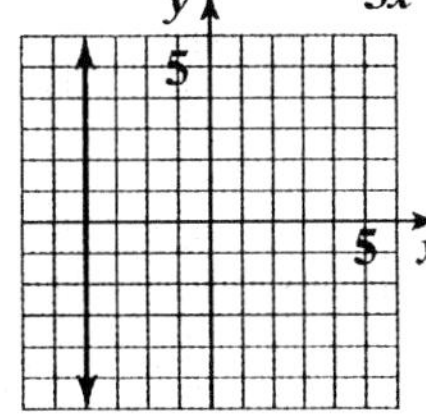

59. **a.** $3x + y - 5 = 0$

$y - 5 = -3x$

$y = -3x + 5$

b. $m = -3;\ b = 5$

c.

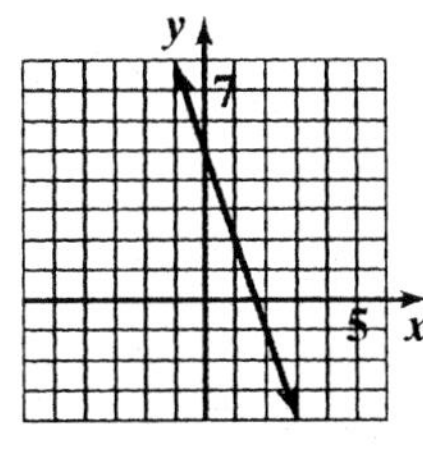

60. **a.** $4x + y - 6 = 0$

$y - 6 = -4x$

$y = -4x + 6$

b. $m = -4; b = 6$

c.

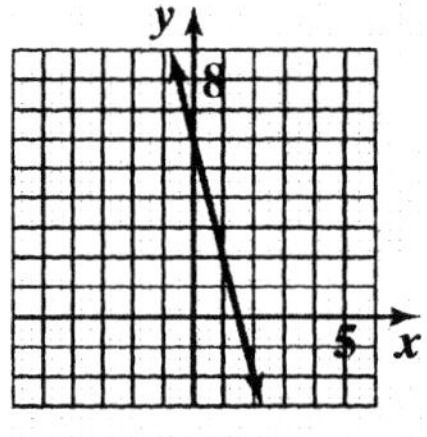

$4x + y - 6 = 0$

61. **a.**

$$\begin{aligned} 2x+3y-18&=0 \\ 2x-18&=-3y \\ -3y&=2x-18 \\ y&=\frac{2}{-3}x-\frac{18}{-3} \\ y&=-\frac{2}{3}x+6 \end{aligned}$$

b. $m=-\frac{2}{3};\ b=6$

c.

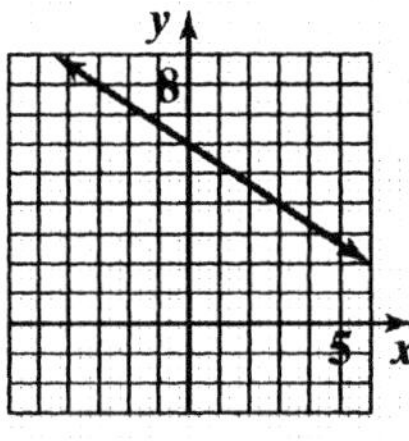

$2x + 3y - 18 = 0$

62. **a.**

$$\begin{aligned} 4x+6y+12&=0 \\ 4x+12&=-6y \\ -6y&=4x+12 \\ y&=\frac{4}{-6}x+\frac{12}{-6} \\ y&=-\frac{2}{3}x-2 \end{aligned}$$

b. $m=-\frac{2}{3};\ b=-2$

c.

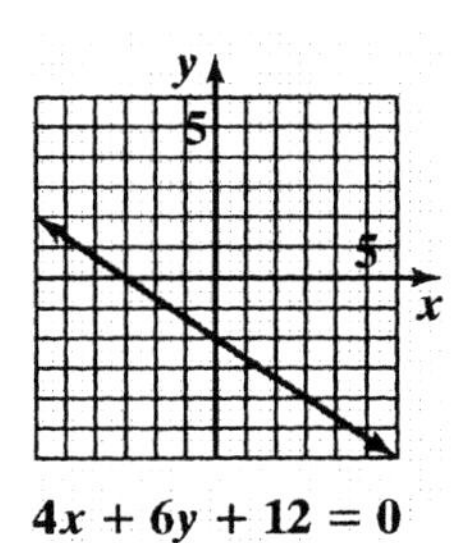

$4x + 6y + 12 = 0$

63. **a.**

$$\begin{aligned} 8x-4y-12&=0 \\ 8x-12&=4y \\ 4y&=8x-12 \\ y&=\frac{8}{4}x-\frac{12}{4} \\ y&=2x-3 \end{aligned}$$

b. $m=2;\ b=-3$

c.

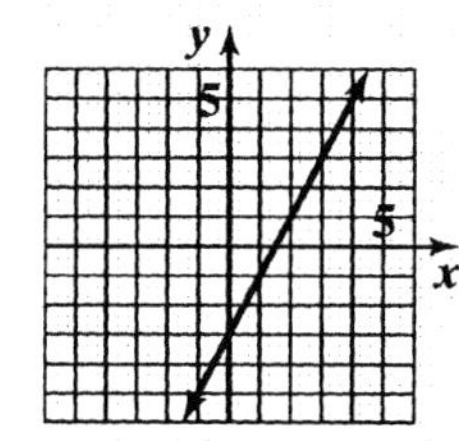

$8x - 4y - 12 = 0$

64. **a.**

$$\begin{aligned} 6x-5y-20&=0 \\ 6x-20&=5y \\ 5y&=6x-20 \\ y&=\frac{6}{5}x-\frac{20}{5} \\ y&=\frac{6}{5}x-4 \end{aligned}$$

b. $m=\frac{6}{5};b=-4$

c.

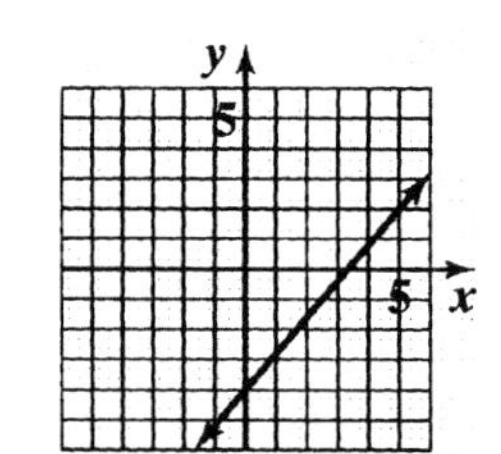

$6x - 5y - 20 = 0$

65. **a.**

$$\begin{aligned} 3y-9&=0 \\ 3y&=9 \\ y&=3 \end{aligned}$$

b. $m=0;\ b=3$

c.

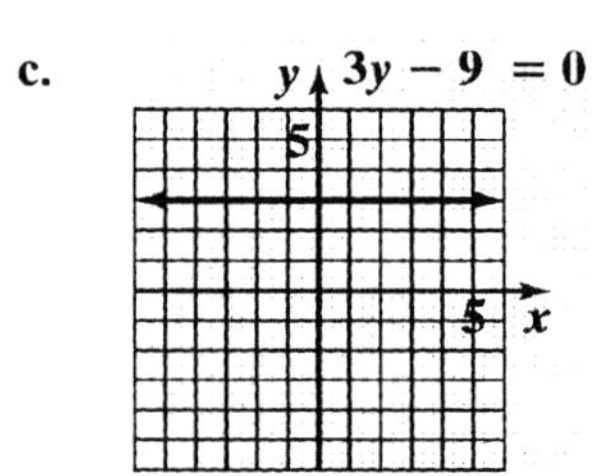

$3y - 9 = 0$

66. a. $4y + 28 = 0$
$4y = -28$
$y = -7$

b. $m = 0$; $b = -7$

c.

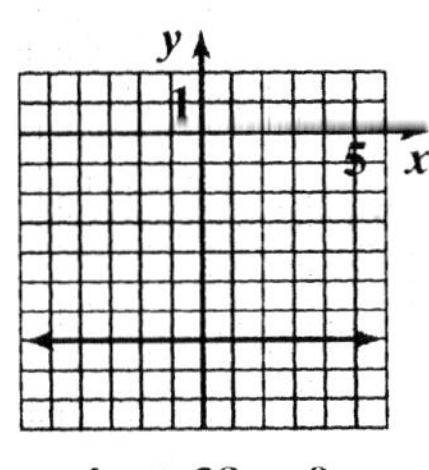

67. Find the x-intercept:
$6x - 2y - 12 = 0$
$6x - 2(0) - 12 = 0$
$6x - 12 = 0$
$6x = 12$
$x = 2$
Find the y-intercept:
$6x - 2y - 12 = 0$
$6(0) - 2y - 12 = 0$
$-2y - 12 = 0$
$-2y = 12$
$y = -6$

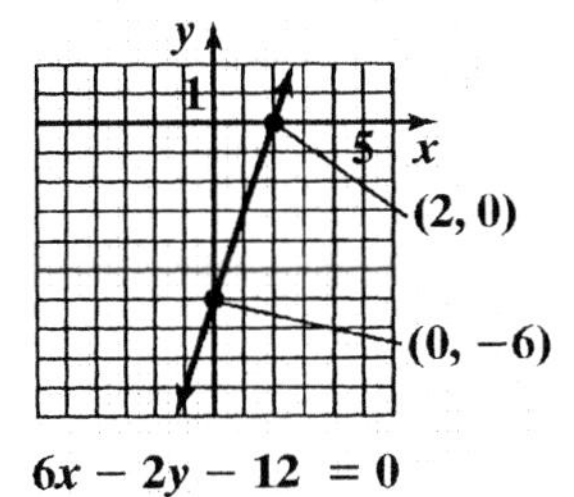

68. Find the x-intercept:
$6x - 9y - 18 = 0$
$6x - 9(0) - 18 = 0$
$6x - 18 = 0$
$6x = 18$
$x = 3$
Find the y-intercept:
$6x - 9y - 18 = 0$
$6(0) - 9y - 18 = 0$
$-9y - 18 = 0$
$-9y = 18$
$y = -2$

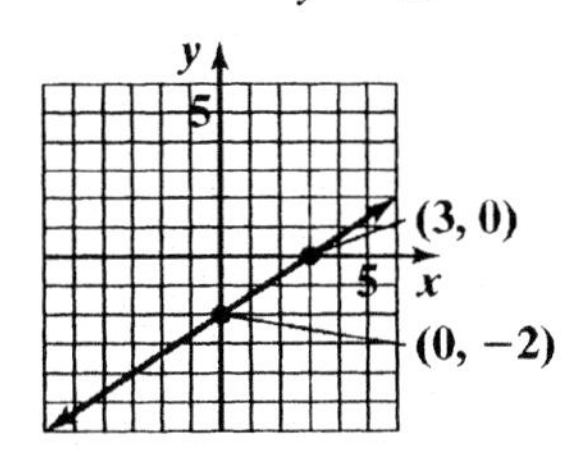

69. Find the x-intercept:
$2x + 3y + 6 = 0$
$2x + 3(0) + 6 = 0$
$2x + 6 = 0$
$2x = -6$
$x = -3$
Find the y-intercept:
$2x + 3y + 6 = 0$
$2(0) + 3y + 6 = 0$
$3y + 6 = 0$
$3y = -6$
$y = -2$

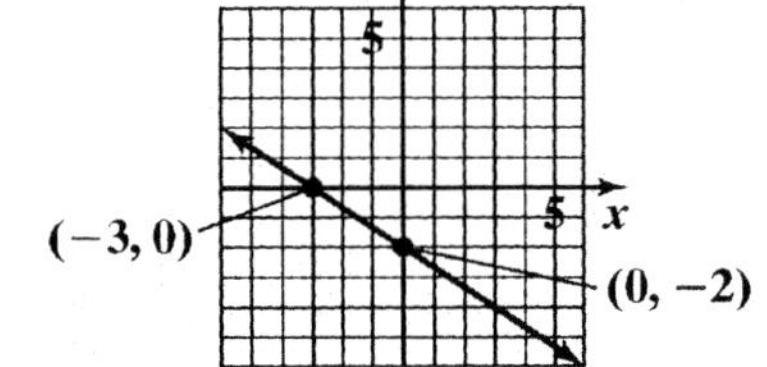

70. Find the x-intercept:

$$3x+5y+15=0$$
$$3x+5(0)+15=0$$
$$3x+15=0$$
$$3x=-15$$
$$x=-5$$

Find the y-intercept:

$$3x+5y+15=0$$
$$3(0)+5y+15=0$$
$$5y+15=0$$
$$5y=-15$$
$$y=-3$$

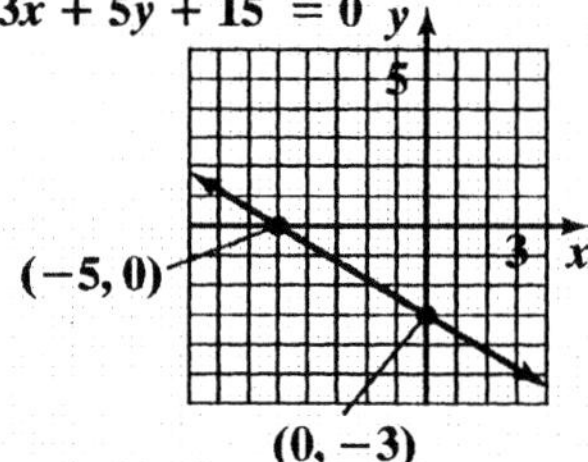

71. Find the x-intercept:

$$8x-2y+12=0$$
$$8x-2(0)+12=0$$
$$8x+12=0$$
$$8x=-12$$
$$\frac{8x}{8}=\frac{-12}{8}$$
$$x=\frac{-3}{2}$$

Find the y-intercept:

$$8x-2y+12=0$$
$$8(0)-2y+12=0$$
$$-2y+12=0$$
$$-2y=-12$$
$$y=-6$$

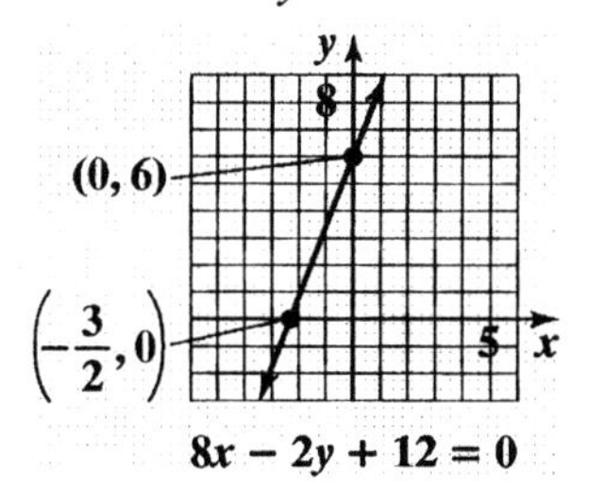

72. Find the x-intercept:

$$6x-3y+15=0$$
$$6x-3(0)+15=0$$
$$6x+15=0$$
$$6x=-15$$
$$\frac{6x}{6}=\frac{-15}{6}$$
$$x=-\frac{5}{2}$$

Find the y-intercept:

$$6x-3y+15=0$$
$$6(0)-3y+15=0$$
$$-3y+15=0$$
$$-3y=-15$$
$$y=5$$

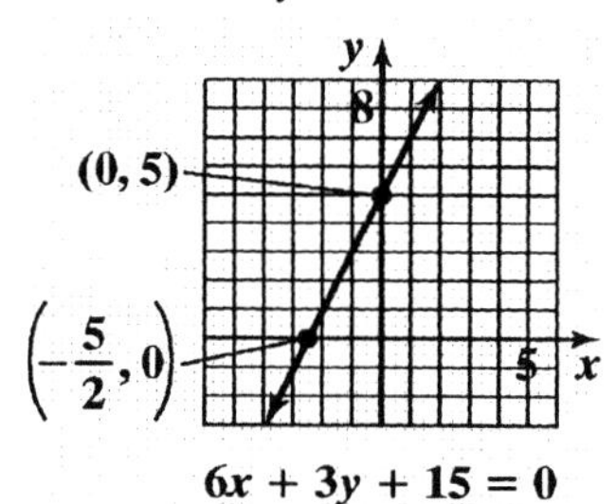

73. $m=\dfrac{0-a}{b-0}=\dfrac{-a}{b}=-\dfrac{a}{b}$

Since a and b are both positive, $-\dfrac{a}{b}$ is negative. Therefore, the line falls.

74. $m=\dfrac{-b-0}{0-(-a)}=\dfrac{-b}{a}=-\dfrac{b}{a}$

Since a and b are both positive, $-\dfrac{b}{a}$ is negative. Therefore, the line falls.

75. $m=\dfrac{(b+c)-b}{a-a}=\dfrac{c}{0}$

The slope is undefined.
The line is vertical.

76. $m=\dfrac{(a+c)-c}{a-(a-b)}=\dfrac{a}{b}$

Since a and b are both positive, $\dfrac{a}{b}$ is positive. Therefore, the line rises.

77. $Ax + By = C$

$By = -Ax + C$

$y = -\frac{A}{B}x + \frac{C}{B}$

The slope is $-\frac{A}{B}$ and the $y-$ intercept is $\frac{C}{B}$.

78. $Ax = By - C$

$Ax + C = By$

$\frac{A}{B}x + \frac{C}{B} = y$

The slope is $\frac{A}{B}$ and the $y-$ intercept is $\frac{C}{B}$.

79. $-3 = \frac{4-y}{1-3}$

$-3 = \frac{4-y}{-2}$

$6 = 4 - y$

$2 = -y$

$-2 = y$

80. $\frac{1}{3} = \frac{-4-y}{4-(-2)}$

$\frac{1}{3} = \frac{-4-y}{4+2}$

$\frac{1}{3} = \frac{-4-y}{6}$

$6 = 3(-4-y)$

$6 = -12 - 3y$

$18 = -3y$

$-6 = y$

81. $3x - 4f(x) = 6$

$-4f(x) = -3x + 6$

$f(x) = \frac{3}{4}x - \frac{3}{2}$

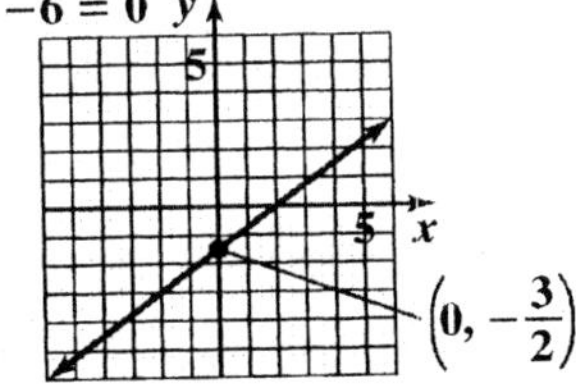

82. $6x - 5f(x) = 20$

$-5f(x) = -6x + 20$

$f(x) = \frac{6}{5}x - 4$

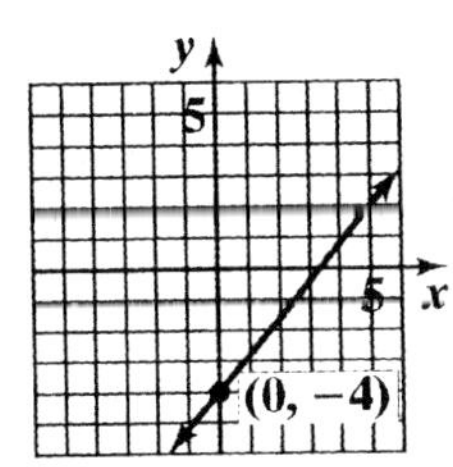

$6x - 5f(x) - 20 = 0$

83. Using the slope-intercept form for the equation of a line:

$-1 = -2(3) + b$

$-1 = -6 + b$

$5 = b$

84. $-6 = -\frac{3}{2}(2) + b$

$-6 = -3 + b$

$-3 = b$

85. m_1, m_3, m_2, m_4

86. b_2, b_1, b_4, b_3

87. **a.** First we must find the slope using $(10,16)$ and $(16,12.7)$.

$m = \frac{12.7-16}{16-10} = -\frac{3.3}{6} = -0.55$

Then use the slope and one of the points to write the equation in point-slope form.

$y - y_1 = m(x - x_1)$

$y - 16 = -0.55(x - 10)$

or

$y - 12.7 = -0.55(x - 16)$

b. $y - 16 = -0.55(x - 10)$

$y - 16 = -0.55x + 5.5$

$y = -0.55x + 21.5$

$f(x) = -0.55x + 21.5$

c. $f(20) = -0.55(20) + 21.5 = 10.5$

The linear function predicts 10.5% of adult women will be on weight-loss diets in 2007.

88. a. First, find the slope.

$$m = \frac{7.1-8}{16-10} = \frac{-0.9}{6} = -0.15$$

Use the slope and point to write the equation in point-slope form.

$y - 8 = -0.15(x - 10)$

b. Solve for y to obtain the slope-intercept form.

$y - 8 = -0.15(x - 10)$

$y - 8 = -0.15x + 1.5$

$y = -0.15x + 9.5$

$f(x) = -0.15x + 9.5$

c. $f(20) = -0.15(20) + 9.5 = 6.5\%$

89. a. points: $(0, 73.7), (5, 74.7), (10, 75.4)$
$(15, 75.8), (19, 76.7), (20, 77.0), (21, 77.2)$

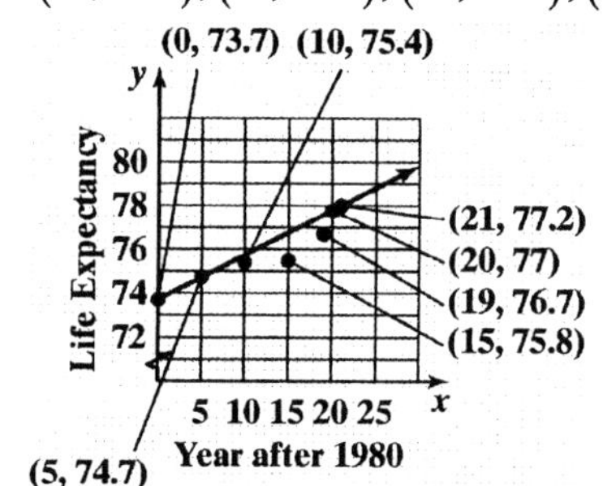

b. $m = \frac{\text{Change in } y}{\text{Change in } x} = \frac{77.0 - 74.7}{20 - 5} \approx 0.15$

$y - y_1 = m(x - x_1)$

$y - 74.7 = 0.15(x - 5)$ [point-slope]

$y - 74.7 = 0.15x - 0.75$

$y = 0.15x + 73.95$ [slope-intercept]

c. $E(x) = 0.15x + 73.95$

$E(40) = 0.15(40) + 73.95$

$= 79.95$

In 2020 the life expectancy is expected to be 79.95.

90. a. points: $(0, 33), (1, 35)$
$(2, 37), (3, 38), (4, 41)$

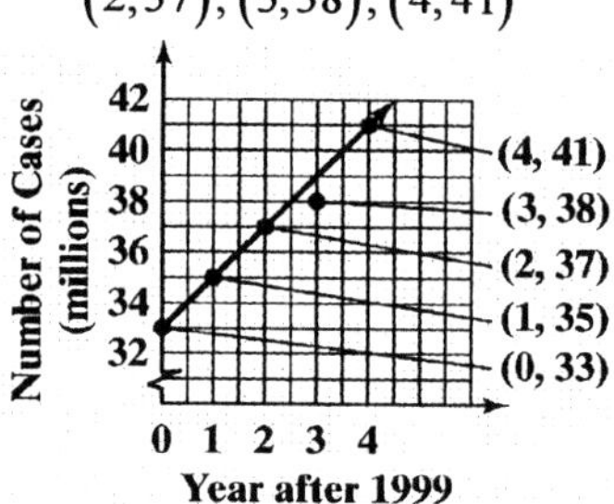

b. $m = \frac{\text{Change in } y}{\text{Change in } x} = \frac{41 - 35}{4 - 1} \approx 2$

$y - y_1 = m(x - x_1)$

$y - 35 = 2(x - 1)$ [point-slope]

$y - 35 = 2x - 2$

$y = 2x + 33$ [slope-intercept]

c. $A(x) = 2x + 33$

$A(11) = 2(11) + 33$

$= 55$

In 2010 there is expected to be 55 million cases.

91. (10, 230) (60, 110) Points may vary.

$$m = \frac{110 - 230}{60 - 10} = -\frac{120}{50} = -2.4$$

$y - 230 = -2.4(x - 10)$

$y - 230 = -2.4x + 24$

$y = -2.4x + 254$

Answers may vary for predictions.

100. Two points are (0,4) and (10,24).

$$m = \frac{24 - 4}{10 - 0} = \frac{20}{10} = 2.$$

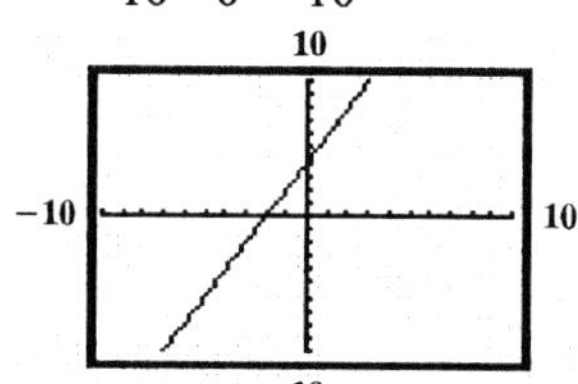

101. Two points are (0, 6) and (10, –24).

$m = \frac{-24-6}{10-0} = \frac{-30}{10} = -3.$

Check: $y = mx + b$: $y = -3x + 6$.

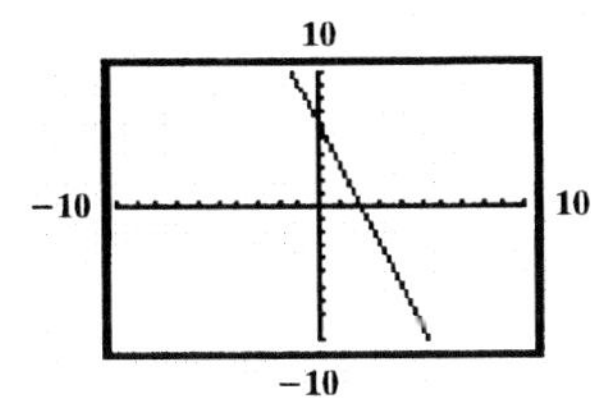

102. Two points are (0,–5) and (10,–10).

$m = \frac{-10-(-5)}{10-0} = \frac{-5}{10} = -\frac{1}{2}.$

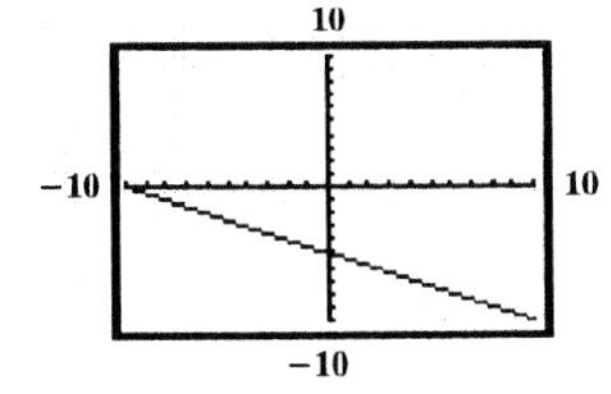

103. Two points are (0, –2) and (10, 5.5).

$m = \frac{5.5-(-2)}{10-0} = \frac{7.5}{10} = 0.75 \text{ or } \frac{3}{4}.$

Check: $y = mx + b$: $y = \frac{3}{4}x - 2$.

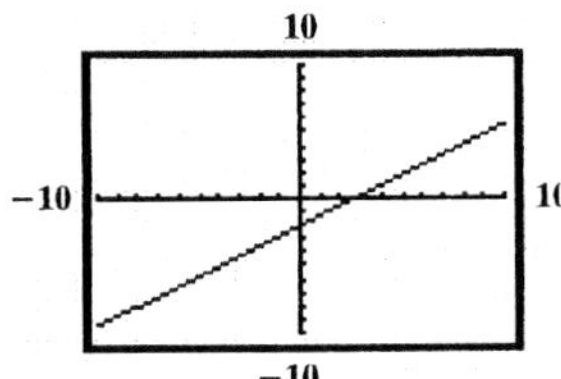

104. a. Enter data from table.

b.

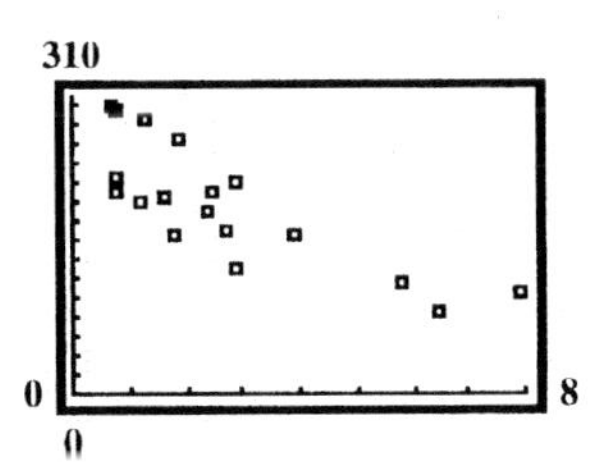

c. $a = -22.96876741$
$b = 260.5633751$
$r = -0.8428126855$

d.

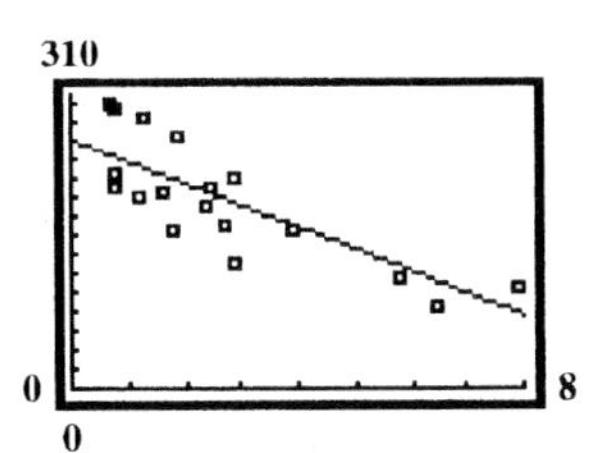

105. Statement **c.** is true.

Statement **a.** is false. One nonnegative slope is 0. A line with slope equal to zero does not rise from left to right.

Statement **b.** is false. Slope-intercept form is $y = mx + b$. Vertical lines have equations of the form $x = a$. Equations of this form have undefined slope and cannot be written in slope-intercept form.

Statement **d.** is false. The graph of $x = 7$ is a vertical line through the point (7, 0).

106. We are given that the $x-$intercept is -2 and the $y-$intercept is 4 . We can use the points $(-2,0)$ and $(0,4)$ to find the slope.

$m = \frac{4-0}{0-(-2)} = \frac{4}{0+2} = \frac{4}{2} = 2$

Using the slope and one of the intercepts, we can write the line in point-slope form.

$$y - y_1 = m(x - x_1)$$
$$y - 0 = 2(x - (-2))$$
$$y = 2(x + 2)$$
$$y = 2x + 4$$
$$-2x + y = 4$$

Find the x– and y–coefficients for the equation of the line with right-hand-side equal to 12. Multiply both sides of $-2x + y = 4$ by 3 to obtain 12 on the right-hand-side.

$$-2x + y = 4$$
$$3(-2x + y) = 3(4)$$
$$-6x + 3y = 12$$

Therefore, the coefficient of x is –6 and the coefficient of y is 3.

107. We are given that the y – intercept is -6 and the slope is $\frac{1}{2}$.

So the equation of the line is $y = \frac{1}{2}x - 6$.

We can put this equation in the form $ax + by = c$ to find the missing coefficients.

$$y = \frac{1}{2}x - 6$$
$$y - \frac{1}{2}x = -6$$
$$2\left(y - \frac{1}{2}x\right) = 2(-6)$$
$$2y - x = -12$$
$$x - 2y = 12$$

Therefore, the coefficient of x is 1 and the coefficient of y is –2.

109. Let (25, 40) and (125, 280) be ordered pairs (M, E) where M is degrees Madonna and E is degrees Elvis. Then $m = \frac{280 - 40}{125 - 25} = \frac{240}{100} = 2.4$. Using $(x_1, y_1) = (25, 40)$, point-slope form tells us that $E - 40 = 2.4\ (M - 25)$ or $E = 2.4\ M - 20$.

Section 1.5

Check Point Exercises

1. The slope of the line $y = 3x + 1$ is 3.

$$y - y_1 = m(x - x_1)$$
$$y - 5 = 3(x - (-2))$$
$$y - 5 = 3(x + 2) \text{ point-slope}$$
$$y - 5 = 3x + 6$$
$$y = 3x + 11 \text{ slope-intercept}$$

2. **a.** Write the equation in slope-intercept form:

$$x + 3y - 12 = 0$$
$$3y = -x + 12$$
$$y = -\frac{1}{3}x + 4$$

The slope of this line is $-\frac{1}{3}$ thus the slope of any line perpendicular to this line is 3.

b. Use $m = 3$ and the point $(-2, -6)$ to write the equation.

$$y - y_1 = m(x - x_1)$$
$$y - (-6) = 3(x - (-2))$$
$$y + 6 = 3(x + 2)$$
$$y + 6 = 3x + 6$$
$$-3x + y = 0$$
$$3x - y = 0 \text{ general form}$$

3. $m = \frac{\text{Change in } y}{\text{Change in } x} = \frac{12 - 10}{2010 - 1995} = \frac{2}{15} \approx 0.13$

The slope indicates that the number of U.S. men living alone is projected to increase by 0.13 million each year.

4. **a.** $\frac{f(x_2) - f(x_1)}{x_2 - x_1} = \frac{1^3 - 0^3}{1 - 0} = 1$

b. $\frac{f(x_2) - f(x_1)}{x_2 - x_1} = \frac{2^3 - 1^3}{2 - 1} = \frac{8 - 1}{1} = 7$

c. $\frac{f(x_2) - f(x_1)}{x_2 - x_1} = \frac{0^3 - (-2)^3}{0 - (-2)} = \frac{8}{2} = 4$

5. $\frac{f(x_2) - f(x_1)}{x_2 - x_1} = \frac{f(3) - f(1)}{3 - 1} = \frac{0.05 - 0.03}{3 - 1} = 0.01$

6. **a.**

$$s(1) = 4(1)^2 = 4$$
$$s(2) = 4(2)^2 = 16$$
$$\frac{\Delta s}{\Delta t} = \frac{16 - 4}{2 - 1} = 12 \text{ feet per second}$$

b.

$$s(1) = 4(1)^2 = 4$$
$$s(1.5) = 4(1.5)^2 = 9$$
$$\frac{\Delta s}{\Delta t} = \frac{9 - 4}{1.5 - 1} = 10 \text{ feet per second}$$

c.

$$s(1) = 4(1)^2 = 4$$
$$s(1.01) = 4(1.01)^2 = 4.0804$$
$$\frac{\Delta s}{\Delta t} = \frac{4.0804 - 4}{1.01 - 1} = 8.04 \text{ feet per second}$$

Exercise Set 1.5

1. Since L is parallel to $y = 2x$, we know it will have slope $m = 2$. We are given that it passes through (4, 2). We use the slope and point to write the equation in point-slope form.

$$y - y_1 = m(x - x_1)$$
$$y - 2 = 2(x - 4)$$

Solve for y to obtain slope-intercept form.

$$y - 2 = 2(x - 4)$$
$$y - 2 = 2x - 8$$
$$y = 2x - 6$$

In function notation, the equation of the line is $f(x) = 2x - 6$.

2. L will have slope $m = -2$. Using the point and the slope, we have $y - 4 = -2(x - 3)$. Solve for y to obtain slope-intercept form.

$$y - 4 = -2x + 6$$
$$y = -2x + 10$$
$$f(x) = -2x + 10$$

3. Since L is perpendicular to $y = 2x$, we know it will have slope $m = -\frac{1}{2}$. We are given that it passes through
(2, 4). We use the slope and point to write the equation in point-slope form.

$$y - y_1 = m(x - x_1)$$
$$y - 4 = -\frac{1}{2}(x - 2)$$

Solve for y to obtain slope-intercept form.

$$y - 4 = -\frac{1}{2}(x - 2)$$
$$y - 4 = -\frac{1}{2}x + 1$$
$$y = -\frac{1}{2}x + 5$$

In function notation, the equation of the line is $f(x) = -\frac{1}{2}x + 5$.

4. L will have slope $m = \frac{1}{2}$. The line passes through (–1, 2). Use the slope and point to write the equation in point-slope form.

$$y - 2 = \frac{1}{2}(x - (-1))$$
$$y - 2 = \frac{1}{2}(x + 1)$$

Solve for y to obtain slope-intercept form.

$$y - 2 = \frac{1}{2}x + \frac{1}{2}$$
$$y = \frac{1}{2}x + \frac{1}{2} + 2$$

$y = \frac{1}{2}x + \frac{5}{2}$

$f(x) = \frac{1}{2}x + \frac{5}{2}$

5. $m = -4$ since the line is parallel to $y = -4x + 3$; $x_1 = -8$, $y_1 = -10$;
point-slope form: $y + 10 = -4(x + 8)$
slope-intercept form: $y + 10 = -4x - 32$
$y = -4x - 42$

6. $m = -5$ since the line is parallel to $y = -5x + 4$; $x_1 = -2$, $y_1 = -7$;
point-slope form: $y + 7 = -5(x + 2)$
slope-intercept form: $y + 7 = -5x - 10$
$y = -5x - 17$

7. $m = -5$ since the line is perpendicular to $y = \frac{1}{5}x + 6$; $x_1 = 2$, $y_1 = -3$;
point-slope form: $y + 3 = -5(x - 2)$
slope-intercept form: $y + 3 = -5x + 10$
$y = -5x + 7$

8. $m = -3$ since the line is perpendicular to $y = \frac{1}{3}x + 7$; $x_1 = -4$, $y_1 = 2$;
point-slope form: $y - 2 = -3(x + 4)$
slope-intercept form: $y - 2 = -3x - 12$
$y = -3x - 10$

9. $2x - 3y - 7 = 0$
$-3y = -2x + 7$
$y = \frac{2}{3}x - \frac{7}{3}$

The slope of the given line is $\frac{2}{3}$, so $m = \frac{2}{3}$ since the lines are parallel.

point-slope form: $y - 2 = \frac{2}{3}(x + 2)$

general form: $2x - 3y + 10 = 0$

10. $3x - 2y - = 0$
$-2y = -3x + 5$
$y = \frac{3}{2}x - \frac{5}{2}$

The slope of the given line is $\frac{3}{2}$, so $m = \frac{3}{2}$ since the lines are parallel.

point-slope form: $y - 3 = \frac{3}{2}(x + 1)$

general form: $3x - 2y + 9 = 0$

11. $x-2y-3=0$

$$-2y=-x+3$$

$$y=\frac{1}{2}x-\frac{3}{2}$$

The slope of the given line is $\frac{1}{2}$, so $m=-2$ since the lines are perpendicular.

point-slope form: $y+7=-2(x-4)$

general form: $2x+y-1=0$

12. $x+7y-12=0$

$$7y=-x+12$$

$$y=\frac{-1}{7}x+\frac{12}{7}$$

The slope of the given line is $-\frac{1}{7}$, so $m=7$ since the lines are perpendicular.

point-slope form: $y+9=7(x-5)$

general form: $7x-y-44=0$

13. $\frac{15-0}{5-0}=\frac{15}{5}=3$

14. $\frac{24-0}{4-0}=\frac{24}{4}=6$

15. $\frac{5^2+2\cdot5-(3^2+2\cdot3)}{5-3}$

$$=\frac{25+10-(9+6)}{2}$$

$$=\frac{20}{2}$$

$$=10$$

16. $\frac{6^2-2(6)-(3^2-2\cdot3)}{6-3}$

$$=\frac{36-12-(9-6)}{3}=\frac{21}{3}=7$$

17. $\frac{\sqrt{9}-\sqrt{4}}{9-4}=\frac{3-2}{5}=\frac{1}{5}$

18. $\frac{\sqrt{16}-\sqrt{9}}{16-9}=\frac{4-3}{7}=\frac{1}{7}$

19. a.

$s(3) = 10(3)^2 = 90$

$s(4) = 10(4)^2 = 160$

$\frac{\Delta s}{\Delta t} = \frac{160-90}{4-3} = 70$ feet per second

b.

$s(3) = 10(3)^2 = 90$

$s(3.5) = 10(3.5)^2 = 122.5$

$\frac{\Delta s}{\Delta t} = \frac{122.5-90}{3.5-3} = 65$ feet per second

c.

$s(3) = 10(3)^2 = 90$

$s(3.01) = 10(3.01)^2 = 90.601$

$\frac{\Delta s}{\Delta t} = \frac{90.601-90}{3.01-3} = 60.1$ feet per second

d.

$s(3) = 10(3)^2 = 90$

$s(3.001) = 10(3.001)^2 = 90.06$

$\frac{\Delta s}{\Delta t} = \frac{90.06-90}{3.001-3} = 60.01$ feet per second

20. a.

$s(3) = 12(3)^2 = 108$

$s(4) = 12(4)^2 = 192$

$\frac{\Delta s}{\Delta t} = \frac{108-192}{4-3} = 84$ feet per second

b.

$s(3) = 12(3)^2 = 108$

$s(3.5) = 12(3.5)^2 = 147$

$\frac{\Delta s}{\Delta t} = \frac{147-108}{3.5-2} = 78$ feet per second

c.

$s(3) = 12(3)^2 = 108$

$s(3.01) = 12(3.01)^2 = 108.7212$

$\frac{\Delta s}{\Delta t} = \frac{108.7212-108}{3.01-3} = 72.12$ feet per second

d.

$s(3) = 12(3)^2 = 108$

$s(3.001) = 12(3.001)^2 = 108.07201$

$\frac{\Delta s}{\Delta t} = \frac{108.07201-108}{3.001-3} = 72.01$ feet per second

21. Since the line is perpendicular to $x = 6$ which is a vertical line, we know the graph of f is a horizontal line with 0 slope. The graph of f passes through $(-1,5)$, so the equation of f is $f(x) = 5$.

22. Since the line is perpendicular to $x = -4$ which is a vertical line, we know the graph of f is a horizontal line with 0 slope. The graph of f passes through $(-2,6)$, so the equation of f is $f(x) = 6$.

23. First we need to find the equation of the line with $x-$intercept of 2 and $y-$intercept of -4. This line will pass through $(2,0)$ and $(0,-4)$. We use these points to find the slope.

$$m = \frac{-4-0}{0-2} = \frac{-4}{-2} = 2$$

Since the graph of f is perpendicular to this line, it will have slope $m = -\frac{1}{2}$.

Use the point $(-6,4)$ and the slope $-\frac{1}{2}$ to find the equation of the line.

$$y - y_1 = m(x - x_1)$$
$$y - 4 = -\frac{1}{2}(x - (-6))$$
$$y - 4 = -\frac{1}{2}(x + 6)$$
$$y - 4 = -\frac{1}{2}x - 3$$
$$y = -\frac{1}{2}x + 1$$
$$f(x) = -\frac{1}{2}x + 1$$

24. First we need to find the equation of the line with $x-$intercept of 3 and $y-$intercept of -9. This line will pass through $(3,0)$ and $(0,-9)$. We use these points to find the slope.

$$m = \frac{-9-0}{0-3} = \frac{-9}{-3} = 3$$

Since the graph of f is perpendicular to this line, it will have slope $m = -\frac{1}{3}$.

Use the point $(-5,6)$ and the slope $-\frac{1}{3}$ to find the equation of the line.

$$y - y_1 = m(x - x_1)$$
$$y - 6 = -\frac{1}{3}(x - (-5))$$
$$y - 6 = -\frac{1}{3}(x + 5)$$
$$y - 6 = -\frac{1}{3}x - \frac{5}{3}$$
$$y = -\frac{1}{3}x + \frac{13}{3}$$
$$f(x) = -\frac{1}{3}x + \frac{13}{3}$$

25. First put the equation $3x-2y-4=0$ in slope-intercept form.

$$3x-2y-4=0$$
$$-2y=-3x+4$$
$$y=\frac{3}{2}x-2$$

The equation of f will have slope $-\frac{2}{3}$ since it is perpendicular to the line above and the same $y-$ intercept -2.

So the equation of f is $f(x)=-\frac{2}{3}x-2$.

26. First put the equation $4x-y-6=0$ in slope-intercept form.

$$4x-y-6=0$$
$$-y=-4x+6$$
$$y=4x-6$$

The equation of f will have slope $-\frac{1}{4}$ since it is perpendicular to the line above and the same $y-$ intercept -6.

So the equation of f is $f(x)=-\frac{1}{4}x-6$.

27. The slope indicates that the global average temperature is projected to increase by 0.01 degrees Fahrenheit each year.

28. The slope indicates that drug industry spending on marketing to doctors increased by 2 billion dollars each year.

29. The slope indicates that the percentage of U.S. adults who smoked cigarettes decreased by 0.52% each year.

30. The slope indicates that the percentage of U.S. taxpayers who were audited by the IRS decreased by 0.28% each year.

31. $f(x)=13x+222$

32. $f(x)=18.50x+135$

33. $f(x)=-2.40x+52.40$

34. $f(x)=-2.80x+46.80$

35. $\frac{f(x_2)-f(x_1)}{x_2-x_1}=\frac{f(2003)-f(1997)}{2003-1997}=\frac{25.2-32.5}{2003-1997}\approx -1.22$

36. $\frac{f(x_2)-f(x_1)}{x_2-x_1}=\frac{f(2003)-f(1997)}{2003-1997}=\frac{13.3-10.1}{2003-1997}\approx 0.53$

43.

$$y=\frac{1}{3}x+1$$
$$y=-3x-2$$

a. The lines are perpendicular because their slopes are negative reciprocals of each other. This is verified because product of their slopes is -1.

b. 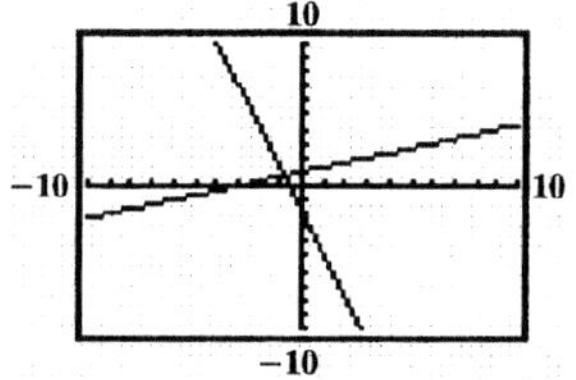

The lines do not appear to be perpendicular.

c. 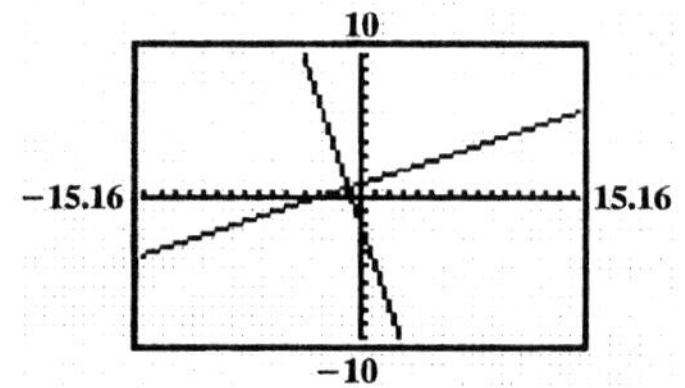

The lines appear to be perpendicular. The calculator screen is rectangular and does not have the same width and height. This causes the scale of the x–axis to differ from the scale on the y–axis despite using the same scale in the window settings. In part (b), this causes the lines not to appear perpendicular when indeed they are. The zoom square feature compensates for this and in part (c), the lines appear to be perpendicular.

44. 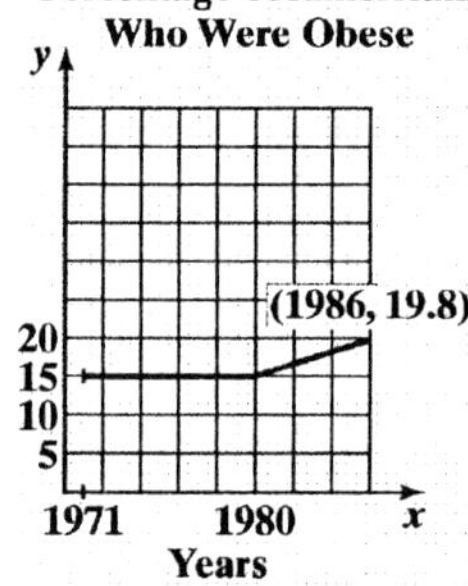

45. 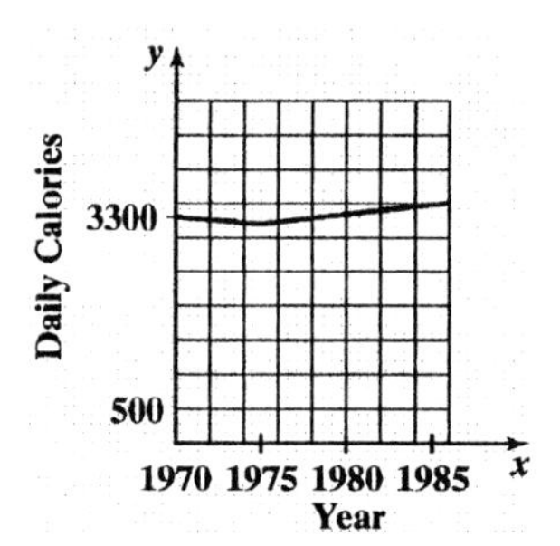

46. Write $Ax + By + C = 0$ in slope-intercept form.

$$Ax + By + C = 0$$
$$By = -Ax - C$$
$$\frac{By}{B} = \frac{-Ax}{B} - \frac{C}{B}$$
$$y = -\frac{A}{B}x - \frac{C}{B}$$

The slope of the given line is $-\frac{A}{B}$.

The slope of any line perpendicular to $Ax + By + C = 0$ is $\frac{B}{A}$.

47. The slope of the line containing $(1,-3)$ and $(-2,4)$ has slope

$m=\frac{4-(-3)}{-2-1}=\frac{4+3}{-3}=\frac{7}{-3}=-\frac{7}{3}$.

Solve $Ax+y-2=0$ for y to obtain slope-intercept form.

$Ax+y-2=0$

$y=-Ax+2$

So the slope of this line is $-A$.

This line is perpendicular to the line above so its slope is $\frac{3}{7}$. Therefore, $-A=\frac{3}{7}$ so $A=-\frac{3}{7}$.

Mid-Chapter 1 Check Point

1. The relation is not a function.
The domain is $\{1,2\}$.
The range is $\{-6,4,6\}$.

2. The relation is a function.
The domain is $\{0,2,3\}$.
The range is $\{1,4\}$.

3. The relation is a function.
The domain is $\{x\,|\,-2\le x<2\}$.
The range is $\{y\,|\,0\le y\le 3\}$.

4. The relation is not a function.
The domain is $\{x\,|\,-3<x\le 4\}$.
The range is $\{y\,|\,-1\le y\le 2\}$.

5. The relation is not a function.
The domain is $\{-2,-1,0,1,2\}$.
The range is $\{-2,-1,1,3\}$.

6. The relation is a function.
The domain is $\{x\,|\,x\le 1\}$.
The range is $\{y\,|\,y\ge -1\}$.

7. $x^2+y=5$

$y=-x^2+5$

For each value of x, there is one and only one value for y, so the equation defines y as a function of x.

8. $x+y^2=5$

$y^2=5-x$

$y=\pm\sqrt{5-x}$

Since there are values of x that give more than one value for y (for example, if $x=4$, then $y=\pm\sqrt{5-4}=\pm1$), the equation does not define y as a function of x.

9. Each value of x corresponds to exactly one value of y.

10. Domain: $(-\infty,\infty)$

11. Range: $(-\infty,4]$

12. x-intercepts: -6 and 2

13. y-intercept: 3

14. increasing: $(-\infty,-2)$

15. decreasing: $(-2,\infty)$

16. $x=-2$

17. $f(-2)=4$

18. $f(-4)=3$

19. $f(-7)=-2$ and $f(3)=-2$

20. $f(-6)=0$ and $f(2)=0$

21. $(-6,2)$

22. $f(100)$ is negative.

23. neither; $f(-x)\ne x$ and $f(-x)\ne -x$

24. $\dfrac{f(x_2)-f(x_1)}{x_2-x_1}=\dfrac{f(4)-f(-4)}{4-(-4)}=\dfrac{-5-3}{4+4}=-1$

25. $y = -2x$

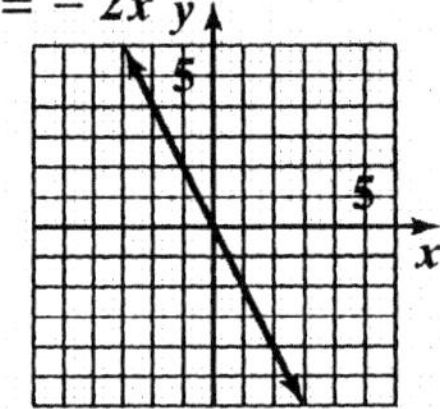

26. $y = -2$

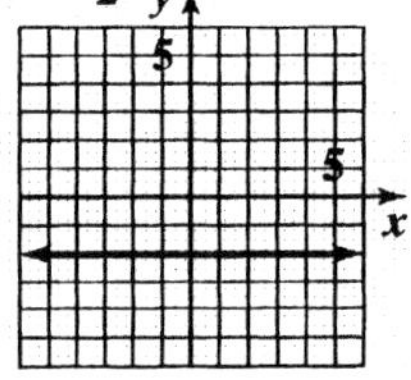

27. $x + y = -2$

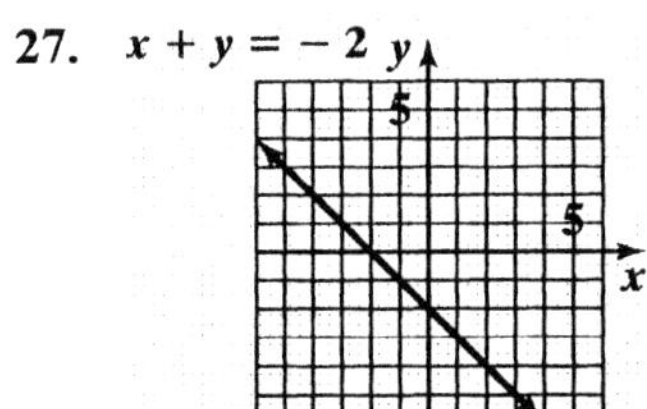

28. $y = \frac{1}{3}x - 2$

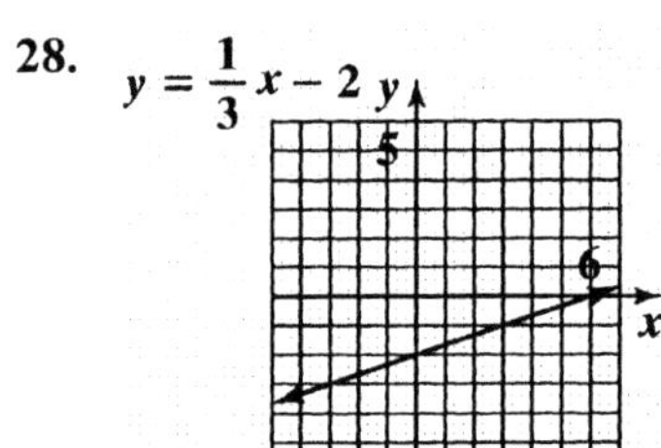

29. $x = 3.5$

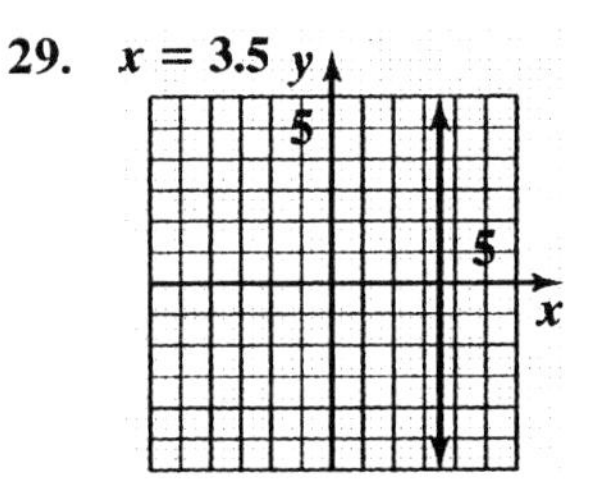

30.

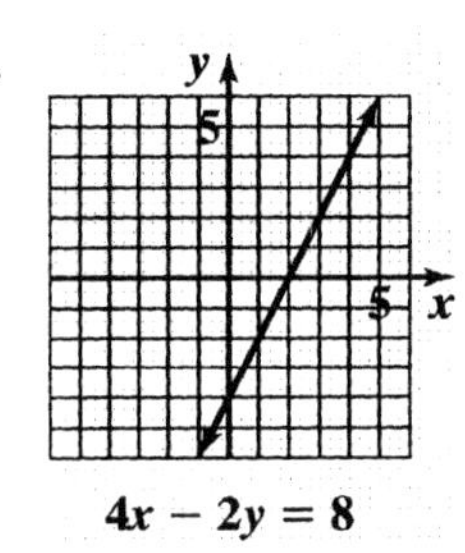

$4x - 2y = 8$

31.

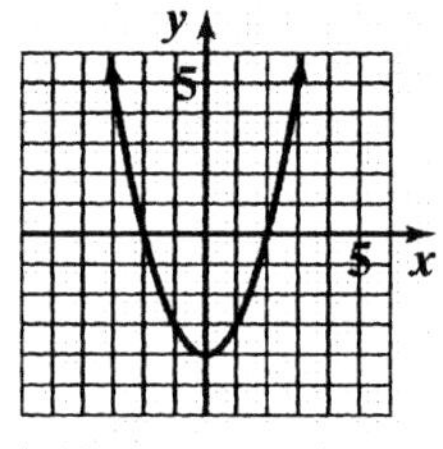

$f(x) = x^2 - 4$

32.

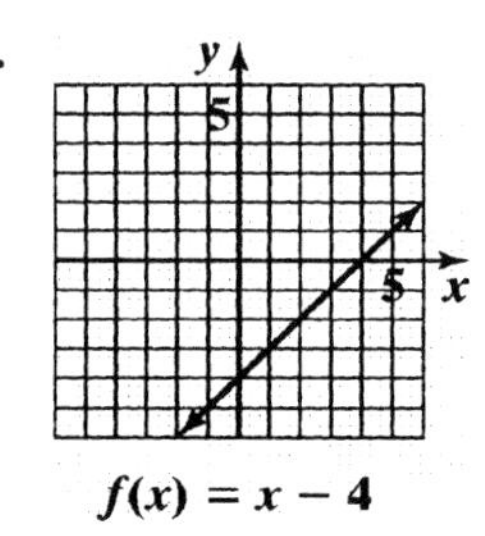

$f(x) = x - 4$

33.

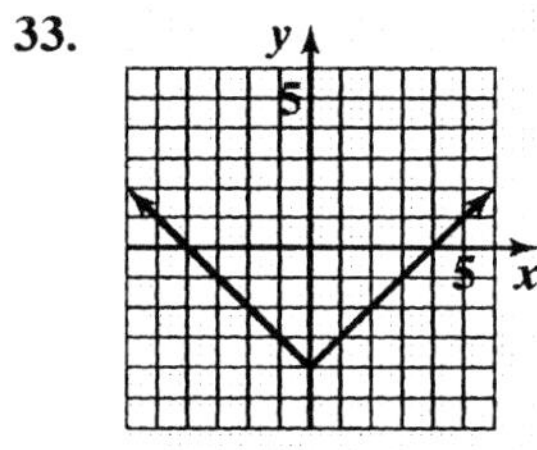

$f(x) = |x| - 4$

34. $5y = -3x$

$y = -\dfrac{3}{5}x$

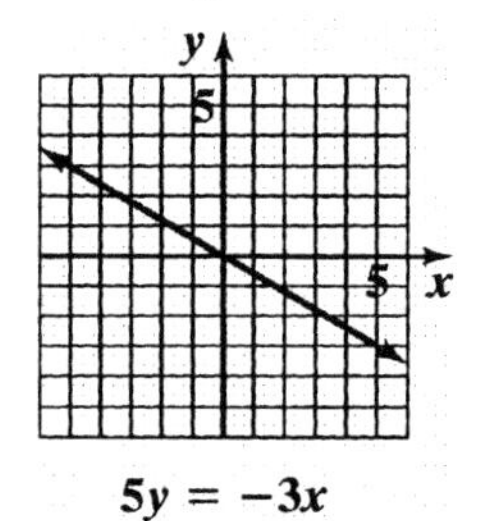

$5y = -3x$

35. $5y = 20$

$y = 4$

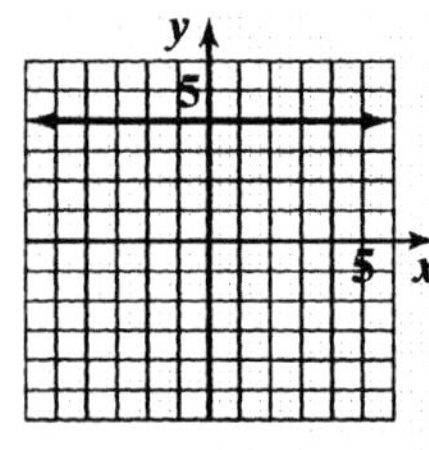

$5y = 20$

36.

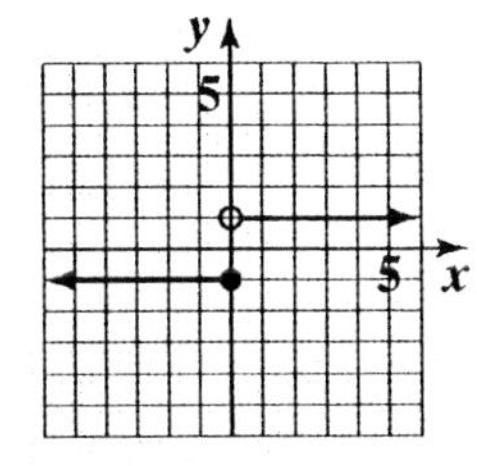

$$f(x) = \begin{cases} -1 \text{ if } x \le 0 \\ 1 \text{ if } x > 0 \end{cases}$$

37. **a.** $f(-x) = -2(-x)^2 - x - 5 = -2x^2 - x - 5$
neither; $f(-x) \ne x$ and $f(-x) \ne -x$

b.
$$\frac{f(x+h) - f(x)}{h}$$
$$= \frac{-2(x+h)^2 + (x+h) - 5 - (-2x^2 + x - 5)}{h}$$
$$= \frac{-2x^2 - 4xh - 2h^2 + x + h - 5 + 2x^2 - x + 5}{h}$$
$$= \frac{-4xh - 2h^2 + h}{h}$$
$$= \frac{h(-4x - 2h + 1)}{h}$$
$$= -4x - 2h + 1$$

38. $C(x) = \begin{cases} 30 & \text{if } 0 \le t \le 200 \\ 30 + 0.40(t - 200) & \text{if } t > 200 \end{cases}$

a. $C(150) = 30$

b. $C(250) = 30 + 0.40(250 - 200) = 50$

39.
$$y - y_1 = m(x - x_1)$$
$$y - 3 = -2(x - (-4))$$
$$y - 3 = -2(x + 4)$$
$$y - 3 = -2x - 8$$
$$y = -2x - 5$$
$$f(x) = -2x - 5$$

40. $m = \dfrac{\text{Change in } y}{\text{Change in } x} = \dfrac{1 - (-5)}{2 - (-1)} = \dfrac{6}{3} = 2$
$$y - y_1 = m(x - x_1)$$
$$y - 1 = 2(x - 2)$$
$$y - 1 = 2x - 4$$
$$y = 2x - 3$$
$$f(x) = 2x - 3$$

41.
$$3x - y - 5 = 0$$
$$-y = -3x + 5$$
$$y = 3x - 5$$
The slope of the given line is 3, and the lines are parallel, so $m = 3$.
$$y - y_1 = m(x - x_1)$$
$$y - (-4) = 3(x - 3)$$
$$y + 4 = 3x - 9$$
$$y = 3x - 13$$
$$f(x) = 3x - 13$$

42.
$$2x - 5y - 10 = 0$$
$$-5y = -2x + 10$$
$$\frac{-5y}{-5} = \frac{-2x}{-5} + \frac{10}{-5}$$
$$y = \frac{2}{5}x - 2$$
The slope of the given line is $\frac{2}{5}$, and the lines are perpendicular, so $m = -\frac{5}{2}$.
$$y - y_1 = m(x - x_1)$$
$$y - (-3) = -\frac{5}{2}(x - (-4))$$
$$y + 3 = -\frac{5}{2}x - 10$$
$$y = -\frac{5}{2}x - 13$$
$$f(x) = -\frac{5}{2}x - 13$$

43. $m_1 = \dfrac{\text{Change in } y}{\text{Change in } x} = \dfrac{0 - (-4)}{7 - 2} = \dfrac{4}{5}$

$m_2 = \dfrac{\text{Change in } y}{\text{Change in } x} = \dfrac{6 - 2}{1 - (-4)} = \dfrac{4}{5}$

The slope of the lines are equal thus the lines are parallel.

44. The slope indicates that the percentage of U.S. colleges offering distance learning is increasing by 7.8% each year.

45.
$$\frac{f(x_2) - f(x_1)}{x_2 - x_1} = \frac{f(2) - f(-1)}{2 - (-1)}$$
$$= \frac{(3(2)^2 - 2) - (3(-1)^2 - (-1))}{2 + 1}$$
$$= 2$$

Section 1.6

Check Point Exercises

1. Shift up vertically 3 units.

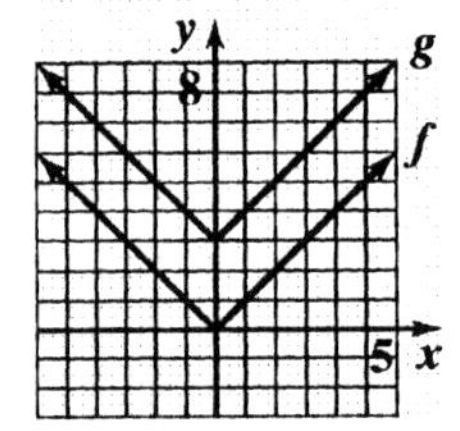

2. Shift to the right 4 units.

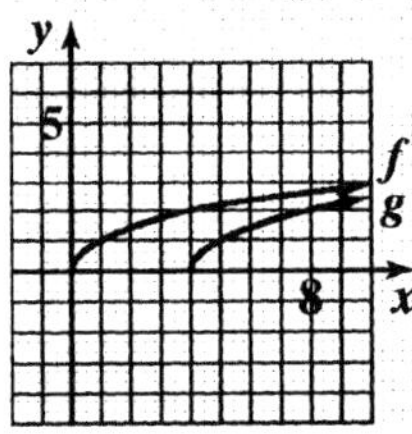

3. Shift to the right 1 unit and down 2 units.

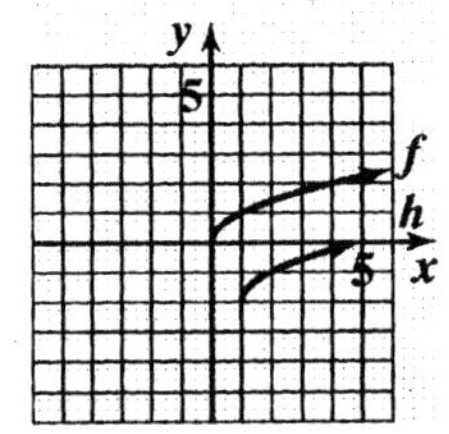

4. Reflect about the x-axis.

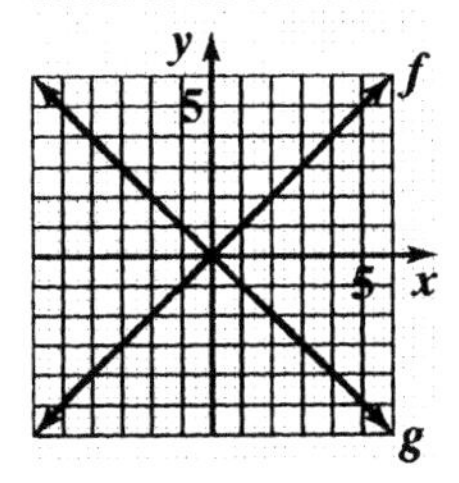

5. Reflect about the y-axis.

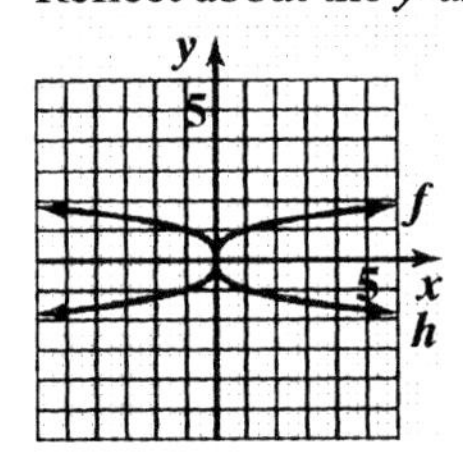

6. Vertically stretch the graph of $f(x) = |x|$.

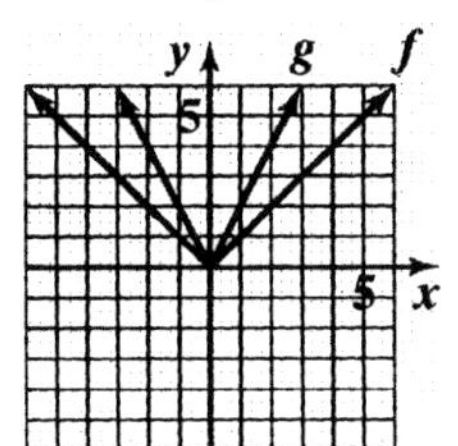

7. **a.** Horizontally shrink the graph of $y = f(x)$.

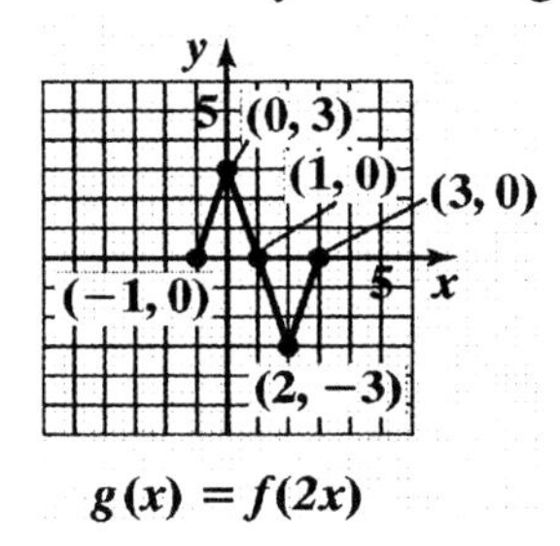

$g(x) = f(2x)$

b. Horizontally stretch the graph of $y = f(x)$.

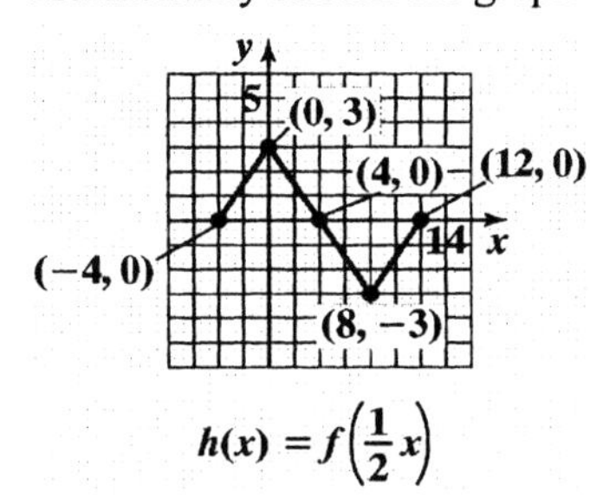

$h(x) = f\left(\frac{1}{2}x\right)$

8. The graph of $y = f(x)$ is shifted 1 unit left, shrunk by a factor of $\frac{1}{3}$, reflected about the x-axis, then shifted down 2 units.

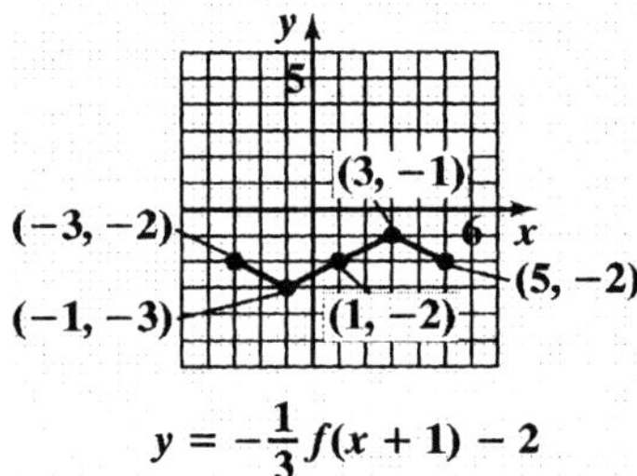

$y = -\frac{1}{3}f(x + 1) - 2$

9. The graph of $f(x) = x^2$ is shifted 1 unit right, stretched by a factor of 2, then shifted up 3 units.

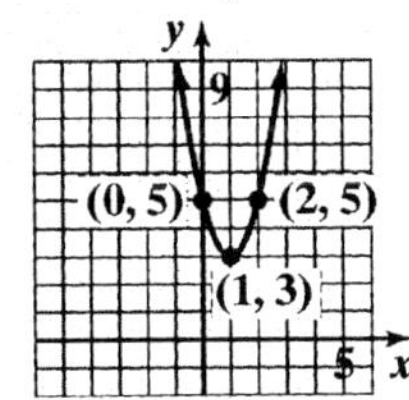

$g(x) = 2(x-1)^2 + 3$

Exercise Set 1.6

1.

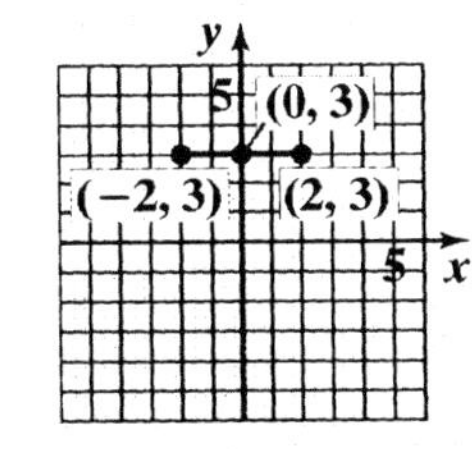

$g(x) = f(x) + 1$

2.

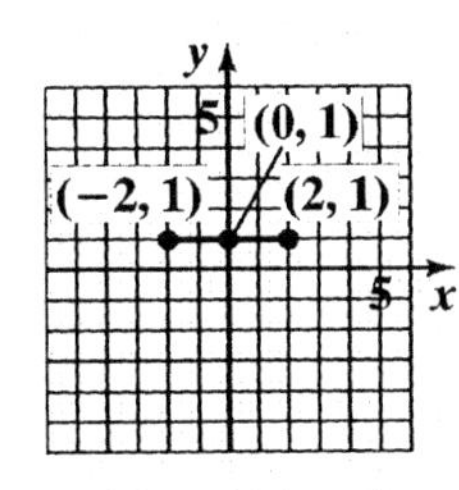

$g(x) = f(x) - 1$

3.

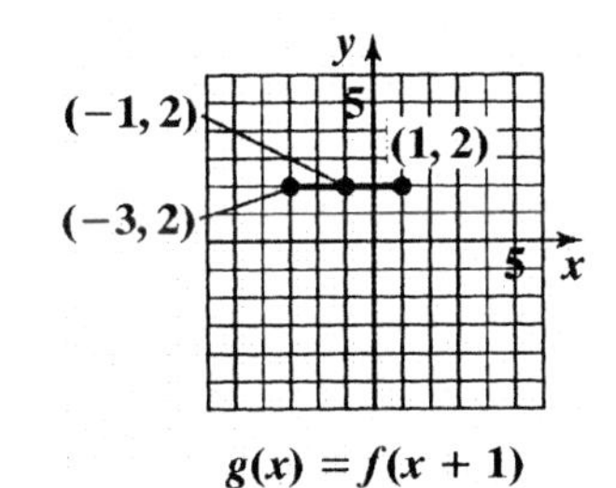

$g(x) = f(x + 1)$

4.

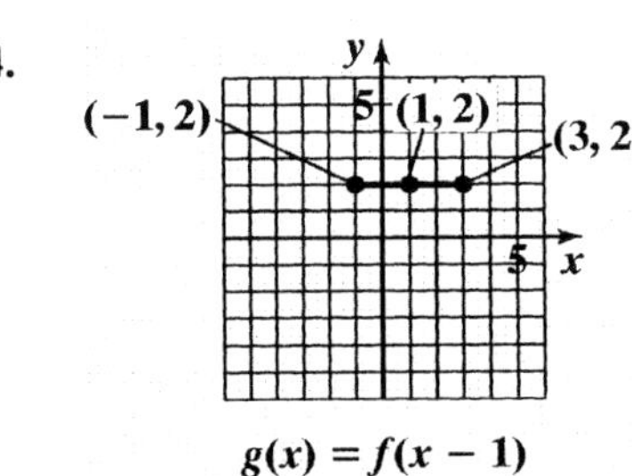

$g(x) = f(x - 1)$

5.

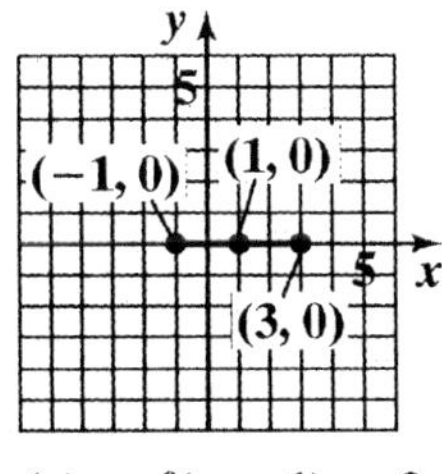

$g(x) = f(x - 1) - 2$

6.

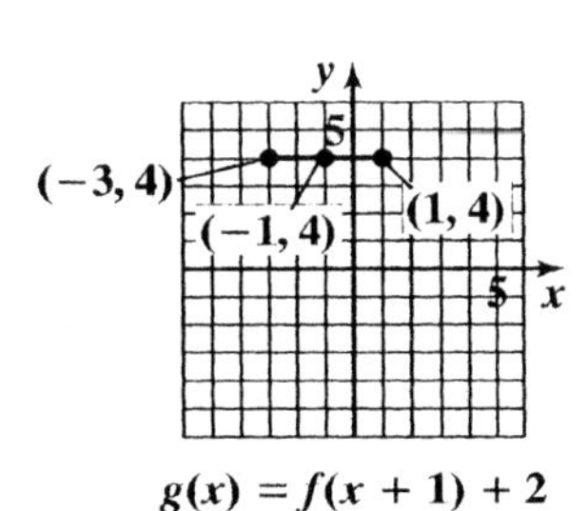

$g(x) = f(x + 1) + 2$

7.

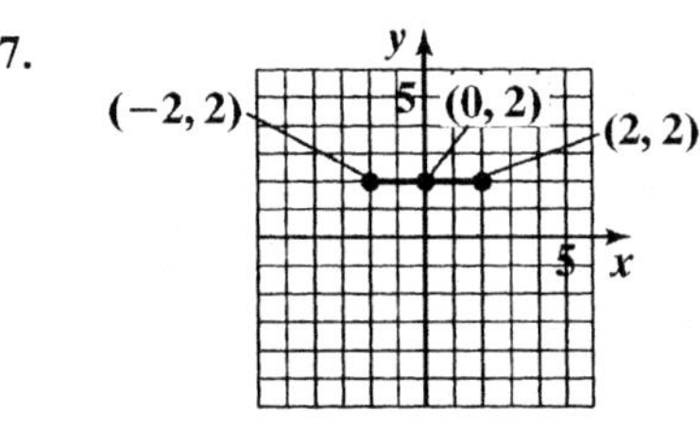

$g(x) = -f(x)$

8.

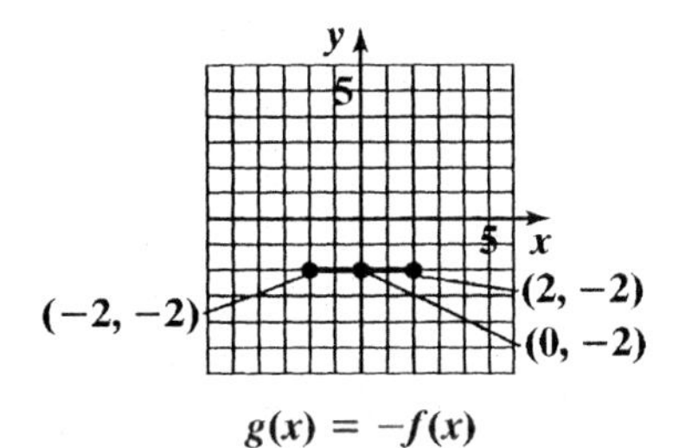

$g(x) = -f(x)$

9.

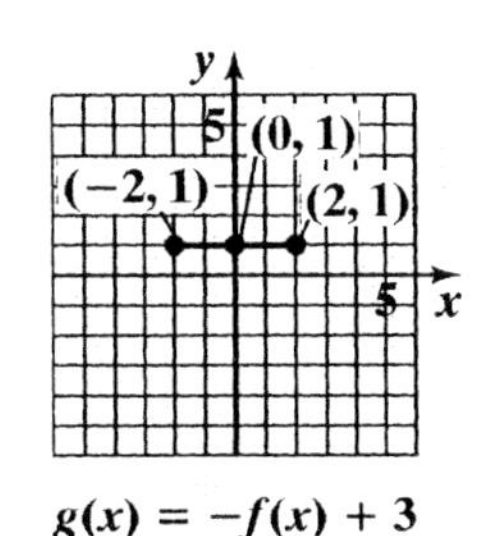

$g(x) = -f(x) + 3$

10.

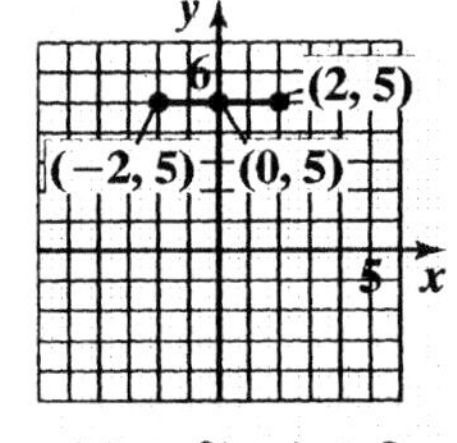

$g(x) = f(-x) + 3$

11.

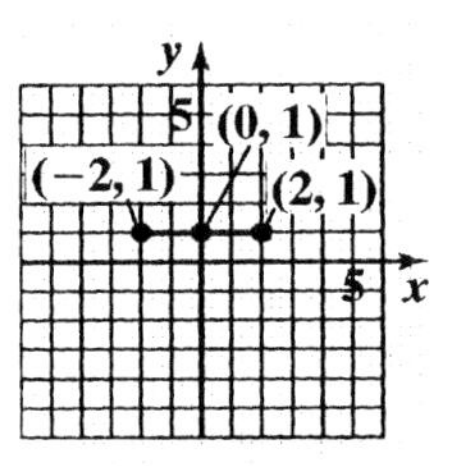

$g(x) = \frac{1}{2}f(x)$

12.

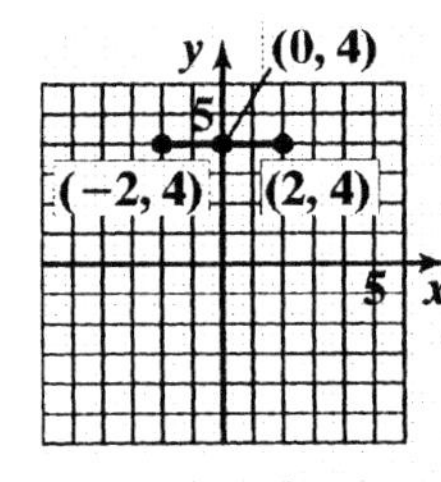

$g(x) = 2f(x)$

13. $g(x) = f\left(\frac{1}{2}x\right)$

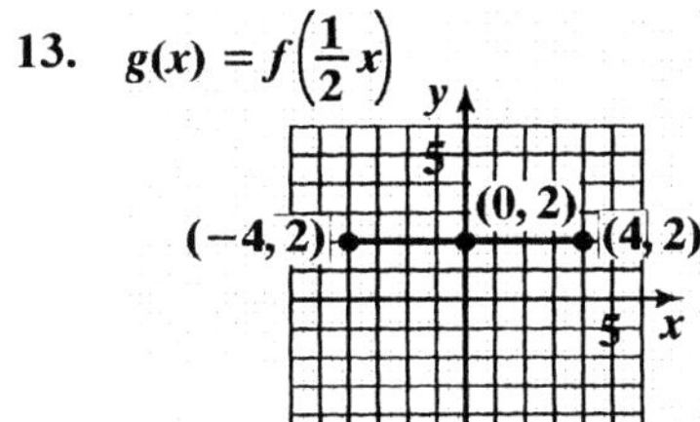

14. $g(x) = f(2x)$

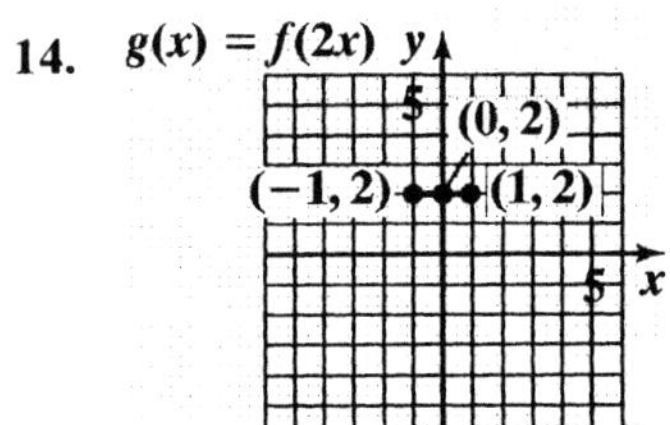

15.

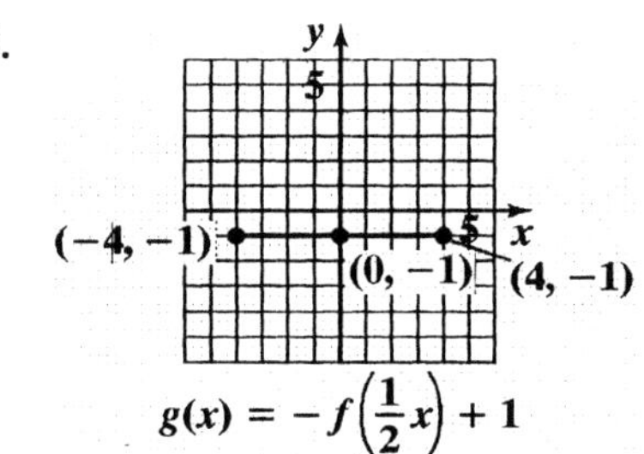

$g(x) = -f\left(\frac{1}{2}x\right) + 1$

16.

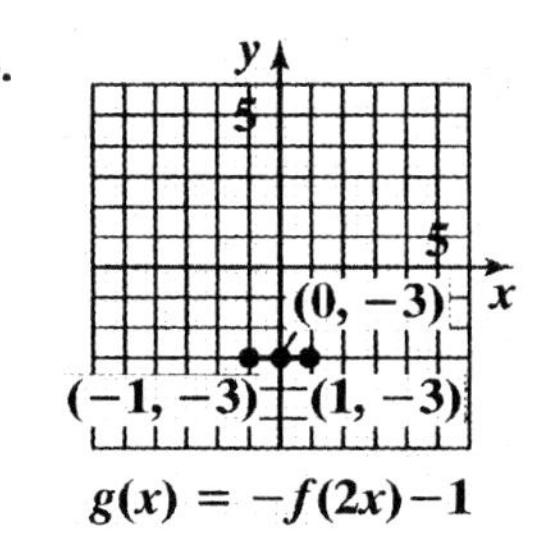

$g(x) = -f(2x) - 1$

17.

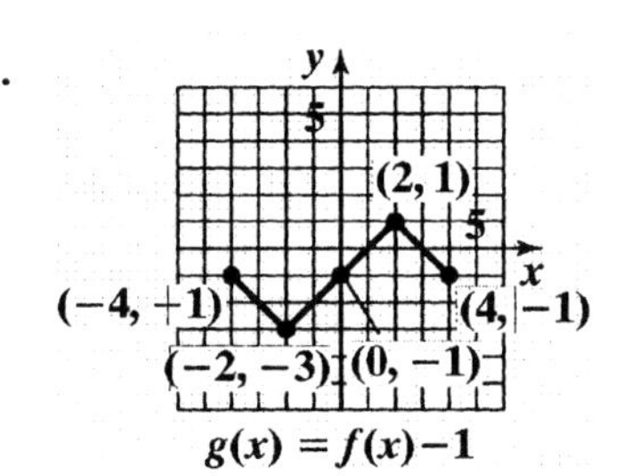

$g(x) = f(x) - 1$

18.

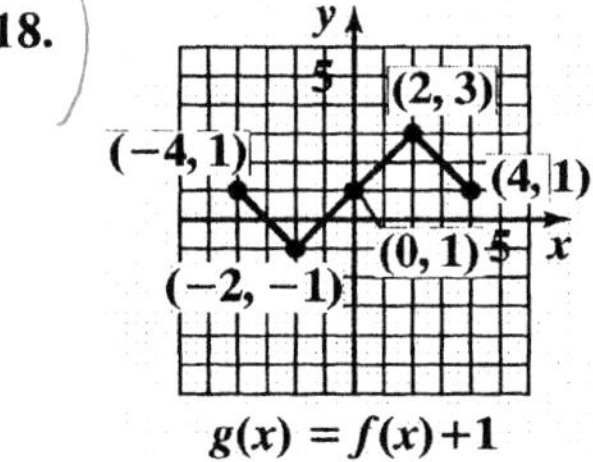

$g(x) = f(x) + 1$

19.

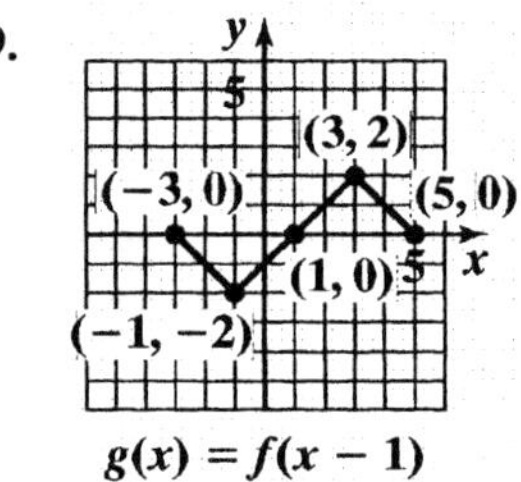

$g(x) = f(x - 1)$

20.

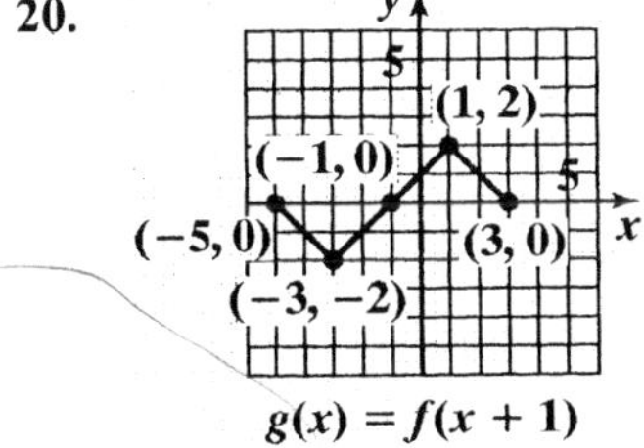

$g(x) = f(x+1)$

21.

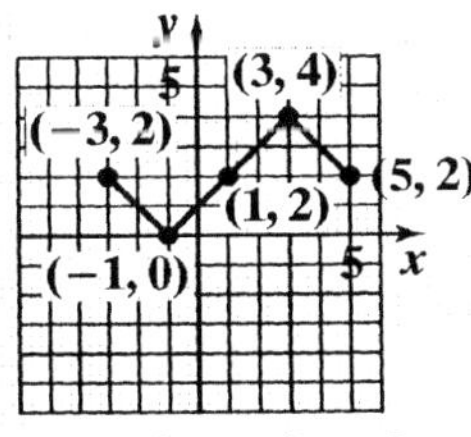

$g(x) = f(x-1)+2$

22.

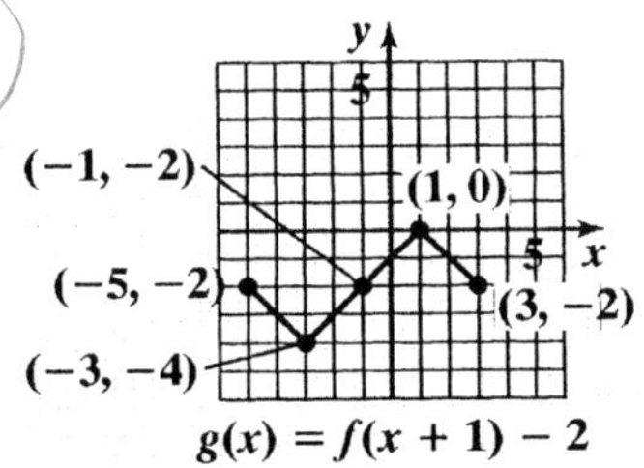

$g(x) = f(x+1)-2$

23.

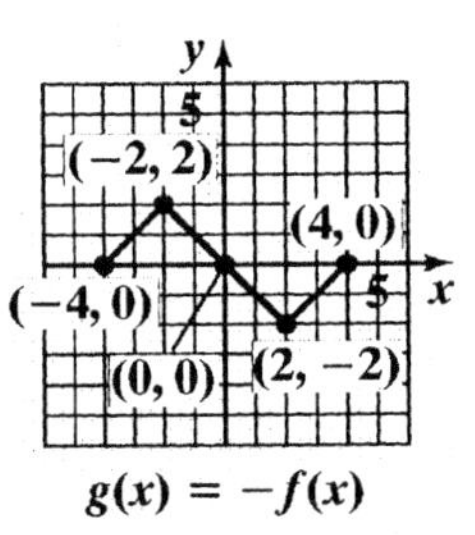

$g(x) = -f(x)$

24.

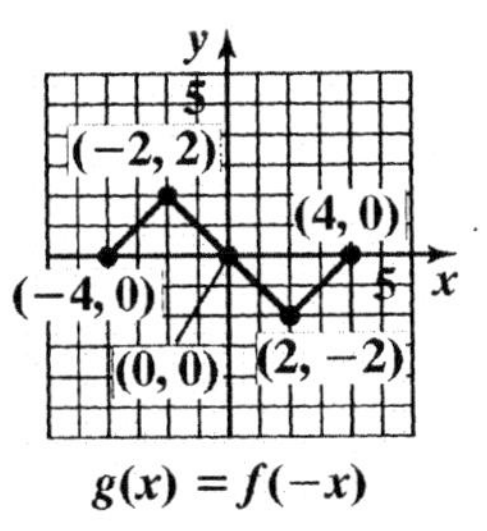

$g(x) = f(-x)$

25.

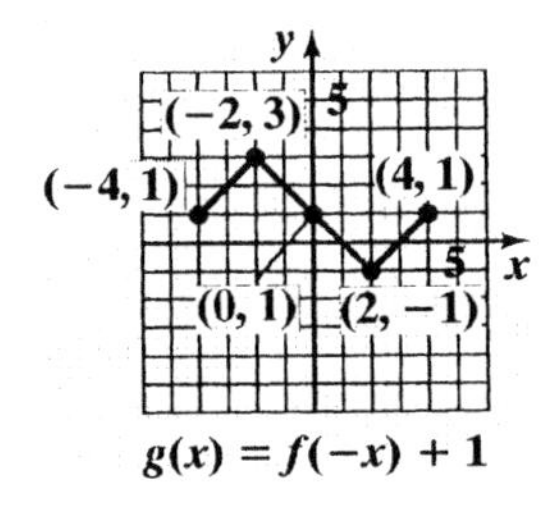

$g(x) = f(-x)+1$

26.

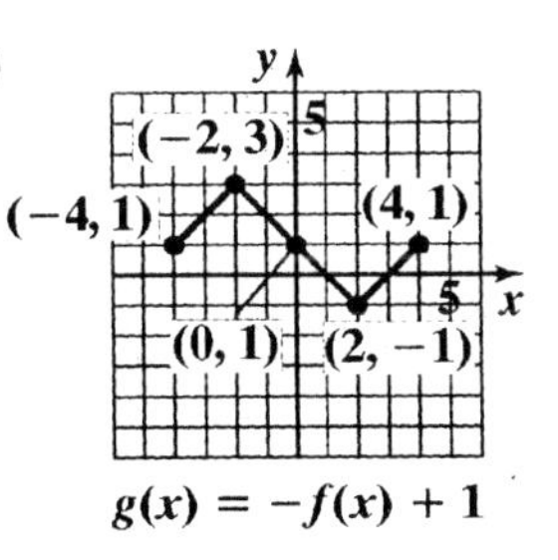

$g(x) = -f(x)+1$

27.

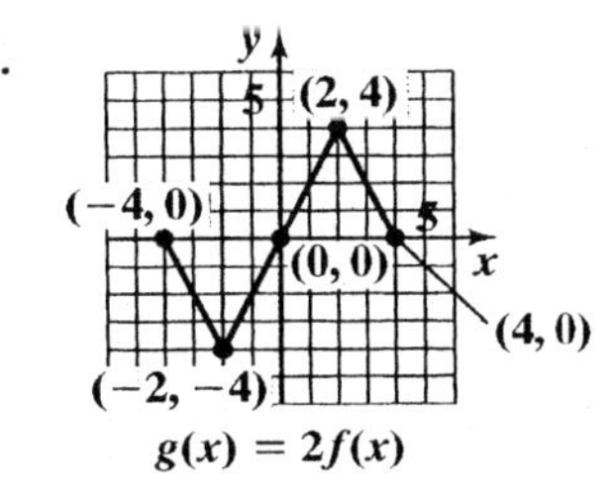

$g(x) = 2f(x)$

28.

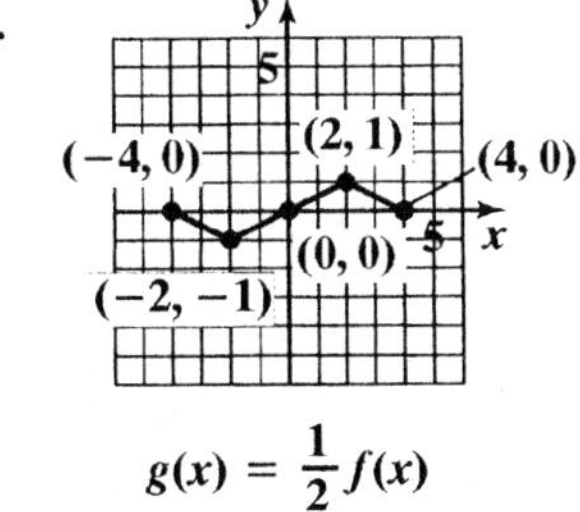

$g(x) = \frac{1}{2}f(x)$

29.

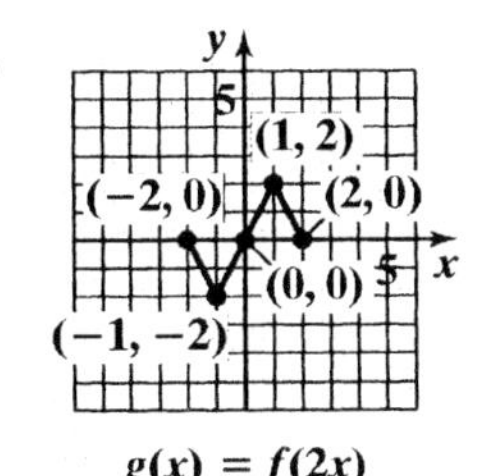

$g(x) = f(2x)$

30.

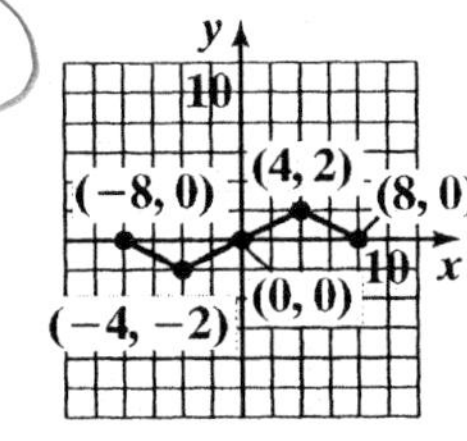

$g(x) = f\left(\frac{1}{2}x\right)$

31.

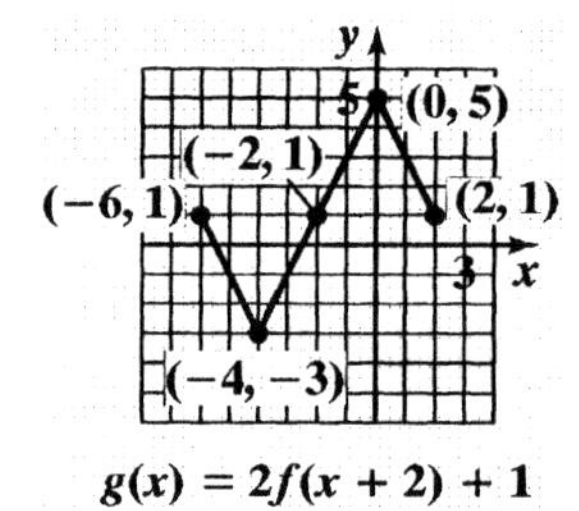

$g(x) = 2f(x + 2) + 1$

32.

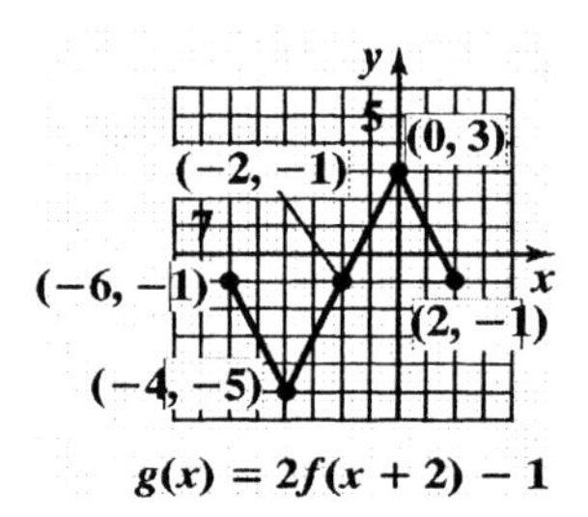

$g(x) = 2f(x + 2) - 1$

33.

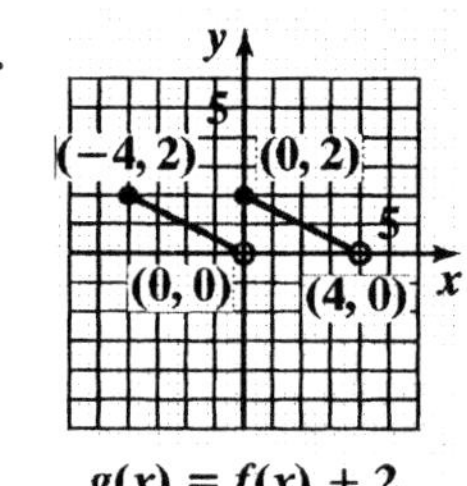

$g(x) = f(x) + 2$

34.

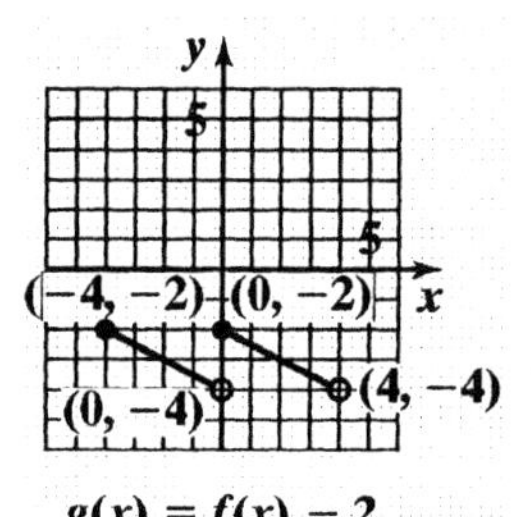

$g(x) = f(x) - 2$

35.

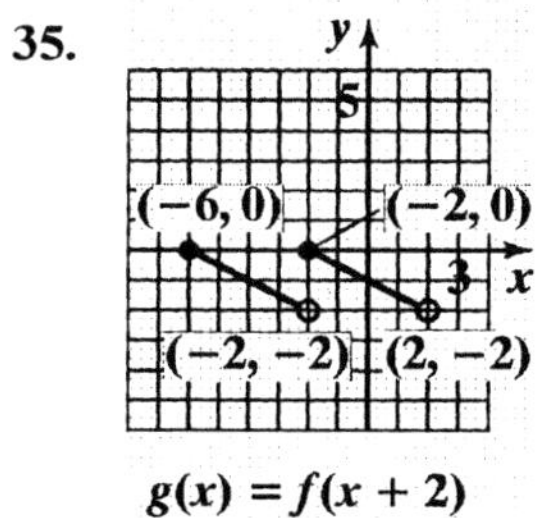

$g(x) = f(x + 2)$

36.

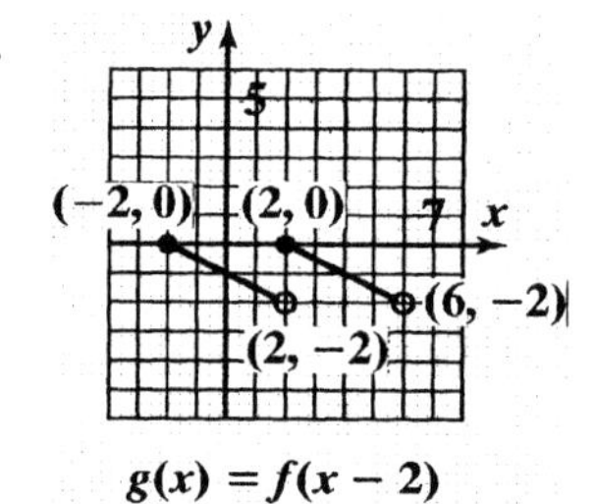

$g(x) = f(x - 2)$

37.

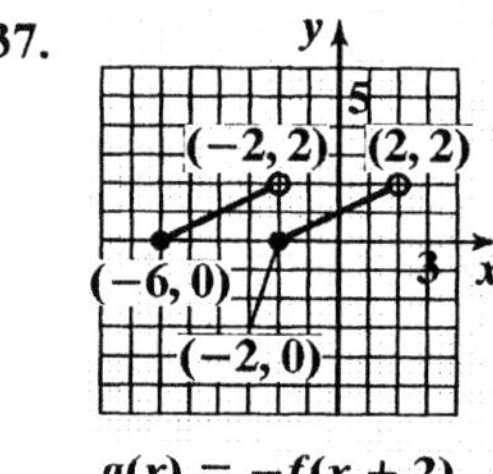

$g(x) = -f(x + 2)$

38.

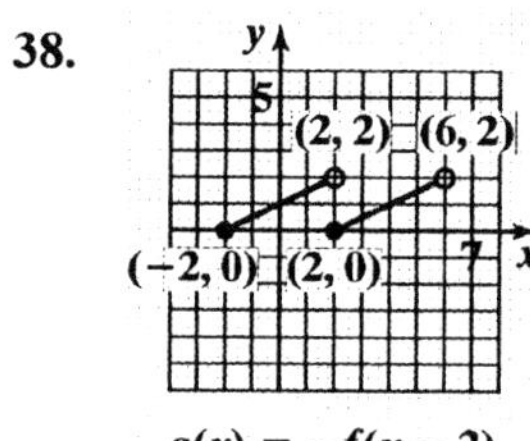

$g(x) = -f(x - 2)$

39.

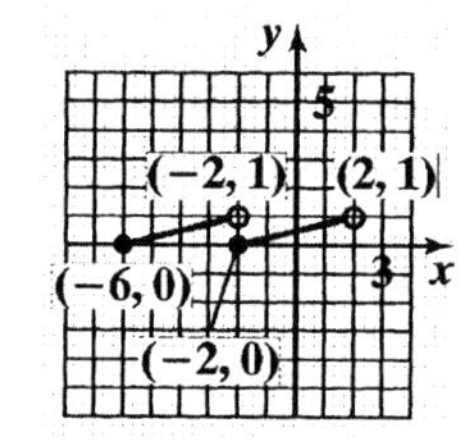

$g(x) = -\frac{1}{2}f(x + 2)$

40.

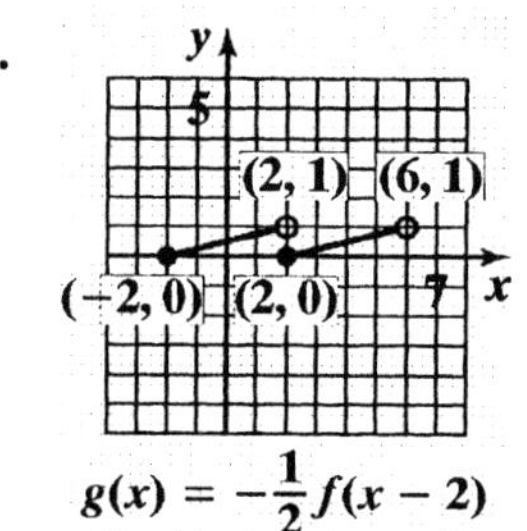

$g(x) = -\frac{1}{2}f(x - 2)$

41.

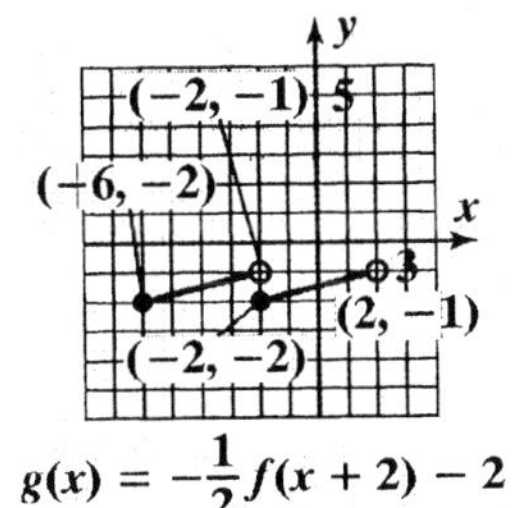

$g(x) = -\frac{1}{2}f(x+2) - 2$

42.

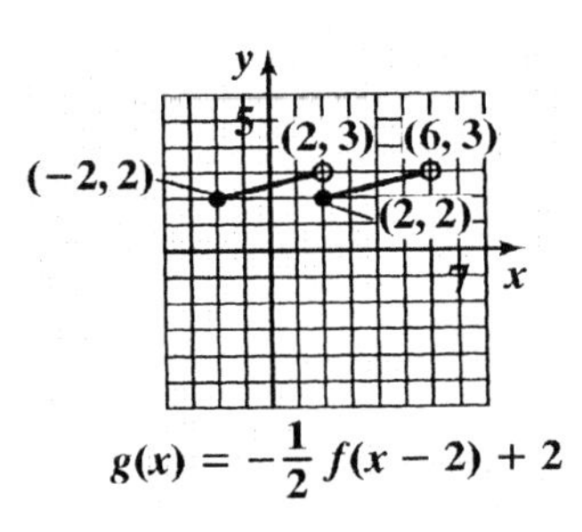

$g(x) = -\frac{1}{2}f(x-2) + 2$

43.

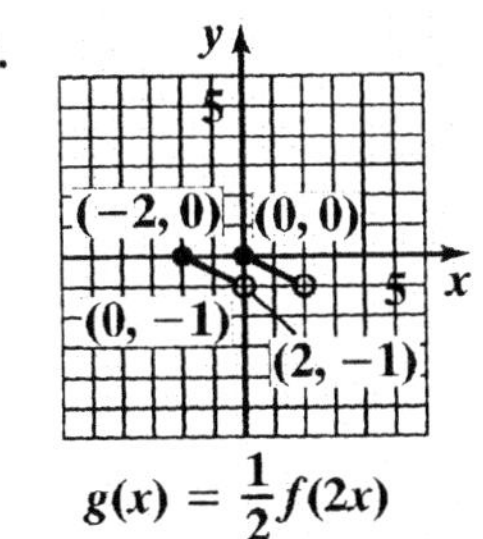

$g(x) = \frac{1}{2}f(2x)$

44.

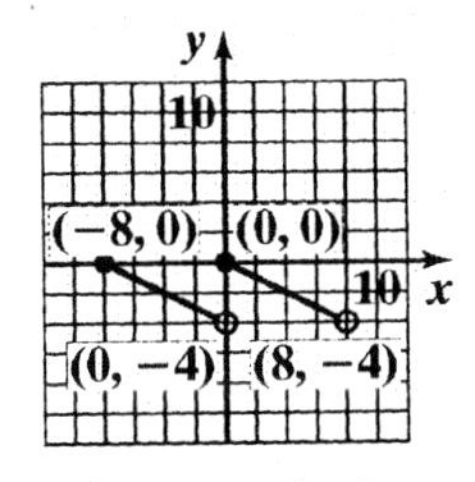

$g(x) = 2f\left(\frac{1}{2}x\right)$

45.

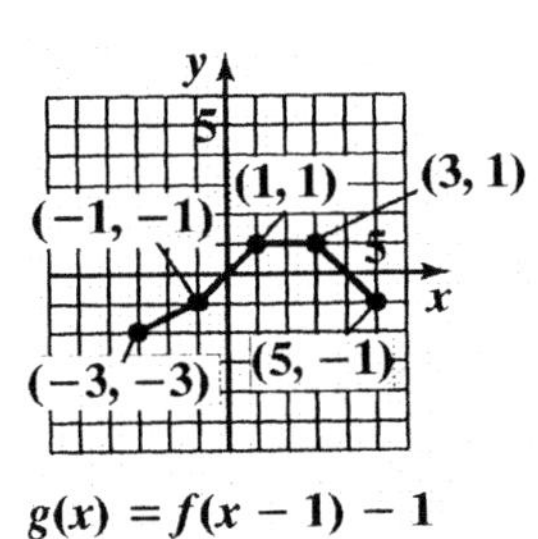

$g(x) = f(x-1) - 1$

46.

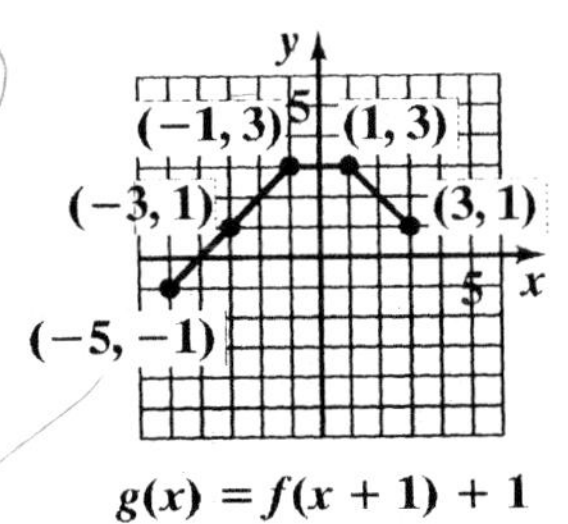

$g(x) = f(x+1) + 1$

47.

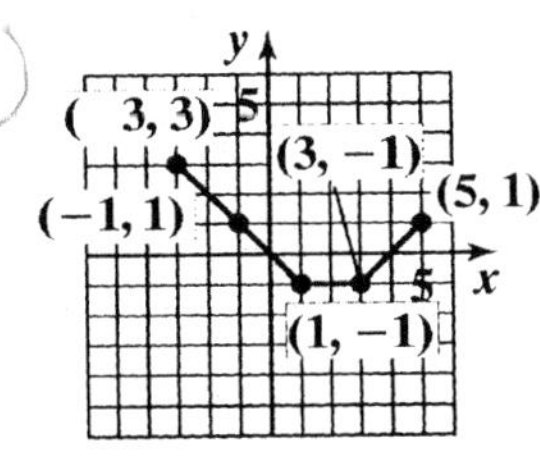

$g(x) = -f(x-1) + 1$

48.

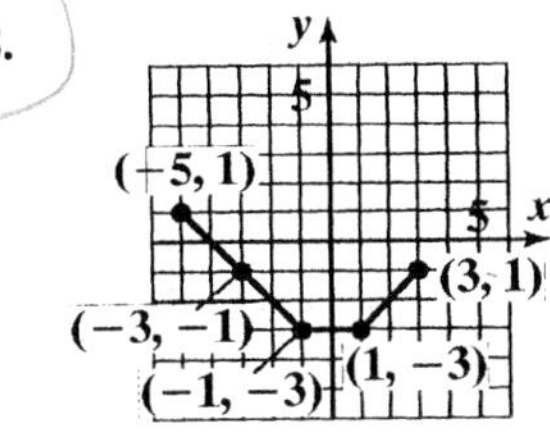

$g(x) = -f(x+1) - 1$

49.

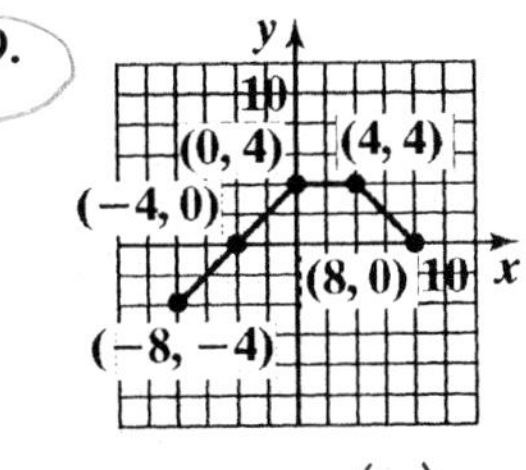

$g(x) = 2f\left(\frac{1}{2}x\right)$

50.

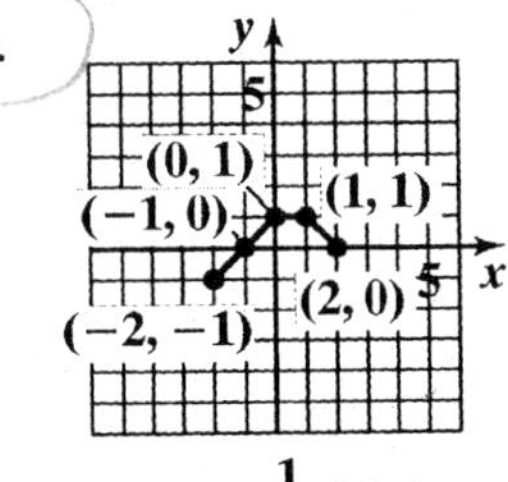

$g(x) = \frac{1}{2}f(2x)$

51.

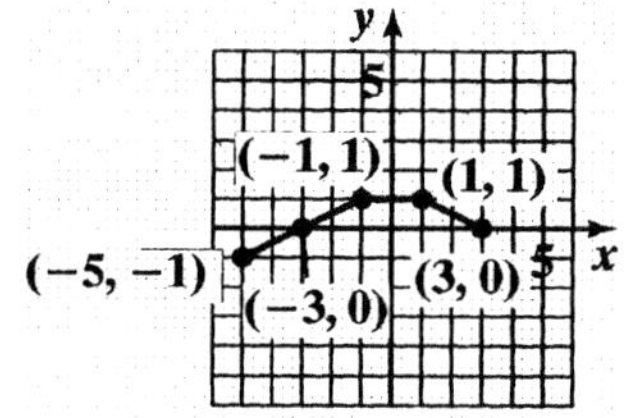

$g(x) = \frac{1}{2}f(x+1)$

52.

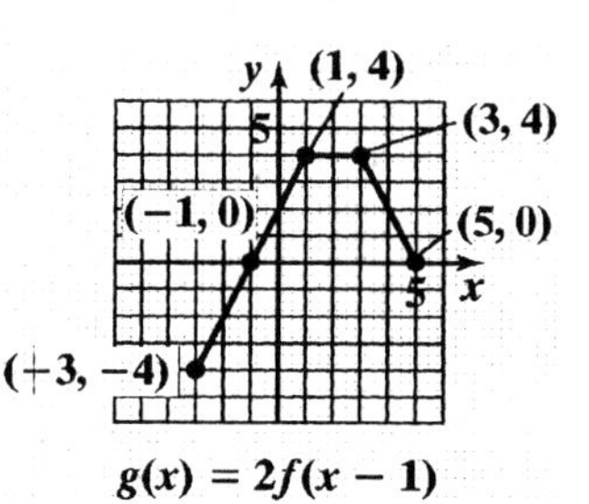

$g(x) = 2f(x-1)$

53.

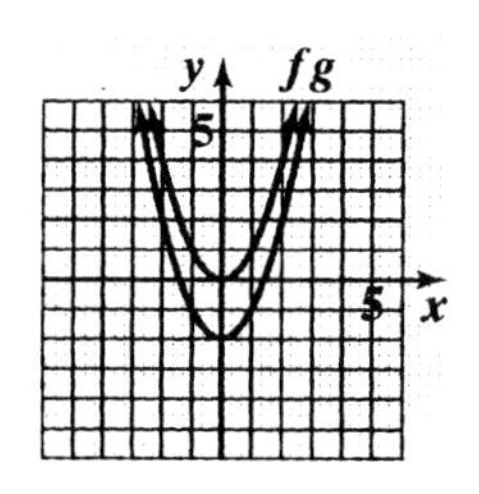

54.

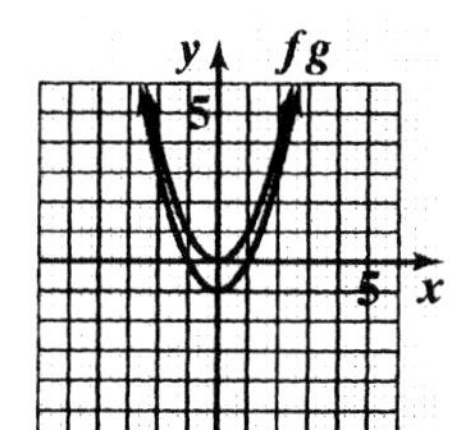

55.

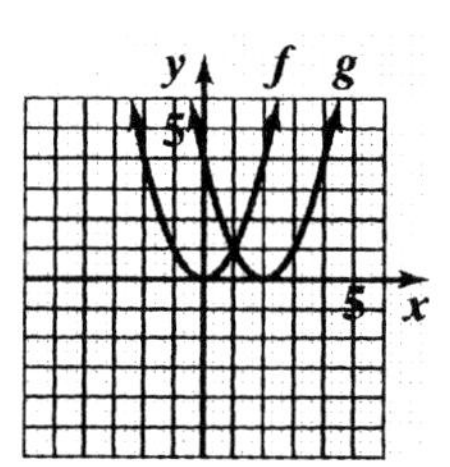

56.

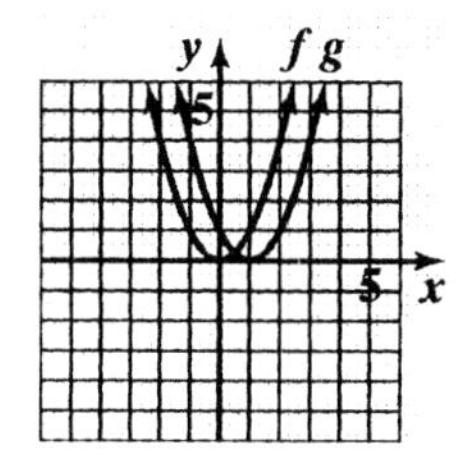

57.

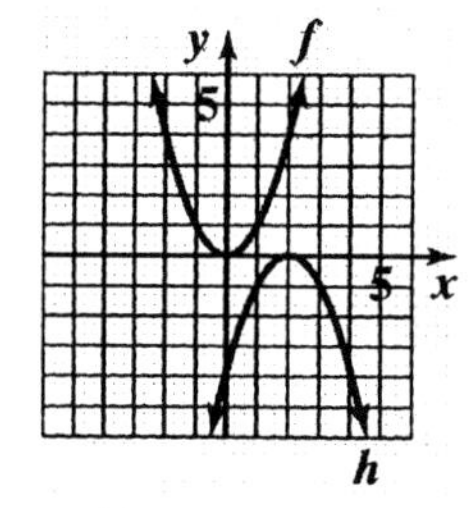

58.

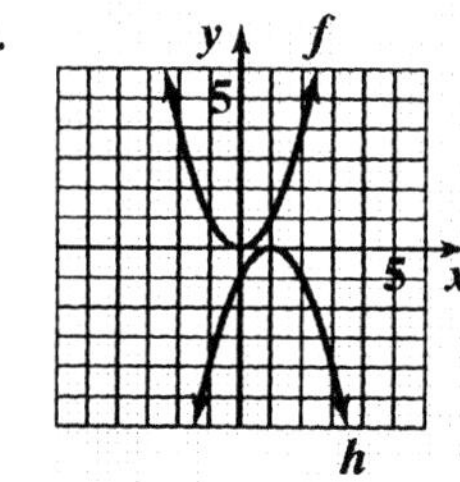

59.

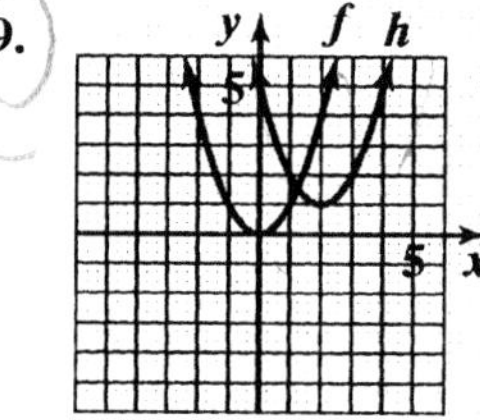

60.

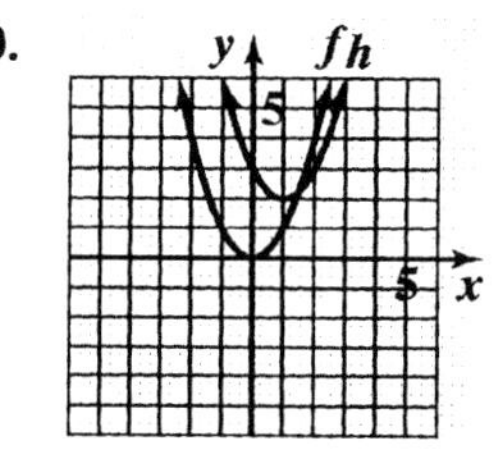

61.

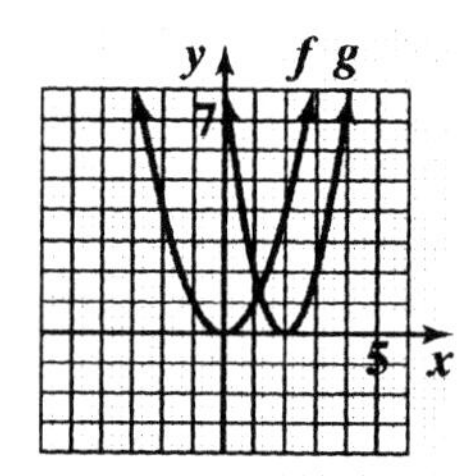

62.

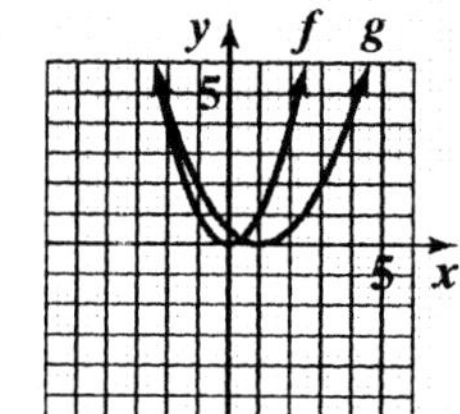

63.

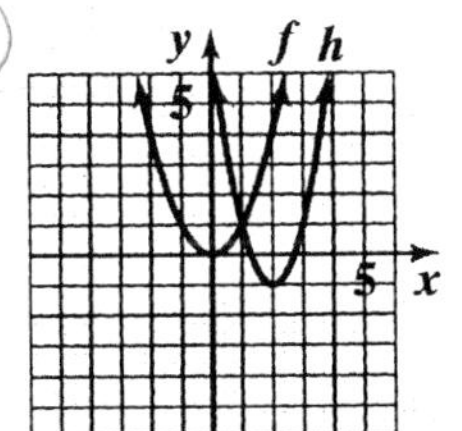

64.

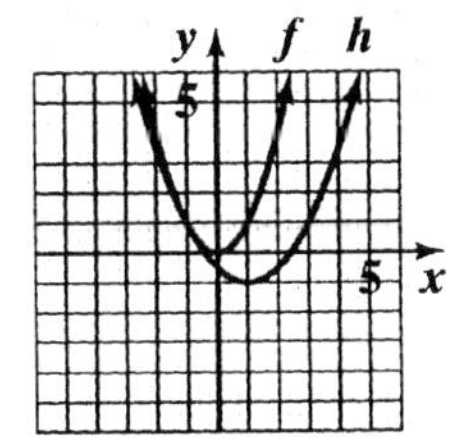

65.

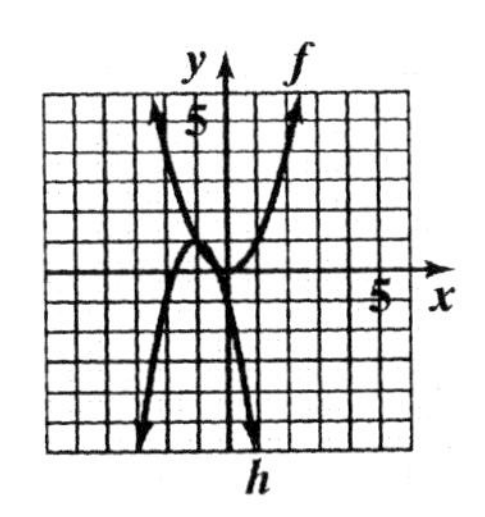

66.

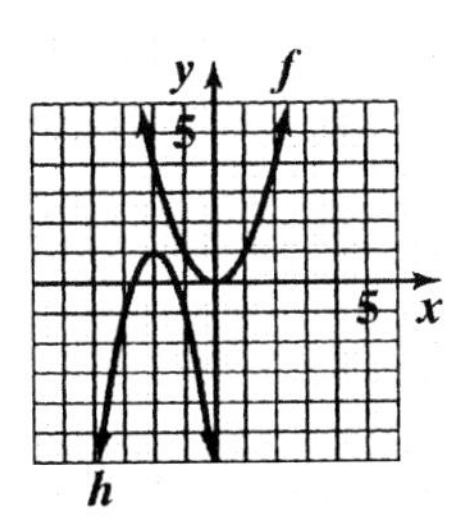

67.

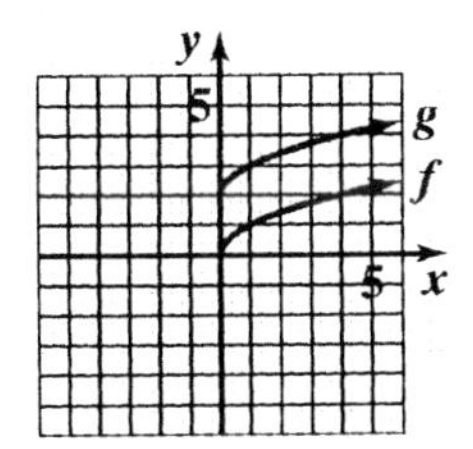

68.

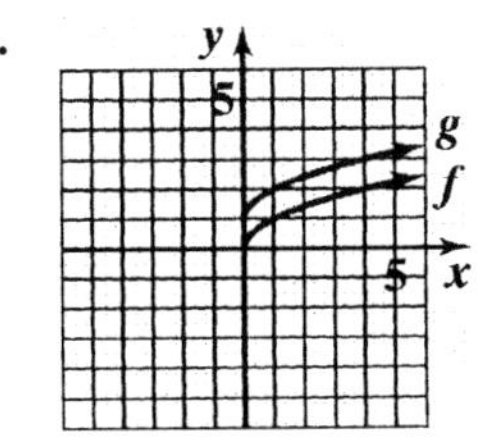

69.

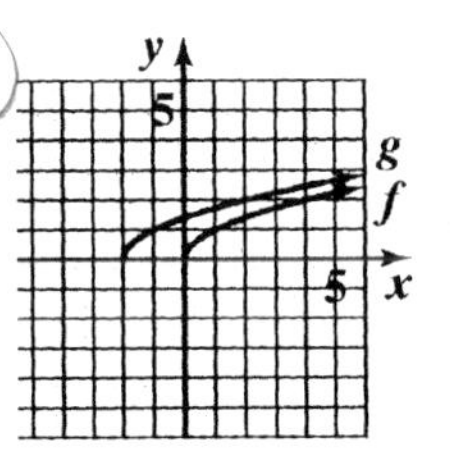

70.

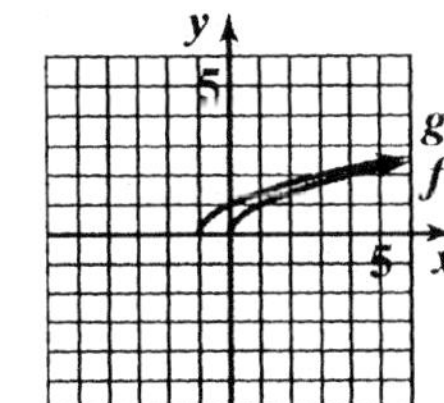

71.

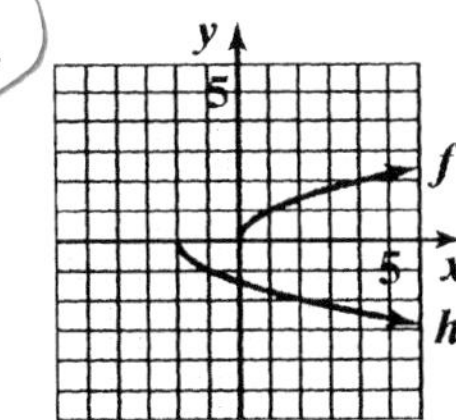

72.

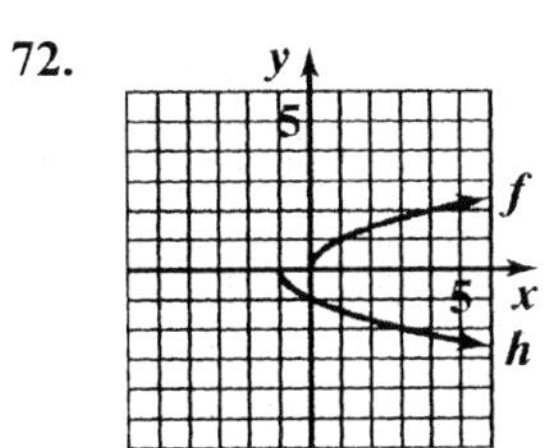

73.

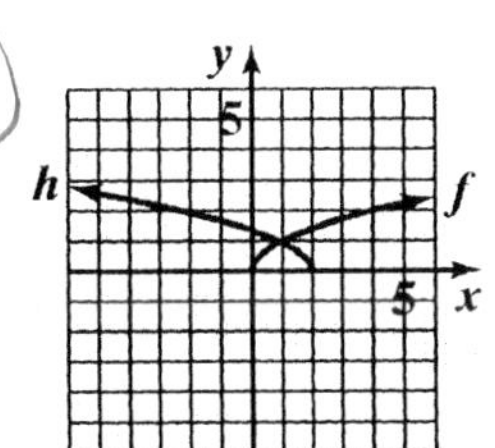

74.

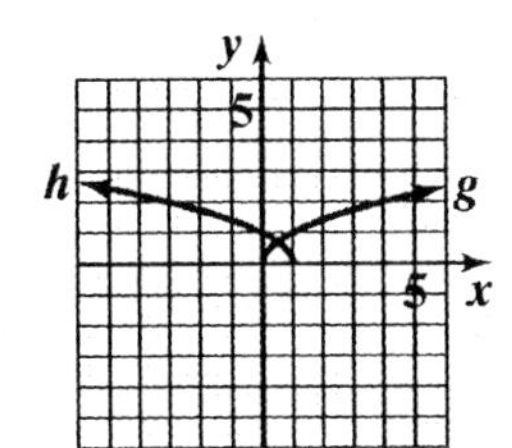

75. 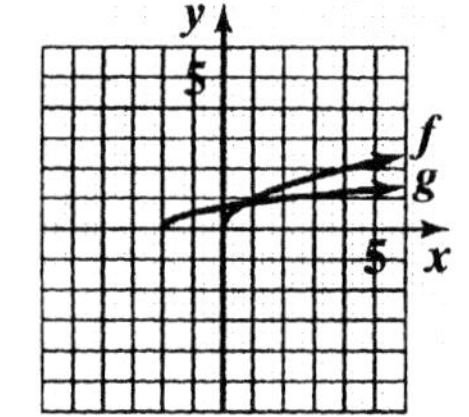

76. 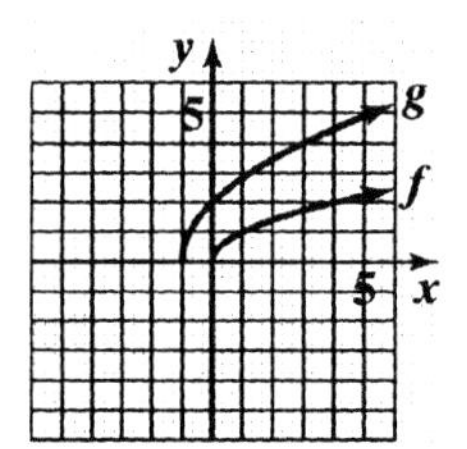

77. 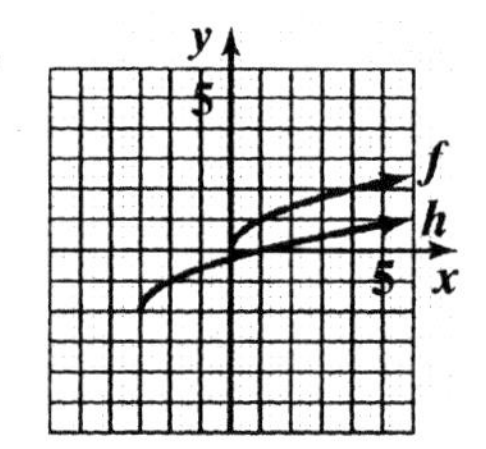

78. 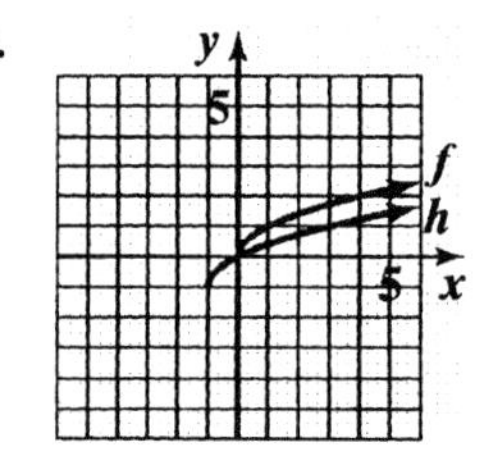

79. 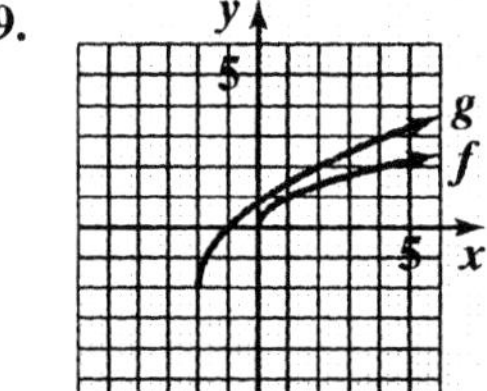

80. 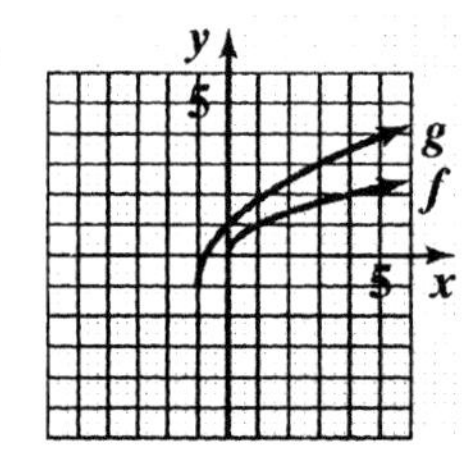

81. 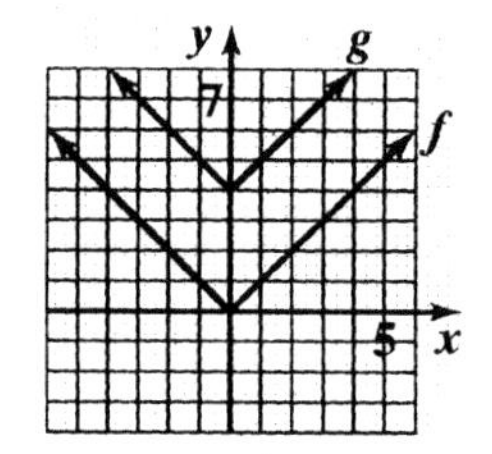

82. 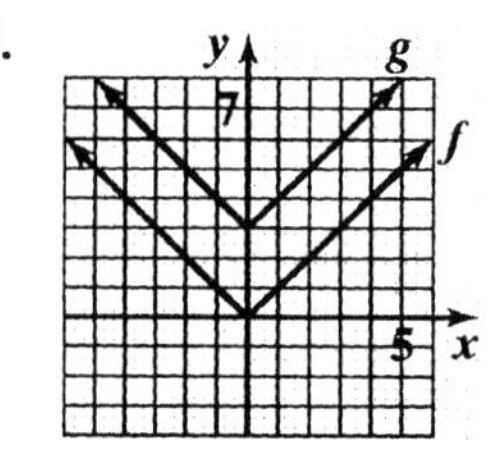

83. 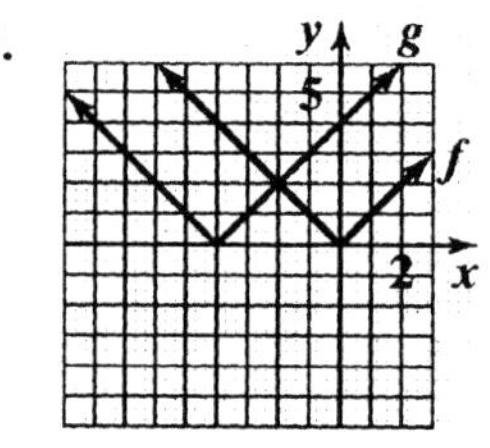

84. 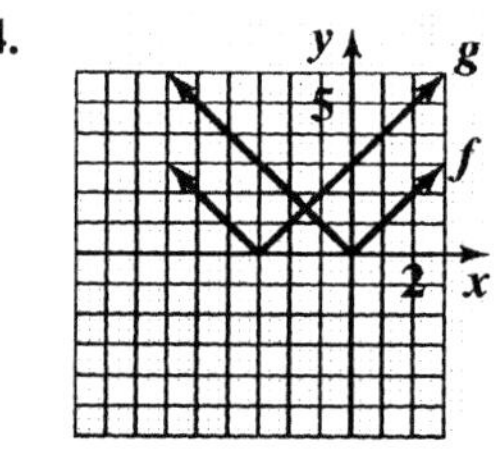

85. 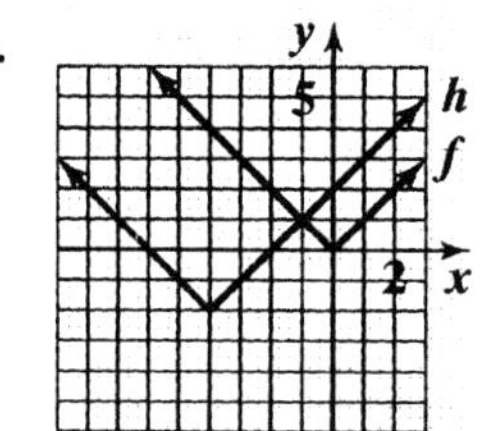

86. 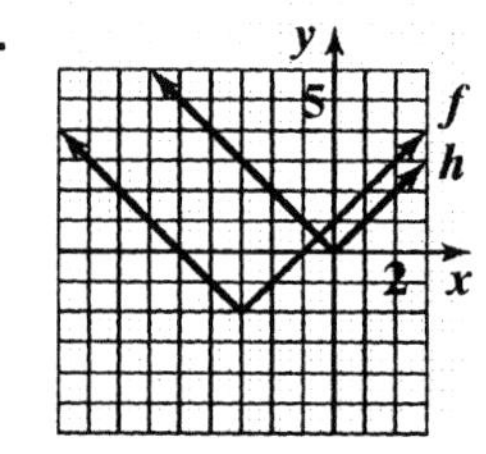

87.

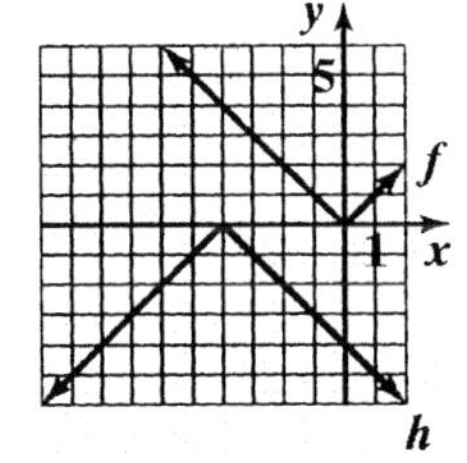

88.

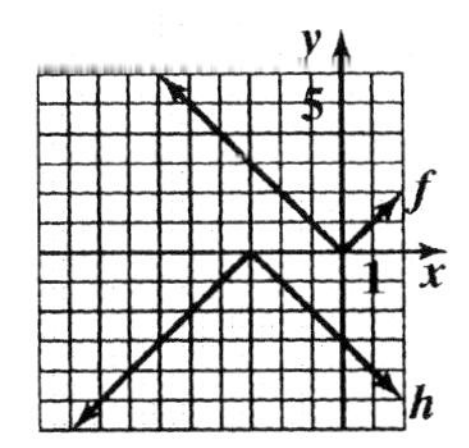

89.

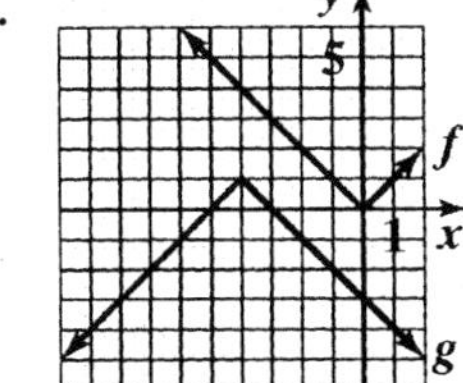

90.

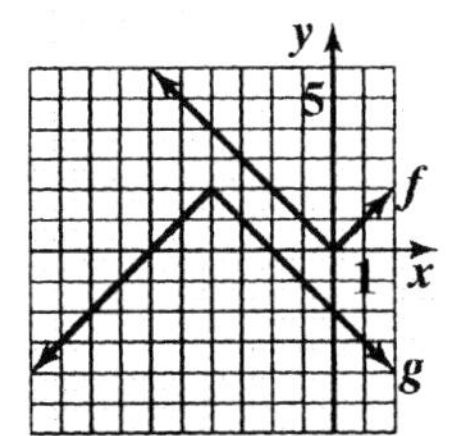

91.

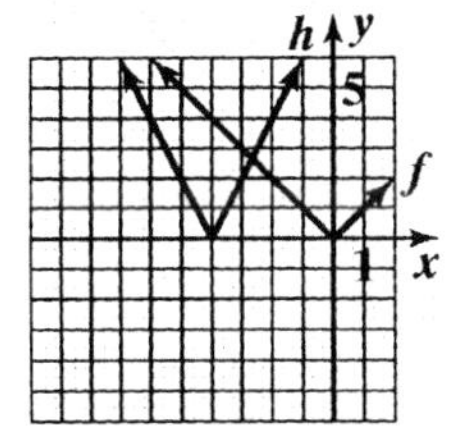

92.

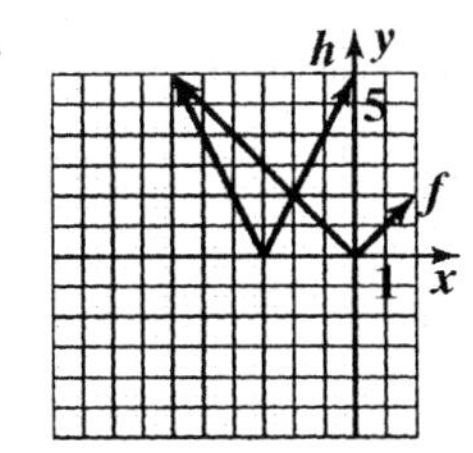

93.

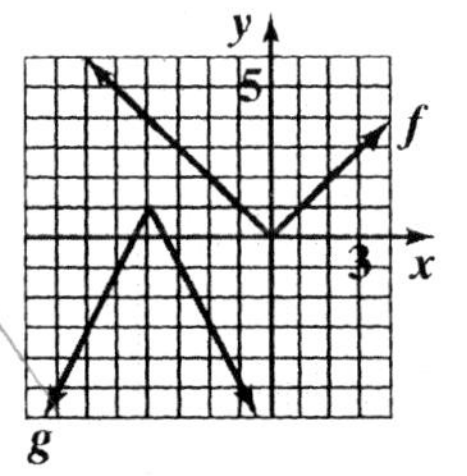

94.

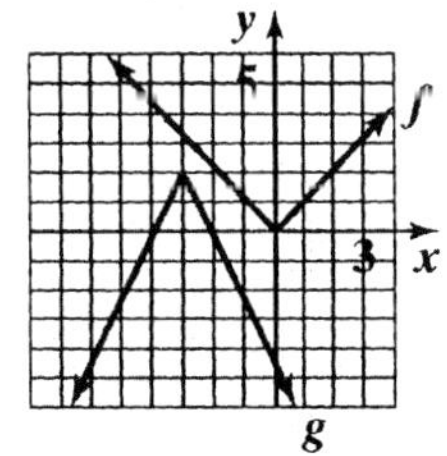

95.

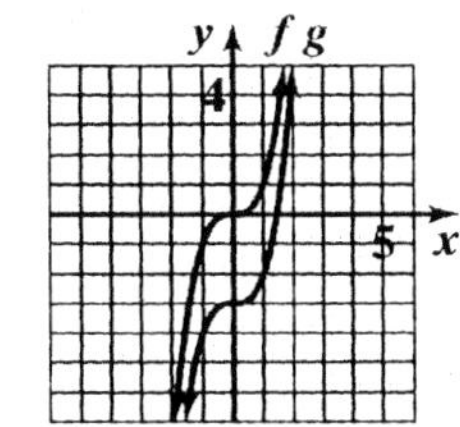

96.

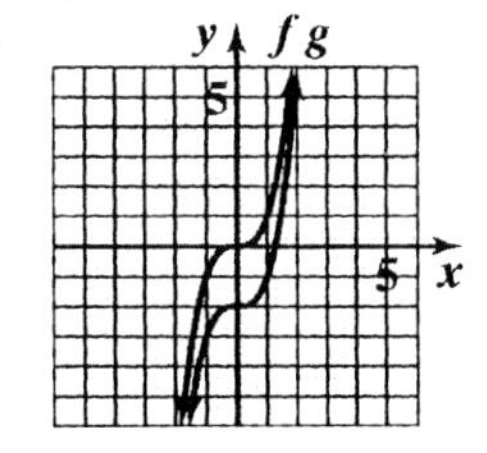

97.

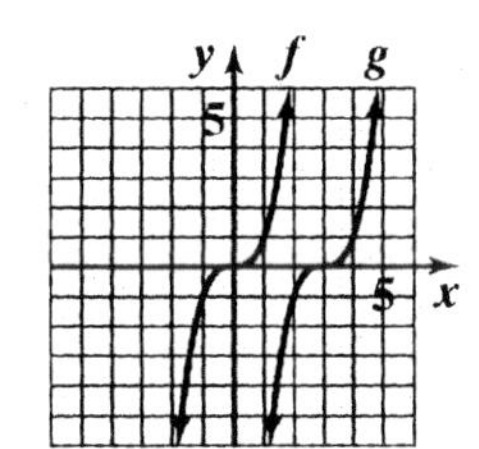

98.

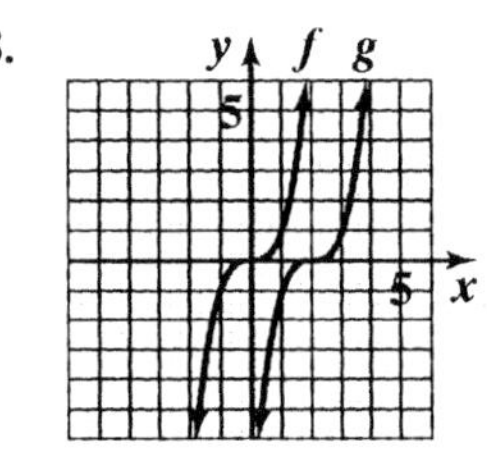

99.

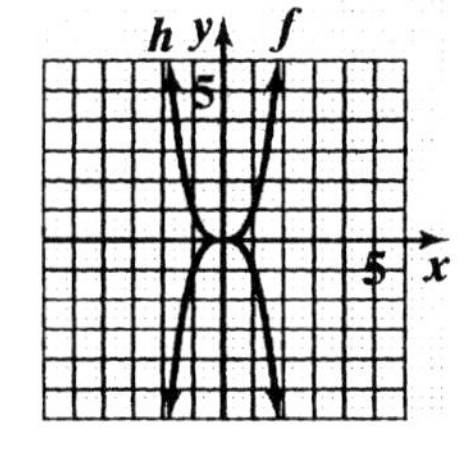

100.

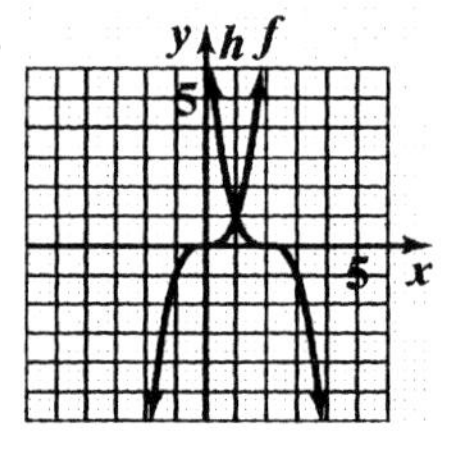

101.

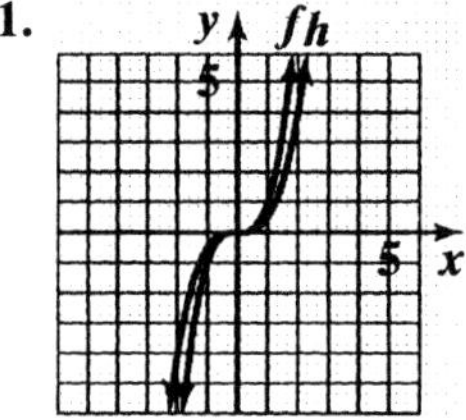

102.

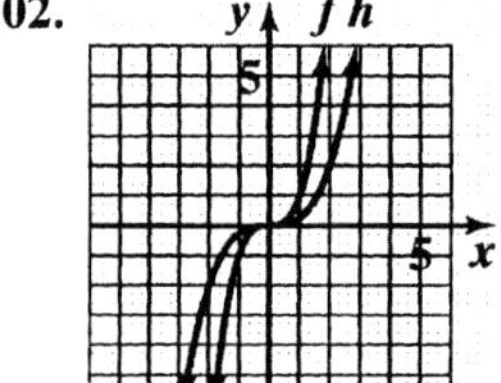

103.

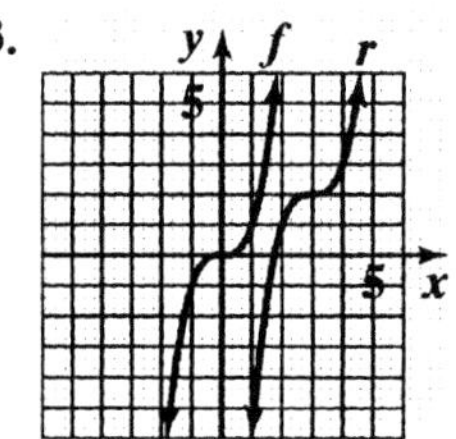

104.

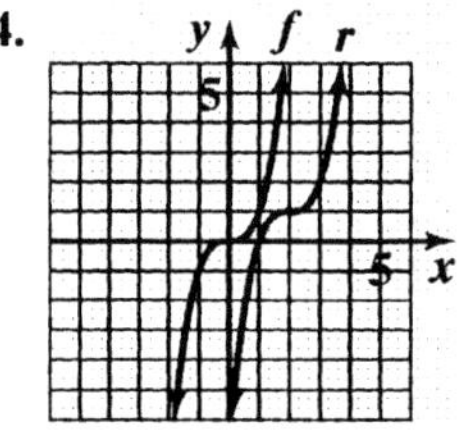

105.

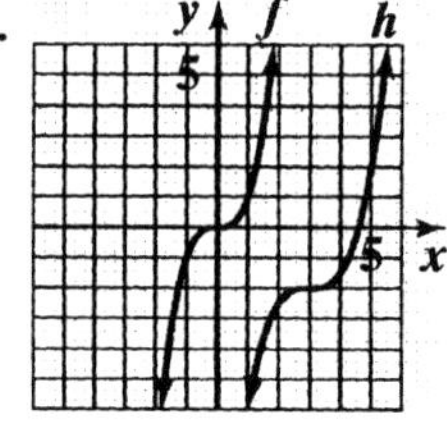

106.

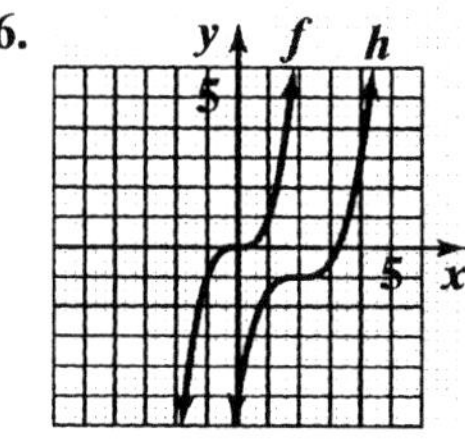

107.

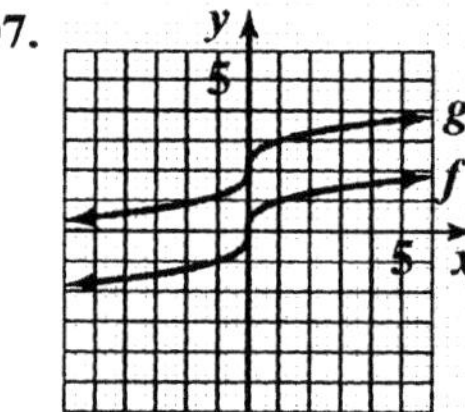

108.

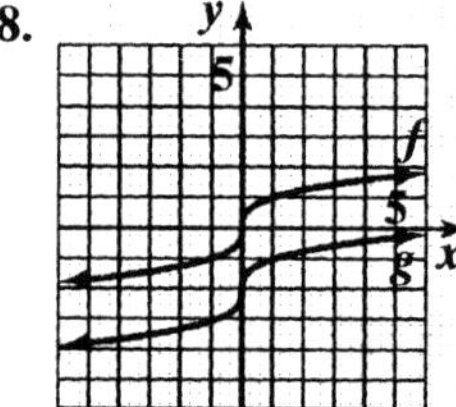

109.

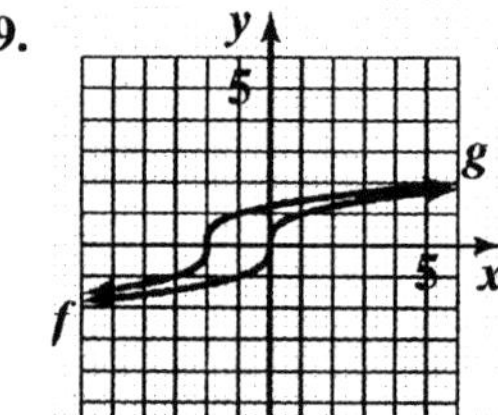

110.

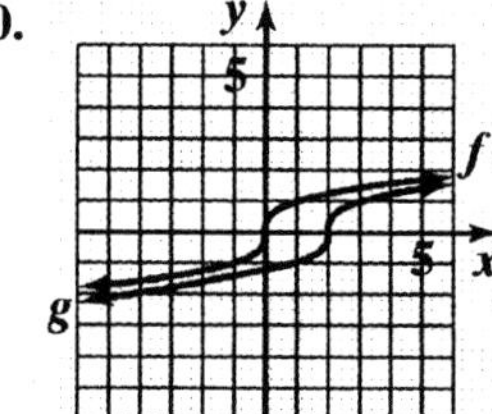

111.
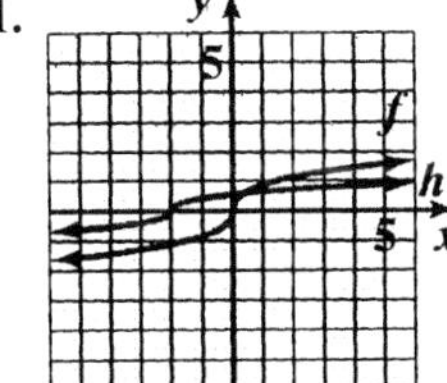

112.
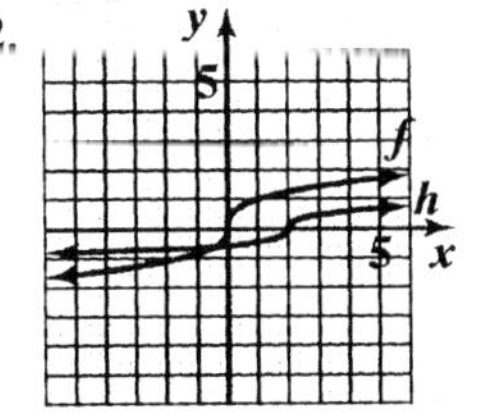

113.
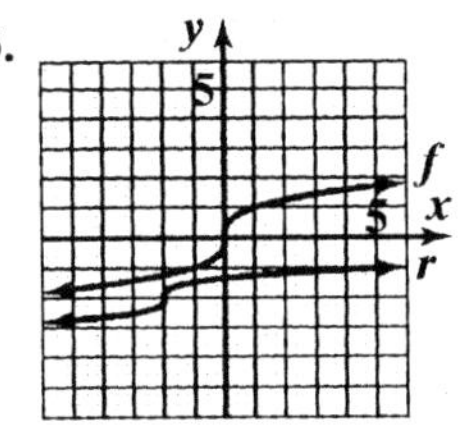

114.
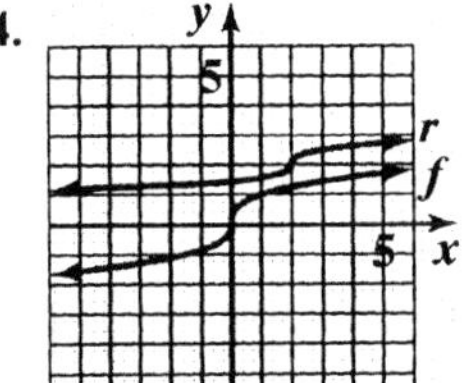

115.
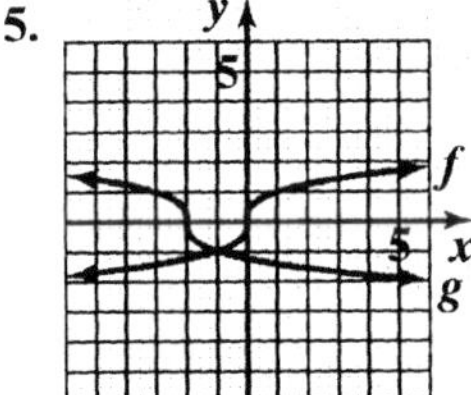

116.
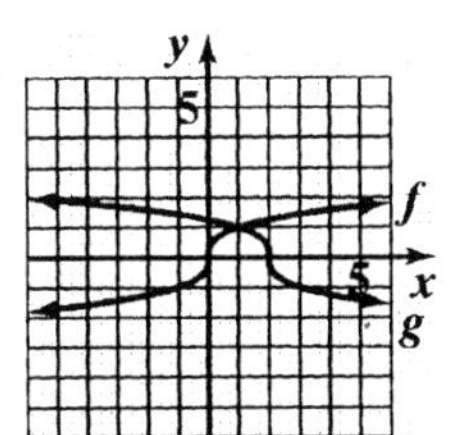

117.
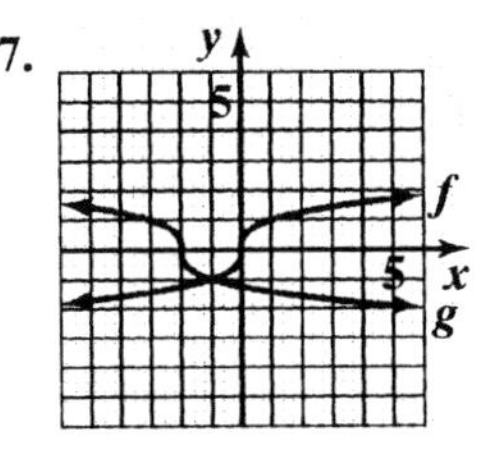

118.
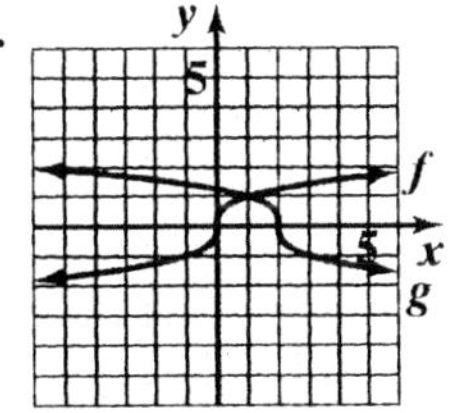

119.
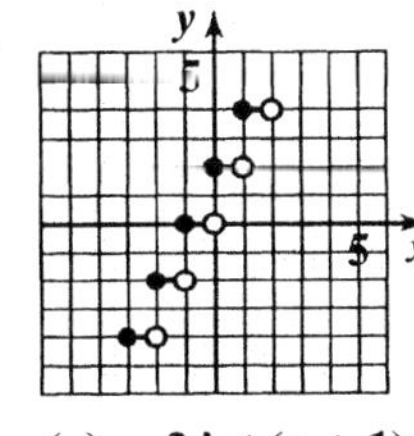

$g(x) = 2\text{ int }(x + 1)$

120.
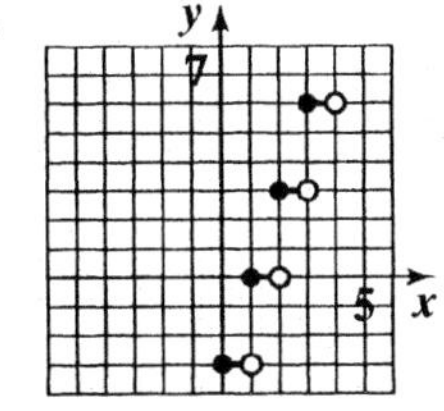

$g(x) = 3\text{ int }(x - 1)$

121.
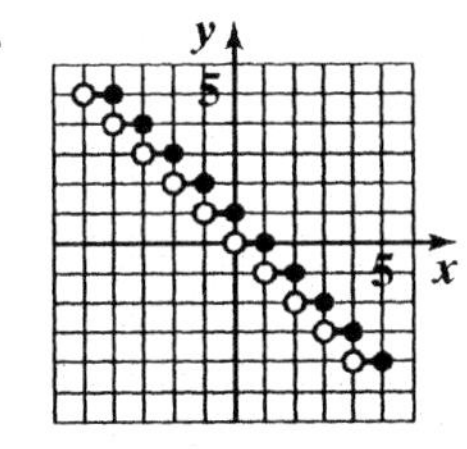

$h(x) = \text{int }(-x) + 1$

122.
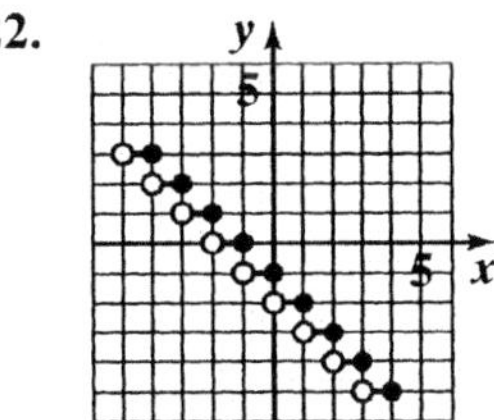

$h(x) = \text{int }(-x) - 1$

123. $y = \sqrt{x-2}$

124. $y = -x^3 + 2$

125. $y = (x+1)^2 - 4$

126. $y = \sqrt{x-2} + 1$

127. a. First, vertically stretch the graph of $f(x)=\sqrt{x}$ by the factor 2.9; then shift the result up 20.1 units.

b. $f(x)=2.9\sqrt{x}+20.1$
$f(48)=2.9\sqrt{48}+20.1\approx 40.2$
The model describes the actual data very well.

c.
$$\frac{f(x_2)-f(x_1)}{x_2-x_1}$$
$$=\frac{f(10)-f(0)}{10-0}$$
$$=\frac{\left(2.9\sqrt{10}+20.1\right)-\left(2.9\sqrt{0}+20.1\right)}{10-0}$$
$$=\frac{29.27-20.1}{10}$$
$$\approx 0.9$$
0.9 inches per month

d.
$$\frac{f(x_2)-f(x_1)}{x_2-x_1}$$
$$=\frac{f(60)-f(50)}{60-50}$$
$$=\frac{\left(2.9\sqrt{60}+20.1\right)-\left(2.9\sqrt{50}+20.1\right)}{60-50}$$
$$=\frac{42.5633-40.6061}{10}$$
$$\approx 0.2$$
This rate of change is lower than the rate of change in part (c). The relative leveling off of the curve shows this difference.

128. a. First, vertically stretch the graph of $f(x)=\sqrt{x}$ by the factor 3.1; then shift the result up 19 units.

b. $f(x)=3.1\sqrt{x}+19$
$f(48)=3.1\sqrt{48}+19\approx 40.5$
The model describes the actual data very well.

c.
$$\frac{f(x_2)-f(x_1)}{x_2-x_1}$$
$$=\frac{f(10)-f(0)}{10-0}$$
$$=\frac{\left(3.1\sqrt{10}+19\right)-\left(3.1\sqrt{0}+19\right)}{10-0}$$
$$=\frac{28.8031-19}{10}$$
$$\approx 1.0$$
1.0 inches per month

d. $\frac{f(x_2)-f(x_1)}{x_2-x_1}$
$= \frac{f(60)-f(50)}{60-50}$
$= \frac{\left(3.1\sqrt{60}+19\right)-\left(3.1\sqrt{50}+19\right)}{60-50}$
$= \frac{43.0125-40.9203}{10}$
≈ 0.2
This rate of change is lower than the rate of change in part (c). The relative leveling off of the curve shows this difference.

135. a.

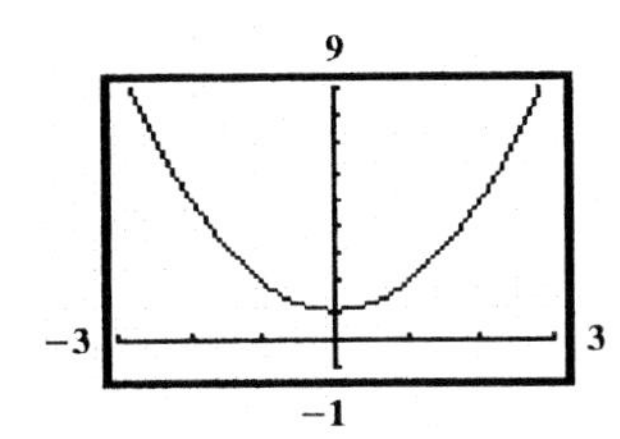

b.

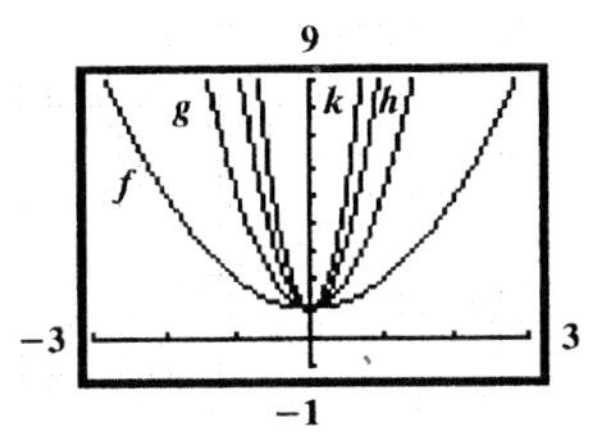

136. a.

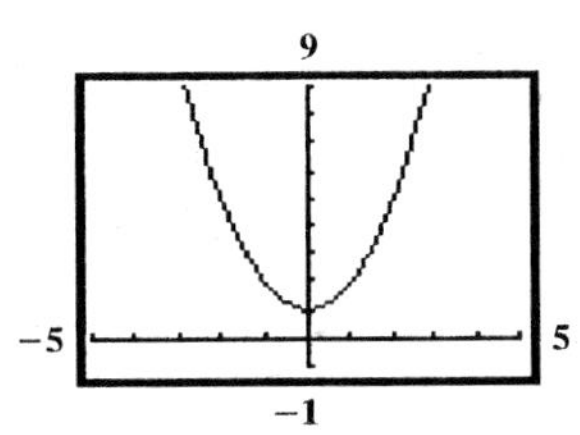

b.

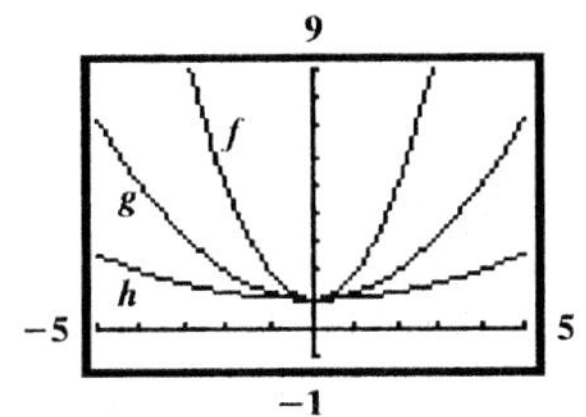

137. a. False; the graph of g is a translation of three units upward and three units to the left of the graph of f.

b. False; the graph of f is a reflection of the graph of $y=\sqrt{x}$ in the x-axis, while the graph of g is a reflection of the graph of $y=\sqrt{x}$ in the y-axis.

c. False; $g(x)=5x^2-10$, so the graph of g can be obtained by stretching f five units followed by a downward shift of ten units.

d. True

(d) is true.

138. $g(x)=-(x+4)^2$

139. $g(x)=-|x-5|+1$

140. $g(x)=-\sqrt{x-2}+2$

141. $g(x)=-\frac{1}{4}\sqrt{16-x^2}-1$

142. $(-a, b)$

143. $(a, 2b)$

144. $(a+3, b)$

145. $(a, b-3)$

Section 1.7

Check Point Exercises

1. **a.** The function $f(x) = x^2 + 3x - 17$ contains neither division nor an even root. The domain of *f* is the set of all real numbers or $(-\infty, \infty)$.

b. The denominator equals zero when $x = 7$ or $x = -7$. These values must be excluded from the domain. Domain of $g = (-\infty, -7) \cup (-7, 7) \cup (7, \infty)$.

c. Since $h(x) = \sqrt{9x - 27}$ contains an even root; the quantity under the radical must be greater than or equal to 0.

$$9x - 27 \geq 0$$
$$9x \geq 27$$
$$x \geq 3$$

Thus, the domain of *h* is $\{x | x \geq 3\}$, or the interval $[3, \infty)$.

2. **a.**
$$\begin{aligned}(f + g)(x) &= f(x) + g(x) \\ &= x - 5 + (x^2 - 1) \\ &= x - 5 + x^2 - 1 \\ &= -x^2 + x - 6\end{aligned}$$

b.
$$\begin{aligned}(f - g)(x) &= f(x) - g(x) \\ &= x - 5 - (x^2 - 1) \\ &= x - 5 - x^2 + 1 \\ &= -x^2 + x - 4\end{aligned}$$

c.
$$\begin{aligned}(fg)(x) &= (x - 5)(x^2 - 1) \\ &= x(x^2 - 1) - 5(x^2 - 1) \\ &= x^3 - x - 5x^2 + 5 \\ &= x^3 - 5x^2 - x + 5\end{aligned}$$

d.
$$\begin{aligned}\left(\frac{f}{g}\right)(x) &= \frac{f(x)}{g(x)} \\ &= \frac{x - 5}{x^2 - 1},\ x \neq \pm 1\end{aligned}$$

3. **a.**
$$\begin{aligned}(f + g)(x) &= f(x) + g(x) \\ &= \sqrt{x - 3} + \sqrt{x + 1}\end{aligned}$$

b. Domain of *f*: $x - 3 \geq 0$

$$x \geq 3$$
$$[3, \infty)$$

Domain of g: $x + 1 \geq 0$

$$x \geq -1$$
$$[-1, \infty)$$

The domain of $f + g$ is the set of all real numbers that are common to the domain of *f* and the domain of *g*. Thus, the domain of $f + g$ is $[3, \infty)$.

4. **a.** $(f \circ g)(x) = f(g(x))$
$= 5(2x^2 - x - 1) + 6$
$= 10x^2 - 5x - 5 + 6$
$= 10x^2 - 5x + 1$

b. $(g \circ f)(x) = g(f(x))$
$= 2(5x + 6)^2 - (5x + 6) - 1$
$= 2(25x^2 + 60x + 36) - 5x - 6 - 1$
$= 50x^2 + 120x + 72 - 5x - 6 - 1$
$= 50x^2 + 115x + 65$

5. **a.** $f \circ g(x) = \dfrac{4}{\frac{1}{x} + 2} = \dfrac{4x}{1 + 2x}$

b. $\left\{x \middle| x \neq 0, x \neq -\frac{1}{2}\right\}$

6. $h(x) = f \circ g$ where $f(x) = \sqrt{x}$; $g(x) = x^2 + 5$

Exercise Set 1.7

1. The function contains neither division nor an even root. The domain $= (-\infty, \infty)$

2. The function contains neither division nor an even root. The domain $= (-\infty, \infty)$

3. The denominator equals zero when $x = 4$. This value must be excluded from the domain.
Domain: $(-\infty, 4) \cup (4, \infty)$.

4. The denominator equals zero when $x = -5$. This value must be excluded from the domain.
Domain: $(-\infty, -5) \cup (-5, \infty)$.

5. The function contains neither division nor an even root. The domain $= (-\infty, \infty)$

6. The function contains neither division nor an even root. The domain $= (-\infty, \infty)$

7. The values that make the denominator equal zero must be excluded from the domain.
Domain: $(-\infty, -3) \cup (-3, 5) \cup (5, \infty)$

8. The values that make the denominator equal zero must be excluded from the domain.
Domain: $(-\infty, -4) \cup (-4, 3) \cup (3, \infty)$

9. The values that make the denominators equal zero must be excluded from the domain.
Domain: $(-\infty, -7) \cup (-7, 9) \cup (9, \infty)$

10. The values that make the denominators equal zero must be excluded from the domain.
Domain: $(-\infty,-8)\cup(-8,10)\cup(10,\infty)$

11. The first denominator cannot equal zero. The values that make the second denominator equal zero must be excluded from the domain.
Domain: $(-\infty,-1)\cup(-1,1)\cup(1,\infty)$

12. The first denominator cannot equal zero. The values that make the second denominator equal zero must be excluded from the domain.
Domain: $(-\infty,-2)\cup(-2,2)\cup(2,\infty)$

13. Exclude x for $x=0$.
Exclude x for $\frac{3}{x}-1=0$.

$$\frac{3}{x}-1=0$$
$$x\left(\frac{3}{x}-1\right)=x(0)$$
$$3-x=0$$
$$-x=-3$$
$$x=3$$

Domain: $(-\infty,0)\cup(0,3)\cup(3,\infty)$

14. Exclude x for $x=0$.
Exclude x for $\frac{4}{x}-1=0$.

$$\frac{4}{x}-1=0$$
$$x\left(\frac{4}{x}-1\right)=x(0)$$
$$4-x=0$$
$$-x=-4$$
$$x=4$$

Domain: $(-\infty,0)\cup(0,4)\cup(4,\infty)$

15. Exclude x for $x-1=0$.

$$x-1=0$$
$$x=1$$

Exclude x for $\frac{4}{x-1}-2=0$.

$$\frac{4}{x-1}-2=0$$
$$(x-1)\left(\frac{4}{x-1}-2\right)=(x-1)(0)$$
$$4-2(x-1)=0$$
$$4-2x+2=0$$
$$-2x+6=0$$
$$-2x=-6$$
$$x=3$$

Domain: $(-\infty,1)\cup(1,3)\cup(3,\infty)$

16. Exclude x for $x-2=0$.

$$x-2=0$$
$$x=2$$

Exclude x for $\frac{4}{x-2}-3=0$.

$$\frac{4}{x-2}-3=0$$
$$(x-2)\left(\frac{4}{x-2}-3\right)=(x-2)(0)$$
$$4-3(x-2)=0$$
$$4-3x+6=0$$
$$-3x+10=0$$
$$-3x=-10$$
$$x=\frac{10}{3}$$

Domain: $(-\infty,2)\cup\left(2,\frac{10}{3}\right)\cup\left(\frac{10}{3},\infty\right)$

17. The expression under the radical must not be negative.
$x-3\geq 0$
$x\geq 3$
Domain: $[3,\infty)$

18. The expression under the radical must not be negative.
$x+2\geq 0$
$x\geq -2$
Domain: $[-2,\infty)$

19. The expression under the radical must be positive.
$x-3>0$
$x>3$
Domain: $(3,\infty)$

20. The expression under the radical must be positive.
$x+2>0$
$x>-2$
Domain: $(-2,\infty)$

21. The expression under the radical must not be negative.
$5x+35\geq 0$
$5x\geq -35$
$x\geq -7$
Domain: $[-7,\infty)$

22. The expression under the radical must not be negative.
$7x-70\geq 0$
$7x\geq 70$
$x\geq 10$
Domain: $[10,\infty)$

23. The expression under the radical must not be negative.
$24-2x\geq 0$
$-2x\geq -24$
$\frac{-2x}{-2}\leq\frac{-24}{-2}$
$x\leq 12$
Domain: $(-\infty,12]$

24. The expression under the radical must not be negative.
$84-6x\geq 0$
$-6x\geq -84$
$\frac{-6x}{-6}\leq\frac{-84}{-6}$
$x\leq 14$
Domain: $(-\infty,14]$

25. The expressions under the radicals must not be negative.

$x-2\geq 0$ and $x+3\geq 0$

$x\geq 2$ $\quad x\geq -3$

To make both inequalities true, $x\geq 2$.

Domain: $[2,\infty)$

26. The expressions under the radicals must not be negative.

$x-3\geq 0$ and $x+4\geq 0$

$x\geq 3$ $\quad x\geq -4$

To make both inequalities true, $x\geq 3$.

Domain: $[3,\infty)$

27. The expression under the radical must not be negative.

$x-2\geq 0$

$x\geq 2$

The denominator equals zero when $x=5$.

Domain: $[2,5)\cup(5,\infty)$.

28. The expression under the radical must not be negative.

$x-3\geq 0$

$x\geq 3$

The denominator equals zero when $x=6$.

Domain: $[3,6)\cup(6,\infty)$.

29. Find the values that make the denominator equal zero and must be excluded from the domain.

$x^3-5x^2-4x+20$

$=x^2(x-5)-4(x-5)$

$=(x-5)(x^2-4)$

$=(x-5)(x+2)(x-2)$

–2, 2, and 5 must be excluded.

Domain: $(-\infty,-2)\cup(-2,2)\cup(2,5)\cup(5,\infty)$

30. Find the values that make the denominator equal zero and must be excluded from the domain.

$x^3-2x^2-9x+18$

$=x^2(x-2)-9(x-2)$

$=(x-2)(x^2-9)$

$=(x-2)(x+3)(x-3)$

–3, 2, and 3 must be excluded.

Domain: $(-\infty,-3)\cup(-3,2)\cup(2,3)\cup(3,\infty)$

31. $(f+g)(x) = 3x+2$
Domain: $(-\infty, \infty)$
$(f-g)(x) = f(x) - g(x)$
$= (2x+3) - (x-1)$
$= x+4$
Domain: $(-\infty, \infty)$
$(fg)(x) = f(x) \cdot g(x)$
$= (2x+3) \cdot (x-1)$
$= 2x^2 + x - 3$
Domain: $(-\infty, \infty)$
$\left(\frac{f}{g}\right)(x) = \frac{f(x)}{g(x)} = \frac{2x+3}{x-1}$
Domain: $(-\infty, 1) \cup (1, \infty)$

32. $(f+g)(x) = 4x-2$
Domain: $(-\infty, \infty)$
$(f-g)(x) = (3x-4) - (x+2) = 2x-6$
Domain: $(-\infty, \infty)$
$(fg)(x) = (3x-4)(x+2) = 3x^2 + 2x - 8$
Domain: $(-\infty, \infty)$
$\left(\frac{f}{g}\right)(x) = \frac{3x-4}{x+2}$
Domain: $(-\infty, -2) \cup (-2, \infty)$

33. $(f+g)(x) = 3x^2 + x - 5$
Domain: $(-\infty, \infty)$
$(f-g)(x) = -3x^2 + x - 5$
Domain: $(-\infty, \infty)$
$(fg)(x) = (x-5)(3x^2) = 3x^3 - 15x^2$
Domain: $(-\infty, \infty)$
$\left(\frac{f}{g}\right)(x) = \frac{x-5}{3x^2}$
Domain: $(-\infty, 0) \cup (0, \infty)$

34. $(f+g)(x) = 5x^2 + x - 6$
Domain: $(-\infty, \infty)$
$(f-g)(x) = -5x^2 + x - 6$
Domain: $(-\infty, \infty)$
$(fg)(x) = (x-6)(5x^2) = 5x^3 - 30x^2$
Domain: $(-\infty, \infty)$
$\left(\frac{f}{g}\right)(x) = \frac{x-6}{5x^2}$
Domain: $(-\infty, 0) \cup (0, \infty)$

35. $(f+g)(x) = 2x^2 - 2$
Domain: $(-\infty, \infty)$
$(f-g)(x) = 2x^2 - 2x - 4$
Domain: $(-\infty, \infty)$
$(fg)(x) = (2x^2 - x - 3)(x+1)$
$= 2x^3 + x^2 - 4x - 3$
Domain: $(-\infty, \infty)$
$\left(\frac{f}{g}\right)(x) = \frac{2x^2 - x - 3}{x+1}$
$= \frac{(2x-3)(x+1)}{(x+1)} = 2x - 3$
Domain: $(-\infty, -1) \cup (-1, \infty)$

36. $(f+g)(x) = 6x^2 - 2$
Domain: $(-\infty, \infty)$
$(f-g)(x) = 6x^2 - 2x$
Domain: $(-\infty, \infty)$
$(fg)(x) = (6x^2 - x - 1)(x-1) = 6x^3 - 7x^2 + 1$
Domain: $(-\infty, \infty)$
$\left(\frac{f}{g}\right)(x) = \frac{6x^2 - x - 1}{x-1}$
Domain: $(-\infty, 1) \cup (1, \infty)$

37. $(f+g)(x) = (3-x^2) + (x^2 + 2x - 15)$
$= 2x - 12$
Domain: $(-\infty, \infty)$
$(f-g)(x) = (3-x^2) - (x^2 + 2x - 15)$
$= -2x^2 - 2x + 18$
Domain: $(-\infty, \infty)$
$(fg)(x) = (3-x^2)(x^2 + 2x - 15)$
$= -x^4 - 2x^3 + 18x^2 + 6x - 45$
Domain: $(-\infty, \infty)$
$\left(\frac{f}{g}\right)(x) = \frac{3-x^2}{x^2 + 2x - 15}$
Domain: $(-\infty, -5) \cup (-5, 3) \cup (3, \infty)$

38. $(f+g)(x) = (5-x^2) + (x^2 + 4x - 12)$
$= 4x - 7$
Domain: $(-\infty, \infty)$
$(f-g)(x) = (5-x^2) - (x^2 + 4x - 12)$
$= -2x^2 - 4x + 17$
Domain: $(-\infty, \infty)$
$(fg)(x) = (5-x^2)(x^2 + 4x - 12)$
$= -x^4 - 4x^3 + 17x^2 + 20x - 60$
Domain: $(-\infty, \infty)$
$\left(\frac{f}{g}\right)(x) = \frac{5-x^2}{x^2 + 4x - 12}$
Domain: $(-\infty, -6) \cup (-6, 2) \cup (2, \infty)$

39. $(f+g)(x)=\sqrt{x}+x-4$
Domain: $[0,\infty)$
$(f-g)(x)=\sqrt{x}-x+4$
Domain: $[0,\infty)$
$(fg)(x)=\sqrt{x}(x-4)$
Domain: $[0,\infty)$
$\left(\frac{f}{g}\right)(x)=\frac{\sqrt{x}}{x-4}$
Domain: $[0,4)\cup(4,\infty)$

40. $(f+g)(x)=\sqrt{x}+x-5$
Domain: $[0,\infty)$
$(f-g)(x)=\sqrt{x}-x+5$
Domain: $[0,\infty)$
$(fg)(x)=\sqrt{x}(x-5)$
Domain: $[0,\infty)$
$\left(\frac{f}{g}\right)(x)=\frac{\sqrt{x}}{x-5}$
Domain: $[0,5)\cup(5,\infty)$

41. $(f+g)(x)=2+\frac{1}{x}+\frac{1}{x}=2+\frac{2}{x}=\frac{2x+2}{x}$
Domain: $(-\infty,0)\cup(0,\infty)$
$(f-g)(x)=2+\frac{1}{x}-\frac{1}{x}=2$
Domain: $(-\infty,0)\cup(0,\infty)$
$(fg)(x)=\left(2+\frac{1}{x}\right)\cdot\frac{1}{x}=\frac{2}{x}+\frac{1}{x^2}=\frac{2x+1}{x^2}$
Domain: $(-\infty,0)\cup(0,\infty)$
$\left(\frac{f}{g}\right)(x)=\frac{2+\frac{1}{x}}{\frac{1}{x}}=\left(2+\frac{1}{x}\right)\cdot x=2x+1$
Domain: $(-\infty,0)\cup(0,\infty)$

42. $(f+g)(x)=6-\frac{1}{x}+\frac{1}{x}=6$
Domain: $(-\infty,0)\cup(0,\infty)$
$(f-g)(x)=6-\frac{1}{x}-\frac{1}{x}=6-\frac{2}{x}=\frac{6x-2}{x}$
Domain: $(-\infty,0)\cup(0,\infty)$
$(fg)(x)=\left(6-\frac{1}{x}\right)\cdot\frac{1}{x}=\frac{6}{x}-\frac{1}{x^2}=\frac{6x-1}{x^2}$
Domain: $(-\infty,0)\cup(0,\infty)$
$\left(\frac{f}{g}\right)(x)=\frac{6-\frac{1}{x}}{\frac{1}{x}}=\left(6-\frac{1}{x}\right)\cdot x=6x-1$
Domain: $(-\infty,0)\cup(0,\infty)$

43. $(f+g)(x)=f(x)+g(x)$
$=\frac{5x+1}{x^2-9}+\frac{4x-2}{x^2-9}$
$=\frac{9x-1}{x^2-9}$
Domain: $(-\infty,-3)\cup(-3,3)\cup(3,\infty)$
$(f-g)(x)=f(x)-g(x)$
$=\frac{5x+1}{x^2-9}-\frac{4x-2}{x^2-9}$
$=\frac{x+3}{x^2-9}$
$=\frac{1}{x-3}$
Domain: $(-\infty,-3)\cup(-3,3)\cup(3,\infty)$
$(fg)(x)=f(x)\cdot g(x)$
$=\frac{5x+1}{x^2-9}\cdot\frac{4x-2}{x^2-9}$
$=\frac{(5x+1)(4x-2)}{\left(x^2-9\right)^2}$
Domain: $(-\infty,-3)\cup(-3,3)\cup(3,\infty)$
$\left(\frac{f}{g}\right)(x)=\frac{\frac{5x+1}{x^2-9}}{\frac{4x-2}{x^2-9}}$
$=\frac{5x+1}{x^2-9}\cdot\frac{x^2-9}{4x-2}$
$=\frac{5x+1}{4x-2}$
The domain must exclude -3, 3, and any values that make $4x-2=0$.
$4x-2=0$
$4x=2$
$x=\frac{1}{2}$
Domain: $(-\infty,-3)\cup(-3,\frac{1}{2})\cup(\frac{1}{2},3)\cup(3,\infty)$

44. $(f+g)(x) = f(x)+g(x)$
$$= \frac{3x+1}{x^2-25} + \frac{2x-4}{x^2-25}$$
$$= \frac{5x-3}{x^2-25}$$
Domain: $(-\infty,-5)\cup(-5,5)\cup(5,\infty)$

$(f-g)(x) = f(x)-g(x)$
$$= \frac{3x+1}{x^2-25} - \frac{2x-4}{x^2-25}$$
$$= \frac{x+5}{x^2-25}$$
$$= \frac{1}{x-5}$$
Domain: $(-\infty,-5)\cup(-5,5)\cup(5,\infty)$

$(fg)(x) = f(x)\cdot g(x)$
$$= \frac{3x+1}{x^2-25}\cdot\frac{2x-4}{x^2-25}$$
$$= \frac{(3x+1)(2x-4)}{\left(x^2-25\right)^2}$$
Domain: $(-\infty,-5)\cup(-5,5)\cup(5,\infty)$

$$\left(\frac{f}{g}\right)(x) = \frac{\frac{3x+1}{x^2-25}}{\frac{2x-4}{x^2-25}}$$
$$= \frac{3x+1}{x^2-25}\cdot\frac{x^2-25}{2x-4}$$
$$= \frac{3x+1}{2x-4}$$
The domain must exclude –5, 5, and any values that make $2x-4=0$.

$2x-4=0$

$2x=4$

$x=2$

Domain: $(-\infty,-5)\cup(-5,2)\cup(2,5)\cup(5,\infty)$

45. $(f+g)(x) = \sqrt{x+4}+\sqrt{x-1}$
Domain: $[1,\infty)$

$(f-g)(x) = \sqrt{x+4}-\sqrt{x-1}$
Domain: $[1,\infty)$

$(fg)(x) = \sqrt{x+4}\cdot\sqrt{x-1} = \sqrt{x^2+3x-4}$
Domain: $[1,\infty)$

$\left(\frac{f}{g}\right)(x) = \frac{\sqrt{x+4}}{\sqrt{x-1}}$
Domain: $(1,\infty)$

46. $(f+g)(x) = \sqrt{x+6}+\sqrt{x-3}$
Domain: $[3,\infty)$

$(f-g)(x) = \sqrt{x+6}-\sqrt{x-3}$
Domain: $[3,\infty)$

$(fg)(x) = \sqrt{x+6}\cdot\sqrt{x-3} = \sqrt{x^2+3x-18}$
Domain: $[3,\infty)$

$\left(\frac{f}{g}\right)(x) = \frac{\sqrt{x+6}}{\sqrt{x-3}}$
Domain: $(3,\infty)$

47. $(f+g)(x) = \sqrt{x-2}+\sqrt{2-x}$
Domain: $\{2\}$

$(f-g)(x) = \sqrt{x-2}-\sqrt{2-x}$
Domain: $\{2\}$

$(fg)(x) = \sqrt{x-2}\cdot\sqrt{2-x} = \sqrt{-x^2+4x-4}$
Domain: $\{2\}$

$\left(\frac{f}{g}\right)(x) = \frac{\sqrt{x-2}}{\sqrt{2-x}}$
Domain: $\varnothing$

48. $(f+g)(x) = \sqrt{x-5}+\sqrt{5-x}$
Domain: $\{5\}$

$(f-g)(x) = \sqrt{x-5}-\sqrt{5-x}$
Domain: $\{5\}$

$(fg)(x) = \sqrt{x-5}\cdot\sqrt{5-x} = \sqrt{-x^2+10x-25}$
Domain: $\{5\}$

$\left(\frac{f}{g}\right)(x) = \frac{\sqrt{x-5}}{\sqrt{5-x}}$
Domain: $\varnothing$

49. $f(x) = 2x$; $g(x) = x+7$

a. $(f\circ g)(x) = 2(x+7) = 2x+14$

b. $(g\circ f)(x) = 2x+7$

c. $(f\circ g)(2) = 2(2)+14 = 18$

50. $f(x) = 3x$; $g(x) = x-5$

a. $(f\circ g)(x) = 3(x-5) = 3x-15$

b. $(g\circ f)(x) = 3x-5$

c. $(f\circ g)(2) = 3(2)-15 = -9$

51. $f(x) = x + 4;\ g(x) = 2x + 1$

a. $(f \circ g)(x) = (2x+1)+4 = 2x+5$

b. $(g \circ f)(x) = 2(x+4)+1 = 2x+9$

c. $(f \circ g)(2) = 2(2)+5 = 9$

52. $f(x) = 5x+2;\ g(x) = 3x-4$

a. $(f \circ g)(x) = 5(3x-4)+2 = 15x-18$

b. $(g \circ f)(x) = 3(5x+2)-4 = 15x+2$

c. $(f \circ g)(2) = 15(2)-18 = 12$

53. $f(x) = 4x-3;\ g(x) = 5x^2-2$

a. $(f \circ g)(x) = 4(5x^2-2)-3$
$= 20x^2-11$

b. $(g \circ f)(x) = 5(4x-3)^2-2$
$= 5(16x^2-24x+9)-2$
$= 80x^2-120x+43$

c. $(f \circ g)(2) = 20(2)^2-11 = 69$

54. $f(x) = 7x+1; g(x) = 2x^2-9$

a. $(f \circ g)(x) = 7(2x^2-9)+1 = 14x^2-62$

b. $(g \circ f)(x) = 2(7x+1)^2-9$
$= 2(49x^2+14x+1)-9$
$= 98x^2+28x-7$

c. $(f \circ g)(2) = 14(2)^2-62 = -6$

55. $f(x) = x^2+2; g(x) = x^2-2$

a. $(f \circ g)(x) = (x^2-2)^2+2$
$= x^4-4x^2+4+2$
$= x^4-4x^2+6$

b. $(g \circ f)(x) = (x^2+2)^2-2$
$= x^4+4x^2+4-2$
$= x^4+4x^2+2$

c. $(f \circ g)(2) = 2^4-4(2)^2+6 = 6$

56. $f(x) = x^2+1; g(x) = x^2-3$

a. $(f \circ g)(x) = (x^2-3)^2+1$
$= x^4-6x^2+9+1$
$= x^4-6x^2+10$

b. $(g \circ f)(x) = (x^2+1)^2-3$
$= x^4+2x^2+1-3$
$= x^4+2x^2-2$

c. $(f \circ g)(2) = 2^4-6(2)^2+10 = 2$

57. $f(x) = 4-x;\ g(x) = 2x^2+x+5$

a. $(f \circ g)(x) = 4-\left(2x^2+x+5\right)$
$= 4-2x^2-x-5$
$= -2x^2-x-1$

b. $(g \circ f)(x) = 2(4-x)^2+(4-x)+5$
$= 2(16-8x+x^2)+4-x+5$
$= 32-16x+2x^2+4-x+5$
$= 2x^2-17x+41$

c. $(f \circ g)(2) = -2(2)^2-2-1 = -11$

58. $f(x) = 5x-2;\ g(x) = -x^2+4x-1$

a. $(f \circ g)(x) = 5\left(-x^2+4x-1\right)-2$
$= -5x^2+20x-5-2$
$= -5x^2+20x-7$

b. $(g \circ f)(x) = -(5x-2)^2+4(5x-2)-1$
$= -(25x^2-20x+4)+20x-8-1$
$= -25x^2+20x-4+20x-8-1$
$= -25x^2+40x-13$

c. $(f \circ g)(2) = -5(2)^2+20(2)-7 = 13$

59. $f(x)=\sqrt{x};\ g(x)=x-1$

a. $(f\circ g)(x)=\sqrt{x-1}$

b. $(g\circ f)(x)=\sqrt{x}-1$

c. $(f\circ g)(2)=\sqrt{2-1}=\sqrt{1}=1$

60. $f(x)=\sqrt{x};g(x)=x+2$

a. $(f\circ g)(x)=\sqrt{x+2}$

b. $(g\circ f)(x)=\sqrt{x}+2$

c. $(f\circ g)(2)=\sqrt{2+2}=\sqrt{4}=2$

61. $f(x)=2x-3;\ g(x)=\dfrac{x+3}{2}$

a. $(f\circ g)(x)=2\left(\dfrac{x+3}{2}\right)-3$

$=x+3-3$

$=x$

b. $(g\circ f)(x)=\dfrac{(2x-3)+3}{2}=\dfrac{2x}{2}=x$

c. $(f\circ g)(2)=2$

62. $f(x)=6x-3;g(x)=\dfrac{x+3}{6}$

a. $(f\circ g)(x)=6\left(\dfrac{x+3}{6}\right)-3=x+3-3=x$

b. $(g\circ f)(x)=\dfrac{6x-3+3}{6}=\dfrac{6x}{6}=x$

c. $(f\circ g)(2)=2$

63. $f(x)=\dfrac{1}{x};\quad g(x)=\dfrac{1}{x}$

a. $(f\circ g)(x)=\dfrac{1}{\frac{1}{x}}=x$

b. $(g\circ f)(x)=\dfrac{1}{\frac{1}{x}}=x$

c. $(f\circ g)(2)=2$

64. $f(x)=\dfrac{2}{x};\quad g(x)=\dfrac{2}{x}$

a. $(f\circ g)(x)=\dfrac{2}{\frac{2}{x}}=x$

b. $(g\circ f)(x)=\dfrac{2}{\frac{2}{x}}=x$

c. $(f\circ g)(2)=2$

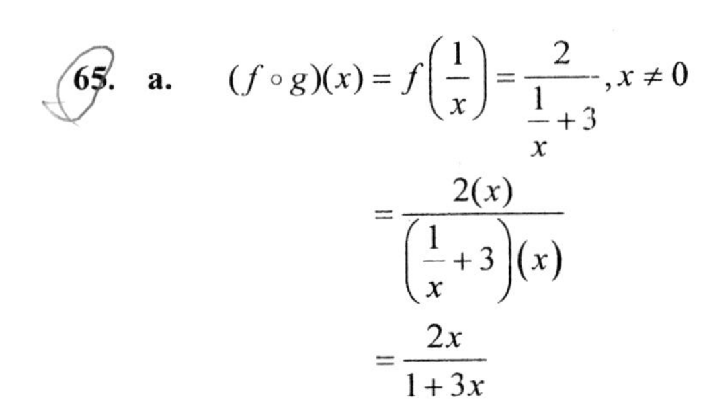

65. a. $(f\circ g)(x)=f\left(\dfrac{1}{x}\right)=\dfrac{2}{\frac{1}{x}+3},x\neq 0$

$=\dfrac{2(x)}{\left(\frac{1}{x}+3\right)(x)}$

$=\dfrac{2x}{1+3x}$

b. We must exclude 0 because it is excluded from g.

We must exclude $-\dfrac{1}{3}$ because it causes the denominator of $f\circ g$ to be 0.

Domain: $\left(-\infty,-\dfrac{1}{3}\right)\cup\left(-\dfrac{1}{3},0\right)\cup(0,\infty)$.

66. **a.** $f \circ g(x) = f\left(\frac{1}{x}\right) = \frac{5}{\frac{1}{x}+4} = \frac{5x}{1+4x}$

b. We must exclude 0 because it is excluded from g.

We must exclude $-\frac{1}{4}$ because it causes the denominator of $f \circ g$ to be 0.

Domain: $\left(-\infty, -\frac{1}{4}\right) \cup \left(-\frac{1}{4}, 0\right) \cup (0, \infty)$.

67. **a.** $(f \circ g)(x) = f\left(\frac{4}{x}\right) = \frac{\frac{4}{x}}{\frac{4}{x}+1}$

$$= \frac{\left(\frac{4}{x}\right)(x)}{\left(\frac{4}{x}+1\right)(x)}$$

$$= \frac{4}{4+x}, x \neq -4$$

b. We must exclude 0 because it is excluded from g.
We must exclude -4 because it causes the denominator of $f \circ g$ to be 0.
Domain: $(-\infty, -4) \cup (-4, 0) \cup (0, \infty)$.

68. **a.** $f \circ g(x) = f\left(\frac{6}{x}\right) = \frac{\frac{6}{x}}{\frac{6}{x}+5} = \frac{6}{6+5x}$

b. We must exclude 0 because it is excluded from g.

We must exclude $-\frac{6}{5}$ because it causes the denominator of $f \circ g$ to be 0.

Domain: $\left(-\infty, -\frac{6}{5}\right) \cup \left(-\frac{6}{5}, 0\right) \cup (0, \infty)$.

69. **a.** $f \circ g(x) = f(x-2) = \sqrt{x-2}$

b. The expression under the radical in $f \circ g$ must not be negative.

$x - 2 \geq 0$

$x \geq 2$

Domain: $[2, \infty)$.

70. **a.** $f \circ g(x) = f(x-3) = \sqrt{x-3}$

b. The expression under the radical in $f \circ g$ must not be negative.

$x - 3 \geq 0$

$x \geq 3$

Domain: $[3, \infty)$.

71. **a.** $(f \circ g)(x) = f(\sqrt{1-x})$
$= \left(\sqrt{1-x}\right)^2 + 4$
$= 1 - x + 4$
$= 5 - x$

b. The domain of $f \circ g$ must exclude any values that are excluded from g.
$1 - x \geq 0$
$-x \geq -1$
$x \leq 1$
Domain: $(-\infty, 1]$.

72. **a.** $(f \circ g)(x) = f(\sqrt{2-x})$
$= \left(\sqrt{2-x}\right)^2 + 1$
$= 2 - x + 1$
$= 3 - x$

b. The domain of $f \circ g$ must exclude any values that are excluded from g.
$2 - x \geq 0$
$-x \geq -2$
$x \leq 2$
Domain: $(-\infty, 2]$.

73. $f(x) = x^4 \quad g(x) = 3x - 1$

74. $f(x) = x^3; g(x) = 2x - 5$

75. $f(x) = \sqrt[3]{x} \quad g(x) = x^2 - 9$

76. $f(x) = \sqrt{x}; g(x) = 5x^2 + 3$

77. $f(x) = |x| \quad g(x) = 2x - 5$

78. $f(x) = |x|; \; g(x) = 3x - 4$

79. $f(x) = \frac{1}{x} \quad g(x) = 2x - 3$

80. $f(x) = \frac{1}{x}; g(x) = 4x + 5$

81. $(f+g)(-3) = f(-3) + g(-3) = 4 + 1 = 5$

82. $(g-f)(-2) = g(-2) - f(-2) = 2 - 3 = -1$

83. $(fg)(2) = f(2)g(2) = (-1)(1) = -1$

84. $\left(\frac{g}{f}\right)(3) = \frac{g(3)}{f(3)} = \frac{0}{-3} = 0$

85. The domain of $f + g$ is $[-4, 3]$.

86. The domain of $\frac{f}{g}$ is $(-4, 3)$

87. The graph of $f + g$

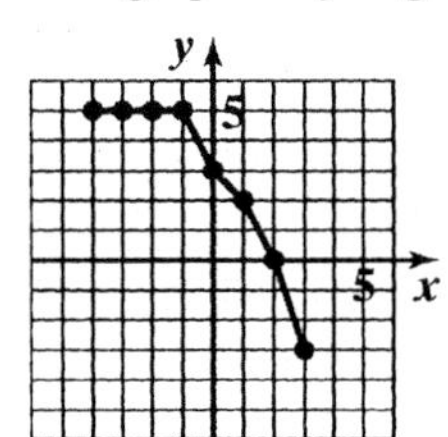

88. The graph of $f - g$

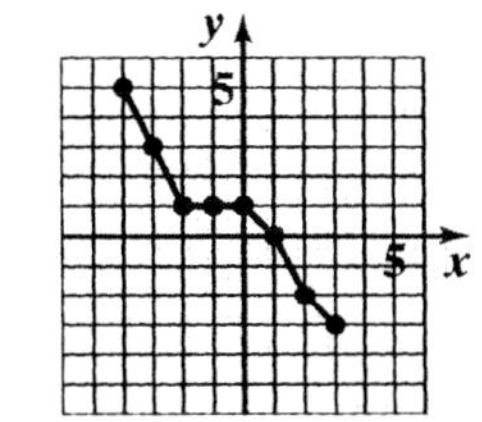

89. $(f \circ g)(-1) = f(g(-1)) = f(-3) = 1$

90. $(f \circ g)(1) = f(g(1)) = f(-5) = 3$

91. $(g \circ f)(0) = g(f(0)) = g(2) = -6$

92. $(g \circ f)(-1) = g(f(-1)) = g(1) = -5$

93.
$$\begin{aligned} (f \circ g)(x) &= 7 \\ 2(x^2 - 3x + 8) - 5 &= 7 \\ 2x^2 - 6x + 16 - 5 &= 7 \\ 2x^2 - 6x + 11 &= 7 \\ 2x^2 - 6x + 4 &= 0 \\ x^2 - 3x + 2 &= 0 \\ (x-1)(x-2) &= 0 \end{aligned}$$
$x - 1 = 0$ or $x - 2 = 0$
$x = 1$ $\quad$ $x = 2$

94.
$$\begin{aligned} (f \circ g)(x) &= -5 \\ 1 - 2(3x^2 + x - 1) &= -5 \\ 1 - 6x^2 - 2x + 2 &= -5 \\ -6x^2 - 2x + 3 &= -5 \\ -6x^2 - 2x + 8 &= 0 \\ 3x^2 + x - 4 &= 0 \\ (3x+4)(x-1) &= 0 \end{aligned}$$
$3x + 4 = 0$ or $x - 1 = 0$
$3x = -4$ $\quad$ $x = 1$
$x = -\frac{4}{3}$

95. Domain: $\{0, 1, 2, 3, 4, 5, 6, 7, 8\}$

96. Domain: $\{0, 1, 2, 3, 4, 5, 6, 7, 8\}$

97. **a.**
$$\begin{aligned} (B - D)(x) &= (26,208x + 3,869,910) - (17,964x + 2,300,198) \\ &= 26,208x + 3,869,910 - 17,964x - 2,300,198 \\ &= 8244x + 1,569,712 \end{aligned}$$
This function represents the net change in population from births and deaths.

b. $(B - D)(x) = 8244x + 1,569,712$
$(B - D)(8) = 8244(8) + 1,569,712 = 1,635,664$
The U.S. population increased by 1,635,664 in 2003.

c. $4,093,000 - 2,423,000 = 1,670,000$
The difference of the functions modeled this value reasonably well.

98. **a.**
$$\begin{aligned} (B - D)(x) &= (26,208x + 3,869,910) - (17,964x + 2,300,198) \\ &= 26,208x + 3,869,910 - 17,964x - 2,300,198 \\ &= 8244x + 1,569,712 \end{aligned}$$
This function represents the net change in population from births and deaths.

b. $(B - D)(x) = 8244x + 1,569,712$
$(B - D)(6) = 8244(6) + 1,569,712 = 1,619,176$
The U.S. population increased by 1,619,176 in 2001.

c. $4,025,933 - 2,416,425 = 1,609,508$
The difference of the functions modeled this value well.

99. $f + g$ represents the total world population in year x.

100. $h - g$ represents the difference between the total world population and the population of the world's less-developed regions. This would be the population of the world's more-developed regions.

101. $(f + g)(2000) \approx 6$ billion people.

102. $(h - g)(2000) = f(2000) = 1.5$ The world's more-developed regions had a population of approximately 1.5 billion in 2000.

103. $(R - C)(20,000)$

$= 65(20,000) - (600,000 + 45(20,000))$

$= -200,000$

The company lost $200,000 since costs exceeded revenues.

$(R - C)(30,000)$

$= 65(30,000) - (600,000 + 45(30,000))$

$= 0$

The company broke even.

104. a. The slope for f is -0.44 This is the decrease in profits for the first store for each year after 1998.

b. The slope of g is 0.51 This is the increase in profits for the second store for each year after 1998.

c. $f + g = -.044x + 13.62 + 0.51x + 11.14$
$= 0.07x + 24.76$
The slope for $f + g$ is 0.07 This is the profit for the two stores combined for each year after 1998.

105. a. f gives the price of the computer after a $400 discount. g gives the price of the computer after a 25% discount.

b. $(f \circ g)(x) = 0.75x - 400$
This models the price of a computer after first a 25% discount and then a $400 discount.

c. $(g \circ f)(x) = 0.75(x - 400)$
This models the price of a computer after first a $400 discount and then a 25% discount.

d. The function $f \circ g$ models the greater discount, since the 25% discount is taken on the regular price first.

106. a. f gives the cost of a pair of jeans for which a $5 rebate is offered.
g gives the cost of a pair of jeans that has been discounted 40%.

b. $(f \circ g)(x) = 0.6x - 5$
The cost of a pair of jeans is 60% of the regular price minus a $5 rebate.

c. $(g \circ f)(x) = 0.6(x - 5)$
$= 0.6x - 3$
The cost of a pair of jeans is 60% of the regular price minus a $3 rebate.

d. $f \circ g$ because of a $5 rebate.

113.

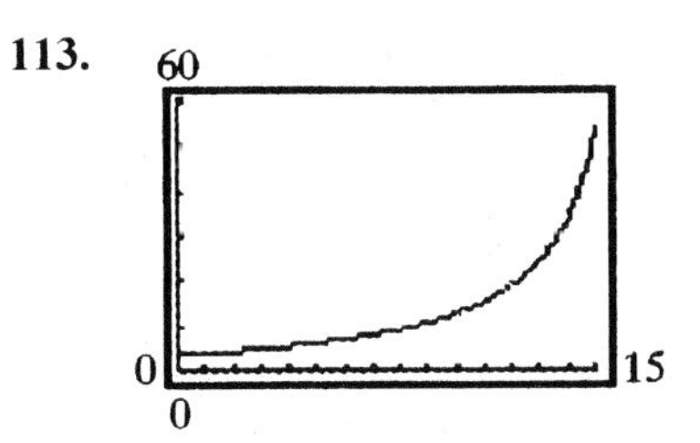

The per capita cost of Medicare is rising.

114. When your trace reaches $x = 0$, the y value disappears because the function is not defined at $x = 0$.

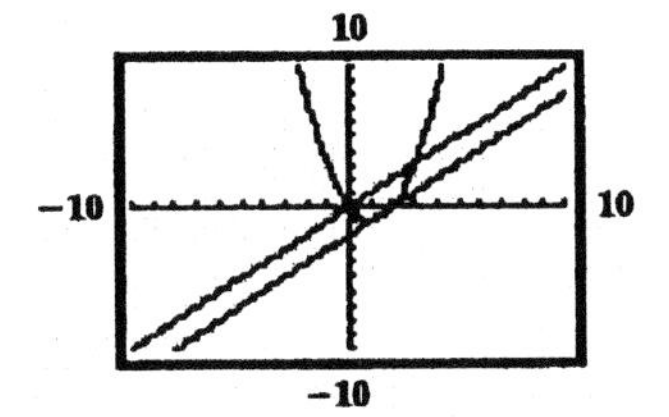

115.

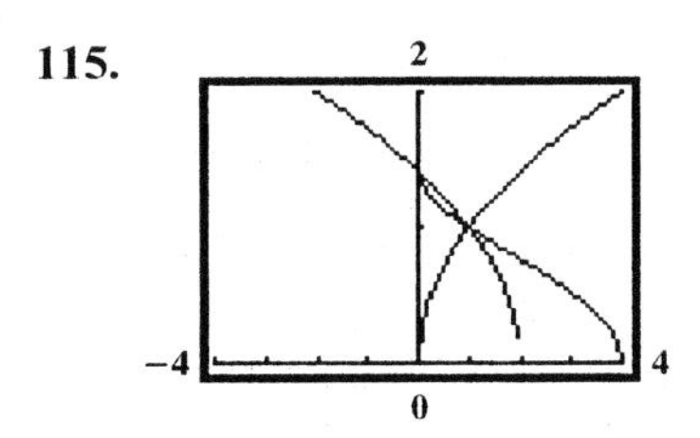

$(f \circ g)(x) = \sqrt{2 - \sqrt{x}}$

The domain of g is $[0, \infty)$.

The expression under the radical in $f \circ g$ must not be negative.

$2 - \sqrt{x} \geq 0$

$-\sqrt{x} \geq -2$

$\sqrt{x} \leq 2$

$x \leq 4$

Domain: $[0, 4]$

116. a. false

$(f \circ g)(x) = f\left(\sqrt{x^2-4}\right)$

$= \left(\sqrt{x^2-4}\right)^2 - 4$

$= x^2 - 4 - 4$

$= x^2 - 8$

b. false

$f(x) = 2x; g(x) = 3x$

$(f \circ g)(x) = f(g(x)) = f(3x) = 2(3x) = 6x$

$(g \circ f)(x) = g(f(x)) = g(f(x)) = 3(2x) = 6x$

c. false

$(f \circ g)(4) = f(g(4)) = f(7) = 5$

d. true

$(f \circ g)(5) = f(g(5)) = f[2(5)-1] = f(\sqrt{9}) = 3$

$g(2) = 2(2) - 1 = 4 - 1 = 3$

$(f \circ g)(5) = g(2)$

(d) is true.

117. $(f \circ g)(x) = (f \circ g)(-x)$

$f(g(x)) = f(g(-x))$ since g is even

$f(g(x)) = f(g(x))$ so $f \circ g$ is even

119.

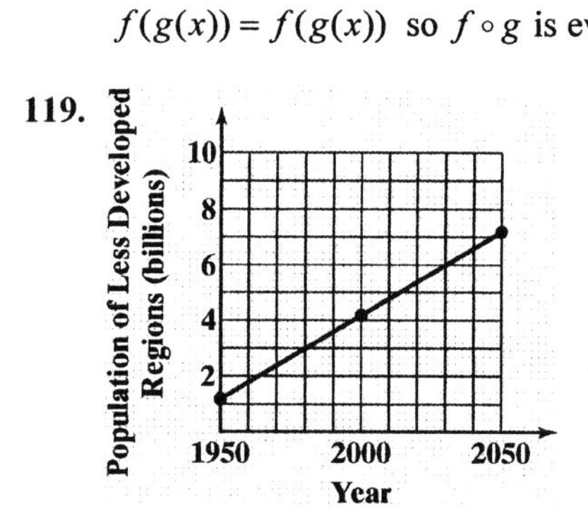

Section 1.8

Check Point Exercises

1. $f\left(g(x)\right)=4\left(\frac{x+7}{4}\right)-7=x$

$g\left(f(x)\right)=\frac{(4x-7)+7}{4}=x$

$f\left(g(x)\right)=g\left(f(x)\right)=x$

2. $f(x)=2x+7$

Replace $f(x)$ with y:

$y=2x+7$

Interchange x and y:

$x=2y+7$

Solve for y:

$x=2y+7$

$x-7=2y$

$\frac{x-7}{2}=y$

Replace y with $f^{-1}(x)$:

$f^{-1}(x)=\frac{x-7}{2}$

3. $f(x)=4x^3-1$

Replace $f(x)$ with y:

$y=4x^3-1$

Interchange x and y:

$x=4y^3-1$

Solve for y:

$x=4y^3-1$

$x+1=4y^3$

$\frac{x+1}{4}=y^3$

$\sqrt[3]{\frac{x+1}{4}}=y$

Replace y with $f^{-1}(x)$:

$f^{-1}(x)=\sqrt[3]{\frac{x+1}{4}}$

Alternative form for answer:

$$f(x)^{-1}=\sqrt[3]{\frac{x+1}{4}}=\frac{\sqrt[3]{x+1}}{\sqrt[3]{4}}$$

$$=\frac{\sqrt[3]{x+1}}{\sqrt[3]{4}}\cdot\frac{\sqrt[3]{2}}{\sqrt[3]{2}}=\frac{\sqrt[3]{2x+2}}{\sqrt[3]{8}}$$

$$=\frac{\sqrt[3]{2x+2}}{2}$$

4. $f(x)=\frac{3}{x}-1$

Replace $f(x)$ with y:

$y=\frac{3}{x}-1$

Interchange x and y:

$x=\frac{3}{y}-1$

Solve for y:

$x=\frac{3}{y}-1$

$xy=3-y$

$xy+y=3$

$y(x+1)=3$

$y=\frac{3}{x+1}$

Replace y with $f^{-1}(x)$:

$f^{-1}(x)=\frac{3}{x+1}$

5. The graphs of (b) and (c) pass the horizontal line test and thus have an inverse.

6. Find points of f^{-1}.

$f(x)$	$f^{-1}(x)$
$(-2,-2)$	$(-2,-2)$
$(-1,0)$	$(0,-1)$
$(1,2)$	$(2,1)$

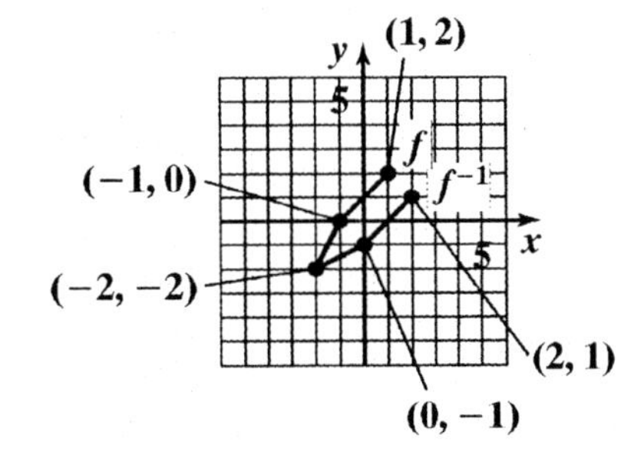

7. $f(x) = x^2 + 1$

Replace $f(x)$ with y:

$y = x^2 + 1$

Interchange x and y:

$x = y^2 + 1$

Solve for y:

$$x = y^2 + 1$$
$$x - 1 = y^2$$
$$\sqrt{x-1} = y$$

Replace y with $f^{-1}(x)$:

$f^{-1}(x) = \sqrt{x-1}$

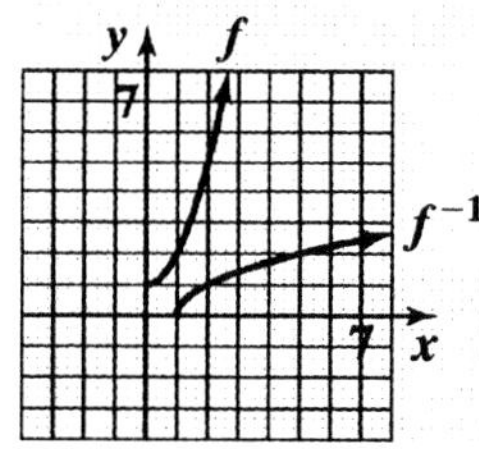

Exercise Set 1.8

1. $f(x) = 4x;\ g(x) = \dfrac{x}{4}$

$$f(g(x)) = 4\left(\frac{x}{4}\right) = x$$

$$g(f(x)) = \frac{4x}{4} = x$$

f and g are inverses.

2. $f(x) = 6x;\ g(x) = \dfrac{x}{6}$

$$f\big(g(x)\big) = 6\left(\frac{x}{6}\right) = x$$

$$g\big(f(x)\big) = \frac{6x}{6} = x$$

f and g are inverses.

3. $f(x) = 3x + 8;\ g(x) = \dfrac{x-8}{3}$

$$f(g(x)) = 3\left(\frac{x-8}{3}\right) + 8 = x - 8 + 8 = x$$

$$g(f(x)) = \frac{(3x+8)-8}{3} = \frac{3x}{3} = x$$

f and g are inverses.

4. $f(x) = 4x + 9;\ g(x) = \dfrac{x-9}{4}$

$$f\big(g(x)\big) = 4\left(\frac{x-9}{4}\right) + 9 = x - 9 + 9 = x$$

$$g\big(f(x)\big) = \frac{(4x+9)-9}{4} = \frac{4x}{4} = x$$

f and g are inverses.

5. $f(x) = 5x - 9;\ g(x) = \dfrac{x+5}{9}$

$$f(g(x)) = 5\left(\frac{x+5}{9}\right) - 9$$
$$= \frac{5x+25}{9} - 9$$
$$= \frac{5x-56}{9}$$

$$g(f(x)) = \frac{5x-9+5}{9} = \frac{5x-4}{9}$$

f and g are not inverses.

6. $f(x) = 3x - 7;\ g(x) = \dfrac{x+3}{7}$

$$f\big(g(x)\big) = 3\left(\frac{x+3}{7}\right) - 7 = \frac{3x+9}{7} - 7 = \frac{3x-40}{7}$$

$$g\big(f(x)\big) = \frac{3x-7+3}{7} = \frac{3x-4}{7}$$

f and g are not inverses.

7. $f(x) = \dfrac{3}{x-4};\ g(x) = \dfrac{3}{x} + 4$

$$f(g(x)) = \frac{3}{\frac{3}{x} + 4 - 4} = \frac{3}{\frac{3}{x}} = x$$

$$g(f(x)) = \frac{3}{\frac{3}{x-4}} + 4$$
$$= 3\cdot\left(\frac{x-4}{3}\right) + 4$$
$$= x - 4 + 4$$
$$= x$$

f and g are inverses.

8. $f(x) = \dfrac{2}{x-5};\ g(x) = \dfrac{2}{x} + 5$

$$f(g(x)) = \frac{2}{\left(\frac{2}{x} + 5\right) - 5} = \frac{2x}{2} = x$$

$$g\big(f(x)\big) = \frac{2}{\frac{2}{x-5}} + 5 = 2\left(\frac{x-5}{2}\right) + 5 = x - 5 + 5 = x$$

f and g are inverses.

9. $f(x) = -x;\ g(x) = -x$
$f(g(x)) = -(-x) = x$
$g(f(x)) = -(-x) = x$
f and g are inverses.

10. $f(x) = \sqrt[3]{x-4};\ g(x) = x^3 + 4$
$f(g(x)) = \sqrt[3]{x^3 + 4 - 4} = \sqrt[3]{x^3} = x$
$g(f(x)) = \left(\sqrt[3]{x-4}\right)^3 + 4 = x - 4 + 4 = x$
f and g are inverses.

11. a.
$f(x) = x + 3$
$y = x + 3$
$x = y + 3$
$y = x - 3$
$f^{-1}(x) = x - 3$

b. $f(f^{-1}(x)) = x - 3 + 3 = x$
$f^{-1}(f(x)) = x + 3 - 3 = x$

12. a.
$f(x) = x + 5$
$y = x + 5$
$x = y + 5$
$y = x - 5$
$f^{-1}(x) = x - 5$

b. $f\left(f^{-1}(x)\right) = x - 5 + 5 = x$
$f^{-1}\left(f(x)\right) = x + 5 - 5 = x$

13. a.
$f(x) = 2x$
$y = 2x$
$x = 2y$
$y = \frac{x}{2}$
$f^{-1}(x) = \frac{x}{2}$

b. $f(f^{-1}(x)) = 2\left(\frac{x}{2}\right) = x$
$f^{-1}(f(x)) = \frac{2x}{2} = x$

14. a.
$f(x) = 4x$
$y = 4x$
$x = 4y$
$y = \frac{x}{4}$
$f^{-1}(x) = \frac{x}{4}$

b. $f\left(f^{-1}(x)\right) = 4\left(\frac{x}{4}\right) = x$
$f^{-1}\left(f(x)\right) = \frac{4x}{4} = x$

15. a.
$f(x) = 2x + 3$
$y = 2x + 3$
$x = 2y + 3$
$x - 3 = 2y$
$y = \frac{x-3}{2}$
$f^{-1}(x) = \frac{x-3}{2}$

b. $f(f^{-1}(x)) = 2\left(\frac{x-3}{2}\right) + 3$
$= x - 3 + 3$
$= x$
$f^{-1}(f(x)) = \frac{2x + 3 - 3}{2} = \frac{2x}{2} = x$

16. a.
$f(x) = 3x - 1$
$y = 3x - 1$
$x = 3y - 1$
$x + 1 = 3y$
$y = \frac{x+1}{3}$
$f^{-1}(x) = \frac{x+1}{3}$

b. $f\left(f^{-1}(x)\right) = 3\left(\frac{x+1}{3}\right) - 1 = x + 1 - 1 = x$
$f^{-1}\left(f(x)\right) = \frac{3x - 1 + 1}{3} = \frac{3x}{3} = x$

17. a.

$$f(x)=x^3+2$$
$$y=x^3+2$$
$$x=y^3+2$$
$$x-2=y^3$$
$$y=\sqrt[3]{x-2}$$
$$f^{-1}(x)=\sqrt[3]{x-2}$$

b. $$f(f^{-1}(x))=\left(\sqrt[3]{x-2}\right)^3+2$$
$$=x-2+2$$
$$=x$$
$$f^{-1}(f(x))=\sqrt[3]{x^3+2-2}=\sqrt[3]{x^3}=x$$

18. a. $$f(x)=x^3-1$$
$$y=x^3-1$$
$$x=y^3-1$$
$$x+1=y^3$$
$$y=\sqrt[3]{x+1}$$
$$f^{-1}(x)=\sqrt[3]{x+1}$$

b. $$f(f^{-1}(x))=\left(\sqrt[3]{x+1}\right)^3-1$$
$$=x+1-1$$
$$=x$$
$$f^{-1}(f(x))=\sqrt[3]{x^3-1+1}=\sqrt[3]{x^3}=x$$

19. a. $$f(x)=(x+2)^3$$
$$y=(x+2)^3$$
$$x=(y+2)^3$$
$$\sqrt[3]{x}=y+2$$
$$y=\sqrt[3]{x}-2$$
$$f^{-1}(x)=\sqrt[3]{x}-2$$

b. $$f(f^{-1}(x))=\left(\sqrt[3]{x}-2+2\right)^3=\left(\sqrt[3]{x}\right)^3=x$$
$$f^{-1}(f(x))=\sqrt[3]{(x+2)^3}-2$$
$$=x+2-2$$
$$=x$$

20. a. $$f(x)=(x-1)^3$$
$$y=(x-1)^3$$
$$x=(y-1)^3$$
$$\sqrt[3]{x}=y-1$$
$$y=\sqrt[3]{x}+1$$

b. $$f\left(f^{-1}(x)\right)=\left(\sqrt[3]{x}+1-1\right)^3=\left(\sqrt[3]{x}\right)^3=x$$
$$f^{-1}\left(f(x)\right)=\sqrt[3]{(x-1^3}+1=x-1+1=x$$

21. a. $$f(x)=\frac{1}{x}$$
$$y=\frac{1}{x}$$
$$x=\frac{1}{y}$$
$$xy=1$$
$$y=\frac{1}{x}$$
$$f^{-1}(x)=\frac{1}{x}$$

b. $$f(f^{-1}(x))=\frac{1}{\frac{1}{x}}=x$$
$$f^{-1}(f(x))=\frac{1}{\frac{1}{x}}=x$$

22. a. $$f(x)=\frac{2}{x}$$
$$y=\frac{2}{x}$$
$$x=\frac{2}{y}$$
$$xy=2$$
$$y=\frac{2}{x}$$
$$f^{-1}(x)=\frac{2}{x}$$

b. $$f(f^{-1}(x))=\frac{2}{\frac{2}{x}}=2\cdot\frac{x}{2}=x$$
$$f^{-1}(f(x))=\frac{2}{\frac{2}{x}}=2\cdot\frac{x}{2}=x$$

23. **a.**
$$f(x) = \sqrt{x}$$
$$y = \sqrt{x}$$
$$x = \sqrt{y}$$
$$y = x^2$$
$$f^{-1}(x) = x^2, x \geq 0$$

b.
$$f(f^{-1}(x)) = \sqrt{x^2} = |x| = x \text{ for } x \geq 0.$$
$$f^{-1}(f(x)) = (\sqrt{x})^2 = x$$

24. **a.**
$$f(x) = \sqrt[3]{x}$$
$$y = \sqrt[3]{x}$$
$$x = \sqrt[3]{y}$$
$$y = x^3$$
$$f^{-1}(x) = x^3$$

b.
$$f\left(f^{-1}(x)\right) = \sqrt[3]{x^3} = x$$
$$f^{-1}\left(f(x)\right) = \left(\sqrt[3]{x}\right)^3 = x$$

25. **a.**
$$f(x) = \frac{7}{x} - 3$$
$$y = \frac{7}{x} - 3$$
$$x = \frac{7}{y} - 3$$
$$xy = 7 - 3y$$
$$xy + 3y = 7$$
$$y(x+3) = 7$$
$$y = \frac{7}{x+3}$$
$$f^{-1}(x) = \frac{7}{x+3}$$

b.
$$f\left(f^{-1}(x)\right) = \frac{7}{\frac{7}{x+3}} - 3 = x$$
$$f^{-1}\left(f(x)\right) = \frac{7}{\frac{7}{x} - 3 + 3} = x$$

26. **a.**
$$f(x) = \frac{4}{x} + 9$$
$$y = \frac{4}{x} + 9$$
$$x = \frac{4}{y} + 9$$
$$xy = 4 + 9y$$
$$xy - 9y = 4$$
$$y(x-9) = 4$$
$$y = \frac{4}{x-9}$$
$$f^{-1}(x) = \frac{4}{x-9}$$

b.
$$f\left(f^{-1}(x)\right) = \frac{4}{\frac{4}{x-9}} + 9 = x$$
$$f^{-1}\left(f(x)\right) = \frac{4}{\frac{4}{x} + 9 - 9} = x$$

27. **a.**
$$f(x) = \frac{2x+1}{x-3}$$
$$y = \frac{2x+1}{x-3}$$
$$x = \frac{2y+1}{y-3}$$
$$x(y-3) = 2y + 1$$
$$xy - 3x = 2y + 1$$
$$xy - 2y = 3x + 1$$
$$y(x-2) = 3x + 1$$
$$y = \frac{3x+1}{x-2}$$
$$f^{-1}(x) = \frac{3x+1}{x-2}$$

b.
$$f(f^{-1}(x)) = \frac{2\left(\frac{3x+1}{x-2}\right) + 1}{\frac{3x+1}{x-2} - 3}$$
$$= \frac{2(3x+1) + x - 2}{3x + 1 - 3(x-2)} = \frac{6x + 2 + x - 2}{3x + 1 - 3x + 6}$$
$$= \frac{7x}{7} = x$$

$$f^{-1}(f(x)) = \frac{3\left(\frac{2x+1}{x-3}\right)+1}{\frac{2x+1}{x-3}-2}$$

$$= \frac{3(2x+1)+x-3}{2x+1-2(x-3)}$$

$$= \frac{6x+3+x-3}{2x+1-2x+6} = \frac{7x}{7} = x$$

28. a. $f(x) = \frac{2x-3}{x+1}$

$y = \frac{2x-3}{x+1}$

$x = \frac{2y-3}{y+1}$

$xy + x = 2y - 3$

$y(x-2) = -x - 3$

$y = \frac{-x-3}{x-2}$

$f^{-1}(x) = \frac{-x-3}{x-2},\quad x \neq 2$

b. $f\left(f^{-1}(x)\right) = \frac{2\left(\frac{-x-3}{x-2}\right)-3}{\frac{-x-3}{x-2}+1}$

$$= \frac{-2x-6-3x+6}{-x-3+x-2} = \frac{-5x}{-5} = x$$

$$f^{-1}(f(x)) = \frac{-\left(\frac{2x-3}{x+1}\right)-3}{\frac{2x-3}{x+1}-2}$$

$$= \frac{-2x+3-3x-3}{2x-3-2x-2} = \frac{-5x}{-5} = x$$

29. The function fails the horizontal line test, so it does not have an inverse function.

30. The function passes the horizontal line test, so it does have an inverse function.

31. The function fails the horizontal line test, so it does not have an inverse function.

32. The function fails the horizontal line test, so it does not have an inverse function.

33. The function passes the horizontal line test, so it does have an inverse function.

34. The function passes the horizontal line test, so it does have an inverse function.

35.

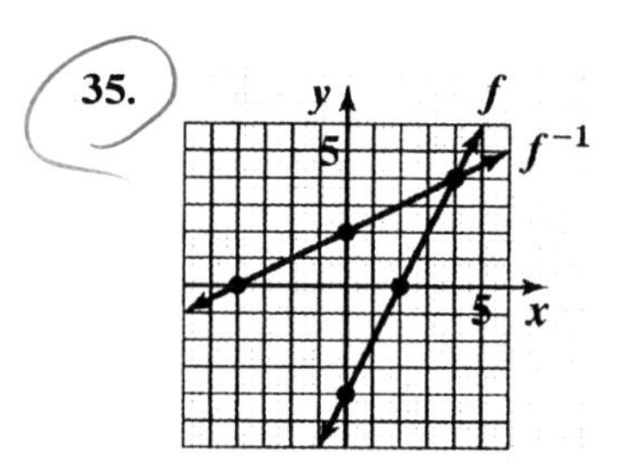

36.

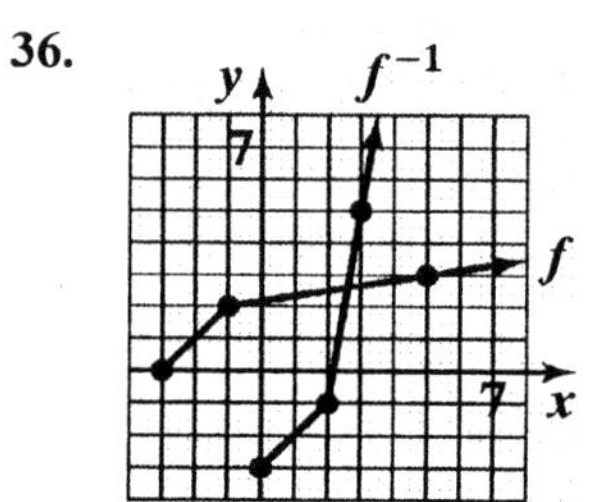

37.

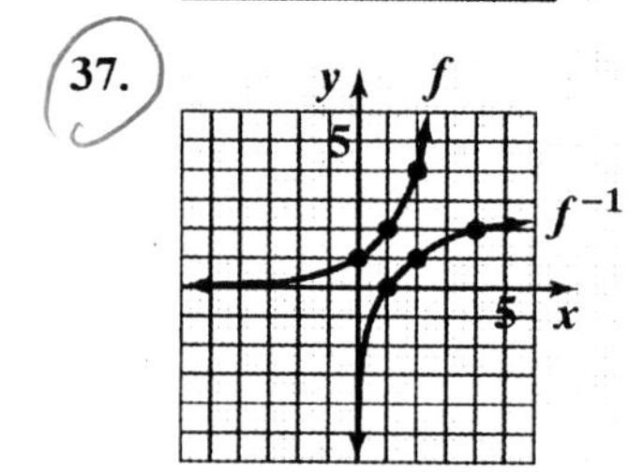

38.

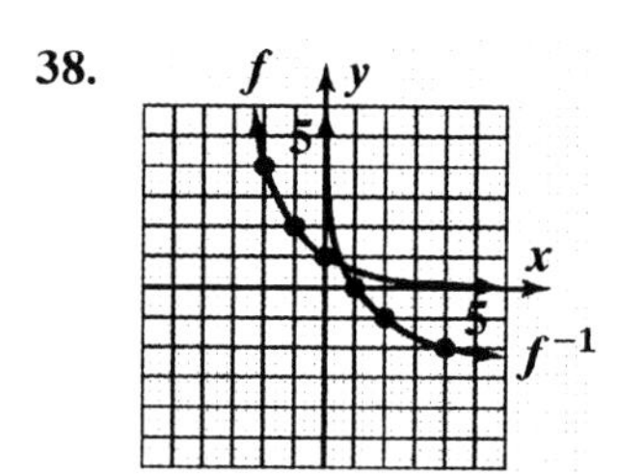

39. a. $f(x) = 2x - 1$

$y = 2x - 1$

$x = 2y - 1$

$x + 1 = 2y$

$\frac{x+1}{2} = y$

$f^{-1}(x) = \frac{x+1}{2}$

b.

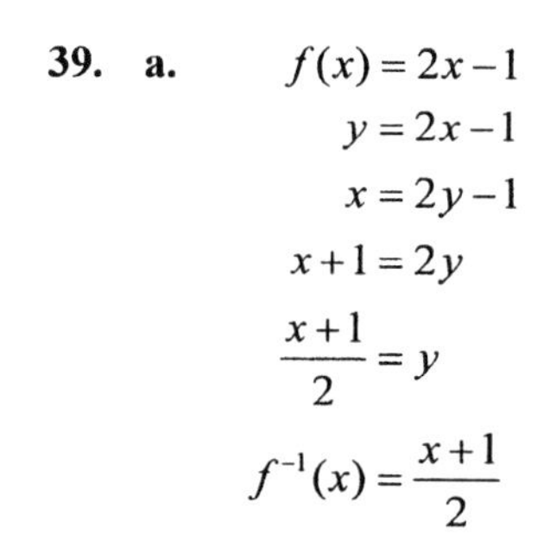

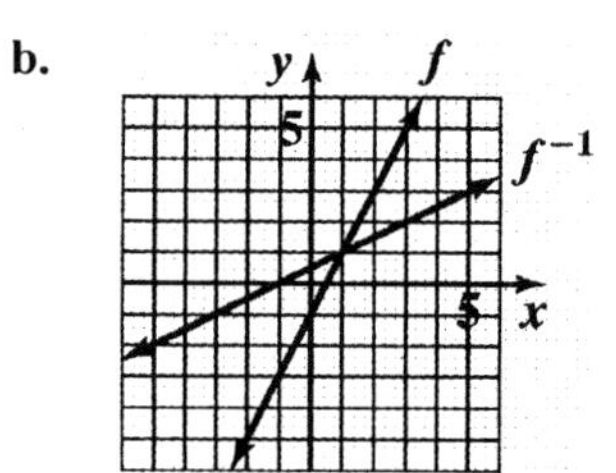

c. Domain of f: $(-\infty, \infty)$
Range of f: $(-\infty, \infty)$
Domain of f^{-1}: $(-\infty, \infty)$
Range of f^{-1}: $(-\infty, \infty)$

40. a.

$$f(x) = 2x - 3$$
$$y = 2x - 3$$
$$x = 2y - 3$$
$$x + 3 = 2y$$
$$\frac{x+3}{2} = y$$
$$f^{-1}(x) = \frac{x+3}{2}$$

b.

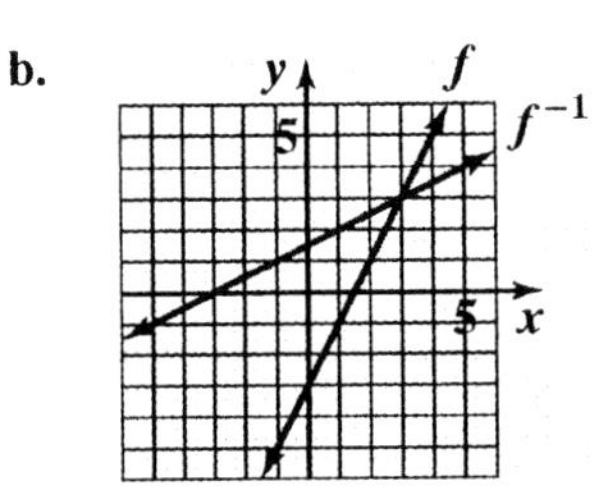

c. Domain of f: $(-\infty, \infty)$
Range of f: $(-\infty, \infty)$
Domain of f^{-1}: $(-\infty, \infty)$
Range of f^{-1}: $(-\infty, \infty)$

41. a.

$$f(x) = x^2 - 4$$
$$y = x^2 - 4$$
$$x = y^2 - 4$$
$$x + 4 = y^2$$
$$\sqrt{x+4} = y$$
$$f^{-1}(x) = \sqrt{x+4}$$

b.

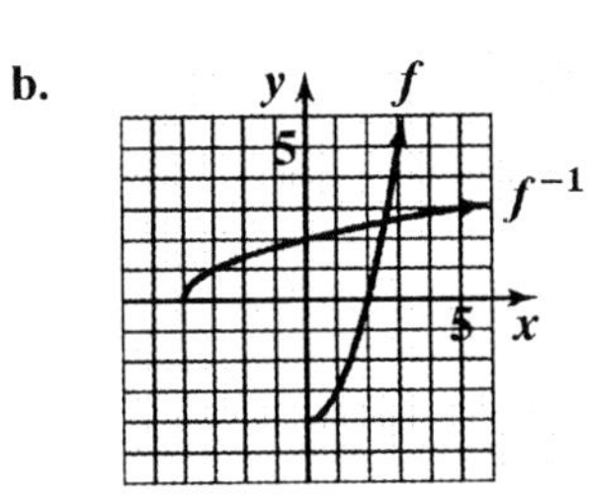

c. Domain of f: $[0, \infty)$
Range of f: $[-4, \infty)$
Domain of f^{-1}: $[-4, \infty)$
Range of f^{-1}: $[0, \infty)$

42. a.

$$f(x) = x^2 - 1$$
$$y = x^2 - 1$$
$$x = y^2 - 1$$
$$x + 1 = y^2$$
$$-\sqrt{x+1} = y$$
$$f^{-1}(x) = -\sqrt{x+1}$$

b.

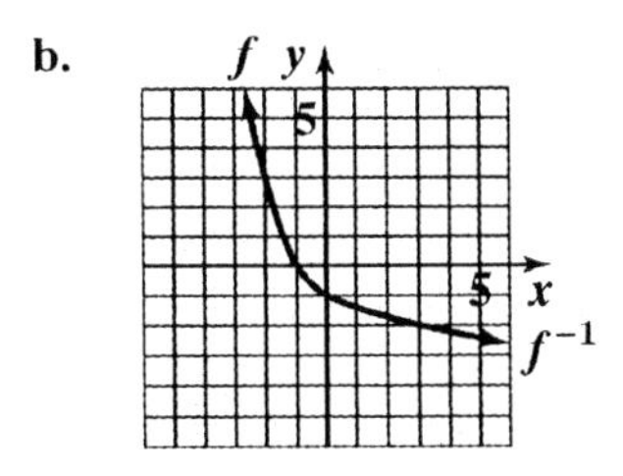

c. Domain of f: $(-\infty, 0]$
Range of f: $[-1, \infty)$
Domain of f^{-1}: $[-1, \infty)$
Range of f^{-1}: $(-\infty, 0]$

43. a.

$$f(x) = (x-1)^2$$
$$y = (x-1)^2$$
$$x = (y-1)^2$$
$$-\sqrt{x} = y - 1$$
$$-\sqrt{x} + 1 = y$$
$$f^{-1}(x) = 1 - \sqrt{x}$$

b.

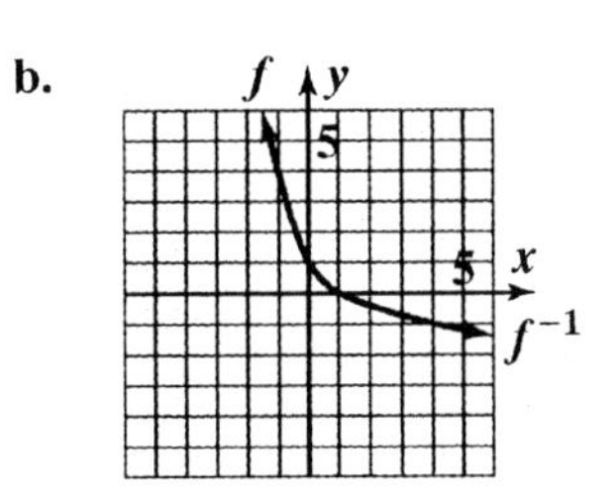

c. Domain of f: $(-\infty, 1]$
Range of f: $[0, \infty)$
Domain of f^{-1}: $[0, \infty)$
Range of f^{-1}: $(-\infty, 1]$

44. a.

$$f(x) = (x-1)^2$$
$$y = (x-1)^2$$
$$x = (y-1)^2$$
$$\sqrt{x} = y-1$$
$$\sqrt{x}+1 = y$$
$$f^{-1}(x) = 1+\sqrt{x}$$

b.

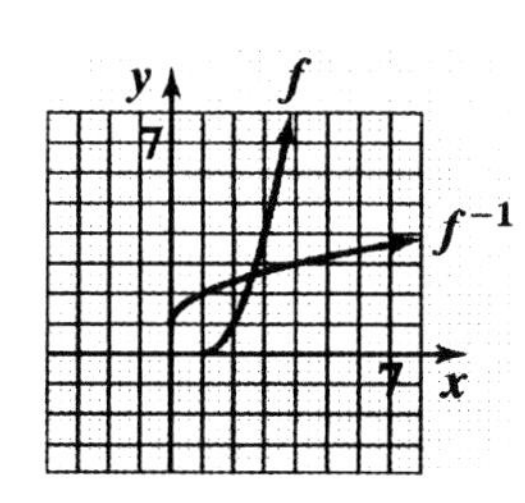

c. Domain of f: $[1,\infty)$

Range of f: $[0,\infty)$

Domain of f^{-1}: $[0,\infty)$

Range of f^{-1}: $[1,\infty)$

45. a.

$$f(x) = x^3-1$$
$$y = x^3-1$$
$$x = y^3-1$$
$$x+1 = y^3$$
$$\sqrt[3]{x+1} = y$$
$$f^{-1}(x) = \sqrt[3]{x+1}$$

b.

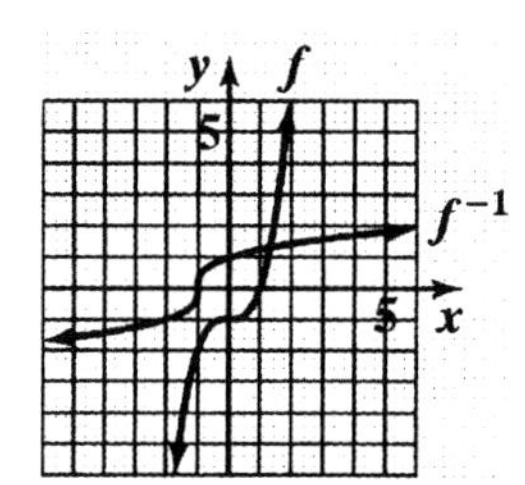

c. Domain of f: $(-\infty,\infty)$

Range of f: $(-\infty,\infty)$

Domain of f^{-1}: $(-\infty,\infty)$

Range of f^{-1}: $(-\infty,\infty)$

46. a.

$$f(x) = x^3+1$$
$$y = x^3+1$$
$$x = y^3+1$$
$$x-1 = y^3$$
$$\sqrt[3]{x-1} = y$$
$$f^{-1}(x) = \sqrt[3]{x-1}$$

b.

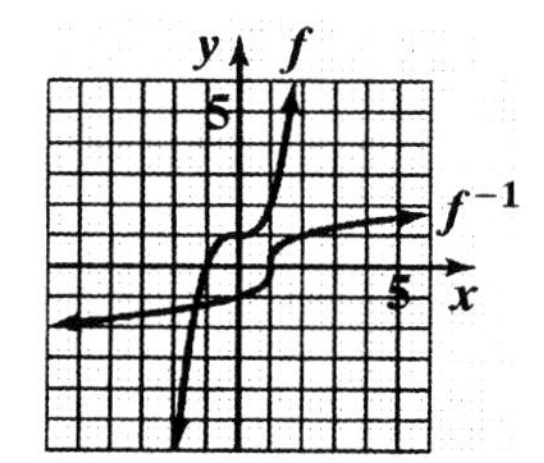

c. Domain of f: $(-\infty,\infty)$

Range of f: $(-\infty,\infty)$

Domain of f^{-1}: $(-\infty,\infty)$

Range of f^{-1}: $(-\infty,\infty)$

47. a.

$$f(x) = (x+2)^3$$
$$y = (x+2)^3$$
$$x = (y+2)^3$$
$$\sqrt[3]{x} = y+2$$
$$\sqrt[3]{x}-2 = y$$
$$f^{-1}(x) = \sqrt[3]{x}-2$$

b.

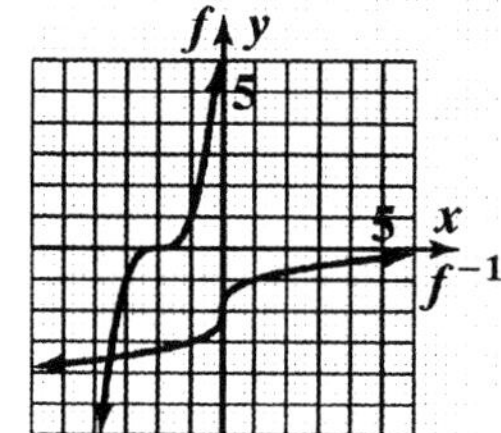

c. Domain of f: $(-\infty,\infty)$

Range of f: $(-\infty,\infty)$

Domain of f^{-1}: $(-\infty,\infty)$

Range of f^{-1}: $(-\infty,\infty)$

48. **a.**
$$f(x) = (x-2)^3$$
$$y = (x-2)^3$$
$$x = (y-2)^3$$
$$\sqrt[3]{x} = y-2$$
$$\sqrt[3]{x} + 2 = y$$
$$f^{-1}(x) = \sqrt[3]{x} + 2$$

b.

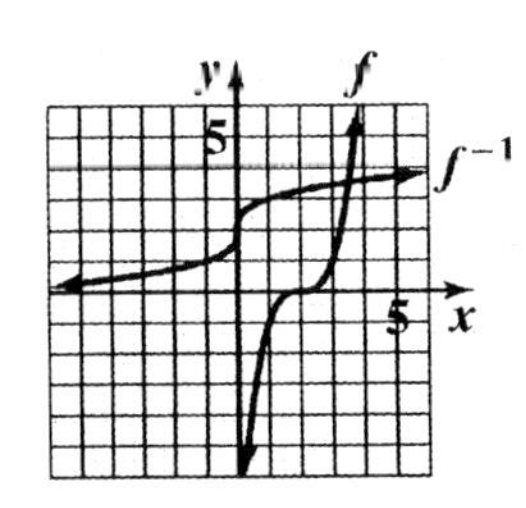

c. Domain of f: $(-\infty, \infty)$

Range of f: $(-\infty, \infty)$

Domain of f^{-1}: $(-\infty, \infty)$

Range of f^{-1}: $(-\infty, \infty)$

49. **a.**
$$f(x) = \sqrt{x-1}$$
$$y = \sqrt{x-1}$$
$$x = \sqrt{y-1}$$
$$x^2 = y-1$$
$$x^2 + 1 = y$$
$$f^{-1}(x) = x^2 + 1$$

b.

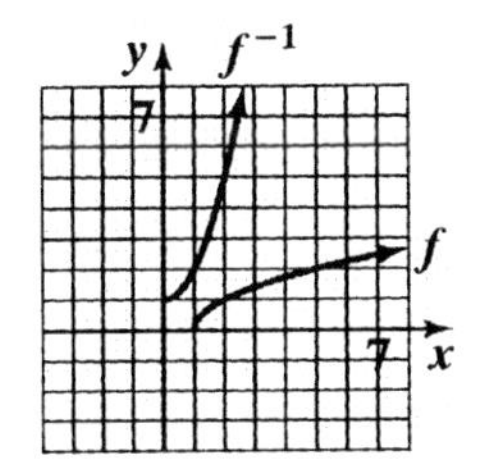

c. Domain of f: $[1, \infty)$

Range of f: $[0, \infty)$

Domain of f^{-1}: $[0, \infty)$

Range of f^{-1}: $[1, \infty)$

50. **a.**
$$f(x) = \sqrt{x} + 2$$
$$y = \sqrt{x} + 2$$
$$x = \sqrt{y} + 2$$
$$x - 2 = \sqrt{y}$$
$$(x-2)^2 = y$$
$$f^{-1}(x) = (x-2)^2$$

b.

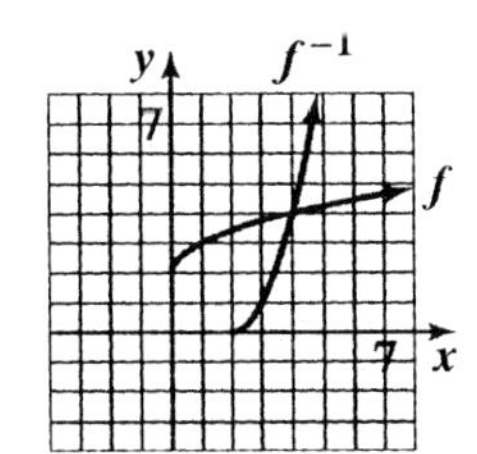

c. Domain of f: $[0, \infty)$

Range of f: $[2, \infty)$

Domain of f^{-1}: $[2, \infty)$

Range of f^{-1}: $[0, \infty)$

51. **a.**
$$f(x) = \sqrt[3]{x} + 1$$
$$y = \sqrt[3]{x} + 1$$
$$x = \sqrt[3]{y} + 1$$
$$x - 1 = \sqrt[3]{y}$$
$$(x-1)^3 = y$$
$$f^{-1}(x) = (x-1)^3$$

b.

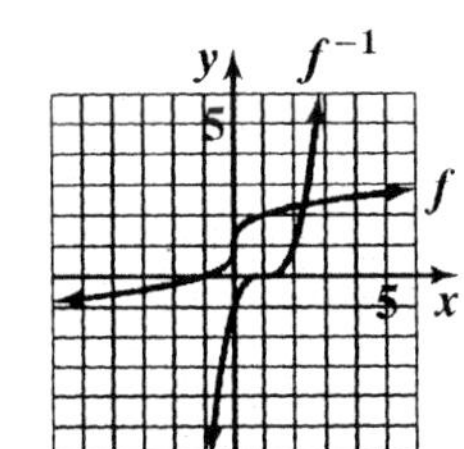

c. Domain of f: $(-\infty, \infty)$

Range of f: $(-\infty, \infty)$

Domain of f^{-1}: $(-\infty, \infty)$

Range of f^{-1}: $(-\infty, \infty)$

52. **a.**
$$\begin{aligned} f(x) &= \sqrt[3]{x-1} \\ y &= \sqrt[3]{x-1} \\ x &= \sqrt[3]{y-1} \\ x^3 &= y-1 \\ x^3+1 &= y \\ f^{-1}(x) &= x^3+1 \end{aligned}$$

b.

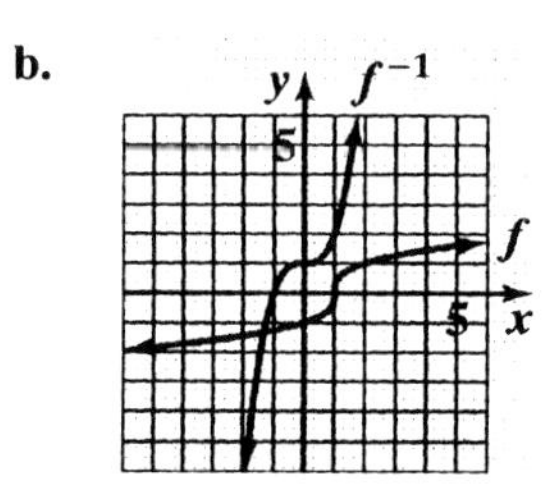

c. Domain of f: $(-\infty, \infty)$
Range of f: $(-\infty, \infty)$
Domain of f^{-1}: $(-\infty, \infty)$
Range of f^{-1}: $(-\infty, \infty)$

53. $f(g(1)) = f(1) = 5$

54. $f(g(4)) = f(2) = -1$

55. $(g \circ f)(-1) = g(f(-1)) = g(1) = 1$

56. $(g \circ f)(0) = g(f(0)) = g(4) = 2$

57. $f^{-1}(g(10)) = f^{-1}(-1) = 2$, since $f(2) = -1$.

58. $f^{-1}(g(1)) = f^{-1}(1) = -1$, since $f(-1) = 1$.

59.
$$\begin{aligned} (f \circ g)(0) &= f(g(0)) \\ &= f(4 \cdot 0 - 1) \\ &= f(-1) = 2(-1) - 5 = -7 \end{aligned}$$

60.
$$\begin{aligned} (g \circ f)(0) &= g(f(0)) \\ &= g(2 \cdot 0 - 5) \\ &= g(-5) = 4(-5) - 1 = -21 \end{aligned}$$

61. Let $f^{-1}(1) = x$. Then
$$\begin{aligned} f(x) &= 1 \\ 2x-5 &= 1 \\ 2x &= 6 \\ x &= 3 \end{aligned}$$
Thus, $f^{-1}(1) = 3$

62. Let $g^{-1}(7) = x$. Then
$$\begin{aligned} g(x) &= 7 \\ 4x-1 &= 7 \\ 4x &= 8 \\ x &= 2 \end{aligned}$$
Thus, $g^{-1}(7) = 2$

63.
$$\begin{aligned} g(f[h(1)]) &= g(f[1^2+1+2]) \\ &= g(f(4)) \\ &= g(2 \cdot 4 - 5) \\ &= g(3) \\ &= 4 \cdot 3 - 1 = 11 \end{aligned}$$

64.
$$\begin{aligned} f(g[h(1)]) &= f(g[1^2+1+2]) \\ &= f(g(4)) \\ &= f(4 \cdot 4 - 1) \\ &= f(15) \\ &= 2 \cdot 15 - 5 = 25 \end{aligned}$$

65. **a.** {(Zambia, −7.3), (Colombia, −4.5), (Poland, −2.8), (Italy, −2.8), (United States, −1.9)}

b. {(−7.3, Zambia), (−4.5, Colombia), (−2.8, Poland), (−2.8, Italy), (−1.9, United States)} This relation is not a function because −2.8 corresponds to two elements in the range.

66. **a.** {(China, 251), (Japan, 243), (Korea, 220), (Israel, 215), (Germany, 210), (Russia, 210)}

b. {(251, China), (243, Japan), (220, Korea), (215, Israel), (210, Germany), (210, Russia)} No; 210 in the domain corresponds to two members of the range, Germany and Russia.

67. **a.** It passes the horizontal line test and is one-to-one.

b. $f^{-1}(0.25)=15$ If there are 15 people in the room, the probability that 2 of them have the same birthday is 0.25.
$f^{-1}(0.5)=21$ If there are 21 people in the room, the probability that 2 of them have the same birthday is 0.5.
$f^{-1}(0.7)=30$ If there are 30 people in the room, the probability that 2 of them have the same birthday is 0.7.

68. **a.** This function fails the horizontal line test. Thus, this function does not have an inverse.

b. The average happiness level is 3 at 12 noon and at 7 p.m. These values can be represented as $(12,3)$ and $(19,3)$.

c. The graph does not represent a one-to-one function. $(12,3)$ and $(19,3)$ are an example of two x-values that correspond to the same y-value.

69.

$$f(g(x))=\frac{9}{5}\left[\frac{5}{9}(x-32)\right]+32$$
$$=x-32+32$$
$$=x$$

f and g are inverses.

76.

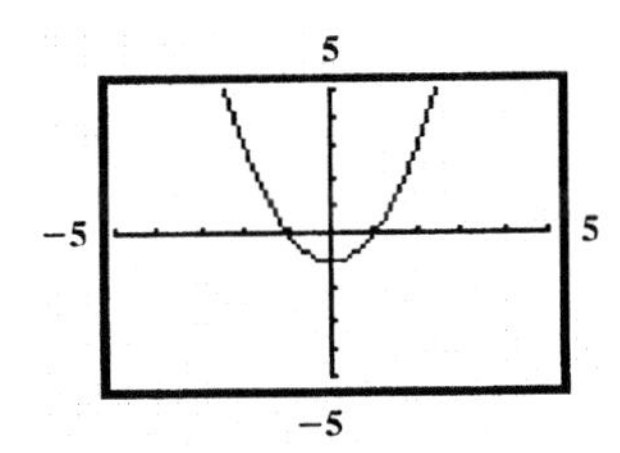

not one-to-one

77.

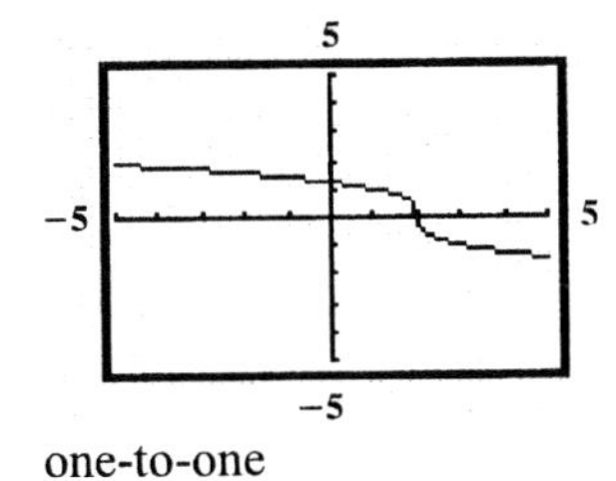

one-to-one

78.

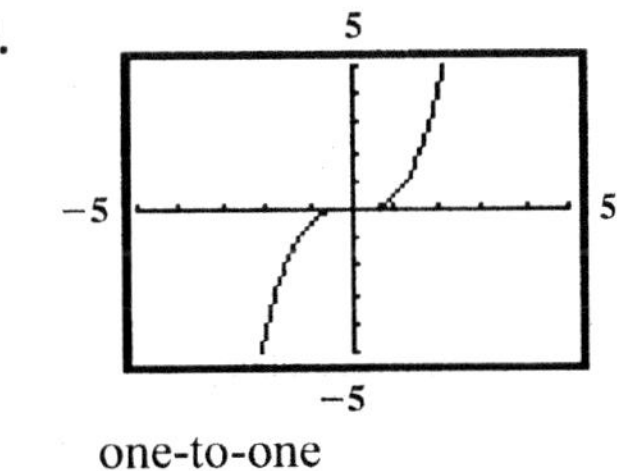

one-to-one

79.

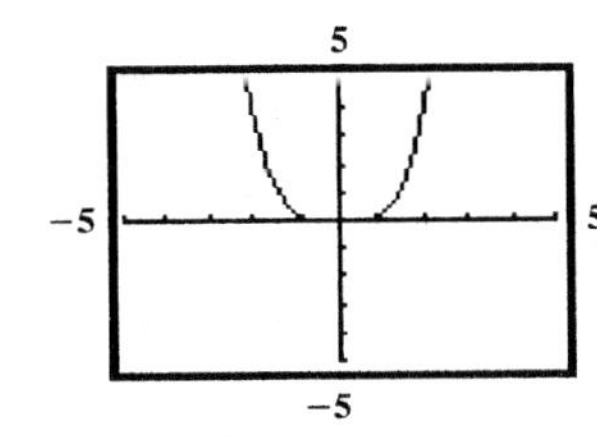

not one-to-one

80.

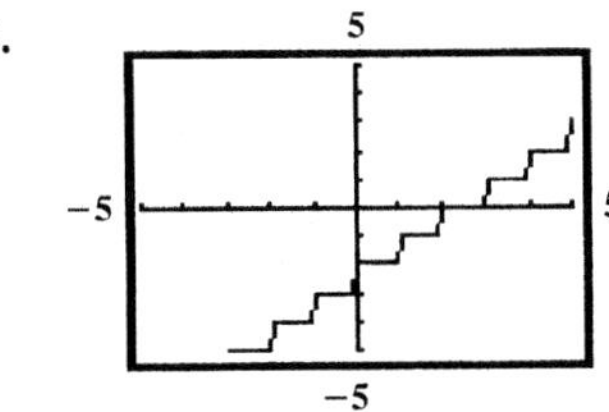

not one-to-one

81.

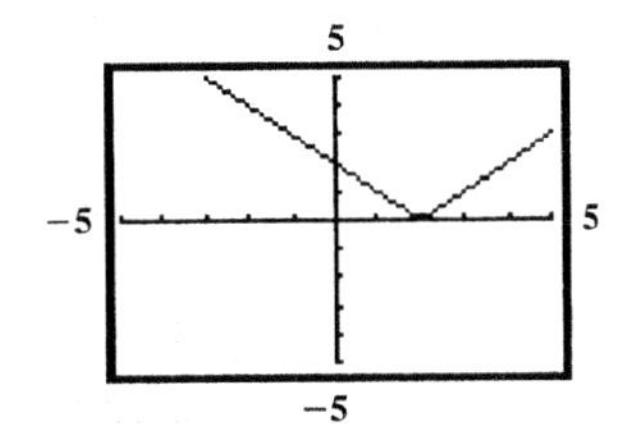

not one-to-one

82.

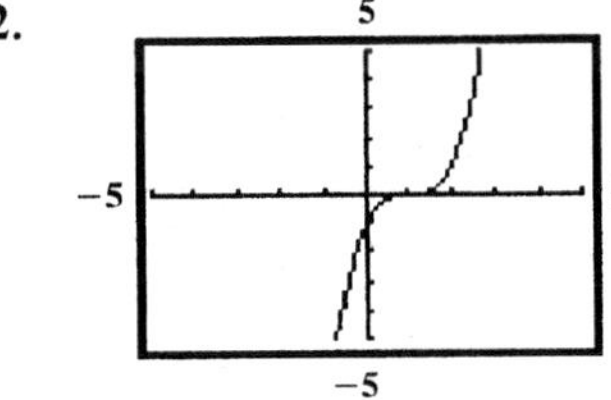

one-to-one

83.

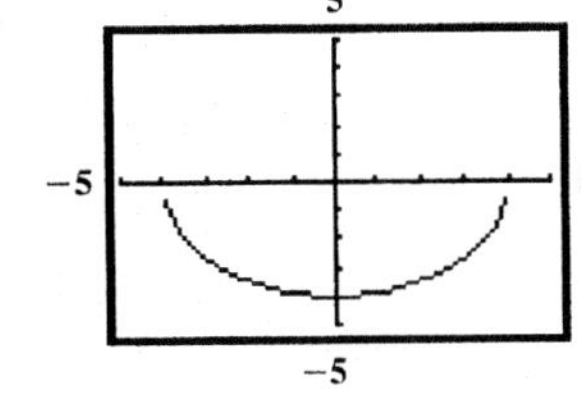

not one-to-one

84.

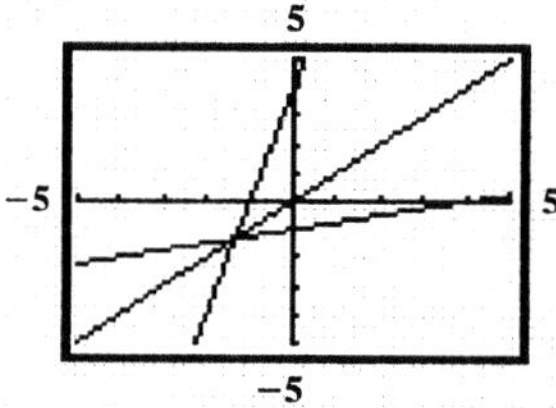

f and g are inverses

85.

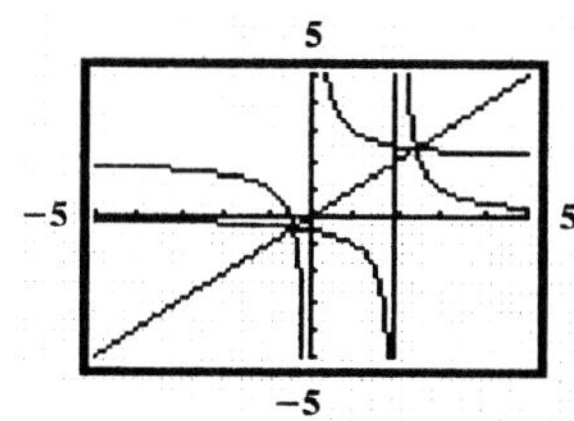

f and g are inverses

86.

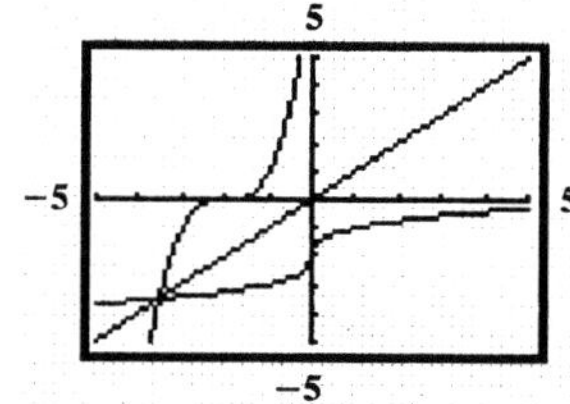

f and g are inverses

87. **a.** False. The inverse is $\{(4,1), (7,2)\}$.

b. False. $f(x) = 5$ is a horizontal line, so it does not pass the horizontal line test.

c. False. $f^{-1}(x) = \frac{x}{3}$.

d. True. The domain of f is the range of f^{-1} and the range of f is the domain of f^{-1}.

(d) is true.

88. $(f \circ g)(x) = 3(x+5) = 3x+15.$

$$y = 3x+15$$
$$x = 3y+15$$
$$y = \frac{x-15}{3}$$
$$(f \circ g)^{-1}(x) = \frac{x-15}{3}$$
$$g(x) = x+5$$
$$y = x+5$$
$$x = y+5$$
$$y = x-5$$
$$g^{-1}(x) = x-5$$
$$f(x) = 3x$$
$$y = 3x$$
$$x = 3y$$
$$y = \frac{x}{3}$$
$$f^{-1}(x) = \frac{x}{3}$$
$$\left(g^{-1} \circ f^{-1}\right)(x) = \frac{x}{3} - 5 = \frac{x-15}{3}$$

89.

$$f(x) = \frac{3x-2}{5x-3}$$
$$y = \frac{3x-2}{5x-3}$$
$$x = \frac{3y-2}{5y-3}$$
$$x(5y-3) = 3y-2$$
$$5xy-3x = 3y-2$$
$$5xy-3y = 3x-2$$
$$y(5x-3) = 3x-2$$
$$y = \frac{3x-2}{5x-3}$$
$$f^{-1}(x) = \frac{3x-2}{5x-3}$$

Note: An alternative approach is to show that $(f \circ f)(x) = x$.

90. No, there will be 2 times when the spacecraft is at the same height, when it is going up and when it is coming down.

91. $8 + f^{-1}(x-1) = 10$

$$f^{-1}(x-1) = 2$$
$$f(2) = x-1$$
$$6 = x-1$$
$$7 = x$$
$$x = 7$$

Section 1.9

Check Point Exercises

1. $d = \sqrt{(x_2 - x_1)^2 + (y_2 - y_1)^2}$

$d = \sqrt{(1-(-4))^2 + (-3-9)^2}$

$= \sqrt{(5)^2 + (-12)^2}$

$= \sqrt{25 + 144}$

$= \sqrt{169}$

$= 13$

2. $\left(\frac{1+7}{2}, \frac{2+(-3)}{2}\right) = \left(\frac{8}{2}, \frac{-1}{2}\right) = \left(4, -\frac{1}{2}\right)$

3. $h = 0,\ k = 0,\ r = 4;$

$(x-0)^2 + (y-0)^2 = 4^2$

$x^2 + y^2 = 16$

4. $h = 5,\ k = -6,\ r = 10;$

$(x-5)^2 + [y-(-6)]^2 = 10^2$

$(x-5)^2 + (y+6)^2 = 100$

5. a. $(x+3)^2 + (y-1)^2 = 4$

$[x-(-3)]^2 + (y-1)^2 = 2^2$

So in the standard form of the circle's equation $(x-h)^2 + (y-k)^2 = r^2$, we have $h = -3,\ k = 1,\ r = 2.$

center: $(h, k) = (-3, 1)$

radius: $r = 2$

b.

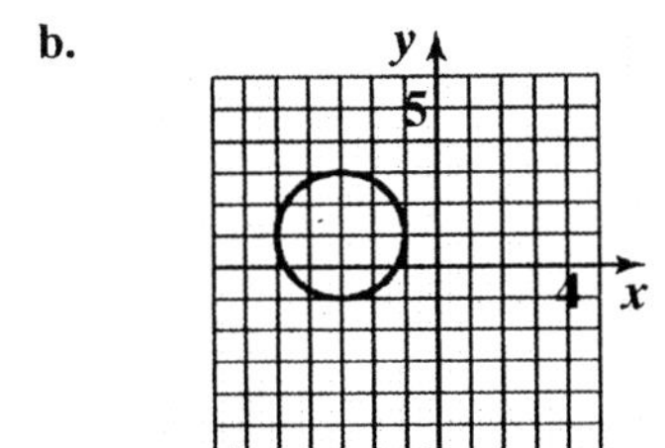

$(x + 3)^2 + (y - 1)^2 = 4$

c. Domain: $[-5, -1]$

Range: $[-1, 3]$

6. $x^2 + y^2 + 4x - 4y - 1 = 0$

$x^2 + y^2 + 4x - 4y - 1 = 0$

$(x^2 + 4x \quad) + (y^2 - 4y \quad) = 0$

$(x^2 + 4x + 4) + (y^2 + 4y + 4) = 1 + 4 + 4$

$(x+2)^2 + (y-2)^2 = 9$

$[x-(-x)]^2 + (y-2)^2 = 3^2$

So in the standard form of the circle's equation $(x-h)^2 + (y-k)^2 = r^2$, we have $h = -2,\ k = 2,\ r = 3$.

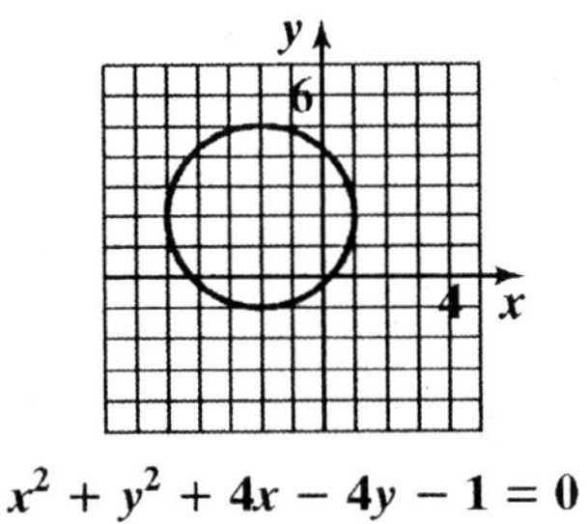

$x^2 + y^2 + 4x - 4y - 1 = 0$

Exercise Set 1.9

1. $d = \sqrt{(14-2)^2 + (8-3)^2}$

$= \sqrt{12^2 + 5^2}$

$= \sqrt{144 + 25}$

$= \sqrt{169}$

$= 13$

2. $d = \sqrt{(8-5)^2 + (5-1)^2}$

$= \sqrt{3^2 + 4^2}$

$= \sqrt{9 + 16}$

$= \sqrt{25}$

$= 5$

3. $d = \sqrt{(-6-4)^2 + (3-(-1))^2}$

$= \sqrt{(-10)^2 + (4)^2}$

$= \sqrt{100 + 16}$

$= \sqrt{116}$

$= 2\sqrt{29}$

≈ 10.77

4. $d=\sqrt{(-1-2)^2+(5-(-3))^2}$
$=\sqrt{(-3)^2+(8)^2}$
$=\sqrt{9+64}$
$=\sqrt{73}$
≈ 8.54

5. $d=\sqrt{(-3-0)^2+(4-0)^2}$
$=\sqrt{3^2+4^2}$
$=\sqrt{9+16}$
$=\sqrt{25}$
$=5$

6. $d=\sqrt{(3-0)^2+(-4-0)^2}$
$=\sqrt{3^2+(-4)^2}$
$=\sqrt{9+16}$
$=\sqrt{25}$
$=5$

7. $d=\sqrt{[3-(-2)]^2+[-4-(-6)]^2}$
$=\sqrt{5^2+2^2}$
$=\sqrt{25+4}$
$=\sqrt{29}$
≈ 5.39

8. $d=\sqrt{[2-(-4)]^2+[-3-(-1)]^2}$
$=\sqrt{6^2+(-2)^2}$
$=\sqrt{36+4}$
$=\sqrt{40}$
$=2\sqrt{10}$
≈ 6.32

9. $d=\sqrt{(4-0)^2+[1-(-3)]^2}$
$=\sqrt{4^2+4^2}$
$=\sqrt{16+16}$
$=\sqrt{32}$
$=4\sqrt{2}$
≈ 5.66

10. $d=\sqrt{(4-0)^2+[3-(-2)]^2}$
$=\sqrt{4^2+[3+2]^2}$
$=\sqrt{16+5^2}$
$=\sqrt{16+25}$
$=\sqrt{41}$
≈ 6.40

11. $d=\sqrt{(-.5-3.5)^2+(6.2-8.2)^2}$
$=\sqrt{(-4)^2+(-2)^2}$
$=\sqrt{16+4}$
$=\sqrt{20}$
$=2\sqrt{5}$
≈ 4.47

12. $d=\sqrt{(1.6-2.6)^2+(-5.7-1.3)^2}$
$=\sqrt{(-1)^2+(-7)^2}$
$=\sqrt{1+49}$
$=\sqrt{50}$
$=5\sqrt{2}$
≈ 7.07

13. $d=\sqrt{(\sqrt{5}-0)^2+[0-(-\sqrt{3})]^2}$
$=\sqrt{(\sqrt{5})^2+(\sqrt{3})^2}$
$=\sqrt{5+3}$
$=\sqrt{8}$
$=2\sqrt{2}$
≈ 2.83

14. $d=\sqrt{(\sqrt{7}-0)^2+[0-(-\sqrt{2})]^2}$
$=\sqrt{(\sqrt{7})^2+[-\sqrt{2}]^2}$
$=\sqrt{7+2}$
$=\sqrt{9}$
$=3$

15. $d=\sqrt{(-\sqrt{3}-3\sqrt{3})^2+(4\sqrt{5}-\sqrt{5})^2}$
$=\sqrt{(-4\sqrt{3})^2+(3\sqrt{5})^2}$
$=\sqrt{16(3)+9(5)}$
$=\sqrt{48+45}$
$=\sqrt{93}$
≈ 9.64

16. $d=\sqrt{\left(-\sqrt{3}-2\sqrt{3}\right)^2+\left(5\sqrt{6}-\sqrt{6}\right)^2}$
$=\sqrt{\left(-3\sqrt{3}\right)^2+\left(4\sqrt{6}\right)^2}$
$=\sqrt{9\cdot 3+16\cdot 6}$
$=\sqrt{27+96}$
$=\sqrt{123}$
≈ 11.09

17. $d=\sqrt{\left(\frac{1}{3}-\frac{7}{3}\right)^2+\left(\frac{6}{5}-\frac{1}{5}\right)^2}$
$=\sqrt{(-2)^2+1^2}$
$=\sqrt{4+1}$
$=\sqrt{5}$
≈ 2.24

18. $d=\sqrt{\left[\frac{3}{4}-\left(-\frac{1}{4}\right)\right]^2+\left[\frac{6}{7}-\left(-\frac{1}{7}\right)\right]^2}$
$=\sqrt{\left(\frac{3}{4}+\frac{1}{4}\right)^2+\left[\frac{6}{7}+\frac{1}{7}\right]^2}$
$=\sqrt{1^2+1^2}$
$=\sqrt{2}$
≈ 1.41

19. $\left(\frac{6+2}{2},\frac{8+4}{2}\right)=\left(\frac{8}{2},\frac{12}{2}\right)=(4,6)$

20. $\left(\frac{10+2}{2},\frac{4+6}{2}\right)=\left(\frac{12}{2},\frac{10}{2}\right)=(6,5)$

21. $\left(\frac{-2+(-6)}{2},\frac{-8+(-2)}{2}\right)$
$=\left(\frac{-8}{2},\frac{-10}{2}\right)=(-4,-5)$

22. $\left(\frac{-4+(-1)}{2},\frac{-7+(-3)}{2}\right)=\left(\frac{-5}{2},\frac{-10}{2}\right)$
$=\left(\frac{-5}{2},-5\right)$

23. $\left(\frac{-3+6}{2},\frac{-4+(-8)}{2}\right)$
$=\left(\frac{3}{2},\frac{-12}{2}\right)=\left(\frac{3}{2},-6\right)$

24. $\left(\frac{-2+(-8)}{2}\frac{-1+6}{2}\right)=\left(\frac{-10}{2},\frac{5}{2}\right)=\left(-5,\frac{5}{2}\right)$

25. $\left(\frac{\frac{-7}{2}+\left(-\frac{5}{2}\right)}{2},\frac{\frac{3}{2}+\left(-\frac{11}{2}\right)}{2}\right)$
$=\left(\frac{\frac{-12}{2}}{2},\frac{\frac{-8}{2}}{2}\right)=\left(-\frac{6}{2},\frac{-4}{2}\right)=(-3,-2)$

26. $\left(\frac{-\frac{2}{5}+\left(-\frac{2}{5}\right)}{2},\frac{\frac{7}{15}+\left(-\frac{4}{15}\right)}{2}\right)=\left(\frac{-\frac{4}{5}}{2},\frac{\frac{3}{15}}{2}\right)$
$=\left(-\frac{4}{5}\cdot\frac{1}{2},\frac{3}{15}\cdot\frac{1}{2}\right)=\left(-\frac{2}{5},\frac{1}{10}\right)$

27. $\left(\frac{8+(-6)}{2},\frac{3\sqrt{5}+7\sqrt{5}}{2}\right)$
$=\left(\frac{2}{2},\frac{10\sqrt{5}}{2}\right)=\left(1,5\sqrt{5}\right)$

28. $\left(\frac{7\sqrt{3}+3\sqrt{3}}{2},\frac{-6+(-2)}{2}\right)=\left(\frac{10\sqrt{3}}{2},\frac{-8}{2}\right)$
$=\left(5\sqrt{3},-4\right)$

29. $\left(\dfrac{\sqrt{18}+\sqrt{2}}{2}, \dfrac{-4+4}{2}\right)$

$=\left(\dfrac{3\sqrt{2}+\sqrt{2}}{2}, \dfrac{0}{2}\right)=\left(\dfrac{4\sqrt{2}}{2}, 0\right)=(2\sqrt{2}, 0)$

30. $\left(\dfrac{\sqrt{50}+\sqrt{2}}{2}, \dfrac{-6+6}{2}\right)=\left(\dfrac{5\sqrt{2}+\sqrt{2}}{2}, \dfrac{0}{2}\right)$

$=\left(\dfrac{6\sqrt{2}}{2}, 0\right)=\left(3\sqrt{2}, 0\right)$

31. $(x-0)^2+(y-0)^2=7^2$

$x^2+y^2=49$

32. $(x-0)^2+(y-0)^2=8^2$

$x^2+y^2=64$

33. $(x-3)^2+(y-2)^2=5^2$

$(x-3)^2+(y-2)^2=25$

34. $(x-2)^2+[y-(-1)]^2=4^2$

$(x-2)^2+(y+1)^2=16$

35. $[x-(-1)]^2+(y-4)^2=2^2$

$(x+1)^2+(y-4)^2=4$

36. $[x-(-3)]^2+(y-5)^2=3^2$

$(x+3)^2+(y-5)^2=9$

37. $[x-(-3)]^2+[y-(-1)]^2=\left(\sqrt{3}\right)^2$

$(x+3)^2+(y+1)^2=3$

38. $[x-(-5)]^2+[y-(-3)]^2=\left(\sqrt{5}\right)^2$

$(x+5)^2+(y+3)^2=5$

39. $[x-(-4)]^2+(y-0)^2=10^2$

$(x+4)^2+(y-0)^2=100$

40. $[x-(-2)]^2+(y-0)^2=6^2$

$(x+2)^2+y^2=36$

41. $x^2+y^2=16$

$(x-0)^2+(y-0)^2=y^2$

$h=0,\ k=0,\ r=4;$

center = (0, 0); radius = 4

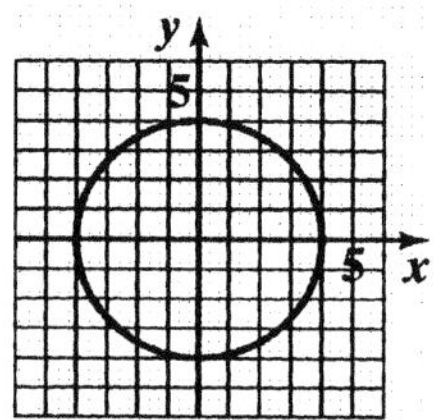

$x^2 + y^2 = 16$

Domain: $[-4,4]$

Range: $[-4,4]$

42. $x^2+y^2=49$

$(x-0)^2+(y-0)^2=7^2$

$h=0, k=0, r=7;$

center = (0, 0); radius = 7

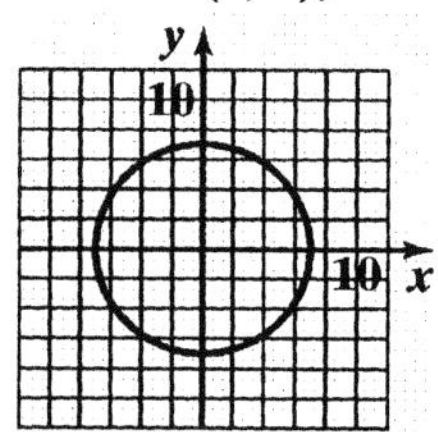

$x^2 + y^2 = 49$

Domain: $[-7,7]$

Range: $[-7,7]$

43. $(x-3)^2+(y-1)^2=36$

$(x-3)^2+(y-1)^2=6^2$

$h=3,\ k=1,\ r=6;$

center = (3, 1); radius = 6

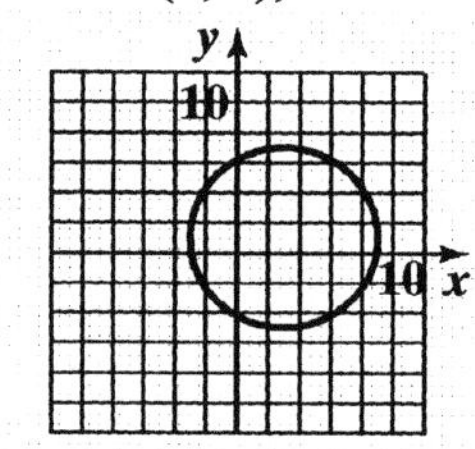

$(x-3)^2 + (y-1)^2 = 36$

Domain: $[-3,9]$

Range: $[-5,7]$

44. $(x-2)^2+(y-3)^2=16$

$(x-2)^2+(y-3)^2=4^2$

$h=2, k=3, r=4;$

center = (2, 3); radius = 4

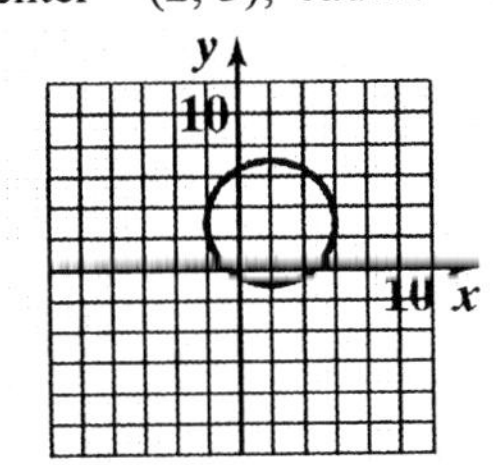

$(x-2)^2+(y-3)^2=16$

Domain: $[-2,6]$

Range: $[-1,7]$

45. $(x+3)^2+(y-2)^2=4$

$[x-(-3)]^2+(y-2)^2=2^2$

$h=-3, k=2, r=2$

center = (–3, 2); radius = 2

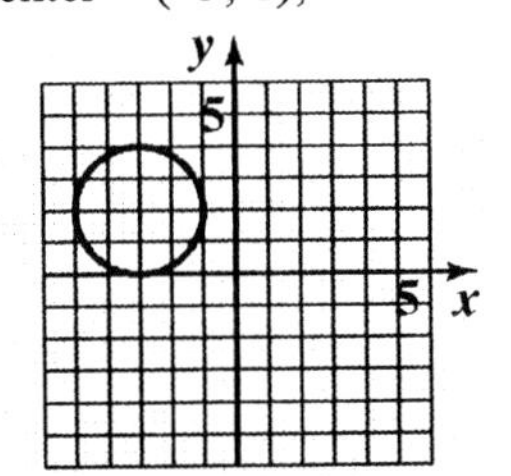

$(x+3)^2+(y-2)^2=4$

Domain: $[-5,-1]$

Range: $[0,4]$

46. $(x+1)^2+(y-4)^2=25$

$[x-(-1)]^2+(y-4)^2=5^2$

$h=-1, k=4, r=5;$

center = (–1, 4); radius = 5

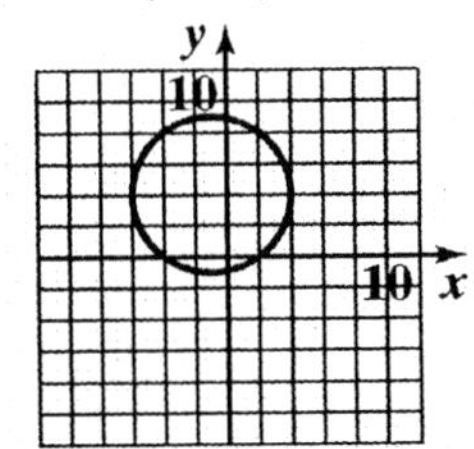

$(x+1)^2+(y-4)^2=25$

Domain: $[-6,4]$

Range: $[-1,9]$

47. $(x+2)^2+(y+2)^2=4$

$[x-(-2)]^2+[y-(-2)]^2=2^2$

$h=-2, k=-2, r=2$

center = (–2, –2); radius = 2

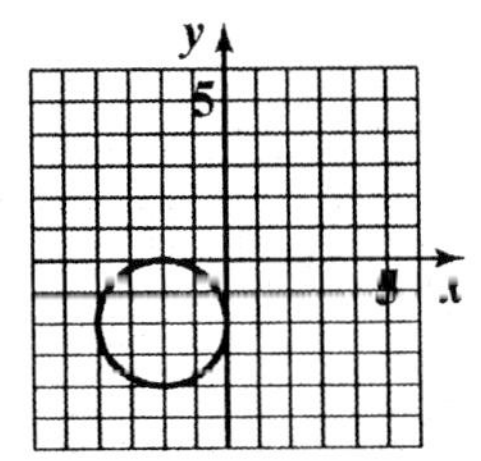

$(x+2)^2+(y+2)^2=4$

Domain: $[-4,0]$

Range: $[-4,0]$

48. $(x+4)^2+(y+5)^2=36$

$[x-(-4)]^2+[y-(-5)]^2=6^2$

$h=-4, k-5, r=6;$

center = (–4, –5); radius = 6

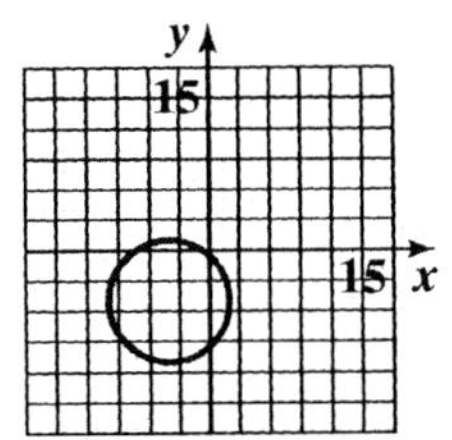

$(x+4)^2+(y+5)^2=36$

Domain: $[-10,2]$

Range: $[-11,1]$

49. $x^2+y^2+6x+2y+6=0$

$(x^2+6x)+(y^2+2y)=-6$

$(x^2+6x+9)+(y^2+2y+1)=9+1-6$

$(x+3)^2+(y+1)^2=4$

$[x-(-3)]^2+[9-(-1)]^2=2^2$

center = (–3, –1); radius = 2

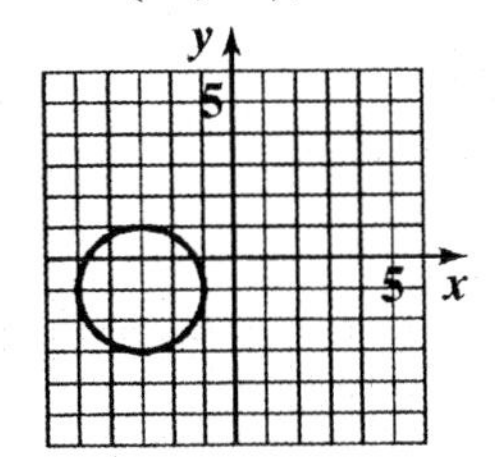

$x^2+y^2+6x+2y+6=0$

50. $x^2+y^2+8x+4y+16=0$

$(x^2+8x)+(y^2+4y)=-16$

$(x^2+8x+16)+(y^2+4y+4)=20-16$

$(x+4)^2+(y+2)^2=4$

$[x-(-4)]^2+[y-(-2)]^2=2^2$

center = (–4, –2); radius = 2

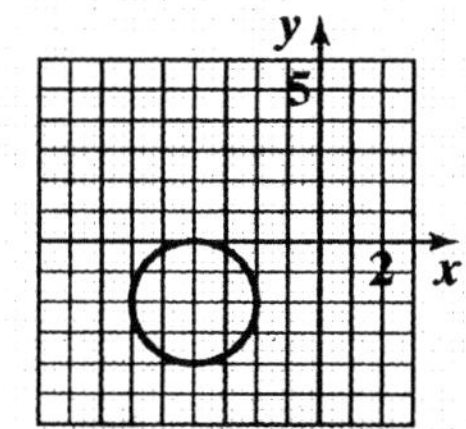

$x^2 + y^2 + 8x + 4y + 16 = 0$

51. $x^2+y^2-10x-6y-30=0$

$(x^2-10x)+(y^2-6y)=30$

$(x^2-10x+25)+(y^2-6y+9)=25+9+30$

$(x-5)^2+(y-3)^2=64$

$(x-5)^2+(y-3)^2=8^2$

center = (5, 3); radius = 8

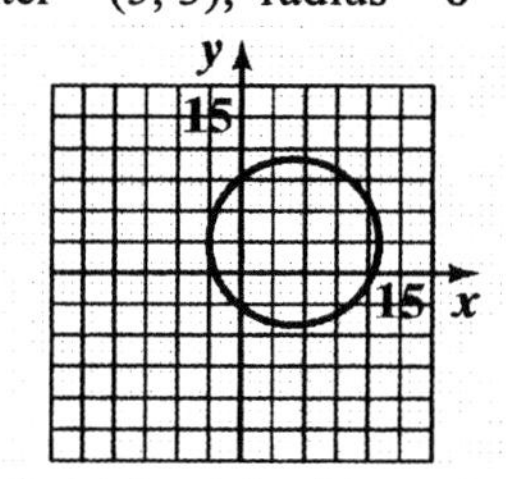

$x^2 + y^2 - 10x - 6y - 30 = 0$

52. $x^2+y^2-4x-12y-9=0$

$(x^2-4x)+(y^2-12y)=9$

$(x^2-4x+4)+(y^2-12y+36)=4+36+9$

$(x-2)^2+(y-6)^2=49$

$(x-2)^2+(y-6)^2=7^2$

center = (2, 6); radius = 7

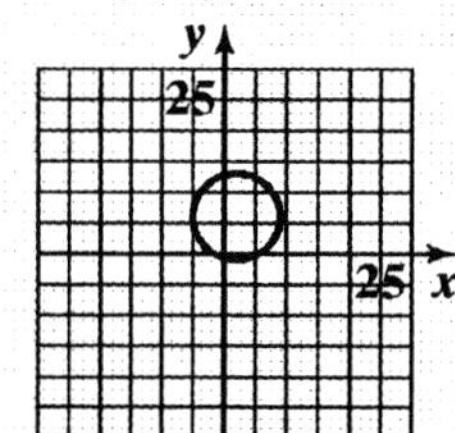

$x^2 + y^2 - 4x - 12y - 9 = 0$

53. $x^2+y^2+8x-2y-8=0$

$(x^2+8x)+(y^2-2y)=8$

$(x^2+8x+16)+(y^2-2y+1)=16+1+8$

$(x+4)^2+(y-1)^2=25$

$[x-(-4)]^2+(y-1)^2=5^2$

center = (–4, 1); radius = 5

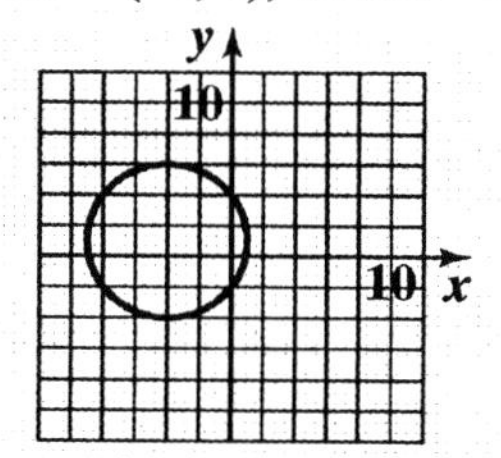

$x^2 + y^2 + 8x - 2y - 8 = 0$

54. $x^2+y^2+12x-6y-4=0$

$(x^2+12x)+(y^2-6y)=4$

$(x^2+12x+36)+(y^2-6y+9)=36+9+4$

$[x-(-6)]^2+(y-3)^2=7^2$

center = (–6, 3); radius = 7

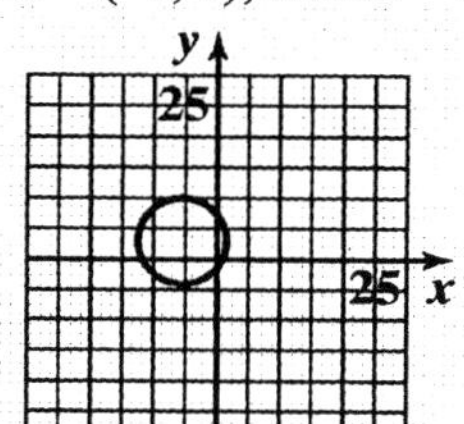

$x^2 + y^2 + 12x - 6y - 4 = 0$

55. $x^2-2x+y^2-15=0$

$(x^2-2x)+y^2=15$

$(x^2-2x+1)+(y-0)^2=1+0+15$

$(x-1)^2+(y-0)^2=16$

$(x-1)^2+(y-0)^2=4^2$

center = (1, 0); radius = 4

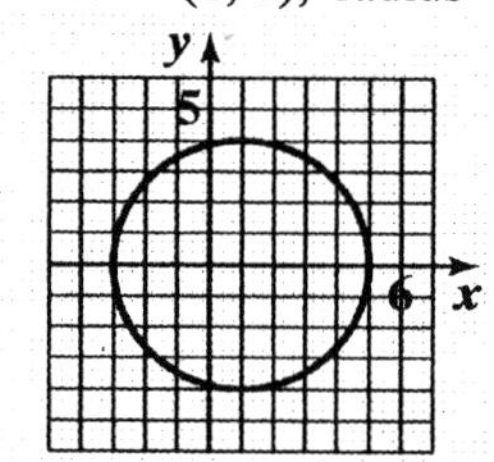

$x^2 - 2x + y^2 - 15 = 0$

56. $x^2 + y^2 - 6y - 7 = 0$

$x^2 + (y^2 - 6y) = 7$

$(x-0)^2 = (y^2 - 6y + 9) = 0 + 9 + 7$

$(x-0)^2 + (y-3)^2 = 16$

$(x-0)^2 + (y-3)^2 = 4^2$

center = (0, 3); radius = 4

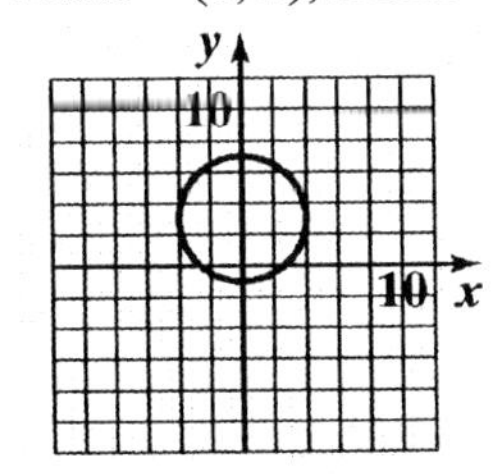

$\mathbf{x^2 + y^2 - 6y - 7 = 0}$

57. $x^2 + y^2 - x + 2y + 1 = 0$

$x^2 - x \quad + y^2 + 2y \quad = -1$

$x^2 - x + \frac{1}{4} + y^2 + 2y + 1 = -1 + \frac{1}{4} + 1$

$\left(x - \frac{1}{2}\right)^2 + (y+1)^2 = \frac{1}{4}$

center = $\left(\frac{1}{2}, -1\right)$; radius = $\frac{1}{2}$

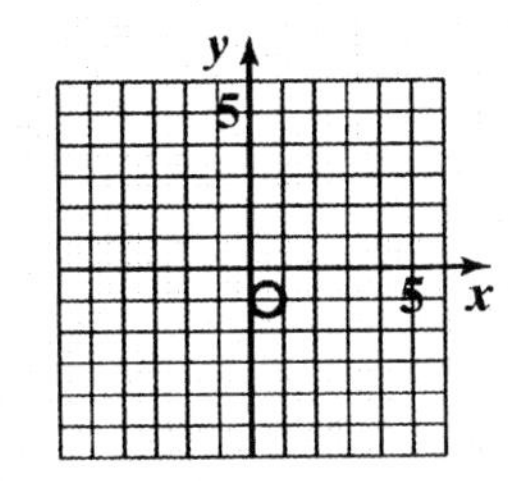

$\mathbf{x^2 + y^2 - x + 2y + 1 = 0}$

58. $x^2 + y^2 + x + y - \frac{1}{2} = 0$

$x^2 + x \quad + y^2 + y \quad = \frac{1}{2}$

$x^2 + x + \frac{1}{4} + y^2 + y + \frac{1}{4} = \frac{1}{2} + \frac{1}{4} + \frac{1}{4}$

$\left(x - \frac{1}{2}\right)^2 + \left(y - \frac{1}{2}\right)^2 = 1$

center = $\left(\frac{1}{2}, \frac{1}{2}\right)$; radius = 1

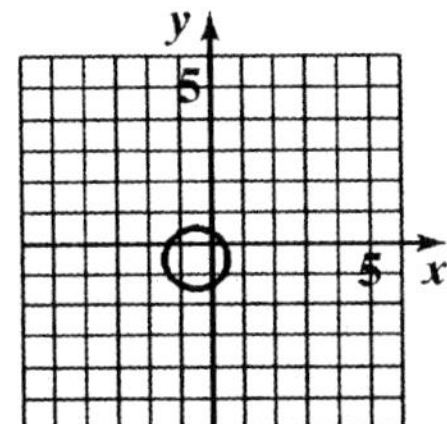

$\mathbf{x^2 + y^2 + x + y - \frac{1}{2} = 0}$

59. $x^2 + y^2 + 3x - 2y - 1 = 0$

$x^2 + 3x \quad + y^2 - 2y \quad = 1$

$x^2 + 3x + \frac{9}{4} + y^2 - 2y + 1 = 1 + \frac{9}{4} + 1$

$\left(x + \frac{3}{2}\right)^2 + (y-1)^2 = \frac{17}{4}$

center = $\left(-\frac{3}{2}, 1\right)$; radius = $\frac{\sqrt{17}}{2}$

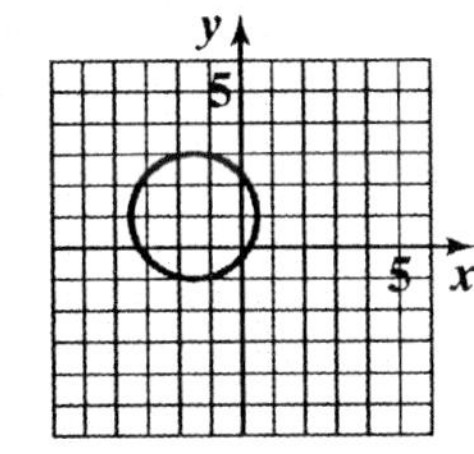

$\mathbf{x^2 + y^2 + 3x - 2y - 1 = 0}$

60.

$$x^2+y^2+3x+5y+\frac{9}{4}=0$$
$$x^2+3x \quad +y^2+5y \quad =-\frac{9}{4}$$
$$x^2+3x+\frac{9}{4}+y^2+5y+\frac{25}{4}=-\frac{9}{4}+\frac{9}{4}+\frac{25}{4}$$
$$\left(x+\frac{3}{2}\right)^2+\left(y+\frac{5}{2}\right)^2=\frac{25}{4}$$

center $=\left(-\frac{3}{2},-\frac{5}{2}\right)$; radius $=\frac{5}{2}$

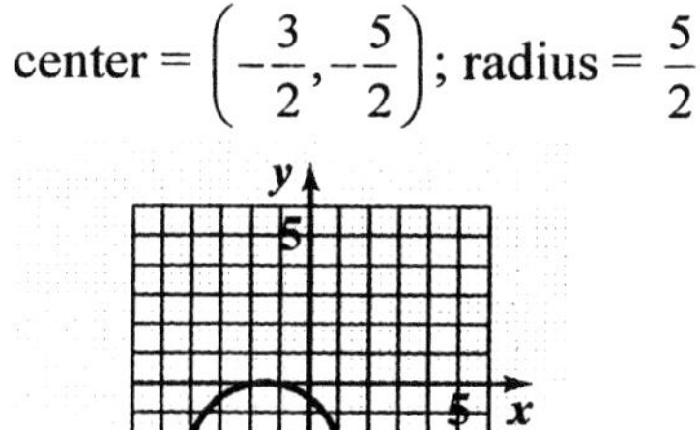

$x^2+y^2+3x+5y+\frac{9}{4}=0$

61. a. Since the line segment passes through the center, the center is the midpoint of the segment.

$$M=\left(\frac{x_1+x_2}{2},\frac{y_1+y_2}{2}\right)$$
$$=\left(\frac{3+7}{2},\frac{9+11}{2}\right)=\left(\frac{10}{2},\frac{20}{2}\right)$$
$$=(5,10)$$

The center is $(5,10)$.

b. The radius is the distance from the center to one of the points on the circle. Using the point $(3,9)$, we get:

$$d=\sqrt{(5-3)^2+(10-9)^2}$$
$$=\sqrt{2^2+1^2}=\sqrt{4+1}$$
$$=\sqrt{5}$$

The radius is $\sqrt{5}$ units.

c.

$$(x-5)^2+(y-10)^2=\left(\sqrt{5}\right)^2$$
$$(x-5)^2+(y-10)^2=5$$

62. a. Since the line segment passes through the center, the center is the midpoint of the segment.

$$M=\left(\frac{x_1+x_2}{2},\frac{y_1+y_2}{2}\right)$$
$$=\left(\frac{3+5}{2},\frac{6+4}{2}\right)=\left(\frac{8}{2},\frac{10}{2}\right)$$
$$=(4,5)$$

The center is $(4,5)$.

b. The radius is the distance from the center to one of the points on the circle. Using the point $(3,6)$, we get:

$$d=\sqrt{(4-3)^2+(5-6)^2}$$
$$=\sqrt{1^2+(-1)^2}=\sqrt{1+1}$$
$$=\sqrt{2}$$

The radius is $\sqrt{2}$ units.

c.

$$(x-4)^2+(y-5)^2=\left(\sqrt{2}\right)^2$$
$$(x-4)^2+(y-5)^2=2$$

63. $x^2+y^2=16$

$x-y=4$

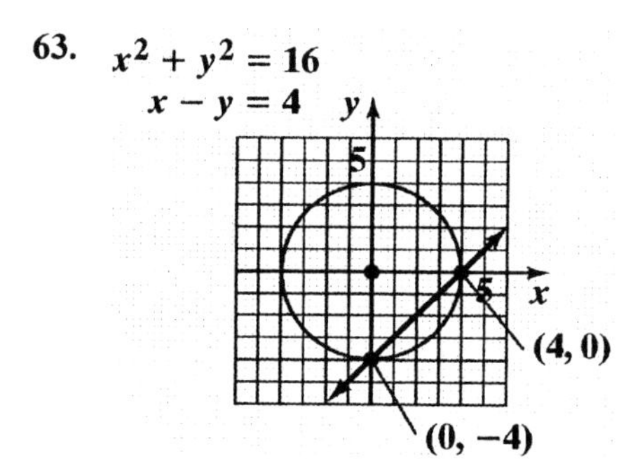

Intersection points: $(0,-4)$ and $(4,0)$

Check $(0,-4)$:

$0^2+(-4)^2=16 \qquad 0-(-4)=4$

$16=16$ true $\qquad 4=4$ true

Check $(4,0)$:

$4^2+0^2=16 \qquad 4-0=4$

$16=16$ true $\qquad 4=4$ true

The solution set is $\{(0,-4),(4,0)\}$.

64. $x^2 + y^2 = 9$
$x - y = 3$

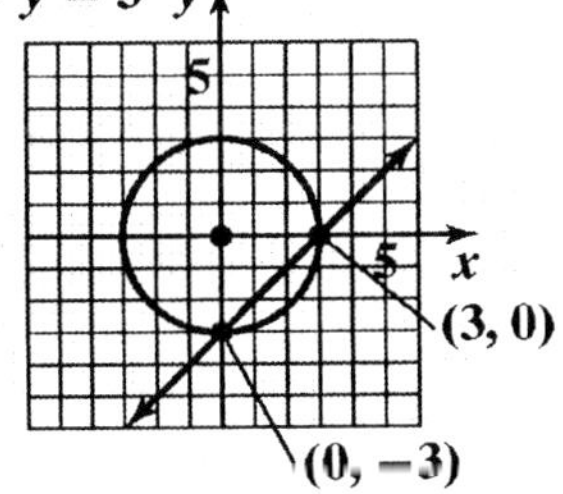

Intersection points: $(0,-3)$ and $(3,0)$

Check $(0,-3)$:

$0^2 + (-3)^2 = 9 \qquad 0-(-3)=3$

$9 = 9$ true $\qquad 3 = 3$ true

Check $(3,0)$:

$3^2 + 0^2 = 9 \qquad 3-0=3$

$9 = 9$ true $\qquad 3 = 3$ true

The solution set is $\{(0,-3),(3,0)\}$.

65. $(x - 2)^2 + (y + 3)^2 = 4$
$y = x - 3$

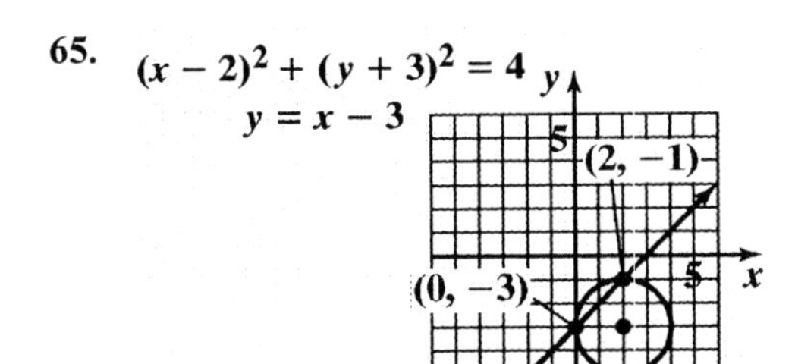

Intersection points: $(0,-3)$ and $(2,-1)$

Check $(0,-3)$:

$(0-2)^2 + (-3+3)^2 = 9 \qquad -3 = 0-3$

$(-2)^2 + 0^2 = 4 \qquad -3 = -3$ true

$4 = 4$

true

Check $(2,-1)$:

$(2-2)^2 + (-1+3)^2 = 4 \qquad -1 = 2-3$

$0^2 + 2^2 = 4 \qquad -1 = -1$ true

$4 = 4$

true

The solution set is $\{(0,-3),(2,-1)\}$.

66. $(x - 3)^2 + (y + 1)^2 = 9$
$y = x - 1$

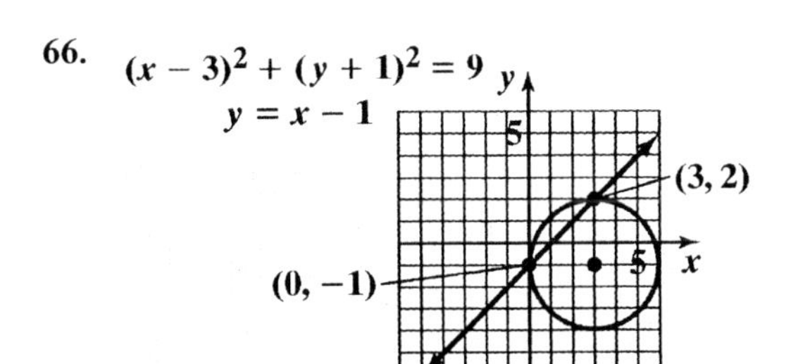

Intersection points: $(0,-1)$ and $(3,2)$

Check $(0,-1)$:

$(0-3)^2 + (-1+1)^2 = 9 \qquad -1 = 0-1$

$(-3)^2 + 0^2 = 9 \qquad -1 = -1$ true

$9 = 9$

true

Check $(3,2)$:

$(3-3)^2 + (2+1)^2 = 9 \qquad 2 = 3-1$

$0^2 + 3^2 = 9 \qquad 2 = 2$ true

$9 = 9$

true

The solution set is $\{(0,-1),(3,2)\}$.

67.
$$d = \sqrt{[65-(-115)]^2 + (70-170)^2}$$
$$d = \sqrt{(65+115)^2 + (-100)^2}$$
$$d = \sqrt{180^2 + 10000}$$
$$d = \sqrt{32400 + 10000}$$
$$d = \sqrt{42400}$$
$$d = 205.9 \text{ miles}$$
$$\frac{205.9 \text{ miles}}{400} = 0.5 \text{ hours or 30 minutes}$$

68. $C(0, 68 + 14) = (0, 82)$
$$(x-0)^2 + (y-82)^2 = 68^2$$
$$x^2 + (y-82)^2 = 4624$$

69. If we place L.A. at the origin, then we want the equation of a circle with center at $(-2.4,-2.7)$ and radius 30.
$$(x-(-2.4))^2 + (y-(-2.7))^2 = 30^2$$
$$(x+2.4)^2 + (y+2.7)^2 = 900$$

77.

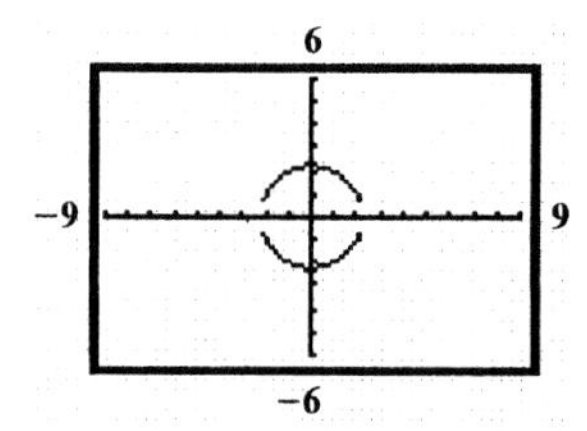

78.

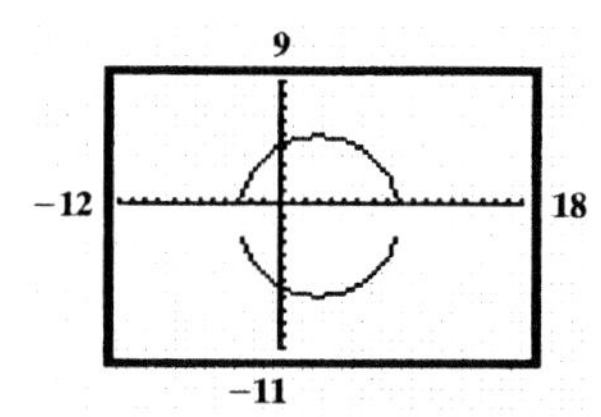

79.

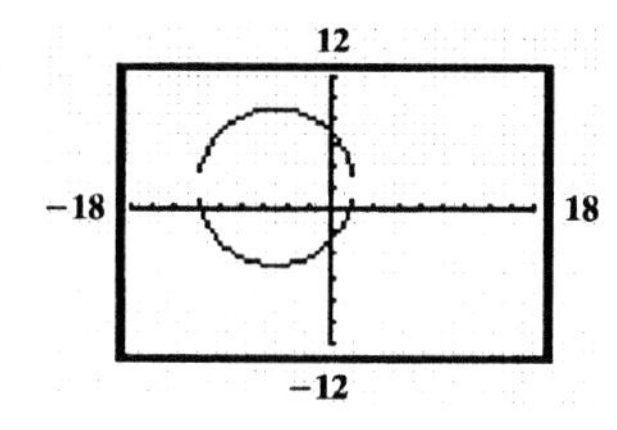

80. **a.** False; the equation should be $x^2 + y^2 = 256$.

b. False; the center is at (3, –5).

c. False; this is not an equation for a circle.

d. True

(d) is true.

81. The distance for A to B:

$$\overline{AB} = \sqrt{(3-1)^2 + [3+d-(1+d)]^2}$$
$$= \sqrt{2^2 + 2^2}$$
$$= \sqrt{4+4}$$
$$= \sqrt{8}$$
$$= 2\sqrt{2}$$

The distance from B to C:

$$\overline{BC} = \sqrt{(6-3)^2 + [3+d-(6+d)]^2}$$
$$= \sqrt{3^2 + (-3)^2}$$
$$= \sqrt{9+9}$$
$$= \sqrt{18}$$
$$= 3\sqrt{2}$$

The distance for A to C:

$$\overline{AC} = \sqrt{(6-1)^2 + [6+d-(1+d)]^2}$$
$$= \sqrt{5^2 + 5^2}$$
$$= \sqrt{25+25}$$
$$= \sqrt{50}$$
$$= 5\sqrt{2}$$

$$\overline{AB} + \overline{BC} = \overline{AC}$$
$$2\sqrt{2} + 3\sqrt{2} = 5\sqrt{2}$$
$$5\sqrt{2} = 5\sqrt{2}$$

82. **a.** d_1 is distance from (x_1, x_2) to midpoint

$$d_1 = \sqrt{\left(\frac{x_1 + x_2}{2} - x_1\right)^2 + \left(\frac{y_1 + y_2}{2} - y_1\right)^2}$$

$$d_1 = \sqrt{\left(\frac{x_1 + x_2 - 2x_1}{2}\right)^2 + \left(\frac{y_1 + y_2 - 2y_1}{2}\right)^2}$$

$$d_1 = \sqrt{\left(\frac{x_2 - x_1}{2}\right)^2 + \left(\frac{y_2 - y_1}{2}\right)^2}$$

$$d_1 = \sqrt{\frac{x_2 - 2x_1x_2 + x_1^2}{4} + \frac{y_2^2 - 2y_2y_1 + y_1^2}{4}}$$

$$d_1 = \sqrt{\frac{1}{4}\left(x_2 - 2x_1x_2 + x_1 + y_2^2 - 2y_2y_1 + y_1^2\right)}$$

$$d_1 = \frac{1}{2}\sqrt{x_2 - 2x_1x_2 + x_1 + y_2^2 - 2y_2y_1 + y_1^2}$$

d_2 is distance from midpoint to (x_2, y_2)

$$d_2 = \sqrt{\left(\frac{x_1+x_2}{2} - x_2\right)^2 + \left(\frac{y_1+y_2}{2} - y_2\right)^2}$$

$$d_2 = \sqrt{\left(\frac{x_1+x_2-2x_2}{2}\right)^2 + \left(\frac{y_1+y_2-2y_2}{2}\right)^2}$$

$$d_2 = \sqrt{\left(\frac{x_1-x_2}{2}\right)^2 + \left(\frac{y_1-y_2}{2}\right)^2}$$

$$d_2 = \sqrt{\frac{x_1^2-2x_1x_2+x_2^2}{4} + \frac{y_1^2-2y_2y_1+y_2^2}{4}}$$

$$d_2 = \sqrt{\frac{1}{4}\left(x_1^2-2x_1x_2+x_2^2+y_1^2-2y_2y_1+y_2^2\right)}$$

$$d_2 = \frac{1}{2}\sqrt{x_1^2-2x_1x_2+x_2^2+y_1^2-2y_2y_1+y_2^2}$$

$$d_1 = d_2$$

b. d_3 is the distance from (x_1, y_1) to $(x_2 y_2)$

$$d_3 = \sqrt{(x_2-x_1)^2+(y_2-y_1)^2}$$

$$d_3 = \sqrt{x_2^2-2x_1x_2+x_1^2+y_2^2-2y_2y_1+y_1^2}$$

$d_1 + d_2 = d_3$ because $\frac{1}{2}\sqrt{a} + \frac{1}{2}\sqrt{a} = \sqrt{a}$

83. Both circles have center (2, –3). The smaller circle has radius 5 and the larger circle has radius 6. The smaller circle is inside of the larger circle. The area between them is given by

$$\begin{aligned}\pi(6)^2 - \pi(5)^2 &= 36\pi - 25\pi \\ &= 11\pi \\ &\approx 34.56 \text{ square units.}\end{aligned}$$

84. The circle is centered at (0,0). The slope of the radius with endpoints (0,0) and (3,–4) is $m = -\frac{-4-0}{3-0} = -\frac{4}{3}$.

The line perpendicular to the radius has slope $\frac{3}{4}$. The tangent line has slope $\frac{3}{4}$ and passes through (3,–4), so its equation is:

$$y + 4 = \frac{3}{4}(x-3).$$

Section 1.10

Check Point Exercises

1. **a.** $f(x) = 15 + 0.08x$

   b. $g(x) = 3 + 0.12x$

   c. $15 + 0.08x = 3 + 0.12x$
   $12 = 0.04x$
   $300 = x$
   The plans cost the same for 300 minutes.

2. **a.** $N(x) = 8000 - 100(x - 100)$
   $= 8000 - 100x + 10000$
   $= 18{,}000 - 100x$

   b. $R(x) = (18{,}000 - 100x)x$
   $= -100x^2 + 18{,}000x$

3. $V(x) = (15 - 2x)(8 - 2x)x$
   $= (120 - 46x + 4x^2)x$
   $= 4x^3 - 46x^2 + 120x$
   Since x represents the inches to be cut off, $x > 0$. The smallest side is 8, so must cut less than 4 off each side. The domain of V is $\{x | 0 < x < 4\}$ or, in interval notation, $(0, 4)$.

4. $2l + 2w = 200$
   $2l = 200 - 2w$
   $l = 100 - w$
   Let x = width, then length = $100 - x$
   $A(x) = x(100 - x)$
   $= 100x - x^2$

5. $V = \pi r^2 h$
   $1000 = \pi r^2 h$
   $\frac{1000}{\pi r^2} = h$

   $A = 2\pi r^2 + 2\pi rh$
   $= 2\pi r^2 + 2\pi r\left(\frac{1000}{\pi r^2}\right)$
   $= 2\pi r^2 + \frac{2000}{r}$

6. $I(x) = 0.07x + 0.09(25{,}000 - x)$

7. $d = \sqrt{(x-0)^2 + (y-0)^2}$
   $= \sqrt{x^2 + y^2}$

   $y = x^3$
   $d = \sqrt{x^2 + (x^3)^2}$
   $= \sqrt{x^2 + x^6}$

Exercise Set 1.10

1. **a.** $f(x) = 200 + 0.15x$

   b. $320 = 200 + 0.15x$
   $120 = 01.5x$
   $800 = x$
   800 miles

2. **a.** $f(x) = 180 + 0.25x$

   b. $395 = 180 + 0.25x$
   $215 = 0.25x$
   $860 = x$
   You drove 860 miles for \$395.

3. **a.** $M(x) = 239.4 - 0.3x$

   b. $180 = 239.4 - 0.3x$
   $0.3x = 59.4$
   $x = 198$
   198 years after 1954, in 2152, someone will run a 3 minute mile.

4. **a.** $P(x) = 28 + 0.6x$

   b. $40 = 28 + 06x$
   $12 = 0.6x$
   $20 = x$
   20 years after 1990, in 2010, 40% of babies born will be out of wedlock.

5. **a.** $f(x) = 1.25x$

   b. $g(x) = 21 + 0.5x$

   c. $1.25x = 21 + 0.5x$
   $0.75x = 21$
   $x = 28$
   $f(28) = 1.25(28) = 35$
   $g(28) = 21 + 0.5(28) = 35$
   If a person crosses the bridge 28 times the cost will be \$35 for both options

6. **a.** $f(x) = 2.5x$

b. $g(x) = 21 + x$

c. $2.5x = 21 + x$
$1.5x = 21$
$x = 14$

$f(14) = 2.5(14) = 35$
$g(14) = 21 + 14 = 35$
To cross the bridge 14 times costs the same, \$35, for either method.

7. **a.** $f(x) = 100 + 0.8x$

b. $g(x) = 40 + 0.9x$

c. $100 + 0.8x = 40 + 0.9x$
$60 = 0.1x$
$600 = x$
For \$600 worth of merchandise, your cost is \$580 for both plans

8. **a.** $f(x) = 300 + 0.7x$

b. $g(x) = 40 + 0.9x$

c. $300 + .7x = 40 + .9x$
$260 = .2x$
$1300 = x$
$f(1300) = 300 + 0.7(1300) = 1210$
$g(1300) = 40 + 0.9(1300) = 1210$
You would have to purchase \$1300 in merchandise at a total cost of \$1210.

9. **a.** $N(x) = 30,000 - 500(x - 20)$
$= 30,000 - 500x + 10000$
$= 40,000 - 500x$

b. $R(x) = (40,000 - 500x)x$
$= -500x^2 + 40,000$

10. **a.** $N(x) = 20,000 - 400(x - 15)$
$= 20,000 - 400x + 6000$
$= 26,000 - 400x$

b. $R(x) = (26,000 - 400x)x$
$= -400x^2 + 26,000x$

11. **a.** $N(x) = 9000 + 50(150 - x)$
$= 9000 - 50x + 7500$
$= 16500 - 50x$

b. $R(x) = (16500 - 50x)x$
$= -50x^2 + 16500x$

12. **a.** $N(x) = 7,000 + 60(90 - x)$
$= 7000 - 60x + 5400$
$= 12400 - 60x$

b. $R(x) = (12400 - 60x)x$
$= -60x^2 + 12400x$

13. **a.** $Y(x) = 320 - 4(x - 50)$
$= 320 - 4x + 200$
$= 520 - 4x$

b. $T(x) = (520 - 4x)x$
$= -4x^2 + 520x$

14. **a.** $Y(x) = 270 - 3(x - 30)$
$= 270 - 3x + 90$
$= 360 - 3x$

b. $T(x) = (360 - 3x)x$
$= -3x^2 + 360x$

15. **a.** $V(x) = (24 - 2x)(24 - 2x)x$
$= (576 - 96x + 4x^2)x$
$= 4x^3 - 96x^2 + 576x$

b. $V(2) = 4(2)^3 - 96(2)^2 + 576(2) = 800$ If 2-inch squares are cut off each corner, the volume will be 800 square inches.

$V(3) = 4(3)^3 - 96(3)^2 + 576(3) = 972$ If 3-inch squares are cut off each corner, the volume will be 972 square inches.

$V(4) = 4(4)^3 - 96(4)^2 + 576(4) = 1024$ If 4-inch squares are cut off each corner, the volume will be 1024 square inches.

$V(5) = 4(5)^3 - 96(5)^2 + 576(5) = 980$ If 5-inch squares are cut off each corner, the volume will be 980 square inches.

$V(6) = 4(6)^3 - 96(6)^2 + 576(6) = 864$ If 6-inch squares are cut off each corner, the volume will be 864 square inches.

c. If x is the inches to be cut off, $x > 0$. Since each side is 24, you must cut less than 12 inches off each end.
$0 < x < 12$

16. a. $V(x) = (30 - 2x)(30 - 2x)x$
$= (900 - 120x + 4x^2)x$
$= 4x^3 - 120x^2 + 900x$

b. $V(3) = 4(3^3) - 120(3^2) + 900(3) = 1728$
If 3 inches are cut from each side, the volume will be 1728 square inches.

$V(4) = 4(4^3) - 120(4^2) + 900(4) = 1936$
If 4 inches are cut from each side, the volume will be 1936 square inches.

$V(5) = 4(5^3) - 120(5^2) + 900(5) = 2000$
If 5 inches are cut from each side, the volume will be 2000 square inches.

$V(6) = 4(6^3) - 120(6^2) + 900(6) = 1944$
If 6 inches are cut from each side, the volume will be 1944 square inches.

$V(7) = 4(7^3) - 120(7^2) + 900(7) = 1792$
If 7 inches are cut from each side, the volume will be 1792 square inches.

c. Since x is the number of inches to be cut from each side, $x > 0$. Since each side is 30 inches, you must cut less than 15 inches from each side.
$0 < x < 15$ or $(0, 15)$

17. $A(x) = x(20 - 2x)$
$= -2x^2 + 20x$

18.
$$A(x) = \left(\frac{x}{4}\right)^2 + \left(\frac{8-x}{4}\right)^2$$
$$= \frac{x^2}{16} + \frac{64 - 16x + x^2}{16}$$
$$= \frac{2x^2 - 16x + 64}{16}$$
$$= \frac{x^2 - 8x + 32}{8}$$

19. $P(x) = x(66 - x)$
$= -x^2 + 66x$

20. $P(x) = x(50 - x)$
$= -x^2 + 50x$

21. $A(x) = x(400 - x)$
$= -x^2 + 400x$

22. $A(x) = x(300 - x)$
$= -x^2 + 300x$

23. $2w + l = 800$
$l = 800 - 2w$
Let $x = w$
$A(x) = x(800 - 2x)$
$= -2x^2 + 800x$

24. $2w + l = 600$
$l = 600 - 2l$
let $x =$ width, $600 - 2x =$ length

$A(x) = (600 - 2x)x$
$= -2x^2 + 600x$

25. $2x + 3y = 1000$
$3y = 1000 - 2x$
$$y = \frac{1000 - 2x}{3}$$
$$A(x) = x\left(\frac{1000 - 2x}{3}\right)$$
$$= \frac{x(10000 - 2x)}{3}$$

26. $2x + 4y = 1200$

$4y = 1200 - 2x$

$y = \frac{1200 - 2x}{4}$

$$A(x) = x\left(\frac{1200 - 2x}{4}\right) = \frac{x(1200 - 2x)}{4} = \frac{2x(600 - x)}{4} = \frac{x(600 - x)}{2}$$

27. $2x$ = distance around 2 straight sides

$\pi 2r$ = distance around 2 curved sides

$2x + 2\pi r = 440$

$2x = 440 - 2\pi r$

$x = 220 - \pi r$

$$A(x) = (220 - \pi r)r + \pi r^2 = 220r - \pi r^2 + \pi r^2 = 220r$$

28. $2x$ = distance around the 2 straight sides

$2\pi r$ = distance around the 2 curved sides

$2x + 2\pi r = 880$

$2x = 880 - 2\pi r$

$x = 440 - \pi r$

$$A(x) = r(440 - \pi r) + \pi r^2 = 440r - \pi r^2 + \pi r^2 = 440r$$

29. $xy = 4000$

$y = \frac{4000}{x}$

$$C(x) = \left[2x + 2\left(\frac{4000}{x}\right)\right]175 + 125x = 350x + \frac{1,400,000}{x} + 125x = 4750x + \frac{1,400,000}{x}$$

30. $125 = lw$

$\frac{125}{l} = w$; let $x = l$

$$C(x) = 20\left(2\left(\frac{125}{x}\right) + x\right) + 9x = \frac{5000}{x} + 20x + 9x = \frac{5000}{x} + 29x$$

31. $10 = x^2 y$

$\frac{10}{x^2} = y$

$$A(x) = x^2 + 4\left(x \cdot \frac{10}{x^2}\right) = x^2 + \frac{40}{x}$$

32. $400 = x^2 y$

$\frac{400}{x^2} = y$

$$A = x^2 + 5\left(\frac{400}{x^2}\right)x = x^2 + \frac{2000}{x}$$

33. $300 = y + 4x$

$300 - 4x = y$

$$A(x) = x^2(300 - 4x) = -4x^3 + 300x^2$$

34. $108 = y + 4x$

$108 - 4x = y$

$$A = x^2(108 - 4x) = -4x^3 + 108x^2$$

35. a. Let x = amount invested at 15%
$50000 - x$ = amount invested at 7%
$I(x) = 0.15x + 0.07(50000 - x)$

b.
$$6000 = 0.15x + 0.07x(50000 - x)$$
$$6000 = 0.15x + 3500 - 0.07x$$
$$2500 = 0.08x$$
$$31250 = x$$
$$50000 - 31250 = 18750$$
Invest $31,250 at 15% and $18,750 at 7%.

36. a. Let x = amount at 10%
$18{,}750 - x$ = amount at 12%
$I(x) = 0.10x + 0.12(18750 - x)$

b. $0.10x + 0.12(18750 - x) = 2117$
$$0.1x + 2250 - 0.12x = 2117$$
$$-0.02x = -133$$
$$x = 6650$$
The amount of money to be invested should be $6650 at 10% and $12100 at 12%.

37. Let x = amount invested at 12%
$8000 - x$ = amount invested at 5% loss
$I(x) = 0.12x - 0.05(8000 - x)$

38. Let x = amount at 14%
$12000 - x$ = amount at 6%
$$I(x) = 0.14x + 0.06(12000 - x)$$
$$= 0.14x + 720 - 0.06x$$
$$= 0.08x + 720$$

39.
$$d = \sqrt{(x-0)^2 + (y-0)^2}$$
$$= \sqrt{x^2 + y^2}$$
$$= \sqrt{x^2 + \left(x^2 - 4\right)^2}$$
$$= \sqrt{x^2 + x^4 - 8x^2 + 16}$$
$$= \sqrt{x^4 - 7x^2 + 16}$$

40.
$$d = \sqrt{(x-0)^2 + (y-0)^2}$$
$$= \sqrt{x^2 + y^2}$$
$$= \sqrt{x^2 + \left(x^2 - 8\right)^2}$$
$$= \sqrt{x^2 + x^4 - 16x^2 + 64}$$
$$= \sqrt{x^4 - 15x^2 + 64}$$

41.
$$d = \sqrt{(x-1)^2 + y^2}$$
$$= \sqrt{x^2 - 2x + 1 + \left(\sqrt{x}\right)^2}$$
$$= \sqrt{x^2 - 2x + 1 + x}$$
$$= \sqrt{x^2 - x + 1}$$

42.
$$d = \sqrt{(x-2)^2 + y^2}$$
$$= \sqrt{x^2 - 4x + 4 + \left(\sqrt{x}\right)^2}$$
$$= \sqrt{x^2 - 3x + 4}$$

43. a.
$$A(x) = 2xy$$
$$= 2x\sqrt{4 - x^2}$$

b.
$$P(x) = 2(2x) + 2y$$
$$= 4x + 2\sqrt{4 - x^2}$$

44. a.
$$A(x) = 2xy$$
$$= 2x\sqrt{9 - x^2}$$

b.
$$P(x) = 2(2x) + 2y$$
$$= 4x + 2\sqrt{9 - x^2}$$

45. 6-foot pole
$$c^2 = 6^2 + x^2$$
$$x = \sqrt{36 + x^2}$$
8-foot pole
$$c^2 = 8^2 + (10 - x)^2$$
$$c = \sqrt{64 + 100 - 20x + x^2}$$
$$c = \sqrt{x^2 - 20x + 164}$$
total length
$$f(x) = \sqrt{36 + x^2} + \sqrt{x^2 - 20x + 164}$$

46. Road from Town A:

$$c^2 = 6^2 + x^2$$
$$c = \sqrt{36 + x^2}$$

Road from Town B:

$$c^2 = 3^2 + (12 - x)^2$$
$$c = \sqrt{9 + 144 - 24x + x^2}$$
$$c = \sqrt{x^2 - 24x + 153}$$

$$f(x) = \sqrt{36 + x^2} + \sqrt{x^2 - 24x + 153}$$

47. $A(x) = \frac{1}{2}x(x-5) + \frac{1}{2}x(x+3) + (x+2)\left[(x-5)+(x+3)\right]$

$$A(x) = \tfrac{1}{2}x^2 - \tfrac{5}{2}x + \tfrac{1}{2}x^2 + \tfrac{3}{2}x + (x+2)\left[2x-2\right]$$
$$A(x) = x^2 - x + 2x^2 + 2x - 4$$
$$A(x) = 3x^2 + x - 4$$

48. $A(x) = \frac{1}{2}x(2x) + \frac{1}{2}(6x - 4x)(x+2) + (4x)(x+2) + 2x(8)$

$$A(x) = x^2 + x(x+2) + 4x^2 + 8x + 16x$$
$$A(x) = x^2 + x^2 + 2x + 4x^2 + 8x + 16x$$
$$A(x) = 6x^2 + 26$$

49. $V(x) = (x+5)(2x+1)(x+2) - (x+5)(3)(x)$

$$V(x) = (x+5)(2x^2 + 5x + 2) - 3x(x+5)$$
$$V(x) = 2x^3 + 15x^2 + 27x + 10 - 3x^2 - 15x$$
$$V(x) = 2x^3 + 12x^2 + 12x + 10$$

50. $V(x) = (x)(2x-1)(x+3) - (x)(x)\left[(2x-1)-(x+1)\right]$

$$V(x) = (x)(2x^2 + 5x - 3) - x^2(x-2)$$
$$V(x) = 2x^3 + 5x^2 - 3x - x^3 + 2x^2$$
$$V(x) = x^3 + 7x^2 - 3x$$

63. Distance and time rowed:

$$d^2 = 2^2 + x^2$$
$$d = \sqrt{4 + x^2}$$
$$rt = d$$
$$2t = \sqrt{4 + x^2}$$
$$t = \frac{\sqrt{4 + x^2}}{2}$$

Distance and time walked:

$$d = 6 - x$$
$$rt = d$$
$$5t = 6 - x$$
$$t = \frac{6 - x}{5}$$

Total time:

$$T(x) = \frac{\sqrt{4 + x^2}}{2} + \frac{6 - x}{5}$$

64. $A(x) = (20 + 2x)(10 + 2x) - 10(20)$

$$= 4x^2 + 60x + 200 - 200$$
$$= 4x^2 + 60x$$

65.

$$P = 2h + 2r + \frac{1}{2}(\pi 2r)$$
$$12 = 2h + 2r + \pi r$$
$$12 - 2r - \pi r = 2h$$
$$\frac{12 - 2r - \pi r}{2} = h$$

$$A = \left(\frac{12 - 2r - \pi r}{2}\right)2r + \frac{1}{2}\left(\pi r^2\right)$$
$$= 12r - 2r^2 - \pi r^2 + \frac{1}{2}\pi r^2$$
$$= 12r - 2r^2 - \frac{1}{2}\pi r^2$$

66.

$$r = \frac{1}{2}h$$

$$V(h) = \frac{1}{3}\pi r^2 h$$
$$= \frac{1}{3}\pi\left(\frac{1}{2}h\right)^2 h$$
$$= \frac{1}{3}\pi\frac{1}{4}h^2 h$$
$$= \frac{\pi}{12}h^3$$

Chapter 1 Review Exercises

1.

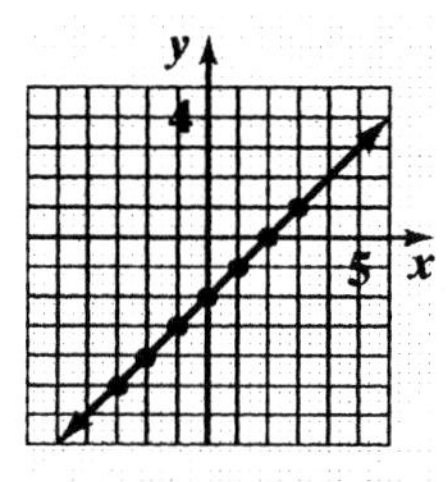

$y = 2x - 2$

$x = -3, y = -8$
$x = -2, y = -6$
$x = -1, y = -4$
$x = 0, y = -2$
$x = 1, y = 0$
$x = 2, y = 2$
$x = 3, y = 4$

2.

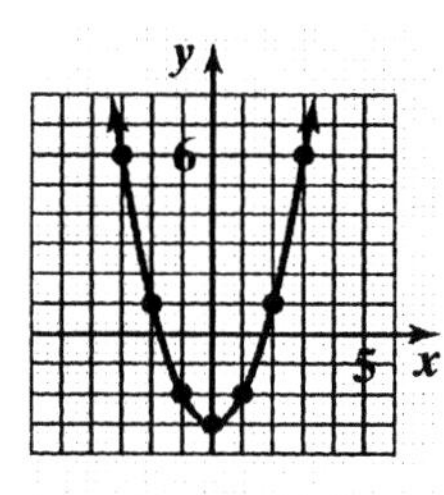

$y = x^2 - 3$

$x = -3, y = 6$
$x = -2, y = 1$
$x = -1, y = -2$
$x = 0, y = -3$
$x = 1, y = -2$
$x = 2, y = 1$
$x = 3, y = 6$

3.

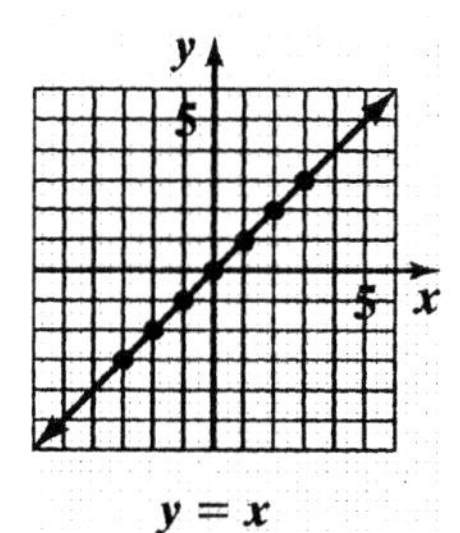

$y = x$

$x = -3, y = -3$
$x = -2, y = -2$
$x = -1, y = -1$
$x = 0, y = 0$
$x = 1, y = 1$
$x = 2, y = 2$
$x = 3, y = 3$

4.

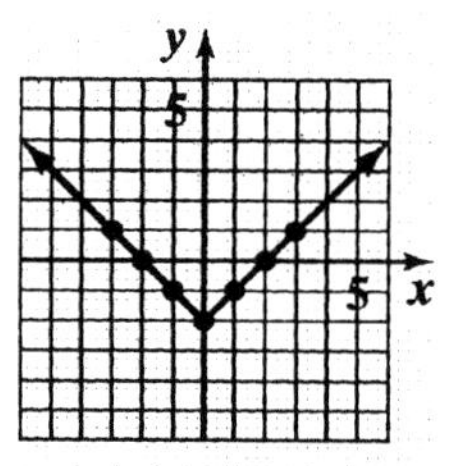

$y = |x| - 2$

$x = -3, y = 1$
$x = -2, y = 0$
$x = -1, y = -1$
$x = 0, y = -2$
$x = 1, y = -1$
$x = 2, y = 0$
$x = 3, y = 1$

5. A portion of Cartesian coordinate plane with minimum x-value equal to –20, maximum x-value equal to 40, x-scale equal to 10 and with minimum y-value equal to –5, maximum y-value equal to 5, and y-scale equal to 1.

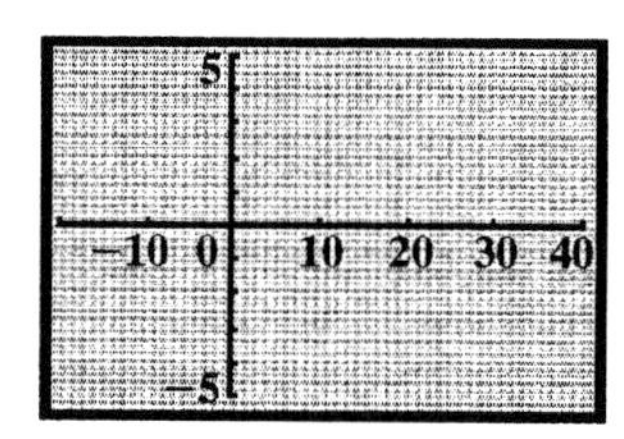

6. x-intercept: –2; The graph intersects the x-axis at (–2, 0).
y-intercept: 2; The graph intersects the y-axis at (0, 2).

7. x-intercepts: 2, –2; The graph intersects the x-axis at (–2, 0) and (2, 0).
y-intercept: –4; The graph intercepts the y-axis at (0, –4).

8. x-intercept: 5; The graph intersects the x-axis at (5, 0).
y-intercept: None; The graph does not intersect the y-axis.

9. Point A is (91, 125). This means that in 1991, 125,000 acres were used for cultivation

10. Opium cultivation was 150,000 acres in 1997.

11. Opium cultivation was at a minimum in 2001 when approximately 25,000 acres were used.

12. Opium cultivation was at a maximum in 2004 when approximately 300,000 acres were used.

13. Opium cultivation did not change between 1991 and 1992.

14. Opium cultivation increased at the greatest rate between 2001 and 2002. The increase in acres used for opium cultivation in this time period was approximately 180,000 – 25,000 = 155,000 acres.

15. function
domain: {2, 3, 5}
range: {7}

16. function
domain: {1, 2, 13}
range: {10, 500, π}

17. not a function
domain: {12, 14}
range: {13, 15, 19}

18. $2x + y = 8$
$y = -2x + 8$
Since only one value of y can be obtained for each value of x, y is a function of x.

19. $3x^2 + y = 14$
$y = -3x^2 + 14$
Since only one value of y can be obtained for each value of x, y is a function of x.

20. $2x + y^2 = 6$
$y^2 = -2x + 6$
$y = \pm\sqrt{-2x+6}$
Since more than one value of y can be obtained from some values of x, y is not a function of x.

21. $f(x) = 5 - 7x$

a. $f(4) = 5 - 7(4) = -23$

b. $$\begin{aligned} f(x+3) &= 5 - 7(x+3) \\ &= 5 - 7x - 21 \\ &= -7x - 16 \end{aligned}$$

c. $f(-x) = 5 - 7(-x) = 5 + 7x$

22. $g(x) = 3x^2 - 5x + 2$

a. $g(0) = 3(0)^2 - 5(0) + 2 = 2$

b. $$\begin{aligned} g(-2) &= 3(-2)^2 - 5(-2) + 2 \\ &= 12 + 10 + 2 \\ &= 24 \end{aligned}$$

c. $$\begin{aligned} g(x-1) &= 3(x-1)^2 - 5(x-1) + 2 \\ &= 3(x^2 - 2x + 1) - 5x + 5 + 2 \\ &= 3x^2 - 11x + 10 \end{aligned}$$

d. $$\begin{aligned} g(-x) &= 3(-x)^2 - 5(-x) + 2 \\ &= 3x^2 + 5x + 2 \end{aligned}$$

23. a. $g(13) = \sqrt{13-4} = \sqrt{9} = 3$

b. $g(0) = 4 - 0 = 4$

c. $g(-3) = 4 - (-3) = 7$

24. a. $f(-2) = \dfrac{(-2)^2 - 1}{-2-1} = \dfrac{3}{-3} = -1$

b. $f(1) = 12$

c. $f(2) = \dfrac{2^2 - 1}{2-1} = \dfrac{3}{1} = 3$

25. The vertical line test shows that this is not the graph of a function.

26. The vertical line test shows that this is the graph of a function.

27. The vertical line test shows that this is the graph of a function.

28. The vertical line test shows that this is not the graph of a function.

29. The vertical line test shows that this is not the graph of a function.

30. The vertical line test shows that this is the graph of a function.

31. $$\begin{aligned} &\frac{8(x+h) - 11 - (8x - 11)}{h} \\ &= \frac{8x + 8h - 11 - 8x + 11}{h} \\ &= \frac{8h}{8} \\ &= 8 \end{aligned}$$

32. $\dfrac{-2(x+h)^2+(x+h)+10-\left(-2x^2+x+10\right)}{h}$

$=\dfrac{-2\left(x^2+2xh+h^2\right)+x+h+10+2x^2-x-10}{h}$

$=\dfrac{-2x^2-4xh-2h^2+x+h+10+2x^2-x-10}{h}$

$=\dfrac{-4xh-2h^2+h}{h}$

$=\dfrac{h\left(-4x-2h+1\right)}{h}$

$-4x-2h+1$

33. **a.** domain: [–3, 5)

b. range: [–5, 0]

c. x-intercept: –3

d. y-intercept: –2

e. increasing: $(-2, 0)$ or $(3, 5)$
decreasing: $(-3, -2)$ or $(0, 3)$

f. $f(-2) = -3$ and $f(3) = -5$

34. **a.** domain: $(-\infty, \infty)$

b. range: $(-\infty, \infty)$

c. x-intercepts: –2 and 3

d. y-intercept: 3

e. increasing: $(-5, 0)$
decreasing: $(-\infty, -5)$ or $(0, \infty)$

f. $f(-2) = 0$ and $f(6) = -3$

35. **a.** domain: $(-\infty, \infty)$

b. range: [–2, 2]

c. x-intercept: 0

d. y-intercept: 0

e. increasing: $(-2, 2)$
constant: $(-\infty, -2)$ or $(2, \infty)$

f. $f(-9) = -2$ and $f(14) = 2$

36. **a.** 0, relative maximum –2

b. –2, 3, relative minimum –3, –5

37. **a.** 0, relative maximum 3

b. –5, relative minimum –6

38. $f(x) = x^3 - 5x$

$f(-x) = (-x)^3 - 5(-x)$

$= -x^3 + 5x$

$= -f(x)$

The function is odd. The function is symmetric with respect to the origin.

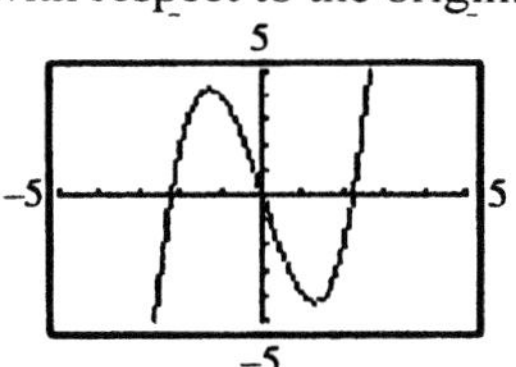

39. $f(x) = x^4 - 2x^2 + 1$

$f(-x) = (-x)^4 - 2(-x)^2 + 1$

$= x^4 - 2x^2 + 1$

$= f(x)$

The function is even. The function is symmetric with respect to the y-axis.

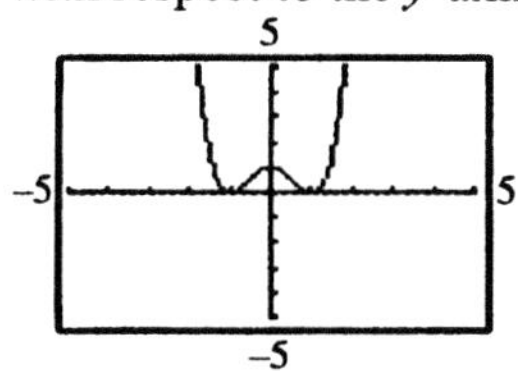

40. $f(x) = 2x\sqrt{1-x^2}$

$f(-x) = 2(-x)\sqrt{1-(-x)^2}$

$= -2x\sqrt{1-x^2}$

$= -f(x)$

The function is odd. The function is symmetric with respect to the origin.

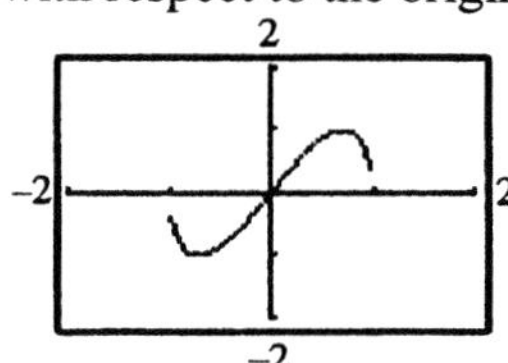

41. **a.** Yes, the eagle's height is a function of time since the graph passes the vertical line test.

b. Decreasing: (3, 12)
The eagle descended.

c. Constant: (0, 3) or (12, 17)
The eagle's height held steady during the first 3 seconds and the eagle was on the ground for 5 seconds.

d. Increasing: (17, 30)
The eagle was ascending.

42.

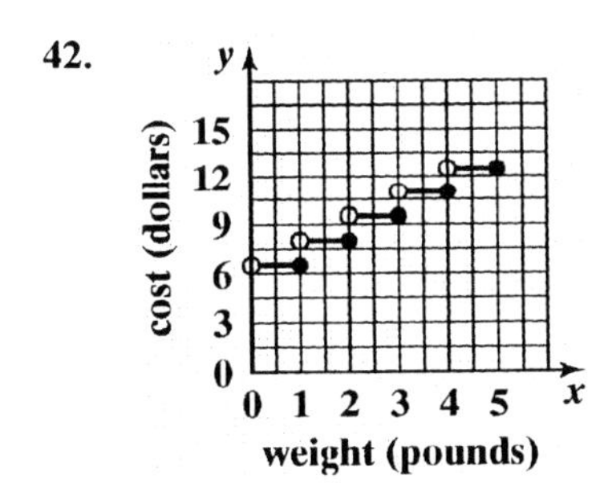

43. $m = \frac{1-2}{5-3} = \frac{-1}{2} = -\frac{1}{2}$; falls

44. $m = \frac{-4-(-2)}{-3-(-1)} = \frac{-2}{-2} = 1$; rises

45. $m = \frac{\frac{1}{4}-\frac{1}{4}}{6-(-3)} = \frac{0}{9} = 0$; horizontal

46. $m = \frac{10-5}{-2-(-2)} = \frac{5}{0}$ undefined; vertical

47. point-slope form: $y - 2 = -6(x + 3)$
slope-intercept form: $y = -6x - 16$

48. $m = \frac{2-6}{-1-1} = \frac{-4}{-2} = 2$
point-slope form: $y - 6 = 2(x - 1)$
or $y - 2 = 2(x + 1)$
slope-intercept form: $y = 2x + 4$

49. $3x + y - 9 = 0$
$y = -3x + 9$
$m = -3$
point-slope form:
$y + 7 = -3(x - 4)$
slope-intercept form:
$y = -3x + 12 - 7$
$y = -3x + 5$

50. perpendicular to $y = \frac{1}{3}x + 4$

$m = -3$
point-slope form:
$y - 6 = -3(x + 3)$
slope-intercept form:
$y = -3x - 9 + 6$
$y = -3x - 3$

51. Write $6x - y - 4 = 0$ in slope intercept form.

$$6x - y - 4 = 0$$
$$-y = -6x + 4$$
$$y = 6x - 4$$

The slope of the perpendicular line is 6, thus the slope of the desired line is $m = -\frac{1}{6}$.

$$y - y_1 = m(x - x_1)$$
$$y - (-1) = -\tfrac{1}{6}(x - (-12))$$
$$y + 1 = -\tfrac{1}{6}(x + 12)$$
$$y + 1 = -\tfrac{1}{6}x - 2$$
$$6y + 6 = -x - 12$$
$$x + 6y + 18 = 0$$

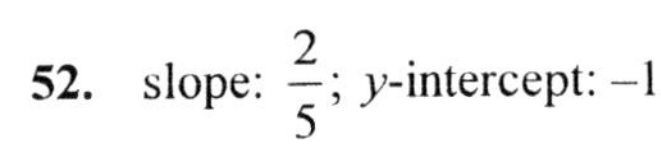

52. slope: $\frac{2}{5}$; y-intercept: -1

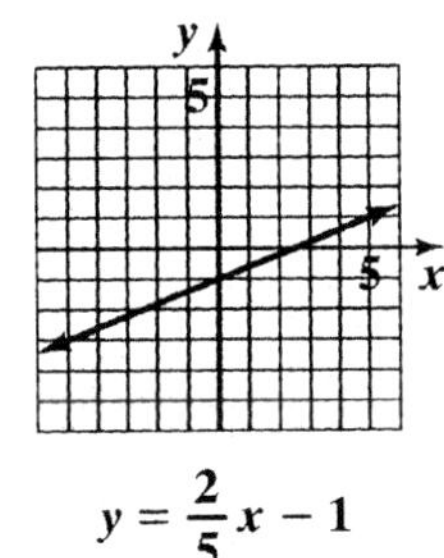

$y = \frac{2}{5}x - 1$

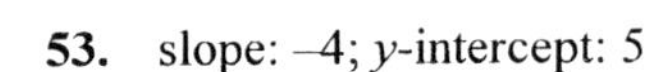

53. slope: -4; y-intercept: 5

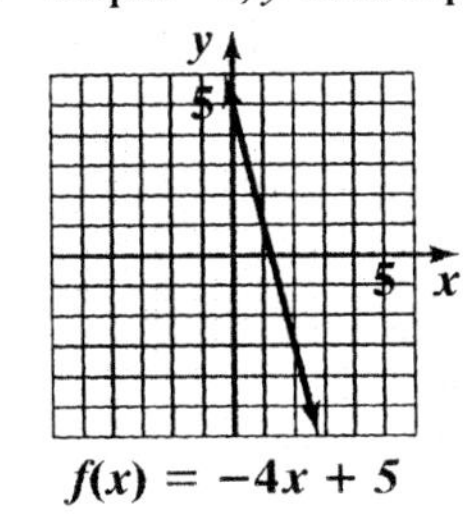

$f(x) = -4x + 5$

54. $2x+3y+6=0$

$$3y=-2x-6$$

$$y=-\frac{2}{3}x-2$$

slope: $-\frac{2}{3}$; y-intercept: -2

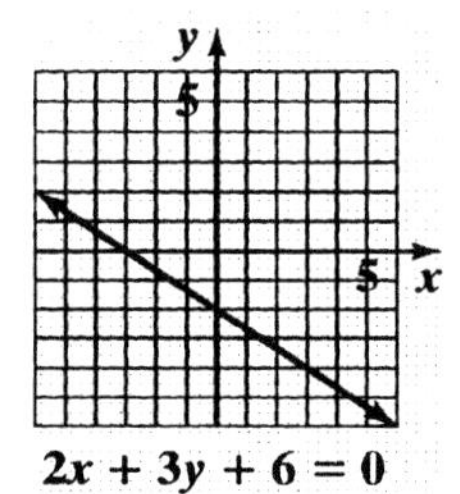

$2x + 3y + 6 = 0$

55. $2y-8=0$

$$2y=8$$

$$y=4$$

slope: 0; y-intercept: 4

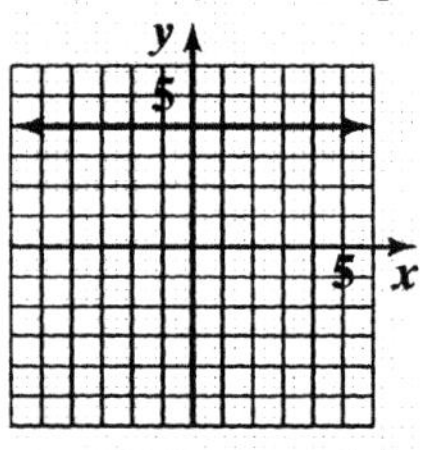

$2y - 8 = 0$

56. $2x-5y-10=0$

Find x-intercept:

$$2x-5(0)-10=0$$

$$2x-10=0$$

$$2x=10$$

$$x=5$$

Find y-intercept:

$$2(0)-5y-10=0$$

$$-5y-10=0$$

$$-5y=10$$

$$y=-2$$

$2x - 5y - 10 = 0$

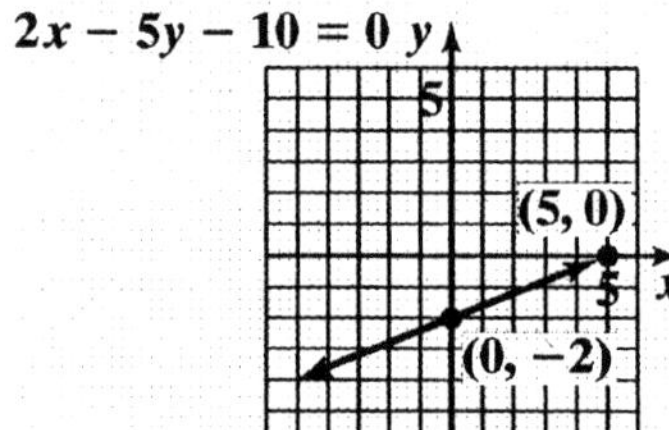

57. $2x-10=0$

$$2x=10$$

$$x=5$$

$2x - 10 = 0$

y 5 6 x

58. **a.** First, find the slope. $(1,1.5)$ and $(3,3.4)$.

$$m=\frac{3.4-1.5}{3-1}=\frac{1.9}{2}=0.95$$

Next, use the slope and one of the points to write the point-slope equation of the line.

$y-1.5=0.95(x-1)$ or $y-3.4=0.95(x-3)$

b. $y-1.5=0.95(x-1)$

$$y-1.5=0.95x-0.95$$

$$y=0.95x-0.55$$

c. Since 2009 is 2009-1999 = 10, let $x = 10$.

$$y=0.95(10)+0.55$$

$$=9.5+0.55=10.05$$

\$10.05 billion in revenue was earned from online gambling in 2009.

59. **a.** $(1999,41315)$ and $(2001,41227)$

$$m=\frac{41227-41315}{2001-1999}=\frac{-88}{2}=-44$$

The number of new AIDS diagnoses decreased at a rate of 44 each year from 1999 to 2001.

b. $(2001,41227)$ and $(2003,43045)$

$$m=\frac{43045-41227}{2003-2001}=\frac{1818}{2}=909$$

The number of new AIDS diagnoses increased at a rate of 909 each year from 2001 to 2003.

c. $(1999,41315)$ and $(2003,43045)$

$$m=\frac{43045-41315}{2003-1999}=\frac{1730}{4}=432.5$$

$$\frac{-44+909}{2}=\frac{865}{2}=432.5$$

Yes, the slope equals the average of the two values.

60. $\dfrac{9^2-4(9)-[4^2-4\cdot 5]}{9-5}=\dfrac{40}{4}=10$

61. **a.** $S(0)=-16(0)^2+64(0)+80=80$

$S(2)=-16(2)^2+64(2)+80=144$

$\dfrac{144-80}{2-0}=32$

b. $S(4)=-16(4)^2+64(4)+80=80$

$\dfrac{80-144}{4-2}=-32$

c. The ball is traveling up until 2 seconds, then it starts to come down.

62. $y=g(x)$

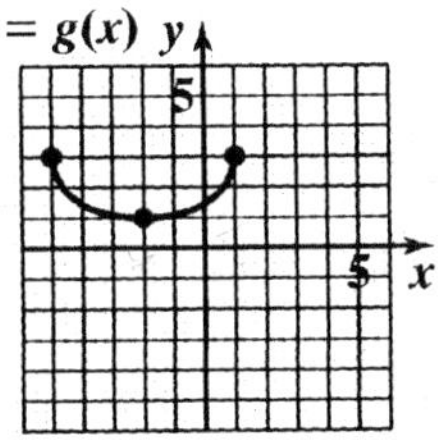

63. $y=g(x)$

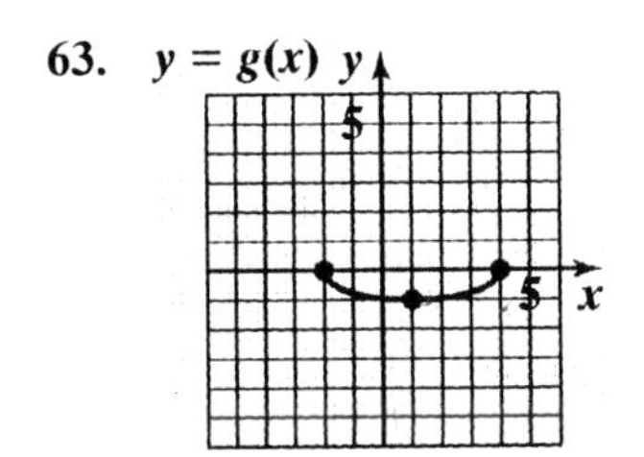

64.

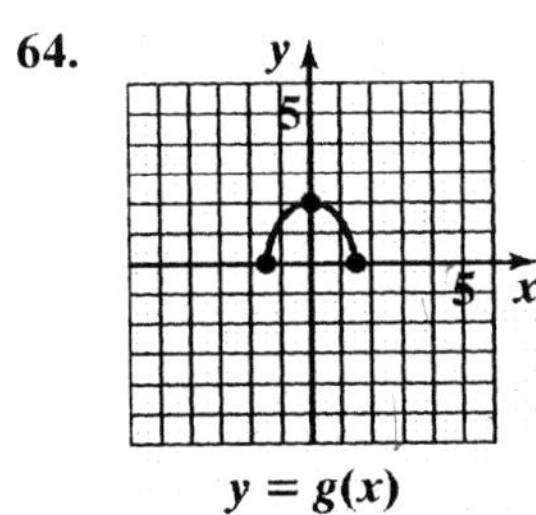

$y=g(x)$

65.

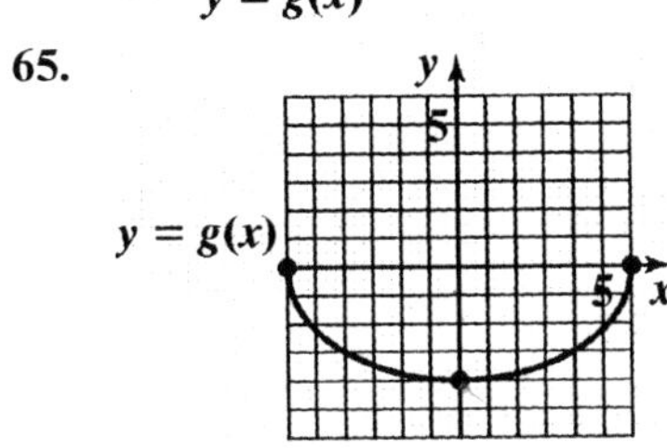

66.

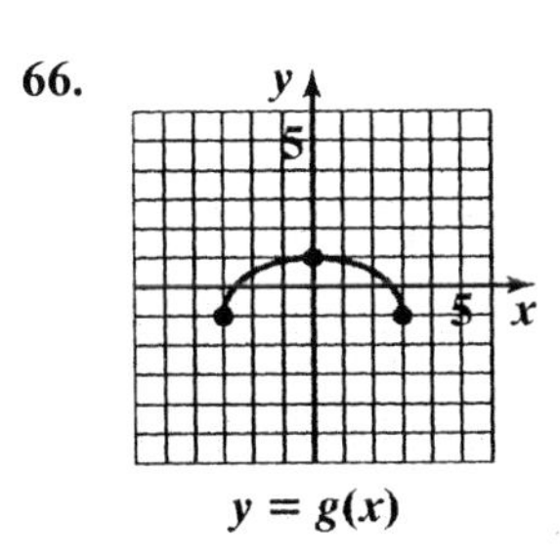

67.

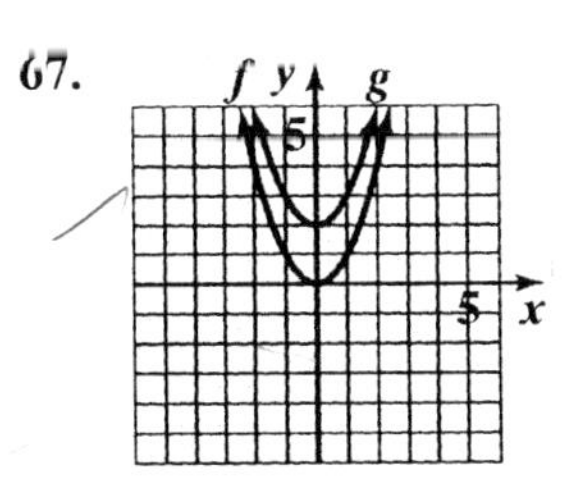

68.

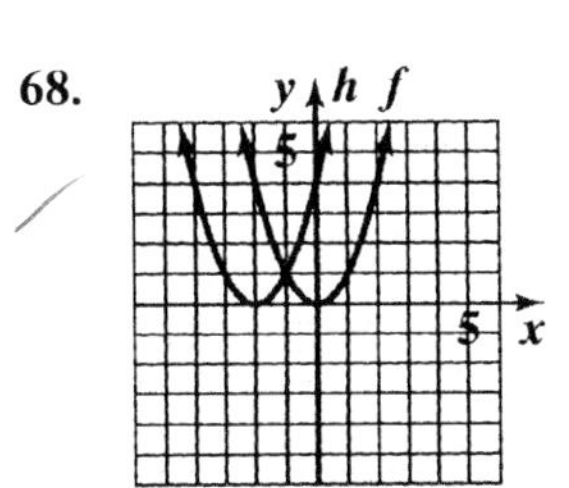

69.

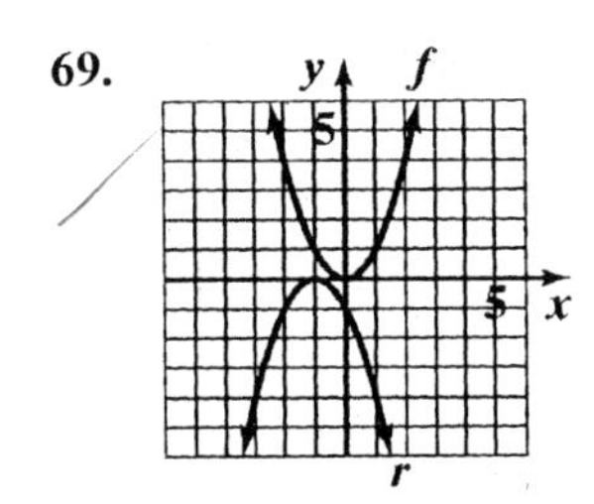

70.

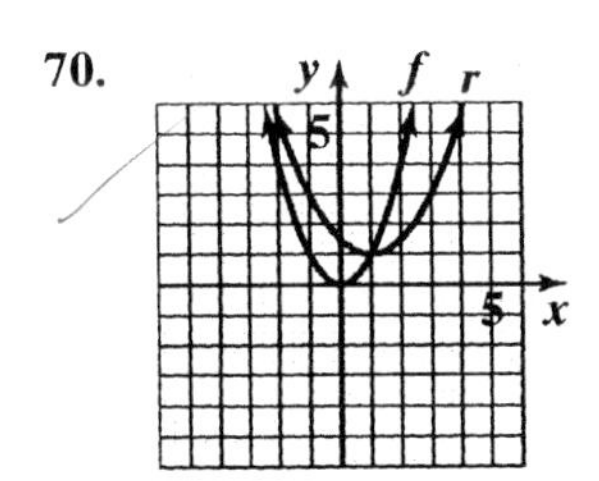

71.

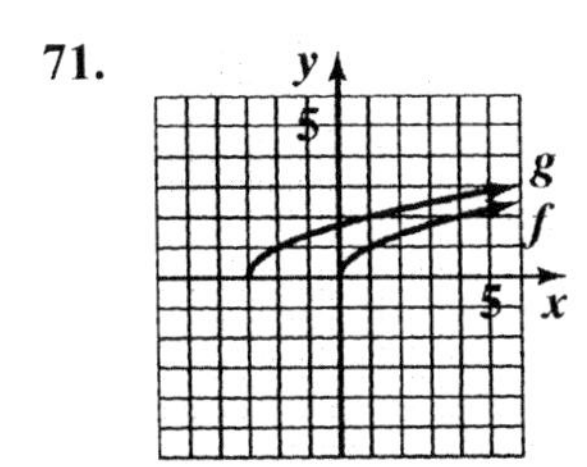

72.

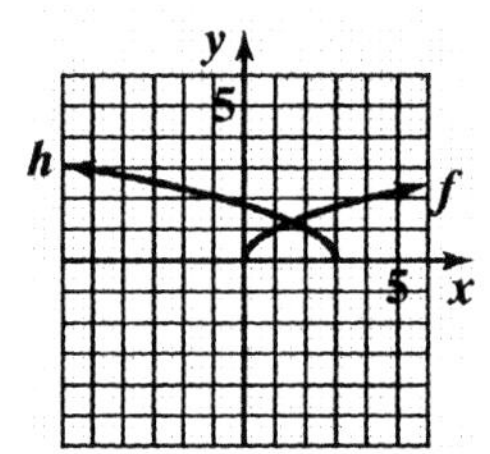

73.

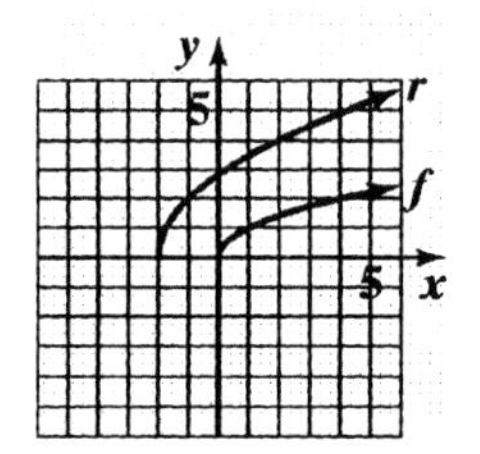

74.

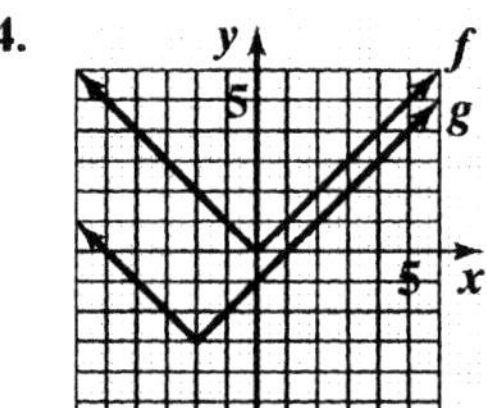

75.

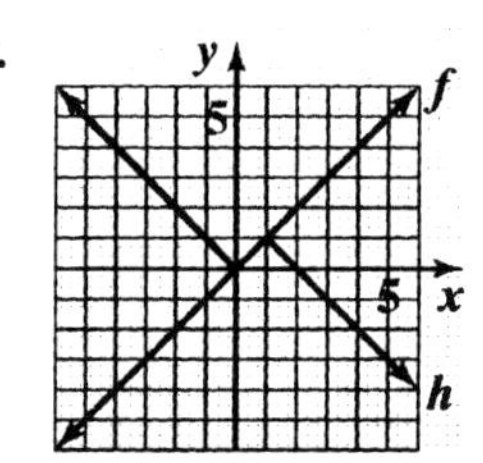

76.

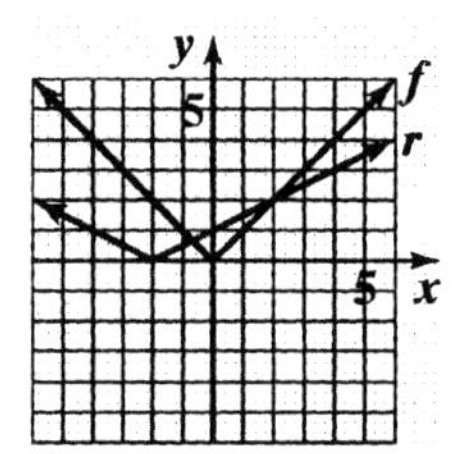

77.

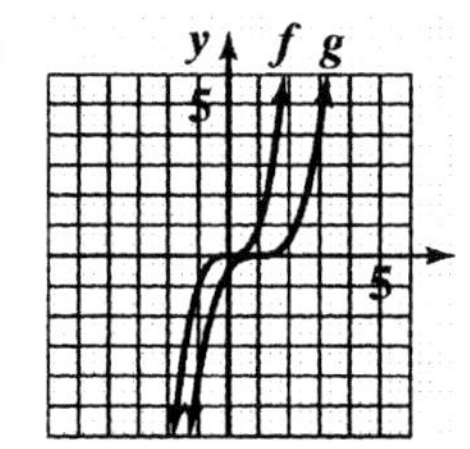

78.

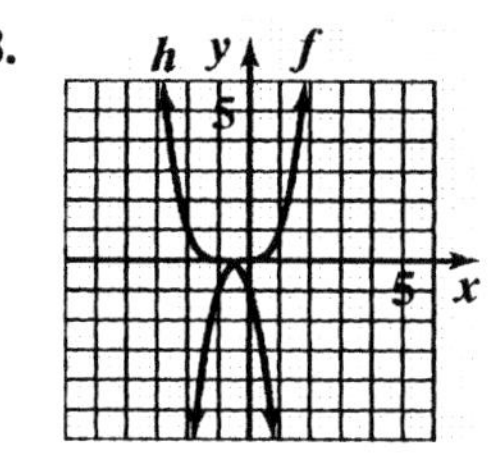

79.

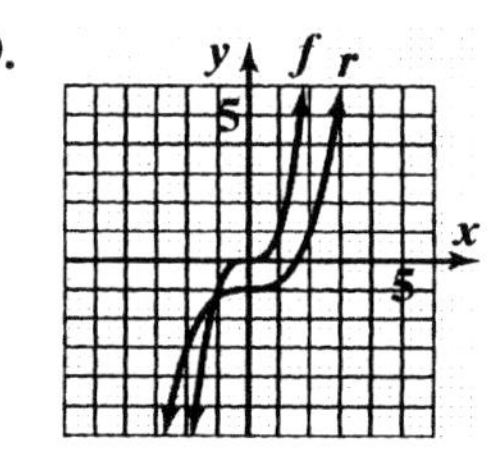

80.

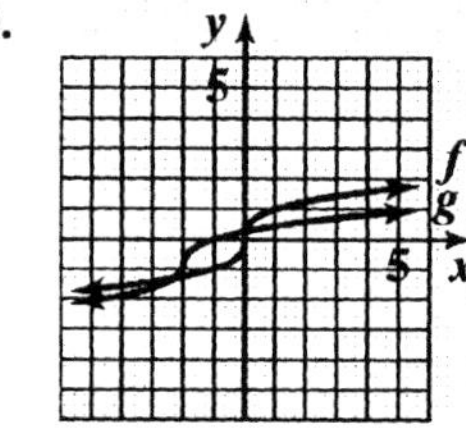

81.

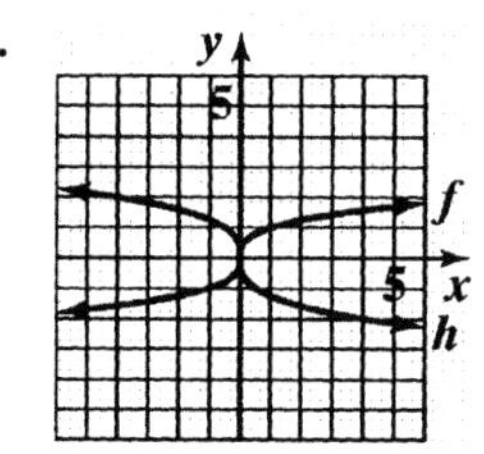

82.

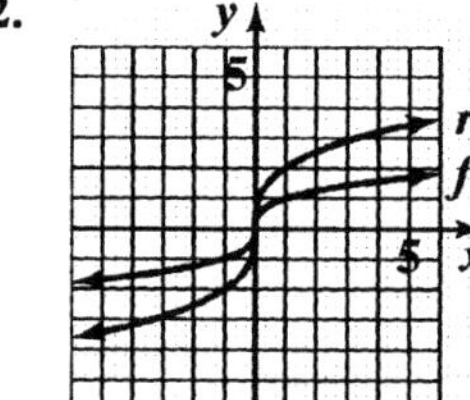

83. domain: $(-\infty, \infty)$

84. The denominator is zero when $x = 7$. The domain is $(-\infty, 7) \cup (7, \infty)$.

85. The expressions under each radical must not be negative.
$8 - 2x \geq 0$
$-2x \geq -8$
$x \leq 4$
Domain: $(-\infty, 4]$.

86. The denominator is zero when $x = -7$ or $x = 3$.
Domain: $(-\infty,-7)\cup(-7,3)\cup(3,\infty)$

87. The expressions under each radical must not be negative. The denominator is zero when $x = 5$.
$x-2\ge 0$
$x\ge 2$
Domain: $[2,5)\cup(5,\infty)$

88. The expressions under each radical must not be negative.
$x-1\ge 0$ and $x+5\ge 0$
$x\ge 1$ $\quad x\ge -5$
Domain: $[1,\infty)$

89. $f(x) = 3x - 1$; $g(x) = x - 5$
$(f+g)(x) = 4x - 6$
Domain: $(-\infty,\infty)$
$(f-g)(x) = (3x-1)-(x-5) = 2x+4$
Domain: $(-\infty,\infty)$
$(fg)(x) = (3x-1)(x-5) = 3x^2-16x+5$
Domain: $(-\infty,\infty)$
$\left(\frac{f}{g}\right)(x) = \frac{3x-1}{x-5}$
Domain: $(-\infty,5)\cup(5,\infty)$

90. $f(x) = x^2+x+1$; $g(x) = x^2-1$
$(f+g)(x) = 2x^2+x$
Domain: $(-\infty,\infty)$
$(f-g)(x) = (x^2+x+1)-(x^2-1) = x+2$
Domain: $(-\infty,\infty)$
$(fg)(x) = (x^2+x+1)(x^2-1)$
$= x^4+x^3-x-1$
$\left(\frac{f}{g}\right)(x) = \frac{x^2+x+1}{x^2-1}$
Domain: $(-\infty,-1)\cup(-1,1)\cup(1,\infty)$

91. $f(x) = \sqrt{x+7}$; $g(x) = \sqrt{x-2}$
$(f+g)(x) = \sqrt{x+7}+\sqrt{x-2}$
Domain: $[2,\infty)$
$(f-g)(x) = \sqrt{x+7}-\sqrt{x-2}$
Domain: $[2,\infty)$
$(fg)(x) = \sqrt{x+7}\cdot\sqrt{x-2}$
$= \sqrt{x^2+5x-14}$
Domain: $[2,\infty)$
$\left(\frac{f}{g}\right)(x) = \frac{\sqrt{x+7}}{\sqrt{x-2}}$
Domain: $(2,\infty)$

92. $f(x) = x^2+3$; $g(x) = 4x-1$

a. $(f\circ g)(x) = (4x-1)^2+3$
$= 16x^2-8x+4$

b. $(g\circ f)(x) = 4(x^2+3)-1$
$= 4x^2+11$

c. $(f\circ g)(3) = 16(3)^2-8(3)+4 = 124$

93. $f(x) = \sqrt{x}$; $g(x) = x+1$

a. $(f\circ g)(x) = \sqrt{x+1}$

b. $(g\circ f)(x) = \sqrt{x}+1$

c. $(f\circ g)(3) = \sqrt{3+1} = \sqrt{4} = 2$

94. a. $(f\circ g)(x) = f\left(\frac{1}{x}\right)$
$= \frac{\frac{1}{x}+1}{\frac{1}{x}-2} = \frac{\left(\frac{1}{x}+1\right)x}{\left(\frac{1}{x}-2\right)x} = \frac{1+x}{1-2x}$

b. $x\ne 0$ $\quad 1-2x\ne 0$
$x\ne\frac{1}{2}$
$(-\infty,0)\cup\left(0,\frac{1}{2}\right)\cup\left(\frac{1}{2},\infty\right)$

95. a. $(f\circ g)(x) = f(x+3) = \sqrt{x+3-1} = \sqrt{x+2}$

b. $x+2\ge 0$
$x\ge -2$
$[-2,\infty)$

96. $f(x) = x^4$ $\quad g(x) = x^2+2x-1$

97. $f(x) = \sqrt[3]{x} \quad g(x) = 7x + 4$

98. $f(x) = \frac{3}{5}x + \frac{1}{2};\ g(x) = \frac{5}{3}x - 2$

$$f(g(x)) = \frac{3}{5}\left(\frac{5}{3}x - 2\right) + \frac{1}{2} = x - \frac{6}{5} + \frac{1}{2} = x - \frac{7}{10}$$

$$g(f(x)) = \frac{5}{3}\left(\frac{3}{5}x + \frac{1}{2}\right) - 2 = x + \frac{5}{6} - 2 = x - \frac{7}{6}$$

f and g are not inverses of each other.

99. $f(x) = 2 - 5x;\ g(x) = \frac{2-x}{5}$

$$f(g(x)) = 2 - 5\left(\frac{2-x}{5}\right) = 2 - (2 - x) = x$$

$$g(f(x)) = \frac{2-(2-5x)}{5} = \frac{5x}{5} = x$$

f and g are inverses of each other.

100. a.

$$f(x) = 4x - 3$$
$$y = 4x - 3$$
$$x = 4y - 3$$
$$y = \frac{x+3}{4}$$
$$f^{-1}(x) = \frac{x+3}{4}$$

b.

$$f(f^{-1}(x)) = 4\left(\frac{x+3}{4}\right) - 3 = x + 3 - 3 = x$$

$$f^{-1}(f(x)) = \frac{(4x-3)+3}{4} = \frac{4x}{4} = x$$

101. a.

$$f(x) = 8x^3 + 1$$
$$y = 8x^3 + 1$$
$$x = 8y^3 + 1$$
$$x - 1 = 8y^3$$
$$\frac{x-1}{8} = y^3$$
$$\sqrt[3]{\frac{x-1}{8}} = y$$
$$\frac{\sqrt[3]{x-1}}{2} = y$$
$$f^{-1}(x) = \frac{\sqrt[3]{x-1}}{2}$$

b.

$$f\left(f^{-1}(x)\right) = 8\left(\frac{\sqrt[3]{x-1}}{2}\right)^3 + 1 = 8\left(\frac{x-1}{8}\right) + 1 = x - 1 + 1 = x$$

$$f^{-1}\left(f(x)\right) = \frac{\sqrt[3]{\left(8x^3+1\right)-1}}{2} = \frac{\sqrt[3]{8x^3}}{2} = \frac{2x}{2} = x$$

102. a.

$$f(x) = \frac{2}{x} + 5$$
$$y = \frac{2}{x} + 5$$
$$x = \frac{2}{y} + 5$$
$$xy = 2 + 5y$$
$$xy - 5y = 2$$
$$y(x-5) = 2$$
$$y = \frac{2}{x-5}$$
$$f^{-1}(x) = \frac{2}{x-5}$$

b. $f\left(f^{-1}(x)\right)=\dfrac{2}{\dfrac{2}{x-5}}+5$

$=\dfrac{2(x-5)}{2}+5$

$=x-5+5$

$=x$

$f^{-1}\left(f(x)\right)=\dfrac{2}{\dfrac{2}{x}+5-5}$

$=\dfrac{2}{\dfrac{2}{x}}$

$=\dfrac{2x}{2}$

$=x$

103. The inverse function exists.

104. The inverse function does not exist since it does not pass the horizontal line test.

105. The inverse function exists.

106. The inverse function does not exist since it does not pass the horizontal line test.

107.

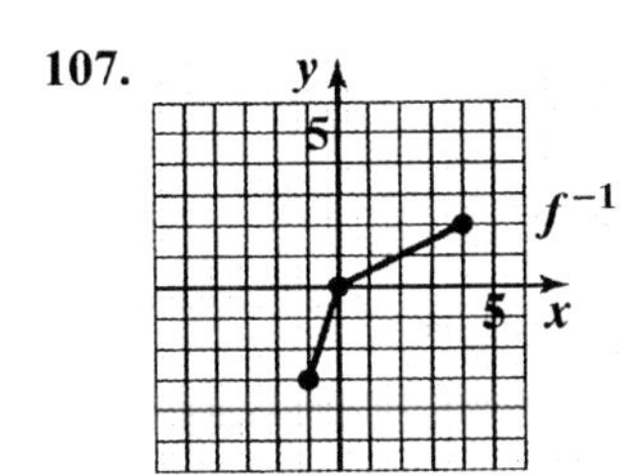

108. $f(x)=1-x^2$

$y=1-x^2$

$x=1-y^2$

$y^2=1-x$

$y=\sqrt{1-x}$

$f^{-1}(x)=\sqrt{1-x}$

109. $f(x)=\sqrt{x}+1$

$y=\sqrt{x}+1$

$x=\sqrt{y}+1$

$x-1=\sqrt{y}$

$(x-1)^2=y$

$f^{-1}(x)=(x-1)^2,\quad x\ge 1$

$\boldsymbol{f(x)=\sqrt{x}+1}$
$\boldsymbol{g(x)=(x-1)^2, x\ge 1}$

110. $d=\sqrt{[3-(-2)]^2+[9-(-3)]^2}$

$=\sqrt{5^2+12^2}$

$=\sqrt{25+144}$

$=\sqrt{169}$

$=13$

111. $d=\sqrt{[-2-(-4)]^2+(5-3)^2}$

$=\sqrt{2^2+2^2}$

$=\sqrt{4+4}$

$=\sqrt{8}$

$=2\sqrt{2}$

≈ 2.83

112. $\left(\dfrac{2+(-12)}{2},\dfrac{6+4}{2}\right)=\left(\dfrac{-10}{2},\dfrac{10}{2}\right)=(-5,5)$

113. $\left(\dfrac{4+(-15)}{2},\dfrac{-6+2}{2}\right)=\left(\dfrac{-11}{2},\dfrac{-4}{2}\right)=\left(\dfrac{-11}{2},-2\right)$

114. $x^2+y^2=3^2$

$x^2+y^2=9$

115. $(x-(-2))^2+(y-4)^2=6^2$

$(x+2)^2+(y-4)^2=36$

116. center: (0, 0); radius: 1

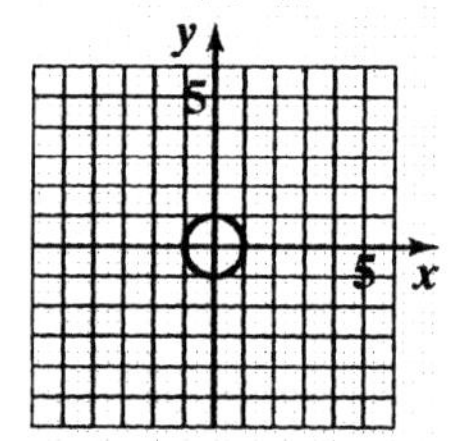

$$x^2 + y^2 = 1$$

Domain: $[-1,1]$

Range: $[-1,1]$

117. center: (–2, 3); radius: 3

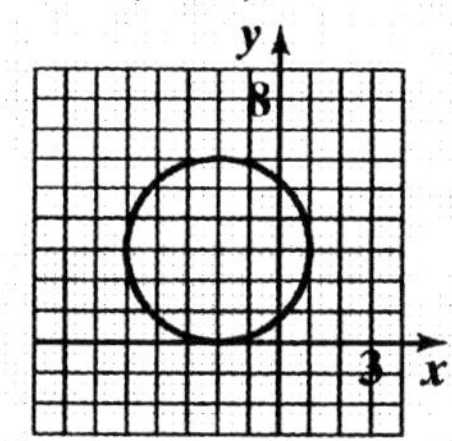

$$(x + 2)^2 + (y - 3)^2 = 9$$

Domain: $[-5,1]$

Range: $[0,6]$

118.
$$x^2 + y^2 - 4x + 2y - 4 = 0$$
$$x^2 - 4x \quad + y^2 + 2y \quad = 4$$
$$x^2 - 4x + 4 + y^2 + 2y + 1 = 4 + 4 + 1$$
$$(x-2)^2 + (y+1)^2 = 9$$

center: (2, –1); radius: 3

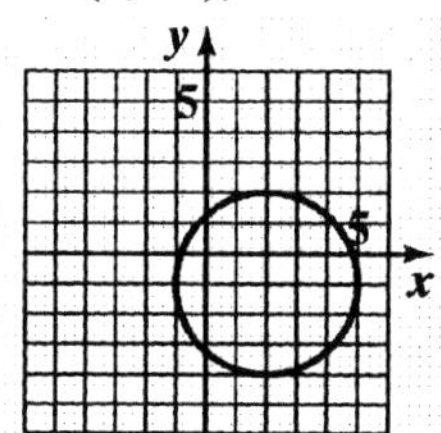

$$x^2 + y^2 - 4x + 2y - 4 = 0$$

Domain: $[-1,5]$

Range: $[-4,2]$

119. a. $W(x) = 567 + 15x$

b.
$$702 = 567 + 15x$$
$$135 = 15x$$
$$9 = x$$

9 years after 2000, in 2009, the average weekly sales will be $702.

120. a. $f(x) = 15 + 0.05x$

b. $g(x) = 5 + 0.07x$

c.
$$15 + 0.05x = 5 + 0.07x$$
$$10 = 0.02x$$
$$500 = x$$

For 500 minutes, the two plans cost the same.

121. a.
$$N(x) = 400 - 2(x - 120)$$
$$= 400 - 2x + 240$$
$$= 640 - 2x$$

b.
$$R(x) = x(640 - 2x)$$
$$= -2x^2 + 640x$$

122. a.
$$w = 16 - 2x \qquad l = 24 - 2x$$
$$V(x) = (16 - 2x)(24 - 2x)x$$

b. $0 < x < 8$

123.
$$2l + 3w = 400$$
$$2l = 400 - 3w$$
$$l = \frac{400 - 3w}{2}$$

Let x = width

$$A(x) = x\left(\frac{400 - 3w}{2}\right)$$
$$= \frac{x(400 - 3w)}{2}$$

124.
$$V = lwh$$
$$8 = x \cdot x \cdot h$$
$$\frac{8}{x^2} = h$$

$$A(x) = 2x \cdot x + 4hx$$
$$= 2x^2 + 4\left(\frac{8}{x^2}\right)x$$
$$= 2x^2 + \frac{32}{x}$$

125. $I = 0.08x + 0.12(10{,}000 - x)$

Chapter 1 Test

1. (b), (c), and (d) are not functions.

2. a. $f(4) - f(-3) = 3 - (-2) = 5$

   b. domain: $(-5, 6]$

   c. range: $[-4, 5]$

   d. increasing: $(-1, 2)$

   e. decreasing: $(-5, -1)$ or $(2, 6)$

   f. $2, f(2) = 5$

   g. $(-1, -4)$

   h. x-intercepts: -4, 1, and 5.

   i. y-intercept: -3

3. a. $-2, 2$

   b. $-1, 1$

   c. 0

   d. even; $f(-x) = f(x)$

   e. no; f fails the horizontal line test

   f. $f(0)$ is a relative minimum.

   g.

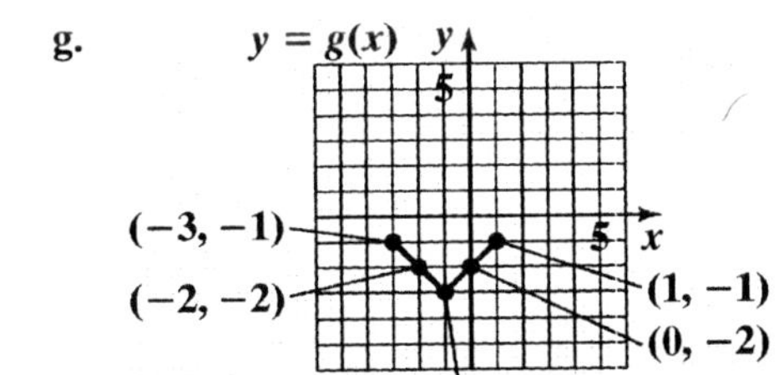

   h.

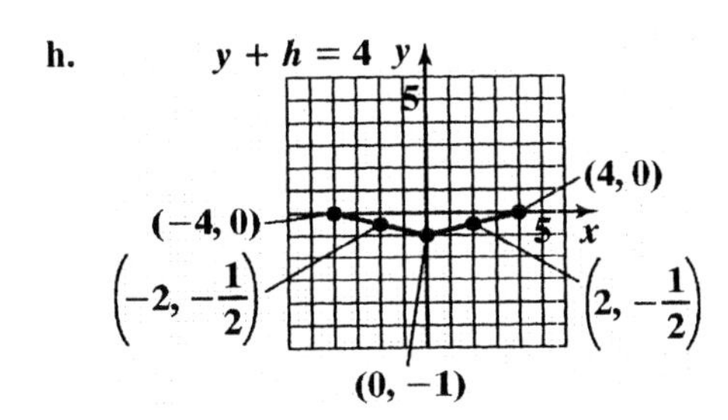

   i.

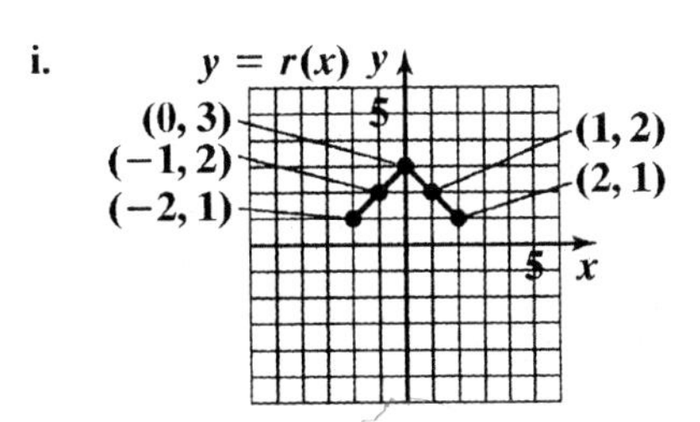

   j.

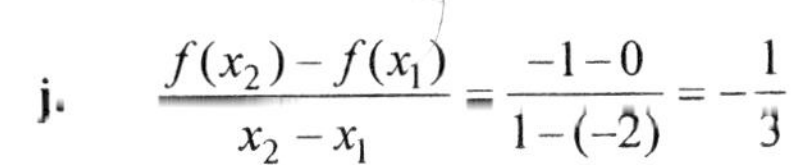

$$\frac{f(x_2) - f(x_1)}{x_2 - x_1} = \frac{-1 - 0}{1 - (-2)} = -\frac{1}{3}$$

4.

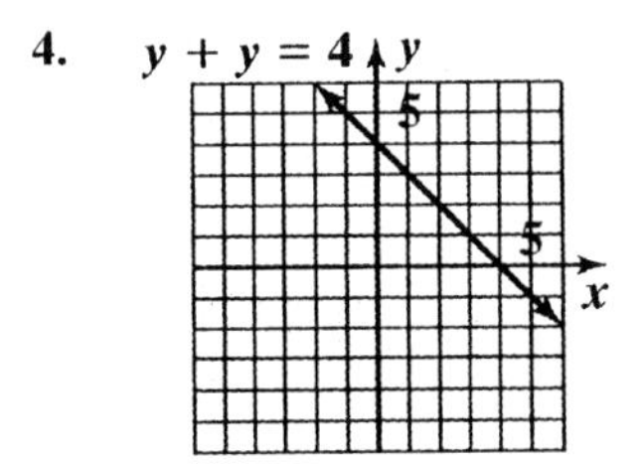

Domain: $(-\infty, \infty)$

Range: $(-\infty, \infty)$

5.

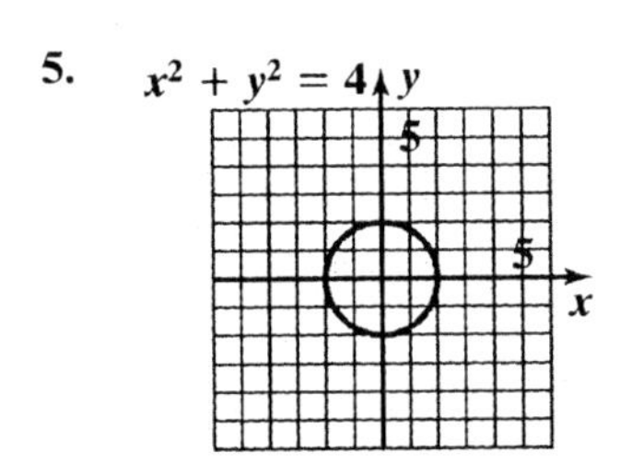

Domain: $[-2, 2]$

Range: $[-2, 2]$

6.

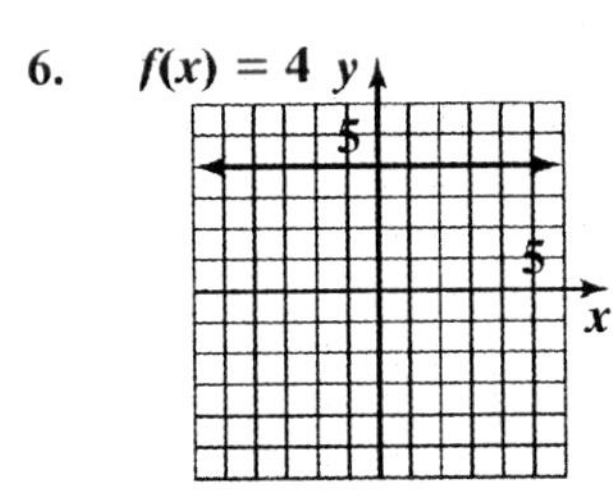

Domain: $(-\infty, \infty)$

Range: $\{4\}$

7.

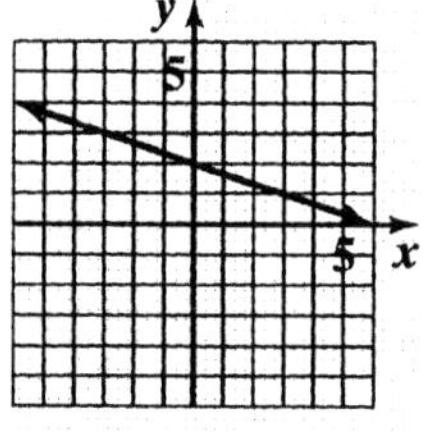

$f(x) = -\frac{1}{3}x + 2$

Domain: $(-\infty, \infty)$

Range: $(-\infty, \infty)$

8.

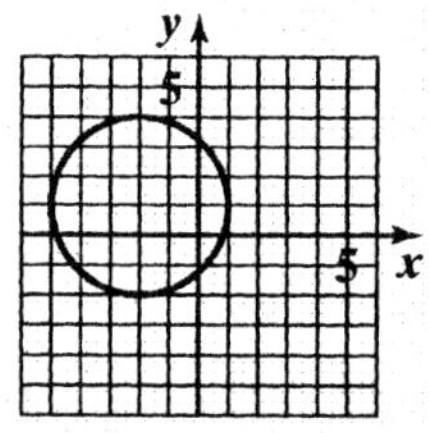

$(x + 2)^2 + (y - 1)^2 = 9$

Domain: $[-5, 1]$

Range: $[-2, 4]$

9.

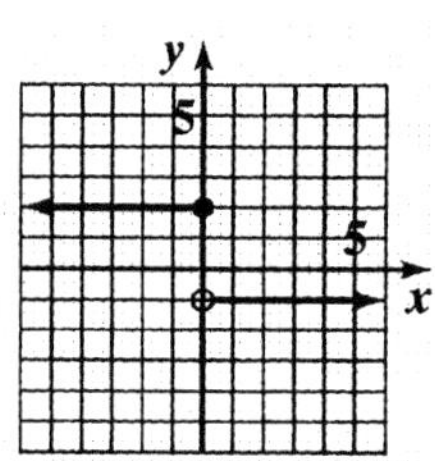

$f(x) = \begin{cases} 2 \text{ if } x \le 0 \\ -1 \text{ if } x > 0 \end{cases}$

Domain: $(-\infty, \infty)$

Range: $\{-1, 2\}$

10.

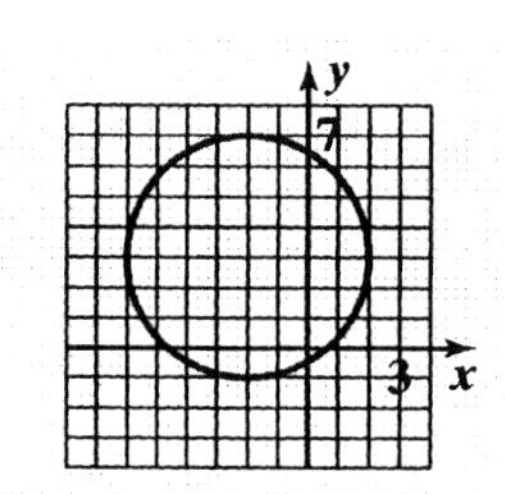

$x^2 + y^2 + 4x + 6y - 3 = 0$

Domain: $[-6, 2]$

Range: $[-1, 7]$

11.

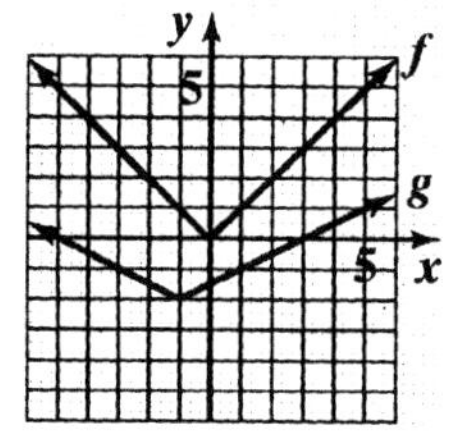

Domain of f: $(-\infty, \infty)$

Range of f: $[0, \infty)$

Domain of g: $(-\infty, \infty)$

Range of g: $[-2, \infty)$

12.

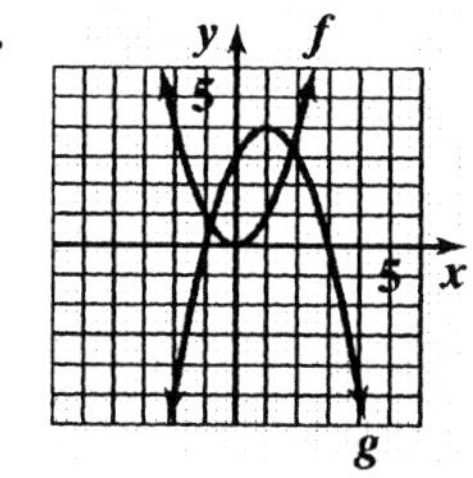

Domain of f: $(-\infty, \infty)$

Range of f: $[0, \infty)$

Domain of g: $(-\infty, \infty)$

Range of g: $(-\infty, 4]$

13.

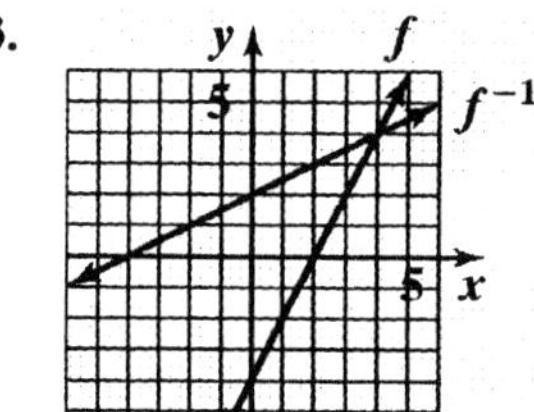

Domain of f: $(-\infty, \infty)$

Range of f: $(-\infty, \infty)$

Domain of f^{-1}: $(-\infty, \infty)$

Range of f^{-1}: $(-\infty, \infty)$

14.

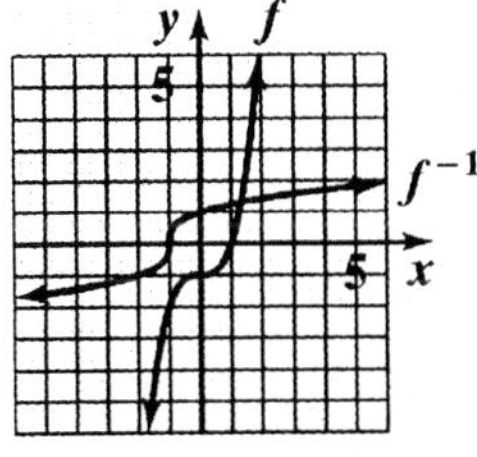

Domain of f: $(-\infty, \infty)$

Range of f: $(-\infty, \infty)$

Domain of f^{-1}: $(-\infty, \infty)$

Range of f^{-1}: $(-\infty, \infty)$

15.

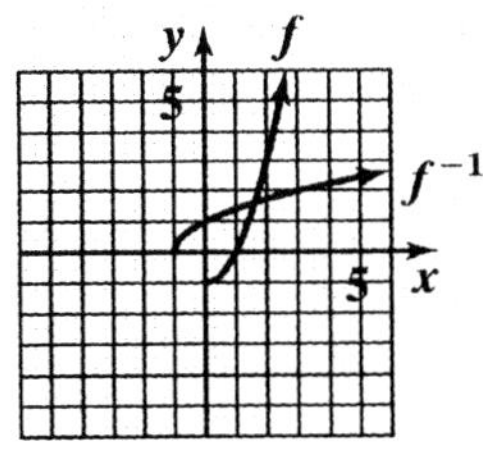

Domain of f: $[0, \infty)$

Range of f: $[-1, \infty)$

Domain of f^{-1}: $[-1, \infty)$

Range of f^{-1}: $[0, \infty)$

16.
$$\begin{aligned} f(x) &= x^2 - x - 4 \\ f(x-1) &= (x-1)^2 - (x-1) - 4 \\ &= x^2 - 2x + 1 - x + 1 - 4 \\ &= x^2 - 3x - 2 \end{aligned}$$

17.
$$\begin{aligned} &\frac{f(x+h) - f(x)}{h} \\ &= \frac{(x+h)^2 - (x+h) - 4 - \left(x^2 - x - 4\right)}{h} \\ &= \frac{x^2 + 2xh + h^2 - x - h - 4 - x^2 + x + 4}{h} \\ &= \frac{2xh + h^2 - h}{h} \\ &= \frac{h\left(2x + h - 1\right)}{h} \\ &= 2x + h - 1 \end{aligned}$$

18.
$$\begin{aligned} (g - f)(x) &= 2x - 6 - \left(x^2 - x - 4\right) \\ &= 2x - 6 - x^2 + x + 4 \\ &= -x^2 + 3x - 2 \end{aligned}$$

19.
$$\left(\frac{f}{g}\right)(x) = \frac{x^2 - x - 4}{2x - 6}$$
Domain: $(-\infty, 3) \cup (3, \infty)$

20.
$$\begin{aligned} (f \circ g)(x) &= f\left(g(x)\right) \\ &= (2x-6)^2 - (2x-6) - 4 \\ &= 4x^2 - 24x + 36 - 2x + 6 - 4 \\ &= 4x^2 - 26x + 38 \end{aligned}$$

21.
$$\begin{aligned} (g \circ f)(x) &= g\left(f(x)\right) \\ &= 2\left(x^2 - x - 4\right) - 6 \\ &= 2x^2 - 2x - 8 - 6 \\ &= 2x^2 - 2x - 14 \end{aligned}$$

22.
$$\begin{aligned} g\left(f(-1)\right) &= 2\left((-1)^2 - (-1) - 4\right) - 6 \\ &= 2\left(1 + 1 - 4\right) - 6 \\ &= 2\left(-2\right) - 6 \\ &= -4 - 6 \\ &= -10 \end{aligned}$$

23.
$$\begin{aligned} f(x) &= x^2 - x - 4 \\ f(-x) &= (-x)^2 - (-x) - 4 \\ &= x^2 + x - 4 \end{aligned}$$
f is neither even nor odd.

24. $m = \dfrac{-8-1}{-1-2} = \dfrac{-9}{-3} = 3$

point-slope form: $y - 1 = 3(x - 2)$
or $y + 8 = 3(x + 1)$
slope-intercept form: $y = 3x - 5$

25. $y = -\dfrac{1}{4}x + 5$ so $m = 4$

point-slope form: $y - 6 = 4(x + 4)$
slope-intercept form: $y = 4x + 22$

26. Write $4x+2y-5=0$ in slope intercept form.

$$4x+2y-5=0$$
$$2y=-4x+5$$
$$y=-2x+\frac{5}{2}$$

The slope of the parallel line is –2, thus the slope of the desired line is $m=-2$.

$$y-y_1=m(x-x_1)$$
$$y-(-10)=-2(x-(-7))$$
$$y+10=-2(x+7)$$
$$y+10=-2x-14$$
$$2x+y+24=0$$

27. **a.** (2,4.85) and (5,4.49)

First, find the slope using the points (2,4.85) and (5,4.49)

$$m=\frac{4.49-4.85}{5-2}=\frac{-0.36}{3}=-0.12$$

Then use the slope and one of the points to write the equation in point-slope form.

$$y-y_1=m(x-x_1)$$
$$y-4.85=-0.12(x-2)$$

or

$$y-4.49=-0.12(x-5)$$

b. Solve for y to obtain slope-intercept form.

$$y-4.85=-0.12(x-2)$$
$$y-4.85=-0.12x+0.24$$
$$y=-0.12x+5.09$$
$$f(x)=-0.12x+5.09$$

c. To predict the minimum hourly inflation-adjusted wages in 2007, let x = 2007 – 1997 = 10.

$$f(10)=-0.12(10)+5.09=3.89$$

The linear function predicts the minimum hourly inflation-adjusted wage in 2007 to \$3.89.

28.

$$\frac{\left[3(10)^2-5\right]-[3(6)^2-5]}{10-6}$$
$$=\frac{205-103}{4}$$
$$=\frac{192}{4}$$
$$=48$$

29. $g(-1)=3-(-1)=4$

$g(7)=\sqrt{7-3}=\sqrt{4}=2$

30. The denominator is zero when $x=1$ or $x=-5$.

Domain: $(-\infty,-5)\cup(-5,1)\cup(1,\infty)$

31. The expressions under each radical must not be negative.

$x+5\ge 0$ and $x-1\ge 0$

$x\ge -5$ $\quad x\ge 1$

Domain: $[1,\infty)$

32.

$$(f\circ g)(x)=\frac{7}{\frac{2}{x}-4}=\frac{7x}{2-4x}$$

$x\ne 0,\quad 2-4x\ne 0$

$$x\ne\frac{1}{2}$$

Domain: $(-\infty,0)\cup\left(0,\frac{1}{2}\right)\cup\left(\frac{1}{2},\infty\right)$

33. $f(x)=x^7 \qquad g(x)=2x+3$

34.

$$d=\sqrt{(x_2-x_1)^2+(y_2-y_1)^2}$$
$$d=\sqrt{(x_2-x_1)^2+(y_2-y_1)^2}$$
$$=\sqrt{(5-2)^2+(2-(-2))^2}$$
$$=\sqrt{3^2+4^2}$$
$$=\sqrt{9+16}$$
$$=\sqrt{25}$$
$$=5$$

$$\left(\frac{x_1+x_2}{2},\frac{y_1+y_2}{2}\right)=\left(\frac{2+5}{2},\frac{-2+2}{2}\right)$$
$$=\left(\frac{7}{2},0\right)$$

The length is 5 and the midpoint is $\left(\frac{7}{2},0\right)$.

35. **a.** $T(x)=41.78-0.19x$

b.

$$37.22=41.78-0.19x$$
$$-4.56=-0.19x$$
$$24=x$$

24 years after 1908, in 2004, the winning time will be 37.22 seconds.

36. **a.**
$$\begin{aligned} Y(x) &= 50 - 1.5(x - 30) \\ &= 50 - 1.5x + 45 \\ &= 95 - 1.5x \end{aligned}$$

b.
$$\begin{aligned} T(x) &= x(95 - 1.5x) \\ &= -1.5x^2 + 95x \end{aligned}$$

37. $2l + 2w - 600$
$$\begin{aligned} 2l &= 600 - 2w \\ l &= 300 - w \end{aligned}$$
Let $x = w$

$$\begin{aligned} A(x) &= x(300 - x) \\ &= -x^2 + 300x \end{aligned}$$

38.
$$\begin{aligned} V &= lwh \\ 8000 &= x \cdot x \cdot h \\ \frac{8000}{x^2} &= h \end{aligned}$$

$$\begin{aligned} A(x) &= 2x^2 + 4x\left(\frac{8000}{x^2}\right) \\ &= 2x^2 + \frac{3200}{x} \end{aligned}$$

Chapter 2

Section 2.1

Check Point Exercises

1. **a.** $(5-2i)+(3+3i)$
$=5-2i+3+3i$
$=(5+3)+(-2+3)i$
$=8+i$

b. $(2+6i)-(12-i)$
$=2+6i-12+i$
$=(2-12)+(6+1)i$
$=-10+7i$

2. **a.** $7i(2-9i)=7i(2)-7i(9i$
$=14i-63i^2$
$=14i-63(-1)$
$=63+14i$

b. $(5+4i)(6-7i)=30-35i+24i-28i^2$
$=30-35i+24i-28(-1)$
$=30+28-35i+24i$
$=58-11i$

3. $\frac{5+4i}{4-i}=\frac{5+4i}{4-i}\cdot\frac{4+i}{4+i}$
$=\frac{20+5i+16i+4i^2}{16+4i-4i-i^2}$
$=\frac{20+21i-4}{16+1}$
$=\frac{16+21i}{17}$
$=\frac{16}{17}+\frac{21}{17}i$

4. **a.** $\sqrt{-27}+\sqrt{-48}=i\sqrt{27}+i\sqrt{48}$
$=i\sqrt{9\cdot 3}+i\sqrt{16\cdot 3}$
$=3i\sqrt{3}+4i\sqrt{3}$
$=7i\sqrt{3}$

b. $(-2+\sqrt{-3})^2=(-2+i\sqrt{3})^2$
$=(-2)^2+2(-2)(i\sqrt{3})+(i\sqrt{3})^2$
$=4-4i\sqrt{3}+3i^2$
$=4-4i\sqrt{3}+3(-1)$
$=1-4i\sqrt{3}$

c. $\frac{-14+\sqrt{-12}}{2}=\frac{-14+i\sqrt{12}}{2}$
$=\frac{-14+2i\sqrt{3}}{2}$
$=\frac{-14}{2}+\frac{2i\sqrt{3}}{2}$
$=-7+i\sqrt{3}$

5. $x^2-2x+2=0$
$a=1, b=-2, c=2$
$x=\frac{-b\pm\sqrt{b^2-4ac}}{2a}$
$x=\frac{-(-2)\pm\sqrt{(-2)^2-4(1)(2)}}{2(1)}$
$x=\frac{2\pm\sqrt{4-8}}{2}$
$x=\frac{2\pm\sqrt{-4}}{2}$
$x=\frac{2\pm 2i}{2}$
$x=1\pm i$
The solution set is $\{1+i, 1-i\}$.

Exercise Set 2.1

1. $(7+2i)+(1-4i)=7+2i+1-4i$
$=7+1+2i-4i$
$=8-2i$

2. $(-2+6i)+(4-i)$
$=-2+6i+4-i$
$=-2+4+6i-i$
$=2+5i$

3. $(3+2i)-(5-7i)=3-5+2i+7i$
$=3+2i-5+7i$
$=-2+9i$

4. $(-7+5i)-(-9-11i) = -7+5i+9+11i$
$= -7+9+5i+11i$
$= 2+16i$

5. $6-(-5+4i)-(-13-i) = 6+5-4i+13+i$
$= 24-3i$

6. $7-(-9+2i)-(-17-i) = 7+9-2i+17+i$
$= 33-i$

7. $8i-(14-9i) = 8i-14+9i$
$= -14+8i+9i$
$= -14+17i$

8. $15i-(12-11i) = 15i-12+11i$
$= -12+15i+11i$
$= -12+26i$

9. $-3i(7i-5) = -21i^2+15i$
$= -21(-1)+15i$
$= 21+15i$

10. $-8i(2i-7) = -16i^2+56i = -16(-1)+56i$
$= 9-25i^2 = 9+25 = 34 = 16+56i$

11. $(-5+4i)(3+i) = -15-5i+12i+4i^2$
$= -15+7i-4$
$= -19+7i$

12. $(-4-8i)(3+i) = -12-4i-24i-8i^2$
$= -12-28i+8$
$= -4-28i$

13. $(7-5i)(-2-3i) = -14-21i+10i+15i^2$
$= -14-15-11i$
$= -29-11i$

14. $(8-4i)(-3+9i) = -24+72i+12i-36i^2$
$= -24+36+84i$
$= 12+84i$

15. $(3+5i)(3-5i)$

16. $(2+7i)(2-7i) = 4-49i^2 = 4+49 = 53$

17. $(-5+i)(-5-i) = 25+5i-5i-i^2$
$= 25+1$
$= 26$

18. $(-7+i)(-7-i) = 49+7i-7i-i^2$
$= 49+1$
$= 50$

19. $(2+3i)^2 = 4+12i+9i^2$
$= 4+12i-9$
$= -5+12i$

20. $(5-2i)^2 = 25-20i+4i^2$
$= 25-20i-4$
$= 21-20i$

21.
$$\frac{2}{3-i} = \frac{2}{3-i}\cdot\frac{3+i}{3+i} = \frac{2(3+i)}{9+1} = \frac{2(3+i)}{10} = \frac{3+i}{5} = \frac{3}{5}+\frac{1}{5}i$$

22.
$$\frac{3}{4+i} = \frac{3}{4+i}\cdot\frac{4-i}{4-i} = \frac{3(4-i)}{16-i^2} = \frac{3(4-i)}{17} = \frac{12}{17}-\frac{3}{17}i$$

23.
$$\frac{2i}{1+i} = \frac{2i}{1+i}\cdot\frac{1-i}{1-i} = \frac{2i-2i^2}{1+1} = \frac{2+2i}{2} = 1+i$$

24.
$$\frac{5i}{2-i} = \frac{5i}{2-i}\cdot\frac{2+i}{2+i} = \frac{10i+5i^2}{4+1} = \frac{-5+10i}{5} = -1+2i$$

25. $\frac{8i}{4-3i}=\frac{8i}{4-3i}\cdot\frac{4+3i}{4+3i}$
$=\frac{32i+24i^2}{16+9}$
$=\frac{-24+32i}{25}$
$=-\frac{24}{25}+\frac{32}{25}i$

26. $\frac{-6i}{3+2i}=\frac{-6i}{3+2i}\cdot\frac{3-2i}{3-2i}=\frac{-18i+12i^2}{9+4}$
$=\frac{-12-18i}{13}=-\frac{12}{13}-\frac{18}{13}i$

27. $\frac{2+3i}{2+i}=\frac{2+3i}{2+i}\cdot\frac{2-i}{2-i}$
$=\frac{4+4i-3i^2}{4+1}$
$=\frac{7+4i}{5}$
$=\frac{7}{5}+\frac{4}{5}i$

28. $\frac{3-4i}{4+3i}=\frac{3-4i}{4+3i}\cdot\frac{4-3i}{4-3i}$
$=\frac{12-25i+12i^2}{16+9}$
$=\frac{-25i}{25}$
$=-i$

29. $\sqrt{-64}-\sqrt{-25}=i\sqrt{64}-i\sqrt{25}$
$=8i-5i=3i$

30. $\sqrt{-81}-\sqrt{-144}=i\sqrt{81}-i\sqrt{144}=9i-12i$
$=-3i$

31. $5\sqrt{-16}+3\sqrt{-81}=5(4i)+3(9i)$
$=20i+27i=47i$

32. $5\sqrt{-8}+3\sqrt{-18}$
$=5i\sqrt{8}+3i\sqrt{18}=5i\sqrt{4\cdot 2}+3i\sqrt{9\cdot 2}$
$=10i\sqrt{2}+9i\sqrt{2}$
$=19i\sqrt{2}$

33. $\left(-2+\sqrt{-4}\right)^2=(-2+2i)^2$
$=4-8i+4i^2$
$=4-8i-4$
$=-8i$

34. $\left(-5-\sqrt{-9}\right)^2=(-5-i\sqrt{9})^2=(-5-3i)^2$
$=25+30i+9i^2$
$=25+30i-9$
$=16+30i$

35. $\left(-3-\sqrt{-7}\right)^2=\left(-3-i\sqrt{7}\right)^2$
$=9+6i\sqrt{7}+i^2(7)$
$=9-7+6i\sqrt{7}$
$=2+6i\sqrt{7}$

36. $\left(-2+\sqrt{-11}\right)^2=\left(-2+i\sqrt{11}\right)^2$
$=4-4i\sqrt{11}+i^2(11)$
$=4-11-4i\sqrt{11}$
$=-7-4i\sqrt{11}$

37. $\frac{-8+\sqrt{-32}}{24}=\frac{-8+i\sqrt{32}}{24}$
$=\frac{-8+i\sqrt{16\cdot 2}}{24}$
$=\frac{-8+4i\sqrt{2}}{24}$
$=-\frac{1}{3}+\frac{\sqrt{2}}{6}i$

38. $\frac{-12+\sqrt{-28}}{32}=\frac{-12+i\sqrt{28}}{32}=\frac{-12+i\sqrt{4\cdot 7}}{32}$
$=\frac{-12+2i\sqrt{7}}{32}=-\frac{3}{8}+\frac{\sqrt{7}}{16}i$

39. $\frac{-6-\sqrt{-12}}{48}=\frac{-6-i\sqrt{12}}{48}$
$=\frac{-6-i\sqrt{4\cdot 3}}{48}$
$=\frac{-6-2i\sqrt{3}}{48}$
$=-\frac{1}{8}-\frac{\sqrt{3}}{24}i$

40. $\frac{-15-\sqrt{-18}}{33}=\frac{-15-i\sqrt{18}}{33}=\frac{-15-i\sqrt{9\cdot 2}}{33}$
$=\frac{-15-3i\sqrt{2}}{33}=-\frac{5}{11}-\frac{\sqrt{2}}{11}i$

41. $\sqrt{-8}\left(\sqrt{-3}-\sqrt{5}\right)=i\sqrt{8}(i\sqrt{3}-\sqrt{5})$
$=2i\sqrt{2}\left(i\sqrt{3}-\sqrt{5}\right)$
$=-2\sqrt{6}-2i\sqrt{10}$

42. $\sqrt{-12}\left(\sqrt{-4}-\sqrt{2}\right)=i\sqrt{12}(i\sqrt{4}-\sqrt{2})$
$=2i\sqrt{3}\left(2i-\sqrt{2}\right)$
$=4i^2\sqrt{3}-2i\sqrt{6}$
$=-4\sqrt{3}-2i\sqrt{6}$

43. $\left(3\sqrt{-5}\right)\left(-4\sqrt{-12}\right)=\left(3i\sqrt{5}\right)\left(-8i\sqrt{3}\right)$
$=-24i^2\sqrt{15}$
$=24\sqrt{15}$

44. $\left(3\sqrt{-7}\right)\left(2\sqrt{-8}\right)$
$=(3i\sqrt{7})(2i\sqrt{8})=(3i\sqrt{7})(2i\sqrt{4\cdot 2})$
$=\left(3i\sqrt{7}\right)\left(4i\sqrt{2}\right)=12i^2\sqrt{14}=-12\sqrt{14}$

45. $x^2-6x+10=0$
$x=\frac{6\pm\sqrt{(-6)^2-4(1)(10)}}{2(1)}$
$x=\frac{6\pm\sqrt{36-40}}{2}$
$x=\frac{6\pm\sqrt{-4}}{2}$
$x=\frac{6\pm 2i}{2}$
$x=3\pm i$

The solution set is $\{3+i, 3-i\}$.

46. $x^2-2x+17=0$
$x=\frac{2\pm\sqrt{(-2)^2-4(1)(17)}}{2(1)}$
$x=\frac{2\pm\sqrt{4-68}}{2}$
$x=\frac{2\pm\sqrt{-64}}{2}$
$x=\frac{2\pm 8i}{2}$
$x=1\pm 4i$

The solution set is $\{1+4i, 1-4i\}$.

47. $4x^2+8x+13=0$
$x=\frac{-8\pm\sqrt{8^2-4(4)(13)}}{2(4)}$
$=\frac{-8\pm\sqrt{64-208}}{8}$
$=\frac{-8\pm\sqrt{-144}}{8}$
$=\frac{-8\pm 12i}{8}$
$=\frac{4(-2\pm 3i)}{8}$
$=\frac{-2\pm 3i}{2}$
$=-1\pm\frac{3}{2}i$

The solution set is $\left\{-1+\frac{3}{2}i, -1-\frac{3}{2}i\right\}$.

48. $2x^2+2x+3=0$

$$x=\frac{-(2)\pm\sqrt{(2)^2-4(2)(3)}}{2(2)}$$
$$=\frac{-2\pm\sqrt{4-24}}{4}$$
$$=\frac{-2\pm\sqrt{-20}}{4}$$
$$=\frac{-2\pm2i\sqrt{5}}{4}$$
$$=\frac{2(-1\pm i\sqrt{5})}{4}$$
$$=\frac{-1\pm i\sqrt{5}}{2}$$
$$=-\frac{1}{2}\pm\frac{\sqrt{5}}{2}i$$

The solution set is $\left\{-\frac{1}{2}+\frac{\sqrt{5}}{2}i,-\frac{1}{2}-\frac{\sqrt{5}}{2}i\right\}$.

49. $3x^2-8x+7=0$

$$x=\frac{-(-8)\pm\sqrt{(-8)^2-4(3)(7)}}{2(3)}$$
$$=\frac{8\pm\sqrt{64-84}}{6}$$
$$=\frac{8\pm\sqrt{-20}}{6}$$
$$=\frac{8\pm2i\sqrt{5}}{6}$$
$$=\frac{2(4\pm i\sqrt{5})}{6}$$
$$=\frac{4\pm i\sqrt{5}}{3}$$
$$=\frac{4}{3}\pm\frac{\sqrt{5}}{3}i$$

The solution set is $\left\{\frac{4}{3}+\frac{\sqrt{5}}{3}i,\frac{4}{3}-\frac{\sqrt{5}}{3}i\right\}$.

50. $3x^2-4x+6=0$

$$x=\frac{-(-4)\pm\sqrt{(-4)^2-4(3)(6)}}{2(3)}$$
$$=\frac{4\pm\sqrt{16-72}}{6}$$
$$=\frac{4\pm\sqrt{-56}}{6}$$
$$=\frac{4\pm2i\sqrt{14}}{6}$$
$$=\frac{2(2\pm i\sqrt{14})}{6}$$
$$=\frac{2\pm i\sqrt{14}}{3}$$
$$=\frac{2}{3}\pm\frac{\sqrt{14}}{3}i$$

The solution set is $\left\{\frac{2}{3}+\frac{\sqrt{14}}{3}i,\frac{2}{3}-\frac{\sqrt{14}}{3}i\right\}$.

51. $(2-3i)(1-i)-(3-i)(3+i)$

$$=(2-2i-3i+3i^2)-(3^2-i^2)$$
$$=2-5i+3i^2-9+i^2$$
$$=-7-5i+4i^2$$
$$=-7-5i+4(-1)$$
$$=-11-5i$$

52. $(8+9i)(2-i)-(1-i)(1+i)$

$$=(16-8i+18i-9i^2)-(1^2-i^2)$$
$$=16+10i-9i^2-1+i^2$$
$$=15+10i-8i^2$$
$$=15+10i-8(-1)$$
$$=23+10i$$

53. $(2+i)^2-(3-i)^2$

$$=(4+4i+i^2)-(9-6i+i^2)$$
$$=4+4i+i^2-9+6i-i^2$$
$$=-5+10i$$

54. $(4-i)^2-(1+2i)^2$
$=(16-8i+i^2)-(1+4i+4i^2)$
$=16-8i+i^2-1-4i-4i^2$
$=15-12i-3i^2$
$=15-12i-3(-1)$
$=18-12i$

55. $5\sqrt{-16}+3\sqrt{-81}$
$=5\sqrt{16}\sqrt{-1}+3\sqrt{81}\sqrt{-1}$
$=5\cdot 4i+3\cdot 9i$
$=20i+27i$
$=47i$ or $0+47i$

56. $5\sqrt{-8}+3\sqrt{-18}$
$=5\sqrt{4}\sqrt{2}\sqrt{-1}+3\sqrt{9}\sqrt{2}\sqrt{-1}$
$=5\cdot 2\sqrt{2}\,i+3\cdot 3\sqrt{2}\,i$
$=10i\sqrt{2}+9i\sqrt{2}$
$=(10+9)i\sqrt{2}$
$=19i\sqrt{2}$ or $0+19i\sqrt{2}$

57. $f(x)=x^2-2x+2$
$f(1+i)=(1+i)^2-2(1+i)+2$
$=1+2i+i^2-2-2i+2$
$=1+i^2$
$=1-1$
$=0$

58. $f(x)=x^2-2x+5$
$f(1-2i)=(1-2i)^2-2(1-2i)+5$
$=1-4i+4i^2-2+4i+5$
$=4+4i^2$
$=4-4$
$=0$

59. $f(x)=\dfrac{x^2+19}{2-x}$
$f(3i)=\dfrac{(3i)^2+19}{2-3i}$
$=\dfrac{9i^2+19}{2-3i}$
$=\dfrac{-9+19}{2-3i}$
$=\dfrac{10}{2-3i}$
$=\dfrac{10}{2-3i}\cdot\dfrac{2+3i}{2+3i}$
$=\dfrac{20+30i}{4-9i^2}$
$=\dfrac{20+30i}{4+9}$
$=\dfrac{20+30i}{13}$
$=\dfrac{20}{13}+\dfrac{30}{13}i$

60. $f(x)=\dfrac{x^2+11}{3-x}$
$f(4i)=\dfrac{(4i)^2+11}{3-4i}=\dfrac{16i^2+11}{3-4i}$
$=\dfrac{-16+11}{3-4i}$
$=\dfrac{-5}{3-4i}$
$=\dfrac{-5}{3-4i}\cdot\dfrac{3+4i}{3+4i}$
$=\dfrac{-15-20i}{9-16i^2}$
$=\dfrac{-15-20i}{9+16}$
$=\dfrac{-15-20i}{25}$
$=\dfrac{-15}{25}-\dfrac{20}{25}i$
$=-\dfrac{3}{5}-\dfrac{4}{5}i$

61. $E=IR=(4-5i)(3+7i)$
$=12+28i-15i-35i^2$
$=12+13i-35(-1)$
$=12+35+13i=47+13i$
The voltage of the circuit is

(47 + 13i) volts.

62. $E = IR = (2-3i)(3+5i)$
$= 6+10i-9i-15i^2 = 6+i-15(-1)$
$= 6+i+15 = 21+i$
The voltage of the circuit is $(21+i)$ volts.

63. Sum:
$(5+i\sqrt{15}\,)+(5-i\sqrt{15}\,)$
$= 5+i\sqrt{15}\,+5-i\sqrt{15}$
$= 5+5$
$= 10$
Product:
$(5+i\sqrt{15}\,)(5-i\sqrt{15})$
$= 25-5i\sqrt{15}+5i\sqrt{15}-15i^2$
$= 25+15$
$= 40$

73. **a.** False; all irrational numbers are complex numbers.

b. False; (3 + 7i)(3 – 7i) = 9 + 49 = 58 is a real number.

c. False; $\frac{7+3i}{5+3i} = \frac{7+3i}{5+3i}\cdot\frac{5-3i}{5-3i}$
$= \frac{44-6i}{34} = \frac{22}{17} - \frac{3}{17}i$

d. True;
$(x+yi)(x-yi) = x^2-(yi)^2 = x^2+y^2$

(d) is true.

74. $\frac{4}{(2+i)(3-i)} = \frac{4}{6-2i+3i-i^2}$
$= \frac{4}{6+i+1}$
$= \frac{4}{7+i}$
$= \frac{4}{7+i}\cdot\frac{7-i}{7-i}$
$= \frac{28-4i}{49-i^2}$
$= \frac{28-4i}{49+1}$
$= \frac{28-4i}{50}$
$= \frac{28}{50}-\frac{4}{50}i$
$= \frac{14}{25}-\frac{2}{25}i$

75. $\frac{1+i}{1+2i}+\frac{1-i}{1-2i}$
$= \frac{(1+i)(1-2i)}{(1+2i)(1-2i)}+\frac{(1-i)(1+2i)}{(1+2i)(1-2i)}$
$= \frac{(1+i)(1-2i)+(1-i)(1+2i)}{(1+2i)(1-2i)}$
$= \frac{1-2i+i-2i^2+1+2i-i-2i^2}{1-4i^2}$
$= \frac{1-2i+i+2+1+2i-i+2}{1+4}$
$= \frac{6}{5}$
$= \frac{6}{5}+0i$

76. $$\frac{8}{1+\frac{2}{i}} = \frac{8}{\frac{i}{i}+\frac{2}{i}} = \frac{8}{\frac{2+i}{i}} = \frac{8i}{2+i} = \frac{8i}{2+i}\cdot\frac{2-i}{2-i} = \frac{16i-8i^2}{4-i^2} = \frac{16i+8}{4+1} = \frac{8+16i}{5} = \frac{8}{5}+\frac{16}{5}i$$

Section 2.2

Check Point Exercises

1. $f(x) = -(x-1)^2 + 4$

$$f(x) = \overset{a=-1}{-}\left(x - \overset{h=1}{1}\right)^2 + \overset{k=4}{4}$$

Step 1: The parabola opens down because $a < 0$.
Step 2: find the vertex: (1, 4)
Step 3: find the x-intercepts:

$0 = -(x-1)^2 + 4$
$(x-1)^2 = 4$
$x - 1 = \pm 2$
$x = 1 \pm 2$
$x = 3$ or $x = -1$

Step 4: find the y-intercept:
$f(0) = -(0-1)^2 + 4 = 3$
Step 5: The axis of symmetry is $x = 1$.

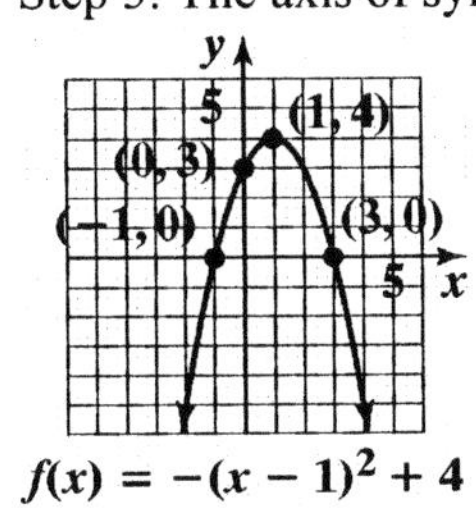

$f(x) = -(x-1)^2 + 4$

2. $f(x) = (x-2)^2 + 1$

Step 1: The parabola opens up because $a > 0$.
Step 2: find the vertex: (2, 1)
Step 3: find the x-intercepts:

$0 = (x-2)^2 + 1$
$(x-2)^2 = -1$
$x - 2 = \sqrt{-1}$
$x = 2 \pm i$

The equation has no real roots, thus the parabola has no x-intercepts.
Step 4: find the y-intercept:
$f(0) = (0-2)^2 + 1 = 5$
Step 5: The axis of symmetry is $x = 2$.

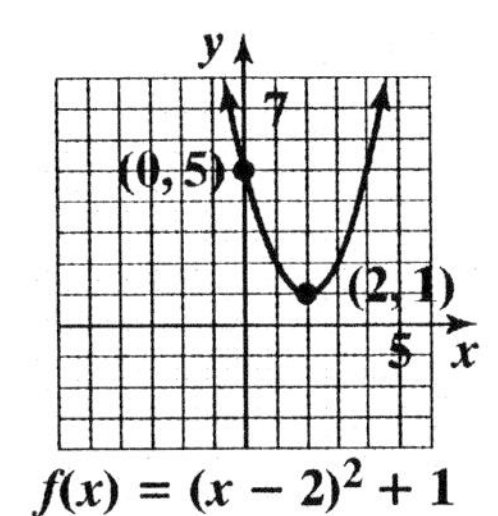

$f(x) = (x-2)^2 + 1$

3. $f(x) = -x^2 + 4x + 1$

Step 1: The parabola opens down because $a < 0$.
Step 2: find the vertex:

$x = -\frac{b}{2a} = -\frac{4}{2(-1)} = 2$

$f(2) = -2^2 + 4(2) + 1 = 5$

The vertex is (2, 5).
Step 3: find the x-intercepts:

$0 = -x^2 + 4x + 1$

$x = \frac{-b \pm \sqrt{b^2 - 4ac}}{2a}$

$x = \frac{-4 \pm \sqrt{4^2 - 4(-1)(1)}}{2(-1)}$

$x = \frac{-4 \pm \sqrt{20}}{-2}$

$x = 2 \pm \sqrt{5}$

The x-intercepts are $x \approx -0.2$ and $x \approx -4.2$.
Step 4: find the y-intercept:
$f(0) = -0^2 + 4(0) + 1 = 1$
Step 5: The axis of symmetry is $x = 2$.

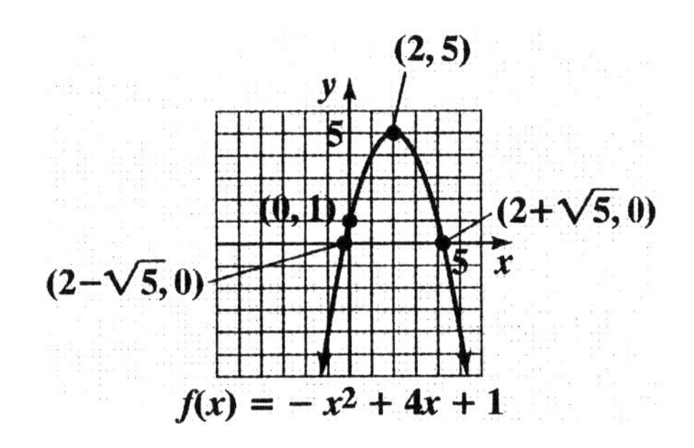

4. $f(x) = 4x^2 - 16x + 1000$
   a. $a = 4$. The parabola opens upward and has a minimum value.
   b. $x = \frac{-b}{2a} = \frac{16}{8} = 2$
      $f(2) = 4(2)^2 - 16(2) + 1000 = 984$
      The minimum point is 984 at $x = 2$.
   c. Domain: $(-\infty, \infty)$ Range: $[984, \infty)$

5. $f(x) = 0.4x^2 - 36x + 1000$
   Because $a > 0$, the function has a minimum value.
   $x = -\frac{b}{2a} = -\frac{-36}{2(0.4)} = 45$
   $f(45) = 0.4(45)^2 - 36(45) + 1000 = 190$
   The age of a driver having the least number of car accidents is 45. The minimum number of accidents per 50 million miles driven is 190.

6. Let x = one of the numbers;
   $x - 8$ = the other number.
   The product is $f(x) = x(x-8) = x^2 - 8x$
   The x-coordinate of the minimum is
   $x = -\frac{b}{2a} = -\frac{-8}{2(1)} = -\frac{-8}{2} = 4.$
   $f(4) = (4)^2 - 8(4)$
   $= 16 - 32 = -16$
   The vertex is $(4, -16)$.
   The minimum product is -16. This occurs when the two number are 4 and $4 - 8 = -4$.

7. Maximize the area of a rectangle constructed with 120 feet of fencing.
   Let x = the length of the rectangle. Let y = the width of the rectangle.
   Since we need an equation in one variable, use the perimeter to express y in terms of x.
   $2x + 2y = 120$
   $2y = 120 - 2x$
   $y = \frac{120 - 2x}{2} = 60 - x$
   We need to maximize $A = xy = x(60 - x)$.
   Rewrite A as a function of x.
   $A(x) = x(60 - x) = -x^2 + 60x$

   Since $a = -1$ is negative, we know the function opens downward and has a maximum at
   $x = -\frac{b}{2a} = -\frac{60}{2(-1)} = -\frac{60}{-2} = 30.$
   When the length x is 30, the width y is
   $y = 60 - x = 60 - 30 = 30.$
   The dimensions of the rectangular region with maximum area are 30 feet by 30 feet. This gives an area of $30 \cdot 30 = 900$ square feet.

Exercise Set 2.2

1. vertex: (1, 1)
   $h(x) = (x-1)^2 + 1$

2. vertex: (–1, 1)
   $g(x) = (x+1)^2 + 1$

3. vertex: (1, –1)
   $j(x) = (x-1)^2 - 1$

4. vertex: (–1, –1)
   $f(x) = (x+1)^2 - 1$

5. The graph is $f(x) = x^2$ translated down one.
   $h(x) = x^2 - 1$

6. The point (–1, 0) is on the graph and $f(-1) = 0$. $f(x) = x^2 + 2x + 1$

7. The point (1, 0) is on the graph and $g(1) = 0$. $g(x) = x^2 - 2x + 1$

8. The graph is $f(x) = -x^2$ translated down one.
   $j(x) = -x^2 - 1$

9. $f(x) = 2(x-3)^2 + 1$
   $h = 3, k = 1$
   The vertex is at (3, 1).

10. $f(x) = -3(x-2)^2 + 12$
$h = 2, k = 12$
The vertex is at (2, 12).

11. $f(x) = -2(x+1)^2 + 5$
$h = -1, k = 5$
The vertex is at (–1, 5).

12. $f(x) = -2(x+4)^2 - 8$
$h = -4, k = -8$
The vertex is at (–4, –8).

13. $f(x) = 2x^2 - 8x + 3$
$x = \frac{-b}{2a} = \frac{8}{4} = 2$
$f(2) = 2(2)^2 - 8(2) + 3$
$= 8 - 16 + 3 = -5$
The vertex is at (2, –5).

14. $f(x) = 3x^2 - 12x + 1$
$x = \frac{-b}{2a} = \frac{12}{6} = 2$
$f(2) = 3(2)^2 - 12(2) + 1$
$= 12 - 24 + 1 = -11$
The vertex is at (2, –11).

15. $f(x) = -x^2 - 2x + 8$
$x = \frac{-b}{2a} = \frac{2}{-2} = -1$
$f(-1) = -(-1)^2 - 2(-1) + 8$
$= -1 + 2 + 8 = 9$
The vertex is at (–1, 9).

16. $f(x) = -2x^2 + 8x - 1$
$x = \frac{-b}{2a} = \frac{-8}{-4} = 2$
$f(2) = -2(2)^2 + 8(2) - 1$
$= -8 + 16 - 1 = 7$
The vertex is at (2, 7).

17. $f(x) = (x-4)^2 - 1$
vertex: (4, –1)
x-intercepts:
$0 = (x-4)^2 - 1$
$1 = (x-4)^2$
$\pm 1 = x - 4$
$x = 3$ or $x = 5$
y-intercept:
$f(0) = (0-4)^2 - 1 = 15$

The axis of symmetry is $x = 4$.

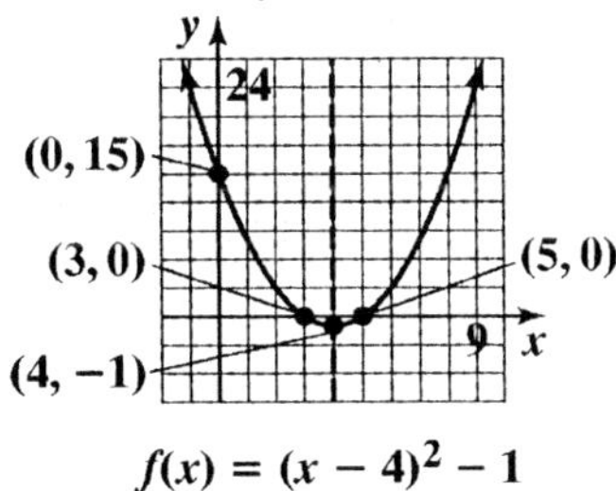

$f(x) = (x-4)^2 - 1$

Domain: $(-\infty, \infty)$
Range: $[-1, \infty)$

18. $f(x) = (x-1)^2 - 2$
vertex: (1, –2)
x-intercepts:
$0 = (x-1)^2 - 2$
$(x-1)^2 = 2$
$x - 1 = \pm\sqrt{2}$
$x = 1 \pm \sqrt{2}$
y-intercept:
$f(0) = (0-1)^2 - 2 = -1$
The axis of symmetry is $x = 1$.

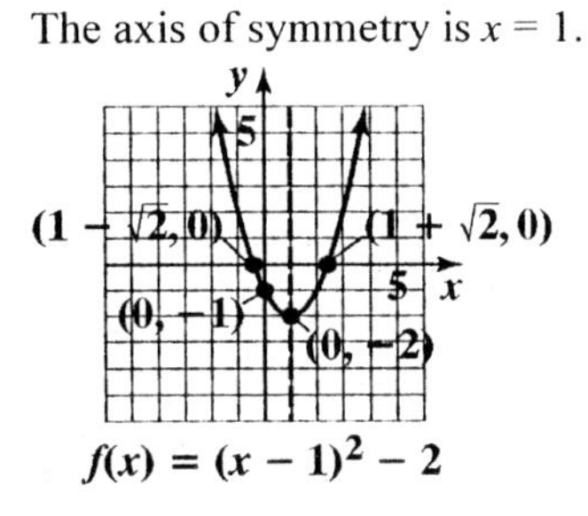

$f(x) = (x-1)^2 - 2$

Domain: $(-\infty, \infty)$
Range: $[-2, \infty)$

19. $f(x) = (x-1)^2 + 2$
vertex: (1, 2)
x-intercepts:
$0 = (x-1)^2 + 2$
$(x-1)^2 = -2$
$x - 1 = \pm\sqrt{-2}$
$x = 1 \pm i\sqrt{2}$
No x-intercepts.
y-intercept:
$f(0) = (0-1)^2 + 2 = 3$

The axis of symmetry is $x = 1$.

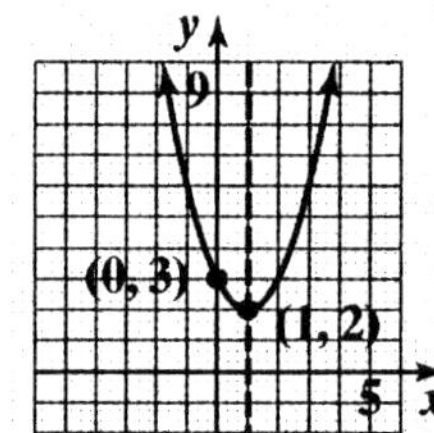

Domain: $(-\infty, \infty)$

Range: $[2, \infty)$

20. $f(x) = (x-3)^2 + 2$
vertex: (3, 2)
x-intercepts:

$0 = (x-3)^2 + 2$

$(x-3)^2 = -2$

$x - 3 = \pm i\sqrt{2}$

$x = 3 \pm i\sqrt{2}$

No x-intercepts.
y-intercept:

$f(0) = (0-3)^2 + 2 = 11$

The axis of symmetry is $x = 3$.

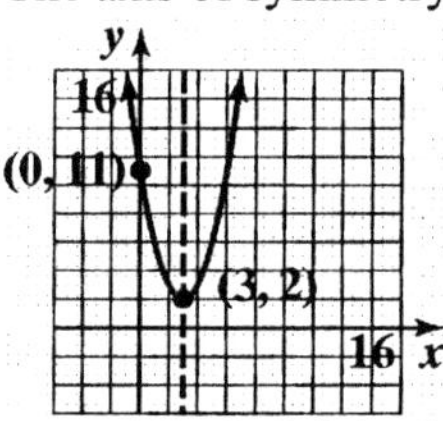

Domain: $(-\infty, \infty)$

Range: $[2, \infty)$

21. $y - 1 = (x-3)^2$

$y = (x-3)^2 + 1$

vertex: (3, 1)
x-intercepts:

$0 = (x-3)^2 + 1$

$(x-3)^2 = -1$

$x - 3 = \pm i$

$x = 3 \pm i$

No x-intercepts.
y-intercept: 10

$y = (0-3)^2 + 1 = 10$

The axis of symmetry is $x = 3$.

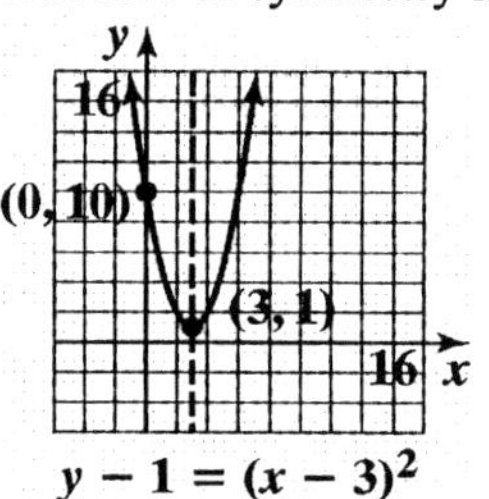

Domain: $(-\infty, \infty)$

Range: $[1, \infty)$

22. $y - 3 = (x-1)^2$

$y = (x-1)^2 + 3$

vertex: (1, 3)
x-intercepts:

$0 = (x-1)^2 + 3$

$(x-1)^2 = -3$

$x - 1 = \pm i\sqrt{3}$

$x = 1 \pm i\sqrt{3}$

No x-intercepts
y-intercept:

$y = (0-1)^2 + 3 = 4$

The axis of symmetry is $x = 1$.

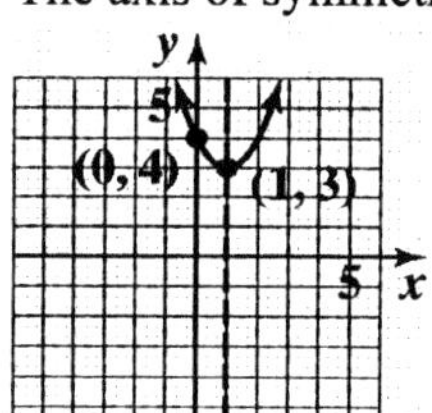

Domain: $(-\infty, \infty)$

Range: $[3, \infty)$

23. $f(x) = 2(x+2)^2 - 1$
vertex: (–2, –1)
x-intercepts:

$0 = 2(x+2)^2 - 1$

$2(x+2)^2 = 1$

$(x+2)^2 = \frac{1}{2}$

$x + 2 = \pm\frac{1}{\sqrt{2}}$

$x = -2 \pm \frac{1}{\sqrt{2}} = -2 \pm \frac{\sqrt{2}}{2}$

y-intercept:

$f(0) = 2(0+2)^2 - 1 = 7$

The axis of symmetry is $x = -2$.

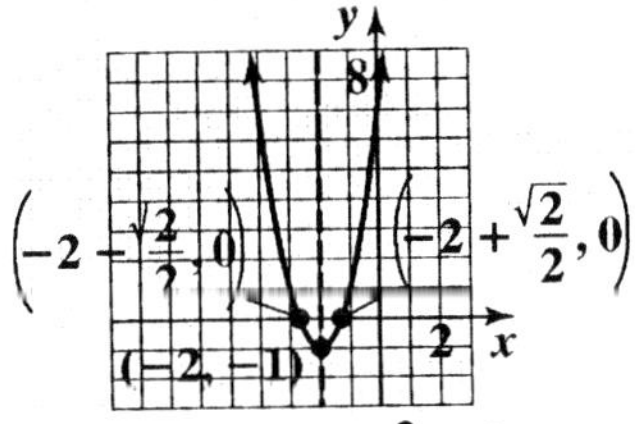

$f(x) = 2(x + 2)^2 - 1$

Domain: $(-\infty, \infty)$

Range: $[-1, \infty)$

24. $f(x) = \frac{5}{4} - \left(x - \frac{1}{2}\right)^2$

$f(x) = -\left(x - \frac{1}{2}\right)^2 + \frac{5}{4}$

vertex: $\left(\frac{1}{2}, \frac{5}{4}\right)$

x-intercepts:

$0 = -\left(x - \frac{1}{2}\right)^2 + \frac{5}{4}$

$\left(x - \frac{1}{2}\right)^2 = \frac{5}{4}$

$x - \frac{1}{2} = \pm\frac{\sqrt{5}}{2}$

$x = \frac{1 \pm \sqrt{5}}{2}$

y-intercept:

$f(0) = -\left(0 - \frac{1}{2}\right)^2 + \frac{5}{4} = 1$

The axis of symmetry is $x = \frac{1}{2}$.

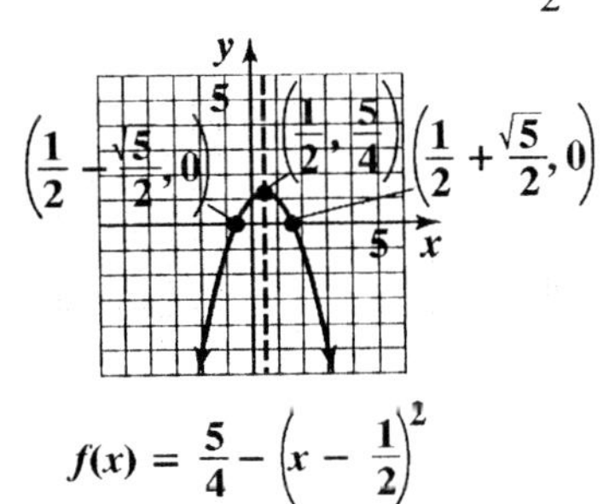

$f(x) = \frac{5}{4} - \left(x - \frac{1}{2}\right)^2$

Domain: $(-\infty, \infty)$

Range: $\left(-\infty, \frac{5}{4}\right]$

25. $f(x) = 4 - (x-1)^2$

$f(x) = -(x-1)^2 + 4$

vertex: $(1, 4)$

x-intercepts:

$0 = -(x-1)^2 + 4$

$(x-1)^2 = 4$

$x - 1 = \pm 2$

$x = -1$ or $x = 3$

y-intercept:

$f(x) = -(0-1)^2 + 4 = 3$

The axis of symmetry is $x = 1$.

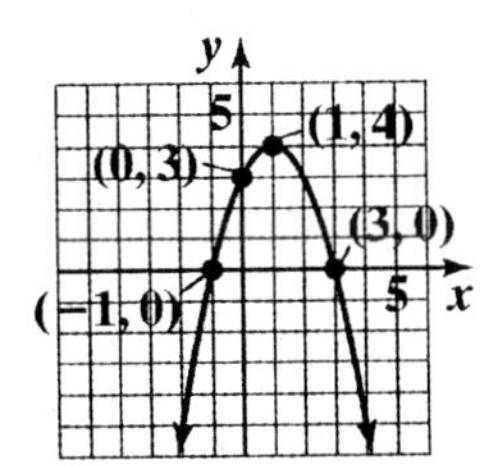

$f(x) = 4 - (x - 1)^2$

Domain: $(-\infty, \infty)$

Range: $(-\infty, 4]$

26. $f(x)=1-(x-3)^2$

$f(x)=-(x-3)^2+1$

vertex: (3, 1)

x-intercepts:

$0=-(x-3)^2+1$

$(x-3)^2=1$

$x-3=\pm 1$

$x=2$ or $x=4$

y-intercept:

$f(0)=-(0-3)^2+1=-8$

The axis of symmetry is $x = 3$.

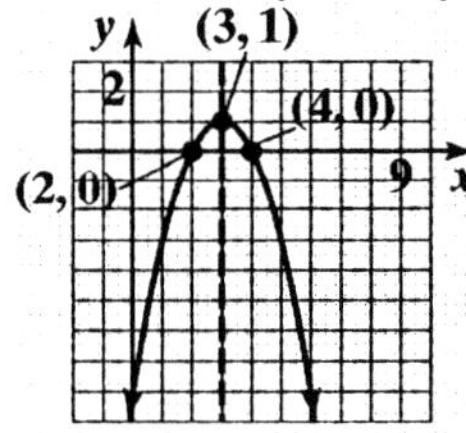

$f(x) = 1 - (x - 3)^2$

Domain: $(-\infty,\infty)$

Range: $(-\infty,1]$

27. $f(x)=x^2-2x-3$

$f(x)=(x^2-2x+1)-3-1$

$f(x)=(x-1)^2-4$

vertex: (1, –4)

x-intercepts:

$0=(x-1)^2-4$

$(x-1)^2=4$

$x-1=\pm 2$

$x=-1$ or $x=3$

y-intercept: –3

$f(0)=0^2-2(0)-3=-3$

The axis of symmetry is $x = 1$.

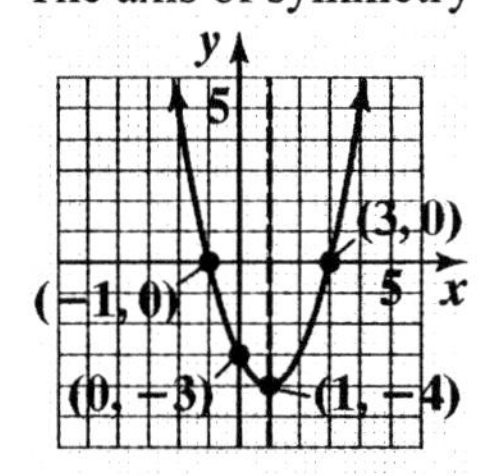

$f(x) = x^2 + 3x - 10$

Domain: $(-\infty,\infty)$

Range: $[-4,\infty)$

28. $f(x)=x^2-2x-15$

$f(x)=(x^2-2x+1)-15-1$

$f(x)=(x-1)^2-16$

vertex: (1, –16)

x-intercepts:

$0=(x-1)^2-16$

$(x-1)^2=16$

$x-1=\pm 4$

$x=-3$ or $x=5$

y-intercept:

$f(0)=0^2-2(0)-15=-15$

The axis of symmetry is $x = 1$.

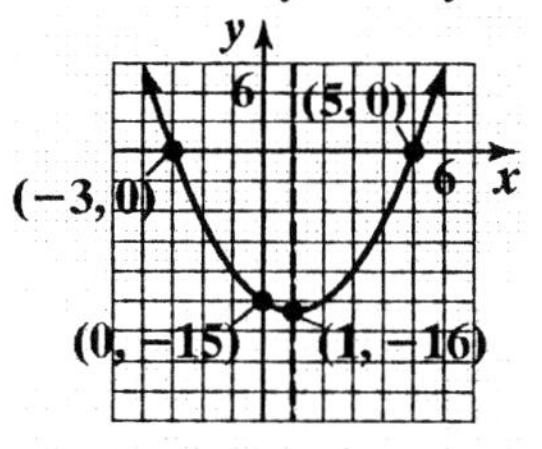

$f(x) = x^2 - 2x - 15$

Domain: $(-\infty,\infty)$

Range: $[-16,\infty)$

29. $f(x)=x^2+3x-10$

$f(x)=\left(x^2+3x+\frac{9}{4}\right)-10-\frac{9}{4}$

$f(x)=\left(x+\frac{3}{2}\right)^2-\frac{49}{4}$

vertex: $\left(-\frac{3}{2},-\frac{49}{4}\right)$

x-intercepts:

$0=\left(x+\frac{3}{2}\right)^2-\frac{49}{4}$

$\left(x+\frac{3}{2}\right)^2=\frac{49}{4}$

$x+\frac{3}{2}=\pm\frac{7}{2}$

$x=-\frac{3}{2}\pm\frac{7}{2}$

$x=2$ or $x=-5$

y-intercept:

$f(x)=0^2+3(0)-10=-10$

The axis of symmetry is $x = -\frac{3}{2}$.

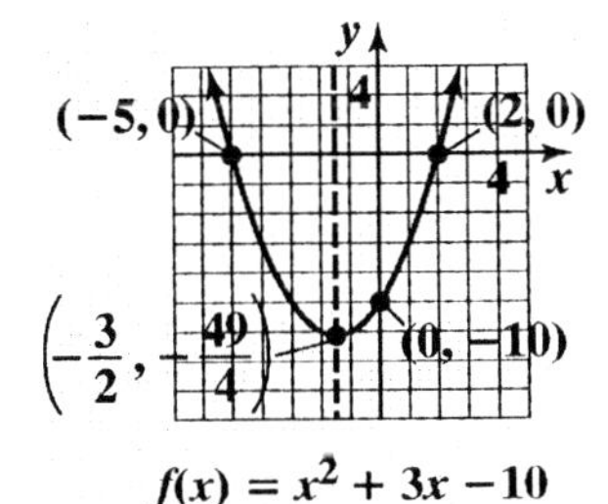

$f(x) = x^2 + 3x - 10$

Domain: $(-\infty, \infty)$

Range: $\left[-\frac{49}{4}, \infty\right)$

30. $f(x) = 2x^2 - 7x - 4$

$f(x) = 2\left(x^2 - \frac{7}{2}x + \frac{49}{16}\right) - 4 - \frac{49}{8}$

$f(x) = 2\left(x - \frac{7}{4}\right)^2 - \frac{81}{8}$

vertex: $\left(\frac{7}{4}, -\frac{81}{8}\right)$

x-intercepts:

$0 = 2\left(x - \frac{7}{4}\right)^2 - \frac{81}{8}$

$2\left(x - \frac{7}{4}\right)^2 = \frac{81}{8}$

$\left(x - \frac{7}{4}\right)^2 = \frac{81}{16}$

$x - \frac{7}{4} = \pm\frac{9}{4}$

$x = \frac{7}{4} \pm \frac{9}{4}$

$x = -\frac{1}{2}$ or $x = 4$

y-intercept:

$f(0) = 2(0)^2 - 7(0) - 4 = -4$

The axis of symmetry is $x = \frac{7}{4}$.

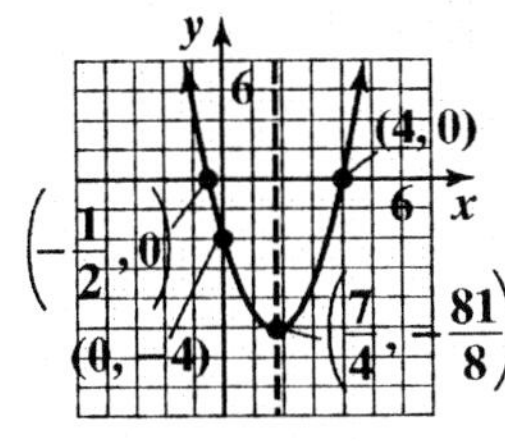

$f(x) = 2x^2 - 7x - 4$

Domain: $(-\infty, \infty)$

Range: $\left[-\frac{81}{8}, \infty\right)$

31. $f(x) = 2x - x^2 + 3$

$f(x) = -x^2 + 2x + 3$

$f(x) = -\left(x^2 - 2x + 1\right) + 3 + 1$

$f(x) = -(x-1)^2 + 4$

vertex: (1, 4)

x-intercepts:

$0 = -(x-1)^2 + 4$

$(x-1)^2 = 4$

$x - 1 = \pm 2$

$x = -1$ or $x = 3$

y-intercept:

$f(0) = 2(0) - (0)^2 + 3 = 3$

The axis of symmetry is $x = 1$.

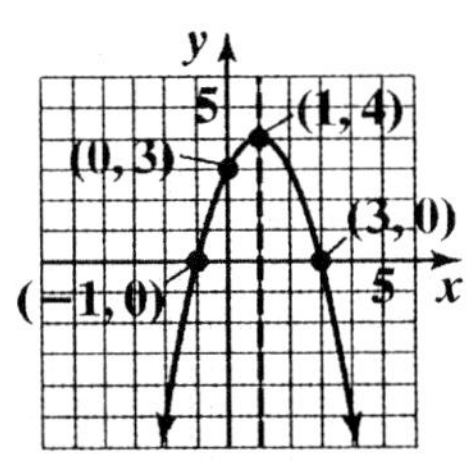

$f(x) = 2x - x^2 + 3$

Domain: $(-\infty, \infty)$

Range: $(-\infty, 4]$

32. $f(x) = 5 - 4x - x^2$

$f(x) = -x^2 - 4x + 5$

$f(x) = -\left(x^2 + 4x + 4\right) + 5 + 4$

$f(x) = -(x+2)^2 + 9$

vertex: (–2, 9)

x-intercepts:

$0 = -(x+2)^2 + 9$

$(x+2)^2 = 9$

$x + 2 = \pm 3$

$x = -5, 1$

y-intercept:

$f(0) = 5 - 4(0) - (0)^2 = 5$

The axis of symmetry is $x = -2$.

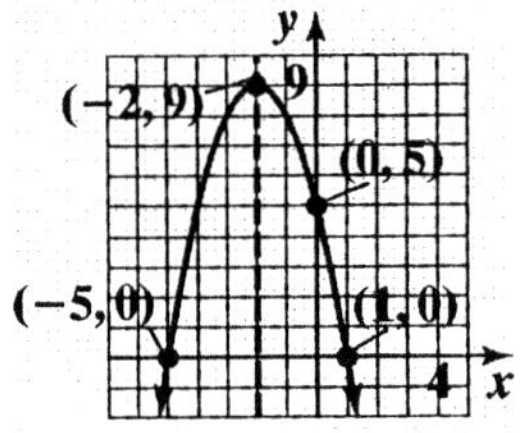

Domain: $(-\infty, \infty)$

Range: $(-\infty, 9]$

33. $f(x) = x^2 + 6x + 3$

$f(x) = (x^2 + 6x + 9) + 3 - 9$

$f(x) = (x+3)^2 - 6$

vertex: $(-3, -6)$

x-intercepts:

$0 = (x+3)^2 - 6$

$(x+3)^2 = 6$

$x + 3 = \pm\sqrt{6}$

$x = -3 \pm \sqrt{6}$

y-intercept:

$f(0) = (0)^2 + 6(0) + 3$

$f(0) = 3$

The axis of symmetry is $x = -3$.

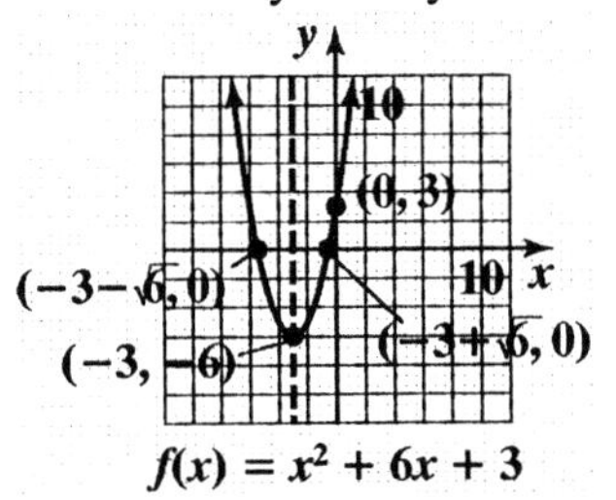

Domain: $(-\infty, \infty)$

Range: $[-6, \infty)$

34. $f(x) = x^2 + 4x - 1$

$f(x) = (x^2 + 4x + 4) - 1 - 4$

$f(x) = (x+2)^2 - 5$

vertex: $(-2, -5)$

x-intercepts:

$0 = (x+2)^2 - 5$

$(x+2)^2 = 5$

$x + 2 = \pm\sqrt{5}$

$x = -2 \pm \sqrt{5}$

y-intercept:

$f(0) = (0)^2 + 4(0) - 1$

$f(0) = -1$

The axis of symmetry is $x = -2$.

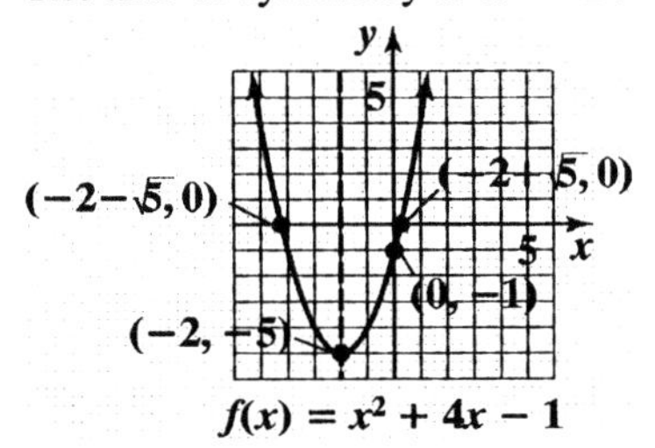

Domain: $(-\infty, \infty)$

Range: $[-5, \infty)$

35. $f(x) = 2x^2 + 4x - 3$

$f(x) = 2(x^2 + 2x \quad) - 3$

$f(x) = 2(x^2 + 2x + 1) - 3 - 2$

$f(x) = 2(x+1)^2 - 5$

vertex: $(-1, -5)$

x-intercepts:

$0 = 2(x+1)^2 - 5$

$2(x+1)^2 = 5$

$(x+1)^2 = \frac{5}{2}$

$x + 1 = \pm\sqrt{\frac{5}{2}}$

$x = -1 \pm \frac{\sqrt{10}}{2}$

y-intercept:

$f(0) = 2(0)^2 + 4(0) - 3$

$f(0) = -3$

The axis of symmetry is $x = -1$.

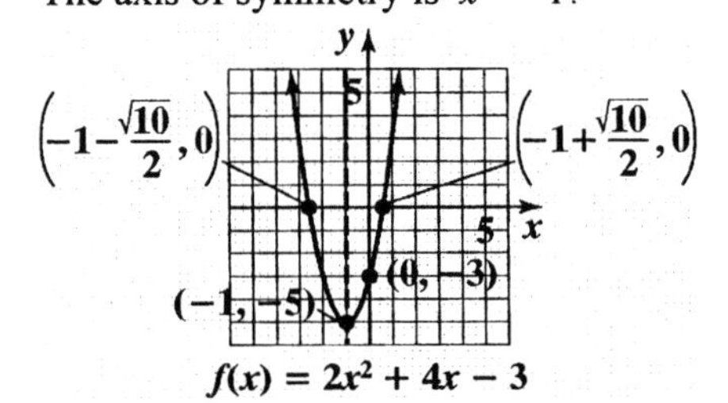

Domain: $(-\infty, \infty)$

Range: $[-5, \infty)$

36. $f(x) = 3x^2 - 2x - 4$

$f(x) = 3\left(x^2 - \frac{2}{3}x \quad\right) - 4$

$f(x) = 3\left(x^2 - \frac{2}{3}x + \frac{1}{9}\right) - 4 - \frac{1}{3}$

$f(x) = 3\left(x - \frac{1}{3}\right)^2 - \frac{13}{3}$

vertex: $\left(\frac{1}{3}, -\frac{13}{3}\right)$

x-intercepts:

$0 = 3\left(x - \frac{1}{3}\right)^2 - \frac{13}{3}$

$3\left(x - \frac{1}{3}\right)^2 = \frac{13}{3}$

$\left(x - \frac{1}{3}\right)^2 = \frac{13}{9}$

$x - \frac{1}{3} = \pm\sqrt{\frac{13}{9}}$

$x = \frac{1}{3} \pm \frac{\sqrt{13}}{3}$

y-intercept:

$f(0) = 3(0)^2 - 2(0) - 4$

$f(0) = -4$

The axis of symmetry is $x = \frac{1}{3}$.

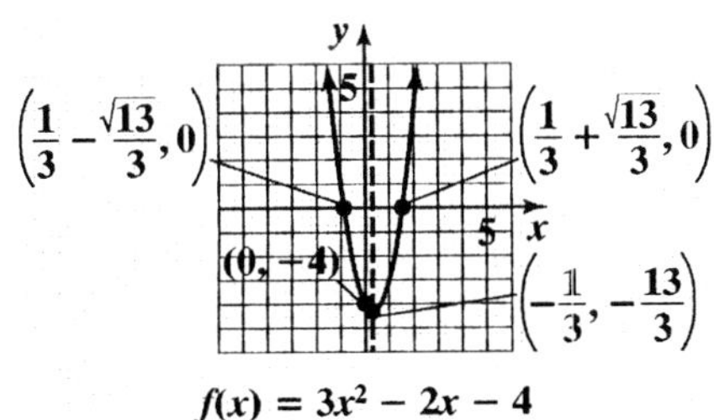

$f(x) = 3x^2 - 2x - 4$

Domain: $(-\infty, \infty)$

Range: $\left[-\frac{13}{3}, \infty\right)$

37. $f(x) = 2x - x^2 - 2$

$f(x) = -x^2 + 2x - 2$

$f(x) = -(x^2 - 2x + 1) - 2 + 1$

$f(x) = -(x-1)^2 - 1$

vertex: $(1, -1)$

x-intercepts:

$0 = -(x-1)^2 - 1$

$(x-1)^2 = -1$

$x - 1 = \pm i$

$x = 1 \pm i$

No x-intercepts.

y-intercept:

$f(0) = 2(0) - (0)^2 - 2 = -2$

The axis of symmetry is $x = 1$.

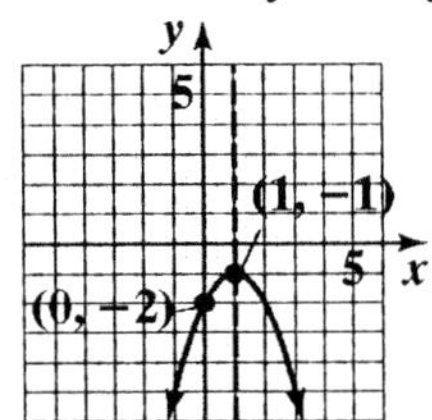

$f(x) = 2x - x^2 - 2$

Domain: $(-\infty, \infty)$

Range: $(-\infty, -1]$

38. $f(x) = 6 - 4x + x^2$

$f(x) = x^2 - 4x + 6$

$f(x) = (x^2 - 4x + 4) + 6 - 4$

$f(x) = (x-2)^2 + 2$

vertex: $(2, 2)$

x-intercepts:

$0 = (x-2)^2 + 2$

$(x-2)^2 = -2$

$x - 2 = \pm i\sqrt{2}$

$x = 2 \pm i\sqrt{2}$

No x-intercepts

y-intercept:

$f(0) = 6 - 4(0) + (0)^2 = 6$

The axis of symmetry is $x = 2$.

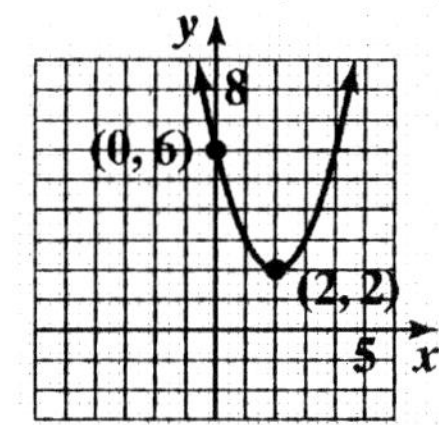

$f(x) = 6 - 4x + x^2$

Domain: $(-\infty, \infty)$

Range: $[2, \infty)$

39. $f(x) = 3x^2 - 12x - 1$

a. $a = 3$. The parabola opens upward and has a minimum value.

b. $x = \frac{-b}{2a} = \frac{12}{6} = 2$

$f(2) = 3(2)^2 - 12(2) - 1$
$= 12 - 24 - 1 = -13$

The minimum is -13 at $x = 2$.

c. Domain: $(-\infty, \infty)$ Range: $[-13, \infty)$

40. $f(x) = 2x^2 - 8x - 3$

a. $a = 2$. The parabola opens upward and has a minimum value.

b. $x = \frac{-b}{2a} = \frac{8}{4} = 2$

$f(2) = 2(2)^2 - 8(2) - 3$
$= 8 - 16 - 3 = -11$

The minimum is -11 at $x = 2$.

c. Domain: $(-\infty, \infty)$ Range: $[-11, \infty)$

41. $f(x) = -4x^2 + 8x - 3$

a. $a = -4$. The parabola opens downward and has a maximum value.

b. $x = \frac{-b}{2a} = \frac{-8}{-8} = 1$

$f(1) = -4(1)^2 + 8(1) - 3$
$= -4 + 8 - 3 = 1$

The maximum is 1 at $x = 1$.

c. Domain: $(-\infty, \infty)$ Range: $(-\infty, 1]$

42. $f(x) = -2x^2 - 12x + 3$

a. $a = -2$. The parabola opens downward and has a maximum value.

b. $x = \frac{-b}{2a} = \frac{12}{-4} = -3$

$f(-3) = -2(-3)^2 - 12(-3) + 3$
$= -18 + 36 + 3 = 21$

The maximum is 21 at $x = -3$.

c. Domain: $(-\infty, \infty)$ Range: $(-\infty, 21]$

43. $f(x) = 5x^2 - 5x$

a. $a = 5$. The parabola opens upward and has a minimum value.

b. $x = \frac{-b}{2a} = \frac{5}{10} = \frac{1}{2}$

$f\left(\frac{1}{2}\right) = 5\left(\frac{1}{2}\right)^2 - 5\left(\frac{1}{2}\right)$

$= \frac{5}{4} - \frac{5}{2} = \frac{5}{4} - \frac{10}{4} = \frac{-5}{4}$

The minimum is $\frac{-5}{4}$ at $x = \frac{1}{2}$.

c. Domain: $(-\infty, \infty)$ Range: $\left[\frac{-5}{4}, \infty\right)$

44. $f(x) = 6x^2 - 6x$

a. $a = 6$. The parabola opens upward and has minimum value.

b. $x = \frac{-b}{2a} = \frac{6}{12} = \frac{1}{2}$

$f\left(\frac{1}{2}\right) = 6\left(\frac{1}{2}\right)^2 - 6\left(\frac{1}{2}\right)$

$= \frac{6}{4} - 3 = \frac{3}{2} - \frac{6}{2} = \frac{-3}{2}$

The minimum is $\frac{-3}{2}$ at $x = \frac{1}{2}$.

c. Domain: $(-\infty, \infty)$ Range: $\left[\frac{-3}{2}, \infty\right)$

45. Since the parabola opens up, the vertex $(-1, -2)$ is a minimum point.
Domain: $(-\infty, \infty)$. Range: $[-2, \infty)$

46. Since the parabola opens down, the vertex $(-3, -4)$ is a maximum point.

Domain: $(-\infty, \infty)$. Range: $(-\infty, -4]$

47. Since the parabola has a maximum, it opens down from the vertex $(10,-6)$.
Domain: $(-\infty,\infty)$. Range: $(-\infty,-6]$

48. Since the parabola has a minimum, it opens up from the vertex $(-6,18)$.
Domain: $(-\infty,\infty)$. Range: $[18,\infty)$

49. $(h,k)=(5,3)$
$f(x)=2(x-h)^2+k=2(x-5)^2+3$

50. $(h,k)=(7,4)$
$f(x)=2(x-h)^2+k=2(x-7)^2+4$

51. $(h,k)=(-10,-5)$

$$\begin{aligned} f(x)&=2(x-h)^2+k\\ &=2[x-(-10)]^2+(-5)\\ &=2(x+10)^2-5 \end{aligned}$$

52. $(h,k)=(-8,-6)$

$$\begin{aligned} f(x)&=2(x-h)^2+k\\ &=2[x-(-8)]^2+(-6)\\ &=2(x+8)^2-6 \end{aligned}$$

53. Since the vertex is a maximum, the parabola opens down and $a=-3$.
$(h,k)=(-2,4)$

$$\begin{aligned} f(x)&=-3(x-h)^2+k\\ &=-3[x-(-2)]^2+4\\ &=-3(x+2)^2+4 \end{aligned}$$

54. Since the vertex is a maximum, the parabola opens down and $a=-3$.
$(h,k)=(5,-7)$

$$\begin{aligned} f(x)&=-3(x-h)^2+k\\ &=-3(x-5)^2+(-7)\\ &=-3(x-5)^2-7 \end{aligned}$$

55. Since the vertex is a minimum, the parabola opens up and $a=3$.
$(h,k)=(11,0)$

$$\begin{aligned} f(x)&=3(x-h)^2+k\\ &=3(x-11)^2+0\\ &=3(x-11)^2 \end{aligned}$$

56. Since the vertex is a minimum, the parabola opens up and $a=3$.
$(h,k)=(9,0)$

$$\begin{aligned} f(x)&=3(x-h)^2+k\\ &=3(x-9)^2+0\\ &=3(x-9)^2 \end{aligned}$$

57. $x=-\dfrac{b}{2a}=-\dfrac{-0.104}{2(0.005)}=10.4$

Wine consumption was at a minimum 10.4 years after 1980, or about 1990.

$$\begin{aligned} f(10)&=0.005(10)^2-0.104(10)+2.626\\ &=0.005(100)-1.04+2.626\\ &=0.5-1.04+2.626\\ &=2.086\\ &\approx 2.1 \end{aligned}$$

In 1990 the function suggests that per capita wine consumption was about 2.1 gallons. This models the graph's data quite well.

58. $x=-\dfrac{b}{2a}=-\dfrac{-255}{2(33)}\approx 4$

Homicides were at a minimum about 4 years after 1995, or 1999.

$$\begin{aligned} f(4)&=33(4)^2-255(4)+1230\\ &=33(16)-1020+1230\\ &=528-1020+1230\\ &=738 \end{aligned}$$

In 1999 the function predicts that there would be 738 gang-related homicides. This exceeds the graph's value of 692.

59. **a.** $t=-\frac{b}{2a}=-\frac{64}{2(-16)}=-\frac{64}{-32}=2$

$s(2)=-16(2)^2+64(2)+200$
$=-16(4)+128+200$
$=-64+128+200=264$

The ball reaches a maximum height of 264 feet 2 seconds after it is thrown.

b. $0=-16t^2+64t+200$

$0=t^2-4t-12.5$

$a=1 \quad b=-4 \quad c=-12.5$

$$t=\frac{-(-4)\pm\sqrt{(-4)^2-4(1)(-12.5)}}{2(1)}$$

$$=\frac{4\pm\sqrt{16+50}}{2}=\frac{4\pm\sqrt{66}}{2}$$

$$\approx\frac{4\pm 8.1}{2}$$

$=\frac{4+8.1}{2}$ or $\frac{4-8.1}{2}$

$=\frac{12.1}{2}$ $\quad =\frac{-4.1}{2}$

≈ 6.1 $\quad \approx -2.1$

We disregard –2.1 because we can't have a negative time measurement. The solution is 6.1 and we conclude that the ball will hit the ground in approximately 6.1 seconds.

c. $s(0)=-16(0)^2+64(0)+200$
$=-16(0)+0+200=200$

At $t=0$, the ball has not yet been thrown and is at a height of 200 feet. This is the height of the building.

d.

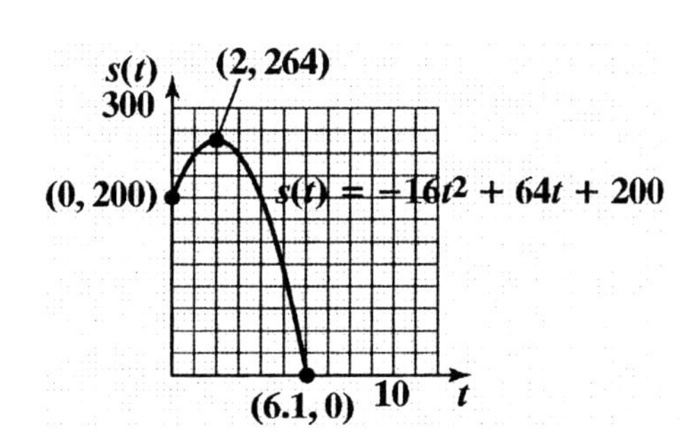

60. $s(t)=-16t^2+64t+160$

a. The t-coordinate of the minimum is $t=-\frac{b}{2a}=-\frac{64}{2(-16)}=-\frac{64}{-32}=2.$ The s-coordinate of the minimum is

$s(2)=-16(2)^2+64(2)+160$
$=-16(4)+128+160$
$=-64+128+160=224$

The ball reaches a maximum height of 224 feet 2 seconds after it is thrown.

b. $0=-16t^2+64t+160$

$0=t^2-4t-10$

$a=1 \quad b=-4 \quad c=-10$

$$t=\frac{-(-4)\pm\sqrt{(-4)^2-4(1)(-10)}}{2(1)}$$

$$=\frac{4\pm\sqrt{16+40}}{2}$$

Evaluate the

$$=\frac{4\pm\sqrt{56}}{2}\approx\frac{4\pm 7.48}{2}$$

expression to obtain two solutions.

$x=\frac{4+7.48}{2}$ or $x=\frac{4-7.48}{2}$

$x=\frac{11.48}{2}$ $\quad x=\frac{-3.48}{2}$

$x=5.74$ $\quad x=-1.74$

We disregard –1.74 because we can't have a negative time
measurement. The solution is 5.74 and we conclude that the ball will hit the ground in approximately 5.7 seconds.

c. $s(0)=-16(0)^2+64(0)+160$
$=-16(0)+0+160=160$

At $t=0$, the ball has not yet been thrown and is at a height of 160 feet. This is the height of the building.

d.

s(t)
250
(2, 224)
s(t) = −16t2 + 64t + 160
(0, 160)
(5.7, 0)
10
t

61. Let x = one of the numbers;
$16-x$ = the other number.

The product is $f(x)=x(16-x)$
$=16x-x^2=-x^2+16x$

The x-coordinate of the maximum is

$$x=-\frac{b}{2a}=-\frac{16}{2(-1)}=-\frac{16}{-2}=8.$$

$f(8)=-8^2+16(8)=-64+128=64$

The vertex is (8, 64). The maximum product is 64. This occurs when the two number are 8 and $16-8=8$.

62. Let x = one of the numbers
Let $20-x$ = the other number
$P(x)=x(20-x)=20x-x^2=-x^2+20x$

$$x=-\frac{b}{2a}=-\frac{20}{2(-1)}=-\frac{20}{-2}=10$$

The other number is $20-x=20-10=10.$

The numbers which maximize the product are 10 and 10. The maximum product is $10\cdot 10=100.$

63. Let x = one of the numbers;
$x-16$ = the other number.
The product is $f(x)=x(x-16)=x^2-16x$
The x-coordinate of the minimum is

$$x=-\frac{b}{2a}=-\frac{-16}{2(1)}=-\frac{-16}{2}=8.$$

$f(8)=(8)^2-16(8)$
$=64-128=-64$

The vertex is $(8,-64)$. The minimum product is -64. This occurs when the two number are 8 and $8-16=-8$.

64. Let x = the larger number. Then $x-24$ is the smaller number. The product of these two numbers is given by
$P(x)=x(x-24)=x^2-24x$
The product is minimized when

$$x=-\frac{b}{2a}=-\frac{(-24)}{2(1)}=12$$

Since $12-(-12)=24$, the two numbers whose difference is 24 and whose product is minimized are 12 and -12.
The minimum product is
$P(12)=12(12-24)=-144$.

65. Maximize the area of a rectangle constructed along a river with 600 feet of fencing.
Let x = the width of the rectangle;
$600-2x$ = the length of the rectangle
We need to maximize.
$A(x)=x(600-2x)$
$=600x-2x^2=-2x^2+600x$
Since $a=-2$ is negative, we know the function opens downward and has a maximum at

$$x=-\frac{b}{2a}=-\frac{600}{2(-2)}=-\frac{600}{-4}=150.$$

When the width is $x=150$ feet, the length is $600-2(150)=600-300=300$ feet.
The dimensions of the rectangular plot with maximum area are 150 feet by 300 feet. This gives an area of $150\cdot 300=45,000$ square feet.

66. From the diagram, we have that x is the width of the rectangular plot and $200-2x$ is the length. Thus, the area of the plot is given by
$A=l\cdot w=(200-2x)(x)=-2x^2+200x$
Since the graph of this equation is a parabola that opens down, the area is maximized at the vertex.

$$x=-\frac{b}{2a}=-\frac{200}{2(-2)}=50$$

$A=-2(50)^2+200(50)=-5000+10,000$
$=5000$
The maximum area is 5000 square feet when the length is 100 feet and the width is 50 feet.

67. Maximize the area of a rectangle constructed with 50 yards of fencing.
Let x = the length of the rectangle. Let y = the width of the rectangle.
Since we need an equation in one variable, use the perimeter to express y in terms of x.
$2x+2y=50$
$2y=50-2x$

$$y=\frac{50-2x}{2}=25-x$$

We need to maximize $A=xy=x(25-x)$. Rewrite A as a function of x.
$A(x)=x(25-x)=-x^2+25x$

Since $a=-1$ is negative, we know the function opens downward and has a maximum at

$$x=-\frac{b}{2a}=-\frac{25}{2(-1)}=-\frac{25}{-2}=12.5.$$

When the length x is 12.5, the width y is $y=25-x=25-12.5=12.5.$

The dimensions of the rectangular region with maximum area are 12.5 yards by 12.5 yards. This gives an area of $12.5\cdot 12.5=156.25$ square yards.

68. Let x = the length of the rectangle
Let y = the width of the rectangle

$$2x+2y=80$$
$$2y=80-2x$$
$$y=\frac{80-2x}{2}$$
$$y=40-x$$
$$A(x)=x(40-x)=-x^2+40x$$
$$x=-\frac{b}{2a}=-\frac{40}{2(-1)}=-\frac{40}{-2}=20.$$

When the length x is 20, the width y is $y=40-x=40-20=20.$

The dimensions of the rectangular region with maximum area are 20 yards by 20 yards. This gives an area of $20\cdot 20=400$ square yards.

69. Maximize the area of the playground with 600 feet of fencing.
Let x = the length of the rectangle. Let y = the width of the rectangle.
Since we need an equation in one variable, use the perimeter to express y in terms of x.

$$2x+3y=600$$
$$3y=600-2x$$
$$y=\frac{600-2x}{3}$$
$$y=200-\frac{2}{3}x$$

We need to maximize $A=xy=x\left(200-\frac{2}{3}x\right)$.

Rewrite A as a function of x.

$$A(x)=x\left(200-\frac{2}{3}x\right)=-\frac{2}{3}x^2+200x$$

Since $a=-\frac{2}{3}$ is negative, we know the function opens downward and has a maximum at

$$x=-\frac{b}{2a}=-\frac{200}{2\left(-\frac{2}{3}\right)}=-\frac{200}{-\frac{4}{3}}=150.$$

When the length x is 150, the width y is

$$y=200-\frac{2}{3}x=200-\frac{2}{3}(150)=100.$$

The dimensions of the rectangular playground with maximum area are 150 feet by 100 feet. This gives an area of $150\cdot 100=15{,}000$ square feet.

70. Maximize the area of the playground with 400 feet of fencing.
Let x = the length of the rectangle. Let y = the width of the rectangle.
Since we need an equation in one variable, use the perimeter to express y in terms of x.

$$2x+3y=400$$
$$3y=400-2x$$
$$y=\frac{400-2x}{3}$$
$$y=\frac{400}{3}-\frac{2}{3}x$$

We need to maximize $A=xy=x\left(\frac{400}{3}-\frac{2}{3}x\right)$.

Rewrite A as a function of x.

$$A(x)=x\left(\frac{400}{3}-\frac{2}{3}x\right)=-\frac{2}{3}x^2+\frac{400}{3}x$$

Since $a=-\frac{2}{3}$ is negative, we know the function opens downward and has a maximum at

$$x=-\frac{b}{2a}=-\frac{\frac{400}{3}}{2\left(-\frac{2}{3}\right)}=-\frac{\frac{400}{3}}{-\frac{4}{3}}=100.$$

When the length x is 100, the width y is

$$y=\frac{400}{3}-\frac{2}{3}x=\frac{400}{3}-\frac{2}{3}(100)=\frac{200}{3}=66\frac{2}{3}.$$

The dimensions of the rectangular playground with maximum area are 100 feet by $66\frac{2}{3}$ feet. This gives an area of $100\cdot 66\frac{2}{3}=6666\frac{2}{3}$ square feet.

71. Maximize the cross-sectional area of the gutter:

$A(x) = x(20 - 2x)$

$= 20x - 2x^2 = -2x^2 + 20x.$

Since $a = -2$ is negative, we know the function opens downward and has a maximum at

$x = -\frac{b}{2a} = -\frac{20}{2(-2)} = -\frac{20}{-4} = 5.$

When the height x is 5, the width is

$20 - 2x = 20 - 2(5) = 20 - 10 = 10.$

$A(5) = -2(5)^2 + 20(5)$

$= -2(25) + 100 = -50 + 100 = 50$

The maximum cross-sectional area is 50 square inches. This occurs when the gutter is 5 inches deep and 10 inches wide.

72. $A(x) = x(12 - 2x) = 12x - 2x^2$

$= -2x^2 + 12x$

$x = -\frac{b}{2a} = -\frac{12}{2(-2)} = -\frac{12}{-4} = 3$

When the height x is 3, the width is

$12 - 2x = 12 - 2(3) = 12 - 6 = 6.$

$A(3) = -2(3)^2 + 12(3) = -2(9) + 36$

$= -18 + 36 = 18$

The maximum cross-sectional area is 18 square inches. This occurs when the gutter is 3 inches deep and 6 inches wide.

73. x = increase

$A = (50 + x)(8000 - 100x)$

$= 400{,}000 + 3000x - 100x^2$

$x = \frac{-b}{2a} = \frac{-3000}{2(-100)} = 15$

The maximum price is 50 + 15 = \$65.

The maximum revenue = 65(800 – 100·15) = \$422,500.

74. Maximize $A = (30 + x)(200 - 5x)$

$= 6000 + 50x - 5x^2$

$x = \frac{-(50)}{2(-5)} = 5$

Maximum rental = 30 + 5 = \$35

Maximum revenue = 35(200 – 5·5) = \$6125

75. x = increase

$A = (20 + x)(60 - 2x)$

$= 1200 + 20x - 2x^2$

$x = \frac{-b}{2a} = \frac{-20}{2(-2)} = 5$

The maximum number of trees is 20 + 5 = 25 trees. The maximum yield is 60 – 2·5=50 pounds per tree, 50 x 25 = 1250 pounds.

76. Maximize $A = (30 + x)(50 - x)$

$= 1500 + 20x - x^2$

$x = \frac{-20}{2(-1)} = 10$

Maximum number of trees = 30 + 10 = 40 trees

Maximum yield = (30 + 10)(50 – 10) = 1600 pounds

84. $y = 2x^2 - 82x + 720$

a.

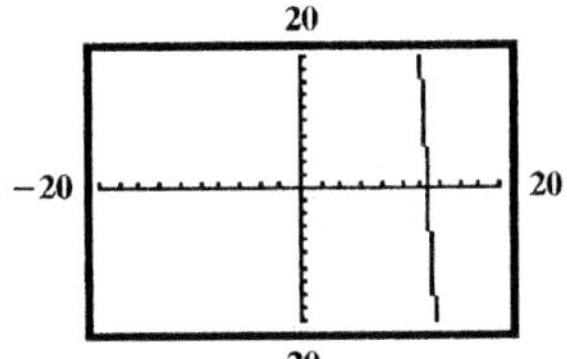

You can only see a little of the parabola.

b. $a = 2$; $b = -82$

$x = -\frac{b}{2a} = -\frac{-82}{4} = 20.5$

$y = 2(20.5)^2 - 82(20.5) + 720$

$= 840.5 - 1681 + 720$

$= -120.5$

vertex: (20.5, –120.5)

c. Ymax = 750

d. You can choose Xmin and Xmax so the x-value of the vertex is in the center of the graph. Choose Ymin to include the y-value of the vertex.

85. $y = -0.25x^2 + 40x$

$x = \frac{-b}{2a} = \frac{-40}{-0.5} = 80$

$y = -0.25(80)^2 + 40(80)$

$= 1600$

vertex: (80, 1600)

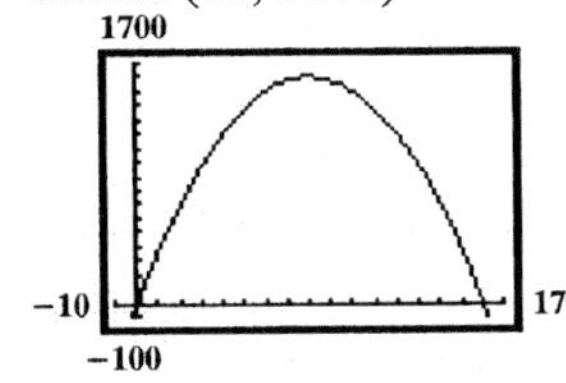

86. $y = -4x^2 + 20x + 160$

$x = \frac{-b}{2a} = \frac{-20}{-8} = 2.5$

$y = -4(2.5)^2 + 20(2.5) + 160$
$= -2.5 + 50 + 160 = 185$

The vertex is at (2.5, 185).

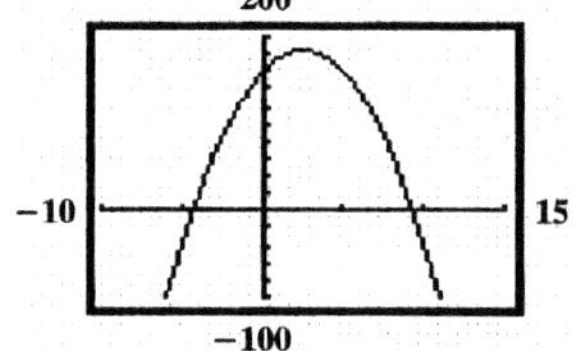

87. $y = 5x^2 + 40x + 600$

$x = \frac{-b}{2a} = \frac{-40}{10} = -4$

$y = 5(-4)^2 + 40(-4) + 600$
$= 80 - 160 + 600 = 520$

vertex: (–4, 520)

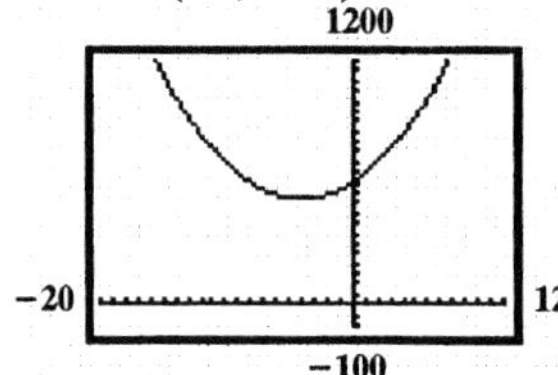

88. $y = 0.01x^2 + 0.6x + 100$

$x = \frac{-b}{2a} = \frac{-0.6}{0.02} = -30$

$y = 0.01(-30)^2 + 0.6(-30) + 100$
$= 9 - 18 + 100 = 91$

The vertex is at (–30, 91).

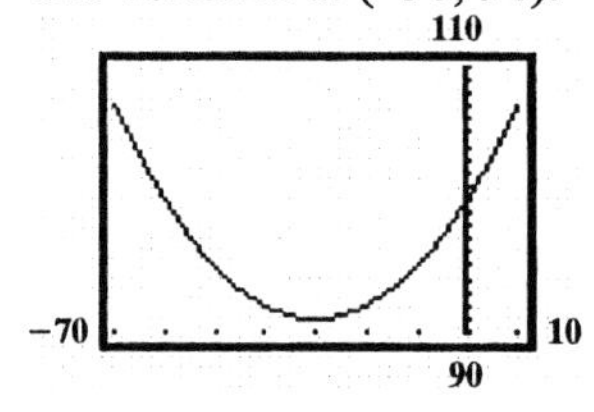

89. **a.** The values of y increase then decrease.

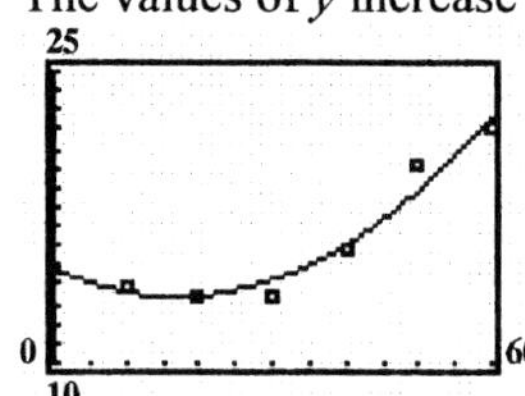

b. $y = 0.005x^2 - 0.170x + 14.817$

c. $x = \frac{-(-0.170)}{2(.005)} = 17; \ 1940 + 17 = 1957$

$y = 0.005(17)^2 - 0.170(17) + 14.817$
≈ 13.372

The worst gas mileage was 13.372 mpg in 1957.

d.

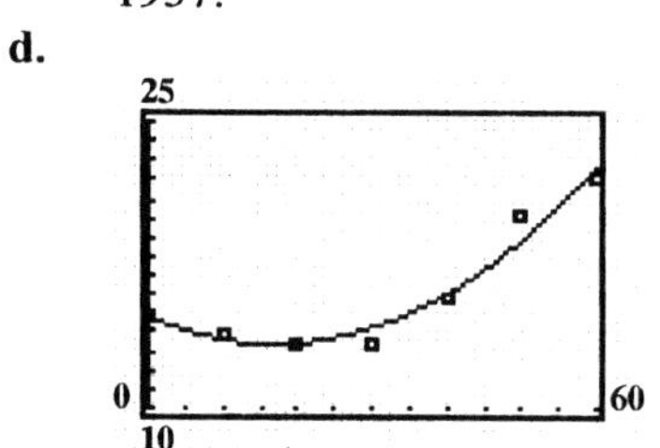

90. Statement **a** is true. Since quadratic functions represent parabolas, we know that the function has a maximum or a minimum. This means that the range cannot be $(-\infty, \infty)$.

Statement **b** is false. The vertex is $(5, -1)$.

Statement **c** is false. The graph has no x–intercepts. To find x–intercepts, set $y = 0$ and solve for x.

$$0 = -2(x+4)^2 - 8$$
$$2(x+4)^2 = -8$$
$$(x+4)^2 = -4$$

Because the solutions to the equation are imaginary, we know that there are no x–intercepts.

Statement **d** is false. The x-coordinate of the maximum is $-\frac{b}{2a} = -\frac{1}{2(-1)} = -\frac{1}{-2} = \frac{1}{2}$ and the y–coordinate of the vertex of the parabola is

$$f\left(-\frac{b}{2a}\right) = f\left(\frac{1}{2}\right)$$
$$= -\left(\frac{1}{2}\right)^2 + \frac{1}{2} + 1$$
$$= -\frac{1}{4} + \frac{1}{2} + 1$$
$$= -\frac{1}{4} + \frac{2}{4} + \frac{4}{4} = \frac{5}{4}.$$

The maximum y–value is $\frac{5}{4}$.

91. $f(x) = 3(x + 2)^2 - 5$; $(-1, -2)$
axis: $x = -2$
$(-1, -2)$ is one unit right of $(-2, -2)$. One unit left of $(-2, -2)$ is $(-3, -2)$.
point: $(-3, -2)$

92. Vertex (3, 2) Axis: $x = 3$
second point (0, 11)

93. We start with the form $f(x) = a(x-h)^2 + k$. Since we know the vertex is $(h,k) = (-3,-4)$, we have $f(x) = a(x+3)^2 - 4$. We also know that the graph passes through the point $(1,4)$, which allows us to solve for a.

$$4 = a(1+3)^2 - 4$$
$$8 = a(4)^2$$
$$8 = 16a$$
$$\frac{1}{2} = a$$

Therefore, the function is

$$f(x) = \frac{1}{2}(x+3)^2 - 4.$$

94. We know $(h,k) = (-3,-4)$, so the equation is of the form

$$f(x) = a(x-h)^2 + k$$
$$= a[x-(-3)]^2 + (-1)$$
$$= a(x+3)^2 - 1$$

We use the point $(-2,-3)$ on the graph to determine the value of a:

$$f(x) = a(x+3)^2 - 1$$
$$-3 = a(-2+3)^2 - 1$$
$$-3 = a(1)^2 - 1$$
$$-3 = a - 1$$
$$-2 = a$$

Thus, the equation of the parabola is

$$f(x) = -2(x+3)^2 - 1.$$

95. $2x + y - 2 = 0$

$y = 2 - 2x$

$$d = \sqrt{x^2 + (2-2x)^2}$$
$$d = \sqrt{x^2 + 4 - 8x + 4x^2}$$
$$d = \sqrt{5x^2 - 8x + 4}$$

Minimize $5x^2 - 8x + 4$

$$x = \frac{-(-8)}{2(5)} = \frac{4}{5}$$
$$y = 2 - 2\left(\frac{4}{5}\right) = \frac{2}{5}$$
$$\left(\frac{4}{5}, \frac{2}{5}\right)$$

96. $f(x) = (80 + x)(300 - 3x) - 10(300 - 3x)$
$= 24000 + 60x - 3x^2 - 3000 + 30x$
$= -3x^2 + 90x + 21000$

$$x = \frac{-b}{2a} = \frac{-90}{2(-3)} = \frac{3}{2} = 15$$

The maximum charge is 80 + 15 = \$95.00. the maximum profit is $-3(15)^2 + 9(15) + 21000 =$ \$21,675.

97.
$$440 = 2x + \pi y$$
$$440 - 2x = \pi y$$
$$\frac{440-2x}{\pi} = y$$

Maximize

$$A = x\left(\frac{440-2x}{\pi}\right) = -\frac{2}{\pi}x^2 + \frac{440}{\pi}x$$

$$x = \frac{-\frac{440}{\pi}}{2\left(-\frac{2}{\pi}\right)} = \frac{\frac{-440}{\pi}}{-\frac{4}{\pi}} = \frac{440}{4} = 110$$

$$\frac{440-2(110)}{\pi} = \frac{220}{\pi}$$

The dimensions are 110 yards by $\frac{220}{\pi}$ yards.

Section 2.3

Check Point Exercises

1. Since n is even and $a_n > 0$, the graph rises to the left and to the right.

2. Since n is odd and the leading coefficient is negative, the function falls to the right. Since the ratio cannot be negative, the model won't be appropriate.

3. The graph does not show the function's end behavior. Since $a_n > 0$ and n is odd, the graph should fall to the left.

4. $f(x) = x^3 + 2x^2 - 4x - 8$

$0 = x^2(x+2) - 4(x+2)$

$0 = (x+2)(x^2 - 4)$

$0 = (x+2)^2(x-2)$

$x = 2$ or $x = -2$

The zeros are 2 and –2.

5. $f(x) = x^4 - 4x^2$

$x^4 - 4x^2 = 0$

$x^2(x^2 - 4) = 0$

$x^2(x+2)(x-2) = 0$

$x = 0$ or $x = -2$ or $x = 2$

The zeros are 0, –2, and 2.

6. $f(x) = -4\left(x + \frac{1}{2}\right)^2 (x-5)^3$

$-4\left(x + \frac{1}{2}\right)^2 (x-5)^3 = 0$

$x = -\frac{1}{2}$ or $x = 5$

The zeros are $-\frac{1}{2}$, with multiplicity 2, and 5, with multiplicity 3.

Because the multiplicity of $-\frac{1}{2}$ is even, the graph touches the x-axis and turns around at this zero.

Because the multiplicity of 5 is odd, the graph crosses the x-axis at this zero.

7. $f(-3) = 3(-3)^3 - 10(-3) + 9 = -42$

$f(-2) = 3(-2)^3 - 10(-2) + 9 = 5$

The sign change shows there is a zero between –3 and –2.

8. $f(x) = x^3 - 3x^2$

Since $a_n > 0$ and n is odd, the graph falls to the left and rises to the right.

$x^3 - 3x^2 = 0$

$x^2(x-3) = 0$

$x = 0$ or $x = 3$

The x-intercepts are 0 and 3.

$f(0) = 0^3 - 3(0)^2 = 0$

The y-intercept is 0.

$f(-x) = (-x)^3 - 3(-x)^2 = -x^3 - 3x^2$

No symmetry.

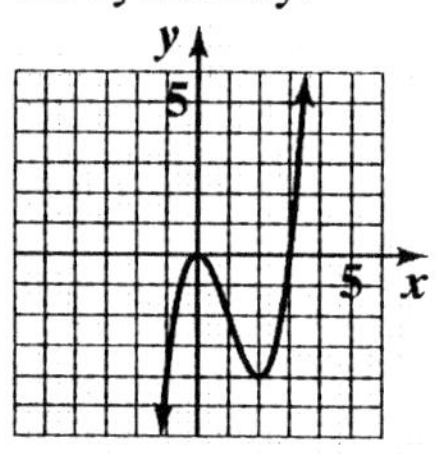

$f(x) = x^3 - 3x^2$

Exercise Set 2.3

1. polynomial function; degree: 3

2. polynomial function; degree: 4

3. polynomial function; degree: 5

4. polynomial function; degree: 7

5. not a polynomial function

6. not a polynomial function

7. not a polynomial function

8. not a polynomial function

9. not a polynomial function

10. polynomial function; degree: 2

11. polynomial function

12. Not a polynomial function because graph is not smooth.

13. Not a polynomial function because graph is not continuous.

14. polynomial function

15. (b)

16. (c)

17. (a)

18. (d)

19. $f(x) = 5x^3 + 7x^2 - x + 9$
Since $a_n > 0$ and n is odd, the graph of $f(x)$ falls to the left and rises to the right.

20. $f(x) = 11x^3 - 6x^2 + x + 3$
Since $a_n > 0$ and n is odd, the graph of $f(x)$ falls to the left and rises to the right.

21. $f(x) = 5x^4 + 7x^2 - x + 9$
Since $a_n > 0$ and n is even, the graph of $f(x)$ rises to the left and to the right.

22. $f(x) = 11x^4 - 6x^2 + x + 3$
Since $a_n > 0$ and n is even, the graph of $f(x)$ rises to the left and to the right.

23. $f(x) = -5x^4 + 7x^2 - x + 9$
Since $a_n < 0$ and n is even, the graph of $f(x)$ falls to the left and to the right.

24. $f(x) = -11x^4 - 6x^2 + x + 3$
Since $a_n < 0$ and n is even, the graph of $f(x)$ falls to the left and to the right.

25. $f(x) = 2(x-5)(x+4)^2$
$x = 5$ has multiplicity 1;
The graph crosses the x-axis.
$x = -4$ has multiplicity 2;
The graph touches the x-axis and turns around.

26. $f(x) = 3(x+5)(x+2)^2$
$x = -5$ has multiplicity 1;
The graph crosses the x-axis.
$x = -2$ has multiplicity 2;
The graph touches the x-axis and turns around.

27. $f(x) = 4(x-3)(x+6)^3$
$x = 3$ has multiplicity 1;
The graph crosses the x-axis.
$x = -6$ has multiplicity 3;
The graph crosses the x-axis.

28. $f(x) = -3\left(x + \frac{1}{2}\right)(x-4)^3$

$x = -\frac{1}{2}$ has multiplicity 1;
The graph crosses the x-axis.
$x = 4$ has multiplicity 3;
The graph crosses the x-axis.

29. $f(x) = x^3 - 2x^2 + x$
$= x\left(x^2 - 2x + 1\right)$
$= x(x-1)^2$
$x = 0$ has multiplicity 1;
The graph crosses the x-axis.
$x = 1$ has multiplicity 2;
The graph touches the x-axis and turns around.

30. $f(x) = x^3 + 4x^2 + 4x$
$= x\left(x^2 + 4x + 4\right)$
$= x(x+2)^2$
$x = 0$ has multiplicity 1;
The graph crosses the x-axis.
$x = -2$ has multiplicity 2;
The graph touches the x-axis and turns around.

31. $f(x) = x^3 + 7x^2 - 4x - 28$
$= x^2(x+7) - 4(x+7)$
$= \left(x^2 - 4\right)(x+7)$
$= (x-2)(x+2)(x+7)$
$x = 2$, $x = -2$ and $x = -7$ have multiplicity 1;
The graph crosses the x-axis.

32. $f(x) = x^3 + 5x^2 - 9x - 45$
$= x^2(x+5) - 9(x+5)$
$= \left(x^2 - 9\right)(x+5)$
$= (x-3)(x+3)(x+5)$
$x = 3$, $x = -3$ and $x = -5$ have multiplicity 1;
The graph crosses the x-axis.

33. $f(x) = x^3 - x - 1$
$f(1) = -1$
$f(2) = 5$
The sign change shows there is a zero between the given values.

34. $f(x) = x^3 - 4x^2 + 2$
$f(0) = 2$
$f(1) = -1$
The sign change shows there is a zero between the given values.

35. $f(x) = 2x^4 - 4x^2 + 1$
$f(-1) = -1$
$f(0) = 1$
The sign change shows there is a zero between the given values.

36. $f(x) = x^4 + 6x^3 - 18x^2$
$f(2) = -8$
$f(3) = 81$
The sign change shows there is a zero between the given values.

37. $f(x) = x^3 + x^2 - 2x + 1$
$f(-3) = -11$
$f(-2) = 1$
The sign change shows there is a zero between the given values.

38. $f(x) = x^5 - x^3 - 1$
$f(1) = -1$
$f(2) = 23$
The sign change shows there is a zero between the given values.

39. $f(x) = 3x^3 - 10x + 9$
$f(-3) = -42$
$f(-2) = 5$
The sign change shows there is a zero between the given values.

40. $f(x) = 3x^3 - 8x^2 + x + 2$
$f(2) = -4$
$f(3) = 14$
The sign change shows there is a zero between the given values.

41. $f(x) = x^3 + 2x^2 - x - 2$

a. Since $a_n > 0$ and n is odd, $f(x)$ rises to the right and falls to the left.

b.
$$x^3 + 2x^2 - x - 2 = 0$$
$$x^2(x+2) - (x+2) = 0$$
$$(x+2)(x^2-1) = 0$$
$$(x+2)(x-1)(x+1) = 0$$
$$x = -2, x = 1, x = -1$$
The zeros at –2, –1, and 1 have odd multiplicity so $f(x)$ crosses the x-axis at these points.

c.
$$f(0) = (0)^3 + 2(0)^2 - 0 - 2$$
$$= -2$$
The y-intercept is –2.

d.
$$f(-x) = (-x) + 2(-x)^2 - (-x) - 2$$
$$= -x^3 + 2x^2 + x - 2$$
$$-f(x) = -x^3 - 2x^2 + x + 2$$
The graph has neither origin symmetry or y-axis symmetry.

e. The graph has 2 turning points and $2 \le 3 - 1$.

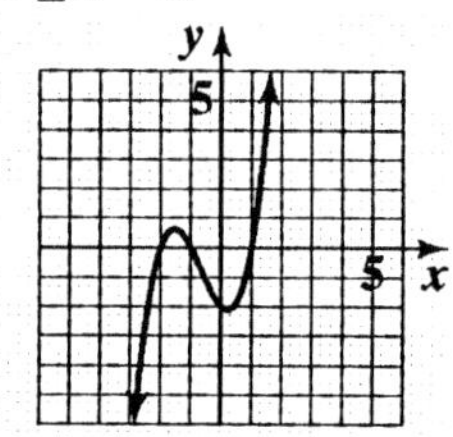

$y = x^3 + 2x^2 - x - 2$

42. $f(x) = x^3 + x^2 - 4x - 4$

a. Since $a_n > 0$ and n is odd, $f(x)$ rises to the right and falls to the left.

b.
$$x^3 + x^2 - 4x - 4 = 0$$
$$x^2(x+1) - 4(x+1) = 0$$
$$(x+1)(x^2-4) = 0$$
$$(x+1)(x-2)(x+2) = 0$$
$x = -1$, or $x = 2$, or $x = -2$
The zeros at –2, –1 and 2 have odd multiplicity, so $f(x)$ crosses the x-axis at these points. The x-intercepts are –2, –1, and 2.

c. $f(0) = 0^3 + (0)^2 - 4(0) - 4 = -4$
The y-intercept is –4.

d.
$$f(-x) = -x^3 + x^2 + 4x - 4$$
$$-f(x) = -x^3 - x^2 + 4x + 4$$
neither symmetry

e. The graph has 2 turning points and $2 \le 3 - 1$.

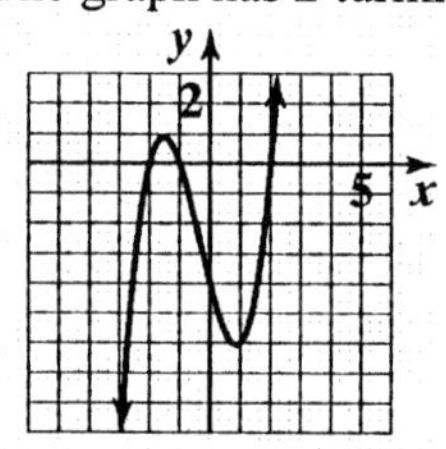

$y = x^3 + x^2 - 4x - 4$

43. $f(x) = x^4 - 9x^2$

a. Since $a_n > 0$ and n is even, $f(x)$ rises to the left and the right.

b.
$$x^4 - 9x^2 = 0$$
$$x^2(x^2 - 9) = 0$$
$$x^2(x-3)(x+3) = 0$$
$$x = 0, x = 3, x = -3$$

The zeros at –3 and 3 have odd multiplicity, so $f(x)$ crosses the x-axis at these points. The root at 0 has even multiplicity, so $f(x)$ touches the x-axis at 0.

c. $f(0) = (0)^4 - 9(0)^2 = 0$
The y-intercept is 0.

d. $f(-x) = x^4 - 9x^2$
$f(-x) = f(x)$
The graph has y-axis symmetry.

e. The graph has 3 turning points and $3 \le 4 - 1$.

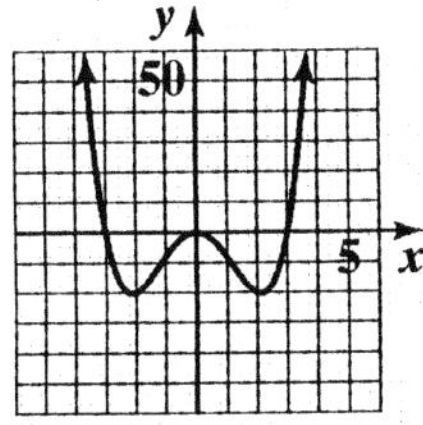

$f(x) = x^4 - 9x^2$

44. $f(x) = x^4 - x^2$

a. Since $a_n > 0$ and n is even, $f(x)$ rises to the left and the right.

b.
$$x^4 - x^2 = 0$$
$$x^2(x^2 - 1) = 0$$
$$x^2(x-1)(x+1) = 0$$
$$x = 0, x = 1, x = -1$$

f touches but does not cross the x-axis at 0.

c. $f(0) = (0)^4 - (0)^2 = 0$
The y-intercept is 0.

d. $f(-x) = x^4 - x^2$
$f(-x) = f(x)$
The graph has y-axis symmetry.

e. The graph has 3 turning points and $3 \le 4 - 1$.

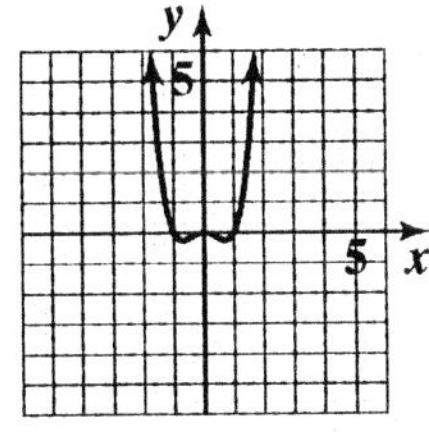

$f(x) = x^4 - x^2$

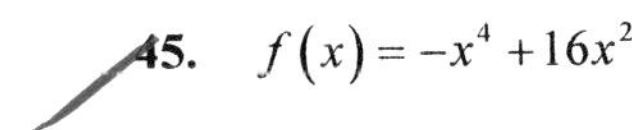

45. $f(x) = -x^4 + 16x^2$

a. Since $a_n < 0$ and n is even, $f(x)$ falls to the left and the right.

b.
$$-x^4 + 16x^2 = 0$$
$$x^2(-x^2 + 16) = 0$$
$$x^2(4-x)(4+x) = 0$$
$$x = 0, x = 4, x = -4$$

The zeros at –4 and 4 have odd multiplicity, so $f(x)$ crosses the x-axis at these points. The root at 0 has even multiplicity, so $f(x)$ touches the x-axis at 0.

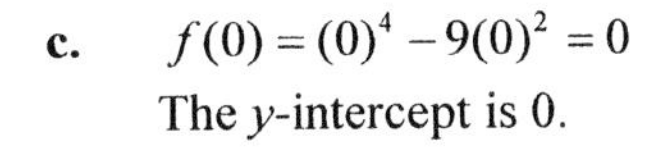

c. $f(0) = (0)^4 - 9(0)^2 = 0$
The y-intercept is 0.

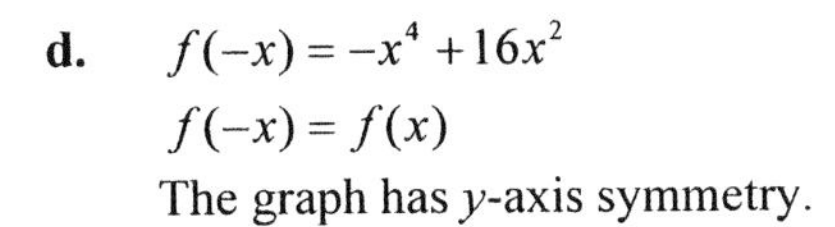

d. $f(-x) = -x^4 + 16x^2$
$f(-x) = f(x)$
The graph has y-axis symmetry.

e. The graph has 3 turning points and $3 \le 4 - 1$.

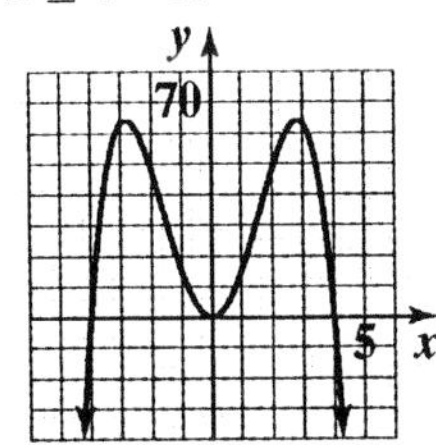

$f(x) = -x^4 + 16x^2$

46. $f(x) = -x^4 + 4x^2$

a. Since $a_n < 0$ and n is even, $f(x)$ falls to the left and the right.

b.
$$-x^4 + 4x^2 = 0$$
$$x^2(4 - x^2) = 0$$
$$x^2(2-x)(2+x) = 0$$
$$x = 0, x = 2, x = -2$$

The x-intercepts are $-2, 0$, and 2. Since f has a double root at 0, it touches but does not cross the x-axis at 0.

c. $f(0) = -(0)^4 + 4(0)^2 = 0$
The y-intercept is 0.

d. $f(-x) = -x^4 + 4x^2$
$f(-x) = f(x)$
The graph has y-axis symmetry.

e. The graph has 3 turning points and $3 \le 4 - 1$.

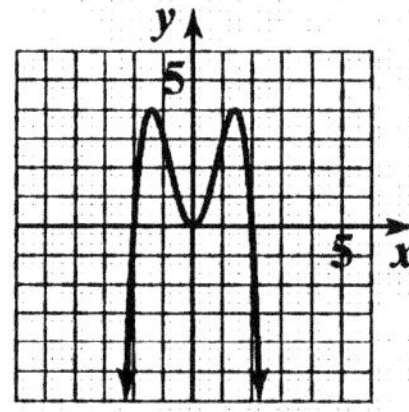

$f(x) = -x^4 + 4x^2$

47. $f(x) = x^4 - 2x^3 + x^2$

a. Since $a_n > 0$ and n is even, $f(x)$ rises to the left and the right.

b. $x^4 - 2x^3 + x^2 = 0$
$x^2(x^2 - 2x + 1) = 0$
$x^2(x-1)(x-1) = 0$
$x = 0, x = 1$
The zeros at 1 and 0 have even multiplicity, so $f(x)$ touches the x-axis at 0 and 1.

c. $f(0) = (0)^4 - 2(0)^3 + (0)^2 = 0$
The y-intercept is 0.

d. $f(-x) = x^4 + 2x^3 + x^2$
The graph has neither y-axis nor origin symmetry.

e. The graph has 3 turning points and $3 \le 4 - 1$.

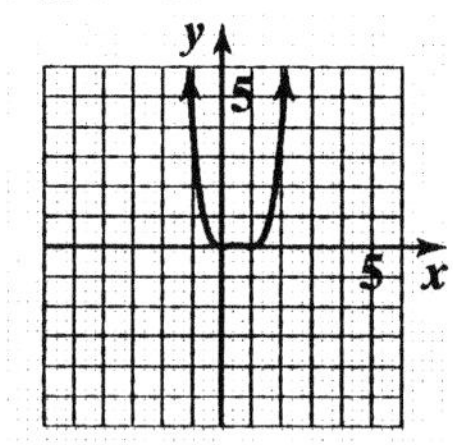

$f(x) = x^4 - 2x^3 + x^2$

48. $f(x) = x^4 - 6x^3 + 9x^2$

a. Since $a_n > 0$ and n is even, $f(x)$ rises to the left and the right.

b. $x^4 - 6x^3 + 9x^2 = 0$
$x^2(x^2 - 6x + 9) = 0$
$x^2(x-3)^2 = 0$
$x = 0, x = 3$
The zeros at 3 and 0 have even multiplicity, so $f(x)$ touches the x-axis at 3 and 0.

c. $f(0) = (0)^4 - 6(0)^3 + 9(0)^2 = 0$
The y-intercept is 0.

d. $f(-x) = x^4 + 6x^3 + 9x^2$
The graph has neither y-axis or origin symmetry.

e. The graph has 3 turning points and $3 \le 4 - 1$.

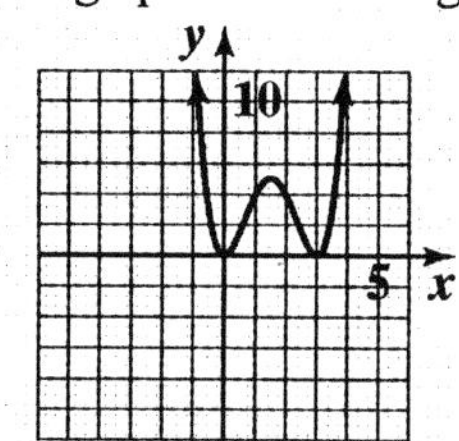

$f(x) = x^4 - 6x^3 + 9x^2$

49. $f(x) = -2x^4 + 4x^3$

a. Since $a_n < 0$ and n is even, $f(x)$ falls to the left and the right.

b. $-2x^4 + 4x^3 = 0$
$x^3(-2x + 4) = 0$
$x = 0, x = 2$
The zeros at 0 and 2 have odd multiplicity, so $f(x)$ crosses the x-axis at these points.

c. $f(0) = -2(0)^4 + 4(0)^3 = 0$
The y-intercept is 0.

d. $f(-x) = -2x^4 - 4x^3$
The graph has neither y-axis nor origin symmetry.

e. The graph has 1 turning point and $1 \le 4 - 1$.

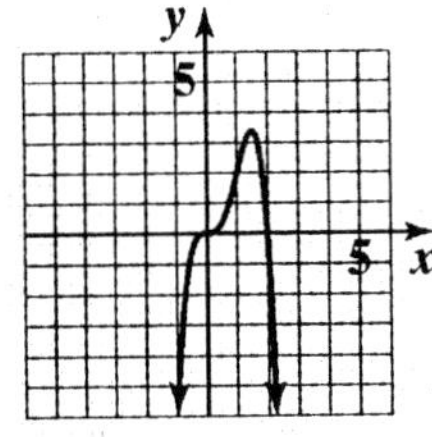

$f(x) = -2x^4 + 4x^3$

50. $f(x) = -2x^4 + 2x^3$

a. Since $a_n < 0$ and n is even, $f(x)$ falls to the left and the right.

b. $-2x^4 + 2x^3 = 0$
$x^3(-2x + 2) = 0$
$x = 0, x = 1$
The zeros at 0 and 1 have odd multiplicity, so $f(x)$ crosses the x-axis at these points.

c. The y-intercept is 0.

d. $f(-x) = -2x^4 - 2x^3$
The graph has neither y-axis or origin symmetry.

e. The graph has 2 turning points and $2 \le 4 - 1$.

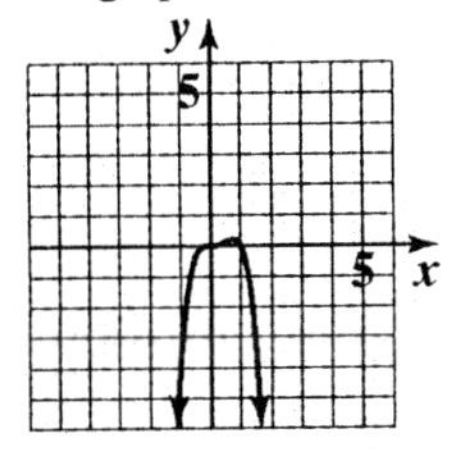

$f(x) = -2x^4 + 2x^3$

51. $f(x) = 6x^3 - 9x - x^5$

a. Since $a_n < 0$ and n is odd, $f(x)$ rises to the left and falls to the right.

b. $-x^5 + 6x^3 - 9x = 0$
$-x(x^4 - 6x^2 + 9) = 0$
$-x(x^2 - 3)(x^2 - 3) = 0$
$x = 0, x = \pm\sqrt{3}$
The root at 0 has odd multiplicity so $f(x)$ crosses the x-axis at (0, 0). The zeros at $-\sqrt{3}$ and $\sqrt{3}$ have even multiplicity so $f(x)$ touches the x-axis at $\sqrt{3}$ and $-\sqrt{3}$.

c. $f(0) = -(0)^5 + 6(0)^3 - 9(0) = 0$
The y-intercept is 0.

d. $f(-x) = x^5 - 6x^3 + 9x$
$f(-x) = -f(x)$
The graph has origin symmetry.

e. The graph has 4 turning point and $4 \le 5 - 1$.

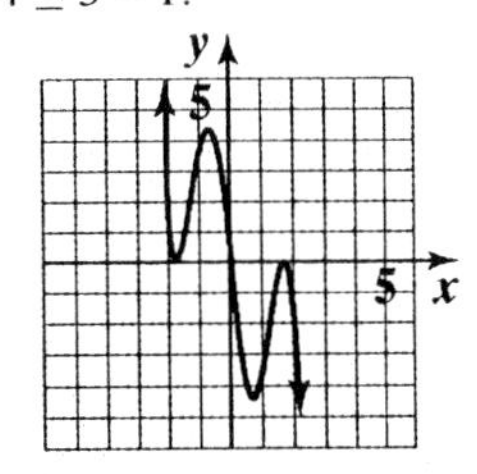

$f(x) = 6x^3 - 9x - x^5$

52. $f(x) = 6x - x^3 - x^5$

a. Since $a_n < 0$ and n is odd, $f(x)$ rises to the left and falls to the right.

b. $-x^5 - x^3 + 6x = 0$
$-x(x^4 + x^2 - 6) = 0$
$-x(x^2 + 3)(x^2 - 2) = 0$
$x = 0, x = \pm\sqrt{2}$
The zeros at $-\sqrt{2}$, 0, and $\sqrt{2}$ have odd multiplicity, so $f(x)$ crosses the x-axis at these points.

c. $f(0) = -(0)^5 - (0)^3 + 6(0) = 0$
The y-intercept is 0.

d. $f(-x) = x^5 + x^3 - 6x$
$f(-x) = -f(x)$
The graph has origin symmetry.

e. The graph has 2 turning points and $2 \le 5 - 1$.

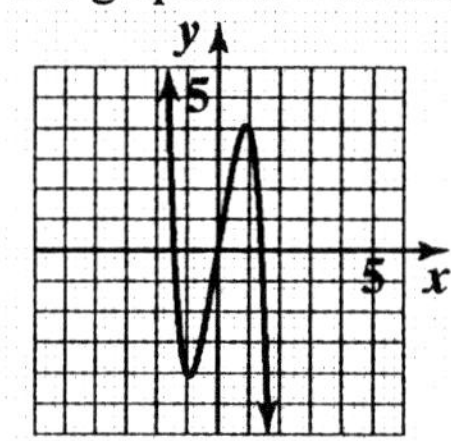

53. $f(x) = 3x^2 - x^3$

a. Since $a_n < 0$ and n is odd, $f(x)$ rises to the left and falls to the right.

b. $-x^3 + 3x^2 = 0$
$-x^2(x-3) = 0$
$x = 0, x = 3$
The zero at 3 has odd multiplicity so $f(x)$ crosses the x-axis at that point. The root at 0 has even multiplicity so $f(x)$ touches the axis at (0, 0).

c. $f(0) = -(0)^3 + 3(0)^2 = 0$
The y-intercept is 0.

d. $f(-x) = x^3 + 3x^2$
The graph has neither y-axis nor origin symmetry.

e. The graph has 2 turning point and $2 \le 3 - 1$.

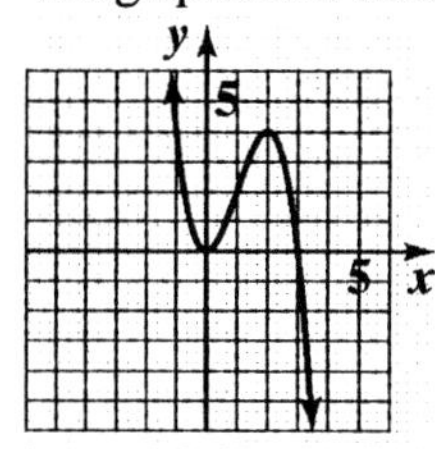

54. $f(x) = \frac{1}{2} - \frac{1}{2}x^4$

a. Since $a_n < 0$ and n is even, $f(x)$ falls to the left and the right.

b. $-\frac{1}{2}x^4 + \frac{1}{2} = 0$
$-\frac{1}{2}(x^4 - 1) = 0$
$-\frac{1}{2}(x^2 + 1)(x^2 - 1) = 0$
$-\frac{1}{2}(x^2 + 1)(x - 1)(x + 1) = 0$
$x = \pm 1$
The zeros at -1 and 1 have odd multiplicity, so $f(x)$ crosses the x-axis at these points.

c. $f(0) = -\frac{1}{2}(0)^4 + \frac{1}{2} = \frac{1}{2}$
The y-intercept is $\frac{1}{2}$.

d. $f(-x) = \frac{1}{2} - \frac{1}{2}x^4$
$f(-x) = f(x)$
The graph has y-axis symmetry.

e. The graph has 1 turning point and $1 \le 4 - 1$.

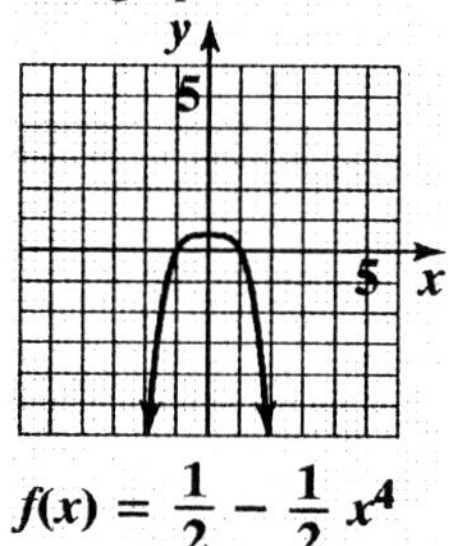

55. $f(x) = -3(x-1)^2(x^2 - 4)$

a. Since $a_n < 0$ and n is even, $f(x)$ falls to the left and the right.

b. $-3(x-1)^2(x^2 - 4) = 0$
$x = 1, x = -2, x = 2$
The zeros at -2 and 2 have odd multiplicity, so $f(x)$ crosses the x-axis at these points. The root at 1 has even multiplicity, so $f(x)$ touches the x-axis at (1, 0).

c. $f(0) = -3(0-1)^2(0^2 - 4)^3$
$= -3(1)(-4) = 12$
The y-intercept is 12.

d. $f(-x) = -3(-x-1)^2(x^2 - 4)$
The graph has neither y-axis nor origin symmetry.

e. The graph has 1 turning point and $1 \le 4 - 1$.

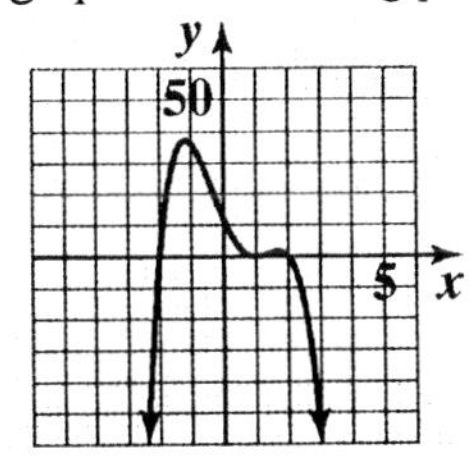

$f(x) = -3(x-1)^2(x^2-4)$

56. $f(x) = -2(x-4)^2(x^2-25)$

a. Since $a_n < 0$ and n is even, $f(x)$ falls to the left and the right.

b. $-2(x-4)^2(x^2-25) = 0$
$x = 4, x = -5, x = 5$
The zeros at –5 and 5 have odd multiplicity so $f(x)$ crosses the x-axis at these points. The root at 4 has even multiplicity so $f(x)$ touches the x-axis at (4, 0).

c. $f(0) = -2(0-4)^2(0^2-25)$
$= -2(16)(-25)$
$= 800$
The y-intercept is 800.

d. $f(-x) = -2(-x-4)^2(x^2-2)$
The graph has neither y-axis nor origin symmetry.

e. The graph has 1 turning point and $1 \le 4 - 1$.

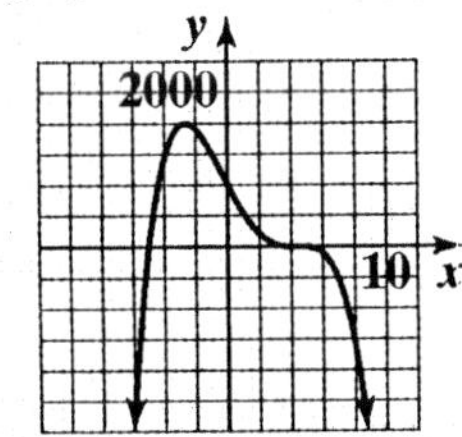

$f(x) = -2(x-4)^2(x^2-25)$

57. $f(x) = x^2(x-1)^3(x+2)$

a. Since $a_n > 0$ and n is even, $f(x)$ rises to the left and the right.

b. $x = 0, x = 1, x = -2$
The zeros at 1 and –2 have odd multiplicity so $f(x)$ crosses the x-axis at those points. The root at 0 has even multiplicity so $f(x)$ touches the axis at (0, 0).

c. $f(0) = 0^2(0-1)^3(0+2) = 0$
The y-intercept is 0.

d. $f(-x) = x^2(-x-1)^3(-x+2)$
The graph has neither y-axis nor origin symmetry.

e. The graph has 2 turning points and $2 \le 6 - 1$.

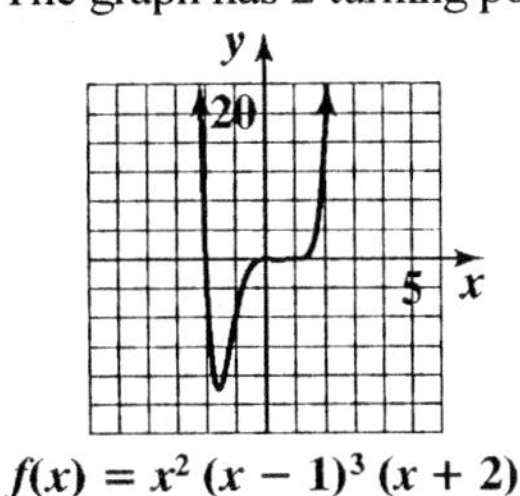

$f(x) = x^2(x-1)^3(x+2)$

58. $f(x) = x^3(x+2)^2(x+1)$

a. Since $a_n > 0$ and n is even, $f(x)$ rises to the left and the right.

b. $x = 0, x = -2, x = -1$
The roots at 0 and –1 have odd multiplicity so $f(x)$ crosses the x-axis at those points. The root at –2 has even multiplicity so $f(x)$ touches the axis at (–2, 0).

c. $f(0) = 0^3(0+2)^2(0+1) = 0$
The y-intercept is 0.

d. $f(-x) = -x^3(-x+2)^2(-x+1)$
The graph has neither y-axis nor origin symmetry.

e. The graph has 3 turning points and $3 \le 6 - 1$.

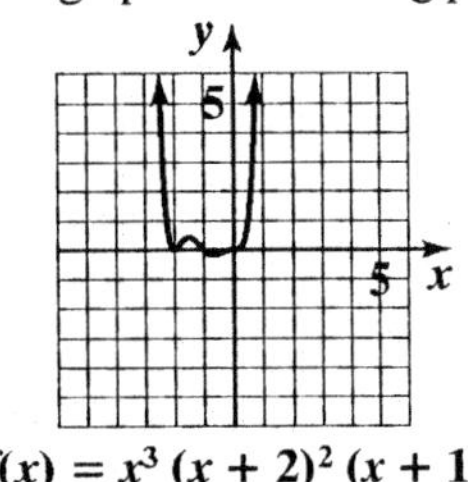

$f(x) = x^3(x+2)^2(x+1)$

59. $f(x) = -x^2(x-1)(x+3)$

a. Since $a_n < 0$ and n is even, $f(x)$ falls to the left and the right.

b. $x = 0, x = 1, x = -3$
The zeros at 1 and –3 have odd multiplicity so $f(x)$ crosses the x-axis at those points. The root at 0 has even multiplicity so $f(x)$ touches the axis at (0, 0).

c. $f(0) = -0^2(0-1)(0+3) = 0$
The y-intercept is 0.

d. $f(-x) = -x^2(-x-1)(-x+3)$
The graph has neither y-axis nor origin symmetry.

e. The graph has 3 turning points and $3 \le 4 - 1$.

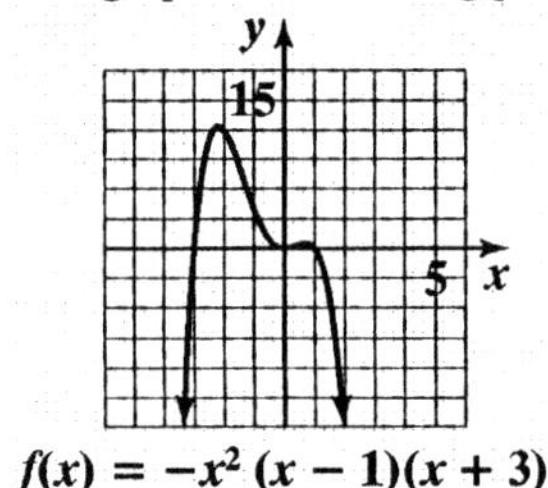

$f(x) = -x^2(x-1)(x+3)$

60. $f(x) = -x^2(x+2)(x-2)$

a. Since $a_n < 0$ and n is even, $f(x)$ falls to the left and the right.

b. $x = 0, x = 2, x = -2$
The zeros at 2 and –2 have odd multiplicity so $f(x)$ crosses the x-axis at those points. The root at 0 has even multiplicity so $f(x)$ touches the axis at (0, 0).

c. $f(0) = -0^2(0+2)(0-2) = 0$
The y-intercept is 0.

d. $f(-x) = -x^2(-x+2)(-x-2)$
$f(-x) = -x^2(-1)(x-2)(-1)(x+2)$
$f(-x) = -x^2(x+2)(x-2)$
$f(-x) = f(x)$
The graph has y-axis symmetry.

e. The graph has 3 turning points and $3 \le 4 - 1$.

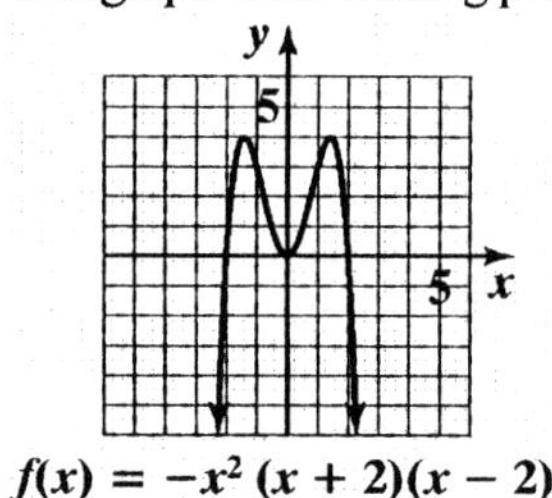

$f(x) = -x^2(x+2)(x-2)$

61. $f(x) = -2x^3(x-1)^2(x+5)$

a. Since $a_n < 0$ and n is even, $f(x)$ falls to the left and the right.

b. $x = 0, x = 1, x = -5$
The roots at 0 and –5 have odd multiplicity so $f(x)$ crosses the x-axis at those points. The root at 1 has even multiplicity so $f(x)$ touches the axis at (1, 0).

c. $f(0) = -2(0)^3(0-1)^2(0+5) = 0$
The y-intercept is 0.

d. $f(-x) = 2x^3(-x-1)^2(-x+5)$
The graph has neither y-axis nor origin symmetry.

e. The graph has 2 turning points and $2 \le 6 - 1$.

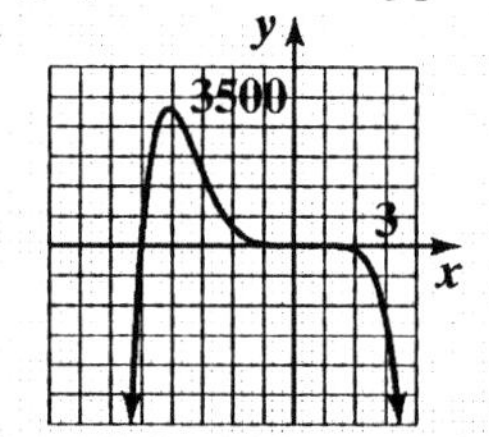

$f(x) = -2x^3(x-1)^2(x+5)$

62. $f(x) = -3x^3(x-1)^2(x+3)$

a. Since $a_n < 0$ and n is even, $f(x)$ falls to the left and the right.

b. $x = 0, x = 1, x = -3$
The roots at 0 and –3 have odd multiplicity so $f(x)$ crosses the x-axis at those points. The root at 1 has even multiplicity so $f(x)$ touches the axis at (1, 0).

c. $f(0) = -3(0)^3(0-1)^2(0+3) - 0$
The y-intercept is 0.

d. $f(-x) = 3x^3(-x-1)^2(-x+3)$
The graph has neither y-axis nor origin symmetry.

e. The graph has 2 turning points and $2 \le 6 - 1$.

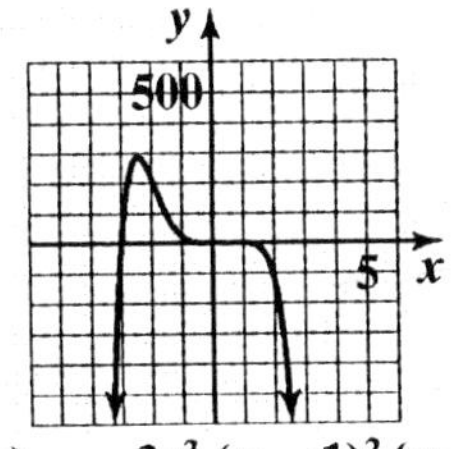

$f(x) = -3x^3(x-1)^2(x+3)$

63. $f(x) = (x-2)^2(x+4)(x-1)$

a. Since $a_n > 0$ and n is even, $f(x)$ rises to the left and rises the right.

b. $x = 2, x = -4, x = 1$
The zeros at –4 and 1 have odd multiplicity so $f(x)$ crosses the x-axis at those points. The root at 2 has even multiplicity so $f(x)$ touches the axis at (2, 0).

c. $f(0) = (0-2)^2(0+4)(0-1) = -16$
The y-intercept is –16.

d. $f(-x) = (-x-2)^2(-x+4)(-x-1)$
The graph has neither y-axis nor origin symmetry.

e. The graph has 3 turning points and $3 \le 4 - 1$.

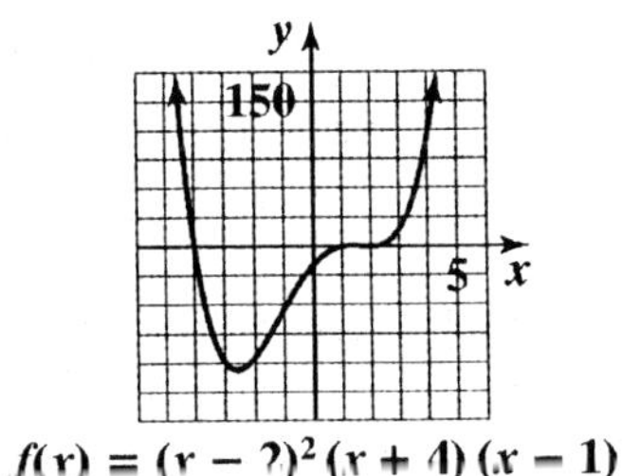

$f(x) = (x-2)^2(x+4)(x-1)$

64. $f(x) = (x+3)(x+1)^3(x+4)$

a. Since $a_n > 0$ and n is odd, $f(x)$ falls to the left and rises to the right.

b. $x = -3, x = -1, x = -4$
The zeros at all have odd multiplicity so $f(x)$ crosses the x-axis at these points.

c. $f(0) = (0+3)(0+1)^3(0+4) = 12$
The y-intercept is 12.

d. $f(-x) = (-x+3)(-x+1)^3(-x+4)$
The graph has neither y-axis nor origin symmetry.

e. The graph has 2 turning points

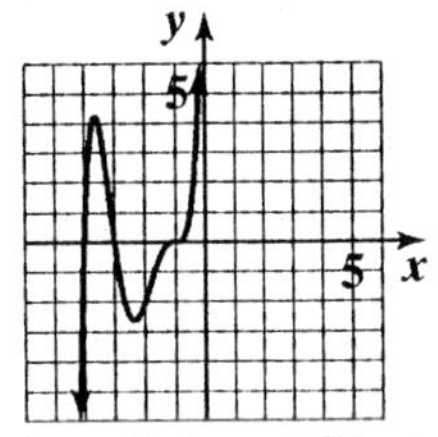

$f(x) = (x+3)(x+1)^3(x+4)$

65. a. The x-intercepts of the graph are -2, 1, and 4, so they are the zeros. Since the graph actually crosses the x-axis at all three places, all three have odd multiplicity.

b. Since the graph has two turning points, the function must be at least of degree 3. Since -2, 1, and 4 are the zeros, $x+2$, $x-1$, and $x-4$ are factors of the function. The lowest odd multiplicity is 1. From the end behavior, we can tell that the leading coefficient must be positive. Thus, the function is $f(x) = (x+2)(x-1)(x-4)$.

c. $f(0) = (0+2)(0-1)(0-4) = 8$

66. **a.** The x-intercepts of the graph are -3, 2, and 5, so they are the zeros. Since the graph actually crosses the x-axis at all three places, all three have odd multiplicity.

b. Since the graph has two turning points, the function must be at least of degree 3. Since -3, 2, and 5 are the zeros, $x+3$, $x-2$, and $x-5$ are factors of the function. The lowest odd multiplicity is 1. From the end behavior, we can tell that the leading coefficient must be positive. Thus, the function is $f(x)=(x+3)(x-2)(x-5)$.

c. $f(0)=(0+3)(0-2)(0-5)=30$

67. **a.** The x-intercepts of the graph are -1 and 3, so they are the zeros. Since the graph crosses the x-axis at -1, it has odd multiplicity. Since the graph touches the x-axis and turns around at 3, it has even multiplicity.

b. Since the graph has two turning points, the function must be at least of degree 3. Since -1 and 3 are the zeros, $x+1$ and $x-3$ are factors of the function. The lowest odd multiplicity is 1, and the lowest even multiplicity is 2. From the end behavior, we can tell that the leading coefficient must be positive. Thus, the function is $f(x)=(x+1)(x-3)^2$.

c. $f(0)=(0+1)(0-3)^2=9$

68. **a.** The x-intercepts of the graph are -2 and 1, so they are the zeros. Since the graph crosses the x-axis at -2, it has odd multiplicity. Since the graph touches the x-axis and turns around at 1, it has even multiplicity.

b. Since the graph has two turning points, the function must be at least of degree 3. Since -2 and 1 are the zeros, $x+2$ and $x-1$ are factors of the function. The lowest odd multiplicity is 1, and the lowest even multiplicity is 2. From the end behavior, we can tell that the leading coefficient must be positive. Thus, the function is $f(x)=(x+2)(x-1)^2$.

c. $f(0)=(0+2)(0-1)^2=2$

69. **a.** The x-intercepts of the graph are -3 and 2, so they are the zeros. Since the graph touches the x-axis and turns around at both -3 and 2, both have even multiplicity.

b. Since the graph has three turning points, the function must be at least of degree 4. Since -3 and 2 are the zeros, $x+3$ and $x-2$ are factors of the function. The lowest even multiplicity is 2. From the end behavior, we can tell that the leading coefficient must be negative. Thus, the function is $f(x)=-(x+3)^2(x-2)^2$.

c. $f(0)=-(0+3)^2(0-2)^2=-36$

70. **a.** The x-intercepts of the graph are -1 and 4, so they are the zeros. Since the graph touches the x-axis and turns around at both -1 and 4, both have even multiplicity.

b. Since the graph has two turning points, the function must be at least of degree 3. Since -1 and 4 are the zeros, $x+1$ and $x-4$ are factors of the function. The lowest even multiplicity is 2. From the end behavior, we can tell that the leading coefficient must be negative. Thus, the function is $f(x)=-(x+1)^2(x-4)^2$.

c. $f(0)=-(0+1)^2(0-4)^2=-16$

71. **a.** The x-intercepts of the graph are -2, -1, and 1, so they are the zeros. Since the graph crosses the x-axis at -1 and 1, they both have odd multiplicity. Since the graph touches the x-axis and turns around at -2, it has even multiplicity.

b. Since the graph has five turning points, the function must be at least of degree 6. Since -2, -1, and 1 are the zeros, $x+2$, $x+1$, and $x-1$ are factors of the function. The lowest even multiplicity is 2, and the lowest odd multiplicity is 1. However, to reach degree 6, one of the odd multiplicities must be 3. From the end behavior, we can tell that the leading coefficient must be positive. The function is $f(x)=(x+2)^2(x+1)(x-1)^3$.

c. $f(0)=(0+2)^2(0+1)(0-1)^3=-4$

72. **a.** The x-intercepts of the graph are -2, -1, and 1, so they are the zeros. Since the graph crosses the x-axis at -2 and 1, they both have odd multiplicity. Since the graph touches the x-axis and turns around at -1, it has even multiplicity.

b. Since the graph has five turning points, the function must be at least of degree 6. Since -2, -1, and 1 are the zeros, $x+2$, $x+1$, and $x-1$ are factors of the function.

The lowest even multiplicity is 2, and the lowest odd multiplicity is 1. However, to reach degree 6, one of the odd multiplicities must be 3. From the end behavior, we can tell that the leading coefficient must be positive. The function is $f(x)=(x+2)(x+1)^2(x-1)^3$.

c. $f(0)=(0+2)(0+1)^2(0-1)^3=-2$

73.

$$f(x)=-2212x^2+57{,}575x+107{,}896$$
$$f(10)=-2212(10)^2+57{,}575(10)+107{,}896=462{,}446$$
$$g(x)=-84x^3-702x^2+50{,}609x+113{,}435$$
$$g(10)=-84(10)^3-702(10)^2+50{,}609(10)+113{,}435=465{,}325$$

The graph indicates that f provides a better description because $f(10)$ is closer to the actual amount of 458,551.

74.

$$f(x)=-2212x^2+57{,}575x+107{,}896$$
$$f(12)=-2212(12)^2+57{,}575(12)+107{,}896=480{,}268$$
$$g(x)=-84x^3-702x^2+50{,}609x+113{,}435$$
$$g(12)=-84(12)^3-702(12)^2+50{,}609(12)+113{,}435=474{,}503$$

The graph indicates that f provides a better description because $f(12)$ is closer to the actual amount of 483,920.

75. The leading coefficient is negative and the degree is even. This means that the graph will fall to the right. This function will not be useful in modeling the number of cumulative AIDS deaths over an extended period of time because the cumulative number of deaths cannot decrease.

76. The leading coefficient is negative and the degree is odd. This means that the graph will fall to the right. This function will not be useful in modeling the number of cumulative AIDS deaths over an extended period of time because the cumulative number of deaths cannot decrease.

77. **a.** The percentage of white classmates was increasing from 1970 through 1980 and from 1985 through 1990.

b. The percentage of white classmates was decreasing from 1980 through 1985 and from 1990 through 2002.

c. Three turning points are shown in the graph.

d. Since there are three turning points, the degree of the polynomial function of best fit would be 4.

e. The leading coefficient would be negative because the graph falls to the left and falls to the right.

78. **a.** From Stage I to Stage IV and from Stage VI to Stage VII marital satisfaction for women is decreasing.

b. From Stage IV to Stage VI and from Stage VII to State VIII marital satisfaction for women is increasing.

c. There are 3 turning points.

d. The degree of the polynomial would be 4.

e. The leading coefficient would be positive since both ends of the graph are rising.

97.

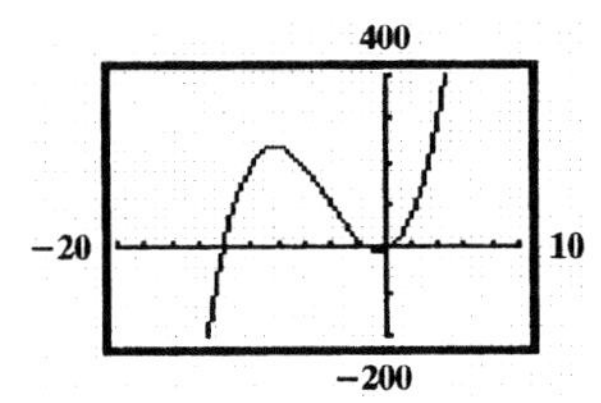

98.

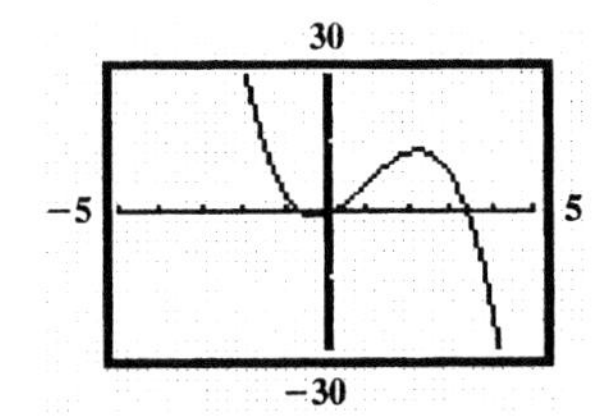

99.

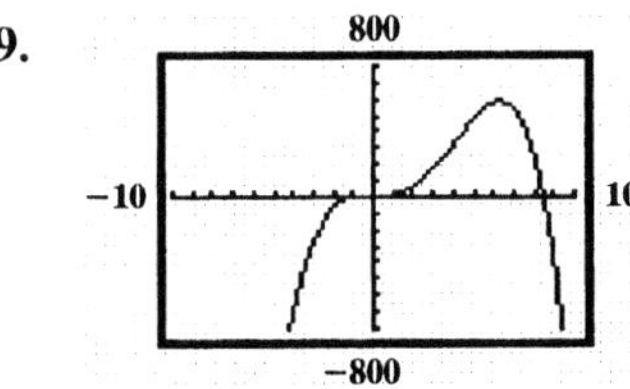

100.

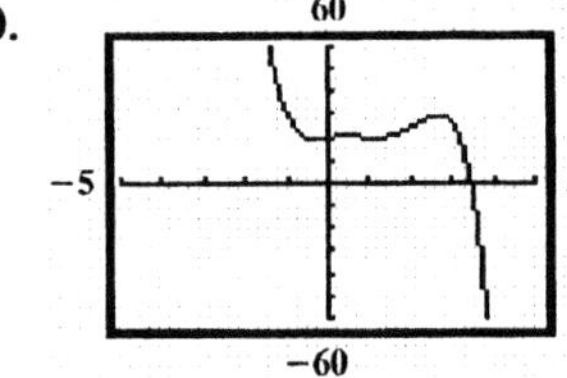

101.

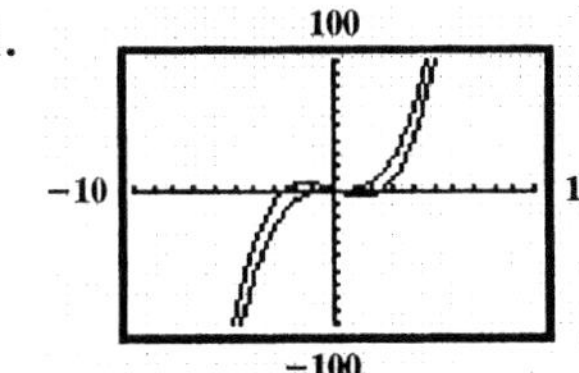

102.

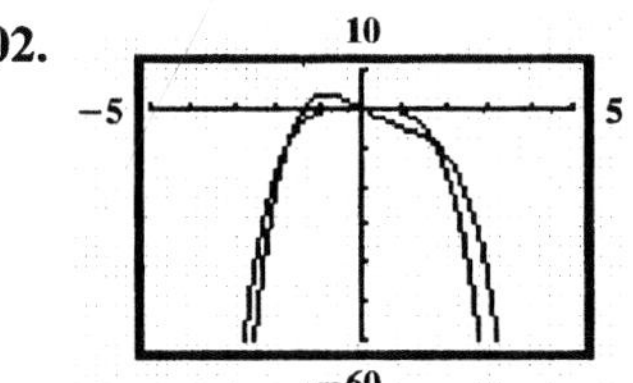

103. **a.** False; $f(x)$ falls to the left and rises to the right.

b. False

c. True; There are many 3rd degree polynomials with the same three x-intercepts.

d. False; a function with origin symmetry either falls to the left and rises to the right, or rises to the left and falls to the right.

(c) is true.

104. $f(x) = x^3 + x^2 - 12x$

105. $f(x) = x^3 - 2x^2$

Section 2.4

Check Point Exercises

1.
$$\begin{array}{r} x+5 \\ x+9\overline{)x^2+14x+45} \\ \underline{x^2+9x} \\ 5x+45 \\ \underline{5x+45} \\ 0 \end{array}$$

The answer is $x + 5$.

2.
$$\begin{array}{r} 2x^2+3x-2 \\ x-3\overline{)2x^3-3x^2-11x+7} \\ \underline{2x^3-6x^2} \\ 3x^2-11x \\ \underline{3x^2-9x} \\ -2x+7 \\ \underline{-2x+6} \\ 1 \end{array}$$

The answer is $2x^2 + 3x - 2 + \dfrac{1}{x-3}$.

3.
$$\begin{array}{r} 2x^2+7x+14 \\ x^2-2x\overline{\smash{\big)}\,2x^4+3x^3+0x^2-7x-10} \\ \underline{2x^4-4x^3} \qquad\qquad\qquad\quad \\ 7x^3+0x^2 \qquad\qquad\quad \\ \underline{7x^3-14x^2} \qquad\qquad\quad \\ 14x^2-7x \qquad\quad \\ \underline{14x^2-28x} \qquad\quad \\ 21x-10 \end{array}$$

The answer is $2x^2+7x+14+\dfrac{21x-10}{x^2-2x}$.

4.

−2	1	0	−7	−6
		−2	4	6
	1	−2	−3	0

The answer is x^2-2x-3.

5.

−4	3	4	−5	3
		−12	32	−108
	3	−8	27	−105

$f(-4)=-105$

6.

−1	15	14	−3	−2
		−15	1	2
	15	−1	−2	0

$15x^2-x-2=0$

$(3x+1)(5x-2)=0$

$x=-\dfrac{1}{3}$ or $x=\dfrac{2}{5}$

The solution set is $\left\{-1,-\dfrac{1}{3},\dfrac{2}{5}\right\}$.

Exercise Set 2.4

1.
$$\begin{array}{r} x+3 \\ x+5\overline{\smash{\big)}\,x^2+8x+15} \\ \underline{x^2+5x} \qquad \\ 3x+15 \\ \underline{3x+15} \\ 0 \end{array}$$

The answer is $x+3$.

2.
$$\begin{array}{r} x+5 \\ x-2\overline{\smash{\big)}\,x^2+3x-10} \\ \underline{x^2-2x} \qquad \\ 5x-10 \\ \underline{5x-10} \\ 0 \end{array}$$

The answer is $x+5$.

3.
$$\begin{array}{r} x^2+3x+1 \\ x+2\overline{\smash{\big)}\,x^3+5x^2+7x+2} \\ \underline{x^3+2x^2} \qquad\qquad \\ 3x^2+7x \qquad \\ \underline{3x^2+6x} \qquad \\ x+2 \\ \underline{x+2} \\ 0 \end{array}$$

The answer is x^2+3x+1.

4.
$$\begin{array}{r} x^2+x-2 \\ x-3\overline{\smash{\big)}\,x^3-2x^2-5x+6} \\ \underline{x^3-3x^2} \qquad\qquad \\ x^2-5x \qquad \\ \underline{x^2-3x} \qquad \\ -2x+6 \\ \underline{-2x+6} \\ 0 \end{array}$$

The answer is x^2+x-2.

5.
$$\begin{array}{r} 2x^2+3x+5 \\ 3x-1\overline{\smash{\big)}\,6x^3+7x^2+12x-5} \\ \underline{6x^3-2x^2} \qquad\qquad \\ 9x^2+12x \qquad \\ \underline{9x^2-3x} \qquad \\ 15x-5 \\ \underline{15x-5} \\ 0 \end{array}$$

The answer is $2x^2+3x+5$.

6.
$$3x+4\overline{)6x^3+17x^2+27x+20}$$
Quotient: $2x^2+3x+5$

$\underline{6x^3+8x^2}$
$9x^2+27x$
$\underline{9x^2+12x}$
$15x+20$
$\underline{15x+20}$
0

The answer is $2x^2+3x+5$.

7.
$$3x-2\overline{)12x^2+x-4}$$
Quotient: $4x+3+\frac{2}{3x-2}$

$\underline{12x^2-8x}$
$9x-4$
$\underline{9x-6}$
2

The answer is $4x+3+\frac{2}{3x-2}$.

8.
$$2x-1\overline{)4x^2-8x+6}$$
Quotient: $2x-3+\frac{3}{2x-1}$

$\underline{4x^2-2x}$
$-6x+6$
$\underline{-6x+6}$
3

The answer is $2x-3+\frac{3}{2x-1}$.

9.
$$x+3\overline{)2x^3+7x^2+9x-20}$$
Quotient: $2x^2+x+6-\frac{38}{x+3}$

$\underline{2x^3+6x^2}$
x^2+9x
$\underline{x^2+3x}$
$6x-20$
$\underline{6x+18}$
-38

The answer is $2x^2+x+6-\frac{38}{x+3}$.

10.
$$x-3\overline{)3x^2-2x+5}$$
Quotient: $3x+7+\frac{26}{x-3}$

$\underline{3x^2-9x}$
$7x+5$
$\underline{7x-21}$
26

The answer is $3x+7+\frac{26}{x-3}$.

11.
$$x-4\overline{)4x^4-4x^2+6x}$$
Quotient: $4x^3+16x^2+60x+246+\frac{984}{x-4}$

$\underline{4x^4-16x^3}$
$16x^3-4x^2$
$\underline{16x^3-64x^2}$
$60x^2+6x$
$\underline{60x^2-240x}$
$246x$
$\underline{246x-984}$
984

The answer is

$$4x^3+16x^2+60x+246+\frac{984}{x-4}.$$

12.
$$x-3\overline{)x^4\qquad\qquad -81}$$
Quotient: $x^3+3x^2+9x+27$

$\underline{x^4-3x^3}$
$3x^3$
$\underline{3x^2-9x^2}$
$9x^2$
$\underline{9x^2-27x}$
$27x-81$
$\underline{27x-81}$
0

The answer is $x^3+3x^2+9x+27$.

13.

$$\begin{array}{r} 2x+5 \\ 3x^2-x-3\overline{)6x^3+13x^2-11x-15} \\ \underline{6x^3-2x^2-6x} \\ 15x^2-5x-15 \\ \underline{15x^2-5x-15} \\ 0 \end{array}$$

The answer is $2x+5$.

14.

$$\begin{array}{r} x^2+x-3 \\ x^2+x-2\overline{)x^4+2x^3-4x^2-5x-6} \\ \underline{x^4+x^3-2x^2} \\ x^3-2x^2-5x \\ \underline{x^3+x^2-2x} \\ -3x^2-3x-6 \\ \underline{-3x^2-3x+6} \\ -12 \end{array}$$

The answer is $x^2+x-3-\dfrac{12}{x^2+x-2}$.

15.

$$\begin{array}{r} 6x^2+3x-1 \\ 3x^2+1\overline{)18x^4+9x^3+3x^2} \\ \underline{18x^4+6x^2} \\ 9x^3-3x^2 \\ \underline{9x^3+3x} \\ -3x^2-3x \\ \underline{-3x^2-1} \\ -3x+1 \end{array}$$

The answer is $6x^2+3x-1-\dfrac{3x-1}{3x^2+1}$.

16.

$$\begin{array}{r} x^2-4x+1 \\ 2x^3+1\overline{)2x^5-8x^4+2x^3+x^2} \\ \underline{2x^5+x^2} \\ -8x^4+2x^3 \\ \underline{-8x^4-4x} \\ 2x^3+4x \\ \underline{2x^3+1} \\ 4x-1 \end{array}$$

The answer is $x^2-4x+1+\dfrac{4x-1}{2x^3+1}$.

17. $(2x^2+x-10)\div(x-2)$

$$\begin{array}{r|rrr} 2 & 2 & 1 & -10 \\ & & 4 & 10 \\ \hline & 2 & 5 & 0 \end{array}$$

The answer is $2x+5$.

18. $(x^2+x-2)\div(x-1)$

$$\begin{array}{r|rrr} 1 & 1 & 1 & -2 \\ & & 1 & 2 \\ \hline & 1 & 2 & 0 \end{array}$$

The answer is $x+2$.

19. $(3x^2+7x-20)\div(x+5)$

$$\begin{array}{r|rrr} -5 & 3 & 7 & -20 \\ & & -15 & 40 \\ \hline & 3 & -8 & 20 \end{array}$$

The answer is $3x-8+\dfrac{20}{x+5}$.

20. $(5x^2-12x-8)\div(x+3)$

$$\begin{array}{r|rrr} -3 & 5 & 12 & -8 \\ & & -15 & 81 \\ \hline & 5 & -27 & 73 \end{array}$$

The answer is $5x-27+\dfrac{73}{x+3}$.

21. $(4x^3-3x^2+3x-1)\div(x-1)$

$$\begin{array}{r|rrrr} 1 & 4 & -3 & 3 & -1 \\ & & 4 & 1 & 4 \\ \hline & 4 & 1 & 4 & 3 \end{array}$$

The answer is $4x^2+x+4+\dfrac{3}{x-1}$.

22. $(5x^3 - 6x^2 + 3x + 11) \div (x - 2)$

2	5	–6	3	11
		10	8	22
	5	4	11	33

The answer is $5x^2 + 4x + 11 + \frac{33}{x-2}$.

23. $(6x^5 - 2x^3 + 4x^2 - 3x + 1) \div (x - 2)$

2	6	0	–2	4	–3	1
		12	24	44	96	186
	6	12	22	48	93	187

The answer is

$6x^4 + 12x^3 + 22x^2 + 48x + 93 + \frac{187}{x-2}$.

24. $(x^5 + 4x^4 - 3x^2 + 2x + 3) \div (x - 3)$

3	1	4	0	–3	2	3
		3	21	63	180	546
	1	7	21	60	182	549

The answer is

$x^4 + 7x^3 + 21x^2 + 60x + 182 + \frac{549}{x-3}$.

25. $(x^2 - 5x - 5x^3 + x^4) \div (5 + x) \Rightarrow$

$(x^4 - 5x^3 + x^2 - 5x) \div (x + 5)$

–5	1	–5	1	–5	0
		–5	50	–255	1300
	1	–10	51	–260	1300

The answer is

$x^3 - 10x^2 + 51x - 260 + \frac{1300}{x+5}$.

26. $(x^2 - 6x - 6x^3 + x^4) \div (6 + x) \Rightarrow$

$(x^4 - 6x^3 + x^2 - 6x) \div (x + 6)$

–6	1	–6	1	–6	0
		–6	72	–438	2664
	1	–12	73	–444	2664

The answer is

$x^3 - 12x^2 + 73x - 444 + \frac{2664}{x+6}$.

27. $\frac{x^5 + x^3 - 2}{x - 1}$

1	1	0	1	0	0	–2
		1	1	2	2	2
	1	1	2	2	2	0

The answer is $x^4 + x^3 + 2x^2 + 2x + 2$.

28. $\frac{x^7 + x^5 - 10x^3 + 12}{x + 2}$

–2	1	0	1	0	–10	0	0	12
		–2	4	–10	20	–20	40	–80
	1	–2	5	–10	10	–20	40	–68

The answer is $x^6 - 2x^5 + 5x^4 - 10x^3 + 10x^2$

$-20x + 40 - \frac{68}{x+2}$.

29. $\frac{x^4 - 256}{x - 4}$

4	1	0	0	0	–256
		4	16	64	256
	1	4	16	64	0

The answer is $x^3 + 4x^2 + 16x + 64$.

30. $\frac{x^7 - 128}{x - 2}$

2	1	0	0	0	0	0	0	–128
		2	4	8	16	32	64	128
	1	2	4	8	16	32	64	0

The answer is

$x^6 + 2x^5 + 4x^4 + 8x^3 + 16x^2 + 32x + 64$.

31. $\dfrac{2x^5 - 3x^4 + x^3 - x^2 + 2x - 1}{x+2}$

−2	2	−3	1	−1	2	−1
		−4	14	−30	62	−128
	2	−7	15	−31	64	−129

The answer is

$2x^4 - 7x^3 + 15x^2 - 31x + 64 - \dfrac{129}{x+2}$.

32. $\dfrac{x^5 - 2x^4 - x^3 + 3x^2 - x + 1}{x-2}$

2	1	−2	−1	3	−1	1
		2	0	−2	2	2
	1	0	−1	1	1	3

The answer is $x^4 - x^2 + x + 1 + \dfrac{3}{x-2}$.

33. $f(x) = 2x^3 - 11x^2 + 7x - 5$

4	2	−11	7	−5
		8	−12	−20
	2	−3	−5	−25

$f(4) = -25$

34.

3	1	−7	5	−6
		3	−12	−21
	1	−4	−7	−27

$f(3) = -27$

35. $f(x) = 3x^3 - 7x^2 - 2x + 5$

−3	3	−7	−2	5
		−9	48	−138
	3	−16	46	−133

$f(-3) = -133$

36.

−2	4	5	−6	−4
		−8	6	0
	4	−3	0	−4

$f(-2) = -4$

37. $f(x) = x^4 + 5x^3 + 5x^2 - 5x - 6$

3	1	5	5	−5	−6
		3	24	87	246
	1	8	29	82	240

$f(3) = 240$

38.

2	1	−5	5	5	−6
		2	−6	−2	6
	1	−3	−1	3	0

$f(2) = 0$

39. $f(x) = 2x^4 - 5x^3 - x^2 + 3x + 2$

$-\frac{1}{2}$	2	−5	−1	3	2
		−1	3	−1	−1
	2	−6	2	2	1

$f\left(-\dfrac{1}{2}\right) = 1$

40.

$-\frac{2}{3}$	6	10	5	1	1
		−4	−4	$-\frac{2}{3}$	$-\frac{2}{9}$
	6	6	1	$\frac{1}{3}$	$\frac{7}{9}$

$f\left(-\frac{2}{3}\right) = \frac{7}{9}$

41. Dividend: $x^3 - 4x^2 + x + 6$
Divisor: $x + 1$

−1	1	−4	1	6
		−1	5	−6
	1	−5	6	0

The quotient is $x^2 - 5x + 6$.

$(x+1)(x^2 - 5x + 6) = 0$

$(x + 1)(x - 2)(x - 3) = 0$

$x = -1, x = 2, x = 3$

The solution set is $\{-1, 2, 3\}$.

42. Dividend: $x^3 - 2x^2 - x + 2$
Divisor: $x + 1$

$$\begin{array}{r|rrrr} -1 & 1 & -2 & -1 & 2 \\ & & -1 & 3 & -2 \\ \hline & 1 & -3 & 2 & 0 \end{array}$$

The quotient is $x^2 - 3x + 2$.
$(x+1)(x^2 - 3x + 2) = 0$
$(x+1)(x-2)(x-1) = 0$
$x = -1, x = 2, x = 1$
The solution set is $\{-1, 2, 1\}$.

43. $2x^3 - 5x^2 + x + 2 = 0$

$$\begin{array}{r|rrrr} 2 & 2 & -5 & 1 & 2 \\ & & 4 & -2 & -2 \\ \hline & 2 & -1 & -1 & 0 \end{array}$$

$(x-2)(2x^2 - x - 1) = 0$
$(x-2)(2x+1)(x-1) = 0$
$x = 2,\ x = -\dfrac{1}{2},\ x = 1$
The solution set is $\left\{-\dfrac{1}{2}, 1, 2\right\}$.

44. $2x^3 - 3x^2 - 11x + 6 = 0$

$$\begin{array}{r|rrrr} -2 & 2 & -3 & -11 & 6 \\ & & -4 & 14 & -6 \\ \hline & 2 & -7 & 3 & 0 \end{array}$$

$(x+2)(2x^2 - 7x + 3) = 0$
$(x+2)(2x-1)(x-3) = 0$
$x = -2,\ x = \dfrac{1}{2},\ x = 3$
The solution set is $\left\{-2, \dfrac{1}{2}, 3\right\}$.

45. $12x^3 + 16x^2 - 5x - 3 = 0$

$$\begin{array}{r|rrrr} -\frac{3}{2} & 12 & 16 & -5 & -3 \\ & & -18 & 3 & 3 \\ \hline & 12 & -2 & -2 & 0 \end{array}$$

$\left(x + \dfrac{3}{2}\right)(12x^2 - 2x - 2) = 0$
$\left(x + \dfrac{3}{2}\right)2\left(6x^2 - x - 1\right) = 0$
$\left(x + \dfrac{3}{2}\right)2(3x+1)(2x-1) = 0$
$x = -\dfrac{3}{2},\ x = -\dfrac{1}{3},\ x = \dfrac{1}{2}$
The solution set is $\left\{-\dfrac{3}{2}, -\dfrac{1}{3}, \dfrac{1}{2}\right\}$.

46. $3x^3 + 7x^2 - 22x - 8 = 0$

$$\begin{array}{r|rrrr} -\frac{1}{3} & 3 & 7 & -22 & -8 \\ & & -1 & -2 & 8 \\ \hline & 3 & 6 & -24 & 0 \end{array}$$

$\left(x + \dfrac{1}{3}\right)3x^2 + 6x - 24 = 0$
$\left(x + \dfrac{1}{3}\right)3(x+4)(x-2) = 0$
$x = -4, x = 2, x = -\dfrac{1}{3}$
The solution set is $\left\{-4, -\dfrac{1}{3}, 2\right\}$.

47. The graph indicates that 2 is a solution to the equation.

$$\begin{array}{r|rrrr} 2 & 1 & 2 & -5 & -6 \\ & & 2 & 8 & 6 \\ \hline & 1 & 4 & 3 & 0 \end{array}$$

The remainder is 0, so 2 is a solution.
$x^3 + 2x^2 - 5x - 6 = 0$
$(x-2)\left(x^2 + 4x + 3\right) = 0$
$(x-2)(x+3)(x+1) = 0$
The solutions are 2, -3, and -1, or $\{-3, -1, 2\}$.

48. The graph indicates that -3 is a solution to the equation.

$$\begin{array}{r|rrrr} -3 & 2 & 1 & -13 & 6 \\ & & -6 & 15 & -6 \\ \hline & 2 & -5 & 2 & 0 \end{array}$$

The remainder is 0, so -3 is a solution.

$2x^3+x^2-13x+6=0$

$(x+3)(2x^2-5x+2)=0$

$(x+3)(2x-1)(x-2)=0$

The solutions are -3, $\frac{1}{2}$, and 2, or $\left\{-3,\frac{1}{2},2\right\}$.

49. The table indicates that 1 is a solution to the equation.

$$\begin{array}{r|rrrr} 1 & 6 & -11 & 6 & -1 \\ & & 6 & -5 & 1 \\ \hline & 6 & -5 & 1 & 0 \end{array}$$

The remainder is 0, so 1 is a solution.

$6x^3-11x^2+6x-1=0$

$(x-1)(6x^2-5x+1)=0$

$(x-1)(3x-1)(2x-1)=0$

The solutions are 1, $\frac{1}{3}$, and $\frac{1}{2}$, or $\left\{\frac{1}{3},\frac{1}{2},1\right\}$.

50. The table indicates that 1 is a solution to the equation.

$$\begin{array}{r|rrrr} 1 & 2 & 11 & -7 & -6 \\ & & 2 & 13 & -6 \\ \hline & 2 & 13 & 6 & 0 \end{array}$$

The remainder is 0, so 1 is a solution.

$2x^3+11x^2-7x-6=0$

$(x-1)(2x^2+13x+6)=0$

$(x-1)(2x+1)(x+6)=0$

The solutions are 1, $-\frac{1}{2}$, and -6, or $\left\{-6,-\frac{1}{2},1\right\}$.

51. **a.** $14x^3-17x^2-16x-177=0$

$$\begin{array}{r|rrrr} 3 & 14 & -17 & -16 & -177 \\ & & 42 & 75 & 177 \\ \hline & 14 & 25 & 59 & 0 \end{array}$$

The remainder is 0 so 3 is a solution.

$14x^3-17x^2-16x-177$

$=(x-3)(14x^2+25x+59)$

b. $f(x)=14x^3-17x^2-16x+34$

We need to find x when $f(x)=211$.

$f(x)=14x^3-17x^2-16x+34$

$211=14x^3-17x^2-16x+34$

$0=14x^3-17x^2-16x-177$

This is the equation obtained in part **a.** One solution is 3. It can be used to find other solutions (if they exist).

$14x^3-17x^2-16x-177=0$

$(x-3)(14x^2+25x+59)=0$

The polynomial $14x^2+25x+59$ cannot be factored, so the only solution is $x=3$. The female moth's abdominal width is 3 millimeters.

52. **a.**

$$\begin{array}{r|rrrr} 2 & 2 & 14 & 0 & -72 \\ & & 4 & 36 & 72 \\ \hline & 2 & 18 & 36 & 0 \end{array}$$

$2h^3+14h^2-72=(h-2)(2h^2+18h+36)$

b. $V=lwh$

$72=(h+7)(2h)(h)$

$72=2h^3+14h^2$

$0=2h^3+14h^2-72$

$0=(h-2)(2h^2+18h+36)$

$0=(h-2)(2(h^2+9h+18))$

$0=(h-2)(2(h+6)(h+3))$

$0=2(h-2)(h+6)(h+3)$

$2(h-2)=0 \quad h+6=0 \quad h+3=0$

$h-2=0 \quad h=-6 \quad h=-3$

$h=2$

The height is 2 inches, the width is $2\cdot 2=4$ inches and the length is $2+7=9$ inches. The dimensions are 2 inches by 4 inches by 9 inches.

53. $A = l \cdot w$ so

$$l = \frac{A}{w} = \frac{0.5x^3 - 0.3x^2 + 0.22x + 0.06}{x + 0.2}$$

$$\begin{array}{r|rrrr} -0.2 & 0.5 & -0.3 & 0.22 & 0.06 \\ & & -0.1 & 0.08 & -0.06 \\ \hline & 0.5 & -0.4 & 0.3 & 0 \end{array}$$

Therefore, the length of the rectangle is $0.5x^2 - 0.4x + 0.3$ units.

54. $A = l \cdot w$ so,

$$l = \frac{A}{w} = \frac{8x^3 - 6x^2 - 5x + 3}{x + \frac{3}{4}}$$

$$\begin{array}{r|rrrr} -\frac{3}{4} & 8 & -6 & -5 & 3 \\ & & -6 & 9 & -3 \\ \hline & 8 & -12 & 4 & 0 \end{array}$$

Therefore, the length of the rectangle is $8x^2 - 12x + 4$ units.

55. a.

$$f(30) = \frac{80(30) - 8000}{30 - 110} = 70$$

(30, 70) At a 30% tax rate, the government tax revenue will be $70 ten billion.

b.

$$\begin{array}{r|rr} 110 & 80 & -8000 \\ & & 8800 \\ \hline & 80 & 800 \end{array}$$

$$f(x) = 80 + \frac{800}{x - 110}$$

$$f(30) = 80 + \frac{800}{80 - 110} = 70$$

(30, 70) same answer as in **a.**

c. $f(x)$ is not a polynomial function. It is a rational function because it is the quotient of two linear polynomials.

56. a.

$$f(40) = \frac{80(40) - 8000}{40 - 110} = 68.57$$

(40, 68.57) At a 40% tax rate, the government's revenue is $68.57 ten billion.

b.

$$\begin{array}{r|rr} 110 & 80 & -8000 \\ & & 8800 \\ \hline & 80 & 800 \end{array}$$

$$f(x) = 80 + \frac{800}{x - 110}$$

$$f(40) = 80 + \frac{800}{40 - 110}$$
$$= 68.57$$

c. $f(x)$ is not a polynomial function. It is a rational function because it is the quotient of two linear polynomials.

65. Exercise 43:

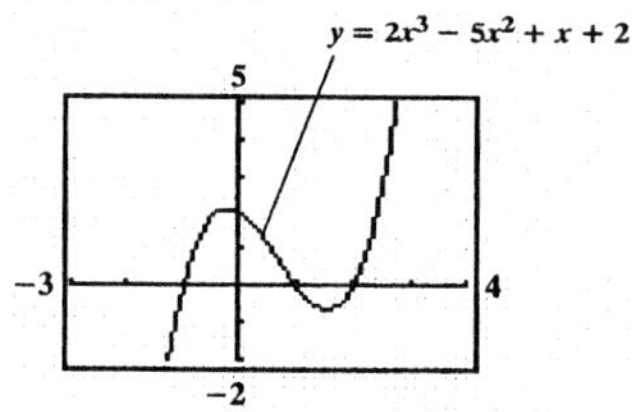

Exercise 44:

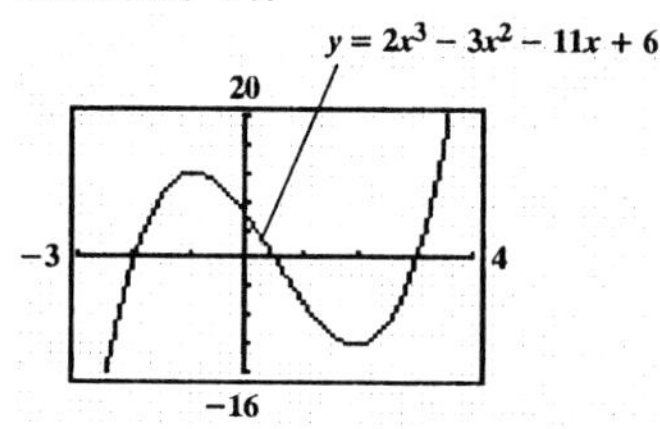

Exercise 45

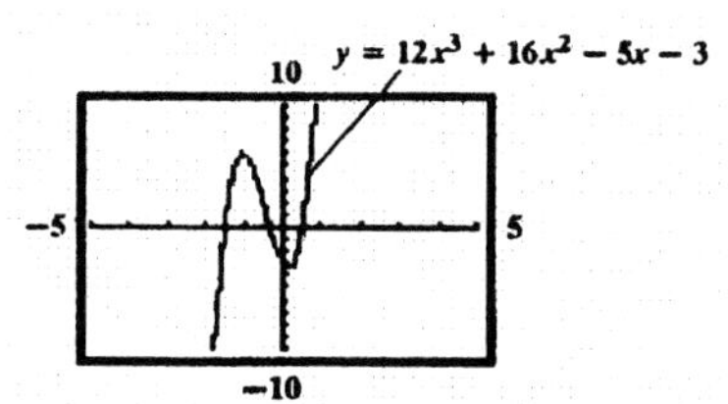

Exercise 46:

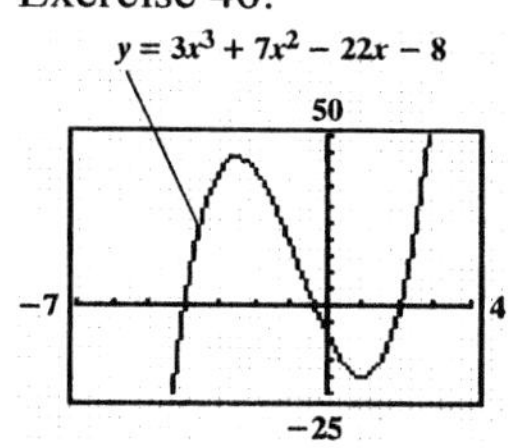

66. a. False. The degree of the quotient is 3, since $\frac{x^6}{x^3} = x^3$.

b. False. Synthetic division could be used since $x - \frac{1}{2}$ is in the form $x - c$.

c. True

d. False. The divisor is a factor of the divided only if the remainder is the whole number 0.

(c) is true.

67.

$$\begin{array}{r} 5x^2 + 2x - 4 \\ 4x+3 \overline{)20x^3 + 23x^2 - 10x + k} \\ \underline{20x^3 + 15x^2} \\ 8x^2 - 10 \\ \underline{8x^2 + 6x} \\ -16x + k \\ \underline{-16x - 12} \end{array}$$

To get a remainder of zero, k must equal −12.
$k = -12$

68. $f(x) = d(x) \cdot q(x) + r(x)$

$2x^2 - 7x + 9 = d(x)(2x - 3) + 3$

$2x^2 - 7x + 6 = d(x)(2x - 3)$

$\frac{2x^2 - 7x + 6}{2x - 3} = d(x)$

$$\begin{array}{r} x - 2 \\ 2x-3 \overline{)2x^2 - 7x + 6} \\ \underline{2x^2 - 3x} \\ -4x + 6 \\ \underline{-4x + 6} \end{array}$$

The polynomial is $x - 2$.

69.

$$\begin{array}{r} x^{2n} - x^n + 1 \\ x^n + 1 \overline{)x^{3n} \qquad\qquad + 1} \\ \underline{x^{3n} + x^{2n}} \\ -x^{2n} \\ \underline{-x^{2n} - x^n} \\ x^n + 1 \\ \underline{x^n + 1} \\ 0 \end{array}$$

70. $2x - 4 = 2(x - 2)$
Use synthetic division to divide by $x - 2$. Then divide the quotient by 2.

71. $x^4 - 4x^3 - 9x^2 + 16x + 20 = 0$

$$\begin{array}{r|rrrrr} 5 & 1 & -4 & -9 & 16 & 20 \\ & & 5 & 5 & -20 & -20 \\ \hline & 1 & 1 & -4 & -4 & 0 \end{array}$$

The remainder is zero and 5 is a solution to the equation.

$x^4 - 4x^3 - 9x^2 + 16x + 20$
$= (x - 5)(x^3 + x^2 - 4x - 4)$

To solve the equation, we set it equal to zero and factor.

$(x - 5)(x^3 + x^2 - 4x - 4) = 0$

$(x - 5)(x^2(x + 1) - 4(x + 1)) = 0$

$(x - 5)(x + 1)(x^2 - 4) = 0$

$(x - 5)(x + 1)(x + 2)(x - 2) = 0$

Apply the zero product principle.

$x - 5 = 0 \qquad x + 1 = 0$
$x = 5 \qquad x = -1$

$x + 2 = 0 \qquad x - 2 = 0$
$x = -2 \qquad x = 2$

The solutions are −2, −1, 2 and 5 and the solution set is $\{-2, -1, 2, 5\}$.

Section 2.5

Check Point Exercises

1. $p: \pm 1, \pm 2, \pm 3, \pm 6$
$q: \pm 1$
$\frac{p}{q}: \pm 1, \pm 2, \pm 3, \pm 6$
are the possible rational zeros.

2. $p: \pm 1, \pm 3$
$q: \pm 1, \pm 2, \pm 4$
$\frac{p}{q}: \pm 1, \pm 3, \pm \frac{1}{2}, \pm \frac{1}{4}, \pm \frac{3}{2}, \pm \frac{3}{4}$
are the possible rational zeros.

3. $\pm 1, \pm 2, \pm 4, \pm 5, \pm 10, \pm 20$ are possible rational zeros

1	1	8	11	–20
		1	9	20
	1	9	20	0

1 is a zero.
$x^2 + 9x + 20 = 0$
$(x+4)(x+5) = 0$
$x = -4$ or $x = -5$
The solution set is $\{1, -4, -5\}$.

4. $\pm 1, \pm 2$ are possible rational zeros

2	1	1	-5	–2
		2	6	2
	1	3	1	0

2 is a zero.
$x^2 + 3x + 1 = 0$

$$x = \frac{-b \pm \sqrt{b^2 - 4ac}}{2a}$$

$$x = \frac{-3 \pm \sqrt{3^2 - 4(1)(1)}}{2(1)}$$

$$= \frac{-3 \pm \sqrt{5}}{2}$$

The solution set is $\left\{2, \frac{-3+\sqrt{5}}{2}, \frac{-3-\sqrt{5}}{2}\right\}$.

5. $\pm 1, \pm 13$ are possible rational zeros.

1	1	–6	22	–30	13
		1	–5	17	–13
	1	–5	17	–13	0

1 is a zero.

1	1	5	17	–13
		1	–4	13
	1	–4	13	0

1 is a double root.
$x^2 - 4x + 13 = 0$

$$x = \frac{4 \pm \sqrt{16-52}}{2} = \frac{4 \pm \sqrt{-36}}{2} = 2 + 3i$$

The solution set is $\{1, 2+3i, 2-3i\}$.

6. $(x+3)(x-i)(x+i) = (x+3)(x^2+1)$
$f(x) = a_n(x+3)(x^2+1)$
$f(1) = a_n(1+3)(1^2+1) = 8a_n = 8$
$a_n = 1$
$f(x) = (x+3)(x^2+1)$ or $x^3 + 3x^2 + x + 3$

7. $f(x) = x^4 - 14x^3 + 71x^2 - 154x + 120$
$f(-x) = x^4 + 14x^3 + 71x^2 + 154x + 120$
Since $f(x)$ has 4 changes of sign, there are 4, 2, or 0 positive real zeros.
Since $f(-x)$ has no changes of sign, there are no negative real zeros.

Exercise Set 2.5

1. $f(x) = x^3 + x^2 - 4x - 4$
$p: \pm 1, \pm 2, \pm 4$
$q: \pm 1$
$\frac{p}{q}: \pm 1, \pm 2, \pm 4$

2. $f(x) = x^3 + 3x^2 - 6x - 8$
$p: \pm 1, \pm 2, \pm 4, \pm 8$
$q: \pm 1$
$\frac{p}{q}: \pm 1, \pm 2, \pm 4, \pm 8$

3. $f(x) = 3x^4 - 11x^3 - x^2 + 19x + 6$

p: ±1, ±2, ±3, ±6

q: ±1, ±3

$\frac{p}{q}: \pm 1, \pm 2, \pm 3, \pm 6, \pm\frac{1}{3}, \pm\frac{2}{3}$

4. $f(x) = 2x^4 + 3x^3 - 11x^2 - 9x + 15$

p: ±1, ±3, ±5, ±15

q: ±1, ±2

$\frac{p}{q}: \pm 1, \pm 3, \pm 5, \pm 15, \pm\frac{1}{2}, \pm\frac{3}{2}, \pm\frac{5}{2}, \pm\frac{15}{2}$

5. $f(x) = 4x^4 - x^3 + 5x^2 - 2x - 6$

p: ±1, ±2, ±3, ±6

q: ±1, ±2, ±4

$\frac{p}{q}: \pm 1, \pm 2, \pm 3, \pm 6, \pm\frac{1}{2}, \pm\frac{1}{4}, \pm\frac{3}{2}, \pm\frac{3}{4}$

6. $f(x) = 3x^4 - 11x^3 - 3x^2 - 6x + 8$

p: ±1, ±2, ±4, ±8

q: ±1, ±3

$\frac{p}{q}: \pm 1, \pm 2, \pm 4, \pm 8, \pm\frac{1}{3}, \pm\frac{2}{3}, \pm\frac{4}{3}, \pm\frac{8}{3}$

7. $f(x) = x^5 - x^4 - 7x^3 + 7x^2 - 12x - 12$

$p: \pm 1, \pm 2, \pm 3 \pm 4 \pm 6 \pm 12$

$q: \pm 1$

$\frac{p}{q}: \pm 1, \pm 2, \pm 3 \pm 4 \pm 6 \pm 12$

8. $f(x) = 4x^5 - 8x^4 - x + 2$

$p: \pm 1, \pm 2$

$q: \pm 1, \pm 2, \pm 4$

$\frac{p}{q}: \pm 1, \pm 2, \pm\frac{1}{2}, \pm\frac{1}{4}$

9. $f(x) = x^3 + x^2 - 4x - 4$

a. $p: \pm 1, \pm 2, \pm 4$

$q: \pm 1$

$\frac{p}{q}: \pm 1, \pm 2, \pm 4$

b.

2	1	1	−4	−4
		2	6	4
	1	3	2	0

2 is a zero.

c. $x^3 + x^2 - 4x - 4 = 0$

$(x-2)(x^2 + 3x + 2) = 0$

$(x-2)(x+2)(x+1) = 0$

$x - 2 = 0 \quad x + 2 = 0 \quad x + 1 = 0$

$x = 2, \ x = -2, \ x = -1$

The solution set is $\{2, -2, -1\}$.

10. a. $f(x) = x^3 - 2x^2 - 11x + 12$

p: ±1, +2, ±3, ±4, ±6, ±12

q: ±1

$\frac{p}{q}: \pm 1, \pm 2, \pm 3, \pm 4, \pm 6, \pm 12$

b.

4	1	−2	−11	12
		4	8	−12
	1	2	−3	0

4 is a zero.

c. $x^3 - 2x^2 - 11x + 12 = 0$

$(x-4)\left(x^2 + 2x - 3\right) = 0$

$(x-4)\ (x+3)(x-1) = 0$

$x = 4, \ x = -3, \ x = 1$

The solution set is $\{4, -3, 1\}$.

11. $f(x) = 2x^3 - 3x^2 - 11x + 6$

a. $p: \pm 1, \pm 2, \pm 3, \pm 6$

$q: \pm 1, \pm 2$

$\frac{p}{q}: \pm 1, \pm 2, \pm 3, \pm 6, \pm\frac{1}{2}, \pm\frac{3}{2}$

b.

3	2	−3	−11	6
		6	9	−6
	2	3	−2	0

3 is a zero.

c. $2x^3 - 3x^2 - 11x + 6 = 0$

$(x-3)(2x^2 + 3x - 2) = 0$

$(x-3)(2x-1)(x+2) = 0$

$x = 3, \ x = \frac{1}{2}, \ x = -2$

The solution set is $\left\{3, \frac{1}{2}, -2\right\}$.

12. a. $f(x) = 2x^3 - 5x^2 + x + 2$
$p: \pm 1, \pm 2$
$q: \pm 1, \pm 2$
$\frac{p}{q}: \pm 1, \pm 2, \pm \frac{1}{2}$

b.

2	2	–5	1	2
		4	–2	–2
	2	–1	–1	0

2 is a zero.

c. $2x^3 - 5x^2 + x + 2 = 0$
$(x-2)\left(2x^2 - x - 1\right) = 0$
$(x-2)(2x+1)(x-1) = 0$
$x = 2,\ x = -\frac{1}{2},\ x = 1$
The solution set is $\left\{2, -\frac{1}{2}, 1\right\}$.

13. a. $f(x) = x^3 + 4x^2 - 3x - 6$
$p: \pm 1, \pm 2, \pm 3, \pm 6$
$q: \pm 1$
$\frac{p}{q}: \pm 1, \pm 2, \pm 3, \pm 6$

b.

–1	1	4	–3	–6
		–1	–3	6
	1	3	–6	0

–1 is a zero.

c. $x^2 + 3x - 6 = 0$
$$x = \frac{-b \pm \sqrt{b^2 - 4ac}}{2a}$$
$$x = \frac{-3 \pm \sqrt{3^2 - 4(1)(-6)}}{2(1)}$$
$$= \frac{-3 \pm \sqrt{33}}{2}$$
The solution set is
$\left\{-1, \frac{-3+\sqrt{33}}{2}, \frac{-3-\sqrt{33}}{2}\right\}$.

14. a. $f(x) = 2x^3 + x^2 - 3x + 1$
$p: \pm 1$
$q: \pm 1, \pm 2$
$\frac{p}{q}: \pm 1, \pm \frac{1}{2}$

b.

$\frac{1}{2}$	2	1	–3	1
		1	1	–1
	2	2	–2	0

$\frac{1}{2}$ is a zero.

c. $2x^2 + 2x - 2 = 0$
$x^2 + x - 1 = 0$
$$x = \frac{-b \pm \sqrt{b^2 - 4ac}}{2a}$$
$$x = \frac{-1 \pm \sqrt{1^2 - 4(1)(-1)}}{2(1)}$$
$$= \frac{-1 \pm \sqrt{5}}{2}$$
The solution set is
$\left\{\frac{1}{2}, \frac{-1+\sqrt{5}}{2}, \frac{-1-\sqrt{5}}{2}\right\}$.

15. a. $f(x) = 2x^3 + 6x^2 + 5x + 2$
$p: \pm 1, \pm 2$
$q: \pm 1, \pm 2$
$\frac{p}{q}: \pm 1, \pm 2, \pm \frac{1}{2}$

b.

–2	2	6	5	2
		–4	–4	–2
	2	2	1	0

–2 is a zero.

c. $2x^2 + 2x + 1 = 0$

$$x = \frac{-b \pm \sqrt{b^2 - 4ac}}{2a}$$

$$x = \frac{-2 \pm \sqrt{2^2 - 4(2)(1)}}{2(2)}$$

$$= \frac{-2 \pm \sqrt{-4}}{4}$$

$$= \frac{-2 \pm 2i}{4}$$

$$= \frac{-1 \pm i}{2}$$

The solution set is $\left\{-2, \frac{-1+i}{2}, \frac{-1-i}{2}\right\}$.

16. a. $f(x) = x^3 - 4x^2 + 8x - 5$

p: ±1, ±5

q: ±1

$\frac{p}{q}$: ±1, ±5

b.

1	1	–4	8	–5
		1	–3	5
	1	–3	5	0

1 is a zero.

c. $x^2 - 3x + 5 = 0$

$$x = \frac{-b \pm \sqrt{b^2 - 4ac}}{2a}$$

$$x = \frac{-(-3) \pm \sqrt{(-3)^2 - 4(1)(5)}}{2(1)}$$

$$= \frac{3 \pm \sqrt{-11}}{2}$$

$$= \frac{3 \pm i\sqrt{11}}{2}$$

The solution set is $\left\{1, \frac{3+i\sqrt{11}}{2}, \frac{3-i\sqrt{11}}{2}\right\}$.

17. $x^3 - 2x^2 - 11x + 12 = 0$

a. p: ±1, ±2, ±3, ±4, ±6, ±12

q: ±1

$\frac{p}{q}$: ±1, ±2, ±3, ±4, ±6, ±12

b.

4	1	–2	–11	12
		4	8	–12
	1	2	–3	0

4 is a root.

c. $x^3 - 2x^2 - 11x + 12$

$(x-4)(x^2 + 2x - 3) = 0$

$(x-4)(x+3)(x-1) = 0$

$x - 4 = 0 \quad x + 3 = 0 \quad x - 1 = 0$

$x = 4 \quad x = -3 \quad x = 1$

The solution set is $\{-3, 1, 4\}$.

18. a. $x^3 - 2x^2 - 7x - 4 = 0$

p: ±1, ±2, ±4

q: ±1

$\frac{p}{q}$: ±1, ±2, ±4

b.

4	1	–2	–7	–4
		4	8	4
	1	2	1	0

4 is a root.

c. $x^3 + 2x^2 - 7x - 4 = 0$

$(x-4)\left(x^2 + 2x + 1\right) = 0$

$(x-4) \quad (x+1)^2$

$x = 4, \quad x = -1$

The solution set is $\{4, -1\}$.

19. $x^3 - 10x - 12 = 0$

a. p: ±1, ±2, ±3, ±4, ±6, ±12

q: ±1

$\frac{p}{q}$: ±1, ±2, ±3, ±4, ±6, ±12

b.

–2	1	0	–10	–12
		–2	4	12
	1	–2	–6	0

–2 is a root.

c. $x^3 - 10x - 12 = 0$

$(x+2)(x^2 - 2x - 6) = 0$

$x = \frac{2 \pm \sqrt{4+24}}{2} = \frac{2 \pm \sqrt{28}}{2}$

$= \frac{2 \pm 2\sqrt{7}}{2} = 1 \pm \sqrt{7}$

The solution set is $\{-2, 1+\sqrt{7}, 1-\sqrt{7}\}$.

20. a. $x^3 - 5x^2 + 17x - 13 = 0$

$p: \pm 1, \pm 13$

$q: \pm 1$

$\frac{p}{q}: \pm 1, \pm 13$

b.

1	1	–5	17	–13
		1	–4	13
	1	–4	13	0

1 is a root.

c. $x^3 - 5x^2 + 17x - 13 = 0$

$(x-1)\left(x^2 - 4x + 13\right) = 0$

$x = \frac{4 + \sqrt{16-52}}{2} = \frac{4 \pm \sqrt{-36}}{2}$

$= \frac{4 \pm 6i}{2} = 2 \pm 3i$

The solution set is $\{1, 2+3i, 2, -3i\}$.

21. $6x^3 + 25x^2 - 24x + 5 = 0$

a. $p: \pm 1, \pm 5$

$q: \pm 1, \pm 2, \pm 3, \pm 6$

$\frac{p}{q}: \pm 1, \pm 5, \pm\frac{1}{2}, \pm\frac{5}{2}, \pm\frac{1}{3}, \pm\frac{5}{3}, \pm\frac{1}{6}, \pm\frac{5}{6}$

b.

–5	6	25	–24	5
		–30	25	–5
	6	–5	1	0

–5 is a root.

c. $6x^3 + 25x^2 - 24x + 5 = 0$

$(x+5)(6x^2 - 5x + 1) = 0$

$(x+5)(2x-1)(3x-1) = 0$

$x+5=0 \quad 2x-1=0 \quad 3x-1=0$

$x=-5, \quad x=\frac{1}{2}, \quad x=\frac{1}{3}$

The solution set is $\left\{-5, \frac{1}{2}, \frac{1}{3}\right\}$.

22. a. $2x^3 - 5x^2 - 6x + 4 = 0$

$p: \pm 1, \pm 2, \pm 4$

$q: \pm 1, \pm 2$

$\frac{p}{q}: \pm 1, \pm 2 \pm 4 \pm \frac{1}{2}$

b.

$\frac{1}{2}$	2	–5	–6	4
		1	–2	–4
	2	–4	–8	0

$\frac{1}{2}$ is a root.

c. $2x^3 - 5x^2 - 6x + 4 = 0$

$(x - \frac{1}{2})\left(2x^2 - 4x - 8\right) = 0$

$2\left(x - \frac{1}{2}\right)\left(x^2 - 2x - 4\right) = 0$

$x = \frac{2 \pm 2\sqrt{5}}{2} = 1 \pm \sqrt{5}$

The solution set is $\left\{\frac{1}{2}, 1+\sqrt{5}, 1-\sqrt{5}\right\}$.

23. $x^4 - 2x^3 - 5x^2 + 8x + 4 = 0$

a. $p: \pm 1, \pm 2, \pm 4$

$q: \pm 1$

$\frac{p}{q}: \pm 1, \pm 2, \pm 4$

b.

2	1	–2	–5	8	4
		2	0	–10	–4
	1	0	–5	–2	0

2 is a root.

c. $x^4 - 2x^3 - 5x^2 + 8x + 4 = 0$

$(x-2)(x^3 - 5x - 2) = 0$

−2	1	0	−5	−2
		−2	4	2
	1	−2	−1	0

2 is a zero of $x^3 - 5x - 2 = 0$.

$(x-2)(x+2)\left(x^2 - 2x - 1\right) = 0$

$x = \frac{2 \pm \sqrt{4+4}}{2} = \frac{2 \pm \sqrt{8}}{2} = \frac{2 \pm 2\sqrt{2}}{2}$

$= 1 \pm \sqrt{2}$

The solution set is

$\left\{-2, 2, 1+\sqrt{2}, 1-\sqrt{2}\right\}$.

24. a. $x^4 - 2x^2 - 16x - 15 = 0$

p: ±1, ±3, ±5, ±15

q: ±1

$\frac{p}{q}$: ±1, ±3 ±5 ±15

b.

3	1	0	−2	−16	−15
		3	9	21	15
	1	3	7	5	0

3 is a root.

c. $x^4 - 2x^2 - 16x - 15 = 0$

$(x-3)\left(x^3 + 3x^2 + 7x + 5\right) = 0$

−1	1	3	7	5
		−1	−2	−5
	1	2	5	0

−1 is a root of $x^3 + 3x^2 + 7x + 5$

$(x-3)\ (x+1)\ \left(x^2 + 2x + 5\right)$

$x = \frac{-2 \pm \sqrt{4-20}}{2} = \frac{-2 \pm \sqrt{-16}}{2}$

$= \frac{-2 \pm 4i}{2} = -1 \pm 2i$

The solution set is $\{3, -1, -1+2i, -1-2i\}$.

25. $(x-1)\ (x+5i)\ (x-5i)$

$= (x-1)\left(x^2 + 25\right)$

$= x^3 + 25x - x^2 - 25$

$= x^3 - x^2 + 25x - 25$

$f(x) = a_n\left(x^3 - x^2 + 25x - 25\right)$

$f(-1) = a_n(-1 - 1 - 25 - 25)$

$-104 = a_n(-52)$

$a_n = 2$

$f(x) = 2\left(x^3 - x^2 + 25x - 25\right)$

$f(x) = 2x^3 - 2x^2 + 50x - 50$

26. $(x-4)\ (x+2i)\ (x-2i)$

$= (x-4)\left(x^2 + 4\right)$

$= x^3 - 4x^2 + 4x - 16$

$f(x) = a_n\left(x^3 - 4x^2 + 4x - 16\right)$

$f(-1) = a_n(-1 - 4 - 4 - 16)$

$-50 = a_n(-25)$

$a_n = 2$

$f(x) = 2\left(x^3 - 4x^2 + 4x - 16\right)$

$f(x) = 2x^3 - 8x^2 + 8x - 32$

27. $(x+5)(x-4-3i)(x-4+3i)$

$= (x+5)(x^2-4x+3ix-4x+16-12i$
$-3ix+12i-9i^2)$
$= (x+5)(x^2-8x+25)$
$= (x^3-8x^2+25x+5x^2-40x+125)$
$= x^3-3x^2-15x+125$

$f(x) = a_n(x^3-3x^2-15x+125)$
$f(2) = a_n(2^3-3(2)^2-15(2)+125)$
$91 = a_n(91)$
$a_n = 1$
$f(x) = 1(x^3-3x^2-15x+125)$
$f(x) = x^3-3x^2-15x+125$

28. $(x-6)(x+5+2i)(x+5-2i)$

$= (x-6)(x^2+5x-2ix+5x+25-10i+2ix+10i-4i^2)$
$= (x-6)(x^2+10x+29)$
$= x^3+10x^2+29x-6x^2-60x-174$
$= x^3+4x^2-31x-174$
$f(x) = a_n(x^3+4x^2-31x-174)$
$f(2) = a_n(8+16-62-174)$
$-636 = a_n(-212)$
$a_n = 3$
$f(x) = 3(x^3+4x^2-31x-174)$
$f(x) = 3x^3+12x^2-93x-522$

29. $(x-i)(x+i)(x-3i)(x+3i)$

$= (x^2-i^2)(x^2-9i^2)$
$= (x^2+1)(x^2+9)$
$= x^4+10x^2+9$
$f(x) = a_n(x^4+10x^2+9)$
$f(-1) = a_n((-1)^4+10(-1)^2+9)$
$20 = a_n(20)$
$a_n = 1$
$f(x) = x^4+10x^2+9$

30. $(x+2)\left(x+\frac{1}{2}\right)(x-i)(x+i)$

$=\left(x^2+\frac{5}{2}x+1\right)\left(x^2+1\right)$

$=x^4+x^2+\frac{5}{2}x^3+\frac{5}{2}x+x^2+1$

$=x^4+\frac{5}{2}x^3+2x^2+\frac{5}{2}x+1$

$f(x)=a_n\left(x^4+\frac{5}{2}x^3+2x^2+\frac{5}{2}x+1\right)$

$f(1)=a_n\left[(1)^4+\frac{5}{2}(1)^3+2(1)^2+\frac{5}{2}(1)+1\right]$

$18=a_n(9)$

$a_n=2$

$f(x)=2\left(x^4+\frac{5}{2}x^3+2x^2+\frac{5}{2}x+1\right)$

$f(x)=2x^4+5x^3+4x^2+5x+2$

31. $(x+2)(x-5)(x-3+2i)(x-3-2i)$

$=\left(x^2-3x-10\right)\left(x^2-3x-2ix-3x+9+6i+2ix-6i-4i\right)$

$=\left(x^2-3x-10\right)\left(x^2-6x+13\right)$

$=x^4-6x+13x^2-3x^3+18x^2-39x-10x^2+60x-130$

$=x^4-9x^3+21x^2+21x-130$

$f(x)=a_n\left(x^4-9x^3+21x^2+21x-130\right)$

$f(1)=a_n(1-9+21+21-130)$

$-96=a_n(-96)$

$a_n=1$

$f(x)=x^4-9x^3+21x^2+21x-130$

32. $(x+4)(3x-1)(x-2+3i)(x-2-3i)$

$=\left(3x^2+11x-4\right)\left(x^2-2x-3ix-2x+4+6i+3ix-6i-9i^2\right)$

$=\left(3x^2+11x-4\right)\left(x^2-4x+13\right)$

$=3x^4-12x^3+39x^2+11x^3-44x^2+143x-4x^2+16x-52$

$=3x^4-x^3-9x^2+159x-52$

$f(x)=a_n\left(3x^4-x^3-9x^2+159x-52\right)$

$f(1)=a_n(3-1-9+159-52)$

$100=a_n(100)$

$a_n=1$

$f(x)=3x^4-x^3-9x^2+159x-52$

33. $f(x) = x^3 + 2x^2 + 5x + 4$
Since $f(x)$ has no sign variations, no positive real roots exist.
$f(-x) = -x^3 + 2x^2 - 5x + 4$
Since $f(-x)$ has 3 sign variations, 3 or 1 negative real roots exist.

34. $f(x) = x^3 + 7x^2 + x + 7$
Since $f(x)$ has no sign variations no positive real roots exist.
$f(-x) = -x^3 + 7x^2 - x + 7$
Since $f(-x)$ has 3 sign variations, 3 or 1 negative real roots exist.

35. $f(x) = 5x^3 - 3x^2 + 3x - 1$
Since $f(x)$ has 3 sign variations, 3 or 1 positive real roots exist.
$f(-x) = -5x^3 - 3x^2 - 3x - 1$
Since $f(-x)$ has no sign variations, no negative real roots exist.

36. $f(x) = -2x^3 + x^2 - x + 7$
Since $f(x)$ has 3 sign variations, 3 or 1 positive real roots exist.
$f(-x) = 2x^3 + x^2 + x + 7$
Since $f(-x)$ has no sign variations, no negative real roots exist.

37. $f(x) = 2x^4 - 5x^3 - x^2 - 6x + 4$
Since $f(x)$ has 2 sign variations, 2 or 0 positive real roots exist.
$f(-x) = 2x^4 + 5x^3 - x^2 + 6x + 4$
Since $f(-x)$ has 2 sign variations, 2 or 0 negative real roots exist.

38. $f(x) = 4x^4 - x^3 + 5x^2 - 2x - 6$
Since $f(x)$ has 3 sign variations, 3 or 1 positive real roots exist.
$f(-x) = 4x^4 + x^3 + 5x^2 + 2x - 6$
Since $f(x)$ has 1 sign variations, 1 negative real roots exist.

39. $f(x) = x^3 - 4x^2 - 7x + 10$
$p: \pm 1, \pm 2, \pm 5, \pm 10$
$q: \pm 1$
$\frac{p}{q}: \pm 1, \pm 2, \pm 5, \pm 10$
Since $f(x)$ has 2 sign variations, 0 or 2 positive real zeros exist.
$f(-x) = -x^3 - 4x^2 + 7x + 10$
Since $f(-x)$ has 1 sign variation, exactly one negative real zeros exists.

–2	1	–4	–7	10
		–2	12	–10
	1	–6	5	0

–2 is a zero.

$$f(x) = (x+2)(x^2 - 6x + 5)$$
$$= (x+2)(x-5)(x-1)$$

$x = -2, x = 5, x = 1$
The solution set is $\{-2, 5, 1\}$.

40. $f(x) = x^3 + 12x^2 + 2x + 10$
$p: \pm 1, \pm 2, \pm 5, \pm 10$
$q: \pm 1,$
$\frac{p}{q}: \pm 1, \pm 2 \pm 5 \pm 10$

Since $f(x)$ has no sign variations, no positive zeros exist.
$f(-x) = -x^3 + 12x^2 - 21x + 10$
Since $f(-x)$ has 3 sign variations, 3 or 1 negative zeros exist.

–1	1	12	21	10
		–1	–11	–10
	1	11	10	0

–1 is a zero.

$$f(x) = (x+1)(x^2 + 11x + 10)$$
$$= (x+1)(x+10)(x+1)$$
$$x = -1, x = -10$$

The solution set is $\{-1, -10\}$.

41. $2x^3 - x^2 - 9x - 4 = 0$

$p: \pm 1, \pm 2, \pm 4$

$q: \pm 1, \pm 2$

$\frac{p}{q}: \pm 1, \pm 2, \pm 4 \pm \frac{1}{2}$

1 positive real root exists.

$f(-x) = -2x^3 - x^2 + 9x - 4$ 2 or no negative real roots exist.

$-\frac{1}{2}$	2	−1	−9	4
		−1	1	4
	2	−2	−8	0

$-\frac{1}{2}$ is a root.

$$\left(x+\frac{1}{2}\right)\left(2x^2 - 2x - 8\right) = 0$$

$$2\left(x+\frac{1}{2}\right)\left(x^2 - x - 4\right) = 0$$

$$x = \frac{1 \pm \sqrt{1+16}}{2} = \frac{1 \pm \sqrt{17}}{2}$$

The solution set is $\left\{-\frac{1}{2}, \frac{1+\sqrt{17}}{2}, \frac{1-\sqrt{17}}{2}\right\}$.

42. $3x^3 - 8x^2 - 8x + 8 = 0$

$p: \pm 1, \pm 2, \pm 4, \pm 8$

$q: \pm 1, \pm 3$

$\frac{p}{q}: \pm 1, \pm 2 \pm 4 \pm 8, \pm \frac{1}{3}, \pm \frac{2}{3}, \pm \frac{4}{3}, \pm \frac{8}{3}$

Since $f(x)$ has 2 sign variations, 2 or no positive real roots exist.

$$f(-x) = -3x^3 - 8x^2 + 8x + 8$$

Since $f(-x)$ has 1 sign changes, exactly 1 negative real zero exists.

$\frac{2}{3}$	3	−8	−8	8
		2	−4	−8
	3	−6	−12	0

$\frac{2}{3}$ is a zero.

$$f(x) = \left(x - \frac{2}{3}\right)\left(3x^2 - 6x - 12\right)$$

$$x = \frac{6 \pm \sqrt{36+144}}{6} = \frac{6 \pm 6\sqrt{5}}{6}$$

$$= 1 \pm \sqrt{5}$$

The solution set is $\left\{\frac{2}{3}, 1+\sqrt{5}, 1-\sqrt{5}\right\}$.

43. $f(x) = x^4 - 2x^3 + x^2 + 12x + 8$

$p: \pm 1, \ \pm 2, \ \pm 4, \ \pm 8$

$q: \pm 1$

$\frac{p}{q}: \pm 1, \ \pm 2, \ \pm 4, \ \pm 8$

Since $f(x)$ has 2 sign changes, 0 or 2 positive roots exist.

$$f(-x) = (-x)^4 - 2(-x)^3 + (-x)^2 - 12x + 8$$
$$= x^4 + 2x^3 + x^2 - 12x + 8$$

Since $f(-x)$ has 2 sign changes, 0 or 2 negative roots exist.

−1	1	−2	1	12	8
		−1	4	−4	−8
	1	−3	4	8	0

−1	1	−3	4	8
		−1	4	−8
	1	−4	8	0

$$0 = x^2 - 4x + 8$$

$$x = \frac{-(-4) \pm \sqrt{(-4)^2 - 4(1)(8)}}{2(1)}$$

$$x = \frac{4 \pm \sqrt{16-32}}{2}$$

$$x = \frac{4 \pm \sqrt{-16}}{2}$$

$$x = \frac{4 \pm 4i}{2}$$

$$x = 2 \pm 2i$$

The solution set is $\{-1, -1, 2+2i, 2-2i\}$.

44. $f(x) = x^4 - 4x^3 - x^2 + 14x + 10$

p: ±1, ±2, ±5, ±10
q: ±1

$\frac{p}{q}$: ±1, ±2, ±5, ±10

–1	1	-4	-1	14	10
		–1	5	-4	-10
	1	-5	4	10	0

–1	1	-5	4	10
		–1	6	–10
	1	-6	10	0

$f(x) = (x-1)(x-1)(x^2 - 6x + 10)$

$x = \frac{-(-6) \pm \sqrt{(-6)^2 - 4(1)(10)}}{2(1)}$ $\quad x = 1$

$x = \frac{6 \pm \sqrt{36 - 40}}{2}$

$x = \frac{6 \pm \sqrt{-4}}{2}$

$x = \frac{6 \pm 2i}{2}$

$x = 3 \pm i$

The solution set is {–1, 3 – i, 3 + i}

45. $x^4 - 3x^3 - 20x^2 - 24x - 8 = 0$

p: ±1, ±2, ±4, ±8

q: ±1

$\frac{p}{q}$: ±1, ±2, ±4 ±8

1 positive real root exists.
3 or 1 negative real roots exist.

–1	1	–3	–20	–24	–8
		–1	4	16	8
	1	–4	–16	–8	0

$(x+1)\left(x^3 - 4x^2 - 16x - 8\right) = 0$

–2	1	–4	–16	–8
		–2	12	8
	1	–6	–4	0

$(x+1)(x+2)\left(x^2 - 6x - 4\right) = 0$

$x = \frac{6 \pm \sqrt{36+16}}{2} = \frac{6 \pm \sqrt{52}}{2}$

$= \frac{6 \pm 2\sqrt{13}}{2} = \frac{3 \pm \sqrt{13}}{2}$

The solution set is
$\left\{-1, -2, 3 \pm \sqrt{13}, 3 - \sqrt{13}\right\}$.

46. $x^4 - x^3 + 2x^2 - 4x - 8 = 0$

p: ±1, ±2, ±4, ±8
q: ±1

$\frac{p}{q}$: ±1, ±2 ±4 ±8

1 negative real root exists.

–1	1	–1	2	–4	–8
		–1	2	–4	8
	1	–2	4	–8	0

$(x+1)\left(x^3 - 2x^2 + 4x - 8\right)$

2	1	–2	4	–8
		2	0	8
	1	0	4	0

$(x+1) \quad (x-2) \quad \left(x^2+4\right)$

$x+1=0 \quad x-2=0 \quad x^2+4=0$

$x=-1 \quad x=2 \quad x^2=-4$

$x = \pm 2i$

The solution set is $\{-1, 2, 2i, -2i\}$.

47. $f(x) = 3x^4 - 11x^3 - x^2 + 19x + 6$

p: ±1, ±2, ±3, ±6

q: ±1, ±3

$\frac{p}{q}$: $\pm 1, \pm 2, \pm 3, \pm 6, \pm\frac{1}{3}, \pm\frac{2}{3}$

2 or no positive real zeros exists.

$f(-x) = 3x^4 + 11x^3 - x^2 - 19x + 6$

2 or no negative real zeros exist.

–1	3	–11	–1	19	6
		–3	14	–13	–6
	3	–14	13	6	0

$f(x) = (x+1)\left(3x^3 - 14x^2 + 13x + 6\right)$

2	3	−14	13	6
		6	−16	−6
	3	−8	−3	0

$f(x)=(x+1)(x-2)(3x^2-8x-3)$
$=(x+1)(x-2)(3x+1)(x-3)$

$x=-1,\ x=2\ x=-\frac{1}{3},\ x=3$

The solution set is $\left\{-1,\ 2,\ -\frac{1}{3},\ 3\right\}$.

48. $f(x)=2x^4+3x^3-11x^2-9x+15$
p: ±1, ±3, ±5, ±15
q: ±1, ±2
$\frac{p}{q}: \pm1, \pm3, \pm5, \pm15, \pm\frac{1}{2}, \pm\frac{3}{2}, \pm\frac{5}{2}, \pm\frac{15}{2}$
2 or no positive real zeros exist.
$f(-x)=2x^4-3x^3-11x^2+9x+15$
2 or no negative real zeros exist.

1	2	3	−11	−9	15
		2	5	−6	−15
	2	5	−6	−15	0

$f(x)=(x-1)(2x^3+5x^2-6x-15)$

$-\frac{5}{2}$	2	5	−6	−15
		−5	0	15
	2	0	−6	0

$f(x)=(x-1)\left(x+\frac{5}{2}\right)(2x^2-6)$
$=2(x-1)\left(x+\frac{5}{2}\right)(x^2-3)$

$x^2-3=0$
$x^2=3$
$x=\pm\sqrt{3}$

$x=1,\ x=-\frac{5}{2},\ x=\sqrt{3},\ x=-\sqrt{3}$

The solution set is $\left\{1, -\frac{5}{2}, \sqrt{3}, -\sqrt{3}\right\}$.

49. $4x^4-x^3+5x^2-2x-6=0$
$p: \pm1, \pm2, \pm3, \pm6$
$q: \pm1, \pm2, \pm4$
$\frac{p}{q}: \pm1, \pm2, \pm3, \pm6, \pm\frac{1}{2}, \pm\frac{3}{2}, \pm\frac{1}{4}, \pm\frac{3}{4}$
3 or 1 positive real roots exists.
1 negative real root exists.

1	4	−1	5	−2	−6
		4	3	8	6
	4	3	8	6	0

$(x-1)(4x^3+3x^2+8x+6)=0$
$4x^3+3x^2+8x+6=0$ has no positive real roots.

$-\frac{3}{4}$	4	3	8	6
		−3	0	−6
	4	0	8	0

$(x-1)\left(x+\frac{3}{4}\right)(4x^2+8)=0$

$4(x-1)\left(x+\frac{3}{4}\right)(x^2+2)=0$

$x^2+2=0$
$x^2=-2$
$x=\pm i\sqrt{2}$

The solution set is $\left\{1, -\frac{3}{4}, i\sqrt{2}, -i\sqrt{2}\right\}$.

50. $3x^4-11x^3-3x^2-6x+8=0$
p: ±1, ±2, ±4, ±8
q: ±1, ±3
$\frac{p}{q}: \pm1, \pm2 \pm4 \pm8, \pm\frac{1}{3}, \pm\frac{2}{3}, \pm\frac{4}{3}, \pm\frac{8}{3}$
2 or no positive real roots exist.
$f(-x)=3x^4+11x^3-3x^2+6x+8$ 2 or no negative real roots exist.

4	3	−11	−3	−6	8
		12	4	4	−8
	3	1	1	−2	0

$(x-4)(3x^3+x^2+x-2)=0$
Another positive real root must exist.

$\frac{2}{3}$	3	1	1	−2
		2	2	2
	3	3	3	0

$$(x-4)\left(x-\frac{2}{3}\right)\left(3x^2+3x+3\right)=0$$

$$3(x-4)\left(x-\frac{2}{3}\right)\left(x^2+x+1\right)=0$$

$$x=\frac{-1\pm\sqrt{1-4}}{2}=\frac{-1\pm i\sqrt{3}}{2}$$

The solution set is $\left\{4, \frac{2}{3}, \frac{-1+i\sqrt{3}}{2}, \frac{-1-i\sqrt{3}}{2}\right\}$.

51. $2x^5+7x^4-18x^2-8x+8=0$

$p: \pm 1, \pm 2, \pm 4, \pm 8$

$q: \pm 1, \pm 2$

$\frac{p}{q}: \pm 1, \pm 2, \pm 4, \pm 8, \pm\frac{1}{2}$

2 or no positive real roots exists.

3 or 1 negative real root exist.

–2	2	7	0	–18	–8	8
		–4	–6	12	12	–8
	2	3	–6	–6	4	0

$(x+2)(2x^4+3x^3-6x^2-6x+4)=0$

$4x^3+3x^2+8x+6=0$ has no positive real roots.

–2	2	3	–6	–6	4
		–4	2	8	–4
	2	–1	–4	2	0

$(x+2)^2(2x^3-x^2-4x+2)$

$\frac{1}{2}$	2	–1	–4	2
		1	0	2
	2	0	–4	0

$$(x+2)^2\left(x-\frac{1}{2}\right)\left(2x^2-4\right)=0$$

$$2(x+2)^2\left(x-\frac{1}{2}\right)\left(x^2-2\right)=0$$

$x^2-2=0$

$x^2=2$

$x=\pm\sqrt{2}$

The solution set is $\left\{-2, \frac{1}{2}, \sqrt{2}, -\sqrt{2}\right\}$.

52. $4x^5+12x^4-41x^3-99x^2+10x+24=0$

p: ±1, ±2, ±3, ±4, ±6, ±8, ±12, ±24

q: ±1, ±2, ±4

$\frac{p}{q}: \pm 1, \pm 2, \pm 3, \pm 4, \pm 6, \pm 8, \pm 12, \pm 24, \pm\frac{1}{2}, \pm\frac{3}{2},$

$\pm\frac{1}{4}, \pm\frac{3}{4}$

2 or no positive real roots exist.

$f(-x)=-4x^5+12x^4+41x^3-99x^2-10x+24$

3 or 1 negative real roots exist.

3	4	12	–41	–99	10	24
		12	72	93	–18	–24
	4	24	31	–6	–8	0

$(x-3)\left(4x^4+24x^3+31x^2-6x-8\right)=0$

–2	4	24	31	–6	–8
		–8	–32	2	8
	4	16	–1	–4	0

$(x-3)(x+2)\left(4x^3+16x^2-x-4\right)=0$

–4	4	16	–1	4
		–16	0	4
	4	0	–1	0

$(x-3)(x+2)(x+4)\left(4x^2-1\right)=0$

$4x^2-1=0$

$4x^2=1$

$x^2=\frac{1}{4}$

$x=\pm\frac{1}{2}$

The solution set is $\left\{3, -2, -4, \frac{1}{2}, -\frac{1}{2}\right\}$.

53. $f(x) = -x^3 + x^2 + 16x - 16$

a. From the graph provided, we can see that -4 is an x-intercept and is thus a zero of the function. We verify this below:

$$\begin{array}{r|rrrr} -4 & -1 & 1 & 16 & -16 \\ & & 4 & -20 & 16 \\ \hline & -1 & 5 & -4 & 0 \end{array}$$

Thus, $-x^3 + x^2 + 16x - 16 = 0$

$$(x+4)(-x^2 + 5x - 4) = 0$$
$$-(x+4)(x^2 - 5x + 4) = 0$$
$$-(x+4)(x-1)(x-4) = 0$$

$x + 4 = 0$ or $x - 1 = 0$ or $x - 4 = 0$

$x = -4$ $\quad$ $x = 1$ $\quad$ $x = 4$

The zeros are -4, 1, and 4.

b.

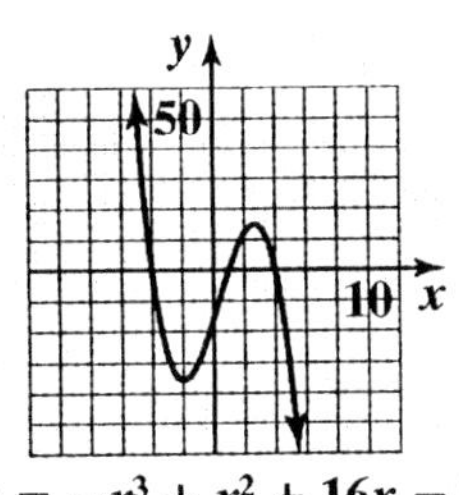

$f(x) = -x^3 + x^2 + 16x - 16$

54. $f(x) = -x^3 + 3x^2 - 4$

a. From the graph provided, we can see that -1 is an x-intercept and is thus a zero of the function. We verify this below:

$$\begin{array}{r|rrrr} -1 & -1 & 3 & 0 & -4 \\ & & 1 & -4 & 4 \\ \hline & -1 & 4 & -4 & 0 \end{array}$$

Thus, $-x^3 + 3x^2 - 4 = 0$

$$(x+1)(-x^2 + 4x - 4) = 0$$
$$-(x+1)(x^2 - 4x + 4) = 0$$
$$-(x+1)(x-2)^2 = 0$$

$x + 1 = 0$ or $(x-2)^2 = 0$

$x = -1$ $\quad$ $x - 2 = 0$

$x = 2$

The zeros are -1 and 2.

b.

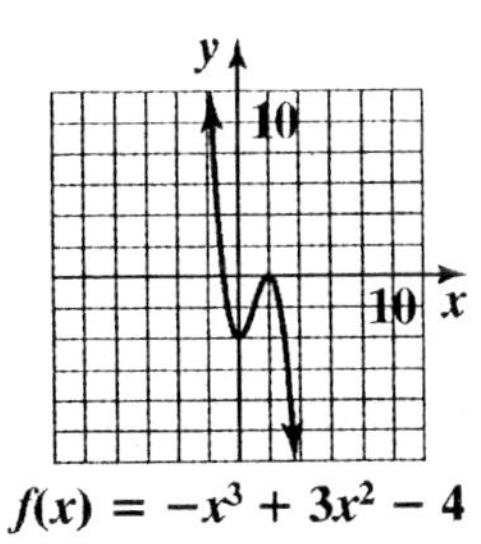

$f(x) = -x^3 + 3x^2 - 4$

55. $f(x) = 4x^3 - 8x^2 - 3x + 9$

a. From the graph provided, we can see that -1 is an x-intercept and is thus a zero of the function. We verify this below:

$$\begin{array}{r|rrrr} -1 & 4 & -8 & -3 & 9 \\ & & -4 & 12 & -9 \\ \hline & 4 & -12 & 9 & 0 \end{array}$$

Thus, $4x^3 - 8x^2 - 3x + 9 = 0$

$$(x+1)(4x^2 - 12x + 9) = 0$$
$$(x+1)(2x-3)^2 = 0$$

$x + 1 = 0$ or $(2x-3)^2 = 0$

$x = -1$ $\quad$ $2x - 3 = 0$

$2x = 3$

$x = \frac{3}{2}$

The zeros are -1 and $\frac{3}{2}$.

b.

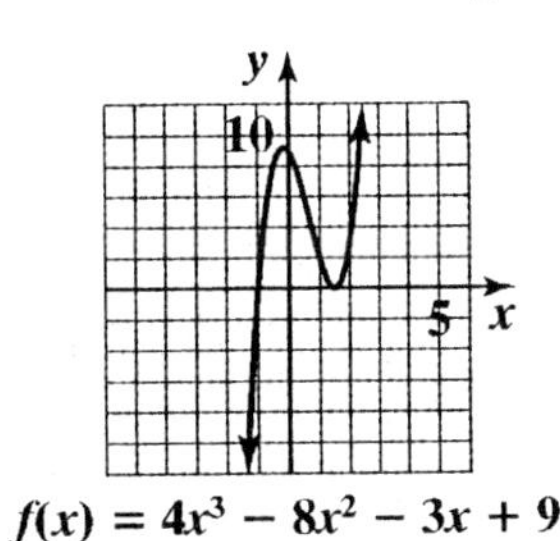

$f(x) = 4x^3 - 8x^2 - 3x + 9$

56. $f(x) = 3x^3 + 2x^2 + 2x - 1$

a. From the graph provided, we can see that $\frac{1}{3}$ is an x-intercept and is thus a zero of the function. We verify this below:

$$\begin{array}{r|rrrr} \frac{1}{3} & 3 & 2 & 2 & -1 \\ & & 1 & 1 & 1 \\ \hline & 3 & 3 & 3 & 0 \end{array}$$

Thus, $3x^3 + 2x^2 + 2x - 1 = 0$

$$\left(x - \frac{1}{3}\right)\left(3x^2 + 3x + 3\right) = 0$$

$$3\left(x - \frac{1}{3}\right)\left(x^2 + x + 1\right) = 0$$

Note that $x^2 + x + 1$ will not factor, so we use the quadratic formula:

$x - \frac{1}{3} = 0$ or $x^2 + x + 1 = 0$

$x = \frac{1}{3}$ $\quad a = 1 \quad b = 1 \quad c = 1$

$$x = \frac{-1 \pm \sqrt{1^2 - 4(1)(1)}}{2(1)}$$

$$= \frac{-1 \pm \sqrt{-3}}{2}$$

$$= \frac{-1 \pm \sqrt{3}i}{2} = -\frac{1}{2} \pm \frac{\sqrt{3}}{2}i$$

The zeros are $\frac{1}{3}$ and $-\frac{1}{2} \pm \frac{\sqrt{3}}{2}i$.

b.

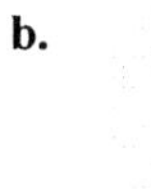

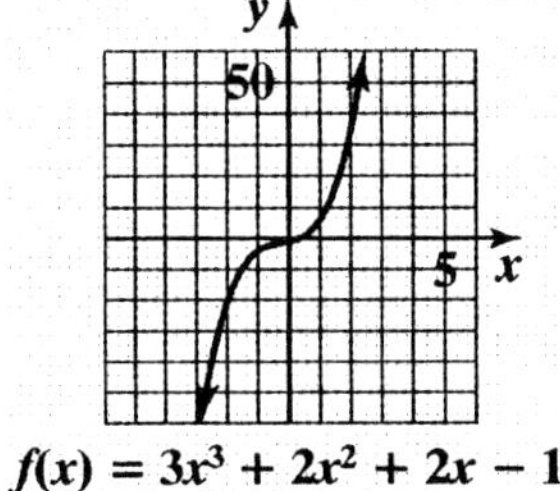

$f(x) = 3x^3 + 2x^2 + 2x - 1$

57. $f(x) = 2x^4 - 3x^3 - 7x^2 - 8x + 6$

a. From the graph provided, we can see that $\frac{1}{2}$ is an x-intercept and is thus a zero of the function. We verify this below:

$$\begin{array}{r|rrrrr} \frac{1}{2} & 2 & -3 & -7 & -8 & 6 \\ & & 1 & -1 & -4 & -6 \\ \hline & 2 & -2 & -8 & -12 & 0 \end{array}$$

Thus, $2x^4 - 3x^3 - 7x^2 - 8x + 6 = 0$

$$\left(x - \frac{1}{2}\right)\left(2x^3 - 2x^2 - 8x - 12\right) = 0$$

$$2\left(x - \frac{1}{2}\right)\left(x^3 - x^2 - 4x - 6\right) = 0$$

To factor $x^3 - x^2 - 4x - 6$, we use the Rational Zero Theorem to determine possible rational zeros.

Factors of the constant term -6: $\pm 1, \pm 2, \pm 3, \pm 6$

Factors of the leading coefficient 1: ± 1

The possible rational zeros are:

$$\frac{\text{Factors of } -6}{\text{Factors of } 1} = \frac{\pm 1, \ \pm 2, \ \pm 3, \ \pm 6}{\pm 1}$$

$= \pm 1, \ \pm 2, \ \pm 3, \ \pm 6$

We test values from above until we find a zero. One possibility is shown next:

Test 3:

$$\begin{array}{r|rrrr} 3 & 1 & -1 & -4 & -6 \\ & & 3 & 6 & 6 \\ \hline & 1 & 2 & 2 & 0 \end{array}$$

The remainder is 0, so 3 is a zero of f.

$$2x^4 - 3x^3 - 7x^2 - 8x + 6 = 0$$

$$\left(x - \frac{1}{2}\right)\left(2x^3 - 2x^2 - 8x - 12\right) = 0$$

$$2\left(x - \frac{1}{2}\right)\left(x^3 - x^2 - 4x - 6\right) = 0$$

$$2\left(x - \frac{1}{2}\right)(x - 3)\left(x^2 + 2x + 2\right) = 0$$

Note that $x^2 + x + 1$ will not factor, so we use the quadratic formula:

$a = 1 \quad b = 2 \quad c = 2$

$$x = \frac{-2 \pm \sqrt{2^2 - 4(1)(2)}}{2(1)}$$

$$= \frac{-2 \pm \sqrt{-4}}{2} = \frac{-2 \pm 2i}{2} = -1 \pm i$$

The zeros are $\frac{1}{2}$, 3, and $-1 \pm i$.

b.

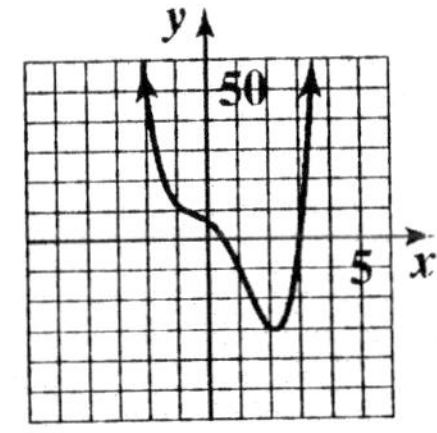

$f(x) = 2x^4 - 3x^3 - 7x^2 - 8x + 6$

58. $f(x) = 2x^4 + 2x^3 - 22x^2 - 18x + 36$

a. From the graph provided, we can see that 1 and 3 are x-intercepts and are thus zeros of the function. We verify this below:

```
1| 2  2  -22  -18   36
      2    4  -18  -36
   ------------------
   2  4  -18  -36    0
```

Thus, $2x^4 + 2x^3 - 22x^2 - 18x + 36$
$= (x-1)(2x^3 + 4x^2 - 18x - 36)$

```
3| 2   4  -18  -36
       6   30   36
   ---------------
   2  10   12    0
```

Thus, $2x^4 + 2x^3 - 22x^2 - 18x + 36 = 0$

$$(x-1)(x-3)(2x^2 + 10x + 12) = 0$$
$$2(x-1)(x-3)(x^2 + 5x + 6) = 0$$
$$2(x-1)(x-3)(x+3)(x+2) = 0$$

$x = 1,\ x = 3,\ x = -3,\ x = -2$

The zeros are -3, -2, 1, and 3.

b.

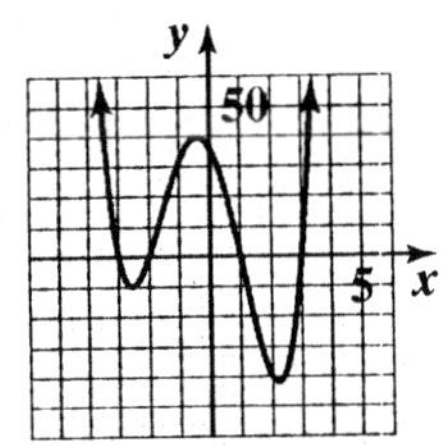

$f(x) = 2x^4 + 2x^3 - 22x^2 - 18x + 36$

59. $f(x) = 3x^5 + 2x^4 - 15x^3 - 10x^2 + 12x + 8$

a. From the graph provided, we can see that 1 and 2 are x-intercepts and are thus zeros of the function. We verify this below:

```
1| 3  2  -15  -10   12   8
      3    5  -10  -20  -8
   -----------------------
   3  5  -10  -20   -8   0
```

Thus, $3x^5 + 2x^4 - 15x^3 - 10x^2 + 12x + 8$
$= (x-1)(3x^4 + 5x^3 - 10x^2 - 20x - 8)$

```
2| 3   5  -10  -20  -8
       6   22   24   8
   -------------------
   3  11   12    4   0
```

Thus, $3x^5 + 2x^4 - 15x^3 - 10x^2 + 12x + 8$
$= (x-1)(3x^4 + 5x^3 - 10x^2 - 20x - 8)$
$= (x-1)(x-2)(3x^3 + 11x^2 + 12x + 4)$

To factor $3x^3 + 11x^2 + 12x + 4$, we use the Rational Zero Theorem to determine possible rational zeros.

Factors of the constant term 4:
$\pm 1,\ \pm 2,\ \pm 4$

Factors of the leading coefficient 3: $\pm 1,\ \pm 3$

The possible rational zeros are:

$$\frac{\text{Factors of 4}}{\text{Factors of 3}} = \frac{\pm 1,\ \pm 2,\ \pm 4}{\pm 1,\ \pm 3}$$
$$= \pm 1,\ \pm 2,\ \pm 4,\ \pm\frac{1}{3},\ \pm\frac{2}{3},\ \pm\frac{4}{3}$$

We test values from above until we find a zero. One possibility is shown next:

Test -1:

```
-1| 3  11  12   4
       -3  -8  -4
    -------------
    3   8   4   0
```

The remainder is 0, so -1 is a zero of f. We can now finish the factoring:

$$3x^5 + 2x^4 - 15x^3 - 10x^2 + 12x + 8 = 0$$
$$(x-1)(3x^4 + 5x^3 - 10x^2 - 20x - 8) = 0$$
$$(x-1)(x-2)(3x^3 + 11x^2 + 12x + 4) = 0$$
$$(x-1)(x-2)(x+1)(3x^2 + 8x + 4) = 0$$
$$(x-1)(x-2)(x+1)(3x+2)(x+2) = 0$$

$x = 1,\ x = 2,\ x = -1,\ x = -\frac{2}{3},\ x = -2$

The zeros are -2, -1, $-\frac{2}{3}$, 1 and 2.

b.

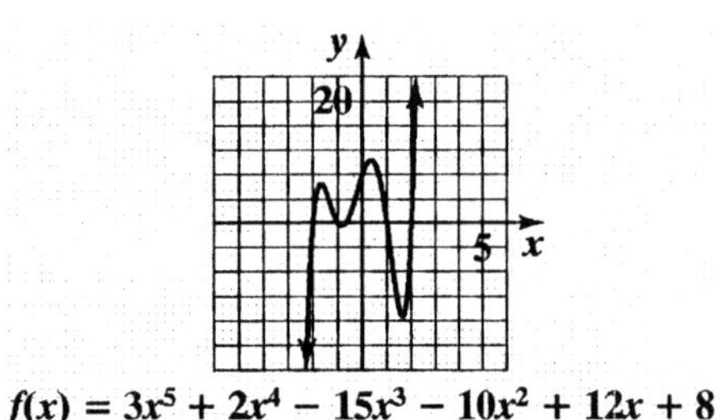

$f(x) = 3x^5 + 2x^4 - 15x^3 - 10x^2 + 12x + 8$

60. $f(x) = -5x^4 + 4x^3 - 19x^2 + 16x + 4$

a. From the graph provided, we can see that 1 is an x-intercept and is thus a zero of the function. We verify this below:

$$\begin{array}{r|rrrrr} 1 & -5 & 4 & -19 & 16 & 4 \\ & & -5 & -1 & -20 & -4 \\ \hline & -5 & -1 & -20 & -4 & 0 \end{array}$$

Thus, $-5x^4 + 4x^3 - 19x^2 + 16x + 4 = 0$

$$(x-1)(-5x^3 - x^2 - 20x - 4) = 0$$

$$-(x-1)(5x^3 + x^2 + 20x + 4) = 0$$

To factor $5x^3 + x^2 + 20x + 4$, we use the Rational Zero Theorem to determine possible rational zeros.

Factors of the constant term 4:
$\pm 1, \pm 2, \pm 4$
Factors of the leading coefficient 5: $\pm 1, \pm 5$

The possible rational zeros are:

$$\frac{\text{Factors of 4}}{\text{Factors of 5}} = \frac{\pm 1, \pm 2, \pm 4}{\pm 1, \pm 5}$$

$$= \pm 1, \pm 2, \pm 4, \pm\frac{1}{5}, \pm\frac{2}{5}, \pm\frac{4}{5}$$

We test values from above until we find a zero. One possibility is shown next:

Test $-\frac{1}{5}$:

$$\begin{array}{r|rrrr} -\frac{1}{5} & 5 & 1 & 20 & 4 \\ & & -1 & 0 & -4 \\ \hline & 5 & 0 & 20 & 0 \end{array}$$

The remainder is 0, so $-\frac{1}{5}$ is a zero of f.

$$-5x^4 + 4x^3 - 19x^2 + 16x + 4 = 0$$

$$(x-1)(-5x^3 - x^2 - 20x - 4) = 0$$

$$-(x-1)(5x^3 + x^2 + 20x + 4) = 0$$

$$-(x-1)\left(x + \frac{1}{5}\right)(5x^2 + 20) = 0$$

$$-5(x-1)\left(x + \frac{1}{5}\right)(x^2 + 4) = 0$$

$$-5(x-1)\left(x + \frac{1}{5}\right)(x + 2i)(x - 2i) = 0$$

$$x = 1, \; x = -\frac{1}{5}, \; x = -2i, \; x = 2i$$

The zeros are $-\frac{1}{5}$, 1, and $\pm 2i$.

b.

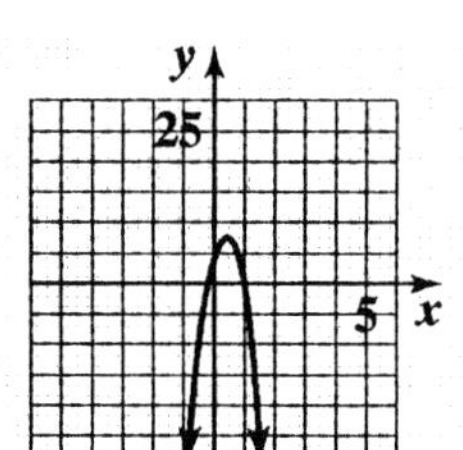

$f(x) = -5x^4 + 4x^3 - 19x^2 + 16x + 4$

61. **a.** $f(x) = 27$, $x = 40$ People in the arts complete 27% of their work in their 40's.

b. The polynomial will have a degree 2 with a negative leading coefficient.

62. **a.** $g(x) = 20$, $x = 60$ Professionals in the sciences completed 20% of their work in their 60's.

b. The function will have degree 2 and a negative coefficient.

63. If you are 25, the equivalent age for dogs is 3 years.

64. About 18 years of age for a dog is equivalent to 90 years of age for a human.

65. Answers may vary.

66. $-x^4+12x^3-58x^2+132x=0$

$x(-x^3+12x^2-58x+132)=0$

$p: \pm1, \pm2, \pm3, \pm4, \pm6, \pm11, \pm12, \pm22, \pm33,$
$\pm44 \pm66, \pm132$

$q: \pm1$

6	−1	12	−58	132
		6	36	−132
	−1	6	−22	0

$x(x-6)(-x^2+6x-22)=0$

$-x(x-6)(x^2-6x+22)=0$

$b^2-4ac=36-88$

0 and 6 are the only real roots.

6 hours

67. $x(10-2x)(8-2x)=48$

$x(80-36x+4x^2)=48$

$4x^3-36x^2+80x-48=0$

$x^3-9x^2+20x-12=0$

2	1	−9	20	−12
		2	−14	12
	1	−7	6	0

$(x-2)(x^2-7x+6)=0$

$(x-2)(x-6)(x-1)=0$

$x=2 \quad x=6 \quad x=1$

If $x=6$, $10-2x<0$

$x=1$ in. or $x=2$ in.

76. $2x^3-15x^2+22x+15=0$

$p: \pm1, \pm3, \pm5, \pm15$

$q: \pm1, \pm2$

$\frac{p}{q}: \pm1, \pm3, \pm5, \pm15, \pm\frac{1}{2}, \pm\frac{3}{2}, \pm\frac{5}{2}, \pm\frac{15}{2}$

From the graph we see that the solutions are $-\frac{1}{2}$, 3 and 5.

77. $6x^3-19x^2+16x-4=0$

$p: \pm1, \pm2, \pm4$

$q: \pm1, \pm2, \pm3, \pm6$

$\frac{p}{q}: \pm1, \pm2, \pm4, \pm\frac{1}{2}, \pm\frac{1}{3}, \pm\frac{2}{3}, \pm\frac{4}{3}, \pm\frac{1}{6}$

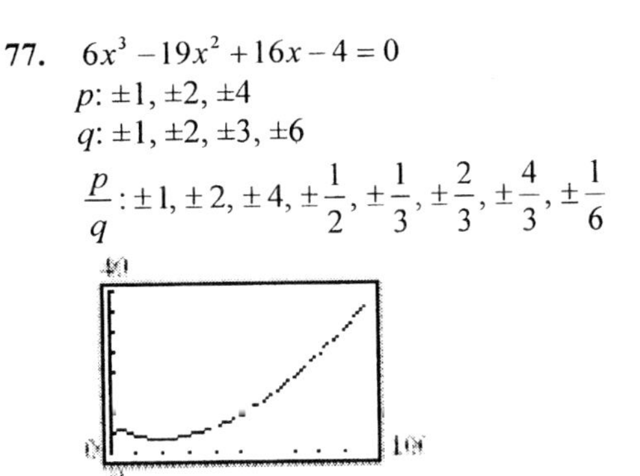

From the graph, we see that the solutions are $\frac{1}{2}, \frac{2}{3}$ and 2.

78. $2x^4+7x^3-4x^2-27x-18=0$

$p: \pm1, \pm2, \pm3, \pm6, \pm9, \pm18$

$q: \pm1, \pm2$

$\frac{p}{q}: \pm1, \pm2, \pm3, \pm6, \pm9, \pm18, \pm\frac{1}{2}, \pm\frac{3}{2}, \pm\frac{9}{2}$

From the graph we see the solutions are $-3, -\frac{3}{2}, -1, 2$.

79. $4x^4+4x^3+7x^2-x-2=0$

$p: \pm1, \pm2$

$q: \pm1, \pm2, \pm4$

$\frac{p}{q}: \pm1, \pm2, \pm\frac{1}{2}, \pm\frac{1}{4}$

30

−5 5

−10

From the graph, we see that the solutions are $-\frac{1}{2}$ and $\frac{1}{2}$.

80. $f(x) = 3x^4 + 5x^2 + 2$

Since $f(x)$ has no sign variations, it has no positive real roots.

$f(-x) = 3x^4 + 5x^2 + 2$

Since $f(-x)$ has no sign variations, no negative roots exist.

The polynomial's graph doesn't intersect the x-axis.

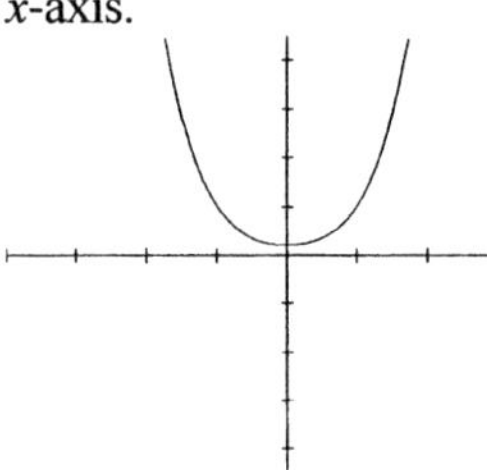

From the graph, we see that there are no real solutions.

81. $f(x) = x^5 - x^4 + x^3 - x^2 + x - 8$

$f(x)$ has 5 sign variations, so either 5, 3, or 1 positive real roots exist.

$f(-x) = -x^5 - x^4 - x^3 - x^2 - x - 8$

$f(-x)$ has no sign variations, so no negative real roots exist.

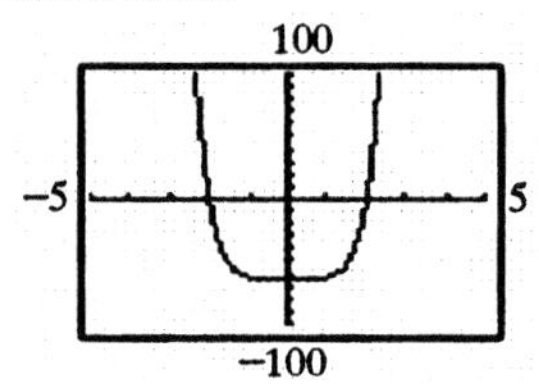

82. Odd functions must have at least one real zero. Even functions do not.

83. $f(x) = x^3 - 6x - 9$

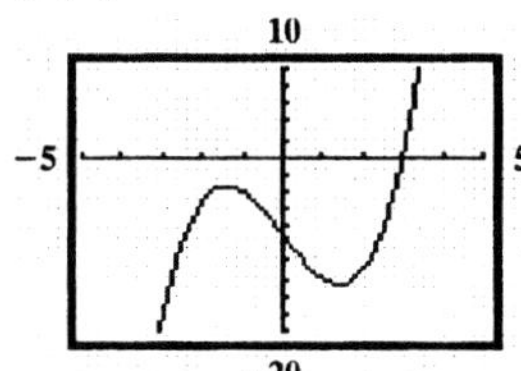

1 real zero

2 nonreal complex zeros

84. $f(x) = 3x^5 - 2x^4 + 6x^3 - 4x^2 - 24x + 16$

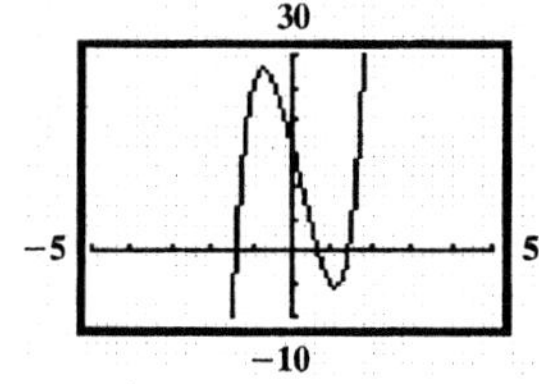

3 real zeros

2 nonreal complex zeros

85. $f(x) = 3x^4 + 4x^3 - 7x^2 - 2x - 3$

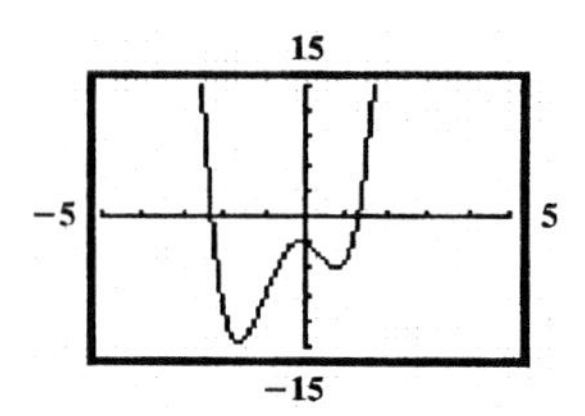

86. $f(x) = x^6 - 64$

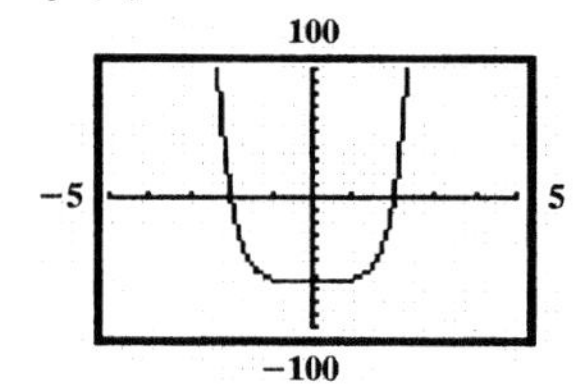

2 real zeros

4 nonreal complex zeros

87. **a.** False; the equation has 0 sign variations, so no positive roots exist.

b. False; Descartes' Rule gives the maximum possible number of real roots.

c. False; every polynomial equation of degree 3 has at least one real root.

d. True

(d) is true.

88. Answers may vary.

89. $(2x+1)(x+5)(x+2) - 3x(x+5) = 208$

$(2x^2 + 11x + 5)(x+2) - 3x^2 - 15x = 208$

$2x^3 + 4x^2 + 11x^2 + 22x + 5x$

$+10 - 3x^2 - 15x = 208$

$2x^3 + 15x^2 + 27x - 3x^2 - 15x - 198 = 0$

$2x^3 + 12x^2 + 12x - 198 = 0$

$2(x^3 + 6x^2 + 6x - 99) = 0$

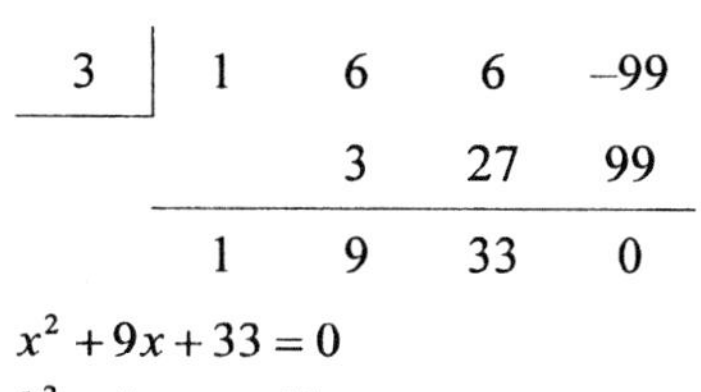

3	1	6	6	−99
		3	27	99
	1	9	33	0

$x^2 + 9x + 33 = 0$

$b^2 - 4ac = -51$

$x = 3$ in.

91. Because the polynomial has two obvious changes of direction; the smallest degree is 3.

92. Because the polynomial has no obvious changes of direction but the graph is obviously not linear, the smallest degree is 3.

93. Because the polynomial has two obvious changes of direction and two roots have multiplicity 2, the smallest degree is 5.

94. Two roots appear twice, the smallest degree is 5.

Mid-Chapter 2 Check Point

1. $(6-2i)-(7-i)=6-2i-7+i=-1-i$

2. $3i(2+i)=6i+3i^2=-3+6i$

3. $(1+i)(4-3i)=4-3i+4i-3i^2$
$=4+i+3$
$=7+i$

4. $\frac{1+i}{1-i}=\frac{1+i}{1-i}\cdot\frac{1+i}{1+i}$
$=\frac{1+i+i+i^2}{1-i^2}$
$=\frac{1+2i-1}{1+1}$
$=\frac{2i}{2}$
$=i$

5. $\sqrt{-75}-\sqrt{-12}=5i\sqrt{3}-2i\sqrt{3}=3i\sqrt{3}$

6. $\left(2-\sqrt{-3}\right)^2=\left(2-i\sqrt{3}\right)^2$
$=4-4i\sqrt{3}+3i^2$
$=4-4i\sqrt{3}-3$
$=1-4i\sqrt{3}$

7. $x(2x-3)=-4$
$2x^2-3x=-4$
$2x^2-3x+4=0$
$x=\frac{-b\pm\sqrt{b^2-4ac}}{2a}$
$x=\frac{-(-3)\pm\sqrt{(-3)^2-4(2)(4)}}{2(2)}$
$x=\frac{3\pm\sqrt{-23}}{4}$
$x=\frac{3}{4}\pm\frac{\sqrt{23}}{4}i$

8. $f(x)=(x-3)^2-4$
The parabola opens up because $a>0$.
The vertex is (3, –4).
x-intercepts:
$0=(x-3)^2-4$
$(x-3)^2=4$
$x-3=\pm\sqrt{4}$
$x=3\pm2$
The equation has x-intercepts at $x=1$ and $x=5$.
y-intercept:
$f(0)=(0-3)^2-4=5$
Domain: $(-\infty,\infty)$ Range: $[-4,\infty)$

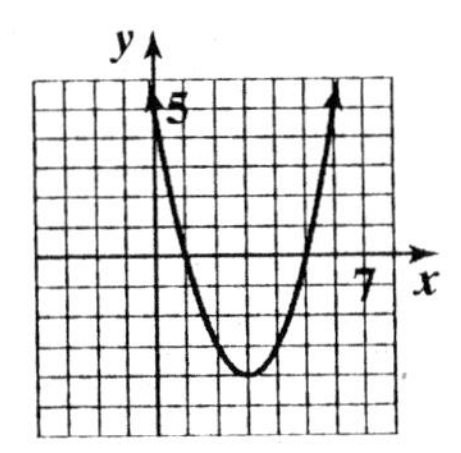

$f(x)=(x-3)^2-4$

9. $f(x)=5-(x+2)^2$
The parabola opens down because $a<0$.
The vertex is (–2, 5).
x-intercepts:
$0=5-(x+2)^2$
$(x+2)^2=5$
$x+2=\pm\sqrt{5}$
$x=-2\pm\sqrt{5}$
y-intercept:
$f(0)=5-(0+2)^2=1$
Domain: $(-\infty,\infty)$ Range: $(-\infty,5]$

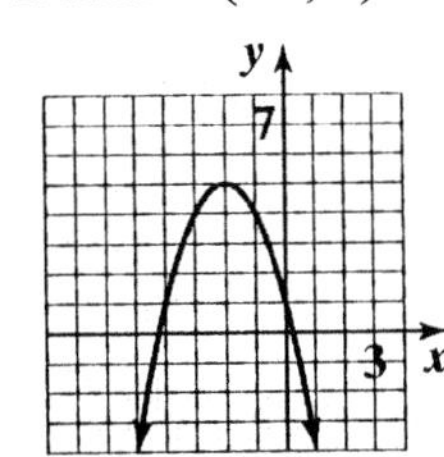

$f(x)=5-(x+2)^2$

10. $f(x) = -x^2 - 4x + 5$

The parabola opens down because $a < 0$.

vertex: $x = -\frac{b}{2a} = -\frac{-4}{2(-1)} = -2$

$f(-2) = -(-2)^2 - 4(-2) + 5 = 9$

The vertex is (–2, 9).

x-intercepts:

$0 = -x^2 - 4x + 5$

$x = \frac{-b \pm \sqrt{b^2 - 4ac}}{2a}$

$x = \frac{-(-4) \pm \sqrt{(-4)^2 - 4(-1)(5)}}{2(-1)}$

$x = \frac{4 \pm \sqrt{36}}{-2}$

$x = -2 \pm 3$

The x-intercepts are $x = 1$ and $x = -5$.

y-intercept:

$f(0) = -0^2 - 4(0) + 5 = 5$

Domain: $(-\infty, \infty)$ Range: $(-\infty, 9]$

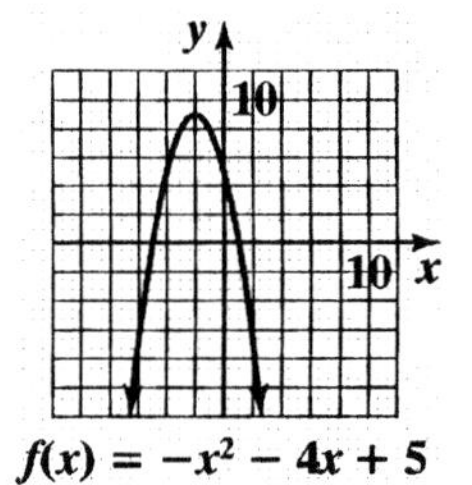

$f(x) = -x^2 - 4x + 5$

11. $f(x) = 3x^2 - 6x + 1$

The parabola opens up because $a > 0$.

vertex: $x = -\frac{b}{2a} = -\frac{-6}{2(3)} = 1$

$f(1) = 3(1)^2 - 6(1) + 1 = -2$

The vertex is (1, –2).

x-intercepts:

$0 = 3x^2 - 6x + 1$

$x = \frac{-b \pm \sqrt{b^2 - 4ac}}{2a}$

$x = \frac{-(-6) \pm \sqrt{(-6)^2 - 4(3)(1)}}{2(3)}$

$x = \frac{6 \pm \sqrt{24}}{6}$

$x = \frac{3 \pm \sqrt{6}}{3}$

y-intercept:

$f(0) = 3(0)^2 - 6(0) + 1 = 1$

Domain: $(-\infty, \infty)$ Range: $[-2, \infty)$

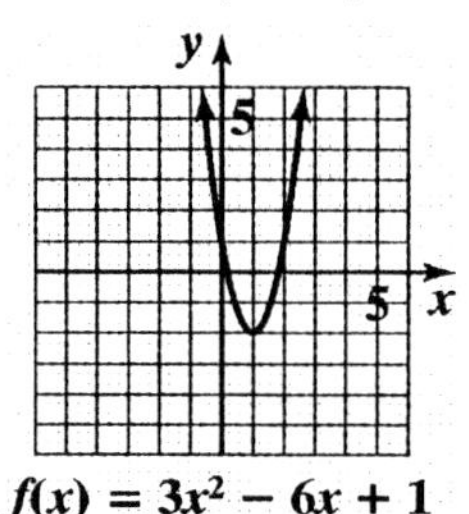

$f(x) = 3x^2 - 6x + 1$

12. $f(x) = (x-2)^2 (x+1)^3$

$(x-2)^2 (x+1)^3 = 0$

Apply the zero-product principle:

$(x-2)^2 = 0$ or $(x+1)^3 = 0$

$x - 2 = 0$ $\quad x + 1 = 0$

$x = 2$ $\quad x = -1$

The zeros are -1 and 2.

The graph of f crosses the x-axis at -1, since the zero has multiplicity 3. The graph touches the x-axis and turns around at 2 since the zero has multiplicity 2.

Since f is an odd-degree polynomial, degree 5, and since the leading coefficient, 1, is positive, the graph falls to the left and rises to the right.

Plot additional points as necessary and construct the graph.

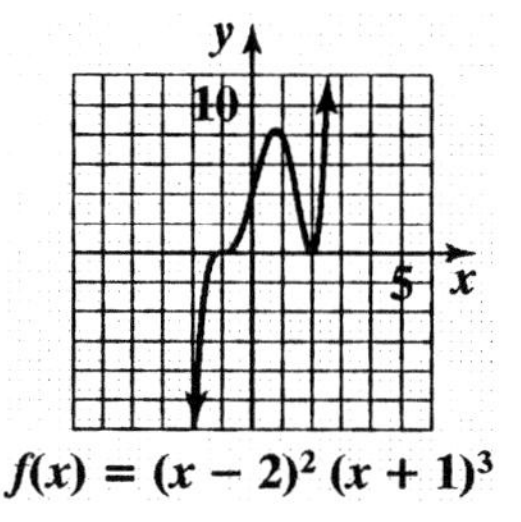

$f(x) = (x-2)^2 (x+1)^3$

13. $f(x) = -(x-2)^2(x+1)^2$

$-(x-2)^2(x+1)^2 = 0$

Apply the zero-product principle:

$(x-2)^2 = 0$ or $(x+1)^2 = 0$

$x-2=0 \qquad x+1=0$

$x=2 \qquad x=-1$

The zeros are -1 and 2.

The graph touches the x-axis and turns around both at -1 and 2 since both zeros have multiplicity 2.

Since f is an even-degree polynomial, degree 4, and since the leading coefficient, -1, is negative, the graph falls to the left and falls to the right.

Plot additional points as necessary and construct the graph.

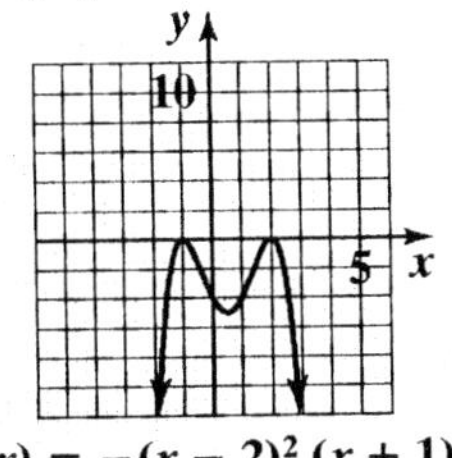

$f(x) = -(x-2)^2(x+1)^2$

14. $f(x) = x^3 - x^2 - 4x + 4$

$x^3 - x^2 - 4x + 4 = 0$

$x^2(x-1) - 4(x-1) = 0$

$(x^2-4)(x-1) = 0$

$(x+2)(x-2)(x-1) = 0$

Apply the zero-product principle:

$x+2=0$ or $x-2=0$ or $x-1=0$

$x=-2 \qquad x=2 \qquad x=1$

The zeros are -2, 1, and 2.

The graph of f crosses the x-axis at all three zeros, -2, 1, and 2, since all have multiplicity 1.

Since f is an odd-degree polynomial, degree 3, and since the leading coefficient, 1, is positive, the graph falls to the left and rises to the right.

Plot additional points as necessary and construct the graph.

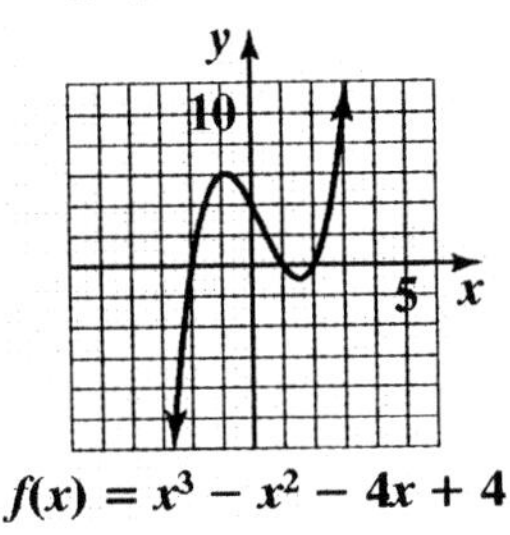

$f(x) = x^3 - x^2 - 4x + 4$

15. $f(x) = x^4 - 5x^2 + 4$

$x^4 - 5x^2 + 4 = 0$

$(x^2-4)(x^2-1) = 0$

$(x+2)(x-2)(x+1)(x-1) = 0$

Apply the zero-product principle,

$x=-2,\ x=2,\ x=-1,\ x=1$

The zeros are -2, -1, 1, and 2.

The graph crosses the x-axis at all four zeros, -2, -1, 1, and 2., since all have multiplicity 1.

Since f is an even-degree polynomial, degree 4, and since the leading coefficient, 1, is positive, the graph rises to the left and rises to the right.

Plot additional points as necessary and construct the graph.

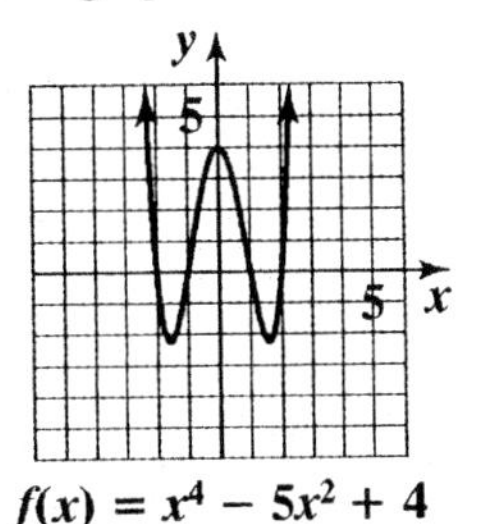

$f(x) = x^4 - 5x^2 + 4$

16. $f(x) = -(x+1)^6$

$-(x+1)^6 = 0$

$(x+1)^6 = 0$

$x+1=0$

$x=-1$

The zero is are -1.

The graph touches the x-axis and turns around at -1 since the zero has multiplicity 6.

Since f is an even-degree polynomial, degree 6, and since the leading coefficient, -1, is negative, the graph falls to the left and falls to the right.

Plot additional points as necessary and construct the graph.

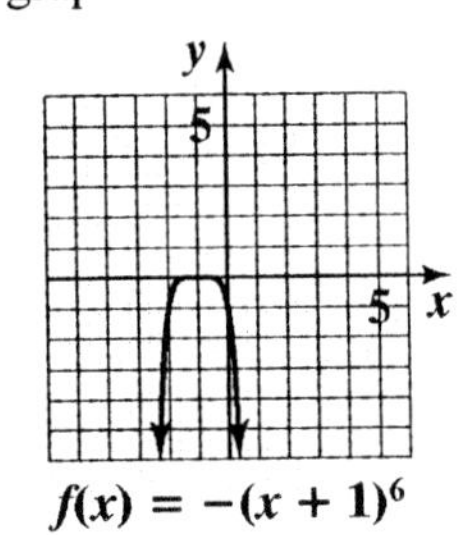

$f(x) = -(x+1)^6$

17. $f(x) = -6x^3 + 7x^2 - 1$

To find the zeros, we use the Rational Zero Theorem:
List all factors of the constant term -1: ± 1
List all factors of the leading coefficient -6: $\pm 1,\ \pm 2,\ \pm 3,\ \pm 6$

The possible rational zeros are:

$$\frac{\text{Factors of } -1}{\text{Factors of } -6} = \frac{\pm 1}{\pm 1,\ \pm 2,\ \pm 3,\ \pm 6}$$

$$= \pm 1,\ \pm\frac{1}{2},\ \pm\frac{1}{3},\ \pm\frac{1}{6}$$

We test values from the above list until we find a zero. One is shown next:

Test 1:

$$\begin{array}{r|rrrr} 1 & -6 & 7 & 0 & -1 \\ & & -6 & 1 & 1 \\ \hline & -6 & 1 & 1 & 0 \end{array}$$

The remainder is 0, so 1 is a zero. Thus,

$$-6x^3 + 7x^2 - 1 = 0$$
$$(x-1)(-6x^2 + x + 1) = 0$$
$$-(x-1)(6x^2 - x - 1) = 0$$
$$-(x-1)(3x+1)(2x-1) = 0$$

Apply the zero-product property:

$$x = 1,\quad x = -\frac{1}{3},\quad x = \frac{1}{2}$$

The zeros are $-\frac{1}{3}$, $\frac{1}{2}$, and 1.

The graph of *f* crosses the *x*-axis at all three zeros, $-\frac{1}{3}$, $\frac{1}{2}$, and 1, since all have multiplicity 1.

Since *f* is an odd-degree polynomial, degree 3, and since the leading coefficient, -6, is negative, the graph rises to the left and falls to the right.

Plot additional points as necessary and construct the graph.

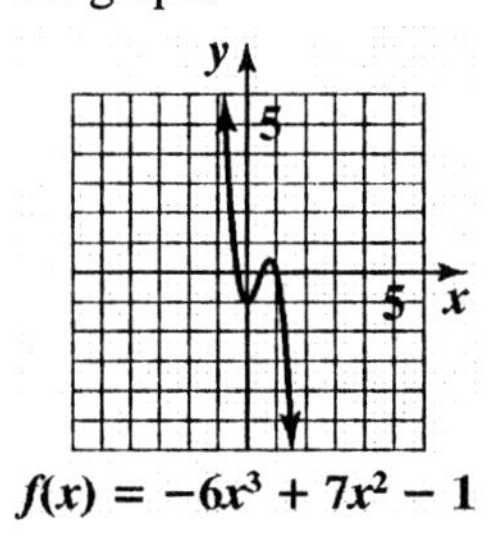

$f(x) = -6x^3 + 7x^2 - 1$

18. $f(x) = 2x^3 - 2x$

$$2x^3 - 2x = 0$$
$$2x(x^2 - 1) = 0$$
$$2x(x+1)(x-1) = 0$$

Apply the zero-product principle:

$x = 0,\qquad x = -1,\qquad x = 1$

The zeros are -1, 0, and 1.

The graph of *f* crosses the *x*-axis at all three zeros, -1, 0, and 1, since all have multiplicity 1.

Since *f* is an odd-degree polynomial, degree 3, and since the leading coefficient, 2, is positive, the graph falls to the left and rises to the right.

Plot additional points as necessary and construct the graph.

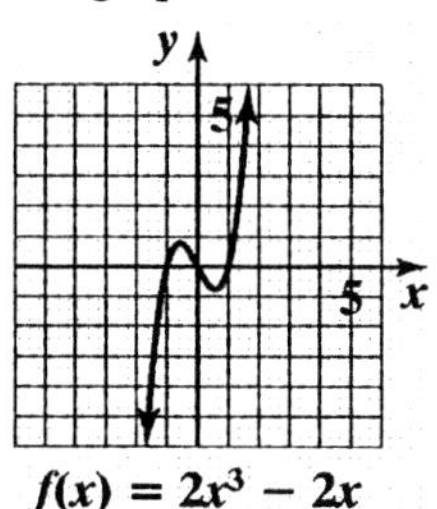

$f(x) = 2x^3 - 2x$

19. $f(x) = x^3 - 2x^2 + 26x$

$$x^3 - 2x^2 + 26x = 0$$
$$x(x^2 - 2x + 26) = 0$$

Note that $x^2 - 2x + 26$ does not factor, so we use the quadratic formula:

$x = 0$ or $x^2 - 2x + 26 = 0$

$$a = 1,\quad b = -2,\quad c = 26$$

$$x = \frac{-(-2) \pm \sqrt{(-2)^2 - 4(1)(26)}}{2(1)}$$

$$= \frac{2 \pm \sqrt{-100}}{2} = \frac{2 \pm 10i}{2} = 1 \pm 5i$$

The zeros are 0 and $1 \pm 5i$.

The graph of *f* crosses the *x*-axis at 0 (the only real zero), since it has multiplicity 1.

Since *f* is an odd-degree polynomial, degree 3, and since the leading coefficient, 1, is positive, the graph falls to the left and rises to the right.

Plot additional points as necessary and construct the graph.

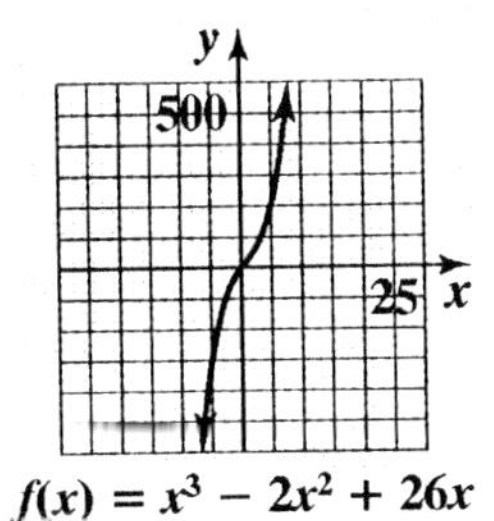

$f(x) = x^3 - 2x^2 + 26x$

20. $f(x) = -x^3 + 5x^2 - 5x - 3$

To find the zeros, we use the Rational Zero Theorem:

List all factors of the constant term -3:
$\pm 1,\ \pm 3$

List all factors of the leading coefficient -1:
± 1

The possible rational zeros are:

$$\frac{\text{Factors of } -3}{\text{Factors of } -1} = \frac{\pm 1,\ \pm 3}{\pm 1} = \pm 1,\ \pm 3$$

We test values from the previous list until we find a zero. One is shown next:

Test 3:

$$\begin{array}{r|rrrr} 3 & -1 & 5 & -5 & -3 \\ & & -3 & 6 & 3 \\ \hline & -1 & 2 & 1 & 0 \end{array}$$

The remainder is 0, so 3 is a zero. Thus,

$$-x^3 + 5x^2 - 5x - 3 = 0$$
$$(x-3)(-x^2 + 2x + 1) = 0$$
$$-(x-3)(x^2 - 2x - 1) = 0$$

Note that $x^2 - 2x - 1$ does not factor, so we use the quadratic formula:

$x - 3 = 0$ or $x^2 - 2x - 1 = 0$

$x = 3$ $\quad a = 1,\ b = -2,\ c = -1$

$$x = \frac{-(-2) \pm \sqrt{(-2)^2 - 4(1)(-1)}}{2(1)}$$

$$= \frac{2 \pm \sqrt{8}}{2} = \frac{2 \pm 2\sqrt{2}}{2} = 1 \pm \sqrt{2}$$

The zeros are 3 and $1 \pm \sqrt{2}$.

The graph of *f* crosses the *x*-axis at all three zeros, 3 and $1 \pm \sqrt{2}$, since all have multiplicity 1.

Since *f* is an odd-degree polynomial, degree 3, and since the leading coefficient, -1, is negative, the graph rises to the left and falls to the right.

Plot additional points as necessary and construct the graph.

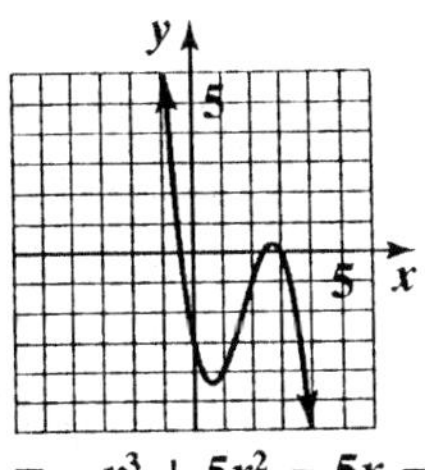

$f(x) = -x^3 + 5x^2 - 5x - 3$

21. $x^3 - 3x + 2 = 0$

We begin by using the Rational Zero Theorem to determine possible rational roots.

Factors of the constant term 2: $\pm 1,\ \pm 2$

Factors of the leading coefficient 1: ± 1

The possible rational zeros are:

$$\frac{\text{Factors of } 2}{\text{Factors of } 1} = \frac{\pm 1,\ \pm 2}{\pm 1} = \pm 1,\ \pm 2$$

We test values from above until we find a root. One is shown next:

Test 1:

$$\begin{array}{r|rrrr} 1 & 1 & 0 & -3 & 2 \\ & & 1 & 1 & -2 \\ \hline & 1 & 1 & -2 & 0 \end{array}$$

The remainder is 0, so 1 is a root of the equation. Thus,

$$x^3 - 3x + 2 = 0$$
$$(x-1)(x^2 + x - 2) = 0$$
$$(x-1)(x+2)(x-1) = 0$$
$$(x-1)^2(x+2) = 0$$

Apply the zero-product property:

$(x-1)^2 = 0$ or $x + 2 = 0$

$x - 1 = 0$ $\quad x = -2$

$x = 1$

The solutions are -2 and 1, and the solution set is $\{-2, 1\}$.

22. $6x^3 - 11x^2 + 6x - 1 = 0$
We begin by using the Rational Zero Theorem to determine possible rational roots.
Factors of the constant term -1: ± 1
Factors of the leading coefficient 6:
$\pm 1, \pm 2, \pm 3, \pm 6$

The possible rational zeros are:

$$\frac{\text{Factors of } -1}{\text{Factors of } 6} = \frac{\pm 1}{\pm 1, \pm 2, \pm 3, \pm 6} = \pm 1, \pm\frac{1}{2}, \pm\frac{1}{3}, \pm\frac{1}{6}$$

We test values from above until we find a root. One is shown next:

Test 1:

$$\begin{array}{r|rrrr} 1 & 6 & -11 & 6 & -1 \\ & & 6 & -5 & 1 \\ \hline & 6 & -5 & 1 & 0 \end{array}$$

The remainder is 0, so 1 is a root of the equation. Thus,

$$6x^3 - 11x^2 + 6x - 1 = 0$$
$$(x-1)(6x^2 - 5x + 1) = 0$$
$$(x-1)(3x-1)(2x-1) = 0$$

Apply the zero-product property:
$x - 1 = 0$ or $3x - 1 = 0$ or $2x - 1 = 0$
$x = 1$ $\quad x = \frac{1}{3}$ $\quad x = \frac{1}{2}$

The solutions are $\frac{1}{3}, \frac{1}{2}$ and 1, and the solution set is $\left\{\frac{1}{3}, \frac{1}{2}, 1\right\}$.

23. $(2x+1)(3x-2)^3(2x-7) = 0$
Apply the zero-product property:
$2x + 1 = 0$ or $(3x-2)^3 = 0$ or $2x - 7 = 0$
$x = -\frac{1}{2}$ $\quad 3x - 2 = 0$ $\quad x = \frac{7}{2}$
$x = \frac{2}{3}$

The solutions are $-\frac{1}{2}, \frac{2}{3}$ and $\frac{7}{2}$, and the solution set is $\left\{-\frac{1}{2}, \frac{2}{3}, \frac{7}{2}\right\}$.

24. $2x^3 + 5x^2 - 200x - 500 = 0$
We begin by using the Rational Zero Theorem to determine possible rational roots.
Factors of the constant term -500:
$\pm 1, \pm 2, \pm 4, \pm 5, \pm 10, \pm 20, \pm 25,$
$\pm 50, \pm 100, \pm 125, \pm 250, \pm 500$
Factors of the leading coefficient 2: $\pm 1, \pm 2$

The possible rational zeros are:

$$\frac{\text{Factors of } 500}{\text{Factors of } 2} = \pm 1, \pm 2, \pm 4, \pm 5, \pm 10, \pm 20, \pm 25, \pm 50, \pm 100, \pm 125, \pm 250, \pm 500, \pm\frac{1}{2}, \pm\frac{5}{2}, \pm\frac{25}{2}, \pm\frac{125}{2}$$

We test values from above until we find a root. One is shown next:

Test 10:

$$\begin{array}{r|rrrr} 10 & 2 & 5 & -200 & -500 \\ & & 20 & 250 & 500 \\ \hline & 2 & 25 & 50 & 0 \end{array}$$

The remainder is 0, so 10 is a root of the equation. Thus,

$$2x^3 + 5x^2 - 200x - 500 = 0$$
$$(x-10)(2x^2 + 25x + 50) = 0$$
$$(x-10)(2x+5)(x+10) = 0$$

Apply the zero-product property:
$x - 10 = 0$ or $2x + 5 = 0$ or $x + 10 = 0$
$x = 10$ $\quad x = -\frac{5}{2}$ $\quad x = -10$

The solutions are -10, $-\frac{5}{2}$, and 10, and the solution set is $\left\{-10, -\frac{5}{2}, 10\right\}$.

25. $x^4 - x^3 - 11x^2 = x + 12$
$x^4 - x^3 - 11x^2 - x - 12 = 0$
We begin by using the Rational Zero Theorem to determine possible rational roots.
Factors of the constant term -12:
$\pm 1, \pm 2, \pm 3, \pm 4, \pm 6, \pm 12$

Factors of the leading coefficient 1: ± 1

The possible rational zeros are:

$$\frac{\text{Factors of } -12}{\text{Factors of } 1} = \frac{\pm1,\ \pm2,\ \pm3,\ \pm4,\ \pm6,\ \pm12}{\pm1}$$
$$= \pm1,\ \pm2,\ \pm3,\ \pm4,\ \pm6,\ \pm12$$

We test values from this list we find a root. One possibility is shown next:

Test -3:

$$\begin{array}{r|rrrrr} -3 & 1 & -1 & -11 & -1 & -12 \\ & & -3 & 12 & -3 & 12 \\ \hline & 1 & -4 & 1 & -4 & 0 \end{array}$$

The remainder is 0, so -3 is a root of the equation. Using the Factor Theorem, we know that $x-1$ is a factor. Thus,

$$x^4 - x^3 - 11x^2 - x - 12 = 0$$
$$(x+3)(x^3 - 4x^2 + x - 4) = 0$$
$$(x+3)\left[x^2(x-4) + 1(x-4)\right] = 0$$
$$(x+3)(x-4)(x^2+1) = 0$$

As this point we know that -3 and 4 are roots of the equation. Note that x^2+1 does not factor, so we use the square-root principle:

$$x^2 + 1 = 0$$
$$x^2 = -1$$
$$x = \pm\sqrt{-1} = \pm i$$

The roots are -3, 4, and $\pm i$, and the solution set is $\{-3,\ 4,\ \pm i\}$.

26. $2x^4 + x^3 - 17x^2 - 4x + 6 = 0$

We begin by using the Rational Zero Theorem to determine possible rational roots.

Factors of the constant term 6: $\pm1,\ \pm2,\ \pm3,\ \pm6$

Factors of the leading coefficient 4: $\pm1,\ \pm2$

The possible rational roots are:

$$\frac{\text{Factors of } 6}{\text{Factors of } 2} = \frac{\pm1,\ \pm2,\ \pm3,\ \pm6}{\pm1,\ \pm2}$$
$$= \pm1,\ \pm2,\ \pm3,\ \pm6,\ \pm\frac{1}{2},\ \pm\frac{3}{2}$$

We test values from above until we find a root. One possibility is shown next:

Test -3:

$$\begin{array}{r|rrrrr} -3 & 2 & 1 & -17 & -4 & 6 \\ & & -6 & 15 & 6 & -6 \\ \hline & 2 & -5 & -2 & 2 & 0 \end{array}$$

The remainder is 0, so -3 is a root. Using the Factor Theorem, we know that $x+3$ is a factor of the polynomial. Thus,

$$2x^4 + x^3 - 17x^2 - 4x + 6 = 0$$
$$(x+3)(2x^3 - 5x^2 - 2x + 2) = 0$$

To solve the equation above, we need to factor $2x^3 - 5x^2 - 2x + 2$. We continue testing potential roots:

Test $\frac{1}{2}$:

$$\begin{array}{r|rrrr} \frac{1}{2} & 2 & -5 & -2 & 2 \\ & & 1 & -2 & -2 \\ \hline & 2 & -4 & -4 & 0 \end{array}$$

The remainder is 0, so $\frac{1}{2}$ is a zero and $x - \frac{1}{2}$ is a factor.

Summarizing our findings so far, we have

$$2x^4 + x^3 - 17x^2 - 4x + 6 = 0$$
$$(x+3)(2x^3 - 5x^2 - 2x + 2) = 0$$
$$(x+3)\left(x - \frac{1}{2}\right)(2x^2 - 4x - 4) = 0$$
$$2(x+3)\left(x - \frac{1}{2}\right)(x^2 - 2x - 2) = 0$$

At this point, we know that -3 and $\frac{1}{2}$ are roots of the equation. Note that $x^2 - 2x - 2$ does not factor, so we use the quadratic formula:

$$x^2 - 2x - 2 = 0$$
$$a = 1,\quad b = -2,\quad c = -2$$

$$x = \frac{-(-2) \pm \sqrt{(-2)^2 - 4(1)(-2)}}{2(1)}$$
$$= \frac{2 \pm \sqrt{4+8}}{2} = \frac{2 \pm \sqrt{12}}{2} = \frac{2 \pm 2\sqrt{3}}{2} = 1 \pm \sqrt{3}$$

The solutions are -3, $\frac{1}{2}$, and $1 \pm \sqrt{3}$, and the solution set is $\left\{-3,\ \frac{1}{2},\ 1 \pm \sqrt{3}\right\}$.

27. $P(x) = -x^2 + 150x - 4425$

Since $a = -1$ is negative, we know the function opens down and has a maximum at

$$x = -\frac{b}{2a} = -\frac{150}{2(-1)} = -\frac{150}{-2} = 75.$$

$$P(75) = -75^2 + 150(75) - 4425$$
$$= -5625 + 11{,}250 - 4425 = 1200$$

The company will maximize its profit by manufacturing and selling 75 cabinets per day. The maximum daily profit is $1200.

28. Let x = one of the numbers;
$-18 - x$ = the other number

The product is $f(x) = x(-18 - x) = -x^2 - 18x$

The x-coordinate of the maximum is

$$x = -\frac{b}{2a} = -\frac{-18}{2(-1)} = -\frac{-18}{-2} = -9.$$

$$f(-9) = -9[-18 - (-9)]$$
$$= -9(-18 + 9) = -9(-9) = 81$$

The vertex is $(-9, 81)$. The maximum product is 81. This occurs when the two number are -9 and $-18 - (-9) = -9$.

29. Let x = height of triangle;
$40 - 2x$ = base of triangle

$$A = \frac{1}{2}bh = \frac{1}{2}x(40 - 2x)$$

$$A(x) = 20x - x^2$$

The height at which the triangle will have maximum area is $x = -\frac{b}{2a} = -\frac{20}{2(-1)} = 10.$

$$A(10) = 20(10) - (10)^2 = 100$$

The maximum area is 100 squares inches.

30.

$$\begin{array}{r} 2x^2 - x - 3 \\ 3x^2 - 1 \overline{)\, 6x^4 - 3x^3 - 11x^2 + 2x + 4} \\ \underline{6x^4 \quad\quad - 2x^2} \\ -3x^3 - 9x^2 + 2x \\ \underline{-3x^3 \quad\quad + x} \\ -9x^2 + x + 4 \\ \underline{-9x^2 \quad\quad + 3} \\ x + 1 \end{array}$$

$$2x^2 - x - 3 + \frac{x + 1}{3x^2 - 1}$$

31. $(2x^4 - 13x^3 + 17x^2 + 18x - 24) \div (x - 4)$

4	2	−13	17	18	−24
		8	-20	−12	24
	2	−5	−3	6	0

The quotient is $2x^3 - 5x^2 - 3x + 6$.

32. $(x - 1)(x - i)(x + i) = (x - 1)(x^2 + 1)$

$f(x) = a_n(x - 1)(x^2 + 1)$

$f(-1) = a_n(-1 - 1)((-1)^2 + 1) = -4a_n = 8$

$a_n = -2$

$f(x) = -2(x - 1)(x^2 + 1)$ or $-2x^3 + 2x^2 - 2x + 2$

33. $(x - 2)(x - 2)(x - 3i)(x + 3i)$

$= (x - 2)(x - 2)(x^2 + 9)$

$f(x) = a_n(x - 2)(x - 2)(x^2 + 9)$

$f(0) = a_n(0 - 2)(0 - 2)(0^2 + 9)$

$36 = 36a_n$

$a_n = 1$

$f(x) = 1(x - 2)(x - 2)(x^2 + 9)$

$f(x) = x^4 - 4x^3 + 13x^2 - 36x + 36$

34. $f(x) = x^3 - x - 5$

$f(1) = 1^3 - 1 - 5 = -5$

$f(2) = 2^3 - 2 - 5 = 1$

Yes, the function must have a real zero between 1 and 2 because $f(1)$ and $f(2)$ have opposite signs.

Section 2.6

Check Point Exercises

1. **a.** $x - 5 = 0$

$x = 5$

$\{x | x \neq 5\}$

b. $x^2 - 25 = 0$

$x^2 = 25$

$x = \pm 5$

$\{x | x \neq 5, x \neq -5\}$

c. The denominator cannot equal zero.
All real numbers.

2. **a.** $x^2 - 1 = 0$

$x^2 = 1$

$x = 1, x = -1$

b. $g(x) = \frac{x-1}{x^2-1} = \frac{x-1}{(x-1)(x+1)} = \frac{1}{x+1}$

$x = -1$

c. The denominator cannot equal zero. No vertical asymptotes.

3. **a.** Since $n = m$, $y = \frac{9}{3} = 3$

$y = 3$ is a horizontal asymptote.

b. Since $n < m$, $y = 0$ is a horizontal asymptote.

c. Since $n > m$, there is no horizontal asymptote.

4. Begin with the graph of $f(x) = \frac{1}{x}$.

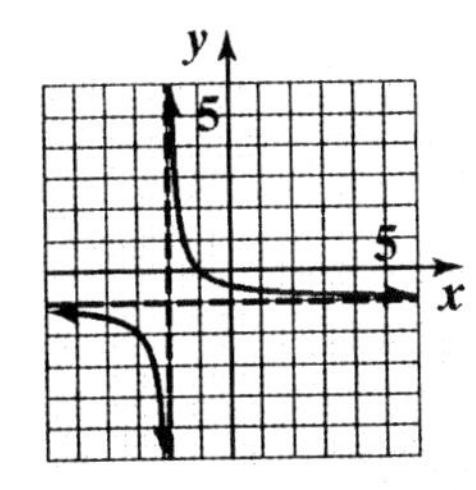

$g(x) = \frac{1}{x+2} - 1$

Shift the graph 2 units to the left by subtracting 2 from each x-coordinate. Shift the graph 1 unit down by subtracting 1 from each y-coordinate.

5. $f(x) = \frac{3x}{x-2}$

$f(-x) = \frac{3(-x)}{-x-2} = \frac{3x}{x+2}$

no symmetry

$f(0) = \frac{3(0)}{0-2} = 0$

The y-intercept is 0.

$3x = 0$

$x = 0$

The x-intercept is 0.

Vertical asymptote:

$x - 2 = 0$

$x = 2$

Horizontal asymptote:

$y = \frac{3}{1} = 3$

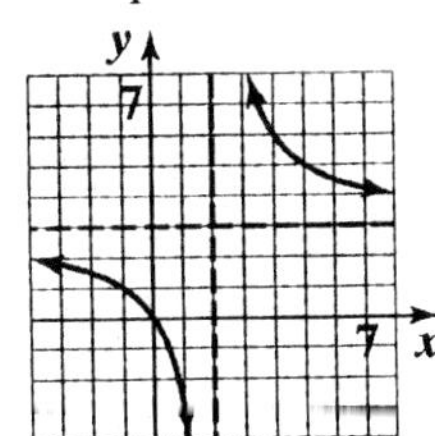

$f(x) = \frac{3x}{x-2}$

6. $f(x) = \frac{2x^2}{x^2-9}$

$f(-x) = \frac{2(-x)^2}{(-x)^2-9} = \frac{2x^2}{x^2-9} = f(x)$

The y-axis symmetry.

$f(0) = \frac{2(0)^2}{0^2-9} = 0$

The y-intercept is 0.

$2x^2 = 0$

$x = 0$

The x-intercept is 0.

vertical asymptotes:

$x^2 - 9 = 0$

$x = 3, x = -3$

horizontal asymptote:

$y = \frac{2}{1} = 2$

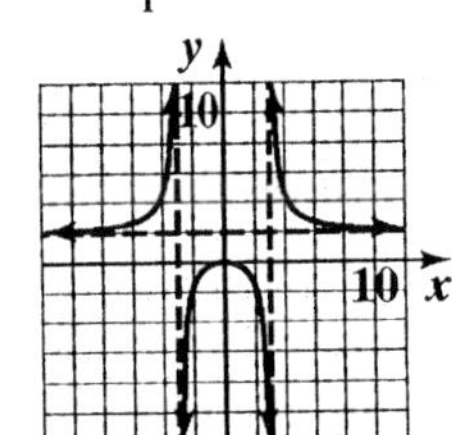

$f(x) = \frac{2x^2}{x^2-9}$

7. $f(x) = \frac{x^4}{x^2 + 2}$

$f(-x) = \frac{(-x)^4}{(-x)^2 + 2} = \frac{x^4}{x^2 + 2} = f(x)$

y-axis symmetry

$f(0) = \frac{0^4}{0^2 + 2} = 0$

The y-intercept is 0.

$x^4 = 0$

$x = 0$

The x-intercept is 0.

vertical asymptotes:

$x^2 + 2 = 0$

$x^2 = -2$

no vertical asymptotes

horizontal asymptote:

Since $n > m$, there is no horizontal asymptote.

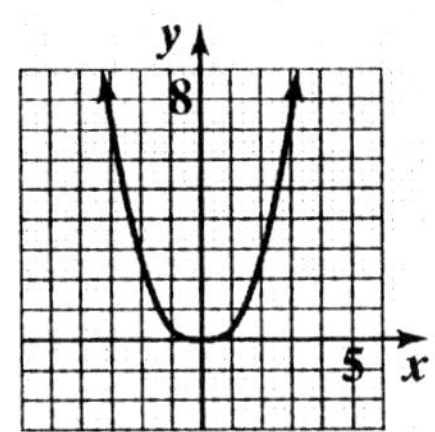

$f(x) = \frac{x^4}{x^2 + 2}$

8.

2	2	–5	7
		4	–2
	2	–1	5

the equation of the slant asymptote is $y = 2x - 1$.

9. **a.** $C(x) = 500x + 600{,}000$

b. $\overline{C}(x) = \frac{500x + 600{,}000}{x}$

c. $\overline{C}(1000) = \frac{500(1000) + 600{,}000}{1000} = 1100$

The cost per computer to replace 1000 computers would be $1100.

$\overline{C}(10000) = \frac{500(10000) + 600{,}000}{10000} = 560$

The cost per computer to replace 10,000 computers would be $560.

$\overline{C}(100{,}000) = \frac{500(100{,}000) + 600{,}000}{100{,}000}$

$= 506$

The cost per computer to replace 100,000 computers would be $506.

d. $y = 500$

The more computers the company replaces, the closer the average cost comes to $500.

10. $x - 10$ = the average velocity on the return trip. The function that expresses the total time required to complete the round trip is

$T(x) = \frac{20}{x} + \frac{20}{x - 10}$.

Exercise Set 2.6

1. $f(x) = \frac{5x}{x - 4}$

$\{x | x \neq 4\}$

2. $f(x) = \frac{7x}{x - 8}$

$\{x | x \neq 8\}$

3. $g(x) = \frac{3x^2}{(x - 5)(x + 4)}$

$\{x | x \neq 5, x \neq -4\}$

4. $g(x) = \frac{2x^2}{(x - 2)(x + 6)}$

$\{x | x \neq 2, x \neq -6\}$

5. $h(x) = \frac{x + 7}{x^2 - 49}$

$x^2 - 49 = (x - 7)(x + 7)$

$\{x | x \neq 7, x \neq -7\}$

6. $h(x) = \frac{x + 8}{x^2 - 64}$

$x^2 - 64 = (x - 8)(x + 8)$

$\{x | x \neq 8, x \neq -8\}$

7. $f(x) = \dfrac{x+7}{x^2+49}$

all real numbers

8. $f(x) = \dfrac{x+8}{x^2+64}$

all real numbers

9. $-\infty$

10. $+\infty$

11. $-\infty$

12. $+\infty$

13. 0

14. 0

15. $+\infty$

16. $-\infty$

17. $-\infty$

18. $+\infty$

19. 1

20. 1

21. $f(x) = \dfrac{x}{x+4}$

$x + 4 = 0$

$x = -4$

vertical asymptote: $x = -4$

22. $f(x) = \dfrac{x}{x-3}$

$x - 3 = 0$

$x = 3$

vertical asymptote: $x = 3$

23. $g(x) = \dfrac{x+3}{x(x+4)}$

$x(x+4) = 0$

$x = 0, x = -4$

vertical asymptotes: $x = 0,\ x = -4$

24. $g(x) = \dfrac{x+3}{x(x-3)}$

$x(x-3) = 0$

$x = 0, x = 3$

vertical asymptotes: $x = 0, x = 3$

25. $h(x) = \dfrac{x}{x(x+4)} = \dfrac{1}{x+4}$

$x + 4 = 0$

$x = 4$

vertical asymptote: $x = -4$

26. $h(x) = \dfrac{x}{x(x-3)} = \dfrac{1}{x-3}$

$x - 3 = 0$

$x = 3$

vertical asymptote: $x = 3$

27. $r(x) = \dfrac{x}{x^2+4}$

$x^2 + 4$ has no real zeros

There are no vertical asymptotes.

28. $r(x) = \dfrac{x}{x^2+3}$

$x^2 + 3$ has no real zeros

There is no vertical asymptotes.

29. $f(x) = \dfrac{12x}{3x^2+1}$

$n < m$

horizontal asymptote: $y = 0$

30. $f(x) = \dfrac{15x}{3x^2+1}$

$n < m$

horizontal asymptote: $y = 0$

31. $g(x) = \dfrac{12x^2}{3x^2+1}$

$n = m,$

horizontal asymptote: $y = \dfrac{12}{3} = 4$

32. $g(x) = \dfrac{15x^2}{3x^2+1}$

$n = m$

horizontal asymptote: $y = \dfrac{15}{3} = 5$

33. $h(x) = \frac{12x^3}{3x^2+1}$

$n > m$
no horizontal asymptote

34. $h(x) = \frac{15x^3}{3x^2+1}$

$n > m$
no horizontal asymptote

35. $f(x) = \frac{-2x+1}{3x+5}$

$n = m$

horizontal asymptote: $y = -\frac{2}{3}$

36. $f(x) = \frac{-3x+7}{5x-2}$

$n = m$ s

horizontal asymptote: $y = -\frac{3}{5}$

37. $g(x) = \frac{1}{x-1}$

Shift the graph of $f(x) = \frac{1}{x}$ 1 unit to the right.

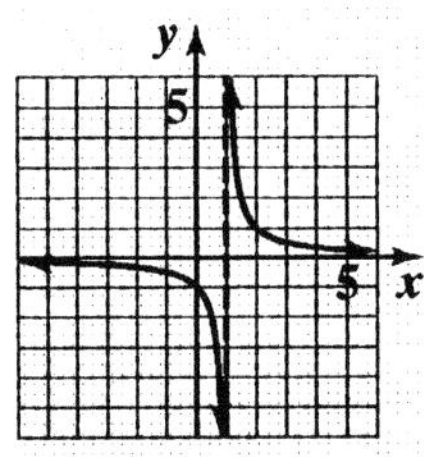

$g(x) = \frac{1}{x-1}$

38. $g(x) = \frac{1}{x-2}$

Shift the graph of $f(x) = \frac{1}{x}$ 2 units to the right.

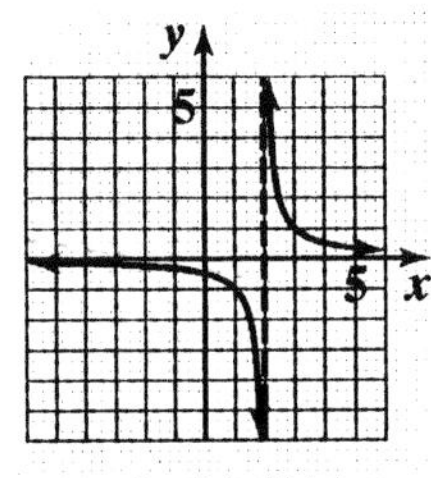

$g(x) = \frac{1}{x-2}$

39. $h(x) = \frac{1}{x} + 2$

Shift the graph of $f(x) = \frac{1}{x}$ 2 units up.

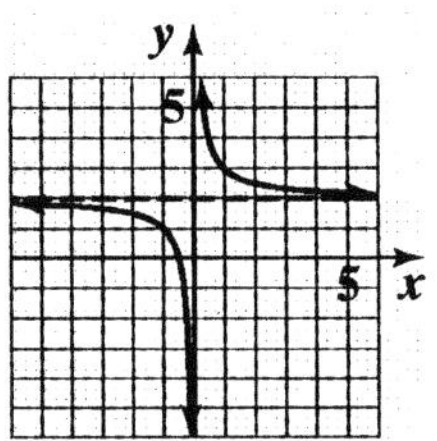

$h(x) = \frac{1}{x} + 2$

40. $h(x) = \frac{1}{x} + 1$

Shift the graph of $f(x) = \frac{1}{x}$ 1 unit up.

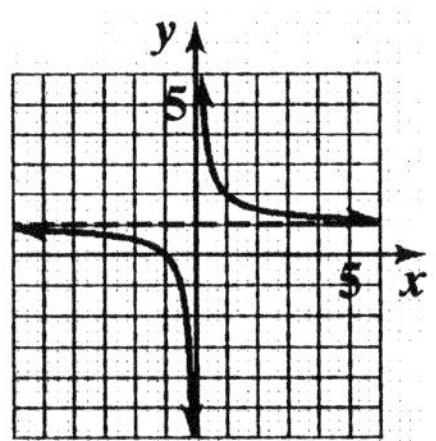

$h(x) = \frac{1}{x} + 1$

41. $g(x) = \frac{1}{x+1} - 2$

Shift the graph of $f(x) = \frac{1}{x}$ 1 unit left and 2 units down.

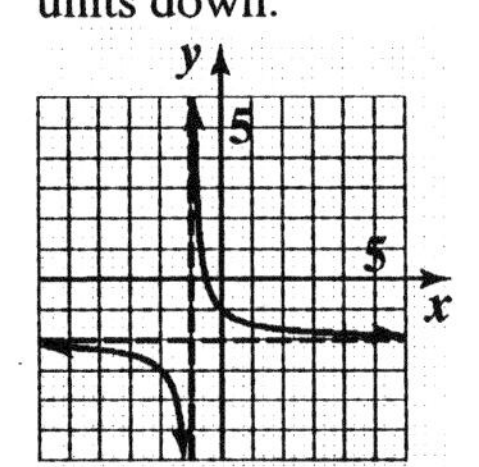

$g(x) = \frac{1}{x+1} - 2$

42. $g(x)=\dfrac{1}{x+2}-2$

Shift the graph of $f(x)=\dfrac{1}{x}$ 2 units left and 2 units down.

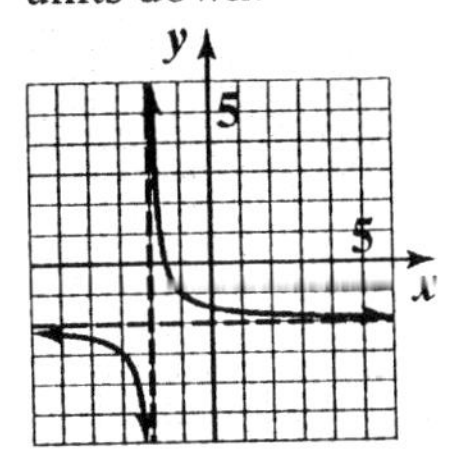

$\mathbf{g(x) = \dfrac{1}{x + 2} - 2}$

43. $g(x)=\dfrac{1}{(x+2)^2}$

Shift the graph of $f(x)=\dfrac{1}{x^2}$ 2 units left.

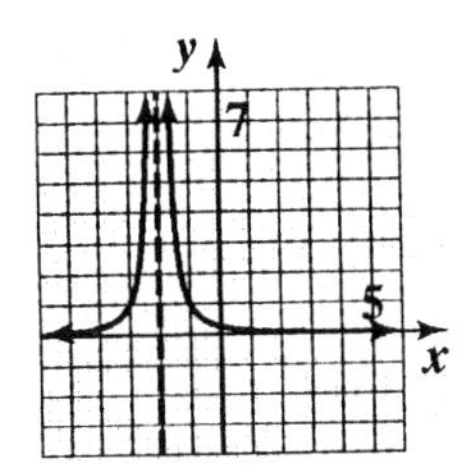

$\mathbf{g(x) = \dfrac{1}{(x + 2)^2}}$

44. $g(x)=\dfrac{1}{(x+1)^2}$

Shift the graph of $f(x)=\dfrac{1}{x^2}$ 1 unit left.

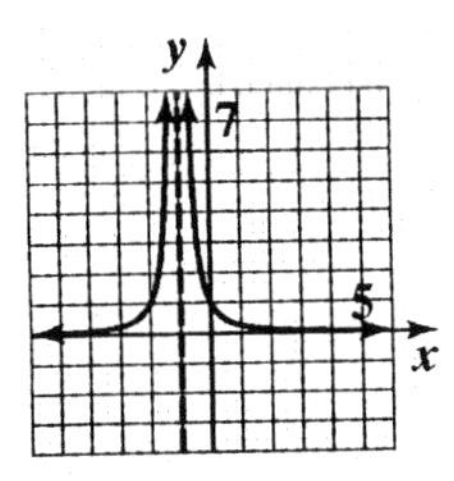

$\mathbf{g(x) = \dfrac{1}{(x + 1)^2}}$

45. $h(x)=\dfrac{1}{x^2}-4$

Shift the graph of $f(x)=\dfrac{1}{x^2}$ 4 units down.

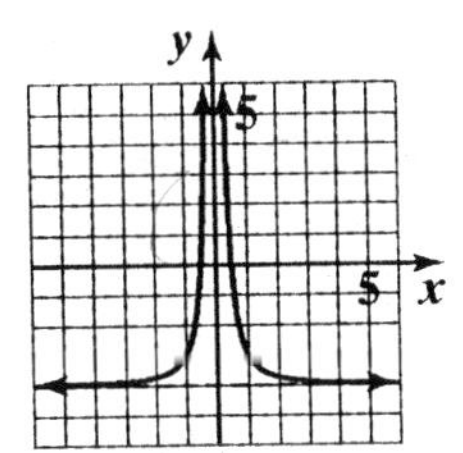

$\mathbf{h(x) = \dfrac{1}{x^2} - 4}$

46. $h(x)=\dfrac{1}{x^2}-3$

Shift the graph of $f(x)=\dfrac{1}{x^2}$ 3 units down.

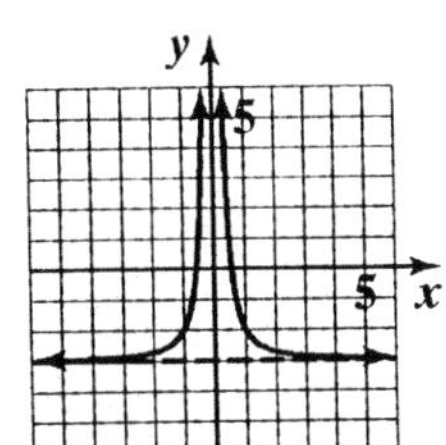

$\mathbf{h(x) = \dfrac{1}{x^2} - 3}$

47. $h(x)=\dfrac{1}{(x-3)^2}+1$

Shift the graph of $f(x)=\dfrac{1}{x^2}$ 3 units right and 1 unit up.

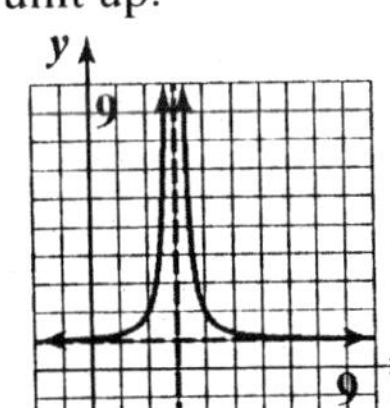

$\mathbf{h(x) = \dfrac{1}{(x - 3)^2} + 1}$

48. $h(x) = \frac{1}{(x-3)^2} + 2$

Shift the graph of $f(x) = \frac{1}{x^2}$ 3 units right and 2 units up.

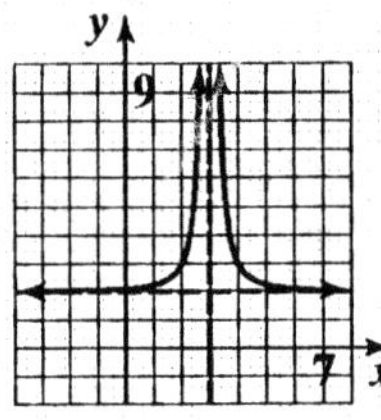

$h(x) = \frac{1}{(x-3)^2} + 2$

49. $f(x) = \frac{4x}{x-2}$

$f(-x) = \frac{4(-x)}{(-x)-2} = \frac{4x}{x+2}$

$f(-x) \neq f(x), f(-x) \neq -f(x)$

no symmetry

y-intercept: $y = \frac{4(0)}{0-2} = 0$

x-intercept: $4x = 0$

$x = 0$

vertical asymptote:

$x - 2 = 0$

$x = 2$

horizontal asymptote:

$n = m$, so $y = \frac{4}{1} = 4$

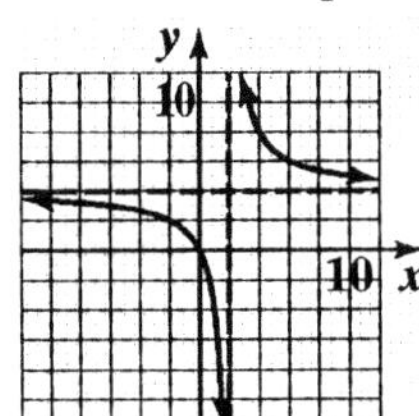

$f(x) = \frac{4x}{x-2}$

50. $f(x) = \frac{3x}{x-1}$

$f(-x) = \frac{3(-x)}{(-x)-1} = \frac{3x}{x+1}$

$f(-x) \neq f(x), f(-x) \neq -f(x)$

no symmetry

y-intercept: $y = \frac{3(0)}{0-1} = 0$

x-intercept: $3x = 0$

$x = 0$

vertical asymptote:

$x - 1 = 0$

$x = 1$

horizontal asymptote:

$n = m$, so $y = \frac{3}{1} = 3$

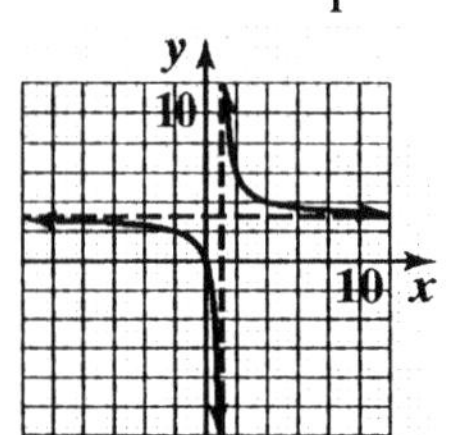

$f(x) = \frac{3x}{x-1}$

51. $f(x) = \frac{2x}{x^2-4}$

$f(-x) = \frac{2(-x)}{(-x)^2-4} = -\frac{2x}{x^2-4} = -f(x)$

Origin symmetry

y-intercept: $\frac{2(0)}{0^2-4} = \frac{0}{-4} = 0$

x-intercept:

$2x = 0$

$x = 0$

vertical asymptotes:

$x^2 - 4 = 0$

$x = \pm 2$

horizontal asymptote:

$n < m$ so $y = 0$

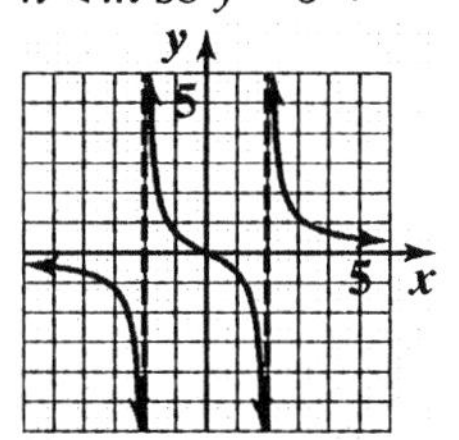

$f(x) = \frac{2x}{x^2-4}$

52. $f(x) = \dfrac{4x}{x^2 - 1}$

$f(-x) = \dfrac{4(-x)}{(-x)^2 - 1} = -\dfrac{4x}{x^2 - 1} = -f(x)$

Origin symmetry

y-intercept: $\dfrac{4(0)}{0^2 - 1} = 0$

x-intercept: $4x = 0$

$x = 0$

vertical asymptotes:

$x^2 - 1 = 0$

$(x - 1)(x + 1) = 0$

$x = \pm 1$

horizontal asymptote:

$n < m$ so $y = 0$

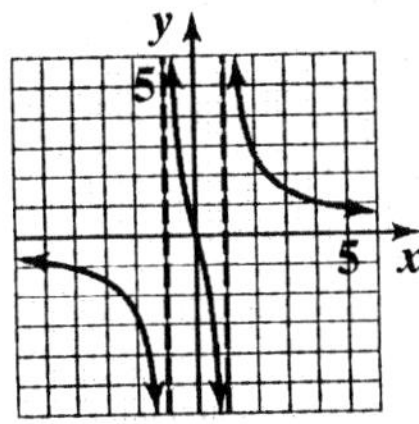

$f(x) = \dfrac{4x}{x^2 - 1}$

53. $f(x) = \dfrac{2x^2}{x^2 - 1}$

$f(-x) = \dfrac{2(-x)^2}{(-x)^2 - 1} = \dfrac{2x^2}{x^2 - 1} = f(x)$

y-axis symmetry

y-intercept: $y = \dfrac{2(0)^2}{0^2 - 1} = \dfrac{0}{1} = 0$

x-intercept:

$2x^2 = 0$

$x = 0$

vertical asymptote:

$x^2 - 1 = 0$

$x^2 = 1$

$x = \pm 1$

horizontal asymptote:

$n = m$, so $y = \dfrac{2}{1} = 2$

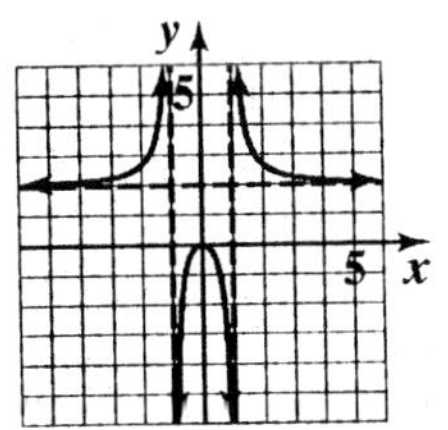

$f(x) = \dfrac{2x^2}{x^2 - 1}$

54. $f(x) = \dfrac{4x^2}{x^2 - 9}$

$f(-x) = \dfrac{4(-x)^2}{(-x)^2 - 9} = \dfrac{4x^2}{x^2 - 9} = f(x)$

y-axis symmetry

y-intercept: $y = \dfrac{4(0)^2}{0^2 - 9} = 0$

x-intercept:

$4x^2 = 0$

$x = 0$

vertical asymptotes:

$x^2 - 9 = 0$

$(x - 3)(x + 3) = 0$

$x = \pm 3$

horizontal asymptote:

$n = m$, so $y = \dfrac{4}{1} = 4$

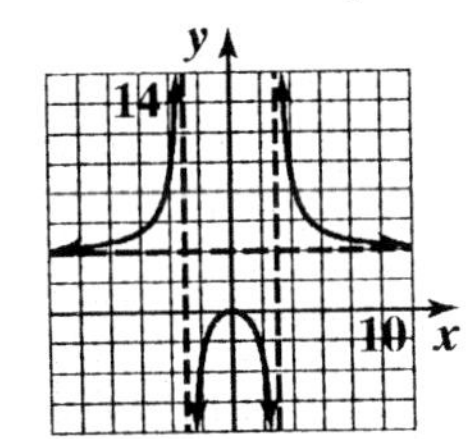

$f(x) = \dfrac{4x^2}{x^2 - 9}$

55. $f(x) = \frac{-x}{x+1}$

$f(-x) = \frac{-(-x)}{(-x)+1} = \frac{x}{-x+1}$

$f(-x) \neq f(x), f(-x) \neq -f(x)$
no symmetry

y-intercept: $y = \frac{-(0)}{0+1} = \frac{0}{1} = 0$

x-intercept:
$-x = 0$
$x = 0$
vertical asymptote:
$x + 1 = 0$
$x = -1$
horizontal asymptote:

$n = m$, so $y = \frac{-1}{1} = -1$

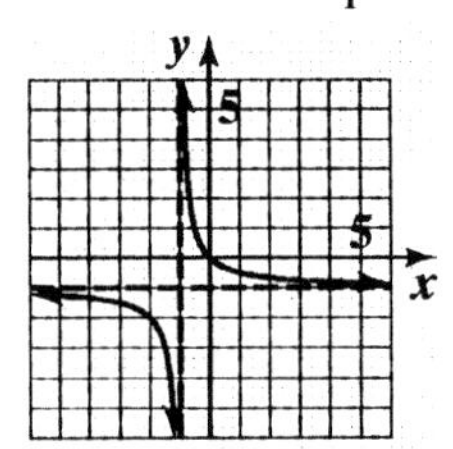

$f(x) = \frac{-x}{x+1}$

56. $f(x) = \frac{-3x}{x+2}$

$f(-x) = \frac{-3(-x)}{(-x)+2} = \frac{3x}{-x+2}$

$f-x) \neq f(x), f(-x) \neq -f(x)$
no symmetry
y-intercept:

$y = \frac{-3(0)}{0+2} = 0$

x-intercept:
$-3x = 0$
$x = 0$
vertical asymptote:
$x + 2 = 0$
$x = -2$

horizontal asymptote:

$n = m$, so $y = \frac{-3}{1} = -3$

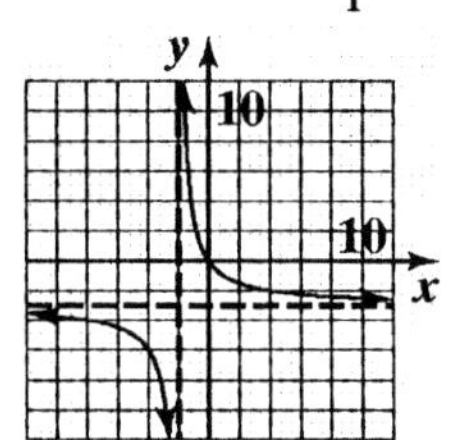

$f(x) = \frac{-3x}{x+2}$

57. $f(x) = -\frac{1}{x^2-4}$

$f(-x) = -\frac{1}{(-x)^2-4} = -\frac{1}{x^2-4} = f(x)$

y-axis symmetry

y-intercept: $y = -\frac{1}{0^2-4} = \frac{1}{4}$

x-intercept: $-1 \neq 0$
no x-intercept
vertical asymptotes:
$x^2 - 4 = 0$
$x^2 = 4$
$x = \pm 2$

horizontal asymptote:
$n < m$ or $y = 0$

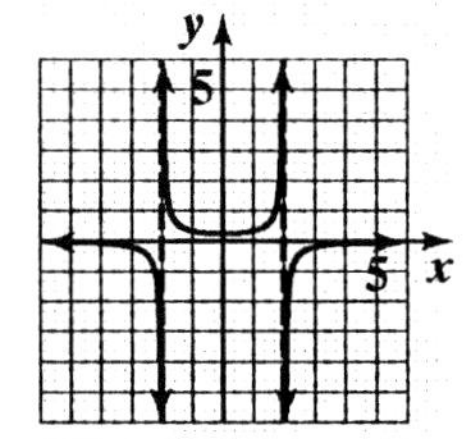

$f(x) = -\frac{1}{x^2-4}$

58. $f(x) = -\frac{2}{x^2-1}$

$f(-x) = -\frac{2}{(-x)^2-1} = -\frac{2}{x^2-1} = f(x)$

y-axis symmetry
y-intercept:

$y = -\frac{2}{0^2-1} = -\frac{2}{-1} = 2$

x-intercept:
$-2 = 0$
no x-intercept

vertical asymptotes:
$x^2 - 1 = 0$
$(x - 1)(x + 1)$
$x = \pm 1$
horizontal asymptote:
$n < m$, so $y = 0$

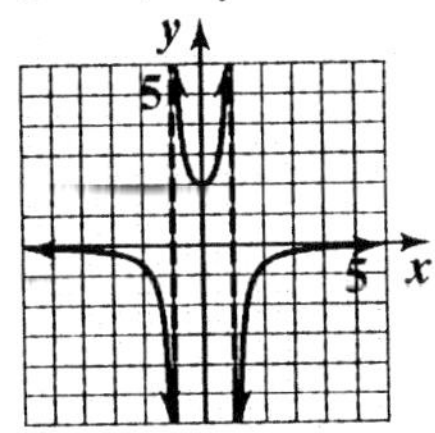

$$f(x) = -\frac{2}{x^2 - 1}$$

59. $f(x) = \dfrac{2}{x^2 + x - 2}$

$$f(-x) = -\frac{2}{(-x)^2 - x - 2} = \frac{2}{x^2 - x - 2}$$

$f(-x) \neq f(x), f(-x) \neq -f(x)$
no symmetry

y-intercept: $y = \dfrac{2}{0^2 + 0 - 2} = \dfrac{2}{-2} = -1$

x-intercept: none
vertical asymptotes:
$x^2 + x - 2 = 0$
$(x + 2)(x - 1) = 0$
$x = -2, x = 1$
horizontal asymptote:
$n < m$ so $y = 0$

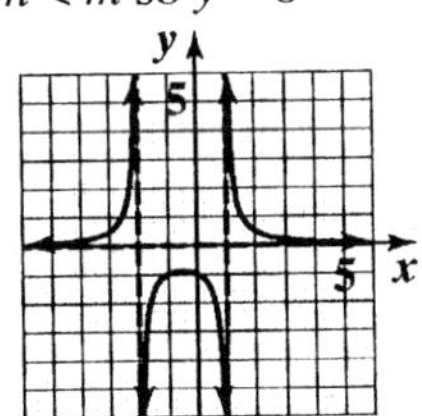

$$f(x) = \frac{2}{x^2 + x - 2}$$

60. $f(x) = \dfrac{-2}{x^2 - x - 2}$

$$f(-x) = \frac{-2}{(-x)^2 - (-x) - 2} = \frac{-2}{x^2 + x - 2}$$

$f(-x) \neq f(x), f(-x) \neq -f(x)$
no symmetry

y-intercept: $y = \dfrac{-2}{0^2 - 0 - 2} = 1$

x-intercept: none
vertical asymptotes:
$x^2 - x - 2 = 0$
$(x - 2)(x + 1) = 0$
$x = 2, x = -1$
horizontal asymptote:
$n < m$ so $y = 0$

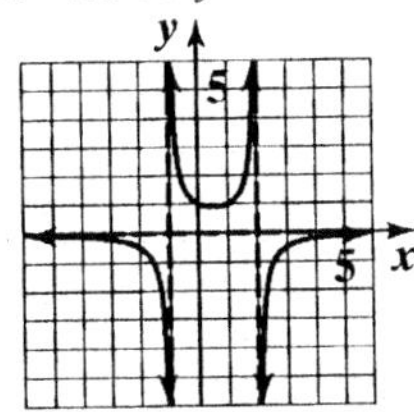

$$f(x) = \frac{-2}{x^2 - x - 2}$$

61. $f(x) = \dfrac{2x^2}{x^2 + 4}$

$$f(-x) = \frac{2(-x)^2}{(-x)^2 + 4} = \frac{2x^2}{x^2 + 4} = f(x)$$

y axis symmetry

y-intercept: $y = \dfrac{2(0)^2}{0^2 + 4} = 0$

x-intercept: $2x^2 = 0$
$x = 0$
vertical asymptote: none
horizontal asymptote:

$n = m$, so $y = \dfrac{2}{1} = 2$

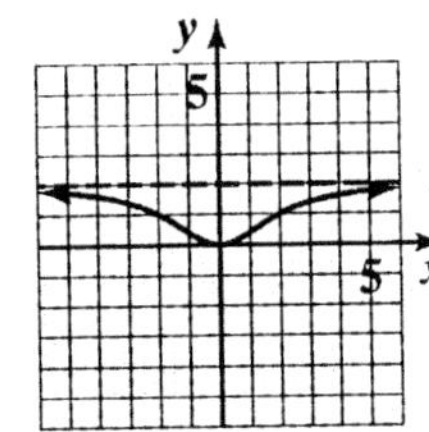

$$f(x) = \frac{2x^2}{x^2 + 4}$$

62. $f(x)=\dfrac{4x^2}{x^2+1}$

$f(-x)=\dfrac{4(-x)^2}{(-x)^2+1}=\dfrac{4x^2}{x^2+1}=f(x)$

y axis symmetry

y-intercept: $y=\dfrac{4(0)^2}{0^2+1}=0$

x-intercept: $4x^2=0$

$x=0$

vertical asymptote: none

horizontal asymptote:

$n=m$, so $y=\dfrac{4}{1}=4$

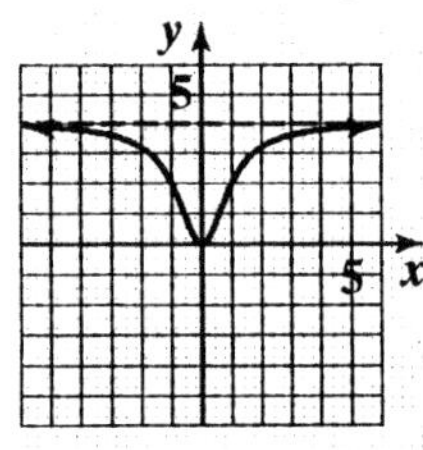

$f(x)=\dfrac{4x^2}{x^2+1}$

63. $f(x)=\dfrac{x+2}{x^2+x-6}$

$f(-x)=\dfrac{-x+2}{(-x)^2-(-x)-6}=\dfrac{-x+2}{x^2+x-6}$

$f(-x)\neq f(x), f(-x)\neq -f(x)$

no symmetry

y-intercept: $y=\dfrac{0+2}{0^2+0-6}=-\dfrac{2}{6}=-\dfrac{1}{3}$

x-intercept:

$x+2=0$

$x=-2$

vertical asymptotes:

$x^2+x-6=0$

$(x+3)(x-2)$

$x=-3, x=2$

horizontal asymptote:

$n<m$, so $y=0$

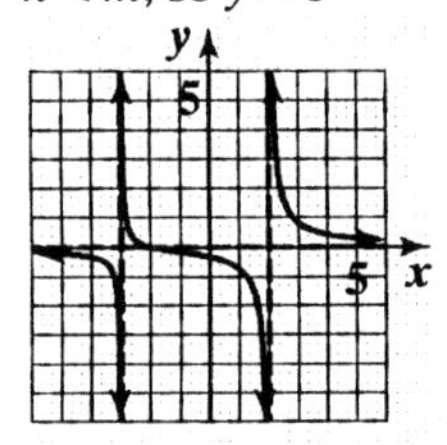

$f(x)=\dfrac{x+2}{x^2+x-6}$

64. $f(x)=\dfrac{x-4}{x^2-x-6}$

$f(-x)=\dfrac{-x-4}{(-x)^2-(-x)-6}=-\dfrac{x+4}{x^2+x-6}$

$f(-x)\neq f(x), f(-x)\neq -f(x)$

no symmetry

y-intercept: $y=\dfrac{0-4}{0^2-0-6}=\dfrac{2}{3}$

x-intercept:

$x-4=0, x=4$

vertical asymptotes:

$x^2-x-6=0$

$(x-3)(x+2)$

$x=3, x=-2$

horizontal asymptote:

$n<m$, so $y=0$

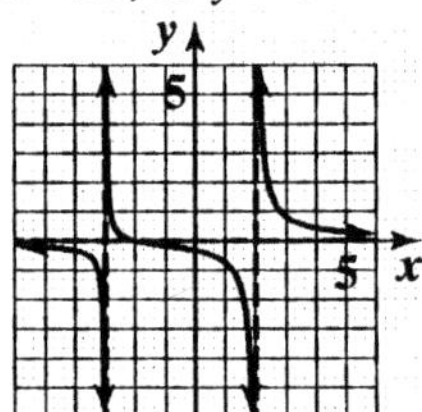

$f(x)=\dfrac{x+2}{x^2-x-6}$

65. $f(x)=\dfrac{x^4}{x^2+2}$

$f(-x)=\dfrac{(-x)^4}{(-x)^2+2}=\dfrac{x^4}{x^2+2}=f(x)$

y-axis symmetry

y-intercept: $y=\dfrac{0^4}{0^2+2}=0$

x-intercept: $x^4=0$

$x=0$

vertical asymptote: none

horizontal asymptote:

$n>m$, so none

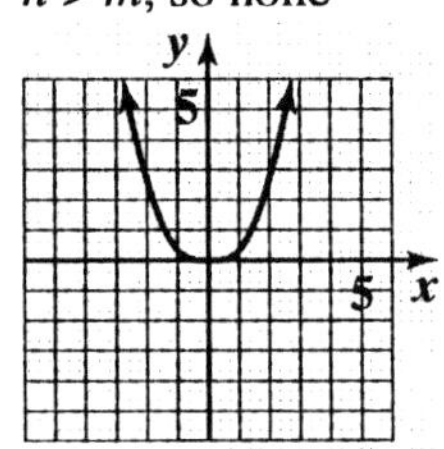

$f(x)=\dfrac{x^4}{x^2+2}$

66. $f(x)=\dfrac{2x^4}{x^2+1}$

$f(-x)=\dfrac{2(-x)^4}{(-x)^2+1}=\dfrac{2x^4}{x^2+1}=f(x)$

y-axis symmetry

y-intercept: $y=\dfrac{2(0^4)}{0^2+2}=0$

x-intercept: $2x^4=0$

$x=0$

vertical asymptote: none

horizontal asymptote:

$n>m$, so none

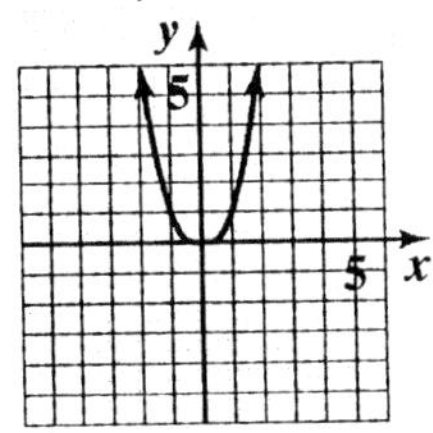

67. $f(x)=\dfrac{x^2+x-12}{x^2-4}$

$f(-x)=\dfrac{(-x)^2-x-12}{(-x)^2-4}=\dfrac{x^2-x-12}{x^2-4}$

$f(-x)\neq f(x), f(-x)\neq -f(x)$

no symmetry

y-intercept: $y=\dfrac{0^2+0-12}{0^2-4}=3$

x-intercept: $x^2+x-12=0$

$(x-3)(x+4)=0$

$x=3, x=-4$

vertical asymptotes:

$x^2-4=0$

$(x-2)(x+2)=0$

$x=2, x=-2$

horizontal asymptote:

$n=m$, so $y=\dfrac{1}{1}=1$

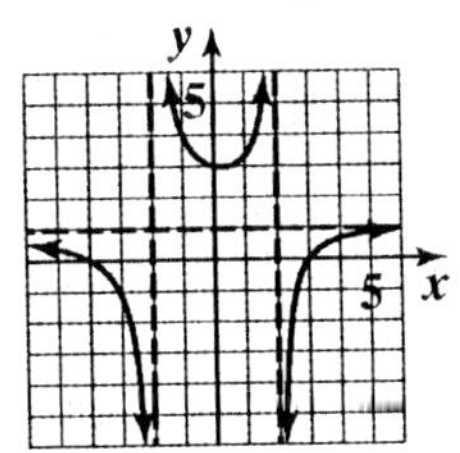

68. $f(x)=\dfrac{x^2}{x^2+x-6}$

$f(-x)=\dfrac{(-x)^2}{(-x)^2-x-6}=\dfrac{x^2}{x^2-x-6}$

$f(-x)\neq f(x), f(-x)\neq -f(x)$

no symmetry

y-intercept: $y=\dfrac{0^2}{0^2+0-6}=0$

x-intercept: $x^2=0, x=0$

vertical asymptotes:

$x^2+x-6=0$

$(x+3)(x-2)=0$

$x=-3, x=2$

horizontal asymptote:

$n=m$, so $y=\dfrac{1}{1}=1$

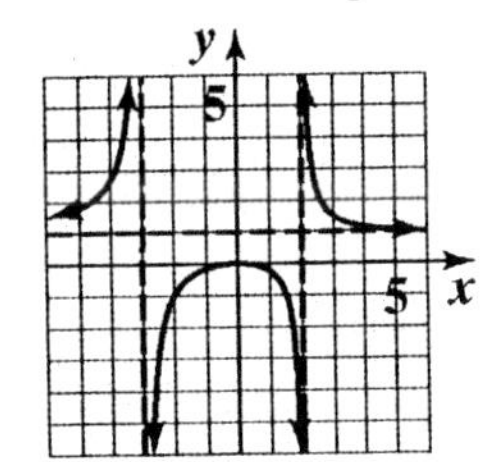

69. $f(x)=\dfrac{3x^2+x-4}{2x^2-5x}$

$f(-x)=\dfrac{3(-x)^2-x-4}{2(-x)^2+5x}=\dfrac{3x^2-x-4}{2x^2+5x}$

$f(-x)\neq f(x), f(-x)\neq -f(x)$

no symmetry

y-intercept: $y=\dfrac{3(0)^2+0-4}{2(0)^2-5(0)}=\dfrac{-4}{0}$

no y-intercept

x-intercepts:

$3x^2+x-4=0$

$(3x+4)(x-1)=0$

$3x+4=0 \quad x-1=0$

$3x=-4$

$x=-\dfrac{4}{3}, x=1$

vertical asymptotes:

$2x^2-5x=0$

$x(2x-5)=0$

$x=0, 2x=5$

$x=\dfrac{5}{2}$

horizontal asymptote:

$n=m$, so $y=\dfrac{3}{2}$

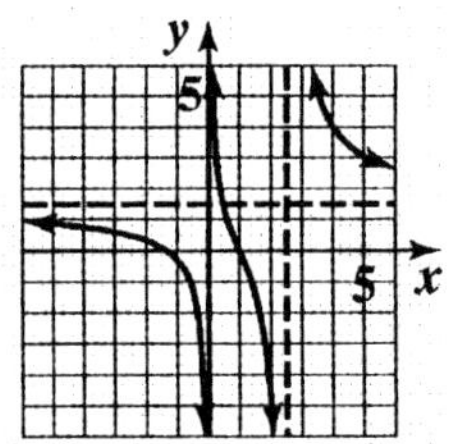

$f(x)=\dfrac{3x^2+x-4}{2x^2-5x}$

70. $f(x)=\dfrac{x^2-4x+3}{(x+1)^2}$

$f(-x)=\dfrac{(-x)^2-4(-x)+3}{(-x+1)^2}=\dfrac{x^2+4x+3}{(-x+1)^2}$

$f(-x)\neq f(x), f(-x)\neq -f(x)$

no symmetry

y-intercept: $y=\dfrac{0^2-4(0)+3}{(0+1)^2}=\dfrac{3}{1}=3$

x-intercept:

$x^2-4x+3=0$

$(x-3)(x-1)=0$

$x=3$ and $x=1$

vertical asymptote:

$(x+1)^2=0$

$x=-1$

horizontal asymptote:

$n=m$, so $y=\dfrac{1}{1}=1$

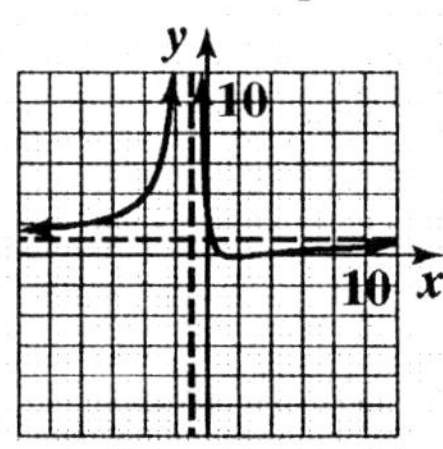

$f(x)=\dfrac{x^2-4x+3}{(x+1)^2}$

71. a. Slant asymptote:

$f(x)=x-\dfrac{1}{x}$

$y=x$

b. $f(x)=\dfrac{x^2-1}{x}$

$f(-x)=\dfrac{(-x)^2-1}{(-x)}=\dfrac{x^2-1}{-x}=-f(x)$

Origin symmetry

y-intercept: $y=\dfrac{0^2-1}{0}=\dfrac{-1}{0}$

no y-intercept

x-intercepts: $x^2-1=0$

$x=\pm 1$

vertical asymptote: $x=0$

horizontal asymptote:

$n<m$, so none exist.

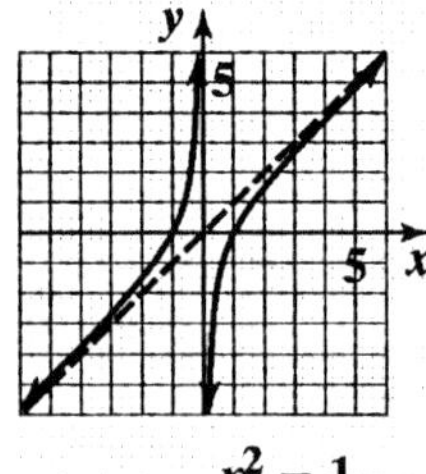

$f(x)=\dfrac{x^2-1}{x}$

72. $f(x) = \frac{x^2 - 4}{x}$

a. slant asymptote:

$f(x) = x - \frac{4}{x}$

$y = x$

b. $f(x) = \frac{x^2 - 4}{x}$

$f(-x) = \frac{(-x)^2 - 4}{-x} = \frac{x^2 - 4}{-x} = -f(x)$

origin symmetry

y-intercept: $y = \frac{0^2 - 4}{0} = -\frac{4}{0}$

no y-intercept

x-intercept:

$x^2 - 4 = 0$

$x = \pm 2$

vertical asymptote: $x = 0$

horizontal asymptote:

$n > m$, so none exist.

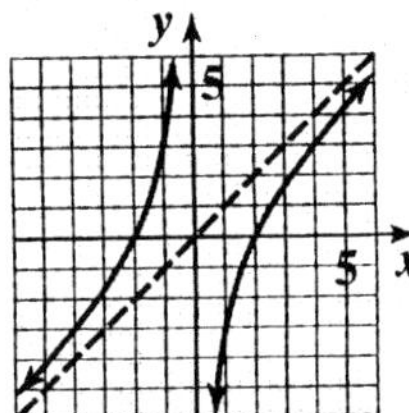

$f(x) = \frac{x^2 - 4}{x}$

73. a. Slant asymptote:

$f(x) = x + \frac{1}{x}$

$y = x$

b. $f(x) = \frac{x^2 + 1}{x}$

$f(-x) = \frac{(-x)^2 + 1}{-x} = \frac{x^2 + 1}{-x} = -f(x)$

Origin symmetry

y-intercept: $y = \frac{0^2 + 1}{0} = \frac{1}{0}$

no y-intercept

x-intercept:

$x^2 + 1 = 0$

$x^2 = -1$

no x-intercept

vertical asymptote: $x = 0$

horizontal asymptote:

$n > m$, so none exist.

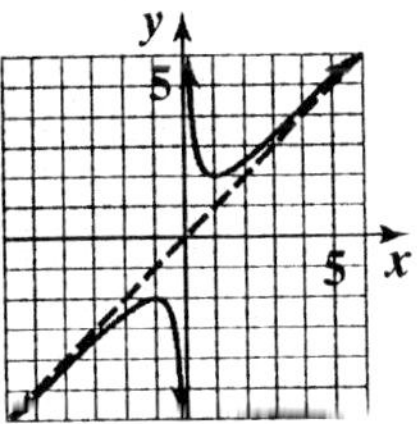

$f(x) = \frac{x^2 + 1}{x}$

74. $f(x) = \frac{x^2 + 4}{x}$

a. slant asymptote:

$g(x) = x + \frac{4}{x}$

$y = x$

b. $f(x) = \frac{x^2 + 4}{x}$

$f(-x) = \frac{(-x)^2 + 4}{-x} = \frac{x^2 + 4}{-x} = -f(x)$

origin symmetry

y-intercept: $y = \frac{0^2 + 4}{0} = \frac{4}{0}$

no y-intercept

$x^2 + 4 = 0$

$x^2 = -4$

no x-intercept

vertical asymptote: $x = 0$

horizontal asymptote:

$n > m$, so none exist.

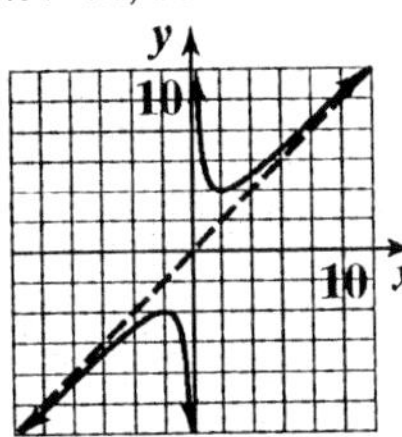

$f(x) = \frac{x^2 + 4}{x}$

75. a. Slant asymptote:

$$f(x) = x + 4 + \frac{6}{x-3}$$

$y = x + 4$

b. $$f(x) = \frac{x^2+x-6}{x-3}$$

$$f(-x) = \frac{(-x)^2 + (-x) - 6}{-x-3} = \frac{x^2 - x - 6}{-x-3}$$

$f(-x) \neq g(x)$, $g(-x) \neq -g(x)$

No symmetry

y-intercept: $y = \frac{0^2+0-6}{0-3} = \frac{-6}{-3} = 2$

x-intercept:

$x^2 + x - 6 = 0$

$(x+3)(x-2) = 0$

$x = -3$ and $x = 2$

vertical asymptote:

$x - 3 = 0$

$x = 3$

horizontal asymptote:

$n > m$, so none exist.

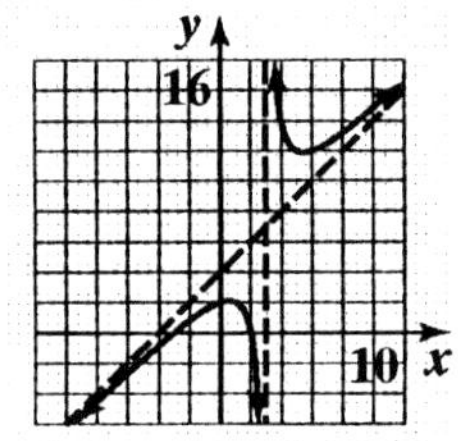

$f(x) = \frac{x^2+x-6}{x-3}$

76. $f(x) = \frac{x^2-x+1}{x-1}$

a. slant asymptote:

$$g(x) = x + \frac{1}{x-1}$$

$y = x$

b. $$f(x) = \frac{x^2-x-1}{x-1}$$

$$f(-x) = \frac{(-x)^2 - (-x) + 1}{-x-1} = \frac{x^2+x+1}{-x-1}$$

no symmetry

$f(-x) \neq f(x), f(-x) \neq -g(x)$

y-intercept: $y = \frac{0^2-0+1}{0-1} = \frac{1}{-1} = -1$

x-intercept:

$x^2 - x + 1 = 0$

no x-intercept

vertical asymptote:

$x - 1 = 0$

$x = 1$

horizontal asymptote:

$n > m$, so none

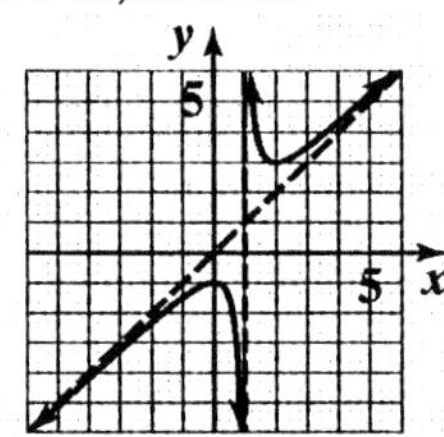

$f(x) = \frac{x^2-x+1}{x-1}$

77. $f(x) = \frac{x^3+1}{x^2+2x}$

a. slant asymptote:

$$\begin{array}{r} x-2 \\ x^2+2x\overline{)x^3 \qquad +1} \\ \underline{x^3+2x^2} \\ -2x^2 \\ \underline{-2x^2+4x} \\ -4x+1 \end{array}$$

$y = x - 2$

b. $$f(-x) = \frac{(-x)^3+1}{(-x)^2+2(-x)} = \frac{-x^3+1}{x^2-2x}$$

$f(-x) \neq f(x)$, $f(-x) \neq -f(x)$

no symmetry

y-intercept: $y = \frac{0^3+1}{0^2+2(0)} = \frac{1}{0}$

no y-intercept

x-intercept: $x^3 + 1 = 0$

$x^3 = -1$

$x = -1$

vertical asymptotes:

$x^2 + 2x = 0$

$x(x+2) = 0$

$x = 0, x = -2$

horizontal asymptote:
$n > m$, so none

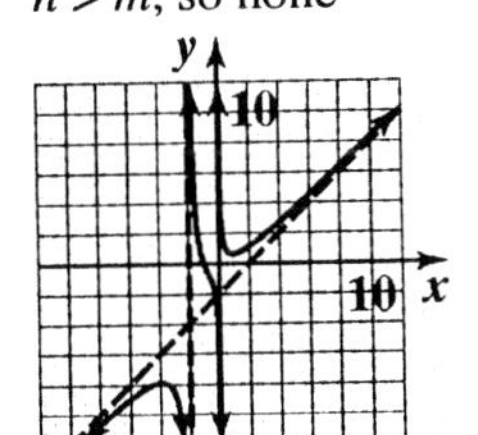

$f(x) = \dfrac{x^3 + 1}{x^2 + 2x}$

78. $f(x) = \dfrac{x^3 - 1}{x^2 - 9}$

a. slant asymptote:

$$\begin{array}{r} x + \dfrac{9x-1}{x^2-9} \\ x^2 - 9 \overline{) x^3 \quad\quad -1} \\ \underline{x^3 - 9x} \\ 9x - 1 \end{array}$$

$y = x$

b. $f(-x) = \dfrac{(-x)^3 - 1}{(-x)^2 - 9} = \dfrac{-x^3 - 1}{x^2 - 9}$

$f(-x) \neq f(x), f(-x) \neq -f(x)$

no symmetry

y-intercept: $y = \dfrac{0^3 - 1}{0^2 - 9} = \dfrac{1}{9}$

x-intercept: $x^3 - 1 = 0$

$x^3 = 1$

$x = 1$

vertical asymptotes:

$x^2 - 9 = 0$

$(x - 3)(x + 3) = 0$

$x = 3, x = -3$

horizontal asymptote:
$n > m$, so none

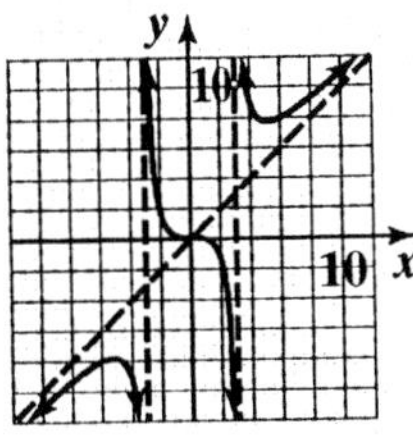

$f(x) = \dfrac{x^3 - 1}{x^2 - 9}$

79. $\dfrac{5x^2}{x^2 - 4} \cdot \dfrac{x^2 + 4x + 4}{10x^3}$

$= \dfrac{\cancel{5}\,\cancel{x^2}}{\cancel{(x+2)}(x-2)} \cdot \dfrac{(x+2)^{\cancel{2}}}{\underset{2}{\cancel{10}}\, x^{\cancel{3}1}}$

$= \dfrac{x+2}{2x(x-2)}$

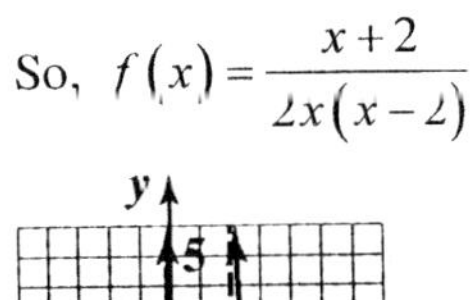

So, $f(x) = \dfrac{x+2}{2x(x-2)}$

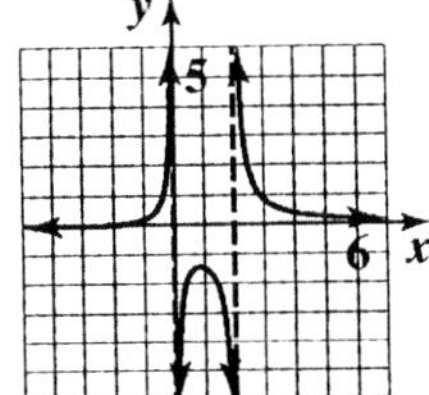

$f(x) = \dfrac{x + 2}{2x(x - 2)}$

80. $\dfrac{x - 5}{10x - 2} \div \dfrac{x^2 - 10x + 25}{25x^2 - 1}$

$= \dfrac{x - 5}{10x - 2} \cdot \dfrac{25x^2 - 1}{x^2 - 10x + 25}$

$= \dfrac{\cancel{x-5}}{2\cancel{(5x-1)}} \cdot \dfrac{(5x+1)\cancel{(5x-1)}}{(x-5)^{\cancel{2}}}$

$= \dfrac{5x+1}{2(x-5)}$

So, $f(x) = \dfrac{5x+1}{2(x-5)}$

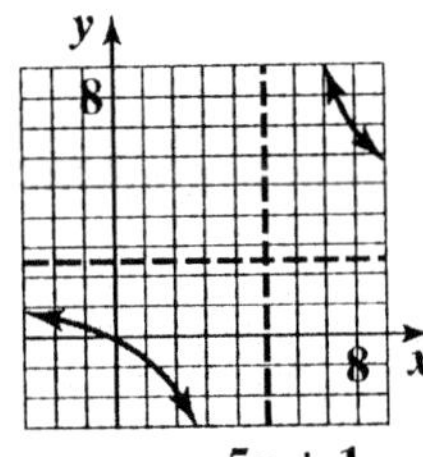

$f(x) = \dfrac{5x + 1}{2(x - 5)}$

81. $\frac{x}{2x+6}-\frac{9}{x^2-9}$

$\frac{x}{2x+6}-\frac{9}{x^2-9}$

$=\frac{x}{2(x+3)}-\frac{9}{(x+3)(x-3)}$

$=\frac{x(x-3)-9(2)}{2(x+3)(x-3)}$

$=\frac{x^2-3x-18}{2(x+3)(x-3)}$

$=\frac{(x-6)\cancel{(x+3)}}{2\cancel{(x+3)}(x-3)}=\frac{x-6}{2(x-3)}$

So, $f(x)=\frac{x-6}{2(x-3)}$

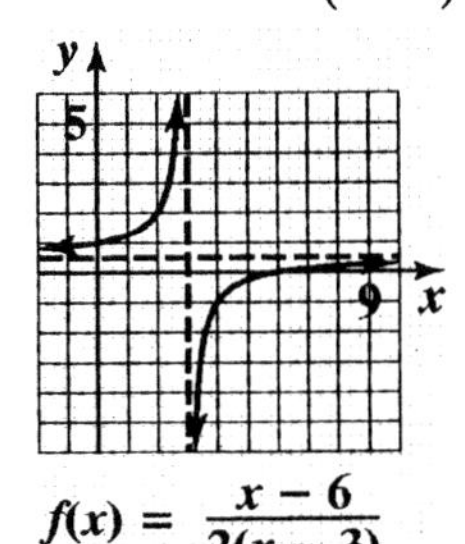

$f(x) = \frac{x-6}{2(x-3)}$

82. $\frac{2}{x^2+3x+2}-\frac{4}{x^2+4x+3}$

$=\frac{2}{(x+2)(x+1)}-\frac{4}{(x+3)(x+1)}$

$=\frac{2(x+3)-4(x+2)}{(x+2)(x+1)(x+3)}$

$=\frac{2x+6-4x-8}{(x+2)(x+1)(x+3)}$

$=\frac{-2x-2}{(x+2)(x+1)(x+3)}$

$=\frac{-2\cancel{(x+1)}}{(x+2)\cancel{(x+1)}(x+3)}=\frac{-2}{(x+2)(x+3)}$

So, $f(x)=\frac{-2}{(x+2)(x+3)}$

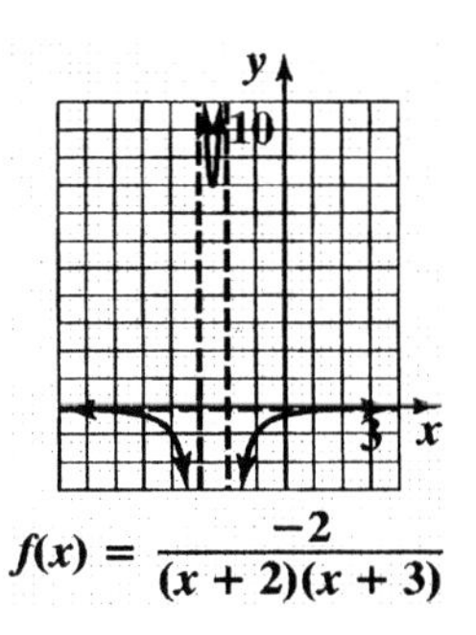

$f(x) = \frac{-2}{(x+2)(x+3)}$

83. $\frac{1-\frac{3}{x+2}}{1+\frac{1}{x-2}}=\frac{1-\frac{3}{x+2}}{1+\frac{1}{x-2}}\cdot\frac{(x+2)(x-2)}{(x+2)(x-2)}$

$=\frac{(x+2)(x-2)-3(x-2)}{(x+2)(x-2)+(x+2)}$

$=\frac{x^2-4-3x+6}{x^2-4+x+2}$

$=\frac{x^2-3x+2}{x^2+x-2}$

$=\frac{(x-2)\cancel{(x-1)}}{(x+2)\cancel{(x-1)}}=\frac{x-2}{x+2}$

So, $f(x)=\frac{x-2}{x+2}$

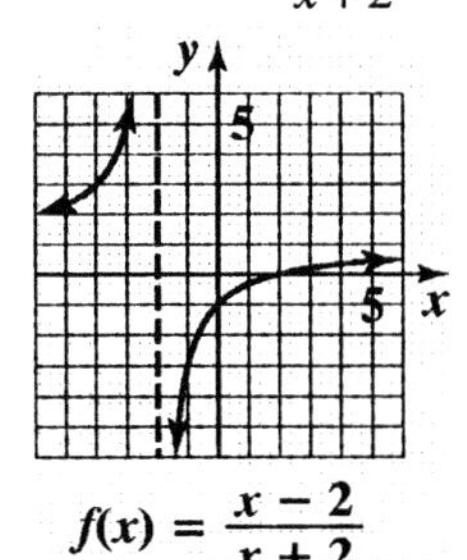

$f(x) = \frac{x-2}{x+2}$

84. $\frac{x-\frac{1}{x}}{x+\frac{1}{x}}\cdot\frac{x}{x}=\frac{x^2-1}{x^2+1}=\frac{(x-1)(x+1)}{x^2+1}$

So, $f(x)=\frac{(x-1)(x+1)}{x^2+1}$

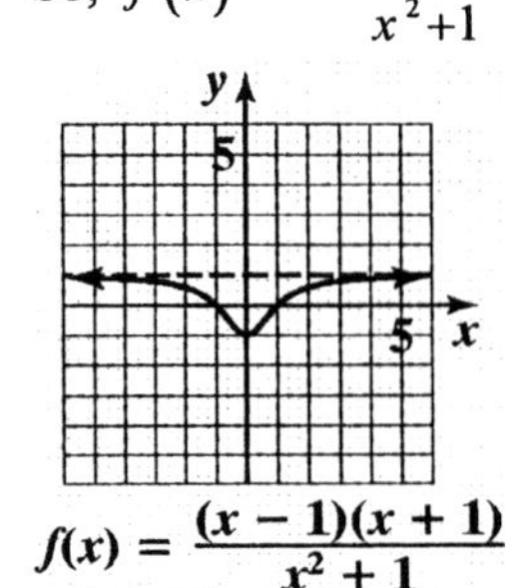

$f(x) = \frac{(x-1)(x+1)}{x^2+1}$

85. $g(x) = \frac{2x+7}{x+3} = \frac{1}{x+3} + 2$

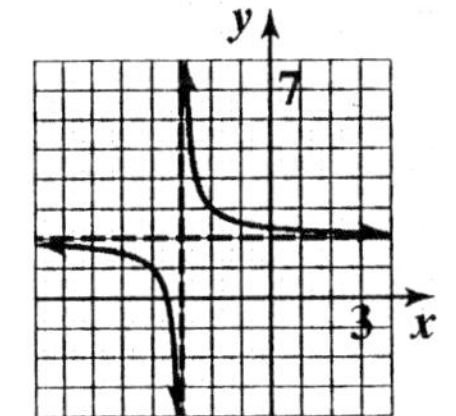

$f(x) = \frac{1}{x+3} + 2$

86. $g(x) = \frac{3x+7}{x+2} = \frac{1}{x+2} + 3$

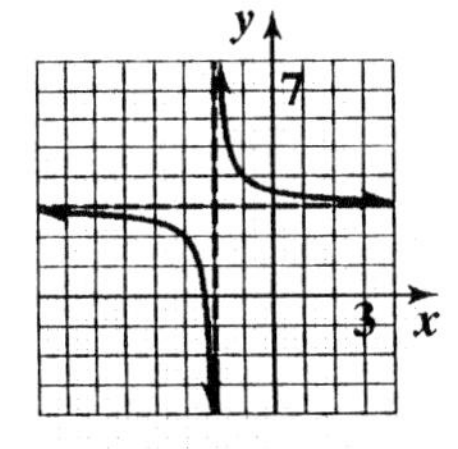

$f(x) = \frac{1}{x+2} + 3$

87. $g(x) = \frac{3x-7}{x-2} = \frac{-1}{x-2} + 3$

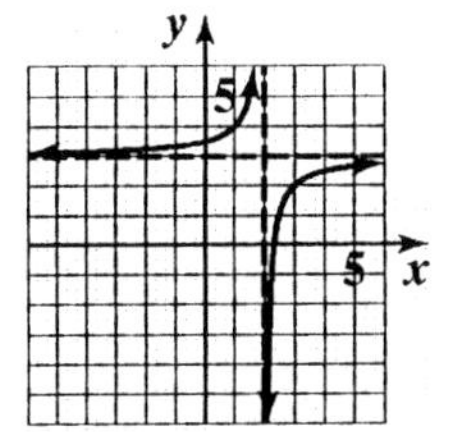

$f(x) = \frac{-1}{x-2} + 3$

88. $g(x) = \frac{2x-9}{x-4} = \frac{-1}{x-4} + 2$

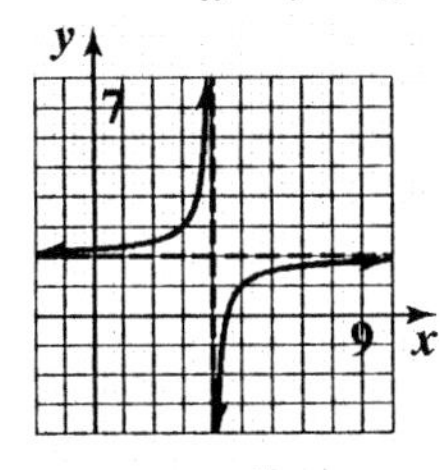

$f(x) = \frac{-1}{x-4} + 2$

89. **a.** $C(x) = 100x + 100{,}000$

b. $\overline{C}(x) = \frac{100x + 100{,}000}{x}$

c. $\overline{C}(500) = \frac{100(500) + 100{,}000}{500} = \$300$

When 500 bicycles are manufactured, it costs \$300 to manufacture each.

$\overline{C}(1000) = \frac{100(1000) + 100{,}000}{1000} = \$200$

When 1000 bicycles are manufactured, it costs \$200 to manufacture each.

$\overline{C}(2000) = \frac{100(2000) + 100{,}000}{2000} = \$150$

When 2000 bicycles are manufactured, it costs \$150 to manufacture each.

$\overline{C}(4000) = \frac{100(4000) + 100{,}000}{4000} = \$125$

When 4000 bicycles are manufactured, it costs \$125 to manufacture each.

The average cost decreases as the number of bicycles manufactured increases.

d. $n = m$, so $y = \frac{100}{1} = 100$.

As greater numbers of bicycles are manufactured, the average cost approaches \$100.

90. **a.** $C(x) = 30x + 300{,}000$

b. $\overline{C} = \frac{300{,}000 + 30x}{x}$

c. $\overline{C}(1000) = \frac{300000 + 30(1000)}{1000} = 330$

When 1000 shoes are manufactured, it costs \$330 to manufacture each.

$\overline{C}(10000) = \frac{300000 + 30(10000)}{10000} = 60$

When 10,000 shoes are manufactured, it costs \$60 to manufacture each.

$\overline{C}(100{,}00) = \frac{300{,}000 + 30(100{,}000)}{100{,}000} = 33$

When 100,000 shoes are manufactured, it costs \$33 to manufacture each.

The average cost decreases as the number of shoes manufactured increases.

d. $n = m$, so $y = \frac{30}{1} = 30$.

As greater numbers of shoes are manufactured, the average cost approaches \$30.

91. **a.** From the graph the pH level of the human mouth 42 minutes after a person eats food containing sugar will be about 6.0.

b. From the graph, the pH level is lowest after about 6 minutes.

$$f(6)=\frac{6.5(6)^2-20.4(6)+234}{6^2+36}$$
$$=4.8$$

The pH level after 6 minutes (i.e. the lowest pH level) is 4.8.

c. From the graph, the pH level appears to approach 6.5 as time goes by. Therefore, the normal pH level must be 6.5.

d. $y=6.5$

Over time, the pH level rises back to the normal level.

e. During the first hour, the pH level drops quickly below normal, and then slowly begins to approach the normal level.

92. **a.** From the graph, the drug's concentration after three hours appears to be about 1.5 milligrams per liter.

$$C(3)=\frac{5(3)}{3^2+1}=\frac{15}{10}=1.5$$

This verifies that the drug's concentration after 3 hours will be 1.5 milligrams per liter.

b. The degree of the numerator, 1, is less than the degree of the denominator, 2, so the the horizontal asymptote is $y=0$.

Over time, the drug's concentration will approach 0 milligrams per liter.

93. $P(10)=\frac{100(10-1)}{10}=90 \quad (10, 90)$

For a disease that smokers are 10 times more likely to contact than non-smokers, 90% of the deaths are smoking related.

94. $P(9)=\frac{100(9-1)}{9}=89 \quad (9, 89)$

For a disease that smokers are 9 times more likely to have than non-smokers, 89% of the deaths are smoking related.

95. $y = 100$ As incidence of the diseases increases, the percent of death approaches, but never gets to be, 100%.

96. No, the percentage approaches 100%, but never reaches 100%.

97. **a.** $\frac{128.3}{134.5}\approx 0.954$

In 1995, there were 954 males per 1000 females.

b. $\frac{141.7}{146.7}\approx 0.966$

In 2002, there were 966 males per 1000 females.

c. $f(x)=\frac{1.256x+74.2}{1.324x+76.71}$

d. 1995 is 45 years after 1950, so we find

$$f(45)=\frac{1.256(45)+74.2}{1.324(45)+76.71}\approx 0.959$$

The model predicts 959 males per 1000 females in 1995. The result from the function modeled the actual number fairly well.

e. 2002 is 52 years after 1950, so we find

$$f(52)=\frac{1.256(52)+74.2}{1.324(52)+76.71}\approx 0.958$$

The model predicts 958 males per 1000 females in 2002. The result from the function modeled the actual number fairly well.

f. $y=\frac{1.256}{1.324}$ or $y=0.949$

This means that over time, the number of males per 1000 females will approach 949.

98. $x-10$ = the average velocity on the return trip. The function that expresses the total time required to complete the round trip is $T(x)=\frac{600}{x}+\frac{600}{x-10}$.

99. $T(x)=\frac{90}{9x}+\frac{5}{x}=\frac{10}{x}+\frac{5}{x}$

The function that expresses the total time for driving and hiking is $T(x)=\frac{10}{x}+\frac{5}{x}$.

100. $A=xy=2500$

$$y=\frac{2500}{x}$$

$$P=2x+2y=2x+2\left(\frac{2500}{x}\right)=2x+\frac{5000}{x}$$

The perimeter of the floor, P, as a function of the width, x is $P(x)=2x+\frac{5000}{x}$.

101. $A = lw$

$xy = 50$

$l = y + 2 = \dfrac{50}{x} + 2$

$w = x + 1$

$A = \left(\dfrac{50}{x} + 2\right)(x + 1)$

$= 50 + \dfrac{50}{x} + 2x + 2$

$= 2x + \dfrac{50}{x} + 52$

The total area of the page is

$A(x) = 2x + \dfrac{50}{x} + 52.$

112.

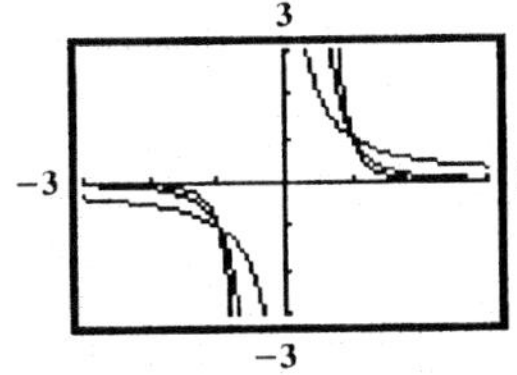

The graph approaches the horizontal asymptote faster and the vertical asymptote slower as n increases.

113.

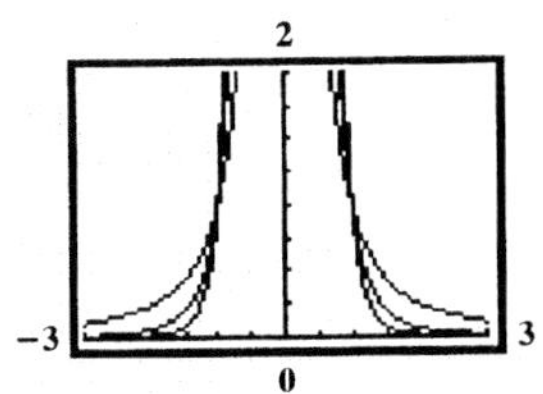

The graph approaches the horizontal asymptote faster and the vertical asymptote slower as n increases.

114.

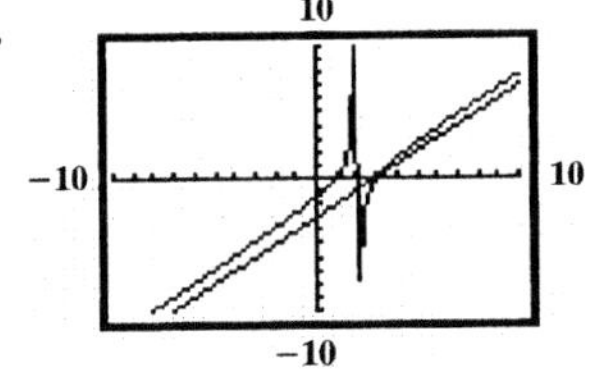

$g(x)$ is the graph of a line where $f(x)$ is the graph of a rational function with a slant asymptote. In $g(x)$, $x - 2$ is a factor of $x^2 - 5x + 6$.

115. a. $f(x) = \dfrac{27725(x-14)}{x^2+9} - 5x$

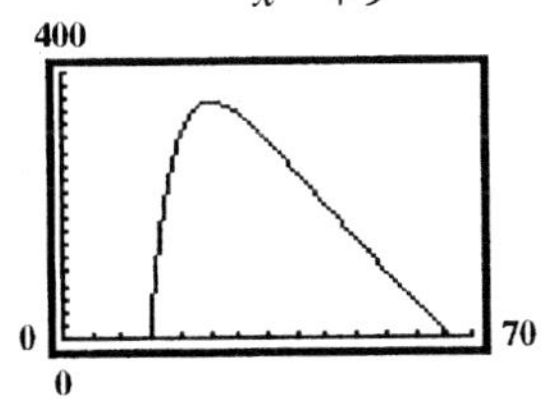

b. The graph increases from late teens until about the age of 25, and then the number of arrests decreases.

c. At age 25 the highest number arrests occurs. There are about 356 arrests for every 100,000 drivers.

116. a. False

b. False; the graph of a rational function may not have a y-intercept when the y-axis is a vertical asymptote.

c. False; the graph can have 1 or no horizontal asymptotes.

d. True; the function is undefined for x values at a vertical asymptote.

(d) is true.

117. a. False, $\sqrt{x-3} = (x-3)^{1/2}$ is not a polynomial function.

b. False, $n = m$ so the horizontal asymptote is $y = \dfrac{4}{1} = 4$

c. False, $n = m$ so the horizontal asymptote is $y = \dfrac{3000}{1} = 3000$, not 30,000.

d. True

(d) is true.

Section 2.7

Check Point Exercises

1. $x^2 - x > 20$

$x^2 - x - 20 > 0$

$(x+4)(x-5) > 0$

Solve the related quadratic equation.

$(x+4)(x-5) = 0$

Apply the zero product principle.

$x + 4 = 0$ or $x - 5 = 0$

$x = -4$ $x = 5$

The boundary points are –2 and 4.

Test Interval	Test Number	Test	Conclusion
$(-\infty, -4)$	–5	$(-5)^2 - (-5) > 20$ $30 > 20$, true	$(-\infty, -4)$ belongs to the solution set.
$(-4, 5)$	0	$(0)^2 - (0) > 20$ $0 > 20$, false	$(-4, 5)$ does not belong to the solution set.
$(5, \infty)$	10	$(10)^2 - (10) > 20$ $90 > 20$, true	$(5, \infty)$ belongs to the solution set.

The solution set is $(-\infty, -4) \cup (5, \infty)$ or $\{x \mid x < -4 \text{ or } x > 5\}$.

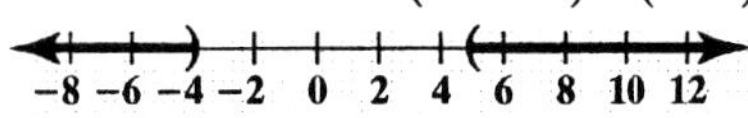

2. $x^3 + 3x^2 \le x + 3$

$x^3 + 3x^2 - x - 3 \le 0$

$(x+1)(x-1)(x+3) \le 0$

$(x+1)(x-1)(x+3) = 0$

$x + 1 = 0$ or $x - 1 = 0$ or $x + 3 = 0$

$x = -1$ $x = 1$ $x = -3$

Test Interval	Test Number	Test	Conclusion
$(-\infty, -3)$	-4	$(-4)^3 + 3(-4)^2 \le (-4) + 3$ $-16 \le -1$ true	$(-\infty, -3)$ belongs to the solution set.
$(-3, -1]$	–2	$(-2)^3 + 3(-2)^2 \le (-2) + 3$ $4 \le 1$ false	$(-3, -1]$ does not belong to the solution set.
$[-1, 1]$	0	$(0)^3 + 3(0)^2 \le (0) + 3$ $0 \le 3$ true	$[-1, 1]$ belongs to the solution set.
$[1, \infty)$	2	$(6+3)(6-5) > 0$ true	$[1, \infty)$ does not belong to the solution set.

The solution set is $(-\infty, -3] \cup [-1, 1]$ or $\{x \mid x \le -3 \text{ or } -1 \le x \le 1\}$.

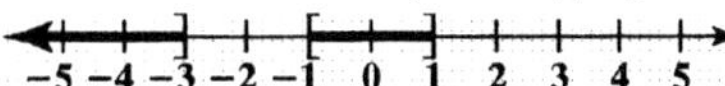

3.

$$\frac{2x}{x+1} \geq 1$$
$$\frac{2x}{x+1} - 1 \geq 0$$
$$\frac{x-1}{x+1} \geq 0$$

$x-1=0$ or $x+1=0$

$x=1$ $\quad x=-1$

Test Interval	Test Number	Test	Conclusion
$(-\infty,-1)$	-2	$\frac{2(-2)}{-2+1} \geq 1$ $4 \geq 1$, true	$(-\infty,-1)$ belongs to the solution set.
$(-1,1]$	0	$\frac{2(0)}{0+1} \geq 1$ $0 \geq 1$, false	$(-1,1]$ does not belong to the solution set.
$[1,\infty)$	2	$\frac{2(2)}{2+1} \geq 1$ $\frac{4}{3} \geq 1$, true	$[1,\infty)$ belongs to the solution set.

The solution set is $(-\infty,-1)\cup[1,\infty)$ or $\{x \mid x<-1 \text{ or } x \geq 1\}$.

−5 −4 −3 −2 −1 0 1 2 3 4 5

4.

$$-16t^2 + 80t > 64$$
$$-16t^2 + 80t - 64 > 0$$
$$-16(t-1)(t-4) > 0$$

$t-1=0$ or $t-4=0$

$t=1$ $\quad t=4$

Test Interval	Test Number	Test	Conclusion
$(-\infty,1)$	0	$-16(0)^2+80(0)>64$ $0>64$, false	$(-\infty,1)$ does not belong to the solution set.
$(1,4)$	2	$-16(2)^2+80(2)>64$ $96>64$, true	$(1,4)$ belongs to the solution set.
$(4,\infty)$	5	$-16(5)^2+80(5)>64$ $0>64$, false	$(4,\infty)$ does not belong to the solution set.

The object will be more than 64 feet above the ground between 1 and 4 seconds.

Exercise Set 2.7

1. $(x-4)(x+2)>0$
$x=4$ or $x=-2$

T	F	T
	−2	4

Test −3: $(-3-4)(-3+2)>0$
$7>0$ True
Test 0: $(0-4)(0+2)>0$
$-8>0$ False
Test 5: $(5-4)(5+2)>0$
$7>0$ True
$(-\infty,-2)$ or $(4,\infty)$

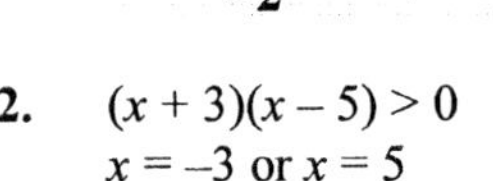

2. $(x+3)(x-5)>0$
$x=-3$ or $x=5$

T	F	T
	−3	5

Test −4: $(-4+3)(-4-5)>0$
$9>0$ True
Test 0: $(0+3)(0-5)>0$
$-15>0$ False
Test 6: $(6+3)(6-5)>0$
$18>0$ True
The solution set is $(-\infty,-3)$ or $(5,\infty)$.

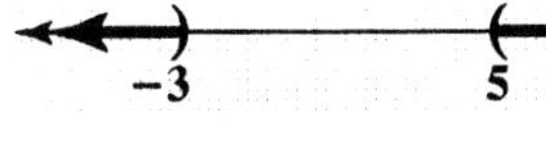

3. $(x-7)(x+3)\le 0$
$x=7$ or $x=-3$

F	T	F
	−3	7

Test −4: $(-4-7)(-4+3)\le 0$
$11\le 0$ False
Test 0: $(0-7)(0+3)\le 0$
$-21\le 0$ True
Test 8: $(8-7)(8+3)\le 0$
$11\le 0$ False
The solution set is $[-3, 7]$.

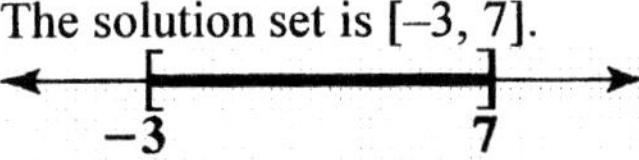

4. $(x+1)(x-7)\le 0$
$x=-1$ or $x=7$

F	T	F
	−1	7

Test −2: $(-2+1)(-2-7)\le 0$
$9\le 0$ False
Test 0: $(0+1)(0-7)\le 0$
$-7\le 0$ True
Test 8: $(8+1)(8-7)\le 0$
$9\le 0$ False
The solution set is $[-1, 7]$.

5. $x^2-5x+4>0$
$(x-4)(x-1)>0$
$x=4$ or $x=1$

T	F	T
	1	4

Test 0: $0^2-5(0)+4>0$
$4>0$ True
Test 2: $2^2-5(2)+4>0$
$-2>0$ False
Test 5: $5^2-5(5)+4>0$
$4>0$ True
The solution set is $(-\infty,1)$ or $(4,\infty)$.

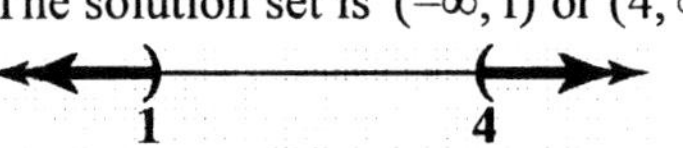

6. $x^2-4x+3<0$
$(x-1)(x-3)<0$
$x=1$ or $x=3$

F	T	F
	1	3

Test 0: $0^2-4(0)+3<0$
$3<0$ False
Test 2: $2^2-4(2)+3<0$
$-1<0$ True
Test 4: $4^2-4(4)+3<0$
$3<0$ False
The solution set is $(1, 3)$.

7. $x^2 + 5x + 4 > 0$

$(x+1)(x+4) > 0$

$x = -1$ or $x = -4$

T	F	T
	−4	−1

Test −5: $(-5)^2 + 5(-5) + 4 > 0$

$4 > 0$ True

Test −3: $(-3)^2 + 5(-3) + 4 > 0$

$-2 > 0$ False

Test 0: $0^2 + 5(0) + 4 > 0$

$4 > 0$ True

The solution set is $(-\infty, -4)$ or $(-1, \infty)$.

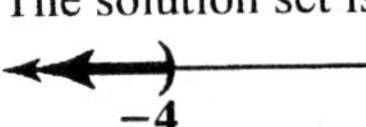

8. $x^2 + x - 6 > 0$

$(x+3)(x-2) > 0$

$x = -3$ or $x = 2$

T	F	T
	−3	2

Test −4: $(-4)^2 - 4 - 6 > 0$

$6 > 0$ True

Test 0: $(0)^2 + 0 - 6 > 0$

$-6 > 0$ False

Test 3: $3^2 + 3 - 6 > 0$

$6 > 0$ True

The solution set is $(-\infty, -3)$ or $(2, \infty)$.

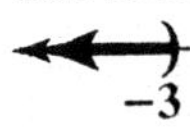

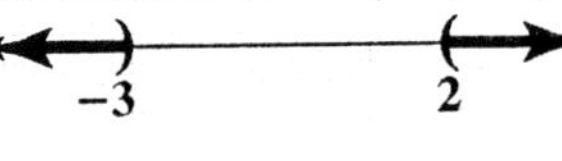

9. $x^2 - 6x + 9 < 0$

$(x-3)(x-3) < 0$

$x = 3$

F	F
	3

Test 0: $0^2 - 6(0) + 9 < 0$

$9 < 0$ False

Test 4: $4^2 - 6(4) + 9 < 0$

$1 < 0$ False

The solution set is the empty set, $\varnothing$.

10. $x^2 - 2x + 1 > 0$

$(x-1)(x-1) > 0$

$x = 1$

T	T
	1

Test 0: $0^2 - 2(0) + 1 > 0$

$1 > 0$ True

Test 2: $2^2 - 2(2) + 1 > 0$

$1 > 0$ True

The solution set is $(-\infty, 1)$ or $(1, \infty)$.

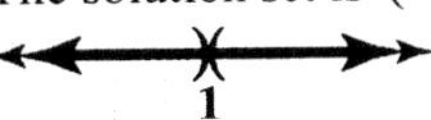

11. $3x^2 + 10x - 8 \le 0$

$(3x-2)(x+4) \le 0$

$x = \frac{2}{3}$ or $x = -4$

F	T	F
	−4	$\frac{2}{3}$

Test −5: $3(-5)^2 + 10(-5) - 8 \le 0$

$17 \le 0$ False

Test 0: $3(0)^2 + 10(0) - 8 \le 0$

$8 \le 0$ True

Test 1: $3(1)^2 + 10(1) - 8 \le 0$

$5 \le 0$ False

The solution set is $\left[-4, \frac{2}{3}\right]$.

12. $9x^2 + 3x - 2 \ge 0$

$(3x-1)(3x+2) \ge 0$

$3x = 1 \quad 3x = -2$

$x = \frac{1}{3} \quad x = \frac{-2}{3}$

T	F	T
	$\frac{-2}{3}$	$\frac{1}{3}$

Test −1: $9(-1)^2 + 3(-1) - 2 \ge 0$

$4 \ge 0$ True

Test 0: $9(0)^2 + 3(0) - 2 \ge 0$

$-2 \ge 0$ False

Test 1: $9(1)^2+3(1)-2\le 0$
$10\ge 0$ True

The solution set is $\left(-\infty,\frac{-2}{3}\right]$ or $\left[\frac{1}{3},\infty\right)$.

$-\frac{2}{3}$ $\frac{1}{3}$

13. $2x^2+x<15$

$2x^2+x-15<0$

$(2x-5)(x+3)<0$

$2x-5=0$ or $x+3=0$

$2x=5$

$x=\frac{5}{2}$ or $x=-3$

F	T	F

-3 $\frac{5}{2}$

Test –4: $2(-4)^2+(-4)<15$
$28<15$ False

Test 0: $2(0)^2+0<15$
$0<15$ True

Test 3: $2(3)^2+3<15$
$21<15$ False

The solution set is $\left(-3,\frac{5}{2}\right)$.

-3 $\frac{5}{2}$

14. $6x^2+x>1$

$6x^2+x-1>0$

$(2x+1)(3x-1)>0$

$2x+1=0$ or $3x-1=0$

$2x=-1$ $3x=1$

$x=-\frac{1}{2}$ $x=\frac{1}{3}$

T	F	T

$-\frac{1}{2}$ $\frac{1}{3}$

Test –1: $6(-1)^2+(-1)>1$
$5>1$ True

Test 0: $6(0)^2+0>1$
$0>1$ False

Test 1: $6(1)^2+1>1$
$7>1$ True

The solution set is $\left(-\infty,-\frac{1}{2}\right)$ or $\left(\frac{1}{3},\infty\right)$.

$-\frac{1}{2}$ $\frac{1}{3}$

15. $4x^2+7x<-3$

$4x^2+7x+3<0$

$(4x+3)(x+1)<0$

$4x+3=0$ or $x+1=0$

$4x-3=0$

$x=-\frac{3}{4}$ or $x=-1$

F	T	F

-1 $-\frac{3}{4}$

Test –2: $4(-2)^2+7(-2)<-3$
$2<-3$ False

Test $-\frac{7}{8}$: $4\left(-\frac{7}{8}\right)^2+7\left(-\frac{7}{8}\right)<-3$

$\frac{49}{16}-\frac{49}{8}<-3$

$-\frac{49}{16}<-3$ True

Test 0: $4(0)^2+7(0)<-3$
$0<-3$ False

The solution set is $\left(-1,-\frac{3}{4}\right)$.

-1 $-\frac{3}{4}$

16. $3x^2+16x<-5$

$3x^2+16x+5<0$

$(3x+1)(x+5)<0$

$3x+1=0$ or $x+5=0$

$3x=-1$

$x=-\frac{1}{3}$ $x=-5$

F	T	F

-5 $-\frac{1}{3}$

Test –6: $3(-6)^2+16(-6)<-5$
$12<-5$ False

Test -2: $3(-2)^2+16(-2)<-5$
$-20<-5$ True
Test 0: $3(0)^2+16(0)<-5$
$0<-5$ False

The solution set is $\left(-5,-\frac{1}{3}\right)$.

17. $5x \le 2-3x^2$
$3x^2+5x-2 \le 0$
$(3x-1)(x+2) \le 0$
$3x-1=0$ or $x+2=0$
$3x=1$
$3x-1=0$ or $x+2=0$
$3x=1$
$x=\frac{1}{3}$ or $x=-2$

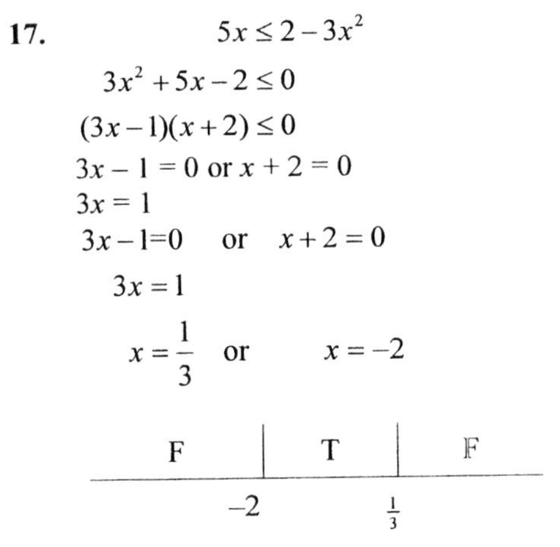

F	T	F
	-2	$\frac{1}{3}$

Test -3: $5(-3) \le 2-3(-3)^2$
$-15 \le -25$ False
Test 0: $5(0) \le 2-3(0)^2$
$0 \le 2$ True
Test 1: $5(1) \le 2-3(1)^2$
$5 \le -1$ False
The solution set is $\left[-2,\frac{1}{3}\right]$.

18.
$4x^2+1 \ge 4x$
$4x^2-4x+1 \ge 0$
$(2x-1)(2x-1) \ge 0$
$2x-1=0$
$x=\frac{1}{2}$

T	T
$\frac{1}{2}$	

Test 0: $4(0)^2+1 \ge 4(0)$
$1 \ge 0$ True
Test 1: $4(1)^2+1 \ge 4(1)$
$5 \ge 4$ True
The solution set is $(-\infty,\infty)$.

19. $x^2-4x \ge 0$
$x(x-4) \ge 0$
$x=0$ or $x-4=0$
$x=4$

T	F	T
	0	4

Test -1: $(-1)^2-4(-1) \ge 0$
$5 \ge 0$ True
Test 1: $(1)^2-4(1) \ge 0$
$-3 \ge 0$ False
$0 \le 2$ True
Test 5: $5^2-4(5) \ge 0$
$5 \ge 0$ True
The solution set is $(-\infty,0]$ or $[4,\infty)$.

20. $x^2+2x<0$
$x(x+2)<0$
$x=0$ or $x=-2$

F	T	F
	-2	0

Test -3: $(-3)^2+2(-3)<0$
$3<0$ False
Test -1: $(-1)^2+2(-1)<0$
$-1<0$ True
Test 1: $(1)^2+2(1)<0$
$3<0$ False
The solution set is $(-2, 0)$.

21. $2x^2 + 3x > 0$

$x(2x+3) > 0$

$x = 0$ or $x = -\frac{3}{2}$

T	F	T

$-\frac{3}{2}$ 0

Test -2: $2(-2)^2 + 3(-2) > 0$

$2 > 0$ True

Test -1: $2(-1)^2 + 3(-1) > 0$

$-1 > 0$ False

Test 1: $2(1)^2 + 3(1) > 0$

$5 > 0$ True

The solution set is $\left(-\infty, -\frac{3}{2}\right)$ or $(0, \infty)$.

$-\frac{3}{2}$ 0

22. $3x^2 - 5x \le 0$

$x(3x-5) \le 0$

$x = 0$ or $x = \frac{5}{3}$

F	T	F

0 $\frac{5}{3}$

Test -1: $3(-1)^2 - 5(-1) \le 0$

$8 \le 0$ False

Test 1: $3(1)^2 - 5(1) \le 0$

$-2 \le 0$ True

Test 2: $3(2)^2 - 5(2) \le 0$

$2 \le 0$ False

The solution set is $\left[0, \frac{5}{3}\right]$.

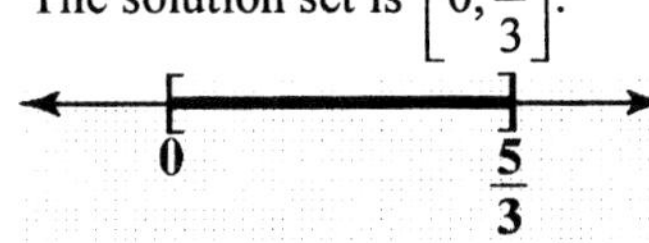

23. $-x^2 + x \ge 0$

$x^2 - x \le 0$

$x(x-1) \le 0$

$x = 0$ or $x = 1$

F	T	F

0 1

Test -1: $-(-1)^2 + (-1) \ge 0$

$-2 \ge 0$ False

Test $\frac{1}{2}$: $-\left(\frac{1}{2}\right)^2 + \left(\frac{1}{2}\right) \ge 0$

$\frac{1}{4} \ge 0$ True

Test 2: $-(2)^2 + 2 \ge 0$

$-2 \ge 0$ False

The solution set is $[0, 1]$.

0 1

24. $-x^2 + 2x \ge 0$

$x(-x+2) \ge 0$

$x = 0$ or $x = 2$

F	T	F

0 2

Test -1: $-(-1)^2 + 2(-1) \ge 0$

$-3 \ge 0$ False

Test 1: $-(1)^2 + 2(1) \ge 0$

$1 \ge 0$ True

Test 3: $-(3)^2 + 2(3) \ge 0$

$-3 \ge 0$ False

The solution set is $[0, 2]$.

0 2

25. $x^2 \le 4x - 2$

$x^2 - 4x + 2 \le 0$

Solve $x^2 - 4x + 2 = 0$

$$x = \frac{-b \pm \sqrt{b^2 - 4ac}}{2a}$$

$$x = \frac{-(-4) \pm \sqrt{(-4)^2 - 4(1)(2)}}{2(1)}$$

$$= \frac{4 \pm \sqrt{8}}{2}$$

$$= 2 \pm \sqrt{2}$$

$x \approx 0.59$ or $x \approx 3.41$

F	T	F

0.59 3.41

The solution set is

$\left[2-\sqrt{2}, 2+\sqrt{2}\right]$ or $[0.59, 3.41]$.

$2-\sqrt{2}$ $2+\sqrt{2}$

26. $x^2 \le 2x+2$

$x^2 - 2x - 2 \le 0$

Solve $x^2 - 2x - 2 = 0$

$x = \dfrac{-b \pm \sqrt{b^2 - 4ac}}{2a}$

$x = \dfrac{-(-2) \pm \sqrt{(-2)^2 - 4(1)(-2)}}{2(1)}$

$= \dfrac{2 \pm \sqrt{12}}{2}$

$= 1 \pm \sqrt{3}$

$x \approx -0.73$ or $x \approx 2.73$

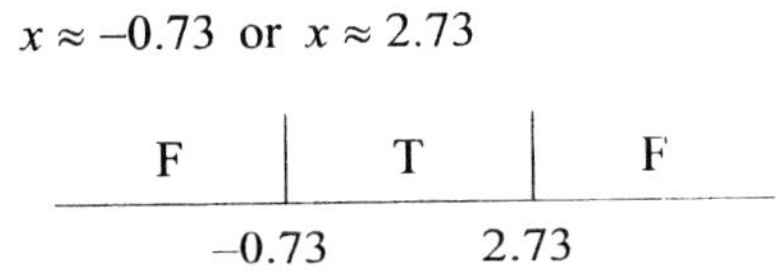

The solution set is

$\left[1-\sqrt{3}, 1+\sqrt{3}\right]$ or $[-0.73, 2.73]$.

$1-\sqrt{3}$ $1+\sqrt{3}$

27. $x^2 - 6x + 9 < 0$

Solve $x^2 - 6x + 9 = 0$

$(x-3)(x-3) = 0$

$(x-3)^2 = 0$

$x = 3$

F	F
3	

The solution set is the empty set, ∅.

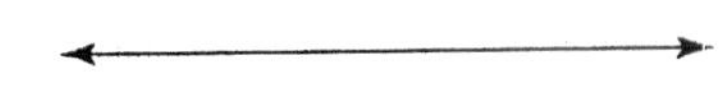

28. $4x^2 - 4x + 1 \ge 0$

Solve $4x^2 - 4x + 1 = 0$

$(2x-1)(2x-1) = 0$

$(2x-1)^2 = 0$

$x = \dfrac{1}{2}$

T	T
$\frac{1}{2}$	

The solution set is $(-\infty, \infty)$.

29. $(x-1)(x-2)(x-3) \ge 0$

Boundary points: 1, 2, and 3

Test one value in each interval.

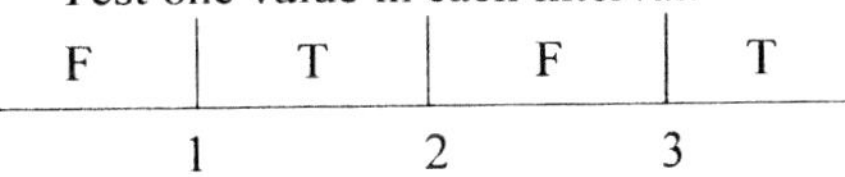

The solution set is $[1, 2] \cup [3, \infty)$.

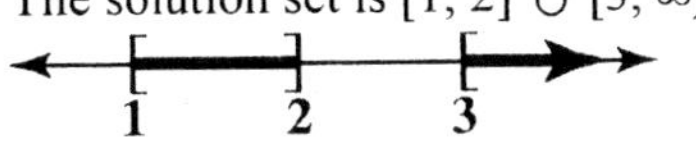

30. $(x+1)(x+2)(x+3) \ge 0$

Boundary points: −1, −2, and −3

Test one value in each interval.

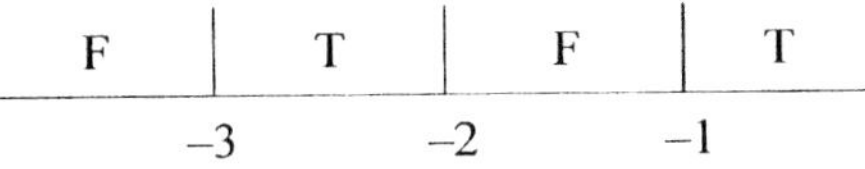

The solution set is $[-3, -2] \cup [-1, \infty)$.

−3 −2 −1

31. $x^3 + 2x^2 - x - 2 \ge 0$

$x^2(x+2) - 1(x+2) \ge 0$

$(x+2)(x^2-1) \ge 0$

$(x+2)(x-1)(x+1) \ge 0$

Boundary points: −2, −1, and 2

Test one value in each interval.

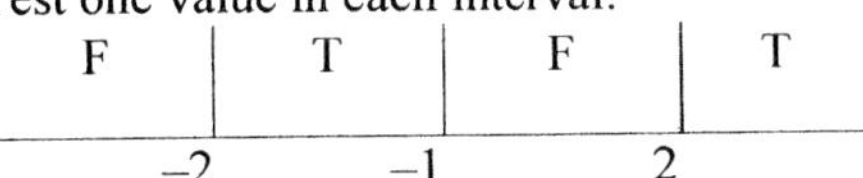

The solution set is $[-2, -1] \cup [1, \infty)$.

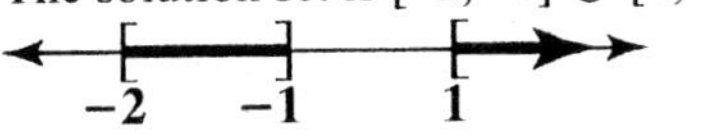

32. $x^3 + 2x^2 - 4x - 8 \ge 0$

$x^2(x+2) - 4(x+1) \ge 0$

$(x+2)(x^2-4) \ge 0$

$(x+2)(x+2)(x-2) \ge 0$

Boundary points: −2, and 2

Test one value in each interval.

The solution set is $[-2, -2] \cup [2, \infty)$.

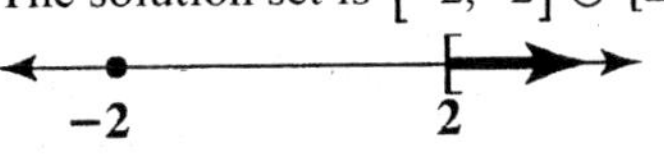

33. $x^3 + 2x^2 - x - 2 \ge 0$
$x^2(x-3) - 9(x-3) \ge 0$
$(x-3)(x^2-9) \ge 0$
$(x-3)(x+3)(x-3) \ge 0$
Boundary points: –3 and 3
Test one value in each interval.

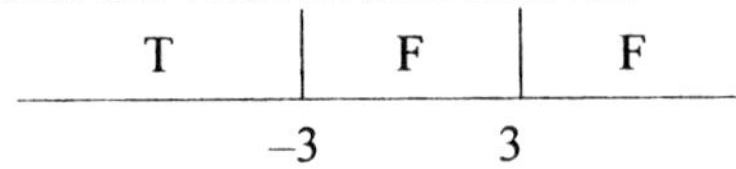

T	F	F

–3 3

The solution set is $(-\infty, -3]$.

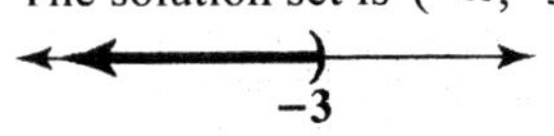

34. $x^3 + 7x^2 - x - 7 < 0$
$x^2(x+7) - (x+7) < 0$
$(x+7)(x^2-1) < 0$
$(x+7)(x+1)(x-1) < 0$
Boundary points: –7, –1 and 1
Test one value in each interval.

T	F	T	F

–7 –1 1

The solution set is $(-\infty, -7) \cup (-1, 1)$.

–7 –1 1

35. $x^3 + x^2 + 4x + 4 > 0$
$x^2(x+1) + 4(x+1) \ge 0$
$(x+1)(x^2+4) \ge 0$
Boundary point: –1
Test one value in each interval.

F	T

–1

The solution set is $(-1, \infty)$.

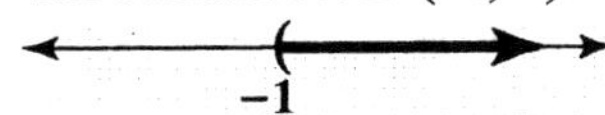

36. $x^3 - x^2 + 9x - 9 > 0$
$x^2(x-1) + 9(x-1) \ge 0$
$(x-1)(x^2+9) \ge 0$
Boundary point: 1.
Test one value in each interval.

F	T

1

The solution set is $(1, \infty)$.

1

37. $x^3 - 9x^2 \ge 0$
$x^2(x-9) \ge 0$
Boundary points: 0 and 9
Test one value in each interval.

F	F	T

0 9

The solution set is $[0, 0] \cup [9, \infty)$.

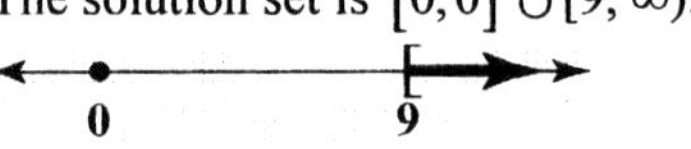

38. $x^3 - 4x^2 \le 0$
$x^2(x-4) \le 0$
Boundary points: 0 and 4.
Test one value in each interval.

T	T	F

0 4

The solution set is $(-\infty, 4]$.

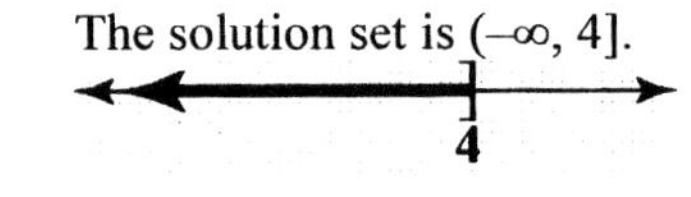

39. $\frac{x-4}{x+3} > 0$
$x - 4 = 0 \quad x + 3 = 0$
$x = 4 \quad x = -3$

T	F	T

–3 4

The solution set is $(-\infty, -3) \cup (4, \infty)$.

40. $\frac{x+5}{x-2} > 0$
$x = -5$ or $x = 2$

T	F	T

–5 2

The solution set is $(-\infty, -5) \cup (2, \infty)$.

–5 2

41. $\frac{x+3}{x+4} < 0$
$x = -3$ or $x = -4$

F	T	F

–4 –3

The solution set is $(-4, -3)$.

–4 –3

42. $\dfrac{x+5}{x+2}<0$

$x=-5$ or $x=-2$

F	T	F
	-5	-2

The solution set is $(-5,-2)$.

-5 -2

43. $\dfrac{-x+2}{x-4}\ge 0$

$x=2$ or $x=4$

F	T	F
	2	4

The solution set is $[2, 4)$.

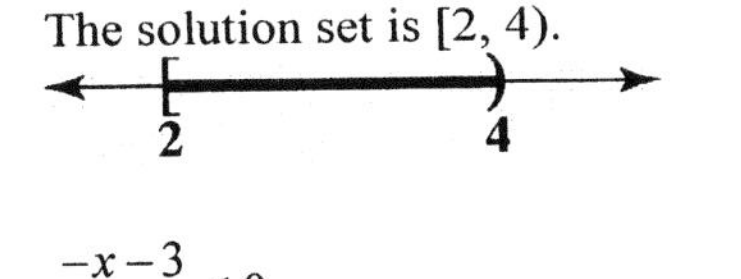

44. $\dfrac{-x-3}{x+2}\le 0$

$x=-3$ or $x=-2$

T	F	T
	-3	-2

The solution set is $(-\infty,-3]\cup(-2,\infty)$.

-3 -2

45. $\dfrac{4-2x}{3x+4}\le 0$

$x=2$ or $x=-\dfrac{4}{3}$

T	F	T
	$-\frac{4}{3}$	2

The solution set is $\left(-\infty,\dfrac{-4}{3}\right)\cup[2,\infty)$.

$-\dfrac{4}{3}$ 2

46. $\dfrac{3x+5}{6-2x}\ge 0$

$x=-\dfrac{5}{3}$ or $x=3$

F	T	F
	$-\frac{5}{3}$	3

The solution set is $\left[-\dfrac{5}{3},3\right)$.

$-\dfrac{5}{3}$ 3

47. $\dfrac{x}{x-3}>0$

$x=0$ or $x=3$

T	F	T
	0	3

The solution set is $(-\infty,0)\cup(3,\infty)$.

0 3

48. $\dfrac{x+4}{x}>0$

$x=-4$ or $x=0$

T	F	T
	-4	0

The solution set is $(-\infty,-4)\cup(0,\infty)$.

-4 0

49. $\dfrac{(x+4)(x-1)}{x+2}\le 0$

$x=-4$ or $x=-2$ or $x=1$.

T	F	T	F
	-4	-2	1

Values of $x=-4$ or $x=1$ result in $f(x)=0$ and, therefore must be included in the solution set.

The solution set is $(-\infty,-4]\cup(-2,1]$

-4 -2 1

50. $\dfrac{(x+3)(x-2)}{x+1}\le 0$

$x=-3$ or $x=-1$ or $x=2$.

T	F	T	F
	-3	-1	2

Values of $x=-3$ or $x=2$ result in $f(x)=0$ and, therefore must be included in the solution set.

The solution set is $(-\infty,-3]\cup(-1,2]$.

-3 -1 2

51.
$$\frac{x+1}{x+3} < 2$$
$$\frac{x+1}{x+3} - 2 < 0$$
$$\frac{x+1-2(x+3)}{x+3} < 0$$
$$\frac{x+1-2x-6}{x+3} < 0$$
$$\frac{-x-5}{x+3} < 0$$

$x =$ or $x = -3$

T	F	T

$-5 \quad -3$

The solution set is $(-\infty, -5) \cup (-3, \infty)$.

$-5 \quad -3$

52.

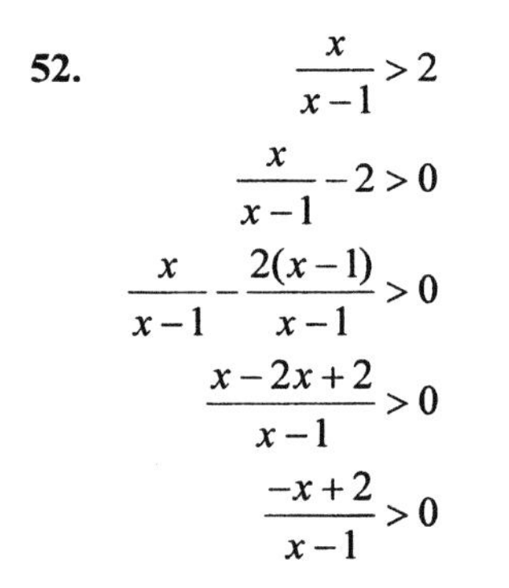

$$\frac{x}{x-1} > 2$$
$$\frac{x}{x-1} - 2 > 0$$
$$\frac{x}{x-1} - \frac{2(x-1)}{x-1} > 0$$
$$\frac{x-2x+2}{x-1} > 0$$
$$\frac{-x+2}{x-1} > 0$$

$x = 2$ or $x = 1$

F	T	F

$1 \quad 2$

The solution set is $(1, 2)$.

$1 \quad 2$

53.
$$\frac{x+4}{2x-1} \le 3$$
$$\frac{x+4}{2x-1} - 3 \le 0$$
$$\frac{x+4-3(2x-1)}{2x-1} \le 0$$
$$\frac{x+4-6x+3}{2x-1} \le 0$$
$$\frac{-5x+7}{2x-1} \le 0$$

$x = \frac{7}{5}$ or $x = \frac{1}{2}$

T	F	T

$\frac{1}{2} \quad \frac{7}{5}$

$\frac{1}{2} \quad \frac{7}{5}$

54.

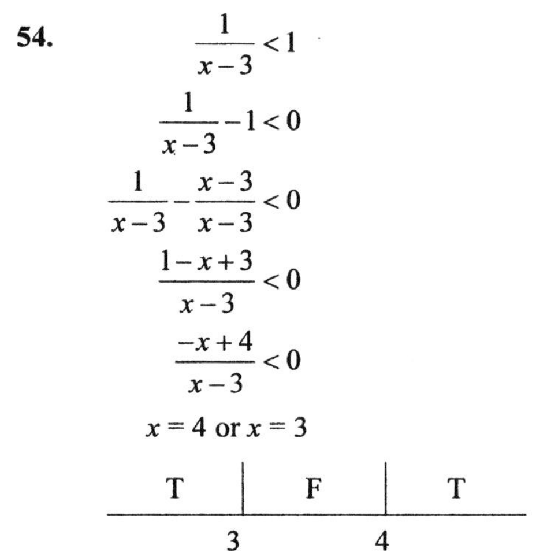

$$\frac{1}{x-3} < 1$$
$$\frac{1}{x-3} - 1 < 0$$
$$\frac{1}{x-3} - \frac{x-3}{x-3} < 0$$
$$\frac{1-x+3}{x-3} < 0$$
$$\frac{-x+4}{x-3} < 0$$

$x = 4$ or $x = 3$

T	F	T

$3 \quad 4$

The solution set is $(-\infty, 3) \cup (4, \infty)$.

$3 \quad 4$

55.
$$\frac{x-2}{x+2} \le 2$$
$$\frac{x-2}{x+2} - 2 \le 0$$
$$\frac{x-2-2(x+2)}{x+2} \le 0$$
$$\frac{x-2-2x-4}{x+2} \le 0$$
$$\frac{-x-6}{x+2} \le 0$$

$x = -6$ or $x = -2$

T	F	T

$-6 \quad -2$

The solution set is $(-\infty, -6] \cup (-2, \infty)$.

$-6 \quad -2$

56. $$\frac{x}{x+2} \ge 2$$
$$\frac{x}{x+2} - 2 \ge 0$$
$$\frac{x}{x+2} - \frac{2(x+2)}{x+2} \ge 0$$
$$\frac{x-2x-4}{x+2} \ge 0$$
$$\frac{-x-4}{x+2} \ge 0$$

$x = -4$ or $x = -2$

F	T	F

$-4 \quad -2$

The solution set is $[-4, -2)$.

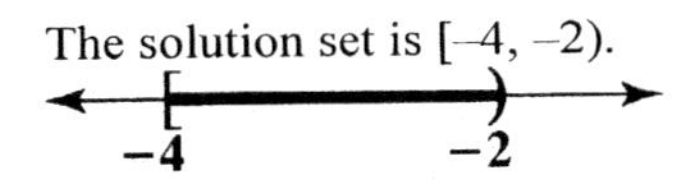

57. $f(x) = \sqrt{2x^2 - 5x + 2}$

The domain of this function requires that

$2x^2 - 5x + 2 \ge 0$

Solve $2x^2 - 5x + 2 = 0$

$(x-2)(2x-1) = 0$

$x = \frac{1}{2}$ or $x = 2$

T	F	T

$\frac{1}{2} \quad 2$

The domain is $\left(-\infty, \frac{1}{2}\right] \cup [2, \infty)$.

58. $f(x) = \dfrac{1}{\sqrt{4x^2 - 9x + 2}}$

The domain of this function requires that

$4x^2 - 9x + 2 > 0$

Solve $4x^2 - 9x + 2 = 0$

$(x-2)(4x-1) = 0$

$x = \frac{1}{4}$ or $x = 2$

T	F	T

$\frac{1}{4} \quad 2$

The domain is $\left(-\infty, \frac{1}{4}\right) \cup (2, \infty)$.

59. $f(x) = \sqrt{\dfrac{2x}{x+1} - 1}$

The domain of this function requires that

$\frac{2x}{x+1} - 1 \ge 0$ or $\frac{x-1}{x+1} \ge 0$

$x = -1$ or $x = 1$

T	F	T

$-1 \quad 1$

The value $x = 1$ results in 0 and, thus, it must be included in the domain.

The domain is $(-\infty, -1) \cup [1, \infty)$.

60. $f(x) = \sqrt{\dfrac{x}{2x-1} - 1}$

The domain of this function requires that

$\frac{x}{2x-1} - 1 \ge 0$ or $\frac{-x+1}{2x-1} \ge 0$

$x = \frac{1}{2}$ or $x = 1$

F	T	F

$\frac{1}{2} \quad 1$

The value $x = 1$ results in 0 and, thus, it must be included in the domain.

The domain is $\left(\frac{1}{2}, 1\right]$.

61. $\left|x^2+2x-36\right|>12$

Express the inequality without the absolute value symbol:

$x^2+2x-36<-12$ or $x^2+2x-36>12$

$x^2+2x-24<0$ $\quad$ $x^2+2x-48>0$

Solve the related quadratic equations.

$x^2+2x-24=0$ or $x^2+2x-48=0$

$(x+6)(x-4)=0$ $\quad$ $(x+8)(x-6)=0$

Apply the zero product principle.

$x+6=0$ or $x-4=0$ or $x+8=0$ or $x-6=0$

$x=-6$ $\quad$ $x=4$ $\quad$ $x=-8$ $\quad$ $x=6$

The boundary points are $-8,\ -6,\ 4$ and 6.

Test Interval	Test Number	Test	Conclusion
$(-\infty,-8)$	-9	$\left\|(-9)^2+2(-9)-36\right\|>12$ $27>12$, True	$(-\infty,-8)$ belongs to the solution set.
$(-8,-6)$	-7	$\left\|(-7)^2+2(-7)-36\right\|>12$ $1>12$, False	$(-8,-6)$ does not belong to the solution set.
$(-6,4)$	0	$\left\|0^2+2(0)-36\right\|>12$ $36>12$, True	$(-6,4)$ belongs to the solution set.
$(4,6)$	5	$\left\|5^2+2(5)-36\right\|>12$ $1>12$, False	$(4,6)$ does not belong to the solution set.
$(6,\infty)$	7	$\left\|7^2+2(7)-36\right\|>12$ $27>12$, True	$(6,\infty)$ belongs to the solution set.

The solution set is $(-\infty,-8)\cup(-6,4)\cup(6,\infty)$ or $\{x\,|\,x<-8 \text{ or } -6<x<4 \text{ or } x>6\}$.

−10 −8 −6 −4 −2 0 2 4 6 8 10

62. $\left|x^2+6x+1\right|>8$

Express the inequality without the absolute value symbol:

$x^2+6x+1<-8$ or $x^2+6x+1>8$

$x^2+6x+9<0$ $\quad$ $x^2+6x-7>0$

Solve the related quadratic equations.

$x^2+6x+9=0$ or $x^2+6x-7=0$

$(x+3)^2=0$ $\quad$ $(x+7)(x-1)=0$

$x+3=\pm\sqrt{0}$ or $x+7=0$ or $x-1=0$

$x+3=0$ $\quad$ $x=-7$ $\quad$ $x=1$

$x=-3$

The boundary points are $-7,-3,$ and 1.

Test Interval	Test Number	Test	Conclusion
$(-\infty,-7)$	-8	$\left\|(-8)^2+6(-8)+1\right\|>8$ $17\ge 8$, True	$(-\infty,-7)$ belongs to the solution set.
$(-7,-3)$	-5	$\left\|(-5)^2+6(-5)+1\right\|>8$ $4\ge 8$, False	$(-7,-3)$ does not belong to the solution set.
$(-3,1)$	0	$\left\|0^2+6(0)+1\right\|>8$ $1\ge 8$, False	$(-3,1)$ does not belong to the solution set.
$(1,\infty)$	2	$\left\|2^2+6(2)+1\right\|>8$ $17\ge 8$, True	$(1,\infty)$ belongs to the solution set.

The solution set is $(-\infty,-7)\cup(1,\infty)$ or $\{x\mid x<-7 \text{ or } x>1\}$.

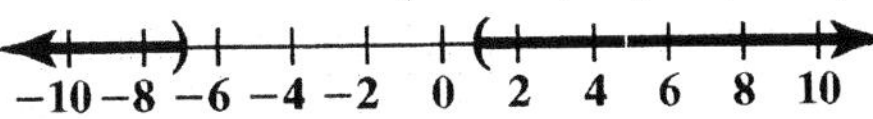

63. $\dfrac{3}{x+3}>\dfrac{3}{x-2}$

Express the inequality so that one side is zero.

$$\frac{3}{x+3}-\frac{3}{x-2}>0$$

$$\frac{3(x-2)}{(x+3)(x-2)}-\frac{3(x+3)}{(x+3)(x-2)}>0$$

$$\frac{3x-6-3x-9}{(x+3)(x-2)}<0$$

$$\frac{-15}{(x+3)(x-2)}<0$$

Find the values of x that make the denominator zero.

$x+3=0 \qquad x-2=0$

$x=-3 \qquad x=2$

The boundary points are -3 and 2.

Test Interval	Test Number	Test	Conclusion
$(-\infty,-3)$	-4	$\frac{3}{-4+3}>\frac{3}{-4-2}$ $-3>\frac{1}{2}$, False	$(-\infty,-3)$ does not belong to the solution set.
$(-3,2)$	0	$\frac{3}{0+3}>\frac{3}{0-2}$ $1>-\frac{3}{2}$, True	$(-3,2)$ belongs to the solution set.
$(2,\infty)$	3	$\frac{3}{3+3}>\frac{3}{3-2}$ $\frac{1}{2}>3$, False	$(2,\infty)$ does not belong to the solution set.

The solution set is $(-3,2)$ or $\{x\mid -3<x<2\}$.

−5 −4 −3 −2 −1 0 1 2 3 4 5

64. $\dfrac{1}{x+1} > \dfrac{2}{x-1}$

Express the inequality so that one side is zero.

$$\frac{1}{x+1} - \frac{2}{x-1} > 0$$

$$\frac{x-1}{(x+1)(x-1)} - \frac{2(x+1)}{(x+1)(x-1)} > 0$$

$$\frac{x-1-2x-2}{(x+1)(x-1)} < 0$$

$$\frac{-x-3}{(x+1)(x-1)} < 0$$

Find the values of x that make the numerator and denominator zero.

$-x-3=0 \qquad x+1=0 \qquad x-1=0$

$-3=x \qquad x=-1 \qquad x=1$

The boundary points are -3, -1, and 1.

Test Interval	Test Number	Test	Conclusion
$(-\infty,-3)$	-4	$\frac{1}{-4+1} > \frac{2}{-3-1}$ $-\frac{1}{3} > -\frac{1}{2}$, True	$(-\infty,-3)$ belongs to the solution set.
$(-3,-1)$	-2	$\frac{1}{-2+1} > \frac{2}{-2-1}$ $-1 > -\frac{2}{3}$, False	$(-3,-1)$ does not belong to the solution set.
$(-1,1)$	0	$\frac{1}{0+1} > \frac{2}{0-1}$ $1 > -2$, True	$(-3,1)$ belongs to the solution set.
$(1,\infty)$	2	$\frac{1}{2+1} > \frac{2}{2-1}$ $\frac{1}{3} > 1$, False	$(1,\infty)$ does not belong to the solution set.

The solution set is $(-\infty,-3)\cup(-1,1)$ or $\{x \mid x<-3 \text{ or } -1<x<1\}$.

−5 −4 −3 −2 −1 0 1 2 3 4 5

65. $\dfrac{x^2-x-2}{x^2-4x+3} > 0$

Find the values of x that make the numerator and denominator zero.

$x^2-x-2=0 \qquad x^2-4x+3=0$

$(x-2)(x+1)=0 \qquad (x-3)(x-1)=0$

Apply the zero product principle.

$x-2=0$ or $x+1=0 \qquad x-3=0$ or $x-1=0$

$x=2 \qquad x=-1 \qquad x=3 \qquad x=1$

The boundary points are -1, 1, 2 and 3.

Test Interval	Test Number	Test	Conclusion
$(-\infty,-1)$	-2	$\frac{(-2)^2-(-2)-2}{(-2)^2-4(-2)+3}>0$ $\frac{4}{15}>0$, True	$(-\infty,-1)$ belongs to the solution set.
$(-1,1)$	0	$\frac{0^2-0-2}{0^2-4(0)+3}>0$ $-\frac{2}{3}>0$, False	$(-1,1)$ does not belong to the solution set.
$(1,2)$	1.5	$\frac{1.5^2-1.5-2}{1.5^2-4(1.5)+3}>0$ $\frac{5}{3}>0$, True	$(1,2)$ belongs to the solution set.
$(2,3)$	2.5	$\frac{2.5^2-2.5-2}{2.5^2-4(2.5)+3}>0$ $-\frac{7}{3}>0$, False	$(2,3)$ does not belong to the solution set.
$(3,\infty)$	4	$\frac{4^2-4-2}{4^2-4(4)+3}>0$ $\frac{10}{3}>0$, True	$(3,\infty)$ belongs to the solution set.

The solution set is $(-\infty,-1)\cup(1,2)\cup(3,\infty)$ or $\{x|x<-1 \text{ or } 1<x<2 \text{ or } x>3\}$.

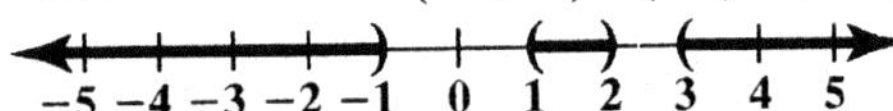

66. $\frac{x^2-3x+2}{x^2-2x-3}>0$

Find the values of x that make the numerator and denominator zero.

$x^2-3x+2=0$ $\quad$ $x^2-2x-3=0$

$(x-2)(x-1)=0$ $\quad$ $(x-3)(x+1)=0$

Apply the zero product principle.

$x-2=0$ or $x-1=0$ $\quad$ $x-3=0$ or $x+1=0$

$x=2$ $\quad$ $x=1$ $\quad$ $x=3$ $\quad$ $x=-1$

The boundary points are -1, 1, 2 and 3.

Test Interval	Test Number	Test	Conclusion
$(-\infty,-1)$	-2 $\frac{x^2-3x+2}{x^2-2x-3}>0$	$\frac{(-2)^2-3(-2)+2}{(-2)^2-2(-2)-3}>0$ $\frac{12}{5}>0$, True	$(-\infty,-1)$ belongs to the solution set.
$(-1,1)$	0	$\frac{0^2-3(0)+2}{0^2-2(0)-3}>0$ $-\frac{2}{3}>0$, False	$(-1,1)$ does not belong to the solution set.
$(1,2)$	1.5	$\frac{1.5^2-3(1.5)+2}{1.5^2-2(1.5)-3}>0$ $\frac{1}{15}>0$, True	$(1,2)$ belongs to the solution set.
$(2,3)$	2.5	$\frac{2.5^2-3(2.5)+2}{2.5^2-2(2.5)-3}>0$ $-\frac{3}{7}>0$, False	$(2,3)$ does not belong to the solution set.
$(3,\infty)$	4	$\frac{4^2-3(4)+2}{4^2-2(4)-3}>0$ $\frac{6}{5}>0$, True	$(3,\infty)$ belongs to the solution set.

The solution set is $(-\infty,-1)\cup(1,2)\cup(3,\infty)$ or $\{x|x<-1 \text{ or } 1<x<2 \text{ or } x>3\}$.

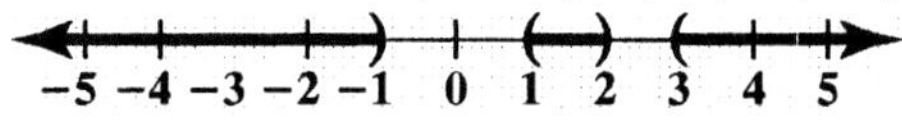

67.

$$2x^3+11x^2\geq 7x+6$$
$$2x^3+11x^2-7x-6\geq 0$$

The graph of $f(x)=2x^3+11x^2-7x-6$ appears to cross the x-axis at -6, $-\frac{1}{2}$, and 1. We verify this numerically by substituting these values into the function:

$$f(-6)=2(-6)^3+11(-6)^2-7(-6)-6=2(-216)+11(36)-(-42)-6=-432+396+42-6=0$$

$$f\left(-\frac{1}{2}\right)=2\left(-\frac{1}{2}\right)^3+11\left(-\frac{1}{2}\right)^2-7\left(-\frac{1}{2}\right)-6=2\left(-\frac{1}{8}\right)+11\left(\frac{1}{4}\right)-\left(-\frac{7}{2}\right)-6=-\frac{1}{4}+\frac{11}{4}+\frac{7}{2}-6=0$$

$$f(1)=2(1)^3+11(1)^2-7(1)-6=2(1)+11(1)-7-6=2+11-7-6=0$$

Thus, the boundaries are -6, $-\frac{1}{2}$, and 1. We need to find the intervals on which $f(x)\geq 0$. These intervals are indicated on the graph where the curve is above the x-axis. Now, the curve is above the x-axis when $-6<x<-\frac{1}{2}$ and when $x>1$. Thus, the solution set is $\left\{x\middle|\ -6\leq x\leq -\frac{1}{2} \text{ or } x\geq 1\right\}$ or $\left[-6,-\frac{1}{2}\right]\cup[1,\infty)$.

68. $2x^3 + 11x^2 < 7x + 6$

$2x^3 + 11x^2 - 7x - 6 < 0$

In Problem 63, we verified that the boundaries are -6, $-\frac{1}{2}$, and 1. We need to find the intervals on which $f(x) < 0$. These intervals are indicated on the graph where the curve is below the x-axis. Now, the curve is below the x-axis when $x < -6$ and when $-\frac{1}{2} < x < 1$. Thus, the solution set is $\left\{x \middle| x < -6 \text{ or } -\frac{1}{2} < x < 1\right\}$ or $(-\infty, -6) \cup \left(-\frac{1}{2}, 1\right)$.

69. $\frac{1}{4(x+2)} \le -\frac{3}{4(x-2)}$

$\frac{1}{4(x+2)} + \frac{3}{4(x-2)} \le 0$

Simplify the left side of the inequality:

$\frac{x-2}{4(x+2)} + \frac{3(x+2)}{4(x-2)} = \frac{x-2+3x+6}{4(x+2)(x-2)} = \frac{4x+4}{4(x+2)(x-2)} = \frac{4(x+1)}{4(x+2)(x-2)} = \frac{x+1}{x^2-4}$.

The graph of $f(x) = \frac{x+1}{x^2-4}$ crosses the x-axis at -1, and has vertical asymptotes at $x = -2$ and $x = 2$. Thus, the boundaries are -2, -1, and 1. We need to find the intervals on which $f(x) \le 0$. These intervals are indicated on the graph where the curve is below the x-axis. Now, the curve is below the x-axis when $x < -2$ and when $-1 < x < 2$. Thus, the solution set is $\{x \mid x < -2 \text{ or } -1 \le x < 2\}$ or $(-\infty, -2) \cup [-1, 2)$.

70. $\frac{1}{4(x+2)} > -\frac{3}{4(x-2)}$

$\frac{1}{4(x+2)} + \frac{3}{4(x-2)} > 0$

$\frac{x+1}{(x+2)(x-2)} > 0$

The boundaries are -2, -1, and 2. We need to find the intervals on which $f(x) > 0$. These intervals are indicated on the graph where the curve is above the x-axis. The curve is above the x-axis when $-2 < x < -1$ and when $x > 2$. Thus, the solution set is $\{x \mid -2 < x < -1 \text{ or } x > 2\}$ or $(-2, -1) \cup (2, \infty)$.

71. $s(t) = -16t^2 + 8t + 87$

The diver's height will exceed that of the cliff when $s(t) > 87$

$-16t^2 + 8t + 87 > 87$

$-16t^2 + 8t > 0$

$-8t(2t - 1) > 0$

The boundaries are 0 and $\frac{1}{2}$. Testing each interval shows that the diver will be higher than the cliff for the first half second after beginning the jump. The interval is $\left(0, \frac{1}{2}\right)$.

72. $s(t) = -16t^2 + 48t + 160$

The ball's height will exceed that of the rooftop when $s(t) > 160$

$-16t^2 + 48t + 160 > 160$

$-16t^2 + 48t > 0$

$-16t(t-3) > 0$

The boundaries are 0 and 3. Testing each interval shows that the ball will be higher than the rooftop for the first three seconds after the throw. The interval is $(0,3)$.

73. $f(8) = 27(8) + 163 = 216 + 163 = 379$

$g(8) = 1.2(8)^2 + 15.2(8) + 181.4 = 1.2(64) + 121.6 + 181.4$

$= 76.8 + 121.6 + 181.4 = 379.8$

Since the graph indicates that Medicare spending will reach \$379 billion, we conclude that both functions model the data quite well.

74. $f(10) = 27(10) + 163 = 270 + 163 = 433$

$g(10) = 1.2(10)^2 + 15.2(10) + 181.4 = 1.2(100) + 152 + 181.4 = 120 + 152 + 181.4 = 453.4$

Since the graph indicates that Medicare spending will reach \$458 billion, we conclude that g is a better model for the data than f.

75. $g(x) = 1.2x^2 + 15.2x + 181.4$

To find when spending exceeds \$536.6 billion, solve the inequality $1.2x^2 + 15.2x + 181.4 > 536.6$.
Solve the related quadratic equation using the quadratic formula.

$1.2x^2 + 15.2x + 181.4 = 536.6$

$1.2x^2 + 15.2x - 355.2 = 0$

$a = 1.2 \quad b = 15.2 \quad c = -355.2$

$$x = \frac{-15.2 \pm \sqrt{15.2^2 - 4(1.2)(-355.2)}}{2(1.2)} = \frac{-15.2 \pm \sqrt{231.04 + 1704.96}}{2.4}$$

$$= \frac{-15.2 \pm \sqrt{1936}}{2.4} = \frac{-15.2 \pm 44}{2.4}$$

$$= \frac{-15.2 - 44}{2.4} \text{ or } \frac{-15.2 + 44}{2.4} = -24\frac{2}{3} \text{ or } 12$$

We disregard $-24\frac{2}{3}$ since x represents the number of years after 1995 and cannot be negative. The boundary point is 12.

Test Interval	Test Number	Test	Conclusion
$(0,12)$	1	$1.2(1)^2 + 15.2(1) + 181.4 > 536.6$ $197.8 > 536.6$, false	$(0,12)$ does not belong to the solution set.
$(12,\infty)$	13	$1.2(13)^2 + 15.2(13) + 181.4 > 536.6$ $581.8 > 536.6$, true	$(12,\infty)$ belongs to the solution set.

The solution set is $(12,\infty)$. This means that spending will exceed \$536.6 billion after 1995 + 12 = 2007.

76. $1.2x^2+15.2x+181.4>629.4$

$1.2x^2+15.2x+181.4=629.4$

$1.2x^2+15.2x-448=0$

$a=1.2 \quad b=15.2 \quad c=-448$

$$x=\frac{-15.2\pm\sqrt{15.2^2-4(1.2)(-448)}}{2(1.2)}=\frac{-15.2\pm\sqrt{231.04+2150.4}}{2.4}=\frac{-15.2\pm\sqrt{2381.44}}{2.4}=\frac{-15.2\pm 48.8}{2.4}$$

$$=\frac{-15.2-48.8}{2.4}\text{ or }\frac{-15.2+48.8}{2.4}\approx -26.7\text{ or }14$$

The disregard –26.7 since x represents the number of years after 1995 and cannot be negative.

Test Interval	Test Number	Test	Conclusion
$(0,14)$	1	$1.2(1)^2+15.2(1)+181.4>629.4$ false	$(0,14)$ does not belong to the solution set.
$(14,\infty)$	15	$1.2(15)^2+15.2(15)+181.4>629.4$ true	$(14,\infty)$ belongs to the solution set.

Spending will exceed \$629.4 billion after 1995 + 14 = 2009.

77. $f(x)=\frac{15}{x}$

If the time must be limited to 3 hours, then $\frac{15}{x}<3$.

$$\frac{15}{x}<3$$

$$\frac{15}{x}-3<0$$

$$\frac{-3(x-5)}{x}<0$$

The boundaries are 0 and 5. Testing each interval shows that $\frac{15}{x}<3$ on the intervals of $(-\infty,0)$ and $(5,\infty)$.

However, a rate below 0 does not fit the constraints of this problem. Thus the solution is $(5,\infty)$.

The graph agrees with this solution.
For rates above 5 mph, the height of the graph is always at or below 3 hours.

78. $f(x)=\frac{15}{x}$

If the time must be limited to 5 hours, then $\frac{15}{x}<5$.

$$\frac{15}{x}<5$$

$$\frac{15}{x}-5<0$$

$$\frac{-5(x-3)}{x}<0$$

The boundaries are 0 and 3. Testing each interval shows that $\frac{15}{x}<5$ on the intervals of $(-\infty,0)$ and $(3,\infty)$.

However, a rate below 0 does not fit the constraints of this problem. Thus the solution is $(3,\infty)$.

The graph agrees with this solution.
For rates above 3 mph, the height of the graph is always at or below 5 hours.

79. As $x \to \infty$, the graph approaches 0. This shows that the higher the running rate, the less time it will take to complete the 95 miles.

80. As $x \to 0^+$, the graph approaches ∞. This shows that the lower the running rate, the more time it will take to complete the 95 miles.

81. To obtain the function that is displayed on the graph, let x represent the hiking rate and, thus, 9x will represent the driving rate.

$$\overbrace{f(x)}^{\text{total time}} = \overbrace{\frac{\text{driving distance}}{\text{driving rate}}}^{\text{driving time}} + \overbrace{\frac{\text{hiking distance}}{\text{hiking rate}}}^{\text{hiking time}}$$

$$\begin{aligned} f(x) &= \frac{90}{9x} + \frac{5}{x} \\ &= \frac{10}{x} + \frac{5}{x} \\ &= \frac{15}{x} \end{aligned}$$

82. Let x = the length of the rectangle.
Since Perimeter $= 2(\text{length}) + 2(\text{width})$, we know

$$\begin{aligned} 50 &= 2x + 2(\text{width}) \\ 50 - 2x &= 2(\text{width}) \\ \text{width} &= \frac{50-2x}{2} = 25 - x \end{aligned}$$

Now, $A = (\text{length})(\text{width})$, so we have that

$$\begin{aligned} A(x) &\le 114 \\ x(25-x) &\le 114 \\ 25x - x^2 &\le 114 \end{aligned}$$

Solve the related equation

$$\begin{aligned} 25x - x^2 &= 114 \\ 0 &= x^2 - 25x + 114 \\ 0 &= (x-19)(x-6) \end{aligned}$$

Apply the zero product principle:

$x - 19 = 0$ or $x - 6 = 0$

$x = 19$ $\quad x = 6$

The boundary points are 6 and 19.

Test Interval	Test Number	Test	Conclusion
$(-\infty, 6)$	0	$25(0) - 0^2 \le 114$ $0 \le 114$, True	$(-\infty, 6)$ belongs to the solution set.
$(6, 19)$	10	$25(10) - 10^2 \le 114$ $150 \le 114$, False	$(6, 19)$ does not belong to the solution set.
$(19, \infty)$	20	$25(20) - 20^2 \le 114$ $100 \le 114$, True	$(19, \infty)$ belongs to the solution set.

If the length is 6 feet, then the width is 19 feet. If the length is less than 6 feet, then the width is greater than 19 feet. Thus, if the area of the rectangle is not to exceed 114 square feet, the length of the shorter side must be 6 feet or less.

83. $2l + 2w = P$

$2l + 2w = 180$

$2l = 180 - 2w$

$l = 90 - w$

We want to restrict the area to 800 square feet. That is,

$A \le 800$

$l \cdot w \le 800$

$(90 - w)w \le 800$

$90w - w^2 \le 800$

$-w^2 + 90w - 800 \le 0$

$w^2 - 90w + 800 \ge 0$

$w^2 - 90w + 800 = 0$

$(w-80)(w-10) = 0$

$w - 80 = 0$ or $w - 10 = 0$

$w = 80$ $\quad$ $w = 10$

Assuming the width is the shorter side, we ignore the larger solution.

Test Interval	Test Number	Test	Conclusion
$(0,10)$	5	$90(5)-(5)^2 \le 800$ true	$(0,10)$ is part of the solution set
$(10,45)$	20	$90(20)-(20)^2 \le 800$ false	$(10,45)$ is not part of the solution set

The solution set is $\{w \mid 0 < w \le 10\}$ or $(0,10]$.

The length of the shorter side cannot exceed 10 feet.

87. Answers will vary.

88. Answers will vary.

89.

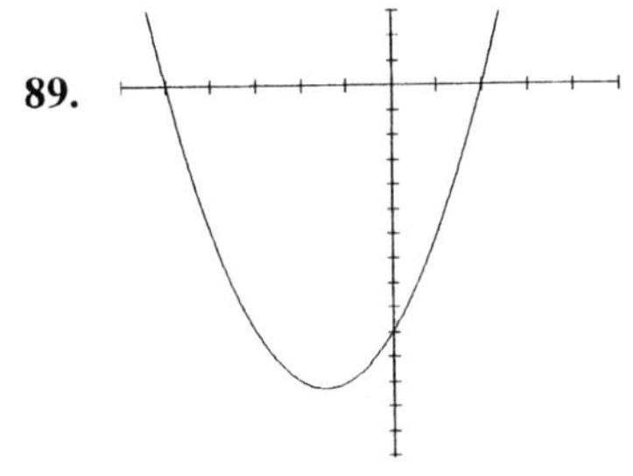

The solution set is (−∞, -5) U(2, ∞).

90. Graph $y_1 = 2x^2 + 5x - 3$ in a standard window. The graph is below or equal to the x-axis for $-3 \le x \le \frac{1}{2}$.

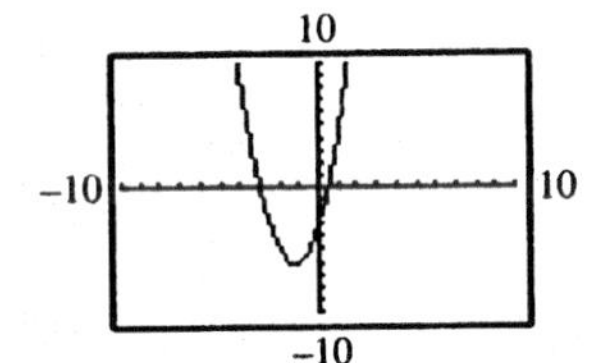

The solution set is $\left\{x \middle| -3 \le x \le \frac{1}{2}\right\}$ or $\left[-3, \frac{1}{2}\right]$.

91.

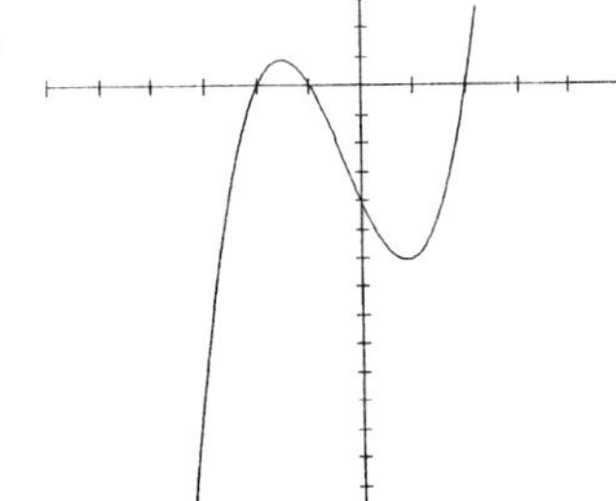

The solution set is (−2, −1) or (2, ∞).

92. Graph $y_1 = \dfrac{x-4}{x-1}$ in a standard viewing window.
The graph is below the x-axis for $1 < x \le 4$.

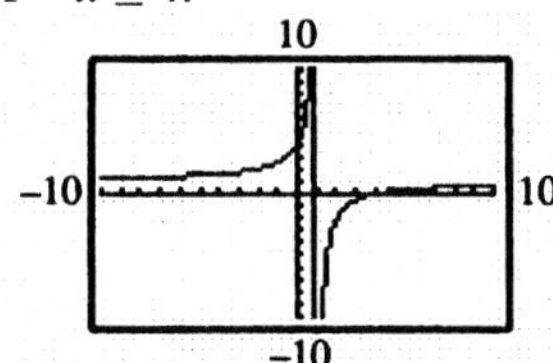

The solution set is (1, 4].

93. Graph $y_1 = \dfrac{x+2}{x-3}$ and $y_2 = 2$
y_1 less than or equal to y_2 for $x < 3$ or $x \ge 8$.
The solution set is $(-\infty, 3) \cup [8, \infty)$

94. Graph $y_1 = \dfrac{1}{x+1}$ and $y_2 = \dfrac{2}{x+4}$
y_1 less than or equal to y_2 for
$-4 < x < -1$ or $x \ge 2$.
The solution set is $(-4, -1) \cup [2, \infty)$

95. **a.** False
$x^2 > 25$
Solution: $(-\infty, -5) \cup (5, \infty)$

b. False
Subtract 2 from both sides and solve.
$\dfrac{x-2-2(x+3)}{x+3} < 0$
Note that when multiplying by $x + 3$, $x + 3$ may be negative.

c. False
The solution set of the first inequality is $(-\infty, -3) \cup (1, \infty)$. The solution set of the second inequality is $(-\infty, -3) \cup (1, \infty)$.

d. True

(d) is true.

96. One possible solution: $x^2 - 2x - 15 \le 0$

97. One possible solution: $\dfrac{x-3}{x+4} \ge 0$

98. Because any non-zero number squared is positive, the solution is all real numbers except 2.

99. Because any number squared other than zero is positive, the solution includes only 2.

100. Because any number squared is positive, the solution is the empty set, $\varnothing$.

101. Because any number squared other than zero is positive, and the reciprocal of zero is undefined, the solution is all real numbers except 2.

102. **a.** The solution set is all real numbers.

b. The solution set is the empty set, $\varnothing$.

c. $4x^2 - 8x + 7 > 0$

$$x = \frac{8 \pm \sqrt{(-8)^2 - 4(4)(7)}}{2(4)}$$

$$x = \frac{8 \pm \sqrt{64 - 112}}{8}$$

$$x = \frac{8 \pm \sqrt{-48}}{8} \Rightarrow \text{imaginary}$$

no critical values
Test 0: $4(0)^2 - 8(0) + 7 > 0$
$7 > 0$ True
The inequality is true for all numbers.

$4x^2 - 8x + 7 < 0$
no critical values
Test 0: $4(0)^2 - 8(0) + 7 = 7 < 0$ False
The solution set is the empty set.

103. $\sqrt{27 - 3x^2} \ge 0$
$27 - 3x^2 \ge 0$
$9 - x^2 \ge 0$
$(3 - x)(3 + x) \ge 0$
$3 - x = 0 \qquad 3 + x = 0$
$x = 3$ or $\qquad x = -3$

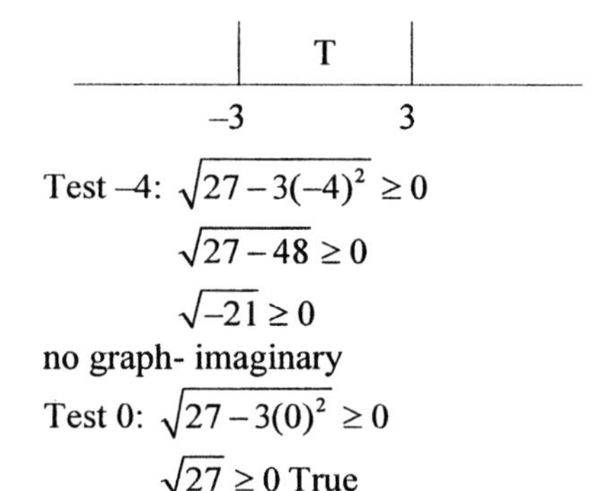

Test −4: $\sqrt{27 - 3(-4)^2} \ge 0$
$\sqrt{27 - 48} \ge 0$
$\sqrt{-21} \ge 0$
no graph- imaginary
Test 0: $\sqrt{27 - 3(0)^2} \ge 0$
$\sqrt{27} \ge 0$ True

est 4: $\sqrt{27-3(4)^2} \geq 0$

$\sqrt{27-48} \geq 0$

$\sqrt{-21} \geq 0$

no graph -imaginary

The solution set is $[-3, 3]$.

Section 2.8

Check Point Exercises

1. y varies directly as x is expressed as $y = kx$.
   The volume of water, W, varies directly as the time, t can be expressed as $W = kt$.
   Use the given values to find k.
   $W = kt$
   $30 = k(5)$
   $6 = k$
   Substitute the value of k into the equation.
   $W = kt$
   $W = 6t$
   Use the equation to find W when $t = 11$.
   $W = 6t$
   $= 6(11)$
   $= 66$
   A shower lasting 11 minutes will use 66 gallons of water.

2. y varies directly as the square of x is expressed as $y = kx$.
   The distance, s, varies directly as the square of the speed, v can be expressed as $s = kv^2$.
   Use the given values to find k.
   $s = kv^2$
   $200 = k(60)^2$
   $0.0556 = k$
   Substitute the value of k into the equation.
   $s = kv^2$
   $s = 0.0556v^2$
   Use the equation to find s when $v = 100$.
   $s = 0.0556v^2$
   $= 0.0556(100)^2$
   $= 556$
   A car traveling 100 mph will require 556 feet to stop.

3. y varies inversely as x is expressed as $y = \frac{k}{x}$.
   The length, L, varies inversely as the frequency, f can be expressed as $L = \frac{k}{f}$.
   Use the given values to find k.
   $L = \frac{k}{f}$
   $8 = \frac{k}{640}$
   $5120 = k$
   Substitute the value of k into the equation.
   $L = \frac{k}{f}$
   $L = \frac{5120}{f}$
   Use the equation to find f when $L = 10$.
   $L = \frac{5120}{f}$
   $10 = \frac{5120}{f}$
   $10f = 5120$
   $f = 512$
   A 10 inch violin string will have a frequency of 512 cycles per second.

4. let M represent the number of minutes
   let Q represent the number of problems
   let P represent the number of people
   M varies directly as Q and inversely as P is expressed as $M = \frac{kQ}{P}$.
   Use the given values to find k.
   $M = \frac{kQ}{P}$
   $32 = \frac{k(16)}{4}$
   $8 = k$
   Substitute the value of k into the equation.
   $M = \frac{kQ}{P}$
   $M = \frac{8Q}{P}$

Use the equation to find M when $P = 8$ and $Q = 24$.

$M = \frac{8Q}{P}$

$M = \frac{8(24)}{8}$

$M = 24$

It will take 24 minutes for 8 people to solve 24 problems.

5. V varies jointly with h and r^2 and can be modeled as $V = khr^2$.

Use the given values to find k.

$V = khr^2$

$120\pi = k(10)(6)^2$

$\frac{\pi}{3} = k$

Therefore, the volume equation is $V = \frac{1}{3}hr^2$.

$V = \frac{\pi}{3}(2)(12)^2 = 96\pi$ cubic feet

Exercise Set 2.8

1. Use the given values to find k.

$y = kx$

$65 = k \cdot 5$

$\frac{65}{5} = \frac{k \cdot 5}{5}$

$13 = k$

The equation becomes $y = 13x$.

When $x = 12$, $y = 13x = 13 \cdot 12 = 156$.

2. $y = kx$

$45 = k \cdot 5$

$9 = k$

$y = 9x = 9 \cdot 13 = 117$

3. Since y varies inversely with x, we have $y = \frac{k}{x}$.

Use the given values to find k.

$y = \frac{k}{x}$

$12 = \frac{k}{5}$

$5 \cdot 12 = 5 \cdot \frac{k}{5}$

$60 = k$

The equation becomes $y = \frac{60}{x}$.

When $x = 2$, $y = \frac{60}{2} = 30$.

4. $y = \frac{k}{x}$

$6 = \frac{k}{3}$

$18 = k$

$y = \frac{18}{9} = 2$

5. Since y varies inversely as x and inversely as the square of z, we have $y = \frac{kx}{z^2}$.

Use the given values to find k.

$y = \frac{kx}{z^2}$

$20 = \frac{k(50)}{5^2}$

$20 = \frac{k(50)}{25}$

$20 = 2k$

$10 = k$

The equation becomes $y = \frac{10x}{z^2}$.

When $x = 3$ and $z = 6$,

$y = \frac{10x}{z^2} = \frac{10(3)}{6^2} = \frac{10(3)}{36} = \frac{30}{36} = \frac{5}{6}$.

6.
$$a = \frac{kb}{c^2}$$
$$7 = \frac{k(9)}{(6)^2}$$
$$7 = \frac{k(9)}{36}$$
$$7 = \frac{k}{4}$$
$$28 = k$$

$$a = \frac{28(4)}{(8)^2} = \frac{28(4)}{64} = \frac{7}{4}$$

7. Since y varies jointly as x and z, we have $y = kxz$.
Use the given values to find k.
$$y = kxz$$
$$25 = k(2)(5)$$
$$25 = k(10)$$
$$\frac{25}{10} = \frac{k(10)}{10}$$
$$\frac{5}{2} = k$$
The equation becomes $y = \frac{5}{2}xz$.

When $x = 8$ and $z = 12$, $y = \frac{5}{2}(8)(12) = 240$.

8.
$$C = kAT$$
$$175 = k(2100)(4)$$
$$175 = k(8400)$$
$$\frac{1}{48} = k$$

$$C = \frac{1}{48}(2400)(6) = \frac{14400}{48} = 300$$

9. Since y varies jointly as a and b and inversely as the square root of c, we have $y = \frac{kab}{\sqrt{c}}$.
Use the given values to find k.
$$y = \frac{kab}{\sqrt{c}}$$
$$12 = \frac{k(3)(2)}{\sqrt{25}}$$
$$12 = \frac{k(6)}{5}$$
$$12(5) = \frac{k(6)}{5}(5)$$
$$60 = 6k$$
$$\frac{60}{6} = \frac{6k}{6}$$
$$10 = k$$
The equation becomes $y = \frac{10ab}{\sqrt{c}}$.

When $a = 5$, $b = 3$, $c = 9$,
$$y = \frac{10ab}{\sqrt{c}} = \frac{10(5)(3)}{\sqrt{9}} = \frac{150}{3} = 50.$$

10.
$$y = \frac{kmn^2}{p}$$
$$15 = \frac{k(2)(1)^2}{6}$$
$$15 = \frac{2k}{6}$$
$$15(6) = \frac{2k}{6}(6)$$
$$90 = 2k$$
$$k = 45$$

$$y = \frac{45mn^2}{p} = \frac{45(3)(4)^2}{10} = \frac{2160}{10} = 216$$

11. $x = kyz$;
Solving for y:
$$x = kyz$$
$$\frac{x}{kz} = \frac{kyz}{yz}.$$
$$y = \frac{x}{kz}$$

12. $x = kyz^2$;
Solving for y:
$$x = kyz^2$$
$$\frac{x}{kz^2} = \frac{kyz^2}{kz^2}$$
$$y = \frac{x}{kz^2}$$

13. $x = \frac{kz^3}{y}$;
Solving for y
$$x = \frac{kz^3}{y}$$
$$xy = y \cdot \frac{kz^3}{y}$$
$$xy = kz^3$$
$$\frac{xy}{x} = \frac{kz^3}{x}$$
$$y = \frac{kz^3}{x}$$

14.
$$x = \frac{k\sqrt[3]{z}}{y}$$
$$yx = y \cdot \frac{k\sqrt[3]{z}}{y}$$
$$yx = k\sqrt[3]{z}$$
$$\frac{yx}{x} = \frac{k\sqrt[3]{z}}{x}$$
$$y = \frac{k\sqrt[3]{z}}{x}$$

15. $x = \frac{kyz}{\sqrt{w}}$;
Solving for y:
$$x = \frac{kyz}{\sqrt{w}}$$
$$x\left(\sqrt{w}\right) = \left(\sqrt{w}\right)\frac{kyz}{\sqrt{w}}$$
$$x\sqrt{w} = kyz$$
$$\frac{x\sqrt{w}}{kz} = \frac{kyz}{kz}$$
$$y = \frac{x\sqrt{w}}{kz}$$

16.
$$x = \frac{kyz}{w^2}$$
$$\left(\frac{w^2}{kz}\right)x = \frac{w^2}{kz}\frac{kyz}{w^2}$$
$$y = \frac{xw^2}{kz}$$

17. $x = kz(y+w)$;
Solving for y:
$$x = kz(y+w)$$
$$x = kzy + kzw$$
$$x - kzw = kzy$$
$$\frac{x-kzw}{kz} = \frac{kzy}{kz}$$
$$y = \frac{x-kzw}{kz}$$

18.
$$x = kz(y-w)$$
$$x = kzy - kzw$$
$$x + kzw = kzy$$
$$\frac{x+kzw}{kz} = \frac{kzy}{kz}$$
$$y = \frac{x+kzw}{kz}$$

19. $x = \frac{kz}{y-w}$;
Solving for y:
$$x = \frac{kz}{y-w}$$
$$(y-w)x = (y-w)\frac{kz}{y-w}$$
$$xy - wx = kz$$
$$xy = kz + wx$$
$$\frac{xy}{x} = \frac{kz+wx}{x}$$
$$y = \frac{xw+kz}{x}$$

20.
$$x = \frac{kz}{y+w}$$
$$(y+w)x = (y+w)\frac{kz}{y+w}$$
$$yx + xw = kz$$
$$yx = kz - xw$$
$$\frac{yx}{x} = \frac{kz - xw}{x}$$
$$y = \frac{kz - xw}{x}$$

21. Since T varies directly as B, we have $T = kB$.
Use the given values to find k.
$$T = kB$$
$$3.6 = k(4)$$
$$\frac{3.6}{4} = \frac{k(4)}{4}$$
$$0.9 = k$$
The equation becomes $T = 0.9B$.
When $B = 6$, $T = 0.9(6) = 5.4$.
The tail length is 5.4 feet.

22.
$$M = kE$$
$$60 = k(360)$$
$$\frac{60}{360} = \frac{k(360)}{360}$$
$$\frac{1}{6} = k$$
$$M = \frac{1}{6}(186) = 31$$
A person who weighs 186 pounds on Earth will weigh 31 pounds on the moon.

23. Since B varies directly as D, we have $B = kD$.
Use the given values to find k.
$$B = kD$$
$$8.4 = k(12)$$
$$\frac{8.4}{12} = \frac{k(12)}{12}$$
$$k = \frac{8.4}{12} = 0.7$$
The equation becomes $B = 0.7D$.
When $B = 56$,
$$56 = 0.7D$$
$$\frac{56}{0.7} = \frac{0.7D}{0.7}$$
$$D = \frac{56}{0.7} = 80$$
It was dropped from 80 inches.

24.
$$d = kf$$
$$9 = k(12)$$
$$\frac{9}{12} = \frac{k(12)}{12}$$
$$0.75 = k$$
$$d = 0.75f$$
$$15 = 0.75f$$
$$\frac{15}{0.75} = \frac{0.75f}{0.75}$$
$$20 = f$$
A force of 20 pounds is needed.

25. Since a man's weight varies directly as the cube of his height, we have $w = kh^3$.
Use the given values to find k.
$$w = kh^3$$
$$170 = k(70)^3$$
$$170 = k(343{,}000)$$
$$\frac{170}{343{,}000} = \frac{k(343{,}000)}{343{,}000}$$
$$0.000496 = k$$
The equation becomes $w = 0.000496h^3$.
When $h = 107$,
$$w = 0.000496(107)^3$$
$$= 0.000496(1{,}225{,}043) \approx 607.$$
Robert Wadlow's weight was approximately 607 pounds.

26.
$$d = ks^2$$
$$67.5 = k(45)^2$$
$$67.5 = k(2025)$$
$$\frac{67.5}{2025} = \frac{k(2025)}{2025}$$
$$\frac{1}{30} = k$$
$$d = \frac{1}{30}s^2 = \frac{1}{30}(60)^2 = \frac{1}{30}(3600) = 120$$
The stopping distance is 120 feet.

27. Since the banking angle varies inversely as the turning radius, we have $B = \frac{k}{r}$.

Use the given values to find k.

$$B = \frac{k}{r}$$
$$28 = \frac{k}{4}$$
$$28(4) = 28\left(\frac{k}{4}\right)$$
$$112 = k$$

The equation becomes $B = \frac{112}{r}$.

When $r = 3.5$, $B = \frac{112}{r} = \frac{112}{3.5} = 32$.

The banking angle is 32° when the turning radius is 3.5 feet.

28.
$$t = \frac{k}{d}$$
$$4.4 = \frac{k}{1000}$$
$$(1000)4.4 = (1000)\frac{k}{1000}$$
$$4400 = k$$

$$t = \frac{4400}{d} = \frac{4400}{5000} = 0.88$$

The water temperature is 0.88° Celsius at a depth of 5000 meters.

29. Since intensity varies inversely as the square of the distance, we have pressure, we have $I = \frac{k}{d}$.

Use the given values to find k.

$$I = \frac{k}{d^2}.$$
$$62.5 = \frac{k}{3^2}$$
$$62.5 = \frac{k}{9}$$
$$9(62.5) = 9\left(\frac{k}{9}\right)$$
$$562.5 = k$$

The equation becomes $I = \frac{562.5}{d^2}$.

When $d = 2.5$, $I = \frac{562.5}{2.5^2} = \frac{562.5}{6.25} = 90$

The intensity is 90 milliroentgens per hour.

30.
$$i = \frac{k}{d^2}$$
$$3.75 = \frac{k}{40^2}$$
$$3.75 = \frac{k}{1600}$$
$$(1600)3.75 = (1600)\frac{k}{1600}$$
$$6000 = k$$

$$i = \frac{6000}{d^2} = \frac{6000}{50^2} = \frac{6000}{2500} = 2.4$$

The illumination is 2.4 foot-candles at a distance of 50 feet.

31. Since index varies directly as weight and inversely as the square of one's height, we have $I = \frac{kw}{h^2}$.

Use the given values to find k.

$$I = \frac{kw}{h^2}$$
$$35.15 = \frac{k(180)}{60^2}$$
$$35.15 = \frac{k(180)}{3600}$$
$$(3600)35.15 = \frac{k(180)}{3600}$$
$$126540 = k(180)$$
$$k = \frac{126540}{180} = 703$$

The equation becomes $I = \frac{703w}{h^2}$.

When $w = 170$ and $h = 70$,

$I = \frac{703(170)}{(70)^2} \approx 24.4.$

This person has a BMI of 24.4 and is not overweight.

32.
$$i = \frac{km}{c}$$
$$125 = \frac{k(25)}{20}$$
$$20(125) = (20)\frac{k(25)}{20}$$
$$2500 = 25k$$
$$\frac{2500}{25} = \frac{25k}{25}$$
$$100 = k$$

$$i = \frac{100m}{c}$$
$$80 = \frac{100(40)}{c}$$
$$80 = \frac{4000}{c}$$
$$80c = c \cdot \frac{4000}{c}$$
$$80c = 4000$$
$$\frac{80c}{80} = \frac{4000}{80}$$
$$c = 50$$

The chronological age is 50.

33. Since heat loss varies jointly as the area and temperature difference, we have $L = kAD$. Use the given values to find *k*.
$$L = kAD$$
$$1200 = k(3 \cdot 6)(20)$$
$$1200 = 360k$$
$$\frac{1200}{360} = \frac{360k}{360}$$
$$k = \frac{10}{3}$$

The equation becomes $L = \frac{10}{3}AD$

When $A = 6 \cdot 9 = 54$, $D = 10$,

$L = \frac{10}{3}(9 \cdot 6)(10) = 1800$.

The heat loss is 1800 Btu .

34.
$$e = kmv^2$$
$$36 = k(8)(3)^2$$
$$36 = k(8)(9)$$
$$36 = 72k$$
$$\frac{36}{72} = \frac{72k}{72}$$
$$k = 0.5$$

$$e = 0.5mv^2 = 0.5(4)(6)^2 = 0.5(4)(36) = 72$$

A mass of 4 grams and velocity of 6 centimeters per second has a kinetic energy of 72 ergs.

35. Since intensity varies inversely as the square of the distance from the sound source, we have $I = \frac{k}{d^2}$. If you move to a seat twice as far, then $d = 2d$. So we have $I = \frac{k}{(2d)^2} = \frac{k}{4d^2} = \frac{1}{4} \cdot \frac{k}{d^2}$. The intensity will be multiplied by a factor of $\frac{1}{4}$. So the sound intensity is $\frac{1}{4}$ of what it was originally.

36.
$$t = \frac{k}{a}$$
$$t = \frac{k}{3a} = \frac{1}{3} \cdot \frac{k}{a}$$

A year will seem to be $\frac{1}{3}$ of a year.

37. **a.** Since the average number of phone calls varies jointly as the product of the populations and inversely as the square of the distance, we have $C = \frac{kP_1P_2}{d^2}$.

b. Use the given values to find *k*.
$$C = \frac{kP_1P_2}{d^2}$$
$$326,000 = \frac{k(777,000)(3,695,000)}{(420)^2}$$
$$326,000 = \frac{k\left(2.87 \times 10^{12}\right)}{176,400}$$
$$326,000 = 16269841.27k$$
$$0.02 \approx k$$

The equation becomes $C = \frac{0.02P_1P_2}{d^2}$.

c.
$$C = \frac{0.02(650,000)(490,000)}{(400)^2}$$
$$\approx 39813$$

The average number of calls is approximately 39,813 daily phone calls.

38.

$$f = kas^2$$
$$150 = k(4 \cdot 5)(30)^2$$
$$150 = k(20)(900)$$
$$150 = 18000k$$
$$\frac{150}{18000} = \frac{18000k}{150}$$
$$\frac{1}{120} = k$$

$$f = \frac{1}{120}as^2 = \frac{1}{120}(3 \cdot 4)(60)^2$$
$$= \frac{1}{120}(12)(3600)$$
$$= 360$$

Yes, the wind will exert a force of 360 pounds on the window.

39. **a.**

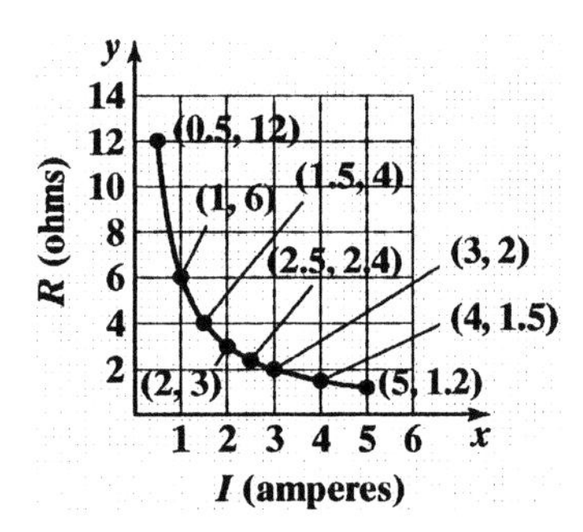

b. Current varies inversely as resistance. Answers will vary.

c. Since the current varies inversely as resistance we have $R = \frac{k}{I}$. Using one of the given ordered pairs to find k.

$$12 = \frac{k}{0.5}$$
$$12(0.5) = \frac{k}{0.5}(0.5)$$
$$k = 6$$

The equation becomes $R = \frac{6}{I}$.

45. z varies directly as the square root of x and inversely as the square of y.

46. z varies jointly as the square of x and the square root of y.

49. Pressure, P, varies directly as the square of wind velocity, v, can be modeled as $P = kv^2$.
If $v = x$ then $P = k(x)^2 = kx^2$
If $v = 2x$ then $P = k(2x)^2 = 4kx^2$
If the wind speed doubles the pressure is 4 times more destructive.

50. Illumination, I, varies inversely as the square of the distance, d, can be modeled as $I = \frac{k}{d^2}$.

If $d = 15$ then $I = \frac{k}{15^2} = \frac{k}{225}$

If $d = 30$ then $I = \frac{k}{30^2} = \frac{k}{900}$

Note that $\frac{900}{225} = 4$

If the distance doubles the illumination is 4 times less intense.

51. The Heat, H, varies directly as the square of the voltage, v, and inversely as the resistance, r.

$$H = \frac{kv^2}{r}$$

If the voltage remains constant, to triple the heat the resistant must be reduced by a multiple of 3.

52. Illumination, I, varies inversely as the square of the distance, d, can be modeled as $I = \frac{k}{d^2}$.

If $I = x$ then $x = \frac{k}{d^2} \Rightarrow d = \sqrt{\frac{k}{x}}$.

If $I = \frac{1}{50}x$ then

$$\frac{1}{50}x = \frac{k}{d^2} \Rightarrow d = \sqrt{\frac{50k}{x}} = \sqrt{50}\sqrt{\frac{k}{x}}.$$

Since $\sqrt{50} \approx 7$, the Hubble telescope is able to see about 7 times farther than a ground-based telescope.

Chapter 2 Review Exercises

1. $(8-3i)-(17-7i) = 8-3i-17+7i$
$= -9+4i$

2. $4i(3i-2) = (4i)(3i)+(4i)(-2)$
$= 12i^2 - 8i$
$= -12-8i$

3. $(7-i)(2+3i)$
$= 7\cdot 2 + 7(3i) + (-i)(2) + (-i)(3i)$
$= 14+21i-2i+3$
$= 17+19i$

4. $(3-4i)^2 = 3^2 + 2\cdot 3(-4i) + (-4i)^2$
$= 9-24i-16$
$= -7-24i$

5. $(7+8i)(7-8i) = 7^2+8^2 = 49+64 = 113$

6. $\frac{6}{5+i} = \frac{6}{5+i}\cdot\frac{5-i}{5-i}$
$= \frac{30-6i}{25+1}$
$= \frac{30-6i}{26}$
$= \frac{15-3i}{13}$
$= \frac{15}{13} - \frac{3}{13}i$

7. $\frac{3+4i}{4-2i} = \frac{3+4i}{4-2i}\cdot\frac{4+2i}{4+2i}$
$= \frac{12+6i+16i+8i^2}{16-4i^2}$
$= \frac{12+22i-8}{16+4}$
$= \frac{4+22i}{20}$
$= \frac{1}{5} + \frac{11}{10}i$

8. $\sqrt{-32} - \sqrt{-18} = i\sqrt{32} - i\sqrt{18}$
$= i\sqrt{16\cdot 2} - i\sqrt{9\cdot 2}$
$= 4i\sqrt{2} - 3i\sqrt{2}$
$= (4i-3i)\sqrt{2}$
$= i\sqrt{2}$

9. $(-2+\sqrt{-100})^2 = (-2+i\sqrt{100})^2$
$= (-2+10i)^2$
$= 4-40i+(10i)^2$
$= 4-40i-100$
$= -96-40i$

10. $\frac{4+\sqrt{-8}}{2} = \frac{4+i\sqrt{8}}{2} = \frac{4+2i\sqrt{2}}{2} = 2+i\sqrt{2}$

11. $x^2 - 2x + 4 = 0$
$x = \frac{-(-2) \pm \sqrt{(-2)^2 - 4(1)(4)}}{2(1)}$
$x = \frac{2 \pm \sqrt{4-16}}{2}$
$x = \frac{2 \pm \sqrt{-12}}{2}$
$x = \frac{2 \pm 2i\sqrt{3}}{2}$
$x = 1 \pm i\sqrt{3}$
The solution set is $\{1-i\sqrt{3},\ 1+i\sqrt{3}\}$

12. $2x^2 - 6x + 5 = 0$
$x = \frac{-(-6) \pm \sqrt{(-6)^2 - 4(2)(5)}}{2(2)}$
$x = \frac{6 \pm \sqrt{36-40}}{4}$
$x = \frac{6 \pm \sqrt{-4}}{4}$
$x = \frac{6 \pm 2i}{4}$
$x = \frac{6}{4} \pm \frac{2i}{4}$
$= \frac{3}{2} \pm \frac{1}{2}i$
The solution set is $\left\{\frac{3}{2} - \frac{1}{2}i,\ \frac{3}{2} + \frac{1}{2}i\right\}$.

13. $f(x) = -(x+1)^2 + 4$
vertex: $(-1, 4)$
x-intercepts:
$0 = -(x+1)^2 + 4$
$(x+1)^2 = 4$
$x+1 = \pm 2$
$x = -1 \pm 2$

$x = -3$ or $x = 1$
y-intercept:
$f(0) = -(0+1)^2 + 4 = 3$
The axis of symmetry is $x = -1$.

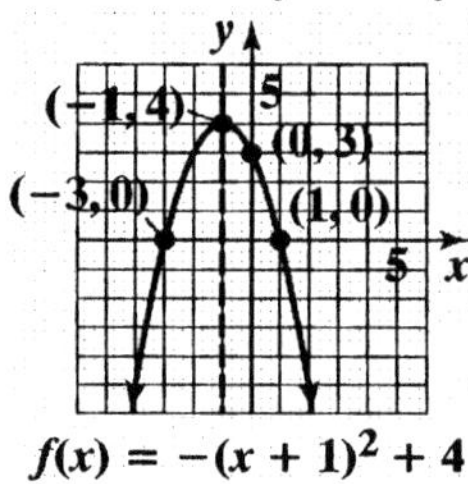

$f(x) = -(x+1)^2 + 4$

Domain: $(-\infty, \infty)$ Range: $(-\infty, 4]$

14. $f(x) = (x+4)^2 - 2$
vertex: $(-4, -2)$
x-intercepts:

$$0 = (x+4)^2 - 2$$
$$(x+4)^2 = 2$$
$$x + 4 = \pm\sqrt{2}$$
$$x = -4 \pm \sqrt{2}$$

y-intercept:
$f(0) = (0+4)^2 - 2 = 14 = -1$

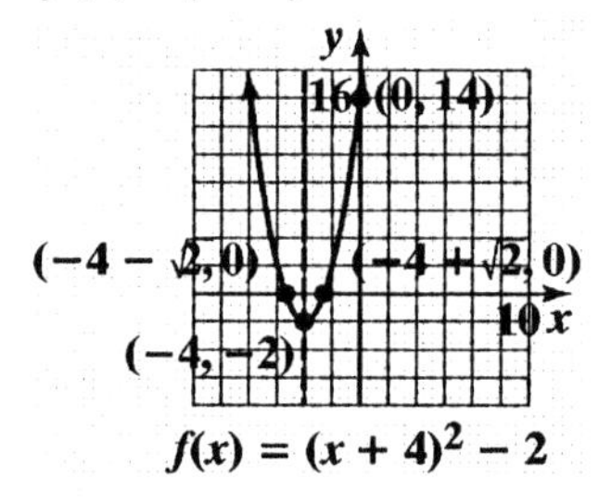

$f(x) = (x+4)^2 - 2$

The axis of symmetry is $x = -4$.
Domain: $(-\infty, \infty)$ Range: $[-2, \infty)$

15. $f(x) = -x^2 + 2x + 3$
$= -(x^2 - 2x + 1) + 3 + 1$
$f(x) = -(x-1)^2 + 4$

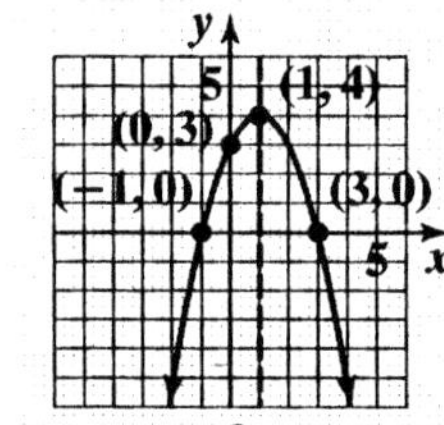

$f(x) = -x^2 + 2x + 3$

Domain: $(-\infty, \infty)$ Range: $(-\infty, 4]$

16. $f(x) = 2x^2 - 4x - 6$
$f(x) = 2(x^2 - 2x + 1) - 6 - 2$
$2(x-1)^2 - 8$

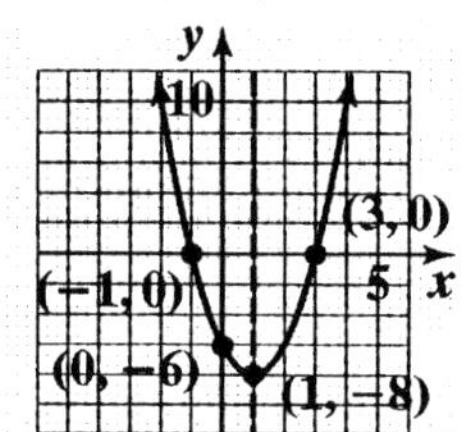

$f(x) = 2x^2 - 4x - 6$
axis of symmetry: $x = 1$
Domain: $(-\infty, \infty)$ Range: $[-8, \infty)$

17. $f(x) = -x^2 + 14x - 106$

a. Since $a < 0$ the parabola opens down with the maximum value occurring at
$$x = -\frac{b}{2a} = -\frac{14}{2(-1)} = 7.$$
The maximum value is $f(7)$.
$$f(7) = -(7)^2 + 14(7) - 106 = -57$$

b. Domain: $(-\infty, \infty)$ Range: $(-\infty, -57]$

18. $f(x) = 2x^2 + 12x + 703$

a. Since $a > 0$ the parabola opens up with the minimum value occurring at
$$x = -\frac{b}{2a} = -\frac{12}{2(2)} = -3.$$
The minimum value is $f(-3)$.
$$f(-3) = 2(-3)^2 + 12(-3) + 703 = 685$$

b. Domain: $(-\infty, \infty)$ Range: $[685, \infty)$

19. $f(x) = -0.02x^2 + x + 1$
Since $a = -0.02$ is negative, we know the function opens downward and has a maximum at $x = -\frac{b}{2a} = -\frac{1}{2(-0.02)} = -\frac{1}{-0.04} = 25$. When 25 inches of rain falls, the maximum growth will occur. The maximum growth is
$$f(25) = -0.02(25)^2 + 25 + 1$$
$$= -0.02(625) + 25 + 1$$
$$= -12.5 + 25 + 1 = 13.5.$$
A maximum yearly growth of 13.5 inches occurs when 25 inches of rain falls per year.

20. According to the graph, the vertex (maximum point) is (25, 5). This means that the maximum divorce rate of 5 divorces per 1000 people occurred in the year 1960 + 25 = 1985.

21. 19

Maximize the area using $A = lw$.

$A(x) = x(1000 - 2x)$

$A(x) = -2x^2 + 1000x$

Since $a = -2$ is negative, we know the function opens downward and has a maximum at

$x = -\frac{b}{2a} = -\frac{1000}{2(-2)} = -\frac{1000}{-4} = 250.$

The maximum area is achieved when the width is 250 yards. The maximum area is

$$A(250) = 250(1000 - 2(250))$$
$$= 250(1000 - 500)$$
$$= 250(500) = 125,000.$$

The area is maximized at 125,000 square yards when the width is 250 yards and the length is $1000 - 2 \cdot 250 = 500$ yards.

22. Let x = one of the numbers

Let $14 + x$ = the other number

We need to minimize the function

$$P(x) = x(14 + x)$$
$$= 14x + x^2$$
$$= x^2 + 14x.$$

The minimum is at

$x = -\frac{b}{2a} = -\frac{14}{2(1)} = -\frac{14}{2} = -7.$

The other number is $14 + x = 14 + (-7) = 7$.

The numbers which minimize the product are 7 and -7. The minimum product is $-7 \cdot 7 = -49$.

23. $3x + 4y = 1000$

$4y = 1000 - 3x$

$y = \frac{1000 - 3x}{4}$

$$A = x\left(\frac{1000 - 3x}{4}\right)$$
$$= -\frac{3}{4}x^2 + 250x$$

$x = \frac{-b}{2a} = \frac{-250}{2\left(-\frac{3}{4}\right)} = 125$

$y = \frac{1000 - 3(125)}{4} = 166.7$

125 feet by 166.7 feet will maximize the area.

24. $y = (35 + x)(150 - 4x)$

$y = 5250 + 10x - 4x^2$

$x = \frac{-b}{2a} = \frac{-10}{2(-4)} = \frac{5}{4} = 1.25$ or 1 tree

The maximum number of trees should be 35 + 1 = 36 trees.

maximum number of trees should be 35 + 1 = 36 trees.

$y = 36(150 - 4x) = 36(150 - 4 \cdot 1) = 5256$

The maximum yield will be 5256 pounds.

25. $f(x) = -x^3 + 12x^2 - x$

The graph rises to the left and falls to the right and goes through the origin, so graph (c) is the best match.

26. $g(x) = x^6 - 6x^4 + 9x^2$

The graph rises to the left and rises to the right, so graph (b) is the best match.

27. $h(x) = x^5 - 5x^3 + 4x$

The graph falls to the left and rises to the right and crosses the y-axis at zero, so graph (a) is the best match.

28. $f(x) = -x^4 + 1$

$f(x)$ falls to the left and to the right so graph (d) is the best match.

29. The leading coefficient is –0.87 and the degree is 3. This means that the graph will fall to the right. This function is not useful in modeling the number of thefts over an extended period of time. The model predicts that eventually, the number of thefts would be negative. This is impossible.

30. In the polynomial, $f(x) = -x^4 + 21x^2 + 100$, the leading coefficient is –1 and the degree is 4. Applying the Leading Coefficient Test, we know that even-degree polynomials with negative leading coefficient will fall to the left and to the right. Since the graph falls to the right, we know that the elk population will die out over time.

31. $f(x) = -2(x-1)(x+2)^2(x+5)^3$
$x = 1$, multiplicity 1, the graph crosses the x-axis
$x = -2$, multiplicity 2, the graph touches the x-axis
$x = -5$, multiplicity 5, the graph crosses the x-axis

32. $f(x) = x^3 - 5x^2 - 25x + 125$

$$= x^2(x-5) - 25(x-5)$$
$$= (x^2 - 25)(x-5)$$
$$= (x+5)(x-5)^2$$

$x = -5$, multiplicity 1, the graph crosses the x-axis
$x = 5$, multiplicity 2, the graph touches the x-axis

33. $f(x) = x^3 - 2x - 1$
$f(1) = (1)^3 - 2(1) - 1 = -2$
$f(2) = (2)^3 - 2(2) - 1 = 3$
The sign change shows there is a zero between the given values.

34. $f(x) = x^3 - x^2 - 9x + 9$

a. Since n is odd and $a_n > 0$, the graph falls to the left and rises to the right.

b. $f(-x) = (-x)^3 - (-x)^2 - 9(-x) + 9$
$= -x^3 - x^2 + 9x + 9$
$f(-x) \neq f(x), f(-x) \neq -f(x)$
no symmetry

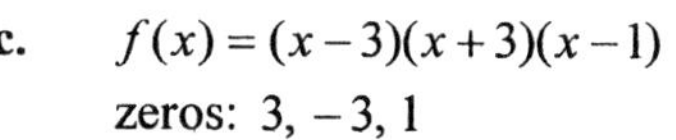

c. $f(x) = (x-3)(x+3)(x-1)$
zeros: 3, –3, 1

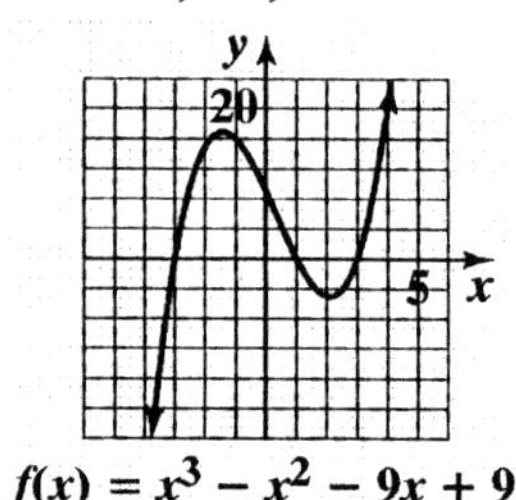

$f(x) = x^3 - x^2 - 9x + 9$

35. $f(x) = 4x - x^3$

a. Since n is odd and $a_n < 0$, the graph rises to the left and falls to the right.

b. $f(-x) = -4x + x^3$
$f(-x) = -f(x)$
origin symmetry

c. $f(x) = x(x^2 - 4) = x(x-2)(x+2)$
zeros: $x = 0, 2, -2$

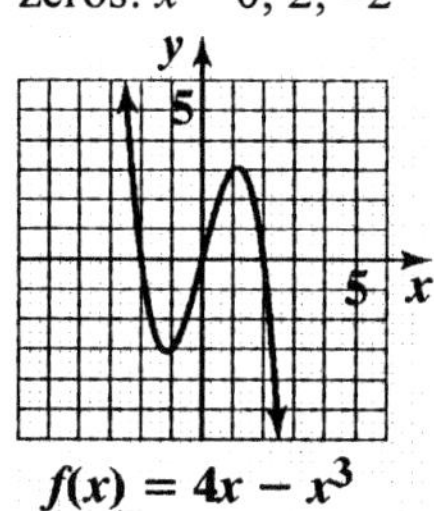

$f(x) = 4x - x^3$

36. $f(x) = 2x^3 + 3x^2 - 8x - 12$

a. Since h is odd and $a_n > 0$, the graph falls to the left and rises to the right.

b. $f(-x) = -2x^3 + 3x^2 + 8x - 12$
$f(-x) \neq f(x),\ f(-x) = -f(x)$
no symmetry

c. $f(x) = (x-2)(x+2)(2x+3)$
zeros: $x = 2, -2, -\frac{3}{2}$

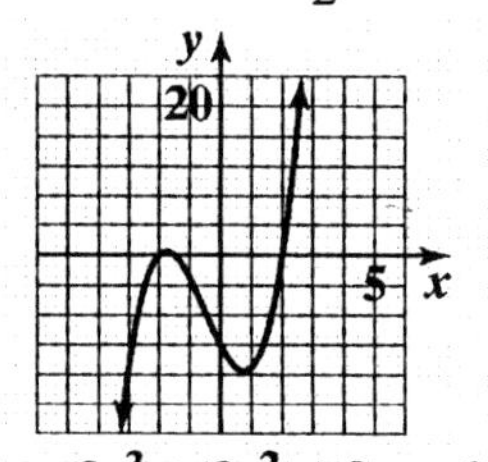

$f(x) = 2x^3 + 3x^2 - 8x - 12$

37. $g(x) = -x^4 + 25x^2$

a. The graph falls to the left and to the right.

b. $f(-x) = -(-x)^4 + 25(-x)^2$
$= -x^4 + 25x^2 = f(x)$
y-axis symmetry

c. $-x^4 + 25x^2 = 0$
$-x^2(x^2 - 25) = 0$
$-x^2(x-5)(x+5) = 0$
zeros: $x = -5, 0, 5$

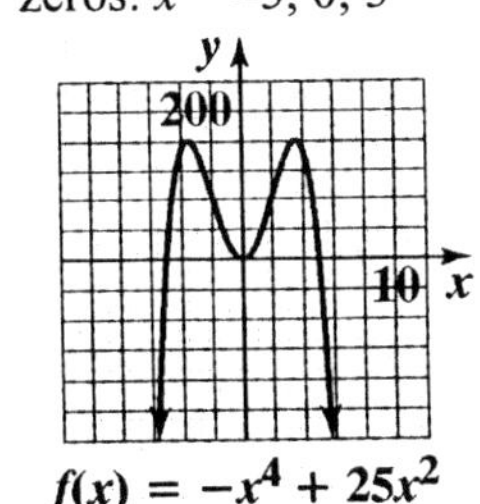

$f(x) = -x^4 + 25x^2$

38. $f(x) = -x^4 + 6x^3 - 9x^2$

a. The graph falls to the left and to the right.

b. $f(-x) = -(-x)^4 + 6(-x)^3 - 9(-x)$
$= -x^4 - 6x^3 - 9x^2 f(-x) \neq f(x)$
$f(-x) \neq -f(x)$
no symmetry

c. $= -x^2(x^2 - 6x + 9) = 0$
$-x^2(x-3)(x-3) = 0$
zeros: $x = 0, 3$

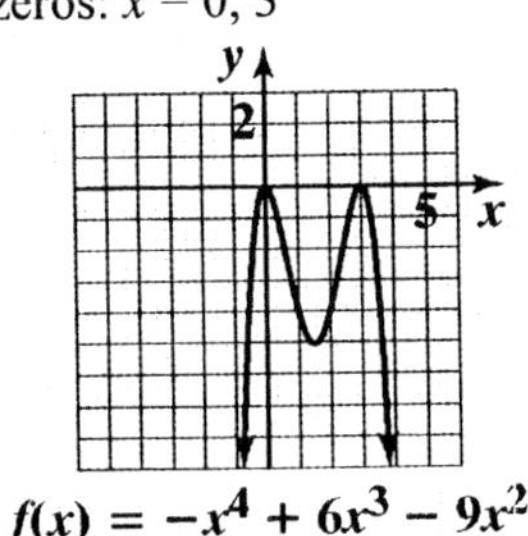

$f(x) = -x^4 + 6x^3 - 9x^2$

39. $f(x) = 3x^4 - 15x^3$

a. The graph rises to the left and to the right.

b. $f(-x) = 3(-x)^4 - 15(-x)^2 = 3x^4 + 15x^3$
$f(-x) \neq f(x),\ f(-x) \neq -f(x)$
no symmetry

c $3x^4 - 15x^3 = 0$
$3x^3(x-5) = 0$
zeros: $x = 0, 5$

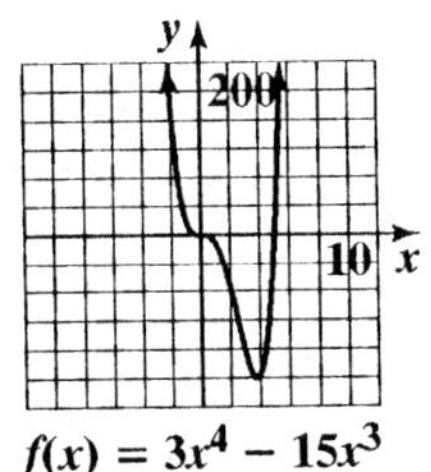

$f(x) = 3x^4 - 15x^3$

40. $f(x) = 2x^2(x-1)^3(x+2)$
Since $a_n > 0$ and n is even, $f(x)$ rises to the left and the right.
$x = 0, x = 1, x = -2$
The zeros at 1 and –2 have odd multiplicity so $f(x)$ crosses the x-axis at those points. The root at 0 has even multiplicity so $f(x)$ touches the axis at (0, 0)
$f(0) = 2(0)^2(0-1)^3(0+2) = 0$
The y-intercept is 0.

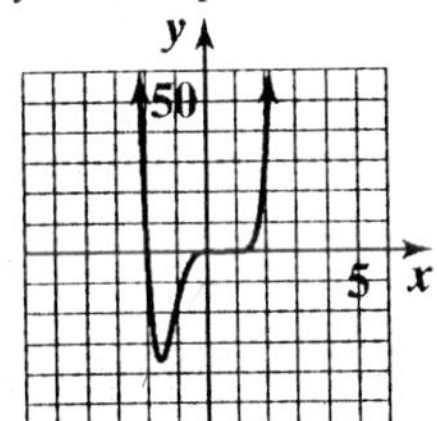

$f(x) = 2x^2 (x - 1)^3 (x + 2)$

41. $f(x) = -x^3(x+4)^2(x-1)$
Since $a_n < 0$ and n is even, $f(x)$ falls to the left and the right.
$x = 0, x = -4, x = 1$
The roots at 0 and 1 have odd multiplicity so $f(x)$ crosses the x-axis at those points. The root at –4 has even multiplicity so $f(x)$ touches the axis at (–4, 0)
$f(0) = -(0)^3(0+4)^2(0-1) = 0$

The y-intercept is 0.

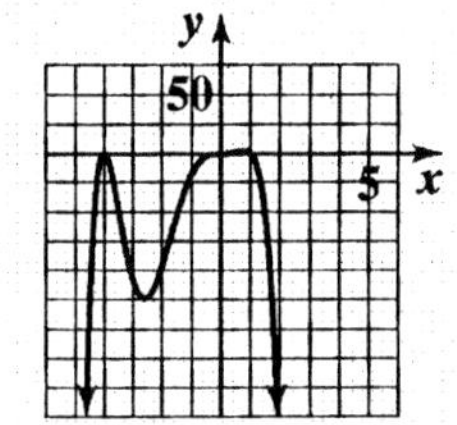

$f(x) = -x^3 (x + 4)^2 (x - 1)$

42.

$$\begin{array}{r} 4x^2 - 7x + 5 \\ x+1\overline{)4x^3 - 3x^2 - 2x + 1} \\ \underline{4x^3 + 4x^2} \\ -7x^2 - 2x \\ \underline{-7x^2 - 7x} \\ 5x + 1 \\ \underline{5x + 5} \\ -4 \end{array}$$

Quotient: $4x^2 - 7x + 5 - \dfrac{4}{x+1}$

43.

$$\begin{array}{r} 2x^2 - 4x + 1 \\ 5x-3\overline{)10x^3 - 26x^2 + 17x - 13} \\ \underline{10x^3 + 6x^2} \\ -20x^2 + 17x \\ \underline{-20x^2 + 12x} \\ 5x - 13 \\ \underline{5x - 3} \\ -10 \end{array}$$

Quotient: $2x^2 - 4x + 1 - \dfrac{10}{5x-3}$

44.

$$\begin{array}{r} 2x^2 + 3x - 1 \\ 2x^2+1\overline{)4x^4 + 6x^3 + 3x - 1} \\ \underline{4x^2 + 2x^2} \\ 6x^3 - 2x^2 + 3x \\ \underline{6x^2 + 3x} \\ -2x^2 - 1 \\ \underline{-2x^2 - 1} \\ 0 \end{array}$$

45. $(3x^4 + 11x^3 - 20x^3 + 7x + 35) \div (x + 5)$

−5	3	11	−20	7	35
		−15	20	0	−35
	3	−4	0	7	0

Quotient: $3x^3 - 4x^2 + 7$

46. $(3x^4 - 2x^2 - 10x) \div (x - 2)$

2	3	0	−2	−10	0
		6	12	20	20
	3	6	10	10	20

Quotient: $3x^3 + 6x^2 + 10x + 10 + \dfrac{20}{x-2}$

47. $f(x) = 2x^3 - 7x^2 + 9x - 3$

−13	2	−7	9	−3
		−26	429	−5694
	2	−33	438	−5697

Quotient: $f(-13) = -5697$

48. $f(x) = 2x^3 + x^2 - 13x + 6$

2	2	1	−13	6
		4	10	−6
	2	5	−3	0

$f(x) = (x-2)(2x^2 + 5x - 3)$
$= (x-2)(2x-1)(x+3)$

Zeros: $x = 2, \frac{1}{2}, -3$

49. $x^3 - 17x + 4 = 0$

4	1	0	−17	4
		4	16	−4
	1	4	−1	0

$(x-4)(x^2 + 4x - 1) = 0$

$x = \dfrac{-4 \pm \sqrt{16+4}}{2} = \dfrac{-4 \pm 2\sqrt{5}}{2} = -2 \pm \sqrt{5}$

The solution set is $\{4, -2+\sqrt{5}, -2-\sqrt{5}\}$.

50. $f(x) = x^4 - 6x^3 + 14x^2 - 14x + 5$

$p: \pm 1, \pm 5$

$q: \pm 1$

$\frac{p}{q}: \pm 1, \pm 5$

51. $f(x) = 3x^5 - 2x^4 - 15x^3 + 10x^2 + 12x - 8$

$p: \pm 1, \pm 2, \pm 4, \pm 8$

$q: \pm 1, \pm 3$

$\frac{p}{q}: \pm 1, \pm 2, \pm 4, \pm 8, \pm \frac{8}{3}, \pm \frac{4}{3}, \pm \frac{2}{3}, \pm \frac{1}{3}$

52. $f(x) = 3x^4 - 2x^3 - 8x + 5$

$f(x)$ has 2 sign variations, so $f(x) = 0$ has 2 or 0 positive solutions.

$f(-x) = 3x^4 + 2x^3 + x + 5$

$f(-x)$ has no sign variations, so $f(x) = 0$ has no negative solutions.

53. $f(x) = 2x^5 - 3x^3 - 5x^2 + 3x - 1$

$f(x)$ has 3 sign variations, so $f(x) = 0$ has 3 or 1 positive real roots.

$f(-x) = -2x^5 + 3x^3 - 5x^2 - 3x - 1$

$f(-x)$ has 2 sign variations, so $f(x) = 0$ has 2 or 0 negative solutions.

54. $f(x) = f(-x) = 2x^4 + 6x^2 + 8$

No sign variations exist for either $f(x)$ or $f(-x)$, so no real roots exist.

55. $f(x) = x^3 + 3x^2 - 4$

a. $p: \pm 1, \pm 2, \pm 4$

$q: \pm 1$

$\frac{p}{q}: \pm 1, \pm 2, \pm 4$

b. 1 sign variation $\Rightarrow$ 1 positive real zero

$f(-x) = -x^3 + 3x^2 - 4$

2 sign variations $\Rightarrow$ 2 or no negative real zeros

c.

1	1	3	0	–4
		1	4	–4
	1	4	4	0

1 is a zero.

d. $(x-1)(x^2 + 4x + 4) = 0$

$(x-1)(x+2)^2 = 0$

$x = 1$ or $x = -2$

The solution set is $\{1, -2\}$.

56. $f(x) = 6x^3 + x^2 - 4x + 1$

a. $p: \pm 1$

$q: \pm 1, \pm 2, \pm 3, \pm 6$

$\frac{p}{q}: \pm 1, \pm \frac{1}{2}, \pm \frac{1}{3}, \pm \frac{1}{6}$

b. $f(x) = 6x^3 + x^2 - 4x + 1$

2 sign variations; 2 or 0 positive real zeros.

$f(-x) = -6x^3 + x^2 + 4x + 1$

1 sign variation; 1 negative real zero.

c.

–1	6	1	–4	1
		–6	5	–1
	6	–5	1	0

–1 is a zero.

d. $6x^3 + x^2 - 4x + 1 = 0$

$(x+1)(6x^2 - 5x + 1) = 0$

$(x+1)(3x-1)(2x-1) = 0$

$x = -1$ or $x = \frac{1}{3}$ or $x = \frac{1}{2}$

The solution set is $\left\{-1, \frac{1}{3}, \frac{1}{2}\right\}$.

57. $f(x) = 8x^3 - 36x^2 + 46x - 15$

a. $p: \pm 1, \pm 3, \pm 5, \pm 15$

$q: \pm 1, \pm 2, \pm 4, \pm 8$

$\frac{p}{q}: \pm 1, \pm 3, \pm 5, \pm 15, \pm \frac{1}{2}, \pm \frac{1}{4}, \pm \frac{1}{8},$

$\pm \frac{3}{2}, \pm \frac{3}{4}, \pm \frac{3}{8}, \pm \frac{5}{2}, \pm \frac{5}{4},$

$\pm \frac{5}{8}, \pm \frac{15}{2}, \pm \frac{15}{4}, \pm \frac{15}{8}$

b. $f(x) = 8x^3 - 36x^2 + 46x - 15$

3 sign variations; 3 or 1 positive real solutions.

$f(-x) = -8x^3 - 36x^2 - 46x - 15$

0 sign variations; no negative real solutions.

c.

$\frac{1}{2}$	8	–36	46	–15
		4	–16	15
	8	–32	30	0

$\frac{1}{2}$ is a zero.

d.

$$8x^3-36x^2+46x-15=0$$
$$\left(x-\frac{1}{2}\right)(8x^2-32x+30)=0$$
$$2\left(x-\frac{1}{2}\right)(4x-16x+15)=0$$
$$2\left(x-\frac{1}{2}\right)(2x-5)(2x-3)=0$$
$$x=\frac{1}{2}\text{ or }x=\frac{5}{2}\text{ or }x=\frac{3}{2}$$

The solution set is $\left\{\frac{1}{2},\frac{3}{2},\frac{5}{2}\right\}$.

58. $2x^3+9x^2-7x+1=0$

a. p: ±1
q: $\pm1, \pm2$
$\frac{p}{q}: \pm1, \pm\frac{1}{2}$

b. $f(x)=2x^3+9x^2-7x+1$
2 sign variations; 2 or 0 positive real zeros.
$f(-x)=-2x^3+9x^2+7x+1$
1 sign variation; 1 negative real zero.

c.

$\frac{1}{2}$	2	9	–7	1
		1	5	–1
	2	10	–2	0

$-\frac{1}{2}$ is a zero.

d.

$$2x^3+9x^2-7x+1=0$$
$$\left(x-\frac{1}{2}\right)(2x^2+10x-2)=0$$
$$2\left(x-\frac{1}{2}\right)(x^2+5x-1)=0$$

Solving $x^2+5x-1=0$ using the quadratic formula gives $x=\frac{-5\pm\sqrt{29}}{2}$

The solution set is $\left\{\frac{1}{2},\frac{-5+\sqrt{29}}{2},\frac{-5-\sqrt{29}}{2}\right\}$.

59. $x^4-x^3-7x^2+x+6=0$

a. $p=\frac{p}{q}: \pm1, \pm2, \pm3, \pm6$

b. $f(x)=x^4-x^3-7x^2+x+6$
2 sign variations; 2 or 0 positive real zeros.
$f(-x)=x^4+x^3-7x^2-x+6$
2 sign variations; 2 or 0 negative real zeros.

c.

1	1	–1	–7	1	6
		1	0	–7	–6
	1	0	–7	–6	0

–1	1	0	–7	–6
		–1	1	6
	1	–1	–6	0

d.

$$x^4-x^3-7x^2+x+6=0$$
$$(x-1)(x+1)(x^2-x+6)=0$$
$$(x-1)(x+1)(x-3)(x+2)=0$$

The solution set is $\{-2,-1,1,3\}$.

60. $4x^4+7x^2-2=0$

a. p: $\pm1, \pm2$
q: $\pm1, \pm2, \pm4$
$\frac{p}{q}: \pm1, \pm2, \pm\frac{1}{2}, \pm\frac{1}{4}$

b. $f(x) = 4x^4 + 7x^2 - 2$
1 sign variation; 1 positive real zero.
$f(-x) = 4x^4 + 7x^2 - 2$
1 sign variation; 1 negative real zero.

c.

$\frac{1}{2}$	4	0	7	0	−2
		2	1	4	2
	4	2	8	4	0

$-\frac{1}{2}$	4	2	8	4
		−2	0	−4
	4	0	8	0

d.
$$4x^4 + 7x^2 - 2 = 0$$
$$\left(x - \frac{1}{2}\right)\left(x + \frac{1}{2}\right)(4x^2 + 8) = 0$$
$$4\left(x - \frac{1}{2}\right)\left(x + \frac{1}{2}\right)(x^2 + 2) = 0$$
Solving $x^2 + 2 = 0$ using the quadratic formula gives $x = \pm 2i$

The solution set is $\left\{ -\frac{1}{2}, \frac{1}{2}, 2i, -2i \right\}$.

61. $f(x) = 2x^4 + x^3 - 9x^2 - 4x + 4$

a. p: ±1, ±2, ±4
q: ±1, ±2
$\frac{p}{q}: \pm 1, \pm 2, \pm 4, \pm \frac{1}{2}$

b. $f(x) = 2x^4 + x^3 - 9x^2 - 4x + 4$
2 sign variations; 2 or 0 positive real zeros.
$f(-x) = 2x^4 - x^3 - 9x^2 + 4x + 4$
2 sign variations; 2 or 0 negative real zeros.

c.

2	2	1	−9	−4	4
		4	10	2	−4
	2	5	1	−2	0

−1	2	5	1	−2
		−2	−3	2
	2	3	−2	0

d.
$$2x^2 + 3x - 2 = 0$$
$$(2x - 1)(x + 2) = 0$$
$$x = -2 \text{ or } x = \frac{1}{2}$$
The solution set is $\left\{ -2, -1, \frac{1}{2}, 2 \right\}$.

62. $f(x) = a_n(x - 2)(x - 2 + 3i)(x - 2 - 3i)$
$f(x) = a_n(x - 2)(x^2 - 4x + 13)$
$f(1) = a_n(1 - 2)\left[1^2 - 4(1) + 13\right]$
$-10 = -10a_n$
$a_n = 1$
$f(x) = 1(x - 2)(x^2 - 4x + 13)$
$f(x) = x^3 - 4x^2 + 13x - 2x^2 + 8x - 26$
$f(x) = x^3 - 6x^2 + 21x - 26$

63. $f(x) = a_n(x - i)(x + i)(x + 3)^2$
$f(x) = a_n(x^2 + 1)(x^2 + 6x + 9)$
$f(-1) = a_n\left[(-1)^2 + 1\right]\left[(-1)^2 + 6(-1) + 9\right]$
$16 = 8a_n$
$a_n = 2$
$f(x) = 2(x^2 + 1)(x^2 + 6x + 9)$
$f(x) = 2(x^4 + 6x^3 + 9x^2 + x^2 + 6x + 9)$
$f(x) = 2x^4 + 12x^3 + 20x^2 + 12x + 18$

64. $f(x) = 2x^4 + 3x^3 + 3x - 2$
p: ±1, ±2
q: ±1, ±2
$\frac{p}{q}: \pm 1, \pm 2, \pm \frac{1}{2}$

−2	2	3	0	3	−2
		−4	2	−4	2
	2	−1	2	−1	0

$$2x^4 + 3x^3 + 3x - 2 = 0$$
$$(x+2)(2x^3 - x^2 + 2x - 1) = 0$$
$$(x+2)[x^2(2x-1) + (2x-1)] = 0$$
$$(x+2)(2x-1)(x^2+1) = 0$$
$$x = -2,\ x = \frac{1}{2} \text{ or } x = \pm i$$

The zeros are $-2,\ \frac{1}{2},\ \pm i$.

$$f(x) = (x-i)(x+i)(x+2)(2x-1)$$

65. $g(x) = x^4 - 6x^3 + x^2 + 24x + 16$

$p: \pm 1, \pm 2, \pm 4, \pm 8, \pm 16$
$q: \pm 1$

$\frac{p}{q}: \pm 1, \pm 2, \pm 4, \pm 8, \pm 16$

−1	1	−6	1	24	16
		−1	7	−8	−16
	1	−7	8	16	0

$$x^4 - 6x^3 + x^2 + 24x + 16 = 0$$
$$(x+1)(x^3 - 7x^2 + 8x + 16) = 0$$

−1	1	−7	8	16
		−1	8	−16
	1	−8	16	0

$$(x+1)^2(x^2 - 8x + 16) = 0$$
$$(x+1)^2(x-4)^2 = 0$$
$$x = -1 \text{ or } x = 4$$
$$g(x) = (x+1)^2(x-4)^2$$

66. 4 real zeros, one with multiplicity two

67. 3 real zeros; 2 nonreal complex zeros

68. 2 real zeros, one with multiplicity two; 2 nonreal complex zeros

69. 1 real zero; 4 nonreal complex zeros

70. $g(x) = \frac{1}{(x+2)^2} - 1$

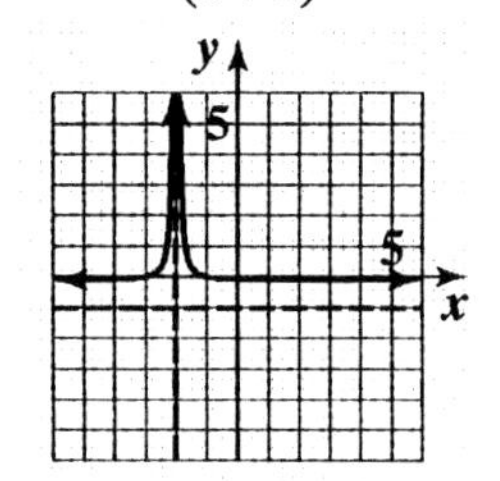

$$g(x) = \frac{1}{(x+2)^2} - 1$$

71. $h(x) = \frac{1}{x-1} + 3$

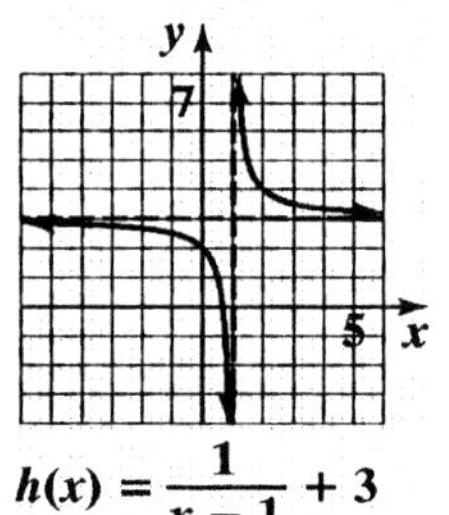

$$h(x) = \frac{1}{x-1} + 3$$

72. $f(x) = \frac{2x}{x^2 - 9}$

Symmetry: $f(-x) = -\frac{2x}{x^2 - 9} = -f(x)$

origin symmetry

x-intercept:

$$0 = \frac{2x}{x^2 - 9}$$
$$2x = 0$$
$$x = 0$$

y-intercept: $y = \frac{2(0)}{0^2 - 9} = 0$

Vertical asymptote:

$$x^2 - 9 = 0$$
$$(x-3)(x+3) = 0$$
$$x = 3 \text{ and } x = -3$$

Horizontal asymptote:
$n < m$, so $y = 0$

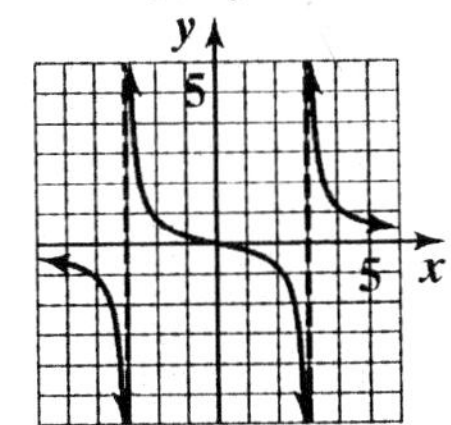

$f(x) = \dfrac{2x}{x^2 - 9}$

73. $g(x) = \dfrac{2x-4}{x+3}$

Symmetry: $g(-x) = \dfrac{-2x-4}{x+3}$
$g(-x) \neq g(x)$, $g(-x) \neq -g(x)$
No symmetry
x-intercept:
$2x - 4 = 0$
$x = 2$

y-intercept: $y = \dfrac{2(0)-4}{(0)+3} = -\dfrac{4}{3}$

Vertical asymptote:
$x + 3 = 0$
$x = -3$
Horizontal asymptote:

$n = m$, so $y = \dfrac{2}{1} = 2$

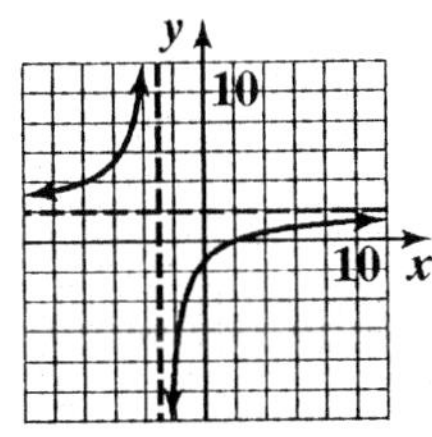

$f(x) = \dfrac{2x - 4}{x + 3}$

74. $h(x) = \dfrac{x^2-3x-4}{x^2-x-6}$

Symmetry: $h(-x) = \dfrac{x^2+3x-4}{x^2+x-6}$
$h(-x) \neq h(x)$, $h(-x) \neq -h(x)$
No symmetry
x-intercepts:
$x^2 - 3x - 4 = 0$
$(x-4)(x+1)$
$x = 4 \quad x = -1$

y-intercept: $y = \dfrac{0^2-3(0)-4}{0^2-0-6} = \dfrac{2}{3}$

Vertical asymptotes:
$x^2 - x - 6 = 0$
$(x-3)(x+2) = 0$
$x = 3, -2$

Horizontal asymptote:

$n = m$, so $y = \dfrac{1}{1} = 1$

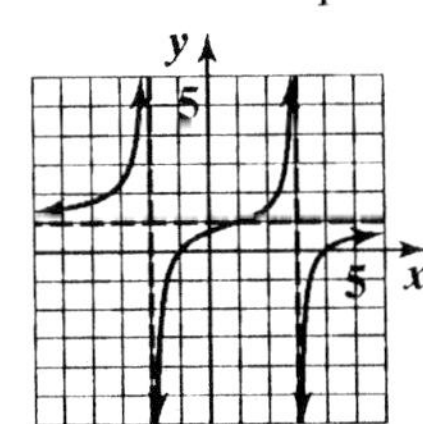

$h(x) = \dfrac{x^2 - 3x - 4}{x^2 - x - 6}$

68 **75.** $r(x) = \dfrac{x^2+4x+3}{(x+2)^2}$

Symmetry: $r(-x) = \dfrac{x^2-4x+3}{(-x+2)^2}$
$r(-x) \neq r(x)$, $r(-x) \neq -r(x)$
No symmetry
x-intercepts:
$x^2 + 4x + 3 = 0$
$(x+3)(x+1) = 0$
$x = -3, -1$

y-intercept: $y = \dfrac{0^2+4(0)+3}{(0+2)^2} = \dfrac{3}{4}$

Vertical asymptote:
$x + 2 = 0$
$x = -2$

Horizontal asymptote:

$n = m$, so $y = \dfrac{1}{1} = 1$

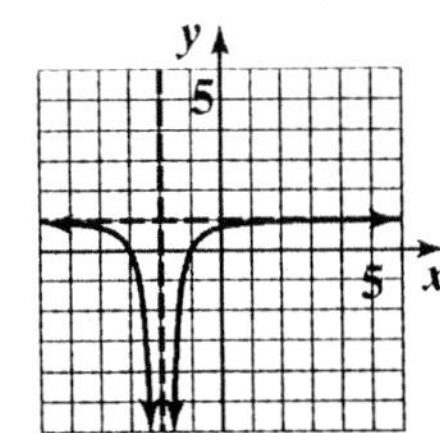

$r(x) = \dfrac{x^2 + 4x + 3}{(x + 4)^2}$

76. $y = \dfrac{x^2}{x+1}$

Symmetry: $f(-x) = \dfrac{x^2}{-x+1}$

$f(-x) \neq f(x), f(-x) \neq -f(x)$

No symmetry

x-intercept:

$x^2 = 0$

$x = 0$

y-intercept: $y = \dfrac{0^2}{0+1} = 0$

Vertical asymptote:

$x + 1 = 0$

$x = -1$

$n > m$, no horizontal asymptote.

Slant asymptote:

$y = x - 1 + \dfrac{1}{x+1}$

$y = x - 1$

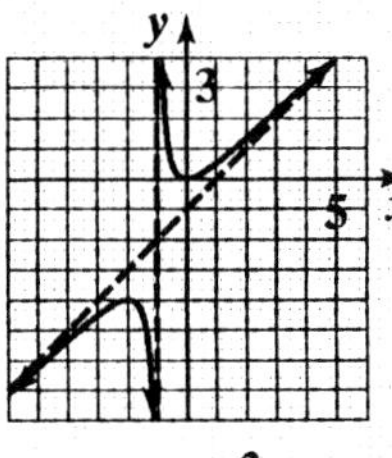

$y = \dfrac{x^2}{x+1}$

77. $y = \dfrac{x^2+2x-3}{x-3}$

Symmetry: $f(-x) = \dfrac{x^2-2x-3}{-x-3}$

$f(-x) \neq f(x), f(-x) \neq -f(x)$

No symmetry

x-intercepts:

$x^2 + 2x - 3 = 0$

$(x+3)(x-1) = 0$

$x = -3, 1$

y-intercept: $y = \dfrac{0^2 + 2(0) - 3}{0-3} = \dfrac{-3}{-3} = 1$

Vertical asymptote:

$x - 3 = 0$

$x = 3$

Horizontal asymptote:

$n > m$, so no horizontal asymptote.

Slant asymptote:

$y = x + 5 + \dfrac{12}{x-3}$

$y = x + 5$

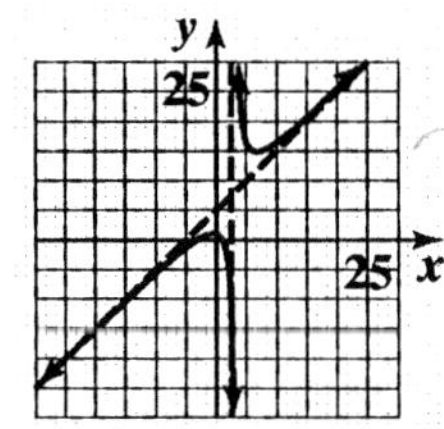

$f(x) = \dfrac{x^2 + 2x - 3}{x - 3}$

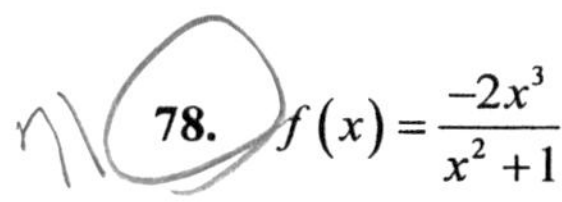

78. $f(x) = \dfrac{-2x^3}{x^2+1}$

Symmetry: $f(-x) = \dfrac{2}{x^2+1} = -f(x)$

Origin symmetry

x-intercept:

$-2x^3 = 0$

$x = 0$

y-intercept: $y = \dfrac{-2(0)^3}{0^2+1} = \dfrac{0}{1} = 0$

Vertical asymptote:

$x^2 + 1 = 0$

$x^2 = -1$

No vertical asymptote.

Horizontal asymptote:

$n > m$, so no horizontal asymptote.

Slant asymptote:

$f(x) = -2x + \dfrac{2x}{x^2+1}$

$y = -2x$

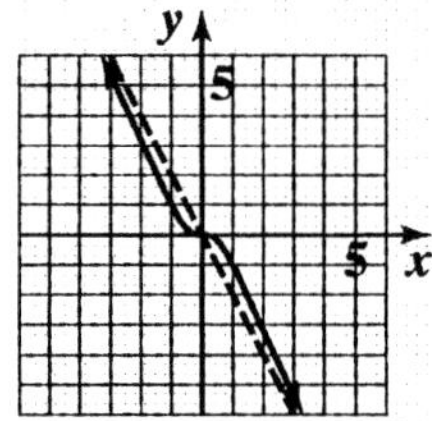

$f(x) = \dfrac{-2x^3}{x^2+1}$

79. $g(x)=\dfrac{4x^2-16x+16}{2x-3}$

Symmetry: $g(-x)=\dfrac{4x^2+16x+16}{-2x-3}$

$g(-x)\neq g(x),\ g(-x)\neq -g(x)$

No symmetry

x-intercept:

$4x^2-16x+16=0$

$4(x-2)^2=0$

$x=2$

y-intercept:

$y=\dfrac{4(0)^2-16(0)+16}{2(0)-3}=-\dfrac{16}{3}$

Vertical asymptote:

$2x-3=0$

$x=\dfrac{3}{2}$

Horizontal asymptote:

$n>m$, so no horizontal asymptote.

Slant asymptote:

$g(x)=2x-5+\dfrac{1}{2x-3}$

$y=2x-5$

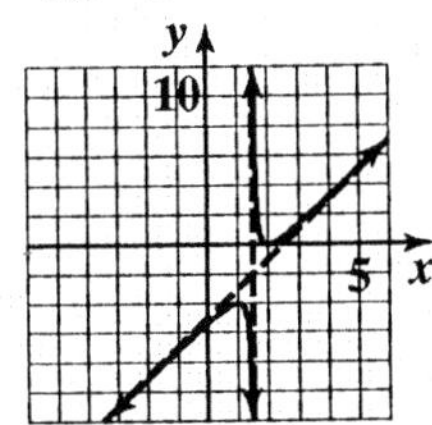

$g(x)=\dfrac{4x^2-16x+16}{2x-3}$

80. **a.** $C(x)=50{,}000+25x$

b. $\overline{C}(x)=\dfrac{25x+50{,}000}{x}$

c. $\overline{C}(50)=\dfrac{25(50)+50{,}000}{50}=1025$

When 50 calculators are manufactured, it costs \$1025 to manufacture each.

$\overline{C}(100)=\dfrac{25(100)+50{,}000}{100}=525$

When 100 calculators are manufactured, it costs \$525 to manufacture each.

$\overline{C}(1000)=\dfrac{25(1000)+50{,}000}{1000}=75$

When 1,000 calculators are manufactured, it costs \$75 to manufacture each.

$\overline{C}(100{,}000)=\dfrac{25(100{,}000)+50{,}000}{100{,}000}=25.5$

When 100,000 calculators are manufactured, it costs \$25.50 to manufacture each.

d. $n=m$, so $y=\dfrac{25}{1}=25$ is the horizontal asymptote. Minimum costs will approach \$25.

81. **a.** $C(90)-C(50)=\dfrac{200(90)}{100-90}-\dfrac{200(50)}{100-50}$

$C(90)-C(50)=1800-200$

$C(90)-C(50)=1600$

The difference in cost of removing 90% versus 50% of the contaminants is 16 million dollars.

b. $x=100$; No amount of money can remove 100% of the contaminants, since $C(x)$ increases without bound as x approaches 100.

82. $f(x)=\dfrac{150x+120}{0.05x+1}$

$n=m$, so $y=\dfrac{150}{0.05}=3000$

The number of fish available in the pond approaches 3000.

83. $P(x)=\dfrac{72{,}900}{100x^2+729}$

$n<m$ so $y=0$

As the number of years of education increases the percentage rate of unemployment approaches zero.

84. **a.** $f(x)=\dfrac{1.96x+3.14}{3.04x+21.79}$

b. $y=\dfrac{1.96}{3.04}=0.645$

The percentage of inmates that are in for violent crimes will approach 64.5%.

c. Answers may vary.

85. $T(x)=\dfrac{4}{x+3}+\dfrac{2}{x}$

86. $1000 = lw$

$\frac{1000}{w} = l$

$P = 2x + 2\left(\frac{1000}{x}\right)$

$P = 2x + \frac{2000}{2}$

87. $2x^2 + 5x - 3 < 0$

Solve the related quadratic equation.

$2x^2 + 5x - 3 = 0$

$(2x-1)(x+3) = 0$

The boundary points are -3 and $\frac{1}{2}$.

Testing each interval gives a solution set of $\left(-3, \frac{1}{2}\right)$

−5 −4 −3 −2 −1 0 1 2 3 4 5

88. $2x^2 + 9x + 4 \geq 0$

Solve the related quadratic equation.

$2x^2 + 9x + 4 = 0$

$(2x+1)(x+4) = 0$

The boundary points are -4 and $-\frac{1}{2}$.

Testing each interval gives a solution set of $(-\infty, -4] \cup \left[-\frac{1}{2}, \infty\right)$

−5 −4 −3 −2 −1 0 1 2 3 4 5

89. $x^3 + 2x^2 > 3x$

Solve the related quadratic equation.

$x^3 + 2x^2 = 3x$

$x^3 + 2x^2 - 3x = 0$

$x\left(x^2 + 2x - 3\right) = 0$

$x(x+3)(x-1) = 0$

The boundary points are -3, 0, and 1.

Testing each interval gives a solution set of $(-3, 0) \cup (1, \infty)$

−4 −3 −2 −1 0 1 2 3 4 5 6

90. $\frac{x-6}{x+2} > 0$

Find the values of x that make the numerator and denominator zero.
The boundary points are -2 and 6.
Testing each interval gives a solution set of $(-\infty, -2) \cup (6, \infty)$.

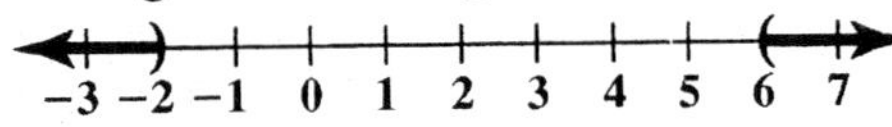

91. $\frac{(x+1)(x-2)}{x-1} \geq 0$

Find the values of x that make the numerator and denominator zero.
The boundary points are -1, 1 and 2. We exclude 1 from the solution set, since this would make the denominator zero.
Testing each interval gives a solution set of $[-1, 1) \cup [2, \infty)$.

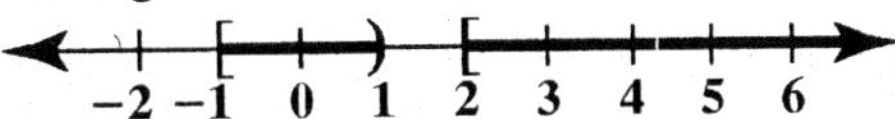

92. $\frac{x+3}{x-4} \leq 5$

Express the inequality so that one side is zero.

$$\frac{x+3}{x-4} - 5 \leq 0$$

$$\frac{x+3}{x-4} - \frac{5(x-4)}{x-4} \leq 0$$

$$\frac{-4x+23}{x-4} \leq 0$$

Find the values of x that make the numerator and denominator zero.

The boundary points are 4 and $\frac{23}{4}$. We exclude 4 from the solution set, since this would make the denominator zero.

Testing each interval gives a solution set of $(-\infty, 4) \cup \left[\frac{23}{4}, \infty\right)$.

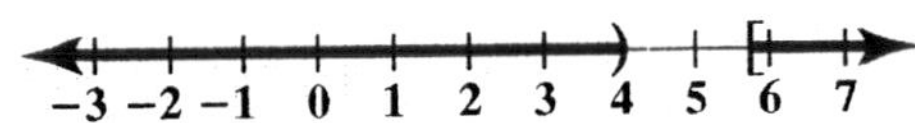

93. $s = -16t^2 + v_0t + s_0$

$32 < -16t^2 + 48t + 0$

$0 < -16t^2 + 48t - 32$

$0 < -16(t^2 - 3t + 2)$

$0 < -16(t-2)(t-1)$

F	T	F
1	2	

The projectile's height exceeds 32 feet during the time period from 1 to 2 seconds.

94. $b = ke$

$98 = k \cdot 1400$

$k = 0.07$

$b = 0.07e$

$b = 0.07(2200) = \$154$

95. $d = kt^2$

$144 = k(3)^2$

$k = 16$

$d = 16t^2$

$d = 16(10)^2 = 1{,}600$ ft

96. $t = \frac{k}{r}$

$4 = \frac{k}{50}$

$k = 200$

$t = \frac{200}{r}$

$t = \frac{200}{40} = 5$ hours

97. $l = \frac{k}{d^2}$

$28 = \frac{k}{8^2}$

$k = 1792$

$l = \frac{1792}{d^2}$

$l = \frac{1792}{4^2} = 112$ decibels

98. $t = \frac{kc}{w}$

$10 = \frac{k \cdot 30}{6}$

$10 = 5h$

$h = 2$

$t = \frac{2c}{w}$

$t = \frac{2(40)}{5} = 16$ hours

99. $V = khB$

$175 = k \cdot 15 \cdot 35$

$k = \frac{1}{3}$

$V = \frac{1}{3}hB$

$V = \frac{1}{3} \cdot 20 \cdot 120 = 800 \text{ ft}^3$

Chapter 2 Test

1. $(6-7i)(2+5i) = 12+30i-14i-35i^2$
$= 12+16i+35$
$= 47+16i$

2. $\frac{5}{2-i} = \frac{5}{2-i} \cdot \frac{2+i}{2+i}$
$= \frac{5(2+i)}{4+1}$
$= \frac{5(2+i)}{5}$
$= 2+i$

3. $2\sqrt{-49} + 3\sqrt{-64} = 2(7i) + 3(8i)$
$= 14i + 24i$
$= 38i$

4. $x^2 = 4x - 8$

$x^2 - 4x + 8 = 0$

$x = \frac{-b \pm \sqrt{b^2 - 4ac}}{2a}$

$x = \frac{-(-4) \pm \sqrt{(-4)^2 - 4(1)(8)}}{2(1)}$

$x = \frac{4 \pm \sqrt{-16}}{2}$

$x = \frac{4 \pm 4i}{2}$

$x = 2 \pm 2i$

5. $f(x) = (x+1)^2 + 4$

vertex: $(-1, 4)$

axis of symmetry: $x = -1$

x-intercepts:

$(x+1)^2 + 4 = 0$

$x^2 + 2x + 5 = 0$

$x = \frac{-2 \pm \sqrt{4-20}}{2} = -1 \pm 2i$

no x-intercepts

y-intercept:

$f(0) = (0+1)^2 + 4 = 5$

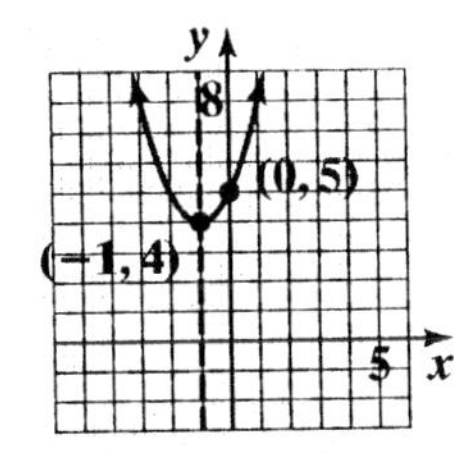

$f(x) = (x+1)^2 + 4$
Domain: $(-\infty, \infty)$; Range: $[4, \infty)$

6. $f(x) - x^2 - 2x - 3$

$x = \frac{-b}{2a} = \frac{2}{2} = 1$

$f(1) = 1^2 - 2(1) - 3 = -4$

vertex: $(1, -4)$
axis of symmetry $x = 1$
x-intercepts:

$x^2 - 2x - 3 = 0$

$(x-3)(x+1) = 0$

$x = 3$ or $x = -1$

y-intercept:
$f(0) = 0^2 - 2(0) - 3 = -3$

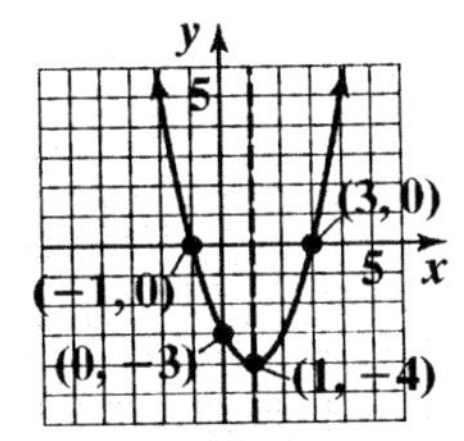

$f(x) = x^2 - 2x - 3$
Domain: $(-\infty, \infty)$; Range: $[-4, \infty)$

7. $f(x) = -2x^2 + 12x - 16$

Since the coefficient of x^2 is negative, the graph of $f(x)$ opens down and $f(x)$ has a maximum point.

$x = \frac{-12}{2(-2)} = 3$

$f(3) = -2(3)^2 + 12(3) - 16$
$= -18 + 36 - 16$
$= 2$

Maximum point: $(3, 2)$
Domain: $(-\infty, \infty)$; Range: $(-\infty, 2]$

8. $f(x) = -x^2 + 46x - 360$

$x = -\frac{b}{2a} = \frac{-46}{-2} = 23$

23 computers will maximize profit.

$f(23) = -(23)^2 + 46(23) - 360 = 169$

Maximum daily profit = \$16,900.

9. Let x = one of the numbers;
$14 - x$ = the other number.
The product is $f(x) = x(14-x)$

$f(x) = x(14-x) = -x^2 + 14x$

The x-coordinate of the maximum is

$x = -\frac{b}{2a} = -\frac{14}{2(-1)} = -\frac{14}{-2} = 7.$

$f(7) = -7^2 + 14(7) = 49$

The vertex is $(7, 49)$. The maximum product is 49. This occurs when the two number are 7 and $14 - 7 = 7$.

10. **a.** $f(x) = x^3 - 5x^2 - 4x + 20$

$x^3 - 5x^2 - 4x + 20 = 0$

$x^2(x-5) - 4(x-5) = 0$

$(x-5)(x-2)(x+2) = 0$

$x = 5, 2, -2$

The solution set is $\{5, 2, -2\}$.

b. The degree of the polynomial is odd and the leading coefficient is positive. Thus the graph falls to the left and rises to the right.

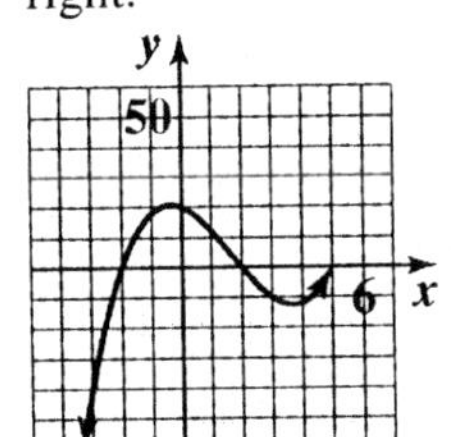

11. $f(x) = x^5 - x$

Since the degree of the polynomial is odd and the leading coefficient is positive, the graph of f should fall to the left and rise to the right. The x-intercepts should be −1 and 1.

12. **a.** The integral root is 2.

b.

2	6	−19	16	−4
		12	−14	4
	6	−7	2	0

$6x^2 - 7x + 2 = 0$

$(3x - 2)(2x - 1) = 0$

$x = \frac{2}{3}$ or $x = \frac{1}{2}$

The other two roots are $\frac{1}{2}$ and $\frac{2}{3}$.

13. $2x^3 + 11x^2 - 7x - 6 = 0$

p: ±1, ±2, ±3, ±6

q: ±1, ±2

$\frac{p}{q}: \pm 1, \pm 2, \pm 3, \pm 6, \pm\frac{1}{2}, \pm\frac{3}{2}$

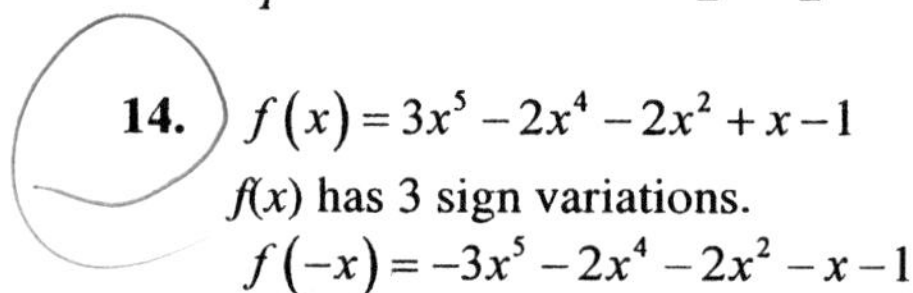

14. $f(x) = 3x^5 - 2x^4 - 2x^2 + x - 1$

$f(x)$ has 3 sign variations.

$f(-x) = -3x^5 - 2x^4 - 2x^2 - x - 1$

$f(-x)$ has no sign variations.

There are 3 or 1 positive real solutions and no negative real solutions.

15. $x^3 + 9x^2 + 16x - 6 = 0$

Since the leading coefficient is 1, the possible rational zeros are the factors of 6

$p = \frac{p}{q}: \pm 1, \pm 2, \pm 3, \pm 6$

−3	1	9	16	−6
		−3	−18	6
	1	6	−2	0

Thus $x = 3$ is a root.

Solve the quotient $x^2 + 6x - 2 = 0$ using the quadratic formula to find the remaining roots.

$$x = \frac{-b \pm \sqrt{b^2 - 4ac}}{2a}$$

$$x = \frac{-(6) \pm \sqrt{(6)^2 - 4(1)(-2)}}{2(1)}$$

$$= \frac{-6 \pm \sqrt{44}}{2}$$

$$= -3 \pm \sqrt{11}$$

The zeros are -3, $-3 + \sqrt{11}$, and $-3 - \sqrt{11}$.

16. $f(x) = 2x^4 - x^3 - 13x^2 + 5x + 15$

a. Possible rational zeros are:

p: ±1, ±3, ±5, ±15

q: ±1, ±2

$\frac{p}{q}: \pm 1, \pm 3, \pm 5, \pm 15, \pm\frac{1}{2}, \pm\frac{3}{2}, \pm\frac{5}{2}, \pm\frac{15}{2}$

b. Verify that -1 and $\frac{3}{2}$ are zeros as it appears in the graph:

−1	2	−1	−13	5	15
		−2	3	10	−15
	2	−3	−10	15	0

$\frac{3}{2}$	2	−3	−10	15
		3	0	−15
	2	0	−10	0

Thus, -1 and $\frac{3}{2}$ are zeros, and the polynomial factors as follows:

$$2x^4 - x^3 - 13x^2 + 5x + 15 = 0$$

$$(x+1)(2x^3 - 3x^2 - 10x + 15) = 0$$

$$(x+1)\left(x - \frac{3}{2}\right)(2x^2 - 10) = 0$$

Find the remaining zeros by solving:

$2x^2 - 10 = 0$

$2x^2 = 10$

$x^2 = 5$

$x = \pm\sqrt{5}$

The zeros are -1, $\frac{3}{2}$, and $\pm\sqrt{5}$.

17. $f(x)$ has zeros at −2 and 1. The zero at −2 has multiplicity of 2.

$x^3 + 3x^2 - 4 = (x-1)(x+2)^2$

18. $f(x) = a_0(x+1)(x-1)(x+i)(x-i)$
$= a_0(x^2-1)(x^2+1)$
$= a_0(x^4-1)$

Since $f(3) = 160$, then

$a_0(3^4-1) = 160$
$a_0(80) = 160$
$a_0 = \frac{160}{80}$
$a_0 = 2$

$f(x) = 2(x^4-1) = 2x^4-2$

19. $f(x) = -3x^3 - 4x^2 + x + 2$

The graph shows a root at $x = -1$.
Use synthetic division to verify this root.

-1	-3	-4	1	2
		3	1	4
	-3	-1	2	0

Factor the quotient to find the remaining zeros.

$-3x^2 - x + 2 = 0$
$-(3x-2)(x+1) = 0$

The zeros (x-intercepts) are -1 and $\frac{2}{3}$.

The y-intercept is $f(0) = 2$

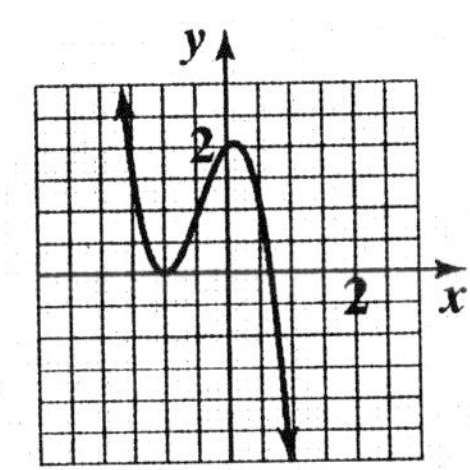

$f(x) = -3x^3 - 4x^2 + x + 2$

20. $f(x) = \frac{1}{(x+3)^2}$

Domain: $\{x \mid x \neq -3\}$ or $(-\infty, -3) \cup (-3, \infty)$

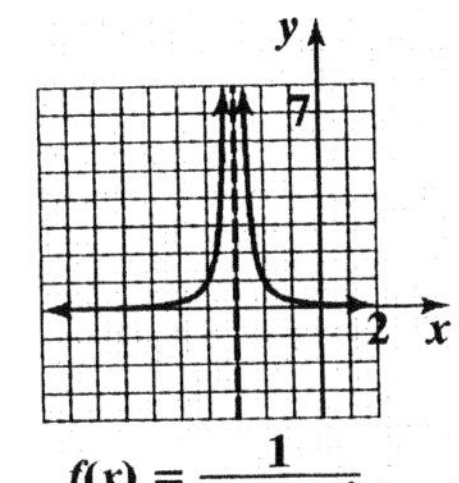

$f(x) = \frac{1}{(x+3)^2}$

21. $f(x) = \frac{1}{x-1} + 2$

Domain: $\{x \mid x \neq 1\}$ or $(-\infty, 1) \cup (1, \infty)$

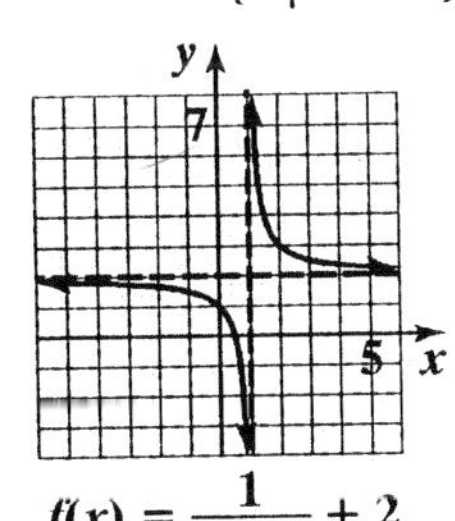

$f(x) = \frac{1}{x-1} + 2$

22. $f(x) = \frac{x}{x^2-16}$

Domain: $\{x \mid x \neq 4, x \neq -4\}$

Symmetry: $f(-x) = \frac{-x}{x^2-16} = -f(x)$

y-axis symmetry
x-intercept: $x = 0$

y-intercept: $y = \frac{0}{0^2-16} = 0$

Vertical asymptotes:
$x^2 - 16 = 0$
$(x-4)(x+4) = 0$
$x = 4, -4$

Horizontal asymptote:
$n < m$, so $y = 0$ is the horizontal asymptote.

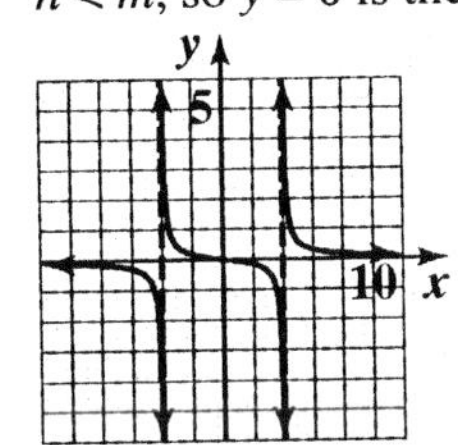

$f(x) = \frac{x}{x^2-16}$

23. $f(x) = \frac{x^2-9}{x-2}$

Domain: $\{x \mid x \neq 2\}$

Symmetry: $f(-x) = \frac{x^2-9}{-x-2}$

$f(-x) \neq f(x), f(-x) \neq -f(x)$
No symmetry

x-intercepts:
$x^2 - 9 = 0$
$(x-3)(x+3) = 0$
$x = 3, -3$

y-intercept: $y = \frac{0^2 - 9}{0 - 2} = \frac{9}{2}$

Vertical asymptote:

$x - 2 = 0$

$x = 2$

Horizontal asymptote:

$n > m$, so no horizontal asymptote exists.

Slant asymptote: $f(x) = x + 2 - \frac{5}{x-2}$

$y = x + 2$

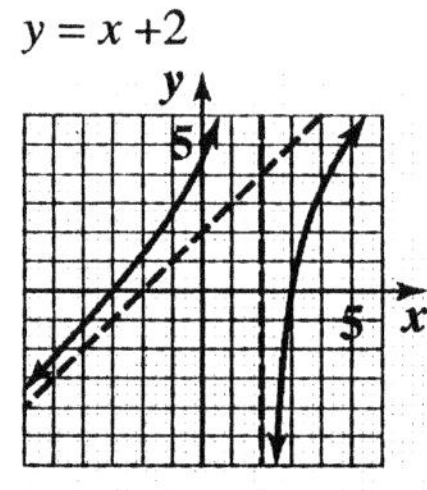

$f(x) = \frac{x^2 - 9}{x - 2}$

24. $f(x) = \frac{x+1}{x^2 + 2x - 3}$

$x^2 + 2x - 3 = (x+3)(x-1)$

Domain: $\{x \mid x \neq -3, x \neq 1\}$

Symmetry: $f(-x) = \frac{-x+1}{x^2 - 2x - 3}$

$f(-x) \neq f(x), f(-x) \neq -f(x)$

No symmetry

x-intercept:

$x + 1 = 0$

$x = -1$

y-intercept: $y = \frac{0+1}{0^2 + 2(0) - 3} = -\frac{1}{3}$

Vertical asymptotes:

$x^2 + 2x - 3 = 0$

$(x + 3)(x - 1) = 0$

x –3, 1

Horizontal asymptote:

$n < m$, so $y = 0$ is the horizontal asymptote.

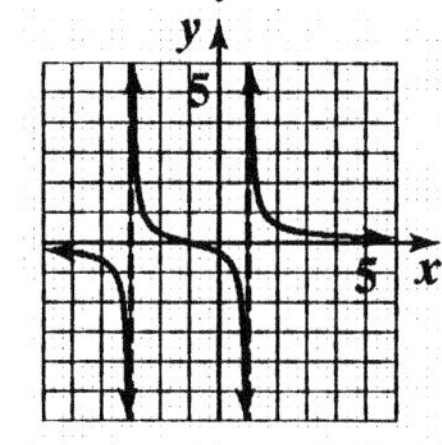

$f(x) = \frac{x + 1}{x^2 + 2x - 3}$

25. $f(x) = \frac{4x^2}{x^2 + 3}$

Domain: all real numbers

Symmetry: $f(-x) = \frac{4x^2}{x^2 + 3} = f(x)$

y-axis symmetry

x-intercept:

$4x^2 = 0$

$x = 0$

y-intercept: $y = \frac{4(0)^2}{0^2 + 3} = 0$

Vertical asymptote:

$x^2 + 3 = 0$

$x^2 = -3$

No vertical asymptote.

Horizontal asymptote:

$n = m$, so $y = \frac{4}{1} = 4$ is the horizontal asymptote.

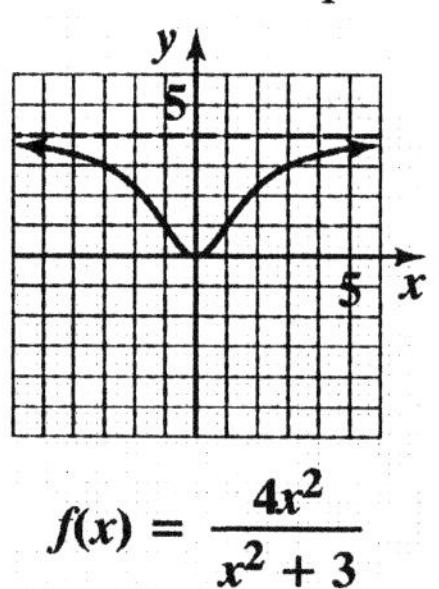

$f(x) = \frac{4x^2}{x^2 + 3}$

26. **a.** $\overline{C}(x) = \frac{300,000 + 10x}{x}$

b. Since the degree of the numerator equals the degree of the denominator, the horizontal asymptote is $x = \frac{10}{1} = 10$.

This represents the fact that as the number of televisions produced increases, the average approaches $10 per unit.

27. **a.** When $x = 5, y = 0.89$

After 5 learning tries, 89% of the responses were correct.

b. When $x = 11, y = 0.95$

After 11 learning tries, 95% of the responses were correct.

c. $y = .9/.9 = 1$

As the number of learning tries increases, the correct responses approaches 100%.

28. $x^2 < x + 12$

$x^2 - x - 12 < 0$

$(x+3)(x-4) < 0$

Boundary values: –3 and 4

Solution set: $(-3, 4)$

29. $\frac{2x+1}{x-3} \le 3$

$\frac{2x+1}{x-3} - 3 \le 0$

$\frac{10-x}{x-3} \le 0$

Boundary values: 3 and 10

Solution set: $(-\infty, 3) \cup [10, \infty)$

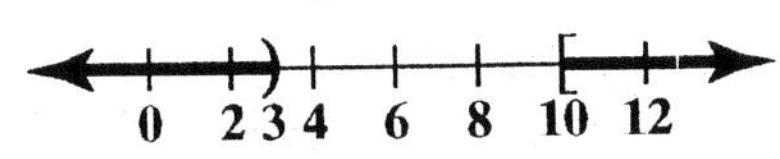

30. $i = \frac{k}{d^2}$

$20 = \frac{k}{15^2}$

$4500 = k$

$i = \frac{4500}{d^2} = \frac{4500}{10^2} = 45$ foot-candles

Cumulative Review Exercises (Chapters P–2)

1. Domain: $(-2, 2)$ Range: $[0, \infty)$

2. The zero at -1 touches the x-axis at turns around so it must have a minimum multiplicity of 2. The zero at 1 touches the x-axis at turns around so it must have a minimum multiplicity of 2.

3. There is a relative maximum at the point (0, 3).

4. $(f \circ f)(-1) = f(f(-1)) = f(0) = 3$

5. $f(x) \to \infty$. as $\underline{x \to -2^+}$ or as $\underline{x \to 2^-}$

6.

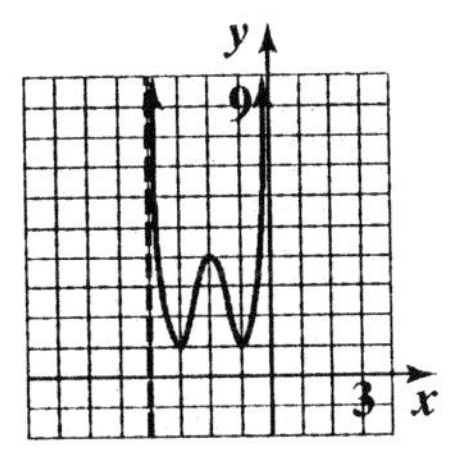

7. $|2x - 1| = 3$

$2x - 1 = 3$

$2x = 4$

$x = 2$

$2x - 1 = -3$

$2x = -2$

$x = -1$

The solution set is $\{2, -1\}$.

8. $3x^2 - 5x + 1 = 0$

$x = \frac{5 \pm \sqrt{25-12}}{6} = \frac{5 \pm \sqrt{13}}{6}$

The solution set is $\left\{\frac{5+\sqrt{13}}{6}, \frac{5-\sqrt{13}}{6}\right\}$.

9. $9 + \frac{3}{x} = \frac{2}{x^2}$

$9x^2 + 3x = 2$

$9x^2 + 3x - 2 = 0$

$(3x-1)(3x+2) = 0$

$3x - 1 = 0 \quad 3x + 2 = 0$

$x = \frac{1}{3}$ or $x = -\frac{2}{3}$

The solution set is $\left\{\frac{1}{3}, -\frac{2}{3}\right\}$.

10. $x^3 + 2x^2 - 5x - 6 = 0$

p: ±1, ±2, ±3, ±6

q: ±1

$\frac{p}{q}$: ±1, ±2, ±3, ±6

–3	1	2	–5	–6
		–3	3	6
	1	–1	–2	0

$x^3 + 2x^2 - 5x - 6 = 0$

$(x+3)(x^2 - x - 2) = 0$

$(x+3)(x+1)(x-2) = 0$

$x = -3$ or $x = -1$ or $x = 2$

The solution set is $\{-3, -1, 2\}$.

11. $|2x-5|>3$

$2x-5>3$

$2x>8$

$x>4$

$2x-5<-3$

$2x<2$

$x<1$

$(-\infty, 1)$ or $(4, \infty)$

12. $3x^2>2x+5$

$3x^2-2x-5>0$

$3x^2-2x-5=0$

$(3x-5)(x+1)=0$

$x=\frac{5}{3}$ or $x=-1$

Test intervals are $(-\infty, -1)$, $\left(-1, \frac{5}{3}\right)$, $\left(\frac{5}{3}, \infty\right)$.

Testing points, the solution is

$(-\infty, -1)$ or $\left(\frac{5}{3}, \infty\right)$.

13. $f(x)=x^3-4x^2-x+4$

x-intercepts:

$x^3-4x^2-x+4=0$

$x^2(x-4)-1(x-4)=0$

$(x-4)(x^2-1)=0$

$(x-4)(x+1)(x-1)=0$

$x=-1, 1, 4$

x-intercepts:

$f(0)=0^3-4(0)^2-0+4=4$

The degree of the polynomial is odd and the leading coefficient is positive. Thus the graph falls to the left and rises to the right.

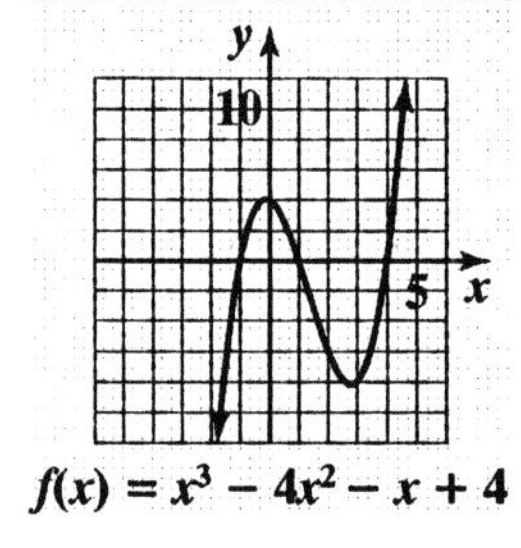

$f(x) = x^3 - 4x^2 - x + 4$

14. $f(x)=x^2+2x-8$

$x=\frac{-b}{2a}=\frac{-2}{2}=-1$

$f(-1)=(-1)^2+2(-1)-8$

$=1-2-8=-9$

vertex: $(-1, -9)$

x-intercepts:

$x^2+2x-8=0$

$(x+4)(x-2)=0$

$x=-4$ or $x=2$

y-intercept: $f(0)=-8$

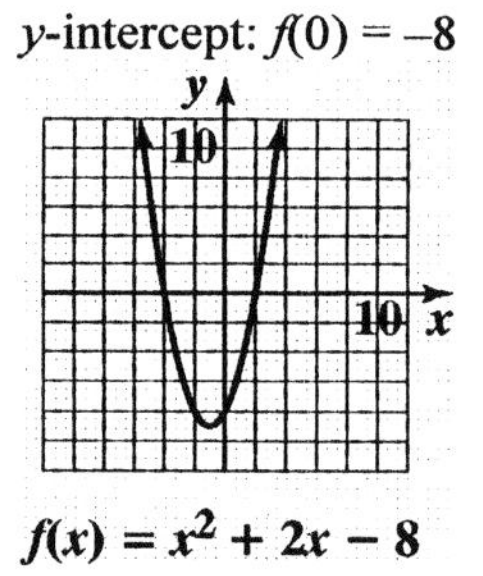

$f(x) = x^2 + 2x - 8$

15. $f(x)=x^2(x-3)$

zeros: $x=0$ (multiplicity 2) and $x=3$

y-intercept: $y=0$

$f(x)=x^3-3x^2$

$n=3, a_n=0$ so the graph falls to the left and rises to the right.

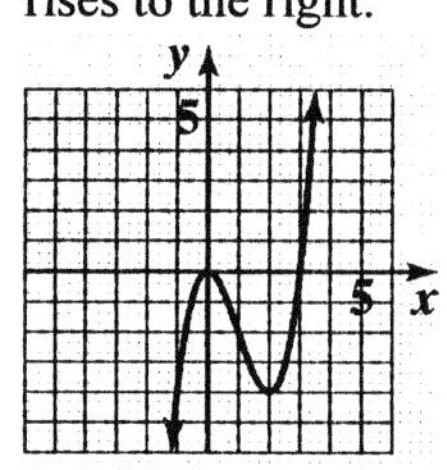

$f(x) = x^2(x - 3)$

16. $f(x)=\frac{x-1}{x-2}$

vertical asymptote: $x=2$

horizontal asymptote: $y=1$

x-intercept: $x=1$

y-intercept: $y=\frac{1}{2}$

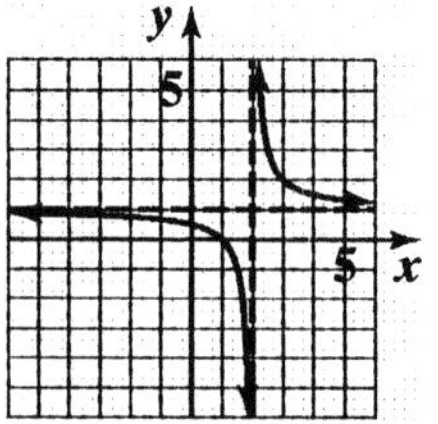

17.

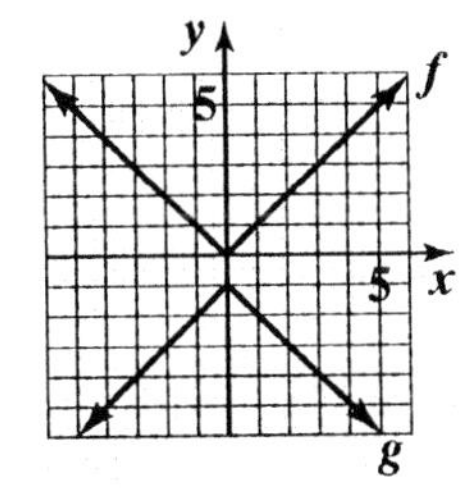

18.

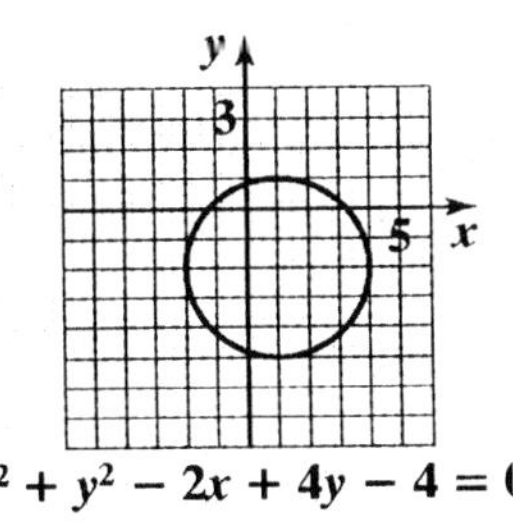

19. $(f \circ g)(x) = f\left(g(x)\right)$

$$= 2(4x-1)^2 - (4x-1) - 1$$
$$= 32x^2 - 20x + 2$$

20. $$\frac{f(x+h)-f(x)}{h} = \frac{\left[2(x+h)^2 - (x+h) - 1\right] - \left[2x^2 - x - 1\right]}{h}$$

$$= \frac{2x^2 + 4hx - x + 2h^2 - h - 1 - 2x^2 + x + 1}{h}$$

$$= \frac{4hx + 2h^2 - h}{h}$$

$$= 4x + 2h - 1$$

Chapter 3

Section 3.1

Check Point Exercises

1. Substitute 60 for x and evaluate the function at 60. $f(60) = 13.49(0.967)^{-60} - 1 \approx 1$
   Thus, one O-ring is expected to fail at a temperature of 60°F.

2. Begin by setting up a table of coordinates.

x	$f(x) = 3^x$
−3	$f(-3) = 3^{-3} = \frac{1}{27}$
−2	$f(-2) = 3^{-2} = \frac{1}{9}$
−1	$f(-1) = 3^{-1} = \frac{1}{3}$
0	$f(0) = 3^0 = 1$
1	$f(1) = 3^1 = 3$
2	$f(2) = 3^2 = 9$
3	$f(3) = 3^3 = 27$

Plot these points, connecting them with a continuous curve.

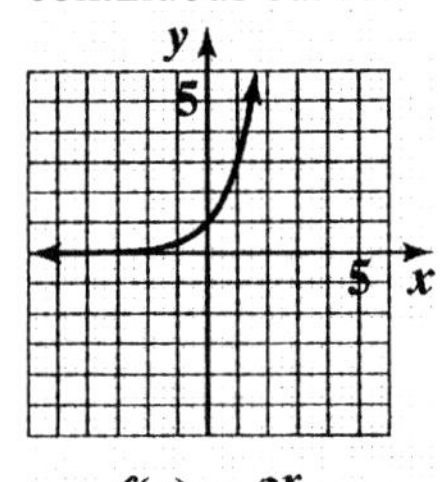

$f(x) = 3^x$

3. Begin by setting up a table of coordinates.

x	$f(x) = \left(\frac{1}{3}\right)^x$
−2	$\left(\frac{1}{3}\right)^{-2} = 9$
−1	$\left(\frac{1}{3}\right)^{-1} = 3$
0	$\left(\frac{1}{3}\right)^{0} = 1$
1	$\left(\frac{1}{3}\right)^{1} = \frac{1}{3}$
2	$\left(\frac{1}{3}\right)^{2} = \frac{1}{9}$

Plot these points, connecting them with a continuous curve.

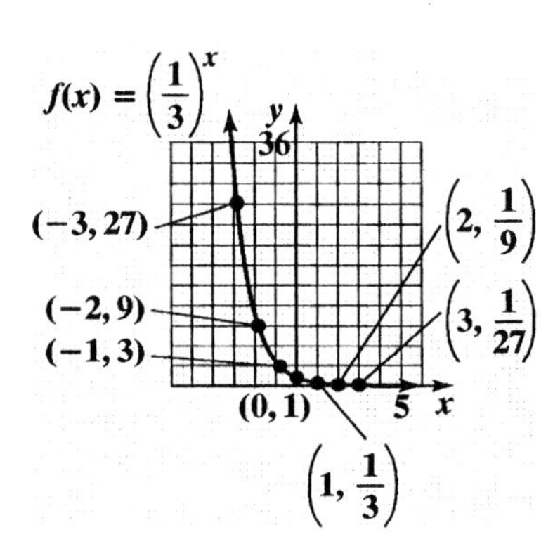

4. Note that the function $g(x) = 3^{x-1}$ has the general form $g(x) = b^{x+c}$ where $c = -1$. Because $c < 0$, we graph $g(x) = 3^{x-1}$ by shifting the graph of $f(x) = 3^x$ one unit to the right. Construct a table showing some of the coordinates for f and g.

x	$f(x) = 3^x$	$g(x) = 3^{x-1}$
−2	$3^{-2} = \frac{1}{9}$	$3^{-2-1} = 3^{-3} = \frac{1}{27}$
−1	$3^{-1} = \frac{1}{3}$	$3^{-1-1} = 3^{-2} = \frac{1}{9}$
0	$3^0 = 1$	$3^{0-1} = 3^{-1} = \frac{1}{3}$
1	$3^1 = 3$	$3^{1-1} = 3^0 = 1$
2	$3^2 = 9$	$3^{2-1} = 3^1 = 3$

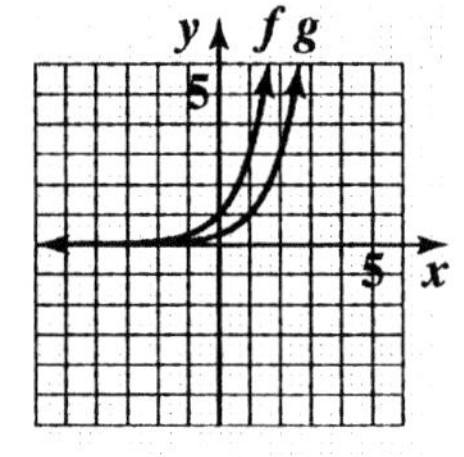

$f(x) = 3^x$
$g(x) = 3^{x-1}$

5. Note that the function $g(x) = 2^x + 1$ has the general form $g(x) = b^x + c$ where $c = 1$. Because $c > 0$, we graph $g(x) = 2^x + 1$ by shifting the graph of $f(x) = 2^x$ up one unit. Construct a table showing some of the coordinates for *f* and *g*.

x	$f(x) = 2^x$	$g(x) = 2^x + 1$
-2	$2^{-2} = \frac{1}{4}$	$2^{-2} + 1 = \frac{1}{4} + 1 = \frac{5}{4}$
-1	$2^{-1} = \frac{1}{2}$	$2^{-1} + 1 = \frac{1}{2} + 1 = \frac{3}{2}$
0	$2^0 = 1$	$2^0 + 1 = 1 + 1 = 2$
1	$2^1 = 2$	$2^1 + 1 = 2 + 1 = 3$
2	$2^2 = 4$	$2^2 + 1 = 4 + 1 = 5$

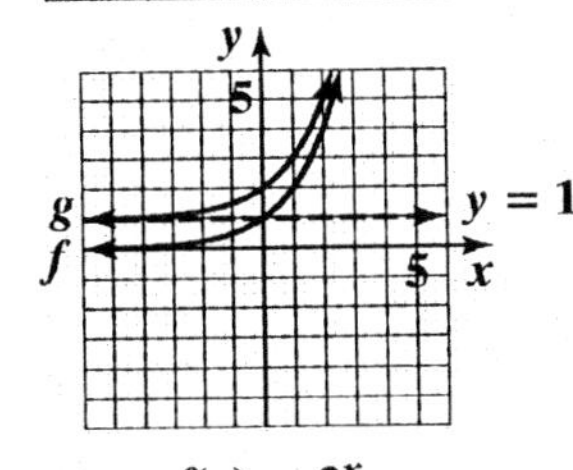

$f(x) = 2^x$
$g(x) = 2^x + 1$

6. $f(x) = 6.4e^{0.0123x}$
The year 2050 is 46 years after 2004.
Find $f(46)$.

$f(x) = 6.4e^{0.0123x}$
$f(46) = 6.4e^{0.0123(46)}$
≈ 11.27

The world population is predicted to be 11.27 billion in 2050.

7. a. $A = P\left(1 + \frac{r}{n}\right)^{nt}$

$A = 10{,}000\left(1 + \frac{0.08}{4}\right)^{4(5)}$
$= \$14{,}859.47$

b. $A = Pe^{rt}$
$A = 10{,}000e^{0.08(5)}$
$= \$14{,}918.25$

Exercise Set 3.1

1. $2^{3.4} \approx 10.556$

2. $3^{2.4} \approx 13.967$

3. $3^{\sqrt{5}} \approx 11.665$

4. $5^{\sqrt{3}} \approx 16.242$

5. $4^{-1.5} = 0.125$

6. $6^{-1.2} \approx 0.116$

7. $e^{2.3} \approx 9.974$

8. $e^{3.4} \approx 29.964$

9. $e^{-0.95} \approx 0.387$

10. $e^{-0.75} \approx 0.472$

11.

x	$f(x) = 4^x$
-2	$4^{-2} = \frac{1}{16}$
-1	$4^{-1} = \frac{1}{4}$
0	$4^0 = 1$
1	$4^1 = 4$
2	$4^2 = 16$

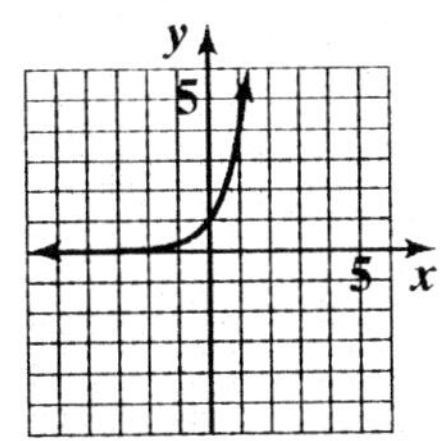

$f(x) = 4^x$

12.

x	$g(x) = 5^x$
-2	$5^{-2} = \frac{1}{25}$
-1	$5^{-1} = \frac{1}{5}$
0	$5^0 = 1$
1	$5^1 = 5$
2	$5^2 = 25$

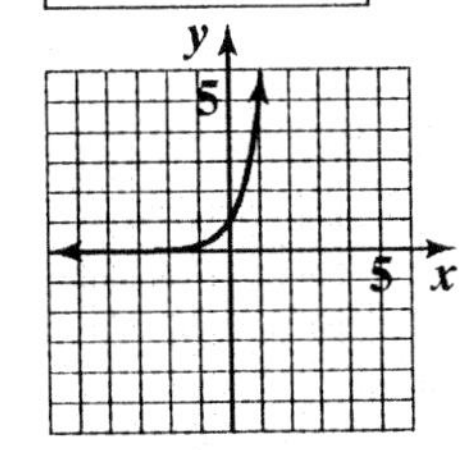

$f(x) = 5^x$

13.

x	$g(x)=\left(\frac{3}{2}\right)^x$
−2	$\left(\frac{3}{2}\right)^{-2}=\frac{4}{9}$
−1	$\left(\frac{3}{2}\right)^{-1}=\frac{2}{3}$
0	$\left(\frac{3}{2}\right)^{0}=1$
1	$\left(\frac{3}{2}\right)^{1}=\frac{3}{2}$
2	$\left(\frac{3}{2}\right)^{2}=\frac{9}{4}$

$g(x)=\left(\frac{3}{2}\right)^x$

14.

x	$g(x)=\left(\frac{4}{3}\right)^x$
−2	$\left(\frac{4}{3}\right)^{-2}=\frac{9}{16}$
−1	$\left(\frac{4}{3}\right)^{-1}=\frac{3}{4}$
0	$\left(\frac{4}{3}\right)^{0}=1$
1	$\left(\frac{4}{3}\right)^{1}=\frac{4}{3}$
2	$\left(\frac{4}{3}\right)^{2}=\frac{16}{9}$

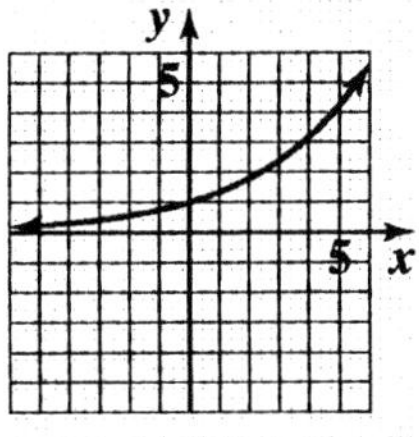

$g(x)=\left(\frac{4}{3}\right)^x$

15.

x	$h(x)=\left(\frac{1}{2}\right)^x$
−2	$\left(\frac{1}{2}\right)^{-2}=4$
−1	$\left(\frac{1}{2}\right)^{-1}=2$
0	$\left(\frac{1}{2}\right)^{0}=1$
1	$\left(\frac{1}{2}\right)^{1}=\frac{1}{2}$
2	$\left(\frac{1}{2}\right)^{2}=\frac{1}{4}$

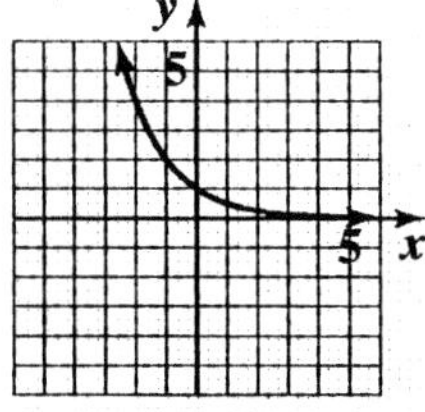

$h(x)=\left(\frac{1}{2}\right)^x$

16.

x	$h(x)=\left(\frac{1}{3}\right)^x$
−2	$\left(\frac{1}{3}\right)^{-2}=9$
−1	$\left(\frac{1}{3}\right)^{-1}=3$
0	$\left(\frac{1}{3}\right)^{0}=1$
1	$\left(\frac{1}{3}\right)^{1}=\frac{1}{3}$
2	$\left(\frac{1}{3}\right)^{2}=\frac{1}{9}$

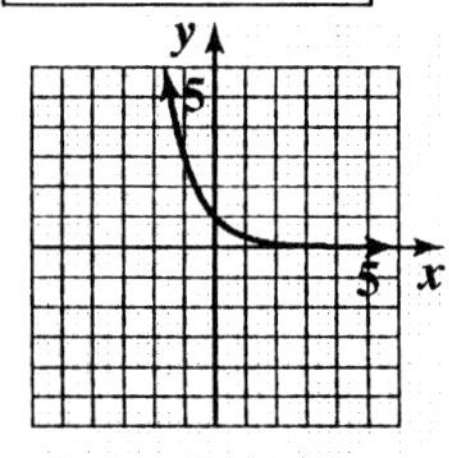

$h(x)=\left(\frac{1}{3}\right)^x$

17.

x	$f(x)=(0.6)^x$
-2	$(0.6)^{-2}=2.\overline{7}$
-1	$(0.6)^{-1}=1.\overline{6}$
0	$(0.6)^0=1$
1	$(0.6)^1=0.6$
2	$(0.6)^2=0.36$

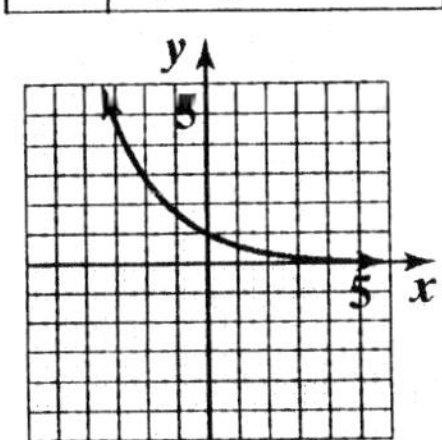

$f(x) = (0.6)^x$

18.

x	$f(x)=(0.8)^x$
-2	$(0.8)^{-2}=1.5625$
-1	$(0.8)^{-1}=1.25$
0	$(0.8)^0=1$
1	$(0.8)^1=0.8$
2	$(0.8)^2=0.64$

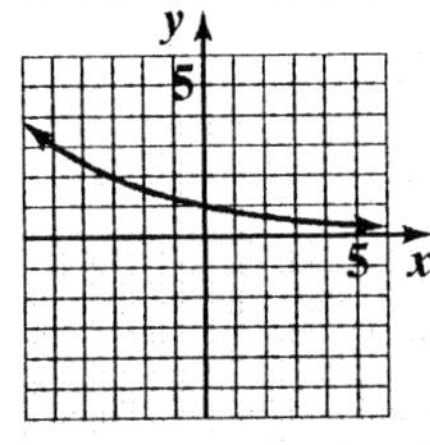

$f(x) = (0.8)^x$

19. This is the graph of $f(x)=3^x$ reflected about the x-axis and about the y-axis, so the function is $H(x)=-3^{-x}$.

20. This is the graph of $f(x)=3^x$ shifted one unit to the right, so the function is $g(x)=3^{x-1}$.

21. This is the graph of $f(x)=3^x$ reflected about the x-axis, so the function is $F(x)=-3^x$.

22. This is the graph of $f(x) = 3^x$.

23. This is the graph of $f(x)=3^x$ shifted one unit downward, so the function is $h(x)=3^x-1$.

24. This is the graph of $f(x)=3^x$ reflected about the y-axis, so the function is $G(x)=3^{-x}$.

25. The graph of $g(x)=2^{x+1}$ can be obtained by shifting the graph of $f(x)=2^x$ one unit to the left.

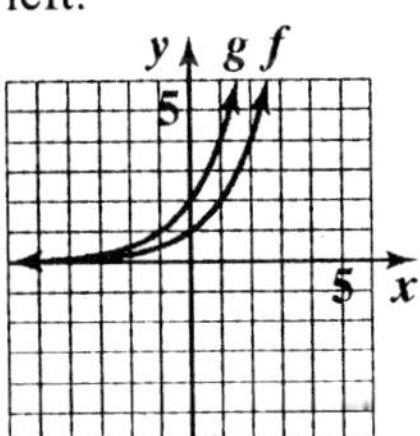

$f(x) = 2^x$
$g(x) = 2^{x+1}$

Asymptote: $y=0$
Domain: $(-\infty,\infty)$; Range: $(0,\infty)$

26. The graph of $g(x)=2^{x+2}$ can be obtained by shifting the graph of $f(x)=2^x$ two units to the left.

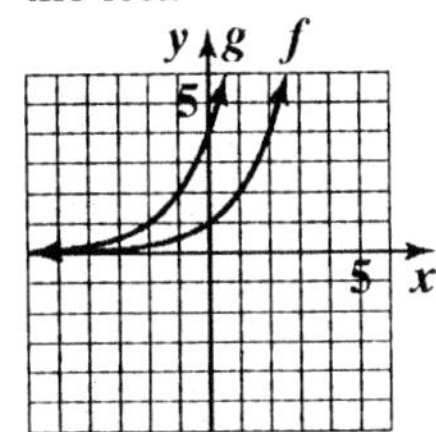

$f(x) = 2^x$
$g(x) = 2^{x+2}$

Asymptote: $y=0$
Domain: $(-\infty,\infty)$; Range: $(0,\infty)$

27. The graph of $g(x)=2^x-1$ can be obtained by shifting the graph of $f(x)=2^x$ downward one unit.

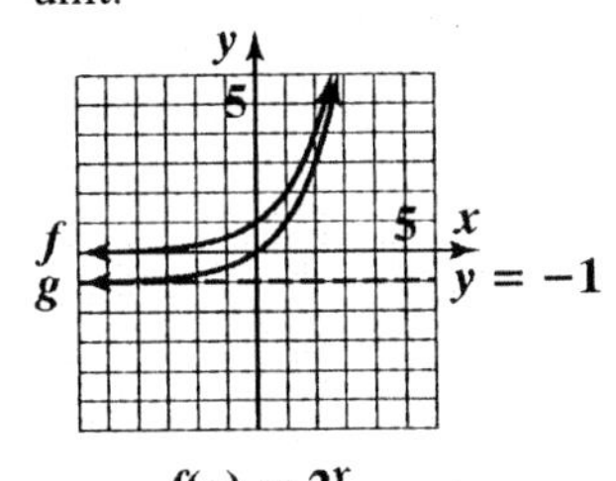

$f(x) = 2^x$
$g(x) = 2^x - 1$

Asymptote: $y=-1$
Domain: $(-\infty,\infty)$; Range: $(-1,\infty)$

28. The graph of $g(x) = 2^x + 2$ can be obtained by shifting the graph of $f(x) = 2^x$ two units upward.

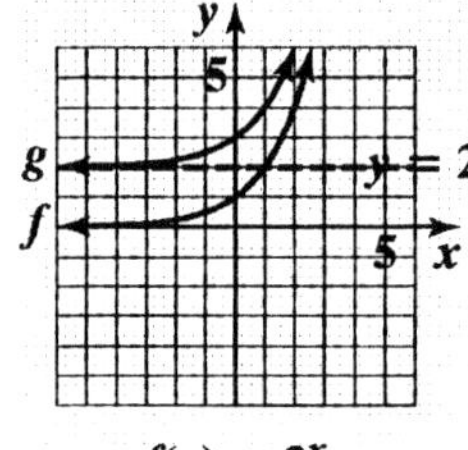

$f(x) = 2^x$
$g(x) = 2^x + 2$

Asymptote: $y = 2$
Domain: $(-\infty, \infty)$; Range: $(2, \infty)$

29. The graph of $h(x) = 2^{x+1} - 1$ can be obtained by shifting the graph of $f(x) = 2^x$ one unit to the left and one unit downward.

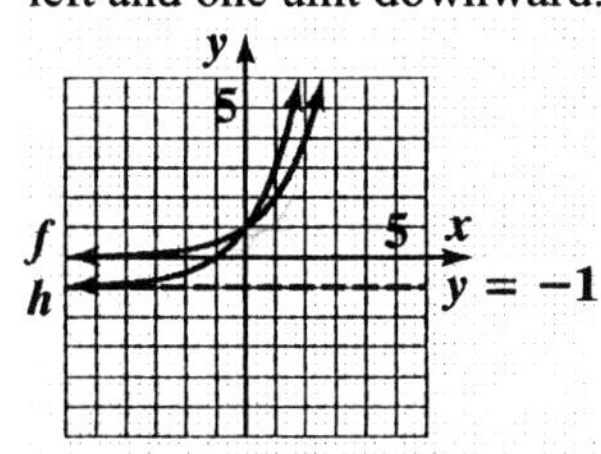

$f(x) = 2^x$
$h(x) = 2^{x+1} - 1$

Asymptote: $y = -1$
Domain: $(-\infty, \infty)$; Range: $(-1, \infty)$

30. The graph of $h(x) = 2^{x+2} - 1$ can be obtained by shifting the graph of $f(x) = 2^x$ two units to the left and one unit downward.

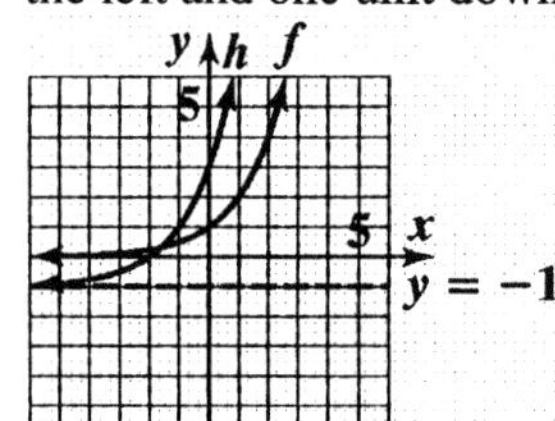

$f(x) = 2^x$
$h(x) = 2^{x+2} - 1$

Asymptote: $y = -1$
Domain: $(-\infty, \infty)$; Range: $(-1, \infty)$

31. The graph of $g(x) = -2^x$ can be obtained by reflecting the graph of $f(x) = 2^x$ about the x-axis.

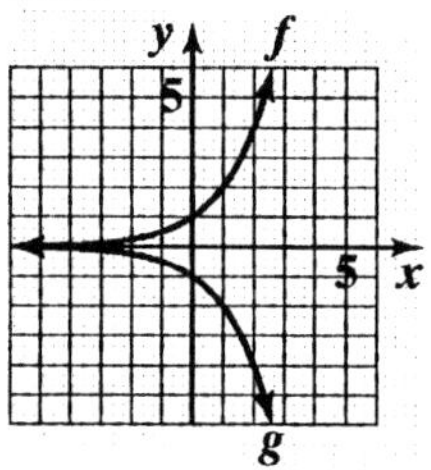

$f(x) = 2^x$
$g(x) = -2^x$

Asymptote: $y = 0$
Domain: $(-\infty, \infty)$; Range: $(-\infty, 0)$

32. The graph of $g(x) = 2^{-x}$ can be obtained by reflecting the graph of $f(x) = 2^x$ about the y-axis.

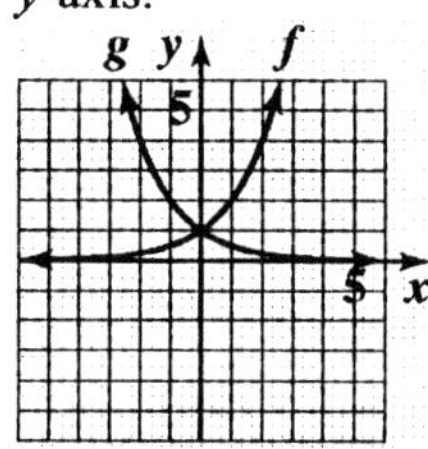

$f(x) = 2^x$
$g(x) = 2^{-x}$

Asymptote: $y = 0$
Domain: $(-\infty, \infty)$; Range: $(0, \infty)$

33. The graph of $g(x) = 2 \cdot 2^x$ can be obtained by vertically stretching the graph of $f(x) = 2^x$ by a factor of two.

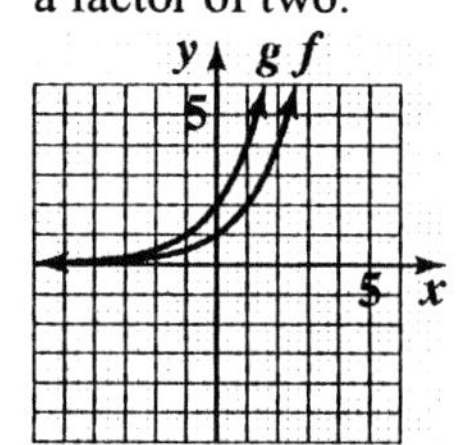

$f(x) = 2^x$
$g(x) = 2 \cdot 2^x$

Asymptote: $y = 0$
Domain: $(-\infty, \infty)$; Range: $(0, \infty)$

34. The graph of $g(x)=\frac{1}{2}\cdot 2^x$ can be obtained by vertically shrinking the graph of $f(x)=2^x$ by a factor of one-half.

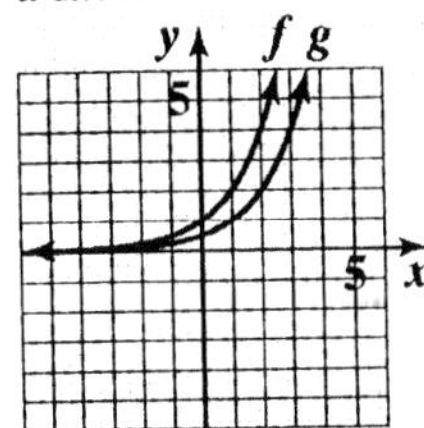

$f(x) = 2^x$
$g(x) = \frac{1}{2}\cdot 2^x$

Asymptote: $y=0$
Domain: $(-\infty,\infty)$; Range: $(0,\infty)$

35. The graph of $g(x)=e^{x-1}$ can be obtained by moving $f(x)=e^x$ 1 unit right.

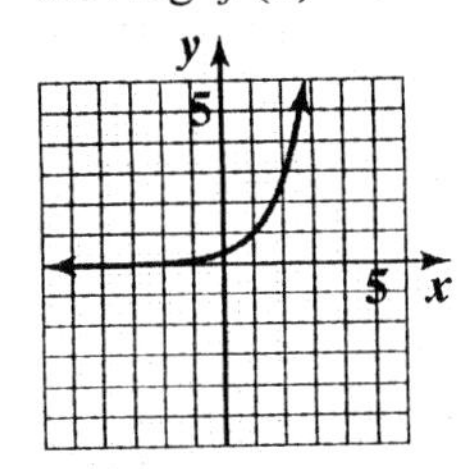

$g(x) = e^{x-1}$

Asymptote: $y=0$
Domain: $(-\infty,\infty)$; Range: $(0,\infty)$

36. The graph of $g(x)=e^{x+1}$ can be obtained by moving $f(x)=e^x$ 1 unit left.

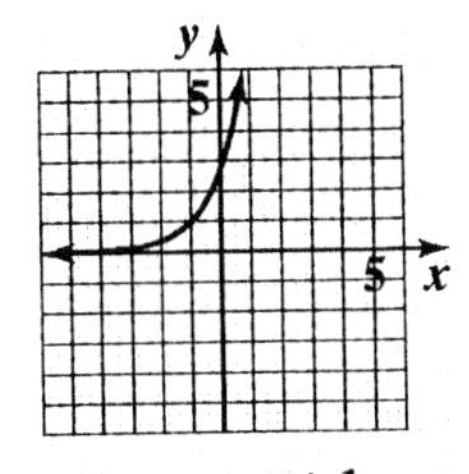

$g(x) = e^{x+1}$

Asymptote: $y=0$
Domain: $(-\infty,\infty)$; Range: $(0,\infty)$

37. The graph of $g(x)=e^x+2$ can be obtained by moving $f(x)=e^x$ 2 units up.

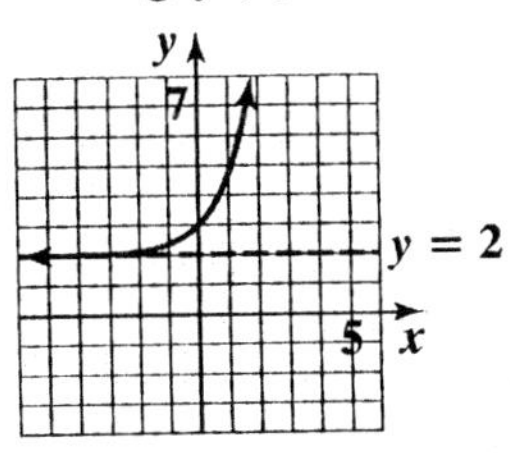

$g(x) = e^x + 2$

Asymptote: $y=2$
Domain: $(-\infty,\infty)$; Range: $(2,\infty)$

38. The graph of $g(x)=e^x-1$ can be obtained by moving $f(x)=e^x$ 1 unit down.

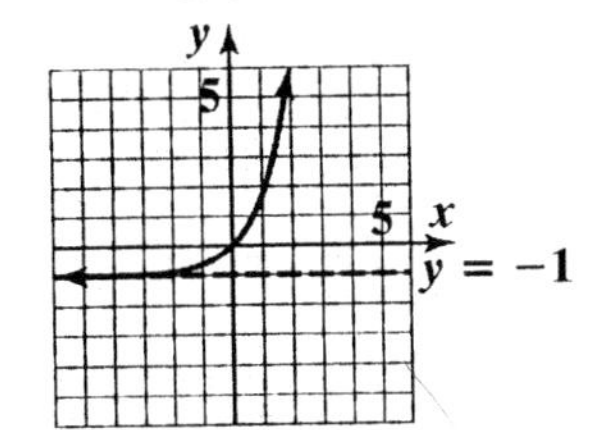

$g(x) = e^x - 1$

Asymptote: $y=-1$
Domain: $(-\infty,\infty)$; Range: $(-1,\infty)$

39. The graph of $h(x)=e^{x-1}+2$ can be obtained by moving $f(x)=e^x$ 1 unit right and 2 units up.

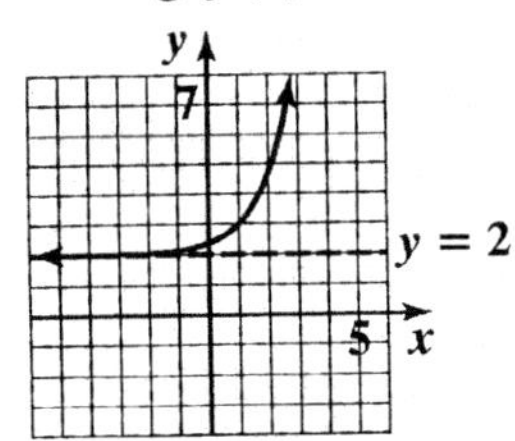

$h(x) = e^{x-1} + 2$

Asymptote: $y=2$
Domain: $(-\infty,\infty)$; Range: $(2,\infty)$

40. The graph of $h(x) = e^{x+1} - 1$ can be obtained by moving $f(x) = e^x$ 1 unit left and 1 unit down.

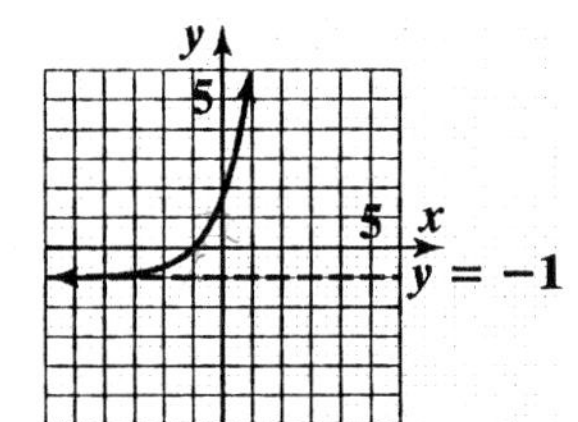

$h(x) = e^{x+1} - 1$

Asymptote: $y = -1$
Domain: $(-\infty, \infty)$; Range: $(-1, \infty)$

41. The graph of $h(x) = e^{-x}$ can be obtained by reflecting $f(x) = e^x$ about the y-axis.

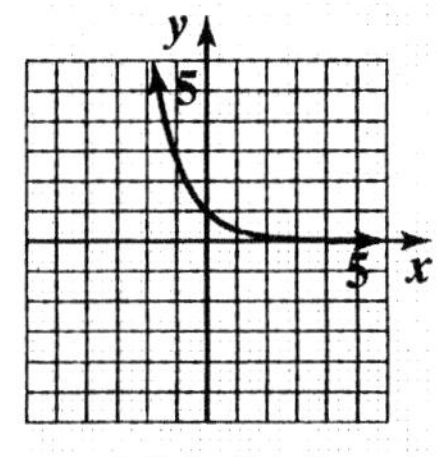

$h(x) = e^{-x}$

Asymptote: $y = 0$
Domain: $(-\infty, \infty)$; Range: $(0, \infty)$

42. The graph of $h(x) = -e^x$ can be obtained by reflecting $f(x) = e^x$ about the x-axis.

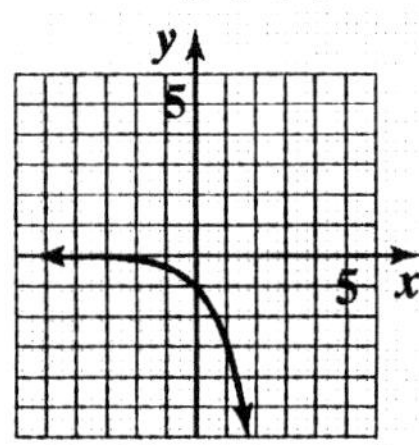

$h(x) = -e^x$

Asymptote: $y = 0$
Domain: $(-\infty, \infty)$; Range: $(-\infty, 0)$

43. The graph of $g(x) = 2e^x$ can be obtained by stretching $f(x) = e^x$ vertically by a factor of 2.

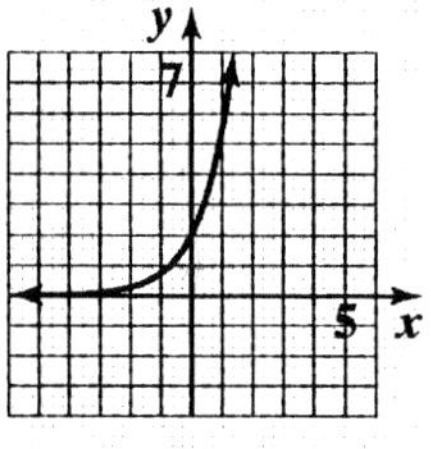

$g(x) = 2e^x$

Asymptote: $y = 0$
Domain: $(-\infty, \infty)$; Range: $(0, \infty)$

44. The graph of $g(x) = \frac{1}{2}e^x$ can be obtained by shrinking $f(x) = e^x$ vertically by a factor of $\frac{1}{2}$.

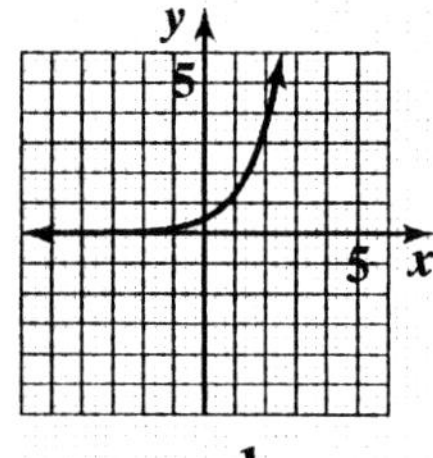

$g(x) = \frac{1}{2}e^x$

Asymptote: $y = 0$
Domain: $(-\infty, \infty)$; Range: $(0, \infty)$

45. The graph of $h(x) = e^{2x} + 1$ can be obtained by stretching $f(x) = e^x$ horizontally by a factor of 2 and then moving the graph up 1 unit.

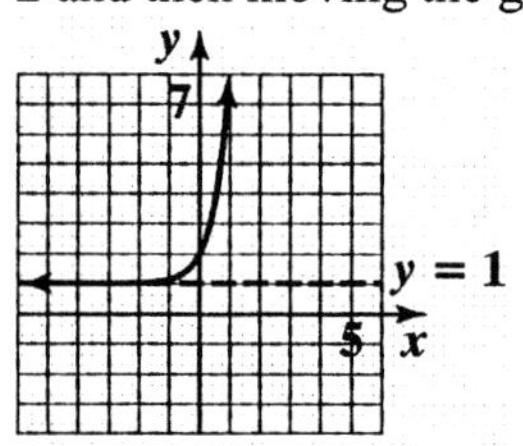

$h(x) = e^{2x} + 1$

Asymptote: $y = 1$
Domain: $(-\infty, \infty)$; Range: $(1, \infty)$

46. The graph of $h(x) = e^{\frac{1}{2}x} + 2$ can be obtained by shrinking $f(x) = e^x$ horizontally by a factor of $\frac{1}{2}$ and then moving the graph up 2 units.

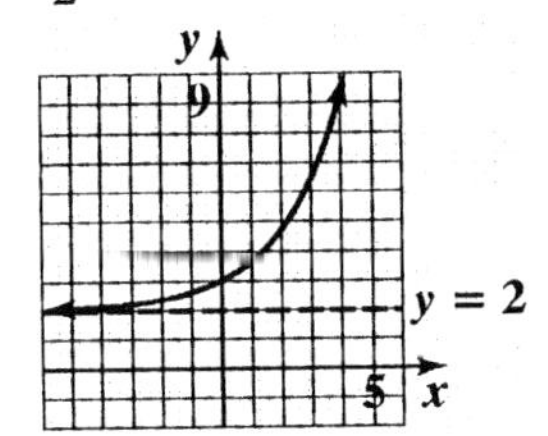

$h(x) = e^{x/2} + 2$

Asymptote: $y = 2$

Domain: $(-\infty, \infty)$; Range: $(2, \infty)$

47. The graph of $g(x)$ can be obtained by reflecting $f(x)$ about the y-axis.

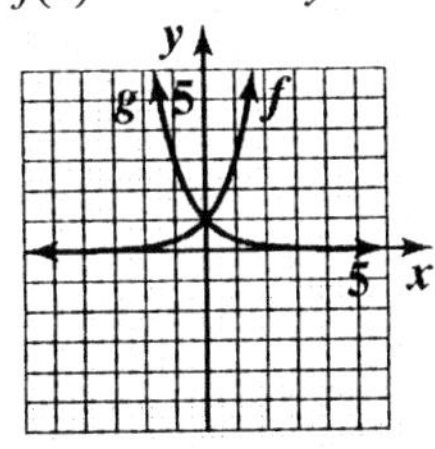

$f(x) = 3^x$

$g(x) = 3^{-x}$

Asymptote of $f(x)$: $y = 0$

Asymptote of $g(x)$: $y = 0$

48. The graph of $g(x)$ can be obtained by reflecting $f(x)$ about the x-axis.

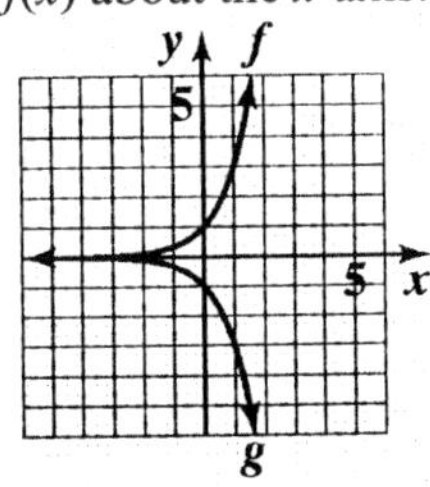

$f(x) = 3^x$

$g(x) = -3^x$

Asymptote of $f(x)$: $y = 0$

Asymptote of $g(x)$: $y = 0$

49. The graph of $g(x)$ can be obtained by vertically shrinking $f(x)$ by a factor of $\frac{1}{3}$.

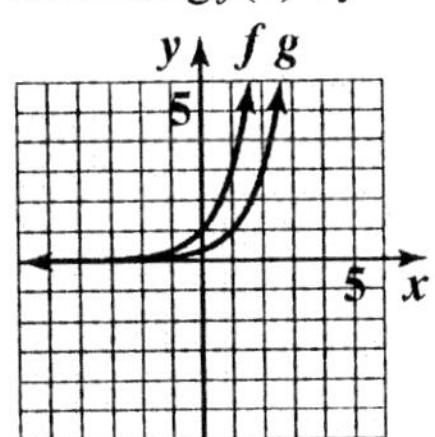

$f(x) = 3^x$

$g(x) = \frac{1}{3} \cdot 3^x$

Asymptote of $f(x)$: $y = 0$

Asymptote of $g(x)$: $y = 0$

50. The graph of $g(x)$ can be obtained by horizontally stretching $f(x)$ by a factor of 3..

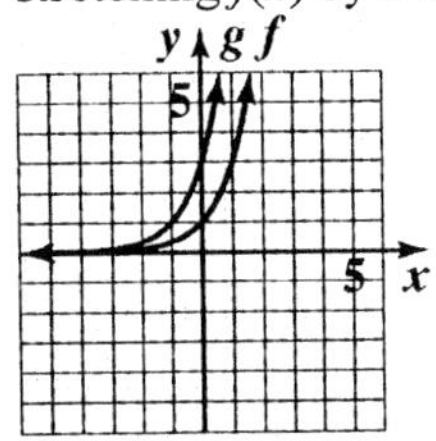

$f(x) = 3^x$

$g(x) = 3 \cdot 3^x$

Asymptote of $f(x)$: $y = 0$

Asymptote of $g(x)$: $y = 0$

51. The graph of $g(x)$ can be obtained by moving the graph of $f(x)$ one space to the right and one space up.

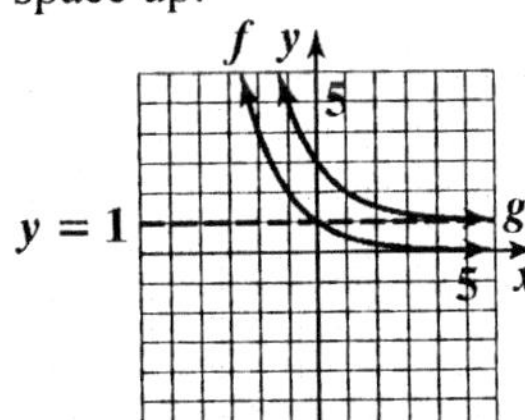

$f(x) = \left(\frac{1}{2}\right)^x$

$g(x) = \left(\frac{1}{2}\right)^{x-1} + 1$

Asymptote of $f(x)$: $y = 0$

Asymptote of $g(x)$: $y = 1$

52. The graph of $g(x)$ can be obtained by moving the graph of $f(x)$ one space to the right and two spaces up.

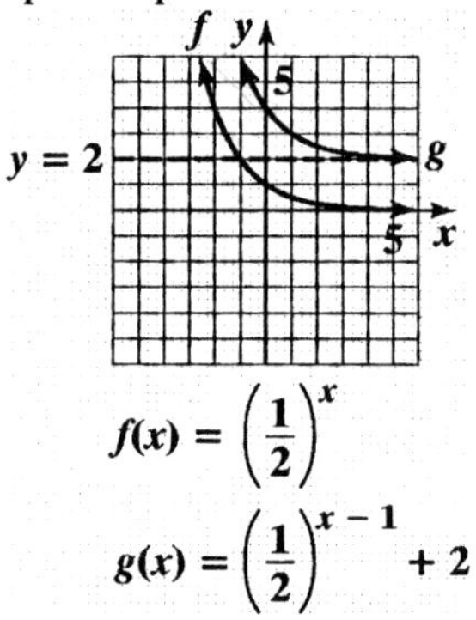

Asymptote of $f(x)$: $y = 0$
Asymptote of $g(x)$: $y = 2$

53. **a.** $A = 10,000\left(1+\frac{0.055}{2}\right)^{2(5)}$
$\approx \$13,116.51$

b. $A = 10,000\left(1+\frac{0.055}{4}\right)^{4(5)}$
$\approx \$13,140.67$

c. $A = 10,000\left(1+\frac{0.055}{12}\right)^{12(5)}$
$\approx \$13,157.04$

d. $A = 10,000e^{0.055(5)}$
$\approx \$13,165.31$

54. **a.** $A = 5000\left(1+\frac{0.065}{2}\right)^{2(10)} \approx \$9479.19$

b. $A = 5000\left(1+\frac{0.065}{4}\right)^{4\cdot 10} \approx \$9527.79$

c. $A = 5000\left(1+\frac{0.065}{12}\right)^{12(10)} \approx = \$9560.92$

d. $A = 5000(e)^{0.065(10)} \approx 9577.70$

55. $A = 12,000\left(1+\frac{0.07}{12}\right)^{12(3)}$
$\approx 14,795.11$ (7% yield)
$A = 12,000e^{0.0685(3)}$
$\approx 14,737.67$ (6.85% yield)
Investing \$12,000 for 3 years at 7% compounded monthly yields the greater return.

56. $A = 6000\left(1+\frac{0.0825}{4}\right)^{4(4)}$
$\approx$ \$8317.84 (8.25% yield)
$A = 6000\left(1+\frac{0.083}{2}\right)^{2(4)}$
$\approx$ \$ 8306.64 (8.3% yield)
Investing \$6000 for 4 years at 8.25% compounded quarterly yields the greater return.

57.

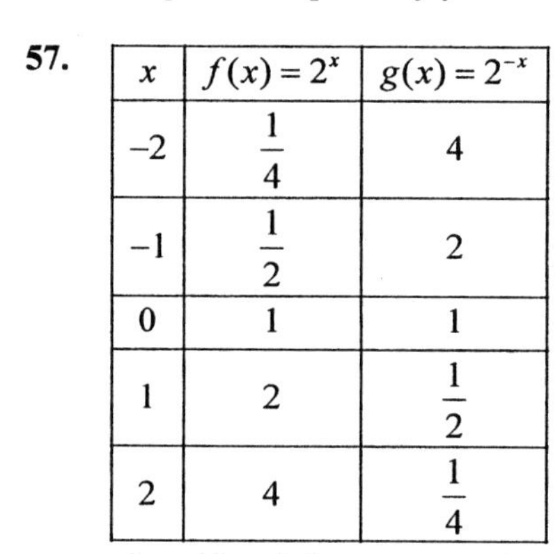

x	$f(x)=2^x$	$g(x)=2^{-x}$
−2	$\frac{1}{4}$	4
−1	$\frac{1}{2}$	2
0	1	1
1	2	$\frac{1}{2}$
2	4	$\frac{1}{4}$

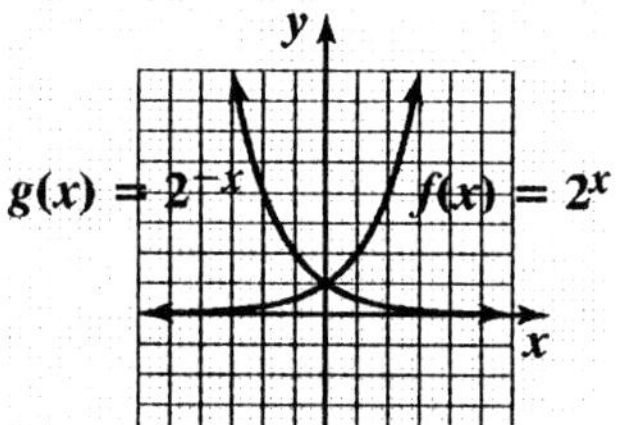

The point of intersection is $(0,1)$.

58.

x	$f(x)=2^{x+1}$	$g(x)=2^{-x+1}$
-2	$\frac{1}{2}$	8
-1	1	4
0	2	2
1	4	1
2	8	$\frac{1}{2}$

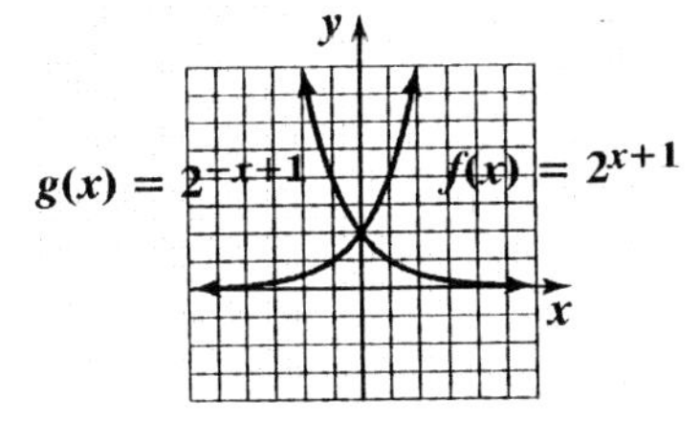

The point of intersection is $(0,2)$.

59

x	$y=2^x$
-2	$\frac{1}{4}$
-1	$\frac{1}{2}$
0	1
1	2
2	4

y	$x=2^y$
-2	$\frac{1}{4}$
-1	$\frac{1}{2}$
0	1
1	2
2	4

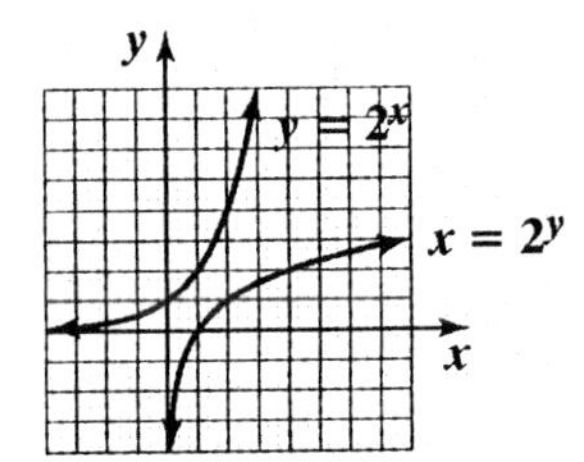

60

x	$y=3^x$
-2	$\frac{1}{9}$
-1	$\frac{1}{3}$
0	1
1	3
2	9

y	$x=3^y$
-2	$\frac{1}{9}$
-1	$\frac{1}{3}$
0	1
1	3
2	9

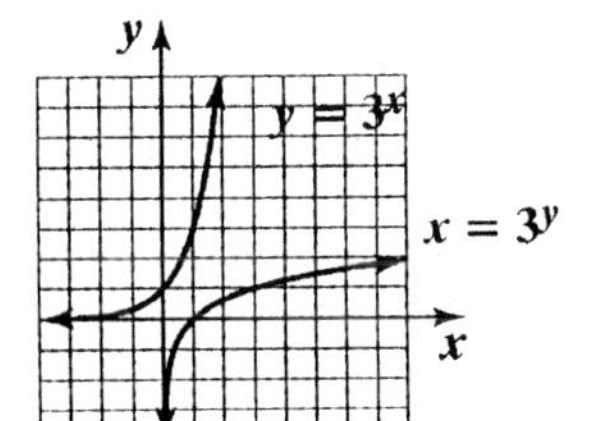

61. The graph is of the form $y=b^x$.
Substitute values from the point (1, 4) to find b.

$y=b^x$

$4=b^1$

$4=b$

The equation of the graph is $y=4^x$

62. The graph is of the form $y=b^x$.
Substitute values from the point (1, 6) to find b.

$y=b^x$

$6=b^1$

$6=b$

The equation of the graph is $y=6^x$

63. The graph is of the form $y=-b^x$.
Substitute values from the point (1, $-e$) to find b.

$y=-b^x$

$-e=-b^1$

$e=b$

The equation of the graph is $y=-e^x$

64. The graph is of the form $y=b^x$.
Substitute values from the point (-1, e) to find b.

$y=b^x$

$e=b^{-1}$

$e=\frac{1}{b}$

$eb=1$

$b=\frac{1}{e}$

The equation of the graph is $y=\left(\frac{1}{e}\right)^x=e^{-x}$

65. **a.** $f(0)=574(1.026)^0$
$=574(1)=574$
India's population in 1974 was 574 million.

b. $f(27)=574(1.026)^{27}\approx 1148$
India's population in 2001 will be 1148 million.

c. Since $2028-1974=54$, find
$f(54)=574(1.026)^{54}\approx 2295$.
India's population in 2028 will be 2295 million.

d. $2055-1974=81$, find
$f(54)=574(1.026)^{81}\approx 4590$.
India's population in 2055 will be 4590 million.

e. India's population appears to be doubling every 27 years.

66. $f(80)=1000(0.5)^{\frac{80}{30}}=157.49$
Chernobyl will not be safe for human habitation by 2066. There will still be 157.5 kilograms of cesium-137 in Chernobyl's atmosphere.

67. $S=465{,}000(1+0.06)^{10}$
$=465{,}000(1.06)^{10}\approx \$832{,}744$

68. $S=510{,}000(1+0.03)^5$
$=510{,}000(1.03)^5$
$\approx \$591{,}230$

69. $2^{1.7}\approx 3.249009585$
$2^{1.73}\approx 3.317278183$
$2^{1.732}\approx 3.321880096$
$2^{1.73205}\approx 3.321995226$
$2^{1.7320508}\approx 3.321997068$
$2^{\sqrt{3}}\approx 3.321997085$
The closer the exponent is to $\sqrt{3}$, the closer the value is to $2^{\sqrt{3}}$.

70. $2^3\approx 8$
$2^{3.1}\approx 8.5741877$
$2^{3.14}\approx 8.815240927$
$2^{3.141}\approx 8.821353305$
$2^{3.1415}\approx 8.824411082$
$2^{3.14159}\approx 8.824961595$
$2^{3.141593}\approx 8.824979946$
$2^{\pi}\approx 8.824977827$
The closer the exponent gets to π, the closer the value is to 2^{π}.

71. $f(45)=0.16(45)+1.43=8.63$
$g(45)=1.8e^{0.04(45)}\approx 10.9$
The linear function is a better model for the graph's value of 8.9.

72. $f(30)=0.16(30)+1.43=6.23$
$g(30)=1.8e^{0.04(30)}\approx 6.0$
The exponential function is a better model for the graph's value of 5.7.

73. **a.** $f(0)=80e^{-0.5(0)}+20$
$=80e^0+20$
$=80(1)+20$
$=100$
100% of the material is remembered at the moment it is first learned.

b. $f(1)=80e^{-0.5(1)}+20\approx 68.5$
68.5% of the material is remembered 1 week after it is first learned.

c. $f(4)=80e^{-0.5(4)}+20\approx 30.8$
30.8% of the material is remembered 4 week after it is first learned.

d. $f(52)=80e^{-0.5(52)}+20\approx 20$
20% of the material is remembered 1 year after it is first learned.

74. **a.** $24\left(1+\frac{0.05}{12}\right)^{12(379)}\approx \$3{,}917{,}360{,}753$

b. $24e^{0.05(379)}\approx \$4{,}074{,}662{,}794$

75. $f(4) = \dfrac{258,051}{1+6.78e^{-1.21(4)}} \approx 244,921$

$g(4) = 55,979.5(4) + 36,217.8 \approx 260,136$

$f(x)$ is a better model for the graph's value of 246,570.

76. $f(3) = \dfrac{258,051}{1+6.78e^{-1.21(3)}} \approx 218,728$

$g(3) = 55,979.5(3) + 36,217.8 \approx 204,156$

$f(x)$ is a better model for the graph's value of 215,093.

83.

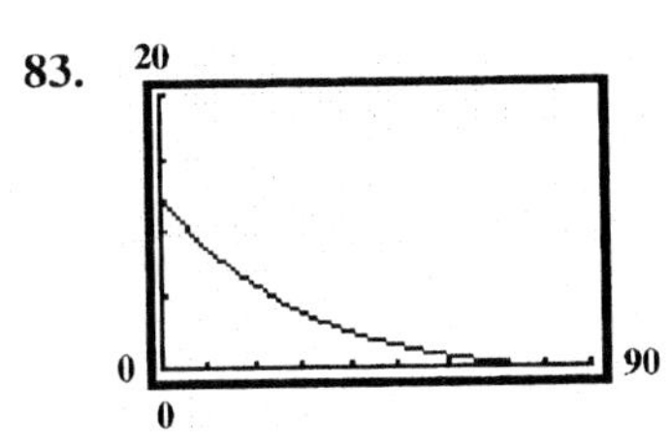

When $x = 31, y \approx 3.77$. NASA would not have launched the *Challenger*, since nearly 4 O-rings are expected to fail.

84. a. $A = 10,000\left(1+\dfrac{0.05}{4}\right)^{4t}$

$A = 10,000\left(1+\dfrac{0.045}{12}\right)^{12t}$

b. 5% compounded quarterly offers the better return.

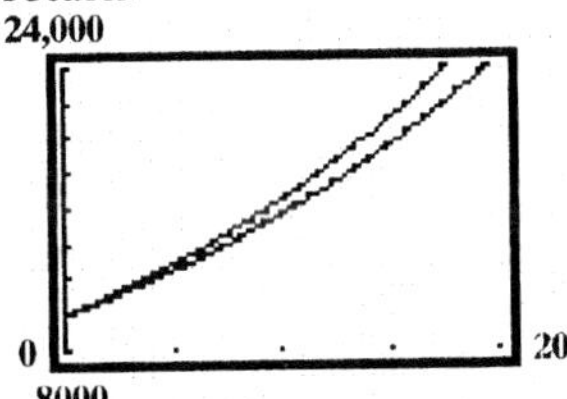

85. a.

5

−5 5

−5

b.

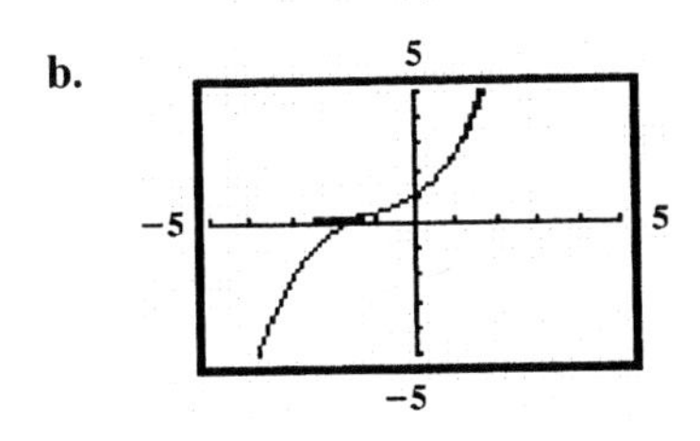

c.

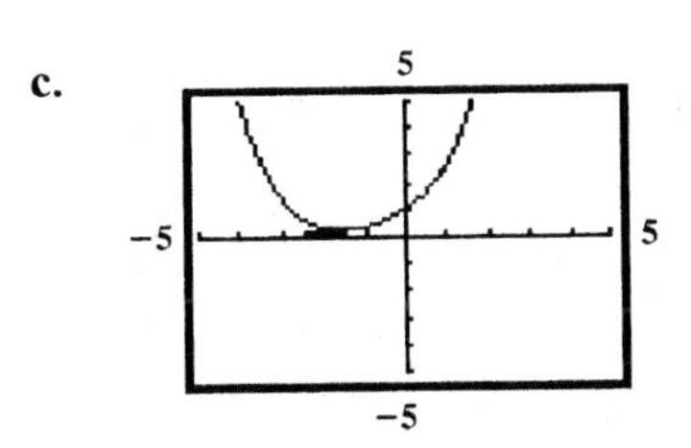

d. Answers may vary.

86. a. False; even continuous compounding has a limit on the amount over a fixed interval of time.

b. False; $f(x) = 3^{-x}$ reflects the graph of $y = 3^x$ about the y-axis while $f(x) = -3^x$ reflects the graph of $y = 3^x$ about the x-axis.

c. False; $e \approx 2.718$, but e is irrational.

d. True; $f(x) = \left(\dfrac{1}{3}\right)^x = \left(3^{-1}\right)^x = (3)^{-x} = g(x)$.

(d) is true.

87. $y = 3^x$ is (d). y increases as x increases, but not as quickly as $y = 5^x$. $y = 5^x$ is (c). $y = \left(\dfrac{1}{3}\right)^x$ is (a).

$y = \left(\dfrac{1}{3}\right)^x$ is the same as $y = 3^{-x}$, so it is (d) reflected about the y-axis. $y = \left(\dfrac{1}{5}\right)^x$ is (b).

$y = \left(\dfrac{1}{5}\right)^x$ is the same as $y = 5^{-x}$, so it is (c) reflected about the y-axis.

88.

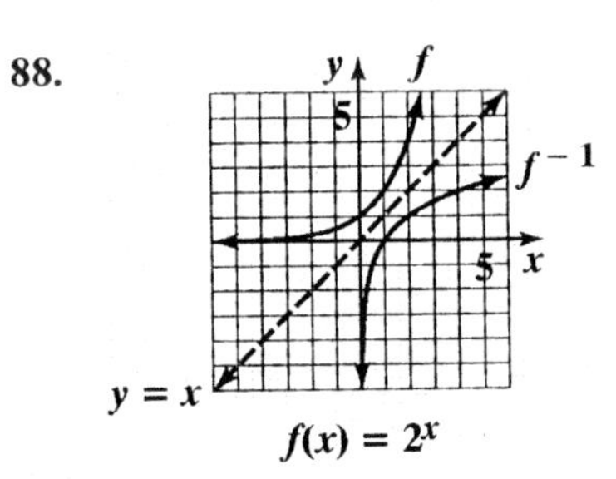

89. a.

$$\cosh(-x) = \frac{e^{-x} + e^{-(-x)}}{2} = \frac{e^{-x} + e^{x}}{2} = \frac{e^{x} + e^{-x}}{2} = \cosh x$$

b.
$$\sinh(-x) = \frac{e^{-x} - e^{-(-x)}}{2} = \frac{e^{-x} - e^{x}}{2} = \frac{-\left(-e^{-x} + e^{x}\right)}{2} = -\frac{e^{x} - e^{-x}}{2} = -\sinh x$$

c.
$$(\cosh x)^2 - (\sinh x)^2 \stackrel{?}{=} 1$$
$$\left(\frac{e^x + e^{-x}}{2}\right)^2 - \left(\frac{e^x - e^{-x}}{2}\right)^2 \stackrel{?}{=} 1$$
$$\frac{e^{2x} + 2 + e^{-2x}}{4} - \frac{e^{2x} - 2 + e^{-2x}}{4} \stackrel{?}{=} 1$$
$$\frac{e^{2x} + 2 + e^{-2x} - e^{2x} + 2 - e^{-2x}}{4} \stackrel{?}{=} 1$$
$$\frac{4}{4} \stackrel{?}{=} 1$$
$$1 = 1$$

Section 3.2

Check Point Exercises

1. **a.** $3 = \log_7 x$ means $7^3 = x$.

b. $2 = \log_b 25$ means $b^2 = 25$.

c. $\log_4 26 = y$ means $4^y = 26$.

2. **a.** $2^5 = x$ means $5 = \log_2 x$.

b. $b^3 = 27$ means $3 = \log_b 27$.

c. $e^y = 33$ means $y = \log_e 33$.

3. **a.** Question: 10 to what power gives 100?
$\log_{10} 100 = 2$ because $10^2 = 100$.

b. Question: 3 to what power gives 3?
$\log_3 3 = 1$ because $3^1 = 3$.

c. Question: 36 to what power gives 6?
$\log_{36} 6 = \frac{1}{2}$ because $36^{1/2} = \sqrt{36} = 6$

4. **a.** Because $\log_b b = 1$, we conclude $\log_9 9 = 1$.

b. Because $\log_b 1 = 0$, we conclude $\log_8 1 = 0$.

5. **a.** Because $\log_b b^x = x$, we conclude $\log_7 7^8 = 8$.

b. Because $b^{\log_b x} = x$, we conclude $3^{\log_3 17} = 17$.

6. First, set up a table of coordinates for $f(x) = 3^x$.

x	–2	–1	0	1	2	3
$f(x) = 3^x$	$\frac{1}{9}$	$\frac{1}{3}$	1	3	9	27

Reversing these coordinates gives the coordinates for the inverse function
$g(x) = \log_3 x$.

x	$\frac{1}{9}$	$\frac{1}{3}$	1	3	9	27
$g(x) = \log_3 x$	–2	–1	0	1	2	3

The graph of the inverse can also be drawn by reflecting the graph of $f(x) = 3^x$ about the line $y = x$.

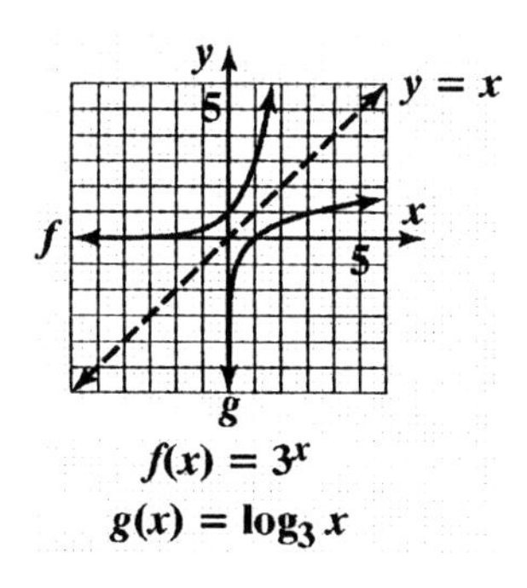

7. The domain of h consists of all x for which $x - 5 > 0$. Solving this inequality for x, we obtain $x > 5$. Thus, the domain of h is $(5, \infty)$.

8. Substitute the boy's age, 10, for x and evaluate the function at 10.
$$f(10) = 29 + 48.8\log(10+1) = 29 + 48.8\log(11) \approx 80$$
Thus, a 10-year-old boy is approximately 80% of his adult height.

9. Because $I = 10{,}000\, I_0$,
$$R = \log\frac{10{,}000I_0}{I_0} = \log 10{,}000 = 4$$
The earthquake registered 4.0 on the Richter scale.

10. a. The domain of f consists of all x for which $4 - x > 0$. Solving this inequality for x, we obtain $x < 4$. Thus, the domain of f is $(-\infty, 4)$

b. The domain of g consists of all x for which $x^2 > 0$. Solving this inequality for x, we obtain $x < 0$ or $x > 0$. Thus the domain of g is $(-\infty, 0) \cup (0, \infty)$.

11. Find the temperature increase after 30 minutes by substituting 30 for x and evaluating the function at 30.
$$f(x) = 13.4\ln x - 11.6$$
$$f(30) = 13.4\ln 30 - 11.6 \approx 34$$
The function models the actual increase shown in the graph quite well.

Exercise Set 3.2

1. $2^4 = 16$

2. $2^6 = 64$

3. $3^2 = x$

4. $9^2 = x$

5. $b^5 = 32$

6. $b^3 = 27$

7. $6^y = 216$

8. $5^y = 125$

9. $\log_2 8 = 3$

10. $\log_5 625 = 4$

11. $\log_2 \frac{1}{16} = -4$

12. $\log_5 \frac{1}{125} = -3$

13. $\log_8 2 = \frac{1}{3}$

14. $\log_{64} 4 = \frac{1}{3}$

15. $\log_{13} x = 2$

16. $\log_{15} x = 2$

17. $\log_b 1000 = 3$

18. $\log_b 343 = 3$

19. $\log_7 200 = y$

20. $\log_8 300 = y$

21. $\log_4 16 = 2$ because $4^2 = 16$.

22. $\log_7 49 = 2$ because $7^2 = 49$.

23. $\log_2 64 = 6$ because $2^6 = 64$.

24. $\log_3 27 = 3$ because $3^3 = 27$.

25. $\log_5 \frac{1}{5} = -1$ because $5^{-1} = \frac{1}{5}$.

26. $\log_6 \frac{1}{6} = -1$ because $6^{-1} = \frac{1}{6}$.

27. $\log_2 \frac{1}{8} = -3$ because $2^{-3} = \frac{1}{8}$.

28. $\log_3 \frac{1}{9} = -2$ because $3^{-2} = \frac{1}{9}$.

29. $\log_7 \sqrt{7} = \frac{1}{2}$ because $7^{\frac{1}{2}} = \sqrt{7}$.

30. $\log_6 \sqrt{6} = \frac{1}{2}$ because $6^{\frac{1}{2}} = \sqrt{6}$.

31. $\log_2 \frac{1}{\sqrt{2}} = -\frac{1}{2}$ because $2^{-\frac{1}{2}} = \frac{1}{\sqrt{2}}$.

32. $\log_3 \frac{1}{\sqrt{3}} = -\frac{1}{2}$ because $3^{-\frac{1}{2}} = \frac{1}{\sqrt{3}}$.

33. $\log_{64} 8 = \frac{1}{2}$ because $64^{1/2} = \sqrt{64} = 8$.

34. $\log_{81} 9 = \frac{1}{2}$ because $81^{1/2} = \sqrt{81} = 9$.

35. Because $\log_b b = 1$, we conclude $\log_5 5 = 1$.

36. Because $\log_b b = 1$, we conclude $\log_{11} 11 = 1$.

37. Because $\log_b 1 = 0$, we conclude $\log_4 1 = 0$.

38. Because $\log_b 1 = 0$, we conclude $\log_6 1 = 0$.

39. Because $\log_b b^x = x$, we conclude $\log_5 5^7 = 7$.

40. Because $\log_b b^x = x$, we conclude $\log_4 4^6 = 6$.

41. Because $b^{\log_b x} = x$, we conclude $8^{\log_8 19} = 19$.

42. Because $b^{\log_b x} = x$, we conclude $7^{\log_7 23} = 23$.

43. First, set up a table of coordinates for $f(x) = 4^x$.

x	-2	-1	0	1	2	3
$f(x) = 4x$	$\frac{1}{16}$	$\frac{1}{4}$	1	4	16	64

Reversing these coordinates gives the coordinates for the inverse function $g(x) = \log_4 x$.

x	$\frac{1}{16}$	$\frac{1}{4}$	1	4	16	64
$g(x) = \log_{4x}$	-2	-1	0	1	2	3

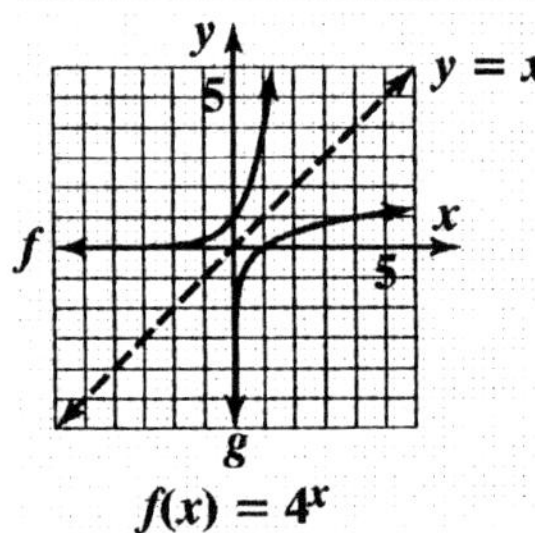

44. First, set up a table of coordinates for $f(x) = 5^x$.

x	-2	-1	0	1	2	3
$f(x) = 5^x$	$\frac{1}{25}$	$\frac{1}{5}$	1	5	25	125

Reversing these coordinates gives the coordinates for the inverse function $g(x) = \log_5 x$.

x	$\frac{1}{25}$	$\frac{1}{5}$	1	5	25	125
$g(x) = \log_5 x$	-2	-1	0	1	2	3

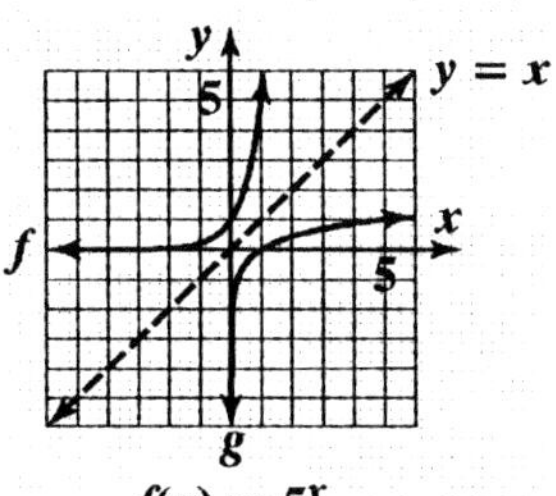

45. First, set up a table of coordinates for $f(x) = \left(\frac{1}{2}\right)^x$.

x	-2	-1	0	1	2	3
$f(x) = \left(\frac{1}{2}\right)^x$	4	2	1	$\frac{1}{2}$	$\frac{1}{4}$	$\frac{1}{8}$

Reversing these coordinates gives the coordinates for the inverse function $g(x) = \log_{1/2} x$.

x	4	2	1	$\frac{1}{2}$	$\frac{1}{4}$	$\frac{1}{8}$
$g(x) = \log_{1/2} x$	-2	-1	0	1	2	3

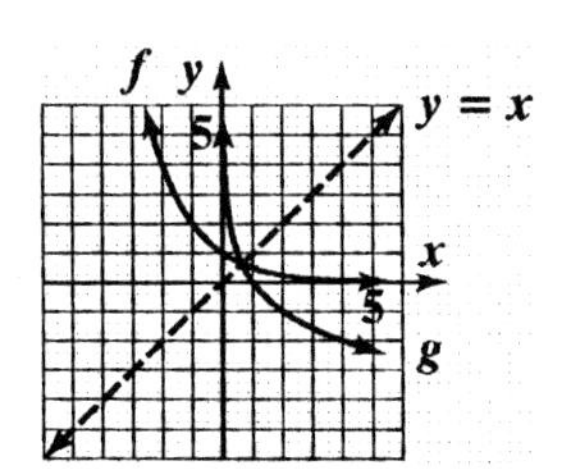

46. First, set up a table of coordinates for $f(x) = \left(\frac{1}{4}\right)^x$.

x	-2	-1	0	1	2	3
$f(x) = \left(\frac{1}{4}\right)^x$	16	4	1	$\frac{1}{4}$	$\frac{1}{16}$	$\frac{1}{64}$

Reversing these coordinates gives the coordinates for the inverse function $g(x) = \log_{1/4} x$.

x	16	4	1	$\frac{1}{4}$	$\frac{1}{16}$	$\frac{1}{64}$
$g(x) = \log_{1/4} x$	-2	-1	0	1	2	3

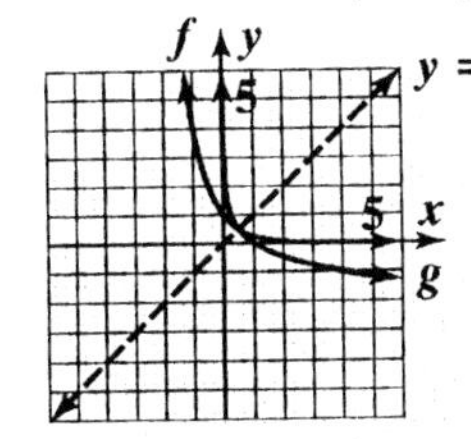

$f(x) = \left(\frac{1}{4}\right)^x$

$g(x) = \log_{1/4} x$

47. This is the graph of $f(x) = \log_3 x$ reflected about the x-axis and shifted up one unit, so the function is $H(x) = 1 - \log_3 x$.

48. This is the graph of $f(x) = \log_3 x$ reflected about the y-axis, so the function is $G(x) = \log_3(-x)$.

49. This is the graph of $f(x) = \log_3 x$ shifted down one unit, so the function is $h(x) = \log_3 x - 1$.

50. This is the graph of $f(x) = \log_3 x$ reflected about the x-axis, so the function is $F(x) = -\log_3 x$.

51. This is the graph of $f(x) = \log_3 x$ shifted right one unit, so the function is $g(x) = \log_3(x-1)$.

52. This is the graph of $f(x) = \log_3 x$.

53.

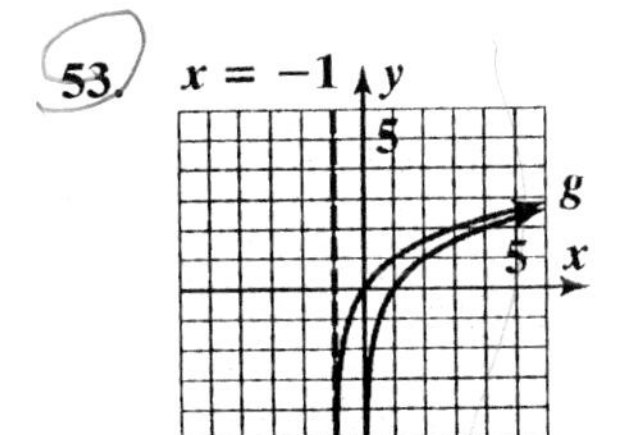

$f(x) = \log_2 x$

$g(x) = \log_2(x+1)$

vertical asymptote: $x = -1$
Domain: $(-1, \infty)$; Range: $(-\infty, \infty)$

54.

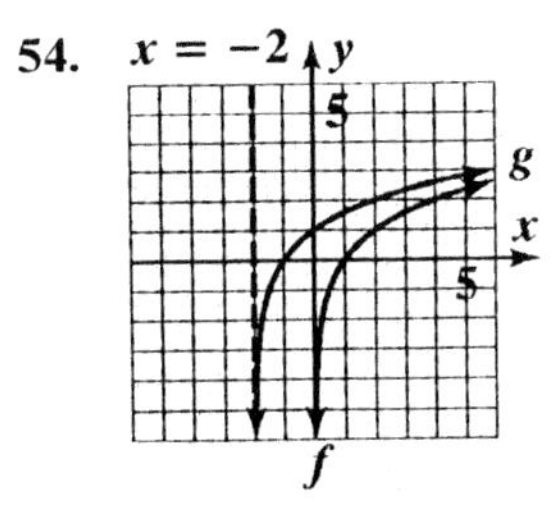

$f(x) = \log_2 x$

$g(x) = \log_2(x+2)$

vertical asymptote: $x = -2$
Domain: $(-2, \infty)$; Range: $(-\infty, \infty)$

55.

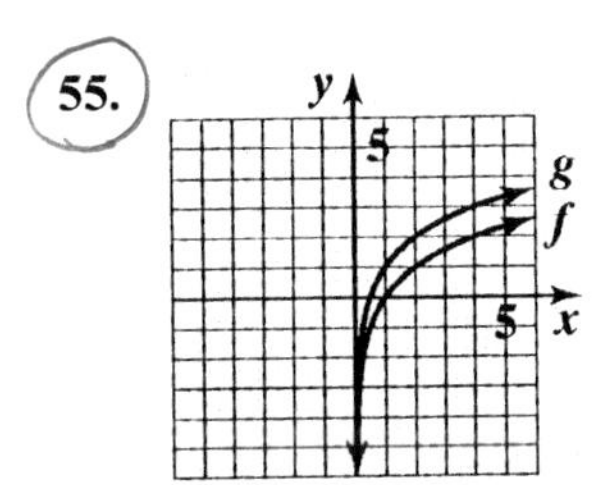

$f(x) = \log_2 x$

$g(x) = 1 + \log_2 x$

vertical asymptote: $x = 0$
Domain: $(0, \infty)$; Range: $(-\infty, \infty)$

56.

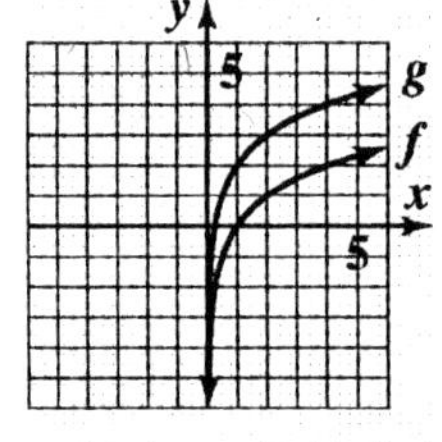

$f(x) = \log_2 x$

$g(x) = 2 + \log_2 x$

vertical asymptote: $x = 0$
Domain: $(0, \infty)$; Range: $(-\infty, \infty)$

57.

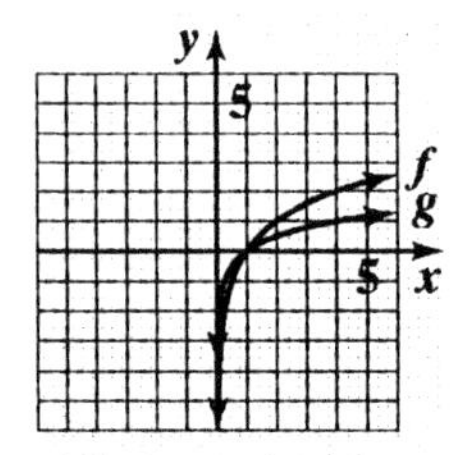

$f(x) = \log_2 x$

$g(x) = \frac{1}{2} \log_2 x$

vertical asymptote: $x = 0$
Domain: $(0, \infty)$; Range: $(-\infty, \infty)$

58.

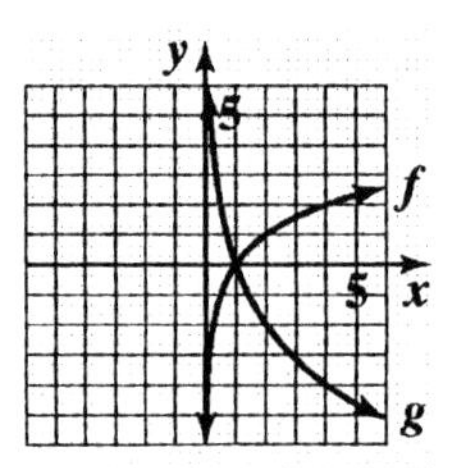

$f(x) = \log_2 x$

$g(x) = -2 \log_2 x$

vertical asymptote: $x = 0$
Domain: $(0, \infty)$; Range: $(-\infty, \infty)$

59.

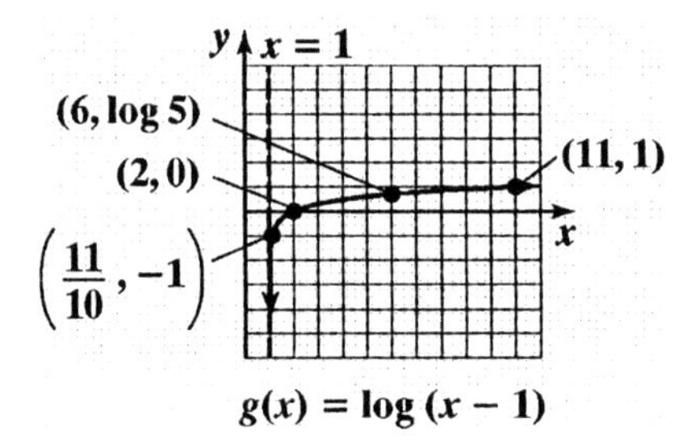

$g(x) = \log(x - 1)$

vertical asymptote: $x = 1$
Domain: $(1, \infty)$; Range: $(-\infty, \infty)$

60.

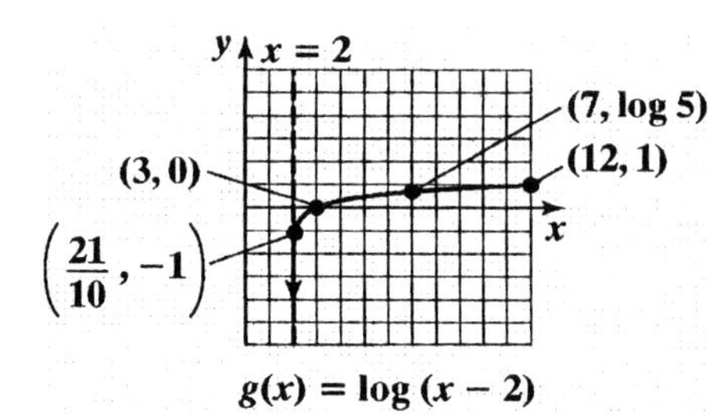

$g(x) = \log(x - 2)$

vertical asymptote: $x = 2$
Domain: $(2, \infty)$; Range: $(-\infty, \infty)$

61.

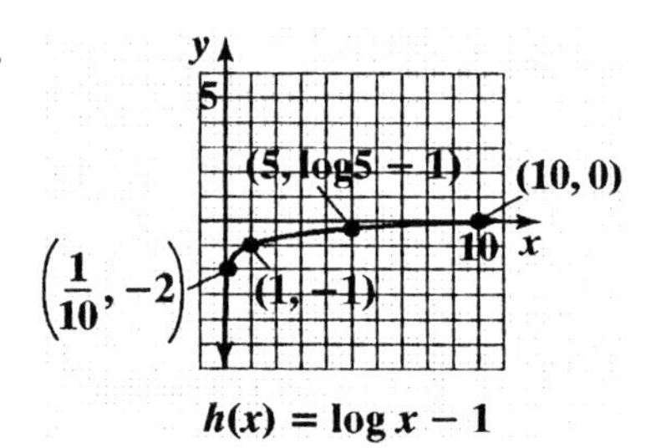

$h(x) = \log x - 1$

vertical asymptote: $x = 0$
Domain: $(0, \infty)$; Range: $(-\infty, \infty)$

62.

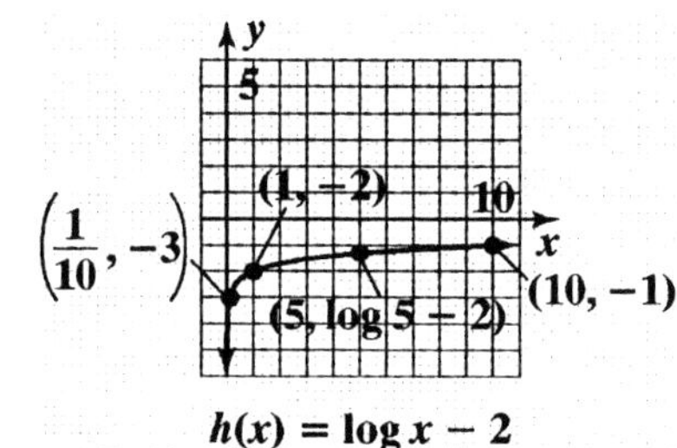

$h(x) = \log x - 2$

vertical asymptote: $x = 0$
Domain: $(0, \infty)$; Range: $(-\infty, \infty)$

63.

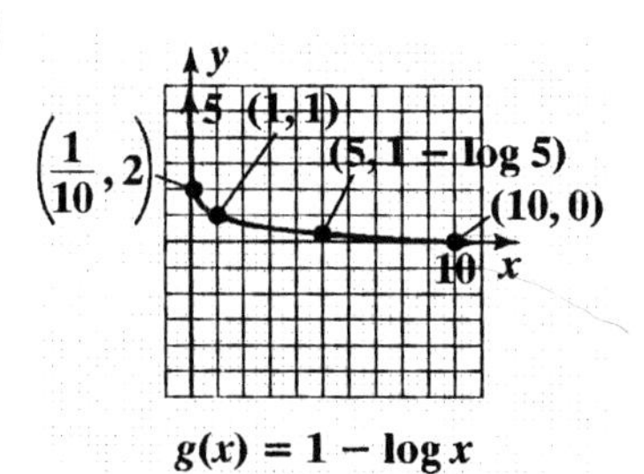

$g(x) = 1 - \log x$

vertical asymptote: $x = 0$
Domain: $(0, \infty)$; Range: $(-\infty, \infty)$

64.

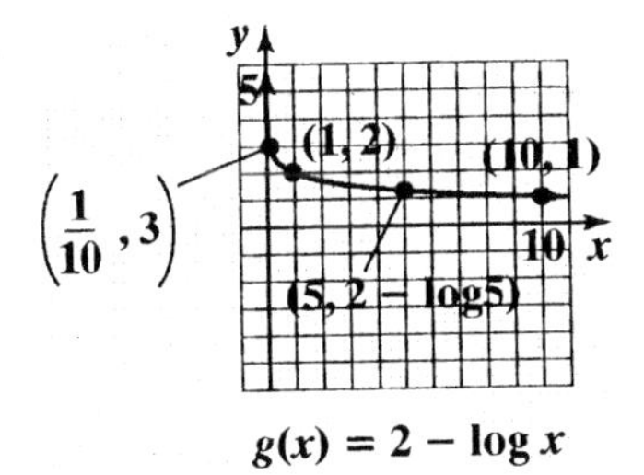

$g(x) = 2 - \log x$

vertical asymptote: $x = 0$
Domain: $(0, \infty)$; Range: $(-\infty, \infty)$

65.

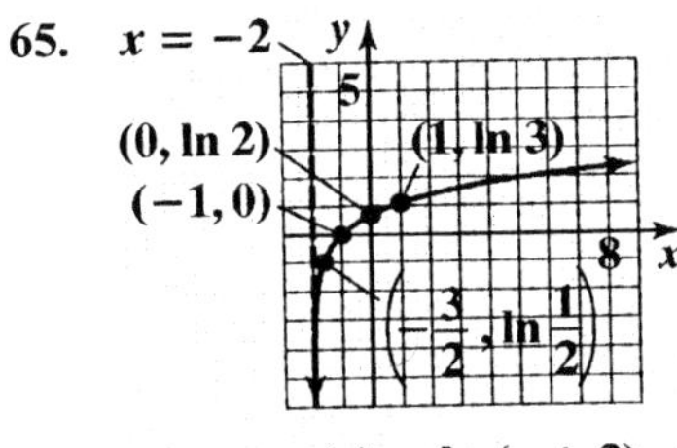

$g(x) = \ln(x + 2)$

vertical asymptote: $x = -2$
Domain: $(-2, \infty)$; Range: $(-\infty, \infty)$

66.

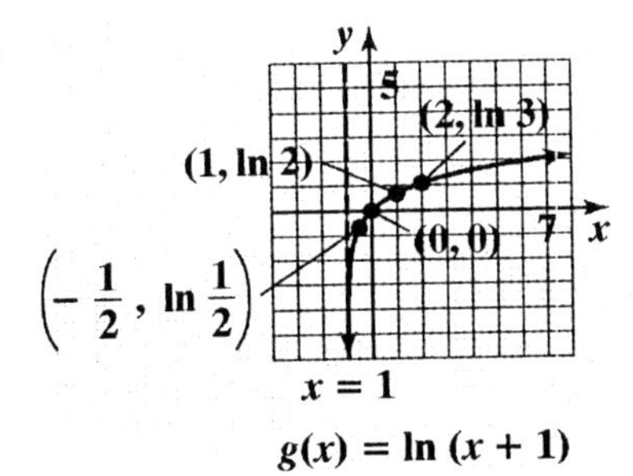

$g(x) = \ln(x + 1)$

vertical asymptote: $x = -1$
Domain: $(-1, \infty)$; Range: $(-\infty, \infty)$

67.

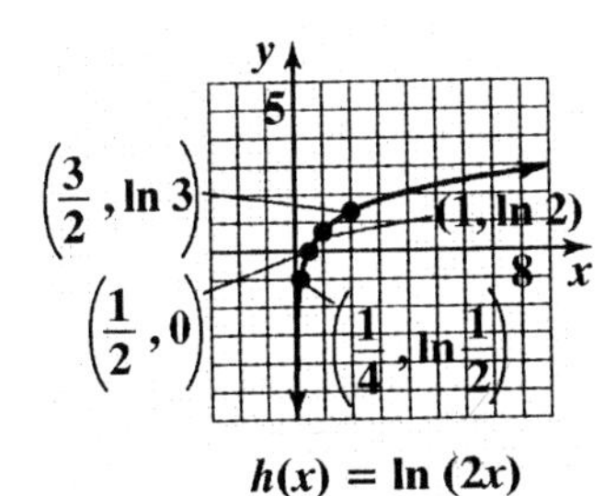

$h(x) = \ln(2x)$

vertical asymptote: $x = 0$
Domain: $(0, \infty)$; Range: $(-\infty, \infty)$

68.

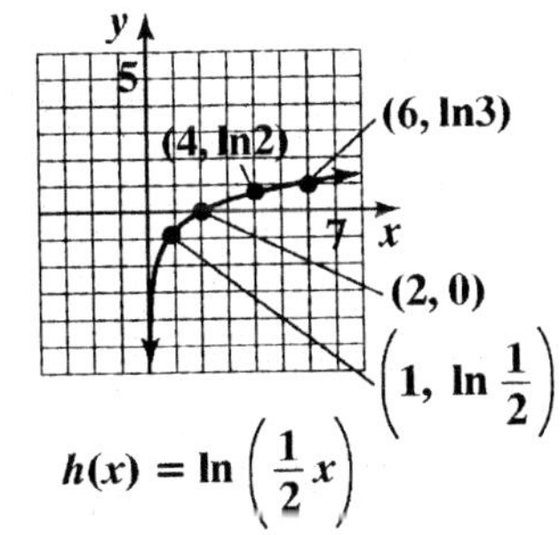

$h(x) = \ln\left(\frac{1}{2}x\right)$

vertical asymptote: $x = 0$
Domain: $(0, \infty)$; Range: $(-\infty, \infty)$

69.

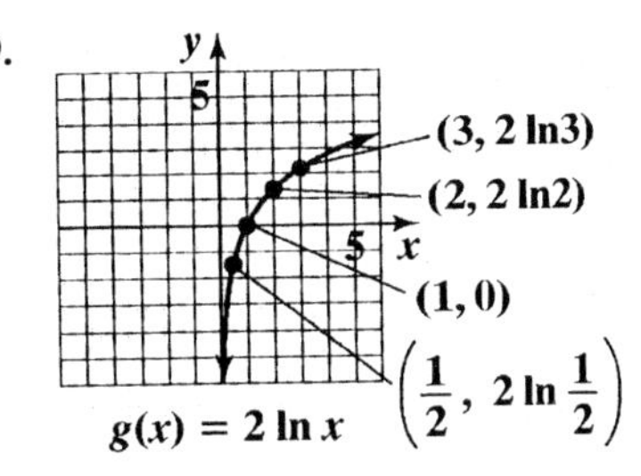

$g(x) = 2\ln x$

vertical asymptote: $x = 0$
Domain: $(0, \infty)$; Range: $(-\infty, \infty)$

70.

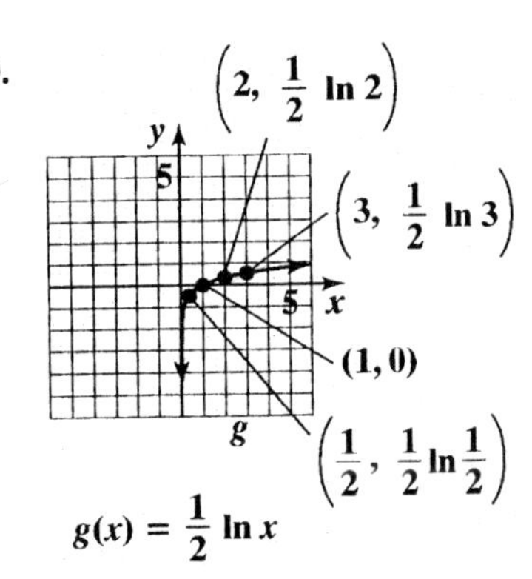

$g(x) = \frac{1}{2}\ln x$

vertical asymptote: $x = 0$
Domain: $(0, \infty)$; Range: $(-\infty, \infty)$

71.

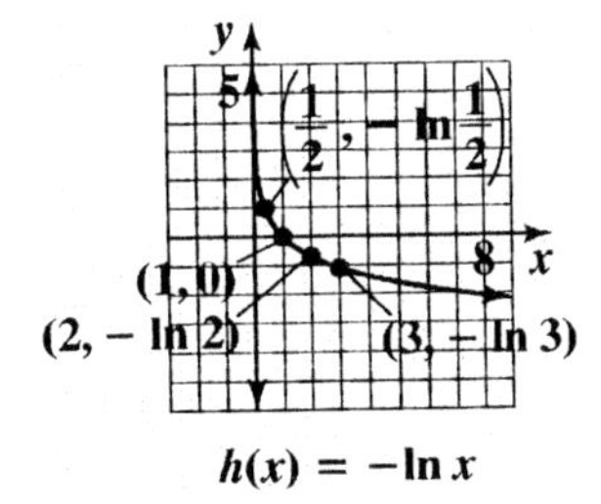

$h(x) = -\ln x$

vertical asymptote: $x = 0$
Domain: $(0, \infty)$; Range: $(-\infty, \infty)$

72.

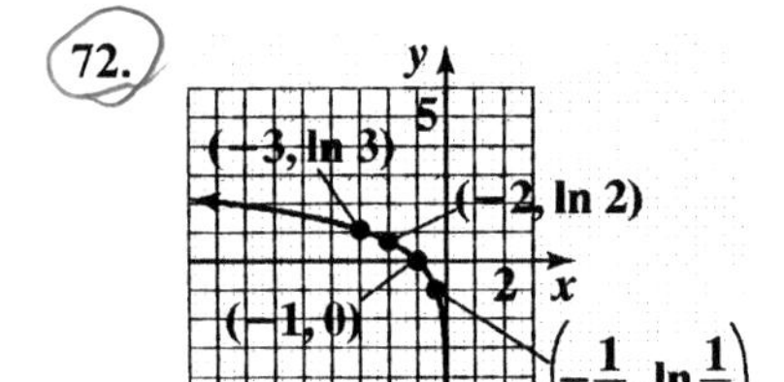

vertical asymptote: $x = 0$
Domain: $(-\infty, 0)$; Range: $(-\infty, \infty)$

73.

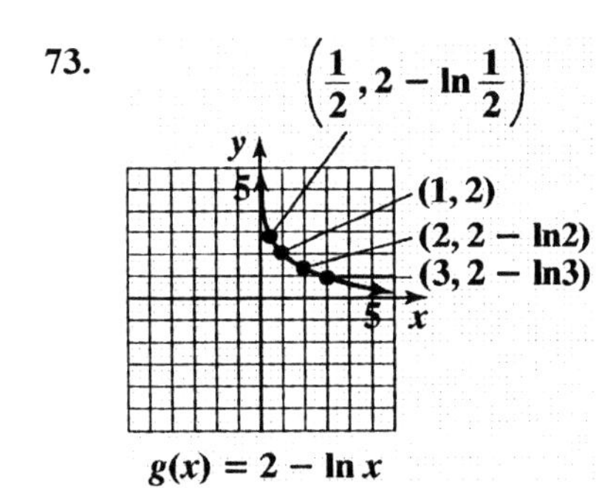

vertical asymptote: $x = 0$
Domain: $(0, \infty)$; Range: $(-\infty, \infty)$

74.

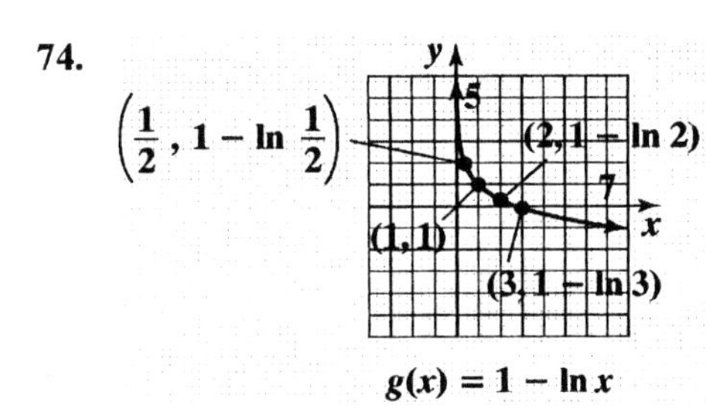

vertical asymptote: $x = 0$
Domain: $(0, \infty)$; Range: $(-\infty, \infty)$

75. The domain of f consists of all x for which $x + 4 > 0$. Solving this inequality for x, we obtain $x > -4$. Thus, the domain of f is $(-4, \infty)$.

76. The domain of f consists of all x for which $x + 6 > 0$. Solving this inequality for x, we obtain $x > -6$. Thus, the domain of f is $(-6, \infty)$.

77. The domain of f consists of all x for which $2 - x > 0$. Solving this inequality for x, we obtain $x < 2$. Thus, the domain of f is $(-\infty, 2)$.

78. The domain of f consists of all x for which $7 - x > 0$. Solving this inequality for x, we obtain $x < 7$. Thus, the domain of f is $(-\infty, 7)$.

79. The domain of f consists of all x for which $(x - 2)^2 > 0$. Solving this inequality for x, we obtain $x < 2$ or $x > 2$. Thus, the domain of f is $(-\infty, 2)$ or $(2, \infty)$.

80. The domain of f consists of all x for which $(x-7)^2 > 0$. Solving this inequality for x, we obtain $x < 7$ or $x > 7$. Thus, the domain of f is $(-\infty, 7)$ or $(7, \infty)$.

81. $\log 100 = \log_{10} 100 = 2$
because $10^2 = 100$.

82. $\log 1000 = \log_{10} 1000 = 3$ because $10^3 = 1000$.

83. Because $\log 10^x = x$, we conclude $\log 10^7 = 7$.

84. Because $\log 10^x = x$, we conclude $\log 10^8 = 8$.

85. Because $10^{\log x} = x$, we conclude $10^{\log 33} = 33$.

86. Because $10^{\log x} = x$, we conclude $10^{\log 53} = 53$.

87. $\ln 1 = 0$ because $e^0 = 1$.

88. $\ln e = \log_e e = 1$ because $e^1 = e$.

89. Because $\ln e^x = x$, we conclude $\ln e^6 = 6$.

90. Because $\ln e^x = x$, we conclude $\ln e^7 = 7$.

91. $\ln \frac{1}{e^6} = \ln e^{-6}$

Because $\ln e^x = x$ we conclude
$\ln e^{-6} = -6$, so $\ln \frac{1}{e^6} = -6$.

92. $\ln \frac{1}{e^7} = \ln e^{-7}$ Because $\ln e^x = x$, we conclude

$\ln e^{-7} = -7$, so $\ln \frac{1}{e^7} = -7$.

93. Because $e^{\ln x} = x$, we conclude $e^{\ln 125} = 125$.

94. Because $e^{\ln x} = x$, we conclude $e^{\ln 300} = 300$.

95. Because $\ln e^x = x$, we conclude $\ln e^{9x} = 9x$.

96. Because $\ln e^x = x$, we conclude $\ln e^{13x} = 13x$.

97. Because $e^{\ln x} = x$, we conclude $e^{\ln 5x^2} = 5x^2$.

98. Because $e^{\ln x} = x$, we conclude $e^{\ln 7x^2} = 7x^2$.

99. Because $10^{\log x} = x$, we conclude $10^{\log\sqrt{x}} = \sqrt{x}$.

100. Because $10^{\log x} = x$, we conclude $10^{\log\sqrt[3]{x}} = \sqrt[3]{x}$.

101. $\log_3(x-1) = 2$

$$3^2 = x-1$$
$$9 = x-1$$
$$10 = x$$

The solution is 10, and the solution set is $\{10\}$.

102. $\log_5(x+4) = 2$

$$5^2 = x+4$$
$$25 = x+4$$
$$21 = x$$

The solution is 21, and the solution set is $\{21\}$.

103. $\log_4 x = -3$

$$4^{-3} = x$$
$$x = \frac{1}{4^3} = \frac{1}{64}$$

The solution is $\frac{1}{64}$, and the solution set is $\left\{\frac{1}{64}\right\}$.

104. $\log_{64} x = \frac{2}{3}$

$$64^{\frac{2}{3}} = x$$
$$x = \left(\sqrt[3]{64}\right)^2 = 4^2 = 16$$

The solution is 16, and the solution set is $\{16\}$.

105. $\log_3(\log_7 7) = \log_3 1 = 0$

106. $\log_5(\log_2 32) = \log_5(\log_2 2^5) = \log_5 5 = 1$

107. $\log_2(\log_3 81) = \log_2(\log_3 3^4)$

$$= \log_2 4 = \log_2 2^2 = 2$$

108. $\log(\ln e) = \log 1 = 0$

109. For $f(x) = \ln(x^2 - x - 2)$ to be real, $x^2 - x - 2 > 0$.

Solve the related equation to find the boundary points:

$$x^2 - x - 2 = 0$$
$$(x+1)(x-2) = 0$$

The boundary points are -1 and 2. Testing each interval gives a domain of $(-\infty,-1)\cup(2,\infty)$.

110. For $f(x) = \ln(x^2 - 4x - 12)$ to be real, $x^2 - 4x - 12 > 0$.

Solve the related equation to find the boundary points:

$$x^2 - 4x - 12 = 0$$
$$(x+2)(x-6) = 0$$

The boundary points are -2 and 6. Testing each interval gives a domain of $(-\infty,-2)\cup(6,\infty)$.

111. For $f(x) = \ln\left(\frac{x+1}{x-5}\right)$ to be real, $\frac{x+1}{x-5} > 0$.

The boundary points are -1 and 5. Testing each interval gives a domain of $(-\infty,-1)\cup(5,\infty)$.

112. For $f(x) = \ln\left(\frac{x-2}{x+5}\right)$ to be real, $\frac{x-2}{x+5} > 0$.

The boundary points are -5 and 2. Testing each interval gives a domain of $(-\infty,-5)\cup(2,\infty)$.

113. $f(13) = 62 + 35\log(13-4) \approx 95.4$

She is approximately 95.4% of her adult height.

114. $f(10) = 62 + 35\log(10-4) \approx 89.2$.

She is approximately 89.2% of her adult height.

115. Since $2003 - 1997 = 6$, we find $f(6)$:

$$f(6) = -4.9\ln 6 + 73.8 \approx 65$$

In 2003, approximately 65% of U.S. companies performed drug tests. The function modeled the actual number very well. It gives the actual percent.

116. Note that $2008 - 1997 = 11$.

$$f(11) = -4.9\ln 11 + 73.8 \approx 62$$

The function predicts that about 62% of U.S. companies will be performing drug tests in 2008.

117. $D = 10\log\left[10^{12}(6.3\times 10^6)\right] \approx 188$

Yes, the sound can rupture the human eardrum.

118. $D = 10\log\left[10^{12}\left(3.2\times10^{-6}\right)\right] \approx 65.05$

A normal conversation is about 65 decibels.

119. a. $f(0) = 88 - 15\ln(0+1) = 88$
The average score on the original exam was 88.

b. $f(2) = 88 - 15\ln(2+1) = 71.5$
$f(4) = 88 - 15\ln(4+1) = 63.9$
$f(6) = 88 - 15\ln(6+1) = 58.8$
$f(8) = 88 - 15\ln(8+1) = 55$
$f(10) = 88 - 15\ln(10+1) = 52$
$f(12) = 88 - 15\ln(12+1) = 49.5$

The average score after 2 months was about 71.5, after 4 months was about 63.9, after 6 months was about 58.8, after 8 months was about 55, after 10 months was about 52, and after one year was about 49.5.

c.

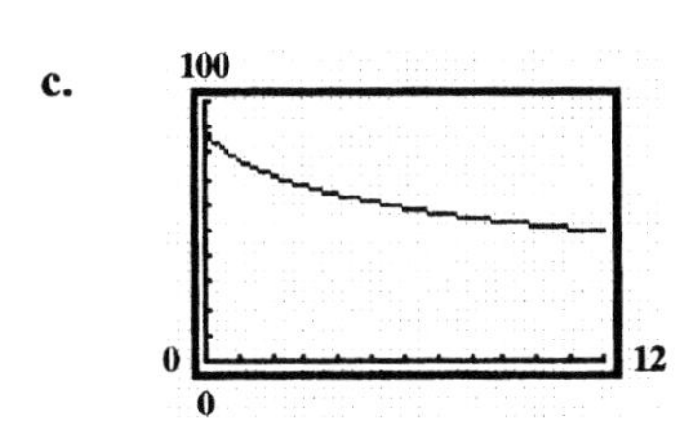

Material retention decreases as time passes.

128.

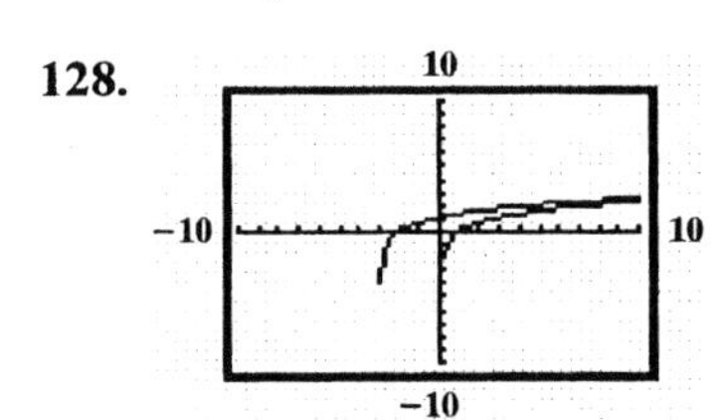

$g(x)$ is $f(x)$ shifted 3 units left.

129.

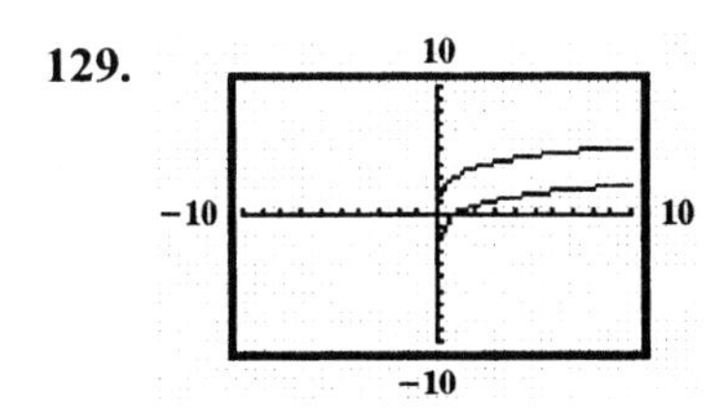

$g(x)$ is $f(x)$ shifted 3 units upward.

130.

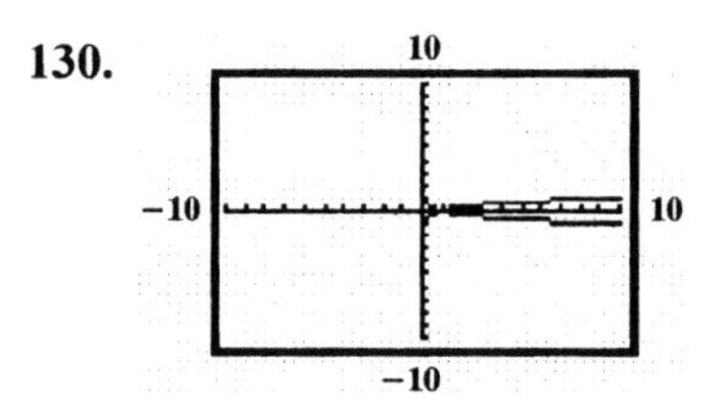

$g(x)$ is $f(x)$ reflected about the x-axis.

131.

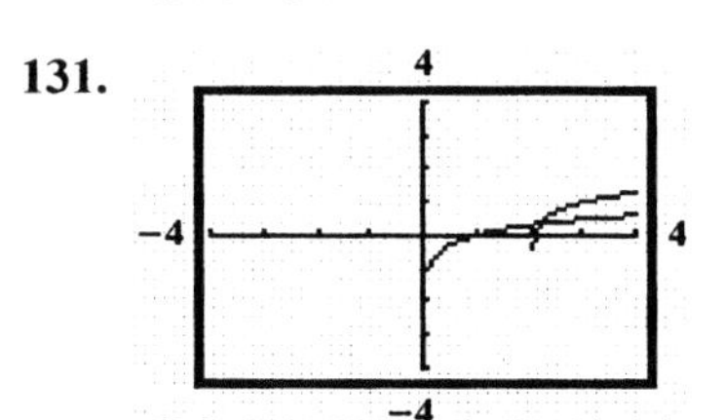

$g(x)$ is $f(x)$ shifted right 2 units and upward 1 unit.

132.

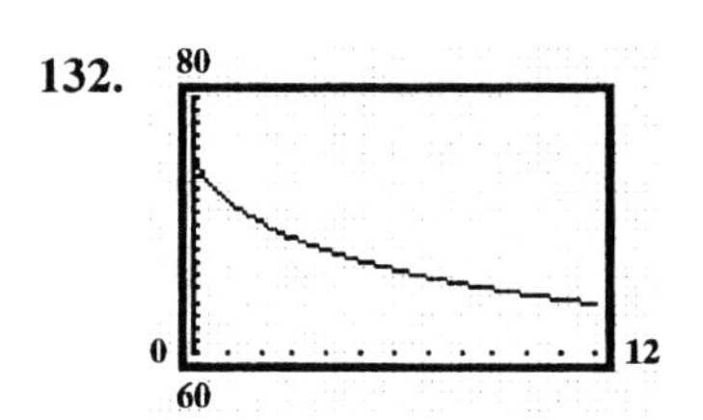

The score falls below 65 after 9 months.

133. a.

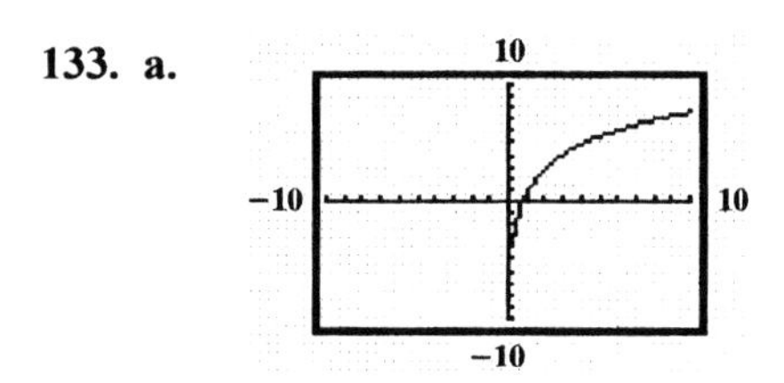

b.

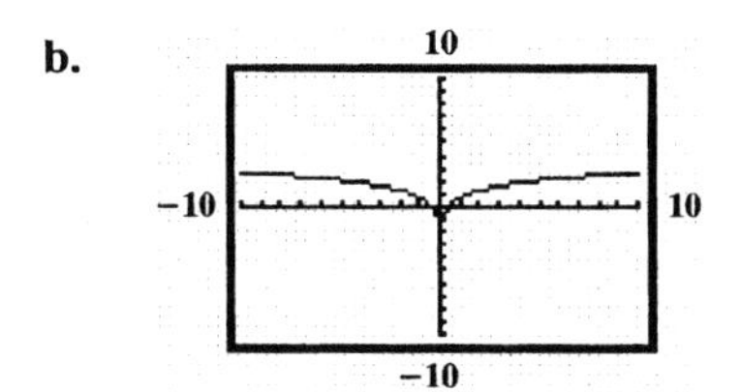

c.

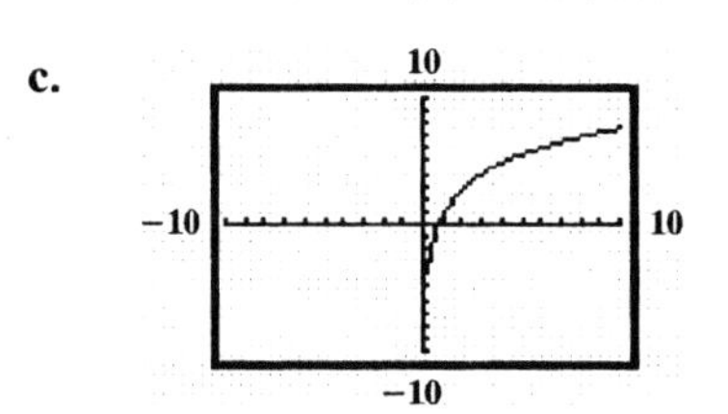

d They are the same.
$\log_b MN = \log_b M + \log_b N$

e. The sum of the logarithms of its factors.

134.

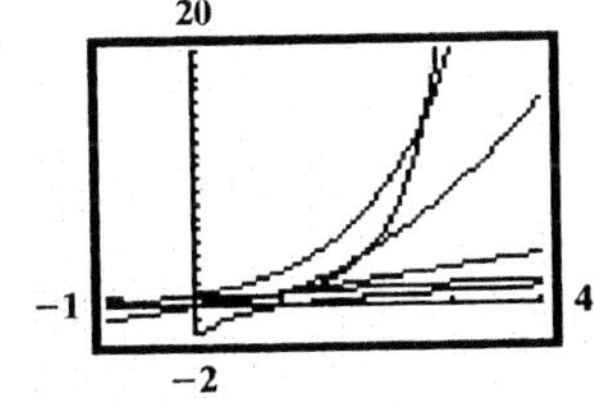

$y = \ln x$, $y = \sqrt{x}$, $y = x$,
$y = x^2$, $y = e^x$, $y = x^x$

135. a. False; $\frac{\log_2 8}{\log_2 4} = \frac{3}{2}$

b. False; $\log x$ is not defined for $x < 0$.

c. False; the domain is $(0, \infty)$.

d. True

(d) is true.

136. $\frac{\log_3 81 - \log_\pi 1}{\log_{2\sqrt{2}} 8 - \log 0.001} = \frac{4-0}{2-(-3)} = \frac{4}{5}$

137. $\log_4 [\log_3 (\log_2 8)]$
$= \log_4 [\log_3 (\log_2 2^3)]$
$= \log_4 [\log_3 3] = \log_4 1 = 0$

138. $\log_4 60 < \log_4 64 = 3$ so $\log_4 60 < 3$.
$\log_3 40 > \log_3 27 = 3$ so $\log_3 40 > 3$.
$\log_4 60 < 3 < \log_3 40$
$\log_3 40 > \log_4 60$

Section 3.3

Check Point Exercises

1. a. $\log_6 (7 \cdot 11) = \log_6 7 + \log_6 11$

b. $\log(100x) = \log 100 + \log x$
$= 2 + \log x$

2. a. $\log_8 \left(\frac{23}{x}\right) = \log_8 23 - \log_8 x$

b. $\ln\left(\frac{e^5}{11}\right) = \ln e^5 - \ln 11$
$= 5 - \ln 11$

3. a. $\log_6 3^9 = 9 \log_6 3$

b. $\ln \sqrt[3]{x} = \ln x^{1/3} = \frac{1}{3} \ln x$

c. $\log(x+4)^2 = 2\log(x+4)$

4. a. $\log_b x^4 \sqrt[3]{y}$
$= \log_b x^4 y^{1/3}$
$= \log_b x^4 + \log_b y^{1/3}$
$= 4\log_b x + \frac{1}{3}\log_b y$

b. $\log_5 \frac{\sqrt{x}}{25y^3}$
$= \log_5 \frac{x^{1/2}}{25y^3}$
$= \log_5 x^{1/2} - \log_5 25y^3$
$= \log_5 x^{1/2} - (\log_5 5^2 + \log_5 y^3)$
$= \frac{1}{2}\log_5 x - \log_5 5^2 - \log_5 y^3$
$= \frac{1}{2}\log_5 x - 2\log_5 5 - 3\log_5 y$
$= \frac{1}{2}\log_5 x - 2 - 3\log_5 y$

5. a. $\log 25 + \log 4 = \log(25 \cdot 4) = \log 100 = 2$

b. $\log(7x+6) - \log x = \log \frac{7x+6}{x}$

6. a. $\ln x^2 + \frac{1}{3}\ln(x+5)$
$= \ln x^2 + \ln(x+5)^{1/3}$
$= \ln x^2 (x+5)^{1/3}$
$= \ln x^2 \sqrt[3]{x+5}$

b. $2\log(x-3) - \log x$
$= \log(x-3)^2 - \log x$
$= \log \frac{(x-3)^2}{x}$

c. $\frac{1}{4}\log_b x - 2\log_b 5 - 10\log_b y$
$= \log_b x^{1/4} - \log_b 5^2 - \log_b y^{10}$
$= \log_b x^{1/4} - \left(\log_b 25 - \log_b y^{10}\right)$
$= \log_b x^{1/4} - \log_b 25y^{10}$
$= \log_b \frac{x^{1/4}}{25y^{10}}$ or $\log_b \frac{\sqrt[4]{x}}{25y^{10}}$

7. $\log_7 2506 = \frac{\log 2506}{\log 7} \approx 4.02$

8. $\log_7 2506 = \frac{\ln 2506}{\ln 7} \approx 4.02$

Exercise Set 3.3

1. $\log_5(7\cdot 3) = \log_5 7 + \log_5 3$

2. $\log_8(13\cdot 7) = \log_8 13 + \log_8 7$

3. $\log_7(7x) = \log_7 7 + \log_7 x = 1 + \log_7 x$

4. $\log_9 9x = \log_9 9 + \log_9 x = 1 + \log_9 x$

5. $\log(1000x) = \log 1000 + \log x = 3 + \log x$

6. $\log(10{,}000x) = \log 10{,}000 + \log x = 4 + \log x$

7. $\log_7\left(\frac{7}{x}\right) = \log_7 7 - \log_7 x = 1 - \log_7 x$

8. $\log_9\left(\frac{9}{x}\right) = \log_9 9 - \log_9 x = 1 - \log_9 x$

9. $\log\left(\frac{x}{100}\right) = \log x - \log 100 = \log_x - 2$

10. $\log\left(\frac{x}{1000}\right) = \log x - \log 1000 = \log x - 3$

11. $\log_4\left(\frac{64}{y}\right) = \log_4 64 - \log_4 y$
$= 3 - \log_4 y$

12. $\log_5\left(\frac{125}{y}\right) = \log_5 125 - \log_5 y = 3 - \log_5 y$

13. $\ln\left(\frac{e^2}{5}\right) = \ln e^2 - \ln 5 = 2\ln e - \ln 5 = 2 - \ln 5$

14. $\ln\left(\frac{e^4}{8}\right) = \ln e^4 - \ln 8 = 4\ln e - \ln 8 = 4 - \ln 8$

15. $\log_b x^3 = 3\log_b x$

16. $\log_b x^7 = 7\log_b x$

17. $\log N^{-6} = -6\log N$

18. $\log M^{-8} = -8\log M$

19. $\ln\sqrt[5]{x} = \ln x^{(1/5)} = \frac{1}{5}\ln x$

20. $\ln\sqrt[7]{x} = \ln x^{\frac{1}{7}} = \frac{1}{7}\ln x$

21. $\log_b x^2 y = \log_b x^2 + \log_b y = 2\log_b x + \log_b y$

22. $\log_b xy^3 = \log_b x + \log_b y^3 = \log_b x + 3\log_b y$

23. $\log_4\left(\frac{\sqrt{x}}{64}\right) = \log_4 x^{1/2} - \log_4 64 = \frac{1}{2}\log_4 x - 3$

24. $\log_5\left(\frac{\sqrt{x}}{25}\right) = \log_5 x^{1/2} - \log_5 25 = \frac{1}{2}\log_5 x - 2$

25. $\log_6\left(\frac{36}{\sqrt{x+1}}\right) = \log_6 36 - \log_6(x+1)^{1/2}$
$= 2 - \frac{1}{2}\log_6(x+1)$

26. $\log_8\left(\frac{64}{\sqrt{x+1}}\right) = \log_8 64 - \log_8(x+1)^{1/2}$
$= 2 - \frac{1}{2}\log_8(x+1)$

27. $\log_b\left(\frac{x^2 y}{z^2}\right) = \log_b\left(x^2 y\right) - \log_b z^2$
$= \log_b x^2 + \log_b y - \log_b z^2$
$= 2\log_b x + \log_b y - 2\log_b z$

28. $\log_b\left(\frac{x^3y}{z^2}\right) = \log_b\left(x^3y\right) - \log_b z^2$

$\log_b\left(\frac{x^3y}{z^2}\right) = \log_b x^3 + \log_b y - \log_b z^2$

$= 3\log_b x + \log_b y - 2\log_b z$

29. $\log\sqrt{100x} = \log(100x)^{1/2}$

$= \frac{1}{2}\log(100x)$

$= \frac{1}{2}(\log 100 + \log x)$

$= \frac{1}{2}(2 + \log x)$

$= 1 + \frac{1}{2}\log x$

30. $\ln\sqrt{ex} = \ln(ex)^{1/2}$

$= \frac{1}{2}\ln(ex)$

$= \frac{1}{2}(\ln e + \ln x)$

$= \frac{1}{2}(1 + \ln x)$

$= \frac{1}{2} + \frac{1}{2}\ln x$

31. $\log\sqrt[3]{\frac{x}{y}} = \log\left(\frac{x}{y}\right)^{1/3}$

$= \frac{1}{3}\left[\log\left(\frac{x}{y}\right)\right]$

$= \frac{1}{3}(\log x - \log y)$

$= \frac{1}{3}\log x - \frac{1}{3}\log y$

32. $\log\sqrt[5]{\frac{x}{y}} = \log\left(\frac{x}{y}\right)^{1/5}$

$= \frac{1}{5}\left[\log\left(\frac{x}{y}\right)\right]$

$= \frac{1}{5}(\log x - \log y)$

$= \frac{1}{5}\log x - \frac{1}{5}\log y$

33. $\log_b \frac{\sqrt{x}y^3}{z^3}$

$= \log_b x^{1/2} + \log_b y^3 - \log_b z^3$

$= \frac{1}{2}\log_b x + 3\log_b y - 3\log_b z$

34. $\log_b \frac{\sqrt[3]{x}y^4}{z^5}$

$= \log_b x^{1/3} + \log_b y^4 - \log_b z^5$

$= \frac{1}{3}\log_b x + 4\log_b y - 5\log_b z$

35. $\log_5 \sqrt[3]{\frac{x^2y}{25}}$

$= \log_5 x^{2/3} + \log_5 y^{1/3} - \log_5 25^{1/3}$

$= \frac{2}{3}\log_5 x + \frac{1}{3}\log_5 y - \log_5 5^{2/3}$

$= \frac{2}{3}\log_5 x + \frac{1}{3}\log_5 y - \frac{2}{3}$

36. $\log_2 \sqrt[5]{\frac{xy^4}{16}}$

$= \log_2 x^{1/5} + \log_2 y^{4/5} - \log_2 16^{1/5}$

$= \frac{1}{5}\log_2 x + \frac{4}{5}\log_2 y - \frac{1}{5}\log_2 16$

$= \frac{1}{5}\log_2 x + \frac{4}{5}\log_2 y - \frac{4}{5}$

37. $\ln\left[\frac{x^3\sqrt{x^2+1}}{(x+1)^4}\right]$

$= \ln x^3 + \ln\sqrt{x^2+1} - \ln(x+1)^4$

$= 3\ln x + \frac{1}{2}\ln(x^2+1) - 4\ln(x+1)$

38. $\ln\left[\frac{x^4\sqrt{x^2+3}}{(x+3)^5}\right]$
$= \ln\left[\frac{x^4(x^2+3)^{1/2}}{(x+3)^5}\right]$
$= \ln x^4 + \ln\left(x^2+3\right)^{1/2} - \ln\left(x+3\right)^5$
$= 4\ln x + \frac{1}{2}\ln\left(x^2+3\right) - 5\ln\left(x+3\right)$

39. $\log\left[\frac{10x^2\sqrt[3]{1-x}}{7(x+1)^2}\right]$
$= \log 10 + \log x^2 + \log\sqrt[3]{1-x} - \log 7 - \log(x+1)^2$
$= 1 + 2\log x + \frac{1}{3}\log(1-x) - \log 7 - 2\log(x+1)$

40. $\log\left[\frac{100x^3\sqrt[3]{5-x}}{3(x+7)^2}\right]$
$= \log 100 + \log x^3 + \log\left(5-x\right)^{\frac{1}{3}} - \log 3 - \log\left(x+7\right)^2$
$= 2 + 3\log x + \frac{1}{3}\log(5-x) - \log 3 - 2\log(x+7)$

41. $\log 5 + \log 2 = \log(5\cdot 2) = \log 10 = 1$

42. $\log 250 + \log 4 = \log 1000 = 3$

43. $\ln x + \ln 7 = \ln(7x)$

44. $\ln x + \ln 3 = \ln(3x)$

45. $\log_2 96 - \log_2 3 = \log_2\left(\frac{96}{3}\right) = \log_2 32 = 5$

46. $\log_3 405 - \log_3 5 = \log_3\left(\frac{405}{5}\right)$
$= \log_3 81$
$= 4$

47. $\log(2x+5) - \log x = \log\left(\frac{2x+5}{x}\right)$

48. $\log(3x+7) - \log x = \log\left(\frac{3x+7}{x}\right)$

49. $\log x + 3\log y = \log x + \log y^3 = \log(xy^3)$

50. $\log x + 7\log y = \log x + \log y^7 = \log(xy^7)$

51. $\frac{1}{2}\ln x + \ln y = \ln x^{1/2} + \ln y$
$= \ln\left(x^{\frac{1}{2}}y\right)$ or $\ln\left(y\sqrt{x}\right)$

52. $\frac{1}{3}\ln x + \ln y = \ln x^{1/3} + \ln y$
$= \ln\left(x^{1/3}y\right)$ or $\ln\left(y\sqrt[3]{x}\right)$

53. $2\log_b x + 3\log_b y = \log_b x^2 + \log_b y^3$
$= \log_b(x^2y^3)$

54. $5\log_b x + 6\log_b y = \log_b x^5 + \log_b y^6$
$= \log_b\left(x^5y^6\right)$

55. $5\ln x - 2\ln y = \ln x^5 - \ln y^2 = \ln\left(\frac{x^5}{y^2}\right)$

56. $7\ln x - 3\ln y = \ln x^7 - \ln y^3 = \ln\left(\frac{x^7}{y^3}\right)$

57. $3\ln x - \frac{1}{3}\ln y = \ln x^3 - \ln y^{1/3}$
$= \ln\left(\frac{x^3}{y^{1/3}}\right)$ or $\ln\left(\frac{x^3}{\sqrt[3]{y}}\right)$

58. $2\ln x - \frac{1}{2}\ln y = \ln x^2 - \ln y^{1/2}$
$= \ln\left(\frac{x^2}{y^{1/2}}\right)$ or $\ln\left(\frac{x^2}{\sqrt{y}}\right)$

59. $4\ln(x+6) - 3\ln x = \ln(x+6)^4 - \ln x^3$
$= \ln\frac{\left(x+6\right)^4}{x^3}$

60. $8\ln(x+9) - 4\ln x = \ln(x+9)^8 - \ln x^4$
$= \ln\frac{(x+9)^8}{x^4}$

61. $3\ln x + 5\ln y - 6\ln z$
$= \ln x^3 + \ln y^5 - \ln z^6$
$= \ln\frac{x^3y^5}{z^6}$

62. $4\ln x + 7\ln y - 3\ln z$
$= \ln x^4 + \ln y^7 - \ln z^3$
$= \ln \dfrac{x^4 y^7}{z^3}$

63. $\frac{1}{2}(\log x + \log y)$
$= \frac{1}{2}(\log xy)$
$= \log(xy)^{1/2}$
$= \log\sqrt{xy}$

64. $\frac{1}{3}(\log_4 x - \log_4 y)$
$= \frac{1}{3}\log_4 \frac{x}{y}$
$= \log_4 \left(\frac{x}{y}\right)^{1/3}$
$= \log_4 \sqrt[3]{\frac{x}{y}}$

65. $\frac{1}{2}(\log_5 x + \log_5 y) - 2\log_5(x+1)$
$= \frac{1}{2}\log_5 xy - \log_5(x+1)^2$
$= \log_5(xy)^{1/2} - \log_5(x+1)^2$
$= \log_5 \dfrac{(xy)^{1/2}}{(x+1)^2}$
$= \log_5 \dfrac{\sqrt{xy}}{(x+1)^2}$

66. $\frac{1}{3}(\log_4 x - \log_4 y) + 2\log_4(x+1)$
$= \frac{1}{3}\log_4 \frac{x}{y} + \log_4(x+1)^2$
$= \log_4 \left[\left(\frac{x}{y}\right)^{1/3} (x+1)^2\right]$
$= \log_4 \left[(x+1)^2 \sqrt[3]{\frac{x}{y}}\right]$

67. $\frac{1}{3}[2\ln(x+5) - \ln x - \ln(x^2-4)]$
$= \frac{1}{3}[\ln(x+5)^2 - \ln x - \ln(x^2-4)]$
$= \frac{1}{3}\left[\ln \dfrac{(x+5)^2}{x(x^2-4)}\right]$
$= \ln\left[\dfrac{(x+5)^2}{x(x^2-4)}\right]^{1/3}$
$= \ln \sqrt[3]{\dfrac{(x+5)^2}{x(x^2-4)}}$

68. $\frac{1}{3}[5\ln(x+6) - \ln x - \ln(x^2-25)]$
$= \frac{1}{3}\ln\left[\dfrac{(x+6)^5}{x(x^2-25)}\right]$
$= \ln\left[\dfrac{(x+6)^5}{x(x^2-25)}\right]^{\frac{1}{3}}$

69. $\log x + \log(x^2-1) - \log 7 - \log(x+1)$
$= \log x + \log(x^2-1) - (\log 7 + \log(x+1))$
$= \log(x(x^2-1)) - \log(7(x+1))$
$= \log \dfrac{x(x^2-1)}{7(x+1)}$
$= \log \dfrac{x(x+1)(x-1)}{7(x+1)}$
$= \log \dfrac{x(x-1)}{7}$

70. $\log x + \log(x^2-4) - \log 15 - \log(x+2)$
$= \log x + \log(x^2-4) - (\log 15 + \log(x+2))$
$= \log(x(x^2-4)) - \log(15(x+2))$
$= \log \dfrac{x(x^2-4)}{15(x+2)}$
$= \log \dfrac{x(x+2)(x-2)}{15(x+2)}$
$= \log \dfrac{x(x-2)}{15}$

71. $\log_5 13 = \dfrac{\log 13}{\log 5} \approx 1.5937$

72. $\log_6 17 = \frac{\log 17}{\log 6} \approx 1.5812$

73. $\log_{14} 87.5 = \frac{\ln 87.5}{\ln 14} \approx 1.6944$

74. $\log_{16} 57.2 = \frac{\ln 57.2}{\ln 16} \approx 1.4595$

75. $\log_{0.1} 17 = \frac{\log 17}{\log 0.1} \approx -1.2304$

76. $\log_{0.3} 19 = \frac{\log 19}{\log 0.3} \approx -2.4456$

77. $\log_\pi 63 = \frac{\ln 63}{\ln \pi} \approx 3.6193$

78. $\log_\pi 400 = \frac{\ln 400}{\ln \pi} \approx 5.2340$

79. $y = \log_3 x = \frac{\log x}{\log 3}$

80. $y = \log_{15} x = \frac{\log x}{\log 15}$

81. $y = \log_2(x+2) = \frac{\log(x+2)}{\log 2}$

82. $y = \log_3(x-2) = \frac{\log(x-2)}{\log 3}$

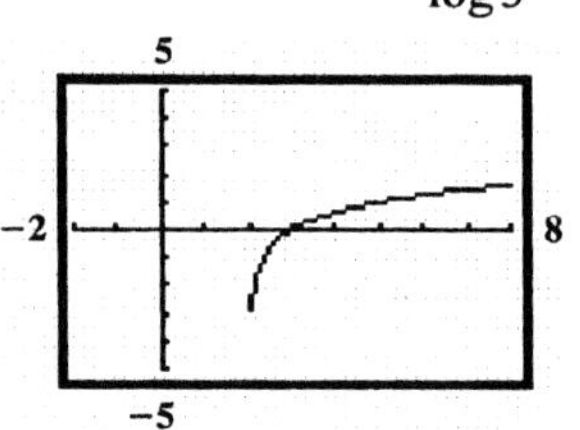

83. $\log_b \frac{3}{2} = \log_b 3 - \log_b 2 = C - A$

84. $\log_b 6 = \log_b (2 \cdot 3)$
$= \log_b 2 + \log_b 3 = A + C$

85. $\log_b 8 = \log_b 2^3 = 3\log_b 2 = 3A$

86. $\log_b 81 = \log_b 3^4 = 4\log_b 3 = 4C$

87.
$$\log_b \sqrt{\frac{2}{27}} = \log_b \left(\frac{2}{27}\right)^{\frac{1}{2}}$$
$$= \frac{1}{2}\log_b\left(\frac{2}{3^3}\right)$$
$$= \frac{1}{2}\left(\log_b 2 - \log_b 3^3\right)$$
$$= \frac{1}{2}\left(\log_b 2 - 3\log_b 3\right)$$
$$= \frac{1}{2}\log_b 2 - \frac{3}{2}\log_b 3$$
$$= \frac{1}{2}A - \frac{3}{2}C$$

88.
$$\log_b \sqrt{\frac{3}{16}} = \log_b \left(\frac{\sqrt{3}}{4}\right)$$
$$= \log_b \sqrt{3} - \log_b 4$$
$$= \log_b 3^{\frac{1}{2}} - \log 2^2$$
$$= \frac{1}{2}\log_b 3 - 2\log 2$$
$$= \frac{1}{2}C - 2A$$

89. false; $\ln e = 1$

90. false; $\ln e^e = 0$

91. false; $\log_4 (2x)^3 = 3\log_4 (2x)$

92. true; $\ln(8x^3) = \ln(2^3x^3) = \ln(2x)^3 = 3\ln(2x)$

93. true; $x\log 10^x = x \cdot x = x^2$

94. false; $\ln(x \cdot 1) = \ln x + \ln 1$

95 true; $\ln(5x) + \ln 1 = \ln 5x + 0 = \ln 5x$

96. false; $\ln x + \ln(2x) = \ln(x \cdot 2x)) = \ln 2x^2$

97. false; $\log(x+3) - \log(2x) = \log\frac{x+3}{2x}$

98. false; $\log\frac{x+2}{x-1} = \log(x+2) - \log(x-1)$

99. true; quotient rule

100. true; product rule

101. true; $\log_3 7 = \frac{\log 7}{\log 3} = \frac{1}{\frac{\log 3}{\log 7}} = \frac{1}{\log_7 3}$

102. false; $e^x = \ln e^{e^x}$

103. a. $D = 10\log\left(\frac{I}{I_0}\right)$

b. $D_1 = 10\log\left(\frac{100I}{I_0}\right)$

$= 10\log(100I - I_0)$

$= 10\log 100 + 10\log I - 10\log I_0$

$= 10(2) + 10\log I - 10\log I_0$

$= 20 + 10\log\left(\frac{I}{I_0}\right)$

This is 20 more than the loudness level of the softer sound. This means that the 100 times louder sound will be 20 decibels louder.

104. a. $t = \frac{1}{c}\ln\left(\frac{A}{A-N}\right)$

b. $t = \frac{1}{0.03}\left[\ln\frac{65}{65-30}\right]$

$t = \frac{1}{0.03}\ln\left(\frac{65}{35}\right)$

$t \approx 20.63$

It will take the chimpanzee a little more than 20.5 weeks to master 30 signs.

113. a. $y = \log_3 x = \frac{\ln x}{\ln 3}$

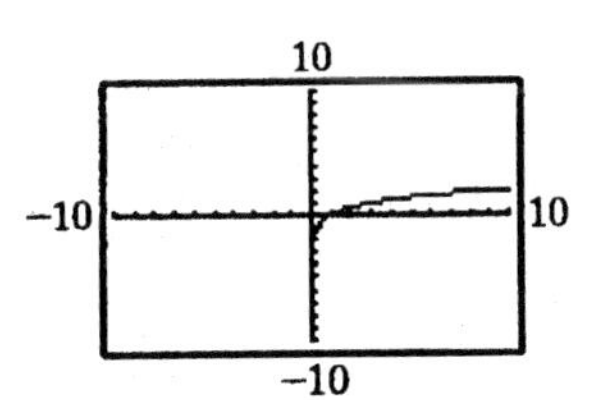

b.

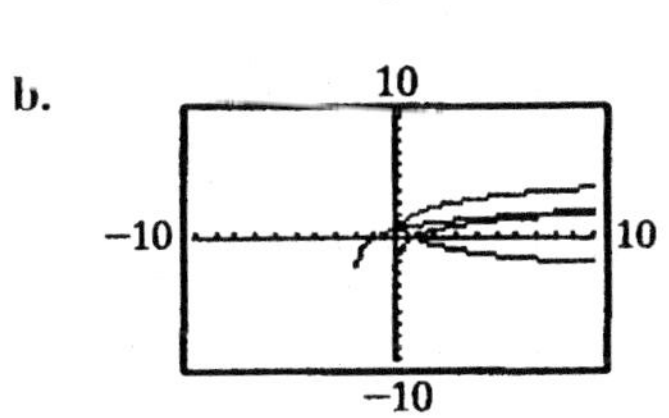

To obtain the graph of $y = 2 + \log_3 x$, shift the graph of $y = \log_3 x$ two units upward. To obtain the graph of $y = \log_3(x + 2)$, shift the graph of $y = \log_3 x$ two units left. To obtain the graph of $y = -\log_3 x$, reflect the graph of $y = \log_3 x$ about the x-axis.

114.

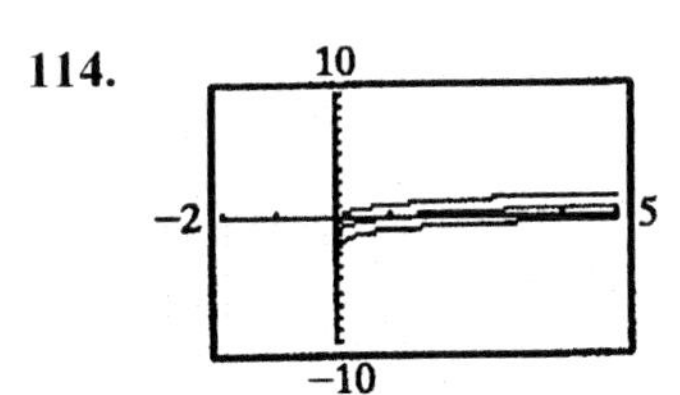

Using the product rule, $\log(10x) = \log x + 1$ and $1\log(0.1x) = \log x - 1$. Hence, these two graphs are just vertical shifts of $y = \log x$.

115. $\log_3 x = \frac{\log x}{\log 3}$;

$\log_{25} x = \frac{\log x}{\log 25}$;

$\log_{100} x = \frac{\log x}{\log 100}$

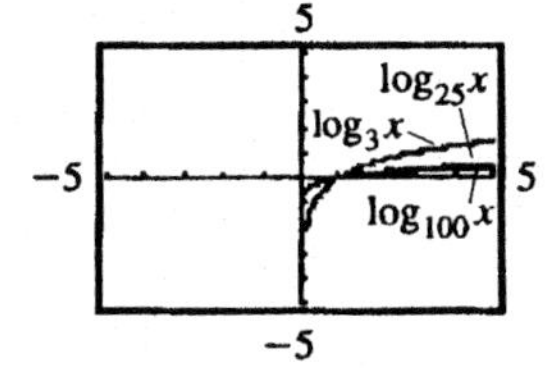

a. top graph: $y = \log_{100} x$

bottom graph: $y = \log_3 x$

b. top graph: $y = \log_3 x$
bottom graph: $y = \log_{100} x$

c. Comparing graphs of $\log_b x$ for $b > 1$, the graph of the equation with the largest b will be on the top in the interval (0, 1) and on the bottom in the interval $(1, \infty)$.

121. a. False;

$$\log_7 49 - \log_7 7 = \log_7 \frac{49}{7} = \log_7 7 = 1$$

$$\frac{\log_7 49}{\log_7 7} = \frac{2}{1} = 2$$

b. False;

$$3\log_b x + 3\log_b y = \log_b (xy)^3 \neq \log_b \left(x^3 + y^3\right)$$

c. False;

$$\log_b (xy)^5 = 5\log_b (xy) = 5(\log_b x + \log_b y) \neq (\log_b x + \log_b y)^5$$

d. True;

$$\ln\sqrt{2} = \ln 2^{1/2} = \frac{1}{2}\ln 2 = \frac{\ln 2}{2}$$

(d) is true.

122. $\log e = \log_{10} e = \dfrac{\ln e}{\ln 10} = \dfrac{1}{\ln 10}$

123. $\log_7 9 = \dfrac{\log 9}{\log 7} = \dfrac{\log 3^2}{\log 7} = \dfrac{2\log 3}{\log 7} = \dfrac{2A}{B}$

124. $e^{\ln 8x^5 - \ln 2x^2} = e^{\ln\left(\frac{8x^5}{2x^2}\right)} = e^{\ln\left(4x^3\right)} = 4x^3$

125.

$$\frac{\log_b (x+h) - \log_b x}{h} = \frac{\log_b \frac{x+h}{x}}{h} = \frac{\log_b\left(1+\frac{h}{x}\right)}{h} = \frac{1}{h}\log_b\left(1+\frac{h}{x}\right) = \log_b\left(1+\frac{x}{h}\right)^{1/h}$$

Mid-Chapter 3 Check Point

1.

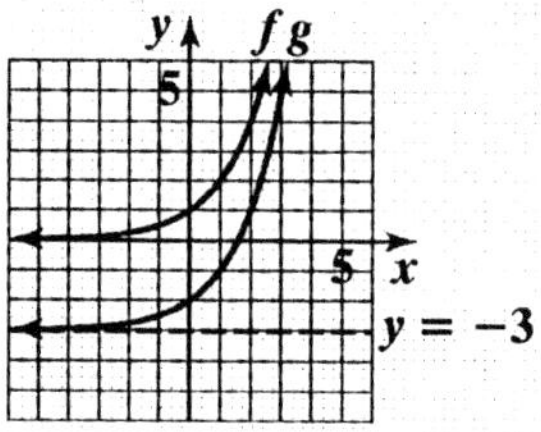

$f(x) = 2^x$
$g(x) = 2^x - 3$

Asymptote of f: $y = 0$
Asymptote of g: $y = -3$
Domain of f = Domain of $g = (-\infty, \infty)$
Range of $f = (0, \infty)$
Range of $g = (-3, \infty)$

2.

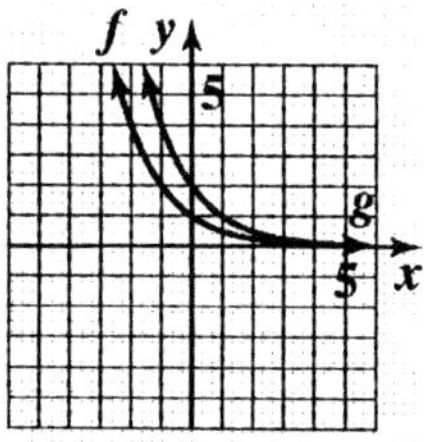

$f(x) = \left(\frac{1}{2}\right)^x$
$g(x) = \left(\frac{1}{2}\right)^{x-1}$

Asymptote of f: $y = 0$
Asymptote of g: $y = 0$
Domain of f = Domain of $g = (-\infty, \infty)$
Range of f = Range of $g = (0, \infty)$

3. 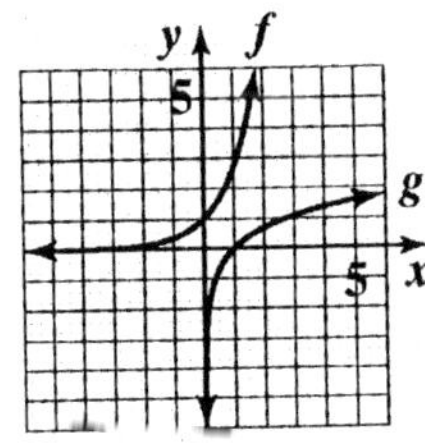

$f(x) = e^x$
$g(x) = \ln x$

Asymptote of f: $y = 0$
Asymptote of g: $x = 0$
Domain of f = Range of g = $(-\infty, \infty)$
Range of f = Domain of g = $(0, \infty)$

4. 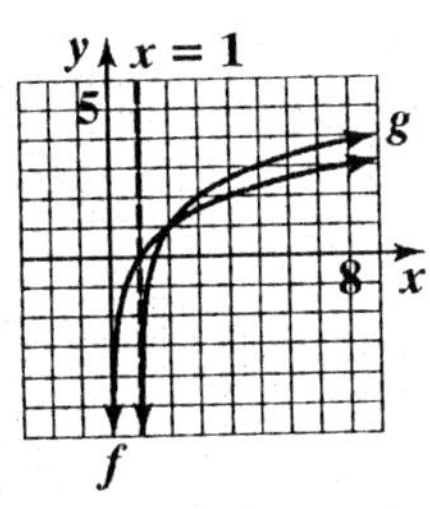

$f(x) = \log_2 x$
$g(x) = \log_2 (x-1) + 1$

Asymptote of f: $x = 0$
Asymptote of g: $x = 1$
Domain of f = $(0, \infty)$
Domain of g = $(1, \infty)$
Range of f = Range of g = $(-\infty, \infty)$

5. 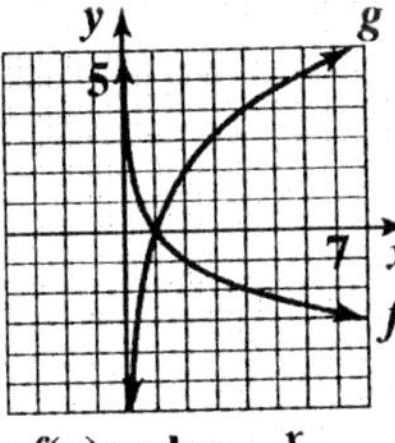

$f(x) = \log_{1/2} x$
$g(x) = -2\log_{1/2} x$

Asymptote of f: $x = 0$
Asymptote of g: $x = 0$
Domain of f = Domain of g = $(0, \infty)$
Range of f = Range of g = $(-\infty, \infty)$

6. $f(x) = \log_3(x+6)$
The argument of the logarithm must be positive:
$x + 6 > 0$
$x > -6$
Domain: $\{x \mid x > -6\}$ or $(-6, \infty)$.

7. $f(x) = \log_3 x + 6$
The argument of the logarithm must be positive:
$x > 0$
Domain: $\{x \mid x > 0\}$ or $(0, \infty)$.

8. $\log_3 (x+6)^2$
The argument of the logarithm must be positive.
Now $(x+6)^2$ is always positive, except when $x = -6$
Domain: $\{x \mid x \neq 0\}$ or $(-\infty, -6) \cup (-6, \infty)$.

9. $f(x) = 3^{x+6}$
Domain: $\{x \mid x$ is a real number$\}$ or $(-\infty, \infty)$.

10. $\log_2 8 + \log_5 25 = \log_2 2^3 + \log_5 5^2$
$= 3 + 2 = 5$

11. $\log_3 \frac{1}{9} = \log_3 \frac{1}{3^2} = \log_3 3^{-2} = -2$

12. Let $\log_{100} 10 = y$
$100^y = 10$
$(10^2)^y = 10^1$
$10^{2y} = 10^1$
$2y = 1$
$y = \frac{1}{2}$

13. $\log \sqrt[3]{10} = \log 10^{\frac{1}{3}} = \frac{1}{3}$

14. $\log_2(\log_3 81) = \log_2(\log_3 3^4)$
$= \log_2 4 = \log_2 2^2 = 2$

15. $\log_3\left(\log_2 \frac{1}{8}\right) = \log_3\left(\log_2 \frac{1}{2^3}\right)$
$= \log_3\left(\log_2 2^{-3}\right)$
$= \log_3(-3)$
$=$ not possible

This expression is impossible to evaluate.

16. $6^{\log_6 5} = 5$

17. $\ln e^{\sqrt{7}} = \sqrt{7}$

18. $10^{\log 13} = 13$

19. $\log_{100} 0.1 = y$
$100^y = 0.1$
$\left(10^2\right)^y = \frac{1}{10}$
$10^{2y} = 10^{-1}$
$2y = -1$
$y = -\frac{1}{2}$

20. $\log_\pi \pi^{\sqrt{\pi}} = \sqrt{\pi}$

21. $\log\left(\frac{\sqrt{xy}}{1000}\right) = \log\left(\sqrt{xy}\right) - \log 1000$
$= \log(xy)^{\frac{1}{2}} - \log 10^3$
$= \frac{1}{2}\log(xy) - 3$
$= \frac{1}{2}(\log x + \log y) - 3$
$= \frac{1}{2}\log x + \frac{1}{2}\log y - 3$

22. $\ln\left(e^{19}x^{20}\right) = \ln e^{19} + \ln x^{20}$
$= 19 + 20\ln x$

23. $8\log_7 x - \frac{1}{3}\log_7 y = \log_7 x^8 - \log_7 y^{\frac{1}{3}}$
$= \log_7\left(\frac{x^8}{y^{\frac{1}{3}}}\right)$
$= \log_7\left(\frac{x^8}{\sqrt[3]{y}}\right)$

24. $7\log_5 x + 2\log_5 x = \log_5 x^7 + \log_5 x^2$
$= \log_5\left(x^7 \cdot x^2\right)$
$= \log_5 x^9$

25. $\frac{1}{2}\ln x - 3\ln y - \ln(z-2)$
$= \ln x^{\frac{1}{2}} - \ln y^3 - \ln(z-2)$
$= \ln\sqrt{x} - \left[\ln y^3 + \ln(z-2)\right]$
$= \ln\sqrt{x} - \ln\left[y^3(z-2)\right]$
$= \ln\left[\frac{\sqrt{x}}{y^3(z-2)}\right]$

26. Continuously: $A = 8000e^{0.08(3)}$
$\approx 10{,}170$

Monthly: $A = 8000\left(1 + \frac{0.08}{12}\right)^{12\cdot 3}$
$\approx 10{,}162$

$10{,}170 - 10{,}162 = 8$

Interest returned will be \$8 more if compounded continuously.

Section 3.4

Check Point Exercises

1. **a.** $5^{3x-6} = 125$
$5^{3x-6} = 5^3$
$3x - 6 = 3$
$3x = 9$
$x = 3$

b. $8^{x+2} = 4^{x-3}$
$\left(2^3\right)^{x+2} = \left(2^2\right)^{x-3}$
$2^{3x+6} = 2^{2x-6}$
$3x + 6 = 2x - 6$
$x = -12$

2.
$$5^x = 134$$
$$\ln 5^x = \ln 134$$
$$x \ln 5 = \ln 134$$
$$x = \frac{\ln 134}{\ln 5} \approx 3.04$$
The solution set is $\left\{\frac{\ln 134}{\ln 5}\right\}$, approximately 3.04.

3.
$$7e^{2x} = 63$$
$$e^{2x} = 9$$
$$\ln e^{2x} = \ln 9$$
$$2x = \ln 9$$
$$x = \frac{\ln 9}{2} \approx 1.10$$
The solution set is $\left\{\frac{\ln 9}{2}\right\}$, approximately 1.10.

4.
$$3^{2x-1} = 7^{x+1}$$
$$\ln 3^{2x-1} = \ln 7^{x+1}$$
$$(2x-1)\ln 3 = (x+1)\ln 7$$
$$2x \ln 3 - \ln 3 = x \ln 7 + \ln 7$$
$$2x \ln 3 - x \ln 7 = \ln 3 + \ln 7$$
$$x(2\ln 3 - \ln 7) = \ln 3 + \ln 7$$
$$x = \frac{\ln 3 + \ln 7}{2\ln 3 - \ln 7}$$
$$x \approx 12.11$$

5.
$$e^{2x} - 8e^x + 7 = 0$$
$$(e^x - 7)(e^x - 1) = 0$$
$e^x - 7 = 0$ or $e^x - 1 = 0$

$e^x = 7$ $\quad$ $e^x = 1$

$\ln e^x = \ln 7$ $\quad$ $\ln e^x = \ln 1$

$x = \ln 7$ $\quad$ $x = 0$

The solution set is $\{0, \ln 7\}$. The solutions are 0 and (approximately) 1.95.

6. a. $\log_2(x-4) = 3$
$$2^3 = x - 4$$
$$8 = x - 4$$
$$12 = x$$
Check:
$$\log_2(x-4) = 3$$
$$\log_2(12-4) = 3$$
$$\log_2 8 = 3$$
$$3 = 3$$
The solution set is $\{12\}$.

b.
$$4\ln 3x = 8$$
$$\ln 3x = 2$$
$$e^{\ln 3x} = e^2$$
$$3x = e^2$$
$$x = \frac{e^2}{3} \approx 2.46$$
Check
$$4\ln 3x = 8$$
$$4\ln 3\left(\frac{e^2}{3}\right) = 8$$
$$4\ln e^2 = 8$$
$$4(2) = 8$$
$$8 = 8$$
The solution set is $\left\{\frac{e^2}{3}\right\}$, approximately 2.46.

7.
$$\log x + \log(x-3) = 1$$
$$\log x(x-3) = 1$$
$$10^1 = x(x-3)$$
$$10 = x^2 - 3x$$
$$0 = x^2 - 3x - 10$$
$$0 = (x-5)(x+2)$$
$x - 5 = 0$ or $x + 2 = 0$

$x = 5$ or $x = -2$

Check

Checking 5:
$$\log 5 + \log(5-3) = 1$$
$$\log 5 + \log 2 = 1$$
$$\log(5 \cdot 2) = 1$$
$$\log 10 = 1$$
$$1 = 1$$
Checking –2:
$$\log x + \log(x-3) = 1$$
$$\log(-2) + \log(-2-3) \; 0 \; 1$$
Negative numbers do not have logarithms so –2 does not check.
The solution set is $\{5\}$.

8.
$$\ln(x-3)=\ln(7x-23)-\ln(x+1)$$
$$\ln(x-3)=\ln\frac{7x-23}{x+1}$$
$$x-3=\frac{7x-23}{x+1}$$
$$(x-3)(x+1)=7x-23$$
$$x^2-2x-3=7x-23$$
$$x^2-9x+20=0$$
$$(x-4)(x-5)=0$$
$x=4$ or $x=5$

Both values produce true statements.
The solution set is {4, 5}

9. For a risk of 7%, let $R=7$ in
$$R=6e^{12.77x}$$
$$6e^{12.77x}=7$$
$$e^{12.77x}=\frac{7}{6}$$
$$\ln e^{12.77x}=\ln\left(\frac{7}{6}\right)$$
$$12.77x=\ln\left(\frac{7}{6}\right)$$
$$x=\frac{\ln\left(\frac{7}{6}\right)}{12.77}\approx 0.01$$
For a blood alcohol concentration of 0.01, the risk of a car accident is 7%.

10. $A=P\left(1+\frac{r}{n}\right)^{nt}$
$$3600=1000\left(1+\frac{0.08}{4}\right)^{4t}$$
$$1000\left(1\_\frac{0.08}{4}\right)^{4t}=3600$$
$$1000(1+0.02)^{4t}=3600$$
$$1000(1.02)^{4t}=3600$$
$$(1.02)^{4t}=\ln 3.6$$
$$4t\ln(1.02)=\ln 3.6$$
$$t=\frac{\ln 3.6}{4\ln 1.02}$$
$$\approx 16.2$$
After approximately 16.2 years, the $1000 will grow to an accumulated value of $3600.

11. $f(x)=34.1\ln x+117.7$

Solve equation when $f(x)=210$.
$$34.1\ln x+117.7=210$$
$$34.1\ln x=92.3$$
$$\ln x=\frac{92.3}{34.1}$$
$$\log_e x=\frac{92.3}{34.1}$$
$$e^{\frac{92.3}{34.1}}=x$$
$$15\approx x$$
Approximately 15 years after 1999, in the year 2014, there will be 210 million Internet users in the United States.

Exercise Set 3.4

1.
$$2^x=64$$
$$2^x=2^6$$
$$x=6$$
The solution is 6, and the solution set is $\{6\}$.

2.
$$3^x=81$$
$$3^x=3^4$$
$$x=4$$
The solution set is {4}.

3.
$$5^x=125$$
$$5^x=5^3$$
$$x=3$$
The solution is 3, and the solution set is $\{3\}$.

4.
$$5^x=625$$
$$5^x=5^4$$
$$x=4$$
The solution set is {4}.

5.
$$2^{2x-1}=32$$
$$2^{2x-1}=2^5$$
$$2x-1=5$$
$$2x=6$$
$$x=3$$
The solution is 3, and the solution set is $\{3\}$.

6.
$$3^{2x+1} = 27$$
$$3^{2x+1} = 3^3$$
$$2x + 1 = 3$$
$$2x = 2$$
$$x = 1$$
The solution set is $\{1\}$

7.
$$4^{2x-1} = 64$$
$$4^{2x-1} = 4^3$$
$$2x - 1 = 3$$
$$2x = 4$$
$$x = 2$$
The solution is 2, and the solution set is $\{2\}$.

8.
$$5^{3x-1} = 125$$
$$5^{3x-1} = 5^3$$
$$3x - 1 = 3$$
$$3x = 4$$
$$x = \frac{4}{3}$$
The solution set is $\left\{\frac{4}{3}\right\}$.

9.
$$32^x = 8$$
$$\left(2^5\right)^x = 2^3$$
$$2^{5x} = 2^3$$
$$5x = 3$$
$$x = \frac{3}{5}$$
The solution is $\frac{3}{5}$, and the solution set is $\left\{\frac{3}{5}\right\}$.

10.
$$4^x = 32$$
$$\left(2^2\right)^x = 2^5$$
$$2^{2x} = 2^5$$
$$2x = 5$$
$$x = \frac{5}{2}$$
The solution set is $\left\{\frac{5}{2}\right\}$.

11.
$$9^x = 27$$
$$\left(3^2\right)^x = 3^3$$
$$3^{2x} = 3^3$$
$$2x = 3$$
$$x = \frac{3}{2}$$
The solution is $\frac{3}{2}$, and the solution set is $\left\{\frac{3}{2}\right\}$.

12.
$$125^x = 625$$
$$\left(5^3\right)^x = 5^4$$
$$5^{3x} = 5^4$$
$$3x = 4$$
$$x = \frac{4}{3}$$
The solution set is $\left\{\frac{4}{3}\right\}$.

13.
$$3^{1-x} = \frac{1}{27}$$
$$3^{1-x} = \frac{1}{3^3}$$
$$3^{1-x} = 3^{-3}$$
$$1 - x = -3$$
$$-x = -4$$
$$x = 4$$
The solution set is $\{4\}$.

14.
$$5^{2-x} = \frac{1}{125}$$
$$5^{2-x} = \frac{1}{5^3}$$
$$5^{2-x} = 5^{-3}$$
$$2 - x = -3$$
$$-x = -5$$
$$x = 5$$
The solution set is $\{5\}$.

15.
$$6^{\frac{x-3}{4}} = \sqrt{6}$$
$$6^{\frac{x-3}{4}} = 6^{\frac{1}{2}}$$
$$\frac{x-3}{4} = \frac{1}{2}$$
$$2(x-3) = 4(1)$$
$$2x-6 = 4$$
$$2x = 10$$
$$x = 5$$
The solution is 5, and the solution set is {5}.

16.
$$7^{\frac{x-2}{6}} = \sqrt{7}$$
$$7^{\frac{x-2}{6}} = 7^{\frac{1}{2}}$$
$$\frac{x-2}{6} = \frac{1}{2}$$
$$2(x-2) = 6(1)$$
$$2x-4 = 6$$
$$2x = 10$$
$$x = 5$$
The solution set is {5}.

17.
$$4^x = \frac{1}{\sqrt{2}}$$
$$\left(2^2\right)^x = \frac{1}{2^{\frac{1}{2}}}$$
$$2^{2x} = 2^{-\frac{1}{2}}$$
$$2x = -\frac{1}{2}$$
$$x = \frac{1}{2}\left(-\frac{1}{2}\right) = -\frac{1}{4}$$
The solution is $-\frac{1}{4}$, and the solution set is $\left\{-\frac{1}{4}\right\}$.

18.
$$9^x = \frac{1}{\sqrt[3]{3}}$$
$$\left(3^2\right)^x = \frac{1}{3^{\frac{1}{3}}}$$
$$3^{2x} = 3^{-\frac{1}{3}}$$
$$2x = -\frac{1}{3}$$
$$x = \frac{1}{2}\left(-\frac{1}{3}\right) = -\frac{1}{6}$$
The solution set is $\left\{-\frac{1}{6}\right\}$.

19.
$$8^{x+3} = 16^{x-1}$$
$$\left(2^3\right)^{x+3} = \left(2^4\right)^{x-1}$$
$$2^{3x+9} = 2^{4x-4}$$
$$3x+9 = 4x-4$$
$$13 = x$$
The solution set is $\{13\}$.

20.
$$8^{1-x} = 4^{x+2}$$
$$\left(2^3\right)^{1-x} = \left(2^2\right)^{x+2}$$
$$2^{3-3x} = 2^{2x+4}$$
$$3-3x = 2x+4$$
$$-5x = 1$$
$$x = -\frac{1}{5}$$
The solution set is $\left\{-\frac{1}{5}\right\}$.

21.
$$e^{x+1} = \frac{1}{e}$$
$$e^{x+1} = e^{-1}$$
$$x+1 = -1$$
$$x = -2$$
The solution set is $\{-2\}$.

22.
$$e^{x+4} = \frac{1}{e^{2x}}$$
$$e^{x+4} = e^{-2x}$$
$$x+4 = -2x$$
$$3x = -4$$
$$x = -\frac{4}{3}$$
The solution set is $\left\{-\frac{4}{3}\right\}$.

23.
$$10^x = 3.91$$
$$\ln 10^x = \ln 3.91$$
$$x \ln 10 = \ln 3.91$$
$$x = \frac{\ln 3.91}{\ln 10} \approx 0.59$$

24.
$$10^x = 8.07$$
$$\ln 10^x = \ln 8.07$$
$$x \ln 10 = \ln 8.07$$
$$x = \frac{\ln 8.07}{\ln 10} \approx 0.91$$

25. $e^x = 5.7$
$\ln e^x = 5.7$
$x = \ln 5.7 \approx 1.74$

26. $e^x = 0.83$
$\ln e^x = \ln 0.83$
$x = \ln 0.83 \approx -0.19$

27. $5^x = 17$
$\ln 5^x = \ln 17$
$x \ln 5 = \ln 17$
$x = \dfrac{\ln 17}{\ln 5} \approx 1.76$

28. $19^x = 143$
$x \ln 19 = \ln 143$
$x = \dfrac{\ln 143}{\ln 19} \approx 1.69$

29. $5e^x = 23$
$e^x = \dfrac{23}{5}$
$\ln e^x = \ln \dfrac{23}{5}$
$x = \ln \dfrac{23}{5} \approx 1.53$

30. $9e^x = 107$
$e^x = \dfrac{107}{9}$
$\ln e^x = \ln \dfrac{107}{9}$
$x = \ln \dfrac{107}{9} \approx 2.48$

31. $3e^{5x} = 1977$
$e^{5x} = 659$
$\ln e^{5x} = \ln 659$
$x = \dfrac{\ln 659}{5} \approx 1.30$

32. $4e^{7x} = 10,273$
$e^{7x} = \dfrac{10,273}{4}$
$\ln e^{7x} = \ln\left(\dfrac{10,273}{4}\right)$
$7x = \ln\left(\dfrac{10,273}{4}\right)$
$x = \dfrac{1}{7}\ln\left(\dfrac{10,273}{4}\right) \approx 1.12$

33. $e^{1-5x} = 793$
$\ln e^{1-5x} = \ln 793$
$(1-5x)(\ln e) = \ln 793$
$1-5x = \ln 793$
$5x = 1 - \ln 793$
$x = \dfrac{1-\ln 793}{5} \approx -1.14$

34. $e^{1-8x} = 7957$
$\ln e^{1-8x} = \ln 7957$
$(1-8x)\ln e = \ln 7957$
$1 - 8x = \ln 7957$
$8x = 1 - \ln 7957$
$x = \dfrac{1-\ln 7957}{8} \approx -1.00$

35. $e^{5x-3} - 2 = 10,476$
$e^{5x-3} = 10,478$
$\ln e^{5x-3} = \ln 10,478$
$(5x-3)\ln e = \ln 10,478$
$5x - 3 = \ln 10,478$
$5x = \ln 10,478 + 3$
$x = \dfrac{\ln 10,478 + 3}{5} \approx 2.45$

36. $e^{4x-5} - 7 = 11,243$
$e^{4x-5} = 11,250$
$\ln e^{4x-5} = \ln 11,250 \; (4x-5)\ln e = \ln 11,250$
$4x - 5 = \ln 11,250$
$x = \dfrac{\ln 11,250 + 5}{4} \approx 3.58$

37.
$$7^{x+2} = 410$$
$$\ln 7^{x+2} = \ln 410$$
$$(x+2)\ln 7 = \ln 410$$
$$x+2 = \frac{\ln 410}{\ln 7}$$
$$x = \frac{\ln 410}{\ln 7} - 2 \approx 1.09$$

38.
$$5^{x-3} = 137$$
$$\ln 5^{x-3} = \ln 137$$
$$(x-3)\ln 5 = \ln 137$$
$$x-3 = \frac{\ln 137}{\ln 5}$$
$$x = 3 + \frac{\ln 137}{\ln 5} \approx 6.06$$

39.
$$7^{0.3x} = 813$$
$$\ln 7^{0.3x} = \ln 813$$
$$0.3x \ln 7 = \ln 813$$
$$x = \frac{\ln 813}{0.3 \ln 7} \approx 11.48$$

40.
$$3^{x/7} = 0.2$$
$$\ln 3^{x/7} = \ln 0.2$$
$$\frac{x}{7}\ln 3 = \ln 0.2$$
$$x \ln 3 = 7 \ln 0.2$$
$$x = \frac{7 \ln 0.2}{\ln 3} \approx -10.25$$

41.
$$5^{2x+3} = 3^{x-1}$$
$$\ln 5^{2x+3} = \ln 3^{x-1}$$
$$(2x+3)\ln 5 = (x-1)\ln 3$$
$$2x \ln 5 + 3\ln 5 = x\ln 3 - \ln 3$$
$$3\ln 5 + \ln 3 = x\ln 3 - 2x\ln 5$$
$$3\ln 5 + \ln 3 = x(\ln 3 - 2\ln 5)$$
$$\frac{3\ln 5 + \ln 3}{\ln 3 - 2\ln 5} = x$$
$$-2.80 \approx x$$

42.
$$7^{2x+1} = 3^{x+2}$$
$$\ln 7^{2x+1} = \ln 3^{x+2}$$
$$(2x+1)\ln 7 = (x+2)\ln 3$$
$$2x+1 = (x+2)\frac{\ln 3}{\ln 7}$$
$$2x+1 = x\frac{\ln 3}{\ln 7} + \frac{2\ln 3}{\ln 7}$$
$$2x - x\frac{\ln 3}{\ln 7} = \frac{2\ln 3}{\ln 7} - 1$$
$$x\left(2 - \frac{\ln 3}{\ln 7}\right) = \frac{2\ln 3}{\ln 7} - 1$$
$$x = \frac{\frac{2\ln 3}{\ln 7} - 1}{2 - \frac{\ln 3}{\ln 7}} \approx 0.09$$

43.
$$e^{2x} - 3e^x + 2 = 0$$
$$(e^x - 2)(e^x - 1) = 0$$
$e^x - 2 = 0$ or $e^x - 1 = 0$

$e^x = 2$ $\quad$ $e^x = 1$

$\ln e^x = \ln 2$ $\quad$ $\ln e^x = \ln 1$

$x = \ln 2$ $\quad$ $x = 0$

The solution set is {0, ln 2). The solutions are 0 and (approximately) 0.69.

44.
$$e^{2x} - 2e^x - 3 = 0$$
$$(e^x - 3)(e^x + 1) = 0$$
$e^x - 3 = 0$ or $e^x + 1 = 0$

$e^x = 3$ $\quad$ $e^x = -1$

$\ln e^x = \ln 3$ $\quad$ $\ln e^x = \ln(-1)$

$x = \ln 3$ $\quad$ no solution

The solution set is {ln 3}. The solutions is (approximately) 1.10.

45.
$$e^{4x} + 5e^{2x} - 24 = 0$$
$$(e^{2x} + 8)(e^{2x} - 3) = 0$$
$e^{2x} + 8 = 0$ or $e^{2x} - 3 = 0$

$e^{2x} = -8$ $\quad$ $e^{2x} = 3$

$\ln e^{2x} = \ln(-8)$ $\quad$ $\ln e^{2x} = \ln 3$

$2x = \ln(-8)$ $\quad$ $2x = \ln 3$

ln(−8) does not exist $\quad$ $x = \frac{\ln 3}{2}$

$$x = \frac{\ln 3}{2} \approx 0.55$$

46. $e^{4x}-3e^{2x}-18=0$

$\left(e^{2x}-6\right)\left(e^{2x}+3\right)=0$

$e^{2x}-6=0$ or $e^{2x}+3=0$

$e^{2x}=6 \qquad e^{2x}=-3$

$\ln e^{2x}=\ln 6 \qquad \ln e^{2x}=\ln(-3)$

$2x=\ln 6 \qquad \ln(-3)$ does not exist.

$x=\frac{\ln 6}{2}\approx 0.90$

47. $3^{2x}+3^x-2=0$

$(3^x+2)(3^x-1)=0$

$3^x+2=0 \qquad 3^x-1=0$

$3^x=-2 \qquad 3^x=1$

$\log 3^x=\log(-2) \qquad \log 3^x=\log 1$

can't do $\qquad x\log 3=0$

$x=\frac{0}{\log 3}$

$x=0$

The solution set is $\{0\}$.

48. $2^{2x}+2^x-12=0$

$(2^x+4)(2^x-3)=0$

$2^x+4=0 \qquad 2^x-3=0$

$2^x=-4 \qquad 2^x=3$

$\ln 2^x=\ln(-4) \qquad \ln 2^x=\ln 3$

can't do $\qquad x\ln 2=\ln 3$

$x=\frac{\ln 3}{\ln 2}$

$x\approx 1.58$

49. $\log_3 x=4$

$3^4=x$

$81=x$

50. $\log_5 x=3$

$5^3=x$

$125=x$

51. $\ln x=2$

$e^2=x$

$7.39\approx x$

52. $\ln x=3$

$e^3=x$

$20.09\approx x$

53. $\log_4(x+5)=3$

$4^3=x+5$

$59=x$

54. $\log_5(x-7)=2$

$5^2=x-7$

$32=x$

55. $\log_3(x-4)=-3$

$3^{-3}=x-4$

$\frac{1}{27}=x-4$

$4\frac{1}{27}=x$

$4.04\approx x$

56. $\log_7(x+2)=-2$

$7^{-2}=x+2$

$\frac{1}{49}=x+2$

$-1\frac{48}{49}=x$

$-1.98\approx x$

57. $\log_4(3x+2)=3$

$4^3=3x+2$

$64=3x+2$

$62=3x$

$\frac{62}{3}=x$

$20.67\approx x$

58. $\log_2(4x+1)=5$

$2^5=4x+1$

$32=4x+1$

$31=4x$

$\frac{31}{4}=x$

$7.75=x$

59. $5\ln 2x=20$

$\ln 2x=4$

$e^{\ln 2x}=e^4$

$2x=e^4$

$x=\frac{e^4}{2}\approx 27.30$

60. $6\ln 2x = 30$
$\ln 2x = 5$
$e^{\ln 2x} = e^5$
$2x = e^5$
$x = \frac{e^5}{2} \approx 74.21$

61. $6 + 2\ln x = 5$
$2\ln x = -1$
$\ln x = -\frac{1}{2}$
$e^{\ln x} = e^{-1/2}$
$x = e^{-1/2} \approx 0.61$

62. $7 + 3\ln x = 6$
$3\ln x = -1$
$\ln x = -\frac{1}{3}$
$e^{\ln x} = e^{-1/3}$
$x = e^{-1/3} \approx 0.72$

63. $\ln\sqrt{x+3} = 1$
$e^{\ln\sqrt{x+3}} = e^1$
$\sqrt{x+3} = e$
$x + 3 = e^2$
$x = e^2 - 3 \approx 4.39$

64. $\ln\sqrt{x+4} = 1$
$e^{\ln\sqrt{x+4}} = e^1$
$\sqrt{x+4} = e$
$x + 4 = e^2$
$x = e^2 - 4 \approx 3.39.$

65. $\log_5 x + \log_5 (4x-1) = 1$
$\log_5 (4x^2 - x) = 1$
$4x^2 - x = 5$
$4x^2 - x - 5 = 0$
$(4x-5)(x+1) = 0$
$x = \frac{5}{4}$ or $x = -1$
$x = -1$ does not check because $\log_5(-1)$ does not exist.
The solution set is $\left\{\frac{5}{4}\right\}$.

66. $\log_6 (x+5) + \log_6 x = 2$
$\log_6 x(x+5) = 2$
$x(x+5) = 6^2$
$x^2 + 5x = 36$
$x^2 + 5x - 36 = 0$
$(x+9)(x-4) = 0$
$x = -9$ or $x = 4$
$x = -9$ does not check because $\log_6(-9+5)$ does not exist.
The solution set is $\{4\}$.

67. $\log_3 (x-5) + \log_3 (x+3) = 2$
$\log_3 [(x-5)(x+3)] = 2$
$(x-5)(x+3) = 3^2$
$x^2 - 2x - 15 = 9$
$x^2 - 2x - 24 = 0$
$(x-6)(x+4) = 0$
$x = 6$ or $x = -4$
$x = -4$ does not check because $\log_3(-4-5)$ does not exist. The solution set is $\{6\}$.

68. $\log_2 (x-1) + \log_2 (x+1) = 3$
$\log_2 [(x-1)(x+1)] = 3$
$(x-1)(x+1) = 2^3$
$x^2 - 1 = 8$
$x^2 = 9$
$x = 3$ or $x = -3$
$x = -3$ does not check because $\log_2(-3-1)$ does not exist.
The solution set is $\{3\}$.

69. $\log_2 (x+2) - \log_2 (x-5) = 3$
$\log_2 \left(\frac{x+2}{x-5}\right) = 3$
$\frac{x+2}{x-5} = 2^3$
$\frac{x+2}{x-5} = 8$
$x + 2 = 8(x-5)$
$x + 2 = 8x - 40$
$7x = 42$
$x = 6$

70. $\log_4(x+2)-\log_4(x-1)=1$

$\log_4\left(\frac{x+2}{x-1}\right)=1$

$\frac{x+2}{x-1}=4^1$

$\frac{x+2}{x-1}=4$

$x+2=4(x-1)$

$x+2=4x-4$

$3x=6$

$x=2$

71. $2\log_3(x+4)=\log_3 9+2$

$2\log_3(x+4)=2+2$

$2\log_3(x+4)=4$

$\log_3(x+4)=2$

$3^2=x+4$

$9=x+4$

$5=x$

72. $3\log_2(x-1)=5-\log_2 4$

$3\log_2(x-1)=5-2$

$3\log_2(x-1)=3$

$\log_2(x-1)=1$

$2^1=x-1$

$3=x$

73. $\log_2(x-6)+\log_2(x-4)-\log_2 x=2$

$\log_2\frac{(x-6)(x-4)}{x}=2$

$\frac{(x-6)(x-4)}{x}=2^2$

$x^2-10x+24=4x$

$x^2-14x+24=0$

$(x-12)(x-2)=0$

$x-12=0 \qquad x-2=0$

$x=12 \qquad x=2$

The solution set is $\{12\}$ since $\log_2(2-6)=\log_2(-4)$ is not possible.

74. $\log_2(x-3)+\log_2 x-\log_2(x+2)=2$

$\log_2\frac{(x-3)x}{(x+2)}=2$

$2^2=\frac{x^2-3x}{x+2}$

$4(x+2)=x^2-3x$

$4x+8=x^2-3x$

$0=x^2-7x-8$

$0=(x+1)(x-8)$

$x+1=0 \qquad x-8=0$

$x=-1 \qquad x=8$

$\log_2(-1-3)=\log_2(-4)$ does not exist, so the solution set is $\{8\}$.

75. $\log(x+4)=\log x+\log 4$

$\log(x+4)=\log 4x$

$x+4=4x$

$4=3x$

$x=\frac{4}{3}$

This value is rejected. The solution set is $\left\{\frac{4}{3}\right\}$.

76. $\log(5x+1)=\log(2x+3)+\log 2$

$\log(5x+1)=\log(4x+6)$

$5x+1=4x+6$

$x=5$

77. $\log(3x-3)=\log(x+1)+\log 4$

$\log(3x-3)=\log(4x+4)$

$3x-3=4x+4$

$-7=x$

This value is rejected. The solution set is $\{\ \}$.

78. $\log(2x-1)=\log(x+3)+\log 3$
$\log(2x-1)=\log(3x+9)$
$2x-1=3x+9$
$-10=x$
This value is rejected. The solution set is $\{\ \}$.

79. $2\log x=\log 25$
$\log x^2=\log 25$
$x^2=25$
$x=\pm 5$
-5 is rejected. The solution set is $\{5\}$.

80. $3\log x=\log 125$
$\log x^3=\log 125$
$x^3=125$
$x=5$

81. $\log(x+4)-\log 2=\log(5x+1)$
$\log\frac{x+4}{2}=\log(5x+1)$
$\frac{x+4}{2}=5x+1$
$x+4=10x+2$
$-9x=-2$
$x=\frac{2}{9}$
$x\approx 0.22$

82. $\log(x+7)-\log 3=\log(7x+1)$
$\log\frac{x+7}{3}=\log(7x+1)$
$\frac{x+7}{3}=7x+1$
$x+7=21x+3$
$-20x=-4$
$x=\frac{1}{5}$
$x\approx 0.2$

83. $2\log x-\log 7=\log 112$
$\log x^2-\log 7=\log 112$
$\log\frac{x^2}{7}=\log 112$
$\frac{x^2}{7}=112$
$x^2=784$
$x=\pm 28$
-28 is rejected. The solution set is $\{28\}$.

84. $\log(x-2)+\log 5=\log 100$
$\log(5x-10)=\log 100$
$5x-10=100$
$5x=110$
$x=22$

85. $\log x+\log(x+3)=\log 10$
$\log(x^2+3x)=\log 10$
$x^2+3x=10$
$x^2+3x-10=0$
$(x+5)(x-2)=0$
$x=-5$ or $x=2$
-5 is rejected. The solution set is $\{2\}$.

86. $\log(x+3)+\log(x-2)=\log 14$
$\log(x^2+x-6)=\log 14$
$x^2+x-6=14$
$x^2+x-20=0$
$(x+5)(x-4)=0$
$x=-5$ or $x=4$
-5 is rejected. The solution set is $\{4\}$.

87. $\ln(x-4)+\ln(x+1)=\ln(x-8)$
$\ln(x^2-3x-4)=\ln(x-8)$
$x^2-3x-4=x-8$
$x^2-4x+4=0$
$(x-2)(x-2)=0$
$x=2$
2 is rejected. The solution set is $\{\ \}$.

88. $\log_2(x-1)-\log_2(x+3)=\log_2\left(\frac{1}{x}\right)$

$$\log_2\frac{x-1}{x+3}=\log_2\left(\frac{1}{x}\right)$$
$$\frac{x-1}{x+3}=\frac{1}{x}$$
$$x^2-x=x+3$$
$$x^2-2x-3=0$$
$$(x+1)(x-3)=0$$
$x=-1$ or $x=3$

-1 is rejected. The solution set is $\{3\}$.

89. $\ln(x-2)-\ln(x+3)=\ln(x-1)-\ln(x+7)$

$$\ln\frac{x-2}{x+3}=\ln\frac{x-1}{x+7}$$
$$\frac{x-2}{x+3}=\frac{x-1}{x+7}$$
$$(x-2)(x+7)=(x+3)(x-1)$$
$$x^2+5x-14=x^2+2x-3$$
$$3x=11$$
$$x=\frac{11}{3}$$
$$x\approx 3.67$$

90. $\ln(x-5)-\ln(x+4)=\ln(x-1)-\ln(x+2)$

$$\ln\frac{x-5}{x+4}=\ln\frac{x-1}{x+2}$$
$$\frac{x-5}{x+4}=\frac{x-1}{x+2}$$
$$(x-5)(x+2)=(x+4)(x-1)$$
$$x^2-3x-10=x^2+3x-4$$
$$-6x=6$$
$$x=-1$$

-1 is rejected. The solution set is $\{\ \}$.

91. $5^{2x}\cdot 5^{4x}=125$

$$5^{2x+4x}=5^3$$
$$5^{6x}=5^3$$
$$6x=3$$
$$x=\frac{1}{2}$$

92. $3^{x+2}\cdot 3^x=81$

$$3^{(x+2)+x}=3^4$$
$$3^{2x+2}=3^4$$
$$2x+2=4$$
$$2x=2$$
$$x=1$$

93. $2|\ln x|-6=0$

$$2|\ln x|=6$$
$$|\ln x|=3$$

$\ln x=3$ or $\ln x=-3$

$x=e^3$ $\qquad$ $x=e^{-3}$

$x\approx 20.09$ $\qquad$ $x\approx 0.05$

94. $3|\log x|-6=0$

$$3|\log x|=6$$
$$|\log x|=2$$

$\log x=2$ or $\log x=-2$

$x=10^2$ $\qquad$ $x=10^{-2}$

$x=100$ $\qquad$ $x=0.01$

95. $3^{x^2}=45$

$$\ln 3^{x^2}=\ln 45$$
$$x^2\ln 3=\ln 45$$
$$x^2=\frac{\ln 45}{\ln 3}$$
$$x=\pm\sqrt{\frac{\ln 45}{\ln 3}}\approx\pm 1.86$$

96. $5^{x^2}=50$

$$\ln 5^{x^2}=\ln 50$$
$$x^2\ln 5=\ln 50$$
$$x^2=\frac{\ln 50}{\ln 5}$$
$$x=\pm\sqrt{\frac{\ln 50}{\ln 5}}\approx\pm 1.56$$

97. $\ln(2x+1)+\ln(x-3)-2\ln x=0$

$$\ln(2x+1)+\ln(x-3)-\ln x^2=0$$
$$\ln\frac{(2x+1)(x-3)}{x^2}=0$$
$$\frac{(2x+1)(x-3)}{x^2}=e^0$$
$$\frac{2x^2-5x-3}{x^2}=1$$
$$2x^2-5x-3=x^2$$
$$x^2-5x-3=0$$
$$x=\frac{-b\pm\sqrt{b^2-4ac}}{2a}$$
$$x=\frac{-(-5)\pm\sqrt{(-5)^2-4(1)(-3)}}{2(1)}$$
$$x=\frac{5\pm\sqrt{37}}{2}$$
$$x=\frac{5+\sqrt{37}}{2}\approx 5.54$$
$$x=\frac{5-\sqrt{37}}{2}\approx -0.54 \text{ (rejected)}$$

The solution set is $\left\{\frac{5+\sqrt{37}}{2}\right\}$.

98. $\ln 3-\ln(x+5)-\ln x=0$

$$\ln\frac{3}{x(x+5)}=0$$
$$e^0=\frac{3}{x(x+5)}$$
$$1=\frac{3}{x(x+5)}$$
$$x(x+5)=3$$
$$x^2+5x=3$$
$$x^2+5x-3=0$$
$$x=\frac{-b\pm\sqrt{b^2-4ac}}{2a}\; x^2+5x-3=0$$
$$x=\frac{-(5)\pm\sqrt{(5)^2-4(1)(-3)}}{2(1)}$$
$$x=\frac{-5\pm\sqrt{37}}{2}$$
$$x=\frac{-5+\sqrt{37}}{2}\approx 0.54$$
$$x=\frac{-5-\sqrt{37}}{2}\approx -5.54 \text{ (rejected)}$$

The solution set is $\left\{\frac{-5+\sqrt{37}}{2}\right\}$.

99.

$$5^{x^2-12}=25^{2x}$$
$$5^{x^2-12}=\left(5^2\right)^{2x}$$
$$5^{x^2-12}=5^{4x}$$
$$x^2-12=4x$$
$$x^2-4x-12=0$$
$$(x-6)(x+2)=0$$

Apply the zero product property:

$x-6=0$ or $x+2=0$

$x=6$ $\quad x=-2$

The solutions are -2 and 6, and the solution set is $\{-2,\ 6\}$.

100.

$$3^{x^2-12}=9^{2x}$$
$$3^{x^2-12}=\left(3^2\right)^{2x}$$
$$3^{x^2-12}=3^{4x}$$
$$x^2-12=4x$$
$$x^2-4x-12=0$$
$$(x-6)(x+2)=0$$

Apply the zero product property:

$x-6=0$ or $x+2=0$

$x=6$ $\quad x=-2$

The solutions are -2 and 6, and the solution set is $\{-2,\ 6\}$.

101. $25 = 6e^{12.77x}$

$\frac{25}{6} = e^{12.77x}$

$\ln\frac{25}{6} = \ln e^{12.77x}$

$\ln\frac{25}{6} = 12.77x$

$\frac{\ln\frac{25}{6}}{12.77} = x$

$0.112 \approx x$

A blood alcohol level of about 0.11 corresponds to a 25% risk of a car accident.

102. $50 = 6e^{12.77x}$

$e^{12.77x} = \frac{25}{3}$

$\ln e^{12.77x} = \ln\left(\frac{25}{3}\right)$

$12.77x = \ln\left(\frac{25}{3}\right)$

$x = \frac{\ln\left(\frac{25}{3}\right)}{12.77} \approx 0.17$

A blood alcohol level of about 0.17 corresponds to a 50% risk of a car accident.

103. a. $A = 18.9e^{0.005(0)}$

$A = 18.9$ million

b. $19.6 = 18.9e^{0.0055t}$

$\frac{19.6}{18.9} = e^{0.0055t}$

$\ln\frac{19.6}{18.9} = \ln e^{0.0055t}$

$\ln\frac{19.6}{18.9} = 0.0055t$

$\frac{\ln\frac{19.6}{18.9}}{0.0055} = t$

$6.6 \approx t$

In 2007 the population of New York will reach 19.6 million.

104. a. In 2000 $t = 0$

$A(0) = 15.9e^{0.0235(0)}$

$A(0) = 15.9e^{0}$

$A(0) = 15.9$

In 2000, the population was 15.9 million.

b. $19.2 = 15.9e^{0.0235t}$

$\frac{19.2}{15.9} = e^{0.0235t}$

$\ln\frac{19.2}{15.9} = \ln e^{0.0235t}$

$\ln\frac{19.2}{15.9} = 0.0235t$

$\frac{\ln\frac{19.2}{15.9}}{0.0235} = t$

$8 \approx t$

The population will reach 19.2 million about 8 years after 2000, in 2008.

105. $20{,}000 = 12{,}500\left(1 + \frac{0.0575}{4}\right)^{4t}$

$12{,}500(1.014375)^{4t} = 20{,}000$

$(1.014375)^{4t} = 1.6$

$\ln(1.014375)^{4t} = \ln 1.6$

$4t\ln(1.014375) = \ln 1.6$

$t = \frac{\ln 1.6}{4\ln 1.014375} \approx 8.2$

8.2 years

106. $15{,}000 = 7250\left(1 + \frac{0.065}{12}\right)^{12t}$

$7250(1.005416667)^{12t} = 15{,}000$

$(1.005416667)^{12t} = \frac{60}{29}$

$\ln(1.005416667)^{12t} = \ln\left(\frac{60}{29}\right)$

$12t\ln(1.00541667) = \ln\left(\frac{60}{29}\right)$

$t = \frac{\ln\left(\frac{60}{29}\right)}{12\ln 1.00541667} \approx 11.2$ years

107.
$$1400 = 1000\left(1+\frac{r}{360}\right)^{360\cdot 2}$$
$$\left(1+\frac{r}{360}\right)^{720} = 1.4$$
$$\ln\left(1+\frac{r}{360}\right)^{720} = \ln 1.4$$
$$720\ln\left(1+\frac{r}{360}\right) = \ln 1.4$$
$$\ln\left(1+\frac{r}{360}\right) = \frac{\ln 1.4}{720}$$
$$e^{\ln(1+r/360)} = e^{(\ln 1.4)/720}$$
$$1+\frac{r}{360} = e^{(\ln 1.4)/720} - 1$$
$$r = 360(e^{(\ln 1.4)/720)} - 1$$
$$\approx 0.168$$
16.8%

108. $9000 = 5000\left(1+\frac{r}{360}\right)^{(360\cdot 4)}$
$$\left(1+\frac{r}{360}\right)^{1440} = 1.8$$
$$\ln\left(1+\frac{r}{360}\right)^{1440} = \ln 1.8$$
$$1440\ln\left(1+\frac{r}{360}\right) = \ln 1.8$$
$$\ln\left(1+\frac{r}{360}\right) = \frac{\ln 1.8}{1440}$$
$$e^{\ln(1+r/360)} = e^{(\ln 1.8)/1440}$$
$$1+\frac{r}{360} = e^{(\ln 1.8)/1440} - 1$$
$$\frac{r}{360} = e^{(\ln 1.8)/1440} - 1$$
$$r = 360(e^{(\ln 1.8)/1440} - 1 \approx 0.147$$
14.7%

109. accumulated amount = 2(8000) = 16,000
$$16{,}000 = 8000e^{0.08t}$$
$$e^{0.08t} = 2$$
$$\ln e^{0.08t} = \ln 2$$
$$0.08t = \ln 2$$
$$t = \frac{\ln 2}{0.08}$$
$$t \approx 8.7$$
The amount would double in 8.7 years.

110. $12{,}000 = 8000e^{r\cdot 2}$
$$e^{2r} = 1.5$$
$$\ln e^{2r} = \ln 1.5$$
$$2r = \ln 1.5$$
$$r = \frac{\ln 1.5}{2} \approx 0.203$$
20.3%

111. accumulated amount = 3(2350) = 7050
$$7050 = 2350e^{r\cdot 7}$$
$$e^{7r} = 3$$
$$\ln e^{7r} = \ln 3$$
$$7r = \ln 3$$
$$r = \frac{\ln 3}{7} \approx 0.157$$
15.7%

112. $25{,}000 = 17{,}425e^{0.0425t}$
$$e^{0.0425t} = \frac{1000}{697}$$
$$\ln e^{0.0425t} = \ln\left(\frac{1000}{697}\right)$$
$$0.0425t = \ln\left(\frac{1000}{697}\right)$$
$$t = \frac{\ln\left(\frac{1000}{697}\right)}{0.0425} \approx 8.5 \text{ years}$$

113. a. $f(x) = 13.4 + 46.3\ln x$
$$f(3) = 13.4 + 46.3\ln 3 \approx 64.3$$
The function result of 64.3% models the actual value of 63% very well.

b. $f(x) = 13.4 + 46.3\ln x$
$$96 = 13.4 + 46.3\ln x$$
$$82.6 = 46.3\ln x$$
$$\frac{82.6}{46.3} = \ln x$$
$$x = e^{\frac{82.6}{46.3}}$$
$$x \approx 6$$
The function predicts that 96% of email will be spam 6 years after 2000, or 2006.

114. a. $f(x) = 34 - 2.6\ln x$

$f(41) = 34 - 2.6\ln 41 \approx 24.3$

The function result of 24.3% models the actual value of 24% very well.

b. $f(x) = 34 - 2.6\ln x$

$23 = 34 - 2.6\ln x$

$-11 = -2.6\ln x$

$\frac{11}{2.6} = \ln x$

$x = e^{\frac{11}{2.6}}$

$x \approx 69$

The function predicts that children under 18 will decline to 23% of the total population 69 years after 1969, or 2038.

115. $30\log_2 x = 45$

$\log_2 x = 1.5$

$x = 2^{1.5} \approx 2.8$

Only half the students recall the important features of the lecture after 2.8 days.
(2.8, 50)

116. $0 = 95 - 30\log_2 x$

$30\log_2 x = 95$

$\log_2 x = \frac{95}{30}$

$2^{\frac{95}{30}} = x$

$9.0 \approx x$

(9.0, 0)

117. $2.4 = -\log x$

$\log x = -2.4$

$x = 10^{-2.4} \approx 0.004$

The hydrogen ion concentration was $10^{-2.4}$, approximately 0.004 moles per liter.

118. $4.2 = -\log x$

$\log x = -4.2$

$x = 10^{-4.2} \approx 0.00006$

The hydrogen ion concentration was $10^{-4.2}$, approximately 0.00006 moles per liter.

123.

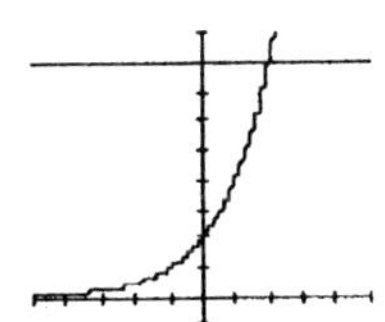

The intersection point is (2, 8).

Verify: $x = 2$

$2^{x+1} = 8$

$2^{2+1} = 2$

$2^3 = 8$

$8 = 8$

The solution set is $\{2\}$.

124.

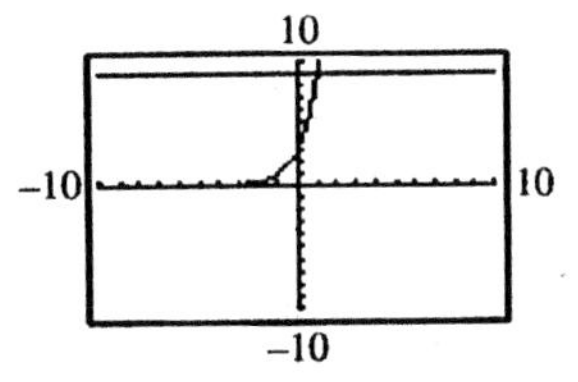

$\{1\}$

The intersection point is (1, 9).

Verify $x = 1$:

$3^{x+1} = 9$

$3^{1+1} = 9$

$3^2 = 9$

$9 = 9$

The solution set is $\{1\}$.

125.

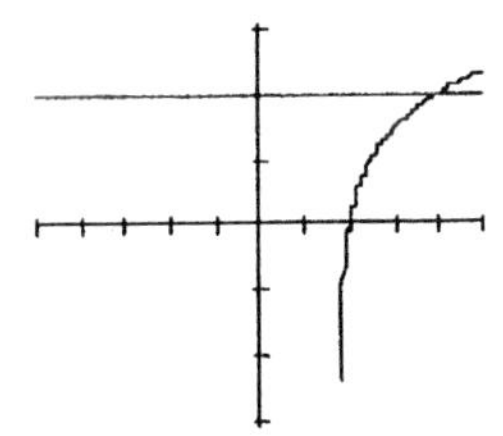

The intersection point is (4, 2).

Verify: $x = 4$

$\log_3(4 \cdot 4 - 7) = 2$

$\log_3 9 = 2$

$2 = 2$

The solution set is $\{4\}$.

126.

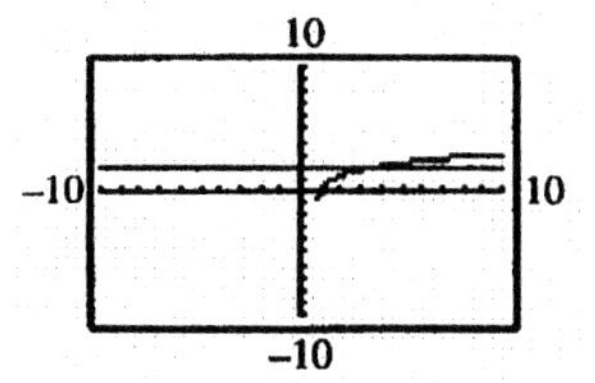

The intersection point is $\left(\frac{11}{3}, 2\right)$.

Verify: $x = \frac{11}{3}$

$$\log_3\left(3\cdot\frac{11}{3}-2\right)=2$$
$$\log_3(11\text{-}2)=2$$
$$\log_3 9=2$$
$$2=2$$

The solution set is $\left\{\frac{11}{3}\right\}$.

127.

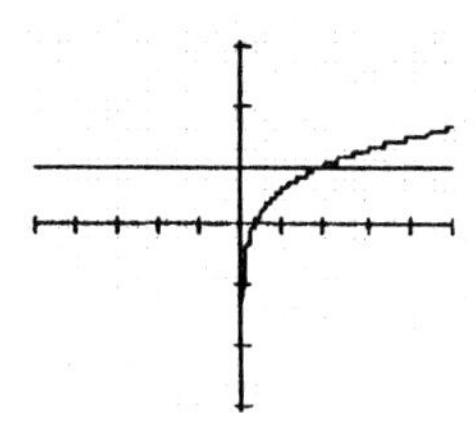

The intersection point is (2, 1).

Verify: $x = 2$

$$\log(2+3)+\log 2=1$$
$$\log 5+\log 2=1$$
$$\log(5\cdot 2)=1$$
$$\log 10=1$$
$$1=1$$

The solution set is {2}.

128.

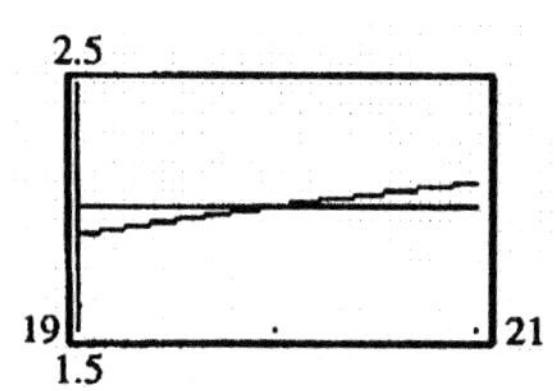

The intersection point is (20, 2).

Verify $x = 20$:

$$\log(x-15)+\log x=2$$
$$\log(20-15)+\log 20=2$$
$$\log 5+\log 20=2$$
$$\log 100=2$$
$$100=10^2$$
$$100=100$$

The solution set is {20}.

129.

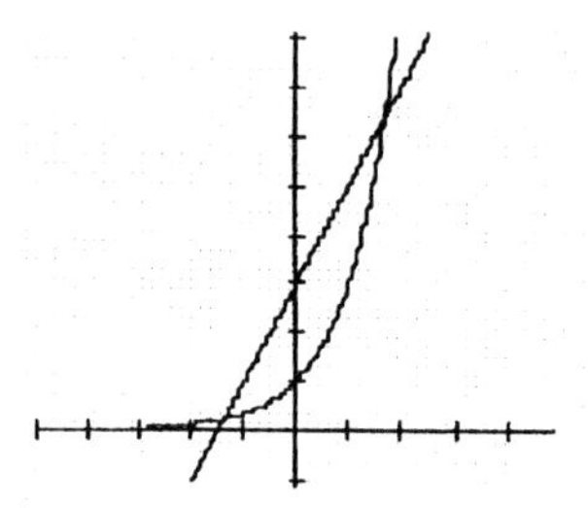

There are 2 points of intersection, approximately (–1.391606, 0.21678798) and (1.6855579, 6.3711158).

Verify $x \approx -1.391606$

$3^x = 2x + 3$

$3^{-1.391606} \approx 2(-1.391606) + 3$

$0.2167879803 \approx 0.216788$

Verify $x \approx 1.6855579$

$3^x = 2x + 3$

$3^{1.6855579} \approx 2(1.6855579) + 3$

$6.37111582 \approx 6.371158$

The solution set is {–1.391606, 1.6855579}.

130.

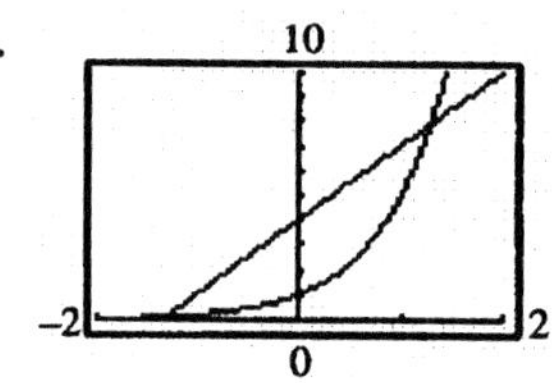

There are 2 points of intersection, approximately (−1.291641, 0.12507831) and (1.2793139, 7.8379416).

Verify:$x \approx -1.291641$

$$5^x = 3x+4$$
$$5^{-1.291641} = 3(-1.291641)+4$$
$$0.1250782178 \approx 0.125077$$

Verify:$x \approx 1.2793139$

$$5^{1.2793139} = 3(1.2793139)+4$$
$$7.837941942 \approx 7.8379417$$

The solution set is {–1.291641, 1,2793139}.

131.

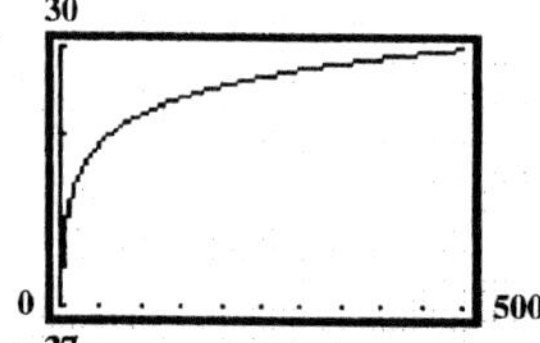

As the distance from the eye increases, barometric air pressure increases, leveling off at about 30 inches of mercury.

132.
$$29 = 0.48\ln(x+1) + 27$$
$$0.48\ln(x+1) = 2$$
$$\ln(x+1) = \frac{1}{0.24}$$
$$e^{\ln(x+1)} = e^{1/0.24}$$
$$x + 1 = e^{1/0.24}$$
$$x = e^{1/0.24} - 1 \approx 63.5$$

The barometric air pressure is 29 inches of mercury at a distance of about 63.5 miles from the eye of a hurricane.

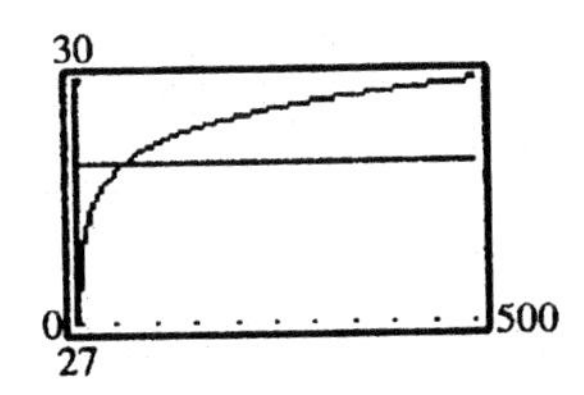

The point of intersection is approximately (63.5, 29).

133.

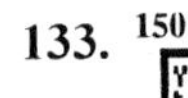

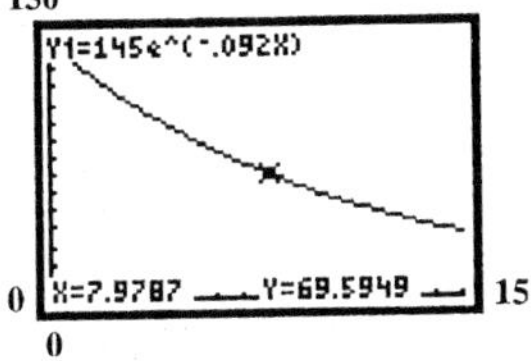

When $P = 70$, $t \approx 7.9$, so it will take about 7.9 minutes.
Verify:
$$70 = 45e^{-0.092(7.9)}$$
$$70 \approx 70.10076749$$
The runner's pulse will be 70 beats per minute after about 7.9 minutes.

134.

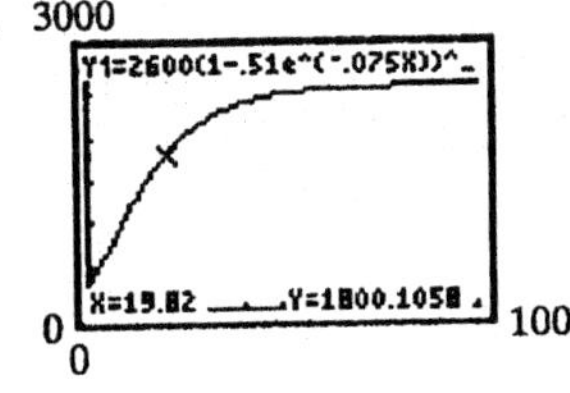

An adult female elephant weighing 1800 kilograms is about 20 years old.

135. a. False; $\log(x+3) = 2$ means $x + 3 = 10^2$

b. False; $\log(7x+3) - \log(2x+5) = 4$ means $\log\frac{7x+3}{2x+5} = 4$ which means $\frac{7x+3}{2x+5} = 10^4$

c. True; $x = \frac{1}{k}\ln y$
$$kx = \ln y$$
$$e^{kx} = e^{\ln y}$$
$$e^{kx} = y$$

d. False; The equation $x^{10} = 5.71$ has no variable in an exponent so is not an exponential equation.

(c) is true

136. Account paying 3% interest:
$$A = 4000\left(1 + \frac{0.03}{1}\right)^{1\cdot t}$$
Account paying 5% interest:
$$A = 2000\left(1 + \frac{0.05}{1}\right)^{1\cdot t}$$
The two accounts will have the same balance when
$$4000(1.03)^t = 2000(1.05)^t$$
$$(1.03)^t = 0.5(1.05)^t$$
$$\left(\frac{1.03}{1.05}\right)^t = 0.5$$
$$\ln\left(\frac{1.03}{1.05}\right)^t = \ln 0.5$$
$$t\ln\left(\frac{1.03}{1.05}\right) = \ln 0.5$$
$$t = \frac{\ln 0.5}{\ln\left(\frac{1.03}{1.05}\right)} \approx 36$$
The accounts will have the same balance in about 36 years.

137.
$$(\ln x)^2 = \ln x^2$$
$$(\ln x)^2 = 2\ln x$$
$$(\ln x)^2 - 2\ln x = 0$$
$$\ln x(\ln x - 2) = 0$$
$$\ln x = 2 \quad \text{or} \quad \ln x = 0$$
$$e^{\ln x} = e^2 \qquad x = 1$$
$$x = e^2$$
The solution set is $\{1, e^2\}$.

Check with graphing utility:

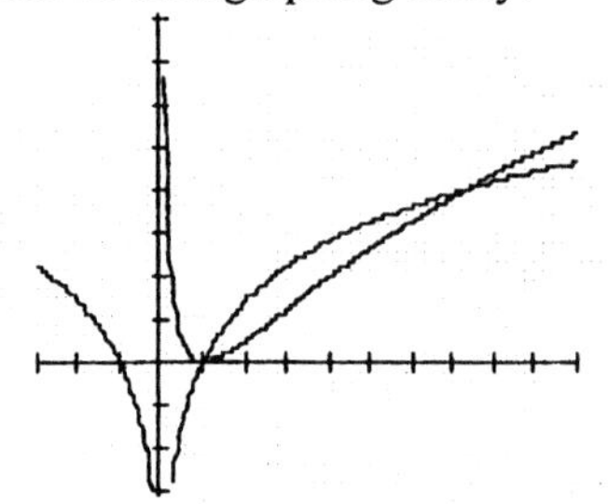

There are two points of intersection: (1, 0) and approximately (7.3890561, 4). Since $e^2 \approx 7.3890566099$, the graph verifies $x = 1$ and $x = e^2$, so the solution set is $\{1, e^2\}$ as determined algebraically.

138. $(\log x)(2\log x + 1) = 6$

$2(\log x)^2 + \log x - 6 = 0$

$(2\log x - 3)(\log x + 2) = 0$

$2\log x - 3 = 0$ or $\log x + 2 = 0$

$2\log x = 3 \qquad \log x = -2$

$\log x = \frac{3}{2} \qquad x = 10^{-2}$

$x = 10^{3/2} \qquad x = \frac{1}{100}$

$x = 10\sqrt{10}$

The solution set is $\left\{\frac{1}{100}, 10\sqrt{10}\right\}$.

Check by direct substitution:

Check: $x = 10\sqrt{10} = 10^{3/2}$

$(\log x)(2\log x + 1) = 6$

$\left(\log 10^{3/2}\right)(2\log 10^{3/2} + 1) = 6$

$\left(\frac{3}{2}\right)\left(2\cdot\frac{3}{2} + 1\right) = 6$

$\left(\frac{3}{2}\right)(3 + 1) = 6$

$\left(\frac{3}{2}\right)(4) = 6$

$6 = 6$

139. $\ln(\ln x) = 0$

$e^{\ln(\ln x)} = e^0$

$\ln x = 1$

$e^{\ln x} = e^1$

$x = e$

The solution set is $\{e\}$.

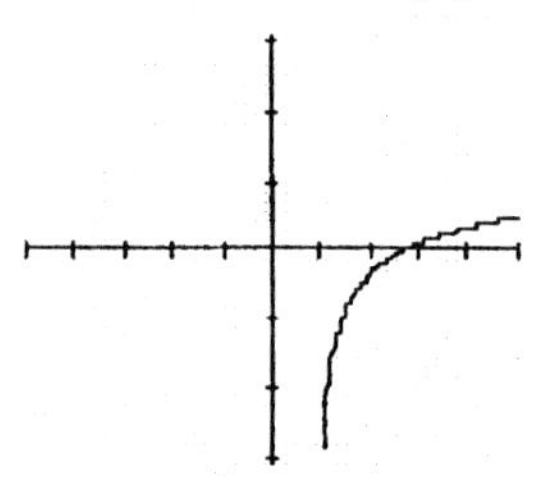

The graph of $\ln(\ln(x))$ crosses the graph $y = 0$ at approximately 2.718.

Section 3.5

Check Point Exercises

1. a. Use the exponential growth model $A = A_0 e^{kt}$ with 1990 corresponding to $t = 0$ when the population was 643 million:

$A = 643e^{kt}$

Substitute $t = 2000 - 1990 = 10$ when the population was 813 million, so $A = 813$, to find k.

$813 = 643e^{k10}$

$\frac{813}{643} = e^{k10}$

$\ln\frac{813}{643} = \ln e^{k10}$

$\ln\frac{813}{643} = 10k$

$\frac{\ln\frac{813}{643}}{10} = k$

$0.023 \approx k$

So the exponential growth function is $A = 643e^{0.023t}$

b. Substitute 2000 for A in the model from part (a) and solve for t.

$$2000 = 643e^{0.023t}$$
$$\frac{2000}{643} = e^{0.023t}$$
$$\ln\frac{2000}{643} = \ln e^{0.023t}$$
$$\ln\frac{2000}{643} = 0.023t$$
$$\frac{\ln\frac{2000}{643}}{0.023} = t$$
$$49 \approx t$$

The population will reach 2000 million, or two billion, about 49 years after 1990, in 2039.

2. a. In the exponential decay model $A = A_0e^{kt}$, substitute $\frac{A_0}{2}$ for A since the amount present after 28 years is half the original amount.

$$\frac{A_0}{2} = A_0e^{k\cdot 28}$$
$$e^{28k} = \frac{1}{2}$$
$$\ln e^{28k} = \ln\frac{1}{2}$$
$$28k = \ln\frac{1}{2}$$
$$k = \frac{\ln^{1/2}}{28} \approx -0.0248$$

So the exponential decay model is $A = A_0e^{-0.0248t}$

b. Substitute 60 for A_0 and 10 for A in the model from part (a) and solve for t.

$$10 = 60e^{-0.0248t}$$
$$e = ^{-0.0248t} = \frac{1}{6}$$
$$\ln e^{-0.0248t} = \ln\frac{1}{6}$$
$$-0.0248t = \ln\frac{1}{6}$$
$$t = \frac{\ln\frac{1}{6}}{-0.0248} \approx 72$$

The strontium-90 will decay to a level of 10 grams about 72 years after the accident.

3. a. The time prior to learning trials corresponds to $t = 0$.

$$f(0) = \frac{0.8}{1+e^{-0.2(0)}} = 0.4$$

The proportion of correct responses prior to learning trials was 0.4.

b. Substitute 10 for t in the model:

$$f(10) = \frac{0.8}{1+e^{-0.2(10)}} \approx 0.7$$

The proportion of correct responses after 10 learning trials was 0.7.

c. In the logistic growth model, $f(t) = \frac{c}{1+ae^{-bt}}$, the constant c represents the limiting size that $f(t)$ can attain. The limiting size of the proportion of correct responses as continued learning trials take place is 0.8.

4. a.

$$T = C + (T_o - C)e^{kt}$$
$$80 = 30 + (100 - 30)e^{k5}$$
$$80 = 30 + 70e^{5k}$$
$$50 = 70e^{5k}$$
$$\frac{5}{7} = e^{5k}$$
$$\ln\frac{5}{7} = \ln e^{5k}$$
$$\ln\frac{5}{7} = 5k$$
$$\frac{\ln\frac{5}{7}}{5} = k$$
$$-0.0673 \approx k$$
$$T = 30 + 70e^{-0.0673t}$$

b. $T = 30 + 70e^{-0.0673(20)} \approx 48^{\circ}$

After 20 minutes, the temperature will be 48°.

c. $35 = 30 + 70e^{-0.0673t}$

$$5 = 70e^{-0.0673t}$$

$$\frac{1}{14} = e^{-0.0673t}$$

$$\ln\frac{1}{14} = \ln e^{-0.0673t}$$

$$\ln\frac{1}{14} = -0.0673t$$

$$\frac{\ln\frac{1}{14}}{-0.0673} = t$$

$$39 \approx t$$

The temperature will reach 35° after 39 min.

5. Scatter plot:

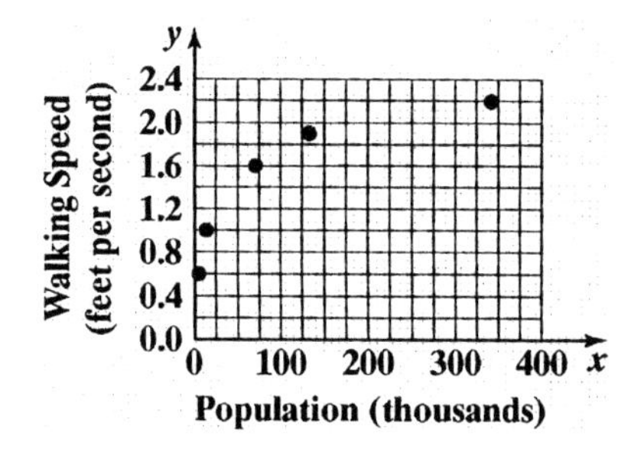

Because the data in the scatter plot increase rapidly at first and then begin to level off, the shape suggests that a logarithmic function is a good choice for modeling the data.

6. Scatter plot:

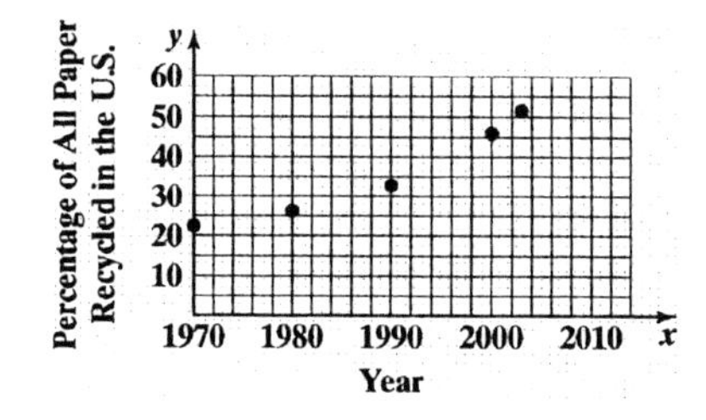

Because the data in the scatter plot appear to increase more and more rapidly, the shape suggests that an exponential function is a good choice for modeling the data.

7. $y = ab^x$ is equivalent to $y = ae^{(\ln b)x}$.

For $y = 4(7.8)^x$, $a = 4$, $b = 7.8$.

Thus, $y = 4(7.8)^x$ is equivalent to $y = 4e^{(\ln 7.8)x}$ in terms of a natural logarithm. Rounded to three decimal places, the model is approximately equivalent to $y = 4e^{2.054x}$.

Exercise Set 3.5

1. Since 2003 is 0 years after 2003, find A when $t = 0$:

$A = 127.2e^{0.001t}$

$A = 127.2e^{0.001(0)}$

$A = 127.2e^{0}$

$A = 127.2(1)$

$A = 127.2$

In 2003, the population was 127.2 million.

2. Since 2003 is 0 years after 2003, find A when $t = 0$. $A = 24.7e^{0.028(0)} = 24.7e^{0} = 24.7$

In 2003, the population of Iraq was about 24.7 million.

3. Iraq has the greatest growth rate at 2.8% per year.

4. Since $k = -0.004$, Russia has a decreasing population. The population is dropping at of 0.4% per year.

5. Substitute $A = 1238$ into the model for India and solve for t:

$$1238 = 1049.7e^{0.015t}$$

$$\frac{1238}{1049.7} = e^{0.015t}$$

$$\ln\frac{1238}{1049.7} = \ln e^{0.015t}$$

$$\ln\frac{1238}{1049.7} = 0.015t$$

$$t = \frac{\ln\frac{1238}{1049.7}}{0.015} \approx 11$$

Now, 2003 + 11 = 2014. The population of India will be 1238 million in approximately the year 2014.

6. Substitute $A = 1416$ into the model for India and solve for t:

$$1416 = 1049.7e^{0.015t}$$
$$\frac{1416}{1049.7} = e^{0.015t}$$
$$\ln\left(\frac{1416}{1049.7}\right) = \ln e^{0.015t}$$
$$\ln\left(\frac{1416}{1049.7}\right) = 0.015t$$
$$t = \frac{\ln\left(\frac{1416}{1049.7}\right)}{0.015} \approx 20$$

Now, 2003 + 20 = 2023. The population of India will be 1416 million in approximately the year 2023.

7. **a.** $A_0 = 6.04$. Since 2050 is 50 years after 2000, when $t = 50$, $A = 10$.

$$A = A_0e^{kt}$$
$$10 = 6.04e^{k(50)}$$
$$\frac{10}{6.04} = e^{50k}$$
$$\ln\left(\frac{10}{6.04}\right) = \ln e^{50k}$$
$$\ln\left(\frac{10}{6.04}\right) = 50k$$
$$k = \frac{\ln\left(\frac{10}{6.04}\right)}{50} \approx 0.01$$

Thus, the growth function is $A = 6.04e^{0.01t}$.

b.

$$9 = 6.04e^{0.01t}$$
$$\frac{9}{6.04} = e^{0.01t}$$
$$\ln\left(\frac{9}{6.04}\right) = \ln e^{0.01t}$$
$$\ln\left(\frac{9}{6.04}\right) = 0.01t$$
$$t = \frac{\ln\left(\frac{9}{6.04}\right)}{0.01} \approx 40$$

Now, 2000 + 40 = 2040, so the population will be 9 million is approximately the year 2040.

8. **a.** $A_0 = 3.2$. Since 2050 is 50 years after 2000, when $t = 50$, $A = 12$.

$$A = A_0e^{kt}$$
$$12 = 3.2e^{k(50)}$$
$$\frac{12}{3.2} = e^{50k}$$
$$\ln\left(\frac{12}{3.2}\right) = \ln e^{50k}$$
$$\ln\left(\frac{12}{3.2}\right) = 50k$$
$$k = \frac{\ln\left(\frac{12}{3.2}\right)}{50} \approx 0.026$$

Thus, the growth function is $A = 3.2e^{0.026t}$.

b.

$$3.2e^{0.026t} = 9$$
$$e^{0.026t} = \frac{9}{3.2}$$
$$\ln e^{0.026t} = \ln\left(\frac{9}{3.2}\right)$$
$$0.026t = \left(\frac{9}{3.2}\right)$$
$$t = \frac{\ln\left(\frac{9}{3.2}\right)}{0.026} \approx 40$$

Now, 2000 + 40 = 2040, so the population will be 9 million is approximately the year 2040.

9.

$$A = 16e^{-0.000121t}$$
$$A = 16e^{-0.000121(5715)}$$
$$A = 16e^{-0.691515}$$
$$A \approx 8.01$$

Approximately 8 grams of carbon-14 will be present in 5715 years.

10.

$$A = 16e^{-0.000121t}$$
$$A = 16e^{-0.000121(11430)}$$
$$A = 16e^{-1.38303}$$
$$A \approx 4.01$$

Approximately 4 grams of carbon-14 will be present in 11,430 years.

11. After 10 seconds, there will be $16 \cdot \frac{1}{2} = 8$ grams present. After 20 seconds, there will be $8 \cdot \frac{1}{2} = 4$ grams present. After 30 seconds, there will be $4 \cdot \frac{1}{2} = 2$ grams present. After 40 seconds, there will be $2 \cdot \frac{1}{2} = 1$ grams present. After 50 seconds, there will be $1 \cdot \frac{1}{2} = \frac{1}{2}$ gram present.

12. After 25,000 years, there will be $16 \cdot \frac{1}{2} = 8$ grams present. After 50,000 years, there will be $8 \cdot \frac{1}{2} = 4$ grams present. After 75,000 years, there will be $4 \cdot \frac{1}{2} = 2$ grams present. After 100,000 years, there will be $2 \cdot \frac{1}{2} = 1$ gram present. After 125,000 years, there will be $1 \cdot \frac{1}{2} = \frac{1}{2}$ gram present.

13.

$$A = A_0 e^{-0.000121t}$$
$$15 = 100e^{-0.000121t}$$
$$\frac{15}{100} = e^{-0.000121t}$$
$$\ln 0.15 = \ln e^{-0.000121t}$$
$$\ln 0.15 = -0.000121t$$
$$t = \frac{\ln 0.15}{-0.000121} \approx 15,679$$

The paintings are approximately 15,679 years old.

14.

$$A = A_0 e^{-0.000121t}$$
$$88 = 100e^{-0.000121t}$$
$$\frac{88}{100} = e^{-0.000121t}$$
$$\ln 0.88 = \ln e^{-0.000121t}$$
$$\ln 0.88 = -0.000121t$$
$$t = \frac{\ln 0.88}{-0.000121} \approx 1056$$

In 1989, the skeletons were approximately 1056 years old.

15. a.

$$\frac{1}{2} = 1e^{k1.31}$$
$$\ln \frac{1}{2} = \ln e^{1.31k}$$
$$\ln \frac{1}{2} = 1.31k$$
$$k = \frac{\ln \frac{1}{2}}{1.31} \approx -0.52912$$

The exponential model is given by $A = A_0 e^{-0.52912t}$.

b.

$$A = A_0 e^{-0.52912t}$$
$$0.945A_0 = A_0 e^{-0.52912t}$$
$$0.945 = e^{-0.52912t}$$
$$\ln 0.945 = \ln e^{-0.52912t}$$
$$\ln 0.945 = -0.52912t$$
$$t = \frac{\ln 0.945}{-0.52912} \approx 0.1069$$

The age of the dinosaur ones is approximately 0.1069 billion or 106,900,000 years old.

16.

$$A = A_0 e^{kt}$$
$$1000 = 1400e^{k5}$$
$$\frac{1000}{1400} = e^{5k}$$
$$\ln \frac{5}{7} = 5k$$
$$k = \frac{\ln \frac{5}{7}}{5} \approx -0.0673$$

The exponential model is given by $A = A_0 e^{-0.0673t}$.

$$100 = 1000e^{-0.0673t}$$
$$\frac{100}{1000} = e^{-0.0673t}$$
$$\ln \frac{1}{10} = \ln e^{-0.0673t}$$
$$\ln \frac{1}{10} = -0.0673t$$
$$t = \frac{\ln \frac{1}{10}}{-0.0673} \approx 34.2$$

The population will drop below 100 birds approximately 34 years from now. (This is 39 years from the time the population was 1400.)

17. $2A_0 = A_0 e^{kt}$

$2 = e^{kt}$

$\ln 2 = \ln e^{kt}$

$\ln 2 = kt$

$t = \dfrac{\ln 2}{k}$

The population will double in $t = \dfrac{\ln 2}{k}$ years.

18. $A = A_0 e^{kt}$

$3A_0 = A_0 e^{kt}$

$3 = e^{kt}$

$\ln 3 = \ln e^{kt}$

$\ln 3 = kt$

$t = \dfrac{\ln 3}{k}$

The population will triple in $t = \dfrac{\ln 3}{k}$ years.

19. $A = e^{0.007t}$

a. $k = 0.007$, so New Zealand's growth rate is 0.7%.

b. $t = \dfrac{\ln 2}{k}$

$t = \dfrac{\ln 2}{0.007} \approx 99$

New Zealand's population will double in approximately 99 years.

20. $A = 104.9e^{0.017t}$

a. $k = 0.017$, so Mexico's growth rate is 1.7%.

b. $t = \dfrac{\ln 2}{k}$

$t = \dfrac{\ln 2}{0.017} \approx 41$

Mexico's population will double in approximately 41 years.

21. a. When the epidemic began, $t = 0$.

$f(0) = \dfrac{100{,}000}{1 + 5000e^0} \approx 20$

Twenty people became ill when the epidemic began.

b. $f(4) = \dfrac{100{,}000}{1 + 5{,}000e^{-4}} \approx 1080$

About 1080 people were ill at the end of the fourth week.

c. In the logistic growth model,

$f(t) = \dfrac{c}{1 + ae^{-bt}}$,

the constant c represents the limiting size that $f(t)$ can attain. The limiting size of the population that becomes ill is 100,000 people.

22. $f(x) = \dfrac{12.85}{1 + 4.21e^{-0.026(x)}}$

$f(51) = \dfrac{12.85}{1 + 4.21e^{-0.026(51)}} \approx 6.1$

The function models the data very well.

23. $f(x) = \dfrac{12.85}{1 + 4.21e^{-0.026(x)}}$

$f(54) = \dfrac{12.85}{1 + 4.21e^{-0.026(54)}} \approx 6.3$

The function models the data very well.

24.

$f(x) = \dfrac{12.85}{1 + 4.21e^{-0.026(x)}}$

$7 = \dfrac{12.85}{1 + 4.21e^{-0.026(x)}}$

$7\left(1 + 4.21e^{-0.026(x)}\right) = 12.85$

$7 + 29.47e^{-0.026(x)} = 12.85$

$29.47e^{-0.026(x)} = 5.85$

$e^{-0.026(x)} = \dfrac{5.85}{29.47}$

$\ln e^{-0.026(x)} = \ln \dfrac{5.85}{29.47}$

$-0.026x = \ln \dfrac{5.85}{29.47}$

$x = \dfrac{\ln \frac{5.85}{29.47}}{-0.026}$

$x \approx 62$

The world population will reach 7 billion 62 years after 1949, or 2011.

25.
$$f(x)=\frac{12.85}{1+4.21e^{-0.026(x)}}$$
$$8=\frac{12.85}{1+4.21e^{-0.026(x)}}$$
$$8\left(1+4.21e^{-0.026(x)}\right)=12.85$$
$$8+33.68e^{-0.026(x)}=12.85$$
$$33.68e^{-0.026(x)}=4.85$$
$$e^{-0.026(x)}=\frac{4.85}{33.68}$$
$$\ln e^{-0.026(x)}=\ln\frac{4.85}{33.68}$$
$$-0.026x=\ln\frac{4.85}{33.68}$$
$$x=\frac{\ln\frac{4.85}{33.68}}{-0.026}$$
$$x\approx 75$$
The world population will reach 8 billion 75 years after 1949, or 2024.

26. $f(x)=\dfrac{12.85}{1+4.21e^{-0.026(x)}}$

As x increases, the exponent of e will decrease. This will make $e^{-0.026(x)}$ become very close to 0 and make the denominator become very close to 1. Thus, the limiting size of this function is 12.85 billion. This population value is somewhat larger than the 10 billion population limit suggested by the U.S. National Academy of Science.

27. $P(20)=\dfrac{90}{1+271e^{-0.122(20)}}\approx 3.7$

The probability that a 20-year-old has some coronary heart disease is about 3.7%.

28. $P(80)=\dfrac{90}{1+271e^{-0.122(80)}}\approx 88.6$

The probability that an 80-year-old has some coronary heart disease is about 88.6%.

29.
$$0.5=\frac{0.9}{1+271e^{-0.122t}}$$
$$0.5\left(1+271e^{-0.122t}\right)=0.9$$
$$1+271e^{-0.122t}=1.8$$
$$271e^{-0.122t}=0.8$$
$$e^{-0.122t}=\frac{0.8}{271}$$
$$\ln e^{-0.122t}=\ln\frac{0.8}{271}$$
$$-0.122t=\ln\frac{0.8}{271}$$
$$t=\frac{\ln\frac{0.8}{271}}{-0.122}\approx 48$$
The probability of some coronary heart disease is 0.5 at about age 48.

30.
$$70=\frac{90}{1+271e^{-0.122x}}$$
$$70(1+271e^{-0.122x})=90$$
$$1+271e^{-0.122x}=\frac{90}{70}$$
$$271e^{-0.122x}=\frac{2}{7}$$
$$e^{-0.122x}=\frac{2}{1897}$$
$$-0.122x=\ln\frac{2}{1897}$$
$$x=\frac{\ln\frac{2}{1897}}{-0.122}$$
$$x\approx 56$$
The probability of some coronary heart disease is 70% at about age 56.

31. a.
$$55=45+(70-45)e^{k10}$$
$$10=25e^{10k}$$
$$\frac{2}{5}=e^{10k}$$
$$\ln\frac{2}{5}=\ln e^{10k}$$
$$\ln\frac{2}{5}=10k$$
$$\frac{\ln\frac{2}{5}}{10}=k$$
$$-0.0916\approx k$$
$$T=45+25e^{-0.0916t}$$

b. $T = 45 + 25e^{-0.0916(15)} \approx 51^{\circ}$
After 15 minutes, the temperature will be 51°.

c.
$$50 = 45 + 25e^{-0.0916t}$$
$$5 = 25e^{-0.0916t}$$
$$\frac{1}{5} = e^{-0.0916t}$$
$$\ln\frac{1}{5} = \ln e^{-0.0916t}$$
$$\ln\frac{1}{5} = -0.0916t$$
$$\frac{\ln\frac{1}{5}}{-0.0916} = t$$
$$18 \approx t$$
The temperature will reach 50° after 18 min.

32. a.
$$T = C + (T_o - C)e^{kt}$$
$$300 = 70 + (450 - 70)e^{k5}$$
$$230 = 380e^{5k}$$
$$\frac{23}{38} = e^{5k}$$
$$\ln\frac{23}{38} = \ln e^{5k}$$
$$\ln\frac{23}{38} = 5k$$
$$\frac{\ln\frac{23}{38}}{5} = k$$
$$-0.1004 \approx k$$
$$T = 70 + 380e^{-0.1004t}$$

b. $T = 70 + 380\text{k}e^{-0.1004(20)} \approx 121^{\circ}$
After 20 minutes, the temperature will be 121°.

c.
$$140 = 70 + 380e^{-0.1004t}$$
$$70 = 380e^{-0.1004t}$$
$$\frac{7}{38} = e^{-0.1004t}$$
$$\ln\frac{7}{38} = \ln e^{-0.1004t}$$
$$\ln\frac{7}{38} = -0.1004t$$
$$\frac{\ln\frac{7}{38}}{-0.1004} = t$$
$$17 \approx t$$
The temperature will reach 140° after 17 min.

33.
$$T = C + (T_o - C)e^{kt}$$
$$38 = 75 + (28 - 75)e^{k10}$$
$$-37 = -47e^{10k}$$
$$\frac{-37}{-47} = e^{10k}$$
$$\ln\frac{37}{47} = \ln e^{10k}$$
$$\ln\frac{37}{47} = 10k$$
$$\frac{\ln\frac{37}{47}}{10} = k$$
$$-0.0239 \approx k$$
$$T = 75 - 47e^{-0.0239t}$$

$$50 = 75 - 47e^{-0.0239t}$$
$$-25 = -47e^{-0.0239t}$$
$$\frac{-25}{-47} = e^{-0.0239t}$$
$$\ln\frac{25}{47} = \ln e^{-0.0239t}$$
$$\ln\frac{25}{47} = -0.0239t$$
$$\frac{\ln\frac{17}{47}}{-0.0239} = t$$
$$26 = t$$
The temperature will reach 50° after 26 min.

34.

$$T = C + (T_o - C)e^{kt}$$

$$30 = 65 + (24 - 65)e^{k10}$$

$$-35 = -41e^{10k}$$

$$\frac{35}{41} = e^{10k}$$

$$\ln\frac{35}{41} = \ln e^{10k}$$

$$\ln\frac{35}{41} = 10k$$

$$\frac{\ln\frac{35}{41}}{10} = k$$

$$-0.0158 \approx k$$

$$T = 65 - 41e^{-0.0158t}$$

$$45 = 65 - 41e^{-0.0158t}$$

$$-20 = -41e^{-0.0158t}$$

$$\frac{20}{41} = e^{-0.0158t}$$

$$\ln\frac{20}{41} = \ln e^{-0.0158t}$$

$$\ln\frac{20}{41} = -0.0158t$$

$$\frac{\ln\frac{20}{41}}{-0.0158} = t$$

$$45 \approx t$$

The temperature will reach 45° after 45 min.

35. **a.**

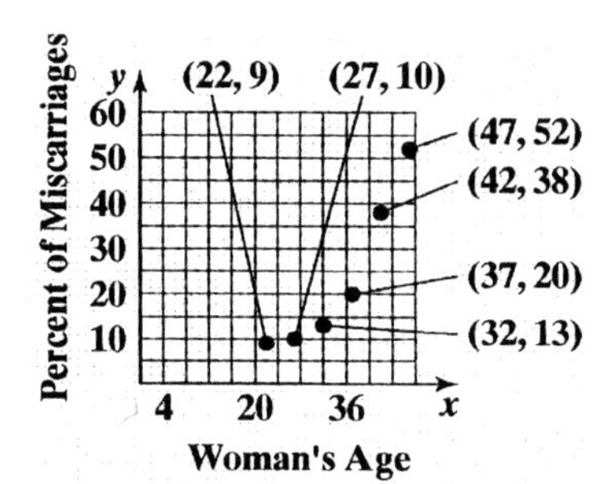

b. An exponential function appears to be the best choice for modeling the data.

36. **a.**

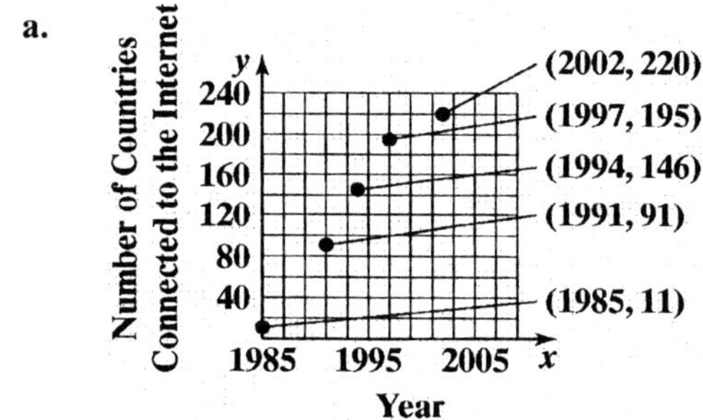

b. A linear function appears to be the best choice for modeling the data.

37. **a.**

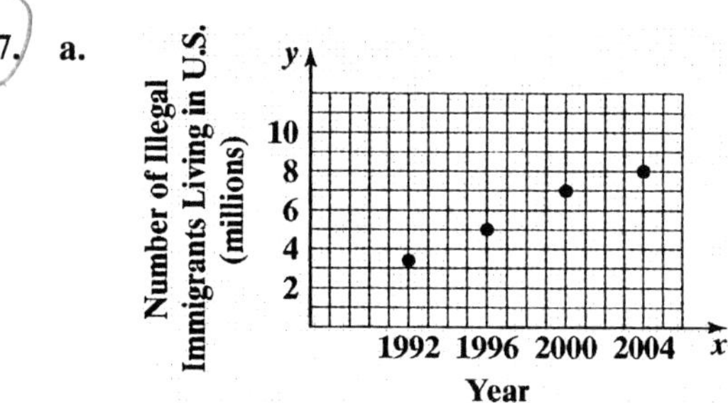

b. A linear function appear to be the best choice for modeling the data.

38. **a.**

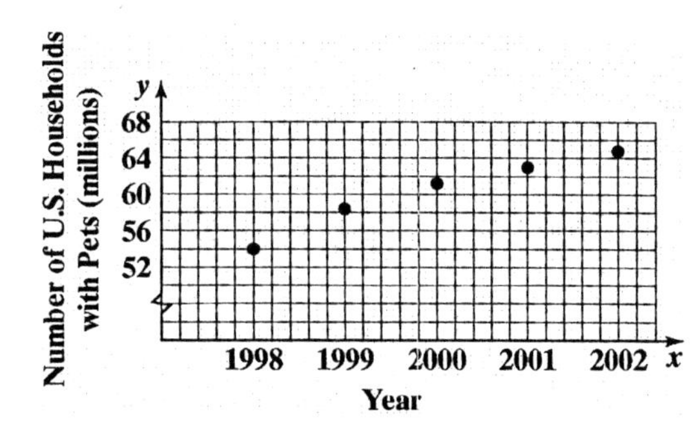

b. An logarithmic function appears to be the best choice for modeling the data.

39. a.

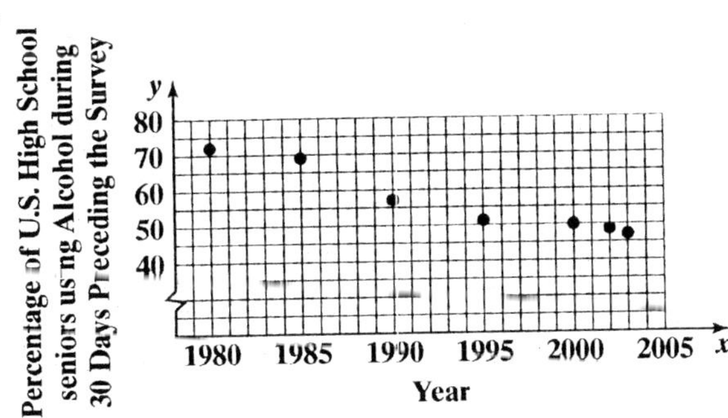

b. An exponential function appears to be the best choice for modeling the data.

40. a.

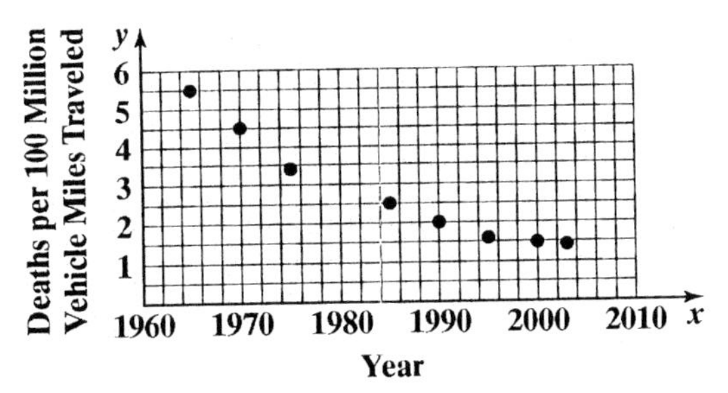

b. An exponential function appears to be the best choice for modeling the data.

41. $y = 100(4.6)^x$ is equivalent to
$y = 100e^{(\ln 4.6)x}$;
Using $\ln 4.6 \approx 1.526$,
$y = 100e^{1.526x}$.

42. $y = 1000(7.3)^x$ is equivalent to
$y = 1000e^{(\ln 7.3)x}$;
Using $\ln 7.3 \approx 1.988$,
$y = 1000e^{1.988x}$.

43. $y = 2.5(0.7)^x$ is equivalent to
$y = 2.5e^{(\ln 0.7)x}$;
Using $\ln 0.7 \approx -0.357$,
$y = 2.5e^{-0.357x}$.

44. $y = 4.5(0.6)^x$ is equivalent to
$y = 4.5e^{(\ln 0.6)x}$;
Using $\ln 0.6 \approx -0.511$,
$y = 4.5e^{-0.511x}$.

For Exercises 54. – 58, enter the data in L1 and L2:

L1	L2	L3 2
1	203.3	------
11	226.5	
21	248.7	
31	281.4	
34	294	

L2(6) =

54. a.

```
ExpReg
 y=a*b^x
 a=200.195411
 b=1.011087157
 r²=.9964717836
 r=.998234333
```

The exponential model is
$y = 200.2(1.011)^x$. Since $r \approx 0.998$ is very close to 1, the model fits the data very well.

b. $y = 200.2(1.011)^x$
$y = 200.2e^{(\ln 1.011)x}$
$y = 200.2e^{0.0109x}$
Since $k = .0109$, the population of the United States is increasing by about 1% each year.

55.

```
LnReg
 y=a+blnx
 a=194.3279438
 b=22.75785045
 r²=.7768947665
 r=.8814163412
```

The logarithmic model is $y = 194.328 + 22.758\ln x$. Since $r = 0.881$ is fairly close to 1, the model fits the data okay, but not great.

56.

```
LinReg
 y=ax+b
 a=2.713988409
 b=197.5858272
 r²=.9890898139
 r=.9945299462
```

The linear model is $y = 2.714x + 197.586$. Since $r \approx 0.995$ is close to 1, the model fits the data well.

57.

```
PwrReg
 y=a*x^b
 a=196.6191762
 b=.0943892895
 r²=.8157235653
 r=.9031741611
```

The power regression model is $y = 196.619x^{0.094}$. Since $r = 0.903$, the model fits the data fairly well.

58. Using r, the model of best fit is the exponential model $y = 200.2(1.011)^x$. The model of second best fit is the linear model $y = 2.714x + 197.586$.

Using the exponential model:

$$315 = 200.2(1.011)^x$$

$$\frac{315}{200.2} = (1.011)^x$$

$$\ln\left(\frac{315}{200.2}\right) = \ln(1.011)^x$$

$$\ln\left(\frac{315}{200.2}\right) = x\ln(1.011)$$

$$x = \frac{\ln\left(\frac{315}{200.2}\right)}{\ln(1.011)} \approx 41$$

$1969 + 41 = 2010$

Using the linear model:

$$315 = 2.714x + 197.586$$

$$117.414 = 2.714x$$

$$x = \frac{117.414}{2.714} \approx 43$$

$1969 + 43 = 2012$

According to the exponential model, the U.S. population will reach 315 million around the year 2010. According to the linear model, the U.S. population will reach 315 million around the year 2012. Both results are reasonably close to the result found in Example 1 (2010).

Explanations may vary.

59.

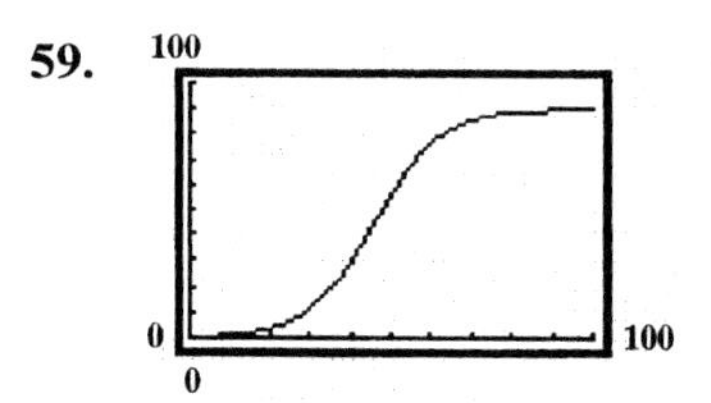

The probability of coronary heart disease starts increasing at a more rapid rate at about age 20. At about age 60, the rate of increase starts to slow down.

60. **a.**

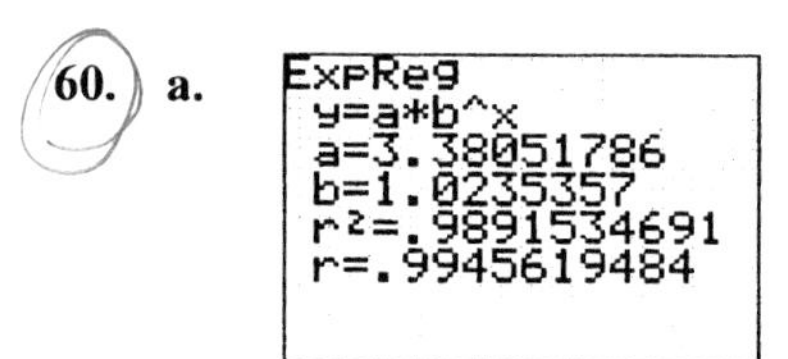

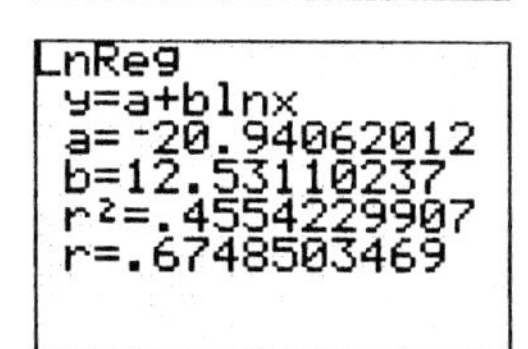

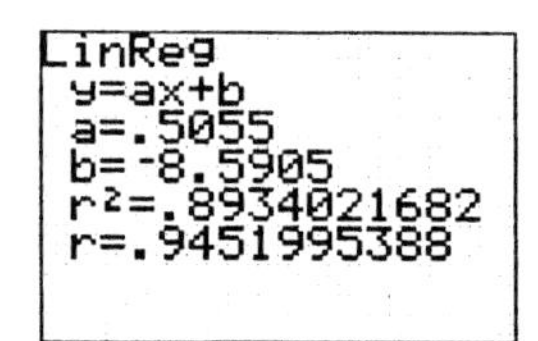

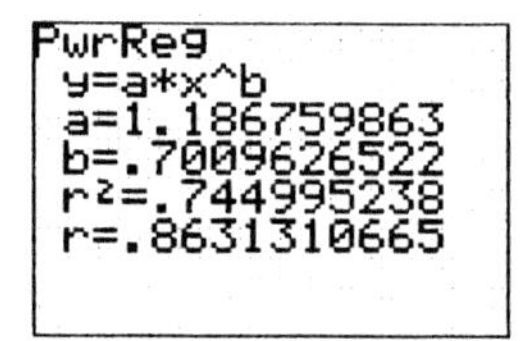

Using r as the determining factor, the model of best fit is the exponential model $y = 3.38(1.02)^x$.

b. $y = 3.38e^{(\ln 1.02)x}$

$y = 3.38e^{0.02x}$

The 65-and-over population is increasing by approximately 2% each year.

62. Statement **c.** is true. In 2012 Uganda's population will exceed Canada's. Uganda's population will be $A = 25.6e^{0.03(9)} \approx 33.5$ million, while Canada's population will be $A = 32.2e^{0.003(9)} \approx 33.1$ million.

Statement **a.** is false. In 2003, Uganda's population was not ten times that of Canada's. In fact, Uganda's population was smaller than Canada's. However, Uganda's rate of growth was ten times that of Canada's.

Statement **b.** is false. In 2003, Canada's population exceeded Uganda's by $32.2 - 25.6 = 6.6$ million, not 660,000.

Statement **d.** is false since statement **c.** is true.

63.

$$827 = 70 + (85.6 - 70)e^{k30}$$
$$12.7 = 15.6e^{30k}$$
$$\frac{12.7}{15.6} = e^{30k}$$
$$\ln\frac{12.7}{15.6} = \ln e^{30k}$$
$$\ln\frac{12.7}{15.6} = 30k$$
$$\frac{\ln\frac{12.7}{15.6}}{30} = k$$
$$-0.0069 \approx k$$
$$85.6 = 70 + (98.6 - 70)e^{-0.0069t}$$
$$15.6 = 28.6e^{-0.0069t}$$
$$\frac{15.6}{28.6} = e^{-0.0069t}$$
$$\ln\frac{15.6}{28.6} = \ln e^{-0.0069t}$$
$$\ln\frac{15.6}{28.6} = -0.0069$$
$$\frac{\ln\frac{15.6}{28.6}}{-0.0069} = t$$
$$88 \approx t$$

The death occurred at 88 minutes before 9:30, or 8:22 am.

Chapter 3 Review Exercises

1. This is the graph of $f(x) = 4^x$ reflected about the y-axis, so the function is $g(x) = 4^{-x}$.

2. This is the graph of $f(x) = 4^x$ reflected about the x-axis and about the y-axis, so the function is $h(x) = -4^{-x}$.

3. This is the graph of $f(x) = 4^x$ reflected about the x-axis and about the y-axis then shifted upward 3 units, so the function is $r(x) = -4^{-x} + 3$.

4. This is the graph of $f(x) = 4^x$.

5. The graph of $g(x)$ shifts the graph of $f(x)$ one unit to the right.

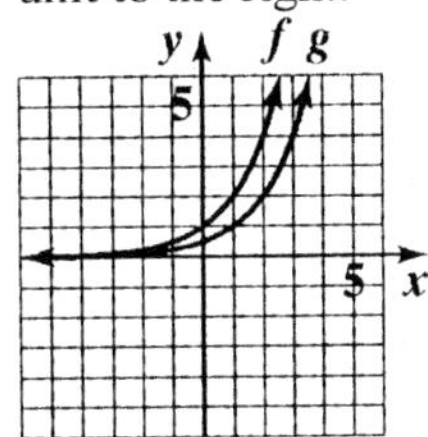

$f(x) = 2^x$
$g(x) = 2^{x-1}$

Asymptote of f: $y = 0$

Asymptote of g: $y = 0$

Domain of f = Domain of g = $(-\infty, \infty)$

Range of f = Range of g = $(0, \infty)$

6. The graph of $g(x)$ shifts the graph of $f(x)$ one unit down.

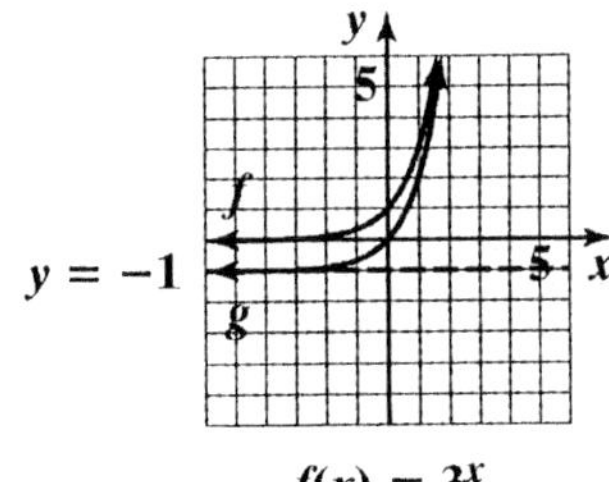

$f(x) = 3^x$
$g(x) = 3^x - 1$

Asymptote of f: $y = 0$

Asymptote of g: $y = -1$

Domain of f = Domain of g = $(-\infty, \infty)$

Range of f = $(0, \infty)$

Range of g = $(-1, \infty)$

7. The graph of $g(x)$ reflects the graph of $f(x)$ about the y – axis.

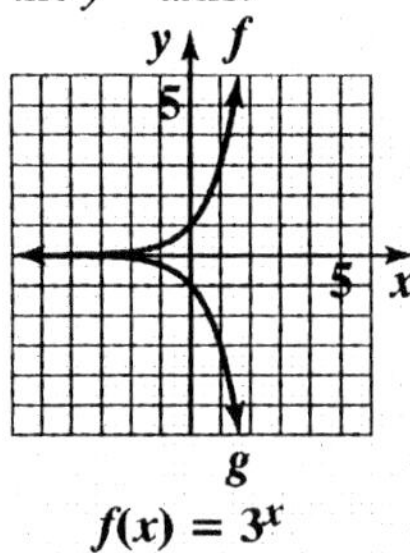

Asymptote of f: $y = 0$

Asymptote of g: $y = 0$

Domain of f = Domain of g = $(-\infty, \infty)$

Range of f = $(0, \infty)$

Range of g = $(-\infty, 0)$

8. The graph of $g(x)$ reflects the graph of $f(x)$ about the x – axis.

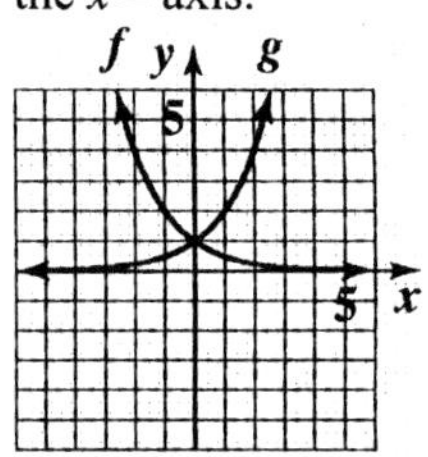

$f(x) = \left(\frac{1}{2}\right)^x$

$g(x) = \left(\frac{1}{2}\right)^{-x}$

Asymptote of f: $y = 0$

Asymptote of g: $y = 0$

Domain of f = Domain of g = $(-\infty, \infty)$

Range of f = Range of g = $(0, \infty)$

9. The graph of $g(x)$ vertically stretches the graph of $f(x)$ by a factor of 2.

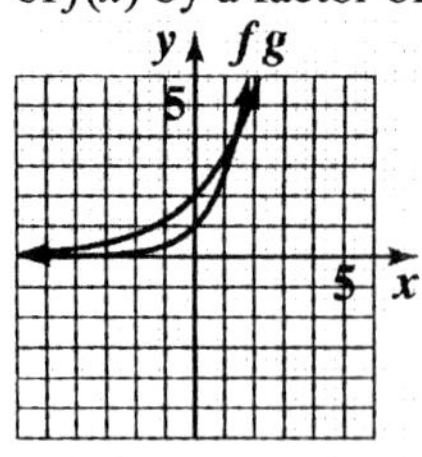

$f(x) = e^x$
$g(x) = 2e^{x/2}$

Asymptote of f: $y = 0$

Asymptote of g: $y = 0$

Domain of f = Domain of g = $(-\infty, \infty)$

Range of f = Range of g = $(0, \infty)$

10. 5.5% compounded semiannually:

$$A = 5000\left(1 + \frac{0.055}{2}\right)^{2\cdot 5} \approx 6558.26$$

5.25% compounded monthly:

$$A = 5000\left(1 + \frac{0.0525}{12}\right)^{12\cdot 5} \approx 6497.16$$

5.5% compounded semiannually yields the greater return.

11. 7% compounded monthly:

$$A = 14,000\left(1 + \frac{0.07}{12}\right)^{12\cdot 10} \approx 28,135.26$$

6.85% compounded continuously:

$A = 14,000e^{0.0685(10)} \approx 27,772.81$

7% compounded monthly yields the greater return.

12. a. When first taken out of the microwave, the temperature of the coffee was 200°.

b. After 20 minutes, the temperature of the coffee was about 120°.

$T = 70 + 130e^{-0.04855(20)} \approx 119.23$

Using a calculator, the temperature is about 119°.

c. The coffee will cool to about 70°; The temperature of the room is 70°.

13. $49^{1/2} = 7$

14. $4^3 = x$

15. $3^y = 81$

16. $\log_6 216 = 3$

17. $\log_b 625 = 4$

18. $\log_{13} 874 = y$

19. $\log_4 64 = 3$ because $4^3 = 64$.

20. $\log_5 \frac{1}{25} = -2$ because $5^{-2} = \frac{1}{25}$.

21. $\log_3(-9)$ cannot be evaluated since $\log_b x$ is defined only for $x > 0$.

22. $\log_{16} 4 = \frac{1}{2}$ because $16^{1/2} = \sqrt{16} = 4$.

23. Because $\log_b b = 1$,
we conclude $\log_{17} 17 = 1$.

24. Because $\log_b b^x = x$,
we conclude $\log_3 3^8 = 8$.

25. Because $\ln e^x = x$,
we conclude $\ln e^5 = 5$.

26. $\log_3 \frac{1}{\sqrt{3}} = \log_3 \frac{1}{3^{\frac{1}{2}}} = \log_3 3^{-\frac{1}{2}} = -\frac{1}{2}$

27. $\ln \frac{1}{e^2} = \ln e^{-2} = -2$

28. $\log \frac{1}{1000} = \log \frac{1}{10^3} = \log 10^{-3} = -3$

29. Because $\log_b = 1$,
we conclude $\log_8 8 = 1$.
So, $\log_3(\log_8 8) = \log_3 1$.
Because $\log_b 1 = 0$
we conclude $\log_3 1 = 0$.
Therefore, $\log_3(\log_8 8) = 0$.

30.

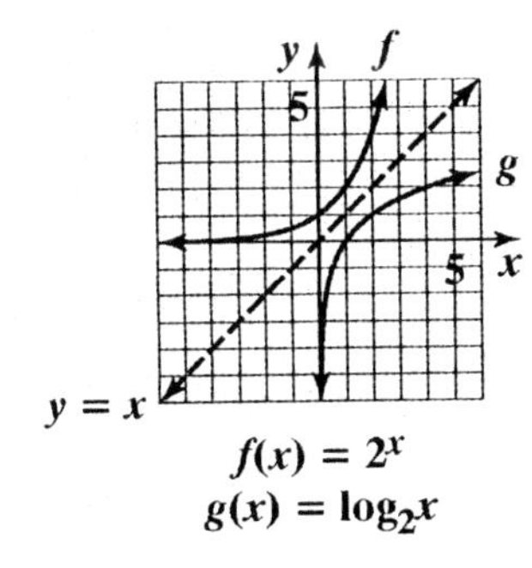

31.

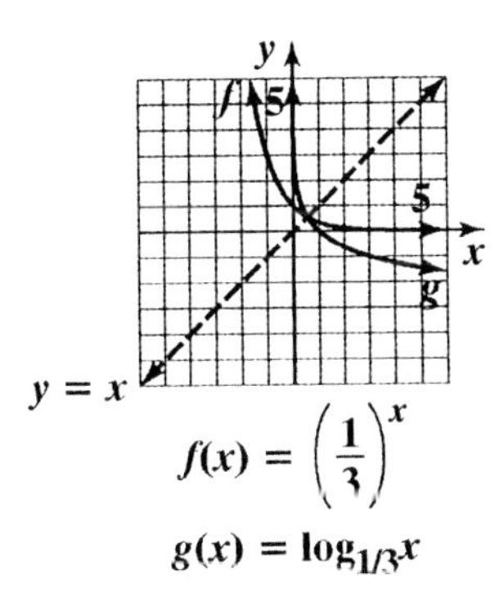

32. This is the graph of $f(x) = \log x$ reflected about the y-axis, so the function is $g(x) = \log(-x)$.

33. This is the graph of $f(x) = \log x$ shifted left 2 units, reflected about the y-axis, then shifted upward one unit, so the function is $r(x) = 1 + \log(2 - x)$.

34. This is the graph of $f(x) = \log x$ shifted left 2 units then reflected about the y-axis, so the function is $h(x) = \log(2 - x)$.

35. This is the graph of $f(x) = \log x$.

36.

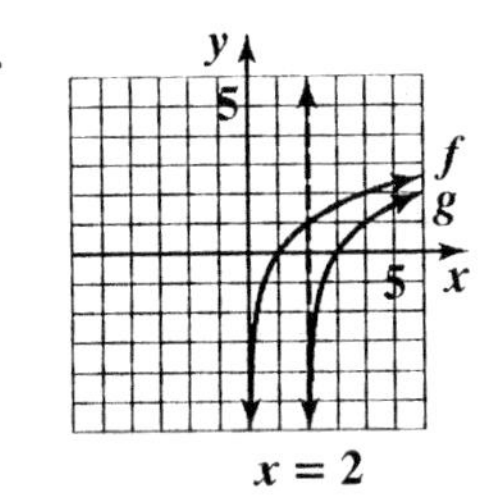

$f(x) = \log_2 x$
$g(x) = \log_2(x - 2)$

x-intercept: (3, 0)
vertical asymptote: $x = 2$
Domain: $(2, \infty)$
Range: $(-\infty, \infty)$

37.

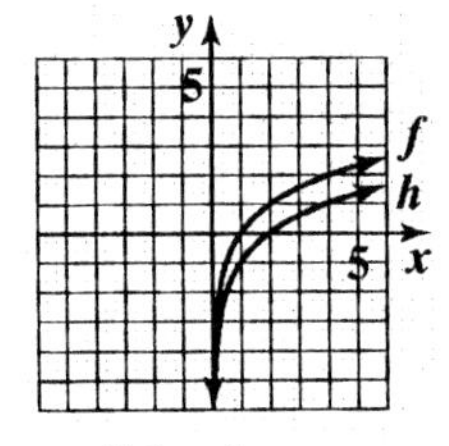

$f(x) = \log_2 x$
$h(x) = -1 + \log_2 x$

x-intercept: (2, 0)
vertical asymptote: $x = 0$
Domain: $(0, \infty)$
Range: $(-\infty, \infty)$

38.

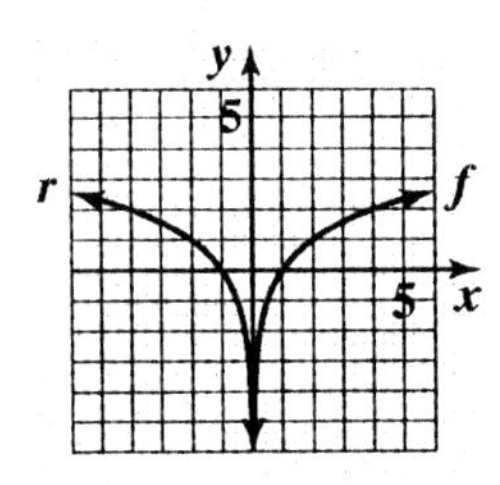

$f(x) = \log_2 x$
$r(x) = \log_2(-x)$

x-intercept: (–1, 0)
vertical asymptote: $x = 0$
Domain: $(-\infty, 0)$
Range: $(-\infty, \infty)$

39.

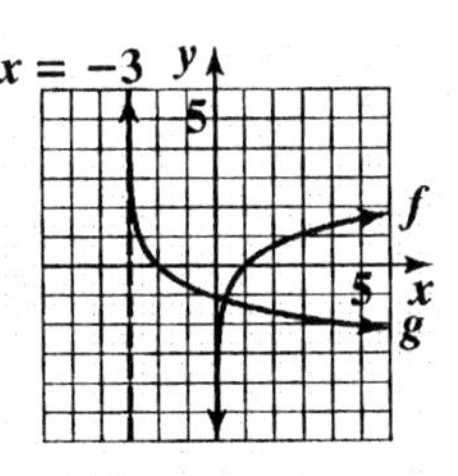

$f(x) = \log x$
$g(x) = -\log(x + 3)$

Asymptote of f: $x = 0$
Asymptote of g: $x = -3$
Domain of f = $(0, \infty)$
Domain of g = $(-3, \infty)$
Range of f = Range of g = $(-\infty, \infty)$

40.

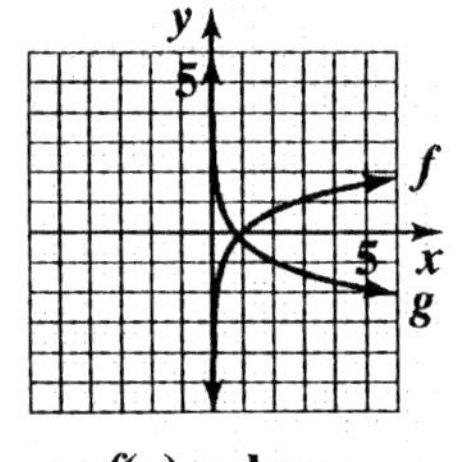

$f(x) = \ln x$
$g(x) = -\ln(2x)$

Asymptote of f: $x = 0$
Asymptote of g: $x = 0$
Domain of f = Domain of g = $(0, \infty)$
Range of f = Range of g = $(-\infty, \infty)$

41. The domain of f consists of all x for which $x + 5 > 0$.
Solving this inequality for x, we obtain $x > -5$.
Thus the domain of f is $(-5, \infty)$

42. The domain of f consists of all x for which $3 - x > 0$.
Solving this inequality for x, we obtain $x < 3$.
Thus, the domain of f is $(-\infty, 3)$.

43. The domain of f consists of all x for which $(x-1)^2 > 0$.
Solving this inequality for x, we obtain $x < 1$ or $x > 1$. Thus, the domain of f is $(-\infty, 1) \cup (1, \infty)$.

44. Because $\ln e^x = x$, we conclude $\ln e^{6x} = 6x$.

45. Because $e^{\ln x} = x$, we conclude $e^{\ln\sqrt{x}} = \sqrt{x}$.

46. Because $10^{\log x} = x$, we conclude $10^{\log 4x^2} = 4x^2$.

47. $R = \log\dfrac{1000I_0}{I_0} = \log 1000 = 3$
The Richter scale magnitude is 3.0.

48. a. $f(0) = 76 - 18\log(0+1) = 76$
When first given, the average score was 76.

b. $f(2) = 76 - 18\log(2+1) \approx 67$
$f(4) = 76 - 18\log(4+1) \approx 63$
$f(6) = 76 - 18\log(6+1) \approx 61$
$f(8) = 76 - 18\log(8+1) \approx 59$
$f(12) = 76 - 18\log(12+1) \approx 56$

After 2, 4, 6, 8, and 12 months, the average scores are about 67, 63, 61, 59, and 56, respectively.

c.

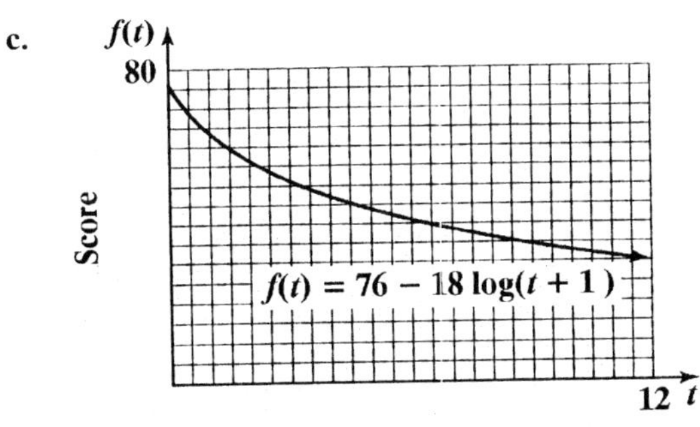

Retention decreases as time passes.

49. $t = \frac{1}{0.06}\ln\left(\frac{12}{12-5}\right) \approx 8.98$

It will take about 9 weeks.

50. $\log_6(36x^3)$
$= \log_6 36 + \log_6 x^3$
$= \log_6 36 + 3\log_6 x$
$= 2 + 3\log_6 x$

51. $\log_4 \frac{\sqrt{x}}{64} = \log_4 x^{1/2} - \log_4 64$
$= \frac{1}{2}\log_4 x - 3$

52. $\log_2 \frac{xy^2}{64} = \log_2 xy^2 - \log_2 64$
$= \log_2 x + \log_2 y^2 - \log_2 64$
$= \log_2 x + 2\log_2^{y} - 6$

53. $\ln \sqrt[3]{\frac{x}{e}}$
$= \ln\left(\frac{x}{e}\right)^{1/3}$
$= \frac{1}{3}[\ln x - \ln e]$
$= \frac{1}{3}\ln x - \frac{1}{3}\ln e$
$= \frac{1}{3}\ln x - \frac{1}{3}$

54. $\log_b 7 + \log_b 3$
$= \log_b(7 \cdot 3)$
$= \log_b 21$

55. $\log 3 - 3\log x$
$= \log 3 - \log x^3$
$= \log \frac{3}{x^3}$

56. $3\ln x + 4\ln y$
$= \ln x^3 + \ln y^4$
$= \ln(x^3 y^4)$

57. $\frac{1}{2}\ln x - \ln y$
$= \ln x^{1/2} - \ln y$
$= \ln \frac{\sqrt{x}}{y}$

58. $\log_6 72{,}348 = \frac{\log 72{,}348}{\log 6} \approx 6.2448$

59. $\log_4 0.863 = \frac{\ln 0.863}{\ln 4} \approx -0.1063$

60. true; $(\ln x)(\ln 1) = (\ln x)(0) = 0$

61. false; $\log(x+9) - \log(x+1) = \log\frac{(x+9)}{(x+1)}$

62. false; $\log_2 x^4 = 4\log_2 x$

63. true; $\ln e^x = x\ln e$

64.
$$\begin{aligned} 2^{4x-2} &= 64 \\ 2^{4x-2} &= 2^6 \\ 4x-2 &= 6 \\ 4x &= 8 \\ x &= 2 \end{aligned}$$

65.
$$\begin{aligned} 125^x &= 25 \\ \left(5^3\right)^x &= 5^2 \\ 5^{3x} &= 5^2 \\ 3x &= 2 \\ x &= \frac{2}{3} \end{aligned}$$

66.
$$\begin{aligned} 9^{x+2} &= 27^{-x} \\ \left(3^2\right)^{x+2} &= \left(3^3\right)^{-x} \\ 3^{2x+4} &= 3^{-3x} \\ 2x+4 &= -3x \\ 5x &= -4 \\ x &= -\frac{4}{5} \end{aligned}$$

67.
$$\begin{aligned} 8^x &= 12{,}143 \\ \ln 8^x &= \ln 12{,}143 \\ x \ln 8 &= \ln 12{,}143 \\ x &= \frac{\ln 12{,}143}{\ln 8} \approx 4.52 \end{aligned}$$

68.
$$\begin{aligned} 9e^{5x} &= 1269 \\ e^{5x} &= 141 \\ \ln e^{5x} &= \ln 141 \\ 5x &= \ln 141 \\ x &= \frac{\ln 141}{5} \approx 0.99 \end{aligned}$$

69.
$$\begin{aligned} e^{12-5x} - 7 &= 123 \\ e^{12-5x} &= 130 \\ \ln e^{12-5x} &= \ln 130 \\ 12-5x &= \ln 130 \\ 5x &= 12 - \ln 130 \\ x &= \frac{12-\ln 130}{5} \approx 1.43 \end{aligned}$$

70.
$$\begin{aligned} 5^{4x+2} &= 37{,}500 \\ \ln 5^{4x+2} &= \ln 37{,}500 \\ (4x+2)\ln 5 &= \ln 37{,}500 \\ 4x\ln 5 + 2\ln 5 &= \ln 37{,}500 \\ 4x \ln 5 &= \ln 37{,}500 - 2\ln 5 \\ x &= \frac{\ln 37{,}500 - 2\ln 5}{4\ln 5} \approx 1.14 \end{aligned}$$

71.
$$\begin{aligned} 3^{x+4} &= 7^{2x-1} \\ \ln 3^{x+4} &= \ln 7^{2x-1} \\ (x+4)\ln 3 &= (2x-1)\ln 7 \\ x\ln 3 + 4\ln 3 &= 2x\ln 7 - \ln 7 \\ x\ln 3 - 2x\ln 7 &= -4\ln 3 - \ln 7 \\ x(\ln 3 - 2\ln 7) &= -4\ln 3 - \ln 7 \\ x &= \frac{-4\ln 3 - \ln 7}{\ln 3 - 2\ln 7} \\ x &= \frac{4\ln 3 + \ln 7}{2\ln 7 - \ln 3} \\ x &\approx 2.27 \end{aligned}$$

72. $e^{2x} - e^x - 6 = 0$

$\left(e^x - 3\right)\left(e^x + 2\right) = 0$

$e^x - 3 = 0$ or $e^x + 2 = 0$

$e^x = 3$ $\quad$ $e^x = -2$

$\ln e^x = \ln 3$ $\quad$ $\ln e^x - \ln(-2)$

$x = \ln 3$ $\quad$ $x = \ln(-2)$

$x = \ln 3 \approx 1.099$ $\quad$ $\ln(-2)$ does not exist.

The solution set is $\{\ln 3\}$, approximately 1.10.

73. $\log_4(3x-5) = 3$

$3x - 5 = 4^3$

$3x - 5 = 64$

$3x = 69$

$x = 23$

The solutions set is $\{23\}$.

74. $3+4\ln(2x)=15$
$4\ln(2x)=12$
$\ln(2x)=3$
$2x=e^3$
$x=\frac{e^3}{2}$
$x\approx 10.04$

The solutions set is $\left\{\frac{e^3}{2}\right\}$.

75. $\log_2(x+3)+\log_2(x-3)=4$
$\log_2(x+3)(x-3)=4$
$\log_2(x^2-9)=4$
$x^2-9=2^4$
$x^2-9=16$
$x^2=25$
$x=\pm 5$
$x=-5$ does not check because $\log_2(-5+3)$ does not exist.
The solution set is $\{5\}$.

76. $\log_3(x-1)-\log_3(x+2)=2$
$\log_3\frac{x-1}{x+2}=2$
$\frac{x-1}{x-2}=3^2$
$\frac{x-1}{x+2}=9$
$x-1=9(x+2)$
$x-1=9x+18$
$8x=-19$
$x=-\frac{19}{8}$
$x=-\frac{19}{8}$ does not check because $\log_3\left(-\frac{19}{8}-1\right)$ does not exist.
The solution set is $\varnothing$.

77. $\ln(x+4)-\ln(x+1)=\ln x$
$\ln\frac{x+4}{x+1}=\ln x$
$\frac{x+4}{x+1}=x$
$x(x+1)=x+4$
$x^2+x=x+4$
$x^2=4$
$x=\pm 2$
$x=-2$ does not check and must be rejected.
The solution set is $\{2\}$.

78. $\log_4(2x+1)=\log_4(x-3)+\log_4(x+5)$
$\log_4(2x+1)=\log_4(x-3)+\log_4(x+5)$
$\log_4(2x+1)=\log_4(x^2+2x-15)$
$2x+1=x^2+2x-15$
$16=x^2$
$x^2=16$
$x=\pm 4$
$x=-4$ does not check and must be rejected.
The solution set is $\{4\}$.

79. $P(x)=14.7e^{-0.21x}$
$4.6=14.7e^{-0.21x}$
$\frac{4.6}{14.7}=e^{-0.21x}$
$\ln\frac{4.6}{14.7}=\ln e^{-0.21x}$
$\ln\frac{4.6}{14.7}=-0.21x$
$t=\frac{\ln\frac{4.6}{14.7}}{-0.21}\approx 5.5$

The peak of Mt. Everest is about 5.5 miles above sea level.

80. $f(t) = 364(1.005)^t$

$560 = 364(1.005)^t$

$\frac{560}{364} = (1.005)^t$

$\ln\frac{560}{364} = \ln(1.005)^t$

$\ln\frac{560}{364} = t\ln 1.005$

$t = \frac{\ln\frac{560}{364}}{\ln 1.005} \approx 86.4$

The carbon dioxide concentration will be double the pre-industrial level approximately 86 years after the year 2000 in the year 2086.

81. $W(x) = 0.37\ln x + 0.05$

$3.38 = 0.37\ln x + 0.05$

$3.33 = 0.37\ln x$

$\frac{3.33}{0.37} = \ln x$

$9 = \ln x$

$e^9 = e^{\ln x}$

$x = e^9 \approx 8103$

The population of New York City is approximately 8103 thousand, or 8,103,000.

82. $20,000 = 12,500\left(1+\frac{0.065}{4}\right)^{4t}$

$12,500(1.01625)^{4t} = 20,000$

$(1.01625)^{4t} = 1.6$

$\ln(1.01625)^{4t} = \ln 1.6$

$4t\ln 1.01625 = \ln 1.6$

$t = \frac{\ln 1.6}{4\ln 1.01625} \approx 7.3$

It will take about 7.3 years.

83. $3\cdot 50,000 = 50,000e^{0.075t}$

$50,000e^{0.075t} = 150,000$

$e^{0.075} = 3$

$\ln e^{0.075t} = \ln 3$

$0.075t = \ln 3$

$t = \frac{\ln 3}{0.075} \approx 14.6$

It will take about 14.6 years.

84. When an investment value triples, $A = 3P$.

$3P = Pe^{5r}$

$e^{5r} = 3$

$\ln e^{5r} = \ln 3$

$5r = \ln 3$

$r = \frac{\ln 3}{5} \approx 0.2197$

The interest rate would need to be about 22%

85. a. $35.3 = 22.4e^{k10}$

$\frac{35.3}{22.4} = e^{10k}$

$\ln\frac{35.3}{22.4} = \ln e^{10k}$

$\ln\frac{35.3}{22.4} = 10k$

$\frac{\ln\frac{35.3}{22.4}}{10} = k$

$0.045 \approx k$

$A = 22.4e^{0.045t}$

b. $A = 22.4e^{0.045(20)} \approx 55.1$

In 2010, the population will be about 55.1 million.

c. $60 = 22.4e^{0.045t}$

$\frac{60}{22.4} = e^{0.045t}$

$\ln\frac{60}{22.4} = \ln e^{0.045t}$

$\ln\frac{60}{22.4} = 0.045t$

$\frac{\ln\frac{60}{22.4}}{0.045} = t$

$22 \approx t$

The population will reach 60 million about 22 years after 1990, in 2012.

86. Use the half-life of 140 days to find k.

$$A = A_0e^{kt}$$
$$\tfrac{1}{2} = e^{k \cdot 140}$$
$$\tfrac{1}{2} = e^{140k}$$
$$\ln\tfrac{1}{2} = \ln e^{140k}$$
$$\ln\tfrac{1}{2} = 140k$$
$$\frac{\ln\frac{1}{2}}{140} = k$$
$$k \approx -0.004951$$

Use $A = A_0e^{kt}$ to find t.

$$A = A_0e^{-0.004951t}$$
$$0.2 = e^{-0.004951t}$$
$$\ln 0.2 = \ln e^{-0.004951t}$$
$$\ln 0.2 = -0.004951t$$
$$t = \frac{\ln 0.2}{-0.004951}$$
$$t \approx 325$$

It will take about 325 days for the substance to decay to 20% of its original amount.

87. **a.** $f(0) = \dfrac{500{,}000}{1+2499e^{-0.92(0)}} = 200$

200 people became ill when the epidemic began.

b. $f(6) = \dfrac{500{,}000}{1+2499e^{-0.92(6)}} = 45{,}411$

45,410 were ill after 6 weeks.

c. 500,000 people

88. **a.**

$$T = C + (T_o - C)e^{kt}$$
$$150 = 65 + (185-65)e^{k2}$$
$$90 = 120e^{2k}$$
$$\frac{90}{120} = e^{2k}$$
$$\ln\frac{3}{4} = \ln e^{2k}$$
$$\ln\frac{3}{4} = 2k$$
$$\frac{\ln\frac{3}{4}}{2} = k$$
$$-0.1438 \approx k$$
$$T = 65 + 120e^{-0.1438t}$$

b.

$$105 = 65 + 120e^{-0.1438t}$$
$$40 = 120e^{-0.1438t}$$
$$\frac{1}{3} = e^{-0.1438t}$$
$$\ln\frac{1}{3} = \ln e^{-0.1438t}$$
$$\ln\frac{1}{3} = -0.1438t$$
$$\frac{\ln\frac{1}{3}}{-0.1438} = t$$
$$7.6 \approx t$$

The temperature will reach 105° after 8 min.

89. **a.**

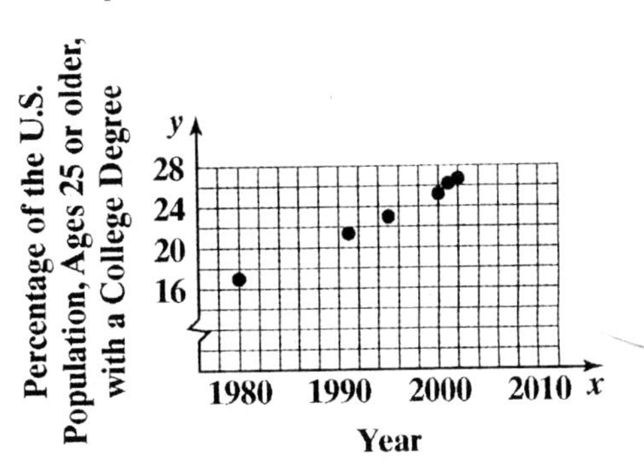

b. An exponential or linear function appears to be the best choice for modeling the data.

90. **a.**

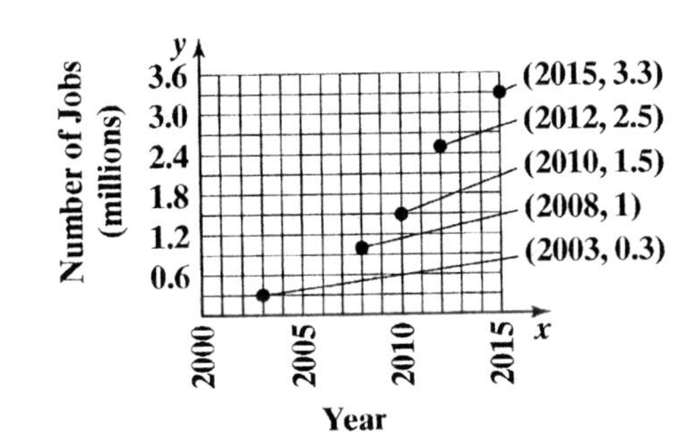

b. An exponential function appears to be the best choice for modeling the data.

91.

$$y = 73(2.6)^x$$
$$y = 73e^{(\ln 2.6)x}$$
$$y = 73e^{0.956x}$$

92.

$$y = 6.5(0.43)^x$$
$$y = 6.5e^{(\ln 0.43)x}$$
$$y = 6.5e^{-0.844x}$$

Chapter 3 Test

1.

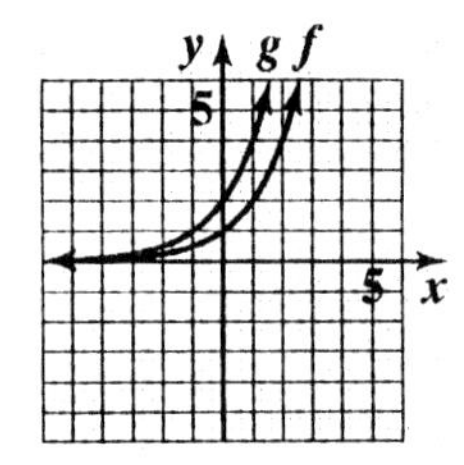

$f(x) = 2^x$
$g(x) = 2^x + 1$

2.

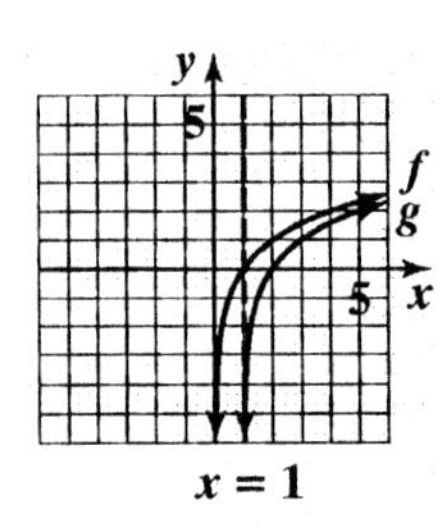

$x = 1$
$f(x) = \log_2 x$
$g(x) = \log_2(x - 1)$

3. $5^3 = 125$

4. $\log_{36} 6 = \frac{1}{2}$

5. The domain of *f* consists of all *x* for which $3 - x > 0$. Solving this inequality for *x*, we obtain $x < 3$.
Thus, the domain of *f* is $(-\infty, 3)$.

6.
$$\log_4\left(64x^5\right) = \log_4 64 + \log_4 x^5$$
$$= 3 + 5\log_4 x$$

7.
$$\log_3 \frac{\sqrt[3]{x}}{81} = \log_3 x^{\frac{1}{3}} - \log_3 81$$
$$= \frac{1}{3}\log_3 x - 4$$

8.
$$6\log x + 2\log y = \log x^6 + \log y^2$$
$$= \log\left(x^6 y^2\right)$$

9.
$$\ln 7 - 3\ln x = \ln 7 - \ln x^3$$
$$= \ln\frac{7}{x^3}$$

10. $\log_{15} 71 = \frac{\log 71}{\log 15} \approx 1.5741$

11.
$$3^{x-2} = 9^{x+4}$$
$$3^{x-2} = \left(3^2\right)^{x+4}$$
$$3^{x-2} = 3^{2x+8}$$
$$x - 2 = 2x + 8$$
$$-x = 10$$
$$x = -10$$

12.
$$5^x = 1.4$$
$$\ln 5^x = \ln 1.4$$
$$x\ln 5 = \ln 1.4$$
$$x = \frac{\ln 1.4}{\ln 5} \approx 0.2091$$

13.
$$400e^{0.005x} = 1600$$
$$e^{0.005x} = 4$$
$$\ln e^{0.005x} = \ln 4$$
$$0.005x = \ln 4$$
$$x = \frac{\ln 4}{0.005} \approx 277.2589$$

14.
$$e^{2x} - 6e^x + 5 = 0$$
$$\left(e^x - 5\right)\left(e^x - 1\right) = 0$$

$e^x - 5 = 0$	or	$e^x - 1 = 0$
$e^x = 5$		$e^x = 1$
$\ln e^x = \ln 5$		$\ln e^x = \ln 1$
$x = \ln 5$		$x = \ln 1$
$x \approx 1.6094$		$x = 0$

The solution set is $\{0, \ln 5\}$; $\ln \approx 1.6094$.

15.
$$\log_6(4x - 1) = 3$$
$$4x - 1 = 6^3$$
$$4x - 1 = 216$$
$$4x = 217$$
$$x = \frac{217}{4} = 54.25$$

16.
$$2\ln 3x = 8$$
$$\ln 3x = 4$$
$$3x = e^4$$
$$x = \frac{e^4}{3} \approx 18.1994$$

17. $\log x + \log(x+15) = 2$

$$\log(x^2+15x) = 2$$
$$x^2+15x = 10^2$$
$$x^2+15x-100 = 0$$
$$(x+20)(x-5) = 0$$
$$x+20 = 0 \text{ or } x-5 = 0$$
$$x = -20 \qquad x = 5$$

$x = -20$ does not check because $\log(-20)$ does not exist.
The solution set is $\{5\}$.

18. $\ln(x-4) - \ln(x+1) = \ln 6$

$$\ln\frac{x-4}{x+1} = \ln 6$$
$$\frac{x-4}{x+1} = 6$$
$$6(x+1) = x-4$$
$$6x+6 = x-4$$
$$5x = -10$$
$$x = -2$$

$x = -2$ does not check and must be rejected.
The solution set is $\{\ \}$.

19. $D = 10\log\frac{10^{12}I_0}{I_0}$

$$= 10\log 10^{12}$$
$$= 10\cdot 12$$
$$= 120$$

The loudness of the sound is 120 decibels.

20. Since $\ln e^x = x$, $\ln e^{5x} = 5x$.

21. $\log_b b = 1$ because $b^1 = b$.

22. $\log_6 1 = 0$ because $6^0 = 1$.

23. 6.5% compounded semiannually:

$$A = 3{,}000\left(1+\frac{0.065}{2}\right)^{2(10)} \approx \$5{,}687.51$$

6% compounded continuously:

$$A = 3{,}000e^{0.06(10)} \approx \$5{,}466.36$$

6.5% compounded semiannually yields about \$221 more than 6% compounded continuously.

24.

$$8000 = 4000\left(1+\frac{0.05}{4}\right)^{4t}$$
$$\frac{8000}{4000} = (1+0.0125)^{4t}$$
$$2 = (1.0125)^{4t}$$
$$\ln 2 = \ln(1.0125)^{4t}$$
$$\ln 2 = 4t\ln(1.0125)$$
$$\frac{\ln 2}{4\ln(1.0125)} = \frac{4t\ln(1.0125)}{4\ln(1.0125)}$$
$$t = \frac{\ln 2}{4\ln(1.0125)} \approx 13.9$$

It will take approximately 13.9 years for the money to grow to \$8000.

25.

$$2 = 1e^{r10}$$
$$2 = e^{10r}$$
$$\ln 2 = \ln e^{10r}$$
$$\ln 2 = 10r$$
$$r = \frac{\ln 2}{10} \approx 0.069$$

The money will double in 10 years with an interest rate of approximately 6.9%.

26. **a.** $P(0) = 82.3e^{-0.002(0)}$

$$= 82.3e^0 = 82.3(1) = 82.3$$

In 2003, the population of Germany was 82.3 million.

b. The population of Germany is decreasing. We can tell the model has a negative, $k = -0.002$.

c.

$$81.5 = 82.3e^{-0.002t}$$
$$\frac{81.5}{82.3} = e^{-0.002t}$$
$$\ln\frac{81.5}{82.3} = \ln e^{-0.002t}$$
$$\ln\frac{81.5}{82.3} = -0.002t$$
$$t = \frac{\ln\frac{81.5}{82.3}}{-0.002} \approx 5$$

The population of Germany will be 81.5 million approximately 5 years after 2003 in the year 2008.

27. In 1990, $t = 0$ and $A_0 = 509$
In 2000, $t = 2000 - 1990 = 10$ and
$A = 729$.

$$729 = 509e^{k10}$$
$$\frac{729}{509} = e^{10k}$$
$$\ln\frac{729}{509} = \ln e^{10k}$$
$$\ln\frac{729}{509} = 10k$$
$$\frac{\ln\frac{729}{509}}{10} = k$$
$$0.036 \approx k$$

The exponential growth function is
$A = 509e^{0.036t}$.

28. When the amount remaining is 5%, $A = 0.05A_0$.

$$0.05A_0 = A_0e^{-0.000121t}$$
$$e^{-0.000121t} = 0.05$$
$$\ln e^{-0.000121t} = \ln 0.05$$
$$-0.000121t = \ln 0.05$$
$$t = \frac{\ln 0.05}{-0.000121} \approx 24{,}758$$

The man died about 24,758 years ago.

29. a. $f(0) = \dfrac{140}{1+9e^{-0.165(0)}} = 14$

Fourteen elk were initially introduced to the habitat.

b. $f(10) = \dfrac{140}{1+9e^{-0.165(10)}} \approx 51$

After 10 years, about 51 elk are expected.

c. In the logistic growth model,

$f(t) = \dfrac{c}{1+ae^{-bt}}$,

the constant c represents the limiting size that $f(t)$ can attain. The limiting size of the elk population is 140 elk.

30. Plot the ordered pairs.

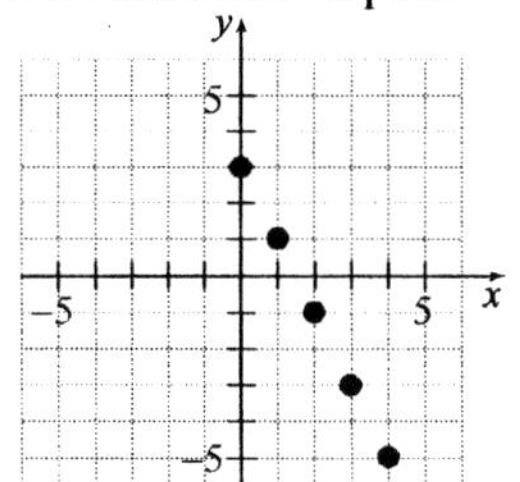

The values appear to belong to a linear function.

31. Plot the ordered pairs.

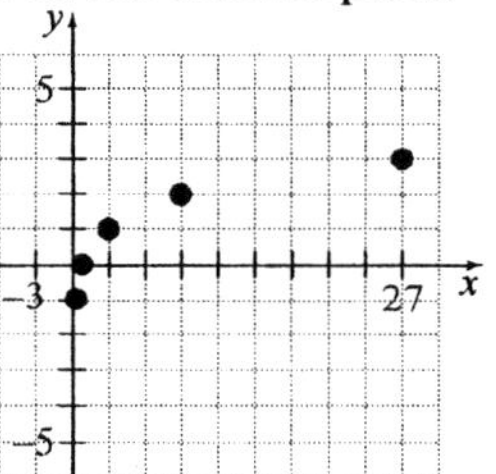

The values appear to belong to a logarithmic function.

32. Plot the ordered pairs.

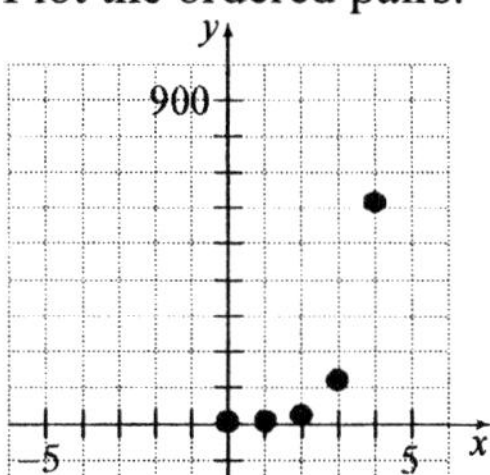

The values appear to belong to an exponential function.

33. Plot the ordered pairs.

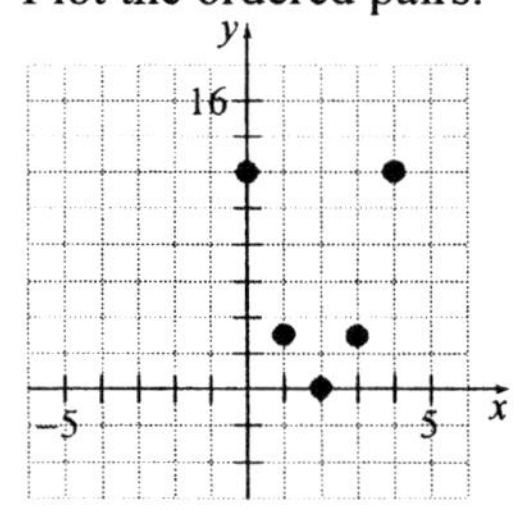

The values appear to belong to a quadratic function.

34.
$$y = 96(0.38)^x$$
$$y = 96e^{(\ln 0.38)x}$$
$$y = 96e^{-0.968x}$$

Cumulative Review Exercises (Chapters P–3)

1. $|3x-4|=2$

$3x-4=2$ or $3x-4=-2$

$3x=6$ $\quad 3x=2$

$x=2$ $\quad x=\frac{2}{3}$

The solution set is $\left\{\frac{2}{3}, 2\right\}$.

2. $x^2+2x+5=0$

$$x=\frac{-b\pm\sqrt{b^2-4ac}}{2a}$$

$$x=\frac{-(2)\pm\sqrt{(2)^2-4(1)(5)}}{2(1)}$$

$$x=\frac{-2\pm\sqrt{-16}}{2}$$

$$x=\frac{-2\pm 4i}{2}$$

$$x=-1\pm 2i$$

3. $x^4+x^3-3x^2-x+2=0$

$p: \pm 1, \pm 2$

$q: \pm 1$

$\frac{p}{q}: \pm 1, \pm 2$

−2	1	1	−3	−1	2
		−2	2	2	−2
	1	−1	−1	1	0

$$(x+2)(x^3-x^2-x+1)=0$$

$$(x+2)[x^2(x-1)-(x-1)]=0$$

$$(x+2)(x^2-1)(x-1)=0$$

$$(x+2)(x+1)(x-1)(x-1)=0$$

$$(x+2)(x+1)(x-1)^2=0$$

$x+2=0$ or $x+1=0$ or $x-1=0$

$x=-2$ $\quad x=-1$ $\quad x=1$

The solution set is $\{-2, -1, 1\}$.

4.

$$e^{5x}-32=96$$

$$e^{5x}=128$$

$$\ln e^{5x}=\ln 128$$

$$5x=\ln 128$$

$$x=\frac{\ln 128}{5}\approx 0.9704$$

The solution set is $\left\{\frac{\ln 128}{5}\right\}$, approximately 0.9704.

5. $\log_2(x+5)+\log_2(x-1)=4$

$$\log_2[(x+5)(x-1)]=4$$

$$(x+5)(x-1)=2^4$$

$$x^2+4x-5=16$$

$$x^2+4x-21=0$$

$$(x+7)(x-3)=0$$

$x+7=0$ or $x-3=0$

$x=-7$ $\quad x=3$

$x=-7$ does not check because $\log_2(-7+5)$ does not exist.

The solution set is $\{3\}$.

6. $\ln(x+4)+\ln(x+1)=2\ln(x+3)$

$$\ln((x+4)(x+1))=\ln(x+3)^2$$

$$(x+4)(x+1)=(x+3)^2$$

$$x^2+5x+4=x^2+6x+9$$

$$5x+4=6x+9$$

$$-x=5$$

$$x=-5$$

$x=-5$ does not check and must be rejected.

The solution set is $\{\ \}$.

7. $14-5x\ge -6$

$-5x\ge -20$

$x\le 4$

The solution set is $(-\infty, 4]$.

8. $|2x-4|\le 2$

$2x-4\le 2$ and $2x-4\ge -2$

$2x\le 6$ $\quad 2x\ge 2$

$x\le 3$ and $x\ge 1$

The solution set is $[1,3]$.

9. Circle with center: (3, –2) and radius of 2

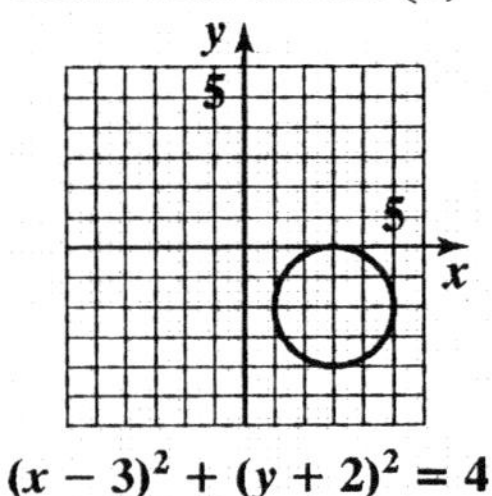

$(x-3)^2+(y+2)^2=4$

10. Parabola with vertex: (2, –1)

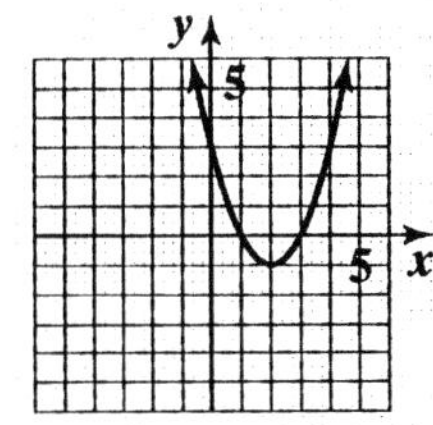

$f(x)=(x-2)^2-1$

11. x-intercepts:

$x^2-1=0$

$x^2=1$

$x=\pm 1$

The x-intercepts are $(1,0)$ and $(-1,0)$.

vertical asymptotes:

$x^2-4=0$

$x^2=4$

$x=\pm 2$

The vertical asymptotes are $x=2$ and $x=-2$.

Horizontal asymptote: y 5 1

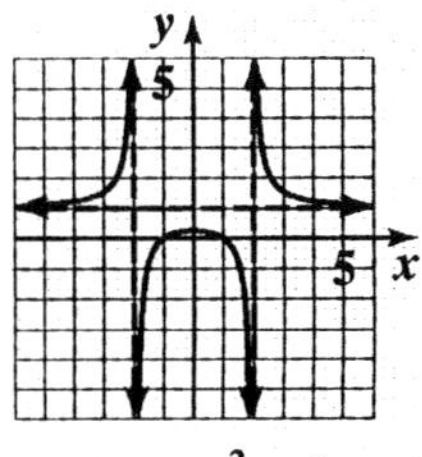

$f(x)=\dfrac{x^2-1}{x^2-4}$

12. x-intercepts:

$x-2=0$ or $x+1=0$

$x=2$ or $x=-1$

The x-intercepts are $(2,0)$ and $(-1,0)$.

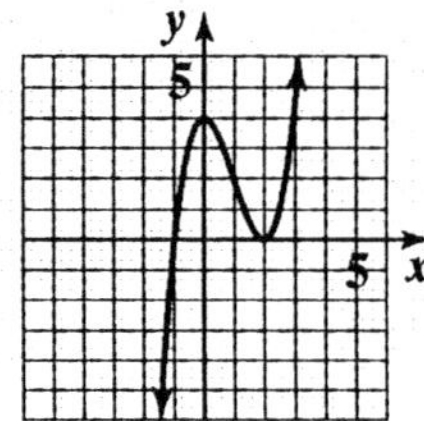

$f(x)=(x-2)^2(x+1)$

13.

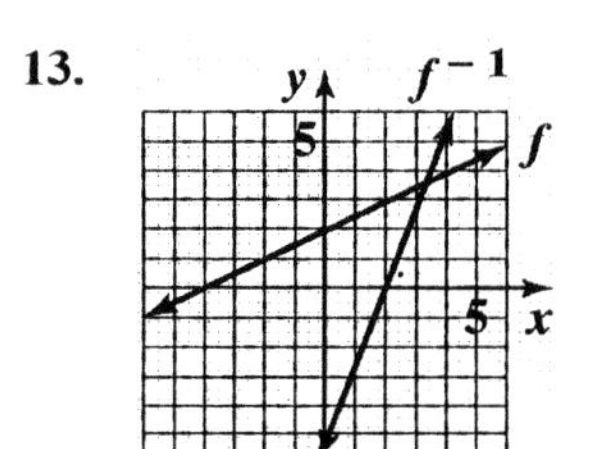

$f(x)=2x-4$

$f^{-1}(x)=\dfrac{x+4}{2}$

14.

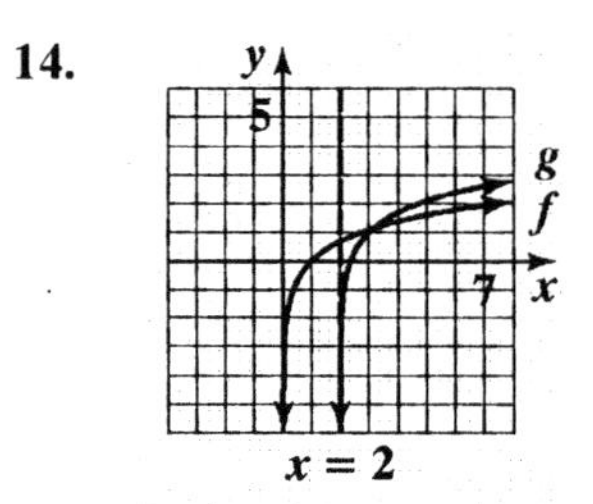

$f(x)=\ln x$

$g(x)=\ln(x-2)+1$

15. $m=\dfrac{3-(-3)}{1-3}=\dfrac{6}{-2}=-3$

Using (1, 3) point-slope form:

$y-3=-3(x-1)$

slope-intercept form:

$y-3=-3(x-1)$

$y-3=-3x+3$

$y=-3x+6$

16. $(f\circ g)(x)=f(x+2)$

$=(x+2)^2$

$=x^2+4x+4$

$(g\circ f)(x)=g(x^2)$

$=x^2+2$

17. y varies inversely as the square of x is expressed as $y = \frac{k}{x^2}$.

The hours, H, vary inversely as the square of the number of cups of coffee, C can be expressed as $H = \frac{k}{C^2}$.

Use the given values to find k.

$$H = \frac{k}{C^2}$$

$$8 = \frac{k}{2^2}$$

$$32 = k$$

Substitute the value of k into the equation.

$$H = \frac{k}{C^2}$$

$$H = \frac{32}{C^2}$$

Use the equation to find H when $C = 4$.

$$H = \frac{32}{C^2}$$

$$H = \frac{32}{4^2}$$

$$H = 2$$

If 4 cups of coffee are consumed you should expect to sleep 2 hours.

18. $s(t) = -16t^2 + 64t + 5$

The ball reaches its maximum height at

$$t = \frac{-b}{2a} = \frac{-(64)}{2(-16)} = 2 \text{ seconds.}$$

The maximum height is $s(2)$.

$$s(2) = -16(2)^2 + 64(2) + 5 = 69 \text{ feet.}$$

19. $s(t) = -16t^2 + 64t + 5$

Let $s(t) = 0$:

$$0 = -16t^2 + 64t + 5$$

Use the quadratic formula to solve.

$$t = \frac{-b \pm \sqrt{b^2 - 4ac}}{2a}$$

$$t = \frac{-(64) \pm \sqrt{(64)^2 - 4(-16)(5)}}{2(-16)}$$

$$t \approx 4.1, \quad t \approx -0.1$$

The negative value is rejected.

The ball hits the ground after about 4.1 seconds.

20. $40x + 10(1.5x) = 660$

$$40x + 15x = 660$$

$$55x = 660$$

$$x = 12$$

Your normal hourly salary is $12 per hour.

Chapter 4

Section 4.1

Check Point Exercises

1. The radian measure of a central angle is the length of the intercepted arc, s, divided by the circle's radius, r. The length of the intercepted arc is 42 feet: $s = 42$ feet. The circle's radius is 12 feet: $r = 12$ feet. Now use the formula for radian measure to find the radian measure of θ.

$$\theta = \frac{s}{r} = \frac{42 \text{ feet}}{12 \text{ feet}} = 3.5$$

Thus, the radian measure of θ is 3.5

2. a. $60° = 60° \cdot \frac{\pi \text{ radians}}{180°} = \frac{60\pi}{180} \text{ radians}$

$= \frac{\pi}{3} \text{ radians}$

b. $270° = 270° \cdot \frac{\pi \text{ radians}}{180°} = \frac{270\pi}{180} \text{ radians}$

$= \frac{3\pi}{2} \text{ radians}$

c. $-300° = -300° \cdot \frac{\pi \text{ radians}}{180°} = \frac{-300\pi}{180} \text{ radians}$

$= -\frac{5\pi}{3} \text{ radians}$

3. a. $\frac{\pi}{4} \text{ radians} = \frac{\pi \text{ radians}}{4} \cdot \frac{180^{\circ}}{\pi \text{ radians}}$

$= \frac{180^{\circ}}{4} = 45^{\circ}$

b. $-\frac{4\pi}{3} \text{ radians} = -\frac{4\pi \text{ radians}}{3} \cdot \frac{180^{\circ}}{\pi}$

$= -\frac{4 \cdot 180^{\circ}}{3} = -240^{\circ}$

c. $6 \text{ radians} = 6 \text{ radians} \cdot \frac{180^{\circ}}{\pi \text{ radians}}$

$= \frac{6 \cdot 180^{\circ}}{\pi} \approx 343.8^{\circ}$

4. a.

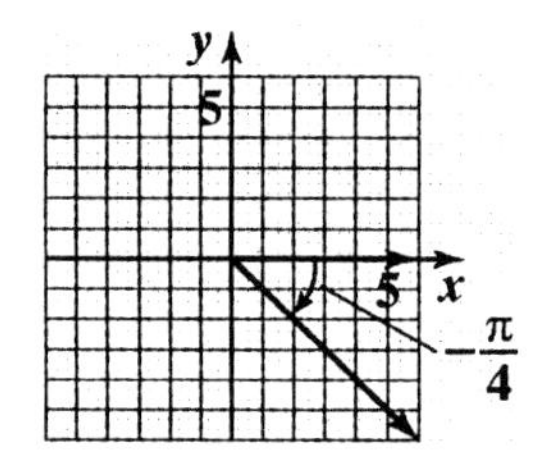

b.

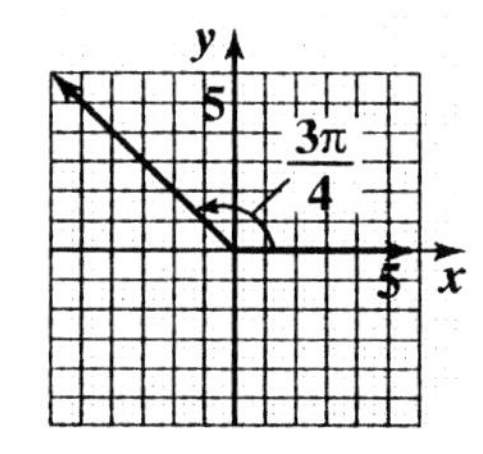

c.

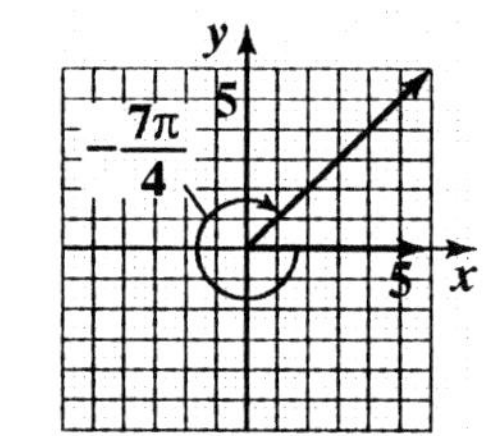

d.

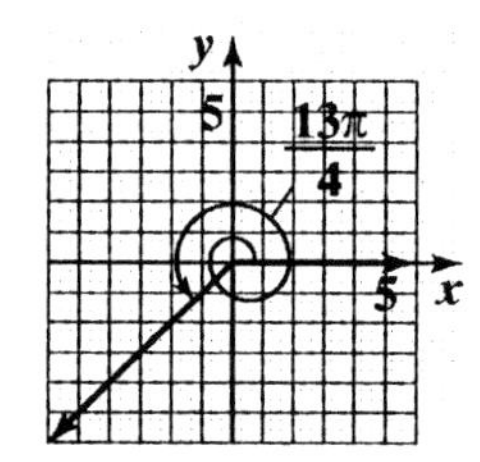

5. a. For a 400° angle, subtract 360° to find a positive coterminal angle.
$400° - 360° = 40°$

b. For a −135° angle, add 360° to find a positive coterminal angle.
$-135° + 360° = 225°$

6. a. $\frac{13\pi}{5} - 2\pi = \frac{13\pi}{5} - \frac{10\pi}{5} = \frac{3\pi}{5}$

b. $-\frac{\pi}{15} + 2\pi = -\frac{\pi}{15} + \frac{30\pi}{15} = \frac{29\pi}{15}$

7. a. $855° - 360° \cdot 2 = 855° - 720° = 135°$

b. $\frac{17\pi}{3} - 2\pi \cdot 2 = \frac{17\pi}{3} - 4\pi$

$= \frac{17\pi}{3} - \frac{12\pi}{3} = \frac{5\pi}{3}$

c. $-\frac{25\pi}{6}+2\pi\cdot 3=-\frac{25\pi}{6}+6\pi$
$=-\frac{25\pi}{6}+\frac{36\pi}{6}=\frac{11\pi}{6}$

8. The formula $s=r\theta$ can only be used when θ is expressed in radians. Thus, we begin by converting 45° to radians. Multiply by $\frac{\pi \text{ radians}}{180°}$

$45°=45°\cdot\frac{\pi \text{ radians}}{180°}=\frac{45}{180}\pi \text{ radians}$
$=\frac{\pi}{4} \text{ radians}$

Now we can use the formula $s=r\theta$ to find the length of the arc. The circle's radius is 6 inches : $r=6$ inches. The measure of the central angle in radians is $\frac{\pi}{4}: \theta=\frac{\pi}{4}$. The length of the arc intercepted by this central angle is

$s=r\theta=(6 \text{ inches})\left(\frac{\pi}{4}\right)=\frac{6\pi}{4} \text{ inches} \approx 4.71 \text{ inches}.$

9. We are given ω, the angular speed.
$\omega=45$ revolutions per minute
We use the formula $v=r\omega$ to find v, the linear speed. Before applying the formula, we must express ω in radians per minute.

$\omega=\frac{45 \text{ revolutions}}{1 \text{ minute}}\cdot\frac{2\pi \text{ radians}}{1 \text{ revolution}}$
$=\frac{90\pi \text{ radians}}{1 \text{ minute}}$

The angular speed of the propeller is 90π radians per minute. The linear speed is

$v=r\omega=1.5 \text{ inches}\cdot\frac{90\pi}{1 \text{ minute}}=\frac{135\pi \text{ inches}}{\text{minute}}$

The linear speed is 135π inches per minute, which is approximately 424 inches per minute.

Exercise Set 4.1

1. obtuse

2. obtuse

3. acute

4. acute

5. straight

6. right

7. $\theta=\frac{s}{r}=\frac{40 \text{ inches}}{10 \text{ inches}}=4 \text{ radians}$

8. $\theta=\frac{s}{r}=\frac{30 \text{ feet}}{5 \text{ feet}}=6 \text{ radians}$

9. $\theta=\frac{s}{r}=\frac{8 \text{ yards}}{6 \text{ yards}}=\frac{4}{3} \text{ radians}$

10. $\theta=\frac{s}{r}=\frac{18 \text{ yards}}{8 \text{ yards}}=2.25 \text{ radians}$

11. $\theta=\frac{s}{r}=\frac{400 \text{ centimeters}}{100 \text{ centimeters}}=4 \text{ radians}$

12. $\theta=\frac{s}{r}=\frac{600 \text{ centimeters}}{100 \text{ centimeters}}=6 \text{ radians}$

13. $45°=45°\cdot\frac{\pi \text{ radians}}{180°}$
$=\frac{45\pi}{180} \text{ radians}$
$=\frac{\pi}{4} \text{ radians}$

14. $18°=18°\cdot\frac{\pi \text{ radians}}{180°}$
$=\frac{18\pi}{180} \text{ radians}$
$=\frac{\pi}{10} \text{ radians}$

15. $135°=135°\cdot\frac{\pi \text{ radians}}{180°}$
$=\frac{135\pi}{180} \text{ radians}$
$=\frac{3\pi}{4} \text{ radians}$

16. $150°=150°\cdot\frac{\pi \text{ radians}}{180°}$
$=\frac{150\pi}{180} \text{ radians}$
$=\frac{5\pi}{6} \text{ radians}$

17. $300° = 300° \cdot \frac{\pi \text{ radians}}{180°} = \frac{300\pi}{180} \text{ radians} = \frac{5\pi}{3} \text{ radians}$

18. $330° = 330° \cdot \frac{\pi \text{ radians}}{180°} = \frac{330\pi}{180} \text{ radians} = \frac{11\pi}{6} \text{ radians}$

19. $-225° = -225° \cdot \frac{\pi \text{ radians}}{180°} = -\frac{225\pi}{180} \text{ radians} = -\frac{5\pi}{4} \text{ radians}$

20. $-270° = -270° \cdot \frac{\pi \text{ radians}}{180°} = -\frac{270\pi}{180} \text{ radians} = -\frac{3\pi}{2} \text{ radians}$

21. $\frac{\pi}{2} \text{ radians} = \frac{\pi \text{ radians}}{2} \cdot \frac{180^\circ}{\pi \text{ radians}} = \frac{180^\circ}{2} = 90^\circ$

22. $\frac{\pi}{9} \text{ radians} = \frac{\pi \text{ radians}}{9} \cdot \frac{180^\circ}{\pi \text{ radians}} = \frac{180^\circ}{9} = 20^\circ$

23. $\frac{2\pi}{3} \text{ radians} = \frac{2\pi \text{ radians}}{3} \cdot \frac{180^\circ}{\pi \text{ radians}} = \frac{2 \cdot 180^\circ}{3} = 120^\circ$

24. $\frac{3\pi \text{ radians}}{4} \cdot \frac{180^\circ}{\pi \text{ radians}} = \frac{3 \cdot 180^\circ}{4} = 135^\circ$

25. $\frac{7\pi}{6} \text{ radians} = \frac{7\pi \text{ radians}}{6} \cdot \frac{180^\circ}{\pi \text{ radians}} = \frac{7 \cdot 180^\circ}{6} = 210^\circ$

26. $\frac{11\pi \text{ radians}}{6} \cdot \frac{180^\circ}{\pi \text{ radians}} = \frac{11 \cdot 180^\circ}{6} = 330^\circ$

27. $-3\pi \text{ radians} = -3\pi \text{ radians} \cdot \frac{180^\circ}{\pi \text{ radians}} = -3 \cdot 180^\circ = -540^\circ$

28. $-4\pi \text{ radians} \cdot \frac{180^\circ}{\pi \text{ radians}} = -4 \cdot 180^\circ = -720^\circ$

29. $18° = 18° \cdot \frac{\pi \text{ radians}}{180°} = \frac{18\pi}{180} \text{ radians} \approx 0.31 \text{ radians}$

30. $76° = 76° \cdot \frac{\pi \text{ radians}}{180°} = \frac{76\pi}{180} \text{ radians} \approx 1.33 \text{ radians}$

31. $-40° = -40° \cdot \frac{\pi \text{ radians}}{180°} = -\frac{40\pi}{180} \text{ radians} \approx -0.70 \text{ radians}$

32. $-50° = -50° \cdot \frac{\pi \text{ radians}}{180°} = -\frac{50\pi}{180} \text{ radians} \approx -0.87 \text{ radians}$

33. $200° = 200° \cdot \frac{\pi \text{ radians}}{180°} = \frac{200\pi}{180} \text{ radians} \approx 3.49 \text{ radians}$

34. $250° = 250° \cdot \dfrac{\pi \text{ radians}}{180°}$
$= \dfrac{250\pi}{180}$ radians
≈ 4.36 radians

35. $2 \text{ radians} = 2 \text{ radians} \cdot \dfrac{180^{\circ}}{\pi \text{ radians}}$
$= \dfrac{2 \cdot 180^{\circ}}{\pi}$
$\approx 114.59^{\circ}$

36. $3 \text{ radians} \cdot \dfrac{180^{\circ}}{\pi \text{ radians}} = \dfrac{3 \cdot 180^{\circ}}{\pi} \approx 171.89^{\circ}$

37. $\dfrac{\pi}{13} \text{ radians} = \dfrac{\pi \text{ radians}}{13} \cdot \dfrac{180^{\circ}}{\pi \text{ radians}}$
$= \dfrac{180^{\circ}}{13}$
$\approx 13.85^{\circ}$

38. $\dfrac{\pi}{17} \text{radians} \cdot \dfrac{180^{\circ}}{\pi \text{radians}} = \dfrac{180^{\circ}}{17} \approx 10.59^{\circ}$

39. $-4.8 \text{ radians} = -4.8 \text{ radians} \cdot \dfrac{180^{\circ}}{\pi \text{ radians}}$
$= \dfrac{-4.8 \cdot 180^{\circ}}{\pi}$
$\approx -275.02^{\circ}$

40. $-5.2 \text{ radians} \cdot \dfrac{180^{\circ}}{\pi \text{radians}} = \dfrac{-5.2 \cdot 180^{\circ}}{\pi}$
$\approx -297.94^{\circ}$

41.

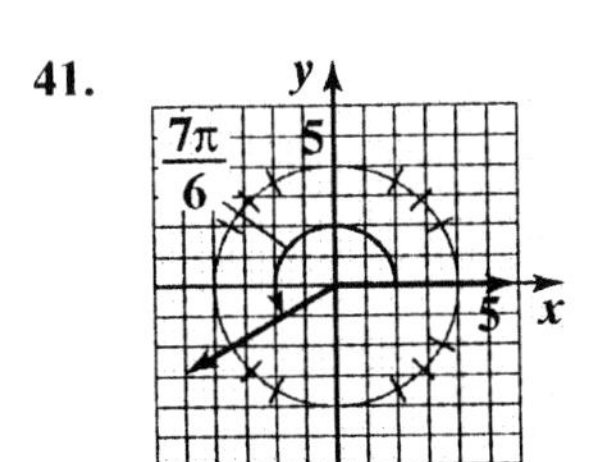

42.

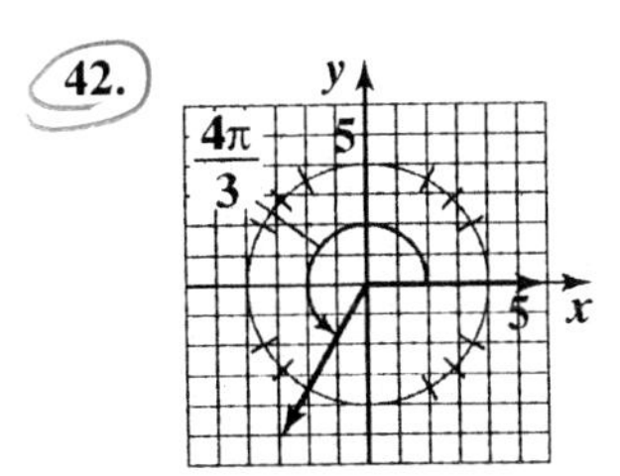

43.

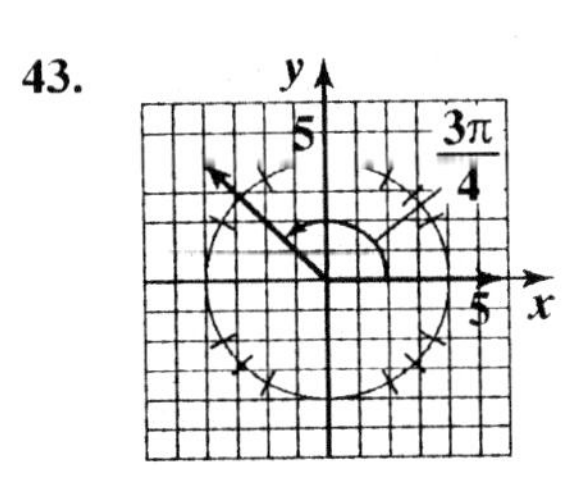

44.

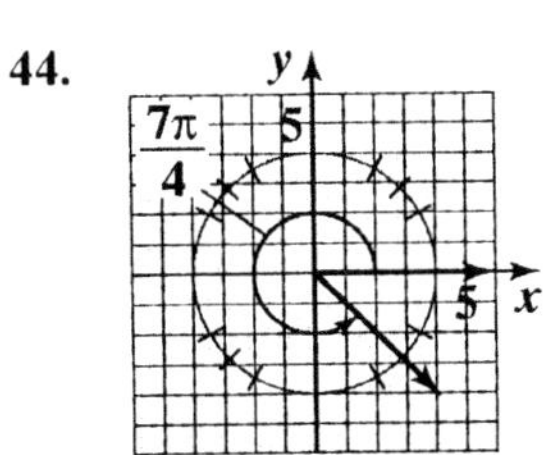

45.

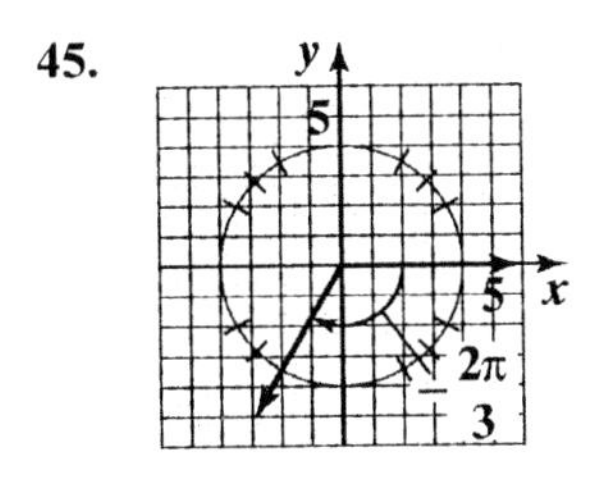

46.

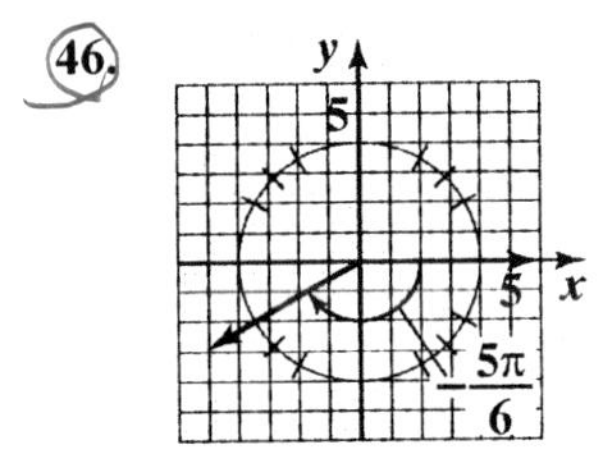

47.

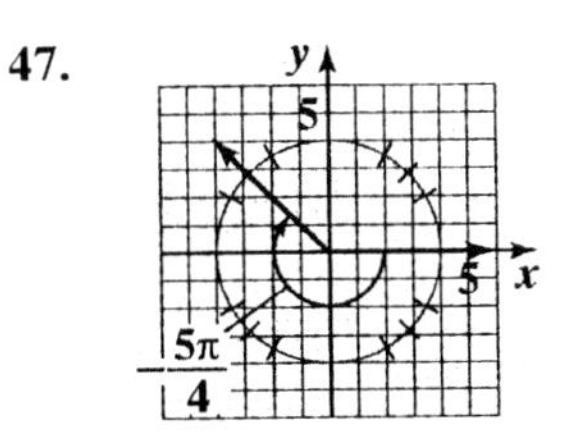

48.

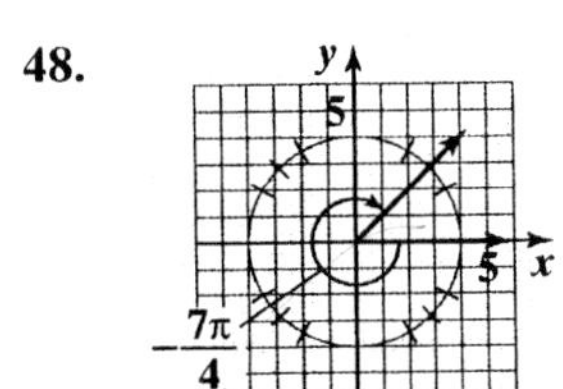

49.

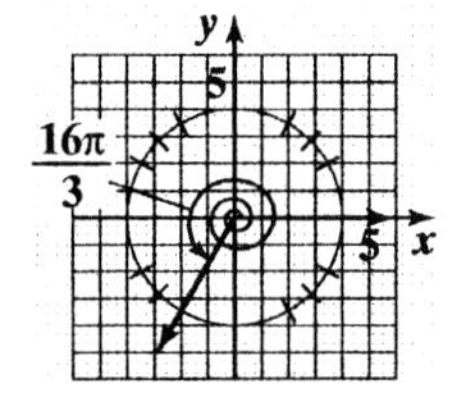

50.

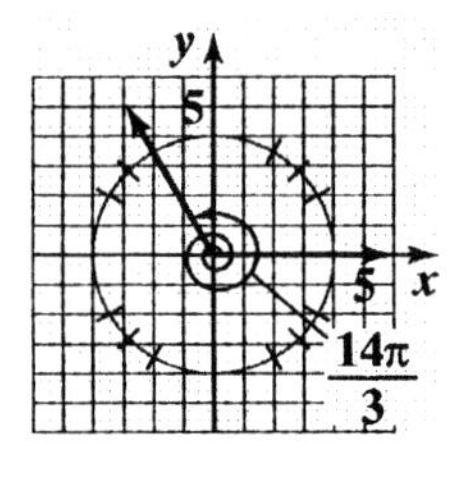

51.

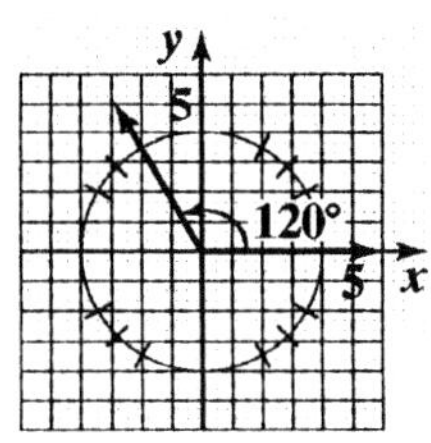

52.

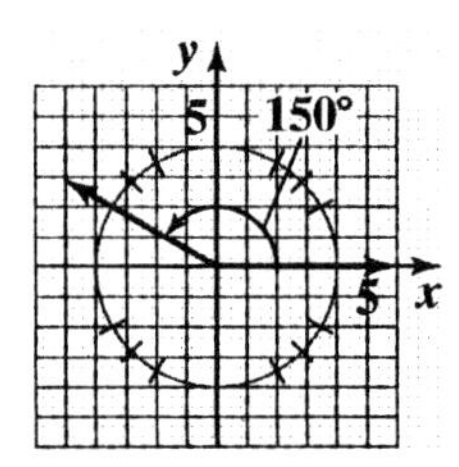

53.

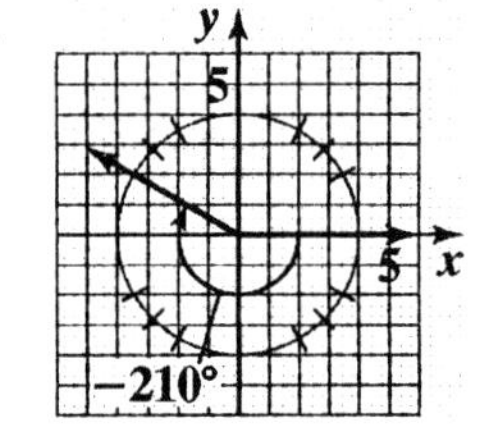

54.

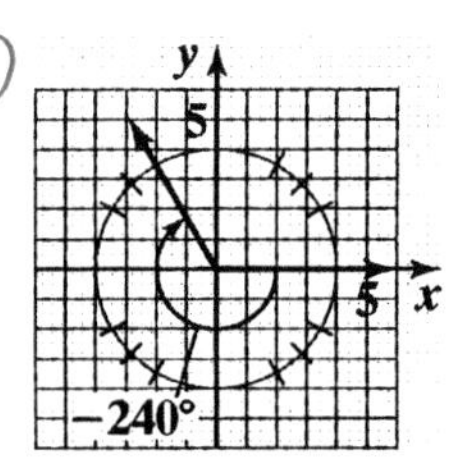

55.

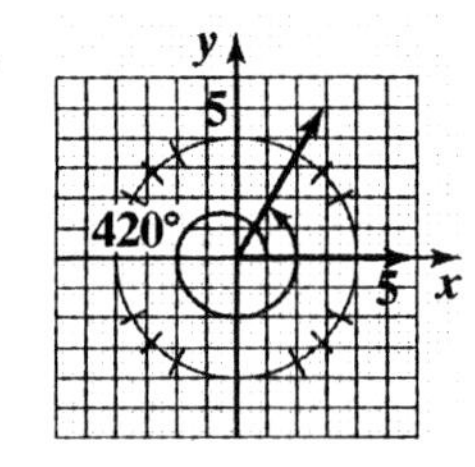

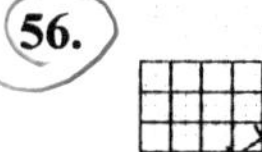

56.

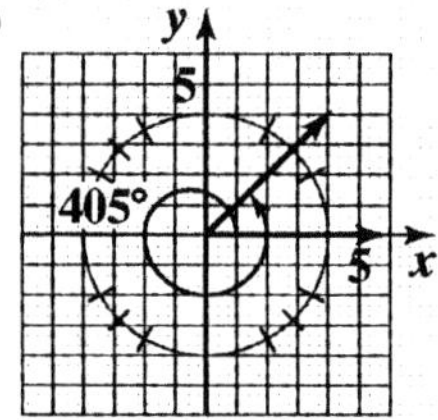

57. $395° - 360° = 35°$

58. $415° - 360° = 55°$

59. $-150° + 360° = 210°$

60. $-160° + 360° = 200°$

61. $-765° + 360° \cdot 3 = -765° + 1080° = 315°$

62. $-760° + 360° \cdot 3 = -760° + 1080° = 320°$

63. $\frac{19\pi}{6} - 2\pi = \frac{19\pi}{6} - \frac{12\pi}{6} = \frac{7\pi}{6}$

64. $\frac{17\pi}{5} - 2\pi = \frac{17\pi}{5} - \frac{10\pi}{5} = \frac{7\pi}{5}$

65. $\frac{23\pi}{5} - 2\pi \cdot 2 = \frac{23\pi}{5} - 4\pi = \frac{23\pi}{5} - \frac{20\pi}{5} = \frac{3\pi}{5}$

66. $\frac{25\pi}{6} - 2\pi \cdot 2 = \frac{25\pi}{6} - 4\pi = \frac{25\pi}{6} - \frac{24\pi}{6} = \frac{\pi}{6}$

67. $-\frac{\pi}{50} + 2\pi = -\frac{\pi}{50} + \frac{100\pi}{50} = \frac{99\pi}{50}$

68. $-\frac{\pi}{40} + 2\pi = -\frac{\pi}{40} + \frac{80\pi}{40} = \frac{79\pi}{40}$

69. $-\frac{31\pi}{7} + 2\pi \cdot 3 = -\frac{31\pi}{7} + 6\pi$

$= -\frac{31\pi}{7} + \frac{42\pi}{7} = \frac{11\pi}{7}$

70. $-\frac{38\pi}{9} + 2\pi \cdot 3 = -\frac{38\pi}{9} + 6\pi$

$= -\frac{38\pi}{9} + \frac{54\pi}{9} = \frac{16\pi}{9}$

71. $r = 12$ inches, $\theta = 45°$

Begin by converting 45° to radians, in order to use the formula $s = r\theta$.

$45° = 45° \cdot \frac{\pi \text{ radians}}{180°} = \frac{\pi}{4}$ radians

Now use the formula $s = r\theta$.

$s = r\theta = 12 \cdot \frac{\pi}{4} = 3\pi$ inches ≈ 9.42 inches

72. $r = 16$ inches, $\theta = 60°$
Begin by converting $60°$ to radians, in order to use the formula $s = r\theta$.

$60° = 60° \cdot \frac{\pi \text{ radians}}{180°} = \frac{\pi}{3}$ radians

Now use the formula $s = r\theta$.

$s = r\theta = 16 \cdot \frac{\pi}{3} = \frac{16\pi}{3}$ inches ≈ 16.76 inches

73. $r = 8$ feet, $\theta = 225°$
Begin by converting $225°$ to radians, in order to use the formula $s = r\theta$.

$225° = 225° \cdot \frac{\pi \text{ radians}}{180°} = \frac{5\pi}{4}$ radians

Now use the formula $s = r\theta$.

$s = r\theta = 8 \cdot \frac{5\pi}{4} = 10\pi$ feet ≈ 31.42 feet

74. $r = 9$ yards, $\theta = 315°$
Begin by converting $315°$ to radians, in order to use the formula $s = r\theta$.

$315° = 315° \cdot \frac{\pi \text{ radians}}{180°} = \frac{7\pi}{4}$ radians

Now use the formula $s = r\theta$.

$s = r\theta = 9 \cdot \frac{7\pi}{4} = \frac{63\pi}{4}$ yards ≈ 49.48 yards

75. 6 revolutions per second

$= \frac{6 \text{ revolutions}}{1 \text{ second}} \cdot \frac{2\pi \text{ radians}}{1 \text{ revolutions}} = \frac{12\pi \text{ radians}}{1 \text{ seconds}}$
$= 12\pi$ radians per second

76. 20 revolutions per second
$= \frac{20 \text{ revolutions}}{1 \text{ second}} \cdot \frac{2\pi \text{ radians}}{1 \text{ revolution}} = \frac{40\pi \text{ radians}}{1 \text{ second}}$
$= 40\pi$ radians per second

77. $-\frac{4\pi}{3}$ and $\frac{2\pi}{3}$

78. $-\frac{7\pi}{6}$ and $\frac{5\pi}{6}$

79. $-\frac{3\pi}{4}$ and $\frac{5\pi}{4}$

80. $-\frac{\pi}{4}$ and $\frac{7\pi}{4}$

81. $-\frac{\pi}{2}$ and $\frac{3\pi}{2}$

82. $-\pi$ and π

83. $\frac{55}{60} \cdot 2\pi = \frac{11\pi}{6}$

84. $\frac{35}{60} \cdot 2\pi = \frac{7\pi}{6}$

85. 3 minutes and 40 seconds equals 220 seconds.
$\frac{220}{60} \cdot 2\pi = \frac{22\pi}{3}$

86. 4 minutes and 25 seconds equals 265 seconds.
$\frac{265}{60} \cdot 2\pi = \frac{53\pi}{6}$

87. First, convert to degrees.
$\frac{1}{6}$ revolution $= \frac{1}{6}$ revolution $\cdot \frac{360°}{1 \text{ revolution}}$
$= \frac{1}{6} \cdot 360° = 60°$

Now, convert $60°$ to radians.
$60° = 60° \cdot \frac{\pi \text{ radians}}{180°} = \frac{60\pi}{180}$ radians
$= \frac{\pi}{3}$ radians

Therefore, $\frac{1}{6}$ revolution is equivalent to $60°$ or $\frac{\pi}{3}$ radians.

88. First, convert to degrees.

$$\frac{1}{3}\text{revolutions} = \frac{1}{3}\text{ revolutions} \cdot \frac{360^\circ}{1\text{ revolution}}$$
$$= \frac{1}{3} \cdot 360^\circ = 120^\circ$$

Now, convert 120° to radians.

$$120° = 120° \cdot \frac{\pi \text{ radians}}{180°} = \frac{120\pi}{180} \text{ radians}$$
$$= \frac{2\pi}{3} \text{ radians}$$

Therefore, $\frac{1}{3}$ revolution is equivalent to 120° or $\frac{2\pi}{3}$ radians.

89. The distance that the tip of the minute hand moves is given by its arc length, s. Since $s = r\theta$, we begin by finding r and θ. We are given that $r = 8$ inches. The minute hand moves from 12 to 2 o'clock, or $\frac{1}{6}$ of a complete revolution. The formula $s = r\theta$ can only be used when θ is expressed in radians. We must convert $\frac{1}{6}$ revolution to radians.

$$\frac{1}{6}\text{ revolution} = \frac{1}{6}\text{ revolution} \cdot \frac{2\pi \text{ radians}}{1\text{ revolution}}$$
$$= \frac{\pi}{3} \text{ radians}$$

The distance the tip of the minute hand moves is

$$s = r\theta = (8 \text{ inches})\left(\frac{\pi}{3}\right) = \frac{8\pi}{3} \text{ inches}$$
$$\approx 8.38 \text{ inches}.$$

90. The distance that the tip of the minute hand moves is given by its arc length, s. Since $s = r\theta$, we begin by finding r and θ. We are given that $r = 6$ inches. The minute hand moves from 12 to 4 o'clock, or $\frac{1}{3}$ of a complete revolution. The formula $s = r\theta$ can only be used when θ is expressed in radians. We must convert $\frac{1}{3}$ revolution to radians.

$$\frac{1}{3}\text{ revolution} = \frac{1}{3}\text{revolution} \cdot \frac{2\pi \text{ radians}}{1\text{ revolution}}$$
$$= \frac{2\pi}{3} \text{ radians}$$

The distance the tip of the minute hand moves is

$$s = r\theta = (6 \text{ inches})\left(\frac{2\pi}{3}\right) = \frac{12\pi}{3} \text{ inches}$$
$$= 4\pi \text{ inches} \approx 12.57 \text{ inches}.$$

91. The length of each arc is given by $s = r\theta$. We are given that $r = 24$ inches and $\theta = 90°$. The formula $s = r\theta$ can only be used when θ is expressed in radians.

$$90° = 90° \cdot \frac{\pi \text{ radians}}{180°} = \frac{90\pi}{180} \text{ radians}$$
$$= \frac{\pi}{2} \text{ radians}$$

The length of each arc is

$$s = r\theta = (24 \text{ inches})\left(\frac{\pi}{2}\right) = 12\pi \text{ inches}$$
$$\approx 37.70 \text{ inches}.$$

92. The distance that the wheel moves is given by $s = r\theta$. We are given that $r = 80$ centimeters and $\theta = 60°$. The formula $s = r\theta$ can only be used when θ is expressed in radians.

$$60° = 60° \cdot \frac{\pi \text{ radians}}{180°} = \frac{60\pi}{180} \text{ radians}$$
$$= \frac{\pi}{3} \text{ radians}$$

The length that the wheel moves is

$$s = r\theta = (80 \text{ centimeters})\left(\frac{\pi}{3}\right) = \frac{80\pi}{3} \text{ centimeters}$$
$$\approx 83.78 \text{ centimeters}.$$

93. Recall that $\theta = \frac{s}{r}$. We are given that $s = 8000$ miles and $r = 4000$ miles.

$$\theta = \frac{s}{r} = \frac{8000 \text{ miles}}{4000 \text{ miles}} = 2 \text{ radians}$$

Now, convert 2 radians to degrees.

$$2 \text{ radians} = 2 \text{ radians} \cdot \frac{180^\circ}{\pi \text{ radians}} \approx 114.59^\circ$$

94. Recall that $\theta = \frac{s}{r}$. We are given that $s = 10{,}000$ miles and $r = 4000$ miles.

$$\theta = \frac{s}{r} = \frac{10{,}000 \text{ miles}}{4000 \text{ miles}} = 2.5 \text{ radians}$$

Now, convert 2.5 radians to degrees.

$$2.5 \text{ radians} \cdot \frac{180^\circ}{2\pi \text{ radians}} \approx 143.24^\circ$$

95. Recall that $s = r\theta$. We are given that $r = 4000$ miles and $\theta = 30°$. The formula $s = r\theta$ can only be used when θ is expressed in radians.

$$30° = 30° \cdot \frac{\pi \text{ radians}}{180°} = \frac{30\pi}{180} \text{ radians}$$

$$= \frac{\pi}{6} \text{ radians}$$

$$s = r\theta = (4000 \text{ miles})\left(\frac{\pi}{6}\right) \approx 2094 \text{ miles}$$

To the nearest mile, the distance from A to B is 2094 miles.

96. Recall that $s = r\theta$. We are given that $r = 4000$ miles and $\theta = 10°$. We can only use the formula $s = r\theta$ when θ is expressed in radians.

$$10° = 10° \cdot \frac{\pi \text{ radians}}{180°} = \frac{10\pi}{180} \text{ radians}$$

$$= \frac{\pi}{18} \text{ radians}$$

$$s = r\theta = (4000 \text{ miles})\left(\frac{\pi}{18}\right) \approx 698 \text{ miles}$$

To the nearest mile, the distance from A to B is 698 miles.

97. Linear speed is given by $v = r\omega$. We are given that $\omega = \frac{\pi}{12}$ radians per hour and $r = 4000$ miles. Therefore,

$$v = r\omega = (4000 \text{ miles})\left(\frac{\pi}{12}\right)$$

$$= \frac{4000\pi}{12} \text{ miles per hour}$$

$$\approx 1047 \text{ miles per hour}$$

The linear speed is about 1047 miles per hour.

98. Linear speed is given by $v = r\omega$. We are given that $r = 25$ feet and the wheel rotates at 3 revolutions per minute. We need to convert 3 revolutions per minute to radians per minute.

3 revolutions per minute

$$= 3 \text{ revolutions per minute} \cdot \frac{2\pi \text{ radians}}{1 \text{ revolution}}$$

$$= 6\pi \text{ radians per minute}$$

$$v = r\omega = (25 \text{ feet})(6\pi) \approx 471 \text{ feet per minute}$$

The linear speed of the Ferris wheel is about 471 feet per minute.

99. Linear speed is given by $v = r\omega$. We are given that $r = 12$ feet and the wheel rotates at 20 revolutions per minute.

20 revolutions per minute

$$= 20 \text{ revolutions per minute} \cdot \frac{2\pi \text{ radians}}{1 \text{ revolution}}$$

$$= 40\pi \text{ radians per minute}$$

$$v = r\omega = (12 \text{ feet})(40\pi)$$

$$\approx 1508 \text{ feet per minute}$$

The linear speed of the wheel is about 1508 feet per minute.

100. Begin by converting 2.5 revolutions per minute to radians per minute.

2.5 revolutions per minute

$$= 2.5 \text{ revolutions per minute} \cdot \frac{2\pi \text{ radians}}{1 \text{ revolution}}$$

$$= 5\pi \text{ radians per minute}$$

The linear speed of the animals in the outer rows is $v = r\omega = (20 \text{ feet})(5\pi) \approx 100$ feet per minute

The linear speed of the animals in the inner rows is $v = r\omega = (10 \text{ feet})(5\pi) \approx 50$ feet per minute

The difference is $100\pi - 50\pi = 50\pi$ feet per minute or about 157.08 feet per minute.

113.

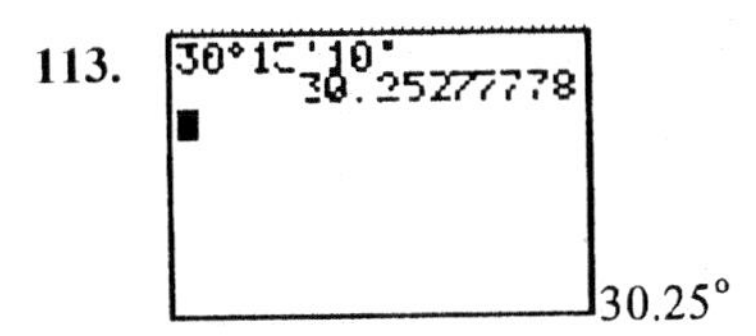

30.25°

114.

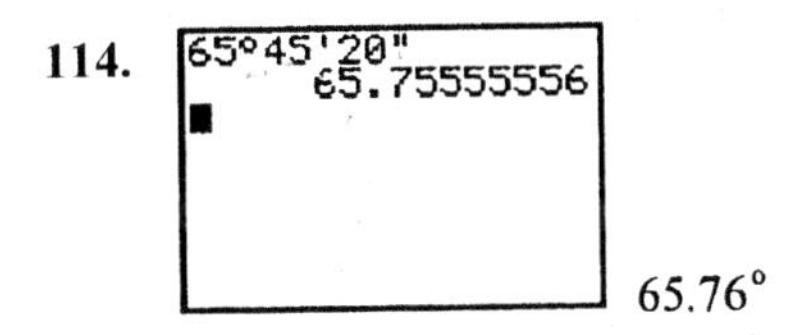

65.76°

115.

30° 25'12"

116.

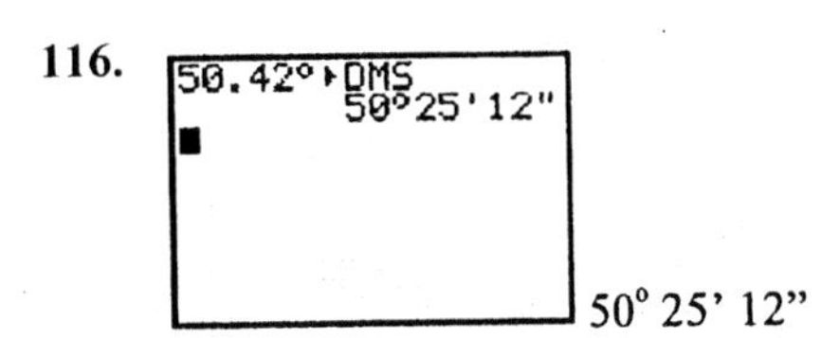

50° 25' 12"

117. A right angle measures 90° and

$90° = \frac{\pi}{2}$ radians ≈ 1.57 radians.

If $\theta = \frac{3}{2}$ radians = 1.5 radians, θ is smaller than a right angle.

118. $s = r\theta$

Begin by changing $\theta = 20°$ to radians.

$$20° = 20° \cdot \frac{\pi}{180°} = \frac{\pi}{9} \text{ radians}$$

$$100 = \frac{\pi}{9} r$$

$$r = \frac{900}{\pi} \approx 286 \text{ miles}$$

To the nearest mile, a radius of 286 miles should be used.

119. $s = r\theta$

Begin by changing $\theta = 26°$ to radians.

$$26° = 26° \cdot \frac{\pi}{180°} = \frac{13\pi}{90} \text{ radians}$$

$$s = 4000 \cdot \frac{13\pi}{90}$$

$$\approx 1815 \text{ miles}$$

To the nearest mile, Miami, Florida is 1815 miles north of the equator.

Section 4.2

Check Point Exercises

1. $P\left(\frac{\sqrt{3}}{2}, \frac{1}{2}\right)$

$$\sin t = y = \frac{1}{2}$$

$$\cos t = x = \frac{\sqrt{3}}{2}$$

$$\tan t = \frac{y}{x} = \frac{\frac{1}{2}}{\frac{\sqrt{3}}{2}} = \frac{\sqrt{3}}{3}$$

$$\csc t = \frac{1}{y} = 2$$

$$\sec t = \frac{1}{x} = \frac{2\sqrt{3}}{3}$$

$$\cot t = \frac{x}{y} = \sqrt{3}$$

2. The point P on the unit circle that corresponds to $t = \pi$ has coordinates (–1, 0). Use $x = -1$ and $y = 0$ to find the values of the trigonometric functions.

$$\sin \pi = y = 0$$

$$\cos \pi = x = -1$$

$$\tan \pi = \frac{y}{x} = \frac{0}{-1} = 0$$

$$\sec \pi = \frac{1}{x} = \frac{1}{-1} = -1$$

$$\cot \pi = \frac{x}{y} = \frac{-1}{0} = \text{undefined}$$

$$\csc \pi = \frac{1}{y} = \frac{1}{0} = \text{undefined}$$

3. $t = \frac{\pi}{4}, P\left(\frac{1}{\sqrt{2}}, \frac{1}{\sqrt{2}}\right)$

$$\csc \frac{\pi}{4} = \frac{1}{y} = \sqrt{2}$$

$$\sec \frac{\pi}{4} = \frac{1}{x} = \sqrt{2}$$

$$\cot \frac{\pi}{4} = \frac{x}{y} = \frac{\frac{1}{y}}{\frac{1}{\sqrt{2}}} = 1$$

4. **a.** $\sec\left(-\frac{\pi}{4}\right) = \sec\left(\frac{\pi}{4}\right) = \sqrt{2}$

b. $\sin\left(-\frac{\pi}{4}\right) = -\sin\left(\frac{\pi}{4}\right) = -\frac{\sqrt{2}}{2}$

5.
$$\tan \theta = \frac{\sin \theta}{\cos \theta} = \frac{\frac{2}{3}}{\frac{\sqrt{5}}{3}}$$

$$= \frac{2}{3} \cdot \frac{3}{\sqrt{5}} = \frac{2}{\sqrt{5}}$$

$$= \frac{2}{\sqrt{5}} \cdot \frac{\sqrt{5}}{\sqrt{5}} = \frac{2\sqrt{5}}{5}$$

$$\csc\theta = \frac{1}{\sin\theta} = \frac{1}{\frac{2}{3}} = \frac{3}{2}$$

$$\sec\theta = \frac{1}{\cos\theta} = \frac{1}{\frac{\sqrt{5}}{3}} = \frac{3}{\sqrt{5}}$$

$$= \frac{3}{\sqrt{5}} \cdot \frac{\sqrt{5}}{\sqrt{5}} = \frac{3\sqrt{5}}{5}$$

$$\cot\theta = \frac{1}{\tan\theta} = \frac{1}{\frac{2}{\sqrt{5}}} = \frac{\sqrt{5}}{2}$$

6. $\sin t = \frac{1}{2}, 0 \le t < \frac{\pi}{2}$

$$\sin^2 t + \cos^2 t = 1$$

$$\left(\frac{1}{2}\right)^2 + \cos^2 t = 1$$

$$\cos^2 t = 1 - \frac{1}{4}$$

$$\cos t = \sqrt{\frac{3}{4}} = \frac{\sqrt{3}}{2}$$

Because $0 \le t < \frac{\pi}{2}$, $\cos t$ is positive.

7. a. $\cot\frac{5\pi}{4} = \cot\left(\frac{\pi}{4} + \pi\right) = \cot\frac{\pi}{4} = 1$

b. $\cos\left(-\frac{9\pi}{4}\right) = \cos\left(-\frac{9\pi}{4} + 4\pi\right)$

$$= \cos\frac{7\pi}{4}$$

$$= \frac{\sqrt{2}}{2}$$

8. a. $\sin\frac{\pi}{4} \approx 0.7071$

b. $\csc 1.5 \approx 1.0025$

Exercise Set 4.2

1. The point P on the unit circle has coordinates $\left(-\frac{15}{17}, \frac{8}{17}\right)$. Use $x = -\frac{15}{17}$ and $y = \frac{8}{17}$ to find the values of the trigonometric functions.

$$\sin t = y = \frac{8}{17}$$

$$\cos t = x = -\frac{15}{17}$$

$$\tan t = \frac{y}{x} = \frac{\frac{8}{17}}{-\frac{15}{17}} = -\frac{8}{15}$$

$$\csc t = \frac{1}{y} = \frac{17}{8}$$

$$\sec t = \frac{1}{x} = -\frac{17}{15}$$

$$\cot t = \frac{x}{y} = -\frac{15}{8}$$

2. The point P on the unit circle has coordinates $\left(-\frac{5}{13}, -\frac{12}{13}\right)$ Use $x = -\frac{5}{13}$ and $y = -\frac{12}{13}$ to find the values of the trigonometric functions.

$$\sin t = y = -\frac{12}{13}$$

$$\cos t = x = -\frac{5}{13}$$

$$\tan t = \frac{y}{x} = \frac{-\frac{12}{13}}{-\frac{5}{13}} = \frac{12}{5}$$

$$\csc t = \frac{1}{y} = -\frac{13}{12}$$

$$\sec t = \frac{1}{x} = -\frac{13}{5}$$

$$\cot t = \frac{x}{y} = \frac{5}{12}$$

3. The point P on the unit circle that corresponds to $t = -\frac{\pi}{4}$ has coordinates $\left(\frac{\sqrt{2}}{2}, -\frac{\sqrt{2}}{2}\right)$. Use $x = \frac{\sqrt{2}}{2}$ and $y = -\frac{\sqrt{2}}{2}$ to find the values of the trigonometric functions.

$\sin t = y = -\frac{\sqrt{2}}{2}$

$\cos t = x = \frac{\sqrt{2}}{2}$

$\tan t = \frac{y}{x} = \frac{-\frac{\sqrt{2}}{2}}{\frac{\sqrt{2}}{2}} = -1$

$\csc t = \frac{1}{y} = -\sqrt{2}$

$\sec t = \frac{1}{x} = \sqrt{2}$

$\cot t = \frac{x}{y} = -1$

4. The point P on the unit circle that corresponds to $t = \frac{3\pi}{4}$ has coordinates $\left(-\frac{\sqrt{2}}{2}, \frac{\sqrt{2}}{2}\right)$. Use $x = -\frac{\sqrt{2}}{2}$ and $y = \frac{\sqrt{2}}{2}$ to find the values of the trigonometric functions.

$\sin t = y = \frac{\sqrt{2}}{2}$

$\cos t = x = -\frac{\sqrt{2}}{2}$

$\tan t = \frac{y}{x} = \frac{\frac{\sqrt{2}}{2}}{-\frac{\sqrt{2}}{2}} = -1$

$\csc t = \frac{1}{y} = \sqrt{2}$

$\sec t = \frac{1}{x} = -\sqrt{2}$

$\cot t = \frac{x}{y} = -1$

5. $\sin\frac{\pi}{6} = \frac{1}{2}$

6. $\sin\frac{\pi}{3} = \frac{\sqrt{3}}{2}$

7. $\cos\frac{5\pi}{6} = -\frac{\sqrt{3}}{2}$

8. $\cos\frac{2\pi}{3} = -\frac{1}{2}$

9. $\tan\pi = \frac{0}{-1} = 0$

10. $\tan 0 = \frac{0}{1} = 0$

11. $\csc\frac{7\pi}{6} = \frac{1}{-\frac{1}{2}} = -2$

12. $\csc\frac{4\pi}{3} = \frac{1}{-\frac{\sqrt{3}}{2}} = \frac{-2\sqrt{3}}{3}$

13. $\sec\frac{11\pi}{6} = \frac{1}{\frac{\sqrt{3}}{2}} = \frac{2\sqrt{3}}{3}$

14. $\sec\frac{5\pi}{3} = \frac{1}{\frac{1}{2}} = 2$

15. $\sin\frac{3\pi}{2} = -1$

16. $\cos\frac{3\pi}{2} = 0$

17. $\sec\frac{3\pi}{2} = \text{undefined}$

18. $\tan\frac{3\pi}{2} = \text{undefined}$

19. a. $\cos\frac{\pi}{6} = \frac{\sqrt{3}}{2}$

b. $\cos\left(-\frac{\pi}{6}\right) = \cos\frac{\pi}{6} = \frac{\sqrt{3}}{2}$

20. a. $\cos\frac{\pi}{3} = \frac{1}{2}$

b. $\cos\left(-\frac{\pi}{3}\right) = \cos\frac{\pi}{3} = \frac{1}{2}$

21. a. $\sin\frac{5\pi}{6} = \frac{1}{2}$

b. $\sin\left(-\frac{5\pi}{6}\right) = -\sin\frac{5\pi}{6} = -\frac{1}{2}$

22. **a.** $\sin\frac{2\pi}{3} = \frac{\sqrt{3}}{2}$

b. $\sin\left(-\frac{2\pi}{3}\right) = -\sin\frac{2\pi}{3} = -\frac{\sqrt{3}}{2}$

23. **a.** $\tan\frac{5\pi}{3} = \frac{-\frac{\sqrt{3}}{2}}{\frac{1}{2}} = -\sqrt{3}$

b. $\tan\left(-\frac{5\pi}{3}\right) = -\tan\frac{5\pi}{3} = \sqrt{3}$

24. **a.** $\tan\frac{11\pi}{6} = \frac{-\frac{1}{2}}{\frac{\sqrt{3}}{2}} = -\frac{\sqrt{3}}{3}$

b. $\tan\left(-\frac{11\pi}{6}\right) = -\tan\frac{11\pi}{6} = \frac{\sqrt{3}}{3}$

25. $\sin t = \frac{8}{17}, \cos t = \frac{15}{17}$

$\tan t = \frac{\frac{8}{17}}{\frac{15}{17}} = \frac{8}{15}$

$\csc t = \frac{17}{8}$

$\sec t = \frac{17}{15}$

$\cot t = \frac{15}{8}$

26. $\sin t = \frac{3}{5}, \cos t = \frac{4}{5}$

$\tan t = \frac{\frac{3}{5}}{\frac{4}{5}} = \frac{3}{4}$

$\csc t = \frac{5}{3}$

$\sec t = \frac{5}{4}$

$\cot t = \frac{4}{3}$

27. $\sin t = \frac{1}{3}, \cos t = \frac{2\sqrt{2}}{3}$

$\tan t = \frac{\frac{1}{3}}{\frac{2\sqrt{2}}{3}} = \frac{\sqrt{2}}{4}$

$\csc t = 3$

$\sec t = \frac{3\sqrt{2}}{4}$

$\cot t = 2\sqrt{2}$

28. $\sin t = \frac{2}{3}, \cos t = \frac{\sqrt{5}}{3}$

$\tan t = \frac{\frac{2}{3}}{\frac{\sqrt{5}}{3}} = \frac{2\sqrt{5}}{5}$

$\csc t = \frac{3}{2}$

$\sec t = \frac{3\sqrt{5}}{5}$

$\cot t = \frac{\sqrt{5}}{2}$

29. $\sin t = \frac{6}{7}, 0 \le t < \frac{\pi}{2}$

$\sin^2 t + \cos^2 t = 1$

$\left(\frac{6}{7}\right)^2 + \cos^2 t = 1$

$\cos^2 t = 1 - \frac{36}{49}$

$\cos t = \sqrt{\frac{13}{49}} = \frac{\sqrt{13}}{7}$

Because $0 \le t < \frac{\pi}{2}$, $\cos t$ is positive.

30. $\sin t = \frac{7}{8}, 0 \le t < \frac{\pi}{2}$

$\sin^2 t + \cos^2 t = 1$

$\left(\frac{7}{8}\right)^2 + \cos^2 t = 1$

$\cos^2 t = 1 - \frac{49}{64}$

$\cos t = \sqrt{\frac{15}{64}} = \frac{\sqrt{15}}{8}$

Because $0 \le t < \frac{\pi}{2}$, $\cos t$ is positive.

31. $\sin t = \frac{\sqrt{39}}{8}, 0 \le t < \frac{\pi}{2}$

$\sin^2 t + \cos^2 t = 1$

$\left(\frac{\sqrt{39}}{8}\right)^2 + \cos^2 t = 1$

$\cos^2 t = 1 - \frac{39}{64}$

$\cos t = \sqrt{\frac{25}{64}} = \frac{5}{8}$

Because $0 \le t < \frac{\pi}{2}$, $\cos t$ is positive.

32. $\sin t = \frac{\sqrt{21}}{5}, 0 \le t < \frac{\pi}{2}$

$\sin^2 t + \cos^2 t = 1$

$\left(\frac{\sqrt{21}}{5}\right)^2 + \cos^2 t = 1$

$\cos^2 t = 1 - \frac{21}{25}$

$\cos t = \sqrt{\frac{4}{25}} = \frac{2}{5}$

Because $0 \le t < \frac{\pi}{2}$, $\cos t$ is positive.

33. $\sin 1.7 \csc 1.7 = \sin 1.7\left(\frac{1}{\sin 1.7}\right) = 1$

34. $\cos 2.3 \sec 2.3 = \cos 2.3\left(\frac{1}{\cos 2.3}\right) = 1$

35. $\sin^2 \frac{\pi}{6} + \cos^2 \frac{\pi}{2} = 1$ by the Pythagorean identity.

36. $\sin^2 \frac{\pi}{3} + \cos^2 \frac{\pi}{3} = 1$ because $\sin^2 t + \cos^2 t = 1$.

37. $\sec^2 \frac{\pi}{3} - \tan^2 \frac{\pi}{3} = 1$ because $1 + \tan^2 t = \sec^2 t$.

38. $\csc^2 \frac{\pi}{6} - \cot^2 \frac{\pi}{6} = 1$ because $1 + \cot^2 t = \csc^2 t$.

39. $\cos \frac{9\pi}{4} = \cos\left(\frac{\pi}{4} + 2\pi\right) = \cos \frac{\pi}{4} = \frac{\sqrt{2}}{2}$

40. $\csc \frac{9\pi}{4} = \csc\left(\frac{\pi}{4} + 2\pi\right) = \csc \frac{\pi}{4} = \sqrt{2}$

41. $\sin\left(-\frac{9\pi}{4}\right) = \sin\left(-\frac{9\pi}{4} + 4\pi\right) = \sin \frac{7\pi}{4} = -\frac{\sqrt{2}}{2}$

42. $\sec\left(-\frac{9\pi}{4}\right) = \sec\left(-\frac{9\pi}{4} + 4\pi\right) = \sec \frac{7\pi}{4} = \sqrt{2}$

43. $\tan \frac{5\pi}{4} = \tan\left(\frac{\pi}{4} + \pi\right) = \tan \frac{\pi}{4} = 1$

44. $\cot \frac{5\pi}{4} = \cot\left(\frac{\pi}{4} + \pi\right) = \cot \frac{\pi}{4} = 1$

45. $\cot\left(-\frac{5\pi}{4}\right) = \cot\left(\frac{3\pi}{4} - 2\pi\right) = \cot \frac{3\pi}{4} = -1$

46. $\tan\left(-\frac{9\pi}{4}\right) = \tan\left(-\frac{9\pi}{4} + 3\pi\right) = \tan \frac{3\pi}{4} = -1$

47. $-\tan\left(\frac{\pi}{4} + 15\pi\right) = -\tan \frac{\pi}{4} = -1$

48. $-\cot\left(\frac{\pi}{4} + 17\pi\right) = -\cot \frac{\pi}{4} = -1$

49. $\sin\left(-\frac{\pi}{4} - 1000\pi\right) = \sin\left(-\frac{\pi}{4} + 2\pi\right)$

$= \sin \frac{7\pi}{4}$

$= -\frac{\sqrt{2}}{2}$

50. $\sin\left(-\frac{\pi}{4} - 2000\pi\right) = \sin\left(-\frac{\pi}{4} + 2\pi\right)$

$= \sin \frac{7\pi}{4}$

$= -\frac{\sqrt{2}}{2}$

51. $\cos\left(-\frac{\pi}{4}-1000\pi\right)=\cos\left(-\frac{\pi}{4}+2\pi\right)$
$=\cos\frac{7\pi}{4}$
$=\frac{\sqrt{2}}{2}$

52. $\cos\left(-\frac{\pi}{4}-2000\pi\right)=\cos\left(-\frac{\pi}{4}+2\pi\right)$
$=\cos\frac{7\pi}{4}$
$=\frac{\sqrt{2}}{2}$

53. a. $\sin\frac{3\pi}{4}=\frac{\sqrt{2}}{2}$

b. $\sin\frac{11\pi}{4}=\sin\left(\frac{3\pi}{4}+2\pi\right)=\sin\frac{3\pi}{4}=\frac{\sqrt{2}}{2}$

54. a. $\cos\frac{3\pi}{4}=-\frac{\sqrt{2}}{2}$

b. $\cos\frac{11\pi}{4}=\cos\left(\frac{3\pi}{4}+2\pi\right)=\cos\frac{3\pi}{4}=-\frac{\sqrt{2}}{2}$

55. a. $\cos\frac{\pi}{2}=0$

b. $\cos\frac{9\pi}{2}=\cos\left(\frac{\pi}{2}+4\pi\right)$
$=\cos\left[\frac{\pi}{2}+2(2\pi)\right]$
$=\cos\frac{\pi}{2}$
$=0$

56. a. $\sin\frac{\pi}{2}=1$

b. $\sin\frac{9\pi}{2}=\sin\left(\frac{\pi}{2}+4\pi\right)=\sin\frac{\pi}{2}=1$

57. a. $\tan\pi=\frac{0}{-1}=0$

b. $\tan 17\pi=\tan(\pi+16\pi)$
$=\tan[\pi+8(2\pi)]$
$=\tan\pi$
$=0$

58. a. $\cot\frac{\pi}{2}=\frac{0}{1}=0$

b. $\cot\frac{15\pi}{2}=\cot\left(\frac{\pi}{2}+7\pi\right)=\cot\frac{\pi}{2}=0$

59. a. $\sin\frac{7\pi}{4}=-\frac{\sqrt{2}}{2}$

b. $\sin\frac{47\pi}{4}=\sin\left(\frac{7\pi}{4}+10\pi\right)$
$=\sin\left[\frac{7\pi}{4}+5(2\pi)\right]$
$=\sin\frac{7\pi}{4}$
$=-\frac{\sqrt{2}}{2}$

60. a. $\cos\frac{7\pi}{4}=\frac{\sqrt{2}}{2}$

b. $\cos\frac{47\pi}{4}=\cos\left(\frac{7\pi}{4}+10\pi\right)=\cos\frac{7\pi}{4}=\frac{\sqrt{2}}{2}$

61. $\sin 0.8\approx 0.7174$

62. $\cos 0.6\approx 0.8253$

63. $\tan 3.4\approx 0.2643$

64. $\tan 3.7\approx 0.6247$

65. $\csc 1\approx 1.1884$

66. $\sec 1\approx 1.8508$

67. $\cos\frac{\pi}{10}\approx 0.9511$

68. $\sin\frac{3\pi}{10}\approx 0.8090$

69. $\cot\frac{\pi}{12}\approx 3.7321$

70. $\cot\frac{\pi}{18} \approx 5.6713$

71. $\sin(-t) - \sin t = -\sin t - \sin t = -2\sin t = -2a$

72. $\tan(-t) - \tan t = -\tan t - \tan t = -2\tan t = -2c$

73. $4\cos(-t) - \cos t = 4\cos t - \cos t = 3\cos t = 3b$

74. $3\cos(-t) - \cos t = 3\cos t - \cos t = 2\cos t = 2b$

75. $\sin(t + 2\pi) - \cos(t + 4\pi) + \tan(t + \pi)$
$= \sin(t) - \cos(t) + \tan(t)$
$= a - b + c$

76. $\sin(t + 2\pi) + \cos(t + 4\pi) - \tan(t + \pi)$
$= \sin(t) + \cos(t) - \tan(t)$
$= a + b - c$

77. $\sin(-t - 2\pi) - \cos(-t - 4\pi) - \tan(-t - \pi)$
$= -\sin(t + 2\pi) - \cos(t + 4\pi) + \tan(t + \pi)$
$= -\sin(t) - \cos(t) + \tan(t)$
$= -a - b + c$

78. $\sin(-t - 2\pi) + \cos(-t - 4\pi) - \tan(-t - \pi)$
$= -\sin(t + 2\pi) + \cos(t + 4\pi) + \tan(t + \pi)$
$= -\sin(t) + \cos(t) + \tan(t)$
$= -a + b + c$

79. $\cos t + \cos(t + 1000\pi) - \tan t - \tan(t + 999\pi) - \sin t + 4\sin(t - 1000\pi)$
$= \cos t + \cos t - \tan t - \tan t - \sin t + 4\sin t$
$= 2\cos t - 2\tan t + 3\sin t$
$= 3a + 2b - 2c$

80. $-\cos t + 7\cos(t + 1000\pi) + \tan t + \tan(t + 999\pi) + \sin t + \sin(t - 1000\pi)$
$= -\cos t + 7\cos t + \tan t + \tan t + \sin t + \sin t$
$= 6\cos t + 2\tan t + 2\sin t$
$= 2a + 6b + 2c$

81. a. $H = 12 + 8.3\sin\left[\frac{2\pi}{365}(80 - 80)\right]$
$= 12 + 8.3\sin 0 = 12 + 8.3(0)$
$= 12$
There are 12 hours of daylight in Fairbanks on March 21.

b. $H = 12 + 8.3\sin\left[\frac{2\pi}{365}(172 - 80)\right]$
$\approx 12 + 8.3\sin 1.5837$
≈ 20.3
There are about 20.3 hours of daylight in Fairbanks on June 21.

c. $H = 12 + 8.3\sin\left[\frac{2\pi}{365}(355 - 80)\right]$
$\approx 12 + 8.3\sin 4.7339$
≈ 3.7
There are about 3.7 hours of daylight in Fairbanks on December 21.

82. a. $H = 12 + 24\sin\left[\frac{2\pi}{365}(80 - 80)\right]$
$12 + 24\sin 0 = 12 + 24(0) = 12$
There are 12 hours of daylight in San Diego on March 21.

b. $H = 12 + 24\sin\left[\frac{2\pi}{365}(172 - 80)\right]$
$\approx 12 + 24\sin 1.5837$
≈ 14.3998
There are about 14.4 hours of daylight in San Diego on June 21.

c. $H = 12 + 24\sin\left[\frac{2\pi}{365}(355 - 80)\right]$
$\approx 12 + 24\sin 4.7339$
≈ 9.6
There are about 9.6 hours of daylight in San Diego on December 21.

83. a. For $t = 7$,
$E = \sin\frac{\pi}{14}\cdot 7 = \sin\frac{\pi}{2} = 1$

For $t = 14$,

$E = \sin\frac{\pi}{14}\cdot 14 = \sin\pi = 0$

For $t = 21$,

$E = \sin\frac{\pi}{14}\cdot 21 = \sin\frac{3\pi}{2} = -1$

For $t = 28$,

$E = \sin\frac{\pi}{14}\cdot 28 = \sin 2\pi = \sin 0 = 0$

For $t = 35$,

$E = \sin\frac{\pi}{14}\cdot 35 = \sin\frac{5\pi}{2} = \sin\frac{\pi}{2} = 1$

Observations may vary.

b. Because $E(35) = E(7) = 1$, the period is $35 - 7 = 28$ or 28 days.

84. a. At 6 A.M., $t = 0$.

$H = 10 + 4\sin\frac{\pi}{6}\cdot 0$
$= 10 + 4\sin 0 = 10 + 4\cdot 0 = 10$

The height is 10 feet.

At 9 A.M., $t = 3$.

$H = 10 + 4\sin\frac{\pi}{6}\cdot 3$
$= 10 + 4\sin\frac{\pi}{2} = 10 + 4(1) = 14$

The height is 14 feet.

At noon, $t = 6$.

$H = 10 + 4\sin\frac{\pi}{6}\cdot 6$
$= 10 + 4\sin\pi = 10 + 4\cdot 0 = 10$

The height is 10 feet.

At 6 P.M., $t = 12$.

$H = 10 + 4\sin\frac{\pi}{6}\cdot 12$
$= 10 + 4\sin 2\pi = 10 + 4\cdot 0 = 10$

The height is 10 feet.

At midnight, $t = 18$.

$H = 10 + 4\sin\frac{\pi}{6}\cdot 18$
$= 10 + 4\sin 3\pi = 10 + 4\sin\pi$
$= 10 + 4\cdot 0 = 10$

The height is 10 feet.

At 3 A.M., $t = 21$.

$H = 10 + 4\sin\frac{\pi}{6}\cdot 21$
$= 10 + 4\sin\frac{7\pi}{2} = 10 + 4\sin\frac{3\pi}{2}$
$= 10 + 4(-1) = 6$

The height is 6 feet.

b. The sine function has a minimum at $\frac{3\pi}{2}$.

Thus, we find a low tide at $\frac{\pi}{6}t = \frac{3\pi}{2}$ or $t = 9$. This value of t corresponds to 3 P.M.

For $t = 9$,

$h = 10 + 4\sin\frac{\pi}{6}\cdot 9$
$= 10 + 4\sin\frac{3\pi}{2} = 10 + 4(-1) = 6$

The height is 6 feet. From part *a*, the height at 3 A.M. is also 6 feet. Thus, low tide is at 3 A.M. and 3 P.M.

The sine function has a maximum at $\frac{\pi}{2}$.

Thus, we find a high tide at $\frac{\pi}{6}t = \frac{\pi}{2}$ or $t = 3$. This value of t corresponds to 9 a.m. From part *a*, the height at 9 A.M. is 14 feet. Because the sine has a period of 2π we also find a maximum at $\frac{5\pi}{2}$. We find another high tide at $\frac{\pi}{6}t = \frac{5\pi}{2}$ or $t = 15$. This value of t corresponds to 9 P.M. Thus, high tide is at 9 A.M. and 9 P.M.

c. The period of the sine function is 2π or on the interval $[0, 2\pi]$. The cycle of the sine function starts at $\frac{\pi}{6}t = \frac{5\pi}{2}$ or $t = 0$, and ends at $\frac{\pi}{6}t = 2\pi$ or $t = 12$. Thus, the period is 12 hours, which means high and low tides occur every 12 hours.

97. t is in the third quadrant therefore $\sin t < 0$, $\tan t > 0$, and $\cot t > 0$.
Thus, only choice (c) is true.

98. $f(x)=\sin x$ and $f(a)=\frac{1}{4}$

$f(a)+f(a+2\pi)+f(a+4\pi)+f(a+6\pi)$

$=4f(a)=4\left(\frac{1}{4}\right)=1$ because $\sin x$ has a period of 2π.

99. $f(x)=\sin x$ and $f(a)=\frac{1}{4}$

$$f(a)+2f(-a)=f(a)-2f(a)=\frac{1}{4}-2\left(\frac{1}{4}\right)=-\frac{1}{4}$$

$f(-a)=-f(a)$ because $\sin(-x)=-\sin x$.
Sine is an odd function.

100. The height is given by
$h=45+40\sin(t-90°)$
$h(765°)=45+40\sin(765°-90°)$
≈ 16.7
You are about 16.7 feet above the ground.

Section 4.3

Checkpoint Exercises

1. Use the Pythagorean Theorem, $c^2=a^2+b^2$, to find c.

$a=3, b=4$

$c^2=a^2+b^2=3^2+4^2=9+16=25$

$c=\sqrt{25}=5$

Referring to these lengths as opposite, adjacent, and hypotenuse, we have

$\sin\theta=\frac{\text{opposite}}{\text{hypotenuse}}=\frac{3}{5}$

$\cos\theta=\frac{\text{adjacent}}{\text{hypotenuse}}=\frac{4}{5}$

$\tan\theta=\frac{\text{opposite}}{\text{adjacent}}=\frac{3}{4}$

$\csc\theta=\frac{\text{hypotenuse}}{\text{opposite}}=\frac{5}{3}$

$\sec\theta=\frac{\text{hypotenuse}}{\text{adjacent}}=\frac{5}{4}$

$\cot\theta=\frac{\text{adjacent}}{\text{opposite}}=\frac{4}{3}$

2. Use the Pythagorean Theorem, $c^2=a^2+b^2$, to find b.

$a^2+b^2=c^2$

$1^2+b^2=5^2$

$1+b^2=25$

$b^2=24$

$b=\sqrt{24}=2\sqrt{6}$

Note that side a is opposite θ and side b is adjacent to θ.

$\sin\theta=\frac{\text{opposite}}{\text{hypotenuse}}=\frac{1}{5}$

$\cos\theta=\frac{\text{adjacent}}{\text{hypotenuse}}=\frac{2\sqrt{6}}{5}$

$\tan\theta=\frac{\text{opposite}}{\text{adjacent}}=\frac{1}{2\sqrt{6}}=\frac{\sqrt{6}}{12}$

$\csc\theta=\frac{\text{hypotenuse}}{\text{opposite}}=\frac{5}{1}=5$

$\sec\theta=\frac{\text{hypotenuse}}{\text{adjacent}}=\frac{5}{2\sqrt{6}}=\frac{5\sqrt{6}}{12}$

$\cot\theta=\frac{\text{adjacent}}{\text{opposite}}=\frac{2\sqrt{6}}{1}=2\sqrt{6}$

3. Apply the definitions of these three trigonometric functions.

$\csc 45°=\frac{\text{length of hypotenuse}}{\text{length of side opposite } 45°}=\frac{\sqrt{2}}{1}=\sqrt{2}$

$\sec 45°=\frac{\text{length of hypotenuse}}{\text{length of side adjacent to } 45°}=\frac{\sqrt{2}}{1}=\sqrt{2}$

$\cot 45°=\frac{\text{length of side adjacent to } 45°}{\text{length of side opposite } 45°}=\frac{1}{1}=1$

4.

$$\tan 60° = \frac{\text{length of side opposite } 60°}{\text{length of side adjacent to } 60°} = \frac{\sqrt{3}}{1} = \sqrt{3}$$

$$\tan 30° = \frac{\text{length of side opposite } 30°}{\text{length of side adjacent to } 30°} = \frac{1}{\sqrt{3}} = \frac{1}{\sqrt{3}} \cdot \frac{\sqrt{3}}{3} = \frac{\sqrt{3}}{3}$$

5. **a.** $\sin 46^{\circ} = \cos(90^{\circ} - 46^{\circ}) = \cos 44^{\circ}$

b.
$$\cot\frac{\pi}{12} = \tan\left(\frac{\pi}{2} - \frac{\pi}{12}\right) = \tan\left(\frac{6\pi}{12} - \frac{\pi}{12}\right) = \tan\frac{5\pi}{12}$$

6. Because we have a known angle, an unknown opposite side, and a known adjacent side, we select the
tangent function.

$$\tan 24^{0} = \frac{a}{750}$$
$$a = 750 \tan 24^{0}$$
$$a \approx 750(0.4452) \approx 334$$

The distance across the lake is approximately 334 yards.

7. $\tan\theta = \frac{\text{side opposite}}{\text{side adjacent}} = \frac{14}{10}$

Use a calculator in degree mode to find θ.

Many Scientific Calculators	Many Graphing Calculators
TAN^{-1} (14 ÷ 10) ENTER	TAN (14 ÷ 10) ENTER

The display should show approximately 54. Thus, the angle of elevation of the sun is approximately 54°.

Exercise Set 4.3

1. $c^2 = 9^2 + 12^2 = 225$

$c = \sqrt{225} = 15$

$$\sin\theta = \frac{\text{opposite}}{\text{hypotenuse}} = \frac{9}{15} = \frac{3}{5}$$
$$\cos\theta = \frac{\text{adjacent}}{\text{hypotenuse}} = \frac{12}{15} = \frac{4}{5}$$
$$\tan\theta = \frac{\text{opposite}}{\text{adjacent}} = \frac{9}{12} = \frac{3}{4}$$
$$\csc\theta = \frac{\text{hypotenuse}}{\text{opposite}} = \frac{15}{9} = \frac{5}{3}$$
$$\sec\theta = \frac{\text{hypotenuse}}{\text{adjacent}} = \frac{15}{12} = \frac{5}{4}$$
$$\cot\theta = \frac{\text{adjacent}}{\text{opposite}} = \frac{12}{9} = \frac{4}{3}$$

2. $c^2 = 6^2 + 8^2 = 100$

$c = \sqrt{100} = 10$

$$\sin\theta = \frac{\text{opposite}}{\text{hypotenuse}} = \frac{6}{10} = \frac{3}{5}$$
$$\cos\theta = \frac{\text{adjacent}}{\text{hypotenuse}} = \frac{8}{10} = \frac{4}{5}$$
$$\tan\theta = \frac{\text{opposite}}{\text{adjacent}} = \frac{6}{8} = \frac{3}{4}$$
$$\csc\theta = \frac{\text{hypotenuse}}{\text{opposite}} = \frac{10}{6} = \frac{5}{3}$$
$$\sec\theta = \frac{\text{hypotenuse}}{\text{adjacent}} = \frac{10}{8} = \frac{5}{4}$$
$$\cot\theta = \frac{\text{adjacent}}{\text{opposite}} = \frac{8}{6} = \frac{4}{3}$$

3. $a^2 + 21^2 = 29^2$

$a^2 = 841 - 441 = 400$

$a = \sqrt{400} = 20$

$\sin\theta = \frac{\text{opposite}}{\text{hypotenuse}} = \frac{20}{29}$

$\cos\theta = \frac{\text{adjacent}}{\text{hypotenuse}} = \frac{21}{29}$

$\tan\theta = \frac{\text{opposite}}{\text{adjacent}} = \frac{20}{21}$

$\csc\theta = \frac{\text{hypotenuse}}{\text{opposite}} = \frac{29}{20}$

$\sec\theta = \frac{\text{hypotenuse}}{\text{adjacent}} = \frac{29}{21}$

$\cot\theta = \frac{\text{adjacent}}{\text{opposite}} = \frac{21}{20}$

4. $a^2 + 15^2 = 17^2$

$a^2 = 289 - 225 = 64$

$a = \sqrt{64} = 8$

$\sin\theta = \frac{\text{opposite}}{\text{hypotenuse}} = \frac{8}{17}$

$\cos\theta = \frac{\text{adjacent}}{\text{hypotenuse}} = \frac{15}{17}$

$\tan\theta = \frac{\text{opposite}}{\text{adjacent}} = \frac{8}{15}$

$\csc\theta = \frac{\text{hypotenuse}}{\text{opposite}} = \frac{17}{8}$

$\sec\theta = \frac{\text{hypotenuse}}{\text{adjacent}} = \frac{17}{15}$

$\cot\theta = \frac{\text{adjacent}}{\text{opposite}} = \frac{15}{8}$

5. $10^2 + b^2 = 26^2$

$b^2 = 676 - 100 = 576$

$b = \sqrt{576} = 24$

$\sin\theta = \frac{\text{opposite}}{\text{hypotenuse}} = \frac{10}{26} = \frac{5}{13}$

$\cos\theta = \frac{\text{adjacent}}{\text{hypotenuse}} = \frac{24}{26} = \frac{12}{13}$

$\tan\theta = \frac{\text{opposite}}{\text{adjacent}} = \frac{10}{24} = \frac{5}{12}$

$\csc\theta = \frac{\text{hypotenuse}}{\text{opposite}} = \frac{26}{10} = \frac{13}{5}$

$\sec\theta = \frac{\text{hypotenuse}}{\text{adjacent}} = \frac{26}{24} = \frac{13}{12}$

$\cot\theta = \frac{\text{adjacent}}{\text{opposite}} = \frac{24}{10} = \frac{12}{5}$

6. $a^2 + 40^2 = 41^2$

$a^2 = 1681 - 1600 = 81$

$a = \sqrt{81} = 9$

$\sin\theta = \frac{\text{opposite}}{\text{hypotenuse}} = \frac{9}{41}$

$\cos\theta = \frac{\text{adjacent}}{\text{hypotenuse}} = \frac{40}{41}$

$\tan\theta = \frac{\text{opposite}}{\text{adjacent}} = \frac{9}{40}$

$\csc\theta = \frac{\text{hypotenuse}}{\text{opposite}} = \frac{41}{9}$

$\sec\theta = \frac{\text{hypotenuse}}{\text{adjacent}} = \frac{41}{40}$

$\cot\theta = \frac{\text{adjacent}}{\text{opposite}} = \frac{40}{9}$

7. $21^2 + b^2 = 35^2$

$b^2 = 1225 - 441 = 784$

$b = \sqrt{784} = 28$

$\sin\theta = \dfrac{\text{opposite}}{\text{hypotenuse}} = \dfrac{28}{35} = \dfrac{4}{5}$

$\cos\theta = \dfrac{\text{adjacent}}{\text{hypotenuse}} = \dfrac{21}{35} = \dfrac{3}{5}$

$\tan\theta = \dfrac{\text{opposite}}{\text{adjacent}} = \dfrac{28}{21} = \dfrac{4}{3}$

$\csc\theta = \dfrac{\text{hypotenuse}}{\text{opposite}} = \dfrac{35}{28} = \dfrac{5}{4}$

$\sec\theta = \dfrac{\text{hypotenuse}}{\text{adjacent}} = \dfrac{35}{21} = \dfrac{5}{3}$

$\cot\theta = \dfrac{\text{adjacent}}{\text{opposite}} = \dfrac{21}{28} = \dfrac{3}{4}$

8. $a^2 + 24^2 = 25^2$

$a^2 = 625 - 576 = 49$

$a = \sqrt{49} = 7$

$\sin\theta = \dfrac{\text{opposite}}{\text{hypotenuse}} = \dfrac{24}{25}$

$\cos\theta = \dfrac{\text{adjacent}}{\text{hypotenuse}} = \dfrac{7}{25}$

$\tan\theta = \dfrac{\text{opposite}}{\text{adjacent}} = \dfrac{24}{7}$

$\csc\theta = \dfrac{\text{hypotenuse}}{\text{opposite}} = \dfrac{25}{24}$

$\sec\theta = \dfrac{\text{hypotenuse}}{\text{adjacent}} = \dfrac{25}{7}$

$\cot\theta = \dfrac{\text{adjacent}}{\text{opposite}} = \dfrac{7}{24}$

9. $\cos 30° = \dfrac{\text{length of side adjacent to } 30°}{\text{length of hypotenuse}}$

$= \dfrac{\sqrt{3}}{2}$

10. $\tan 30° = \dfrac{\text{length of side opposite } 30°}{\text{length of side adjacent to } 30°}$

$= \dfrac{1}{\sqrt{3}} = \dfrac{1}{\sqrt{3}} \cdot \dfrac{\sqrt{3}}{\sqrt{3}} = \dfrac{\sqrt{3}}{3}$

11. $\sec 45° = \dfrac{\text{length of hypotenuse}}{\text{length of side adjacent to } 45°}$

$= \dfrac{\sqrt{2}}{1} = \sqrt{2}$

12. $\csc 45° = \dfrac{\text{length of hypotenuse}}{\text{length of side opposite } 45°}$

$= \dfrac{\sqrt{2}}{1} = \sqrt{2}$

13. $\tan\dfrac{\pi}{3} = \tan 60°$

$= \dfrac{\text{length of side opposite } 60°}{\text{length of side adjacent to } 60°}$

$= \dfrac{\sqrt{3}}{1} = \sqrt{3}$

14. $\cot\dfrac{\pi}{3} = \cot 60° = \dfrac{\text{length of side adjacent to } 60°}{\text{length of side opposite } 60°}$

$= \dfrac{1}{\sqrt{3}} = \dfrac{1}{\sqrt{3}} \cdot \dfrac{\sqrt{3}}{\sqrt{3}} = \dfrac{\sqrt{3}}{3}$

15. $\sin\dfrac{\pi}{4} - \cos\dfrac{\pi}{4} = \sin 45° - \cos 45°$

$= \dfrac{1}{\sqrt{2}} - \dfrac{1}{\sqrt{2}} = 0$

16. $\tan\dfrac{\pi}{4} + \csc\dfrac{\pi}{6} = \tan 45° + \csc 30°$

$= \dfrac{1}{1} + \dfrac{2}{1} = 1 + 2 = 3$

17. $\sin\dfrac{\pi}{3}\cos\dfrac{\pi}{4} - \tan\dfrac{\pi}{4} = \left(\dfrac{\sqrt{3}}{2}\right)\left(\dfrac{\sqrt{2}}{2}\right) - 1$

$= \dfrac{\sqrt{6}}{4} - 1$

$= \dfrac{\sqrt{6} - 4}{4}$

18. $\cos\dfrac{\pi}{3}\sec\dfrac{\pi}{3} - \cot\dfrac{\pi}{3} = 1 - \dfrac{\sqrt{3}}{3} = \dfrac{3 - \sqrt{3}}{3}$

19. $2\tan\frac{\pi}{3}+\cos\frac{\pi}{4}\tan\frac{\pi}{6}=2\left(\sqrt{3}\right)+\left(\frac{\sqrt{2}}{2}\right)\left(\frac{\sqrt{3}}{3}\right)$
$=2\sqrt{3}+\frac{\sqrt{6}}{6}$
$=\frac{12\sqrt{3}+\sqrt{6}}{6}$

20. $6\tan\frac{\pi}{4}+\sin\frac{\pi}{3}\sec\frac{\pi}{6}=6(1)+\left(\frac{\sqrt{3}}{2}\right)\left(\frac{2\sqrt{3}}{3}\right)$
$=6+\frac{6}{6}$
$=7$

21. $\sin 7^\circ=\cos(90^\circ-7^\circ)=\cos 83^\circ$

22. $\sin 19^\circ=\cos\left(90^\circ-19^\circ\right)=\cos 71^\circ$

23. $\csc 25^\circ=\sec(90^\circ-25^\circ)=\sec 65^\circ$

24. $\csc 35^\circ=\sec(90^\circ-35^\circ)=\sec 55^\circ$

25. $\tan\frac{\pi}{9}=\cot\left(\frac{\pi}{2}-\frac{\pi}{9}\right)$
$=\cot\left(\frac{9\pi}{18}-\frac{2\pi}{18}\right)$
$=\cot\frac{7\pi}{18}$

26. $\tan\frac{\pi}{7}=\cot\left(\frac{\pi}{2}-\frac{\pi}{7}\right)=\cot\left(\frac{7\pi}{14}-\frac{2\pi}{14}\right)=\cot\frac{5\pi}{14}$

27. $\cos\frac{2\pi}{5}=\sin\left(\frac{\pi}{2}-\frac{2\pi}{5}\right)$
$=\sin\left(\frac{5\pi}{10}-\frac{4\pi}{10}\right)$
$=\sin\frac{\pi}{10}$

28. $\cos\frac{3\pi}{8}=\sin\left(\frac{\pi}{2}-\frac{3\pi}{8}\right)=\sin\left(\frac{4\pi}{8}-\frac{3\pi}{8}\right)=\sin\frac{\pi}{8}$

29. $\tan 37^\circ=\frac{a}{250}$
$a=250\tan 37^\circ$
$a\approx 250(0.7536)\approx 188$ cm

30. $\tan 61^\circ=\frac{a}{10}$
$a=10\tan 61^\circ$
$a\approx 10(1.8040)\approx 18$ cm

31. $\cos 34^\circ=\frac{b}{220}$
$b=220\cos 34^\circ$
$b\approx 220(0.8290)\approx 182$ in.

32. $\sin 34^\circ=\frac{a}{13}$
$a=13\sin 34^\circ$
$a\approx 13(0.5592)\approx 7$ m

33. $\sin 23^\circ=\frac{16}{c}$
$c=\frac{16}{\sin 23^\circ}\approx\frac{16}{0.3907}\approx 41$ m

34. $\tan 44^\circ=\frac{23}{b}$
$b=\frac{23}{\tan 44^\circ}\approx\frac{23}{0.9657}\approx 24$ yd

35.

Scientific Calculator	Graphing Calculator	Display (rounded to the nearest degree)
.2974 [SIN^{-1}]	[SIN^{-1}] .2974 [ENTER]	17

If $\sin\theta=0.2974$, then $\theta\approx 17^\circ$.

36.

Scientific Calculator	Graphing Calculator	Display (rounded to the nearest degree)
.877 [COS^{-1}]	[COS^{-1}] .877 [ENTER]	29

If $\cos\theta=0.877$, then $\theta\approx 29^\circ$.

37.

Scientific Calculator	Graphing Calculator	Display (rounded to the nearest degree)
4.6252 [TAN⁻¹]	[TAN⁻¹] 4.6252 [ENTER]	78

If $\tan\theta = 4.6252$, then $\theta \approx 78°$.

38.

Scientific Calculator	Graphing Calculator	Display (rounded to the nearest degree)
26.0307 [TAN⁻¹]	[TAN⁻¹] 26.0307 [ENTER]	88

If $\tan\theta = 26.0307$, then $\theta \approx 88°$.

39.

Scientific Calculator	Graphing Calculator	Display (rounded to three places)
.4112 [COS⁻¹]	[COS⁻¹] .4112 [ENTER]	1.147

If $\cos\theta = 0.4112$, then $\theta \approx 1.147$ radians.

40.

Scientific Calculator	Graphing Calculator	Display (rounded to three places)
.9499 [SIN⁻¹]	[SIN⁻¹] .9499 [ENTER]	1.253

If $\sin\theta = 0.9499$, then $\theta = 1.253$ radians.

41.

Scientific Calculator	Graphing Calculator	Display (rounded to three places)
.4169 [TAN⁻¹]	[TAN⁻¹] .4169 [ENTER]	.395

If $\tan\theta = 0.4169$, then $\theta \approx 0.395$ radians.

42.

Scientific Calculator	Graphing Calculator	Display (rounded to three places)
.5117 [TAN⁻¹]	[TAN⁻¹] .5117 [ENTER]	.473

If $\tan\theta = 0.5117$, then $\theta = 0.473$

43. $\dfrac{\tan\frac{\pi}{3}}{2} - \dfrac{1}{\sec\frac{\pi}{6}} = \dfrac{\sqrt{3}}{2} - \dfrac{1}{\frac{1}{\cos\frac{\pi}{6}}}$

$= \dfrac{\sqrt{3}}{2} - \dfrac{1}{\frac{1}{\frac{\sqrt{3}}{2}}}$

$= \dfrac{\sqrt{3}}{2} - \dfrac{\sqrt{3}}{2}$

$= 0$

44. $\dfrac{1}{\cot\frac{\pi}{4}} - \dfrac{2}{\csc\frac{\pi}{6}} = \dfrac{1}{\frac{1}{\tan\frac{\pi}{4}}} - \dfrac{2}{\frac{1}{\sin\frac{\pi}{6}}}$

$= \dfrac{1}{\frac{1}{1}} - \dfrac{2}{\frac{1}{1/2}}$

$= \dfrac{1}{1} - \dfrac{2}{2}$

$= 1 - 1$

$= 0$

45. $1 + \sin^2 40° + \sin^2 50°$

$= 1 + \sin^2(90° - 50°) + \sin^2 50°$

$= 1 + \cos^2 50° + \sin^2 50°$

$= 1 + 1$

$= 2$

46. $1 - \tan^2 10° + \csc^2 80°$
$= 1 - \cot^2 80° + \csc^2 80°$
$= 1 + \csc^2 80° - \cot^2 80°$
$= 1 + 1$
$= 2$

47. $\csc 37° \sec 53° - \tan 53° \cot 37°$
$= \sec 53° \sec 53° - \tan 53° \tan 53°$
$= \sec^2 53° - \tan^2 53°$
$= 1$

48. $\cos 12° \sin 78° + \cos 78° \sin 12°$
$= \sin 78° \sin 78° + \cos 78° \cos 78°$
$= \sin^2 78° + \cos^2 78°$
$= 1$

49. $f(\theta) = 2\cos\theta - \cos 2\theta$

$$f\left(\frac{\pi}{6}\right) = 2\cos\frac{\pi}{6} - \cos\left(2\cdot\frac{\pi}{6}\right)$$
$$= 2\left(\frac{\sqrt{3}}{2}\right) - \cos\left(\frac{\pi}{3}\right)$$
$$= \frac{2\sqrt{3}}{2} - \frac{1}{2}$$
$$= \frac{2\sqrt{3} - 1}{2}$$

50. $f(\theta) = 2\sin\theta - \sin\frac{\theta}{2}$

$$f\left(\frac{\pi}{3}\right) = 2\sin\frac{\pi}{3} - \sin\frac{\frac{\pi}{3}}{2}$$
$$= 2\left(\frac{\sqrt{3}}{2}\right) - \sin\left(\frac{\pi}{6}\right)$$
$$= \frac{2\sqrt{3}}{2} - \frac{1}{2}$$
$$= \frac{2\sqrt{3} - 1}{2}$$

51. $\tan\left(\frac{\pi}{2} - \theta\right) = \cot\theta = \frac{1}{4}$

52. $\csc\left(\frac{\pi}{2} - \theta\right) = \sec\theta = \frac{1}{\cos\theta} = \frac{1}{\frac{1}{3}} = 3$

53. $\tan 40° = \frac{a}{630}$
$a = 630 \tan 40°$
$a \approx 630(0.8391) \approx 529$
The distance across the lake is approximately 529 yards.

54. $\tan 40° = \frac{h}{35}$
$h = 35 \tan 40°$
$h \approx 35(0.8391) \approx 29$
The tree's height is approximately 29 feet.

55. $\tan\theta = \frac{125}{172}$

Use a calculator in degree mode to find θ.

Many Scientific Calculators	Many Graphing Calculators
125 [÷] 172 [=] [TAN⁻¹]	[TAN⁻¹] [(] 125 [÷] 172 [)] [ENTER]

The display should show approximately 36. Thus, the angle of elevation of the sun is approximately 36°.

56. $\tan\frac{555}{1320}$

Use a calculator in degree mode to find θ.

Many Scientific Calculators	Many Graphing Calculators
555 ÷ 1320 [=] [TAN⁻¹]	[TAN⁻¹] [(] 555 ÷ 1320 [)] [ENTER]

The display should show approximately 23. Thus, the angle of elevation is approximately 23°.

57. $\sin 10° = \dfrac{500}{c}$

$c = \dfrac{500}{\sin 10°} \approx \dfrac{500}{0.1736} \approx 2880$

The plane has flown approximately 2880 feet.

58. $\sin 5° = \dfrac{a}{5000}$

$a = 5000 \sin 5° \approx 5000(0.0872) = 436$

The driver's increase in altitude was approximately 436 feet.

59. $\cos\theta = \dfrac{60}{75}$

Use a calculator in degree mode to find θ.

Many Scientific Calculators	Many Graphing Calculators
60 [÷] 75 [=] [COS^{-1}]	[COS^{-1}] [(] 60 [÷] 75 [)] [ENTER]

The display should show approximately 37. Thus, the angle between the wire and the pole is approximately 37°.

60. $\cos\theta = \dfrac{55}{80}$

Use a calculator in degree mode to find θ.

Many Scientific Calculators	Many Graphing Calculators
55 ÷ 80 [=] [COS^{-1}]	[COS^{-1}] [(] 55 ÷ 80 [)] [ENTER]

The display should show approximately 47. Thus, the angle between the wire and the pole is approximately 47°.

68.

θ	0.4	0.3	0.2	0.1	0.01	0.001	0.0001	0.00001
$\sin\theta$	0.3894	0.2955	0.1987	0.0998	0.0099998	9.999998×10^{-4}	9.99999998×10^{-5}	1×10^{-5}
$\dfrac{\sin\theta}{\theta}$	0.9736	0.9851	0.9933	0.9983	0.99998	0.9999998	0.999999998	1

$\dfrac{\sin\theta}{\theta}$ approaches 1 as θ approaches 0.

69.

θ	0.4	0.3	0.2	0.1	0.01	0.001	0.0001	0.00001
$\cos\theta$	0.92106	0.95534	0.98007	0.99500	0.99995	0.9999995	0.999999995	1
$\dfrac{\cos\theta-1}{\theta}$	−0.19735	−0.148878	−0.099667	−0.04996	−0.005	−0.0005	−0.00005	0

$\dfrac{\cos\theta-1}{\theta}$ approaches 0 as θ approaches 0.

70. **a.** False;

$\frac{\tan 45°}{\tan 15°} \approx \frac{1}{0.2679} \approx 3.7327$

$\tan 3° \approx 0.0524$

b. True;

$1+\tan^2\theta = \sec^2\theta$ for any θ so $1+\tan^2\theta - \sec^2\theta = 0$ and $\tan^2\theta - \sec^2\theta = -1$

c. False;

$\sin 45° + \cos 45° = \frac{1}{\sqrt{2}} + \frac{1}{\sqrt{2}} = \frac{2}{\sqrt{2}} \neq 1$

d. False;

$\tan^2 5° \approx (0.0875)^2 \approx 0.0077$

$\tan 25° \approx 0.4663$

(b) is true.

71. In a right triangle, the hypotenuse is greater than either other side. Therefore both $\frac{\text{opposite}}{\text{hypotenuse}}$ and $\frac{\text{adjacent}}{\text{hypotenuse}}$ must be less than 1 for an acute angle in a right triangle.

72. Use a calculator in degree mode to generate the following table. Then use the table to describe what happens to the tangent of an acute angle as the angle gets close to 90°.

θ	60	70	80	89	89.9	89.99	89.999	89.9999
$\tan\theta$	1.7321	2.7475	5.6713	57	573	5730	57,296	572,958

As θ approaches 90°, $\tan\theta$ increases without bound. At 90°, $\tan\theta$ is undefined.

73. **a.** Let a = distance of the ship from the lighthouse.

$\tan 35° = \frac{250}{a}$

$a = \frac{250}{\tan 35°} \approx \frac{250}{0.7002} \approx 357$

The ship is approximately 357 feet from the lighthouse.

b. Let b = the plane's height above the lighthouse.

$\tan 22° = \frac{b}{357}$

$b = 357\tan 22° \approx 357(0.4040) \approx 144$

$144 + 250 = 394$

The plane is approximately 394 feet above the water.

Section 4.4

Checkpoint Exercises

1. $r=\sqrt{x^2+y^2}$

$r=\sqrt{1^2+(-3)^2}=\sqrt{1+9}=\sqrt{10}$

Now that we know *x, y,* and *r,* we can find the six trigonometric functions of θ.

$\sin\theta=\frac{y}{r}=\frac{-3}{\sqrt{10}}=-\frac{3\sqrt{10}}{10}$

$\cos\theta=\frac{x}{r}=\frac{1}{\sqrt{10}}=\frac{\sqrt{10}}{10}$

$\tan\theta=\frac{y}{x}=\frac{-3}{1}=-3$

$\csc\theta=\frac{r}{y}=\frac{\sqrt{10}}{-3}=-\frac{\sqrt{10}}{3}$

$\sec\theta=\frac{r}{x}=\frac{\sqrt{10}}{1}=\sqrt{10}$

$\cot\theta=\frac{x}{y}=\frac{1}{-3}=-\frac{1}{3}$

2. **a.** $\theta=0°=0$ radians

The terminal side of the angle is on the positive *x*-axis. Select the point $P=(1,0)$: $x=1, y=0, r=1$

Apply the definitions of the cosine and cosecant functions.

$\cos 0°=\cos 0=\frac{x}{r}=\frac{1}{1}=1$

$\csc 0°=\csc 0=\frac{r}{y}=\frac{1}{0}$, undefined

b. $\theta=90°=\frac{\pi}{2}$ radians

The terminal side of the angle is on the positive *y*-axis. Select the point $P=(0,1)$: $x=0, y=1, r=1$

Apply the definitions of the cosine and cosecant functions.

$\cos 90°=\cos\frac{\pi}{2}=\frac{x}{r}=\frac{0}{1}=0$

$\csc 90°=\csc\frac{\pi}{2}=\frac{r}{y}=\frac{1}{1}=1$

c. $\theta=180°=\pi$ radians

The terminal side of the angle is on the negative *x*-axis. Select the point $P=(-1,0)$: $x=-1, y=0, r=1$

Apply the definitions of the cosine and cosecant functions.

$\cos 180°=\cos\pi=\frac{x}{r}=\frac{-1}{1}=-1$

$\csc 180°=\csc\pi=\frac{r}{y}=\frac{1}{0}$, undefined

d. $\theta=270°=\frac{3\pi}{2}$ radians

The terminal side of the angle is on the negative *y*-axis. Select the point $P=(0,-1)$: $x=0, y=-1, r=1$

Apply the definitions of the cosine and cosecant functions.

$\cos 270°=\cos\frac{3\pi}{2}=\frac{x}{r}=\frac{0}{1}=0$

$\csc 270°=\csc\frac{3\pi}{2}=\frac{r}{y}=\frac{1}{-1}=-1$

3. Because $\sin\theta<0$, θ cannot lie in quadrant I; all the functions are positive in quadrant I. Furthermore, θ cannot lie in quadrant II; $\sin\theta$ is positive in quadrant II. Thus, with $\sin\theta<0$, θ lies in quadrant III or quadrant IV. We are also given that $\cos\theta<0$. Because quadrant III is the only quadrant in which cosine is negative and the sine is negative, we conclude that θ lies in quadrant III.

4. Because the tangent is negative and the cosine is negative, θ lies in quadrant II. In quadrant II, *x* is negative and *y* is positive. Thus,

$\tan\theta=-\frac{1}{3}=\frac{y}{x}=\frac{1}{-3}$

$x=-3,\ y=1$

Furthermore,

$r=\sqrt{x^2+y^2}=\sqrt{(-3)^2+1^2}=\sqrt{9+1}=\sqrt{10}$

Now that we know *x, y,* and *r,* we can find $\sin\theta$ and $\sec\theta$.

$\sin\theta=\frac{y}{r}=\frac{1}{\sqrt{10}}=\frac{1}{\sqrt{10}}\cdot\frac{\sqrt{10}}{\sqrt{10}}=\frac{\sqrt{10}}{10}$

$\sec\theta=\frac{r}{x}=\frac{\sqrt{10}}{-3}=-\frac{\sqrt{10}}{3}$

5. a. Because 210° lies between 180° and 270°, it is in quadrant III. The reference angle is $\theta' = 210° - 180° = 30°$.

b. Because $\frac{7\pi}{4}$ lies between $\frac{3\pi}{2} = \frac{6\pi}{4}$ and $2\pi = \frac{8\pi}{4}$, it is in quadrant IV. The reference angle is
$\theta' = 2\pi - \frac{7\pi}{4} = \frac{8\pi}{4} - \frac{7\pi}{4} = \frac{\pi}{4}$.

c. Because –240° lies between –180° and –270°, it is in quadrant II. The reference angle is $\theta = 240 - 180 = 60°$.

d. Because 3.6 lies between $\pi \approx 3.14$ and $\frac{3\pi}{2} \approx 4.71$, it is in quadrant III. The reference angle is $\theta' = 3.6 - \pi \approx 0.46$.

6. a. $665° - 360° = 305°$
This angle is in quadrant IV, thus the reference angle is $\theta' = 360° - 305° = 55°$.

b. $\frac{15\pi}{4} - 2\pi = \frac{15\pi}{4} - \frac{8\pi}{4} = \frac{7\pi}{4}$
This angle is in quadrant IV, thus the reference angle is
$\theta' = 2\pi - \frac{7\pi}{4} = \frac{8\pi}{4} - \frac{7\pi}{4} = \frac{\pi}{4}$.

c. $-\frac{11\pi}{3} + 2 \cdot 2\pi = -\frac{11\pi}{3} + \frac{12\pi}{3} = \frac{\pi}{3}$
This angle is in quadrant I, thus the reference angle is $\theta' = \frac{\pi}{3}$.

7. a. 300° lies in quadrant IV. The reference angle is $\theta' = 360° - 300° = 60°$.
$\sin 60° = \frac{\sqrt{3}}{2}$
Because the sine is negative in quadrant IV, $\sin 300° = -\sin 60° = -\frac{\sqrt{3}}{2}$.

b. $\frac{5\pi}{4}$ lies in quadrant III. The reference angle is $\theta' = \frac{5\pi}{4} - \pi = \frac{5\pi}{4} - \frac{4\pi}{4} = \frac{\pi}{4}$.

$\tan\frac{\pi}{4} = 1$

Because the tangent is positive in quadrant III, $\tan\frac{5\pi}{4} = +\tan\frac{\pi}{4} = 1$.

c. $-\frac{\pi}{6}$ lies in quadrant IV. The reference angle is $\theta' = \frac{\pi}{6}$.

$\sec\frac{\pi}{6} = \frac{2\sqrt{3}}{3}$
Because the secant is positive in quadrant IV, $\sec\left(-\frac{\pi}{6}\right) = +\sec\frac{\pi}{6} = \frac{2\sqrt{3}}{3}$.

8. a. $\frac{17\pi}{6} - 2\pi = \frac{17\pi}{6} - \frac{12\pi}{6} = \frac{5\pi}{6}$ lies in quadrant II. The reference angle is
$\theta' = \pi - \frac{5\pi}{6} = \frac{\pi}{6}$.
The function value for the reference angle is $\cos\frac{\pi}{6} = \frac{\sqrt{3}}{2}$.
Because the cosine is negative in quadrant II, $\cos\frac{17\pi}{6} = \cos\frac{5\pi}{6} = -\cos\frac{\pi}{6} = -\frac{\sqrt{3}}{2}$.

b. $\frac{-22\pi}{3} + 8\pi = \frac{-22\pi}{3} + \frac{24\pi}{3} = \frac{2\pi}{3}$ lies in quadrant II. The reference angle is
$\theta' = \pi - \frac{2\pi}{3} = \frac{\pi}{3}$.
The function value for the reference angle is $\sin\frac{\pi}{3} = \frac{\sqrt{3}}{2}$.
Because the sine is positive in quadrant II, $\sin\frac{-22\pi}{3} = \sin\frac{2\pi}{3} = \sin\frac{\pi}{3} = \frac{\sqrt{3}}{2}$.

Exercise Set 4.4

1. We need values for x, y, and r. Because $P = (-4, 3)$ is a point on the terminal side of θ, $x = -4$ and $y = 3$. Furthermore,

$$r = \sqrt{x^2 + y^2} = \sqrt{(-4)^2 + 3^2} = \sqrt{16+9} = \sqrt{25} = 5$$

Now that we know x, y, and r, we can find the six trigonometric functions of θ.

$$\sin\theta = \frac{y}{r} = \frac{3}{5}$$

$$\cos\theta = \frac{x}{r} = \frac{-4}{5} = -\frac{4}{5}$$

$$\tan\theta = \frac{y}{x} = \frac{3}{-4} = -\frac{3}{4}$$

$$\csc\theta = \frac{r}{y} = \frac{5}{3}$$

$$\sec\theta = \frac{r}{x} = \frac{5}{-4} = -\frac{5}{4}$$

$$\cot\theta = \frac{x}{y} = \frac{-4}{3} = -\frac{4}{3}$$

2. We need values for x, y, and r. Because $P = (-12, 5)$ is a point on the terminal side of θ, $x = -12$ and $y = 5$. Furthermore,

$$r = \sqrt{x^2 + y^2} = \sqrt{(-12)^2 + 5^2} = \sqrt{144+25}$$
$$= \sqrt{169} = 13$$

Now that we know x, y, and r, we can find the six trigonometric functions of θ.

$$\sin\theta = \frac{y}{r} = \frac{5}{13}$$

$$\cos\theta = \frac{x}{r} = \frac{-12}{13} = -\frac{12}{13}$$

$$\tan\theta = \frac{y}{x} = \frac{5}{-12} = -\frac{5}{12}$$

$$\csc\theta = \frac{r}{y} = \frac{13}{5}$$

$$\sec\theta = \frac{r}{x} = \frac{13}{-12} = -\frac{13}{12}$$

$$\cot\theta = \frac{x}{y} = \frac{-12}{5} - \frac{12}{5}$$

3. We need values for x, y, and r. Because $P = (2, 3)$ is a point on the terminal side of θ, $x = 2$ and $y = 3$. Furthermore,

$$r = \sqrt{x^2 + y^2} = \sqrt{2^2 + 3^2} = \sqrt{4+9} = \sqrt{13}$$

Now that we know x, y, and r, we can find the six trigonometric functions of θ.

$$\sin\theta = \frac{y}{r} = \frac{3}{\sqrt{13}} = \frac{3}{\sqrt{13}} \cdot \frac{\sqrt{13}}{\sqrt{13}} = \frac{3\sqrt{13}}{13}$$

$$\cos\theta = \frac{x}{r} = \frac{2}{\sqrt{13}} = \frac{2}{\sqrt{13}} \cdot \frac{\sqrt{13}}{\sqrt{13}} = \frac{2\sqrt{13}}{13}$$

$$\tan\theta = \frac{y}{x} = \frac{3}{2}$$

$$\csc\theta = \frac{r}{y} = \frac{\sqrt{13}}{3}$$

$$\sec\theta = \frac{r}{x} = \frac{\sqrt{13}}{2}$$

$$\cot\theta = \frac{x}{y} = \frac{2}{3}$$

4. We need values for x, y, and r. Because $P = (3, 7)$ is a point on the terminal side of θ, $x = 3$ and $y = 7$. Furthermore,

$$r = \sqrt{x^2 + y^2} = \sqrt{3^2 + 7^2} = \sqrt{9+49} = \sqrt{58}$$

Now that we know x, y, and r, we can find the six trigonometric functions of θ.

$$\sin\theta = \frac{y}{r} = \frac{7}{\sqrt{58}} = \frac{7}{\sqrt{58}} \cdot \frac{\sqrt{58}}{\sqrt{58}} = \frac{7\sqrt{58}}{58}$$

$$\cos\theta = \frac{x}{r} = \frac{3}{\sqrt{58}} = \frac{3}{\sqrt{58}} \cdot \frac{\sqrt{58}}{\sqrt{58}} = \frac{3\sqrt{58}}{58}$$

$$\tan\theta = \frac{y}{x} = \frac{7}{3}$$

$$\csc\theta = \frac{r}{y} = \frac{\sqrt{58}}{7}$$

$$\sec\theta = \frac{r}{x} = \frac{\sqrt{58}}{3}$$

$$\cot\theta = \frac{x}{y} = \frac{3}{7}$$

5. We need values for *x, y,* and *r*. Because $P = (3, -3)$ is a point on the terminal side of θ, $x = 3$ and $y = -3$. Furthermore,

$$r = \sqrt{x^2 + y^2} = \sqrt{3^2 + (-3)^2} = \sqrt{9+9}$$
$$= \sqrt{18} = 3\sqrt{2}$$

Now that we know *x, y,* and *r,* we can find the six trigonometric functions of θ.

$$\sin\theta = \frac{y}{r} = \frac{-3}{3\sqrt{2}} = -\frac{-1}{\sqrt{2}} \cdot \frac{\sqrt{2}}{\sqrt{2}} = -\frac{\sqrt{2}}{2}$$

$$\cos\theta = \frac{x}{r} = \frac{3}{3\sqrt{2}} = \frac{1}{\sqrt{2}} \cdot \frac{\sqrt{2}}{\sqrt{2}} = \frac{\sqrt{2}}{2}$$

$$\tan\theta = \frac{y}{x} = \frac{-3}{3} = -1$$

$$\csc\theta = \frac{r}{y} = \frac{3\sqrt{2}}{-3} = -\sqrt{2}$$

$$\sec\theta = \frac{r}{x} = \frac{3\sqrt{2}}{3} = \sqrt{2}$$

$$\cot\theta = \frac{x}{y} = \frac{3}{-3} = -1$$

6. We need values for *x, y,* and *r,* Because $P = (5, -5)$ is a point on the terminal side of θ, $x = 5$ and $y = -5$. Furthermore,

$$r = \sqrt{x^2 + y^2} = \sqrt{5 + (-5)^2} = \sqrt{25+25} = \sqrt{50}$$
$$= 5\sqrt{2}$$

Now that we know *x, y,* and *r,* we can find the six trigonometric functions of θ.

$$\sin\theta = \frac{y}{r} = \frac{-5}{5\sqrt{2}} = \frac{-1}{\sqrt{2}} \cdot \frac{\sqrt{2}}{\sqrt{2}} = -\frac{\sqrt{2}}{2}$$

$$\cos\theta = \frac{x}{r} = \frac{5}{5\sqrt{2}} = \frac{1}{\sqrt{2}} \cdot \frac{\sqrt{2}}{\sqrt{2}} = \frac{\sqrt{2}}{2}$$

$$\tan\theta = \frac{y}{x} = \frac{-5}{5} = -1$$

$$\csc\theta = \frac{r}{y} = \frac{5\sqrt{2}}{-5} = -\sqrt{2}$$

$$\sec\theta = \frac{r}{x} = \frac{5\sqrt{2}}{5} = \sqrt{2}$$

$$\cot\theta = \frac{x}{y} = \frac{5}{-5} = -1$$

7. We need values for *x, y,* and *r*. Because $P = (-2, -5)$ is a point on the terminal side of θ, $x = -2$ and $y = -5$. Furthermore,

$$r = \sqrt{x^2 + y^2} = \sqrt{(-2)^2 + (-5)^2} = \sqrt{4+25} = \sqrt{29}$$

Now that we know *x, y,* and *r,* we can find the six trigonometric functions of θ.

$$\sin\theta = \frac{y}{r} = \frac{-5}{\sqrt{29}} = \frac{-5}{\sqrt{29}} \cdot \frac{\sqrt{29}}{\sqrt{29}} = -\frac{5\sqrt{29}}{29}$$

$$\cos\theta = \frac{x}{r} = \frac{-2}{\sqrt{29}} = \frac{-2}{\sqrt{29}} \cdot \frac{\sqrt{29}}{\sqrt{29}} = -\frac{2\sqrt{29}}{29}$$

$$\tan\theta = \frac{y}{x} = \frac{-5}{-2} = \frac{5}{2}$$

$$\csc\theta = \frac{r}{y} = \frac{\sqrt{29}}{-5} = -\frac{\sqrt{29}}{5}$$

$$\sec\theta = \frac{r}{x} = \frac{\sqrt{29}}{-2} = -\frac{\sqrt{29}}{2}$$

$$\cot\theta = \frac{x}{y} = \frac{-2}{-5} = \frac{2}{5}$$

8. We need values for *x, y,* and *r,* Because $P = (-1, -3)$ is a point on the terminal side of θ, $x = -1$ and $y = -3$. Furthermore,

$$r = \sqrt{x^2 + y^2} = \sqrt{(-1)^2 + (-3)^2} = \sqrt{1+9} = \sqrt{10}$$

Now that we know *x, y,* and *r,* we can find the six trigonometric functions of θ.

$$\sin\theta = \frac{y}{r} = \frac{-3}{\sqrt{10}} = \frac{-3}{\sqrt{10}} \cdot \frac{\sqrt{10}}{\sqrt{10}} = -\frac{3\sqrt{10}}{10}$$

$$\cos\theta = \frac{x}{r} = \frac{-1}{\sqrt{10}} = \frac{-1}{\sqrt{10}} \cdot \frac{\sqrt{10}}{\sqrt{10}} = -\frac{\sqrt{10}}{10}$$

$$\tan\theta = \frac{y}{x} = \frac{-3}{-1} = 3$$

$$\csc\theta = \frac{r}{y} = \frac{\sqrt{10}}{-3} = -\frac{\sqrt{10}}{3}$$

$$\sec\theta = \frac{r}{x} = \frac{\sqrt{10}}{-1} = -\sqrt{10}$$

$$\cot\theta = \frac{x}{y} = \frac{-1}{-3} = \frac{1}{3}$$

9. $\theta = \pi$ radians

The terminal side of the angle is on the negative *x*-axis. Select the point $P = (-1, 0)$:

$x = -1$, $y = 0$, $r = 1$ Apply the definition of the cosine function.

$$\cos\pi = \frac{x}{r} = \frac{-1}{1} = -1$$

10. $\theta = \pi$ radians
The terminal side of the angle is on the negative x-axis. Select the point $P = (-1, 0)$:
$x = -1, y = 0, r = 1$
Apply the definition of the tangent function.
$\tan \pi = \frac{y}{x} = \frac{0}{-1} = 0$

11. $\theta = \pi$ radians
The terminal side of the angle is on the negative x-axis. Select the point $P = (-1, 0)$:
$x = -1,\ y = 0,\ r = 1$ Apply the definition of the secant function.
$\sec \pi = \frac{r}{x} = \frac{1}{-1} = -1$

12. $\theta = \pi$ radians
The terminal side of the angle is on the negative x-axis. Select the point $P = (-1, 0)$:
$x = -1, y = 0, r = 1$
Apply the definition of the cosecant function.
$\csc \pi = \frac{r}{y} = \frac{1}{0}$, undefined

13. $\theta = \frac{3\pi}{2}$ radians
The terminal side of the angle is on the negative y-axis. Select the point $P = (0, -1)$:
$x = 0,\ y = -1,\ r = 1$ Apply the definition of the tangent function. $\tan \frac{3\pi}{2} = \frac{y}{x} = \frac{-1}{0}$, undefined

14. $\theta = \frac{3\pi}{2}$ radians
The terminal side of the angle is on the negative y-axis. Select the point $P = (0, -1)$:
$x = 0, y = -1, r = 1$
Apply the definition of the cosine function.
$\cos \frac{3\pi}{2} = \frac{x}{r} = \frac{0}{1} = 0$

15. $\theta = \frac{\pi}{2}$ radians
The terminal side of the angle is on the positive y-axis. Select the point $P = (0, 1)$:
$x = 0,\ y = 1,\ r = 1$ Apply the definition of the cotangent function. $\cot \frac{\pi}{2} = \frac{x}{y} = \frac{0}{1} = 0$

16. $\theta = \frac{\pi}{2}$ radians
The terminal side of the angle is on the positive y-axis. Select the point $P = (0, 1)$:
$x = 0, y = 1, r = 1$
Apply the definition of the tangent function.
$\tan \frac{\pi}{2} = \frac{y}{x} = \frac{1}{0}$, undefined

17. Because $\sin \theta > 0$, θ cannot lie in quadrant III or quadrant IV; the sine function is negative in those quadrants. Thus, with $\sin \theta > 0$, θ lies in quadrant I or quadrant II. We are also given that $\cos \theta > 0$. Because quadrant I is the only quadrant in which the cosine is positive and sine is positive, we conclude that θ lies in quadrant I.

18. Because $\sin \theta < 0$, θ cannot lie in quadrant I or quadrant II; the sine function is positive in those two quadrants. Thus, with $\sin \theta < 0$, θ lies in quadrant III or quadrant IV. We are also given that $\cos \theta > 0$. Because quadrant IV is the only quadrant in which the cosine is positive and the sine is negative, we conclude that θ lies in quadrant IV.

19. Because $\sin \theta < 0$, θ cannot lie in quadrant I or quadrant II; the sine function is positive in those two quadrants. Thus, with $\sin \theta < 0$, θ lies in quadrant III or quadrant IV. We are also given that $\cos \theta < 0$. Because quadrant III is the only quadrant in which the cosine is positive and the sine is negative, we conclude that θ lies in quadrant III.

20. Because $\tan \theta < 0$, θ cannot lie in quadrant I or quadrant III; the tangent function is positive in those two quadrants. Thus, with $\tan \theta < 0$, θ lies in quadrant II or quadrant IV. We are also given that $\sin \theta < 0$. Because quadrant IV is the only quadrant in which the sine is negative and the tangent is negative, we conclude that θ lies in quadrant IV.

21. Because $\tan \theta < 0$, θ cannot lie in quadrant I or quadrant III; the tangent function is positive in those quadrants. Thus, with $\tan \theta < 0$, θ lies in quadrant II or quadrant IV. We are also given that $\cos \theta < 0$. Because quadrant II is the only quadrant in which the cosine is negative and the tangent is negative, we conclude that θ lies in quadrant II.

22. Because $\cot\theta > 0$, θ cannot lie in quadrant II or quadrant IV; the cotangent function is negative in those two quadrants. Thus, with $\cot\theta > 0$, θ lies in quadrant I or quadrant III. We are also given that $\sec\theta < 0$. Because quadrant III is the only quadrant in which the secant is negative and the cotangent is positive, we conclude that θ lies in quadrant III.

23. In quadrant III x is negative and y is negative.

Thus, $\cos\theta = -\frac{3}{5} = \frac{x}{r} = \frac{-3}{5}$, $x = -3$, $r = 5$.

Furthermore,

$r^2 = x^2 + y^2$

$5^2 = (-3)^2 + y^2$

$y^2 = 25 - 9 = 16$

$y = -\sqrt{16} = -4$

Now that we know x, y, and r, we can find the remaining trigonometric functions of θ.

$\sin\theta = \frac{y}{r} = \frac{-4}{5} = -\frac{4}{5}$

$\tan\theta = \frac{y}{x} = \frac{-4}{-3} = \frac{4}{3}$

$\csc\theta = \frac{r}{y} = \frac{5}{-4} = -\frac{5}{4}$

$\sec\theta = \frac{r}{x} = \frac{5}{-3} = -\frac{5}{3}$

$\cot\theta = \frac{x}{y} = \frac{-3}{-4} = \frac{3}{4}$

24. In quadrant III, x is negative and y is negative.

Thus, $\sin\theta = -\frac{12}{13} = \frac{y}{r} = \frac{-12}{13}$, $y = -12$, $r = 13$.

Furthermore,

$x^2 + y^2 = r^2$

$x^2 + (-12)^2 = 13^2$

$x^2 = 169 - 144 = 25$

$x = -\sqrt{25} = -5$

Now that we know x, y, and r, we can find the remaining trigonometric functions of θ.

$\cos\theta = \frac{x}{r} = \frac{-5}{13} = -\frac{5}{13}$

$\tan\theta = \frac{y}{x} = \frac{-12}{-5} = \frac{12}{5}$

$\csc\theta = \frac{r}{y} = \frac{13}{-12} = -\frac{13}{12}$

$\sec\theta = \frac{r}{x} = \frac{13}{-5} = -\frac{13}{5}$

$\cot\theta = \frac{x}{y} = \frac{-5}{-12} = \frac{5}{12}$

25. In quadrant II x is negative and y is positive.

Thus, $\sin\theta = \frac{5}{13} = \frac{y}{r}$, $y = 5$, $r = 13$.

Furthermore,

$x^2 + y^2 = r^2$

$x^2 + 5^2 = 13^2$

$x^2 = 169 - 25 = 144$

$x = -\sqrt{144} = -12$

Now that we know x, y, and r, we can find the remaining trigonometric functions of θ.

$\cos\theta = \frac{x}{r} = \frac{-12}{13} = -\frac{12}{13}$

$\tan\theta = \frac{y}{x} = \frac{5}{-12} = -\frac{5}{12}$

$\csc\theta = \frac{r}{y} = \frac{13}{5}$

$\sec\theta = \frac{r}{x} = \frac{13}{-12} = -\frac{13}{12}$

$\cot\theta = \frac{x}{y} = \frac{-12}{5} = -\frac{12}{5}$

26. In quadrant IV, x is positive and y is negative.

Thus, $\cos\theta = \frac{4}{5} = \frac{x}{r}$, $x = 4$, $r = 5$. Furthermore,

$$x^2 + y^2 = r^2$$
$$4^2 + y^2 = 5^2$$
$$y^2 = 25 - 16 = 9$$
$$y = -\sqrt{9} = -3$$

Now that we know x, y, and r, we can find the remaining trigonometric functions of θ.

$$\sin\theta = \frac{y}{r} = \frac{-3}{5} = -\frac{3}{5}$$
$$\tan\theta = \frac{y}{x} = \frac{-3}{4} = -\frac{3}{4}$$
$$\csc\theta = \frac{r}{y} = \frac{5}{-3} = -\frac{5}{3}$$
$$\sec\theta = \frac{r}{x} = \frac{5}{4}$$
$$\cot\theta = \frac{x}{y} = \frac{4}{-3} = -\frac{4}{3}$$

27. Because $270° < \theta < 360°$, θ is in quadrant IV. In quadrant IV x is positive and y is negative.

Thus, $\cos\theta = \frac{8}{17} = \frac{x}{r}$, $x = 8$,

$r = 17$. Furthermore

$$x^2 + y^2 = r^2$$
$$8^2 + y^2 = 17^2$$
$$y^2 = 289 - 64 = 225$$
$$y = -\sqrt{225} = -15$$

Now that we know x, y, and r, we can find the remaining trigonometric functions of θ.

$$\sin\theta = \frac{y}{r} = \frac{-15}{17} = -\frac{15}{17}$$
$$\tan\theta = \frac{y}{x} = \frac{-15}{8} = -\frac{15}{8}$$
$$\csc\theta = \frac{r}{y} = \frac{17}{-15} = -\frac{17}{15}$$
$$\sec\theta = \frac{r}{x} = \frac{17}{8}$$
$$\cot\theta = \frac{x}{y} = \frac{8}{-15} = -\frac{8}{15}$$

28. Because $270° < \theta < 360°$, θ is in quadrant IV. In quadrant IV, x is positive and y is negative.

Thus, $\cos\theta = \frac{1}{3} = \frac{x}{r}$, $x = 1$, $r = 3$. Furthermore,

$$x^2 + y^2 = r^2$$
$$1^2 + y^2 = 3^2$$
$$y^2 = 9 - 1 = 8$$
$$y = -\sqrt{8} = -2\sqrt{2}$$

Now that we know x, y, and r, we can find the remaining trigonometric functions of θ.

$$\sin\theta = \frac{y}{r} = \frac{-2\sqrt{2}}{3} = -\frac{2\sqrt{2}}{3}$$
$$\tan\theta = \frac{y}{x} = \frac{-2\sqrt{2}}{1} = -2\sqrt{2}$$
$$\csc\theta = \frac{r}{y} = \frac{3}{-2\sqrt{2}} = \frac{3}{-2\sqrt{2}} \cdot \frac{\sqrt{2}}{\sqrt{2}} = -\frac{3\sqrt{2}}{4}$$
$$\sec\theta = \frac{r}{x} = \frac{3}{1} = 3$$
$$\cot\theta = \frac{x}{y} = \frac{1}{-2\sqrt{2}} = \frac{1}{-2\sqrt{2}} \cdot \frac{\sqrt{2}}{\sqrt{2}} = -\frac{\sqrt{2}}{4}$$

29. Because the tangent is negative and the sine is positive, θ lies in quadrant II. In quadrant II, x is negative and y is positive. Thus,

$\tan\theta = -\frac{2}{3} = \frac{y}{x} = \frac{2}{-3}$, $x = -3$, $y = 2$.

Furthermore,

$$r = \sqrt{x^2 + y^2} = \sqrt{(-3)^2 + 2^2} = \sqrt{9 + 4} = \sqrt{13}$$

Now that we know x, y, and r, we can find the remaining trigonometric functions of θ.

$$\sin\theta = \frac{y}{r} = \frac{2}{\sqrt{13}} = \frac{2}{\sqrt{13}} \cdot \frac{\sqrt{13}}{\sqrt{13}} = \frac{2\sqrt{13}}{13}$$
$$\cos\theta = \frac{x}{r} = \frac{-3}{\sqrt{13}} = \frac{-3}{\sqrt{13}} \cdot \frac{\sqrt{13}}{\sqrt{13}} = -\frac{3\sqrt{13}}{13}$$
$$\csc\theta = \frac{r}{y} = \frac{\sqrt{13}}{2}$$
$$\sec\theta = \frac{r}{x} = \frac{\sqrt{13}}{-3} = -\frac{\sqrt{13}}{3}$$
$$\cot\theta = \frac{x}{y} = \frac{-3}{2} = -\frac{3}{2}$$

30. Because the tangent is negative and the sine is positive, θ lies in quadrant II. In quadrant II, x is negative and y is positive. Thus,

$\tan\theta = -\frac{1}{3} = \frac{y}{x} = \frac{1}{-3}$, $y = 1$, $x = -3$.

Furthermore,

$r = \sqrt{x^2+y^2} = \sqrt{(-3)^2+1^2} = \sqrt{9+1} = \sqrt{10}$

Now that we know x, y, and r, we can find the remaining trigonometric functions of θ.

$\sin\theta = \frac{y}{r} = \frac{1}{\sqrt{10}} = \frac{1}{\sqrt{10}}\cdot\frac{\sqrt{10}}{\sqrt{10}} = \frac{\sqrt{10}}{10}$

$\cos\theta = \frac{x}{r} = \frac{-3}{\sqrt{10}} = \frac{-3}{\sqrt{10}}\cdot\frac{\sqrt{10}}{\sqrt{10}} = -\frac{3\sqrt{10}}{10}$

$\csc\theta = \frac{r}{y} = \frac{\sqrt{10}}{1} = \sqrt{10}$

$\sec\theta = \frac{r}{x} = \frac{\sqrt{10}}{-3} = -\frac{\sqrt{10}}{3}$

$\cot\theta = \frac{x}{y} = \frac{-3}{1} = -3$

31. Because the tangent is positive and the cosine is negative, θ lies in quadrant III. In quadrant III, x is negative and y is negative. Thus,

$\tan\theta = \frac{4}{3} = \frac{y}{x} = \frac{-4}{-3}$, $x = -3, y = -4$.

Furthermore,

$r = \sqrt{x^2+y^2} = \sqrt{(-3)^2+(-4)^2} = \sqrt{9+16}$
$= \sqrt{25} = 5$

Now that we know x, y, and r, we can find the remaining trigonometric functions of θ.

$\sin\theta = \frac{y}{r} = \frac{-4}{5} = -\frac{4}{5}$

$\cos\theta = \frac{x}{r} = \frac{-3}{5} = -\frac{3}{5}$

$\csc\theta = \frac{r}{y} = \frac{5}{-4} = -\frac{5}{4}$

$\sec\theta = \frac{r}{x} = \frac{5}{-3} = -\frac{5}{3}$

$\cot\theta = \frac{x}{y} = \frac{-3}{-4} = \frac{3}{4}$

32. Because the tangent is positive and the cosine is negative, θ lies in quadrant III. In quadrant III, x is negative and y is negative. Thus,

$\tan\theta = \frac{5}{12} = \frac{y}{x} = \frac{-5}{-12}$, $x = -12$, $y = -5$.

Furthermore,

$r = \sqrt{x^2+y^2} = \sqrt{(-12)^2+(-5)^2} = \sqrt{144+25}$
$= \sqrt{169} = 13$

Now that we know x, y, and r, we can find the remaining trigonometric functions of θ.

$\sin\theta = \frac{y}{r} = \frac{-5}{13} = -\frac{5}{13}$

$\cos\theta = \frac{x}{r} = \frac{-12}{13} = -\frac{12}{13}$

$\csc\theta = \frac{r}{y} = \frac{13}{-5} = -\frac{13}{5}$

$\sec\theta = \frac{r}{x} = \frac{13}{-12} = -\frac{13}{12}$

$\cot\theta = \frac{x}{y} = \frac{-12}{-5} = \frac{12}{5}$

33. Because the secant is negative and the tangent is positive, θ lies in quadrant III. In quadrant III, x is negative and y is negative. Thus,

$\sec\theta = -3 = \frac{r}{x} = \frac{3}{-1}$, $x = -1$, $r = 3$.

Furthermore,

$x^2 + y^2 = r^2$
$(-1)^2 + y^2 = 3^2$
$y^2 = 9 - 1 = 8$
$y = -\sqrt{8} = -2\sqrt{2}$

Now that we know x, y, and r, we can find the remaining trigonometric functions of θ.

$\sin\theta = \frac{y}{r} = \frac{-2\sqrt{2}}{3} = -\frac{2\sqrt{2}}{3}$

$\cos\theta = \frac{x}{r} = \frac{-1}{3} = -\frac{1}{3}$

$\tan\theta = \frac{y}{x} = \frac{-2\sqrt{2}}{-1} = 2\sqrt{2}$

$\csc\theta = \frac{r}{y} = \frac{3}{-2\sqrt{2}} = \frac{3}{-2\sqrt{2}}\cdot\frac{\sqrt{2}}{\sqrt{2}} = -\frac{3\sqrt{2}}{4}$

$\cot\theta = \frac{x}{y} = \frac{-1}{-2\sqrt{2}} = \frac{1}{2\sqrt{2}}\cdot\frac{\sqrt{2}}{\sqrt{2}} = \frac{\sqrt{2}}{4}$

34. Because the cosecant is negative and the tangent is positive, θ lies in quadrant III. In quadrant III, x is negative and y is negative. Thus,

$\csc\theta = -4 = \frac{r}{y} = \frac{4}{-1}$, $y = -1$, $r = 4$.

Furthermore,

$$x^2 + y^2 = r^2$$
$$x^2 + (-1)^2 = 4^2$$
$$x^2 = 16 - 1 = 15$$
$$x = -\sqrt{15}$$

Now that we know x, y, and r, we can find the remaining trigonometric functions of θ.

$$\sin\theta = \frac{y}{r} = \frac{-1}{4} = -\frac{1}{4}$$
$$\cos\theta = \frac{x}{r} = \frac{-\sqrt{15}}{4} = -\frac{\sqrt{15}}{4}$$
$$\tan\theta = \frac{y}{x} = \frac{-1}{-\sqrt{15}} = \frac{1}{\sqrt{15}} \cdot \frac{\sqrt{15}}{\sqrt{15}} = \frac{\sqrt{15}}{15}$$
$$\sec\theta = \frac{r}{x} = \frac{4}{-\sqrt{15}} = -\frac{4}{\sqrt{15}} \cdot \frac{\sqrt{15}}{\sqrt{15}} = -\frac{4\sqrt{15}}{15}$$
$$\cot\theta = \frac{x}{y} = \frac{-\sqrt{15}}{-1} = \sqrt{15}$$

35. Because 160° lies between 90° and 180°, it is in quadrant II. The reference angle is $\theta' = 180° - 160° = 20°$.

36. Because 170° lies between 90° and 180°, it is in quadrant II. The reference angle is $\theta' = 180° - 170° = 10°$.

37. Because 205° lies between 180° and 270°, it is in quadrant III. The reference angle is $\theta' = 205° - 180° = 25°$.

38. Because 210° lies between 180° and 270°, it is in quadrant III. The reference angle is $\theta' = 210° - 180° = 30°$.

39. Because 355° lies between 270° and 360°, it is in quadrant IV. The reference angle is $\theta' = 360° - 355° = 5°$.

40. Because 351° lies between 270° and 360°, it is in quadrant IV. The reference angle is $\theta' = 360° - 351° = 9°$.

41. Because $\frac{7\pi}{4}$ lies between $\frac{3\pi}{2} = \frac{6\pi}{4}$ and $2\pi = \frac{8\pi}{4}$, it is in quadrant IV. The reference angle is $\theta' = 2\pi - \frac{7\pi}{4} = \frac{8\pi}{4} - \frac{7\pi}{4} = \frac{\pi}{4}$.

42. Because $\frac{5\pi}{4}$ lies between $\pi = \frac{4\pi}{4}$ and $\frac{3\pi}{2} = \frac{6\pi}{4}$, it is in quadrant III. The reference angle is $\theta' = \frac{5\pi}{4} - \pi = \frac{5\pi}{4} - \frac{4\pi}{4} = \frac{\pi}{4}$.

43. Because $\frac{5\pi}{6}$ lies between $\frac{\pi}{2} = \frac{3\pi}{6}$ and $\pi = \frac{6\pi}{6}$, it is in quadrant II. The reference angle is $\theta' = \pi - \frac{5\pi}{6} = \frac{6\pi}{6} - \frac{5\pi}{6} = \frac{\pi}{6}$.

44. Because $\frac{5\pi}{7} = \frac{10\pi}{14}$ lies between $\frac{\pi}{2} = \frac{7\pi}{14}$ and $\pi = \frac{14\pi}{14}$, it is in quadrant II. The reference angle is $\theta' = \pi - \frac{5\pi}{7} = \frac{7\pi}{7} - \frac{5\pi}{7} = \frac{2\pi}{7}$.

45. $-150° + 360° = 210°$
Because the angle is in quadrant III, the reference angle is $\theta' = 210° - 180° = 30°$.

46. $-250° + 360° = 110°$
Because the angle is in quadrant II, the reference angle is $\theta' = 180° - 110° = 70°$.

47. $-335° + 360° = 25°$
Because the angle is in quadrant I, the reference angle is $\theta' = 25°$.

48. $-359° + 360° = 1°$
Because the angle is in quadrant I, the reference angle is $\theta' = 1°$.

49. Because 4.7 lies between $\pi \approx 3.14$ and $\frac{3\pi}{2} \approx 4.71$, it is in quadrant III. The reference angle is $\theta' = 4.7 - \pi \approx 1.56$.

50. Because 5.5 lies between $\frac{3\pi}{2} \approx 4.71$ and $2\pi \approx 6.28$, it is in quadrant IV. The reference angle is $\theta' = 2\pi - 5.5 \approx 0.78$.

51. $565° - 360° = 205°$
Because the angle is in quadrant III, the reference angle is $\theta' = 205° - 180° = 25°$.

52. $553° - 360° = 193°$
Because the angle is in quadrant III, the reference angle is $\theta' = 193° - 180° = 13°$.

53. $\frac{17\pi}{6} - 2\pi = \frac{17\pi}{6} - \frac{12\pi}{6} = \frac{5\pi}{6}$
Because the angle is in quadrant II, the reference angle is $\theta' = \pi - \frac{5\pi}{6} = \frac{\pi}{6}$.

54. $\frac{11\pi}{4} - 2\pi = \frac{11\pi}{4} - \frac{8\pi}{4} = \frac{3\pi}{4}$
Because the angle is in quadrant II, the reference angle is $\theta' = \pi - \frac{3\pi}{4} = \frac{\pi}{4}$.

55. $\frac{23\pi}{4} - 4\pi = \frac{23\pi}{4} - \frac{16\pi}{4} = \frac{7\pi}{4}$
Because the angle is in quadrant IV, the reference angle is $\theta' = 2\pi - \frac{7\pi}{4} = \frac{\pi}{4}$.

56. $\frac{17\pi}{3} - 4\pi = \frac{17\pi}{3} - \frac{12\pi}{3} = \frac{5\pi}{3}$
Because the angle is in quadrant IV, the reference angle is $\theta' = 2\pi - \frac{5\pi}{3} = \frac{\pi}{3}$.

57. $-\frac{11\pi}{4} + 4\pi = -\frac{11\pi}{4} + \frac{16\pi}{4} = \frac{5\pi}{4}$
Because the angle is in quadrant III, the reference angle is $\theta' = \frac{5\pi}{4} - \pi = \frac{\pi}{4}$.

58. $-\frac{17\pi}{6} + 4\pi = -\frac{17\pi}{6} + \frac{24\pi}{6} = \frac{7\pi}{6}$
Because the angle is in quadrant III, the reference angle is $\theta' = \frac{7\pi}{6} - \pi = \frac{\pi}{6}$.

59. $-\frac{25\pi}{6} + 6\pi = -\frac{25\pi}{6} + \frac{36\pi}{6} = \frac{11\pi}{6}$
Because the angle is in quadrant IV, the reference angle is $\theta' = 2\pi - \frac{11\pi}{6} = \frac{\pi}{6}$.

60. $-\frac{13\pi}{3} + 6\pi = -\frac{13\pi}{3} + \frac{18\pi}{3} = \frac{5\pi}{3}$
Because the angle is in quadrant IV, the reference angle is $\theta' = 2\pi - \frac{5\pi}{3} = \frac{\pi}{3}$.

61. 225° lies in quadrant III. The reference angle is $\theta' = 225° - 180° = 45°$.
$\cos 45° = \frac{\sqrt{2}}{2}$
Because the cosine is negative in quadrant III, $\cos 225° = -\cos 45° = -\frac{\sqrt{2}}{2}$.

62. 300° lies in quadrant IV. The reference angle is $\theta' = 360° - 300° = 60°$.
$\sin 60° = \frac{\sqrt{3}}{2}$
Because the sine is negative in quadrant IV, $\sin 300° = -\sin 60° = -\frac{\sqrt{3}}{2}$.

63. 210° lies in quadrant III. The reference angle is $\theta' = 210° - 180° = 30°$.
$\tan 30° = \frac{\sqrt{3}}{3}$
Because the tangent is positive in quadrant III, $\tan 210° = \tan 30° = \frac{\sqrt{3}}{3}$.

64. 240° lies in quadrant III. The reference angle is $\theta' = 240° - 180° = 60°$.
$\sec 60° = 2$
Because the secant is negative in quadrant III, $\sec 240° = -\sec 60° - 2$.

65. 420° lies in quadrant I. The reference angle is $\theta' = 420° - 360° = 60°$.
$\tan 60° = \sqrt{3}$
Because the tangent is positive in quadrant I, $\tan 420° = \tan 60° = \sqrt{3}$.

66. 405° lies in quadrant I. The reference angle is $\theta' = 405° - 360° = 45°$.

$\tan 45° = 1$
Because the tangent is positive in quadrant I, $\tan 405° = \tan 45° = 1$.

67. $\frac{2\pi}{3}$ lies in quadrant II. The reference angle is

$\theta' = \pi - \frac{2\pi}{3} = \frac{3\pi}{3} - \frac{2\pi}{3} = \frac{\pi}{3}$.

$\sin\frac{\pi}{3} = \frac{\sqrt{3}}{2}$
Because the sine is positive in quadrant II,

$\sin\frac{2\pi}{3} = \sin\frac{\pi}{3} = \frac{\sqrt{3}}{2}$.

68. $\frac{3\pi}{4}$ lies in quadrant II. The reference angle is

$\theta' = \pi - \frac{3\pi}{4} = \frac{4\pi}{4} - \frac{3\pi}{4} = \frac{\pi}{4}$.

$\cos\frac{\pi}{4} = \frac{\sqrt{2}}{2}$
Because the cosine is negative in quadrant II,

$\cos\frac{3\pi}{4} = -\cos\frac{\pi}{4} = -\frac{\sqrt{2}}{2}$.

69. $\frac{7\pi}{6}$ lies in quadrant III. The reference angle is

$\theta' = \frac{7\pi}{6} - \pi = \frac{7\pi}{6} - \frac{6\pi}{6} = \frac{\pi}{6}$.

$\csc\frac{\pi}{6} = 2$
Because the cosecant is negative in quadrant III,

$\csc\frac{7\pi}{6} = -\csc\frac{\pi}{6} = -2$.

70. $\frac{7\pi}{4}$ lies in quadrant IV. The reference angle is

$\theta' = 2\pi - \frac{7\pi}{4} = \frac{8\pi}{4} - \frac{7\pi}{4} = \frac{\pi}{4}$.

$\cot\frac{\pi}{4} = 1$
Because the cotangent is negative in quadrant IV,

$\cot\frac{7\pi}{4} = -\cot\frac{\pi}{4} = -1$.

71. $\frac{9\pi}{4}$ lies in quadrant I. The reference angle is

$\theta' = \frac{9\pi}{4} - 2\pi = \frac{9\pi}{4} - \frac{8\pi}{4} = \frac{\pi}{4}$.

$\tan\frac{\pi}{4} = 1$
Because the tangent is positive in quadrant I,

$\tan\frac{9\pi}{4} = \tan\frac{\pi}{4} = 1$

72. $\frac{9\pi}{2}$ lies on the positive y-axis. The reference angle is $\theta' = \frac{9\pi}{2} - 4\pi = \frac{9\pi}{2} - \frac{8\pi}{2} = \frac{\pi}{2}$.

Because $\tan\frac{\pi}{2}$ is undefined, $\tan\frac{9\pi}{2}$ is also undefined.

73. −240° lies in quadrant II. The reference angle is $\theta' = 240° - 180° = 60°$.

$\sin 60° = \frac{\sqrt{3}}{2}$
Because the sine is positive in quadrant II,

$\sin(-240°) = \sin 60° = \frac{\sqrt{3}}{2}$.

74. −225° lies in quadrant II. The reference angle is $\theta' = 225° - 180° = 45°$.

$\sin 45° = \frac{\sqrt{2}}{2}$
Because the sine is positive in quadrant II,

$\sin(-225°) = \sin 45° = \frac{\sqrt{2}}{2}$.

75. $-\frac{\pi}{4}$ lies in quadrant IV. The reference angle is

$\theta' = \frac{\pi}{4}$.

$\tan\frac{\pi}{4} = 1$
Because the tangent is negative in quadrant IV,

$\tan\left(-\frac{\pi}{4}\right) = -\tan\frac{\pi}{4} = -1$

76. $-\frac{\pi}{6}$ lies in quadrant IV. The reference angle is $\theta = \frac{\pi}{6}$. $\tan\frac{\pi}{6} = \frac{\sqrt{3}}{3}$

Because the tangent is negative in quadrant IV, $\tan\left(-\frac{\pi}{6}\right) = -\frac{\sqrt{3}}{3}$.

77. $\sec 495° = \sec 135° = -\sqrt{2}$

78. $\sec 510° = \sec 150° = -\frac{2\sqrt{3}}{3}$

79. $\cot\frac{19\pi}{6} = \cot\frac{7\pi}{6} = \sqrt{3}$

80. $\cot\frac{13\pi}{3} = \cot\frac{\pi}{3} = \frac{\sqrt{3}}{3}$

81. $\cos\frac{23\pi}{4} = \cos\frac{7\pi}{4} = \frac{\sqrt{2}}{2}$

82. $\cos\frac{35\pi}{6} = \cos\frac{11\pi}{6} = \frac{\sqrt{3}}{2}$

83. $\tan\left(-\frac{17\pi}{6}\right) = \tan\frac{7\pi}{6} = \frac{\sqrt{3}}{3}$

84. $\tan\left(-\frac{11\pi}{4}\right) = \tan\frac{\pi}{4} = 1$

85. $\sin\left(-\frac{17\pi}{3}\right) = \sin\frac{\pi}{3} = \frac{\sqrt{3}}{2}$

86. $\sin\left(-\frac{35\pi}{6}\right) = \sin\frac{\pi}{6} = \frac{1}{2}$

87. $\sin\frac{\pi}{3}\cos\pi - \cos\frac{\pi}{3}\sin\frac{3\pi}{2}$

$$= \left(\frac{\sqrt{3}}{2}\right)(-1) - \left(\frac{1}{2}\right)(-1)$$

$$= -\frac{\sqrt{3}}{2} + \frac{1}{2}$$

$$= \frac{1-\sqrt{3}}{2}$$

88. $\sin\frac{\pi}{4}\cos 0 - \sin\frac{\pi}{6}\cos\pi$

$$= \left(\frac{\sqrt{2}}{2}\right)(1) - \left(\frac{1}{2}\right)(-1)$$

$$= \frac{\sqrt{2}}{2} + \frac{1}{2}$$

$$= \frac{\sqrt{2}+1}{2}$$

89. $\sin\frac{11\pi}{4}\cos\frac{5\pi}{6} + \cos\frac{11\pi}{4}\sin\frac{5\pi}{6}$

$$= \left(\frac{\sqrt{2}}{2}\right)\left(-\frac{\sqrt{3}}{2}\right) + \left(-\frac{\sqrt{2}}{2}\right)\left(\frac{1}{2}\right)$$

$$= -\frac{\sqrt{6}}{4} - \frac{\sqrt{2}}{4}$$

$$= -\frac{\sqrt{6}+\sqrt{2}}{4}$$

90. $\sin\frac{17\pi}{3}\cos\frac{5\pi}{4} + \cos\frac{17\pi}{3}\sin\frac{5\pi}{4}$

$$= \left(-\frac{\sqrt{3}}{2}\right)\left(-\frac{\sqrt{2}}{2}\right) + \left(\frac{1}{2}\right)\left(-\frac{\sqrt{2}}{2}\right)$$

$$= \frac{\sqrt{6}}{4} - \frac{\sqrt{2}}{4}$$

$$= \frac{\sqrt{6}-\sqrt{2}}{4}$$

91. $\sin\frac{3\pi}{2}\tan\left(-\frac{15\pi}{4}\right) - \cos\left(-\frac{5\pi}{3}\right)$

$$= (-1)(1) - \left(\frac{1}{2}\right)$$

$$= -1 - \frac{1}{2}$$

$$= -\frac{2}{2} - \frac{1}{2}$$

$$= -\frac{3}{2}$$

92. $\sin\frac{3\pi}{2}\tan\left(-\frac{8\pi}{3}\right)+\cos\left(-\frac{5\pi}{6}\right)$

$=(-1)(\sqrt{3})+\left(-\frac{\sqrt{3}}{2}\right)$

$=-\sqrt{3}-\frac{\sqrt{3}}{2}$

$=-\frac{2\sqrt{3}}{2}-\frac{\sqrt{3}}{2}$

$=-\frac{3\sqrt{3}}{2}$

93. $f\left(\frac{4\pi}{3}+\frac{\pi}{6}\right)+f\left(\frac{4\pi}{3}\right)+f\left(\frac{\pi}{6}\right)$

$=\sin\left(\frac{4\pi}{3}+\frac{\pi}{6}\right)+\sin\frac{4\pi}{3}+\sin\frac{\pi}{6}$

$=\sin\frac{3\pi}{2}+\sin\frac{4\pi}{3}+\sin\frac{\pi}{6}$

$=(-1)+\left(-\frac{\sqrt{3}}{2}\right)+\left(\frac{1}{2}\right)$

$=-\frac{\sqrt{3}+1}{2}$

94. $g\left(\frac{5\pi}{6}+\frac{\pi}{6}\right)+g\left(\frac{5\pi}{6}\right)+g\left(\frac{\pi}{6}\right)$

$=\cos\left(\frac{5\pi}{6}+\frac{\pi}{6}\right)+\cos\frac{5\pi}{6}+\cos\frac{\pi}{6}$

$=\cos\pi+\cos\frac{5\pi}{6}+\cos\frac{\pi}{6}$

$=(-1)+\left(-\frac{\sqrt{3}}{2}\right)+\left(\frac{\sqrt{3}}{2}\right)$

$=-1$

95. $(h\circ g)\left(\frac{17\pi}{3}\right)=h\left(g\left(\frac{17\pi}{3}\right)\right)$

$=2\left(\cos\left(\frac{17\pi}{3}\right)\right)$

$=2\left(\frac{1}{2}\right)$

$=1$

96. $(h\circ f)\left(\frac{11\pi}{4}\right)=h\left(f\left(\frac{11\pi}{4}\right)\right)$

$=2\left(\sin\left(\frac{11\pi}{4}\right)\right)$

$=2\left(\frac{\sqrt{2}}{2}\right)$

$=\sqrt{2}$

97. The average rate of change is the slope of the line through the points $(x_1, f(x_1))$ and $(x_2, f(x_2))$

$m=\frac{f(x_2)-f(x_1)}{x_2-x_1}$

$=\frac{\sin\left(\frac{3\pi}{2}\right)-\sin\left(\frac{5\pi}{4}\right)}{\frac{3\pi}{2}-\frac{5\pi}{4}}$

$=\frac{-1-\left(-\frac{\sqrt{2}}{2}\right)}{\frac{\pi}{4}}$

$=\frac{-1+\frac{\sqrt{2}}{2}}{\frac{\pi}{4}}$

$=\frac{4\left(-1+\frac{\sqrt{2}}{2}\right)}{4\left(\frac{\pi}{4}\right)}$

$=\frac{2\sqrt{2}-4}{\pi}$

98. The average rate of change is the slope of the line through the points $(x_1, g(x_1))$ and $(x_2, g(x_2))$

$$m = \frac{g(x_2) - g(x_1)}{x_2 - x_1}$$
$$= \frac{\cos(\pi) - \cos\left(\frac{3\pi}{4}\right)}{\pi - \frac{3\pi}{4}}$$
$$= \frac{-1 - \left(-\frac{\sqrt{2}}{2}\right)}{\frac{\pi}{4}}$$
$$= \frac{4\left(-1 + \frac{\sqrt{2}}{2}\right)}{4\left(\frac{\pi}{4}\right)}$$
$$= \frac{2\sqrt{2} - 4}{\pi}$$

99. $\sin\theta = \frac{\sqrt{2}}{2}$ when the reference angle is $\frac{\pi}{4}$ and θ is in quadrants I or II.

QI: $\theta = \frac{\pi}{4}$

QII: $\theta = \pi - \frac{\pi}{4} = \frac{3\pi}{4}$

$\theta = \frac{\pi}{4}, \frac{3\pi}{4}$

100. $\cos\theta = \frac{1}{2}$ when the reference angle is $\frac{\pi}{3}$ and θ is in quadrants I or IV.

QI: $\theta = \frac{\pi}{3}$

QIV: $\theta = 2\pi - \frac{\pi}{3} = \frac{5\pi}{3}$

$\theta = \frac{\pi}{3}, \frac{5\pi}{3}$

101. $\sin\theta = -\frac{\sqrt{2}}{2}$ when the reference angle is $\frac{\pi}{4}$ and θ is in quadrants III or IV.

QIII: $\theta = \pi + \frac{\pi}{4} = \frac{5\pi}{4}$

QIV: $\theta = 2\pi - \frac{\pi}{4} = \frac{7\pi}{4}$

$\theta = \frac{5\pi}{4}, \frac{7\pi}{4}$

102. $\cos\theta = -\frac{1}{2}$ when the reference angle is $\frac{\pi}{3}$ and θ is in quadrants II or III.

QII: $\theta = \pi - \frac{\pi}{3} = \frac{2\pi}{3}$

QIII: $\theta = \pi + \frac{\pi}{3} = \frac{4\pi}{3}$

$\theta = \frac{2\pi}{3}, \frac{4\pi}{3}$

103. $\tan\theta = -\sqrt{3}$ when the reference angle is $\frac{\pi}{3}$ and θ is in quadrants II or IV.

QII: $\theta = \pi - \frac{\pi}{3} = \frac{2\pi}{3}$

QIV: $\theta = 2\pi - \frac{\pi}{3} = \frac{5\pi}{3}$

$\theta = \frac{2\pi}{3}, \frac{5\pi}{3}$

104. $\tan\theta = -\frac{\sqrt{3}}{3}$ when the reference angle is $\frac{\pi}{6}$ and θ is in quadrants II or IV.

QII: $\theta = \pi - \frac{\pi}{6} = \frac{5\pi}{6}$

QIV: $\theta = 2\pi - \frac{\pi}{6} = \frac{11\pi}{6}$

$\theta = \frac{5\pi}{6}, \frac{11\pi}{6}$

Mid-Chapter 4 Check Point

1. $10° = 10° \cdot \frac{\pi \text{ radians}}{180°} = \frac{10\pi}{180} \text{ radians}$

$= \frac{\pi}{18} \text{ radians}$

2. $-105° = -105° \cdot \frac{\pi \text{ radians}}{180°} = -\frac{105\pi}{180} \text{ radians}$

$= -\frac{7\pi}{12} \text{ radians}$

3. $\frac{5\pi}{12} \text{ radians} = \frac{5\pi \text{ radians}}{12} \cdot \frac{180^{\text{o}}}{\pi \text{ radians}} = 75^{\text{o}}$

4. $-\frac{13\pi}{20} \text{ radians} = -\frac{13\pi \text{ radians}}{20} \cdot \frac{180^{\text{o}}}{\pi \text{ radians}}$

$= -117^{\text{o}}$

5. **a.** $\frac{11\pi}{3} - 2\pi = \frac{11\pi}{3} - \frac{6\pi}{3} = \frac{5\pi}{3}$

b.

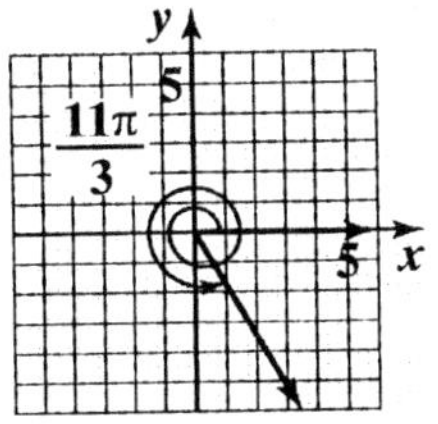

c. Since $\frac{5\pi}{3}$ is in quadrant IV, the reference angle is $2\pi - \frac{5\pi}{3} = \frac{6\pi}{3} - \frac{5\pi}{3} = \frac{\pi}{3}$

6. **a.** $-\frac{19\pi}{4} + 6\pi = -\frac{19\pi}{4} + \frac{24\pi}{4} = \frac{5\pi}{4}$

b.

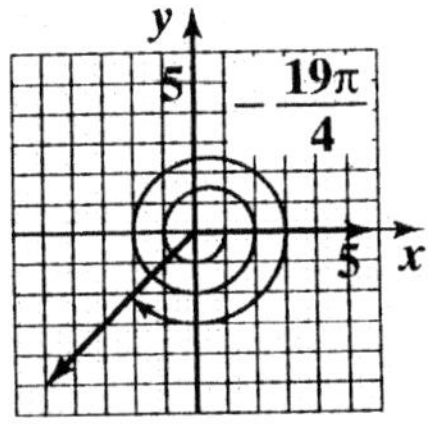

c. Since $\frac{5\pi}{4}$ is in quadrant III, the reference angle is $\frac{5\pi}{4} - \pi = \frac{5\pi}{4} - \frac{4\pi}{4} = \frac{\pi}{4}$

7. **a.** $510° - 360° = 150°$

b.

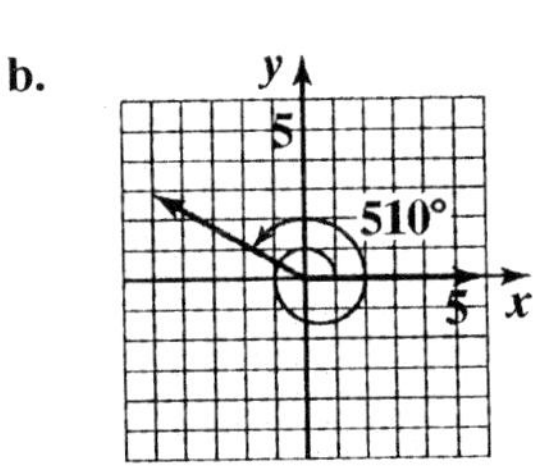

c. Since $150°$ is in quadrant II, the reference angle is $180° - 150° = 30°$

8. $r = \sqrt{x^2 + y^2}$

$r = \sqrt{\left(-\frac{3}{5}\right)^2 + \left(-\frac{4}{5}\right)^2} = \sqrt{\frac{9}{25} + \frac{16}{25}} = \sqrt{\frac{25}{25}} = 1$

Now that we know x, y, and r, we can find the six trigonometric functions of θ.

$\sin\theta = \frac{y}{r} = \frac{-\frac{4}{5}}{1} = -\frac{4}{5}$

$\cos\theta = \frac{x}{r} = \frac{-\frac{3}{5}}{1} = -\frac{3}{5}$

$\tan\theta = \frac{y}{x} = \frac{-\frac{4}{5}}{-\frac{3}{5}} = \frac{4}{3}$

$\csc\theta = \frac{r}{y} = \frac{1}{-\frac{4}{5}} = -\frac{5}{4}$

$\sec\theta = \frac{r}{x} = \frac{1}{-\frac{3}{5}} = -\frac{5}{3}$

$\cot\theta = \frac{x}{y} = \frac{-\frac{3}{5}}{-\frac{4}{5}} = \frac{3}{4}$

9. Use the Pythagorean theorem to find b.

$a^2 + b^2 = c^2$

$5^2 + b^2 = 6^2$

$25 + b^2 = 36$

$b^2 = 11$

$b = \sqrt{11}$

$\sin\theta = \dfrac{\text{opposite}}{\text{hypotenuse}} = \dfrac{5}{6}$

$\cos\theta = \dfrac{\text{adjacent}}{\text{hypotenuse}} = \dfrac{\sqrt{11}}{6}$

$\tan\theta = \dfrac{\text{opposite}}{\text{adjacent}} = \dfrac{5\sqrt{11}}{11}$

$\csc\theta = \dfrac{\text{hypotenuse}}{\text{opposite}} = \dfrac{6}{5}$

$\sec\theta = \dfrac{\text{hypotenuse}}{\text{adjacent}} = \dfrac{6}{\sqrt{11}} = \dfrac{6\sqrt{11}}{11}$

$\cot\theta = \dfrac{\text{adjacent}}{\text{opposite}} = \dfrac{\sqrt{11}}{5}$

10. $r = \sqrt{x^2 + y^2}$

$r = \sqrt{3^2 + (-2)^2} = \sqrt{9+4} = \sqrt{13}$

Now that we know x, y, and r, we can find the six trigonometric functions of θ.

$\sin\theta = \dfrac{y}{r} = \dfrac{-2}{\sqrt{13}} = -\dfrac{2\sqrt{13}}{13}$

$\cos\theta = \dfrac{x}{r} = \dfrac{3}{\sqrt{13}} = \dfrac{3\sqrt{13}}{13}$

$\tan\theta = \dfrac{y}{x} = \dfrac{-2}{3} = -\dfrac{2}{3}$

$\csc\theta = \dfrac{r}{y} = \dfrac{\sqrt{13}}{-2} = -\dfrac{\sqrt{13}}{2}$

$\sec\theta = \dfrac{r}{x} = \dfrac{\sqrt{13}}{3}$

$\cot\theta = \dfrac{x}{y} = \dfrac{3}{-2} = -\dfrac{3}{2}$

11. Because the tangent is negative and the cosine is negative, θ is in quadrant II. In quadrant II, x is negative and y is positive. Thus,

$\tan\theta = -\dfrac{3}{4} = \dfrac{x}{y}$, $x = -4$, $y = 3$. Furthermore,

$r^2 = x^2 + y^2$

$r^2 = (-3)^2 + 4^2$

$r^2 = 9 + 16 = 25$

$r = 5$

Now that we know x, y, and r, we can find the remaining trigonometric functions of θ.

$\sin\theta = \dfrac{y}{r} = \dfrac{3}{5}$

$\cos\theta = \dfrac{x}{r} = \dfrac{-4}{5} = -\dfrac{4}{5}$

$\csc\theta = \dfrac{r}{y} = \dfrac{5}{3}$

$\sec\theta = \dfrac{r}{x} = \dfrac{5}{-3} = -\dfrac{5}{4}$

$\cot\theta = \dfrac{x}{y} = \dfrac{-3}{4} = -\dfrac{4}{3}$

12. Since $\cos\theta = \dfrac{3}{7} = \dfrac{x}{r}$, $x = 3$, $r = 7$. Furthermore,

$x^2 + y^2 = r^2$

$3^2 + y^2 = 7^2$

$9 + y^2 = 49$

$y^2 = 40$

$y = \pm\sqrt{40} = \pm 2\sqrt{10}$

Because the cosine is positive and the sine is negative, θ is in quadrant IV. In quadrant IV, x is positive and y is negative.

Therefore $y = -2\sqrt{10}$

Use x, y, and r to find the remaining trigonometric functions of θ.

$\sin\theta = \dfrac{y}{r} = \dfrac{-2\sqrt{10}}{7} = -\dfrac{2\sqrt{10}}{7}$

$\tan\theta = \dfrac{y}{x} = \dfrac{-2\sqrt{10}}{3} = -\dfrac{2\sqrt{10}}{3}$

$\csc\theta = \dfrac{r}{y} = \dfrac{7}{-2\sqrt{10}} = -\dfrac{7\sqrt{10}}{20}$

$\sec\theta = \dfrac{r}{x} = \dfrac{7}{3}$

$\cot\theta = \dfrac{x}{y} = \dfrac{3}{-2\sqrt{10}} = -\dfrac{3\sqrt{10}}{20}$

13. $\tan\theta = \dfrac{\text{side opposite } \theta}{\text{side adjacent } \theta}$

$\tan 41° = \dfrac{a}{60}$

$a = 60\tan 41°$

$a \approx 52$ cm

14. $\cos\theta = \dfrac{\text{side adjacent } \theta}{\text{hypotenuse}}$

$\cos 72° = \dfrac{250}{c}$

$c = \dfrac{250}{\cos 72°}$

$c \approx 809$ m

15. Since $\cos\theta = \dfrac{1}{6} = \dfrac{x}{r}$, $x = 1$, $r = 6$. Furthermore,

$x^2 + y^2 = r^2$

$1^2 + y^2 = 6^2$

$1 + y^2 = 36$

$y^2 = 35$

$y = \pm\sqrt{35}$

Since θ is acute, $y = +\sqrt{35} = \sqrt{35}$

$\cot\left(\dfrac{\pi}{2} - \theta\right) = \tan\theta = \dfrac{y}{x} = \dfrac{\sqrt{35}}{1} = \sqrt{35}$

16. $\tan 30° = \dfrac{\sqrt{3}}{3}$

17. $\cot 120° = \dfrac{1}{\tan 120°} = \dfrac{1}{-\tan 60°} = \dfrac{1}{-\sqrt{3}} = -\dfrac{\sqrt{3}}{3}$

18. $\cos 240° = -\cos 60° = -\dfrac{1}{2}$

19. $\sec\dfrac{11\pi}{6} = \dfrac{1}{\cos\dfrac{11\pi}{6}} = \dfrac{1}{\cos\dfrac{\pi}{6}} = \dfrac{1}{\dfrac{\sqrt{3}}{2}} = \dfrac{2}{\sqrt{3}} = \dfrac{2\sqrt{3}}{3}$

20. $\sin^2\dfrac{\pi}{7} + \cos^2\dfrac{\pi}{7} = 1$

21. $\sin\left(-\dfrac{2\pi}{3}\right) = \sin\left(-\dfrac{2\pi}{3} + 2\pi\right)$

$= \sin\dfrac{4\pi}{3} = -\sin\dfrac{\pi}{3}$

$= -\dfrac{\sqrt{3}}{2}$

22. $\csc\left(\dfrac{22\pi}{3}\right) = \csc\left(\dfrac{22\pi}{3} - 6\pi\right) = \csc\dfrac{4\pi}{3}$

$= \dfrac{1}{\sin\dfrac{4\pi}{3}} = \dfrac{1}{-\sin\dfrac{\pi}{3}} = \dfrac{1}{-\dfrac{\sqrt{3}}{2}}$

$= -\dfrac{2}{\sqrt{3}} = -\dfrac{2\sqrt{3}}{3}$

23. $\cos 495° = \cos(495° - 360°) = \cos 135°$

$= -\cos 45° = -\dfrac{\sqrt{2}}{2}$

24. $\tan\left(-\dfrac{17\pi}{6}\right) = \tan\left(-\dfrac{17\pi}{6} + 4\pi\right) = \tan\dfrac{7\pi}{6}$

$= \tan\dfrac{\pi}{6} = \dfrac{\sqrt{3}}{3}$

25. $\sin^2\dfrac{\pi}{2} - \cos\pi = (1)^2 - (-1) = 1 + 1 = 2$

26. $\cos\left(\dfrac{5\pi}{6} + 2\pi n\right) + \tan\left(\dfrac{5\pi}{6} + n\pi\right)$

$= \cos\dfrac{5\pi}{6} + \tan\dfrac{5\pi}{6} = -\cos\dfrac{\pi}{6} - \tan\dfrac{\pi}{6}$

$= -\dfrac{\sqrt{3}}{2} - \dfrac{\sqrt{3}}{3} = -\dfrac{3\sqrt{3}}{6} - \dfrac{2\sqrt{3}}{6}$

$= -\dfrac{5\sqrt{3}}{6}$

27. Begin by converting from degrees to radians.

$36° = 36° \cdot \dfrac{\pi \text{ radians}}{180°} = \dfrac{\pi}{5}$ radians

$s = r\theta = 40 \cdot \dfrac{\pi}{5} = 8\pi \approx 25.13$ cm

28. Linear speed is given by $v = r\omega$. It is given that $r = 10$ feet and the merry-go-round rotates at 8 revolutions per minute. Convert 8 revolutions per minute to radians per minute.

8 revolutions per minute

$= 8 \text{ revolutions per minute} \cdot \frac{2\pi \text{ radians}}{1 \text{ revolution}}$

$= 16\pi$ radians per minute

$v = r\omega = (10)(16\pi) = 160\pi \approx 502.7$ feet per minute

The linear speed of the horse is about 502.7 feet per minute.

29. $\sin\theta = \frac{\text{side opposite } \theta}{\text{hypotenuse}}$

$\sin 6° = \frac{h}{5280}$

$h = 5280 \sin 6°$

$h \approx 551.9$ feet

30. $\tan\theta = \frac{\text{side opposite } \theta}{\text{side adjacent } \theta}$

$\tan\theta = \frac{50}{60}$

$\theta = \tan^{-1}\left(\frac{50}{60}\right)$

$\theta \approx 40°$

Section 4.5

Checkpoint Exercises

1. The equation $y = 3\sin x$ is of the form $y = A\sin x$ with $A = 3$. Thus, the amplitude is $|A| = |3| = 3$ The period for both $y = 3\sin x$ and $y = \sin x$ is 2π. We find the three x–intercepts, the maximum point, and the minimum point on the interval [0, 2π] by dividing the period, 2π, by 4, $\frac{\text{period}}{4} = \frac{2\pi}{4} = \frac{\pi}{2}$, then by adding quarter-periods to generate x-values for each of the key points. The five x-values are

$x = 0$

$x = 0 + \frac{\pi}{2} = \frac{\pi}{2}$

$x = \frac{\pi}{2} + \frac{\pi}{2} = \pi$

$x = \pi + \frac{\pi}{2} = \frac{3\pi}{2}$

$x = \frac{3\pi}{2} + \frac{\pi}{2} = 2\pi$

Evaluate the function at each value of x.

x	$y = 3\sin x$	coordinates
0	$y = 3\sin 0 = 3 \cdot 0 = 0$	(0, 0)
$\frac{\pi}{2}$	$y = 3\sin\frac{\pi}{2} = 3 \cdot 1 = 3$	$\left(\frac{\pi}{2}, 3\right)$
π	$y = 3\sin x = 3 \cdot 0 = 0$	$(\pi, 0)$
$\frac{3\pi}{2}$	$y = 3\sin\frac{3\pi}{2}$ $= 3(-1) = -3$	$\left(\frac{3\pi}{2}, -3\right)$
2π	$y = 3\sin 2\pi = 3 \cdot 0 = 0$	$(2\pi, 0)$

Connect the five points with a smooth curve and graph one complete cycle of the given function with the graph of $y = \sin x$.

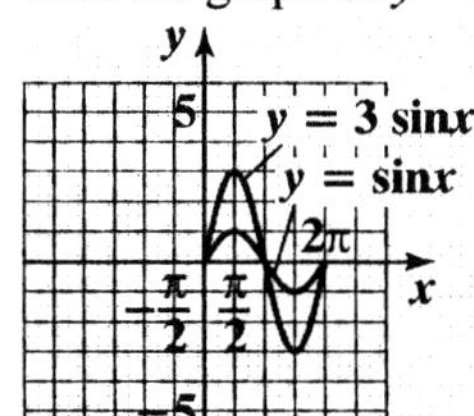

2. The equation $y = -\frac{1}{2}\sin x$ is of the form $y = A\sin x$ with $A = -\frac{1}{2}$. Thus, the amplitude is $|A| = \left|-\frac{1}{2}\right| = \frac{1}{2}$. The period for both $y = -\frac{1}{2}\sin x$ and $y = \sin x$ is 2π.

Find the x–values for the five key points by dividing the period, 2π, by 4, $\frac{\text{period}}{4} = \frac{2\pi}{4} = \frac{\pi}{2}$, then by adding quarter-periods. The five x-values are

$x = 0$

$x = 0 + \frac{\pi}{2} = \frac{\pi}{2}$

$x = \frac{\pi}{2} + \frac{\pi}{2} = \pi$

$x = \pi + \frac{\pi}{2} = \frac{3\pi}{2}$

$x = \frac{3\pi}{2} + \frac{\pi}{2} = 2\pi$

Evaluate the function at each value of x.

x	$y = -\frac{1}{2}\sin x$	coordinates
0	$y = -\frac{1}{2}\sin 0$ $= -\frac{1}{2}\cdot 0 = 0$	(0, 0)
$\frac{\pi}{2}$	$y = -\frac{1}{2}\sin\frac{\pi}{2}$ $= -\frac{1}{2}\cdot 1 = -\frac{1}{2}$	$\left(\frac{\pi}{2}, -\frac{1}{2}\right)$
π	$y = -\frac{1}{2}\sin\pi$ $= -\frac{1}{2}\cdot 0 = 0$	$(\pi, 0)$
$\frac{3\pi}{2}$	$y = -\frac{1}{2}\sin\frac{3\pi}{2}$ $= -\frac{1}{2}(-1) = \frac{1}{2}$	$\left(\frac{3\pi}{2}, \frac{1}{2}\right)$
2π	$y = -\frac{1}{2}\sin 2\pi$ $= -\frac{1}{2}\cdot 0 = 0$	$(2\pi, 0)$

Connect the five key points with a smooth curve and graph one complete cycle of the given function with the graph of $y = \sin x$. Extend the pattern of each graph to the left and right as desired.

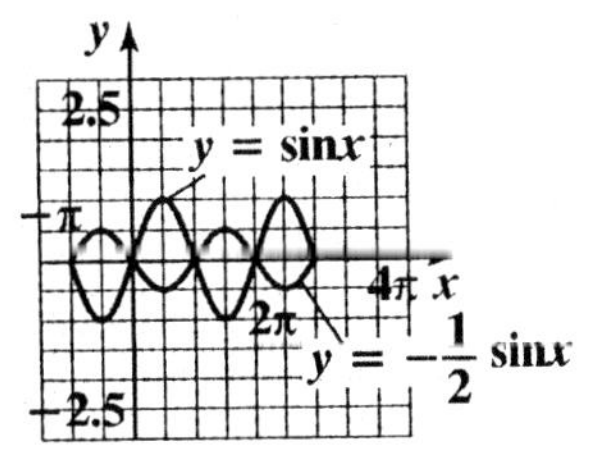

3. The equation $y = 2\sin\frac{1}{2}x$ is of the form $y = A\sin Bx$ with $A = 2$ and $B = \frac{1}{2}$.

The amplitude is $|A| = |2| = 2$.

The period is $\frac{2\pi}{B} = \frac{2\pi}{\frac{1}{2}} = 4\pi$.

Find the x–values for the five key points by dividing the period, 4π, by 4, $\frac{\text{period}}{4} = \frac{4\pi}{4} = \pi$, then by adding quarter-periods.

The five x-values are

$x = 0$

$x = 0 + \pi = \pi$

$x = \pi + \pi = 2\pi$

$x = 2\pi + \pi = 3\pi$

$x = 3\pi + \pi = 4\pi$

Evaluate the function at each value of x.

x	$y = 2\sin\frac{1}{2}x$	coordinates
0	$y = 2\sin\left(\frac{1}{2}\cdot 0\right)$ $= 2\sin 0$ $= 2\cdot 0 = 0$	(0, 0)
π	$y = 2\sin\left(\frac{1}{2}\cdot\pi\right)$ $= 2\sin\frac{\pi}{2} = 2\cdot 1 = 2$	$(\pi, 2)$
2π	$y = 2\sin\left(\frac{1}{2}\cdot 2\pi\right)$ $= 2\sin\pi = 2\cdot 0 = 0$	$(2\pi, 0)$

3π	$y = 2\sin\left(\frac{1}{2}\cdot 3\pi\right)$ $= 2\sin\frac{3\pi}{2}$ $= 2\cdot(-1) = -2$	$(3\pi, -2)$
4π	$y = 2\sin\left(\frac{1}{2}\cdot 4\pi\right)$ $= 2\sin 2\pi = 2\cdot 0 = 0$	$(4\pi, 0)$

Connect the five key points with a smooth curve and graph one complete cycle of the given function. Extend the pattern of the graph another full period to the right.

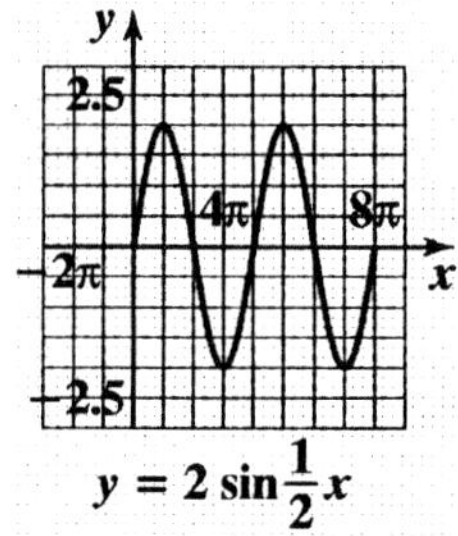

4. The equation $y = 3\sin\left(2x - \frac{\pi}{3}\right)$ is of the form $y = A\sin(Bx - C)$ with $A = 3$, $B = 2$, and $C = \frac{\pi}{3}$.
The amplitude is $|A| = |3| = 3$.
The period is $\frac{2\pi}{B} = \frac{2\pi}{2} = \pi$.
The phase shift is $\frac{C}{B} = \frac{\frac{\pi}{3}}{2} = \frac{\pi}{3}\cdot\frac{1}{2} = \frac{\pi}{6}$.
Find the x-values for the five key points by dividing the period, π, by 4, $\frac{\text{period}}{4} = \frac{\pi}{4}$, then by adding quarter-periods to the value of x where the cycle begins, $x = \frac{\pi}{6}$.
The five x-values are

$x = \frac{\pi}{6}$

$x = \frac{\pi}{6} + \frac{\pi}{4} = \frac{2\pi}{12} + \frac{3\pi}{12} = \frac{5\pi}{12}$

$x = \frac{5\pi}{12} + \frac{\pi}{4} = \frac{5\pi}{12} + \frac{3\pi}{12} = \frac{8\pi}{12} = \frac{2\pi}{3}$

$x = \frac{2\pi}{3} + \frac{\pi}{4} = \frac{8\pi}{12} + \frac{3\pi}{12} = \frac{11\pi}{12}$

$x = \frac{11\pi}{12} + \frac{\pi}{4} = \frac{11\pi}{12} + \frac{3\pi}{12} = \frac{14\pi}{12} = \frac{7\pi}{6}$

Evaluate the function at each value of x.

x	$y = 3\sin\left(2x - \frac{\pi}{3}\right)$	coordinates
$\frac{\pi}{6}$	$y = 3\sin\left(2\cdot\frac{\pi}{6} - \frac{\pi}{3}\right)$ $= 3\sin 0 = 3\cdot 0 = 0$	$\left(\frac{\pi}{6}, 0\right)$
$\frac{5\pi}{12}$	$y = 3\sin\left(2\cdot\frac{5\pi}{12} - \frac{\pi}{3}\right)$ $= 3\sin\frac{3\pi}{6} = 3\sin\frac{\pi}{2}$ $= 3\cdot 1 = 3$	$\left(\frac{5\pi}{12}, 3\right)$
$\frac{2\pi}{3}$	$y = 3\sin\left(2\cdot\frac{2\pi}{3} - \frac{\pi}{3}\right)$ $= 3\sin\frac{3\pi}{3} = 3\sin\pi$ $= 3\cdot 0 = 0$	$\left(\frac{2\pi}{3}, 0\right)$
$\frac{11\pi}{12}$	$y = 3\sin\left(2\cdot\frac{11\pi}{12} - \frac{\pi}{3}\right)$ $= 3\sin\frac{9\pi}{6} = 3\sin\frac{3\pi}{2}$ $= 3(-1) = -3$	$\left(\frac{11\pi}{12}, -3\right)$
$\frac{7\pi}{6}$	$y = 3\sin\left(2\cdot\frac{7\pi}{6} - \frac{\pi}{3}\right)$ $= 3\sin\frac{6\pi}{3} = 3\sin 2\pi$ $= 3\cdot 0 = 0$	$\left(\frac{7\pi}{6}, 0\right)$

Connect the five key points with a smooth curve and graph one complete cycle of the given graph.

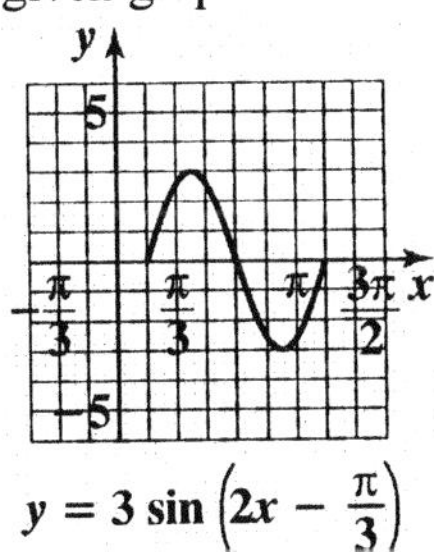

$y = 3\sin\left(2x - \frac{\pi}{3}\right)$

5. The equation $y = -4\cos \pi x$ is of the form $y = A\cos Bx$ with $A = -4$, and $B = \pi$.
Thus, the amplitude is $|A| = |-4| = 4$.
The period is $\frac{2\pi}{B} = \frac{2\pi}{\pi} = 2$.
Find the x-values for the five key points by dividing the period, 2, by 4, $\frac{\text{period}}{4} = \frac{2}{4} = \frac{1}{2}$, then by adding quarter periods to the value of x where the cycle begins. The five x-values are

$x = 0$

$x = 0 + \frac{1}{2} = \frac{1}{2}$

$x = \frac{1}{2} + \frac{1}{2} = 1$

$x = 1 + \frac{1}{2} = \frac{3}{2}$

$x = \frac{3}{2} + \frac{1}{2} = 2$

Evaluate the function at each value of x.

x	$y = -4\cos \pi x$	coordinates
0	$y = -4\cos(\pi \cdot 0)$ $= -4\cos 0 = -4$	$(0, -4)$
$\frac{1}{2}$	$y = -4\cos\left(\pi \cdot \frac{1}{2}\right)$ $= -4\cos\frac{\pi}{2} = 0$	$\left(\frac{1}{2}, 0\right)$
1	$y = -4\cos(\pi \cdot 1)$ $= -4\cos \pi = 4$	$(1, 4)$
$\frac{3}{2}$	$y = -4\cos\left(\pi \cdot \frac{3}{2}\right)$ $= -4\cos\frac{3\pi}{2} = 0$	$\left(\frac{3}{2}, 0\right)$
2	$y = -4\cos(\pi \cdot 2)$ $= -4\cos 2\pi = -4$	$(2, -4)$

Connect the five key points with a smooth curve and graph one complete cycle of the given function. Extend the pattern of the graph another full period to the left.

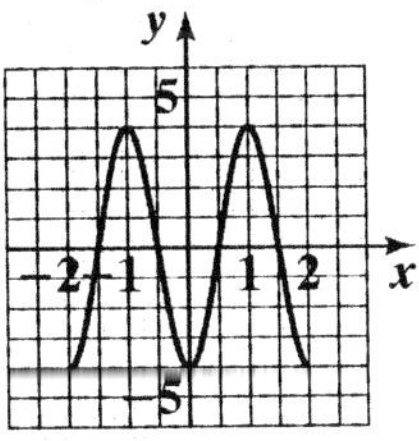

$y = -4\cos \pi x$

6. $y = \frac{3}{2}\cos(2x + \pi) = \frac{3}{2}\cos(2x - (-\pi))$
The equation is of the form $y = A\cos(Bx - C)$ with $A = \frac{3}{2}$, $B = 2$, and $C = -\pi$.

Thus, the amplitude is $|A| = \left|\frac{3}{2}\right| = \frac{3}{2}$.

The period is $\frac{2\pi}{B} = \frac{2\pi}{2} = \pi$.

The phase shift is $\frac{C}{B} = \frac{-\pi}{2} = -\frac{\pi}{2}$.

Find the x-values for the five key points by dividing the period, π, by 4, $\frac{\text{period}}{4} = \frac{\pi}{4}$, then by adding quarter-periods to the value of x where the cycle begins, $x = -\frac{\pi}{2}$.

The five x-values are

$x = -\frac{\pi}{2}$

$x = -\frac{\pi}{2} + \frac{\pi}{4} = -\frac{\pi}{4}$

$x = -\frac{\pi}{4} + \frac{\pi}{4} = 0$

$x = 0 + \frac{\pi}{4} = \frac{\pi}{4}$

$x = \frac{\pi}{4} + \frac{\pi}{4} = \frac{\pi}{2}$

Evaluate the function at each value of x.

x	$y = \frac{3}{2}\cos(2x + \pi)$	coordinates
$-\frac{\pi}{2}$	$y = \frac{3}{2}\cos(-\pi + \pi)$ $= \frac{3}{2} \cdot 1 = \frac{3}{2}$	$\left(-\frac{\pi}{2}, \frac{3}{2}\right)$

$-\frac{\pi}{4}$	$y=\frac{3}{2}\cos\left(-\frac{\pi}{2}+\pi\right)$ $=\frac{3}{2}\cdot 0=0$	$\left(-\frac{\pi}{4},0\right)$
0	$y=\frac{3}{2}\cos(0+\pi)$ $=\frac{3}{2}\cdot -1=-\frac{3}{2}$	$\left(0,-\frac{3}{2}\right)$
$\frac{\pi}{4}$	$y=\frac{3}{2}\cos\left(\frac{\pi}{2}+\pi\right)$ $=\frac{3}{2}\cdot 0=0$	$\left(\frac{\pi}{4},0\right)$
$\frac{\pi}{2}$	$y=\frac{3}{2}\cos(\pi+\pi)$ $=\frac{3}{2}\cdot 1=\frac{3}{2}$	$\left(\frac{\pi}{2},\frac{3}{2}\right)$

Connect the five key points with a smooth curve and graph one complete cycle of the given graph.

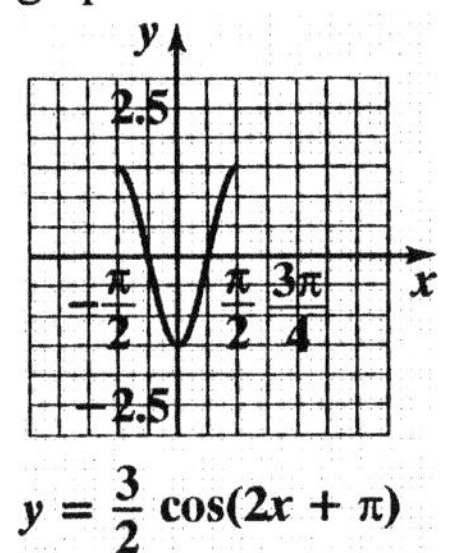

$y=\frac{3}{2}\cos(2x+\pi)$

7. The graph of $y=2\cos x+1$ is the graph of $y=2\cos x$ shifted one unit upwards. The period for both functions is 2π. The quarter-period is $\frac{2\pi}{4}$ or $\frac{\pi}{2}$. The cycle begins at $x = 0$. Add quarter-periods to generate x-values for the key points.

$x=0$

$x=0+\frac{\pi}{2}=\frac{\pi}{2}$

$x=\frac{\pi}{2}+\frac{\pi}{2}=\pi$

$x=\pi+\frac{\pi}{2}=\frac{3\pi}{2}$

$x=\frac{3\pi}{2}+\frac{\pi}{2}=2\pi$

Evaluate the function at each value of x.

x	$y=2\cos x+1$	coordinates
0	$y=2\cos 0+1$ $=2\cdot 1+1=3$	$(0, 3)$
$\frac{\pi}{2}$	$y=2\cos\frac{\pi}{2}+1$ $=2\cdot 0+1=1$	$\left(\frac{\pi}{2},1\right)$
π	$y=2\cos\pi+1$ $=2\cdot(-1)+1=-1$	$(\pi,-1)$
$\frac{3\pi}{2}$	$y=2\cos\frac{3\pi}{2}+1$ $=2\cdot 0+1=1$	$\left(\frac{3\pi}{2},1\right)$
2π	$y=2\cos 2\pi+1$ $=2\cdot 1+1=3$	$(2\pi,3)$

By connecting the points with a smooth curve, we obtain one period of the graph.

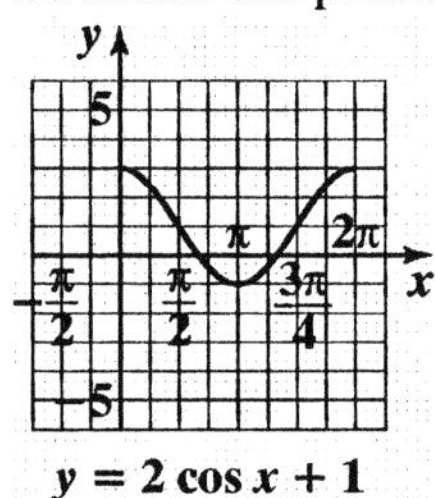

$y=2\cos x+1$

8. A, the amplitude, is the maximum value of y. The graph shows that this maximum value is 4, Thus, $A=4$. The period is $\frac{\pi}{2}$, and period $=\frac{2\pi}{B}$.

Thus, $\frac{\pi}{2}=\frac{2\pi}{B}$

$\pi B=4\pi$

$B=4$

Substitute these values into $y=A\sin Bx$. The graph is modeled by $y=4\sin 4x$.

9. Because the hours of daylight ranges from a minimum of 10 hours to a maximum of 14 hours, the curve oscillates about the middle value, 12 hours. Thus, $D = 12$. The maximum number of hours is 2 hours above 12 hours. Thus, $A = 2$. The graph shows that one complete cycle occurs in 12–0, or 12 months. The period is 12. Thus,

$$12 = \frac{2\pi}{B}$$

$$12B = 2\pi$$

$$B = \frac{2\pi}{12} = \frac{\pi}{6}$$

The graph shows that the starting point of the cycle is shifted from 0 to 3. The phase shift, $\frac{C}{B}$, is 3.

$$3 = \frac{C}{B}$$

$$3 = \frac{C}{\frac{\pi}{6}}$$

$$\frac{\pi}{2} = C$$

Substitute these values into $y = A\sin(Bx - C) + D$. The number of hours of daylight is modeled by $y = 2\sin\left(\frac{\pi}{6}x - \frac{\pi}{2}\right) + 12$.

Exercise Set 4.5

1. The equation $y = 4\sin x$ is of the form $y = A\sin x$ with $A = 4$. Thus, the amplitude is $|A| = |4| = 4$. The period is 2π. The quarter-period is $\frac{2\pi}{4}$ or $\frac{\pi}{2}$. The cycle begins at $x = 0$. Add quarter-periods to generate x-values for the key points.

$$x = 0$$

$$x = 0 + \frac{\pi}{2} = \frac{\pi}{2}$$

$$x = \frac{\pi}{2} + \frac{\pi}{2} = \pi$$

$$x = \pi + \frac{\pi}{2} = \frac{3\pi}{2}$$

$$x = \frac{3\pi}{2} + \frac{\pi}{2} = 2\pi$$

Evaluate the function at each value of x.

x	$y = 4\sin x$	coordinates
0	$y = 4\sin 0 = 4\cdot 0 = 0$	$(0, 0)$
$\frac{\pi}{2}$	$y = 4\sin\frac{\pi}{2} = 4\cdot 1 = 4$	$\left(\frac{\pi}{2}, 4\right)$
π	$y = 4\sin\pi = 4\cdot 0 = 0$	$(\pi, 0)$
$\frac{3\pi}{2}$	$y = 4\sin\frac{3\pi}{2}$ $= 4(-1) = -4$	$\left(\frac{3\pi}{2}, -4\right)$
2π	$y = 4\sin 2\pi = 4\cdot 0 = 0$	$(2\pi, 0)$

Connect the five key points with a smooth curve and graph one complete cycle of the given function with the graph of $y = \sin x$.

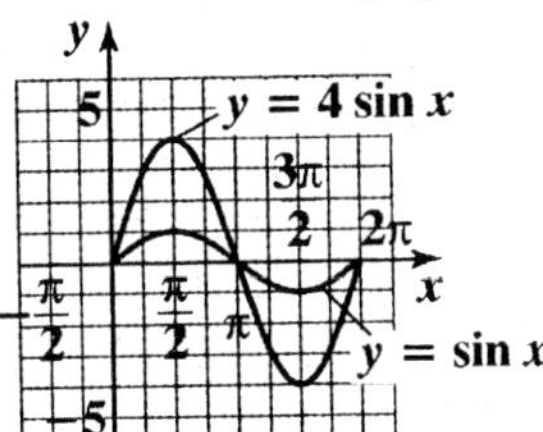

2. The equation $y = 5\sin x$ is of the form $y = A\sin x$ with $A = 5$. Thus, the amplitude is $|A| = |5| = 5$. The period is 2π. The quarter-period is $\frac{2\pi}{4}$ or $\frac{\pi}{2}$. The cycle begins at $x = 0$.
Add quarter-periods to generate x-values for the key points.
$x = 0$

$x = 0 + \frac{\pi}{2} = \frac{\pi}{2}$

$x = \frac{\pi}{2} + \frac{\pi}{2} = \pi$

$x = \pi + \frac{\pi}{2} = \frac{3\pi}{2}$

$x = \frac{3\pi}{2} + \frac{\pi}{2} = 2\pi$

Evaluate the function at each value of x.

x	$y = 5\sin x$	coordinates
0	$y = 5\sin 0 = 5\cdot 0 = 0$	(0, 0)
$\frac{\pi}{2}$	$y = 5\sin\frac{\pi}{2} = 5\cdot 1 = 5$	$\left(\frac{\pi}{2}, 5\right)$
π	$y = 5\sin\pi = 5\cdot 0 = 0$	$(\pi, 0)$
$\frac{3\pi}{2}$	$y = 5\sin\frac{3\pi}{2} = 5(-1) = -5$	$\left(\frac{3\pi}{2}, -5\right)$
2π	$y = 5\sin 2\pi = 5\cdot 0 = 0$	$(2\pi, 0)$

Connect the five key points with a smooth curve and graph one complete cycle of the given function with the graph of $y = \sin x$.

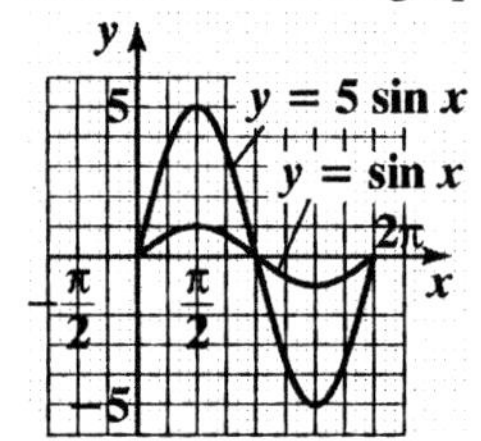

3. The equation $y = \frac{1}{3}\sin x$ is of the form $y = A\sin x$ with $A = \frac{1}{3}$. Thus, the amplitude is $|A| = \left|\frac{1}{3}\right| = \frac{1}{3}$. The period is 2π. The quarter-period is $\frac{2\pi}{4}$ or $\frac{\pi}{2}$. The cycle begins at $x = 0$.
Add quarter-periods to generate x-values for the key points.
$x = 0$

$x = 0 + \frac{\pi}{2} = \frac{\pi}{2}$

$x = \frac{\pi}{2} + \frac{\pi}{2} = \pi$

$x = \pi + \frac{\pi}{2} = \frac{3\pi}{2}$

$x = \frac{3\pi}{2} + \frac{\pi}{2} = 2\pi$

Evaluate the function at each value of x.

x	$y = \frac{1}{3}\sin x$	coordinates
0	$y = \frac{1}{3}\sin 0 = \frac{1}{3}\cdot 0 = 0$	(0, 0)
$\frac{\pi}{2}$	$y = \frac{1}{3}\sin\frac{\pi}{2} = \frac{1}{3}\cdot 1 = \frac{1}{3}$	$\left(\frac{\pi}{2}, \frac{1}{3}\right)$
π	$y = \frac{1}{3}\sin\pi = \frac{1}{3}\cdot 0 = 0$	$(\pi, 0)$
$\frac{3\pi}{2}$	$y = \frac{1}{3}\sin\frac{3\pi}{2}$ $= \frac{1}{3}(-1) = -\frac{1}{3}$	$\left(\frac{3\pi}{2}, -\frac{1}{3}\right)$
2π	$y = \frac{1}{3}\sin 2\pi = \frac{1}{3}\cdot 0 = 0$	$(2\pi, 0)$

Connect the five key points with a smooth curve and graph one complete cycle of the given function with the graph of $y = \sin x$.

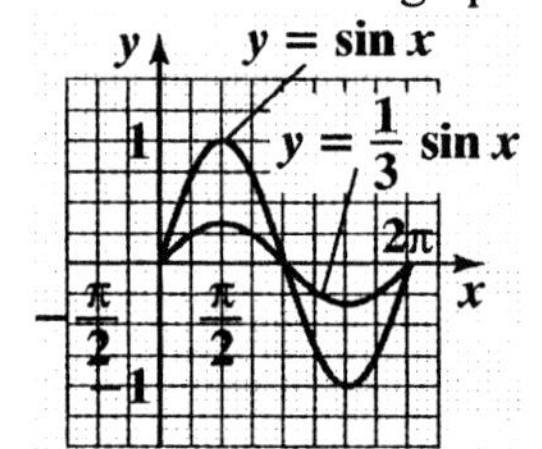

4. The equation $y = \frac{1}{4}\sin x$ is of the form $y = A\sin x$ with $A = \frac{1}{4}$. Thus, the amplitude is $|A| = \left|\frac{1}{4}\right| = \frac{1}{4}$. The period is 2π. The quarter-period is $\frac{2\pi}{4}$ or $\frac{\pi}{2}$. The cycle begins at $x = 0$.
Add quarter-periods to generate x-values for the key points.

$x - 0$

$x = 0 + \frac{\pi}{2} = \frac{\pi}{2}$

$x = \frac{\pi}{2} + \frac{\pi}{2} = \pi$

$x = \pi + \frac{\pi}{2} = \frac{3\pi}{2}$

$x = \frac{3\pi}{2} + \frac{\pi}{2} = 2\pi$

Evaluate the function at each value of x.

x	$y = \frac{1}{4}\sin x$	coordinates
0	$y = \frac{1}{4}\sin 0 = \frac{1}{4} \cdot 0 = 0$	(0, 0)
$\frac{\pi}{2}$	$y = \frac{1}{4}\sin\frac{\pi}{2} = \frac{1}{4} \cdot 1 = \frac{1}{4}$	$\left(\frac{\pi}{2}, \frac{1}{4}\right)$
π	$y = \frac{1}{4}\sin \pi = \frac{1}{4} \cdot 0 = 0$	$(\pi, 0)$
$\frac{3\pi}{2}$	$y = \frac{1}{4}\sin\frac{3\pi}{2} = \frac{1}{4}(-1) = -\frac{1}{4}$	$\left(\frac{3\pi}{2}, -\frac{1}{4}\right)$
2π	$y = \frac{1}{4}\sin 2\pi = \frac{1}{4} \cdot 0 = 0$	$(2\pi, 0)$

Connect the five key points with a smooth curve and graph one complete cycle of the given function with the graph of $y = \sin x$.

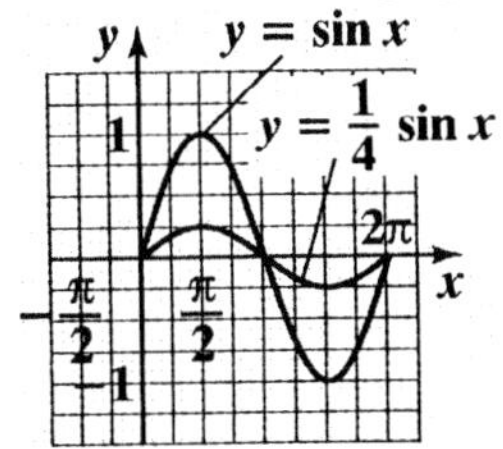

5. The equation $y = -3\sin x$ is of the form $y = A\sin x$ with $A = -3$. Thus, the amplitude is $|A| = |-3| = 3$. The period is 2π. The quarter-period is $\frac{2\pi}{4}$ or $\frac{\pi}{2}$. The cycle begins at $x = 0$.
Add quarter-periods to generate x-values for the key points.

$x = 0$

$x = 0 + \frac{\pi}{2} = \frac{\pi}{2}$

$x = \frac{\pi}{2} + \frac{\pi}{2} = \pi$

$x = \pi + \frac{\pi}{2} = \frac{3\pi}{2}$

$x = \frac{3\pi}{2} + \frac{\pi}{2} = 2\pi$

Evaluate the function at each value of x.

x	$y = -3\sin x$	coordinates
0	$y = -3\sin x$ $= -3 \cdot 0 = 0$	(0, 0)
$\frac{\pi}{2}$	$y = -3\sin\frac{\pi}{2}$ $= -3 \cdot 1 = -3$	$\left(\frac{\pi}{2}, -3\right)$
π	$y = -3\sin \pi$ $= -3 \cdot 0 = 0$	$(\pi, 0)$
$\frac{3\pi}{2}$	$y = -3\sin\frac{3\pi}{2}$ $= -3(-1) = 3$	$\left(\frac{3\pi}{2}, 3\right)$
2π	$y = -3\sin 2\pi$ $= -3 \cdot 0 = 0$	$(2\pi, 0)$

Connect the five key points with a smooth curve and graph one complete cycle of the given function with the graph of $y = \sin x$.

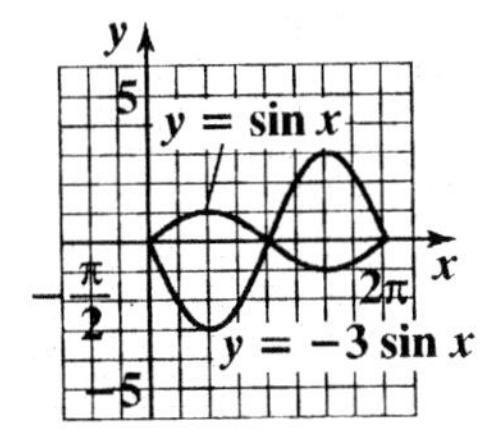

6. The equation $y = -4\sin x$ is of the form $y = A\sin x$ with $A = -4$. Thus, the amplitude is $|A| = |-4| = 4$. The period is 2π. The quarter-period is $\frac{2\pi}{4}$ or $\frac{\pi}{2}$. The cycle begins at $x = 0$. Add quarter-periods to generate x-values for the key points.

$x = 0$

$x = 0 + \frac{\pi}{2} = \frac{\pi}{2}$

$x = \frac{\pi}{2} + \frac{\pi}{2} = \pi$

$x = \pi + \frac{\pi}{2} = \frac{3\pi}{2}$

$x = \frac{3\pi}{2} + \frac{\pi}{2} = 2\pi$

Evaluate the function at each value of x.

x	$y = -4\sin x$	coordinates
0	$y = -4\sin 0 = -4\cdot 0 = 0$	$(0, 0)$
$\frac{\pi}{2}$	$y = -4\sin\frac{\pi}{2} = -4\cdot 1 = -4$	$\left(\frac{\pi}{2}, -4\right)$
π	$y = -4\sin\pi = -4\cdot 0 = 0$	$(\pi, 0)$
$\frac{3\pi}{2}$	$y = -4\sin\frac{3\pi}{2} = -4(-1) = 4$	$\left(\frac{3\pi}{2}, 4\right)$
2π	$y = -4\sin 2\pi = -4\cdot 0 = 0$	$(2\pi, 0)$

Connect the five key points with a smooth curve and graph one complete cycle of the given function with the graph of $y = \sin x$.

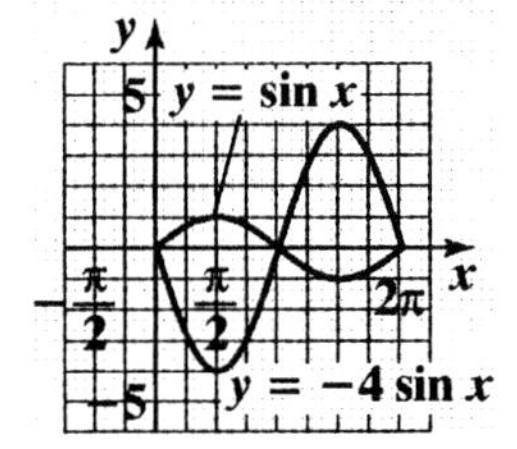

7. The equation $y = \sin 2x$ is of the form $y = A\sin Bx$ with $A = 1$ and $B = 2$. The amplitude is $|A| = |1| = 1$. The period is $\frac{2\pi}{B} = \frac{2\pi}{2} = \pi$. The quarter-period is $\frac{\pi}{4}$. The cycle begins at $x = 0$. Add quarter-periods to generate x-values for the key points.

$x = 0$

$x = 0 + \frac{\pi}{4}$

$x = \frac{\pi}{4} + \frac{\pi}{4} = \frac{\pi}{2}$

$x = \frac{\pi}{2} + \frac{\pi}{4} = \frac{3\pi}{4}$

$x = \frac{3\pi}{4} + \frac{\pi}{4} = \pi$

Evaluate the function at each value of x.

x	$y = \sin 2x$	coordinates
0	$y = \sin 2\cdot 0 = \sin 0 = 0$	$(0, 0)$
$\frac{\pi}{4}$	$y = \sin\left(2\cdot\frac{\pi}{4}\right) = \sin\frac{\pi}{2} = 1$	$\left(\frac{\pi}{4}, 1\right)$
$\frac{\pi}{2}$	$y = \sin\left(2\cdot\frac{\pi}{2}\right) = \sin\pi = 0$	$\left(\frac{\pi}{2}, 0\right)$
$\frac{3\pi}{4}$	$y = \sin\left(2\cdot\frac{3\pi}{4}\right) = \sin\frac{3\pi}{2} = -1$	$\left(\frac{3\pi}{4}, -1\right)$
π	$y = \sin(2\cdot\pi) = \sin 2\pi = 0$	$(\pi, 0)$

Connect the five key points with a smooth curve and graph one complete cycle of the given function.

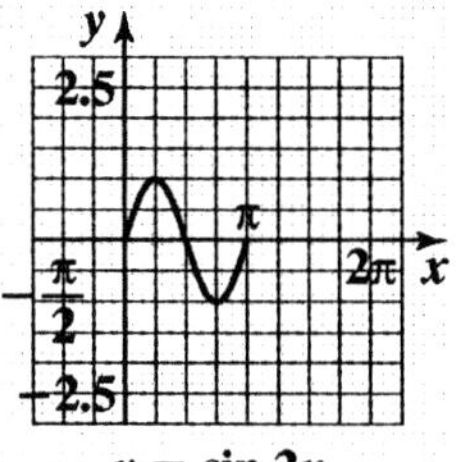

8. The equation $y = \sin 4x$ is of the form $y = A\sin Bx$ with $A = 1$ and $B = 4$. Thus, the amplitude is $|A| = |1| = 1$. The period is $\frac{2\pi}{B} = \frac{2\pi}{4} = \frac{\pi}{2}$. The quarter-period is $\frac{\frac{\pi}{2}}{4} = \frac{\pi}{2}\cdot\frac{1}{4} = \frac{\pi}{8}$. The cycle begins at $x = 0$. Add quarter-periods to generate x-values for the key points.

$x = 0$

$x = 0 + \frac{\pi}{8} = \frac{\pi}{8}$

$x = \frac{\pi}{8} + \frac{\pi}{8} = \frac{\pi}{4}$

$x = \frac{\pi}{4} + \frac{\pi}{8} = \frac{3\pi}{8}$

$x = \frac{3\pi}{8} + \frac{\pi}{8} = \frac{\pi}{2}$

Evaluate the function at each value of x.

x	$y = \sin 4x$	coordinates
0	$y = \sin(4\cdot 0) = \sin 0 = 0$	(0, 0)
$\frac{\pi}{8}$	$y = \sin\left(4\cdot\frac{\pi}{8}\right) = \sin\frac{\pi}{2} = 1$	$\left(\frac{\pi}{8}, 1\right)$
$\frac{\pi}{4}$	$y = \sin\left(4\cdot\frac{\pi}{4}\right) = \sin\pi = 0$	$\left(\frac{\pi}{4}, 0\right)$
$\frac{3\pi}{8}$	$y = \sin\left(4\cdot\frac{3\pi}{8}\right)$ $= \sin\frac{3\pi}{2} = -1$	$\left(\frac{3\pi}{8}, -1\right)$
$\frac{\pi}{2}$	$y = \sin 2\pi = 0$	$\left(\frac{\pi}{2}, 0\right)$

Connect the five key points with a smooth curve and graph one complete cycle of the given function.

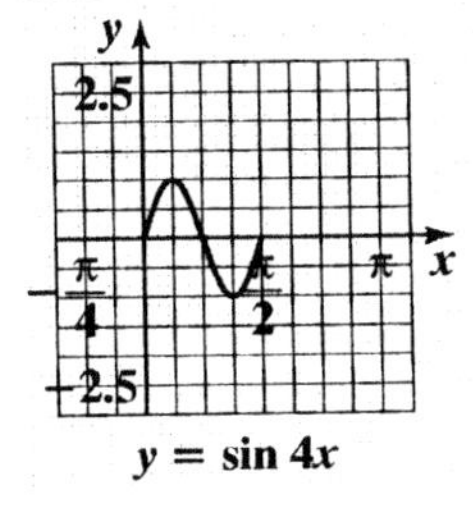

$y = \sin 4x$

9. The equation $y = 3\sin\frac{1}{2}x$ is of the form $y = A\sin Bx$ with $A = 3$ and $B = \frac{1}{2}$. The amplitude is $|A| = |3| = 3$. The period is $\frac{2\pi}{B} = \frac{2\pi}{\frac{1}{2}} = 2\pi\cdot 2 = 4\pi$. The quarter-period is $\frac{4\pi}{4} = \pi$. The cycle begins at $x = 0$. Add quarter-periods to generate x-values for the key points.

$x = 0$

$x = 0 + \pi = \pi$

$x = \pi + \pi = 2\pi$

$x = 2\pi + \pi = 3\pi$

$x = 3\pi + \pi = 4\pi$

Evaluate the function at each value of x.

x	$y = 3\sin\frac{1}{2}x$	coordinates
0	$y = 3\sin\left(\frac{1}{2}\cdot 0\right)$ $= 3\sin 0 = 3\cdot 0 = 0$	(0, 0)
π	$y = 3\sin\left(\frac{1}{2}\cdot\pi\right)$ $= 3\sin\frac{\pi}{2} = 3\cdot 1 = 3$	$(\pi, 3)$
2π	$y = 3\sin\left(\frac{1}{2}\cdot 2\pi\right)$ $= 3\sin\pi = 3\cdot 0 = 0$	$(2\pi, 0)$
3π	$y = 3\sin\left(\frac{1}{2}\cdot 3\pi\right)$ $= 3\sin\frac{3\pi}{2}$ $= 3(-1) = -3$	$(3\pi, -3)$
4π	$y = 3\sin\left(\frac{1}{2}\cdot 4\pi\right)$ $= 3\sin 2\pi = 3\cdot 0 = 0$	$(4\pi, 0)$

Connect the five points with a smooth curve and graph one complete cycle of the given function.

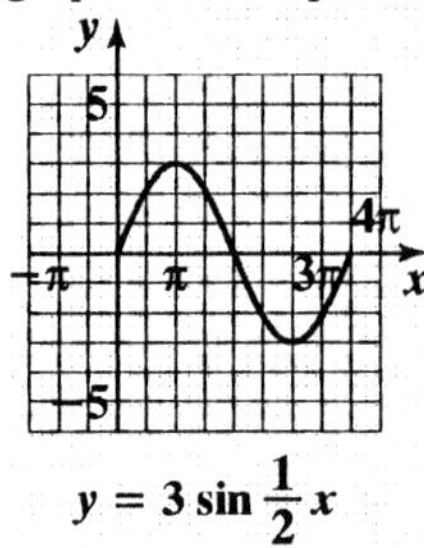

10. The equation $y = 2\sin\frac{1}{4}x$ is of the form $y = A\sin Bx$ with $A = 2$ and $B = \frac{1}{4}$. Thus, the amplitude is $|A| = |2| = 2$. The period is $\frac{2\pi}{B} = \frac{2\pi}{\frac{1}{4}} = 2\pi \cdot 4 = 8\pi$. The quarter-period is $\frac{8\pi}{4} = 2\pi$. The cycle begins at $x = 0$. Add quarter-periods to generate x-values for the key points.

$x = 0$

$x = 0 + 2\pi = 2\pi$

$x = 2\pi + 2\pi = 4\pi$

$x = 4\pi + 2\pi = 6\pi$

$x = 6\pi + 2\pi = 8\pi$

Evaluate the function at each value of x.

x	$y = 2\sin\frac{1}{4}x$	coordinates
0	$y = 2\sin\left(\frac{1}{4}\cdot 0\right)$ $= 2\sin 0 = 2\cdot 0 = 0$	(0, 0)
2π	$y = 2\sin\left(\frac{1}{4}\cdot 2\pi\right)$ $= 2\sin\frac{\pi}{2} = 2\cdot 1 = 2$	$(2\pi, 2)$
4π	$y = 2\sin\pi = 2\cdot 0 = 0$	$(4\pi, 0)$
6π	$y = 2\sin\frac{3\pi}{2} = 2(-1) = -2$	$(6\pi, -2)$
8π	$y = 2\sin 2\pi = 2\cdot 0 = 0$	$(8\pi, 0)$

Connect the five key points with a smooth curve and graph one complete cycle of the given function.

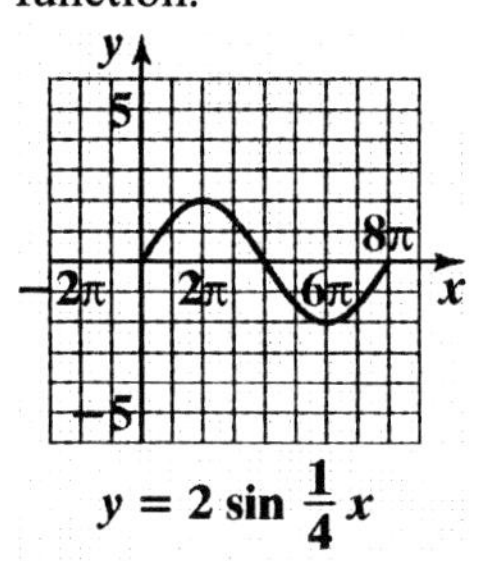

11. The equation $y = 4\sin\pi x$ is of the form $y = A\sin Bx$ with $A = 4$ and $B = \pi$. The amplitude is $|A| = |4| = 4$. The period is $\frac{2\pi}{B} = \frac{2\pi}{\pi} = 2$. The quarter-period is $\frac{2}{4} = \frac{1}{2}$. The cycle begins at $x = 0$. Add quarter-periods to generate x-values for the key points.

$x = 0$

$x = 0 + \frac{1}{2} = \frac{1}{2}$

$x = \frac{1}{2} + \frac{1}{2} = 1$

$x = 1 + \frac{1}{2} = \frac{3}{2}$

$x = \frac{3}{2} + \frac{1}{2} = 2$

Evaluate the function at each value of x.

x	$y = 4\sin\pi x$	coordinates
0	$y = 4\sin(\pi\cdot 0)$ $= 4\sin 0 = 4\cdot 0 = 0$	(0, 0)
$\frac{1}{2}$	$y = 4\sin\left(\pi\cdot\frac{1}{2}\right)$ $= 4\sin\frac{\pi}{2} = 4(1) = 4$	$\left(\frac{1}{2}, 4\right)$
1	$y = 4\sin(\pi\cdot 1)$ $= 4\sin\pi = 4\cdot 0 = 0$	(1, 0)
$\frac{3}{2}$	$y = 4\sin\left(\pi\cdot\frac{3}{2}\right)$ $= 4\sin\frac{3\pi}{2}$ $= 4(-1) = -4$	$\left(\frac{3}{2}, -4\right)$
2	$y = 4\sin(\pi\cdot 2)$ $= 4\sin 2\pi = 4\cdot 0 = 0$	(2, 0)

Connect the five points with a smooth curve and graph one complete cycle of the given function.

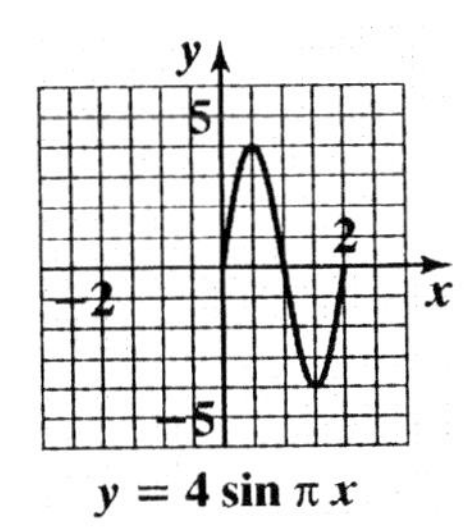

$y = 4 \sin \pi x$

12. The equation $y = 3\sin 2\pi x$ is of the form $y = A\sin Bx$ with $A = 3$ and $B = 2\pi$. The amplitude is $|A| = |3| = 3$. The period is $\frac{2\pi}{B} = \frac{2\pi}{2\pi} = 1$. The quarter-period is $\frac{1}{4}$. The cycle begins at $x = 0$. Add quarter-periods to generate x-values for the key points.

$x = 0$

$x = 0 + \frac{1}{4} = \frac{1}{4}$

$x = \frac{1}{4} + \frac{1}{4} = \frac{1}{2}$

$x = \frac{1}{2} + \frac{1}{4} = \frac{3}{4}$

$x = \frac{3}{4} + \frac{1}{4} = 1$

Evaluate the function at each value of x.

x	$y = 3\sin 2\pi x$	coordinates
0	$y = 3\sin(2\pi \cdot 0)$ $= 3\sin 0 = 3 \cdot 0 = 0$	$(0, 0)$
$\frac{1}{4}$	$y = 3\sin\left(2\pi \cdot \frac{1}{4}\right)$ $= 3\sin\frac{\pi}{2} = 3 \cdot 1 = 3$	$\left(\frac{1}{4}, 3\right)$
$\frac{1}{2}$	$y = 3\sin\left(2\pi \cdot \frac{1}{2}\right)$ $= 3\sin \pi = 3 \cdot 0 = 0$	$\left(\frac{1}{2}, 0\right)$
$\frac{3}{4}$	$y = 3\sin\left(2\pi \cdot \frac{3}{4}\right)$ $= 3\sin\frac{3\pi}{2} = 3(-1) = -3$	$\left(\frac{3}{4}, -3\right)$
1	$y = 3\sin(2\pi \cdot 1)$ $= 3\sin 2\pi = 3 \cdot 0 = 0$	$(1, 0)$

Connect the five key points with a smooth curve and graph one complete cycle of the given function.

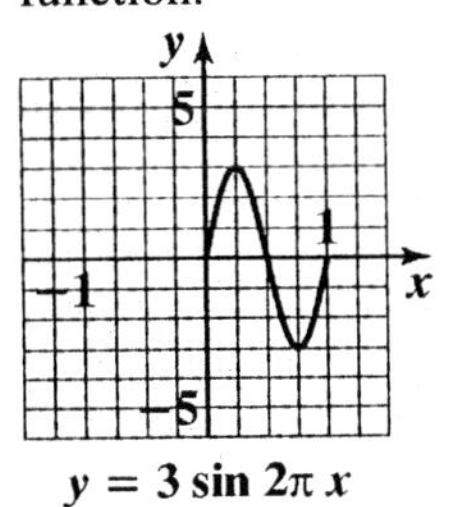

$y = 3 \sin 2\pi x$

13. The equation $y = -3\sin 2\pi x$ is of the form $y = A\sin Bx$ with $A = -3$ and $B = 2\pi$. The amplitude is $|A| = |-3| = 3$. The period is $\frac{2\pi}{B} = \frac{2\pi}{2\pi} = 1$. The quarter-period is $\frac{1}{4}$. The cycle begins at $x = 0$. Add quarter-periods to generate x-values for the key points.

$x = 0$

$x = 0 + \frac{1}{4} = \frac{1}{4}$

$x = \frac{1}{4} + \frac{1}{4} = \frac{1}{2}$

$x = \frac{1}{2} + \frac{1}{4} = \frac{3}{4}$

$x = \frac{3}{4} + \frac{1}{4} = 1$

Evaluate the function at each value of x.

x	$y = -3\sin 2\pi x$	coordinates
0	$y = -3\sin(2\pi \cdot 0)$ $= -3\sin 0$ $= -3 \cdot 0 = 0$	$(0, 0)$
$\frac{1}{4}$	$y = -3\sin\left(2\pi \cdot \frac{1}{4}\right)$ $= -3\sin\frac{\pi}{2}$ $= -3 \cdot 1 = -3$	$\left(\frac{1}{4}, -3\right)$
$\frac{1}{2}$	$y = -3\sin\left(2\pi \cdot \frac{1}{2}\right)$ $= -3\sin \pi$ $= -3 \cdot 0 = 0$	$\left(\frac{1}{2}, 0\right)$

$\frac{3}{4}$	$y=-3\sin\left(2\pi\cdot\frac{3}{4}\right)$ $=-3\sin\frac{3\pi}{2}$ $=-3(-1)=3$	$\left(\frac{3}{4},3\right)$
1	$y=-3\sin(2\pi\cdot 1)$ $=-3\sin 2\pi$ $=-3\cdot 0=0$	(1, 0)

Connect the five points with a smooth curve and graph one complete cycle of the given function.

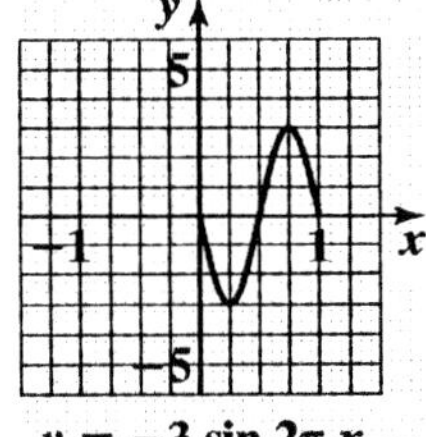

14. The equation $y=-2\sin\pi x$ is of the form $y=A\sin Bx$ with $A=-2$ and $B=\pi$. The amplitude is $|A|=|-2|=2$. The period is $\frac{2\pi}{B}=\frac{2\pi}{\pi}=2$. The quarter-period is $\frac{2}{4}=\frac{1}{2}$.
The cycle begins at $x = 0$. Add quarter-periods to generate x-values for the key points.

$x=0$

$x=0+\frac{1}{2}=\frac{1}{2}$

$x=\frac{1}{2}+\frac{1}{2}=1$

$x=1+\frac{1}{2}=\frac{3}{2}$

$x=\frac{3}{2}+\frac{1}{2}=2$

Evaluate the function at each value of x.

x	$y=-2\sin\pi x$	coordinates
0	$y=-2\sin(\pi\cdot 0)$ $=-2\sin 0=-2\cdot 0=0$	(0, 0)
$\frac{1}{2}$	$y=-2\sin\left(\pi\cdot\frac{1}{2}\right)$ $=-2\sin\frac{\pi}{2}=-2\cdot 1=-2$	$\left(\frac{1}{2},-2\right)$
1	$y=-2\sin(\pi\cdot 1)$ $=-2\sin\pi=-2\cdot 0=0$	(1, 0)
$\frac{3}{2}$	$y=-2\sin\left(\pi\cdot\frac{3}{2}\right)$ $=-2\sin\frac{3\pi}{2}=-2(-1)=2$	$\left(\frac{3}{2},2\right)$
2	$y=-2\sin(\pi\cdot 2)$ $=-2\sin 2\pi=-2\cdot 0=0$	(2, 0)

Connect the five key points with a smooth curve and graph one complete cycle of the given function.

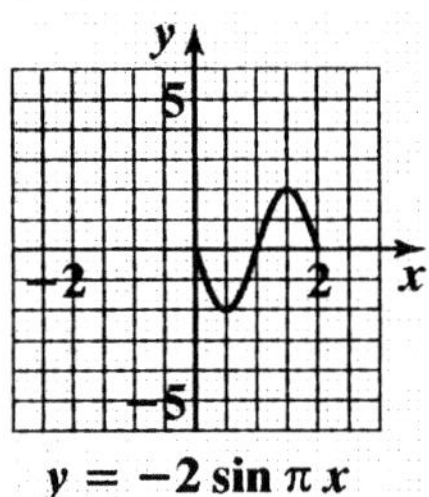

15. The equation $y=-\sin\frac{2}{3}x$ is of the form $y=A\sin Bx$ with $A=-1$ and $B=\frac{2}{3}$. The amplitude is $|A|=|-1|=1$. The period is $\frac{2\pi}{B}=\frac{2\pi}{\frac{2}{3}}=2\pi\cdot\frac{3}{2}=3\pi$. The quarter-period is $\frac{3\pi}{4}$. The cycle begins at $x = 0$. Add quarter-periods to generate x-values for the key points.

$x = 0$

$x = 0 + \frac{3\pi}{4} = \frac{3\pi}{4}$

$x = \frac{3\pi}{4} + \frac{3\pi}{4} = \frac{3\pi}{2}$

$x = \frac{3\pi}{2} + \frac{3\pi}{4} = \frac{9\pi}{4}$

$x = \frac{9\pi}{4} + \frac{3\pi}{4} = 3\pi$

Evaluate the function at each value of x.

x	$y = -\sin\frac{2}{3}x$	coordinates
0	$y = -\sin\left(\frac{2}{3}\cdot 0\right)$ $= -\sin 0 = 0$	$(0, 0)$
$\frac{3\pi}{4}$	$y = -\sin\left(\frac{2}{3}\cdot\frac{3\pi}{4}\right)$ $= -\sin\frac{\pi}{2} = -1$	$\left(\frac{3\pi}{4}, -1\right)$
$\frac{3\pi}{2}$	$y = -\sin\left(\frac{2}{3}\cdot\frac{3\pi}{2}\right)$ $= -\sin\pi = 0$	$\left(\frac{3\pi}{2}, 0\right)$
$\frac{9\pi}{4}$	$y = -\sin\left(\frac{2}{3}\cdot\frac{9\pi}{4}\right)$ $= -\sin\frac{3\pi}{2}$ $= -(-1) = 1$	$\left(\frac{9\pi}{4}, 1\right)$
3π	$y = -\sin\left(\frac{2}{3}\cdot 3\pi\right)$ $= -\sin 2\pi = 0$	$(3\pi, 0)$

Connect the five points with a smooth curve and graph one complete cycle of the given function.

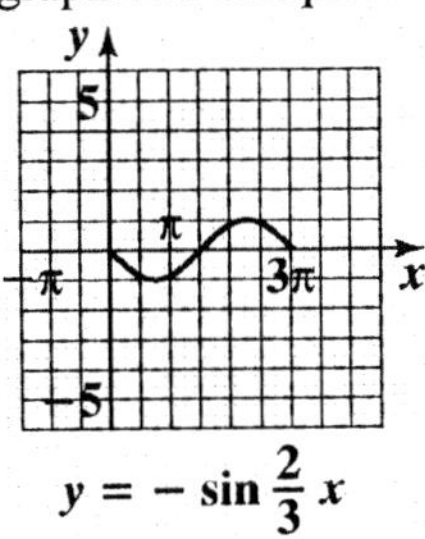

$y = -\sin\frac{2}{3}x$

16. The equation $y = -\sin\frac{4}{3}x$ is of the form $y = A\sin Bx$ with $A = -1$ and $B = \frac{4}{3}$. The amplitude is $|A| = |-1| = 1$. The period is $\frac{2\pi}{B} = \frac{2\pi}{\frac{4}{3}} = 2\pi\cdot\frac{3}{4} = \frac{3\pi}{2}$. The quarter-period is $\frac{\frac{3\pi}{2}}{4} = \frac{3\pi}{2}\cdot\frac{1}{4} = \frac{3\pi}{8}$. The cycle begins at $x = 0$. Add quarter-periods to generate x-values for the key points.

$x = 0$

$x = 0 + \frac{3\pi}{8} = \frac{3\pi}{8}$

$x = \frac{3\pi}{8} + \frac{3\pi}{8} = \frac{3\pi}{4}$

$x = \frac{3\pi}{4} + \frac{3\pi}{8} = \frac{9\pi}{8}$

$x = \frac{9\pi}{8} + \frac{3\pi}{8} = \frac{3\pi}{2}$

Evaluate the function at each value of x.

x	$y = -\sin\frac{4}{3}x$	coordinates
0	$y = -\sin\left(\frac{4}{3}\cdot 0\right)$ $= -\sin 0 = 0$	$(0, 0)$
$\frac{3\pi}{8}$	$y = -\sin\left(\frac{4}{3}\cdot\frac{3\pi}{8}\right)$ $= -\sin\frac{\pi}{2} = -1$	$\left(\frac{3\pi}{8}, -1\right)$
$\frac{3\pi}{4}$	$y = -\sin\left(\frac{4}{3}\cdot\frac{3\pi}{4}\right)$ $= -\sin\pi = 0$	$\left(\frac{3\pi}{4}, 0\right)$
$\frac{9\pi}{8}$	$y = -\sin\left(\frac{4}{3}\cdot\frac{9\pi}{8}\right)$ $= -\sin\frac{3\pi}{2} = -(-1) = 1$	$\left(\frac{9\pi}{8}, 1\right)$
$\frac{3\pi}{2}$	$y = -\sin\left(\frac{4}{3}\cdot\frac{3\pi}{2}\right)$ $= -\sin 2\pi = 0$	$\left(\frac{3\pi}{2}, 0\right)$

Connect the five key points with a smooth curve and graph one complete cycle of the given function.

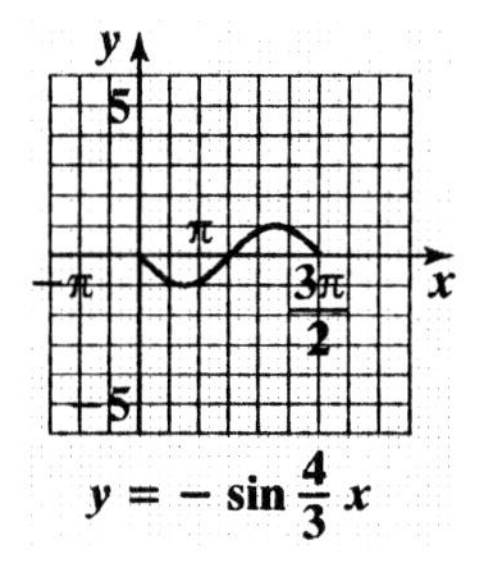

Connect the five points with a smooth curve and graph one complete cycle of the given function.

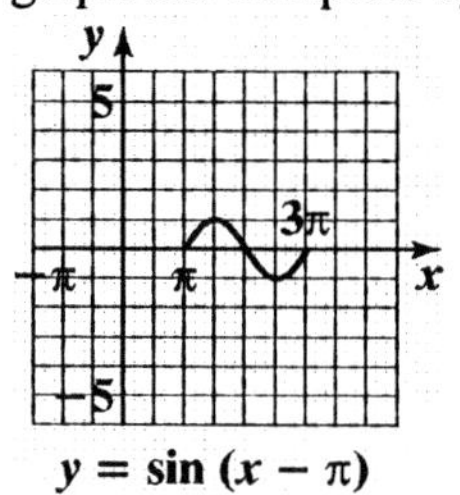

17. The equation $y = \sin(x - \pi)$ is of the form $y = A\sin(Bx - C)$ with $A = 1$, $B = 1$, and $C = \pi$. The amplitude is $|A| = |1| = 1$. The period is $\frac{2\pi}{B} = \frac{2\pi}{1} = 2\pi$. The phase shift is $\frac{C}{B} = \frac{\pi}{1} = \pi$. The quarter-period is $\frac{2\pi}{4} = \frac{\pi}{2}$. The cycle begins at $x = \pi$. Add quarter-periods to generate x-values for the key points.

$x = \pi$

$x = \pi + \frac{\pi}{2} = \frac{3\pi}{2}$

$x = \frac{3\pi}{2} + \frac{\pi}{2} = 2\pi$

$x = 2\pi + \frac{\pi}{2} = \frac{5\pi}{2}$

$x = \frac{5\pi}{2} + \frac{\pi}{2} = 3\pi$

Evaluate the function at each value of x.

x	$y = \sin(x - \pi)$	coordinates
π	$y = \sin(\pi - \pi)$ $= \sin 0 = 0$	$(\pi, 0)$
$\frac{3\pi}{2}$	$y = \sin\left(\frac{3\pi}{2} - \pi\right)$ $= \sin\frac{\pi}{2} = 1$	$\left(\frac{3\pi}{2}, 1\right)$
2π	$y = \sin(2\pi - \pi)$ $= \sin\pi = 0$	$(2\pi, 0)$
$\frac{5\pi}{2}$	$y = \sin\left(\frac{5\pi}{2} - \pi\right)$ $= \sin\frac{3\pi}{2} = -1$	$\left(\frac{5\pi}{2}, -1\right)$
3π	$y = \sin(3\pi - \pi)$ $= \sin 2\pi = 0$	$(3\pi, 0)$

18. The equation $y = \sin\left(x - \frac{\pi}{2}\right)$ is of the form $y = A\sin(Bx - C)$ with $A = 1$, $B = 1$, and $C = \frac{\pi}{2}$. The amplitude is $|A| = |1| = 1$. The period is $\frac{2\pi}{B} = \frac{2\pi}{1} = 2\pi$. The phase shift is $\frac{C}{B} = \frac{\frac{\pi}{2}}{1} = \frac{\pi}{2}$. The quarter-period is $\frac{2\pi}{4} = \frac{\pi}{2}$. The cycle begins at $x = \frac{\pi}{2}$. Add quarter-periods to generate x-values for the key points.

$x = \frac{\pi}{2}$

$x = \frac{\pi}{2} + \frac{\pi}{2} = \pi$

$x = \pi + \frac{\pi}{2} = \frac{3\pi}{2}$

$x = \frac{3\pi}{2} + \frac{\pi}{2} = 2\pi$

$x = 2\pi + \frac{\pi}{2} = \frac{5\pi}{2}$

Evaluate the function at each value of x.

x	$y=\sin\left(x-\frac{\pi}{2}\right)$	coordinates
$\frac{\pi}{2}$	$y=\sin\left(\frac{\pi}{2}-\frac{\pi}{2}\right)=\sin 0=0$	$\left(\frac{\pi}{2},0\right)$
π	$y=\sin\left(\pi-\frac{\pi}{2}\right)=\sin\frac{\pi}{2}=1$	$(\pi, 1)$
$\frac{3\pi}{2}$	$y=\sin\left(\frac{3\pi}{2}-\frac{\pi}{2}\right)$ $=\sin\pi=0$	$\left(\frac{3\pi}{2},0\right)$
2π	$y=\sin\left(2\pi-\frac{\pi}{2}\right)$ $=\sin\frac{3\pi}{2}=-1$	$(2\pi, -1)$
$\frac{5\pi}{2}$	$y=\sin\left(\frac{5\pi}{2}-\frac{\pi}{2}\right)$ $=\sin 2\pi=0$	$\left(\frac{5\pi}{2},0\right)$

Connect the five key points with a smooth curve and graph one complete cycle of the given function.

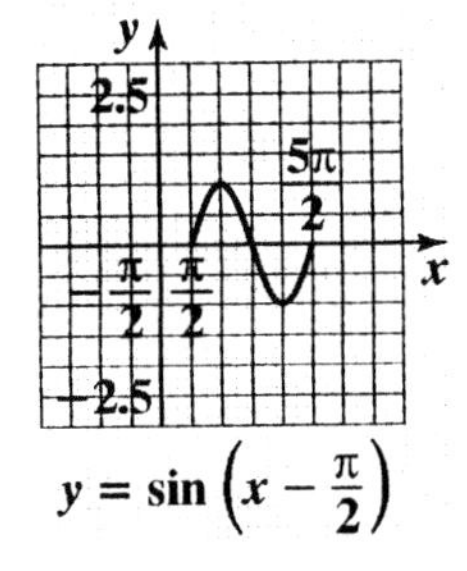

$y=\sin\left(x-\frac{\pi}{2}\right)$

19. The equation $y=\sin(2x-\pi)$ is of the form $y=A\sin(Bx-C)$ with $A=1$, $B=2$, and $C=\pi$. The amplitude is $|A|=|1|=1$. The period is $\frac{2\pi}{B}=\frac{2\pi}{2}=\pi$. The phase shift is $\frac{C}{B}=\frac{\pi}{2}$. The quarter-period is $\frac{\pi}{4}$. The cycle begins at $x=\frac{\pi}{2}$. Add quarter-periods to generate x-values for the key points.

$x=\frac{\pi}{2}$

$x=\frac{\pi}{2}+\frac{\pi}{4}=\frac{3\pi}{4}$

$x=\frac{3\pi}{4}+\frac{\pi}{4}=\pi$

$x=\pi+\frac{\pi}{4}=\frac{5\pi}{4}$

$x=\frac{5\pi}{4}+\frac{\pi}{4}=\frac{3\pi}{2}$

Evaluate the function at each value of x.

x	$y=\sin(2x-\pi)$	coordinates
$\frac{\pi}{2}$	$y=\sin\left(2\cdot\frac{\pi}{2}-\pi\right)$ $=\sin(\pi-\pi)$ $=\sin 0=0$	$\left(\frac{\pi}{2},0\right)$
$\frac{3\pi}{4}$	$y=\sin\left(2\cdot\frac{3\pi}{4}-\pi\right)$ $=\sin\left(\frac{3\pi}{2}-\pi\right)$ $=\sin\frac{\pi}{2}=1$	$\left(\frac{3\pi}{4},1\right)$
π	$y=\sin(2\cdot\pi-\pi)$ $=\sin(2\pi-\pi)$ $=\sin\pi=0$	$(\pi, 0)$
$\frac{5\pi}{4}$	$y=\sin\left(2\cdot\frac{5\pi}{4}-\pi\right)$ $=\sin\left(\frac{5\pi}{2}-\pi\right)$ $=\sin\frac{3\pi}{2}=-1$	$\left(\frac{5\pi}{4},-1\right)$
$\frac{3\pi}{2}$	$y=\sin\left(2\cdot\frac{3\pi}{2}-\pi\right)$ $=\sin(3\pi-\pi)$ $=\sin 2\pi=0$	$\left(\frac{3\pi}{2},0\right)$

Connect the five points with a smooth curve and graph one complete cycle of the given function.

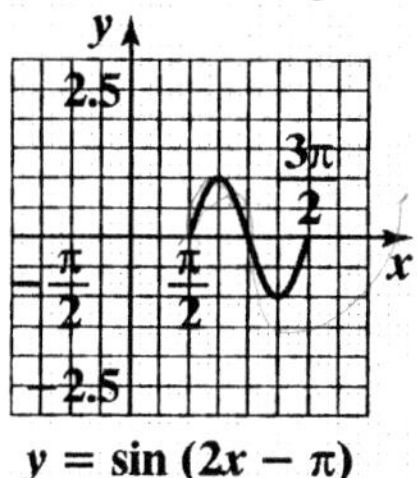

$y = \sin(2x - \pi)$

20. The equation $y = \sin\left(2x - \frac{\pi}{2}\right)$ is of the form $y = A\sin(Bx - C)$ with $A = 1$, $B = 2$, and $C = \frac{\pi}{2}$.

The amplitude is $|A| = |1| = 1$.

The period is $\frac{2\pi}{B} = \frac{2\pi}{2} = \pi$.

The phase shift is $\frac{C}{B} = \frac{\frac{\pi}{2}}{2} = \frac{\pi}{2} \cdot \frac{1}{2} = \frac{\pi}{4}$.

The quarter-period is $\frac{\pi}{4}$.

The cycle begins at $x = \frac{\pi}{4}$. Add quarter-periods to generate x-values for the key points.

$x = \frac{\pi}{4}$

$x = \frac{\pi}{4} + \frac{\pi}{4} = \frac{\pi}{2}$

$x = \frac{\pi}{2} + \frac{\pi}{4} = \frac{3\pi}{4}$

$x = \frac{3\pi}{4} + \frac{\pi}{4} = \pi$

$x = \pi + \frac{\pi}{4} = \frac{5\pi}{4}$

Evaluate the function at each value of x.

x	$y = \sin\left(2x - \frac{\pi}{2}\right)$	coordinates
$\frac{\pi}{4}$	$y = \sin\left(2 \cdot \frac{\pi}{4} - \frac{\pi}{2}\right)$ $= \sin\left(\frac{\pi}{2} - \frac{\pi}{2}\right) = \sin 0 = 0$	$\left(\frac{\pi}{4}, 0\right)$
$\frac{\pi}{2}$	$y = \sin\left(2 \cdot \frac{\pi}{2} - \frac{\pi}{2}\right)$ $= \sin\left(\pi - \frac{\pi}{2}\right) = \sin\frac{\pi}{2} = 1$	$\left(\frac{\pi}{2}, 1\right)$
$\frac{3\pi}{4}$	$y = \sin\left(2 \cdot \frac{3\pi}{4} - \frac{\pi}{2}\right)$ $= \sin\left(\frac{3\pi}{2} - \frac{\pi}{2}\right)$ $= \sin\pi = 0$	$\left(\frac{3\pi}{4}, 0\right)$
π	$y = \sin\left(2 \cdot \pi - \frac{\pi}{2}\right)$ $= \sin\left(2\pi - \frac{\pi}{2}\right)$ $= \sin\frac{3\pi}{2} = -1$	$(\pi, -1)$
$\frac{5\pi}{4}$	$y = \sin\left(2 \cdot \frac{5\pi}{4} - \frac{\pi}{2}\right)$ $= \sin\left(\frac{5\pi}{2} - \frac{\pi}{2}\right)$ $= \sin 2\pi = 0$	$\left(\frac{5\pi}{4}, 0\right)$

Connect the five key points with a smooth curve and graph one complete cycle of the given function.

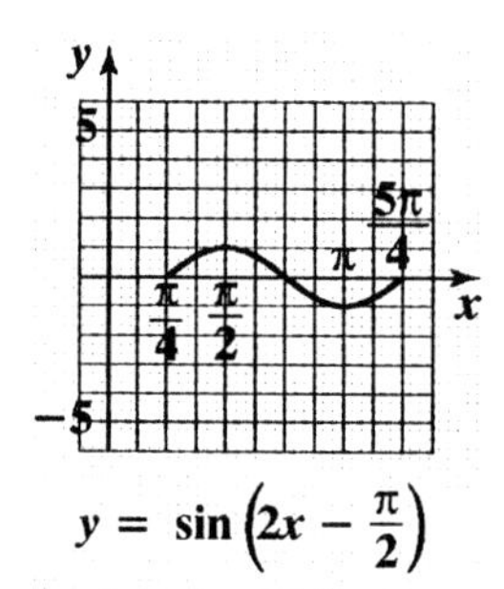

$y = \sin\left(2x - \frac{\pi}{2}\right)$

21. The equation $y = 3\sin(2x - \pi)$ is of the form $y = A\sin(Bx - C)$ with $A = 3$, $B = 2$, and $C = \pi$. The amplitude is $|A| = |3| = 3$. The period is $\frac{2\pi}{B} = \frac{2\pi}{2} = \pi$. The phase shift is $\frac{C}{B} = \frac{\pi}{2}$. The quarter-period is $\frac{\pi}{4}$. The cycle begins at $x = \frac{\pi}{2}$. Add quarter-periods to generate x-values for the key points.

$x = \frac{\pi}{2}$

$x = \frac{\pi}{2} + \frac{\pi}{4} = \frac{3\pi}{4}$

$x = \frac{3\pi}{4} + \frac{\pi}{4} = \pi$

$x = \pi + \frac{\pi}{4} = \frac{5\pi}{4}$

$x = \frac{5\pi}{4} + \frac{\pi}{4} = \frac{3\pi}{2}$

Evaluate the function at each value of x.

x	$y = 3\sin(2x - \pi)$	coordinates
$\frac{\pi}{2}$	$y = 3\sin\left(2 \cdot \frac{\pi}{2} - \pi\right)$ $= 3\sin(\pi - \pi)$ $= 3\sin 0 = 3 \cdot 0 = 0$	$\left(\frac{\pi}{2}, 0\right)$
$\frac{3\pi}{4}$	$y = 3\sin\left(2 \cdot \frac{3\pi}{4} - \pi\right)$ $= 3\sin\left(\frac{3\pi}{2} - \pi\right)$ $= 3\sin\frac{\pi}{2} = 3 \cdot 1 = 3$	$\left(\frac{3\pi}{4}, 3\right)$
π	$y = 3\sin(2 \cdot \pi - \pi)$ $= 3\sin(2\pi - \pi)$ $= 3\sin\pi = 3 \cdot 0 = 0$	$(\pi, 0)$
$\frac{5\pi}{4}$	$y = 3\sin\left(2 \cdot \frac{5\pi}{4} - \pi\right)$ $= 3\sin\left(\frac{5\pi}{2} - \pi\right)$ $= 3\sin\frac{3\pi}{2}$ $= 3(-1) = -3$	$\left(\frac{5\pi}{4}, -3\right)$
$\frac{3\pi}{2}$	$y = 3\sin\left(2 \cdot \frac{3\pi}{2} - \pi\right)$ $= 3\sin(3\pi - \pi)$ $= 3\sin 2\pi = 3 \cdot 0 = 0$	$\left(\frac{3\pi}{2}, 0\right)$

Connect the five points with a smooth curve and graph one complete cycle of the given function.

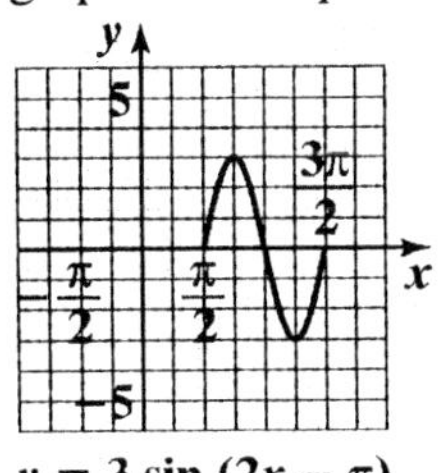

$y = 3\sin(2x - \pi)$

22. The equation $y = 3\sin\left(2x - \frac{\pi}{2}\right)$ is of the form $y = A\sin(Bx - C)$ with $A = 3$, $B = 2$, and $C = \frac{\pi}{2}$.
The amplitude is $|A| = |3| = 3$. The period is $\frac{2\pi}{B} = \frac{2\pi}{2} = \pi$. The phase shift is $\frac{C}{B} = \frac{\frac{\pi}{2}}{2} = \frac{\pi}{2} \cdot \frac{1}{2} = \frac{\pi}{4}$. The quarter-period is $\frac{\pi}{4}$.
The cycle begins at $x = \frac{\pi}{4}$. Add quarter-periods to generate x-values for the key points.

$x = \frac{\pi}{4}$

$x = \frac{\pi}{4} + \frac{\pi}{4} = \frac{\pi}{2}$

$x = \frac{\pi}{2} + \frac{\pi}{4} = \frac{3\pi}{4}$

$x = \frac{3\pi}{4} + \frac{\pi}{4} = \pi$

$x = \pi + \frac{\pi}{4} = \frac{5\pi}{4}$

Evaluate the function at each value of x.

x	$y = 3\sin\left(2x - \frac{\pi}{2}\right)$	coordinates
$\frac{\pi}{4}$	$y = 3\sin\left(2 \cdot \frac{\pi}{4} - \frac{\pi}{2}\right)$ $= \sin\left(\frac{\pi}{2} - \frac{\pi}{2}\right)$ $= 3\sin 0 = 3 \cdot 0 = 0$	$\left(\frac{\pi}{4}, 0\right)$

$\frac{\pi}{2}$	$y = 3\sin\left(2\cdot\frac{\pi}{2}-\frac{\pi}{2}\right)$ $= 3\sin\left(\pi-\frac{\pi}{2}\right)$ $= 3\sin\frac{\pi}{2} = 3\cdot 1 = 3$	$\left(\frac{\pi}{2}, 3\right)$
$\frac{3\pi}{4}$	$y = 3\sin\left(2\cdot\frac{3\pi}{4}-\frac{\pi}{2}\right)$ $= 3\sin\left(\frac{3\pi}{2}-\frac{\pi}{2}\right)$ $= 3\sin\pi = 3\cdot 0 = 0$	$\left(\frac{3\pi}{4}, 0\right)$
π	$y = 3\sin\left(2\cdot\pi-\frac{\pi}{2}\right)$ $= 3\sin\left(2\pi-\frac{\pi}{2}\right)$ $= 3\sin\frac{3\pi}{2} = 3\cdot(-1) = -3$	$(\pi, -3)$
$\frac{5\pi}{4}$	$y = 3\sin\left(2\cdot\frac{5\pi}{4}-\frac{\pi}{2}\right)$ $= 3\sin\left(\frac{5\pi}{2}-\frac{\pi}{2}\right)$ $= 3\sin 2\pi = 3\cdot 0 = 0$	$\left(\frac{5\pi}{4}, 0\right)$

Connect the five key points with a smooth curve and graph one complete cycle of the given function.

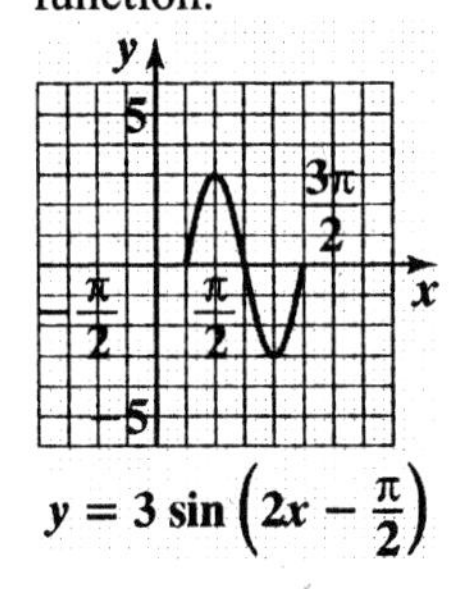

$y = 3\sin\left(2x-\frac{\pi}{2}\right)$

23. $y = \frac{1}{2}\sin\left(x+\frac{\pi}{2}\right) = \frac{1}{2}\sin\left(x-\left(-\frac{\pi}{2}\right)\right)$

The equation $y = \frac{1}{2}\sin\left(x-\left(-\frac{\pi}{2}\right)\right)$ is of the form $y = A\sin(Bx-C)$ with $A = \frac{1}{2}$, $B = 1$, and $C = -\frac{\pi}{2}$. The amplitude is $|A| = \left|\frac{1}{2}\right| = \frac{1}{2}$. The period is $\frac{2\pi}{B} = \frac{2\pi}{1} = 2\pi$. The phase shift is $\frac{C}{B} = \frac{-\frac{\pi}{2}}{1} = -\frac{\pi}{2}$. The quarter-period is $\frac{2\pi}{4} = \frac{\pi}{2}$.

The cycle begins at $x = -\frac{\pi}{2}$. Add quarter-periods to generate x-values for the key points.

$x = -\frac{\pi}{2}$

$x = -\frac{\pi}{2}+\frac{\pi}{2} = 0$

$x = 0+\frac{\pi}{2} = \frac{\pi}{2}$

$x = \frac{\pi}{2}+\frac{\pi}{2} = \pi$

$x = \pi+\frac{\pi}{2} = \frac{3\pi}{2}$

Evaluate the function at each value of x.

x	$y = \frac{1}{2}\sin\left(x+\frac{\pi}{2}\right)$	coordinates
$-\frac{\pi}{2}$	$y = \frac{1}{2}\sin\left(-\frac{\pi}{2}+\frac{\pi}{2}\right)$ $= \frac{1}{2}\sin 0 = \frac{1}{2}\cdot 0 = 0$	$\left(-\frac{\pi}{2}, 0\right)$
0	$y = \frac{1}{2}\sin\left(0+\frac{\pi}{2}\right)$ $= \frac{1}{2}\sin\frac{\pi}{2} = \frac{1}{2}\cdot 1 = \frac{1}{2}$	$\left(0, \frac{1}{2}\right)$
$\frac{\pi}{2}$	$y = \frac{1}{2}\sin\left(\frac{\pi}{2}+\frac{\pi}{2}\right)$ $= \frac{1}{2}\sin\pi = \frac{1}{2}\cdot 0 = 0$	$\left(\frac{\pi}{2}, 0\right)$
π	$y = \frac{1}{2}\sin\left(\pi+\frac{\pi}{2}\right)$ $= \frac{1}{2}\sin\frac{3\pi}{2}$ $= \frac{1}{2}\cdot(-1) = -\frac{1}{2}$	$\left(\pi, -\frac{1}{2}\right)$

$\frac{3\pi}{2}$	$y=\frac{1}{2}\sin\left(\frac{3\pi}{2}+\frac{\pi}{2}\right)$ $=\frac{1}{2}\sin 2\pi$ $=\frac{1}{2}\cdot 0=0$	$\left(\frac{3\pi}{2},0\right)$

Connect the five points with a smooth curve and graph one complete cycle of the given function.

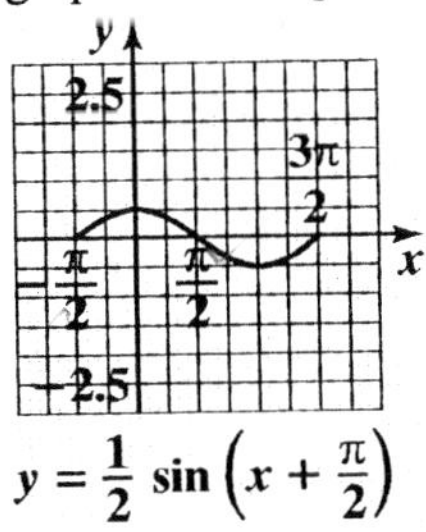

$y=\frac{1}{2}\sin\left(x+\frac{\pi}{2}\right)$

24. $y=\frac{1}{2}\sin(x+\pi)=\frac{1}{2}\sin(x-(-\pi))$

The equation $y=\frac{1}{2}\sin(x-(-\pi))$ is of the form $y=A\sin(Bx-C)$ with $A=\frac{1}{2}$, $B=1$, and $C=-\pi$. The amplitude is $|A|=\left|\frac{1}{2}\right|=\frac{1}{2}$. The period is $\frac{2\pi}{B}=\frac{2\pi}{1}=2\pi$. The phase shift is $\frac{C}{B}=\frac{-\pi}{1}=-\pi$. The quarter-period is $\frac{2\pi}{4}=\frac{\pi}{2}$.
The cycle begins at $x=-\pi$. Add quarter-periods to generate x-values for the key points.

$x=-\pi$

$x=-\pi+\frac{\pi}{2}=-\frac{\pi}{2}$

$x=-\frac{\pi}{2}+\frac{\pi}{2}=0$

$x=0+\frac{\pi}{2}=\frac{\pi}{2}$

$x=\frac{\pi}{2}+\frac{\pi}{2}=\pi$

Evaluate the function at each value of x.

x	$y=\frac{1}{2}\sin(x+\pi)$	coordinates
$-\pi$	$y=\frac{1}{2}\sin(-\pi+\pi)$ $=\frac{1}{2}\sin 0=\frac{1}{2}\cdot 0=0$	$(-\pi,0)$
$-\frac{\pi}{2}$	$y=\frac{1}{2}\sin\left(-\frac{\pi}{2}+\pi\right)$ $=\frac{1}{2}\sin\frac{\pi}{2}=\frac{1}{2}\cdot 1=\frac{1}{2}$	$\left(-\frac{\pi}{2},\frac{1}{2}\right)$
0	$y=\frac{1}{2}\sin(0+\pi)$ $=\frac{1}{2}\sin\pi=\frac{1}{2}\cdot 0=0$	$(0,0)$
$\frac{\pi}{2}$	$y=\frac{1}{2}\sin\left(\frac{\pi}{2}+\pi\right)$ $=\frac{1}{2}\sin\frac{3\pi}{2}=\frac{1}{2}\cdot(-1)=-\frac{1}{2}$	$\left(\frac{\pi}{2},-\frac{1}{2}\right)$
π	$y=\frac{1}{2}\sin(\pi+\pi)$ $=\frac{1}{2}\sin 2\pi=\frac{1}{2}\cdot 0=0$	$(\pi,0)$

Connect the five key points with a smooth curve and graph one complete cycle of the given function.

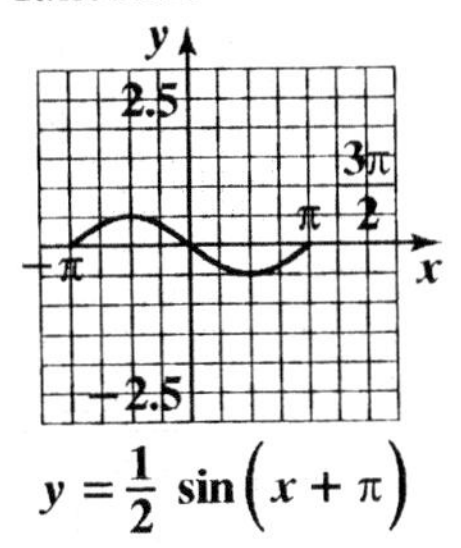

$y=\frac{1}{2}\sin(x+\pi)$

25. $y=-2\sin\left(2x+\frac{\pi}{2}\right)=-2\sin\left(2x-\left(-\frac{\pi}{2}\right)\right)$

The equation $y=-2\sin\left(2x-\left(-\frac{\pi}{2}\right)\right)$ is of the form $y=A\sin(Bx-C)$ with $A=-2$, $B=2$, and $C=-\frac{\pi}{2}$. The amplitude is $|A|=|-2|=2$. The period is $\frac{2\pi}{B}=\frac{2\pi}{2}=\pi$.

The phase shift is $\frac{C}{B}=\frac{-\frac{\pi}{2}}{2}=-\frac{\pi}{2}\cdot\frac{1}{2}=-\frac{\pi}{4}$. The quarter-period is $\frac{\pi}{4}$. The cycle begins at $x=-\frac{\pi}{4}$. Add quarter-periods to generate x-values for the key points.

$x = -\frac{\pi}{4}$

$x = -\frac{\pi}{4} + \frac{\pi}{4} = 0$

$x = 0 + \frac{\pi}{4} = \frac{\pi}{4}$

$x = \frac{\pi}{4} + \frac{\pi}{4} = \frac{\pi}{2}$

$x = \frac{\pi}{2} + \frac{\pi}{4} = \frac{3\pi}{4}$

Evaluate the function at each value of x.

x	$y = -2\sin\left(2x + \frac{\pi}{2}\right)$	coordinates
$-\frac{\pi}{4}$	$y = -2\sin\left(2\cdot\left(-\frac{\pi}{4}\right) + \frac{\pi}{2}\right)$ $= -2\sin\left(-\frac{\pi}{2} + \frac{\pi}{2}\right)$ $= -2\sin 0 = -2\cdot 0 = 0$	$\left(-\frac{\pi}{4}, 0\right)$
0	$y = -2\sin\left(2\cdot 0 + \frac{\pi}{2}\right)$ $= -2\sin\left(0 + \frac{\pi}{2}\right)$ $= -2\sin\frac{\pi}{2}$ $= -2\cdot 1 = -2$	(0, –2)
$\frac{\pi}{4}$	$y = -2\sin\left(2\cdot\frac{\pi}{4} + \frac{\pi}{2}\right)$ $= -2\sin\left(\frac{\pi}{2} + \frac{\pi}{2}\right)$ $= -2\sin\pi$ $= -2\cdot 0 = 0$	$\left(\frac{\pi}{4}, 0\right)$
$\frac{\pi}{2}$	$y = -2\sin\left(2\cdot\frac{\pi}{2} + \frac{\pi}{2}\right)$ $= -2\sin\left(\pi + \frac{\pi}{2}\right)$ $= -2\sin\frac{3\pi}{2}$ $= -2(-1) = 2$	$\left(\frac{\pi}{2}, 2\right)$
$\frac{3\pi}{4}$	$y = -2\sin\left(2\cdot\frac{3\pi}{4} + \frac{\pi}{2}\right)$ $= -2\sin\left(\frac{3\pi}{2} + \frac{\pi}{2}\right)$ $= -2\sin 2\pi$ $= -2\cdot 0 = 0$	$\left(\frac{3\pi}{4}, 0\right)$

Connect the five points with a smooth curve and graph one complete cycle of the given function.

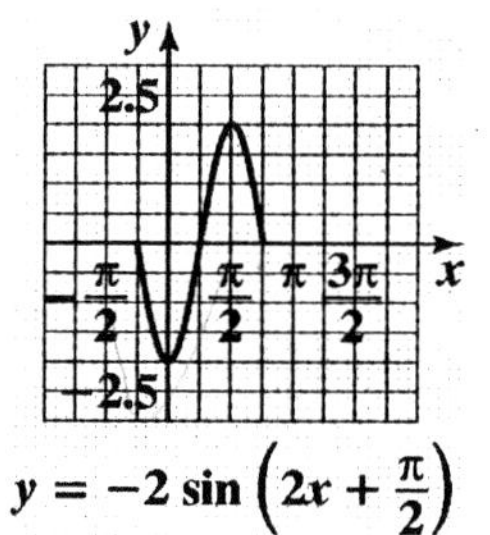

$y = -2\sin\left(2x + \frac{\pi}{2}\right)$

26. $y = -3\sin\left(2x + \frac{\pi}{2}\right) = -3\sin\left(2x - \left(-\frac{\pi}{2}\right)\right)$

The equation $y = -3\sin\left(2x - \left(-\frac{\pi}{2}\right)\right)$ is of the form $y = A\sin(Bx - C)$ with $A = -3$, $B = 2$, and $C = -\frac{\pi}{2}$. The amplitude is $|A| = |-3| = 3$. The period is $\frac{2\pi}{B} = \frac{2\pi}{2} = \pi$. The phase shift is $\frac{C}{B} = \frac{-\frac{\pi}{2}}{2} = -\frac{\pi}{2}\cdot\frac{1}{2} = -\frac{\pi}{4}$. The quarter-period is $\frac{\pi}{4}$. The cycle begins at $x = -\frac{\pi}{4}$. Add quarter-periods to generate x-values for the key points.

$x = -\frac{\pi}{4}$

$x = -\frac{\pi}{4} + \frac{\pi}{4} = 0$

$x = 0 + \frac{\pi}{4} = \frac{\pi}{4}$

$x = \frac{\pi}{4} + \frac{\pi}{4} = \frac{\pi}{2}$

$x = \frac{\pi}{2} + \frac{\pi}{4} = \frac{3\pi}{4}$

Evaluate the function at each value of x.

x	$y=-3\sin\left(2x+\frac{\pi}{2}\right)$	coordinates
$-\frac{\pi}{4}$	$y=-3\sin\left(2\cdot\left(-\frac{\pi}{4}\right)+\frac{\pi}{2}\right)$ $=-3\sin\left(-\frac{\pi}{2}+\frac{\pi}{2}\right)$ $=-3\sin 0=-3\cdot 0=0$	$\left(-\frac{\pi}{4},0\right)$
0	$y=-3\sin\left(2\cdot 0+\frac{\pi}{2}\right)$ $=-3\sin\left(0+\frac{\pi}{2}\right)$ $=-3\sin\frac{\pi}{2}=-3\cdot 1=-3$	(0, –3)
$\frac{\pi}{4}$	$y=-3\sin\left(2\cdot\frac{\pi}{4}+\frac{\pi}{2}\right)$ $=-3\sin\left(\frac{\pi}{2}+\frac{\pi}{2}\right)$ $=-3\sin\pi=-3\cdot 0=0$	$\left(\frac{\pi}{4},0\right)$
$\frac{\pi}{2}$	$y=-3\sin\left(2\cdot\frac{\pi}{2}+\frac{\pi}{2}\right)$ $=-3\sin\left(\pi+\frac{\pi}{2}\right)$ $=-3\sin\frac{3\pi}{2}=-3\cdot(-1)=3$	$\left(\frac{\pi}{2},3\right)$
$\frac{3\pi}{4}$	$y=-3\sin\left(2\cdot\frac{3\pi}{4}+\frac{\pi}{2}\right)$ $=-3\sin\left(\frac{3\pi}{2}+\frac{\pi}{2}\right)$ $=-3\sin 2\pi=-3\cdot 0=0$	$\left(\frac{3\pi}{4},0\right)$

Connect the five key points with a smooth curve and graph one complete cycle of the given function.

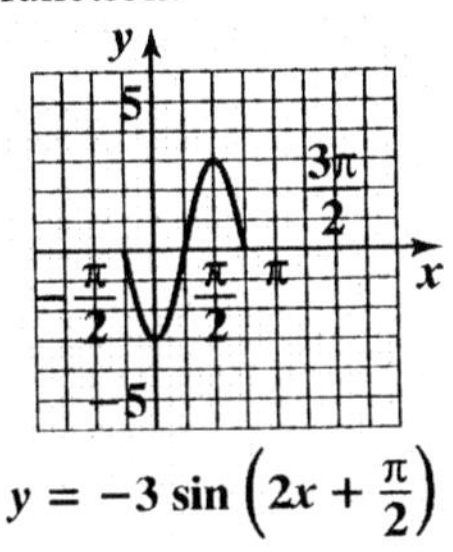

$y=-3\sin\left(2x+\frac{\pi}{2}\right)$

27. $y=3\sin(\pi x+2)$

The equation $y=3\sin(\pi x-(-2))$ is of the form $y=A\sin(Bx-C)$ with $A=3$, $B=\pi$, and $C=-2$. The amplitude is $|A|=|3|=3$. The period is $\frac{2\pi}{B}=\frac{2\pi}{\pi}=2$. The phase shift is $\frac{C}{B}=\frac{-2}{\pi}=-\frac{2}{\pi}$. The quarter-period is $\frac{2}{4}=\frac{1}{2}$.

The cycle begins at $x=-\frac{2}{\pi}$. Add quarter-periods to generate x-values for the key points.

$x=-\frac{2}{\pi}$

$x=-\frac{2}{\pi}+\frac{1}{2}=\frac{\pi-4}{2\pi}$

$x=\frac{\pi-4}{2\pi}+\frac{1}{2}=\frac{\pi-2}{\pi}$

$x=\frac{\pi-2}{\pi}+\frac{1}{2}=\frac{3\pi-4}{2\pi}$

$x=\frac{3\pi-4}{2\pi}+\frac{1}{2}=\frac{2\pi-2}{\pi}$

Evaluate the function at each value of x.

x	$y=3\sin(\pi x+2)$	coordinates
$-\frac{2}{\pi}$	$y=3\sin\left(\pi\left(-\frac{2}{\pi}\right)+2\right)$ $=3\sin(-2+2)$ $=3\sin 0=3\cdot 0=0$	$\left(-\frac{2}{\pi},0\right)$
$\frac{\pi-4}{2\pi}$	$y=3\sin\left(\pi\left(\frac{\pi-4}{2\pi}\right)+2\right)$ $=3\sin\left(\frac{\pi-4}{2}+2\right)$ $=3\sin\left(\frac{\pi}{2}-2+2\right)$ $=3\sin\frac{\pi}{2}$ $=3\cdot 1=3$	$\left(\frac{\pi-4}{2\pi},3\right)$
$\frac{\pi-2}{\pi}$	$y=3\sin\left(\pi\left(\frac{\pi-2}{\pi}\right)+2\right)$ $=3\sin(\pi-2+2)$ $=3\sin\pi=3\cdot 0=0$	$\left(\frac{\pi-2}{\pi},0\right)$

$\frac{3\pi-4}{2\pi}$	$y=3\sin\left(\pi\left(\frac{3\pi-4}{2\pi}\right)+2\right)$ $=3\sin\left(\frac{3\pi-4}{2}+2\right)$ $=3\sin\left(\frac{3\pi}{2}-2+2\right)$ $=3\sin\frac{3\pi}{2}$ $=3(-1)=-3$	$\left(\frac{5\pi}{4},-3\right)$
$\frac{2\pi-2}{\pi}$	$y=3\sin\left(\pi\left(\frac{2\pi-2}{\pi}\right)+2\right)$ $=3\sin(2\pi-2+2)$ $=3\sin 2\pi=3\cdot 0=0$	$\left(\frac{2\pi-2}{\pi},0\right)$

Connect the five points with a smooth curve and graph one complete cycle of the given function.

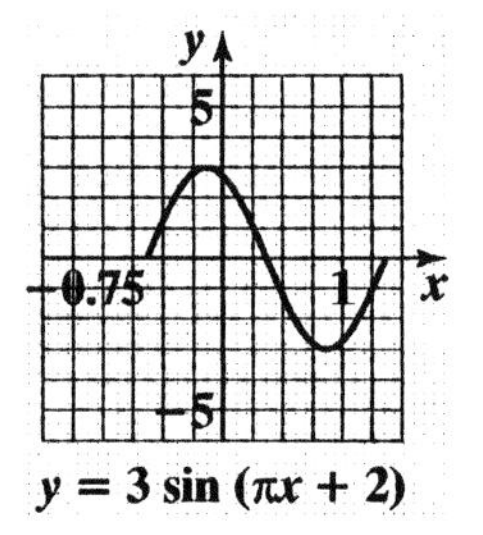

$y = 3\sin(\pi x + 2)$

28. $y=3\sin(2\pi x+4)=3\sin(2\pi x-(-4))$
The equation $y=3\sin(2\pi x-(-4))$ is of the form $y=A\sin(Bx-C)$ with $A=3$, $B=2\pi$, and $C=-4$. The amplitude is $|A|=|3|=3$. The period is $\frac{2\pi}{B}=\frac{2\pi}{2\pi}=1$. The phase shift is $\frac{C}{B}=\frac{-4}{2\pi}=-\frac{2}{\pi}$. The quarter-period is $\frac{1}{4}$. The cycle begins at $x=-\frac{2}{\pi}$. Add quarter-periods to generate x-values for the key points.

$x=-\frac{2}{\pi}$

$x=-\frac{2}{\pi}+\frac{1}{4}=\frac{\pi-8}{4\pi}$

$x=\frac{\pi-8}{4\pi}+\frac{1}{4}=\frac{\pi-4}{2\pi}$

$x=\frac{\pi-4}{2\pi}+\frac{1}{4}=\frac{3\pi-8}{4\pi}$

$x=\frac{3\pi-8}{4\pi}+\frac{1}{4}=\frac{\pi-2}{\pi}$

Evaluate the function at each value of x.

x	$y=3\sin(2\pi x+4)$	coordinates
$-\frac{2}{\pi}$	$y=3\sin\left(2\pi\left(-\frac{2}{\pi}\right)+4\right)$ $=3\sin(-4+4)$ $=3\sin 0=3\cdot 0=0$	$\left(-\frac{2}{\pi},0\right)$
$\frac{\pi-8}{4\pi}$	$y=3\sin\left(2\pi\left(\frac{\pi-8}{4\pi}\right)+4\right)$ $=3\sin\left(\frac{\pi-8}{2}+4\right)$ $=3\sin\left(\frac{\pi}{2}-4+4\right)$ $=3\sin\frac{\pi}{2}=3\cdot 1=3$	$\left(\frac{\pi-8}{4\pi},3\right)$
$\frac{\pi-4}{2\pi}$	$y=3\sin\left(2\pi\left(\frac{\pi-4}{2\pi}\right)+4\right)$ $=3\sin(\pi-4+4)$ $=3\sin\pi=3\cdot 0=0$	$\left(\frac{\pi-4}{2\pi},0\right)$
$\frac{3\pi-8}{4\pi}$	$y=3\sin\left(2\pi\left(\frac{3\pi-8}{4\pi}\right)+4\right)$ $=3\sin\left(\frac{3\pi-8}{2}+4\right)$ $=3\sin\left(\frac{3\pi}{2}-4+4\right)$ $=3\sin\frac{3\pi}{2}=3(-1)=-3$	$\left(\frac{3\pi-8}{4\pi},-3\right)$
$\frac{\pi-2}{\pi}$	$y=3\sin\left(2\pi\left(\frac{\pi-2}{\pi}\right)+4\right)$ $=3\sin(2\pi-4+4)$ $=3\sin 2\pi=3\cdot 0=0$	$\left(\frac{\pi-2}{\pi},0\right)$

Connect the five key points with a smooth curve and graph one complete cycle of the given function.

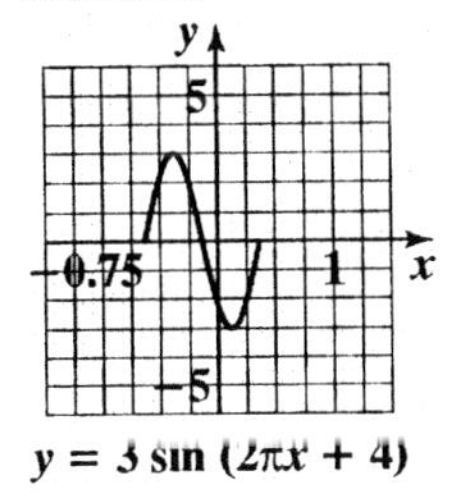

29. $y = -2\sin(2\pi x + 4\pi) = -2\sin(2\pi x - (-4\pi))$
The equation $y = -2\sin(2\pi x - (-4\pi))$ is of the form $y = A\sin(Bx - C)$ with $A = -2$, $B = 2\pi$, and $C = -4\pi$. The amplitude is $|A| = |-2| = 2$. The period is $\frac{2\pi}{B} = \frac{2\pi}{2\pi} = 1$.

The phase shift is $\frac{C}{B} = \frac{-4\pi}{2\pi} = -2$. The quarter-period is $\frac{1}{4}$. The cycle begins at $x = -2$. Add quarter-periods to generate x-values for the key points.

$x = -2$

$x = -2 + \frac{1}{4} = -\frac{7}{4}$

$x = -\frac{7}{4} + \frac{1}{4} = -\frac{3}{2}$

$x = -\frac{3}{2} + \frac{1}{4} = -\frac{5}{4}$

$x = -\frac{5}{4} + \frac{1}{4} = -1$

Evaluate the function at each value of x.

x	$y = -2\sin(2\pi x + 4\pi)$	coordinates
-2	$y = -2\sin(2\pi(-2) + 4\pi)$ $= -2\sin(-4\pi + 4\pi)$ $= -2\sin 0$ $= -2 \cdot 0 = 0$	$(-2, 0)$
$-\frac{7}{4}$	$y = -2\sin\left(2\pi\left(-\frac{7}{4}\right) + 4\pi\right)$ $= -2\sin\left(-\frac{7\pi}{2} + 4\pi\right)$ $= -2\sin\frac{\pi}{2} = -2 \cdot 1 = -2$	$\left(-\frac{7}{4}, -2\right)$
$-\frac{3}{2}$	$y = -2\sin\left(2\pi\left(-\frac{3}{2}\right) + 4\pi\right)$ $= -2\sin(-3\pi + 4\pi)$ $= -2\sin\pi = -2 \cdot 0 = 0$	$\left(-\frac{3}{2}, 0\right)$
$-\frac{5}{4}$	$y = -2\sin\left(2\pi\left(-\frac{5}{4}\right) + 4\pi\right)$ $= -2\sin\left(-\frac{5\pi}{2} + 4\pi\right)$ $= -2\sin\frac{3\pi}{2}$ $= -2(-1) = 2$	$\left(-\frac{5}{4}, 2\right)$
-1	$y = -2\sin(2\pi(-1) + 4\pi)$ $= -2\sin(-2\pi + 4\pi)$ $= -2\sin 2\pi$ $= -2 \cdot 0 = 0$	$(-1, 0)$

Connect the five points with a smooth curve and graph one complete cycle of the given function.

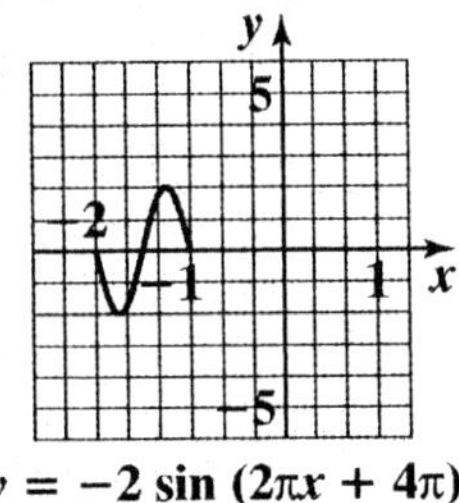

30. $y = -3\sin(2\pi x + 4\pi) = -3\sin(2\pi x - (-4\pi))$
The equation $y = -3\sin(2\pi x - (-4\pi))$ is of the form $y = A\sin(Bx - C)$ with $A = -3$, $B = 2\pi$, and $C = -4\pi$. The amplitude is $|A| = |-3| = 3$.
The period is $\frac{2\pi}{B} = \frac{2\pi}{2\pi} = 1$. The phase shift is

$\frac{C}{B} = \frac{-4\pi}{2\pi} = -2$. The quarter-period is $\frac{1}{4}$. The cycle begins at $x = -2$. Add quarter-periods to generate x-values for the key points.

$x = -2$

$x = -2 + \frac{1}{4} = -\frac{7}{4}$

$x = -\frac{7}{4} + \frac{1}{4} = -\frac{3}{2}$

$x = -\frac{3}{2} + \frac{1}{4} = -\frac{5}{4}$

$x = -\frac{5}{4} + \frac{1}{4} = -1$

Evaluate the function at each value of x.

x	$y = -3\sin(2\pi x + 4\pi)$	coordinates
-2	$y = -3\sin(2\pi(-2) + 4\pi)$ $= -3\sin(-4\pi + 4\pi)$ $= -3\sin 0 = -3 \cdot 0 = 0$	$(-2, 0)$
$-\frac{7}{4}$	$y = -3\sin\left(2\pi\left(-\frac{7}{4}\right) + 4\pi\right)$ $= -3\sin\left(-\frac{7\pi}{2} + 4\pi\right)$ $= -3\sin\frac{\pi}{2} = -3 \cdot 1 = -3$	$\left(-\frac{7}{4}, -3\right)$
$-\frac{3}{2}$	$y = -3\sin\left(2\pi\left(-\frac{3}{2}\right) + 4\pi\right)$ $= -3\sin(-3\pi + 4\pi)$ $= -3\sin\pi = -3 \cdot 0 = 0$	$\left(-\frac{3}{2}, 0\right)$
$-\frac{5}{4}$	$y = -3\sin\left(2\pi\left(-\frac{5}{4}\right) + 4\pi\right)$ $= -3\sin\left(-\frac{5\pi}{2} + 4\pi\right)$ $= -3\sin\frac{3\pi}{2} = -3(-1) = 3$	$\left(-\frac{5}{4}, 3\right)$
-1	$y = -3\sin(2\pi(-1) + 4\pi)$ $= -3\sin(-2\pi + 4\pi)$ $= -3\sin 2\pi = -3 \cdot 0 = 0$	$(-1, 0)$

Connect the five key points with a smooth curve and graph one complete cycle of the given function.

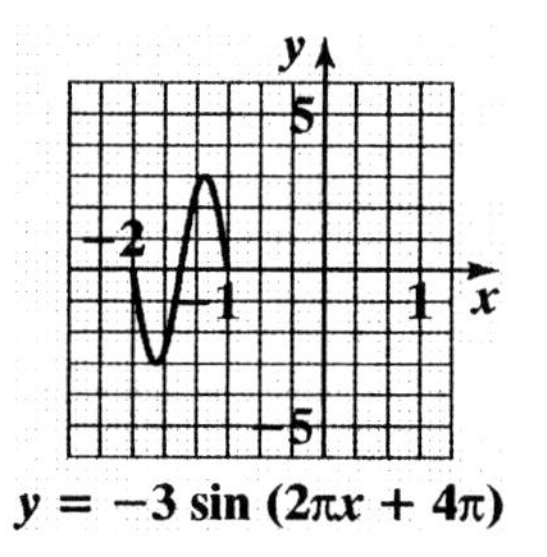

$y = -3\sin(2\pi x + 4\pi)$

31. The equation $y = 2\cos x$ is of the form $y = A\cos x$ with $A = 2$. Thus, the amplitude is $|A| = |2| = 2$. The period is 2π. The quarter-period is $\frac{2\pi}{4}$ or $\frac{\pi}{2}$. The cycle begins at $x = 0$. Add quarter-periods to generate x-values for the key points.

$x = 0$

$x = 0 + \frac{\pi}{2} = \frac{\pi}{2}$

$x = \frac{\pi}{2} + \frac{\pi}{2} = \pi$

$x = \pi + \frac{\pi}{2} = \frac{3\pi}{2}$

$x = \frac{3\pi}{2} + \frac{\pi}{2} = 2\pi$

Evaluate the function at each value of x.

x	$y = 2\cos x$	coordinates
0	$y = 2\cos 0$ $= 2 \cdot 1 = 2$	$(0, 2)$
$\frac{\pi}{2}$	$y = 2\cos\frac{\pi}{2}$ $= 2 \cdot 0 = 0$	$\left(\frac{\pi}{2}, 0\right)$
π	$y = 2\cos\pi$ $= 2 \cdot (-1) = -2$	$(\pi, -2)$
$\frac{3\pi}{2}$	$y = 2\cos\frac{3\pi}{2}$ $= 2 \cdot 0 = 0$	$\left(\frac{3\pi}{2}, 0\right)$
2π	$y = 2\cos 2\pi$ $= 2 \cdot 1 = 2$	$(2\pi, 2)$

Connect the five points with a smooth curve and graph one complete cycle of the given function with the graph of $y = 2\cos x$.

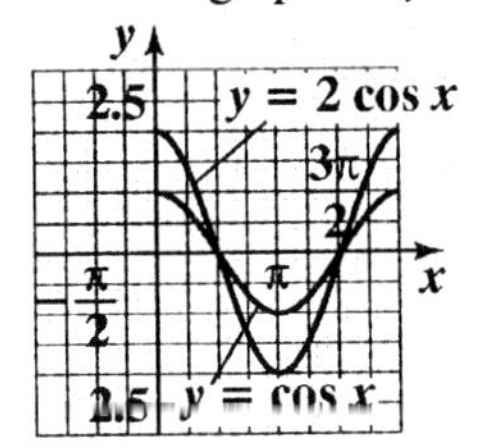

32. The equation $y = 3\cos x$ is of the form $y = A\cos x$ with $A = 3$. Thus, the amplitude is $|A| = |3| = 3$. The period is 2π. The quarter-period is $\frac{2\pi}{4}$ or $\frac{\pi}{2}$. The cycle begins at $x = 0$. Add quarter-periods to generate x-values for the key points.

$x = 0$

$x = 0 + \frac{\pi}{2} = \frac{\pi}{2}$

$x = \frac{\pi}{2} + \frac{\pi}{2} = \pi$

$x = \pi + \frac{\pi}{2} = \frac{3\pi}{2}$

$x = \frac{3\pi}{2} + \frac{\pi}{2} = 2\pi$

Evaluate the function at each value of x.

x	$y = 3\cos x$	coordinates
0	$y = 3\cos 0 = 3\cdot 1 = 3$	(0, 3)
$\frac{\pi}{2}$	$y = 3\cos\frac{\pi}{2} = 3\cdot 0 = 0$	$\left(\frac{\pi}{2}, 0\right)$
π	$y = 3\cos\pi = 3\cdot(-1) = -3$	$(\pi, -3)$
$\frac{3\pi}{2}$	$y = 3\cos\frac{3\pi}{2} = 3\cdot 0 = 0$	$\left(\frac{3\pi}{2}, 0\right)$
2π	$y = 3\cos 2\pi = 3\cdot 1 = 3$	$(2\pi, 3)$

Connect the five key points with a smooth curve and graph one complete cycle of the given function with the graph of $y = \cos x$.

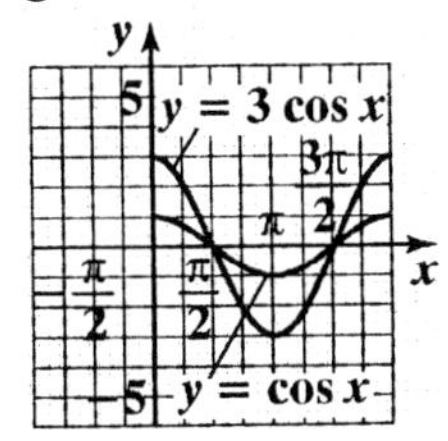

33. The equation $y = -2\cos x$ is of the form $y = A\cos x$ with $A = -2$. Thus, the amplitude is $|A| = |-2| = 2$. The period is 2π. The quarter-period is $\frac{2\pi}{4}$ or $\frac{\pi}{2}$. The cycle begins at $x = 0$. Add quarter-periods to generate x-values for the key points.

$x = 0$

$x = 0 + \frac{\pi}{2} = \frac{\pi}{2}$

$x = \frac{\pi}{2} + \frac{\pi}{2} = \pi$

$x = \pi + \frac{\pi}{2} = \frac{3\pi}{2}$

$x = \frac{3\pi}{2} + \frac{\pi}{2} = 2\pi$

Evaluate the function at each value of x.

x	$y = -2\cos x$	coordinates
0	$y = -2\cos 0$ $= -2\cdot 1 = -2$	(0, –2)
$\frac{\pi}{2}$	$y = -2\cos\frac{\pi}{2}$ $= -2\cdot 0 = 0$	$\left(\frac{\pi}{2}, 0\right)$
π	$y = -2\cos\pi$ $= -2\cdot(-1) = 2$	$(\pi, 2)$
$\frac{3\pi}{2}$	$y = -2\cos\frac{3\pi}{2}$ $= -2\cdot 0 = 0$	$\left(\frac{3\pi}{2}, 0\right)$
2π	$y = -2\cos 2\pi$ $= -2\cdot 1 = -2$	$(2\pi, -2)$

Connect the five points with a smooth curve and graph one complete cycle of the given function with the graph of $y = \cos x$.

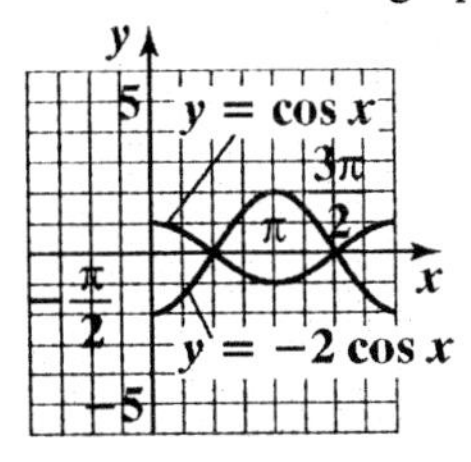

34. The equation $y = -3\cos x$ is of the form $y = A\cos x$ with $A = -3$. Thus, the amplitude is $|A| = |-3| = 3$. The period is 2π. The quarter-period is $\frac{2\pi}{4}$ or $\frac{\pi}{2}$. The cycle begins at $x = 0$. Add quarter-periods to generate x-values for the key points.

$x = 0$

$x = 0 + \frac{\pi}{2} = \frac{\pi}{2}$

$x = \frac{\pi}{2} + \frac{\pi}{2} = \pi$

$x = \pi + \frac{\pi}{2} = \frac{3\pi}{2}$

$x = \frac{3\pi}{2} + \frac{\pi}{2} = 2\pi$

Evaluate the function at each value of x.

x	$y = -3\cos x$	coordinates
0	$y = -3\cos 0 = -3\cdot 1 = -3$	$(0, -3)$
$\frac{\pi}{2}$	$y = -3\cos\frac{\pi}{2} = -3\cdot 0 = 0$	$\left(\frac{\pi}{2}, 0\right)$
π	$y = -3\cos\pi = -3\cdot(-1) = 3$	$(\pi, 3)$
$\frac{3\pi}{2}$	$y = -3\cos\frac{3\pi}{2} = -3\cdot 0 = 0$	$\left(\frac{3\pi}{2}, 0\right)$
2π	$y = -3\cos 2\pi = -3\cdot 1 = -3$	$(2\pi, -3)$

Connect the five key points with a smooth curve and graph one complete cycle of the given function with the graph of $y = \cos x$.

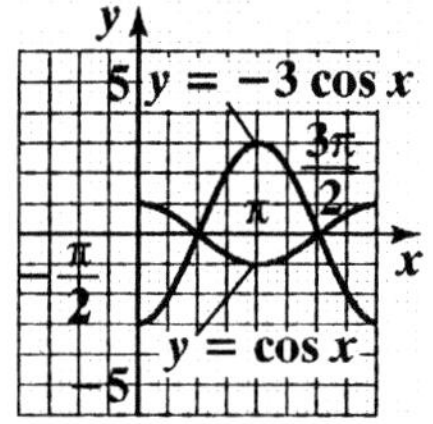

35. The equation $y = \cos 2x$ is of the form $y = A\cos Bx$ with $A = 1$ and $B = 2$. Thus, the amplitude is $|A| = |1| = 1$. The period is $\frac{2\pi}{B} = \frac{2\pi}{2} = \pi$. The quarter-period is $\frac{\pi}{4}$. The cycle begins at $x = 0$. Add quarter-periods to generate x-values for the key points.

$x = 0$

$x = 0 + \frac{\pi}{4} = \frac{\pi}{4}$

$x = \frac{\pi}{4} + \frac{\pi}{4} = \frac{\pi}{2}$

$x = \frac{\pi}{2} + \frac{\pi}{4} = \frac{3\pi}{4}$

$x = \frac{3\pi}{4} + \frac{\pi}{4} = \pi$

Evaluate the function at each value of x.

x	$y = \cos 2x$	coordinates
0	$y = \cos(2\cdot 0) = \cos 0 = 1$	$(0, 1)$
$\frac{\pi}{4}$	$y = \cos\left(2\cdot\frac{\pi}{4}\right) = \cos\frac{\pi}{2} = 0$	$\left(\frac{\pi}{4}, 0\right)$
$\frac{\pi}{2}$	$y = \cos\left(2\cdot\frac{\pi}{2}\right) = \cos\pi = -1$	$\left(\frac{\pi}{2}, -1\right)$
$\frac{3\pi}{4}$	$y = \cos\left(2\cdot\frac{3\pi}{4}\right) = \cos\frac{3\pi}{2} = 0$	$\left(\frac{3\pi}{4}, 0\right)$
π	$y = \cos(2\cdot\pi) = \cos 2\pi = 1$	$(\pi, 1)$

Connect the five points with a smooth curve and graph one complete cycle of the given function.

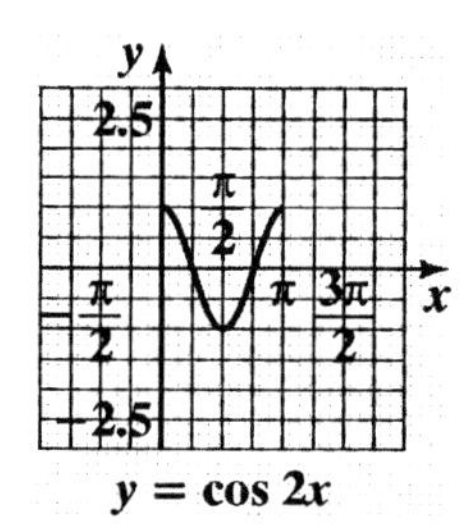

36. The equation $y = \cos 4x$ is of the form $y = A\cos Bx$ with $A = 1$ and $B = 4$. Thus, the amplitude is $|A| = |1| = 1$. The period is $\frac{2\pi}{B} = \frac{2\pi}{4} = \frac{\pi}{2}$. The quarter-period is $\frac{\frac{\pi}{2}}{4} = \frac{\pi}{2}\cdot\frac{1}{4} = \frac{\pi}{8}$. The cycle begins at $x = 0$. Add quarter-periods to generate x-values for the key points.

$x = 0$

$x = 0 + \frac{\pi}{8} = \frac{\pi}{8}$

$x = \frac{\pi}{8} + \frac{\pi}{8} = \frac{\pi}{4}$

$x = \frac{\pi}{4} + \frac{\pi}{8} = \frac{3\pi}{8}$

$x = \frac{3\pi}{8} + \frac{\pi}{8} = \frac{\pi}{2}$

Evaluate the function at each value of x.

x	$y = \cos 4x$	coordinates
0	$y = \cos(4 \cdot 0) = \cos 0 = 1$	(0, 1)
$\frac{\pi}{8}$	$y = \cos\left(4 \cdot \frac{\pi}{8}\right) = \cos\frac{\pi}{2} = 0$	$\left(\frac{\pi}{8}, 0\right)$
$\frac{\pi}{4}$	$y = \cos\left(4 \cdot \frac{\pi}{4}\right) = \cos \pi = -1$	$\left(\frac{\pi}{4}, -1\right)$
$\frac{3\pi}{8}$	$y = \cos\left(4 \cdot \frac{3\pi}{8}\right)$ $= \cos\frac{3\pi}{2} = 0$	$\left(\frac{3\pi}{8}, 0\right)$
$\frac{\pi}{2}$	$y = \cos\left(4 \cdot \frac{\pi}{2}\right) = \cos 2\pi = 1$	$\left(\frac{\pi}{2}, 1\right)$

Connect the five key points with a smooth curve and graph one complete cycle of the given function.

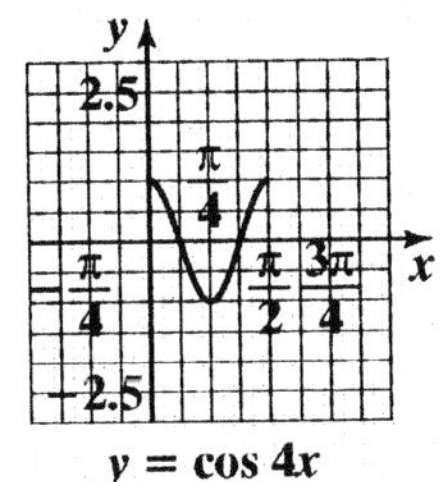

y* = cos 4*x

37. The equation $y = 4\cos 2\pi x$ is of the form $y = A\cos Bx$ with $A = 4$ and $B = 2\pi$. Thus, the amplitude is $|A| = |4| = 4$. The period is $\frac{2\pi}{B} = \frac{2\pi}{2\pi} = 1$. The quarter-period is $\frac{1}{4}$. The cycle begins at $x = 0$. Add quarter-periods to generate x-values for the key points.

$x = 0$

$x = 0 + \frac{1}{4} = \frac{1}{4}$

$x = \frac{1}{4} + \frac{1}{4} = \frac{1}{2}$

$x = \frac{1}{2} + \frac{1}{4} = \frac{3}{4}$

$x = \frac{3}{4} + \frac{1}{4} = 1$

Evaluate the function at each value of x.

x	$y = 4\cos 2\pi x$	coordinates
0	$y = 4\cos(2\pi \cdot 0)$ $= 4\cos 0$ $= 4 \cdot 1 = 4$	(0, 4)
$\frac{1}{4}$	$y = 4\cos\left(2\pi \cdot \frac{1}{4}\right)$ $= 4\cos\frac{\pi}{2}$ $= 4 \cdot 0 = 0$	$\left(\frac{1}{4}, 0\right)$
$\frac{1}{2}$	$y = 4\cos\left(2\pi \cdot \frac{1}{2}\right)$ $= 4\cos \pi$ $= 4 \cdot (-1) = -4$	$\left(\frac{1}{2}, -4\right)$
$\frac{3}{4}$	$y = 4\cos\left(2\pi \cdot \frac{3}{4}\right)$ $= 4\cos\frac{3\pi}{2}$ $= 4 \cdot 0 = 0$	$\left(\frac{3}{4}, 0\right)$
1	$y = 4\cos(2\pi \cdot 1)$ $= 4\cos 2\pi$ $= 4 \cdot 1 = 4$	(1, 4)

Connect the five points with a smooth curve and graph one complete cycle of the given function.

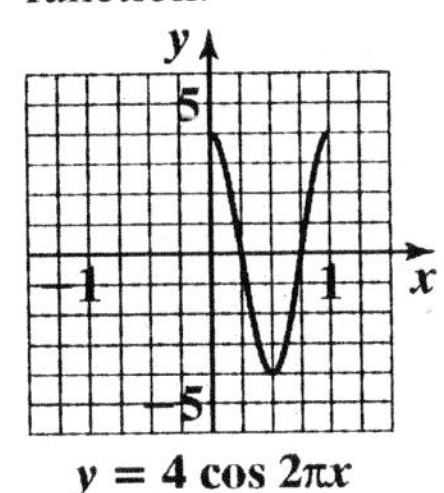

y* = 4 cos 2π*x

38. The equation $y = 5\cos 2\pi x$ is of the form $y = A\cos Bx$ with $A = 5$ and $B = 2\pi$. Thus, the amplitude is $|A| = |5| = 5$. The period is $\frac{2\pi}{B} = \frac{2\pi}{2\pi} = 1$. The quarter-period is $\frac{1}{4}$. The cycle begins at $x = 0$. Add quarter-periods to generate x-values for the key points.

$x = 0$

$x = 0 + \frac{1}{4} = \frac{1}{4}$

$x = \frac{1}{4} + \frac{1}{4} = \frac{1}{2}$

$x = \frac{1}{2} + \frac{1}{4} = \frac{3}{4}$

$x = \frac{3}{4} + \frac{1}{4} = 1$

Evaluate the function at each value of x.

x	$y = 5\cos 2\pi x$	coordinates
0	$y = 5\cos(2\pi \cdot 0)$ $= 5\cos 0 = 5 \cdot 1 = 5$	(0, 5)
$\frac{1}{4}$	$y = 5\cos\left(2\pi \cdot \frac{1}{4}\right)$ $= 5\cos\frac{\pi}{2} = 5 \cdot 0 = 0$	$\left(\frac{1}{4}, 0\right)$
$\frac{1}{2}$	$y = 5\cos\left(2\pi \cdot \frac{1}{2}\right)$ $= 5\cos\pi = 5 \cdot (-1) = -5$	$\left(\frac{1}{2}, -5\right)$
$\frac{3}{4}$	$y = 5\cos\left(2\pi \cdot \frac{3\pi}{4}\right)$ $= 5\cos\frac{3\pi}{2} = 5 \cdot 0 = 0$	$\left(\frac{3}{4}, 0\right)$
1	$y = 5\cos(2\pi \cdot 1)$ $= 5\cos 2\pi = 5 \cdot 1 = 5$	(1, 5)

Connect the five key points with a smooth curve and graph one complete cycle of the given function.

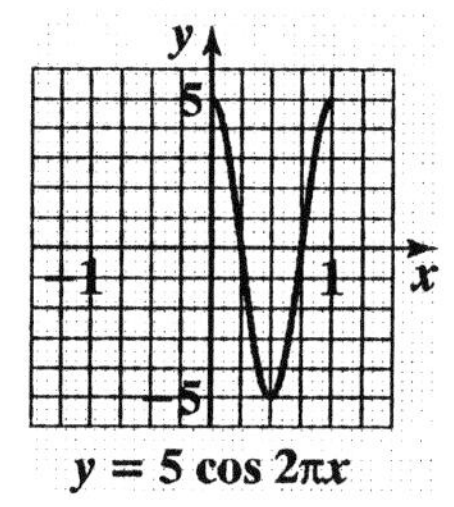

$y = 5\cos 2\pi x$

39. The equation $y = -4\cos\frac{1}{2}x$ is of the form

$y = A\cos Bx$ with $A = -4$ and $B = \frac{1}{2}$. Thus, the amplitude is $|A| = |-4| = 4$. The period is $\frac{2\pi}{B} = \frac{2\pi}{\frac{1}{2}} = 2\pi \cdot 2 = 4\pi$. The quarter-period is $\frac{4\pi}{4} = \pi$. The cycle begins at $x = 0$. Add quarter-periods to generate x-values for the key points.

$x = 0$

$x = 0 + \pi = \pi$

$x = \pi + \pi = 2\pi$

$x = 2\pi + \pi = 3\pi$

$x = 3\pi + \pi = 4\pi$

Evaluate the function at each value of x.

x	$y = -4\cos\frac{1}{2}x$	coordinates
0	$y = -4\cos\left(\frac{1}{2} \cdot 0\right)$ $= -4\cos 0$ $= -4 \cdot 1 = -4$	(0, –4)
π	$y = -4\cos\left(\frac{1}{2} \cdot \pi\right)$ $= -4\cos\frac{\pi}{2}$ $= -4 \cdot 0 = 0$	(π, 0)
2π	$y = -4\cos\left(\frac{1}{2} \cdot 2\pi\right)$ $= -4\cos\pi$ $= -4 \cdot (-1) = 4$	(2π, 4)
3π	$y = -4\cos\left(\frac{1}{2} \cdot 3\pi\right)$ $= -4\cos\frac{3\pi}{2}$ $= -4 \cdot 0 = 0$	(3π, 0)
4π	$y = -4\cos\left(\frac{1}{2} \cdot 4\pi\right)$ $= -4\cos 2\pi$ $= -4 \cdot 1 = -4$	(4π, – 4)

Connect the five points with a smooth curve and graph one complete cycle of the given function.

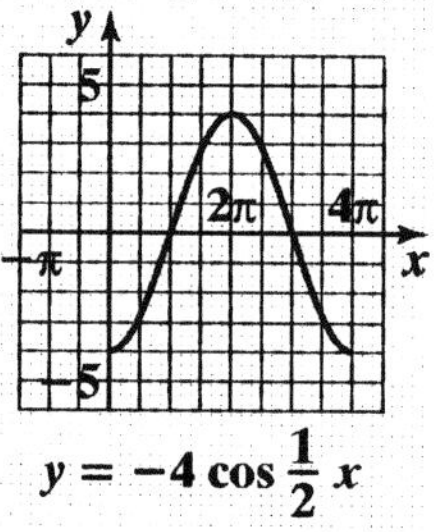

$y = -4\cos\frac{1}{2}x$

40. The equation $y = -3\cos\frac{1}{3}x$ is of the form

$y = A\cos Bx$ with $A = -3$ and $B = \frac{1}{3}$. Thus, the amplitude is $|A| = |-3| = 3$. The period is $\frac{2\pi}{B} = \frac{2\pi}{\frac{1}{3}} = 2\pi \cdot 3 = 6\pi$. The quarter-period is $\frac{6\pi}{4} = \frac{3\pi}{2}$. The cycle begins at $x = 0$. Add quarter-periods to generate x-values for the key points.

$x = 0$

$x = 0 + \frac{3\pi}{2} = \frac{3\pi}{2}$

$x = \frac{3\pi}{2} + \frac{3\pi}{2} = 3\pi$

$x = 3\pi + \frac{3\pi}{2} = \frac{9\pi}{2}$

$x = \frac{9\pi}{2} + \frac{3\pi}{2} = 6\pi$

Evaluate the function at each value of x.

x	$y = -3\cos\frac{1}{3}x$	coordinates
0	$y = -3\cos\left(\frac{1}{3}\cdot 0\right)$ $= -3\cos 0 = -3\cdot 1 = -3$	$(0, -3)$
$\frac{3\pi}{2}$	$y = -3\cos\left(\frac{1}{3}\cdot\frac{3\pi}{2}\right)$ $= -3\cos\frac{\pi}{2} = -3\cdot 0 = 0$	$\left(\frac{3\pi}{2}, 0\right)$
3π	$y = -3\cos\left(\frac{1}{3}\cdot 3\pi\right)$ $= -3\cos\pi = -3\cdot(-1) = 3$	$(3\pi, 3)$
$\frac{9\pi}{2}$	$y = -3\cos\left(\frac{1}{3}\cdot\frac{9\pi}{2}\right)$ $= -3\cos\frac{3\pi}{2} = -3\cdot 0 = 0$	$\left(\frac{9\pi}{2}, 0\right)$
6π	$y = -3\cos\left(\frac{1}{3}\cdot 6\pi\right)$ $= -3\cos 2\pi = -3\cdot 1 = -3$	$(6\pi, -3)$

Connect the five key points with a smooth curve and graph one complete cycle of the given function.

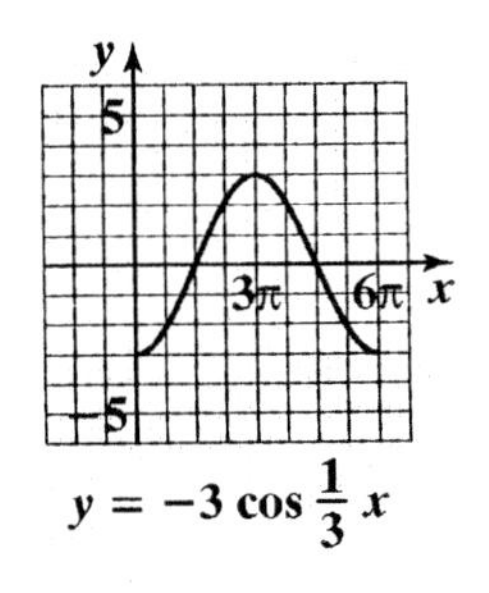

$y = -3\cos\frac{1}{3}x$

41. The equation $y = -\frac{1}{2}\cos\frac{\pi}{3}x$ is of the form $y = A\cos Bx$ with $A = -\frac{1}{2}$ and $B = \frac{\pi}{3}$. Thus, the amplitude is $|A| = \left|-\frac{1}{2}\right| = \frac{1}{2}$. The period is $\frac{2\pi}{B} = \frac{2\pi}{\frac{\pi}{3}} = 2\pi\cdot\frac{3}{\pi} = 6$. The quarter-period is $\frac{6}{4} = \frac{3}{2}$. The cycle begins at $x = 0$. Add quarter-periods to generate x-values for the key points.

$x = 0$

$x = 0 + \frac{3}{2} = \frac{3}{2}$

$x = \frac{3}{2} + \frac{3}{2} = 3$

$x = 3 + \frac{3}{2} = \frac{9}{2}$

$x = \frac{9}{2} + \frac{3}{2} = 6$

Evaluate the function at each value of x.

x	$y = -\frac{1}{2}\cos\frac{\pi}{3}x$	coordinates
0	$y = -\frac{1}{2}\cos\left(\frac{\pi}{3}\cdot 0\right)$ $= -\frac{1}{2}\cos 0$ $= -\frac{1}{2}\cdot 1 = -\frac{1}{2}$	$\left(0, -\frac{1}{2}\right)$
$\frac{3}{2}$	$y = -\frac{1}{2}\cos\left(\frac{\pi}{3}\cdot\frac{3}{2}\right)$ $= -\frac{1}{2}\cos\frac{\pi}{2}$ $= -\frac{1}{2}\cdot 0 = 0$	$\left(\frac{3}{2}, 0\right)$
3	$y = -\frac{1}{2}\cos\left(\frac{\pi}{3}\cdot 3\right)$ $= -\frac{1}{2}\cos\pi$ $= -\frac{1}{2}\cdot(-1) = \frac{1}{2}$	$\left(3, \frac{1}{2}\right)$

$\frac{9}{2}$	$y=-\frac{1}{2}\cos\left(\frac{\pi}{3}\cdot\frac{9}{2}\right)$ $=-\frac{1}{2}\cos\frac{3\pi}{2}$ $=-\frac{1}{2}\cdot 0=0$	$\left(\frac{9}{2}, 0\right)$
6	$y=-\frac{1}{2}\cos\left(\frac{\pi}{3}\cdot 6\right)$ $=-\frac{1}{2}\cos 2\pi$ $=-\frac{1}{2}\cdot 1=-\frac{1}{2}$	$\left(6, -\frac{1}{2}\right)$

Connect the five points with a smooth curve and graph one complete cycle of the given function.

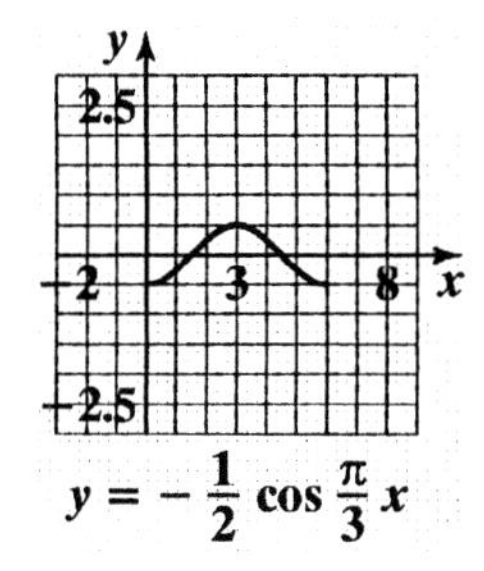

42. The equation $y=-\frac{1}{2}\cos\frac{\pi}{4}x$ is of the form $y=A\cos Bx$ with $A=-\frac{1}{2}$ and $B=\frac{\pi}{4}$. Thus, the amplitude is $|A|=\left|-\frac{1}{2}\right|=\frac{1}{2}$. The period is $\frac{2\pi}{B}=\frac{2\pi}{\frac{\pi}{4}}=2\pi\cdot\frac{4}{\pi}=8$. The quarter-period is $\frac{8}{4}=2$. The cycle begins at $x=0$. Add quarter-periods to generate x-values for the key points.

$x=0$

$x=0+2=2$

$x=2+2=4$

$x=4+2=6$

$x=6+2=8$

Evaluate the function at each value of x.

x	$y=-\frac{1}{2}\cos\frac{\pi}{4}x$	coordinates
0	$y=-\frac{1}{2}\cos\left(\frac{\pi}{4}\cdot 0\right)$ $=-\frac{1}{2}\cos 0=-\frac{1}{2}\cdot 1=-\frac{1}{2}$	$\left(0, -\frac{1}{2}\right)$
2	$y=-\frac{1}{2}\cos\left(\frac{\pi}{4}\cdot 2\right)$ $=-\frac{1}{2}\cos\frac{\pi}{2}=-\frac{1}{2}\cdot 0=0$	(2, 0)
4	$y=-\frac{1}{2}\cos\left(\frac{\pi}{4}\cdot 4\right)$ $=-\frac{1}{2}\cos\pi=-\frac{1}{2}\cdot(-1)=\frac{1}{2}$	$\left(4, \frac{1}{2}\right)$
6	$y=-\frac{1}{2}\cos\left(\frac{\pi}{4}\cdot 6\right)$ $=-\frac{1}{2}\cos\left(\frac{3\pi}{2}\right)=-\frac{1}{2}\cdot 0=0$	(6, 0)
8	$y=-\frac{1}{2}\cos\left(\frac{\pi}{4}\cdot 8\right)$ $=-\frac{1}{2}\cos 2\pi=-\frac{1}{2}\cdot 1=-\frac{1}{2}$	$\left(8, -\frac{1}{2}\right)$

Connect the five key points with a smooth curve and graph one complete cycle of the given function.

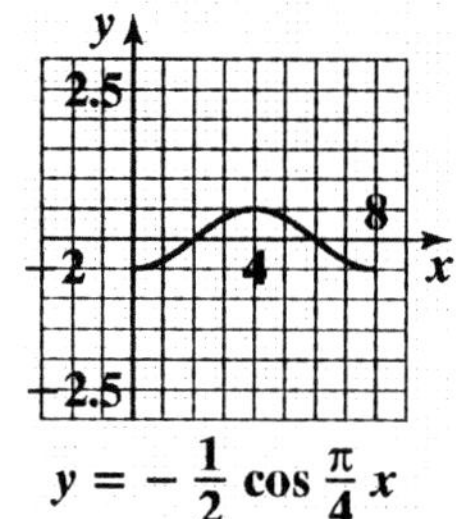

43. The equation $y=\cos\left(x-\frac{\pi}{2}\right)$ is of the form $y=A\cos(Bx-C)$ with $A=1$, and $B=1$, and $C=\frac{\pi}{2}$. Thus, the amplitude is $|A|=|1|=1$. The period is $\frac{2\pi}{B}=\frac{2\pi}{1}=2\pi$. The phase shift is $\frac{C}{B}=\frac{\frac{\pi}{2}}{1}=\frac{\pi}{2}$. The quarter-period is $\frac{2\pi}{4}=\frac{\pi}{2}$.

The cycle begins at $x = \frac{\pi}{2}$. Add quarter-periods to generate x-values for the key points.

$x = \frac{\pi}{2}$

$x = \frac{\pi}{2} + \frac{\pi}{2} = \pi$

$x = \pi + \frac{\pi}{2} = \frac{3\pi}{2}$

$x = \frac{3\pi}{2} + \frac{\pi}{2} = 2\pi$

$x = 2\pi + \frac{\pi}{2} = \frac{5\pi}{2}$

Evaluate the function at each value of x.

x	coordinates
$\frac{\pi}{2}$	$\left(\frac{\pi}{2}, 1\right)$
π	$(\pi, 0)$
$\frac{3\pi}{2}$	$\left(\frac{3\pi}{2}, -1\right)$
2π	$(2\pi, 0)$
$\frac{5\pi}{2}$	$\left(\frac{5\pi}{2}, 1\right)$

Connect the five points with a smooth curve and graph one complete cycle of the given function

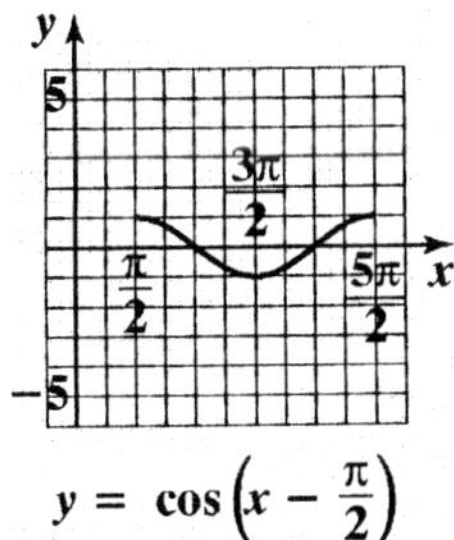

44. The equation $y = \cos\left(x + \frac{\pi}{2}\right)$ is of the form $y = A\cos(Bx - C)$ with $A = 1$, and $B = 1$, and $C = -\frac{\pi}{2}$. Thus, the amplitude is $|A| = |1| = 1$. The period is $\frac{2\pi}{B} = \frac{2\pi}{1} = 2\pi$. The phase shift is $\frac{C}{B} = \frac{-\frac{\pi}{2}}{1} = -\frac{\pi}{2}$. The quarter-period is $\frac{2\pi}{4} = \frac{\pi}{2}$. The cycle begins at $x = -\frac{\pi}{2}$. Add quarter-periods to generate x-values for the key points.

$x = -\frac{\pi}{2}$

$x = -\frac{\pi}{2} + \frac{\pi}{2} = 0$

$x = 0 + \frac{\pi}{2} = \frac{\pi}{2}$

$x = \frac{\pi}{2} + \frac{\pi}{2} = \pi$

$x = \pi + \frac{\pi}{2} = \frac{3\pi}{2}$

Evaluate the function at each value of x.

x	coordinates
$-\frac{\pi}{2}$	$\left(-\frac{\pi}{2}, 1\right)$
0	$(0, 0)$
$\frac{\pi}{2}$	$\left(\frac{\pi}{2}, -1\right)$
π	$(\pi, 0)$
$\frac{3\pi}{2}$	$\left(\frac{3\pi}{2}, 1\right)$

Connect the five points with a smooth curve and graph one complete cycle of the given function

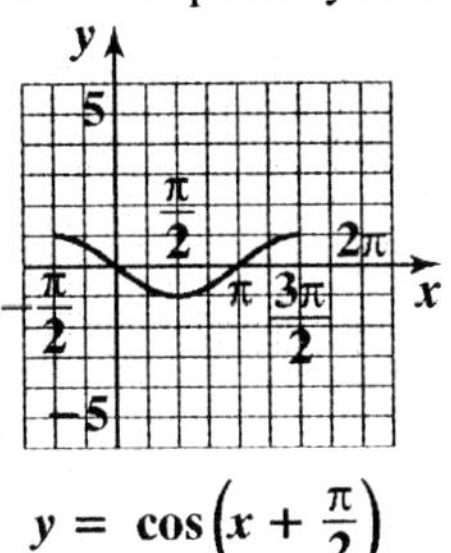

45. The equation $y = 3\cos(2x - \pi)$ is of the form $y = A\cos(Bx - C)$ with $A = 3$, and $B = 2$, and $C = \pi$. Thus, the amplitude is $|A| = |3| = 3$. The period is $\frac{2\pi}{B} = \frac{2\pi}{2} = \pi$. The phase shift is $\frac{C}{B} = \frac{\pi}{2}$. The quarter-period is $\frac{\pi}{4}$. The cycle begins at $x = \frac{\pi}{2}$. Add quarter-periods to generate x-values for the key points.

$x = \frac{\pi}{2}$

$x = \frac{\pi}{2} + \frac{\pi}{4} = \frac{3\pi}{4}$

$x = \frac{3\pi}{4} + \frac{\pi}{4} = \pi$

$x = \pi + \frac{\pi}{4} = \frac{5\pi}{4}$

$x = \frac{5\pi}{4} + \frac{\pi}{4} = \frac{3\pi}{2}$

Evaluate the function at each value of x.

x	coordinates
$\frac{\pi}{2}$	$\left(\frac{\pi}{2}, 3\right)$
$\frac{3\pi}{4}$	$\left(\frac{3\pi}{4}, 0\right)$
π	$(\pi, -3)$
$\frac{5\pi}{4}$	$\left(\frac{5\pi}{4}, 0\right)$
$\frac{3\pi}{2}$	$\left(\frac{3\pi}{2}, 3\right)$

Connect the five points with a smooth curve and graph one complete cycle of the given function

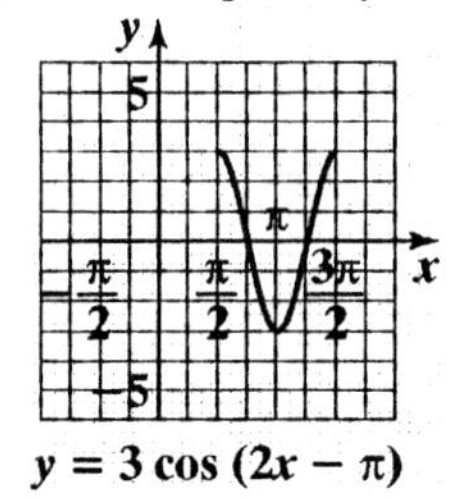

$y = 3\cos(2x - \pi)$

46. The equation $y = 4\cos(2x - \pi)$ is of the form $y = A\cos(Bx - C)$ with $A = 4$, and $B = 2$, and $C = \pi$. Thus, the amplitude is $|A| = |4| = 4$. The period is $\frac{2\pi}{B} = \frac{2\pi}{2} = \pi$. The phase shift is $\frac{C}{B} = \frac{\pi}{2}$. The quarter-period is $\frac{\pi}{4}$. The cycle begins at $x = \frac{\pi}{2}$. Add quarter-periods to generate x-values for the key points.

$x = \frac{\pi}{2}$

$x = \frac{\pi}{2} + \frac{\pi}{4} = \frac{3\pi}{4}$

$x = \frac{3\pi}{4} + \frac{\pi}{4} = \pi$

$x = \pi + \frac{\pi}{4} = \frac{5\pi}{4}$

$x = \frac{5\pi}{4} + \frac{\pi}{4} = \frac{3\pi}{2}$

Evaluate the function at each value of x.

x	coordinates
$\frac{\pi}{2}$	$\left(\frac{\pi}{2}, 4\right)$
$\frac{3\pi}{4}$	$\left(\frac{3\pi}{4}, 0\right)$
π	$(\pi, -4)$
$\frac{5\pi}{4}$	$\left(\frac{5\pi}{4}, 0\right)$
$\frac{3\pi}{2}$	$\left(\frac{3\pi}{2}, 4\right)$

Connect the five key points with a smooth curve and graph one complete cycle of the given function.

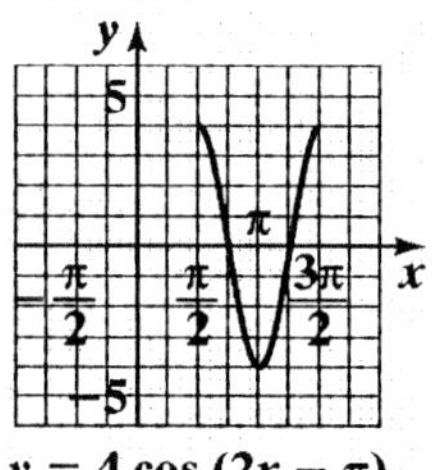

$y = 4\cos(2x - \pi)$

47. $y=\frac{1}{2}\cos\left(3x+\frac{\pi}{2}\right)=\frac{1}{2}\cos\left(3x-\left(-\frac{\pi}{2}\right)\right)$

The equation $y=\frac{1}{2}\cos\left(3x-\left(-\frac{\pi}{2}\right)\right)$ is of the form $y=A\cos(Bx-C)$ with $A=\frac{1}{2}$, and $B=3$, and $C=-\frac{\pi}{2}$. Thus, the amplitude is $|A|=\left|\frac{1}{2}\right|=\frac{1}{2}$. The period is $\frac{2\pi}{B}=\frac{2\pi}{3}$. The phase shift is $\frac{C}{B}=\frac{-\frac{\pi}{2}}{3}=-\frac{\pi}{2}\cdot\frac{1}{3}=-\frac{\pi}{6}$. The quarter-period is $\frac{\frac{2\pi}{3}}{4}=\frac{2\pi}{3}\cdot\frac{1}{4}=\frac{\pi}{6}$. The cycle begins at $x=-\frac{\pi}{6}$. Add quarter-periods to generate x-values for the key points.

$x=-\frac{\pi}{6}$

$x=-\frac{\pi}{6}+\frac{\pi}{6}=0$

$x=0+\frac{\pi}{6}=\frac{\pi}{6}$

$x=\frac{\pi}{6}+\frac{\pi}{6}=\frac{\pi}{3}$

$x=\frac{\pi}{3}+\frac{\pi}{6}=\frac{\pi}{2}$

Evaluate the function at each value of x.

x	coordinates
$-\frac{\pi}{6}$	$\left(-\frac{\pi}{6},\frac{1}{2}\right)$
0	(0, 0)
$\frac{\pi}{6}$	$\left(\frac{\pi}{6},-\frac{1}{2}\right)$
$\frac{\pi}{3}$	$\left(\frac{\pi}{3},0\right)$
$\frac{\pi}{2}$	$\left(\frac{\pi}{2},\frac{1}{2}\right)$

Connect the five points with a smooth curve and graph one complete cycle of the given function

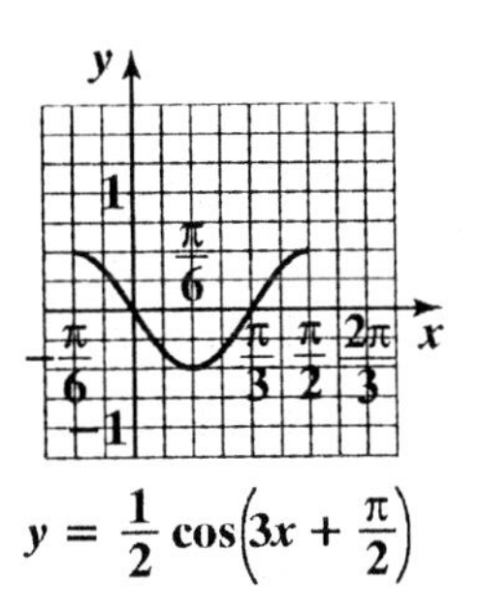

$y=\frac{1}{2}\cos\left(3x+\frac{\pi}{2}\right)$

48. $y=\frac{1}{2}\cos(2x+\pi)=\frac{1}{2}\cos(2x-(-\pi))$

The equation $y=\frac{1}{2}\cos(2x-(-\pi))$ is of the form $y=A\cos(Bx-C)$ with $A=\frac{1}{2}$, and $B=2$, and $C=-\pi$. Thus, the amplitude is $|A|=\left|\frac{1}{2}\right|=\frac{1}{2}$. The period is $\frac{2\pi}{B}=\frac{2\pi}{2}=\pi$. The phase shift is $\frac{C}{B}=\frac{-\pi}{2}=-\frac{\pi}{2}$. The quarter-period is $\frac{\pi}{4}$. The cycle begins at $x=-\frac{\pi}{2}$. Add quarter-periods to generate x-values for the key points.

$x=-\frac{\pi}{2}$

$x=-\frac{\pi}{2}+\frac{\pi}{4}=-\frac{\pi}{4}$

$x=-\frac{\pi}{4}+\frac{\pi}{4}=0$

$x=0+\frac{\pi}{4}=\frac{\pi}{4}$

$x=\frac{\pi}{4}+\frac{\pi}{4}=\frac{\pi}{2}$

Evaluate the function at each value of x.

x	coordinates
$-\frac{\pi}{2}$	$\left(-\frac{\pi}{2},\frac{1}{2}\right)$
$-\frac{\pi}{4}$	$\left(-\frac{\pi}{4},0\right)$
0	$\left(0,-\frac{1}{2}\right)$
$\frac{\pi}{4}$	$\left(\frac{\pi}{4},0\right)$

$\frac{\pi}{2}$	$\left(\frac{\pi}{2}, \frac{1}{2}\right)$

Connect the five key points with a smooth curve and graph one complete cycle of the given function.

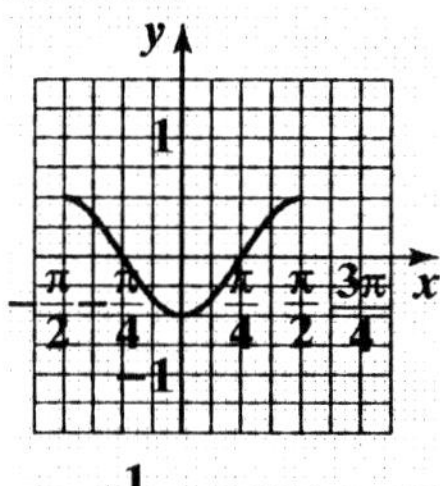

49. The equation $y = -3\cos\left(2x - \frac{\pi}{2}\right)$ is of the form $y = A\cos(Bx - C)$ with $A = -3$, and $B = 2$, and $C = \frac{\pi}{2}$. Thus, the amplitude is $|A| = |-3| = 3$. The period is $\frac{2\pi}{B} = \frac{2\pi}{2} = \pi$. The phase shift is $\frac{C}{B} = \frac{\frac{\pi}{2}}{2} = \frac{\pi}{2} \cdot \frac{1}{2} = \frac{\pi}{4}$. The quarter-period is $\frac{\pi}{4}$. The cycle begins at $x = \frac{\pi}{4}$. Add quarter-periods to generate x-values for the key points.

$x = \frac{\pi}{4}$

$x = \frac{\pi}{4} + \frac{\pi}{4} = \frac{\pi}{2}$

$x = \frac{\pi}{2} + \frac{\pi}{4} = \frac{3\pi}{4}$

$x = \frac{3\pi}{4} + \frac{\pi}{4} = \pi$

$x = \pi + \frac{\pi}{4} = \frac{5\pi}{4}$

Evaluate the function at each value of x.

x	coordinates
$\frac{\pi}{4}$	$\left(\frac{\pi}{4}, -3\right)$
$\frac{\pi}{2}$	$\left(\frac{\pi}{2}, 0\right)$
$\frac{3\pi}{4}$	$\left(\frac{3\pi}{4}, 3\right)$
π	$(\pi, 0)$
$\frac{5\pi}{4}$	$\left(\frac{5\pi}{4}, -3\right)$

Connect the five points with a smooth curve and graph one complete cycle of the given function

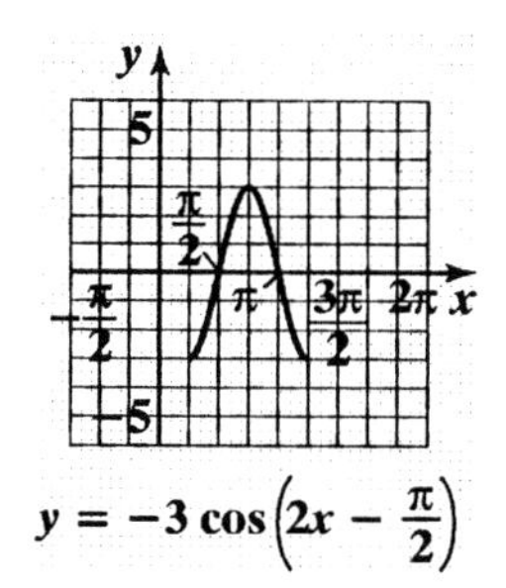

50. The equation $y = -4\cos\left(2x - \frac{\pi}{2}\right)$ is of the form $y = A\cos(Bx - C)$ with $A = -4$, and $B = 2$, and $C = \frac{\pi}{2}$. Thus, the amplitude is $|A| = |-4| = 4$. The period is $\frac{2\pi}{B} = \frac{2\pi}{2} = \pi$. The phase shift is $\frac{C}{B} = \frac{\frac{\pi}{2}}{2} = \frac{\pi}{2} \cdot \frac{1}{2} = \frac{\pi}{4}$. The quarter-period is $\frac{\pi}{4}$. The cycle begins at $x = \frac{\pi}{4}$. Add quarter-periods to generate x-values for the key points.

$x = \frac{\pi}{4}$

$x = \frac{\pi}{4} + \frac{\pi}{4} = \frac{\pi}{2}$

$x = \frac{\pi}{2} + \frac{\pi}{4} = \frac{3\pi}{4}$

$x = \frac{3\pi}{4} + \frac{\pi}{4} = \pi$

$x = \pi + \frac{\pi}{4} = \frac{5\pi}{4}$

Evaluate the function at each value of x.

x	coordinates
$\frac{\pi}{4}$	$\left(\frac{\pi}{4}, -4\right)$
$\frac{\pi}{2}$	$\left(\frac{\pi}{2}, 0\right)$
$\frac{3\pi}{4}$	$\left(\frac{3\pi}{4}, 4\right)$
π	$(\pi, 0)$
$\frac{5\pi}{4}$	$\left(\frac{5\pi}{4}, -4\right)$

Connect the five key points with a smooth curve and graph one complete cycle of the given function.

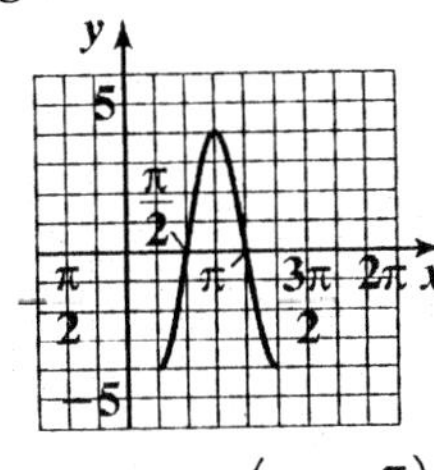

$y = -4\cos\left(2x - \frac{\pi}{2}\right)$

51. $y = 2\cos(2\pi x + 8\pi) = 2\cos(2\pi x - (-8\pi))$
The equation $y = 2\cos(2\pi x - (-8\pi))$ is of the form $y = A\cos(Bx - C)$ with $A = 2$, $B = 2\pi$, and $C = -8\pi$. Thus, the amplitude is $|A| = |2| = 2$. The period is $\frac{2\pi}{B} = \frac{2\pi}{2\pi} = 1$. The phase shift is $\frac{C}{B} = \frac{-8\pi}{2\pi} = -4$. The quarter-period is $\frac{1}{4}$. The cycle begins at $x = -4$. Add quarter-periods to generate x-values for the key points.

$x = -4$

$x = -4 + \frac{1}{4} = -\frac{15}{4}$

$x = -\frac{15}{4} + \frac{1}{4} = -\frac{7}{2}$

$x = -\frac{7}{2} + \frac{1}{4} = -\frac{13}{4}$

$x = -\frac{13}{4} + \frac{1}{4} = -3$

Evaluate the function at each value of x.

x	coordinates
-4	$(-4, 2)$
$-\frac{15}{4}$	$\left(-\frac{15}{4}, 0\right)$
$-\frac{7}{2}$	$\left(-\frac{7}{2}, -2\right)$
$-\frac{13}{4}$	$\left(-\frac{13}{4}, 0\right)$
-3	$(-3, 2)$

Connect the five points with a smooth curve and graph one complete cycle of the given function

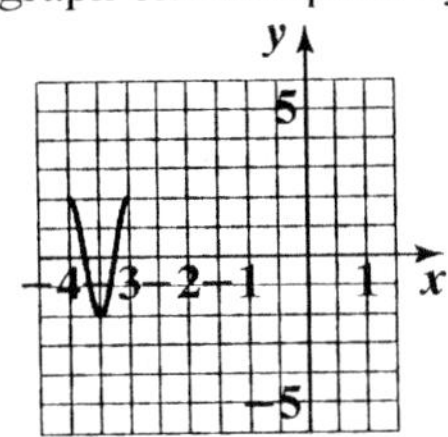

$y = 2\cos(2\pi x + 8\pi)$

52. $y = 3\cos(2\pi x + 4\pi) = 3\cos(2\pi x - (-4\pi))$
The equation $y = 3\cos(2\pi x - (-4\pi))$ is of the form $y = A\cos(Bx - C)$ with $A = 3$, and $B = 2\pi$, and $C = -4\pi$. Thus, the amplitude is $|A| = |3| = 3$. The period is $\frac{2\pi}{B} = \frac{2\pi}{2\pi} = 1$. The phase shift is $\frac{C}{B} = \frac{-4\pi}{2\pi} = -2$. The quarter-period is $\frac{1}{4}$. The cycle begins at $x = -2$. Add quarter-periods to generate x-values for the key points.

$x = -2$

$x = -2 + \frac{1}{4} = -\frac{7}{4}$

$x = -\frac{7}{4} + \frac{1}{4} = -\frac{3}{2}$

$x = -\frac{3}{2} + \frac{1}{4} = -\frac{5}{4}$

$x = -\frac{5}{4} + \frac{1}{4} = -1$

Evaluate the function at each value of x.

x	coordinates
-2	$(-2, 3)$
$-\frac{7}{4}$	$\left(-\frac{7}{4}, 0\right)$
$-\frac{3}{2}$	$\left(-\frac{3}{2}, -3\right)$
$-\frac{5}{4}$	$\left(-\frac{5\pi}{4}, 0\right)$
-1	$(-1, 3)$

Connect the five key points with a smooth curve and graph one complete cycle of the given function.

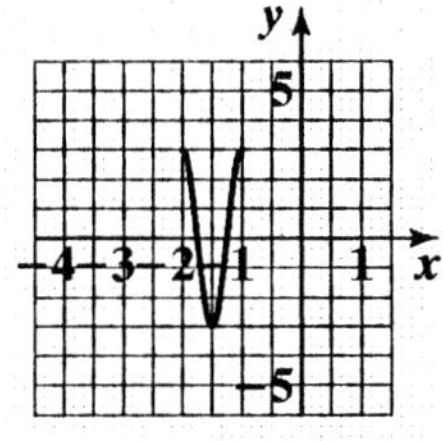

$y = 3\cos(2\pi x + 4\pi)$

53. The graph of $y = \sin x + 2$ is the graph of $y = \sin x$ shifted up 2 units upward. The period for both functions is 2π. The quarter-period is $\frac{2\pi}{4}$ or $\frac{\pi}{2}$. The cycle begins at $x = 0$. Add quarter-periods to generate x-values for the key points.

$x = 0$

$x = 0 + \frac{\pi}{2} = \frac{\pi}{2}$

$x = \frac{\pi}{2} + \frac{\pi}{2} = \pi$

$x = \pi + \frac{\pi}{2} = \frac{3\pi}{2}$

$x = \frac{3\pi}{2} + \frac{\pi}{2} = 2\pi$

Evaluate the function at each value of x.

x	$y = \sin x + 2$	coordinates
0	$y = \sin 0 + 2$ $= 0 + 2 = 2$	$(0, 2)$
$\frac{\pi}{2}$	$y = \sin\frac{\pi}{2} + 2$ $= 1 + 2 = 3$	$\left(\frac{\pi}{2}, 3\right)$
π	$y = \sin \pi + 2$ $= 0 + 2 = 2$	$(\pi, 2)$
$\frac{3\pi}{2}$	$y = \sin\frac{3\pi}{2} + 2$ $= -1 + 2 = 1$	$\left(\frac{3\pi}{2}, 1\right)$
2π	$y = \sin 2\pi + 2$ $= 0 + 2 = 2$	$(2\pi, 2)$

By connecting the points with a smooth curve we obtain one period of the graph.

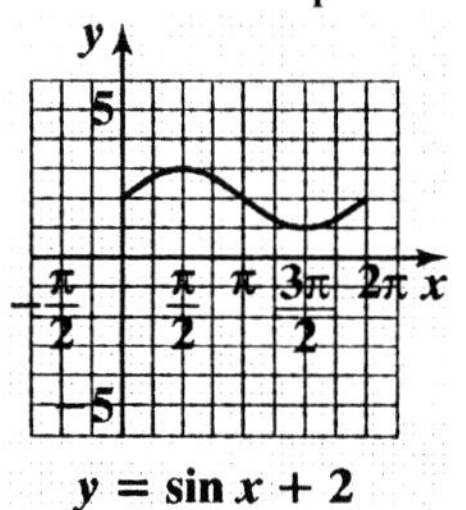

$y = \sin x + 2$

54. The graph of $y = \sin x - 2$ is the graph of $y = \sin x$ shifted 2 units downward. The period for both functions is 2π. The quarter-period is $\frac{2\pi}{4}$ or $\frac{\pi}{2}$. The cycle begins at $x = 0$. Add quarter-periods to generate x-values for the key points.

$x = 0$

$x = 0 + \frac{\pi}{2} = \frac{\pi}{2}$

$x = \frac{\pi}{2} + \frac{\pi}{2} = \pi$

$x = \pi + \frac{\pi}{2} = \frac{3\pi}{2}$

$x = \frac{3\pi}{2} + \frac{\pi}{2} = 2\pi$

Evaluate the function at each value of x.

x	$y = \sin x - 2$	coordinates
0	$y = \sin 0 - 2 = 0 - 2 = -2$	$(0, -2)$
$\frac{\pi}{2}$	$y = \sin\frac{\pi}{2} - 2 = 1 - 2 = -1$	$\left(\frac{\pi}{4}, -1\right)$
π	$y = \sin \pi - 2 = 0 - 2 = -2$	$(\pi, -2)$
$\frac{3\pi}{2}$	$y = \sin\frac{3\pi}{2} - 2 = -1 - 2 = -3$	$\left(\frac{3\pi}{2}, -3\right)$
2π	$y = \sin 2\pi - 2 = 0 - 2 = -2$	$(2\pi, -2)$

By connecting the points with a smooth curve we obtain one period of the graph.

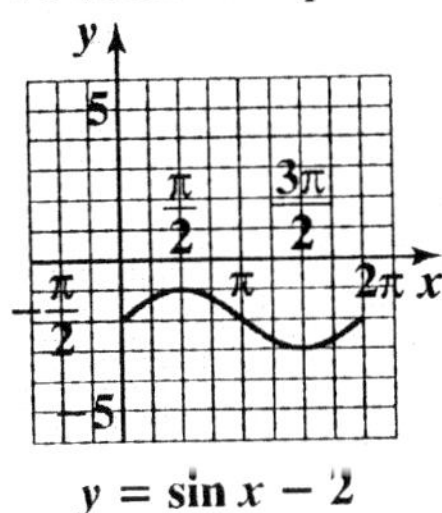

$y = \sin x - 2$

55. The graph of $y = \cos x - 3$ is the graph of $y = \cos x$ shifted 3 units downward. The period for both functions is 2π. The quarter-period is $\frac{2\pi}{4}$ or $\frac{\pi}{2}$. The cycle begins at $x = 0$. Add quarter-periods to generate x-values for the key points.

$x = 0$

$x = 0 + \frac{\pi}{2} = \frac{\pi}{2}$

$x = \frac{\pi}{2} + \frac{\pi}{2} = \pi$

$x = \pi + \frac{\pi}{2} = \frac{3\pi}{2}$

$x = \frac{3\pi}{2} + \frac{\pi}{2} = 2\pi$

Evaluate the function at each value of x.

x	$y = \cos x - 3$	coordinates
0	$y = \cos 0 - 3$ $= 1 - 3 = -2$	$(0, -2)$
$\frac{\pi}{2}$	$y = \cos\frac{\pi}{2} - 3$ $= 0 - 3 = -3$	$\left(\frac{\pi}{2}, -3\right)$
π	$y = \cos\pi - 3$ $= -1 - 3 = -4$	$(\pi, -4)$
$\frac{3\pi}{2}$	$y = \cos\frac{3\pi}{2} - 3$ $= 0 - 3 = -3$	$\left(\frac{3\pi}{2}, -3\right)$
2π	$y = \cos 2\pi - 3$ $= 1 - 3 = -2$	$(2\pi, -2)$

By connecting the points with a smooth curve we obtain one period of the graph.

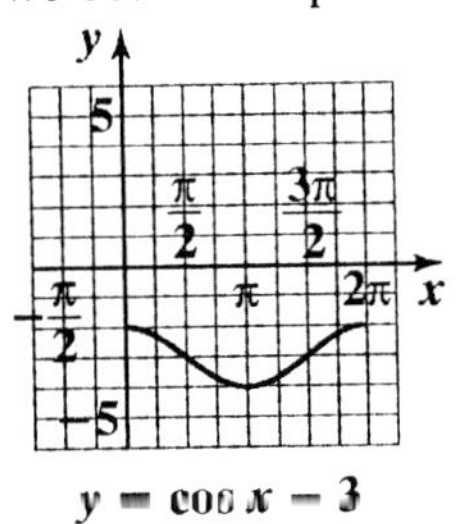

$y = \cos x - 3$

56. The graph of $y = \cos x + 3$ is the graph of $y = \cos x$ shifted 3 units upward. The period for both functions is 2π. The quarter-period is $\frac{2\pi}{4}$ or $\frac{\pi}{2}$. The cycle begins at $x = 0$. Add quarter-periods to generate x-values for the key points.

$x = 0$

$x = 0 + \frac{\pi}{2} = \frac{\pi}{2}$

$x = \frac{\pi}{2} + \frac{\pi}{2} = \pi$

$x = \pi + \frac{\pi}{2} = \frac{3\pi}{2}$

$x = \frac{3\pi}{2} + \frac{\pi}{2} = 2\pi$

Evaluate the function at each value of x.

x	$y = \cos x + 3$	coordinates
0	$y = \cos 0 + 3 = 1 + 3 = 4$	$(0, 4)$
$\frac{\pi}{2}$	$y = \cos\frac{\pi}{2} + 3 = 0 + 3 = 3$	$\left(\frac{\pi}{2}, 3\right)$
π	$y = \cos\pi + 3 = -1 + 3 = 2$	$(\pi, 2)$
$\frac{3\pi}{2}$	$y = \cos\frac{3\pi}{2} + 3 = 0 + 3 = 3$	$\left(\frac{3\pi}{2}, 3\right)$
2π	$y = \cos 2\pi + 3 = 1 + 3 = 4$	$(2\pi, 4)$

By connecting the points with a smooth curve we obtain one period of the graph.

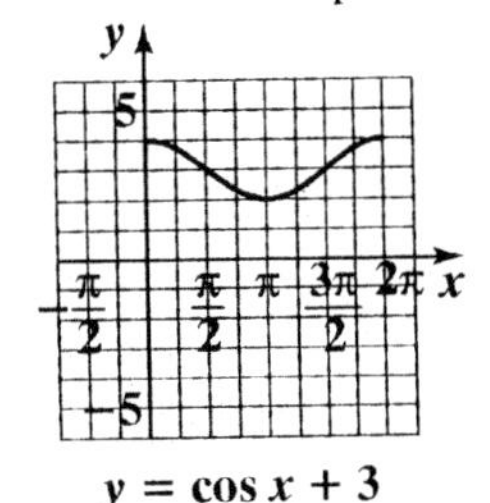

$y = \cos x + 3$

57. The graph of $y = 2\sin\frac{1}{2}x + 1$ is the graph of $y = 2\sin\frac{1}{2}x$ shifted one unit upward. The amplitude for both functions is $|2| = 2$. The period for both functions is $\frac{2\pi}{\frac{1}{2}} = 2\pi \cdot 2 = 4\pi$.

The quarter-period is $\frac{4\pi}{4} = \pi$. The cycle begins at $x = 0$. Add quarter-periods to generate x-values for the key points.

$x = 0$

$x = 0 + \pi = \pi$

$x = \pi + \pi = 2\pi$

$x = 2\pi + \pi = 3\pi$

$x = 3\pi + \pi = 4\pi$

Evaluate the function at each value of x.

x	$y = 2\sin\frac{1}{2}x + 1$	coordinates
0	$y = 2\sin\left(\frac{1}{2}\cdot 0\right) + 1$ $= 2\sin 0 + 1$ $= 2\cdot 0 + 1 = 0 + 1 = 1$	(0, 1)
π	$y = 2\sin\left(\frac{1}{2}\cdot \pi\right) + 1$ $= 2\sin\frac{\pi}{2} + 1$ $= 2\cdot 1 + 1 = 2 + 1 = 3$	$(\pi, 3)$
2π	$y = 2\sin\left(\frac{1}{2}\cdot 2\pi\right) + 1$ $= 2\sin \pi + 1$ $= 2\cdot 0 + 1 = 0 + 1 = 1$	$(2\pi, 1)$
3π	$y = 2\sin\left(\frac{1}{2}\cdot 3\pi\right) + 1$ $= 2\sin\frac{3\pi}{2} + 1$ $= 2\cdot(-1) + 1$ $= -2 + 1 = -1$	$(3\pi, -1)$
4π	$y = 2\sin\left(\frac{1}{2}\cdot 4\pi\right) + 1$ $= 2\sin 2\pi + 1$ $= 2\cdot 0 + 1 = 0 + 1 = 1$	$(4\pi, 1)$

By connecting the points with a smooth curve we obtain one period of the graph.

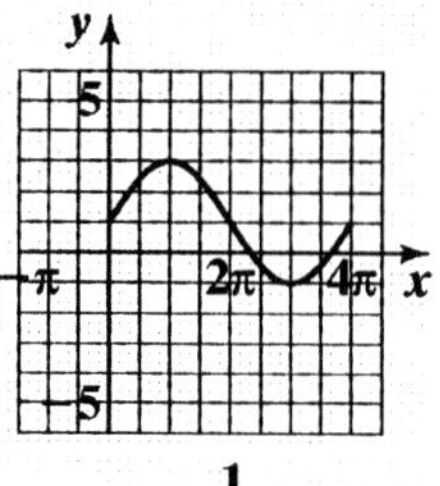

$y = 2\sin\frac{1}{2}x + 1$

58. The graph of $y = 2\cos\frac{1}{2}x + 1$ is the graph of $y = 2\cos\frac{1}{2}x$ shifted one unit upward. The amplitude for both functions is $|2| = 2$. The period for both functions is $\frac{2\pi}{\frac{1}{2}} = 2\pi \cdot 2 = 4\pi$.

The quarter-period is $\frac{4\pi}{4} = \pi$. The cycle begins at $x = 0$. Add quarter-periods to generate x-values for the key points.

$x = 0$

$x = 0 + \pi = \pi$

$x = \pi + \pi = 2\pi$

$x = 2\pi + \pi = 3\pi$

$x = 3\pi + \pi = 4\pi$

Evaluate the function at each value of x.

x	$y = 2\cos\frac{1}{2}x + 1$	coordinates
0	$y = 2\cos\left(\frac{1}{2}\cdot 0\right) + 1$ $= 2\cos 0 + 1$ $= 2\cdot 1 + 1 = 2 + 1 = 3$	(0, 3)
π	$y = 2\cos\left(\frac{1}{2}\cdot \pi\right) + 1$ $= 2\cos\frac{\pi}{2} + 1$ $= 2\cdot 0 + 1 = 0 + 1 = 1$	$(\pi, 1)$
2π	$y = 2\cos\left(\frac{1}{2}\cdot 2\pi\right) + 1$ $= 2\cos \pi + 1$ $= 2\cdot(-1) + 1 = -2 + 1 = -1$	$(2\pi, -1)$

3π	$y = 2\cos\left(\frac{1}{2}\cdot 3\pi\right)+1$ $= 2\cdot 0+1 = 0+1 = 1$	$(3\pi, 1)$
4π	$y = 2\cos\left(\frac{1}{2}\cdot 4\pi\right)+1$ $= 2\cos 2\pi + 1$ $= 2\cdot 1+1 = 2+1 = 3$	$(4\pi, 3)$

By connecting the points with a smooth curve we obtain one period of the graph.

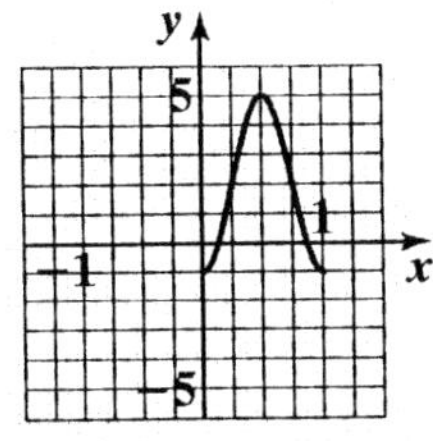

$y = -3\cos 2\pi x + 2$

59. The graph of $y = -3\cos 2\pi x + 2$ is the graph of $y = -3\cos 2\pi x$ shifted 2 units upward. The amplitude for both functions is $|-3| = 3$. The period for both functions is $\frac{2\pi}{2\pi} = 1$. The quarter-period is $\frac{1}{4}$. The cycle begins at $x = 0$. Add quarter-periods to generate x-values for the key points.

$x = 0$

$x = 0 + \frac{1}{4} = \frac{1}{4}$

$x = \frac{1}{4} + \frac{1}{4} = \frac{1}{2}$

$x = \frac{1}{2} + \frac{1}{4} = \frac{3}{4}$

$x = \frac{3}{4} + \frac{1}{4} = 1$

Evaluate the function at each value of x.

x	$y = -3\cos 2\pi x + 2$	coordinates
0	$y = -3\cos(2\pi\cdot 0)+2$ $= -3\cos 0 + 2$ $= -3\cdot 1+2$ $= -3+2 = -1$	$(0, -1)$
$\frac{1}{4}$	$y = -3\cos\left(2\pi\cdot\frac{1}{4}\right)+2$ $= -3\cos\frac{\pi}{2}+2$ $= -3\cdot 0+2$ $= 0+2 = 2$	$\left(\frac{1}{4}, 2\right)$
$\frac{1}{2}$	$y = -3\cos\left(2\pi\cdot\frac{1}{2}\right)+2$ $= -3\cos\pi+2$ $= -3\cdot(-1)+2$ $= 3+2 = 5$	$\left(\frac{1}{2}, 5\right)$
$\frac{3}{4}$	$y = -3\cos\left(2\pi\cdot\frac{3}{4}\right)+2$ $= -3\cos\frac{3\pi}{2}+2$ $= -3\cdot 0+2$ $= 0+2 = 2$	$\left(\frac{3}{4}, 2\right)$
1	$y = -3\cos(2\pi\cdot 1)+2$ $= -3\cos 2\pi+2$ $= -3\cdot 1+2$ $= -3+2 = -1$	$(1, -1)$

By connecting the points with a smooth curve we obtain one period of the graph.

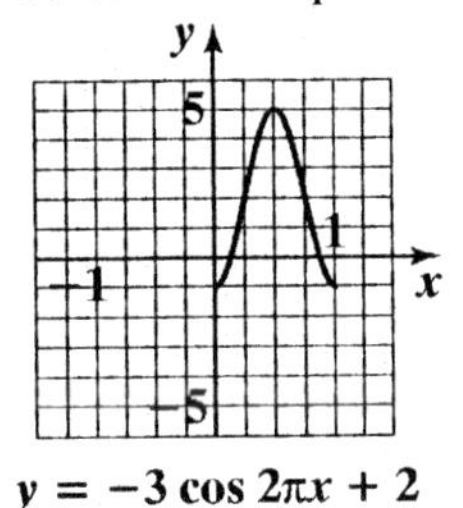

$y = -3\cos 2\pi x + 2$

60. The graph of $y = -3\sin 2\pi x + 2$ is the graph of $y = -3\sin 2\pi x$ shifted two units upward. The amplitude for both functions is $|A| = |-3| = 3$. The period for both functions is $\frac{2\pi}{2\pi} = 1$. The quarter-period is $\frac{1}{4}$. The cycle begins at $x = 0$. Add quarter–periods to generate x-values for the key points.

$x = 0$

$x = 0 + \frac{1}{4} = \frac{1}{4}$

$x = \frac{1}{4} + \frac{1}{4} = \frac{1}{2}$

$x = \frac{1}{2} + \frac{1}{4} = \frac{3}{4}$

$x = \frac{3}{4} + \frac{1}{4} = 1$

Evaluate the function at each value of x.

x	$y = -3\sin 2\pi x + 2$	coordinates
0	$y = -3\sin(2\pi \cdot 0) + 2$ $= -3\sin 0 + 2$ $= -3 \cdot 0 + 2 = 0 + 2 = 2$	(0, 2)
$\frac{1}{4}$	$y = -3\sin\left(2\pi \cdot \frac{1}{4}\right) + 2$ $= -3\sin\frac{\pi}{2} + 2$ $= -3 \cdot 1 + 2 = -3 + 2 = -1$	$\left(\frac{1}{4}, -1\right)$
$\frac{1}{2}$	$y = -3\sin\left(2\pi \cdot \frac{1}{2}\right) + 2$ $= -3\sin \pi + 2$ $= -3 \cdot 0 + 2 = 0 + 2 = 2$	$\left(\frac{1}{2}, 2\right)$
$\frac{3}{4}$	$y = -3\sin\left(2\pi \cdot \frac{3}{4}\right) + 2$ $= -3\sin\frac{3\pi}{2} + 2$ $= -3 \cdot (-1) + 2 = 3 + 2 = 5$	$\left(\frac{3}{4}, 5\right)$
1	$y = -3\sin(2\pi \cdot 1) + 2$ $= -3\sin 2\pi + 2$ $= -3 \cdot 0 + 2 = 0 + 2 = 2$	(1, 2)

By connecting the points with a smooth curve we obtain one period of the graph.

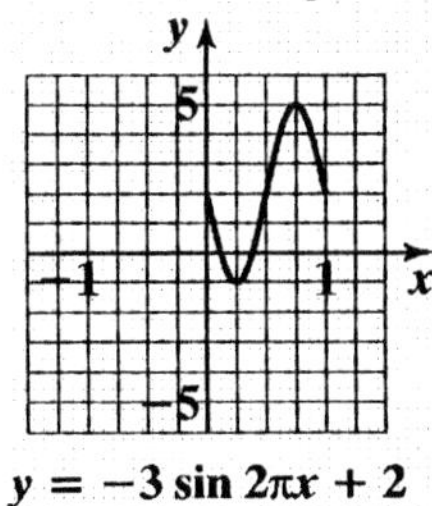

$y = -3\sin 2\pi x + 2$

61. Using $y = A\cos Bx$ the amplitude is 3 and $A = 3$, The period is 4π and thus

$B = \frac{2\pi}{\text{period}} = \frac{2\pi}{4\pi} = \frac{1}{2}$

$y = A\cos Bx$

$y = 3\cos\frac{1}{2}x$

62. Using $y = A\sin Bx$ the amplitude is 3 and $A = 3$, The period is 4π and thus

$B = \frac{2\pi}{\text{period}} = \frac{2\pi}{4\pi} = \frac{1}{2}$

$y = A\sin Bx$

$y = 3\sin\frac{1}{2}x$

63. Using $y = A\sin Bx$ the amplitude is 2 and $A = -2$, The period is π and thus

$B = \frac{2\pi}{\text{period}} = \frac{2\pi}{\pi} = 2$

$y = A\sin Bx$

$y = -2\sin 2x$

64. Using $y = A\cos Bx$ the amplitude is 2 and $A = -2$, The period is 4π and thus

$B = \frac{2\pi}{\text{period}} = \frac{2\pi}{\pi} = 2$

$y = A\cos Bx$

$y = -2\cos 2x$

65. Using $y = A\sin Bx$ the amplitude is 2 and $A = 2$, The period is 4 and thus

$B = \frac{2\pi}{\text{period}} = \frac{2\pi}{4} = \frac{\pi}{2}$

$y = A\sin Bx$

$y = 2\sin\left(\frac{\pi}{2}x\right)$

66. Using $y = A\cos Bx$ the amplitude is 2 and $A = 2$, The period is 4 and thus

$B = \frac{2\pi}{\text{period}} = \frac{2\pi}{4} = \frac{\pi}{2}$

$y = A\cos Bx$

$y = 2\cos\left(\frac{\pi}{2}x\right)$

67.

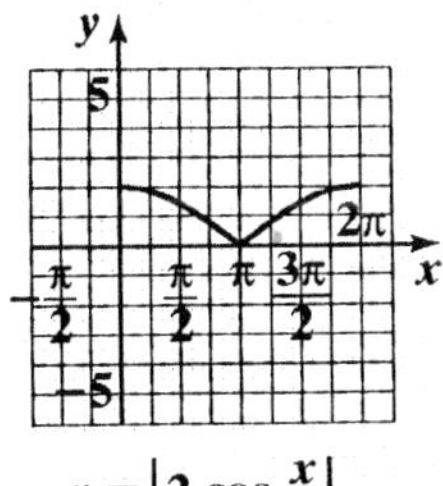

$y = \left|2\cos\frac{x}{2}\right|$

68.

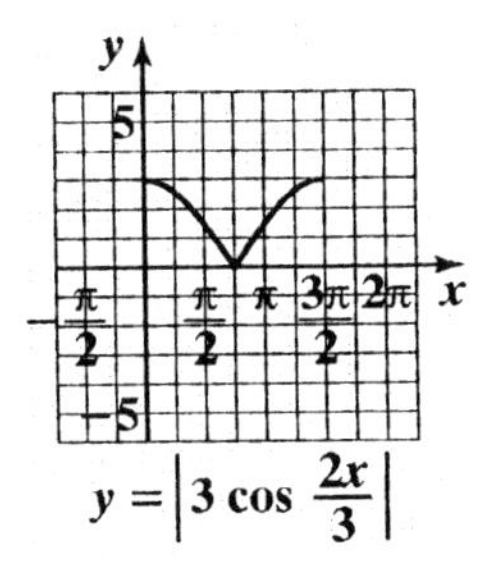

$y = \left|3\cos\frac{2x}{3}\right|$

69.

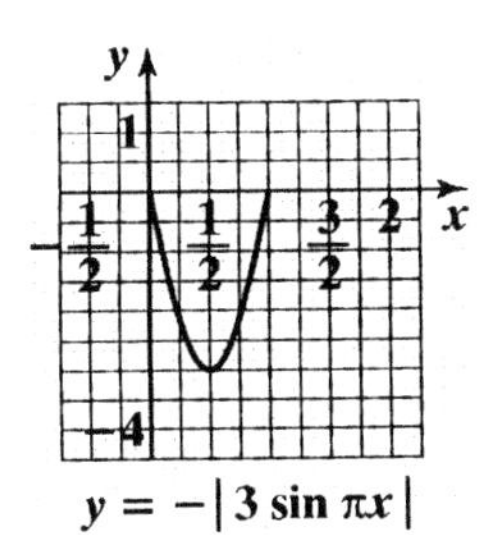

$y = -|3\sin \pi x|$

70.

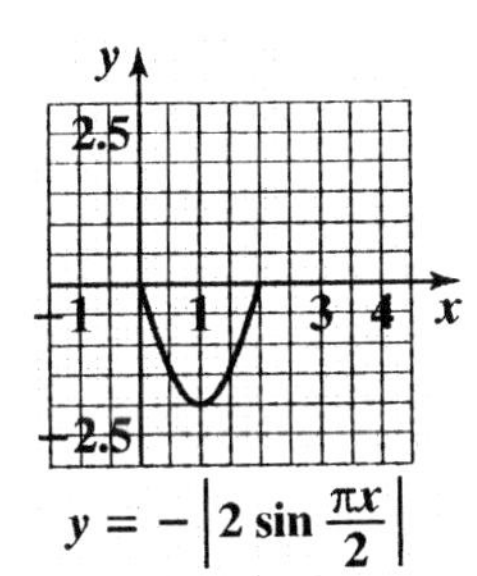

$y = -\left|2\sin\frac{\pi x}{2}\right|$

71.

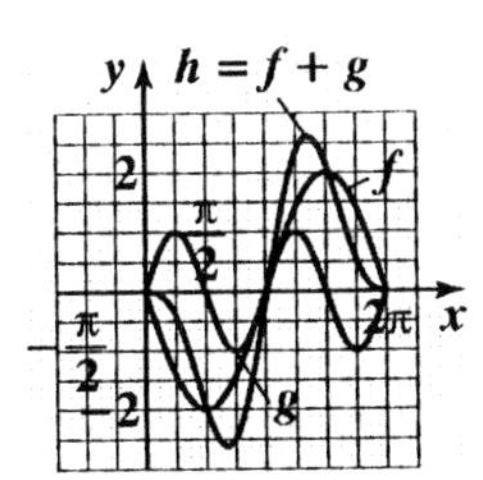

72.

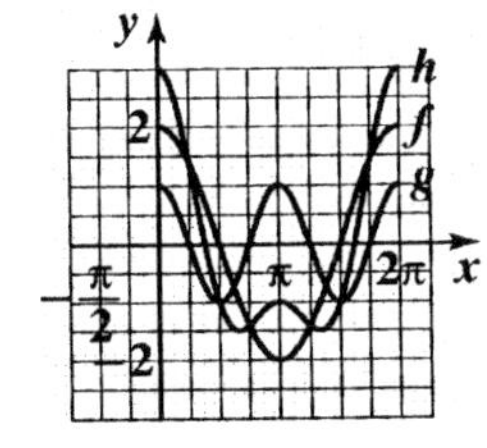

73.

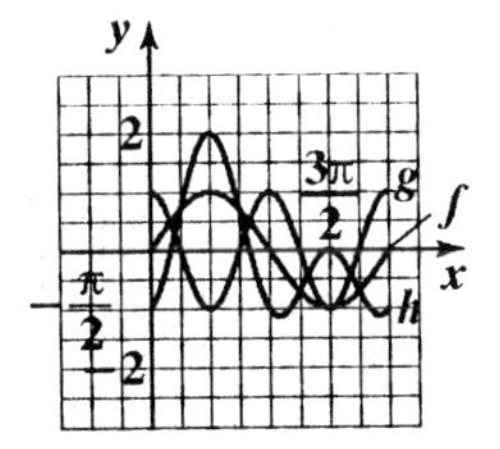

74.

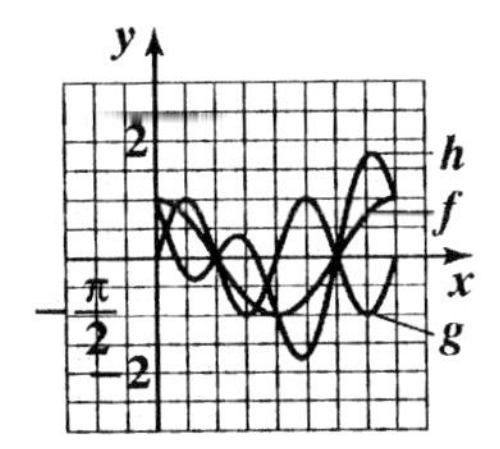

75. The period of the physical cycle is 33 days.

76. The period of the emotional cycle is 28 days.

77. The period of the intellectual cycle is 23 days.

78. In the month of February, the physical cycle is at a minimum on February 18. Thus, the author should not run in a marathon on February 18.

79. In the month of March, March 21 would be the best day to meet an on-line friend for the first time, because the emotional cycle is at a maximum.

80. In the month of February, the intellectual cycle is at a maximum on February 11. Thus, the author should begin writing the on February 11.

81. Answers may vary.

82. Answers may vary.

83. The information gives the five key point of the graph.
(0, 14) corresponds to June,
(3, 12) corresponds to September,
(6, 10) corresponds to December,
(9, 12) corresponds to March,
(12, 14) corresponds to June
By connecting the five key points with a smooth curve we graph the information from June of one year to June of the following year.

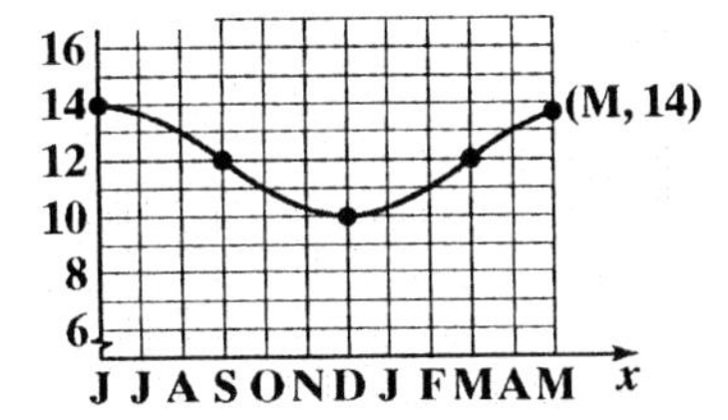

84. The information gives the five key points of the graph.
(0, 23) corresponds to Noon,
(3, 38) corresponds to 3 P.M.,
(6, 53) corresponds to 6 P.M.,
(9, 38) corresponds to 9 P.M.,
(12, 23) corresponds to Midnight.
By connecting the five key points with a smooth curve we graph information from noon to midnight. Extend the graph one cycle to the right to graph the information for $0 \le x \le 24$.

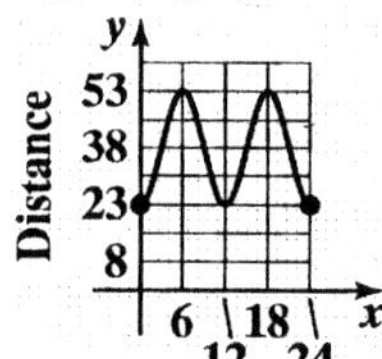

85. The function $y = 3\sin\frac{2\pi}{365}(x-79)+12$ is of the

form $y = A\sin B\left(x - \frac{C}{B}\right) + D$ with

$A = 3$ and $B = \frac{2\pi}{365}$.

a. The amplitude is $|A| = |3| = 3$.

b. The period is $\frac{2\pi}{B} = \frac{2\pi}{\frac{2\pi}{365}} = 2\pi \cdot \frac{365}{2\pi} = 365$.

c. The longest day of the year will have the most hours of daylight. This occurs when the sine function equals 1.

$y = 3\sin\frac{2\pi}{365}(x-79)+12$
$y = 3(1)+12$
$y = 15$
There will be 15 hours of daylight.

d. The shortest day of the year will have the least hours of daylight. This occurs when the sine function equals –1.

$y = 3\sin\frac{2\pi}{365}(x-79)+12$
$y = 3(-1)+12$
$y = 9$
There will be 9 hours of daylight.

e. The amplitude is 3. The period is 365. The phase shift is $\frac{C}{B} = 79$. The quarter-period is $\frac{365}{4} = 91.25$. The cycle begins at $x =$ 79. Add quarter-periods to find the x-values of the key points.

$x = 79$
$x = 79 + 91.25 = 170.25$
$x = 170.25 + 91.25 = 261.5$
$x = 261.5 + 91.25 = 352.75$
$x = 352.75 + 91.25 = 444$

Because we are graphing for $0 \le x \le 365$, we will evaluate the function for the first four x-values along with $x = 0$ and $x = 365$. Using a calculator we have the following points.
(0, 9.07) (79, 12) (170.25, 15)
(261.5, 12) (352.75, 9) (365, 9.07)

By connecting the points with a smooth curve we obtain one period of the graph, starting on January 1.

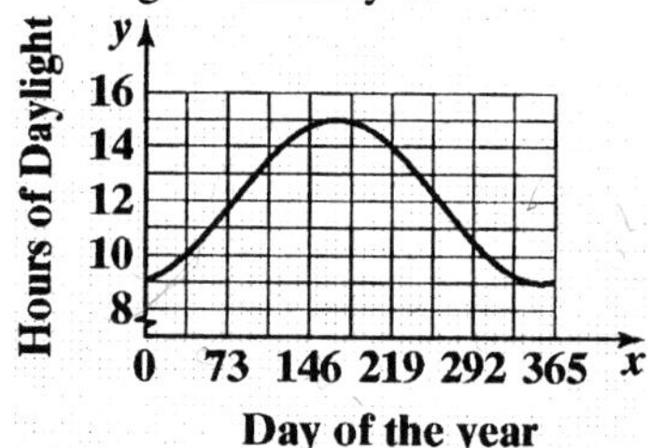

86. The function $y = 16\sin\left(\frac{\pi}{6}x - \frac{2\pi}{3}\right) + 40$ is in the form $y = A\sin(Bx - C) + D$ with $A = 16$, $B = \frac{\pi}{6}$, and $C = \frac{2\pi}{3}$. The amplitude is $|A| = |16| = 16$. The period is $\frac{2\pi}{B} = \frac{2\pi}{\frac{\pi}{6}} = 2\pi \cdot \frac{6}{\pi} = 12$. The phase shift is $\frac{C}{B} = \frac{\frac{2\pi}{3}}{\frac{\pi}{6}} = \frac{2\pi}{3} \cdot \frac{6}{\pi} = 4$. The quarter-period is $\frac{12}{4} = 3$. The cycle begins at $x = 4$. Add quarter-periods to find the x-values for the key points.

$x = 4$
$x = 4 + 3 = 7$
$x = 7 + 3 = 10$
$x = 10 + 3 = 13$
$x = 13 + 3 = 16$

Because we are graphing for $1 \le x \le 12$, we will evaluate the function for the three x-values between 1 and 12, along with $x = 1$ and $x = 12$. Using a calculator we have the following points. $(1, 24)$ $(4, 40)$ $(7, 56)$ $(10, 40)$ $(12, 26.1)$ By connecting the points with a smooth curve we obtain the graph for $1 \le x \le 12$.

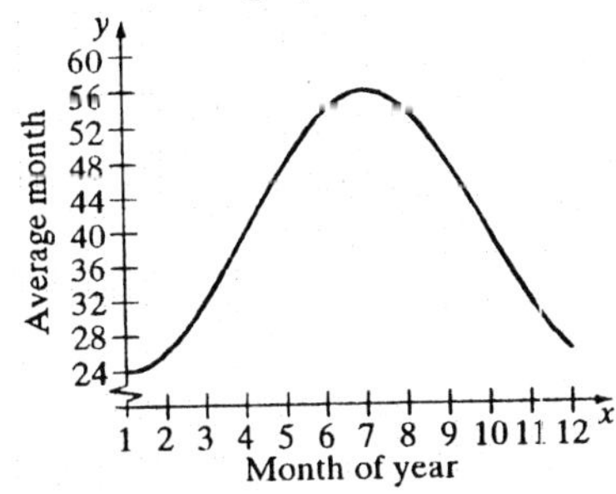

The highest average monthly temperature is 56° in July.

87. Because the depth of the water ranges from a minimum of 6 feet to a maximum of 12 feet, the curve oscillates about the middle value, 9 feet. Thus, $D = 9$. The maximum depth of the water is 3 feet above 9 feet. Thus, $A = 3$. The graph shows that one complete cycle occurs in 12-0, or 12 hours. The period is 12. Thus,

$$12 = \frac{2\pi}{B}$$

$$12B = 2\pi$$

$$B = \frac{2\pi}{12} = \frac{\pi}{6}$$

Substitute these values into $y = A\cos Bx + D$. The depth of the water is modeled by

$$y = 3\cos\frac{\pi x}{6} + 9.$$

88. Because the depth of the water ranges from a minimum of 3 feet to a maximum of 5 feet, the curve oscillates about the middle value, 4 feet. Thus, $D = 4$. The maximum depth of the water is 1 foot above 4 feet. Thus, $A = 1$. The graph shows that one complete cycle occurs in 12–0, or 12 hours. The period is 12. Thus,

$$12 = \frac{2\pi}{B}$$

$$12B = 2\pi$$

$$B = \frac{2\pi}{12} = \frac{\pi}{6}$$

Substitute these values into $y = A\cos Bx + D$.

The depth of the water is modeled by

$$y = \cos\frac{\pi x}{6} + 4.$$

101. The function $y = 3\sin(2x + \pi) = 3\sin(2x - (-\pi))$ is of the form $y = A\sin(Bx - C)$ with $A = 3$, $B = 2$, and $C = -\pi$. The amplitude is $|A| = |3| = 3$. The period is $\frac{2\pi}{B} = \frac{2\pi}{2} = \pi$.

The cycle begins at $x = \frac{C}{B} = \frac{-\pi}{2} = -\frac{\pi}{2}$. We choose $-\frac{\pi}{2} \le x \le \frac{3\pi}{2}$, and $-4 \le y \le 4$ for our graph.

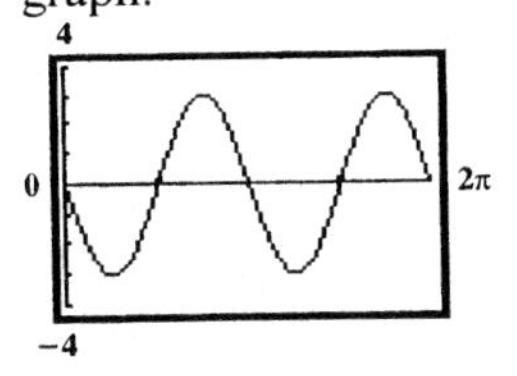

102. The function $y = -2\cos\left(2\pi x - \frac{\pi}{2}\right)$ is of the form $y = A\cos(Bx - C)$ with $A = -2$, $B = 2\pi$, and $C = \frac{\pi}{2}$. The amplitude is $|A| = |-2| = 2$.

The period is $\frac{2\pi}{B} = \frac{2\pi}{2\pi} = 1$. The cycle begins at $x = \frac{C}{B} = \frac{\frac{\pi}{2}}{2\pi} = \frac{\pi}{2} \cdot \frac{1}{2\pi} = \frac{1}{4}$. We choose $\frac{1}{4} \le x \le \frac{9}{4}$, and $-3 \le y \le 3$ for our graph.

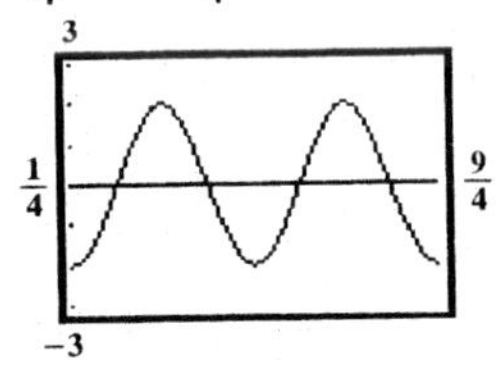

103. The function

$$y = 0.2\sin\left(\frac{\pi}{10}x + \pi\right) = 0.2\sin\left(\frac{\pi}{10}x - (-\pi)\right)$$ is of the form $y = A\sin(Bx - C)$ with $A = 0.2$, $B = \frac{\pi}{10}$, and $C = -\pi$. The amplitude is $|A| = |0.2| = 0.2$. The period is $\frac{2\pi}{B} = \frac{2\pi}{\frac{\pi}{10}} = 2\pi \cdot \frac{10}{\pi} = 20$. The cycle begins at

$x = \frac{C}{B} = \frac{-\pi}{\frac{\pi}{10}} = -\pi \cdot \frac{10}{\pi} = -10$. We choose $-10 \le x \le 30$, and $-1 \le y \le 1$ for our graph.

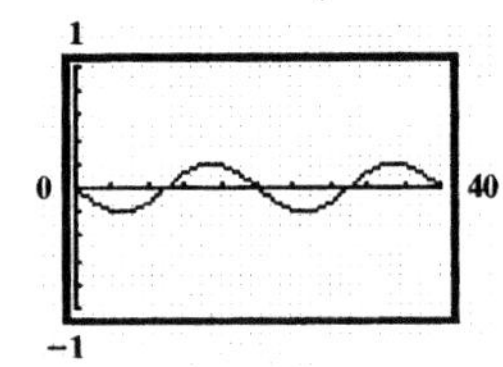

104. The function $y = 3\sin(2x - \pi) + 5$ is of the form $y = A\cos(Bx - C) + D$ with $A = 3$, $B = 2$, $C = \pi$, and $D = 5$. The amplitude is $|A| = |3| = 3$. The period is $\frac{2\pi}{B} = \frac{2\pi}{2} = \pi$.

The cycle begins at $x = \frac{C}{B} = \frac{\pi}{2}$. Because $D = 5$, the graph has a vertical shift 5 units upward. We choose $\frac{\pi}{2} \le x \le \frac{5\pi}{2}$, and $0 \le y \le 10$ for our graph.

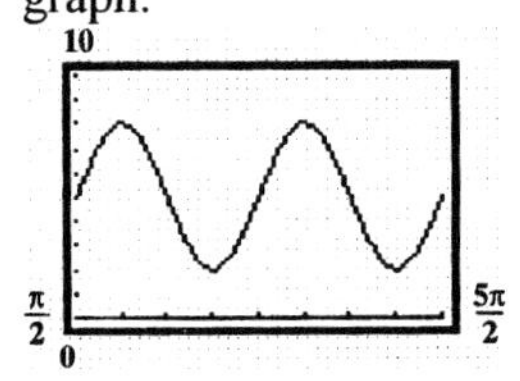

105.

The graphs appear to be the same from $-\frac{\pi}{2}$ to $\frac{\pi}{2}$.

106.

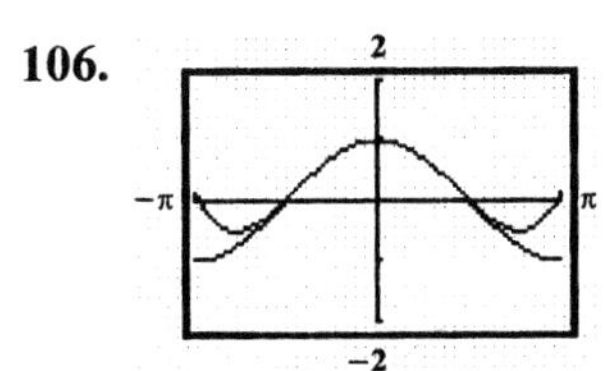

The graphs appear to be the same from $-\frac{\pi}{2}$ to $\frac{\pi}{2}$.

107.

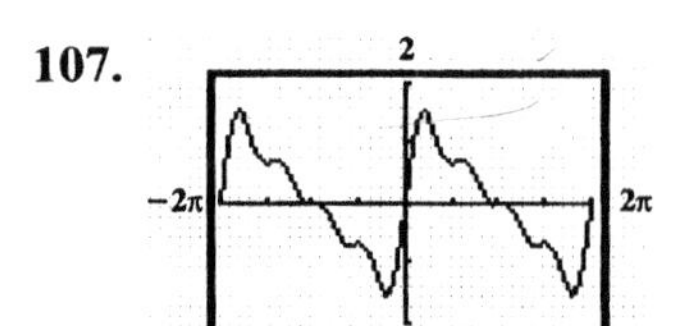

The graph is similar to $y = \sin x$, except the amplitude is greater and the curve is less smooth.

108.

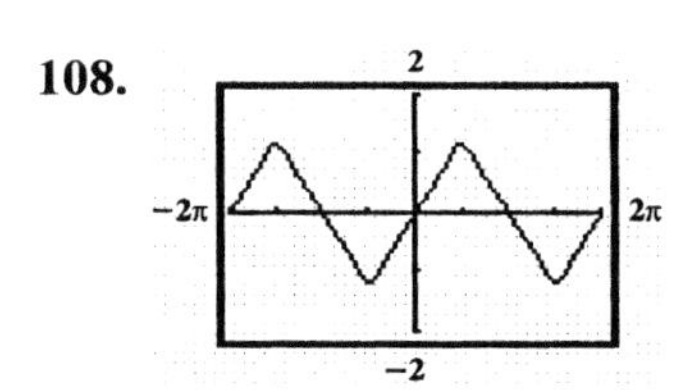

The graph is very similar to $y = \sin x$, except not smooth.

109. a.

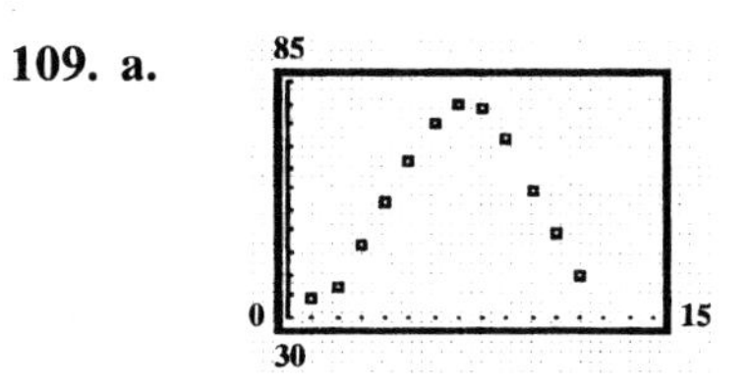

b. $y = 22.61\sin(0.50x - 2.04) + 57.17$

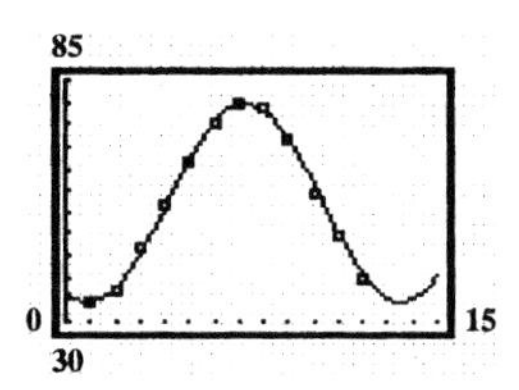

111. a. Since $A = 3$ and $D = -2$, the maximum will occur at $3 - 2 = 1$ and the minimum will occur at $-3 - 2 = -5$. Thus the range is $[-5,1]$

Viewing rectangle: $\left[-\frac{\pi}{6}, \frac{23\pi}{6}, \frac{\pi}{6}\right]$ by $[-5,1,1]$

b. Since $A = 1$ and $D = -2$, the maximum will occur at $1 - 2 = -1$ and the minimum will occur at $-1 - 2 = -3$. Thus the range is $[-3,-1]$

112. $A = \pi$

$B = \dfrac{2\pi}{\text{period}} = \dfrac{2\pi}{1} = 2\pi$

$\dfrac{C}{B} = \dfrac{C}{2\pi} = -2$

$C = -4\pi$

$y = A\cos(Bx - C)$

$y = \pi\cos(2\pi x + 4\pi)$

or

$y = \pi\cos\left[2\pi(x+2)\right]$

113. $y = \sin^2 x = \dfrac{1}{2} - \dfrac{1}{2}\cos 2x$

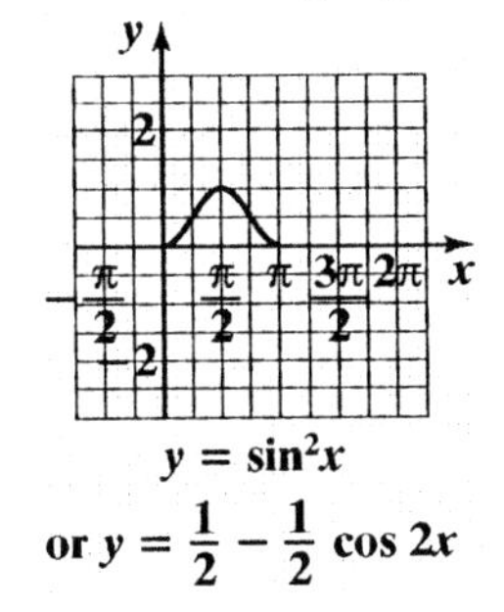

$y = \sin^2 x$

or $y = \dfrac{1}{2} - \dfrac{1}{2}\cos 2x$

114. $y = \cos^2 x = \dfrac{1}{2} + \dfrac{1}{2}\cos 2x$

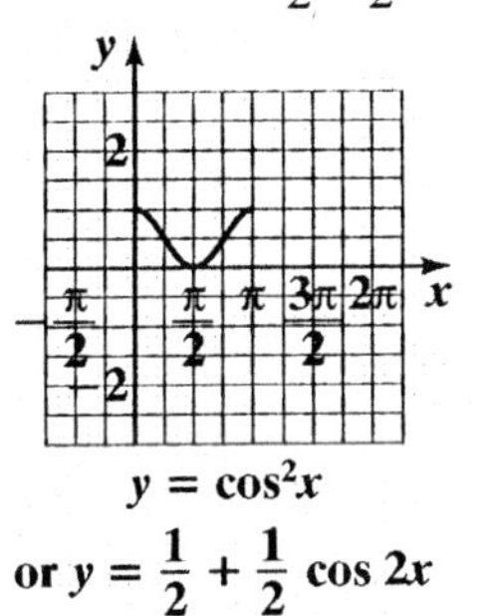

$y = \cos^2 x$

or $y = \dfrac{1}{2} + \dfrac{1}{2}\cos 2x$

Section 4.6

Check Point Exercises

1. Solve the equations

$2x = -\dfrac{\pi}{2}$ and $2x = \dfrac{\pi}{2}$

$x = -\dfrac{\pi}{4}$ $\quad x = \dfrac{\pi}{4}$

Thus, two consecutive asymptotes occur at $x = -\dfrac{\pi}{4}$ and $x = \dfrac{\pi}{4}$. Midway between these asymptotes is $x = 0$. An x-intercept is 0 and the graph passes through (0, 0). Because the coefficient of the tangent is 3, the points on the graph midway between an x-intercept and the asymptotes have y-coordinates of -3 and 3. Use the two asymptotes, the x-intercept, and the points midway between to graph one period of $y = 3\tan 2x$ from $-\dfrac{\pi}{4}$ to $\dfrac{\pi}{4}$. In order to graph for $-\dfrac{\pi}{4} < x < \dfrac{3\pi}{4}$, Continue the pattern and extend the graph another full period to the right.

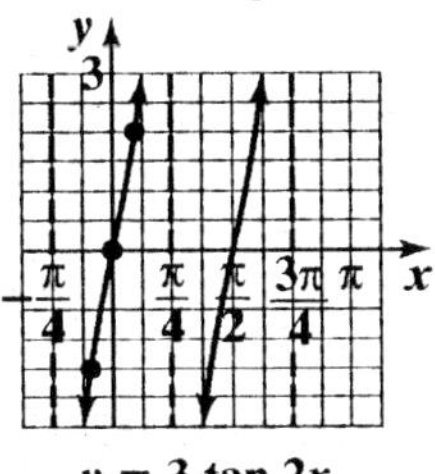

$y = 3\tan 2x$

2. Solve the equations

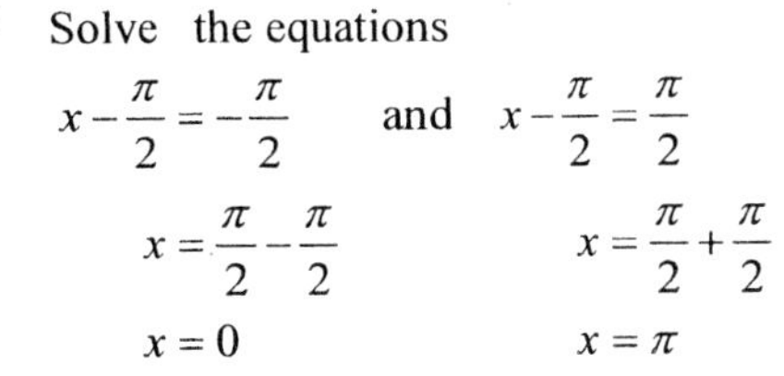

$x - \dfrac{\pi}{2} = -\dfrac{\pi}{2}$ and $x - \dfrac{\pi}{2} = \dfrac{\pi}{2}$

$x = \dfrac{\pi}{2} - \dfrac{\pi}{2}$ $\quad x = \dfrac{\pi}{2} + \dfrac{\pi}{2}$

$x = 0$ $\quad x = \pi$

Thus, two consecutive asymptotes occur at $x = 0$ and $x = \pi$.

x-intercept $= \dfrac{0+\pi}{2} = \dfrac{\pi}{2}$

An x-intercept is $\dfrac{\pi}{2}$ and the graph passes through $\left(\dfrac{\pi}{2}, 0\right)$. Because the coefficient of the tangent is 1, the points on the graph midway between an x-intercept and the asymptotes have y-coordinates of -1 and 1. Use the two consecutive asymptotes, $x = 0$ and $x = \pi$, to graph one full period of $y = \tan\left(x - \dfrac{\pi}{2}\right)$ from 0 to π. Continue the pattern and extend the graph another full period to the right.

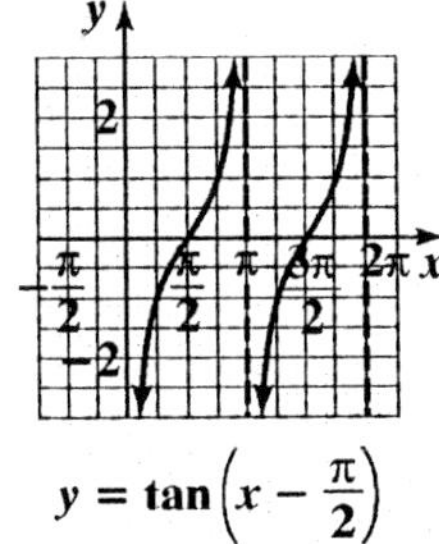

$y = \tan\left(x - \dfrac{\pi}{2}\right)$

3. Solve the equations

$$\frac{\pi}{2}x = 0 \quad \text{and} \quad \frac{\pi}{2}x = \pi$$

$$x = 0 \qquad x = \frac{\pi}{\frac{\pi}{2}}$$

$$x = 2$$

Two consecutive asymptotes occur at $x = 0$ and $x = 2$. Midway between $x = 0$ and $x = 2$ is $x = 1$. An x-intercept is 1 and the graph passes through (1, 0). Because the coefficient of the cotangent is $\frac{1}{2}$, the points on the graph midway between an x-intercept and the asymptotes have y-coordinates of $-\frac{1}{2}$ and $\frac{1}{2}$. Use the two consecutive asymptotes, $x = 0$ and $x = 2$, to graph one full period of $y = \frac{1}{2}\cot\frac{\pi}{2}x$. The curve is repeated along the x-axis one full period as shown.

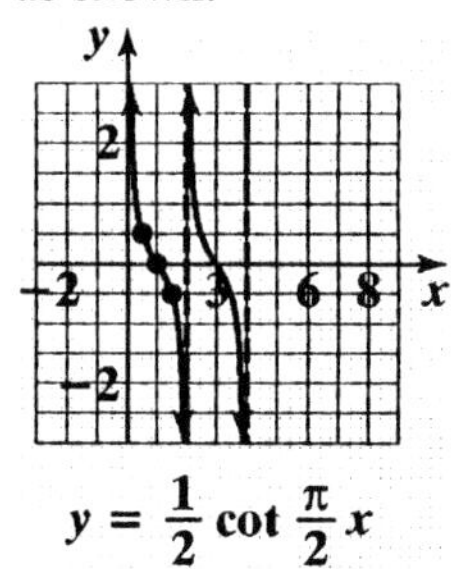

$y = \frac{1}{2}\cot\frac{\pi}{2}x$

4. The x-intercepts of $y = \sin\left(x + \frac{\pi}{4}\right)$ correspond to vertical asymptotes of $y = \csc\left(x + \frac{\pi}{4}\right)$.

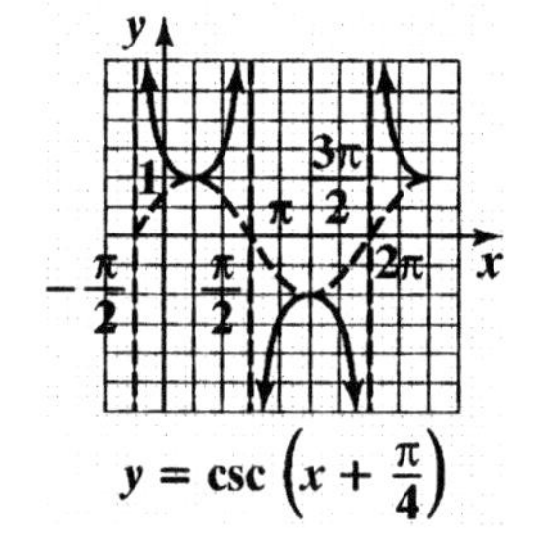

$y = \csc\left(x + \frac{\pi}{4}\right)$

5. Graph the reciprocal cosine function, $y = 2\cos 2x$. The equation is of the form $y = A\cos Bx$ with $A = 2$ and $B = 2$.

amplitude: $|A| = |2| = 2$

period: $\frac{2\pi}{B} = \frac{2\pi}{2} = \pi$

Use quarter-periods, $\frac{\pi}{4}$, to find x-values for the five key points. Starting with $x = 0$, the x-values are $0, \frac{\pi}{4}, \frac{\pi}{2}, \frac{3\pi}{4}$, and π. Evaluating the function at each value of x, the key points are $(0,2), \left(\frac{\pi}{4},0\right), \left(\frac{\pi}{2},-2\right), \left(\frac{3\pi}{4}, 0\right), (\pi,2)$. In order to graph for $-\frac{3\pi}{4} \le x \le \frac{3\pi}{4}$, Use the first four points and extend the graph $-\frac{3\pi}{4}$ units to the left. Use the graph to obtain the graph of the reciprocal function. Draw vertical asymptotes through the x-intercepts, and use them as guides to graph $y = 2\sec 2x$.

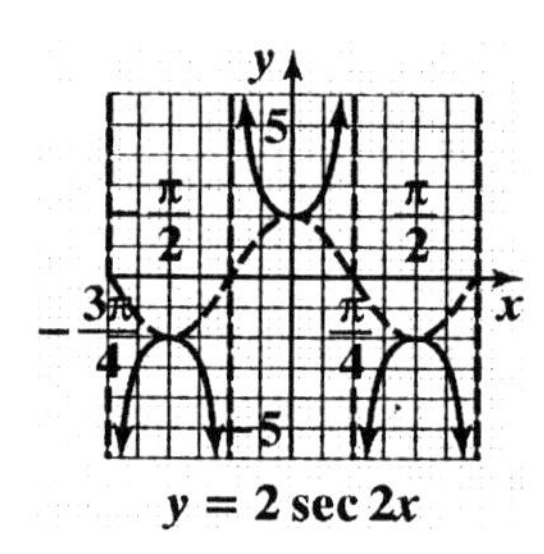

$y = 2\sec 2x$

Exercise Set 4.6

1. The graph has an asymptote at $x = -\frac{\pi}{2}$.

The phase shift, $\frac{C}{B}$, from $\frac{\pi}{2}$ to $-\frac{\pi}{2}$ is $-\pi$ units. Thus, $\frac{C}{B} = \frac{C}{1} = -\pi$

$$C = -\pi$$

The function with $C = -\pi$ is $y = \tan(x + \pi)$.

2. The graph has an asymptote at $x = 0$.

The phase shift, $\frac{C}{B}$, from $\frac{\pi}{2}$ to 0 is $-\frac{\pi}{2}$ units.

Thus, $\frac{C}{B} = \frac{C}{1} = -\frac{\pi}{2}$

$C = -\frac{\pi}{2}$

The function with $C = -\frac{\pi}{2}$ is $y = \tan\left(x + \frac{\pi}{2}\right)$.

3. The graph has an asymptote at $x = \pi$.

$\pi = \frac{\pi}{2} + C$

$C = \frac{\pi}{2}$

The function is $y = -\tan\left(x - \frac{\pi}{2}\right)$.

4. The graph has an asymptote at $\frac{\pi}{2}$.

There is no phase shift. Thus, $\frac{C}{B} = \frac{C}{1} = 0$

$C = 0$

The function with $C = 0$ is $y = -\tan x$.

5. Solve the equations

$\frac{x}{4} = -\frac{\pi}{2}$ and $\frac{x}{4} = \frac{\pi}{2}$

$x = \left(-\frac{\pi}{2}\right)4$ $\qquad x = \left(\frac{\pi}{2}\right)4$

$x = -2\pi$ $\qquad x = 2\pi$

Thus, two consecutive asymptotes occur at $x = -2\pi$ and $x = 2\pi$.

$x\text{-intercept} = \frac{-2\pi + 2\pi}{2} = \frac{0}{2} = 0$

An x-intercept is 0 and the graph passes through (0, 0). Because the coefficient of the tangent is 3, the points on the graph midway between an x-intercept and the asymptotes have y-coordinates of –3 and 3. Use the two consecutive asymptotes, $x = -2\pi$ and $x = 2\pi$, to graph one full period of $y = 3\tan\frac{x}{4}$ from -2π to 2π.

Continue the pattern and extend the graph another full period to the right.

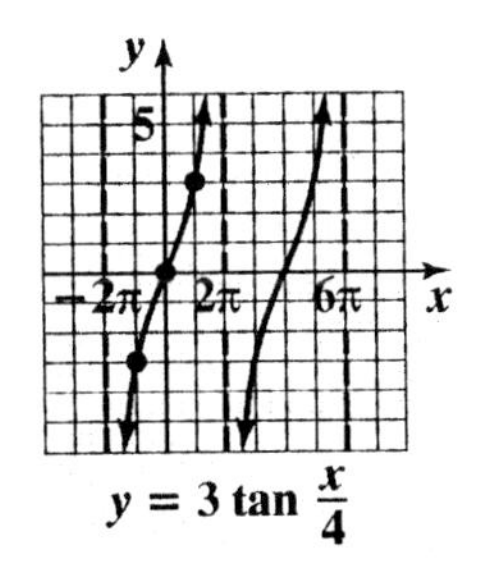

$y = 3\tan\frac{x}{4}$

6. Solve the equations

$\frac{x}{4} = -\frac{\pi}{2}$ and $\frac{x}{4} = \frac{\pi}{2}$

$x = \left(-\frac{\pi}{2}\right)4$ $\qquad x = \left(\frac{\pi}{2}\right)4$

$x = -2\pi$ $\qquad x = 2\pi$

Thus, two consecutive asymptotes occur at $x = -2\pi$ and $x = 2\pi$.

$x\text{-intercept} = \frac{-2\pi + 2\pi}{2} = \frac{0}{2} = 0$

An x-intercept is 0 and the graph passes through (0, 0). Because the coefficient of the tangent is 2, the points on the graph midway between an x-intercept and the asymptotes have y-coordinates of –2 and 2. Use the two consecutive asymptotes, $x = -2\pi$ and $x = 2\pi$, to graph one full period of $y = 2\tan\frac{x}{4}$ from -2π to 2π. Continue the pattern and extend the graph another full period to the right.

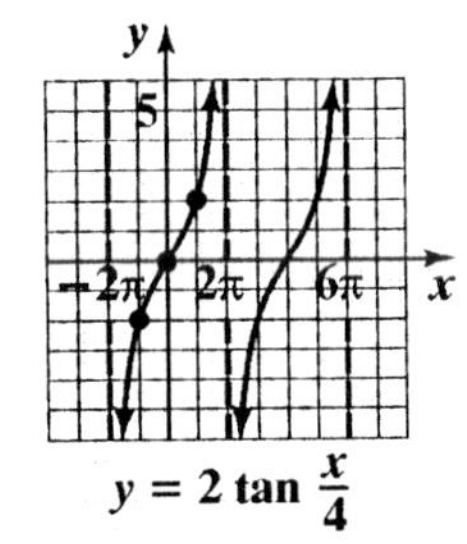

$y = 2\tan\frac{x}{4}$

7. Solve the equations $2x = -\frac{\pi}{2}$ and $2x = \frac{\pi}{2}$

$x = \frac{-\frac{\pi}{2}}{2}$ $\quad x = \frac{\frac{\pi}{2}}{2}$

$x = -\frac{\pi}{4}$ $\quad x = \frac{\pi}{4}$

Thus, two consecutive asymptotes occur at $x = -\frac{\pi}{4}$ and $x = \frac{\pi}{4}$.

$x\text{-intercept} = \frac{-\frac{\pi}{4} + \frac{\pi}{4}}{2} = \frac{0}{2} = 0$

An x-intercept is 0 and the graph passes through (0, 0). Because the coefficient of the tangent is $\frac{1}{2}$, the points on the graph midway between an x-intercept and the asymptotes have y-coordinates of $-\frac{1}{2}$ and $\frac{1}{2}$. Use the two consecutive asymptotes, $x = -\frac{\pi}{4}$ and $x = \frac{\pi}{4}$, to graph one full period of $y = \frac{1}{2}\tan 2x$ from $-\frac{\pi}{4}$ to $\frac{\pi}{4}$. Continue the pattern and extend the graph another full period to the right.

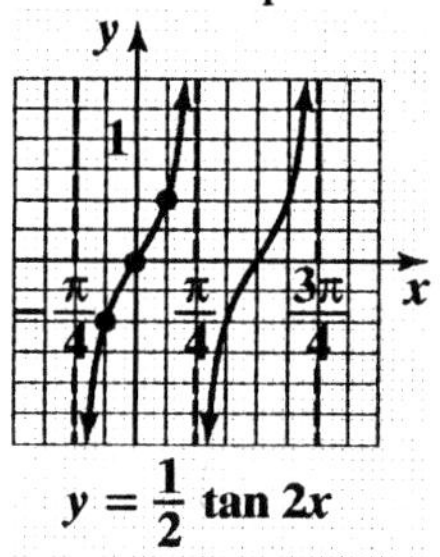

$y = \frac{1}{2}\tan 2x$

8. Solve the equations

$2x = -\frac{\pi}{2}$ and $2x = \frac{\pi}{2}$

$x = \frac{-\frac{\pi}{2}}{2}$ $\quad x = \frac{\frac{\pi}{2}}{2}$

$x = -\frac{\pi}{4}$ $\quad x = \frac{\pi}{4}$

Thus, two consecutive asymptotes occur at $x = -\frac{\pi}{4}$ and $x = \frac{\pi}{4}$.

$x\text{-intercept} = \frac{-\frac{\pi}{4} + \frac{\pi}{4}}{2} = \frac{0}{2} = 0$

An x-intercept is 0 and the graph passes through (0, 0). Because the coefficient of the tangent is 2, the points on the graph midway between an x-intercept and the asymptotes have y-coordinates of –2 and 2. Use the two consecutive asymptotes, $x = -\frac{\pi}{4}$ and $x = \frac{\pi}{4}$, to graph one full period of $y = 2\tan 2x$ from $-\frac{\pi}{4}$ to $\frac{\pi}{4}$.

Continue the pattern and extend the graph another full period to the right.

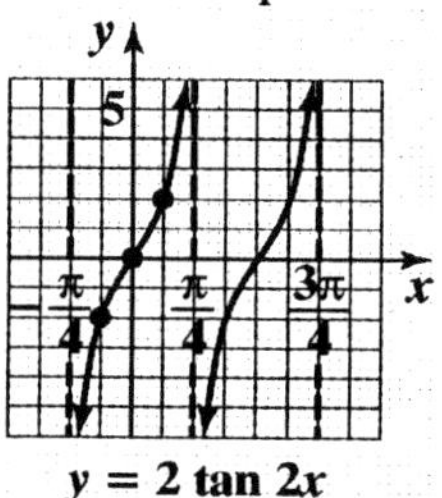

$y = 2\tan 2x$

9. Solve the equations

$\frac{1}{2}x = -\frac{\pi}{2}$ and $\frac{1}{2}x = \frac{\pi}{2}$

$x = \left(-\frac{\pi}{2}\right)2$ $\quad x = \left(\frac{\pi}{2}\right)2$

$x = -\pi$ $\quad x = \pi$

Thus, two consecutive asymptotes occur at $x = -\pi$ and $x = \pi$.

$x\text{-intercept} = \frac{-\pi + \pi}{2} = \frac{0}{2} = 0$

An x-intercept is 0 and the graph passes through (0, 0). Because the coefficient of the tangent is –2, the points on the graph midway between an x-intercept and the asymptotes have y-coordinates of 2 and –2. Use the two consecutive asymptotes, $x = -\pi$ and $x = \pi$, to graph one full period of $y = -2\tan\frac{1}{2}x$ from $-\pi$ to π.

Continue the pattern and extend the graph another full period to the right.

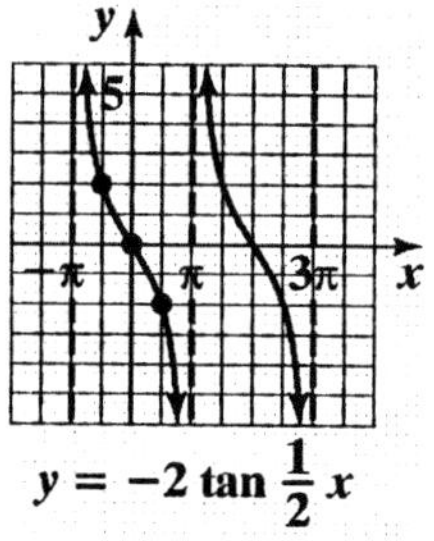

$y = -2\tan\frac{1}{2}x$

10. Solve the equations

$\frac{1}{2}x = -\frac{\pi}{2}$ and $\frac{1}{2}x = \frac{\pi}{2}$

$x = \left(-\frac{\pi}{2}\right)2$ $\quad x = \left(\frac{\pi}{2}\right)2$

$x = -\pi$ $\quad x = \pi$

Thus, two consecutive asymptotes occur at $x = -\pi$ and $x = \pi$.

$x\text{-intercept} = \frac{-\pi + \pi}{2} = \frac{0}{2} = 0$

An x-intercept is 0 and the graph passes through (0, 0). Because the coefficient of the tangent is –3, the points on the graph midway between an x-intercept and the asymptotes have y-coordinates of 3 and –3. Use the two consecutive asymptotes, $x = -\pi$ and $x = \pi$, to graph one full period of $y = -3\tan\frac{1}{2}x$ from $-\pi$ to π. Continue the pattern and extend the graph another full period to the right.

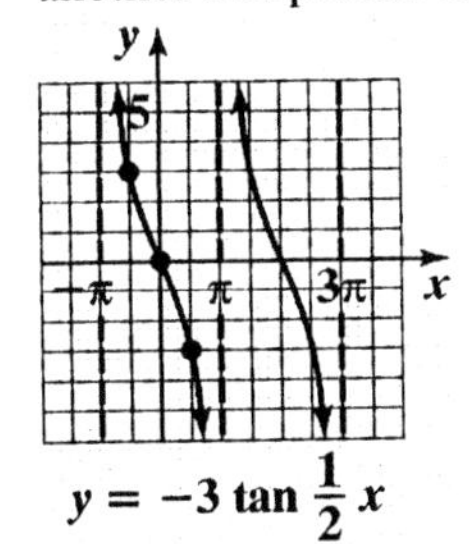

$y = -3\tan\frac{1}{2}x$

11. Solve the equations

$x - \pi = -\frac{\pi}{2}$ and $x - \pi = \frac{\pi}{2}$

$x = -\frac{\pi}{2} + \pi$ $\quad x = \frac{\pi}{2} + \pi$

$x = \frac{\pi}{2}$ $\quad x = \frac{3\pi}{2}$

Thus, two consecutive asymptotes occur at $x = \frac{\pi}{2}$ and $x = \frac{3\pi}{2}$.

$x\text{-intercept} = \frac{\frac{\pi}{2} + \frac{3\pi}{2}}{2} = \frac{\frac{4\pi}{2}}{2} = \frac{4\pi}{4} = \pi$

An x-intercept is π and the graph passes through $(\pi, 0)$. Because the coefficient of the tangent is 1, the points on the graph midway between an x-intercept and the asymptotes have y-coordinates of –1 and 1. Use the two consecutive asymptotes, $x = \frac{\pi}{2}$ and $x = \frac{3\pi}{2}$, to graph one full period of $y = \tan(x - \pi)$ from $\frac{\pi}{2}$ to $\frac{3\pi}{2}$. Continue the pattern and extend the graph another full period to the right.

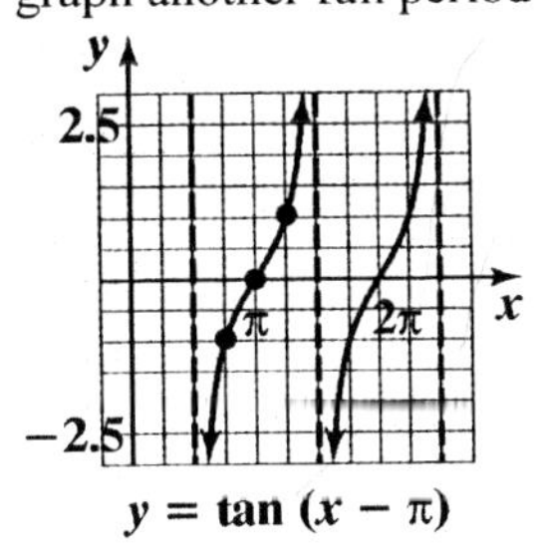

$y = \tan(x - \pi)$

12. Solve the equations

$x - \frac{\pi}{4} = -\frac{\pi}{2}$ and $x - \frac{\pi}{4} = \frac{\pi}{2}$

$x = -\frac{2\pi}{4} + \frac{\pi}{4}$ $\quad x = \frac{2\pi}{4} + \frac{\pi}{4}$

$x = -\frac{\pi}{4}$ $\quad x = \frac{3\pi}{4}$

Thus, two consecutive asymptotes occur at $x = -\frac{\pi}{4}$ and $x = \frac{3\pi}{4}$.

$x\text{-intercept} = \frac{-\frac{\pi}{4} + \frac{3\pi}{4}}{2} = \frac{\frac{2\pi}{4}}{2} = \frac{\pi}{4}$

An x-intercept is $\frac{\pi}{4}$ and the graph passes through $\left(\frac{\pi}{4}, 0\right)$. Because the coefficient of the tangent is 1, the points on the graph midway between an x-intercept and the asymptotes have y-coordinates of –1 and 1. Use the two consecutive asymptotes, $x = -\frac{\pi}{4}$ and $x = \frac{3\pi}{4}$, to graph one full period of $y = \tan\left(x - \frac{\pi}{4}\right)$ from 0 to π. Continue the pattern and extend the graph another full period to the right.

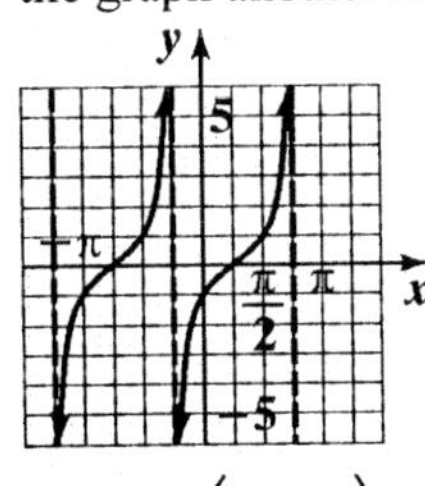

$y = \tan\left(x - \frac{\pi}{4}\right)$

13. There is no phase shift. Thus,

$\frac{C}{B} = \frac{C}{1} = 0$

$C = 0$

Because the points on the graph midway between an x-intercept and the asymptotes have y-coordinates of -1 and 1, $A = -1$. The function with $C = 0$ and $A = -1$ is $y = -\cot x$.

14. The graph has an asymptote at $\frac{\pi}{2}$. The phase shift, $\frac{C}{B}$, from 0 to $\frac{\pi}{2}$ is $\frac{\pi}{2}$ units.

Thus, $\frac{C}{B} = \frac{C}{1} = \frac{\pi}{2}$

$C = \frac{\pi}{2}$

The function with $C = \frac{\pi}{2}$ is $y = -\cot\left(x - \frac{\pi}{2}\right)$.

15. The graph has an asymptote at $-\frac{\pi}{2}$. The phase shift, $\frac{C}{B}$, from 0 to $-\frac{\pi}{2}$ is $-\frac{\pi}{2}$ units. Thus,

$\frac{C}{B} = \frac{C}{1} = -\frac{\pi}{2}$

$C = -\frac{\pi}{2}$

The function with $C = -\frac{\pi}{2}$ is $y = \cot\left(x + \frac{\pi}{2}\right)$.

16. The graph has an asymptote at $-\pi$. The phase shift, $\frac{C}{B}$, from 0 to $-\pi$ is $-\pi$ units.

Thus, $\frac{C}{B} = \frac{C}{1} = -\pi$

$C = -\pi$

The function with $C = -\pi$ is $y = \cot(x + \pi)$.

17. Solve the equations $x = 0$ and $x = \pi$. Two consecutive asymptotes occur at $x = 0$ and $x = \pi$.

$x\text{-intercept} = \frac{0 + \pi}{2} = \frac{\pi}{2}$

An x-intercept is $\frac{\pi}{2}$ and the graph passes through $\left(\frac{\pi}{2}, 0\right)$. Because the coefficient of the cotangent is 2, the points on the graph midway between an x-intercept and the asymptotes have y-coordinates of 2 and -2. Use the two consecutive asymptotes, $x = 0$ and $x = \pi$, to graph one full period of $y = 2\cot x$. The curve is repeated along the x-axis one full period as shown.

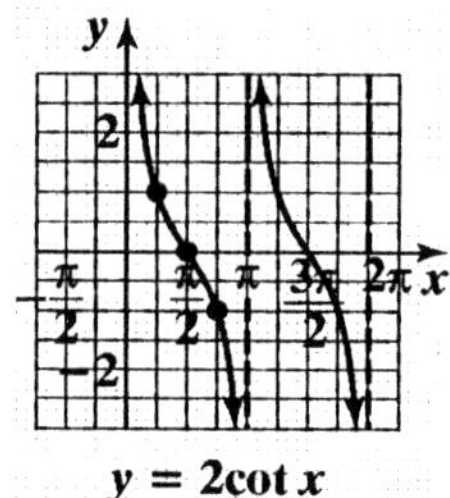

$y = 2\cot x$

18. Solve the equations

$x = 0$ and $x = \pi$

Two consecutive asymptotes occur at $x = 0$ and $x = \pi$.

$x\text{-intercept} = \frac{0 + \pi}{2} = \frac{\pi}{2}$

An x-intercept is $\frac{\pi}{2}$ and the graph passes through $\left(\frac{\pi}{2}, 0\right)$. Because the coefficient of the cotangent is $\frac{1}{2}$, the points on the graph midway between an x-intercept and the asymptotes have y-coordinates of $\frac{1}{2}$ and $-\frac{1}{2}$. Use the two consecutive asymptotes, $x = 0$ and $x = \pi$, to graph one full period of $y = \frac{1}{2}\cot x$.

The curve is repeated along the x-axis one full period as shown.

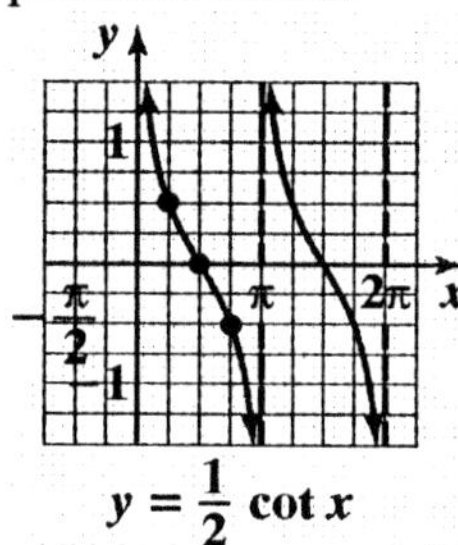

$y = \frac{1}{2}\cot x$

19. Solve the equations

$$2x = 0 \quad \text{and} \quad 2x = \pi$$
$$x = 0 \qquad x = \frac{\pi}{2}$$

Two consecutive asymptotes occur at $x = 0$ and $x = \frac{\pi}{2}$.

$x\text{-intercept} = \dfrac{0 + \frac{\pi}{2}}{2} = \dfrac{\frac{\pi}{2}}{2} = \dfrac{\pi}{4}$

An x-intercept is $\frac{\pi}{4}$ and the graph passes through $\left(\frac{\pi}{4}, 0\right)$. Because the coefficient of the cotangent is $\frac{1}{2}$, the points on the graph midway between an x-intercept and the asymptotes have y-coordinates of $\frac{1}{2}$ and $-\frac{1}{2}$. Use the two consecutive asymptotes, $x = 0$ and $x = \frac{\pi}{2}$, to graph one full period of $y = \frac{1}{2}\cot 2x$. The curve is repeated along the x-axis one full period as shown.

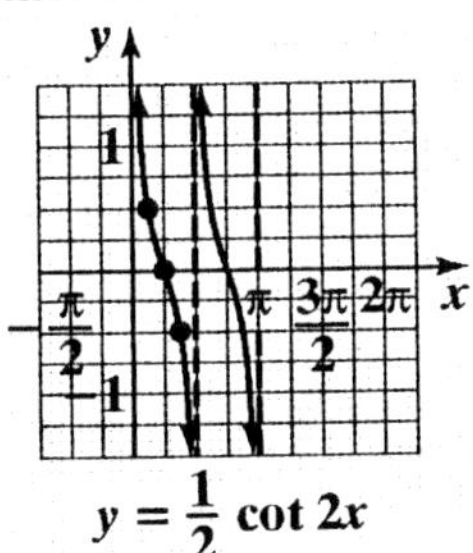

$y = \frac{1}{2}\cot 2x$

20. Solve the equations

$$2x = 0 \quad \text{and} \quad 2x = \pi$$
$$x = 0 \qquad x = \frac{\pi}{2}$$

Two consecutive asymptotes occur at $x = 0$ and $x = \frac{\pi}{2}$.

$x\text{-intercept} = \dfrac{0 + \frac{\pi}{2}}{2} = \dfrac{\frac{\pi}{2}}{2} = \dfrac{\pi}{4}$

An x-intercept is $\frac{\pi}{4}$ and the graph passes through $\left(\frac{\pi}{4}, 0\right)$. Because the coefficient of the cotangent is 2, the points on the graph midway between an x-intercept and the asymptotes have y-coordinates of 2 and –2. Use the two consecutive asymptotes, $x = 0$ and $x = \frac{\pi}{2}$, to graph one full period of $y = 2\cot 2x$. The curve is repeated along the x-axis one full period as shown.

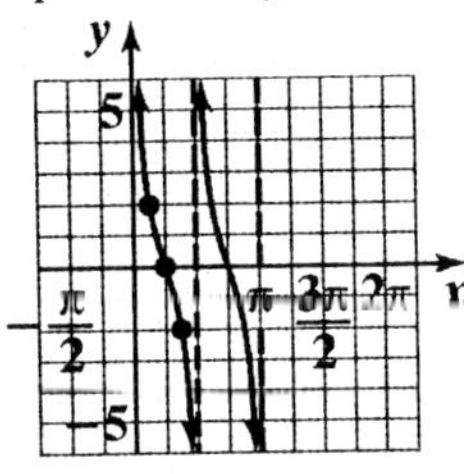

$y = 2\cot 2x$

21. Solve the equations

$$\frac{\pi}{2}x = 0 \quad \text{and} \quad \frac{\pi}{2}x = \pi$$
$$x = 0 \qquad x = \frac{\pi}{\frac{\pi}{2}}$$
$$x = 2$$

Two consecutive asymptotes occur at $x = 0$ and $x = 2$.

$x\text{-intercept} = \dfrac{0+2}{2} = \dfrac{2}{2} = 1$

An x-intercept is 1 and the graph passes through (1, 0). Because the coefficient of the cotangent is –3, the points on the graph midway between an x-intercept and the asymptotes have y-coordinates of –3 and 3. Use the two consecutive asymptotes, $x = 0$ and $x = 2$, to graph one full period of $y = -3\cot \frac{\pi}{2}x$. The curve is repeated along the x-axis one full period as shown.

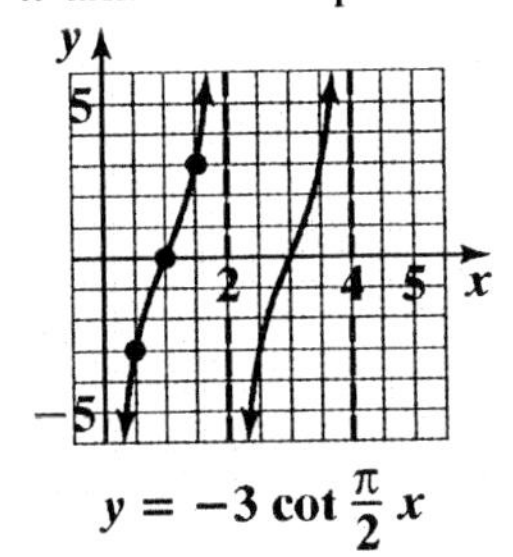

$y = -3\cot \frac{\pi}{2}x$

22. Solve the equations

$$\frac{\pi}{4}x = 0 \quad \text{and} \quad \frac{\pi}{4}x = \pi$$

$$x = 0 \qquad x = \frac{\pi}{\frac{\pi}{4}}$$

$$x = 4$$

Two consecutive asymptotes occur at $x = 0$ and $x = 4$.

$$x\text{-intercept} = \frac{0+4}{2} = \frac{4}{2} = 2$$

An x-intercept is 2 and the graph passes through $(2, 0)$. Because the coefficient of the cotangent is –2, the points on the graph midway between an x-intercept and the asymptotes have y-coordinates of –2 and 2. Use the two consecutive asymptotes, $x = 0$ and $x = 4$, to graph one full period of $y = -2\cot\frac{\pi}{4}x$. The curve is repeated along the x-axis one full period as shown.

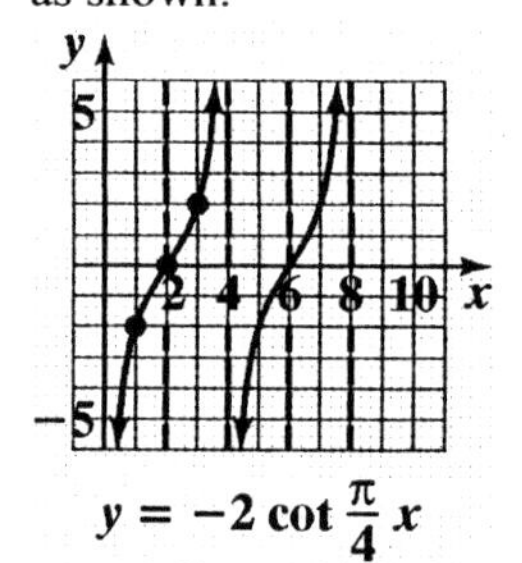

$y = -2\cot\frac{\pi}{4}x$

23. Solve the equations

$$x + \frac{\pi}{2} = 0 \quad \text{and} \quad x + \frac{\pi}{2} = \pi$$

$$x = 0 - \frac{\pi}{2} \qquad x = \pi - \frac{\pi}{2}$$

$$x = -\frac{\pi}{2} \qquad x = \frac{\pi}{2}$$

Two consecutive asymptotes occur at $x = -\frac{\pi}{2}$ and $x = \frac{\pi}{2}$.

$$x\text{-intercept} = \frac{-\frac{\pi}{2} + \frac{\pi}{2}}{2} = \frac{0}{2} = 0$$

An x-intercept is 0 and the graph passes through (0, 0). Because the coefficient of the cotangent is 3, the points on the graph midway between an x-intercept and the asymptotes have y-coordinates of 3 and –3. Use the two consecutive asymptotes, $x = -\frac{\pi}{2}$ and $x = \frac{\pi}{2}$, to graph one full period of $y = 3\cot\left(x + \frac{\pi}{2}\right)$. The curve is repeated along the x-axis one full period as shown.

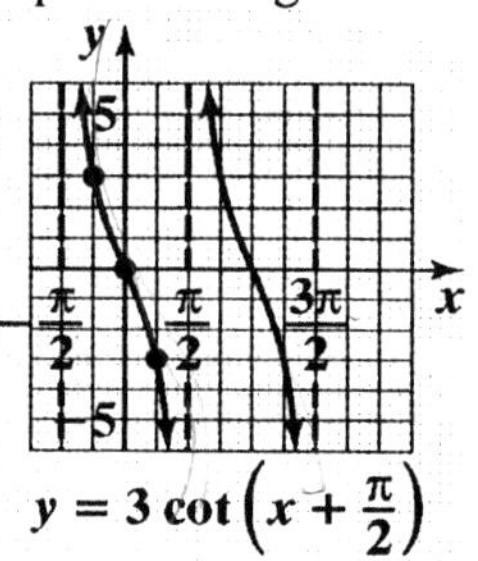

$y = 3\cot\left(x + \frac{\pi}{2}\right)$

24. Solve the equations

$$x + \frac{\pi}{4} = 0 \quad \text{and} \quad x + \frac{\pi}{4} = \pi$$

$$x = 0 - \frac{\pi}{4} \qquad x = \pi - \frac{\pi}{4}$$

$$x = -\frac{\pi}{4} \qquad x = \frac{3\pi}{4}$$

Two consecutive asymptotes occur at $x = -\frac{\pi}{4}$ and $x = \frac{3\pi}{4}$.

$$x\text{-intercept} = \frac{-\frac{\pi}{4} + \frac{3\pi}{4}}{2} = \frac{\frac{2\pi}{4}}{2} = \frac{\pi}{4}$$

An x-intercept is $\frac{\pi}{4}$ and the graph passes through $\left(\frac{\pi}{4}, 0\right)$. Because the coefficient of the cotangent is 3, the points on the graph midway between an x-intercept and the asymptotes have y-coordinates of 3 and –3. Use the two consecutive asymptotes, $x = -\frac{\pi}{4}$ and $x = \frac{3\pi}{4}$, to graph one full period of $y = 3\cot\left(x + \frac{\pi}{4}\right)$.

The curve is repeated along the x-axis one full period as shown.

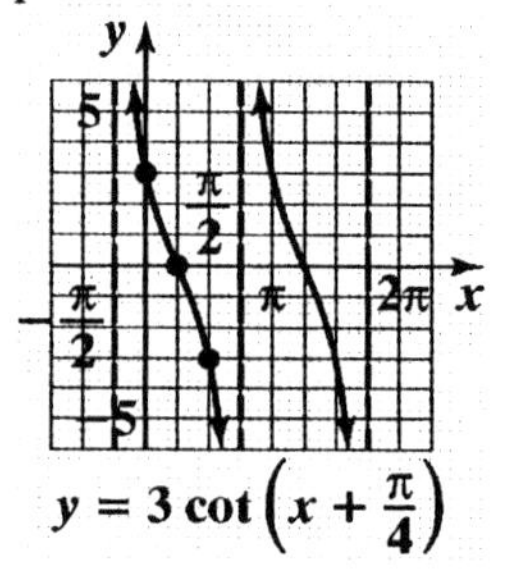

$y = 3\cot\left(x + \frac{\pi}{4}\right)$

25. The x-intercepts of $y = -\frac{1}{2}\sin\frac{x}{2}$ corresponds to vertical asymptotes of $y = -\frac{1}{2}\csc\frac{x}{2}$. Draw the vertical asymptotes, and use them as a guide to sketch the graph of $y = -\frac{1}{2}\csc\frac{x}{2}$.

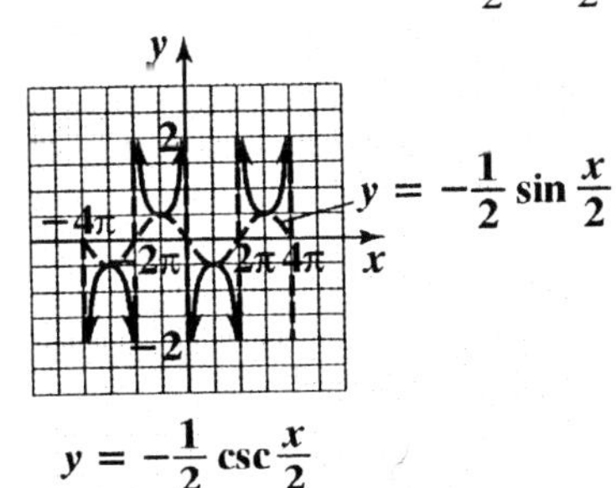

$y = -\frac{1}{2}\csc\frac{x}{2}$

26. The x-intercepts of $y = 3\sin 4x$ correspond to vertical asymptotes of $y = 3\csc 4x$. Draw the vertical asymptotes, and use them as a guide to sketch the graph of $y = 3\csc 4x$.

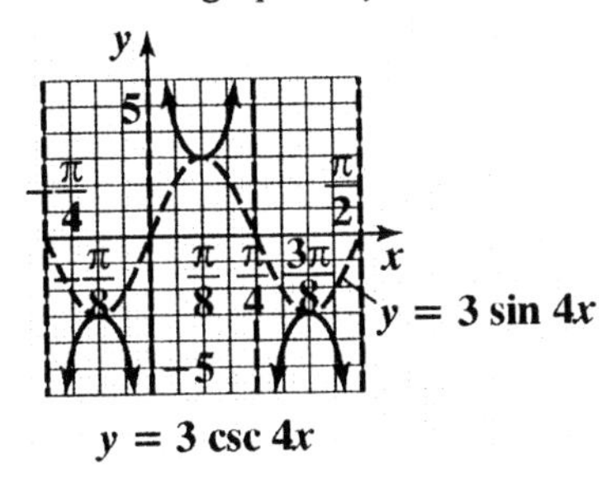

$y = 3\csc 4x$

27. The x-intercepts of $y = \frac{1}{2}\cos 2\pi x$ corresponds to vertical asymptotes of $y = \frac{1}{2}\sec 2\pi x$. Draw the vertical asymptotes, and use them as a guide to sketch the graph of $y = \frac{1}{2}\sec 2\pi x$.

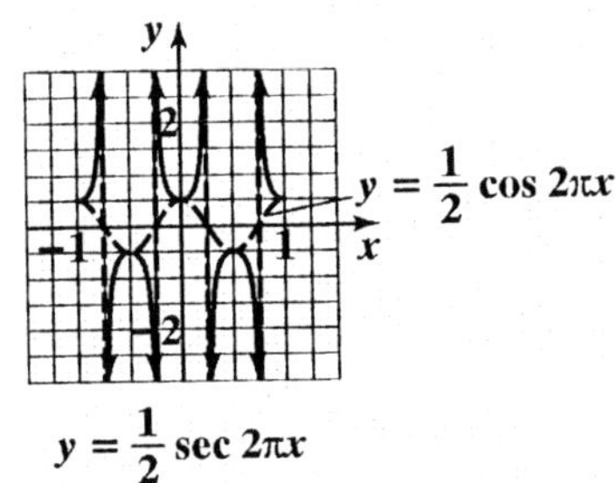

$y = \frac{1}{2}\sec 2\pi x$

28. The x-intercepts of $y = -3\cos\frac{\pi}{2}x$ correspond to vertical asymptotes of $y = -3\sec\frac{\pi}{2}x$. Draw the vertical asymptotes, and use them as a guide to sketch the graph of $y = -3\sec\frac{\pi}{2}x$.

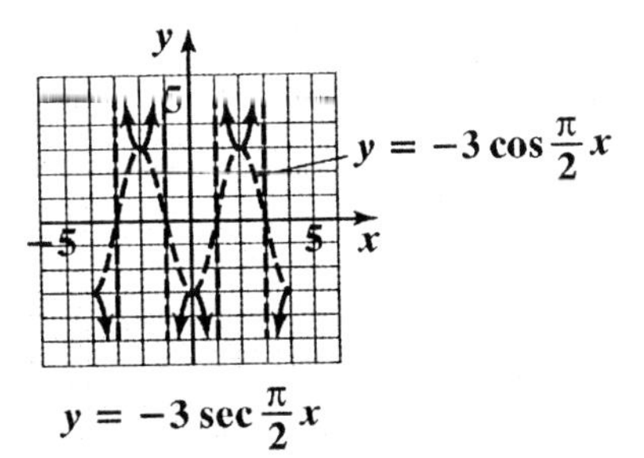

$y = -3\sec\frac{\pi}{2}x$

29. Graph the reciprocal sine function, $y = 3\sin x$. The equation is of the form $y = A\sin Bx$ with $A = 3$ and $B = 1$.
amplitude: $|A| = |3| = 3$

period: $\frac{2\pi}{B} = \frac{2\pi}{1} = 2\pi$

Use quarter-periods, $\frac{\pi}{2}$, to find x-values for the five key points. Starting with $x = 0$, the x-values are $0, \frac{\pi}{2}, \pi, \frac{3\pi}{2}$, and 2π. Evaluating the function at each value of x, the key points are $(0, 0)$, $\left(\frac{\pi}{2}, 3\right)$, $(\pi, 0)$, $\left(\frac{3\pi}{2}, -3\right)$, and $(2\pi, 0)$. Use these key points to graph $y = 3\sin x$ from 0 to 2π. Extend the graph one cycle to the right.
Use the graph to obtain the graph of the reciprocal function. Draw vertical asymptotes through the x-intercepts, and use them as guides to graph $y = 3\csc x$.

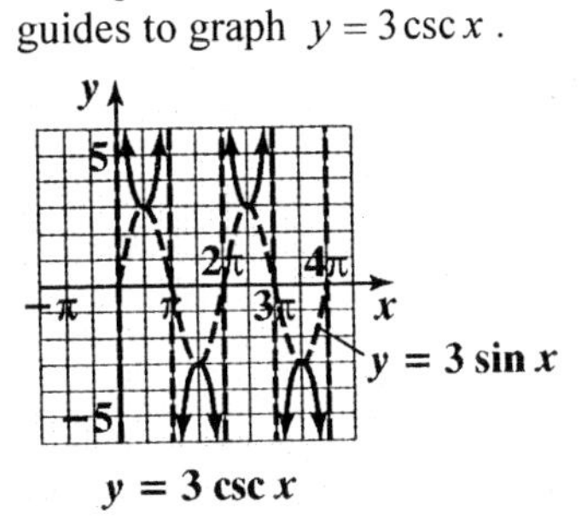

$y = 3\csc x$

30. Graph the reciprocal sine function, $y = 2\sin x$.
The equation is of the form $y = A\sin Bx$ with $A = 2$ and $B = 1$.
amplitude: $|A| = |2| = 2$

period: $\frac{2\pi}{B} = \frac{2\pi}{1} = 2\pi$

Use quarter-periods, $\frac{\pi}{2}$, to find x-values for the five key points. Starting with $x = 0$, the x-values are $0, \frac{\pi}{2}, \pi, \frac{3\pi}{2}$, and 2π. Evaluating the function at each value of x, the key points are $(0, 0), \left(\frac{\pi}{2}, 2\right), (\pi, 0), \left(\frac{3\pi}{2}, -2\right)$, and $(2\pi, 0)$.
Use these key points to graph $y = 2\sin x$ from 0 to 2π. Extend the graph one cycle to the right. Use the graph to obtain the graph of the reciprocal function. Draw vertical asymptotes through the x-intercepts, and use them as guides to graph $y = 2\csc x$.

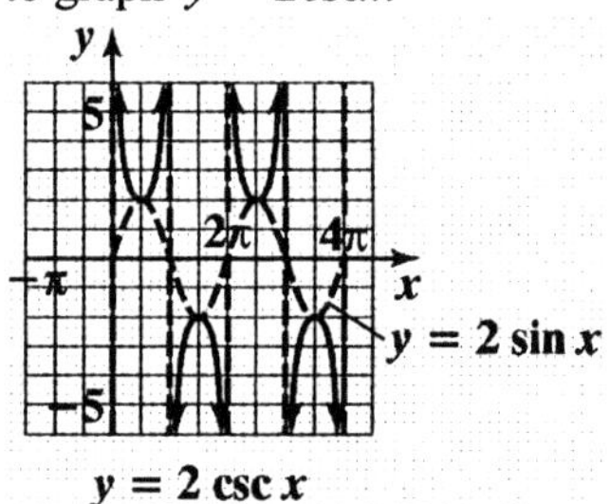

31. Graph the reciprocal sine function, $y = \frac{1}{2}\sin\frac{x}{2}$.
The equation is of the form $y = A\sin Bx$ with $A = \frac{1}{2}$ and $B = \frac{1}{2}$.

amplitude: $|A| = \left|\frac{1}{2}\right| = \frac{1}{2}$

period: $\frac{2\pi}{B} = \frac{2\pi}{\frac{1}{2}} = 2\pi \cdot 2 = 4\pi$

Use quarter-periods, π, to find x-values for the five key points. Starting with $x = 0$, the x-values are $0, \pi, 2\pi, 3\pi$, and 4π. Evaluating the function at each value of x, the key points are $(0, 0), \left(\pi, \frac{1}{2}\right), (2\pi, 0), \left(3\pi, -\frac{1}{2}\right)$, and $(4\pi, 0)$.

Use these key points to graph $y = \frac{1}{2}\sin\frac{x}{2}$ from 0 to 4π. Extend the graph one cycle to the right. Use the graph to obtain the graph of the reciprocal function. Draw vertical asymptotes through the x-intercepts, and use them as guides to graph $y = \frac{1}{2}\csc\frac{x}{2}$.

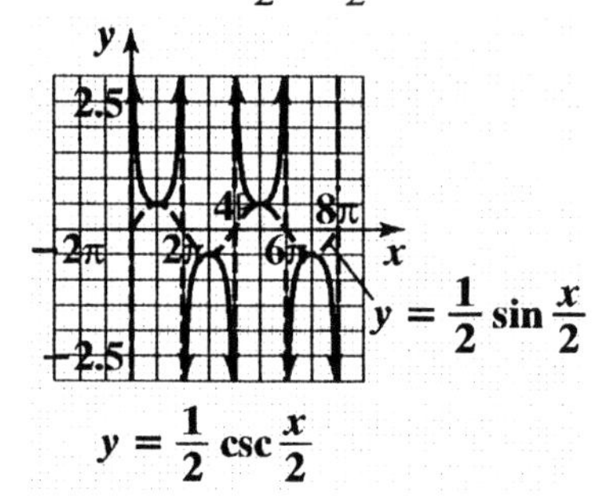

32. Graph the reciprocal sine function, $y = \frac{3}{2}\sin\frac{x}{4}$.
The equation is of the form $y = A\sin Bx$ with $A = \frac{3}{2}$ and $B = \frac{1}{4}$.

amplitude: $|A| = \left|\frac{3}{2}\right| = \frac{3}{2}$

period: $\frac{2\pi}{B} = \frac{2\pi}{\frac{1}{4}} = 2\pi \cdot 4 = 8\pi$

Use quarter-periods, 2π, to find x-values for the five key points. Starting with $x = 0$, the x-values are $0, 2\pi, 4\pi, 6\pi$, and 8π. Evaluating the function at each value of x, the key points are $(0, 0), \left(2\pi, \frac{3}{2}\right), (4\pi, 0), \left(6\pi, -\frac{3}{2}\right)$, and $(8\pi, 0)$

.

Use these key points to graph $y = \frac{3}{2}\sin\frac{x}{4}$ from 0 to 8π. Extend the graph one cycle to the right.

Use the graph to obtain the graph of the reciprocal function. Draw vertical asymptotes through the x-intercepts, and use them as guides to graph $y = \frac{3}{2}\csc\frac{x}{4}$.

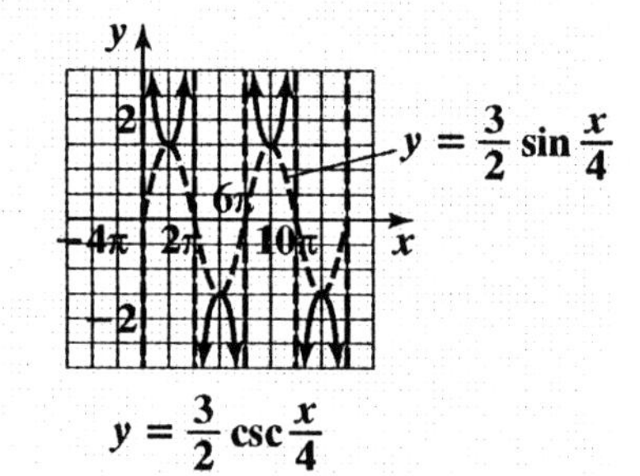

33. Graph the reciprocal cosine function, $y = 2\cos x$. The equation is of the form $y = A\cos Bx$ with $A = 2$ and $B = 1$.

amplitude: $|A| = |2| = 2$

period: $\frac{2\pi}{B} = \frac{2\pi}{1} = 2\pi$

Use quarter-periods, $\frac{\pi}{2}$, to find x-values for the five key points. Starting with $x = 0$, the x-values are $0, \frac{\pi}{2}, \pi, \frac{3\pi}{2}, 2\pi$. Evaluating the function at each value of x, the key points are $(0, 2)$, $\left(\frac{\pi}{2}, 0\right)$, $(\pi, -2)$, $\left(\frac{3\pi}{2}, 0\right)$, and $(2\pi, 2)$. Use these key points to graph $y = 2\cos x$ from 0 to 2π. Extend the graph one cycle to the right. Use the graph to obtain the graph of the reciprocal function. Draw vertical asymptotes through the x-intercepts, and use them as guides to graph $y = 2\sec x$.

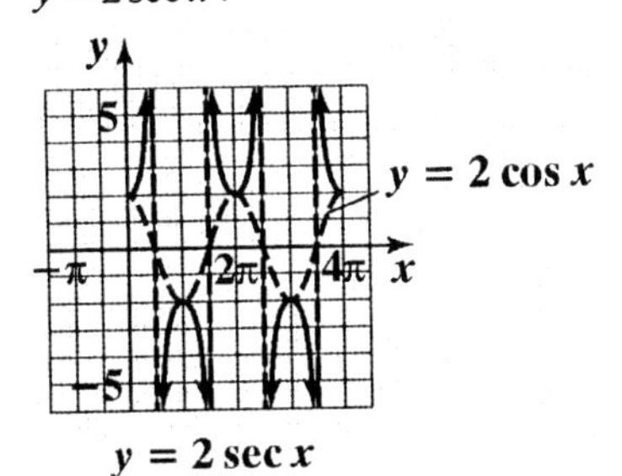

$y = 2\sec x$

34. Graph the reciprocal cosine function, $y = 3\cos x$. The equation is of the form $y = A\cos Bx$ with $A = 3$ and $B = 1$.

amplitude: $|A| = |3| = 3$

period: $\frac{2\pi}{B} = \frac{2\pi}{1} = 2\pi$

Use quarter-periods, $\frac{\pi}{2}$, to find x-values for the five key points. Starting with $x = 0$, the x-values are $0, \frac{\pi}{2}, \pi, \frac{3\pi}{2}$, and 2π. Evaluating the function at each value of x, the key points are $(0, 3), \left(\frac{\pi}{2}, 0\right), (\pi, -3), \left(\frac{3\pi}{2}, 0\right), (2\pi, 3)$.

Use these key points to graph $y = 3\cos x$ from 0 to 2π. Extend the graph one cycle to the right. Use the graph to obtain the graph of the reciprocal function. Draw vertical asymptotes through the x-intercepts, and use them as guides to graph $y = 3\sec x$.

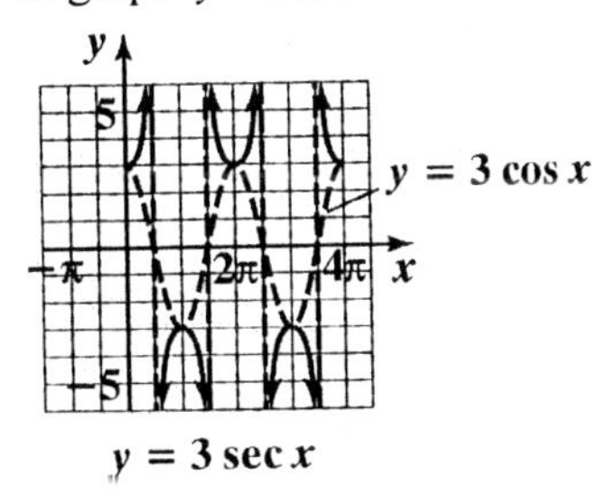

$y = 3\sec x$

35. Graph the reciprocal cosine function, $y = \cos\frac{x}{3}$.

The equation is of the form $y = A\cos Bx$ with $A = 1$ and $B = \frac{1}{3}$.

amplitude: $|A| = |1| = 1$

period: $\frac{2\pi}{B} = \frac{2\pi}{\frac{1}{3}} = 2\pi \cdot 3 = 6\pi$

Use quarter-periods, $\frac{6\pi}{4} = \frac{3\pi}{2}$, to find x-values for the five key points. Starting with $x = 0$, the x-values are $0, \frac{3\pi}{2}, 3\pi, \frac{9\pi}{2}$, and 6π.

Evaluating the function at each value of x, the key points are $(0, 1)$, $\left(\frac{3\pi}{2}, 0\right)$, $(3\pi, -1)$, $\left(\frac{9\pi}{2}, 0\right)$, and $(6\pi, 1)$. Use these key points to graph $y = \cos\frac{x}{3}$ from 0 to 6π. Extend the graph one cycle to the right. Use the graph to obtain the graph of the reciprocal function. Draw vertical asymptotes through the x-intercepts, and use them as guides to graph $y = \sec\frac{x}{3}$.

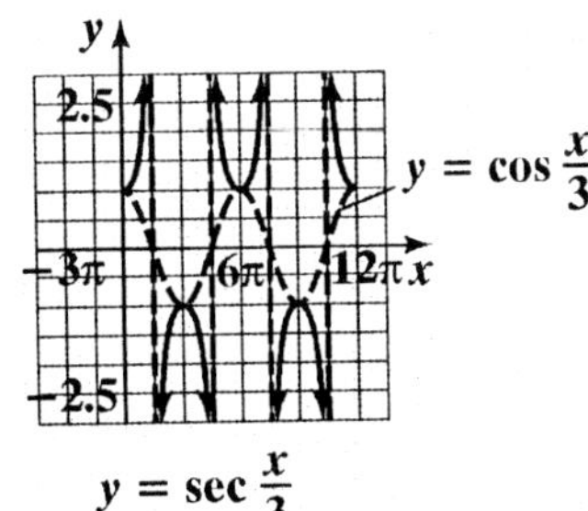

$y = \sec\frac{x}{3}$

36. Graph the reciprocal cosine function, $y = \cos\frac{x}{2}$.
The equation is of the form $y = A\cos Bx$ with $A = 1$ and $B = \frac{1}{2}$.
amplitude: $|A| = |1| = 1$
period: $\frac{2\pi}{B} = \frac{2\pi}{\frac{1}{2}} = 2\pi \cdot 2 = 4\pi$
Use quarter-periods, π, to find x-values for the five key points. Starting with $x = 0$, the x-values are $0, \pi, 2\pi, 3\pi,$ and 4π. Evaluating the function at each value of x, the key points are $(0, 1), (\pi, 0), (2\pi, -1), (3\pi, 0),$ and $(4\pi, 1)$.
Use these key points to graph $y = \cos\frac{x}{2}$ from 0 to 4π. Extend the graph one cycle to the right. Use the graph to obtain the graph of the reciprocal function. Draw vertical asymptotes through the x-intercepts, and use them as guides to graph $y = \sec\frac{x}{2}$.

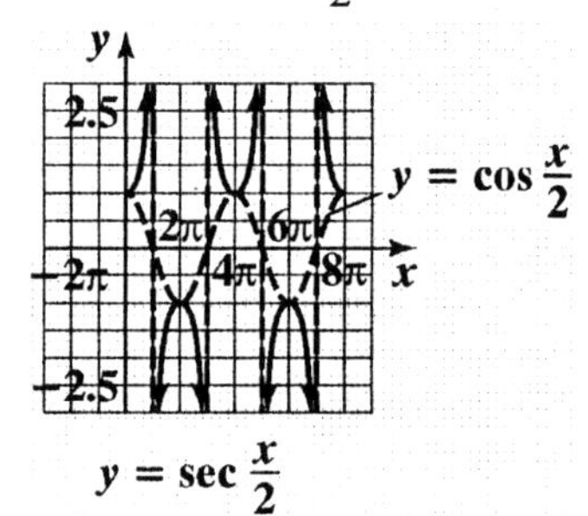

37. Graph the reciprocal sine function, $y = -2\sin\pi x$. The equation is of the form $y = A\sin Bx$ with $A = -2$ and $B = \pi$.
amplitude: $|A| = |-2| = 2$
period: $\frac{2\pi}{B} = \frac{2\pi}{\pi} = 2$
Use quarter-periods, $\frac{2}{4} = \frac{1}{2}$, to find x-values for the five key points. Starting with $x = 0$, the x-values are $0, \frac{1}{2}, 1, \frac{3}{2},$ and 2. Evaluating the function at each value of x, the key points are $(0, 0), \left(\frac{1}{2}, -2\right), (1, 0), \left(\frac{3}{2}, 2\right),$ and $(2, 0)$. Use these key points to graph $y = -2\sin\pi x$ from 0 to 2. Extend the graph one cycle to the right. Use the graph to obtain the graph of the reciprocal function. Draw vertical asymptotes through the x-intercepts, and use them as guides to graph $y = -2\csc\pi x$.

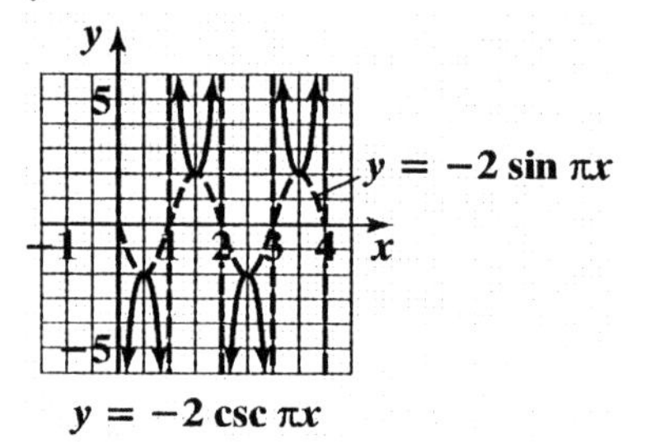

38. Graph the reciprocal sine function, $y = -\frac{1}{2}\sin\pi x$. The equation is of the form $y = A\sin Bx$ with $A = -\frac{1}{2}$ and $B = \pi$.
amplitude: $|A| = \left|-\frac{1}{2}\right| = \frac{1}{2}$
period: $\frac{2\pi}{B} = \frac{2\pi}{\pi} = 2$
Use quarter-periods, $\frac{2}{4} = \frac{1}{2}$, to find x-values for the five key points. Starting with $x = 0$, the x-values are $0, \frac{1}{2}, 1, \frac{3}{2},$ and 2. Evaluating the function at each value of x, the key points are $(0, 0), \left(\frac{1}{2}, -\frac{1}{2}\right), (1, 0), \left(\frac{3}{2}, \frac{1}{2}\right),$ and $(2, 0)$.
Use these key points to graph $y = -\frac{1}{2}\sin\pi x$ from 0 to 2. Extend the graph one cycle to the right. Use the graph to obtain the graph of the reciprocal function. Draw vertical asymptotes through the x-intercepts, and use them as guides to graph $y = -\frac{1}{2}\csc\pi x$.

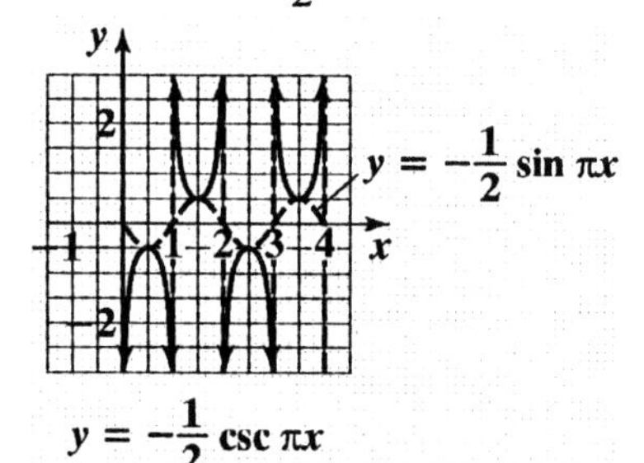

39. Graph the reciprocal cosine function, $y = -\frac{1}{2}\cos \pi x$. The equation is of the form $y = A\cos Bx$ with $A = -\frac{1}{2}$ and $B = \pi$.

amplitude: $|A| = \left|-\frac{1}{2}\right| = \frac{1}{2}$

period: $\frac{2\pi}{B} = \frac{2\pi}{\pi} = 2$

Use quarter-periods, $\frac{2}{4} = \frac{1}{2}$, to find x-values for the five key points. Starting with $x = 0$, the x-values are 0, $\frac{1}{2}$, 1, $\frac{3}{2}$, and 2. Evaluating the function at each value of x, the key points are $\left(0, -\frac{1}{2}\right)$, $\left(\frac{1}{2}, 0\right)$, $\left(1, \frac{1}{2}\right)$, $\left(\frac{3}{2}, 0\right)$, $\left(2, -\frac{1}{2}\right)$.

Use these key points to graph $y = -\frac{1}{2}\cos \pi x$ from 0 to 2. Extend the graph one cycle to the right. Use the graph to obtain the graph of the reciprocal function. Draw vertical asymptotes through the x-intercepts, and use them as guides to graph $y = -\frac{1}{2}\sec \pi x$.

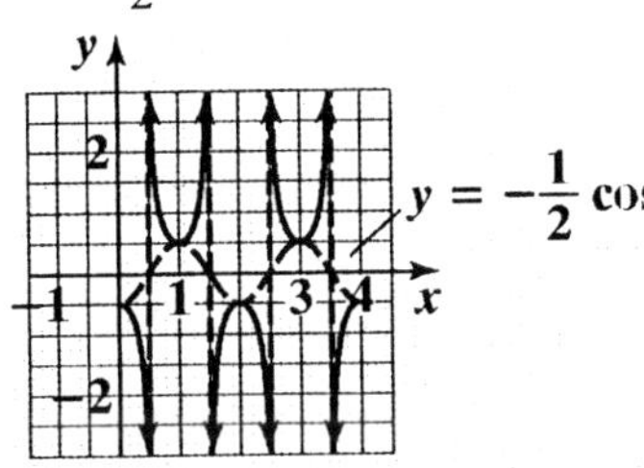

40. Graph the reciprocal cosine function, $y = -\frac{3}{2}\cos \pi x$. The equation is of the form $y = A\cos Bx$ with $A = -\frac{3}{2}$ and $B = \pi$.

amplitude: $|A| = \left|-\frac{3}{2}\right| = \frac{3}{2}$

period: $\frac{2\pi}{B} = \frac{2\pi}{\pi} = 2$

Use quarter-periods, $\frac{2}{4} = \frac{1}{2}$, to find x-values for the five key points. Starting with $x = 0$, the x-values are 0, $\frac{1}{2}$, 1, $\frac{3}{2}$, and 2. Evaluating the function at each value of x, the key points are $\left(0, -\frac{3}{2}\right)$, $\left(\frac{1}{2}, 0\right)$, $\left(1, \frac{3}{2}\right)$, $\left(\frac{3}{2}, 0\right)$, $\left(2, -\frac{3}{2}\right)$.

Use these key points to graph $y = -\frac{3}{2}\cos \pi x$ from 0 to 2. Extend the graph one cycle to the right. Use the graph to obtain the graph of the reciprocal function. Draw vertical asymptotes through the x-intercepts, and use them as guides to graph $y = -\frac{3}{2}\sec \pi x$.

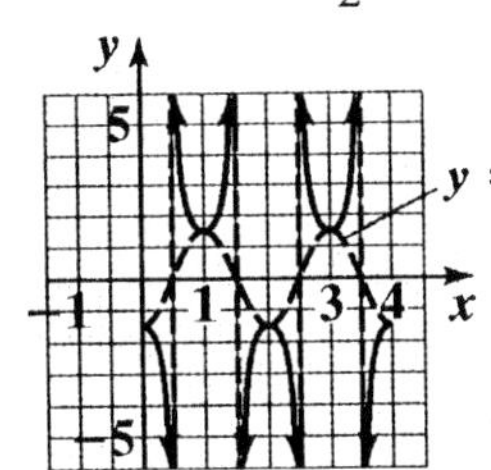

41. Graph the reciprocal sine function, $y = \sin(x - \pi)$. The equation is of the form $y = A\sin(Bx - C)$ with $A = 1$, and $B = 1$, and $C = \pi$.

amplitude: $|A| = |1| = 1$

period: $\frac{2\pi}{B} = \frac{2\pi}{1} = 2\pi$

phase shift: $\frac{C}{B} = \frac{\pi}{1} = \pi$

Use quarter-periods, $\frac{2\pi}{4} = \frac{\pi}{2}$, to find x-values for the five key points. Starting with $x = \pi$, the x-values are π, $\frac{3\pi}{2}$, 2π, $\frac{5\pi}{2}$, and 3π. Evaluating the function at each value of x, the key points are $(\pi, 0)$, $\left(\frac{3\pi}{2}, 1\right)$, $(2\pi, 0)$, $\left(\frac{5\pi}{2}, -1\right)$, $(3\pi, 0)$. Use these key points to graph $y = \sin(x - \pi)$ from π to 3π. Extend the graph one cycle to the right. Use the graph to obtain the graph of the reciprocal function. Draw vertical asymptotes through the x-intercepts, and use them as guides to graph $y = \csc(x - \pi)$.

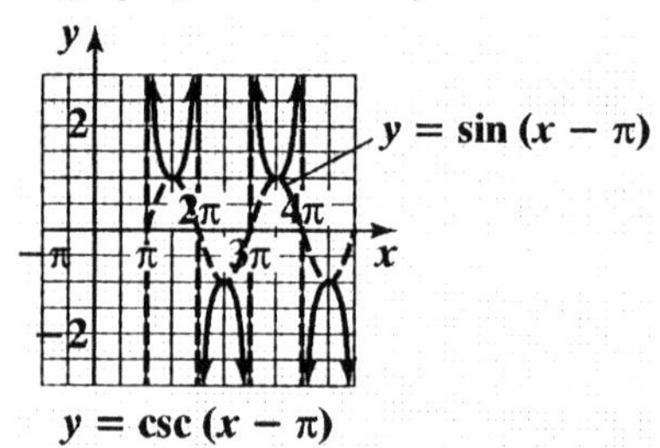

42. Graph the reciprocal sine function, $y = \sin\left(x - \frac{\pi}{2}\right)$. The equation is of the form $y = A\sin(Bx - C)$ with $A = 1$, $B = 1$, and $C = \frac{\pi}{2}$.

amplitude: $|A| = |1| = 1$

period: $\frac{2\pi}{B} = \frac{2\pi}{1} = 2\pi$

phase shift: $\frac{C}{B} = \frac{\frac{\pi}{2}}{1} = \frac{\pi}{2}$

Use quarter-periods, $\frac{\pi}{2}$, to find x-values for the five key points. Starting with $x = \frac{\pi}{2}$, the x-values are $\frac{\pi}{2}, \pi, \frac{3\pi}{2}, 2\pi$, and $\frac{5\pi}{2}$.
Evaluating the function at each value of x, the key points are
$\left(\frac{\pi}{2}, 0\right), (\pi, 1), \left(\frac{3\pi}{2}, 0\right), (2\pi, -1)$, and $\left(\frac{5\pi}{2}, 0\right)$.

Use these key points to graph $y = \sin\left(x - \frac{\pi}{2}\right)$ from $\frac{\pi}{2}$ to $\frac{5\pi}{2}$. Extend the graph one cycle to the right. Use the graph to obtain the graph of the reciprocal function. Draw vertical asymptotes through the x-intercepts, and use them as guides to graph $y = \csc\left(x - \frac{\pi}{2}\right)$.

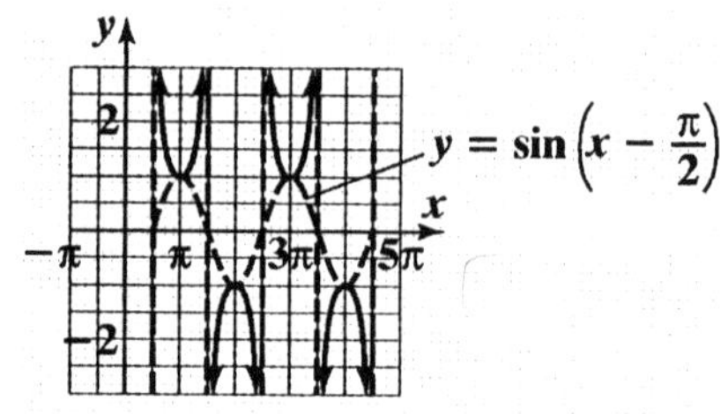

43. Graph the reciprocal cosine function, $y = 2\cos(x + \pi)$. The equation is of the form $y = A\cos(Bx + C)$ with $A = 2$, $B = 1$, and $C = -\pi$.

amplitude: $|A| = |2| = 2$

period: $\frac{2\pi}{B} = \frac{2\pi}{1} = 2\pi$

phase shift: $\frac{C}{B} = \frac{-\pi}{1} = -\pi$

Use quarter-periods, $\frac{2\pi}{4} = \frac{\pi}{2}$, to find x-values for the five key points. Starting with $x = -\pi$, the x-values are $-\pi, -\frac{\pi}{2}$, $0, \frac{\pi}{2}$, and π. Evaluating the function at each value of x, the key points are $(-\pi, 2)$, $\left(-\frac{\pi}{2}, 0\right), (0, -2), \left(\frac{\pi}{2}, 0\right)$, and $(\pi, 2)$. Use these key points to graph $y = 2\cos(x + \pi)$ from $-\pi$ to π. Extend the graph one cycle to the right. Use the graph to obtain the graph of the reciprocal function. Draw vertical asymptotes through the x-intercepts, and use them as guides to graph $y = 2\sec(x + \pi)$.

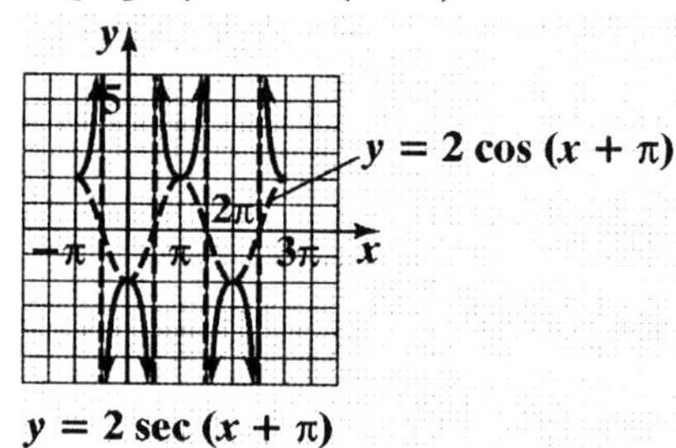

44. Graph the reciprocal cosine function, $y = 2\cos\left(x + \dfrac{\pi}{2}\right)$. The equation is of the form $y = A\cos(Bx + C)$ with $A = 2$ and $B = 1$, and $C = -\dfrac{\pi}{2}$.

amplitude: $|A| = |2| = 2$

period: $\dfrac{2\pi}{B} = \dfrac{2\pi}{1} = 2\pi$

phase shift: $\dfrac{C}{B} = \dfrac{-\frac{\pi}{2}}{1} = -\dfrac{\pi}{2}$

Use quarter-periods, $\dfrac{\pi}{2}$, to find x-values for the five key points. Starting with $x = -\dfrac{\pi}{2}$, the x-values are $-\dfrac{\pi}{2}$, 0, $\dfrac{\pi}{2}$, π, and $\dfrac{3\pi}{2}$. Evaluating the function at each value of x, the key points are

$\left(-\dfrac{\pi}{2}, 2\right), (0, 0), \left(\dfrac{\pi}{2}, -2\right), (\pi, 0), \left(\dfrac{3\pi}{2}, 2\right)$.

Use these key points to graph $y = 2\cos\left(x + \dfrac{\pi}{2}\right)$ from $-\dfrac{\pi}{2}$ to $\dfrac{3\pi}{2}$. Extend the graph one cycle to the right. Use the graph to obtain the graph of the reciprocal function. Draw vertical asymptotes through the x-intercepts, and use them as guides to graph $y = 2\sec\left(x + \dfrac{\pi}{2}\right)$.

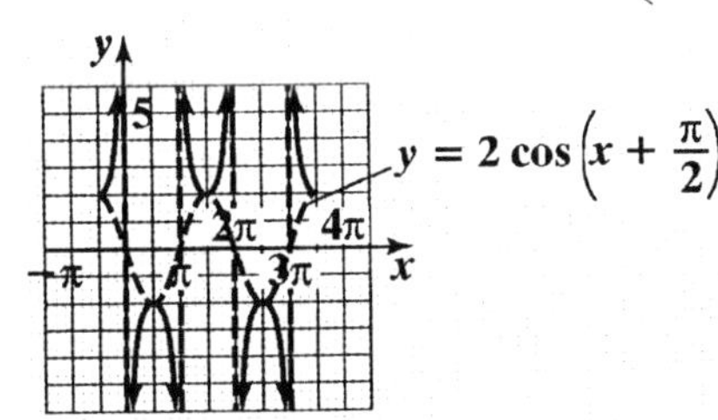

$y = 2\sec\left(x + \dfrac{\pi}{2}\right)$

45.

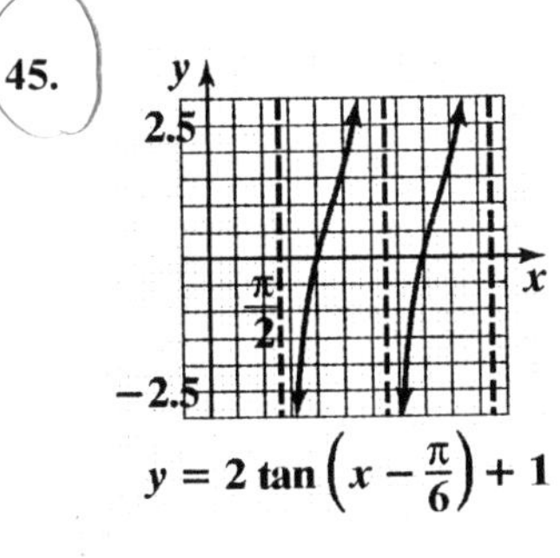

$y = 2\tan\left(x - \dfrac{\pi}{6}\right) + 1$

46.

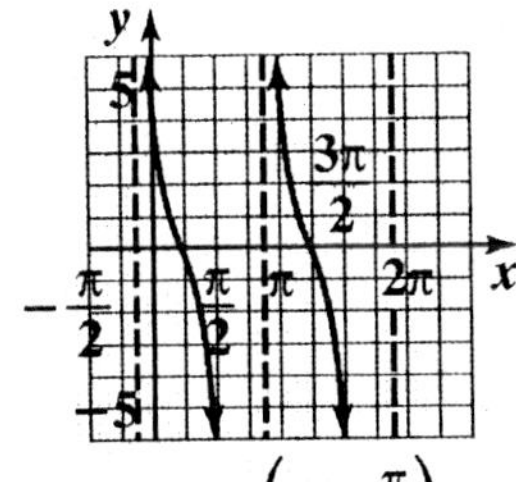

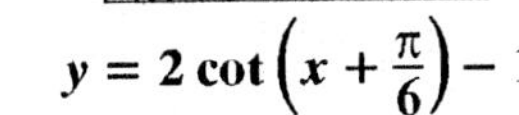

$y = 2\cot\left(x + \dfrac{\pi}{6}\right) - 1$

47.

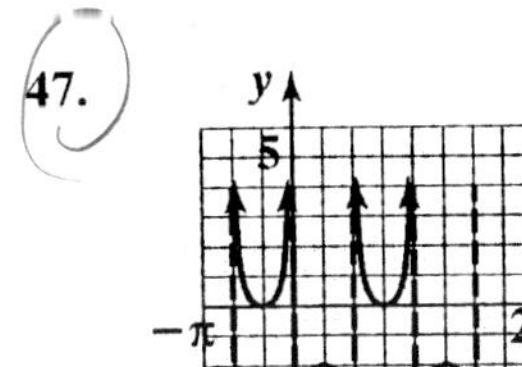

$y = \sec\left(2x + \dfrac{\pi}{2}\right) - 1$

48.

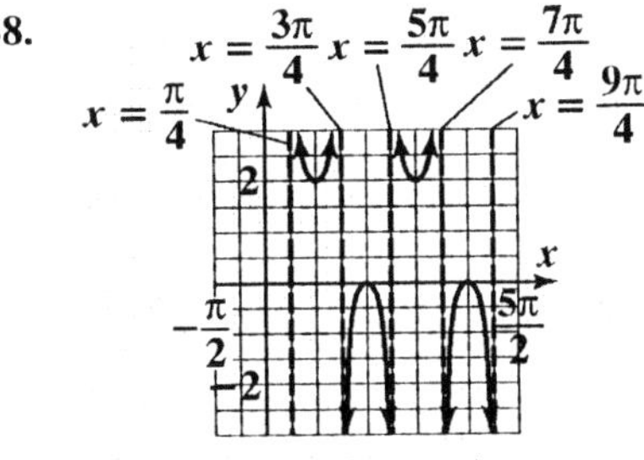

$y = \csc\left(2x - \dfrac{\pi}{2}\right) + 1$

49.

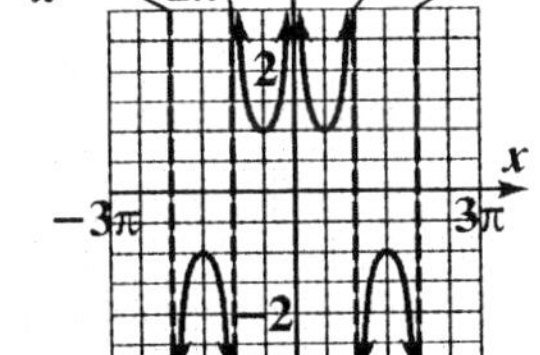

50.

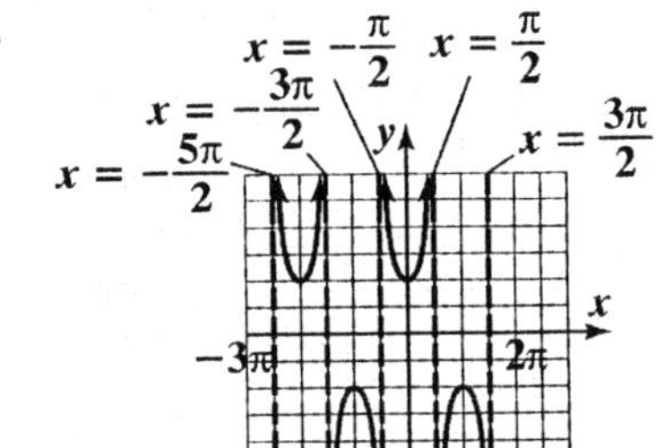

51.

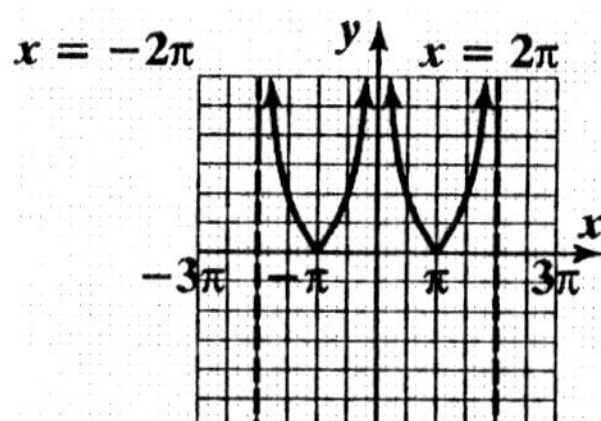

52.

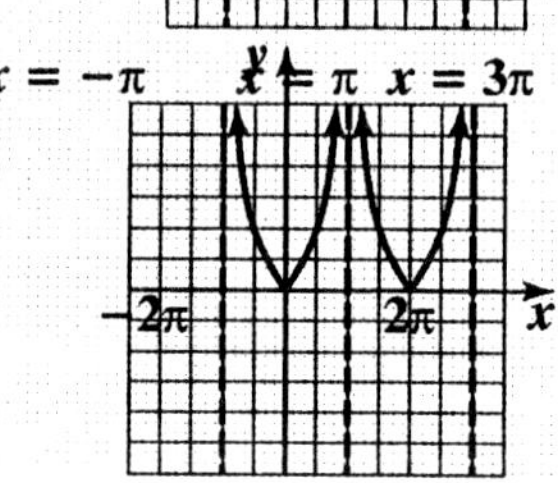

53. $y = (f \circ h)(x) = f(h(x)) = 2\sec\left(2x - \frac{\pi}{2}\right)$

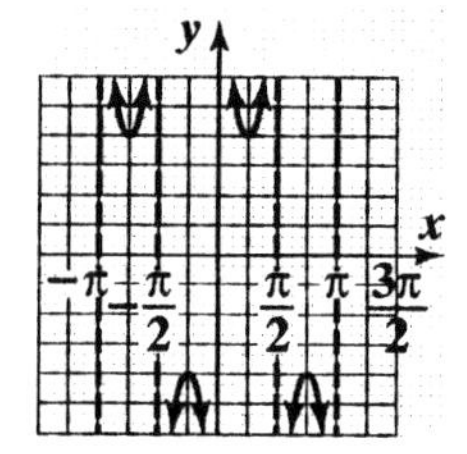

54. $y = (g \circ h)(x) = g(h(x)) = -2\tan\left(2x - \frac{\pi}{2}\right)$

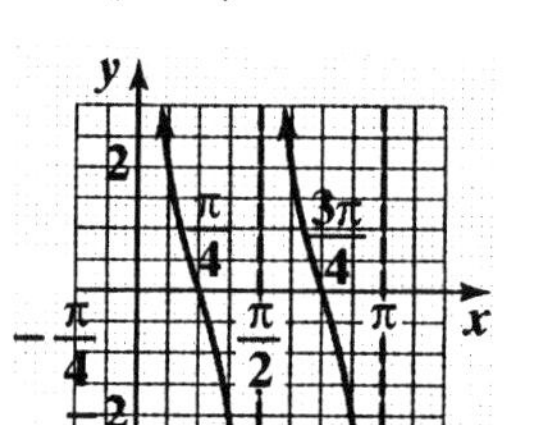

55. Use a graphing utility with $y_1 = \tan x$ and $y_2 = -1$. For the window use $\text{Xmin} = -2\pi$, $\text{Xmax} = 2\pi$, $\text{Ymin} = -2$, and $\text{Ymax} = 2$.

$x = -\frac{5\pi}{4}, -\frac{\pi}{4}, \frac{3\pi}{4}, \frac{7\pi}{4}$

$x \approx -3.93, -0.79, 2.36, 5.50$

56. Use a graphing utility with $y_1 = 1/\tan x$ and $y_2 = -1$. For the window use $\text{Xmin} = -2\pi$, $\text{Xmax} = 2\pi$, $\text{Ymin} = -2$, and $\text{Ymax} = 2$.

$x = -\frac{5\pi}{4}, -\frac{\pi}{4}, \frac{3\pi}{4}, \frac{7\pi}{4}$

$x \approx -3.93, -0.79, 2.36, 5.50$

57. Use a graphing utility with $y_1 = 1/\sin x$ and $y_2 = 1$. For the window use $\text{Xmin} = -2\pi$, $\text{Xmax} = 2\pi$, $\text{Ymin} = -2$, and $\text{Ymax} = 2$.

$x = -\frac{3\pi}{2}, \frac{\pi}{2}$

$x \approx -4.71, 1.57$

58. Use a graphing utility with $y_1 = 1/\cos x$ and $y_2 = 1$. For the window use $\text{Xmin} = -2\pi$, $\text{Xmax} = 2\pi$, $\text{Ymin} = -2$, and $\text{Ymax} = 2$.

$x = -2\pi, 0, 2\pi$

$x \approx -6.28, 0, 6.28$

59. $d = 12\tan 2\pi t$

a. Solve the equations

$2\pi t = -\frac{\pi}{2}$ and $2\pi t = \frac{\pi}{2}$

$t = \frac{-\frac{\pi}{2}}{2\pi}$ $\quad t = \frac{\frac{\pi}{2}}{2\pi}$

$t = -\frac{1}{4}$ $\quad t = \frac{1}{4}$

Thus, two consecutive asymptotes occur at $x = -\frac{1}{4}$ and $x = \frac{1}{4}$.

$x\text{-intercept} = \frac{-\frac{1}{4} + \frac{1}{4}}{2} = \frac{0}{2} = 0$

An x-intercept is 0 and the graph passes through (0, 0). Because the coefficient of the tangent is 12, the points on the graph midway between an x-intercept and the asymptotes have y-coordinates of -12 and 12. Use the two consecutive asymptotes, $x = -\frac{1}{4}$ and $x = \frac{1}{4}$, to graph one full period of $d = 12\tan 2\pi t$. To graph on [0, 2], continue the pattern and extend the graph to 2. (Do not use the left hand side of the first period of the graph on [0, 2].)

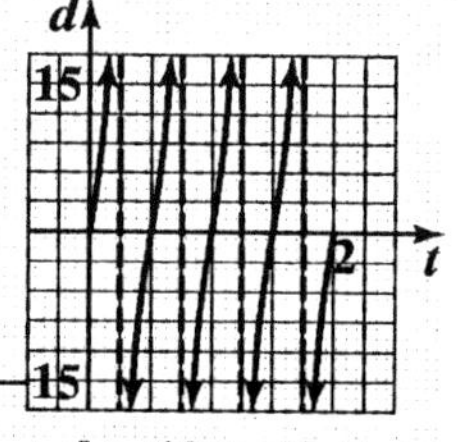

$d = 12\tan 2\pi t$

b. The function is undefined for $t = 0.25$, 0.75, 1.25, and 1.75.
The beam is shining parallel to the wall at these times.

60. In a right triangle the angle of elevation is one of the acute angles, the adjacent leg is the distance *d*, and the opposite leg is 2 mi. Use the cotangent function.

$\cot x = \frac{d}{2}$

$d = 2\cot x$

Use the equations

$x = 0$ and $x = \pi$

Two consecutive asymptotes occur at $x = 0$ and $x = \pi$. Midway between $x = 0$ and $x = \pi$ is $x = \frac{\pi}{2}$. An x-intercept is $\frac{\pi}{2}$ and the graph passes through $\left(\frac{\pi}{2}, 0\right)$. Because the coefficient of the cotangent is 2, the points on the graph midway between an x-intercept and the asymptotes have y-coordinates of –2 and 2. Use the two consecutive asymptotes, $x = 0$ and $x = \pi$, to graph $y = 2\cot x$ for $0 < x < \pi$.

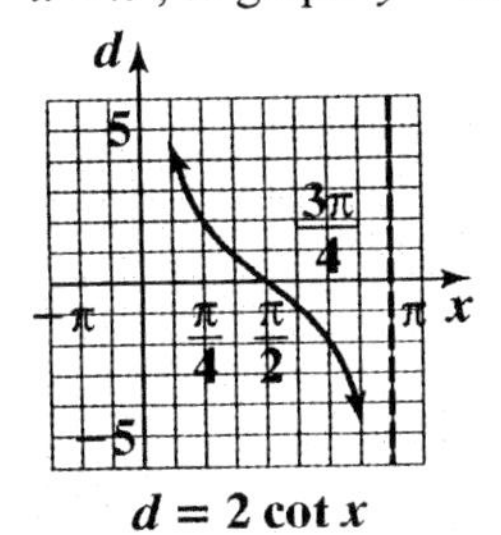

$d = 2\cot x$

61. Use the function that relates the acute angle with the hypotenuse and the adjacent leg, the secant function.

$\sec x = \frac{d}{10}$

$d = 10\sec x$

Graph the reciprocal cosine function, $y = 10\cos x$. The equation is of the form $y = A\cos Bx$ with $A = 10$ and $B = 1$.

amplitude: $|A| = |10| = 10$

period: $\frac{2\pi}{B} = \frac{2\pi}{1} = 2\pi$

For $-\frac{\pi}{2} < x < \frac{\pi}{2}$, use the x-values $-\frac{\pi}{2}$, 0, and $\frac{\pi}{2}$ to find the key points $\left(-\frac{\pi}{2}, 0\right)$, (0, 10), and $\left(\frac{\pi}{2}, 0\right)$. Connect these points with a smooth curve, then draw vertical asymptotes through the x-intercepts, and use them as guides to graph $d = 10\sec x$ on $\left[-\frac{\pi}{2}, \frac{\pi}{2}\right]$.

62.

Hours of Daylight

Number of Years

63.

Height of board

Seconds after dive

64. Graphs will vary

Distance (feet)

time (sec)

77. period: $\frac{\pi}{B} = \frac{\pi}{\frac{1}{4}} = \pi \cdot 4 = 4\pi$

Graph $y = \tan\frac{x}{4}$ for $0 \le x \le 8\pi$.

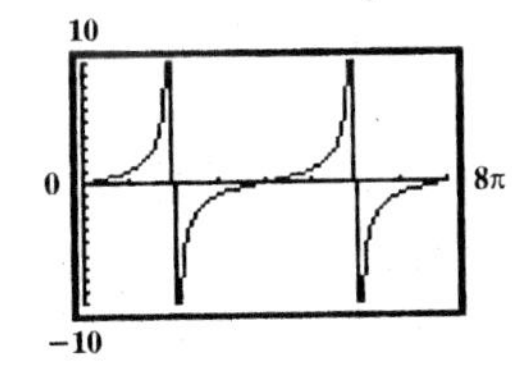

78. period: $\frac{\pi}{B} = \frac{\pi}{4}$

Graph $y = \tan 4x$ for $0 \le x \le \frac{\pi}{2}$.

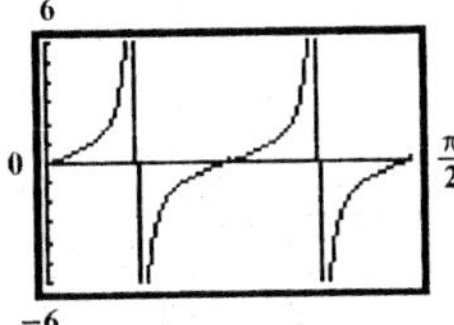

79. period: $\frac{\pi}{B} = \frac{\pi}{2}$

Graph $y = \cot 2x$ for $0 \le x \le \pi$.

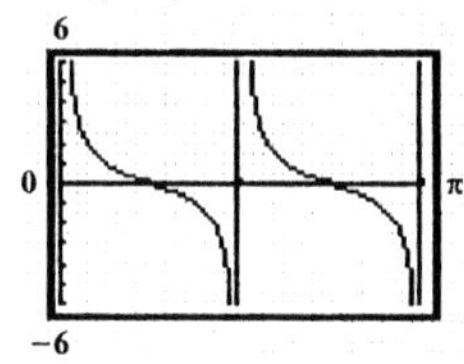

80. period: $\frac{\pi}{B} = \frac{\pi}{\frac{1}{2}} = \pi \cdot 2 = 2\pi$

Graph $y = \cot \frac{x}{2}$ for $0 \le x \le 4\pi$.

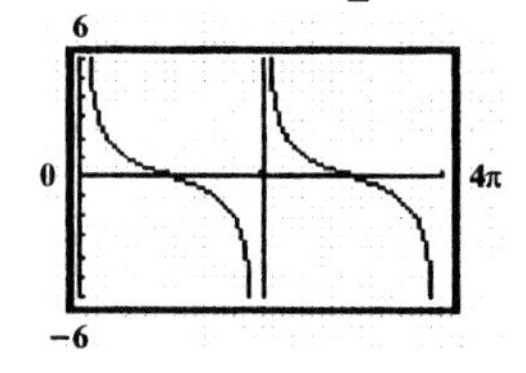

81. period: $\frac{\pi}{B} = \frac{\pi}{\pi} = 1$

Graph $y = \frac{1}{2} \tan \pi x$ for $0 \le x \le 2$.

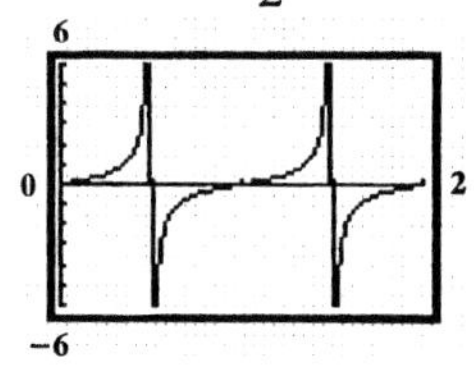

82. Solve the equations

$$\pi x + 1 = -\frac{\pi}{2} \quad \text{and} \quad \pi x + 1 = \frac{\pi}{2}$$

$$\pi x = -\frac{\pi}{2} - 1 \qquad \pi x = \frac{\pi}{2} - 1$$

$$x = \frac{\frac{-\pi}{2} - 1}{\pi} \qquad x = \frac{\frac{\pi}{2} - 1}{\pi}$$

$$x = \frac{-\pi - 2}{2\pi} \qquad x = \frac{\pi - 2}{2\pi}$$

$$x \approx -0.82 \qquad x \approx 0.18$$

period: $\frac{\pi}{B} = \frac{\pi}{\pi} = 1$

Thus, we include $-0.82 \le x \le 1.18$ in our graph of $y = \frac{1}{2} \tan(\pi x + 1)$, and graph for $-0.85 \le x \le 1.2$.

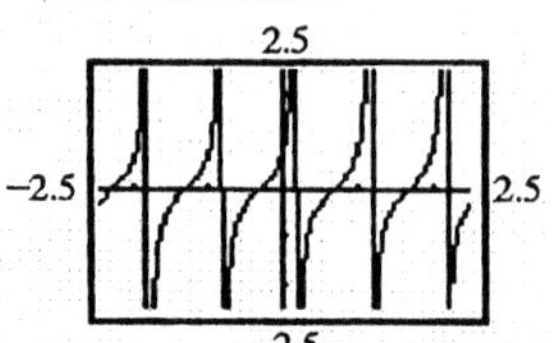

83. period: $\frac{2\pi}{B} = \frac{2\pi}{\frac{1}{2}} = 2\pi \cdot 2 = 4\pi$

Graph the functions for $0 \le x \le 8\pi$.

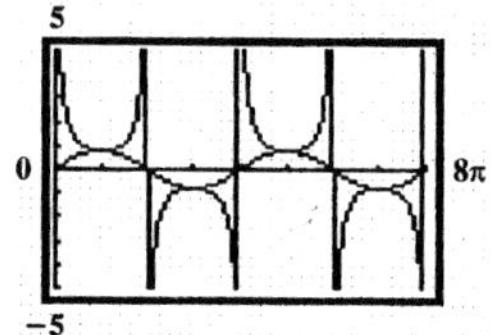

84. period: $\frac{2\pi}{B} = \frac{2\pi}{\frac{\pi}{3}} = 2\pi \cdot \frac{3}{\pi} = 6$

Graph the functions for $0 \le x \le 12$.

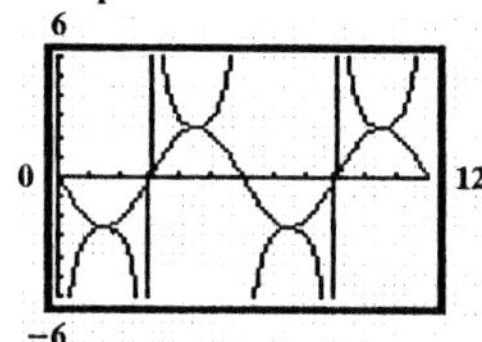

85. period: $\frac{2\pi}{B} = \frac{2\pi}{2} = \pi$

phase shift: $\frac{C}{B} = \frac{\frac{\pi}{6}}{2} = \frac{\pi}{12}$

Thus, we include $\frac{\pi}{12} \le x \le \frac{25\pi}{12}$ in our graph, and graph for $0 \le x \le \frac{5\pi}{2}$.

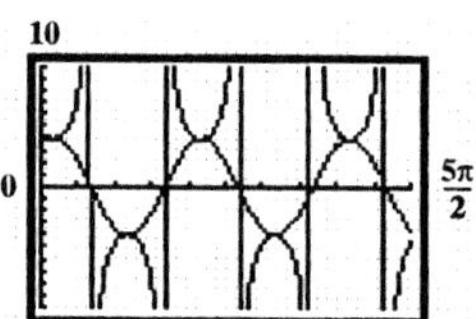

86. period: $\frac{2\pi}{B} = \frac{2\pi}{\pi} = 2$

phase shift: $\frac{C}{B} = \frac{\frac{\pi}{6}}{\pi} = \frac{\pi}{6} \cdot \frac{1}{\pi} = \frac{1}{6}$

Thus, we include $\frac{1}{6} \le x \le \frac{25}{6}$ in our graph, and graph for $0 \le x \le \frac{9}{2}$.

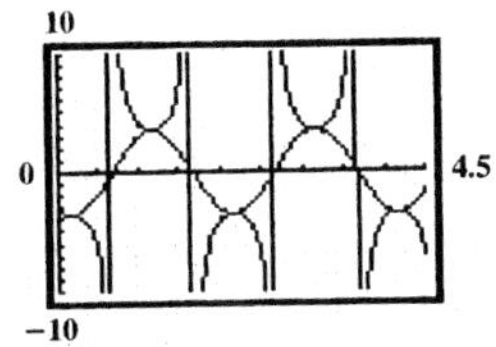

87.

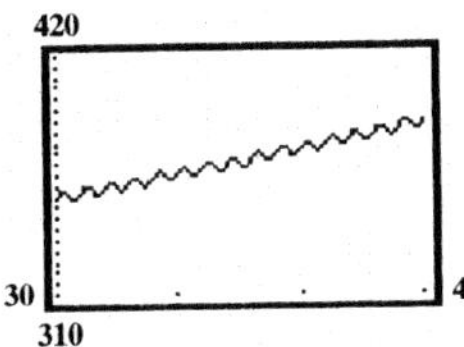

The graph shows that carbon dioxide concentration rises and falls each year, but over all the concentration increased from 1990 to 2005.

88. $y = \sin\frac{1}{x}$

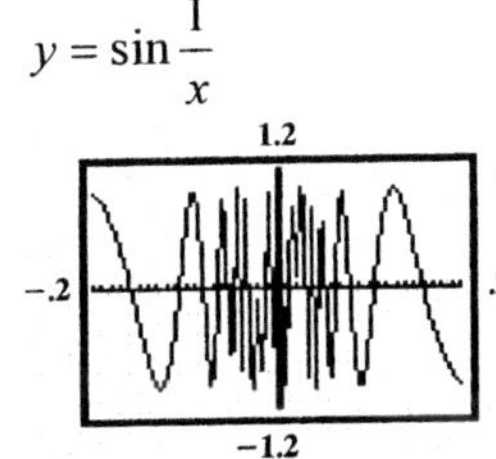

The graph is oscillating between 1 and −1. The oscillation is faster as x gets closer to 0. Explanations may vary.

89. The graph has the shape of a cotangent function with consecutive asymptotes at $x = 0$ and $x = \frac{2\pi}{3}$. The period is $\frac{2\pi}{3} - 0 = \frac{2\pi}{3}$.

Thus,

$$\frac{\pi}{B} = \frac{2\pi}{3}$$
$$2\pi B = 3\pi$$
$$B = \frac{3\pi}{2\pi} = \frac{3}{2}$$

The points on the graph midway between an x-intercept and the asymptotes have y-coordinates of 1 and −1. Thus, $A = 1$. There is no phase shift. Thus, $C = 0$. An equation for this graph is $y = \cot\frac{3}{2}x$.

90. The graph has the shape of a secant function. The reciprocal function has amplitude $|A| = 1$. The period is $\frac{8\pi}{3}$. Thus,

$$\frac{2\pi}{B} = \frac{8\pi}{3}$$
$$8\pi B = 6\pi$$
$$B = \frac{6\pi}{8\pi} = \frac{3}{4}$$

There is no phase shift. Thus, $C = 0$. An equation for the reciprocal function is $y = \cos\frac{3}{4}x$. Thus, an equation for this graph is $y = \sec\frac{3}{4}x$.

91. The range shows that $A = 2$.
Since the period is 3π, the coefficient of x is given by B where $\frac{2\pi}{B} = 3\pi$

$$\frac{2\pi}{B} = 3\pi$$
$$3B\pi = 2\pi$$
$$B = \frac{2}{3}$$

Thus, $y = 2\csc\frac{2x}{3}$

92. The range shows that $A = \pi$.
Since the period is 2, the coefficient of x is given by B where $\frac{2\pi}{B} = 2$

$$\frac{2\pi}{B} = 2$$
$$2B = 2\pi$$
$$B = \pi$$

Thus, $y = \pi\csc\pi x$

93. **a.** Since A=1, the range is $(-\infty,-1]\cup[1,\infty)$

Viewing rectangle: $\left[-\frac{\pi}{6},\pi,\frac{7\pi}{6}\right]$ by $[-3,3,1]$

b. Since A=3, the range is $(-\infty,-3]\cup[3,\infty)$

Viewing rectangle: $\left[-\frac{1}{2},\frac{7}{2},1\right]$ by $[-6,6,1]$

94. $y=2^{-x}\sin x$

2^{-x} decreases the amplitude as x gets larger. Examples may vary.

Section 4.7

Check Point Exercises

1. Let $\theta=\sin^{-1}\frac{\sqrt{3}}{2}$, then $\sin\theta=\frac{\sqrt{3}}{2}$.

The only angle in the interval $\left[-\frac{\pi}{2},\frac{\pi}{2}\right]$ that satisfies $\sin\theta=\frac{\sqrt{3}}{2}$ is $\frac{\pi}{3}$. Thus, $\theta=\frac{\pi}{3}$, or $\sin^{-1}\frac{\sqrt{3}}{2}=\frac{\pi}{3}$.

2. Let $\theta=\sin^{-1}\left(-\frac{\sqrt{2}}{2}\right)$, then $\sin\theta=-\frac{\sqrt{2}}{2}$.

The only angle in the interval $\left[-\frac{\pi}{2},\frac{\pi}{2}\right]$ that satisfies $\cos\theta=-\frac{\sqrt{2}}{2}$ is $-\frac{\pi}{4}$. Thus $\theta=-\frac{\pi}{4}$, or $\sin^{-1}\left(-\frac{\sqrt{2}}{2}\right)=-\frac{\pi}{4}$.

3. Let $\theta=\cos^{-1}\left(-\frac{1}{2}\right)$, then $\cos\theta=-\frac{1}{2}$. The only angle in the interval $[0,\pi]$ that satisfies $\cos\theta=-\frac{1}{2}$ is $\frac{2\pi}{3}$. Thus, $\theta=\frac{2\pi}{3}$, or $\cos^{-1}\left(-\frac{1}{2}\right)=\frac{2\pi}{3}$.

4. Let $\theta=\tan^{-1}(-1)$, then $\tan\theta=-1$. The only angle in the interval $\left(-\frac{\pi}{2},\frac{\pi}{2}\right)$ that satisfies $\tan\theta=-1$ is $-\frac{\pi}{4}$. Thus $\theta=-\frac{\pi}{4}$ or $\tan^{-1}\theta=-\frac{\pi}{4}$.

5.

Scientific Calculator Solution

	Function	Mode	Keystrokes	Display (rounded to four places)
a.	$\cos^{-1}\left(\frac{1}{3}\right)$	Radian	1 [÷] 3 [=] [COS^{-1}]	1.2310
b.	$\tan^{-1}(-35.85)$	Radian	35.85 [+/−] [TAN^{-1}]	−1.5429

Graphing Calculator Solution

	Function	Mode	Keystrokes	Display (rounded to four places)
a.	$\cos^{-1}\left(\frac{1}{3}\right)$	Radian	[COS^{-1}] [(] 1 [÷] 3 [)] [ENTER]	1.2310
b.	$\tan^{-1}(-35.85)$	Radian	[TAN^{-1}] [−] 35.85 [ENTER]	−1.5429

6. **a.** $\cos\left(\cos^{-1} 0.7\right)$

$x = 0.7$, x is in $[-1,1]$ so $\cos(\cos^{-1} 0.7) = 0.7$

b. $\sin^{-1}(\sin \pi)$

$x = \pi$, x is not in $\left[-\frac{\pi}{2}, \frac{\pi}{2}\right]$ x is in the domain of $\sin x$, so $\sin^{-1}(\sin \pi) = \sin^{-1}(0) = 0$

c. $\cos\left(\cos^{-1} \pi\right)$

$x = \pi$, x is not in $[-1,1]$ so $\cos\left(\cos^{-1} \pi\right)$ is not defined.

7. Let $\theta = \tan^{-1}\left(\frac{3}{4}\right)$, then $\tan\theta = \frac{3}{4}$. Because $\tan\theta$ is positive, θ is in the first quadrant.

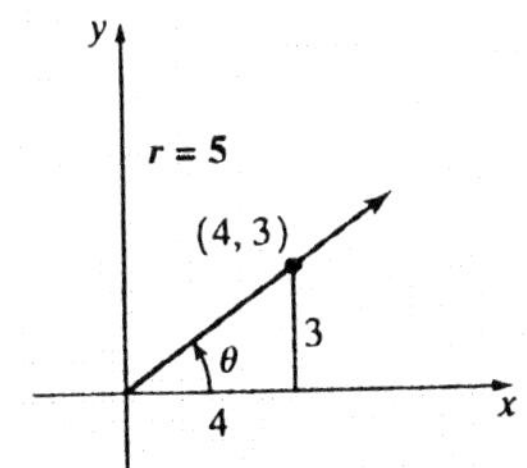

Use the Pythagorean Theorem to find r.

$r^2 = 3^2 + 4^2 = 9 + 16 = 25$

$r = \sqrt{25} = 5$

Use the right triangle to find the exact value.

$$\sin\left(\tan^{-1}\frac{3}{4}\right) = \sin\theta = \frac{\text{side opposite } \theta}{\text{hypotenuse}} = \frac{3}{5}$$

8. Let $\theta = \sin^{-1}\left(-\frac{1}{2}\right)$, then $\sin\theta = -\frac{1}{2}$. Because $\sin\theta$ is negative, θ is in quadrant IV.

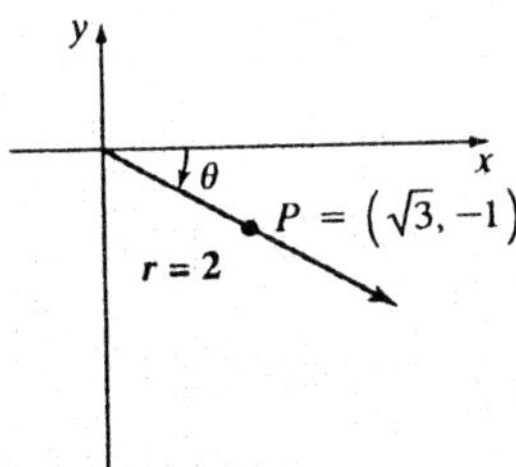

Use the Pythagorean Theorem to find x.

$x^2 + (-1)^2 = 2^2$

$x^2 + 1 = 4$

$x^2 = 3$

$x = \sqrt{3}$

Use values for x and r to find the exact value.

$$\cos\left[\sin^{-1}\left(-\frac{1}{2}\right)\right] = \cos\theta = \frac{x}{r} = \frac{\sqrt{3}}{2}$$

9. Let $\theta = \tan^{-1} x$, then $\tan\theta = x = \frac{x}{1}$.

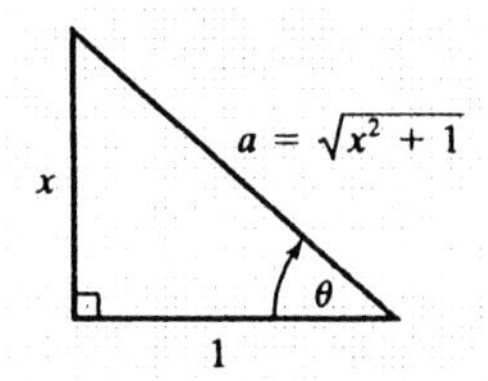

Use the Pythagorean Theorem to find the third side, a.

$a^2 = x^2 + 1^2$

$a = \sqrt{x^2 + 1}$

Use the right triangle to write the algebraic expression.

$\sec(\tan^{-1} x) = \sec\theta = \frac{\sqrt{x^2 + 1}}{1} = \sqrt{x^2 + 1}$

Exercise Set 4.7

1. Let $\theta = \sin^{-1}\frac{1}{2}$, then $\sin\theta = \frac{1}{2}$. The only angle in the interval $\left[-\frac{\pi}{2}, \frac{\pi}{2}\right]$ that satisfies $\sin\theta = \frac{1}{2}$ is $\frac{\pi}{6}$. Thus, $\theta = \frac{\pi}{6}$, or $\sin^{-1}\frac{1}{2} = \frac{\pi}{6}$.

2. Let $\theta = \sin^{-1} 0$, then $\sin\theta = 0$. The only angle in the interval $\left[-\frac{\pi}{2}, \frac{\pi}{2}\right]$ that satisfies $\sin\theta = 0$ is 0. Thus $\theta = 0$, or $\sin^{-1} 0 = 0$.

3. Let $\theta = \sin^{-1}\frac{\sqrt{2}}{2}$, then $\sin\theta = \frac{\sqrt{2}}{2}$. The only angle in the interval $\left[-\frac{\pi}{2}, \frac{\pi}{2}\right]$ that satisfies $\sin\theta = \frac{\sqrt{2}}{2}$ is $\frac{\pi}{4}$. Thus $\theta = \frac{\pi}{4}$, or $\sin^{-1}\frac{\sqrt{2}}{2} = \frac{\pi}{4}$.

4. Let $\theta = \sin^{-1}\frac{\sqrt{3}}{2}$, then $\sin\theta = \frac{\sqrt{3}}{2}$. The only angle in the interval $\left[-\frac{\pi}{2}, \frac{\pi}{2}\right]$ that satisfies $\sin\theta = \frac{\sqrt{3}}{2}$ is $\frac{\pi}{3}$. Thus $\theta = \frac{\pi}{3}$, or $\sin^{-1}\frac{\sqrt{3}}{2} = \frac{\pi}{3}$.

5. Let $\theta = \sin^{-1}\left(-\frac{1}{2}\right)$, then $\sin\theta = -\frac{1}{2}$. The only angle in the interval $\left[-\frac{\pi}{2}, \frac{\pi}{2}\right]$ that satisfies $\sin\theta = -\frac{1}{2}$ is $-\frac{\pi}{6}$. Thus $\theta = -\frac{\pi}{6}$, or $\sin^{-1}\left(-\frac{1}{2}\right) = -\frac{\pi}{6}$.

6. Let $\theta = \sin^{-1}\left(-\frac{\sqrt{3}}{2}\right)$, then $\sin\theta = -\frac{\sqrt{3}}{2}$.

The only angle in the interval $\left[-\frac{\pi}{2}, \frac{\pi}{2}\right]$ that satisfies $\sin\theta = -\frac{\sqrt{3}}{2}$ is $-\frac{\pi}{3}$. Thus $\theta = -\frac{\pi}{3}$, or $\sin^{-1}\left(-\frac{\sqrt{3}}{2}\right) = -\frac{\pi}{3}$.

7. Let $\theta = \cos^{-1}\frac{\sqrt{3}}{2}$, then $\cos\theta = \frac{\sqrt{3}}{2}$. The only angle in the interval $[0, \pi]$ that satisfies $\cos\theta = \frac{\sqrt{3}}{2}$ is $\frac{\pi}{6}$. Thus $\theta = \frac{\pi}{6}$, or $\cos^{-1}\frac{\sqrt{3}}{2} = \frac{\pi}{6}$.

8. Let $\theta = \cos^{-1}\frac{\sqrt{2}}{2}$, then $\cos\theta = \frac{\sqrt{2}}{2}$. The only angle in the interval $[0, \pi]$ that satisfies $\cos\theta = \frac{\sqrt{2}}{2}$ is $\frac{\pi}{4}$. Thus $\theta = \frac{\pi}{4}$, or $\cos^{-1}\frac{\sqrt{2}}{2} = \frac{\pi}{4}$.

9. Let $\theta = \cos^{-1}\left(-\frac{\sqrt{2}}{2}\right)$, then $\cos\theta = -\frac{\sqrt{2}}{2}$. The only angle in the interval $[0, \pi]$ that satisfies $\cos\theta = -\frac{\sqrt{2}}{2}$ is $\frac{3\pi}{4}$. Thus $\theta = \frac{3\pi}{4}$, or $\cos^{-1}\left(-\frac{\sqrt{2}}{2}\right) = \frac{3\pi}{4}$.

10. Let $\theta = \cos^{-1}\left(-\frac{\sqrt{3}}{2}\right)$, then $\cos\theta = -\frac{\sqrt{3}}{2}$.
The only angle in the interval $[0, \pi]$ that satisfies $\cos\theta = -\frac{\sqrt{3}}{2}$ is $\frac{5\pi}{6}$. Thus $\theta = \frac{5\pi}{6}$, or $\cos^{-1}\left(-\frac{\sqrt{3}}{2}\right) = \frac{5\pi}{6}$.

11. Let $\theta = \cos^{-1} 0$, then $\cos\theta = 0$. The only angle in the interval $[0, \pi]$ that satisfies $\cos\theta = 0$ is $\frac{\pi}{2}$. Thus $\theta = \frac{\pi}{2}$, or $\cos^{-1} 0 = \frac{\pi}{2}$.

12. Let $\theta = \cos^{-1} 1$, then $\cos\theta = 1$. The only angle in the interval $[0, \pi]$ that satisfies $\cos\theta = 1$ is 0. Thus $\theta = 0$, or $\cos^{-1} 1 = 0$.

13. Let $\theta = \tan^{-1}\frac{\sqrt{3}}{3}$, then $\tan\theta = \frac{\sqrt{3}}{3}$. The only angle in the interval $\left(-\frac{\pi}{2}, \frac{\pi}{2}\right)$ that satisfies $\tan\theta = \frac{\sqrt{3}}{3}$ is $\frac{\pi}{6}$. Thus $\theta = \frac{\pi}{6}$, or $\tan^{-1}\frac{\sqrt{3}}{3} = \frac{\pi}{6}$.

14. Let $\theta = \tan^{-1} 1$, then $\tan\theta = 1$. The only angle in the interval $\left(-\frac{\pi}{2}, \frac{\pi}{2}\right)$ that satisfies $\tan\theta = 1$ is $\frac{\pi}{4}$. Thus $\theta = \frac{\pi}{4}$, or $\tan^{-1} 1 = \frac{\pi}{4}$.

15. Let $\theta = \tan^{-1} 0$, then $\tan\theta = 0$. The only angle in the interval $\left(-\frac{\pi}{2}, \frac{\pi}{2}\right)$ that satisfies $\tan\theta = 0$ is 0. Thus $\theta = 0$, or $\tan^{-1} 0 = 0$.

16. Let $\theta = \tan^{-1}(-1)$, then $\tan\theta = -1$. The only angle in the interval $\left(-\frac{\pi}{2}, \frac{\pi}{2}\right)$ that satisfies $\tan\theta = -1$ is $-\frac{\pi}{4}$. Thus $\theta = -\frac{\pi}{4}$, or $\tan^{-1}(-1) = -\frac{\pi}{4}$.

17. Let $\theta = \tan^{-1}\left(-\sqrt{3}\right)$, then $\tan\theta = -\sqrt{3}$. The only angle in the interval $\left(-\frac{\pi}{2}, \frac{\pi}{2}\right)$ that satisfies $\tan\theta = -\sqrt{3}$ is $-\frac{\pi}{3}$. Thus $\theta = -\frac{\pi}{3}$, or $\tan^{-1}\left(-\sqrt{3}\right) = -\frac{\pi}{3}$.

18. Let $\theta = \tan^{-1}\left(-\frac{\sqrt{3}}{3}\right)$, then $\tan\theta = -\frac{\sqrt{3}}{3}$. The only angle in the interval $\left(-\frac{\pi}{2}, \frac{\pi}{2}\right)$ that satisfies $\tan\theta = -\frac{\sqrt{3}}{3}$ is $-\frac{\pi}{6}$. Thus $\theta = -\frac{\pi}{6}$, or $\tan^{-1}\left(-\frac{\sqrt{3}}{3}\right) = -\frac{\pi}{6}$.

19.

Scientific Calculator Solution			
Function	**Mode**	**Keystrokes**	**Display** (rounded to two places)
$\sin^{-1} 0.3$	Radian	0.3 [SIN⁻¹]	0.30

Graphing Calculator Solution			
Function	**Mode**	**Keystrokes**	**Display** (rounded to two places)
$\sin^{-1} 0.3$	Radian	[SIN⁻¹] 0.3 [ENTER]	0.30

20.

Scientific Calculator Solution			
Function	**Mode**	**Keystrokes**	**Display** (rounded to two places)
$\sin^{-1} 0.47$	Radian	0.47 [SIN⁻¹]	0.49

Graphing Calculator Solution			
Function	**Mode**	**Keystrokes**	**Display** (rounded to two places)
$\sin^{-1} 0.47$	Radian	[SIN⁻¹] 0.47 [ENTER]	0.49

21.

Scientific Calculator Solution			
Function	**Mode**	**Keystrokes**	**Display** (rounded to two places)
$\sin^{-1}(-0.32)$	Radian	0.32 [+/−] [SIN⁻¹]	–0.33

Graphing Calculator Solution			
Function	**Mode**	**Keystrokes**	**Display** (rounded to two places)
$\sin^{-1}(-0.32)$	Radian	[SIN⁻¹] [−] 0.32 [ENTER]	–0.33

22.

Scientific Calculator Solution			
Function	**Mode**	**Keystrokes**	**Display** (rounded to two places)
$\sin^{-1}(-0.625)$	Radian	0.625 [+/-] [SIN⁻¹]	–0.68

Graphing Calculator Solution			
Function	**Mode**	**Keystrokes**	**Display** (rounded to two places)
$\sin^{-1}(-0.625)$	Radian	[SIN⁻¹] [–] 0.625 [ENTER]	–0.68

23.

Scientific Calculator Solution			
Function	**Mode**	**Keystrokes**	**Display** (rounded to two places)
$\cos^{-1}\left(\frac{3}{8}\right)$	Radian	3 [÷] 8 [=] [COS⁻¹]	1.19

Graphing Calculator Solution			
Function	**Mode**	**Keystrokes**	**Display** (rounded to two places)
$\cos^{-1}\left(\frac{3}{8}\right)$	Radian	[COS⁻¹] [(] 3 [÷] 8 [)] [ENTER]	1.19

24.

Scientific Calculator Solution			
Function	**Mode**	**Keystrokes**	**Display** (rounded to two places)
$\cos^{-1}\left(\frac{4}{9}\right)$	Radian	4 [÷] 9 [=] [COS⁻¹]	1.11

Graphing Calculator Solution			
Function	**Mode**	**Keystrokes**	**Display** (rounded to two places)
$\cos^{-1}\left(\frac{4}{9}\right)$	Radian	[COS⁻¹] [(] 4 [÷] 9 [)] [ENTER]	1.11

25.

Scientific Calculator Solution			
Function	**Mode**	**Keystrokes**	**Display** (rounded to two places)
$\cos^{-1}\frac{\sqrt{5}}{7}$	Radian	5 [√] [÷] 7 [=] [COS⁻¹]	1.25

Graphing Calculator Solution			
Function	**Mode**	**Keystrokes**	**Display** (rounded to two places)
$\cos^{-1}\frac{\sqrt{5}}{7}$	Radian	[COS⁻¹] [(] [√] 5 [÷] 7 [)] [ENTER]	1.25

26.

Scientific Calculator Solution			
Function	**Mode**	**Keystrokes**	**Display** (rounded to two places)
$\cos^{-1}\frac{\sqrt{7}}{10}$	Radian	7 [√] [÷] 10 [=] [COS⁻¹]	1.30

Graphing Calculator Solution			
Function	**Mode**	**Keystrokes**	**Display** (rounded to two places)
$\cos^{-1}\frac{\sqrt{7}}{10}$	Radian	[COS⁻¹] [(] [√] 7 [÷] 10 [)] [ENTER]	1.30

27.

Scientific Calculator Solution			
Function	**Mode**	**Keystrokes**	**Display** (rounded to two places)
$\tan^{-1}(-20)$	Radian	20 [+/−] [TAN⁻¹]	−1.52

Graphing Calculator Solution			
Function	**Mode**	**Keystrokes**	**Display** (rounded to two places)
$\tan^{-1}(-20)$	Radian	[TAN⁻¹] [−] 20 [ENTER]	−1.52

28.

Scientific Calculator Solution			
Function	**Mode**	**Keystrokes**	**Display** (rounded to two places)
$\tan^{-1}(-30)$	Radian	30 [+/−] [TAN⁻¹]	−1.54

Graphing Calculator Solution			
Function	**Mode**	**Keystrokes**	**Display** (rounded to two places)
$\tan^{-1}(-30)$	Radian	[TAN⁻¹] [−] 30 [ENTER]	−1.54

29.

Scientific Calculator Solution			
Function	**Mode**	**Keystrokes**	**Display** (rounded to two places)
$\tan^{-1}\left(-\sqrt{473}\right)$	Radian	473 [√] [+/−] [TAN⁻¹]	−1.52

Graphing Calculator Solution			
Function	**Mode**	**Keystrokes**	**Display** (rounded to two places)
$\tan^{-1}\left(-\sqrt{473}\right)$	Radian	[TAN⁻¹] [(] [−] [√] 473 [)] [ENTER]	−1.52

30.

Scientific Calculator Solution			
Function	**Mode**	**Keystrokes**	**Display** (rounded to two places)
$\tan^{-1}\left(-\sqrt{5061}\right)$	Radian	5061 [√] [+/-] [TAN⁻¹]	–1.56

Graphing Calculator Solution			
Function	**Mode**	**Keystrokes**	**Display** (rounded to two places)
$\tan^{-1}\left(-\sqrt{5061}\right)$	Radian	[TAN⁻¹] [(] [–] [√] 5061 [)] [ENTER]	–1.56

31. $\sin\left(\sin^{-1} 0.9\right)$

$x = 0.9$, x is in $[-1, 1]$, so $\sin(\sin^{-1} 0.9) = 0.9$

32. $\cos(\cos^{-1} 0.57)$

$x = 0.57$, x is in $[-1, 1]$,

so $\cos(\cos^{-1} 0.57) = 0.57$

33. $\sin^{-1}\left(\sin\frac{\pi}{3}\right)$

$x = \frac{\pi}{3}$, x is in $\left[-\frac{\pi}{2}, \frac{\pi}{2}\right]$, so $\sin^{-1}\left(\sin\frac{\pi}{3}\right) = \frac{\pi}{3}$

34. $\cos^{-1}\left(\cos\frac{2\pi}{3}\right)$

$x = \frac{2\pi}{3}$, x is in $[0, \pi]$,

so $\cos^{-1}\left(\cos\frac{2\pi}{3}\right) = \frac{2\pi}{3}$

35. $\sin^{-1}\left(\sin\frac{5\pi}{6}\right)$

$x = \frac{5\pi}{6}$, x is not in $\left[-\frac{\pi}{2}, \frac{\pi}{2}\right]$, x is in the domain of $\sin x$, so

$\sin^{-1}\left(\sin\frac{5\pi}{6}\right) = \sin^{-1}\left(\frac{1}{2}\right) = \frac{\pi}{6}$

36. $\cos^{-1}\left(\cos\frac{4\pi}{3}\right)$

$x = \frac{4\pi}{3}$, x is not in $[0, \pi]$,

x is in the domain of $\cos x$,

so $\cos^{-1}\left(\cos\frac{4\pi}{3}\right) = \cos^{-1}\left(-\frac{1}{2}\right) = \frac{2\pi}{3}$

37. $\tan\left(\tan^{-1} 125\right)$

$x = 125$, x is a real number, so

$\tan\left(\tan^{-1} 125\right) = 125$

38. $\tan(\tan^{-1} 380)$

$x = 380$, x is a real number,

so $\tan(\tan^{-1} 380) = 380$

39. $\tan^{-1}\left[\tan\left(-\frac{\pi}{6}\right)\right]$

$x=-\frac{\pi}{6}$, x is in $\left(-\frac{\pi}{2},\frac{\pi}{2}\right)$, so

$\tan^{-1}\left[\tan\left(-\frac{\pi}{6}\right)\right]=-\frac{\pi}{6}$

40. $\tan^{-1}\left[\tan\left(-\frac{\pi}{3}\right)\right]$

$x=-\frac{\pi}{3}$, x is in $\left(-\frac{\pi}{2},\frac{\pi}{2}\right)$,

so $\tan^{-1}\left[\tan\left(-\frac{\pi}{3}\right)\right]=-\frac{\pi}{3}$

41. $\tan^{-1}\left(\tan\frac{2\pi}{3}\right)$

$x=\frac{2\pi}{3}$, x is not in $\left(-\frac{\pi}{2},\frac{\pi}{2}\right)$, x is in the domain of $\tan x$, so

$\tan^{-1}\left(\tan\frac{2\pi}{3}\right)=\tan^{-1}\left(-\sqrt{3}\right)=-\frac{\pi}{3}$

42. $\tan^{-1}\left(\tan\frac{3\pi}{4}\right)$

$x=\frac{3\pi}{4}$, x is not in $\left(-\frac{\pi}{2},\frac{\pi}{2}\right)$,

x is in the domain of $\tan x$

so $\tan^{-1}\left(\tan\frac{3\pi}{4}\right)=\tan^{-1}(-1)=-\frac{\pi}{4}$

43. $\sin^{-1}(\sin\pi)$

$x=\pi$, x is not in $\left[-\frac{\pi}{2},\frac{\pi}{2}\right]$,

x is in the domain of $\sin x$, so

$\sin^{-1}(\sin\pi)=\sin^{-1}0=0$

44. $\cos^{-1}(\cos 2\pi)$

$x=2\pi$, x is not in $[0,\pi]$,

x is in the domain of $\cos x$,

so $\cos^{-1}(\cos 2\pi)=\cos^{-1}1=0$

45. $\sin\left(\sin^{-1}\pi\right)$

$x=\pi$, x is not in $[-1, 1]$, so $\sin\left(\sin^{-1}\pi\right)$ is not defined.

46. $\cos(\cos^{-1}3\pi)$

$x=3\pi$, x is not in $[-1, 1]$

so $\cos(\cos^{-1}3\pi)$ is not defined.

47. Let $\theta=\sin^{-1}\frac{4}{5}$, then $\sin\theta=\frac{4}{5}$. Because $\sin\theta$ is positive, θ is in the first quadrant.

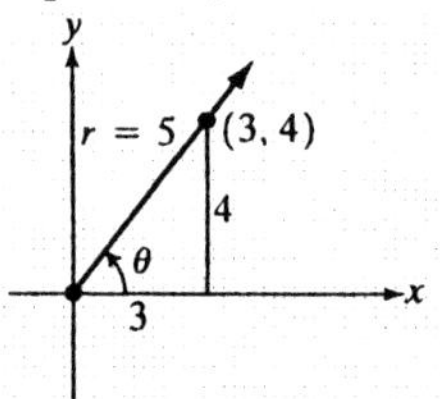

$x^2+y^2=r^2$

$x^2+4^2=5^2$

$x^2=25-16=9$

$x=3$

$\cos\left(\sin^{-1}\frac{4}{5}\right)=\cos\theta=\frac{x}{r}=\frac{3}{5}$

48. Let $\theta=\tan^{-1}\frac{7}{24}$, then $\tan\theta=\frac{7}{24}$.

Because $\tan\theta$ is positive, θ is in the first quadrant.

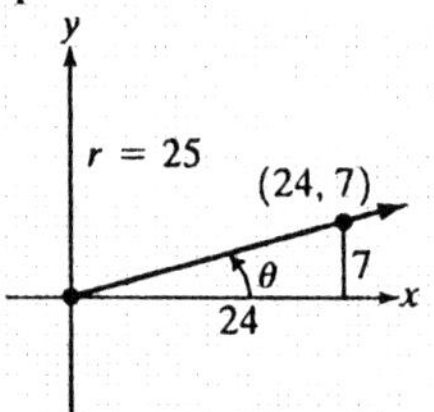

$r^2=x^2+y^2$

$r^2=7^2+24^2$

$r^2=625$

$r=25$

$\sin\left(\tan^{-1}\frac{7}{24}\right)=\sin\theta=\frac{y}{r}=\frac{7}{25}$

49. Let $\theta = \cos^{-1}\frac{5}{13}$, then $\cos\theta = \frac{5}{13}$. Because $\cos\theta$ is positive, θ is in the first quadrant.

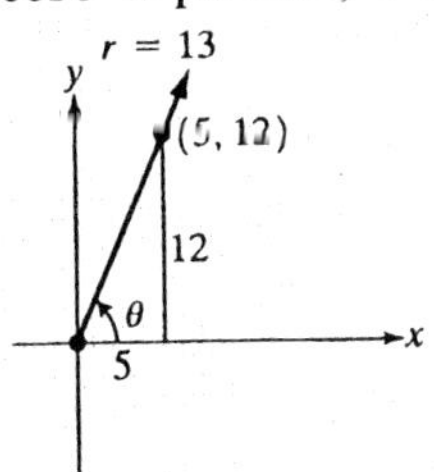

$$x^2 + y^2 = r^2$$
$$5^2 + y^2 = 13^2$$
$$y^2 = 169 - 25$$
$$y^2 = 144$$
$$y = 12$$
$$\tan\left(\cos^{-1}\frac{5}{13}\right) = \tan\theta = \frac{y}{x} = \frac{12}{5}$$

50. Let $\theta = \sin^{-1}\frac{5}{13}$ then $\sin\theta = \frac{5}{13}$.

because $\sin\theta$ is positive, θ is in the first quadrant.

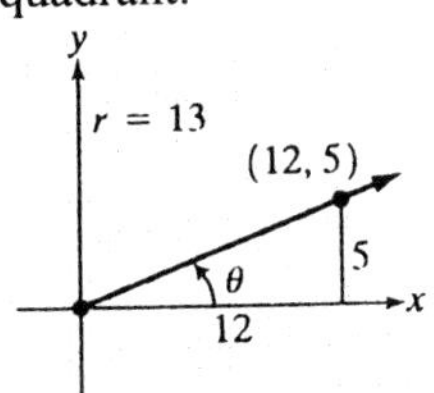

$$x^2 + y^2 = r^2$$
$$x^2 + 5^2 = 13^2$$
$$x^2 = 144$$
$$x = 12$$
$$\cot\left(\sin^{-1}\frac{5}{13}\right) = \cot\theta = \frac{x}{y} = \frac{12}{5}$$

51. Let $\theta = \sin^{-1}\left(-\frac{3}{5}\right)$, then $\sin\theta = -\frac{3}{5}$. Because $\sin\theta$ is negative, θ is in quadrant IV.

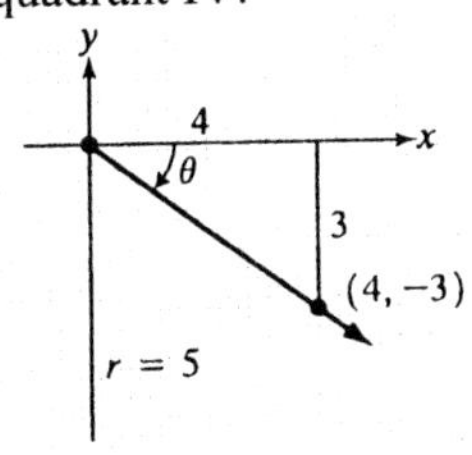

$$x^2 + y^2 = r^2$$
$$x^2 + (-3)^2 = 5^2$$
$$x^2 = 16$$
$$x = 4$$
$$\tan\left[\sin^{-1}\left(-\frac{3}{5}\right)\right] = \tan\theta = \frac{y}{x} = -\frac{3}{4}$$

52. Let $\theta = \sin^{-1}\left(-\frac{4}{5}\right)$, then $\sin\theta = -\frac{4}{5}$.

Because $\sin\theta$ is negative, θ is in quadrant IV.

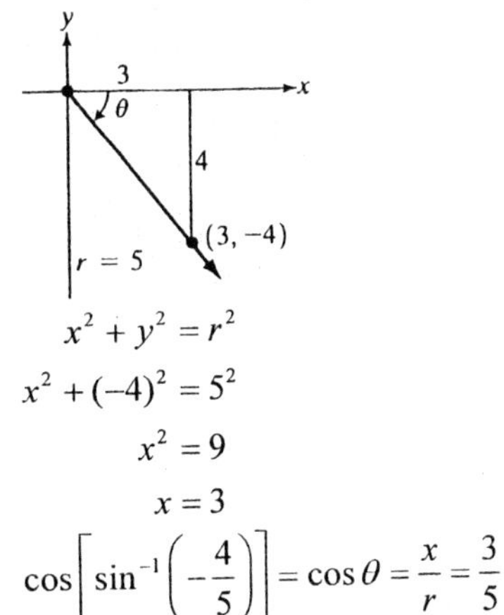

$$x^2 + y^2 = r^2$$
$$x^2 + (-4)^2 = 5^2$$
$$x^2 = 9$$
$$x = 3$$
$$\cos\left[\sin^{-1}\left(-\frac{4}{5}\right)\right] = \cos\theta = \frac{x}{r} = \frac{3}{5}$$

53. Let, $\theta = \cos^{-1}\frac{\sqrt{2}}{2}$, then $\cos\theta = \frac{\sqrt{2}}{2}$. Because $\cos\theta$ is positive, θ is in the first quadrant.

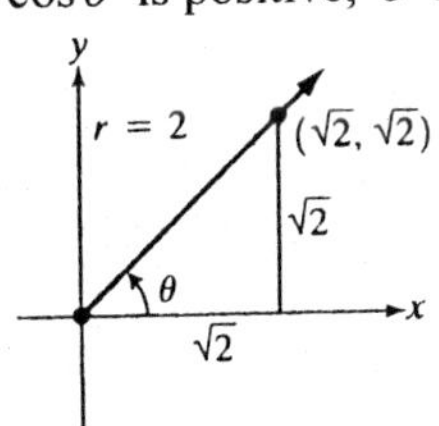

$$x^2 + y^2 = r^2$$
$$\left(\sqrt{2}\right)^2 + y^2 = 2^2$$
$$y^2 = 2$$
$$y = \sqrt{2}$$
$$\sin\left(\cos^{-1}\frac{\sqrt{2}}{2}\right) = \sin\theta = \frac{y}{r} = \frac{\sqrt{2}}{2}$$

54. Let $\theta = \sin^{-1}\frac{1}{2}$, then $\sin\theta = \frac{1}{2}$.
Because $\sin\theta$ is positive, θ is in the first quadrant.

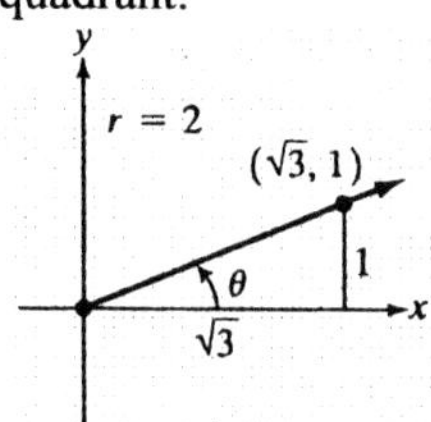

$$x^2 + y^2 = r^2$$
$$x^2 + 1^2 = 2^2$$
$$x^2 = 3$$
$$x = \sqrt{3}$$
$$\cos\left(\sin^{-1}\frac{1}{2}\right) = \cos\theta = \frac{x}{r} = \frac{\sqrt{3}}{2}$$

55. Let $\theta = \sin^{-1}\left(-\frac{1}{4}\right)$, then $\sin\theta = -\frac{1}{4}$. Because $\sin\theta$ is negative, θ is in quadrant IV.

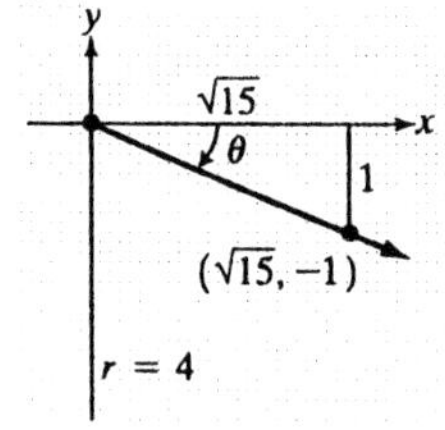

$$x^2 + y^2 = r^2$$
$$x^2 + (-1)^2 = 4^2$$
$$x^2 = 15$$
$$x = \sqrt{15}$$
$$\sec\left[\sin^{-1}\left(-\frac{1}{4}\right)\right] = \sec\theta = \frac{r}{x} = \frac{4}{\sqrt{15}} = \frac{4\sqrt{15}}{15}$$

56. Let $\theta = \sin^{-1}\left(-\frac{1}{2}\right)$, then $\sin\theta = -\frac{1}{2}$.
Because $\sin\theta$ is negative, θ is in quadrant IV.

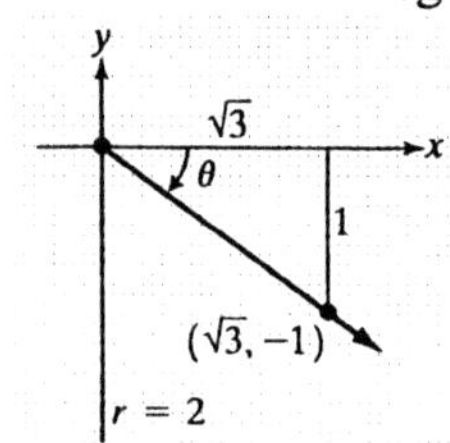

$$x^2 + y^2 = r^2$$
$$x^2 + (-1)^2 = 2^2$$
$$x^2 = 3$$
$$x = \sqrt{3}$$
$$\sec\left[\sin^{-1}\left(-\frac{1}{2}\right)\right] = \sec\theta = \frac{r}{x} = \frac{2}{\sqrt{3}} = \frac{2\sqrt{3}}{3}$$

57. Let $\theta = \cos^{-1}\left(-\frac{1}{3}\right)$, then $\cos\theta = -\frac{1}{3}$. Because $\cos\theta$ is negative, θ is in quadrant II.

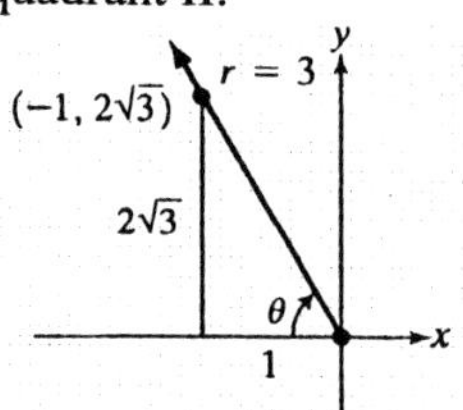

$$x^2 + y^2 = r^2$$
$$(-1)^2 + y^2 = 3^2$$
$$y^2 = 8$$
$$y = \sqrt{8}$$
$$y = 2\sqrt{2}$$
Use the right triangle to find the exact value.
$$\tan\left[\cos^{-1}\left(-\frac{1}{3}\right)\right] = \tan\theta = \frac{y}{x} = \frac{2\sqrt{2}}{-1} = -2\sqrt{2}$$

58. Let $\theta = \cos^{-1}\left(-\frac{1}{4}\right)$, then $\cos\theta = -\frac{1}{4}$.
Because $\cos\theta$ is negative, θ is in quadrant II.

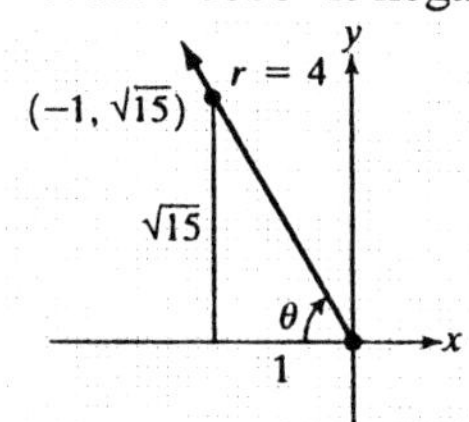

$$x^2 + y^2 = r^2$$
$$(-1)^2 + y^2 = 4^2$$
$$y^2 = 15$$
$$y = \sqrt{15}$$
$$\tan\left[\cos^{-1}\left(-\frac{1}{4}\right)\right] = \tan\theta = \frac{y}{x} = \frac{\sqrt{15}}{-1} = -\sqrt{15}$$

59. Let $\theta = \cos^{-1}\left(-\frac{\sqrt{3}}{2}\right)$, then $\cos\theta = -\frac{\sqrt{3}}{2}$.

Because $\cos\theta$ is negative, θ is in quadrant II.

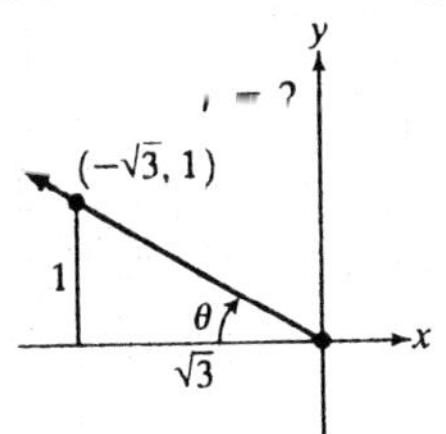

$$x^2 + y^2 = r^2$$
$$\left(-\sqrt{3}\right)^2 + y^2 = 2^2$$
$$y^2 = 1$$
$$y = 1$$
$$\csc\left[\cos^{-1}\left(-\frac{\sqrt{3}}{2}\right)\right] = \csc\theta = \frac{r}{y} = \frac{2}{1} = 2$$

60. Let $\theta = \sin^{-1}\left(-\frac{\sqrt{2}}{2}\right)$, then $\sin\theta = -\frac{\sqrt{2}}{2}$.

Because $\sin\theta$ is negative, θ is in quadrant IV.

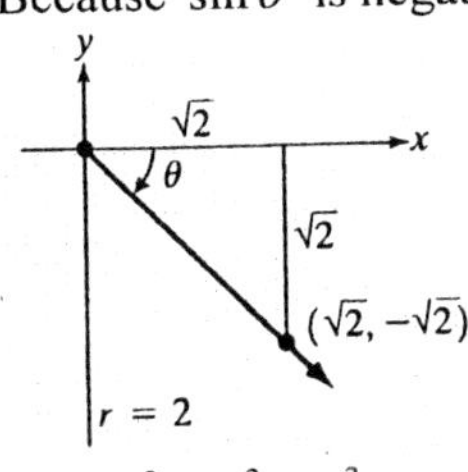

$$x^2 + y^2 = r^2$$
$$x^2 + \left(-\sqrt{2}\right)^2 = 2^2$$
$$x^2 = 2$$
$$x = \sqrt{2}$$
$$\sec\left[\sin^{-1}\left(-\frac{\sqrt{2}}{2}\right)\right] = \sec\theta = \frac{r}{x} = \frac{2}{\sqrt{2}} = \sqrt{2}$$

61. Let $\theta = \tan^{-1}\left(-\frac{2}{3}\right)$, then $\tan\theta = -\frac{2}{3}$.

Because $\tan\theta$ is negative, θ is in quadrant IV.

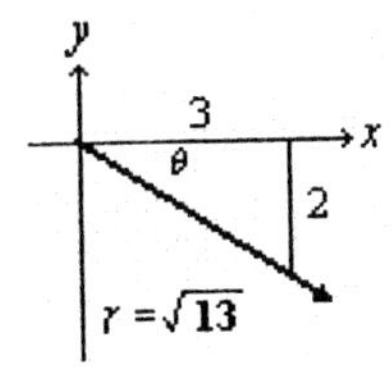

$$r^2 = x^2 + y^2$$
$$r^2 = 3^2 + (-2)^2$$
$$r^2 = 9 + 4$$
$$r^2 = 13$$
$$r = \sqrt{13}$$
$$\cos\left[\tan^{-1}\left(-\frac{2}{3}\right)\right] = \cos\theta = \frac{x}{r} = \frac{3}{\sqrt{13}} = \frac{3\sqrt{13}}{13}$$

62. Let $\theta = \tan^{-1}\left(-\frac{3}{4}\right)$, then $\tan\theta = -\frac{3}{4}$.

Because $\tan\theta$ is negative, θ is in quadrant IV.

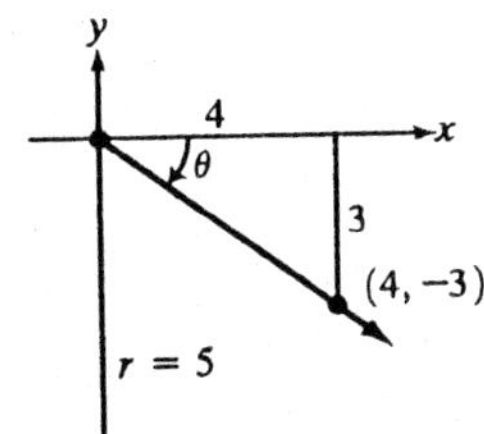

$$r^2 = x^2 + y^2$$
$$r^2 = 4^2 + (-3)^2$$
$$r^2 = 16 + 9$$
$$r^2 = 25$$
$$r = 5$$
$$\sin\left[\tan^{-1}\left(-\frac{3}{4}\right)\right] = \sin\theta = \frac{y}{r} = \frac{-3}{5} = -\frac{3}{5}$$

63. Let $\theta = \cos^{-1} x$, then $\cos\theta = x = \frac{x}{1}$.

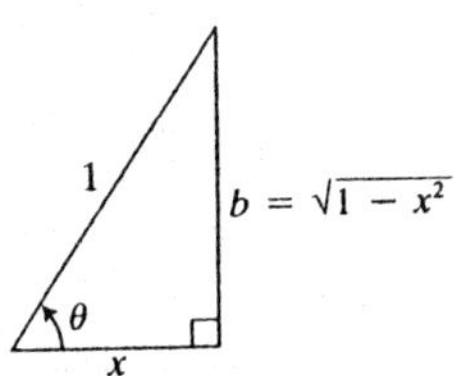

Use the Pythagorean Theorem to find the third side, b.

$$x^2 + b^2 = 1^2$$
$$b^2 = 1 - x^2$$
$$b = \sqrt{1 - x^2}$$

Use the right triangle to write the algebraic expression.

$$\tan\left(\cos^{-1} x\right) = \tan\theta = \frac{\sqrt{1-x^2}}{x}$$

64. Let $\theta = \tan^{-1} x$, then $\tan\theta = x = \frac{x}{1}$.

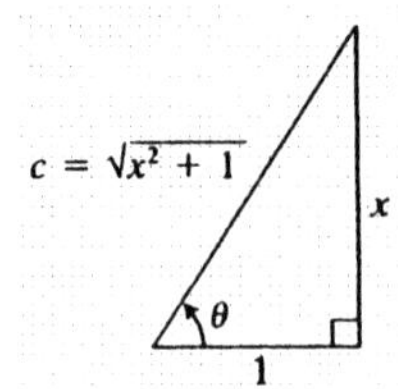

Use the Pythagorean Theorem to find the third side, c.

$c^2 = x^2 + 1^2$

$c = \sqrt{x^2+1}$

Use the right triangle to write the algebraic expression.

$$\sin(\tan^{-1}) = \sin\theta = \frac{x}{\sqrt{x^2+1}} = \frac{x}{\sqrt{x^2+1}}\cdot\frac{\sqrt{x^2+1}}{\sqrt{x^2+1}} = \frac{x\sqrt{x^2+1}}{x^2+1}$$

65. Let $\theta = \sin^{-1} 2x$, then $\sin\theta = 2x$

$y = 2x,\ r = 1$

Use the Pythagorean Theorem to find x.

$x^2 + (2x)^2 = 1^2$

$x^2 = 1 - 4x^2$

$x = \sqrt{1-4x^2}$

$\cos(\sin^{-1} 2x) = \sqrt{1-4x^2}$

66. Let $\theta = \cos^{-1} 2x$.

Use the Pythagorean Theorem to find the third side, b.

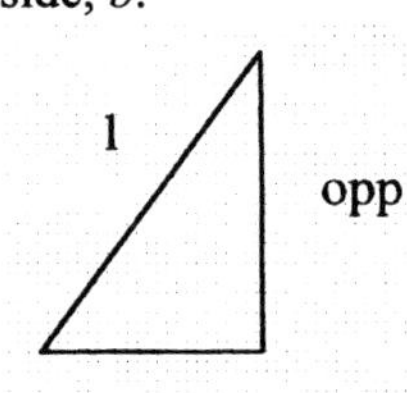

$2x$

$(2x)^2 + b^2 = 1^2$

$b^2 = 1 - 4x^2$

$b = \sqrt{1-4x^2}$

$\sin\left(\cos^{-1} 2x\right) = \frac{\sqrt{1-4x^2}}{1} = \sqrt{1-4x^2}$

67. Let $\theta = \sin^{-1}\frac{1}{x}$, then $\sin\theta = \frac{1}{x}$.

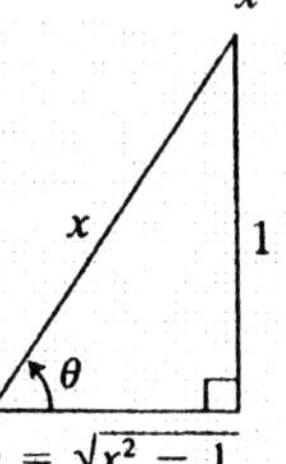

$a = \sqrt{x^2 - 1}$

Use the Pythagorean Theorem to find the third side, a.

$a^2 + 1^2 = x^2$

$a^2 = x^2 - 1$

$a = \sqrt{x^2-1}$

Use the right triangle to write the algebraic expression.

$$\cos\left(\sin^{-1}\frac{1}{x}\right) = \cos\theta = \frac{\sqrt{x^2-1}}{x}$$

68. Let $\theta = \cos^{-1}\frac{1}{x}$, then $\cos\theta = \frac{1}{x}$.

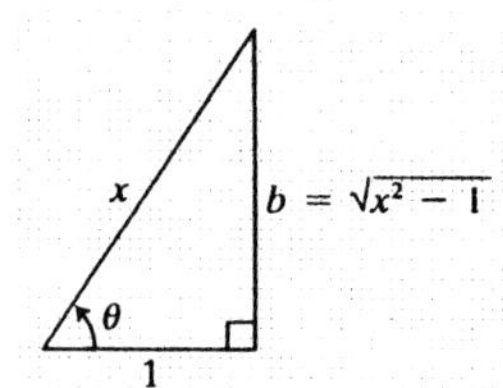

Use the Pythagorean Theorem to find the third side, b.

$1^2 + b^2 = x^2$

$b^2 = x^2 - 1$

$b = \sqrt{x^2-1}$

Use the right triangle to write the algebraic expression.

$$\sec\left(\cos^{-1}\frac{1}{x}\right) = \sec\theta = \frac{x}{1} = x$$

69. $\cot\left(\tan^{-1}\frac{x}{\sqrt{3}}\right) = \frac{\sqrt{3}}{x}$

70. $\cot\left(\tan^{-1}\frac{x}{\sqrt{2}}\right) = \frac{\sqrt{2}}{x}$

71. Let $\theta = \sin^{-1}\dfrac{x}{\sqrt{x^2+4}}$, then $\sin\theta = \dfrac{x}{\sqrt{x^2+4}}$.

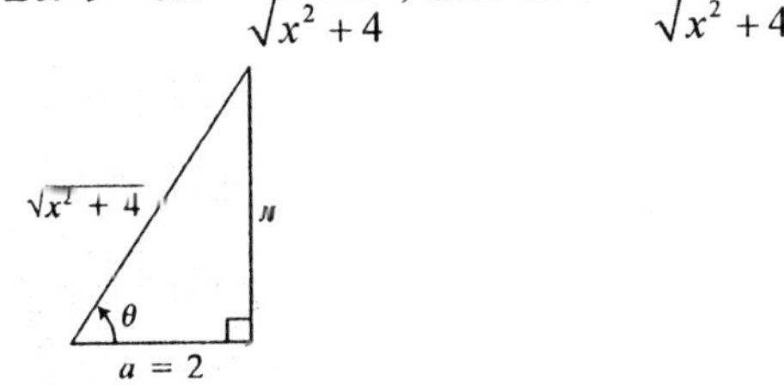

Use the Pythagorean Theorem to find the third side, a.

$$a^2 + x^2 = \left(\sqrt{x^2+4}\right)^2$$
$$a^2 = x^2 + 4 - x^2 = 4$$
$$a = 2$$

Use the right triangle to write the algebraic expression.

$$\sec\left(\sin^{-1}\frac{x}{\sqrt{x^2+4}}\right) = \sec\theta = \frac{\sqrt{x^2+4}}{2}$$

72. Let $\theta = \sin^{-1}\dfrac{\sqrt{x^2-9}}{x}$, then $\sin\theta = \dfrac{\sqrt{x^2-9}}{x}$.

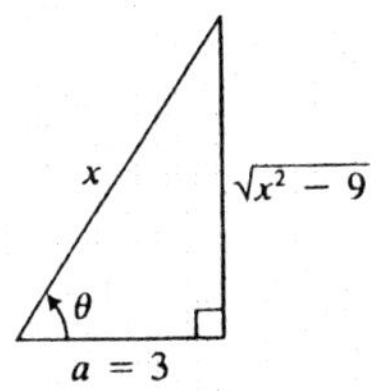

Use the Pythagorean Theorem to find the third side, a.

$$a^2 + \left(\sqrt{x^2-9}\right)^2 = x^2$$
$$a^2 = x^2 - x^2 + 9 = 9$$
$$a = 3$$

Use the right triangle to write the algebraic expression.

$$\cot\left(\sin^{-1}\frac{\sqrt{x^2-9}}{x}\right) = \frac{3}{\sqrt{x^2-9}}$$
$$= \frac{3}{\sqrt{x^2-9}} \cdot \frac{\sqrt{x^2-9}}{\sqrt{x^2-9}} = \frac{3\sqrt{x^2-9}}{x^2-9}$$

73. a. $y = \sec x$ is the reciprocal of $y = \cos x$. The x-values for the key points in the interval $[0, \pi]$ are $0, \dfrac{\pi}{4}, \dfrac{\pi}{2}, \dfrac{3\pi}{4}$, and π.

The key points are $(0, 1)$, $\left(\dfrac{\pi}{4}, \dfrac{\sqrt{2}}{2}\right)$, $\left(\dfrac{\pi}{2}, 0\right)$, $\left(\dfrac{3\pi}{4}, -\dfrac{\sqrt{2}}{2}\right)$, and $(\pi, -1)$,

Draw a vertical asymptote at $x = \dfrac{\pi}{2}$. Now draw our graph from $(0, 1)$ through $\left(\dfrac{\pi}{4}, \sqrt{2}\right)$ to ∞ on the left side of the asymptote. From $-\infty$ on the right side of the asymptote through $\left(\dfrac{3\pi}{4}, -\sqrt{2}\right)$ to $(\pi, -1)$.

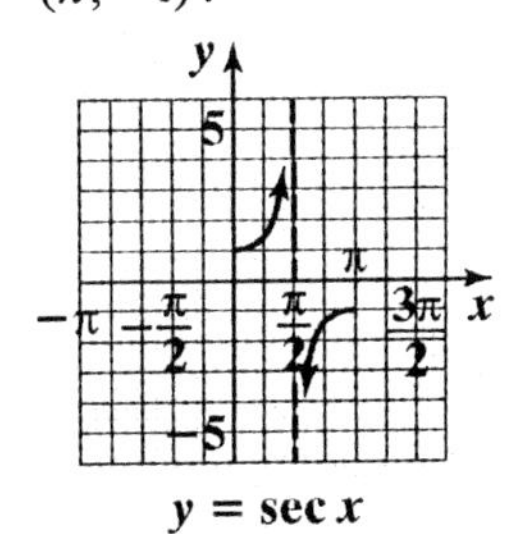

$y = \sec x$

b. With this restricted domain, no horizontal line intersects the graph of $y = \sec x$ more than once, so the function is one-to-one and has an inverse function.

c. Reflecting the graph of the restricted secant function about the line $y = x$, we get the graph of $y = \sec^{-1} x$.

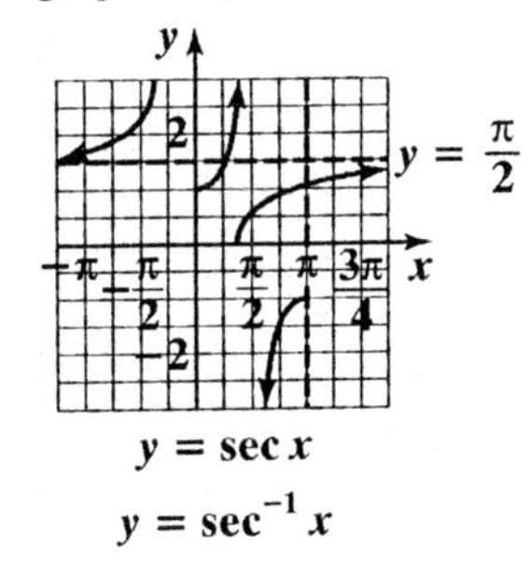

$y = \sec x$

$y = \sec^{-1} x$

74. a. Two consecutive asymptotes occur at $x = 0$ and $x = \pi$. Midway between $x = 0$ and $x = \pi$ is $x = 0$ and $x = \pi$. An x-intercept for the graph is $\left(\frac{\pi}{2}, 0\right)$. The graph goes through the points $\left(\frac{\pi}{4}, 1\right)$ and $\left(\frac{3\pi}{4}, -1\right)$. Now graph the function through these points and using the asymptotes.

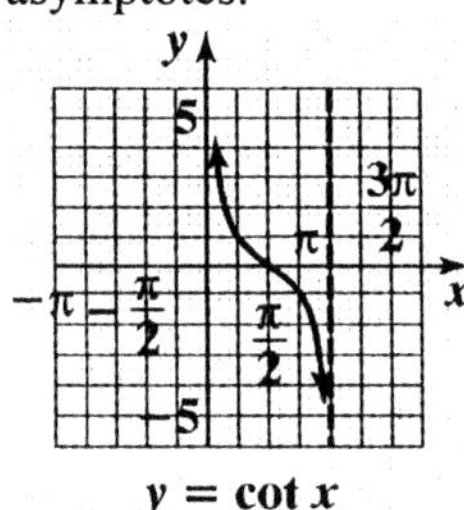

$y = \cot x$

b. With this restricted domain no horizontal line intersects the graph of $y = \cot x$ more than once, so the function is one-to-one and has an inverse function. Reflecting the graph of the restricted cotangent function about the line $y = x$, we get the graph of $y = \cot^{-1} x$.

c.

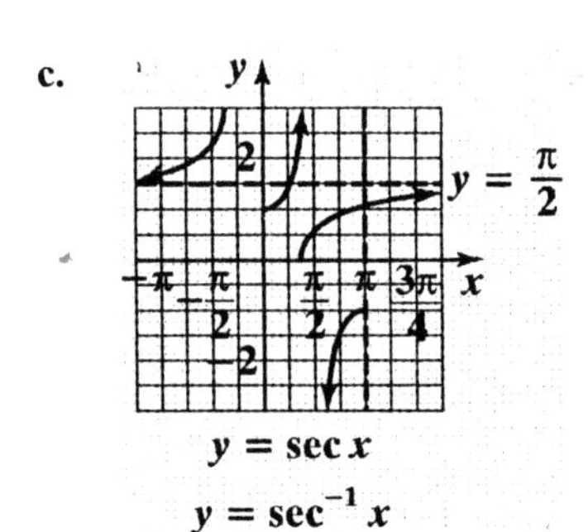

$y = \sec x$

$y = \sec^{-1} x$

75.

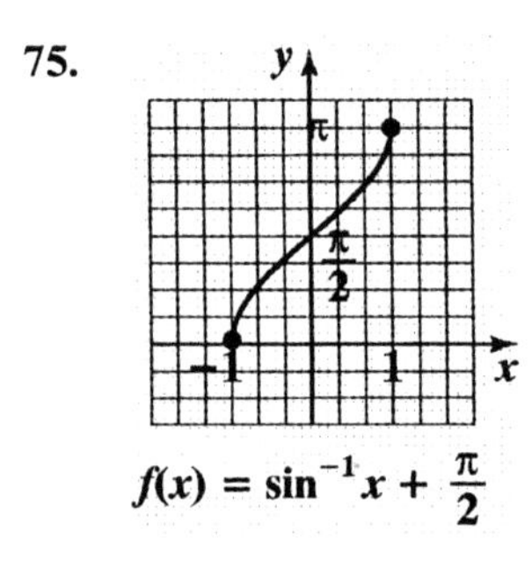

$f(x) = \sin^{-1} x + \frac{\pi}{2}$

Domain: $[-1, 1]$;
Range: $[0, \pi]$

76.

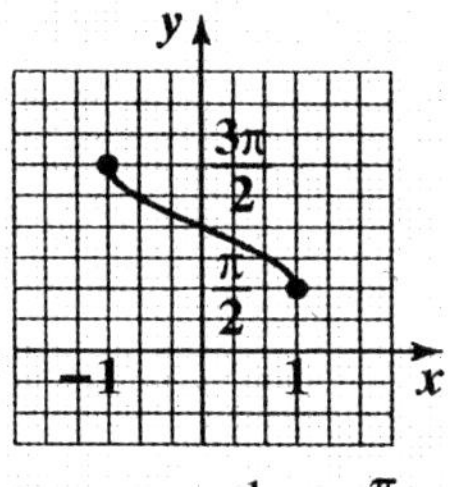

$f(x) = \cos^{-1} x + \frac{\pi}{2}$

Domain: $[-1, 1]$;
Range: $\left[\frac{\pi}{2}, \frac{3\pi}{2}\right]$

77.

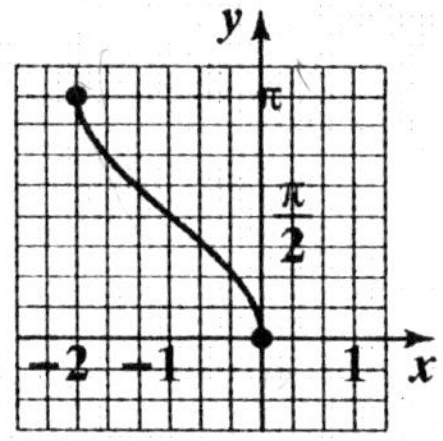

$g(x) = \cos^{-1}(x + 1)$

Domain: $[-2, 0]$;
Range: $[0, \pi]$

78.

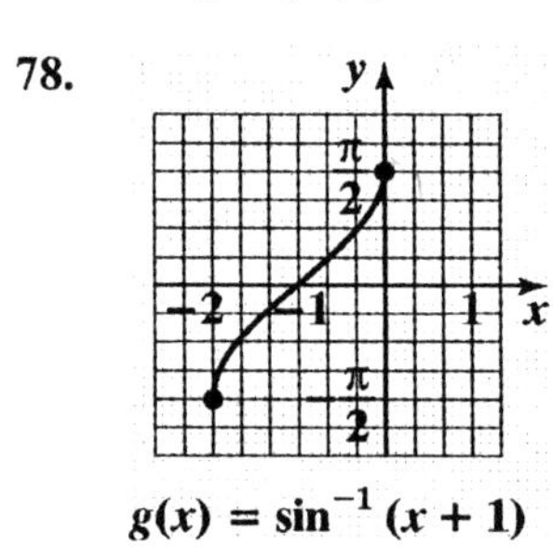

$g(x) = \sin^{-1}(x + 1)$

Domain: $[-2, 0]$;
Range: $\left[-\frac{\pi}{2}, \frac{\pi}{2}\right]$

79.

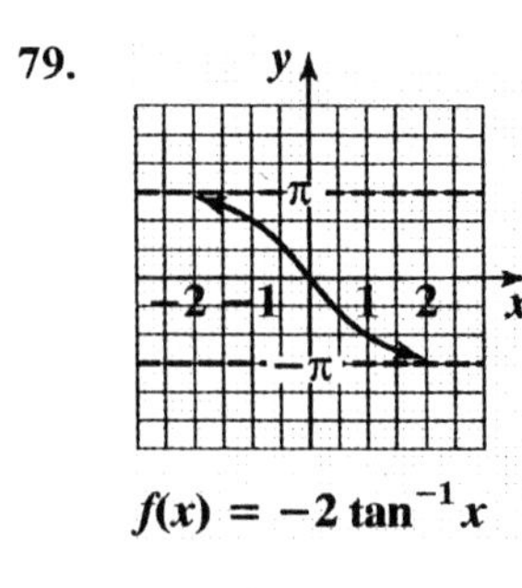

$f(x) = -2\tan^{-1} x$

Domain: $(-\infty, \infty)$;
Range: $(-\pi, \pi)$

80.

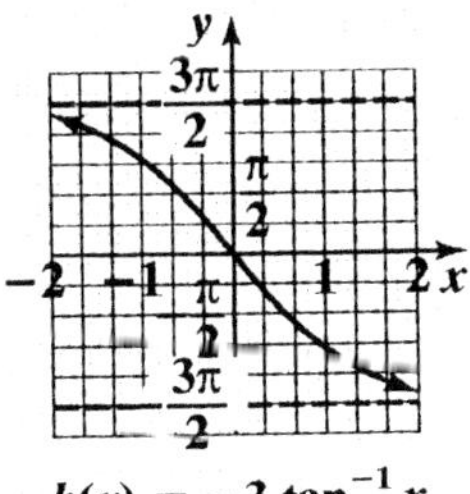

$h(x) = -3\tan^{-1} x$

Domain: $(-\infty, \infty)$;
Range: $\left(-\frac{3\pi}{2}, \frac{3\pi}{2}\right)$

81.

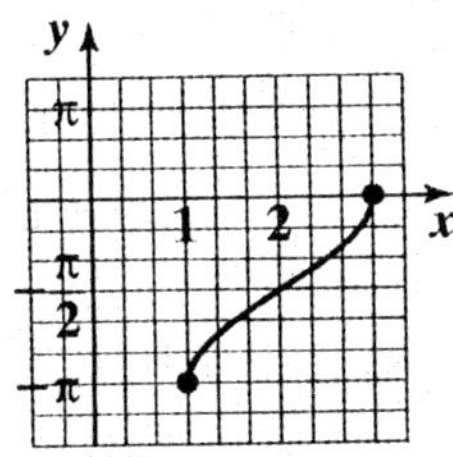

$f(x) = \sin^{-1}(x-2) - \frac{\pi}{2}$

Domain: $(1, 3]$;
Range: $[-\pi, 0]$

82.

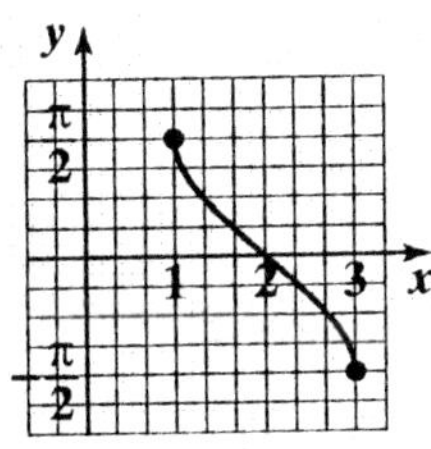

$f(x) = \cos^{-1}(x-2) - \frac{\pi}{2}$

Domain: $[1, 3]$;
Range: $\left[-\frac{\pi}{2}, \frac{\pi}{2}\right]$

83.

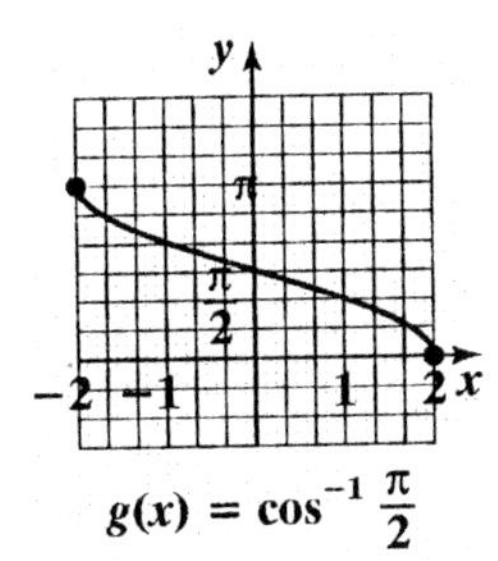

$g(x) = \cos^{-1}\frac{\pi}{2}$

Domain: $[-2, 2]$;
Range: $[0, \pi]$

84.

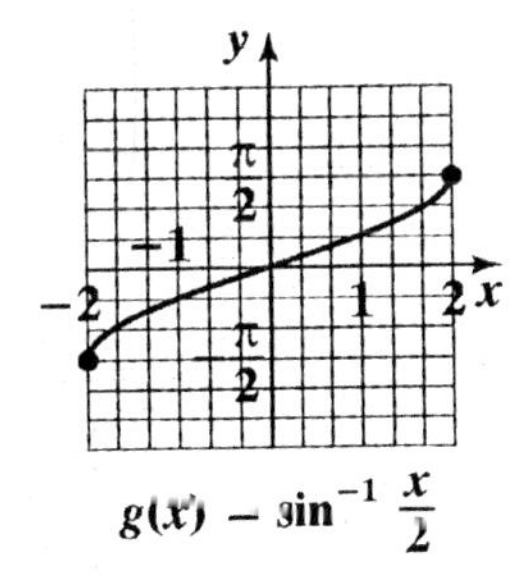

$g(x) = \sin^{-1}\frac{x}{2}$

Domain: $[-2, 2]$;
Range: $\left[-\frac{\pi}{2}, \frac{\pi}{2}\right]$

85. The inner function, $\sin^{-1} x$, accepts values on the interval $[-1,1]$. Since the inner and outer functions are inverses of each other, the domain and range are as follows.
Domain: $[-1,1]$; Range: $[-1,1]$

86. The inner function, $\cos^{-1} x$, accepts values on the interval $[-1,1]$. Since the inner and outer functions are inverses of each other, the domain and range are as follows.
Domain: $[-1,1]$; Range: $[-1,1]$

87. The inner function, $\cos x$, accepts values on the interval $(-\infty,\infty)$. The outer function returns values on the interval $[0,\pi]$
Domain: $(-\infty,\infty)$; Range: $[0,\pi]$

88. The inner function, $\sin x$, accepts values on the interval $(-\infty,\infty)$. The outer function returns values on the interval $\left[-\frac{\pi}{2},\frac{\pi}{2}\right]$
Domain: $(-\infty,\infty)$; Range: $\left[-\frac{\pi}{2},\frac{\pi}{2}\right]$

89. The inner function, $\cos x$, accepts values on the interval $(-\infty,\infty)$. The outer function returns values on the interval $\left[-\frac{\pi}{2},\frac{\pi}{2}\right]$
Domain: $(-\infty,\infty)$; Range: $\left[-\frac{\pi}{2},\frac{\pi}{2}\right]$

90. The inner function, $\sin x$, accepts values on the interval $(-\infty,\infty)$. The outer function returns values on the interval $[0,\pi]$
Domain: $(-\infty,\infty)$; Range: $[0,\pi]$

91. The functions $\sin^{-1} x$ and $\cos^{-1} x$ accept values on the interval $[-1,1]$. The sum of these values is always $\frac{\pi}{2}$.
Domain: $[-1,1]$
Range: $\left\{\frac{\pi}{2}\right\}$

92. The functions $\sin^{-1} x$ and $\cos^{-1} x$ accept values on the interval $[-1,1]$. The difference of these values range from $-\frac{\pi}{2}$ to $\frac{3\pi}{2}$
Domain: $[-1,1]$
Range: $\left[-\frac{\pi}{2},\frac{3\pi}{2}\right]$

93. $\theta=\tan^{-1}\frac{33}{x}-\tan^{-1}\frac{8}{x}$

x	θ
5	$\tan^{-1}\frac{33}{5}-\tan^{-1}\frac{8}{5}\approx 0.408$ radians
10	$\tan^{-1}\frac{33}{10}-\tan^{-1}\frac{8}{10}\approx 0.602$ radians
15	$\tan^{-1}\frac{33}{15}-\tan^{-1}\frac{8}{15}\approx 0.654$ radians
20	$\tan^{-1}\frac{33}{20}-\tan^{-1}\frac{8}{20}\approx 0.645$ radians
25	$\tan^{-1}\frac{33}{25}-\tan^{-1}\frac{8}{25}\approx 0.613$ radians

94. The viewing angle increases rapidly up to about 16 feet, then it decreases less rapidly; about 16 feet; when $x = 15$, $\theta = 0.6542$ radians; when $x = 17$, $\theta = 0.6553$ radians.

95. $\theta=2\tan^{-1}\frac{21.634}{28}\approx 1.3157$ radians;
$1.3157\left(\frac{180}{\pi}\right)\approx 75.4°$

96. $\theta=2\tan^{-1}\frac{21.634}{300}\approx 0.1440$ radians;
$0.1440\left(\frac{180}{\pi}\right)\approx 8.3°$

97. $\tan^{-1} b-\tan^{-1} a=\tan^{-1} 2-\tan^{-1} 0$
≈ 1.1071 square units

98. $\tan^{-1} b-\tan^{-1} a=\tan^{-1} 1-\tan^{-1}(-2)$
≈ 1.8925 square units

110. $y=\sin^{-1} x$
$y=\sin^{-1} x+2$

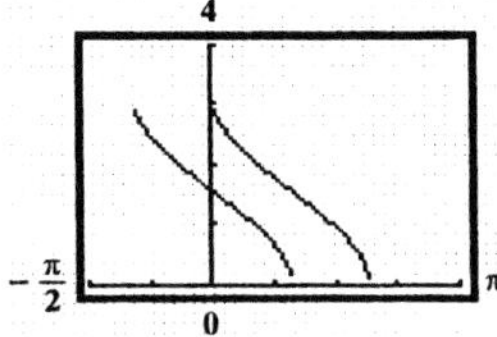

The graph of the second equation is the graph of the first equation shifted up 2 units.

111. The domain of $y=\cos^{-1} x$ is the interval $[-1, 1]$, and the range is the interval $[0, \pi]$. Because the second equation is the first equation with 1 subtracted from the variable, we will move our x max to π, and graph in a $\left[-\frac{\pi}{2}, \pi, \frac{\pi}{4}\right]$ by $[0, 4, 1]$ viewing rectangle.

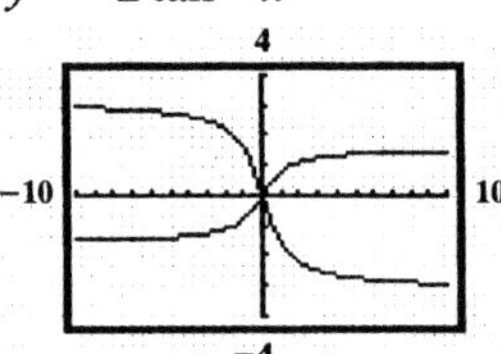

The graph of the second equation is the graph of the first equation shifted right 1 unit.

112. $y=\tan^{-1} x$
$y=-2\tan^{-1} x$

The graph of the second equation is the graph of the first equation reversed and stretched.

113. The domain of $y = \sin^{-1} x$ is the interval $[-1, 1]$, and the range is $\left[-\frac{\pi}{2}, \frac{\pi}{2}\right]$. Because the second equation is the first equation plus 1, and with 2 added to the variable, we will move our y max to 3, and move our x min to $-\pi$, and graph in a $\left[-\pi, \frac{\pi}{2}, \frac{\pi}{2}\right]$ by $[-2, 3, 1]$ viewing rectangle.

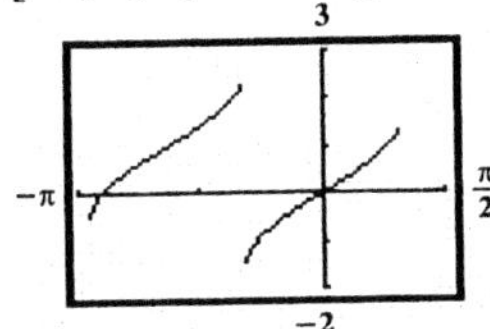

The graph of the second equation is the graph of the first equation shifted left 2 units and up 1 unit.

114. $y = \tan^{-1} x$

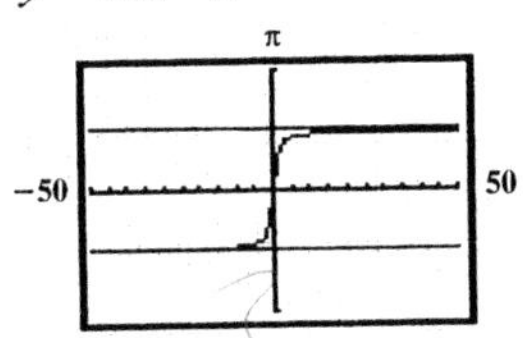

Observations may vary.

115.

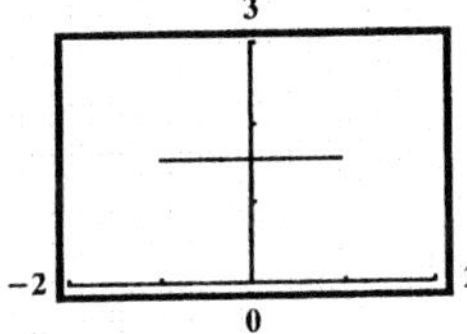

It seems $\sin^{-1} x + \cos^{-1} x = \frac{\pi}{2}$ for $-1 \le x \le 1$.

116.

$$y = 2\sin^{-1}(x-5)$$

$$\frac{y}{2} = \sin^{-1}(x-5)$$

$$\sin\frac{y}{2} = x-5$$

$$x = \sin\frac{y}{2} + 5$$

117. $2\sin^{-1} x = \frac{\pi}{4}$

$$\sin^{-1} x = \frac{\pi}{8}$$

$$x = \sin\frac{\pi}{8}$$

118. Prove: If $x > 0$, $\tan^{-1} x + \tan^{-1}\frac{1}{x} = \frac{\pi}{2}$

Since $x > 0$, there is an angle θ with $0 < \theta < \frac{\pi}{2}$ as shown in the figure.

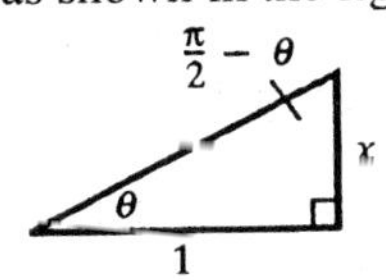

$\tan\theta = x$ and $\tan\left(\frac{\pi}{2} - \theta\right) = \frac{1}{x}$ thus

$\tan^{-1} x = \theta$ and $\tan^{-1}\left(\frac{1}{x}\right) = \frac{\pi}{2} - \theta$ so

$$\tan^{-1} x + \tan^{-1}\frac{1}{x} = \theta + \frac{\pi}{2} - \theta = \frac{\pi}{2}$$

119. Let α equal the acute angle in the smaller right triangle.

$$\tan\alpha = \frac{8}{x}$$

so $\tan^{-1}\frac{8}{x} = \alpha$

$$\tan(\alpha + \theta) = \frac{33}{x}$$

so $\tan^{-1}\frac{33}{x} = \alpha + \theta$

$$\theta = \alpha + \theta - \alpha = \tan^{-1}\frac{33}{x} - \tan^{-1}\frac{8}{x}$$

Section 4.8

Check Point Exercises

1. We begin by finding the measure of angle B. Because $C = 90°$ and the sum of a triangle's angles is 180°, we see that $A + B = 90°$. Thus, $B = 90° - A = 90° - 62.7° = 27.3°$.
Now we find b. Because we have a known angle, a known opposite side, and an unknown adjacent side, use the tangent function.

$$\tan 62.7° = \frac{8.4}{b}$$

$$b = \frac{8.4}{\tan 62.7°} \approx 4.34$$

Finally, we need to find c. Because we have a known angle, a known opposite side and an unknown hypotenuse, use the sine function.

$\sin 62.7° = \frac{8.4}{c}$

$c = \frac{8.4}{\sin 62.7} \approx 9.45$

In summary, $B = 27.3°$, $b \approx 4.34$, and $c \approx 9.45$.

2. Using a right triangle, we have a known angle, an unknown opposite side, a, and a known adjacent side. Therefore, use the tangent function.

$\tan 85.4° = \frac{a}{80}$

$a = 80 \tan 85.4° \approx 994$

The Eiffel tower is approximately 994 feet high.

3. Using a right triangle, we have an unknown angle, A, a known opposite side, and a known hypotenuse. Therefore, use the sine function.

$\sin A = \frac{6.7}{13.8}$

$A = \sin^{-1} \frac{6.7}{13.8} \approx 29.0°$

The wire makes an angle of approximately 29.0° with the ground.

4. Using two right triangles, a smaller right triangle corresponding to the smaller angle of elevation drawn inside a larger right triangle corresponding to the larger angle of elevation, we have a known angle, an unknown opposite side, a in the smaller triangle, b in the larger triangle, and a known adjacent side in each triangle. Therefore, use the tangent function.

$\tan 32° = \frac{a}{800}$

$a = 800 \tan 32° \approx 499.9$

$\tan 35° = \frac{b}{800}$

$b = 800 \tan 35° \approx 560.2$

The height of the sculpture of Lincoln's face is $560.2 - 499.9$, or approximately 60.3 feet.

5. **a.** We need the acute angle between ray OD and the north-south line through O. The measurement of this angle is given to be 25°. The angle is measured from the south side of the north-south line and lies east of the north-south line. Thus, the bearing from O to D is S 25°E.

b. We need the acute angle between ray OC and the north-south line through O. This angle measures $90° - 75° = 15°$. This angle is measured from the south side of the north-south line and lies west of the north-south line. Thus the bearing from O to C is S 15° W.

6. **a.** Your distance from the entrance to the trail system is represented by the hypotenuse, c, of a right triangle. Because we know the length of the two sides of the right triangle, we find c using the Pythagorean Theorem. We have

$c^2 = a^2 + b^2 = (2.3)^2 + (3.5)^2 = 17.54$

$c = \sqrt{17.54} \approx 4.2$

You are approximately 4.2 miles from the entrance to the trail system.

b. To find your bearing from the entrance to the trail system, consider a north-south line passing through the entrance. The acute angle from this line to the ray on which you lie is $31° + \theta$. Because we are measuring the angle from the south side of the line and you are west of the entrance, your bearing from the entrance is $\text{S}(31° + \theta)\text{ W}$. To find θ, Use a right triangle and the tangent function.

$\tan \theta = \frac{3.5}{2.3}$

$\theta = \tan^{-1} \frac{3.5}{2.3} \approx 56.7°$

Thus, $31° + \theta = 31° + 56.7° = 87.7°$. Your bearing from the entrance to the trail system is S 87.7° W.

7. When the object is released $(t = 0)$, the ball's distance, d, from its rest position is 6 inches down. Because it is down, d is negative: when $t = 0$, $d = -6$. Notice the greatest distance from rest position occurs at $t = 0$. Thus, we will use the equation with the cosine function, $y = a \cos \omega t$, to model the ball's motion. Recall that $|a|$ is the maximum distance. Because the ball's initial position is down, $a = -6$. The value of ω can be found using the formula for the period.

$\text{period} = \frac{2\pi}{\omega} = 4$

$2\pi = 4\omega$

$\omega = \frac{2\pi}{4} = \frac{\pi}{2}$

Substitute these values into $d = a\cos wt$. The equation for the ball's simple harmonic motion is $d = -6\cos\frac{\pi}{2}t$.

8. We begin by identifying values for a and ω.

$d = 12\cos\frac{\pi}{4}t$, $a = 12$ and $\omega = \frac{\pi}{4}$.

a. The maximum displacement from the rest position is the amplitude. Because $a = 12$, the maximum displacement is 12 centimeters.

b. The frequency, f, is

$f = \frac{\omega}{2\pi} = \frac{\frac{\pi}{4}}{2\pi} = \frac{\pi}{4}\cdot\frac{1}{2\pi} = \frac{1}{8}$

The frequency is $\frac{1}{8}$ cycle per second.

c. The time required for one cycle is the period.

$\text{period} = \frac{2\pi}{\omega} = \frac{2\pi}{\frac{\pi}{4}} = 2\pi\cdot\frac{4}{\pi} = 8$

The time required for one cycle is 8 seconds.

Exercise Set 4.8

1. Find the measure of angle B. Because $C = 90°$, $A + B = 90°$. Thus, $B = 90° - A = 90° - 23.5° = 66.5°$.
Because we have a known angle, a known adjacent side, and an unknown opposite side, use the tangent function.

$\tan 23.5° = \frac{a}{10}$

$a = 10\tan 23.5° \approx 4.35$

Because we have a known angle, a known adjacent side, and an unknown hypotenuse, use the cosine function.

$\cos 23.5° = \frac{10}{c}$

$c = \frac{10}{\cos 23.5°} \approx 10.90$

In summary, $B = 66.5°$, $a \approx 4.35$, and $c \approx 10.90$.

2. Find the measure of angle B. Because $C = 90°$, $A + B = 90°$.
Thus, $B = 90° - A = 90° - 41.5° = 48.5°$.
Because we have a known angle, a known adjacent side, and an unknown opposite side, use the tangent function.

$\tan 41.5° = \frac{a}{20}$

$a = 20\tan 41.5° \approx 17.69$

Because we have a known angle, a known adjacent side, and an unknown hypotenuse, use the cosine function.

$\cos 41.5° = \frac{20}{c}$

$c = \frac{20}{\cos 41.5°} \approx 26.70$

In summary, $B = 48.5°$, $a \approx 17.69$, and $c \approx 26.70$.

3. Find the measure of angle B. Because $C = 90°$, $A + B = 90°$.
Thus, $B = 90° - A = 90° - 52.6° = 37.4°$.
Because we have a known angle, a known hypotenuse, and an unknown opposite side, use the sine function.

$\sin 52.6 = \frac{a}{54}$

$a = 54\sin 52.6° \approx 42.90$

Because we have a known angle, a known hypotenuse, and an unknown adjacent side, use the cosine function.

$\cos 52.6° = \frac{b}{54}$

$b = 54\cos 52.6° \approx 32.80$

In summary, $B = 37.4°$, $a \approx 42.90$, and $b \approx 32.80$.

4. Find the measure of angle B. Because $C = 90°$, $A + B = 90°$.
Thus, $B = 90° - A = 90° - 54.8° = 35.2°$.
Because we have a known angle, a known hypotenuse, and an unknown opposite side, use the sine function.

$\sin 54.8° = \frac{a}{80}$

$a = 80\sin 54.8° \approx 65.37$

Because we have a known angle, a known hypotenuse, and an unknown adjacent side, use the cosine function.

$\cos 54.8 = \frac{b}{80}$

$b = 80\cos 54.8° \approx 46.11$

In summary, $B = 35.2°$, $a \approx 65.37$, and $c \approx 46.11$.

5. Find the measure of angle A. Because $C = 90°$, $A + B = 90°$.
Thus, $A = 90° - B = 90° - 16.8° = 73.2°$.
Because we have a known angle, a known opposite side and an unknown adjacent side, use the tangent function.

$$\tan 16.8° = \frac{30.5}{a}$$

$$a = \frac{30.5}{\tan 16.8°} \approx 101.02$$

Because we have a known angle, a known opposite side, and an unknown hypotenuse, use the sine function.

$$\sin 16.8° = \frac{30.5}{c}$$

$$c = \frac{30.5}{\sin 16.8°} \approx 105.52$$

In summary, $A = 73.2°$, $a \approx 101.02$, and $c \approx 105.52$.

6. Find the measure of angle A. Because $C = 90°$, $A + B = 90°$.
Thus, $A = 90° - B = 90° - 23.8° = 66.2°$.
Because we have a known angle, a known opposite side, and an unknown adjacent side, use the tangent function.

$$\tan 23.8° = \frac{40.5}{a}$$

$$a = \frac{40.5}{\tan 23.8°} \approx 91.83$$

Because we have a known angle, a known opposite side, and an unknown hypotenuse, use the sine function.

$$\sin 23.8° = \frac{40.5}{c}$$

$$c = \frac{40.5}{\sin 23.8°} \approx 100.36$$

In summary, $A = 66.2°$, $a \approx 91.83$, and $c \approx 100.36$.

7. Find the measure of angle A. Because we have a known hypotenuse, a known opposite side, and an unknown angle, use the sine function.

$$\sin A = \frac{30.4}{50.2}$$

$$A = \sin^{-1}\left(\frac{30.4}{50.2}\right) \approx 37.3°$$

Find the measure of angle B. Because $C = 90°$, $A + B = 90°$. Thus,
$B = 90° - A \approx 90° - 37.3° = 52.7°$.
Because we have a known hypotenuse, a known opposite side, and an unknown adjacent side, use the Pythagorean Theorem.

$$a^2 + b^2 = c^2$$

$$(30.4)^2 + b^2 = (50.2)^2$$

$$b^2 = (50.2)^2 - (30.4)^2 = 1595.88$$

$$b = \sqrt{1595.88} \approx 39.95$$

In summary, $A \approx 37.3°$, $B \approx 52.7°$, and $b \approx 39.95$.

8. Find the measure of angle A. Because we have a known hypotenuse, a known opposite side, and an unknown angle, use the sine function.

$$\sin A = \frac{11.2}{65.8}$$

$$A = \sin^{-1}\left(\frac{11.2}{65.8}\right) \approx 9.8°$$

Find the measure of angle B. Because $C = 90°$, $A + B = 90°$.
Thus, $B = 90° - A \approx 90° - 9.8° = 80.2°$.
Because we have a known hypotenuse, a known opposite side, and an unknown adjacent side, use the Pythagorean Theorem.

$$a^2 + b^2 = c^2$$

$$(11.2)^2 + b^2 = (65.8)^2$$

$$b^2 = (65.8)^2 - (11.2)^2 = 4204.2$$

$$b = \sqrt{4204.2} \approx 64.84$$

9. Find the measure of angle A. Because we have a known opposite side, a known adjacent side, and an unknown angle, use the tangent function.

$$\tan A = \frac{10.8}{24.7}$$

$$A = \tan^{-1}\left(\frac{10.8}{24.7}\right) \approx 23.6°$$

Find the measure of angle B. Because $C = 90°$, $A + B = 90°$.
Thus, $B = 90° - A \approx 90° - 23.6° = 66.4°$.
Because we have a known opposite side, a known adjacent side, and an unknown hypotenuse, use the Pythagorean Theorem.

$$c^2 = a^2 + b^2 = (10.8)^2 + (24.7)^2 = 726.73$$

$$c = \sqrt{726.73} \approx 26.96$$

In summary, $A \approx 23.6°$, $B \approx 66.4°$, and $c \approx 26.96$.

10. Find the measure of angle A. Because we have a known opposite side, a known adjacent side, and an unknown angle, use the tangent function.

$$\tan A = \frac{15.3}{17.6}$$

$$A = \tan^{-1}\left(\frac{15.3}{17.6}\right) \approx 41.0°$$

Find the measure of angle B. Because $C = 90°$, $A + B = 90°$.
Thus, $B = 90° - A \approx 90° - 41.0° = 49.0°$.
Because we have a known opposite side, a known adjacent side, and an unknown hypotenuse, use the Pythagorean Theorem.

$$c^2 = a^2 + b^2 = (15.3)^2 + (17.6)^2 = 543.85$$

$$c = \sqrt{543.85} \approx 23.32$$

In summary, $A \approx 41.0°$, $B \approx 49.0°$, and $c \approx 23.32$.

11. Find the measure of angle A. Because we have a known hypotenuse, a known adjacent side, and unknown angle, use the cosine function.

$$\cos A = \frac{2}{7}$$

$$A = \cos^{-1}\left(\frac{2}{7}\right) \approx 73.4°$$

Find the measure of angle B. Because $C = 90°$, $A + B = 90°$.
Thus, $B = 90° - A \approx 90° - 73.4° = 16.6°$.
Because we have a known hypotenuse, a known adjacent side, and an unknown opposite side, use the Pythagorean Theorem.

$$a^2 + b^2 = c^2$$

$$a^2 + (2)^2 = (7)^2$$

$$a^2 = (7)^2 - (2)^2 = 45$$

$$a = \sqrt{45} \approx 6.71$$

In summary, $A \approx 73.4°$, $B \approx 16.6°$, and $a \approx 6.71$.

12. Find the measure of angle A. Because we have a known hypotenuse, a known adjacent side, and an unknown angle, use the cosine function.

$$\cos A = \frac{4}{9}$$

$$A = \cos^{-1}\left(\frac{4}{9}\right) \approx 63.6°$$

Find the measure of angle B. Because $C = 90°$, $A + B = 90°$.
Thus, $B = 90° - A \approx 90° - 63.6° = 26.4°$.
Because we have a known hypotenuse, a known adjacent side and an unknown opposite side, use the Pythagorean Theorem.

$$a^2 + b^2 = c^2$$

$$a^2 + (4)^2 = (9)^2$$

$$a^2 = (9)^2 - (4)^2 = 65$$

$$a = \sqrt{65} \approx 8.06$$

In summary, $A \approx 63.6°$, $B \approx 26.4°$, and $a \approx 8.06$.

13. We need the acute angle between ray OA and the north-south line through O. This angle measure $90° - 75° = 15°$. This angle is measured from the north side of the north-south line and lies east of the north-south line. Thus, the bearing from O and A is N 15° E.

14. We need the acute angle between ray OB and the north-south line through O. This angle measures $90° - 60° = 30°$. This angle is measured from the north side of the north-south line and lies west of the north-south line. Thus, the bearing from O to B is N 30° W.

15. The measurement of this angle is given to be 80°. The angle is measured from the south side of the north-south line and lies west of the north-south line. Thus, the bearing from O to C is S 80° W.

16. We need the acute angle between ray OD and the north-south line through O. This angle measures $90° - 35° = 55°$. This angle is measured from the south side of the north-south line and lies east of the north-south line. Thus, the bearing from O to D is S 55° E.

17. When the object is released ($t = 0$), the object's distance, d, from its rest position is 6 centimeters down. Because it is down, d is negative: When $t = 0$, $d = -6$. Notice the greatest distance from rest position occurs at $t = 0$. Thus, we will use the equation with the cosine function, $y = a\cos\omega t$ to model the object's motion.

Recall that $|a|$ is the maximum distance.

Because the object's initial position is down, $a = -6$. The value of ω can be found using the formula for the period.

$$\text{period} = \frac{2\pi}{\omega} = 4$$
$$2\pi = 4\omega$$
$$\omega = \frac{2\pi}{4} = \frac{\pi}{2}$$

Substitute these values into $d = a\cos\omega t$. The equation for the object's simple harmonic motion is $d = -6\cos\frac{\pi}{2}t$.

18. When the object is released ($t = 0$), the object's distance, d, from its rest position is 8 inches down. Because it is down, d, is negative: When $t = 0$, $d = -8$. Notice the greatest distance from rest position occurs at $t = 0$. Thus, we will use the equation with the cosine function, $y = a\cos\omega t$, to model the object's motion.

Recall that $|a|$ is the maximum distance.

Because the object's initial position is down, $a = -8$. The value of ω can be found using the formula for the period.

$$\text{period} = \frac{2\pi}{\omega} = 2$$
$$2\pi = 2\omega$$
$$\omega = \frac{2\pi}{2} = \pi$$

Substitute these values into $d = a\cos\omega t$. The equation for the object's simple harmonic motion is $d = -8\cos\pi t$.

19. When the object is released ($t = 0$), the object's distance, d, from its rest position is 0 inches: When $t = 0$, $d = 0$. Therefore, we will use the equation with the sine function, $y = a\sin\omega t$, to model the object's motion.

Recall that $|a|$ is the maximum distance.

Because the object's initial position is down, and has an amplitude of 3 inches, $a = -3$. The value of ω can be found using the formula for the period.

$$\text{period} = \frac{2\pi}{\omega} = 1.5$$
$$2\pi = 1.5\omega$$
$$\omega = \frac{2\pi}{1.5} = \frac{4\pi}{3}$$

Substitute these values into $d = a\sin\omega t$. The equation for the object's simple harmonic motion is $d = -3\sin\frac{4\pi}{3}t$.

20. When the object is released ($t = 0$), the object's distance, d, from its rest position is 0 centimeters: When $t = 0$, $d = 0$. Therefore, we will use the equation with the sine function, $y = a\sin\omega t$, to model the object's motion.

Recall that $|a|$ is the maximum distance.

Because the object's initial position is down,, and has an amplitude of 5 centimeters, $a = -5$. The value of ω can be found using the formula for the period.

$$\text{period} = \frac{2\pi}{\omega} = 2.5$$
$$2\pi = 2.5\omega$$
$$\omega = \frac{2\pi}{2.5} = \frac{4\pi}{5}$$

Substitute these values into $d = a\sin\omega t$. The equation for the object's simple harmonic motion is $d = -5\sin\frac{4\pi}{5}t$.

21. We begin by identifying values for a and ω.

$d = 5\cos\frac{\pi}{2}t$, $a = 5$ and $\omega = \frac{\pi}{2}$

a. The maximum displacement from the rest position is the amplitude. Because $a = 5$, the maximum displacement is 5 inches.

b. The frequency, *f*, is

$$f = \frac{\omega}{2\pi} = \frac{\frac{\pi}{2}}{2\pi} = \frac{\pi}{2} \cdot \frac{1}{2\pi} = \frac{1}{4}.$$

The frequency is $\frac{1}{4}$ inch per second.

c. The time required for one cycle is the period.

$$\text{period} = \frac{2\pi}{\omega} = \frac{2\pi}{\frac{\pi}{2}} = 2\pi \cdot \frac{2}{\pi} = 4$$

The time required for one cycle is 4 seconds.

22. We begin by identifying values for a and ω.
$d = 10\cos 2\pi t$, $a = 10$ and $\omega = 2\pi$

a. The maximum displacement from the rest position is the amplitude.
Because $a = 10$, the maximum displacement is 10 inches.

b. The frequency, *f*, is $f = \frac{\omega}{2\pi} = \frac{2\pi}{2\pi} = 1$.

The frequency is 1 inch per second.

c. The time required for one cycle is the period.

$$\text{period} = \frac{2\pi}{\omega} = \frac{2\pi}{2\pi} = 1$$

The time required for one cycle is 1 second.

23. We begin by identifying values for a and ω.
$d = -6\cos 2\pi t$, $a = -6$ and $\omega = 2\pi$

a. The maximum displacement from the rest position is the amplitude.
Because $a = -6$, the maximum displacement is 6 inches.

b. The frequency, *f*, is

$$f = \frac{\omega}{2\pi} = \frac{2\pi}{2\pi} = 1.$$

The frequency is 1 inch per second.

c. The time required for one cycle is the period.

$$\text{period} = \frac{2\pi}{\omega} = \frac{2\pi}{2\pi} = 1$$

The time required for one cycle is 1 second.

24. We begin by identifying values for a and ω.

$d = -8\cos\frac{\pi}{2}t$, $a = -8$ and $\omega = \frac{\pi}{2}$

a. The maximum displacement from the rest position is the amplitude.
Because $a = -8$, the maximum displacement is 8 inches.

b. The frequency, *f*, is $f = \frac{\omega}{2\pi} = \frac{\frac{\pi}{2}}{2\pi} = \frac{1}{4}$.

The frequency is $\frac{1}{4}$ inch per second.

c. The time required for one cycle is the period.

$$\text{period} = \frac{2\pi}{\omega} = \frac{2\pi}{\frac{\pi}{2}} = 2\pi \cdot \frac{2}{\pi} = 4$$

The time required for one cycle is 4 seconds.

25. We begin by identifying values for a and ω.

$d = \frac{1}{2}\sin 2t$, $a = \frac{1}{2}$ and $\omega = 2$

a. The maximum displacement from the rest position is the amplitude.

Because $a = \frac{1}{2}$, the maximum displacement is $\frac{1}{2}$ inch.

b. The frequency, *f*, is

$$f = \frac{\omega}{2\pi} = \frac{2}{2\pi} = \frac{1}{\pi} \approx 0.32.$$

The frequency is approximately 0.32 cycle per second.

c. The time required for one cycle is the period.

$$\text{period} = \frac{2\pi}{\omega} = \frac{2\pi}{2} = \pi \approx 3.14$$

The time required for one cycle is approximately 3.14 seconds.

26. We begin by identifying values for a and ω.

$d = \frac{1}{3}\sin 2t,\ a = \frac{1}{3}$ and $\omega = 2$

a. The maximum displacement from the rest position is the amplitude.

Because $a = \frac{1}{3}$, the maximum

displacement is $\frac{1}{3}$ inch.

b. The frequency, f, is

$f = \frac{\omega}{2\pi} = \frac{2}{2\pi} = \frac{1}{\pi} \approx 0.32$.

The frequency is approximately 0.32 cycle per second.

c. The time required for one cycle is the period.

$\text{period} = \frac{2\pi}{\omega} = \frac{2\pi}{2} = \pi \approx 3.14$

The time required for one cycle is approximately 3.14 seconds.

27. We begin by identifying values for a and ω.

$d = -5\sin\frac{2\pi}{3}t,\ a = -5$ and $\omega = \frac{2\pi}{3}$

a. The maximum displacement from the rest position is the amplitude. Because $a = -5$, the maximum displacement is 5 inches.

b. The frequency, f, is

$f = \frac{\omega}{2\pi} = \frac{\frac{2\pi}{3}}{2\pi} = \frac{2\pi}{3}\cdot\frac{1}{2\pi} = \frac{1}{3}$.

The frequency is $\frac{1}{3}$ cycle per second.

c. The time required for one cycle is the period.

$\text{period} = \frac{2\pi}{\omega} = \frac{2\pi}{\frac{2\pi}{3}} = 2\pi\cdot\frac{3}{2\pi} = 3$

The time required for one cycle is 3 seconds.

28. We begin by identifying values for a and ω.

$d = -4\sin\frac{3\pi}{2}t,\ a = -4$ and $\omega = \frac{3\pi}{2}$

a. The maximum displacement from the rest position is the amplitude.
Because $a = -4$, the maximum displacement is 4 inches.

b. The frequency, f, is

$f = \frac{\omega}{2\pi} = \frac{\frac{3\pi}{2}}{2\pi} = \frac{3\pi}{2}\cdot\frac{1}{2\pi} = \frac{3}{4}$.

The frequency is $\frac{3}{4}$ cycle per second.

c. The time required for one cycle is the period.

$\text{period} = \frac{2\pi}{\omega} = \frac{2\pi}{\frac{3\pi}{2}} = 2\pi\cdot\frac{2}{3\pi} = \frac{4}{3}$

The required time for one cycle is $\frac{4}{3}$ seconds.

29. $x = 500\tan 40° + 500\tan 25°$

$x \approx 653$

30. $x = 100\tan 20° + 100\tan 8°$

$x \approx 50$

31. $x = 600\tan 28° - 600\tan 25°$

$x \approx 39$

32. $x = 400\tan 40° - 400\tan 28°$

$x \approx 123$

33. $x = \frac{300}{\tan 34°} - \frac{300}{\tan 64°}$

$x \approx 298$

34. $x = \frac{500}{\tan 20°} - \frac{500}{\tan 48°}$

$x \approx 924$

35. $x = \frac{400\tan 40°\tan 20°}{\tan 40° - \tan 20°}$

$x \approx 257$

36. $x = \frac{100\tan 43°\tan 38°}{\tan 43° - \tan 38°}$

$x \approx 482$

37. $d = 4\cos\left(\pi t - \frac{\pi}{2}\right)$

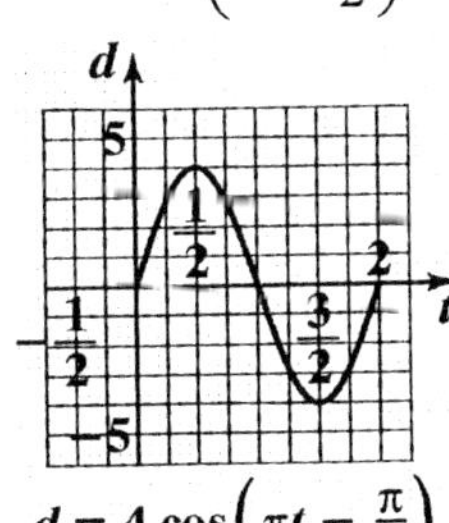

$d = 4\cos\left(\pi t - \frac{\pi}{2}\right)$

a. 4 in.

b. $\frac{1}{2}$ in. per sec

c. 2 sec

d. $\frac{1}{2}$

38. $d = 3\cos\left(\pi t + \frac{\pi}{2}\right)$

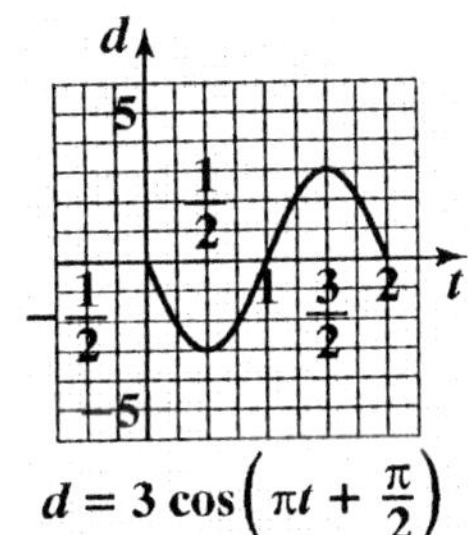

$d = 3\cos\left(\pi t + \frac{\pi}{2}\right)$

a. 3 in.

b. $\frac{1}{2}$ in. per sec

c. 2 sec

d. $-\frac{1}{2}$

39. $d = -2\sin\left(\frac{\pi t}{4} + \frac{\pi}{2}\right)$

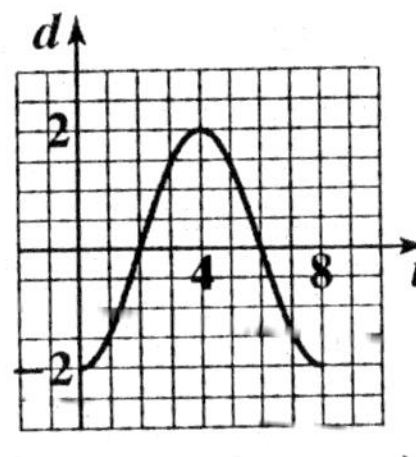

$d = -2\sin\left(\frac{\pi}{4}t + \frac{\pi}{2}\right)$

a. 2 in.

b. $\frac{1}{8}$ in. per sec

c. 8 sec

d. -2

40. $d = -\frac{1}{2}\sin\left(\frac{\pi t}{4} - \frac{\pi}{2}\right)$

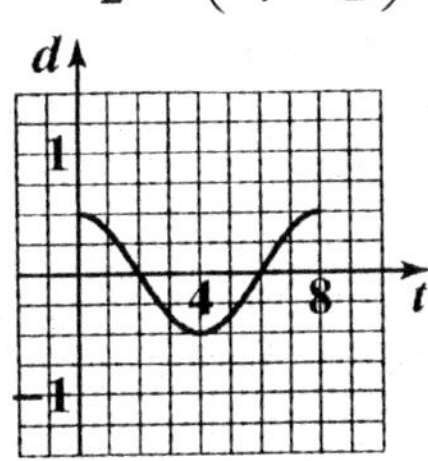

$d = -\frac{1}{2}\sin\left(\frac{\pi t}{4} - \frac{\pi}{2}\right)$

a. $\frac{1}{2}$ in.

b. $\frac{1}{8}$ in. per sec

c. 8 sec

d. 2

41. Using a right triangle, we have a known angle, an unknown opposite side, a, and a known adjacent side. Therefore, use tangent function.

$$\tan 21.3° = \frac{a}{5280}$$
$$a = 5280\tan 21.3° \approx 2059$$

The height of the tower is approximately 2059 feet.

42. $30 \text{ yd} \cdot \frac{3 \text{ ft}}{1 \text{ yd}} = 90 \text{ ft}$

Using a right triangle, we have a known angle, an unknown opposite side, *a*, and a known adjacent side. Therefore, use the tangent function.

$$\tan 38.7° = \frac{a}{90}$$
$$a = 90 \tan 38.7° \approx 72$$

The height of the building is approximately 72 feet.

43. Using a right triangle, we have a known angle, a known opposite side, and an unknown adjacent side, *a*. Therefore, use the tangent function.

$$\tan 23.7° = \frac{305}{a}$$
$$a = \frac{305}{\tan 23.7°} \approx 695$$

The ship is approximately 695 feet from the statue's base.

44. Using a right triangle, we have a known angle, a known opposite side, and an unknown adjacent side, *a*. Therefore, use the tangent function.

$$\tan 22.3° = \frac{200}{a}$$
$$a = \frac{200}{\tan 22.3°} \approx 488$$

The ship is about 488 feet offshore.

45. The angle of depression from the helicopter to point *P* is equal to the angle of elevation from point *P* to the helicopter. Using a right triangle, we have a known angle, a known opposite side, and an unknown adjacent side, *d*. Therefore, use the tangent function.

$$\tan 36° = \frac{1000}{d}$$
$$d = \frac{1000}{\tan 36°} \approx 1376$$

The island is approximately 1376 feet off the coast.

46. The angle of depression from the helicopter to the stolen car is equal to the angle of elevation from the stolen car to the helicopter. Using a right triangle, we have a known angle, a known opposite side, and an unknown adjacent side, *d*. Therefore, use the tangent function.

$$\tan 72° = \frac{800}{d}$$
$$d = \frac{800}{\tan 72°} \approx 260$$

The stolen car is approximately 260 feet from a point directly below the helicopter.

47. Using a right triangle, we have an unknown angle, *A*, a known opposite side, and a known hypotenuse. Therefore, use the sine function.

$$\sin A = \frac{6}{23}$$
$$A = \sin^{-1}\left(\frac{6}{23}\right) \approx 15.1°$$

The ramp makes an angle of approximately 15.1° with the ground.

48. Using a right triangle, we have an unknown angle, *A*, a known opposite side, and a known adjacent side. Therefore, use the tangent function.

$$\tan A = \frac{250}{40}$$
$$A = \tan^{-1}\left(\frac{250}{40}\right) \approx 80.9°$$

The angle of elevation of the sun is approximately 80.9°.

49. Using the two right triangles, we have a known angle, an unknown opposite side, *a* in the smaller triangle, *b* in the larger triangle, and a known adjacent side in each triangle. Therefore, use the tangent function.

$$\tan 19.2° = \frac{a}{125}$$
$$a = 125 \tan 19.2° \approx 43.5$$
$$\tan 31.7° = \frac{b}{125}$$
$$b = 125 \tan 31.7° \approx 77.2$$

The balloon rises approximately 77.2 – 43.5 or 33.7 feet.

50. Using two right triangles, a smaller right triangle corresponding to the smaller angle of elevation drawn inside a larger right triangle corresponding to the larger angle of elevation, we have a known angle, an unknown opposite side, a in the smaller triangle, b in the larger triangle, and a known adjacent side in each triangle. Therefore, use the tangent function.

$$\tan 53° = \frac{a}{330}$$
$$a = 330 \tan 53° = 437.9$$
$$\tan 63° = \frac{b}{330}$$
$$b = 330 \tan 63° \approx 647.7$$

The height of the flagpole is approximately 647.7 – 437.9, or 209.8 feet.

51. Using a right triangle, we have a known angle, a known hypotenuse, and unknown sides. To find the opposite side, a, use the sine function.

$$\sin 53° = \frac{a}{150}$$
$$a = 150 \sin 53° \approx 120$$

To find the adjacent side, b, use the cosine function.

$$\cos 53° = \frac{b}{150}$$
$$b = 150 \cos 53° \approx 90$$

The boat has traveled approximately 90 miles north and 120 miles east.

52. Using a right triangle, we have a known angle, a known hypotenuse, and unknown sides. To find the opposite side, a, use the sine function.

$$\sin 64° = \frac{a}{40}$$
$$a = 40 \sin 64° \approx 36$$

To find the adjacent side, b, use the cosine function.

$$\cos 64° = \frac{b}{40}$$
$$b = 40 \cos 64° \approx 17.5$$

The boat has traveled about 17.5 mi south and 36 mi east.

53. The bearing from the fire to the second ranger is N 28° E. Using a right triangle, we have a known angle, a known opposite side, and an unknown adjacent side, b. Therefore, use the tangent function.

$$\tan 28° = \frac{7}{b}$$
$$b = \frac{7}{\tan 28°} \approx 13.2$$

The first ranger is 13.2 miles from the fire, to the nearest tenth of a mile.

54. The bearing from the lighthouse to the second ship is N 34° E. Using a right triangle, we have a known angle, a known opposite side, and an unknown adjacent side, b. Therefore, use the tangent function.

$$\tan 34° = \frac{9}{b}$$
$$b = \frac{9}{\tan 34°} \approx 13.3$$

The first ship is about 13.3 miles from the lighthouse, to the nearest tenth of a mile.

55. Using a right triangle, we have a known adjacent side, a known opposite side, and an unknown angle, A. Therefore, use the tangent function.

$$\tan A = \frac{1.5}{2}$$
$$A = \tan\left(\frac{1.5}{2}\right) \approx 37°$$

We need the acute angle between the ray that runs from your house through your location, and the north-south line through your house. This angle measures approximately $90° - 37° = 53°$. This angle is measured from the north side of the north-south line and lies west of the north-south line. Thus, the bearing from your house to you is N 53° W.

56. Using a right triangle, we have a known adjacent side, a known opposite side, and an unknown angle, A. Therefore, use the tangent function.

$$\tan A = \frac{6}{9}$$
$$A = \tan^{-1}\left(\frac{6}{9}\right) \approx 34°$$

We need the acute angle between the ray that runs from the ship through the harbor, and the north-south line through the ship. This angle measures $90° - 34° = 56°$. This angle is measured from the north side of the north-south line and lies west of the north-south line. Thus, the bearing from the ship to the harbor is N 56° W. The ship should use a bearing of N 56° W to sail directly to the harbor.

57. To find the jet's bearing from the control tower, consider a north-south line passing through the tower. The acute angle from this line to the ray on which the jet lies is $35° + \theta$. Because we are measuring the angle from the north side of the line and the jet is east of the tower, the jet's bearing from the tower is N $(35° + \theta)$ E. To find θ, use a right triangle and the tangent function.

$\tan\theta = \frac{7}{5}$

$\theta = \tan^{-1}\left(\frac{7}{5}\right) \approx 54.5°$

Thus, $35° + \theta = 35° + 54.5° = 89.5°$.
The jet's bearing from the control tower is N 89.5° E.

58. To find the ship's bearing from the port, consider a north-south line passing through the port. The acute angle from this line to the ray on which the ship lies is $40° + \theta$. Because we are measuring the angle from the south side of the line and the ship is west of the port, the ship's bearing from the port is S $(40° + \theta)$ W. To find θ, use a right triangle and the tangent function.

$\tan\theta = \frac{11}{7}$

$\theta = \tan^{-1}\left(\frac{11}{7}\right) \approx 57.5°$

Thus, $40° + \theta = 40° + 57.5 = 97.5°$. Because this angle is over 90° we subtract this angle from 180° to find the bearing from the north side of the north-south line. The bearing of the ship from the port is N 82.5° W.

59. The frequency, *f*, is $f = \frac{\omega}{2\pi}$, so

$\frac{1}{2} = \frac{\omega}{2\pi}$

$\omega = \frac{1}{2} \cdot 2\pi = \pi$

Because the amplitude is 6 feet, $a = 6$. Thus, the equation for the object's simple harmonic motion is $d = 6\sin \pi t$.

60. The frequency, *f*, is $f = \frac{\omega}{2\pi}$, so

$\frac{1}{4} = \frac{\omega}{2\pi}$

$\omega = \frac{1}{4} \cdot 2\pi = \frac{\pi}{2}$

Because the amplitude is 8 feet, $a = 8$. Thus, the equation for the object's simple harmonic motion is $d = 8\sin\frac{\pi}{2}t$.

61. The frequency, *f*, is $f = \frac{\omega}{2\pi}$, so

$264 = \frac{\omega}{2\pi}$

$\omega = 264 \cdot 2\pi = 528\pi$

Thus, the equation for the tuning fork's simple harmonic motion is $d = \sin 528\pi t$.

62. The frequency, *f*, is $f = \frac{\omega}{2\pi}$, so

$98,100,000 = \frac{\omega}{2\pi}$

$\omega = 98,100,000 \cdot 2\pi = 196,200,000\pi$

Thus, the equation for the radio waves' simple harmonic motion is $d = \sin 196,200,000\pi t$.

70. $y = 4e^{-0.1x}\cos 2x$

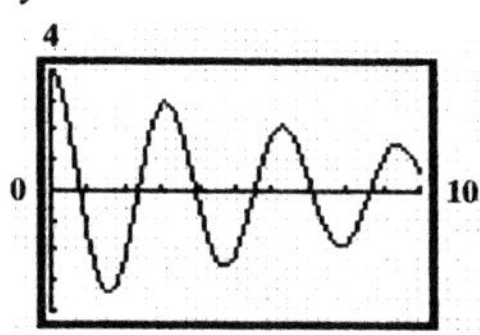

3 complete oscillations occur.

71. $y = -6e^{-0.09x}\cos 2\pi x$

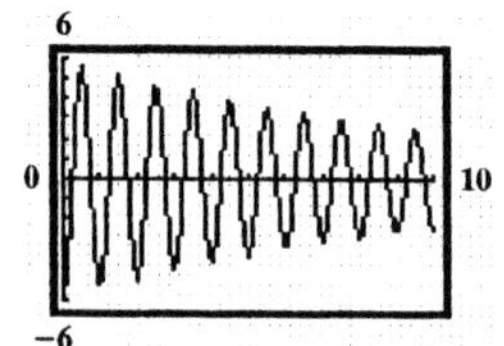

10 complete oscillations occur.

72. Using the right triangle, we have a known angle, an unknown opposite side, r, and an unknown hypotenuse, $r + 112$. Because both sides are in terms of the variable r, we can find r by using the sine function.

$$\sin 76.6° = \frac{r}{r+112}$$
$$\sin 76.6°(r+112) = r$$
$$r\sin 76.6° + 112\sin 76.6° = r$$
$$r - r\sin 76.6° = 112\sin 76.6°$$
$$r(1-\sin 76.6°) = 112\sin 76.6°$$
$$r = \frac{112\sin 76.6°}{1-\sin 76.6°} \approx 4002$$

The Earth's radius is approximately 4002 miles.

73. Let d be the adjacent side to the 40° angle. Using the right triangles, we have a known angle and unknown sides in both triangles. Use the tangent function.

$$\tan 20° = \frac{h}{75+d}$$
$$h = (75+d)\tan 20°$$

Also, $\tan 40° = \frac{h}{d}$

$$h = d\tan 40°$$

Using the transitive property we have

$$(75+d)\tan 20° = d\tan 40°$$
$$75\tan 20° + d\tan 20° = d\tan 40°$$
$$d\tan 40° - d\tan 20° = 75\tan 20°$$
$$d(\tan 40° - \tan 20°) = 75\tan 20°$$
$$d = \frac{75\tan 20°}{\tan 40° - \tan 20°}$$

Thus, $h = d\tan 40°$

$$= \frac{75\tan 20°}{\tan 40° - \tan 20°}\tan 40° \approx 48$$

The height of the building is approximately 48 feet.

Chapter 4 Review Exercises

1. The radian measure of a central angle is the length of the intercepted arc divided by the circle's radius.

$$\theta = \frac{27}{6} = 4.5 \text{ radians}$$

2. $15° = 15° \cdot \frac{\pi \text{ radians}}{180°} = \frac{15\pi}{180}$ radian

$= \frac{\pi}{12}$ radian

3. $120° = 120° \cdot \frac{\pi \text{ radians}}{180°} = \frac{120\pi}{180}$ radians

$= \frac{2\pi}{3}$ radians

4. $315° = 315° \cdot \frac{\pi \text{ radians}}{180°} = \frac{315\pi}{180}$ radians

$= \frac{7\pi}{4}$ radians

5. $\frac{5\pi}{3}$ radians $= \frac{5\pi}{3}$ radians $\cdot \frac{180°}{\pi \text{ radians}}$

$= \frac{5 \cdot 180°}{3} = 300°$

6. $\frac{7\pi}{5}$ radians $= \frac{7\pi}{5}$ radians $\cdot \frac{180°}{\pi \text{ radians}}$

$= \frac{7 \cdot 180°}{5} = 252°$

7. $-\frac{5\pi}{6}$ radians $= -\frac{5\pi}{6}$ radians $\cdot \frac{180°}{\pi \text{ radians}}$

$= -\frac{5 \cdot 180°}{6} = -150°$

8.

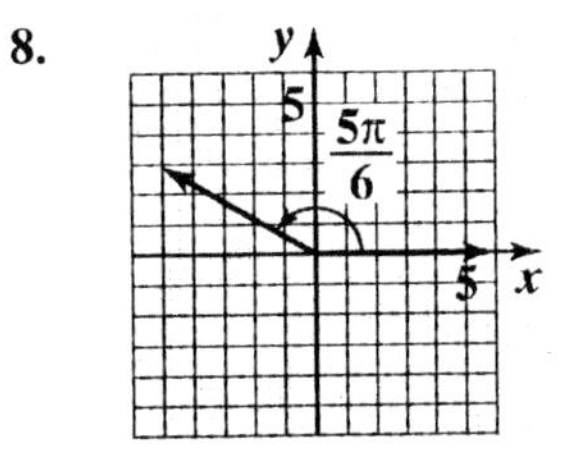

9.

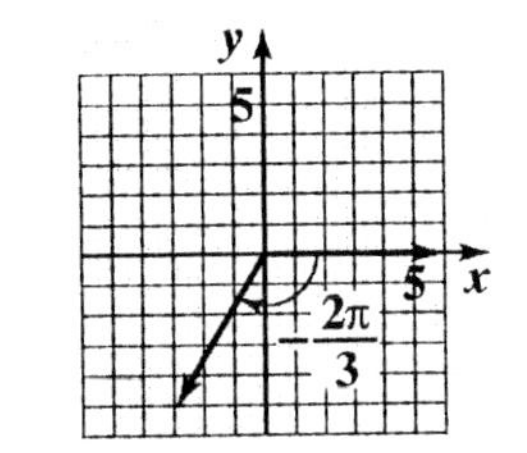

10.

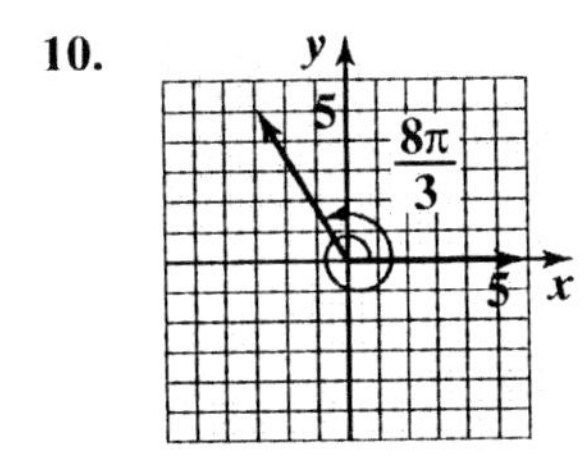

11.

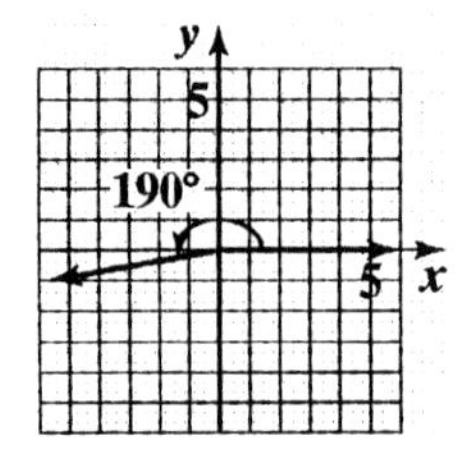

12.

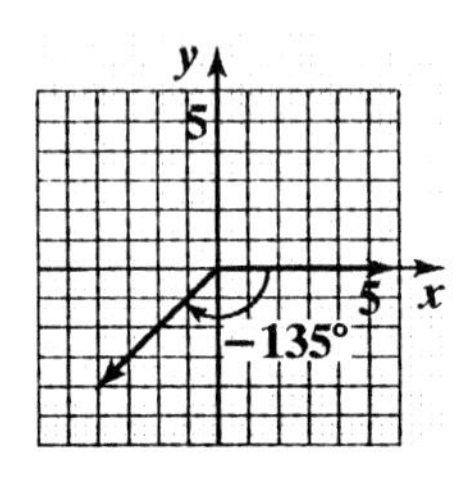

13. $400° - 360° = 40°$

14. $-445° + (2)360° = 275°$

15. $\frac{13\pi}{4} - 2\pi = \frac{13\pi}{4} - \frac{8\pi}{4} = \frac{5\pi}{4}$

16. $\frac{31\pi}{6} - (2)2\pi = \frac{31\pi}{6} - \frac{24\pi}{6} = \frac{7\pi}{6}$

17. $-\frac{8\pi}{3} + (2)2\pi = -\frac{8\pi}{3} + \frac{12\pi}{3} = \frac{4\pi}{3}$

18. $135° = 135° \cdot \frac{\pi \text{ radians}}{180°} = \frac{135 \cdot \pi}{180} \text{ radians}$

$= \frac{3\pi}{4} \text{ radians}$

$s = r\theta$

$s = (10 \text{ ft})\left(\frac{3\pi}{4}\right) = \frac{15\pi}{2} \text{ ft} \approx 23.56 \text{ ft}$

19. $\frac{10.3 \text{ revolutions}}{1 \text{ minute}} \cdot \frac{2\pi \text{ radians}}{1 \text{ revolution}}$

$= \frac{20.6\pi \text{ radians}}{1 \text{ minute}} = 20.6\pi \text{ radians per minute}$

20. Use $v = r\omega$ where v is the linear speed and ω is the angular speed in radians per minute.

$\omega = \frac{2250 \text{ revolutions}}{1 \text{ minute}} \cdot \frac{2\pi \text{ radians}}{1 \text{ revolution}}$

$= 4500\pi \text{ radians per minute}$

$v = 3 \text{ feet} \frac{4500\pi}{\text{minute}} = \frac{13,500\pi \text{ feet}}{\text{min}}$

$\approx 42,412 \text{ ft per min}$

21. $P\left(-\frac{4}{5}, -\frac{3}{5}\right)$

$\sin t = y = -\frac{3}{5}$

$\cos t = x = -\frac{4}{5}$

$\tan t = \frac{y}{x} = \frac{-\frac{3}{5}}{-\frac{4}{5}} = \frac{3}{4}$

$\csc t = \frac{1}{y} = -\frac{5}{3}$

$\sec t = \frac{1}{x} = -\frac{5}{4}$

$\cot t = \frac{x}{y} = \frac{4}{3}$

22. $P\left(\frac{8}{17}, -\frac{15}{17}\right)$

$\sin t = y = -\frac{15}{17}$

$\cos t = x = \frac{8}{17}$

$\tan t = \frac{y}{x} = \frac{-\frac{15}{17}}{\frac{8}{17}} = -\frac{15}{8}$

$\csc t = \frac{1}{y} = -\frac{17}{15}$

$\sec t = \frac{1}{x} = \frac{17}{8}$

$\cot t = \frac{x}{y} = -\frac{8}{15}$

23. $\sec\frac{5\pi}{6} = \frac{1}{-\frac{\sqrt{3}}{2}} = -\frac{2\sqrt{3}}{3}$

24. $\tan\frac{4\pi}{3} = \frac{-\frac{\sqrt{3}}{2}}{-\frac{1}{2}} = \sqrt{3}$

25. $\sec\frac{\pi}{2}$ is undefined.

26. $\cot \pi$ is undefined.

27. $\sin t = \frac{2}{\sqrt{7}}, 0 \le t < \frac{\pi}{2}$

$\sin^2 t + \cos^2 t = 1$

$\left(\frac{2}{\sqrt{7}}\right)^2 + \cos^2 t = 1$

$\cos^2 t = 1 - \frac{4}{7}$

$\cos t = \sqrt{\frac{3}{7}} = \frac{\sqrt{21}}{7}$

Because $0 \le t < \frac{\pi}{2}$, $\cos t$ is positive.

$\tan t = \frac{\frac{2}{\sqrt{7}}}{\sqrt{\frac{3}{7}}} = \frac{2\sqrt{3}}{3}$

$\csc t = \frac{\sqrt{7}}{2}$

$\sec t = \frac{\sqrt{21}}{3}$

$\cot t = \frac{\sqrt{3}}{2}$

28. $\tan 4.7 \cot 4.7 = \tan 4.7\left(\frac{1}{\tan 4.7}\right) = 1$

29. $\sin^2 \frac{\pi}{17} + \cos^2 \frac{\pi}{17} = 1$ because $\sin^2 t + \cos^2 t = 1$.

30. $\tan^2 1.4 - \sec^2 1.4 = -\left(\sec^2 1.4 - \tan^2 1.4\right)$
$= -1$

31. Use the Pythagorean Theorem to find the hypotenuse, c.
$c^2 = a^2 + b^2$

$c = \sqrt{8^2 + 5^2} = \sqrt{64 + 25} = \sqrt{89}$

$\sin\theta = \frac{5}{\sqrt{89}} = \frac{5\sqrt{89}}{\sqrt{89}}$

$\cos\theta = \frac{8}{\sqrt{89}} = \frac{8\sqrt{89}}{\sqrt{89}}$

$\tan\theta = \frac{5}{8}$

$\csc\theta = \frac{\sqrt{89}}{5}$

$\sec\theta = \frac{\sqrt{89}}{8}$

$\cot\theta = \frac{3}{5}$

32. $\sin\frac{\pi}{6} + \tan^2\frac{\pi}{3} = \frac{1}{2} - \left(\sqrt{3}\right)^2$
$= \frac{1}{2} - 3$
$= -\frac{5}{2}$

33. $\cos^2\frac{\pi}{4} + \tan^2\frac{\pi}{4} = \left(\frac{\sqrt{2}}{2}\right)^2 - (1)^2$
$= \frac{1}{2} - 1$
$= -\frac{1}{2}$

34. $4\cot\frac{\pi}{4} + \cos\frac{\pi}{3}\csc\frac{\pi}{6} = 4(1) + \left(\frac{1}{2}\right)(2)$
$= 4 + 1$
$= 5$

35. $\cos\left(-\frac{\pi}{6}\right)\sin\left(-\frac{\pi}{4}\right) - \tan\frac{\pi}{4}$
$= \left(\frac{\sqrt{3}}{2}\right)\left(-\frac{\sqrt{2}}{2}\right) - 1$
$= -\frac{\sqrt{6}}{4} - 1$

36. $\sin 70° = \cos(90° - 70°) = \cos 20°$

37. $\cos\frac{\pi}{2} = \sin\left(\frac{\pi}{2} - \frac{\pi}{2}\right) = \sin 0$

38. $\tan 23° = \frac{a}{100}$

$a = 100\tan 23°$

$a \approx 100(0.4245) \approx 42\,\text{mm}$

39. $\sin 61° = \frac{20}{c}$

$c = \frac{20}{\sin 61°}$

$c \approx \frac{20}{0.8746} \approx 23\,\text{cm}$

40. $\sin 48° = \frac{a}{50}$

$a = 50\sin 48°$

$a \approx 50(0.7431) \approx 37\,\text{in.}$

41. $\sin\theta = \frac{y}{r} = \frac{1}{4}$

$x^2 + y^2 = r^2$

$x^2 + 1^2 = 4^2$

$x^2 = 15$

$x = \sqrt{15}$

$\tan\left(\frac{\pi}{2} - \theta\right) = \cot\theta = \frac{x}{y} = \frac{\sqrt{15}}{1} = \sqrt{15}$

42. $\frac{1}{2}\text{ mi.} = \frac{1}{2}\cdot 5280\text{ ft} = 2640\text{ ft}$

$\sin 17° = \frac{a}{2640}$

$a = 2640\cdot\sin 17°$

$a \approx 2640(0.2924) \approx 772$

The hiker gains 772 feet of altitude.

43. $\tan 32° = \frac{d}{50}$

$d = 50\tan 32°$

$d \approx 50(0.6249) \approx 31$

The distance across the lake is about 31 meters.

44. $\tan\theta = \frac{6}{4}$

Use a calculator in degree mode to find θ.

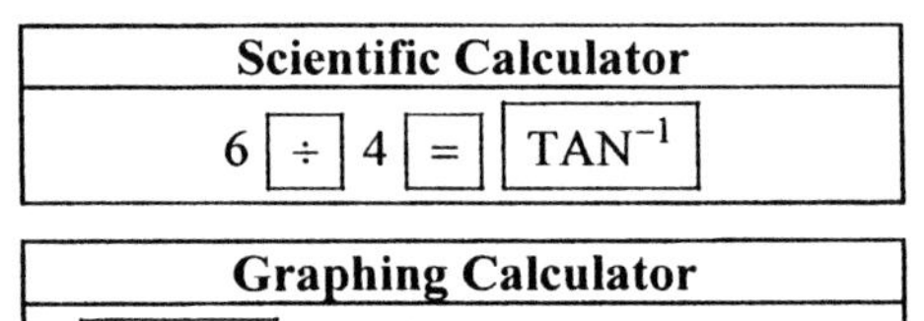

Scientific Calculator
6 [÷] 4 [=] [TAN⁻¹]

Graphing Calculator
[TAN⁻¹] [(] 6 [÷] 4 [)] [ENTER]

The display should show approximately 56. Thus, the angle of elevation of the sun is approximately 56°.

45. We need values for x, y, and r. Because $P = (-1, -5)$ is a point on the terminal side of θ, $x = -1$ and $y = -5$. Furthermore,

$r = \sqrt{(-1)^2 + (-5)^2}$

$= \sqrt{1+25} = \sqrt{26}$

Now that we know x, y, and r, we can find the six trigonometric functions of θ.

$\sin\theta = \frac{y}{r} = \frac{-5}{\sqrt{26}} = \frac{-5\sqrt{26}}{\sqrt{26}\cdot\sqrt{26}} = -\frac{5\sqrt{26}}{26}$

$\cos\theta = \frac{x}{r} = \frac{-1}{\sqrt{26}} = \frac{-1\sqrt{26}}{\sqrt{26}\cdot\sqrt{26}} = -\frac{\sqrt{26}}{26}$

$\tan\theta = \frac{y}{x} = \frac{-5}{-1} = 5$

$\csc\theta = \frac{r}{y} = \frac{\sqrt{26}}{-5} = -\frac{\sqrt{26}}{5}$

$\sec\theta = \frac{r}{x} = \frac{\sqrt{26}}{-1} = -\sqrt{26}$

$\cot\theta = \frac{x}{y} = \frac{-1}{-5} = \frac{1}{5}$

46. We need values for x, y, and r. Because $P = (0, -1)$ is a point on the terminal side of θ, $x = 0$ and $y = -1$. Furthermore,

$$r = \sqrt{x^2 + y^2} = \sqrt{0^2 + (-1)^2} = \sqrt{0+1} = \sqrt{1} = 1$$

Now that we know x, y, and r, we can find the six trigonometric functions of θ.

$$\sin\theta = \frac{y}{r} = \frac{-1}{1} = -1$$

$$\cos\theta = \frac{x}{r} = \frac{0}{1} = 0$$

$$\tan\theta = \frac{y}{x} = \frac{-1}{0}, \text{ undefined}$$

$$\csc\theta = \frac{r}{y} = \frac{1}{-1} = -1$$

$$\sec\theta = \frac{r}{x} = \frac{1}{0}, \text{ undefined}$$

$$\cot\theta = \frac{x}{y} = \frac{0}{-1} = 0$$

47. Because $\tan\theta > 0$, θ cannot lie in quadrant II and quadrant IV; the tangent function is negative in those two quadrants. Thus, with $\tan\theta > 0$, θ lies in quadrant I or quadrant III. We are also given that $\sec\theta > 0$. Because quadrant I is the only quadrant in which the tangent is positive and the secant is positive, we conclude that θ lies in quadrant I.

48. Because $\tan\theta > 0$, θ cannot lie in quadrant II and quadrant IV; the tangent function is negative in those two quadrants. Thus, with $\tan\theta > 0$, θ lies in quadrant I or quadrant III. We are also given that $\cos\theta < 0$. Because quadrant III is the only quadrant in which the tangent is positive and the cosine is negative, we conclude that θ lies in quadrant III.

49. Because the cosine is positive and the sine is negative, θ lies in quadrant IV. In quadrant IV, x is positive and y is negative. Thus,

$\cos\theta = \frac{2}{5} = \frac{x}{r}$, $x = 2$, $r = 5$. Furthermore,

$$x^2 + y^2 = r^2$$

$$2^2 + y^2 = 5^2$$

$$y^2 = 25 - 4 = 21$$

$$y = -\sqrt{21}$$

Now that we know x, y, and r, we can find the six trigonometric functions of θ.

$$\sin\theta = \frac{y}{r} = \frac{-\sqrt{21}}{5} = -\frac{\sqrt{21}}{5}$$

$$\tan\theta = \frac{y}{x} = \frac{-\sqrt{21}}{2} = -\frac{\sqrt{21}}{2}$$

$$\csc\theta = \frac{r}{y} = \frac{5}{-\sqrt{21}} = -\frac{5\cdot\sqrt{21}}{\sqrt{21}\cdot\sqrt{21}} = -\frac{5\sqrt{21}}{21}$$

$$\sec\theta = \frac{r}{x} = \frac{5}{2}$$

$$\cot\theta = \frac{x}{y} = \frac{2}{-\sqrt{21}} = -\frac{2\sqrt{21}}{\sqrt{21}\cdot\sqrt{21}} = -\frac{2\sqrt{21}}{21}$$

50. Because the tangent is negative and the sine is positive, θ lies in quadrant II. In quadrant II x is negative and y is positive. Thus,

$\tan\theta = -\frac{1}{3} = \frac{y}{x} = \frac{1}{-3}$, $x = -3$, $y = 1$.

Furthermore,

$$r = \sqrt{x^2 + y^2} = \sqrt{(-3)^2 + 1^2} = \sqrt{9+1} = \sqrt{10}$$

Now that we know x, y, and r, we can find the six trigonometric functions of θ.

$$\sin\theta = \frac{y}{r} = \frac{1}{\sqrt{10}} = \frac{1\cdot\sqrt{10}}{\sqrt{10}\cdot\sqrt{10}} = \frac{\sqrt{10}}{10}$$

$$\cos\theta = \frac{x}{r} = \frac{-3}{\sqrt{10}} = -\frac{3\sqrt{10}}{\sqrt{10}\cdot\sqrt{10}} = -\frac{3\sqrt{10}}{10}$$

$$\csc\theta = \frac{r}{y} = \frac{\sqrt{10}}{1} = \sqrt{10}$$

$$\sec\theta = \frac{r}{x} = \frac{\sqrt{10}}{-3} = -\frac{\sqrt{10}}{3}$$

$$\cot\theta = \frac{x}{y} = \frac{-3}{1} = -3$$

51. Because the cotangent is positive and the cosine is negative, θ lies in quadrant III. In quadrant III x and y are both negative. Thus,

$\cot\theta = \frac{3}{1} = \frac{x}{y} = \frac{-3}{-1}$, $x = -3, y = -1$.

Furthermore,

$r = \sqrt{x^2 + y^2} = \sqrt{(-3)^2 + (-1)^2} = \sqrt{9+1} = \sqrt{10}$

Now that we know x, y, and r, we can find the six trigonometric functions of θ.

$\sin\theta = \frac{y}{r} = \frac{-1}{\sqrt{10}} = -\frac{\sqrt{10}}{10}$

$\cos\theta = \frac{x}{r} = \frac{-3}{\sqrt{10}} = -\frac{3\sqrt{10}}{10}$

$\tan\theta = \frac{y}{x} = \frac{-1}{-3} = \frac{1}{3}$

$\csc\theta = \frac{r}{y} = \frac{\sqrt{10}}{-1} = -\sqrt{10}$

$\sec\theta = \frac{r}{x} = \frac{\sqrt{10}}{-3} = -\frac{\sqrt{10}}{3}$

52. Because 265° lies between 180° and 270°, it is in quadrant III.
The reference angle is $\theta' = 265° - 180° = 85°$.

53. Because $\frac{5\pi}{8}$ lies between $\frac{\pi}{2} = \frac{4\pi}{8}$ and $\pi = \frac{8\pi}{8}$, it is in quadrant II.
The reference angle is
$\theta' = \pi - \frac{5\pi}{8} = \frac{8\pi}{8} - \frac{5\pi}{8} = \frac{3\pi}{8}$.

54. Find the coterminal angle:
$-410° + (2)360° = 310°$
Find the reference angle: $360° - 310° = 50°$

55. Find the coterminal angle: $\frac{17\pi}{6} - 2\pi = \frac{5\pi}{6}$

Find the reference angle: $2\pi - \frac{5\pi}{6} = \frac{\pi}{6}$

56. Find the coterminal angle: $-\frac{11\pi}{3} + 4\pi = \frac{\pi}{3}$

Find the reference angle: $\frac{\pi}{3}$

57. 240° lies in quadrant III.
The reference angle is
$\theta' = 240° - 180° = 60°$.
$\sin 60° = \frac{\sqrt{3}}{2}$
In quadrant III, $\sin\theta < 0$, so
$\sin 240° = -\sin 60° = -\frac{\sqrt{3}}{2}$.

58. 120° lies in quadrant II.
The reference angle is
$\theta' = 180° - 120° = 60°$.
$\tan 60° = \sqrt{3}$
In quadrant II, $\tan\theta < 0$, so
$\tan 120° = -\tan 60° = -\sqrt{3}$.

59. $\frac{7\pi}{4}$ lies in quadrant IV.
The reference angle is
$\theta' = 2\pi - \frac{7\pi}{4} = \frac{8\pi}{4} - \frac{7\pi}{4} = \frac{\pi}{4}$.
$\sec\frac{\pi}{4} = \sqrt{2}$
In quadrant IV, $\sec\theta > 0$, so
$\sec\frac{7\pi}{4} = \sec\frac{\pi}{4} = \sqrt{2}$.

60. $\frac{11\pi}{6}$ lies in quadrant IV.
The reference angle is
$\theta' = 2\pi - \frac{11\pi}{6} = \frac{12\pi}{6} - \frac{11\pi}{6} = \frac{\pi}{6}$.
$\cos\frac{\pi}{6} = \frac{\sqrt{3}}{2}$
In quadrant IV, $\cos\theta > 0$, so
$\cos\frac{11\pi}{6} = \cos\frac{\pi}{6} = \frac{\sqrt{3}}{2}$.

61. –210° lies in quadrant II.
The reference angle is
$\theta' = 210° - 180° = 30°$.
$\cot 30° = \sqrt{3}$
In quadrant II, $\cot\theta < 0$, so
$\cot(-210°) = -\cot 30° = -\sqrt{3}$.

62. $-\frac{2\pi}{3}$ lies in quadrant III.

The reference angle is

$\theta' = \pi + \frac{-2\pi}{3} = \frac{3\pi}{3} - \frac{2\pi}{3} = \frac{\pi}{3}$.

$\csc\left(\frac{\pi}{3}\right) = \frac{2\sqrt{3}}{3}$

In quadrant III, $\csc\theta < 0$, so

$\csc\left(-\frac{2\pi}{3}\right) = -\csc\left(\frac{\pi}{3}\right) = -\frac{2\sqrt{3}}{3}$.

63. $-\frac{\pi}{3}$ lies in quadrant IV.

The reference angle is

$\theta' = \frac{\pi}{3}$.

$\sin\left(\frac{\pi}{3}\right) = \frac{\sqrt{3}}{2}$

In quadrant IV, $\sin\theta < 0$, so

$\sin\left(-\frac{\pi}{3}\right) = -\sin\left(\frac{\pi}{3}\right) = -\frac{\sqrt{3}}{2}$.

64. 495° lies in quadrant II.

$495° - 360° = 135°$

The reference angle is

$\theta' = 180° - 135° = 45°$.

$\sin 45° = \frac{\sqrt{2}}{2}$

In quadrant II, $\sin\theta > 0$, so

$\sin 495° = \sin 45° = \frac{\sqrt{2}}{2}$.

65. $\frac{13\pi}{4}$ lies in quadrant III.

$\frac{13\pi}{4} - 2\pi = \frac{13\pi}{4} - \frac{8\pi}{4} = \frac{5\pi}{4}$

The reference angle is

$\theta' = \frac{5\pi}{4} - \pi = \frac{5\pi}{4} - \frac{4\pi}{4} = \frac{\pi}{4}$.

$\tan\frac{\pi}{4} = 1$

In quadrant III, $\tan\theta > 0$, so

$\tan\frac{13\pi}{4} = \tan\frac{\pi}{4} = 1$.

66. $\sin\frac{22\pi}{3} = \sin\left(\frac{22\pi}{3} - 6\pi\right)$

$= \sin\frac{4\pi}{3}$

$= -\sin\frac{\pi}{3}$

$= -\frac{\sqrt{3}}{2}$

67. $\cos\left(-\frac{35\pi}{6}\right) = \cos\left(-\frac{35\pi}{6} + 6\pi\right)$

$= \cos\frac{\pi}{6}$

$= \frac{\sqrt{3}}{2}$

68. The equation $y = 3\sin 4x$ is of the form $y = A\sin Bx$ with $A = 3$ and $B = 4$. The amplitude is $|A| = |3| = 3$. The period is $\frac{2\pi}{B} = \frac{2\pi}{4} = \frac{\pi}{2}$. The quarter-period is $\frac{\frac{\pi}{2}}{4} = \frac{\pi}{2} \cdot \frac{1}{4} = \frac{\pi}{8}$. The cycle begins at $x = 0$. Add quarter-periods to generate x-values for the key points.

$x = 0$

$x = 0 + \frac{\pi}{8} = \frac{\pi}{8}$

$x = \frac{\pi}{8} + \frac{\pi}{8} = \frac{\pi}{4}$

$x = \frac{\pi}{4} + \frac{\pi}{8} = \frac{3\pi}{8}$

$x = \frac{3\pi}{8} + \frac{\pi}{8} = \frac{\pi}{2}$

Evaluate the function at each value of x.

x	coordinates
0	(0, 0)
$\frac{\pi}{8}$	$\left(\frac{\pi}{8}, 3\right)$
$\frac{\pi}{4}$	$\left(\frac{\pi}{4}, 0\right)$
$\frac{3\pi}{8}$	$\left(\frac{3\pi}{8}, -3\right)$
$\frac{\pi}{2}$	$(2\pi, 0)$

Connect the five key points with a smooth curve and graph one complete cycle of the given function.

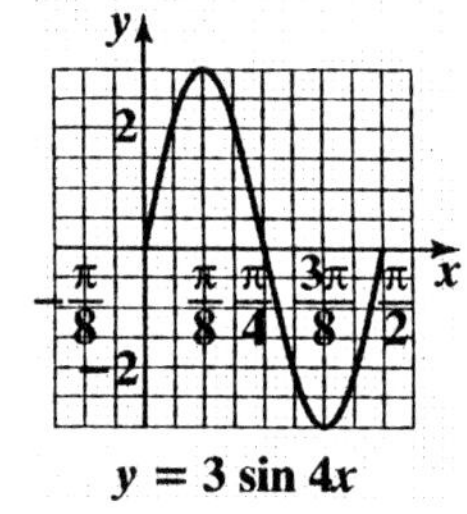

x	coordinates
0	(0, –2)
$\frac{\pi}{4}$	$\left(\frac{\pi}{4}, 0\right)$
$\frac{\pi}{2}$	$\left(\frac{\pi}{2}, 2\right)$
$\frac{3\pi}{4}$	$\left(\frac{3\pi}{4}, 0\right)$
π	$(\pi, -2)$

Connect the five key points with a smooth curve and graph one complete cycle of the given function.

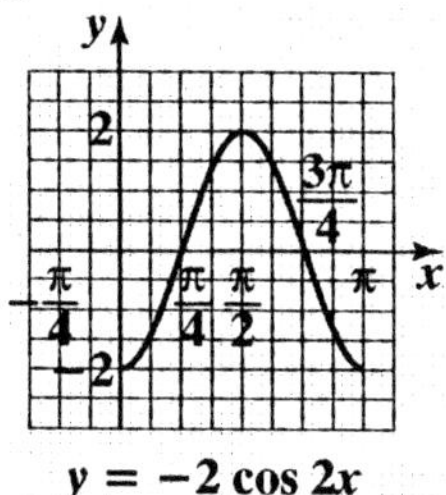

69. The equation $y = -2\cos 2x$ is of the form $y = A\cos Bx$ with $A = -2$ and $B = 2$. The amplitude is $|A| = |-2| = 2$. The period is $\frac{2\pi}{B} = \frac{2\pi}{2} = \pi$. The quarter-period is $\frac{\pi}{4}$. The cycle begins at $x = 0$. Add quarter-periods to generate x-values for the key points.

$x = 0$

$x = 0 + \frac{\pi}{4} = \frac{\pi}{4}$

$x = \frac{\pi}{4} + \frac{\pi}{4} = \frac{\pi}{2}$

$x = \frac{\pi}{2} + \frac{\pi}{4} = \frac{3\pi}{4}$

$x = \frac{3\pi}{4} + \frac{\pi}{4} = \pi$

Evaluate the function at each value of x.

70. The equation $y = 2\cos\frac{1}{2}x$ is of the form $y = A\cos Bx$ with $A = 2$ and $B = \frac{1}{2}$. The amplitude is $|A| = |2| = 2$. The period is $\frac{2\pi}{B} = \frac{2\pi}{\frac{1}{2}} = 2\pi \cdot 2 = 4\pi$. The quarter-period is $\frac{4\pi}{4} = \pi$. The cycle begins at $x = 0$. Add quarter-periods to generate x-values for the key points.

$x = 0$

$x = 0 + \pi = \pi$

$x = \pi + \pi = 2\pi$

$x = 2\pi + \pi = 3\pi$

$x = 3\pi + \pi = 4\pi$

Evaluate the function at each value of x.

x	coordinates
0	$(0, 2)$
π	$(\pi, 0)$
2π	$(2\pi, -2)$
3π	$(3\pi, 0)$
4π	$(4\pi, 2)$

Connect the five key points with a smooth curve and graph one complete cycle of the given function.

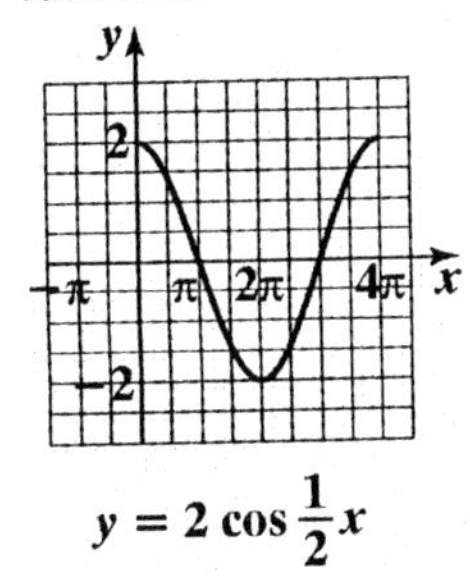

$y = 2\cos\frac{1}{2}x$

x	coordinates
0	$(0, 0)$
$\frac{3}{2}$	$\left(\frac{3}{2}, \frac{1}{2}\right)$
3	$(3, 0)$
$\frac{9}{2}$	$\left(\frac{9}{2}, -\frac{1}{2}\right)$
6	$(6, 0)$

Connect the five key points with a smooth curve and graph one complete cycle of the given function.

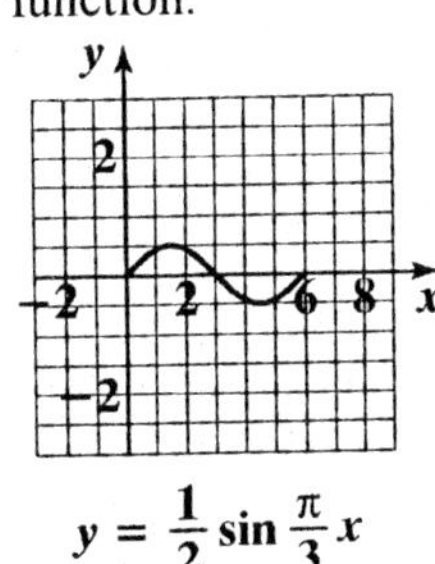

$y = \frac{1}{2}\sin\frac{\pi}{3}x$

71. The equation $y = \frac{1}{2}\sin\frac{\pi}{3}x$ is of the form $y = A\sin Bx$ with $A = \frac{1}{2}$ and $B = \frac{\pi}{3}$. The amplitude is $|A| = \left|\frac{1}{2}\right| = \frac{1}{2}$. The period is $\frac{2\pi}{B} = \frac{2\pi}{\frac{\pi}{3}} = 2\pi \cdot \frac{3}{\pi} = 6$. The quarter-period is $\frac{6}{4} = \frac{3}{2}$. The cycle begins at $x = 0$. Add quarter-periods to generate x-values for the key points.

$x = 0$

$x = 0 + \frac{3}{2} = \frac{3}{2}$

$x = \frac{3}{2} + \frac{3}{2} = 3$

$x = 3 + \frac{3}{2} = \frac{9}{2}$

$x = \frac{9}{2} + \frac{3}{2} = 6$

Evaluate the function at each value of x.

72. The equation $y = -\sin\pi x$ is of the form $y = A\sin Bx$ with $A = -1$ and $B = \pi$. The amplitude is $|A| = |-1| = 1$. The period is $\frac{2\pi}{B} = \frac{2\pi}{\pi} = 2$. The quarter-period is $\frac{2}{4} = \frac{1}{2}$. The cycle begins at $x = 0$. Add quarter-periods to generate x-values for the key points.

$x = 0$

$x = 0 + \frac{1}{2} = \frac{1}{2}$

$x = \frac{1}{2} + \frac{1}{2} = 1$

$x = 1 + \frac{1}{2} = \frac{3}{2}$

$x = \frac{3}{2} + \frac{1}{2} = 2$

Evaluate the function at each value of x.

x	coordinates
0	(0, 0)
$\frac{1}{2}$	$\left(\frac{1}{2}, -1\right)$
1	(1, 0)
$\frac{3}{2}$	$\left(\frac{3}{2}, 1\right)$
2	(2, 0)

Connect the five key points with a smooth curve and graph one complete cycle of the given function.

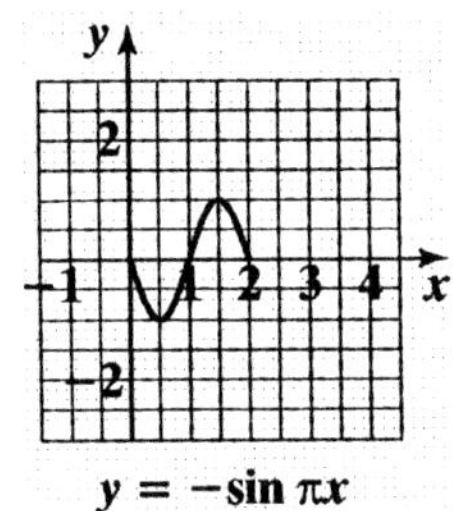

73. The equation $y = 3\cos\frac{x}{3}$ is of the form $y = A\cos Bx$ with $A = 3$ and $B = \frac{1}{3}$. The amplitude is $|A| = |3| = 3$. The period is $\frac{2\pi}{B} = \frac{2\pi}{\frac{1}{3}} = 2\pi \cdot 3 = 6\pi$. The quarter-period is $\frac{6\pi}{4} = \frac{3\pi}{2}$. The cycle begins at $x = 0$. Add quarter-periods to generate x-values for the key points.

$x = 0$

$x = 0 + \frac{3\pi}{2} = \frac{3\pi}{2}$

$x = \frac{3\pi}{2} + \frac{3\pi}{2} = 3\pi$

$x = 3\pi + \frac{3\pi}{2} = \frac{9\pi}{2}$

$x = \frac{9\pi}{2} + \frac{3\pi}{2} = 6\pi$

Evaluate the function at each value of x.

x	coordinates
0	(0, 3)
$\frac{3\pi}{2}$	$\left(\frac{3\pi}{2}, 0\right)$
3π	$(3\pi, -3)$
$\frac{9\pi}{2}$	$\left(\frac{9\pi}{2}, 0\right)$
6π	$(6\pi, 3)$

Connect the five key points with a smooth curve and graph one complete cycle of the given function.

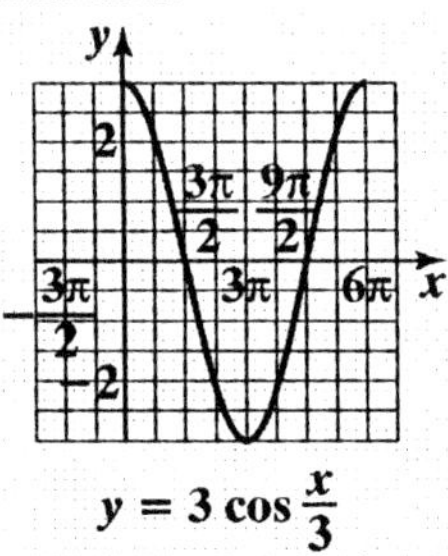

74. The equation $y = 2\sin(x - \pi)$ is of the form $y = A\sin(Bx - C)$ with $A = 2$, $B = 1$, and $C = \pi$. The amplitude is $|A| = |2| = 2$. The period is $\frac{2\pi}{B} = \frac{2\pi}{1} = 2\pi$. The phase shift is $\frac{C}{B} = \frac{\pi}{1} = \pi$. The quarter-period is $\frac{2\pi}{4} = \frac{\pi}{2}$. The cycle begins at $x = \pi$. Add quarter-periods to generate x-values for the key points.

$x = \pi$

$x = \pi + \frac{\pi}{2} = \frac{3\pi}{2}$

$x = \frac{3\pi}{2} + \frac{\pi}{2} = 2\pi$

$x = 2\pi + \frac{\pi}{2} = \frac{5\pi}{2}$

$x = \frac{5\pi}{2} + \frac{\pi}{2} = 3\pi$

Evaluate the function at each value of x.

x	coordinates
π	$(\pi, 0)$
$\frac{3\pi}{2}$	$\left(\frac{3\pi}{2}, 2\right)$
2π	$(2\pi, 0)$
$\frac{5\pi}{2}$	$\left(\frac{5\pi}{2}, -2\right)$
3π	$(3\pi, 0)$

Connect the five key points with a smooth curve and graph one complete cycle of the given function.

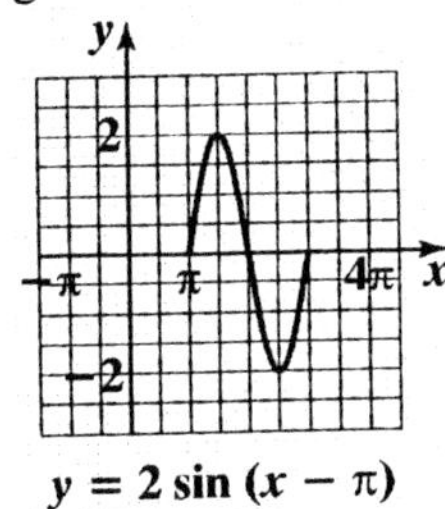

$y = 2\sin(x - \pi)$

75. $y = -3\cos(x+\pi) = -3\cos(x-(-\pi))$
The equation $y = -3\cos(x-(-\pi))$ is of the form $y = A\cos(Bx - C)$ with $A = -3$, $B = 1$, and $C = -\pi$. The amplitude is $|A| = |-3| = 3$.
The period is $\frac{2\pi}{B} = \frac{2\pi}{1} = 2\pi$. The phase shift is $\frac{C}{B} = \frac{-\pi}{1} = -\pi$. The quarter-period is $\frac{2\pi}{4} = \frac{\pi}{2}$.
The cycle begins at $x = -\pi$. Add quarter-periods to generate x-values for the key points.
$x = -\pi$
$x = -\pi + \frac{\pi}{2} = -\frac{\pi}{2}$
$x = -\frac{\pi}{2} + \frac{\pi}{2} = 0$
$x = 0 + \frac{\pi}{2} = \frac{\pi}{2}$
$x = \frac{\pi}{2} + \frac{\pi}{2} = \pi$
Evaluate the function at each value of x.

x	coordinates
$-\pi$	$(-\pi, -3)$
$-\frac{\pi}{2}$	$\left(-\frac{\pi}{2}, 0\right)$
0	$(0, 3)$
$\frac{\pi}{2}$	$\left(\frac{\pi}{2}, 0\right)$
π	$(\pi, -3)$

Connect the five key points with a smooth curve and graph one complete cycle of the given function.

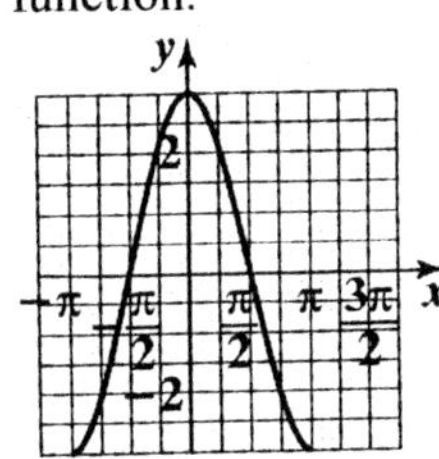

$y = -3\cos(x + \pi)$

76. $y = \frac{3}{2}\cos\left(2x + \frac{\pi}{4}\right) = \frac{3}{2}\cos\left(2x - \left(-\frac{\pi}{4}\right)\right)$

The equation $y = \frac{3}{2}\cos\left(2x - \left(-\frac{\pi}{4}\right)\right)$ is of the form $y = A\cos(Bx - C)$ with $A = \frac{3}{2}$, $B = 2$, and $C = -\frac{\pi}{4}$. The amplitude is $|A| = \left|\frac{3}{2}\right| = \frac{3}{2}$.

The period is $\frac{2\pi}{B} = \frac{2\pi}{2} = \pi$. The phase shift is $\frac{C}{B} = \frac{\frac{-\pi}{4}}{2} = -\frac{\pi}{4}\cdot\frac{1}{2} = -\frac{\pi}{8}$. The quarter-period is $\frac{\pi}{4}$. The cycle begins at $x = -\frac{\pi}{8}$. Add quarter-periods to generate x-values for the key points.

$x = -\frac{\pi}{8}$

$x = -\frac{\pi}{8} + \frac{\pi}{4} = \frac{\pi}{8}$

$x = \frac{\pi}{8} + \frac{\pi}{4} = \frac{3\pi}{8}$

$x = \frac{3\pi}{8} + \frac{\pi}{4} = \frac{5\pi}{8}$

$x = \frac{5\pi}{8} + \frac{\pi}{4} = \frac{7\pi}{8}$

Evaluate the function at each value of x.

x	coordinates
$-\frac{\pi}{8}$	$\left(-\frac{\pi}{8}, \frac{3}{2}\right)$
$\frac{\pi}{8}$	$\left(\frac{\pi}{8}, 0\right)$
$\frac{3\pi}{8}$	$\left(\frac{3\pi}{8}, -\frac{3}{2}\right)$
$\frac{5\pi}{8}$	$\left(\frac{5\pi}{8}, 0\right)$
$\frac{7\pi}{8}$	$\left(\frac{7\pi}{8}, \frac{3}{2}\right)$

Connect the five key points with a smooth curve and graph one complete cycle of the given function.

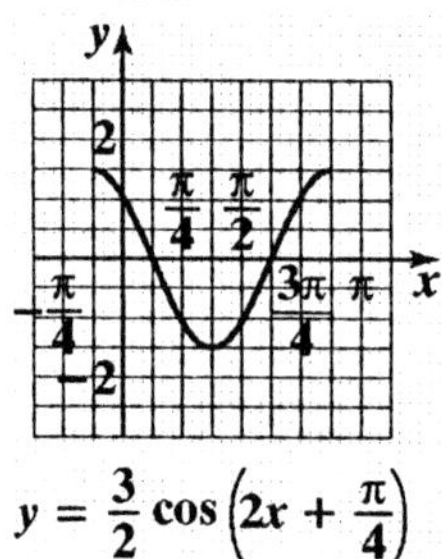

$y = \frac{3}{2}\cos\left(2x + \frac{\pi}{4}\right)$

77. $y = \frac{5}{2}\sin\left(2x + \frac{\pi}{2}\right) = \frac{5}{2}\sin\left(2x - \left(-\frac{\pi}{2}\right)\right)$

The equation $y = \frac{5}{2}\sin\left(2x - \left(-\frac{\pi}{2}\right)\right)$ is of the form $y = A\sin(Bx - C)$ with $A = \frac{5}{2}$, $B = 2$, and $C = -\frac{\pi}{2}$. The amplitude is $|A| = \left|\frac{5}{2}\right| = \frac{5}{2}$.

The period is $\frac{2\pi}{B} = \frac{2\pi}{2} = \pi$. The phase shift is $\frac{C}{B} = \frac{-\frac{\pi}{2}}{2} = -\frac{\pi}{2}\cdot\frac{1}{2} = -\frac{\pi}{4}$. The quarter-period is $\frac{\pi}{4}$. The cycle begins at $x = -\frac{\pi}{4}$. Add quarter-periods to generate x-values for the key points.

$x = -\frac{\pi}{4}$

$x = -\frac{\pi}{4} + \frac{\pi}{4} = 0$

$x = 0 + \frac{\pi}{4} = \frac{\pi}{4}$

$x = \frac{\pi}{4} + \frac{\pi}{4} = \frac{\pi}{2}$

$x = \frac{\pi}{2} + \frac{\pi}{4} = \frac{3\pi}{4}$

Evaluate the function at each value of x.

x	coordinates
$-\frac{\pi}{4}$	$(-\frac{\pi}{4}, 0)$
0	$\left(0, \frac{5}{2}\right)$
$\frac{\pi}{4}$	$\left(\frac{\pi}{4}, 0\right)$
$\frac{\pi}{2}$	$\left(\frac{\pi}{2}, -\frac{5}{2}\right)$
$\frac{3\pi}{4}$	$\left(\frac{3\pi}{4}, 0\right)$

Connect the five key points with a smooth curve and graph one complete cycle of the given function.

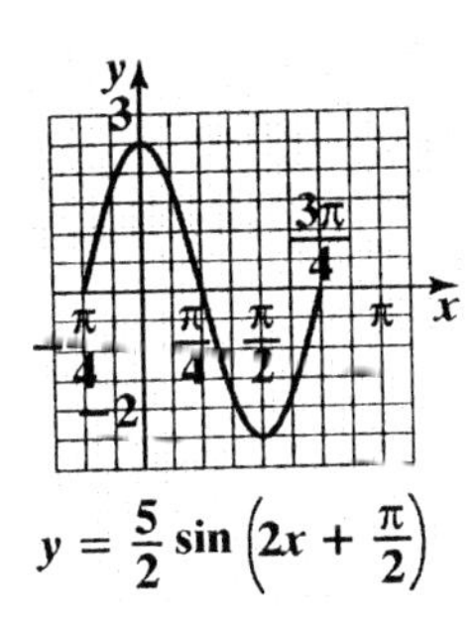

$y = \frac{5}{2}\sin\left(2x + \frac{\pi}{2}\right)$

78. The equation $y = -3\sin\left(\frac{\pi}{3}x - 3\pi\right)$ is of the form $y = A\sin(Bx - C)$ with $A = -3$, $B = \frac{\pi}{3}$, and $C = 3\pi$. The amplitude is $|A| = |-3| = 3$.

The period is $\frac{2\pi}{B} = \frac{2\pi}{\frac{\pi}{3}} = 2\pi \cdot \frac{3}{\pi} = 6$. The phase shift is $\frac{C}{B} = \frac{3\pi}{\frac{\pi}{3}} = 3\pi \cdot \frac{3}{\pi} = 9$. The quarter-period is $\frac{6}{4} = \frac{3}{2}$. The cycle begins at $x = 9$. Add quarter-periods to generate x-values for the key points.

$x = 9$

$x = 9 + \frac{3}{2} = \frac{21}{2}$

$x = \frac{21}{2} + \frac{3}{2} = 12$

$x = 12 + \frac{3}{2} = \frac{27}{2}$

$x = \frac{27}{2} + \frac{3}{2} = 15$

Evaluate the function at each value of x.

x	coordinates
9	$(9, 0)$
$\frac{21}{2}$	$\left(\frac{21}{2}, -3\right)$
12	$(12, 0)$
$\frac{27}{2}$	$\left(\frac{27}{2}, 3\right)$
15	$(15, 0)$

Connect the five key points with a smooth curve and graph one complete cycle of the given function.

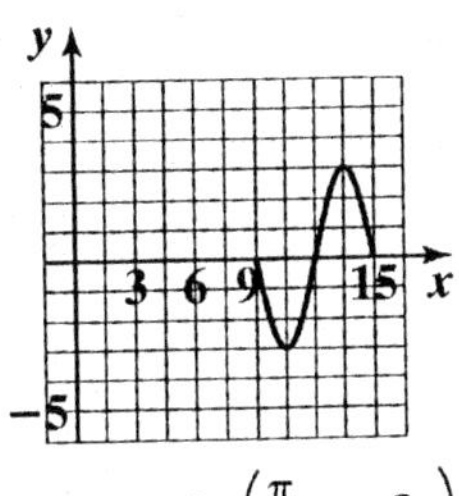

$y = -3\sin\left(\frac{\pi}{3}x - 3\pi\right)$

79. The graph of $y = \sin 2x + 1$ is the graph of $y = \sin 2x$ shifted one unit upward. The period for both functions is $\frac{2\pi}{2} = \pi$. The quarter-period is $\frac{\pi}{4}$. The cycle begins at $x = 0$. Add quarter-periods to generate x-values for the key points.

$x = 0$

$x = 0 + \frac{\pi}{4} = \frac{\pi}{4}$

$x = \frac{\pi}{4} + \frac{\pi}{4} = \frac{\pi}{2}$

$x = \frac{\pi}{2} + \frac{\pi}{4} = \frac{3\pi}{4}$

$x = \frac{3\pi}{4} + \frac{\pi}{4} = \pi$

Evaluate the function at each value of x.

x	coordinates
0	(0, 1)
$\frac{\pi}{4}$	$\left(\frac{\pi}{4}, 2\right)$
$\frac{\pi}{2}$	$\left(\frac{\pi}{2}, 1\right)$
$\frac{3\pi}{4}$	$\left(\frac{3\pi}{4}, 0\right)$
π	$(\pi, 1)$

By connecting the points with a smooth curve we obtain one period of the graph.

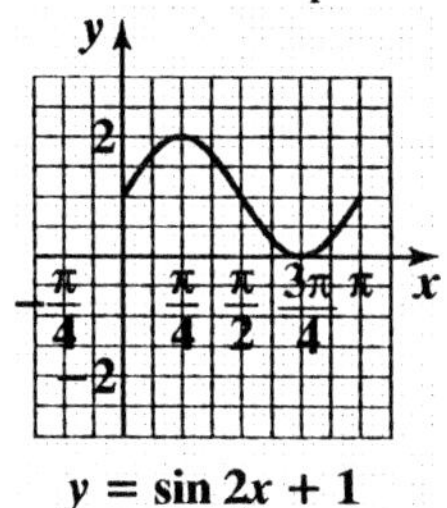

$y = \sin 2x + 1$

x	coordinates
0	(0, 0)
$\frac{3\pi}{2}$	$\left(\frac{3\pi}{2}, -2\right)$
3π	$(3\pi, -4)$
$\frac{9\pi}{2}$	$\left(\frac{9\pi}{2}, -2\right)$
6π	$(6\pi, 0)$

By connecting the points with a smooth curve we obtain one period of the graph.

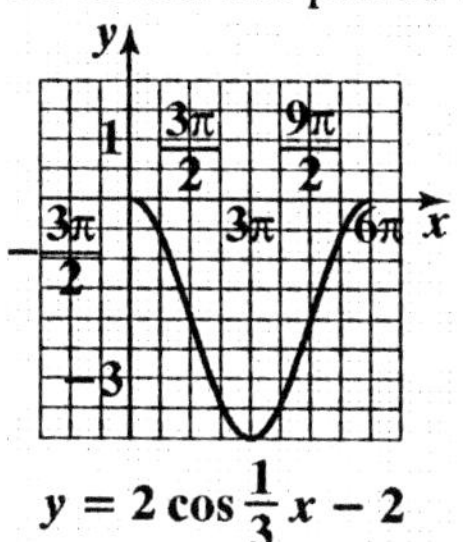

$y = 2\cos\frac{1}{3}x - 2$

80. The graph of $y = 2\cos\frac{1}{3}x - 2$ is the graph of $y = 2\cos\frac{1}{3}x$ shifted two units downward. The period for both functions is $\frac{2\pi}{\frac{1}{3}} = 2\pi \cdot 3 = 6\pi$.

The quarter-period is $\frac{6\pi}{4} = \frac{3\pi}{2}$. The cycle begins at $x = 0$. Add quarter-periods to generate x-values for the key points.

$x = 0$

$x = 0 + \frac{3\pi}{2} = \frac{3\pi}{2}$

$x = \frac{3\pi}{2} + \frac{3\pi}{2} = 3\pi$

$x = 3\pi + \frac{3\pi}{2} = \frac{9\pi}{2}$

$x = \frac{9\pi}{2} + \frac{3\pi}{2} = 6\pi$

Evaluate the function at each value of x.

81. a. At midnight $x = 0$. Thus,

$$y = 98.6 + 0.3\sin\left(\frac{\pi}{12}\cdot 0 - \frac{11\pi}{12}\right)$$
$$= 98.6 + 0.3\sin\left(-\frac{11\pi}{12}\right)$$
$$\approx 98.6 + 0.3(-0.2588) \approx 98.52$$

The body temperature is about 98.52°F.

b. period: $\frac{2\pi}{B} = \frac{2\pi}{\frac{\pi}{12}} = 2\pi \cdot \frac{12}{\pi} = 24$ hours

c. Solve the equation

$$\frac{\pi}{12}x - \frac{11\pi}{12} = \frac{\pi}{2}$$
$$\frac{\pi}{12}x = \frac{\pi}{2} + \frac{11\pi}{12} = \frac{6\pi}{12} + \frac{11\pi}{12} = \frac{17\pi}{12}$$
$$x = \frac{17\pi}{12}\cdot\frac{12}{\pi} = 17$$

The body temperature is highest for $x = 17$.

$y = 98.6 + 0.3\sin\left(\frac{\pi}{12} \cdot 17 - \frac{11\pi}{12}\right)$

$= 98.6 + 0.3\sin\frac{\pi}{2} = 98.6 + 0.3 = 98.9$

17 hours after midnight, which is 5 P.M., the body temperature is 98.9°F.

d. Solve the equation

$\frac{\pi}{12}x - \frac{11\pi}{12} = \frac{3\pi}{2}$

$\frac{\pi}{12}x = \frac{3\pi}{2} + \frac{11\pi}{12} = \frac{18\pi}{12} + \frac{11\pi}{12} = \frac{29\pi}{12}$

$x = \frac{29\pi}{12} \cdot \frac{12}{\pi} = 29$

The body temperature is lowest for $x = 29$.

$y = 98.6 + 0.3\sin\left(\frac{\pi}{12} \cdot 29 - \frac{11\pi}{12}\right)$

$= 98.6 + 0.3\sin\left(\frac{3\pi}{2}\right)$

$= 98.6 + 0.3(-1) = 98.3°$

29 hours after midnight or 5 hours after midnight, at 5 A.M., the body temperature is 98.3°F.

e. The graph of

$y = 98.6 + 0.3\sin\left(\frac{\pi}{12}x - \frac{11\pi}{12}\right)$ is of the form $y = D + A\sin(Bx - C)$ with $A = 0.3$, $B = \frac{\pi}{12}$, $C = \frac{11\pi}{12}$, and $D = 98.6$. The amplitude is $|A| = |0.3| = 0.3$. The period from part (b) is 24. The quarter-period is $\frac{24}{4} = 6$. The phase shift is

$\frac{C}{B} = \frac{\frac{11\pi}{12}}{\frac{\pi}{12}} = \frac{11\pi}{12} \cdot \frac{12}{\pi} = 11$. The cycle begins at $x = 11$. Add quarter-periods to generate x-values for the key points.

$x = 11$

$x = 11 + 6 = 17$

$x = 17 + 6 = 23$

$x = 23 + 6 = 29$

$x = 29 + 6 = 35$

Evaluate the function at each value of x. The key points are (11, 98.6), (17, 98.9), (23, 98.6), (29, 98.3), (35, 98.6). Extend the pattern to the left, and graph the function for $0 \le x \le 24$.

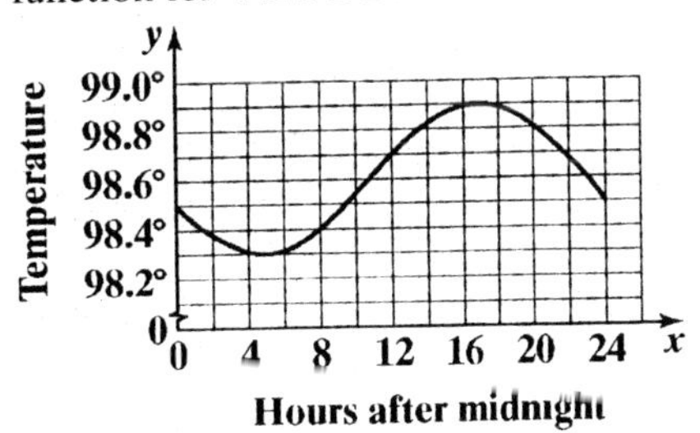

82. Solve the equations

$2x = -\frac{\pi}{2}$ and $2x = \frac{\pi}{2}$

$x = \frac{-\frac{\pi}{2}}{2}$ $\quad x = \frac{\frac{\pi}{2}}{2}$

$x = -\frac{\pi}{4}$ $\quad x = \frac{\pi}{4}$

Thus, two consecutive asymptotes occur at $x = -\frac{\pi}{4}$ and $x = \frac{\pi}{4}$.

$x\text{-intercept} = \frac{-\frac{\pi}{4} + \frac{\pi}{4}}{2} = \frac{0}{2} = 0$

An x-intercept is 0 and the graph passes through (0, 0). Because the coefficient of the tangent is 4, the points on the graph midway between an x-intercept and the asymptotes have y-coordinates of –4 and 4. Use the two consecutive asymptotes. $x = -\frac{\pi}{4}$ and $x = \frac{\pi}{4}$, to graph one full period of $y = 4\tan 2x$ from $-\frac{\pi}{4}$ to $\frac{\pi}{4}$.

Continue the pattern and extend the graph another full period to the right.

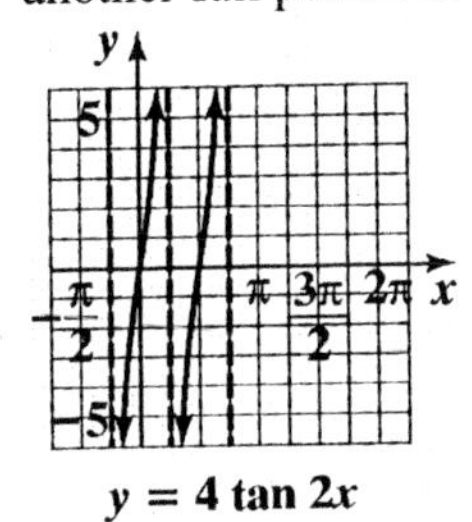

$y = 4\tan 2x$

83. Solve the equations

$\frac{\pi}{4}x = -\frac{\pi}{2}$ and $\frac{\pi}{4}x = \frac{\pi}{2}$

$x = -\frac{\pi}{2}\cdot\frac{4}{\pi}$ $\qquad x = \frac{\pi}{2}\cdot\frac{4}{\pi}$

$x = -2$ $\qquad x = 2$

Thus, two consecutive asymptotes occur at $x = -2$ and $x = 2$.

$x\text{-intercept} = \frac{-2+2}{2} = \frac{0}{2} = 0$

An x-intercept is 0 and the graph passes through (0, 0). Because the coefficient of the tangent is –2, the points on the graph midway between an x-intercept and the asymptotes have y-coordinates of 2 and –2. Use the two consecutive asymptotes, $x = -2$ and $x = 2$, to graph one full period of $y = -2\tan\frac{\pi}{4}x$ from –2 to 2. Continue the pattern and extend the graph another full period to the right.

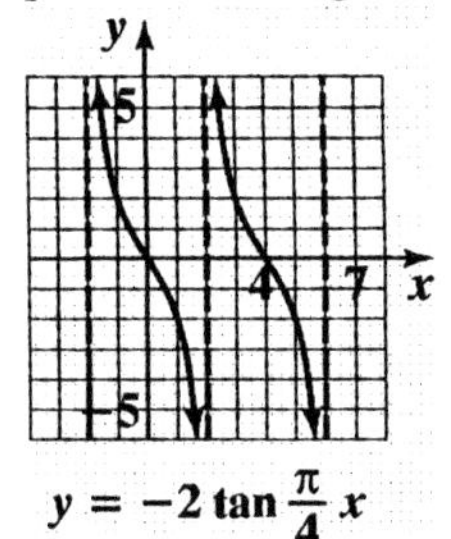

$y = -2\tan\frac{\pi}{4}x$

84. Solve the equations

$x+\pi = -\frac{\pi}{2}$ and $x+\pi = \frac{\pi}{2}$

$x = -\frac{\pi}{2} - \pi$ $\qquad x = \frac{\pi}{2} - \pi$

$x = -\frac{3\pi}{2}$ $\qquad x = -\frac{\pi}{2}$

Thus, two consecutive asymptotes occur at $x = -\frac{3\pi}{2}$ and $x = -\frac{\pi}{2}$.

$x\text{-intercept} = \frac{-\frac{3\pi}{2} - \frac{\pi}{2}}{2} = \frac{-2\pi}{2} = -\pi$

An x-intercept is $-\pi$ and the graph passes through $(-\pi, 0)$. Because the coefficient of the tangent is 1, the points on the graph midway between an x-intercept and the asymptotes have y-coordinates of –1 and 1. Use the two consecutive asymptotes,

$x = -\frac{3\pi}{2}$ and $x = -\frac{\pi}{2}$, to graph one full period of $y = \tan(x+\pi)$ from $-\frac{3\pi}{2}$ to $-\frac{\pi}{2}$. Continue the pattern and extend the graph another full period to the right.

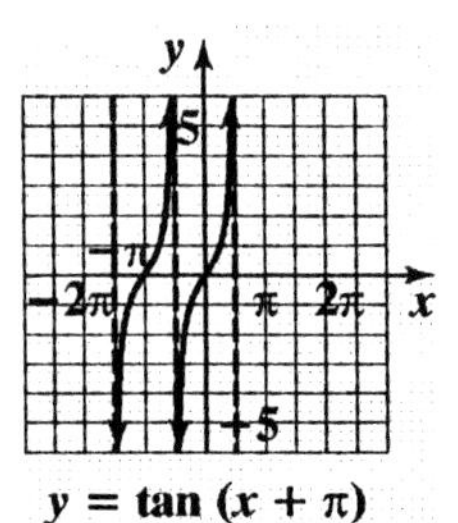

$y = \tan(x + \pi)$

85. Solve the equations

$x - \frac{\pi}{4} = -\frac{\pi}{2}$ and $x - \frac{\pi}{4} = \frac{\pi}{2}$

$x = -\frac{\pi}{2} + \frac{\pi}{4}$ $\qquad x = \frac{\pi}{2} + \frac{\pi}{4}$

$x = -\frac{\pi}{4}$ $\qquad x = \frac{3\pi}{4}$

Thus, two consecutive asymptotes occur at $x = -\frac{\pi}{4}$ and $x = -\frac{3\pi}{4}$.

$x\text{-intercept} = \frac{-\frac{\pi}{4} - \frac{3\pi}{4}}{2} = \frac{\frac{\pi}{2}}{2} = \frac{\pi}{4}$

An x-intercept is $\frac{\pi}{4}$ and the graph passes through $\left(\frac{\pi}{4}, 0\right)$. Because the coefficient of the tangent is –1, the points on the graph midway between an x-intercept and the asymptotes have y-coordinates of 1 and –1. Use the two consecutive asymptotes,

$x = -\frac{\pi}{4}$ and $x = \frac{3\pi}{4}$, to graph one full period of $y = -\tan\left(x - \frac{\pi}{4}\right)$ from $-\frac{\pi}{4}$ to $\frac{3\pi}{4}$.

Continue the pattern and extend the graph another full period to the right.

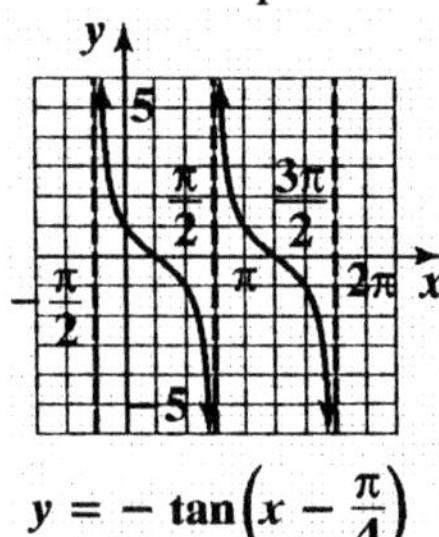

$y = -\tan\left(x - \frac{\pi}{4}\right)$

86. Solve the equations

$3x = 0$ and $3x = \pi$

$x = 0$ $\quad x = \frac{\pi}{3}$

Thus, two consecutive asymptotes occur at $x = 0$ and $x = \frac{\pi}{3}$.

$x\text{-intercept} = \frac{0 + \frac{\pi}{3}}{2} = \frac{\frac{\pi}{3}}{2} = \frac{\pi}{6}$

An x-intercept is $\frac{\pi}{6}$ and the graph passes through $\left(\frac{\pi}{6}, 0\right)$.

Because the coefficient of the tangent is 2, the points on the graph midway between an x-intercept and the asymptotes have y-coordinates of 2 and –2. Use the two consecutive asymptotes, $x = 0$ and $x = \frac{\pi}{3}$, to graph one full period of $y = 2\cot 3x$ from 0 to $\frac{\pi}{3}$. Continue the pattern and extend the graph another full period to the right.

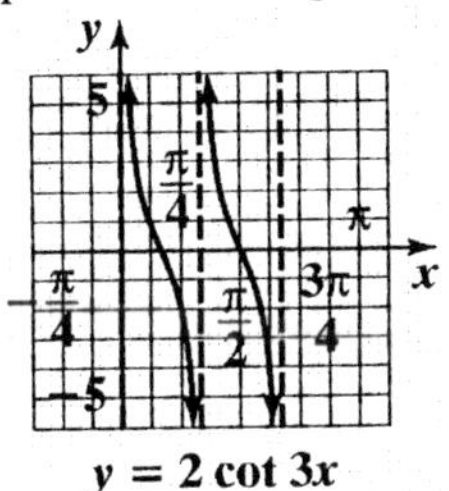

$y = 2 \cot 3x$

87. Solve the equations

$\frac{\pi}{2}x = 0$ and $\frac{\pi}{2}x = \pi$

$x = 0$ $\quad x = \pi \cdot \frac{2}{\pi}$

$x = 2$

Thus, two consecutive asymptotes occur at $x = 0$ and $x = 2$.

$x\text{-intercept} = \frac{0+2}{2} = \frac{2}{2} = 1$

An x-intercept is 1 and the graph passes through (1, 0). Because the coefficient of the cotangent is $-\frac{1}{2}$, the points on the graph midway between an x-intercept and the asymptotes have y-coordinates of $-\frac{1}{2}$ and $\frac{1}{2}$. Use the two consecutive asymptotes, $x = 0$ and $x = 2$, to graph one full period of $y = -\frac{1}{2}\cot\frac{\pi}{2}x$ from 0 to 2. Continue the pattern and extend the graph another full period to the right.

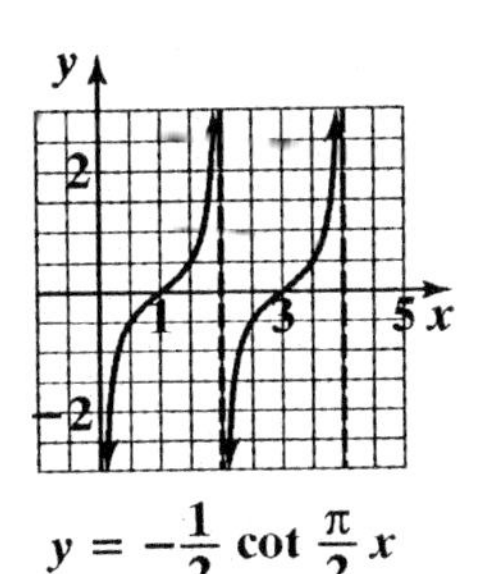

$y = -\frac{1}{2} \cot \frac{\pi}{2} x$

88. Solve the equations

$x + \frac{\pi}{2} = 0$ and $x + \frac{\pi}{2} = \pi$

$x = 0 - \frac{\pi}{2}$ $\quad x = \pi - \frac{\pi}{2}$

$x = -\frac{\pi}{2}$ $\quad x = \frac{\pi}{2}$

Thus, two consecutive asymptotes occur at $x = -\frac{\pi}{2}$ and $x = \frac{\pi}{2}$.

$x\text{-intercept} = \frac{-\frac{\pi}{2} + \frac{\pi}{2}}{2} = \frac{0}{2} = 0$

An x-intercept is 0 and the graph passes through (0, 0). Because the coefficient of the cotangent is 2, the points on the graph midway between an x-intercept and the asymptotes have y-coordinates of 2 and –2. Use the two consecutive asymptotes, $x = -\frac{\pi}{2}$ and $x = \frac{\pi}{2}$, to graph one full period of $y = 2\cot\left(x + \frac{\pi}{2}\right)$ from $-\frac{\pi}{2}$ to $\frac{\pi}{2}$. Continue the pattern and extend the graph another full period to the right.

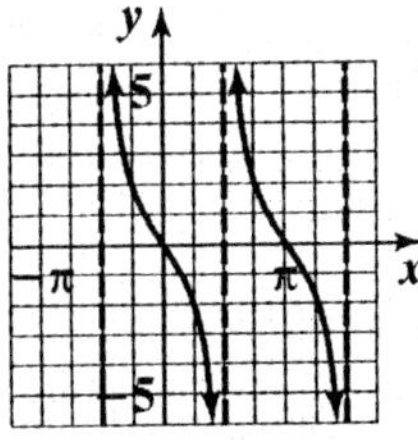

$y = 2 \cot\left(x + \frac{\pi}{2}\right)$

89. Graph the reciprocal cosine function, $y = 3\cos 2\pi x$. The equation is of the form $y = A\cos Bx$ with $A = 3$ and $B = 2\pi$.
amplitude: $|A| = |3| = 3$

period: $\frac{2\pi}{B} = \frac{2\pi}{2\pi} = 1$

Use quarter-periods, $\frac{1}{4}$, to find x-values for the five key points. Starting with $x = 0$, the x-values are 0, $\frac{1}{4}$, $\frac{1}{2}$, $\frac{3}{4}$, 1. Evaluating the function at each value of x, the key points are (0, 3), $\left(\frac{1}{4}, 0\right)$, $\left(\frac{1}{2}, -3\right)$, $\left(\frac{3}{4}, 0\right)$, (1, 3) Use these key points to graph $y = 3\cos 2\pi x$ from 0 to 1. Extend the graph one cycle to the right. Use the graph to obtain the graph of the reciprocal function. Draw vertical asymptotes through the x-intercepts, and use them as guides to graph $y = 3\sec 2\pi x$.

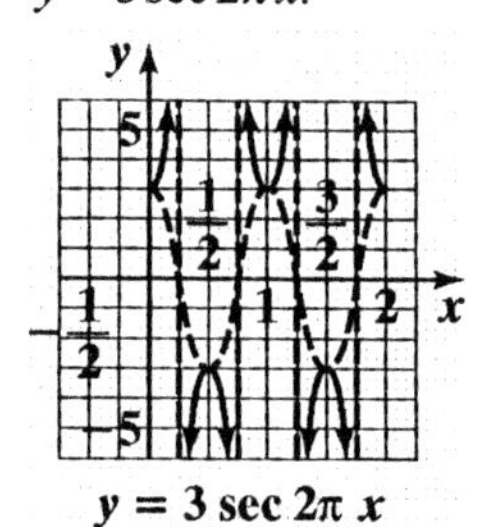

$y = 3 \sec 2\pi x$

90. Graph the reciprocal sine function, $y = -2\sin \pi x$. The equation is of the form $y = A\sin Bx$ with $A = -2$ and $B = \pi$.
amplitude: $|A| = |-2| = 2$

period: $\frac{2\pi}{B} = \frac{2\pi}{\pi} = 2$

Use quarter-periods, $\frac{2}{4} = \frac{1}{2}$, to find x-values for the five key points. Starting with $x = 0$, the x-values are 0, $\frac{1}{2}$, 1, $\frac{3}{2}$, 2. Evaluating the function at each value of x, the key points are (0, 0), $\left(\frac{1}{2}, -2\right)$, (1, 0), $\left(\frac{3}{2}, 2\right)$, (2,0). Use these key points to graph $y = -2\sin \pi x$ from 0 to 2. Extend the graph one cycle to the right. Use the graph to obtain the graph of the reciprocal function. Draw vertical asymptotes through the x-intercepts, and use them as guides to graph $y = -2\csc \pi x$.

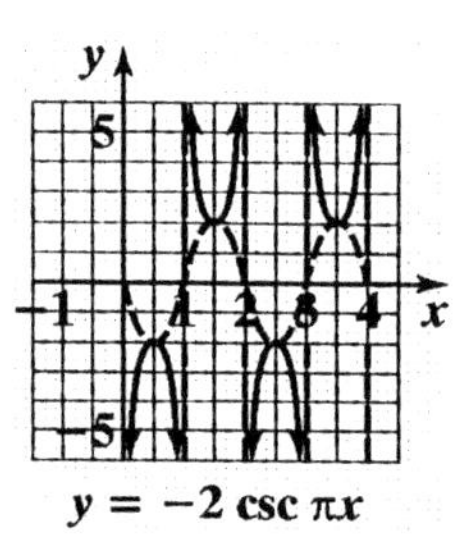

$y = -2 \csc \pi x$

91. Graph the reciprocal cosine function, $y = 3\cos(x + \pi)$. The equation is of the form $y = A\cos(Bx - C)$ with $A = 3$, $B = 1$, and $C = -\pi$.
amplitude: $|A| = |3| = 3$

period: $\frac{2\pi}{B} = \frac{2\pi}{1} = 2\pi$

phase shift: $\frac{C}{B} = \frac{-\pi}{1} = -\pi$

Use quarter-periods, $\frac{2\pi}{4} = \frac{\pi}{2}$, to find x-values for the five key points. Starting with $x = -\pi$, the x-values are $-\pi$, $-\frac{\pi}{2}$, 0, $\frac{\pi}{2}$, π. Evaluating the function at each value of x, the key points are $(-\pi, 3)$, $\left(-\frac{\pi}{2}, 0\right)$, $(0, -3)$, $\left(\frac{\pi}{2}, 0\right)$, $(\pi, 3)$. Use these key points to graph $y = 3\cos(x + \pi)$ from $-\pi$ to π. Extend the graph one cycle to the right. Use the graph to obtain the graph of the reciprocal function. Draw vertical asymptotes through the x-intercepts, and use them as guides to graph $y = 3\sec(x + \pi)$.

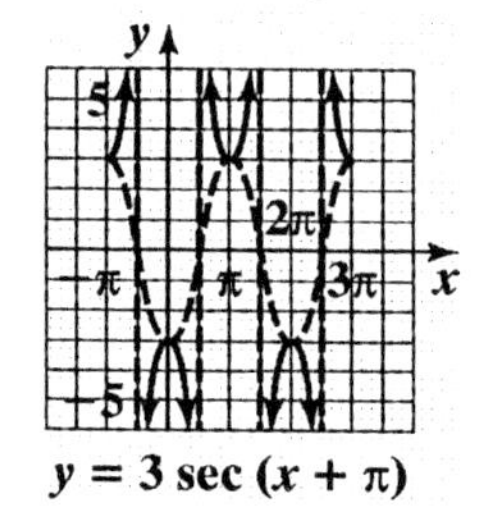

$y = 3 \sec (x + \pi)$

92. Graph the reciprocal sine function, $y = \frac{5}{2}\sin(x - \pi)$. The equation is of the form $y = A\sin(Bx - C)$ with $A = \frac{5}{2}$, $B = 1$, and $C = \pi$.

amplitude: $|A| = \left|\frac{5}{2}\right| = \frac{5}{2}$

period: $\frac{2\pi}{B} = \frac{2\pi}{1} = 2\pi$

phase shift: $\frac{C}{B} = \frac{\pi}{1} = \pi$

Use quarter-periods, $\frac{2\pi}{4} = \frac{\pi}{2}$, to find x-values for the five key points. Starting with $x = \pi$, the x-values are $\pi, \frac{3\pi}{2}, 2\pi, \frac{5\pi}{2}, 3\pi$. Evaluating the function at each value of x, the key points are $(\pi, 0)$, $\left(\frac{3\pi}{2}, \frac{5}{2}\right)$, $(2\pi, 0)$, $\left(\frac{5\pi}{2}, -\frac{5}{2}\right)$, $(3\pi, 0)$. Use these key points to graph $y = \frac{5}{2}\sin(x - \pi)$ from π to 3π. Extend the graph one cycle to the right. Use the graph to obtain the graph of the reciprocal function. Draw vertical asymptotes through the x-intercepts, and use them as guides to graph $y = \frac{5}{2}\csc(x - \pi)$.

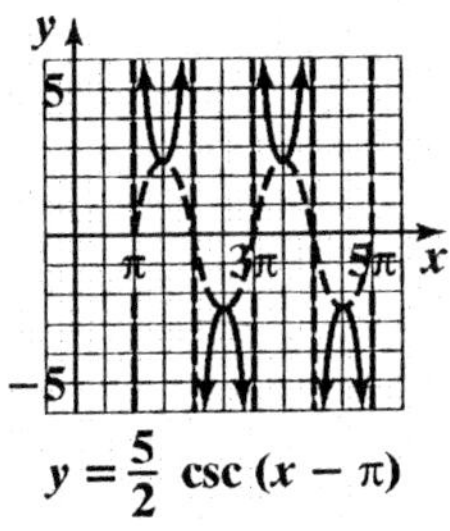

$y = \frac{5}{2}\csc(x - \pi)$

93. Let $\theta = \sin^{-1} 1$, then $\sin\theta = 1$.
The only angle in the interval $\left[-\frac{\pi}{2}, \frac{\pi}{2}\right]$ that satisfies $\sin\theta = 1$ is $\frac{\pi}{2}$. Thus $\theta = \frac{\pi}{2}$, or $\sin^{-1} 1 = \frac{\pi}{2}$.

94. Let $\theta = \cos^{-1} 1$, then $\cos\theta = 1$.
The only angle in the interval $[0, \pi]$ that satisfies $\cos\theta = 1$ is 0. Thus $\theta = 0$, or $\cos^{-1} 1 = 0$.

95. Let $\theta = \tan^{-1} 1$, then $\tan\theta = 1$.
The only angle in the interval $\left(-\frac{\pi}{2}, \frac{\pi}{2}\right)$ that satisfies $\tan\theta = 1$ is $\frac{\pi}{4}$. Thus $\theta = \frac{\pi}{4}$, or $\tan^{-1} 1 = \frac{\pi}{4}$.

96. Let $\theta = \sin^{-1}\left(-\frac{\sqrt{3}}{2}\right)$, then $\sin\theta = -\frac{\sqrt{3}}{2}$.
The only angle in the interval $\left(-\frac{\pi}{2}, \frac{\pi}{2}\right)$ that satisfies $\sin\theta = -\frac{\sqrt{3}}{2}$ is $-\frac{\pi}{3}$. Thus $\theta = -\frac{\pi}{3}$, or $\sin^{-1}\left(-\frac{\sqrt{3}}{2}\right) = -\frac{\pi}{3}$.

97. Let $\theta = \cos^{-1}\left(-\frac{1}{2}\right)$, then $\cos\theta = -\frac{1}{2}$.
The only angle in the interval $[0, \pi]$ that satisfies $\cos\theta = -\frac{1}{2}$ is $\frac{2\pi}{3}$. Thus $\theta = \frac{2\pi}{3}$, or $\cos^{-1}\left(-\frac{1}{2}\right) = \frac{2\pi}{3}$.

98. Let $\theta = \tan^{-1}\left(-\frac{\sqrt{3}}{3}\right)$, then $\tan\theta = -\frac{\sqrt{3}}{3}$.
The only angle in the interval $\left(-\frac{\pi}{2}, \frac{\pi}{2}\right)$ that satisfies $\tan\theta = -\frac{\sqrt{3}}{3}$ is $-\frac{\pi}{6}$.
Thus $\theta = -\frac{\pi}{6}$, or $\tan^{-1}\left(-\frac{\sqrt{3}}{3}\right) = -\frac{\pi}{6}$.

99. Let $\theta = \sin^{-1}\frac{\sqrt{2}}{2}$, then $\sin\theta = \frac{\sqrt{2}}{2}$. The only angle in the interval $\left[-\frac{\pi}{2}, \frac{\pi}{2}\right]$ that satisfies $\sin\theta = \frac{\sqrt{2}}{2}$ is $\frac{\pi}{4}$.

Thus, $\cos\left(\sin^{-1}\frac{\sqrt{2}}{2}\right) = \cos\frac{\pi}{4} = \frac{\sqrt{2}}{2}$.

100. Let $\theta = \cos^{-1}0$, then $\cos\theta = 0$. The only angle in the interval $[0, \pi]$ that satisfies $\cos\theta = 0$ is $\frac{\pi}{2}$. Thus, $\sin\left(\cos^{-1}0\right) = \sin\frac{\pi}{2} = 1$.

101. Let $\theta = \sin^{-1}\left(-\frac{1}{2}\right)$, then $\sin\theta = -\frac{1}{2}$. The only angle in the interval $\left[-\frac{\pi}{2}, \frac{\pi}{2}\right]$ that satisfies $\sin\theta = -\frac{1}{2}$ is $-\frac{\pi}{6}$.

Thus, $\tan\left[\sin^{-1}\left(-\frac{1}{2}\right)\right] = \tan\left(-\frac{\pi}{6}\right) = -\frac{\sqrt{3}}{3}$.

102. Let $\theta = \cos^{-1}\left(-\frac{\sqrt{3}}{2}\right)$, then $\cos\theta = -\frac{\sqrt{3}}{2}$. The only angle in the interval $[0, \pi]$ that satisfies $\cos\theta = -\frac{\sqrt{3}}{2}$ is $\frac{5\pi}{6}$.

Thus, $\tan\left[\cos^{-1}\left(-\frac{\sqrt{3}}{2}\right)\right] = \tan\frac{5\pi}{6} = -\frac{\sqrt{3}}{3}$.

103. Let $\theta = \tan^{-1}\frac{\sqrt{3}}{3}$, then $\tan\theta = \frac{\sqrt{3}}{3}$.

The only angle in the interval $\left(-\frac{\pi}{2}, \frac{\pi}{2}\right)$ that satisfies $\tan\theta = \frac{\sqrt{3}}{3}$ is $\frac{\pi}{6}$.

Thus $\csc\left(\tan^{-1}\frac{\sqrt{3}}{3}\right) = \csc\frac{\pi}{6} = 2$.

104. Let $\theta = \tan^{-1}\frac{3}{4}$, then $\tan\theta = \frac{3}{4}$

Because $\tan\theta$ is positive, θ is in the first quadrant.

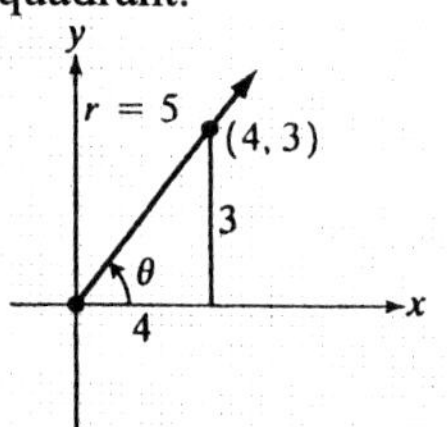

$r^2 = x^2 + y^2$

$r^2 = 4^2 + 3^2$

$r^2 = 25$

$r = 5$

$\cos\left(\tan^{-1}\frac{3}{4}\right) = \cos\theta = \frac{x}{r} = \frac{4}{5}$

105. Let $\theta = \cos^{-1}\frac{3}{5}$, then $\cos\theta = \frac{3}{5}$.

Because $\cos\theta$ is positive, θ is in the first quadrant.

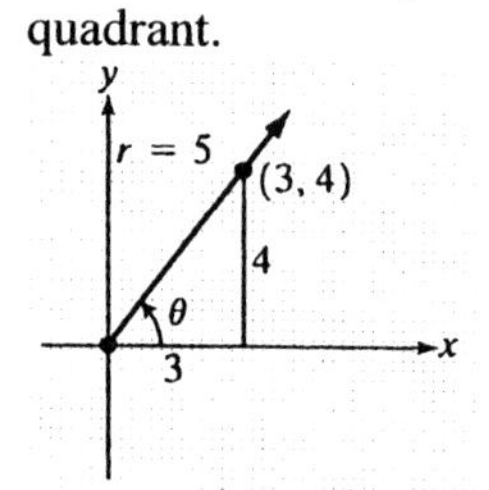

$x^2 + y^2 = r^2$

$3^2 + y^2 = 5^2$

$y^2 = 25 - 9 = 16$

$y = \sqrt{16} = 4$

$\sin\left(\cos^{-1}\frac{3}{5}\right) = \sin\theta = \frac{y}{r} = \frac{4}{5}$

106. Let $\theta = \sin^{-1}\left(-\frac{3}{5}\right)$, then $\sin\theta = -\frac{3}{5}$.

Because $\sin\theta$ is negative, θ is in quadrant IV.

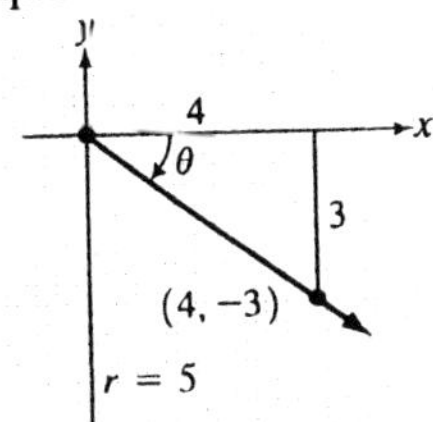

$$x^2 + (-3)^2 = 5^2$$
$$x^2 + y^2 = r^2$$
$$x^2 = 25 - 9 = 16$$
$$x = \sqrt{16} = 4$$
$$\tan\left[\sin^{-1}\left(-\frac{3}{5}\right)\right] = \tan\theta = \frac{y}{x} = -\frac{3}{4}$$

107. Let $\theta = \cos^{-1}\left(-\frac{4}{5}\right)$, then $\cos\theta = -\frac{4}{5}$.

Because $\cos\theta$ is negative, θ is in quadrant II.

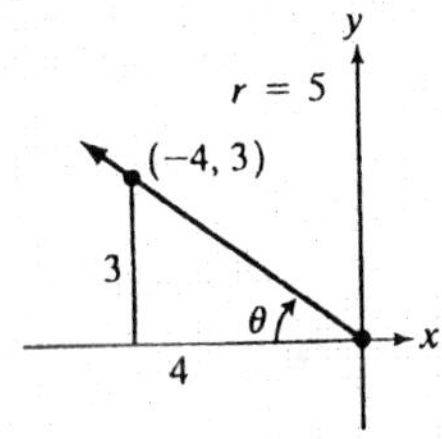

$$x^2 + y^2 = r^2$$
$$(-4)^2 + y^2 = 5^2$$
$$y^2 = 25 - 16 = 9$$
$$y = \sqrt{9} = 3$$

Use the right triangle to find the exact value.

$$\tan\left[\cos^{-1}\left(-\frac{4}{5}\right)\right] = \tan\theta = -\frac{3}{4}$$

108. Let $\theta = \tan^{-1}\left(-\frac{1}{3}\right)$,

Because $\tan\theta$ is negative, θ is in quadrant IV and $x = 3$ and $y = -1$.

$$r^2 = x^2 + y^2$$
$$r^2 = 3^2 + (-1)^2$$
$$r^2 = 10$$
$$r = \sqrt{10}$$
$$\sin\left[\tan^{-1}\left(-\frac{4}{5}\right)\right] = \sin\theta = \frac{y}{r} = \frac{-1}{\sqrt{10}} = -\frac{\sqrt{10}}{10}$$

109. $x = \frac{\pi}{3}$, x is in $\left[-\frac{\pi}{2}, \frac{\pi}{2}\right]$, so $\sin^{-1}\left(\sin\frac{\pi}{3}\right) = \frac{\pi}{3}$

110. $x = \frac{2\pi}{3}$, x is not in $\left[-\frac{\pi}{2}, \frac{\pi}{2}\right]$. x is in the domain of $\sin x$, so

$$\sin^{-1}\left(\sin\frac{2\pi}{3}\right) = \sin^{-1}\frac{\sqrt{3}}{2} = \frac{\pi}{3}$$

111. $\sin^{-1}\left(\cos\frac{2\pi}{3}\right) = \sin^{-1}\left(-\frac{1}{2}\right)$

Let $\theta = \sin^{-1}\left(-\frac{1}{2}\right)$, then $\sin\theta = -\frac{1}{2}$. The only angle in the interval $\left[-\frac{\pi}{2}, \frac{\pi}{2}\right]$ that satisfies $\sin\theta = -\frac{1}{2}$ is $-\frac{\pi}{6}$. Thus, $\theta = -\frac{\pi}{6}$, or

$$\sin^{-1}\left(\cos\frac{2\pi}{3}\right) = \sin^{-1}\left(-\frac{1}{2}\right) = -\frac{\pi}{6}.$$

112. Let $\theta = \tan^{-1}\frac{x}{2}$, then $\tan\theta = \frac{x}{2}$.

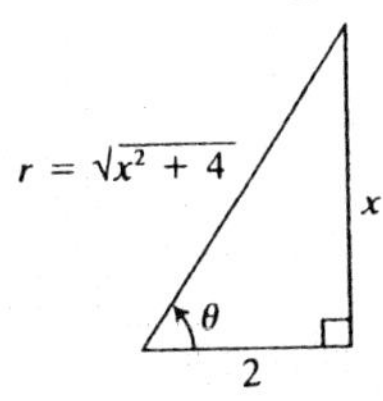

$$r^2 = x^2 + 2^2$$
$$r^2 = x^2 + y^2$$
$$r = \sqrt{x^2 + 4}$$

Use the right triangle to write the algebraic expression.

$$\cos\left(\tan^{-1}\frac{x}{2}\right) = \cos\theta = \frac{2}{\sqrt{x^2+4}} = \frac{2\sqrt{x^2+4}}{x^2+4}$$

113. Let $\theta = \sin^{-1}\frac{1}{x}$, then $\sin\theta = \frac{1}{x}$.

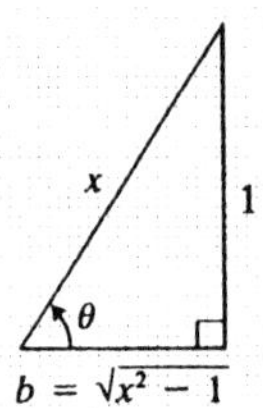

Use the Pythagorean theorem to find the third side, b.

$$1^2 + b^2 = x^2$$
$$b^2 = x^2 - 1$$
$$b = \sqrt{x^2 - 1}$$

Use the right triangle to write the algebraic expression.

$$\sec\left(\sin^{-1}\frac{1}{x}\right) = \sec\theta = \frac{x}{\sqrt{x^2-1}} = \frac{x\sqrt{x^2-1}}{x^2-1}$$

114. Find the measure of angle B. Because $C = 90°$, $A + B = 90°$. Thus,
$B = 90° - A = 90° - 22.3° = 67.7°$
We have a known angle, a known hypotenuse, and an unknown opposite side. Use the sine function.

$$\sin 22.3° = \frac{a}{10}$$
$$a = 10\sin 22.3° \approx 3.79$$

We have a known angle, a known hypotenuse, and an unknown adjacent side. Use the cosine function.

$$\cos 22.3° = \frac{b}{10}$$
$$b = 10\cos 22.3° \approx 9.25$$

In summary, $B = 67.7°$, $a \approx 3.79$, and $b \approx 9.25$.

115. Find the measure of angle A. Because $C = 90°$, $A + B = 90°$. Thus,
$A = 90° - B = 90° - 37.4° = 52.6°$
We have a known angle, a known opposite side, and an unknown adjacent side. Use the tangent function.

$$\tan 37.4° = \frac{6}{a}$$
$$a = \frac{6}{\tan 37.4°} \approx 7.85$$

We have a known angle, a known opposite side, and an unknown hypotenuse. Use the sine function.

$$\sin 37.4° = \frac{6}{c}$$
$$c = \frac{6}{\sin 37.4°} \approx 9.88$$

In summary, $A = 52.6°$, $a \approx 7.85$, and $c \approx 9.88$.

116. Find the measure of angle A. We have a known hypotenuse, a known opposite side, and an unknown angle. Use the sine function.

$$\sin A = \frac{2}{7}$$
$$A = \sin^{-1}\left(\frac{2}{7}\right) \approx 16.6°$$

Find the measure of angle B. Because $C = 90°$, $A + B = 90°$. Thus,
$B = 90° - A \approx 90° - 16.6° = 73.4°$
We have a known hypotenuse, a known opposite side, and an unknown adjacent side. Use the Pythagorean theorem.

$$a^2 + b^2 = c^2$$
$$2^2 + b^2 = 7^2$$
$$b^2 = 7^2 - 2^2 = 45$$
$$b = \sqrt{45} \approx 6.71$$

In summary, $A \approx 16.6°$, $B \approx 73.4°$, and $b \approx 6.71$.

117. Find the measure of angle A. We have a known opposite side, a known adjacent side, and an unknown angle. Use the tangent function.

$$\tan A = \frac{1.4}{3.6}$$
$$A = \tan^{-1}\left(\frac{1.4}{3.6}\right) \approx 21.3°$$

Find the measure of angle B. Because $C = 90°$, $A + B = 90°$. Thus,
$B = 90° - A \approx 90° - 21.3° = 68.7°$
We have a known opposite side, a known adjacent side, and an unknown hypotenuse.

Use the Pythagorean theorem.

$$c^2 = a^2 + b^2 = (1.4)^2 + (3.6)^2 = 14.92$$
$$c = \sqrt{14.92} \approx 3.86$$

In summary, $A \approx 21.3°$, $B \approx 68.7°$, and $c \approx 3.86$.

118. Using a right triangle, we have a known angle, an unknown opposite side, h, and a known adjacent side. Therefore, use the tangent function.

$$\tan 25.6° = \frac{h}{80}$$
$$h = 80 \tan 25.6°$$
$$\approx 38.3$$

The building is about 38 feet high.

119. Using a right triangle, we have a known angle, an unknown opposite side, h, and a known adjacent side. Therefore, use the tangent function.

$$\tan 40° = \frac{h}{60}$$
$$h = 60 \tan 40° \approx 50 \text{ yd}$$

The second building is 50 yds taller than the first. Total height $= 40 + 50 = 90$ yd.

120. Using two right triangles, a smaller right triangle corresponding to the smaller angle of elevation drawn inside a larger right triangle corresponding to the larger angle of elevation, we have a known angle, a known opposite side, and an unknown adjacent side, d, in the smaller triangle. Therefore, use the tangent function.

$$\tan 68° = \frac{125}{d}$$
$$d = \frac{125}{\tan 68°} \approx 50.5$$

We now have a known angle, a known adjacent side, and an unknown opposite side, h, in the larger triangle. Again, use the tangent function.

$$\tan 71° = \frac{h}{50.5}$$
$$h = 50.5 \tan 71° \approx 146.7$$

The height of the antenna is $146.7 - 125$, or 21.7 ft, to the nearest tenth of a foot.

121. We need the acute angle between ray OA and the north-south line through O. This angle measures $90° - 55° = 35°$. This angle measured from the north side of the north-south line and lies east of the north-south line. Thus the bearing from O to A is N35°E.

122. We need the acute angle between ray OA and the north-south line through O. This angle measures $90° - 55° = 35°$. This angle measured from the south side of the north-south line and lies west of the north-south line. Thus the bearing from O to A is S35°W.

123. Using a right triangle, we have a known angle, a known adjacent side, and an unknown opposite side, d. Therefore, use the tangent function.

$$\tan 64° = \frac{d}{12}$$
$$d = 12 \tan 64° \approx 24.6$$

The ship is about 24.6 miles from the lighthouse.

124.

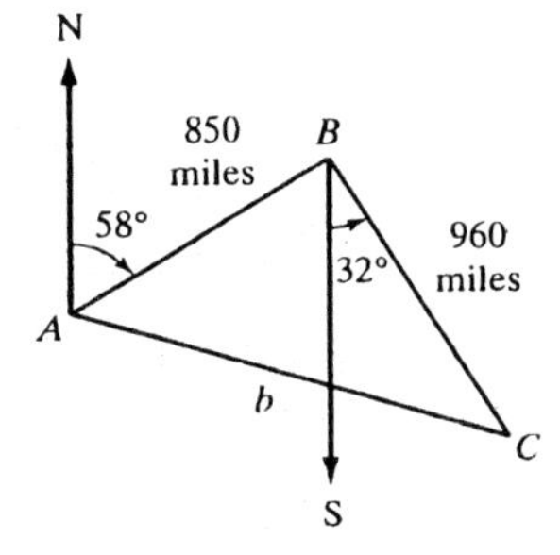

a. Using the figure,
$B = 58° + 32° = 90°$
Thus, use the Pythagorean Theorem to find the distance from city A to city C.

$$850^2 + 960^2 = b^2$$
$$b^2 = 722{,}500 + 921{,}600$$
$$b^2 = 1{,}644{,}100$$
$$b = \sqrt{1{,}644{,}100} \approx 1282.2$$

The distance from city A to city B is about 1282.2 miles.

b. Using the figure,

$$\tan A = \frac{\text{opposite}}{\text{adjacent}} = \frac{960}{850} \approx 1.1294$$
$$A \approx \tan^{-1}(1.1294) \approx 48°$$
$$180° - 58° - 48° = 74°$$

The bearing from city A to city C is S74°E.

125. $d = 20 \cos \frac{\pi}{4} t$

$a = 20$ and $\omega = \frac{\pi}{4}$

a. maximum displacement:
$|a| = |20| = 20\,\text{cm}$

b. $f = \frac{\omega}{2\pi} = \frac{\frac{\pi}{4}}{2\pi} = \frac{\pi}{4} \cdot \frac{1}{2\pi} = \frac{1}{8}$

frequency: $\frac{1}{8}$ cycle per second

c. period: $\frac{2\pi}{\omega}=\frac{2\pi}{\frac{\pi}{4}}=2\pi\cdot\frac{4}{\pi}=8$
The time required for one cycle is 8 seconds.

126. $d=\frac{1}{2}\sin 4t$

$a=\frac{1}{2}$ and $\omega=4$

a. maximum displacement:
$|a|=\left|\frac{1}{2}\right|=\frac{1}{2}\text{ cm}$

b. $f=\frac{\omega}{2\pi}=\frac{4}{2\pi}=\frac{2}{\pi}\approx 0.64$
frequency: 0.64 cycle per second

c. period: $\frac{2\pi}{\omega}=\frac{2\pi}{4}=\frac{\pi}{2}\approx 1.57$
The time required for one cycle is about 1.57 seconds.

127. Because the distance of the object from the rest position at $t=0$ is a maximum, use the form $d=a\cos\omega t$. The period is $\frac{2\pi}{\omega}$ so,

$2=\frac{2\pi}{\omega}$

$\omega=\frac{2\pi}{2}=\pi$

Because the amplitude is 30 inches, $|a|=30$. because the object starts below its rest position $a=-30$. the equation for the object's simple harmonic motion is $d=-30\cos\pi t$.

128. Because the distance of the object from the rest position at $t=0$ is 0, use the form $d=a\sin\omega t$. The period is $\frac{2\pi}{\omega}$ so

$5=\frac{2\pi}{\omega}$

$\omega=\frac{2\pi}{5}$

Because the amplitude is $\frac{1}{4}$ inch, $|a|=\frac{1}{4}$. a is negative since the object begins pulled down. The equation for the object's simple harmonic motion is $d=-\frac{1}{4}\sin\frac{2\pi}{5}t$.

Chapter 4 Test

1. $135°=135°\cdot\frac{\pi\text{ radians}}{180°}$
$=\frac{135\pi}{180}\text{ radians}$
$=\frac{3\pi}{4}\text{ radians}$

2. $75°=75°\cdot\frac{\pi\text{ radians}}{180°}=\frac{75\pi}{180}\text{ radians}$
$=\frac{5\pi}{12}\text{ radians}$
$s=r\theta$
$s=20\left(\frac{5\pi}{12}\right)=\frac{25\pi}{3}\text{ ft}\approx 26.18\text{ ft}$

3. a. $\frac{16\pi}{3}-4\pi=\frac{16\pi}{3}-\frac{12\pi}{3}=\frac{4\pi}{3}$

b. $\frac{16\pi}{3}$ is coterminal with $\frac{4\pi}{3}$.
$\frac{4\pi}{3}-\pi=\frac{4\pi}{3}-\frac{3\pi}{3}=\frac{\pi}{3}$

4. P = (–2, 5) is a point on the terminal side of θ, $x=-2$ and $y=5$. Furthermore,
$r=\sqrt{x^2+y^2}=\sqrt{(-2)^2+(5)^2}$
$=\sqrt{4+25}=\sqrt{29}$
Use x, y, and r, to find the six trigonometric functions of θ.

$\sin\theta=\frac{y}{r}=\frac{5}{\sqrt{29}}=\frac{5\sqrt{29}}{\sqrt{29}\sqrt{29}}=\frac{5\sqrt{29}}{29}$

$\cos\theta=\frac{x}{r}=\frac{-2}{\sqrt{29}}=-\frac{2\sqrt{29}}{\sqrt{29}\sqrt{29}}=-\frac{2\sqrt{29}}{29}$

$\tan\theta=\frac{y}{x}=\frac{5}{-2}=-\frac{5}{2}$

$\csc\theta=\frac{r}{y}=\frac{\sqrt{29}}{5}$

$\sec\theta=\frac{r}{x}=\frac{\sqrt{29}}{-2}=-\frac{\sqrt{29}}{2}$

$\cot\theta=\frac{x}{y}=\frac{-2}{5}=-\frac{2}{5}$

5. Because $\cos\theta < 0$, θ cannot lie in quadrant I and quadrant IV; the cosine function is positive in those two quadrants. Thus, with $\cos\theta < 0$, θ lies in quadrant II or quadrant III. We are also given that $\cot\theta > 0$. Because quadrant III is the only quadrant in which the cosine is negative and the cotangent is positive, θ lies in quadrant III.

6. Because the cosine is positive and the tangent is negative, θ lies in quadrant IV. In quadrant IV x is positive and y is negative. Thus,

$\cos\theta = \frac{1}{3} = \frac{x}{r}$, $x = 1$, $r = 3$. Furthermore,

$$x^2 + y^2 = r^3$$
$$1^2 + y^2 = 3^2$$
$$y^2 = 9 - 1 = 8$$
$$y = -\sqrt{8} = -2\sqrt{2}$$

Use x, y, and r, to find the six trigonometric functions of θ.

$$\sin\theta = \frac{y}{r} = \frac{-2\sqrt{2}}{3} = -\frac{2\sqrt{2}}{3}$$

$$\tan\theta = \frac{y}{x} = \frac{-2\sqrt{2}}{1} = -2\sqrt{2}$$

$$\csc\theta = \frac{r}{y} = \frac{3}{-2\sqrt{2}} = -\frac{3\sqrt{2}}{2\sqrt{2}\cdot\sqrt{2}} = -\frac{3\sqrt{2}}{4}$$

$$\sec\theta = \frac{r}{x} = \frac{3}{1} = 3$$

$$\cot\theta = \frac{x}{y} = \frac{1}{-2\sqrt{2}} = -\frac{1\cdot\sqrt{2}}{2\sqrt{2}\sqrt{2}} = -\frac{\sqrt{2}}{4}$$

7. $\tan\frac{\pi}{6}\cos\frac{\pi}{3} - \cos\frac{\pi}{2} = \frac{\sqrt{3}}{3}\cdot\frac{1}{2} - 0 = \frac{\sqrt{3}}{6}$

8. 300° lies in quadrant IV.
The reference angle is

$$\theta' = 360° - 300° = 60°$$

$\tan 60° = \sqrt{3}$

In quadrant IV, $\tan\theta < 0$, so

$\tan 300° = -\tan 60 = -\sqrt{3}$.

9. $\frac{7\pi}{4}$ lies in quadrant IV.
The reference angle is

$$\theta' = 2\pi - \frac{7\pi}{4} = \frac{8\pi}{4} - \frac{7\pi}{4} = \frac{\pi}{4}$$

$$\sin\frac{\pi}{4} = \frac{\sqrt{2}}{2}$$

In quadrant IV, $\sin\theta < 0$, so

$$\sin\frac{7\pi}{4} = -\sin\frac{\pi}{4} = -\frac{\sqrt{2}}{2}.$$

10. $\sec\frac{22\pi}{3} = \sec\frac{4\pi}{3} = -\sec\frac{\pi}{3}$

$$= \frac{1}{-\cos\frac{\pi}{3}} = \frac{1}{-\frac{1}{2}} = -2$$

11. $\cot\left(-\frac{8\pi}{3}\right) = \cot\left(\frac{4\pi}{3}\right) = \cot\frac{\pi}{3}$

$$= \frac{1}{\tan\frac{\pi}{3}} = \frac{1}{\sqrt{3}} = \frac{\sqrt{3}}{3}$$

12. $\tan\left(\frac{7\pi}{3} + n\pi\right) = \tan\frac{7\pi}{3} = \tan\frac{\pi}{3} = \sqrt{3}$

13. a. $\sin(-\theta) + \cos(-\theta) = -\sin(\theta) + \cos(\theta)$

$$= -a + b$$

b. $\tan\theta - \sec\theta = \frac{\sin\theta}{\cos\theta} - \frac{1}{\cos\theta}$

$$= \frac{a}{b} - \frac{1}{b}$$
$$= \frac{a-1}{b}$$

14. The equation $y = 3\sin 2x$ is of the form $y = A\sin Bx$ with $A = 3$ and $B = 2$. The amplitude is $|A| = |3| = 3$. The period is $\frac{2\pi}{B} = \frac{2\pi}{2} = \pi$. The quarter-period is $\frac{\pi}{4}$.
The cycle begins at $x = 0$. Add quarter-periods to generate x-values for the key points.

$$x = 0$$
$$x = 0 + \frac{\pi}{4} = \frac{\pi}{4}$$
$$x = \frac{\pi}{4} + \frac{\pi}{4} = \frac{\pi}{2}$$
$$x = \frac{\pi}{2} + \frac{\pi}{4} = \frac{3\pi}{4}$$
$$x = \frac{3\pi}{4} + \frac{\pi}{4} = \pi$$

Evaluate the function at each value of x.

x	coordinates
0	(0, 0)
$\frac{\pi}{4}$	$\left(\frac{\pi}{4}, 3\right)$
$\frac{\pi}{2}$	$\left(\frac{\pi}{2}, 0\right)$
$\frac{3\pi}{4}$	$\left(\frac{3\pi}{4}, -3\right)$
π	$(\pi, 0)$

Connect the five key points with a smooth curve and graph one complete cycle of the given function.

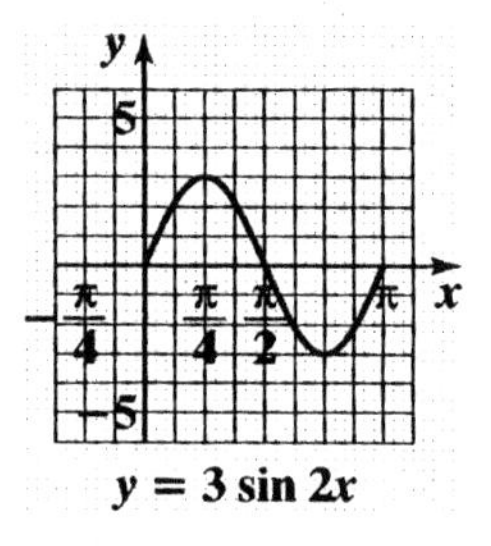

$y = 3 \sin 2x$

15. The equation $y = -2\cos\left(x - \frac{\pi}{2}\right)$ is of the form $y = A\cos(Bx - C)$ with $A = -2$, $B = 1$, and $C = \frac{\pi}{2}$. The amplitude is $|A| = |-2| = 2$.

The period is $\frac{2\pi}{B} = \frac{2\pi}{1} = 2\pi$. The phase shift is $\frac{C}{B} = \frac{\frac{\pi}{2}}{1} = \frac{\pi}{2}$. The quarter-period is $\frac{2\pi}{4} = \frac{\pi}{2}$.

The cycle begins at $x = \frac{\pi}{2}$. Add quarter-periods to generate x-values for the key points.

$x = \frac{\pi}{2}$

$x = \frac{\pi}{2} + \frac{\pi}{2} = \pi$

$x = \pi + \frac{\pi}{2} = \frac{3\pi}{2}$

$x = \frac{3\pi}{2} + \frac{\pi}{2} = 2\pi$

$x = 2\pi + \frac{\pi}{2} = \frac{5\pi}{2}$

Evaluate the function at each value of x.

x	coordinates
$\frac{\pi}{2}$	$\left(\frac{\pi}{2}, -2\right)$
π	$(\pi, 0)$
$\frac{3\pi}{2}$	$\left(\frac{3\pi}{2}, 2\right)$
2π	$(2\pi, 0)$
$\frac{5\pi}{2}$	$\left(\frac{5\pi}{2}, -2\right)$

Connect the five key points with a smooth curve and graph one complete cycle of the given function.

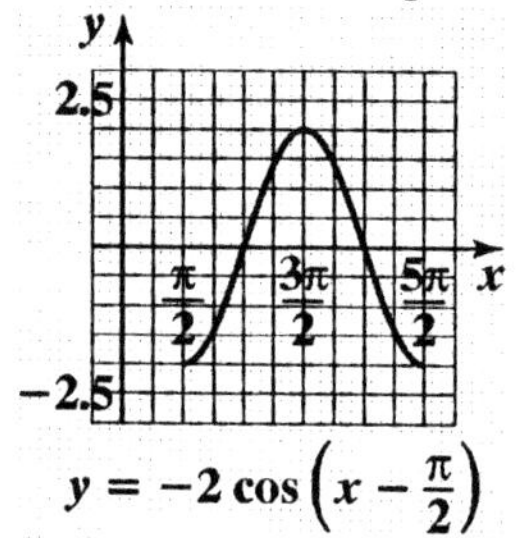

$y = -2\cos\left(x - \frac{\pi}{2}\right)$

16. Solve the equations

$\frac{x}{2} = -\frac{\pi}{2}$ and $\frac{x}{2} = \frac{\pi}{2}$

$x = -\frac{\pi}{2} \cdot 2$ $\quad$ $x = \frac{\pi}{2} \cdot 2$

$x = -\pi$ $\quad$ $x = \pi$

Thus, two consecutive asymptotes occur at $x = -\pi$ and $x = \pi$.

x-intercept $= \frac{-\pi + \pi}{2} = \frac{0}{2} = 0$

An x-intercept is 0 and the graph passes through (0, 0). Because the coefficient of the tangent is 2, the points on the graph midway between an x-intercept and the asymptotes have y-coordinates of -2 and 2. Use the two consecutive asymptotes, $x = -\pi$ and $x = \pi$, to graph one full period of $y = 2\tan\frac{x}{2}$ from $-\pi$ to π.

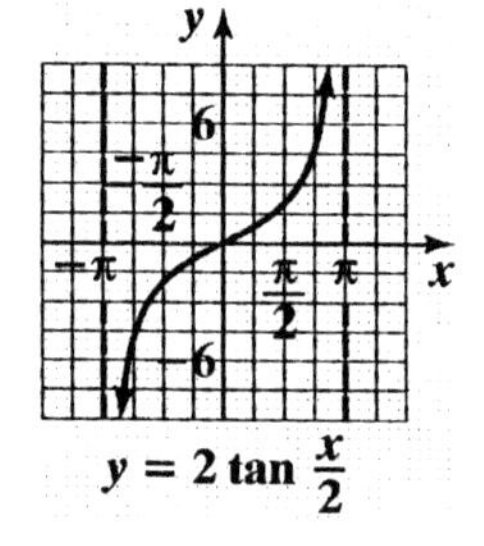

$y = 2\tan\frac{x}{2}$

17. Graph the reciprocal sine function, $y=-\frac{1}{2}\sin\pi x.$ The equation is of the form $y=A\sin Bx$ with $A=-\frac{1}{2}$ and $B=\pi.$

amplitude: $|A|=\left|-\frac{1}{2}\right|=\frac{1}{2}$

period: $\frac{2\pi}{B}=\frac{2\pi}{\pi}=2$

Use quarter-periods, $\frac{2}{4}=\frac{1}{2}$, to find x-values for the five key points. Starting with $x=0$, the x-values are 0, $\frac{1}{2}$, 1, $\frac{3}{2}$, 2. Evaluating the function at each value of x, the key points are $(0, 3)$, $\left(\frac{1}{2}, -\frac{1}{2}\right)$, $(1, 0)$, $\left(\frac{3}{2}, \frac{1}{2}\right)$, $(2, 0)$.

Use these key points to graph $y=-\frac{1}{2}\sin\pi x$ from 0 to 2. Use the graph to obtain the graph of the reciprocal function. Draw vertical asymptotes through the x-intercepts, and use them as guides to graph $y=-\frac{1}{2}\csc\pi x.$

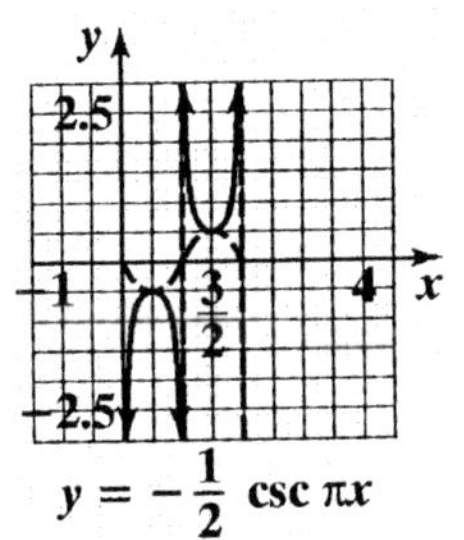

$y=-\frac{1}{2}\csc\pi x$

18. Let $\theta=\cos^{-1}\left(-\frac{1}{2}\right)$, then $\cos\theta=-\frac{1}{2}.$ Because $\cos\theta$ is negative, θ is in quadrant II.

$$x^2+y^2=r^2$$
$$(-1)^2+y^2=2^2$$
$$y^2=4-1=3$$
$$y=\sqrt{3}$$
$$\tan\left[\cos^{-1}\left(-\frac{1}{2}\right)\right]=\tan\theta=\frac{y}{x}=\frac{\sqrt{3}}{-1}=-\sqrt{3}$$

19. Let $\theta=\cos^{-1}\left(\frac{x}{3}\right)$, then $\cos\theta=\frac{x}{3}.$ Because $\cos\theta$ is positive, θ is in quadrant I.

$$x^2+y^2=r^2$$
$$x^2+y^2=3^2$$
$$y^2=9-x^2$$
$$y=\sqrt{9-x^2}$$
$$\sin\left[\cos^{-1}\left(\frac{x}{3}\right)\right]=\sin\theta=\frac{y}{r}=\frac{\sqrt{9-x^2}}{3}$$

20. Find the measure of angle B. Because $C=90°$, $A+B=90°$.
Thus, $B=90°-A=90°-21°=69°$.
We have a known angle, a known hypotenuse, and an unknown opposite side. Use the sine function.

$$\sin 21°=\frac{a}{13}$$
$$a=13\sin 21°\approx 4.7$$

We have a known angle, a known hypotenuse, and an unknown adjacent side. Use the cosine function.

$$\cos 21°=\frac{b}{13}$$
$$b=13\cos 21°\approx 12.1$$

In summary, $B=69°$, $a\approx 4.7$, and $b\approx 12.1$.

21. Using a right triangle, we have a known angle, an unknown opposite side, h, and a known adjacent side. Therefore, use the tangent function.

$$\tan 37°=\frac{h}{30}$$
$$h=30\tan 37°\approx 23$$

The building is about 23 yards high.

22. Using a right triangle, we have a known hypotenuse, a known opposite side, and an unknown angle. Therefore, use the sine function.

$$\sin\theta=\frac{43}{73}$$
$$\theta=\sin^{-1}\left(\frac{43}{73}\right)\approx 36.1°$$

The rope makes an angle of about 36.1° with the pole.

23. We need the acute angle between ray OP and the north-south line through O. This angle measures 90° – 10°. This angle is measured from the north side of the north-south line and lies west of the north-south line. Thus the bearing from O to P is N80°W.

24. $d = -6\cos \pi t$

$a = -6$ and $\omega = \pi$

a. maximum displacement:
$|a| = |-6| = 6$ in.

b. $f = \dfrac{\omega}{2\pi} = \dfrac{\pi}{2\pi} = \dfrac{1}{2}$

frequency: $\dfrac{1}{2}$ cycle per second

c. period $= \dfrac{2\pi}{\omega} = \dfrac{2\pi}{\pi} = 2$

The time required for one cycle is 2 seconds.

25. Trigonometric functions are periodic.

Cumulative Review Exercises (Chapters P-4)

1. $x^2 = 18 + 3x$

$x^2 - 3x - 18 = 0$

$(x-6)(x+3) = 0$

$x - 6 = 0$ or $x + 3 = 0$

$x = 6$ $\quad x = -3$

The solution set is $\{-3, 6\}$.

2. $x^3 + 5x^2 - 4x - 20 = 0$

$x^2(x+5) - 4(x+5) = 0$

$(x^2 - 4)(x+5) = 0$

$(x-2)(x+2)(x+5) = 0$

$x - 2 = 0$ or $x + 2 = 0$ or $x + 5 = 0$

$x = 2$ $\quad x = -2$ $\quad x = -5$

The solution set is $\{-5, -2, 2\}$.

3. $\log_2 x + \log_2(x-2) = 3$

$\log_2 x(x-2) = 3$

$x(x-2) = 2^3$

$x^2 - 2x = 2^3$

$x^2 - 2x - 8 = 0$

$(x-4)(x+2) = 0$

$x - 4 = 0$ or $x + 2 = 0$

$x = 4$ $\quad x = -2$

$x = -2$ is extraneous

The solution set is $\{4\}$

4. $\sqrt{x-3} + 5 = x$

$\sqrt{x-3} = x - 5$

$\left(\sqrt{x-3}\right)^2 = (x-5)^2$

$x - 3 = x^2 - 10x + 25$

$x^2 - 11x + 28 = 0$

$(x-4)(x-7) = 0$

$x - 4 = 0$ or $x - 7 = 0$

$x = 4$ $\quad x = 7$

$\sqrt{4-3} + 5 = 4$

$\sqrt{1} + 5 = 4$

$1 + 5 = 4$ false

$x = 4$ is not a solution

$\sqrt{7-3} + 5 = 7$

$\sqrt{4} + 5 = 7$

$2 + 5 = 7$ true

The solution set is $\{7\}$.

5. $x^3 - 4x^2 + x + 6 = 0$

$p: \pm 1, \pm 2, \pm 3, \pm 6$

$q: \pm 1$

$\dfrac{p}{q}: \pm 1, \pm 2, \pm 3, \pm 6$

$$\begin{array}{r|rrrr} 2 & 1 & -4 & 1 & 6 \\ & & 2 & -4 & -6 \\ \hline & 1 & -2 & -3 & 0 \end{array}$$

$x^3 - 4x^2 + x + 6 = (x-2)\left(x^2 - 2x - 3\right)$

Thus,

$x^3 - 4x^2 + x + 6 = 0$

$(x-2)\left(x^2 - 2x - 3\right) = 0$

$(x-2)(x-3)(x+1) = 0$

$x - 2 = 0$ or $x - 3 = 0$ or $x + 1 = 0$

$x = 2$ $\quad x = 3$ $\quad x = -1$

The solution set is $\{-1, 2, 3\}$

6. $|2x - 5| \le 11$

$-11 \le 2x - 5 \le 11$

$-6 \le 2x \le 16$

$-3 \le x \le 8$

The solution set is $\{x \mid -3 \le x \le 8\}$

7. $f(x)=\sqrt{x-6}$

$x=\sqrt{y-6}$

$x^2=y-6$

$y=x^2+6$

$f^{-1}(x)=x^2+6$

8.
$$\begin{array}{r} 4x^2-\frac{14}{5}x-\frac{17}{25} \\ 5x+2\overline{\big)20x^3-6x^2-9x+10} \\ \underline{20x^3+8x^2} \\ -14x^2-9x \\ \underline{-14x^2-\frac{28}{5}x} \\ -\frac{17}{5}x+10 \\ \underline{-\frac{17}{5}x-\frac{34}{25}} \\ \frac{284}{25} \end{array}$$

The quotient is $4x^2-\frac{14}{5}x-\frac{17}{25}+\frac{284}{125x+50}$.

9. $\log 25+\log 40=\log(25\cdot 40)$
$=\log 1000$
$=\log 10^3$
$=3$

10. $\frac{14\pi}{9}$ radians $=\frac{14\pi}{9}$ radians $\cdot\frac{180°}{\pi \text{ radians}}$
$=\frac{14\cdot 180°}{9}=280°$

11. $3x^4-2x^3+5x^2+x-9=0$
The sign changes 3 times so the equation has at most 3 positive real roots;
$f(-x)=3x^4+2x^3+5x^2-x-9$
The sign changes 1 time, so the equation has at most 1 negative real root.

12. $f(x)=\frac{x}{x^2-1}$
vertical asymptotes: $x^2-1=0$, $x=1$ and $x=-1$
horizontal asymptote: $m=1$ and $n=2$ so $m<n$ and the x-axis is a horizontal asymptote.

x-intercept: (0,0)

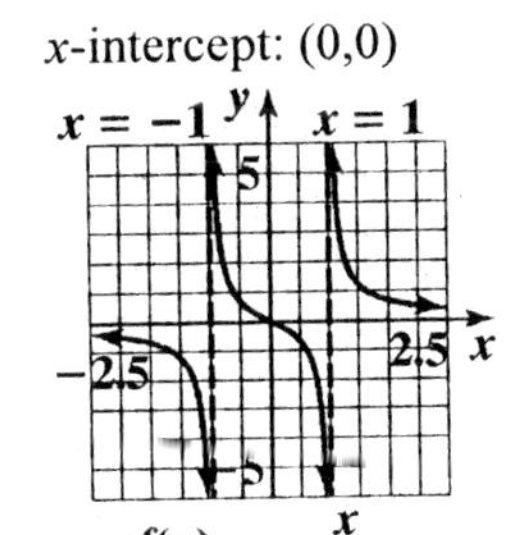

13. $(x-2)^2+y^2=1$
The graph is a circle with center (2,0) and $r=1$.

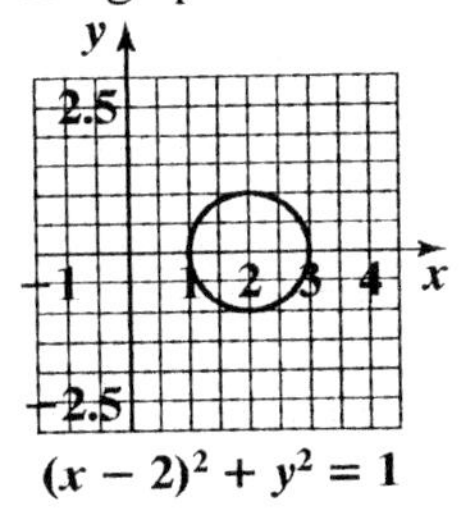

14. $y=(x-1)(x+2)^2$
x-intercepts: (1,0) and (–2,0)
y-intercept: $y=(-1)(2)^2=-4$
(0,–4)

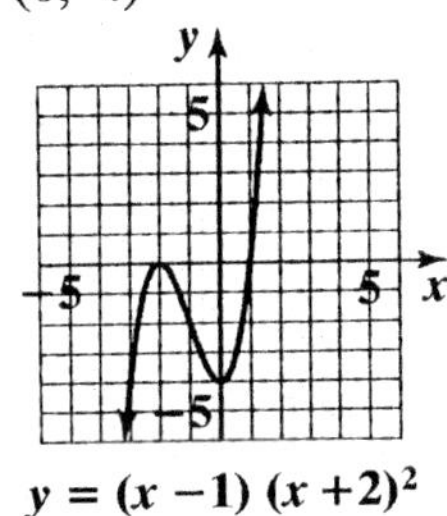

15. $y=\sin\left(2x+\frac{\pi}{2}\right)=\sin\left(2x-\left(-\frac{\pi}{2}\right)\right)$

The equation $y=\sin\left(2x-\left(-\frac{\pi}{2}\right)\right)$ is of the form $y=A\sin(Bx-C)$ with $A=1$, $B=2$, and $C=-\frac{\pi}{2}$. The amplitude is $|A|=|1|=1$
The period is $\frac{2\pi}{B}=\frac{2\pi}{2}=\pi$. The phase shift is $\frac{C}{B}=\frac{-\frac{\pi}{2}}{2}=-\frac{\pi}{2}\cdot\frac{1}{2}=-\frac{\pi}{4}$. The quarter-period is $\frac{\pi}{4}$. The cycle begins at $x=-\frac{\pi}{4}$. Add quarter-periods to generate x-values for the key points.
$x=-\frac{\pi}{4}$, $x=-\frac{\pi}{4}+\frac{\pi}{4}=0$, $x=0+\frac{\pi}{4}=\frac{\pi}{4}$,

$x = \frac{\pi}{4} + \frac{\pi}{4} = \frac{\pi}{2}$, $x = \frac{\pi}{2} + \frac{\pi}{4} = \frac{3\pi}{4}$ To graph from 0 to π, evaluate the function at the last four key points and at $x = \pi$.

x	coordinates
0	(0, 1)
$\frac{\pi}{4}$	$\left(\frac{\pi}{4}, 0\right)$
$\frac{\pi}{2}$	$\left(\frac{\pi}{2}, -1\right)$
$\frac{3\pi}{4}$	$\left(\frac{3\pi}{4}, 0\right)$
π	$(\pi, 1)$

Connect the points with a smooth curve and extend the graph one cycle to the right to graph from 0 to 2π.

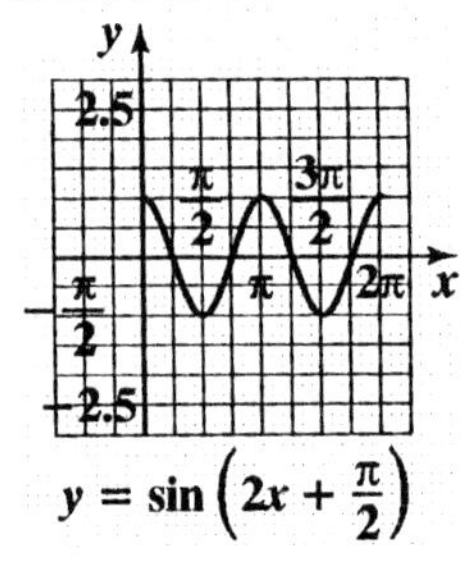

$y = \sin\left(2x + \frac{\pi}{2}\right)$

16. Solve the equations

$$3x = -\frac{\pi}{2} \quad \text{and} \quad 3x = \frac{\pi}{2}$$

$$x = \frac{-\frac{\pi}{2}}{3} \qquad x = \frac{\frac{\pi}{2}}{3}$$

$$x = -\frac{\pi}{6} \qquad x = \frac{\pi}{6}$$

Thus, two consecutive asymptotes occur at $x = -\frac{\pi}{6}$ and $x = \frac{\pi}{6}$.

$$x\text{-intercept} = \frac{-\frac{\pi}{6} + \frac{\pi}{6}}{2} = \frac{0}{2} = 0$$

An x-intercept is 0 and the graph passes through (0, 0). Because the coefficient of the tangent is 2, the points on the graph midway between an x-intercept and the asymptotes have y-coordinates of –2 and 2. Use the two consecutive asymptotes, $x = -\frac{\pi}{6}$ and $x = \frac{\pi}{6}$, to graph one full period of $y = 2\tan 3x$ from $-\frac{\pi}{6}$ to $\frac{\pi}{6}$. Extend the pattern to the right to graph two complete cycles.

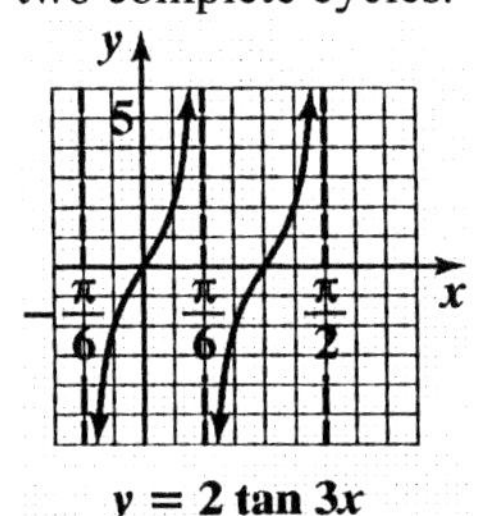

$y = 2 \tan 3x$

17.
$$C(p) = 30{,}000 + 2500p$$
$$R(p) = 3125p$$
$$30{,}000 + 2500p = 3125p$$
$$30{,}000 = 625p$$
$$p = 48$$

48 performances must be played for you to break even.

18. a. Let t be the number of years after 1984. 2003 is 19 years after 1984.

$$A = A_0e^{kt}$$
$$A = 25000e^{kt}$$
$$70500000 = 25000e^{k(19)}$$
$$\frac{70500000}{25000} = e^{k(19)}$$
$$2820 = e^{k(19)}$$
$$\ln 2820 = \ln e^{k(19)}$$
$$\ln 2820 = 19k$$
$$\frac{\ln 2820}{19} = k$$
$$0.418 \approx k$$

Thus, $A = 25000e^{0.418t}$

b.

$$A = 25,000e^{0.418t}$$
$$150,000,000 = 25,000e^{0.418t}$$
$$\frac{150,000,000}{25,000} = e^{0.418t}$$
$$6000 = e^{0.418t}$$
$$\ln 6000 = \ln e^{0.418t}$$
$$\ln 6000 = 0.418t$$
$$\frac{\ln 6000}{0.418} = t$$
$$21 \approx t$$

150,000,000 cell phones will be sold in the United States 21 years after 1984, or 2005

19. $2200 = \frac{k}{3.5}$

$k = 7700$

$h = \frac{7700}{5} = 1540$

The rate of heat loss is 1540 Btu per hour.

20. Using a right triangle, we have a known opposite side, a known adjacent side, and an unknown angle. Therefore, use the tangent function.

$\tan\theta = \frac{200}{50} = 4$

$\theta = \tan^{-1}(4) \approx 76°$

The angle of elevation is about $76°$.